Die Grundlehren der mathematischen Wissenschaften

in Einzeldarstellungen
mit besonderer Berücksichtigung
der Anwendungsgebiete

Band 188

Garth Warner

Harmonic Analysis on Semi-Simple Lie Groups I

Springer-Verlag Berlin Heidelberg GmbH

Garth Warner

Department of Mathematics, University of Washington
Seattle, Washington

AMS Subject Classifications (1970)
22 E 45, 22 D 10, 43 A 90, 43 A 85, 22 D 30

ISBN 978-3-642-50277-4 ISBN 978-3-642-50275-0 (eBook)
DOI 10.1007/978-3-642-50275-0

 Library of Congress Catalog Card Number 70-160590.
setting: Dai Nippon Printing Co., Ltd., Tokyo.

For Helen and Wyeth

Foreword

The representation theory of locally compact groups has been vigorously developed in the past twenty-five years or so; of the various branches of this theory, one of the most attractive (and formidable) is the representation theory of semi-simple Lie groups which, to a great extent, is the creation of a single man: Harish-Chandra. The chief objective of the present volume and its immediate successor is to provide a reasonably self-contained introduction to Harish-Chandra's theory. Granting certain basic prerequisites (cf. infra), we have made an effort to give full details and complete proofs of the theorems on which the theory rests.

The structure of this volume and its successor is as follows. Each book is divided into chapters; each chapter is divided into sections; each section into numbers. We then use the decimal system of reference; for example, 1.3.2 refers to the second number in the third section of the first chapter. Theorems, Propositions, Lemmas, and Corollaries are listed consecutively throughout any given number. Numbers which are set in fine print may be omitted at a first reading. There are a variety of Examples scattered throughout the text; the reader, if he is so inclined, can view them as exercises ad libitum. The Appendices to the text collect certain ancillary results which will be used on and off in the systematic exposition; a reference of the form A2.4, say, would mean the fourth number in the second appendix. Numbers in square brackets after an individual's name designate works in the Bibliography found at the end of the volume; a list of notational conventions can be found there too (the reader is also advised to glance through Appendices 1 and 2 where additional commonly employed notations are introduced).

The basic prerequisites for reading this work (aside from elementary algebra and analysis) are as follows. The reader who is interested in the structure theory of semi-simple Lie groups and algebras, as presented in Chapters 1 and 2, will have to have background commensurate with Helgason [2], Hochschild [1] and Jacobson [3]. A little sheaf theory (cf. Hirzebruch [3]) is needed in Chapter 3 where the Borel-Weil Theorem is proved. Chapters 4, 5, 6, and 7 comprise an introduction to general group representation theory and spherical functions (with applications to semi-simple Lie groups); the reader who would like to put off the study of the first three chapters and instead move directly to representation theory could easily begin with Chapter 4 – the only prerequisites for this would be functional analysis in general along with distribution theory (cf. Schwartz [1], [2]) and elementary harmonic analysis (cf. Hewitt and Ross [1]) in particular (a familiarity with the contents of the book of Dixmier [15] would be helpful for Chapter 7). Chapters 8, 9, and 10 deal exclusively with analysis on semi-simple Lie groups and all the prerequisites cited above are needed for their study.

In preparing this work, I have drawn freely on the published and unpublished literature. Identification of sources is ordinarily not made in the text proper; however, in the Guide to the Literature at the end of the volume, the reader will find the works listed on which the exposition in any particular number is based. We apologize in advance for possible incompleteness and omissions in the list. In addition, the author makes no claim whatsoever to originality in the proofs or mode of presentation and assumes, of course, full responsibility for any errors or mistakes.

I have lectured on this material for a number of years at the University of Washington. I want to thank my students Pat Donaly, Helaman Ferguson, Harry Glover, Scott Osborne, and Cary Rader for helping me read the manuscript in its various stages of preparation. My colleagues here at the University, Professors Ramesh Gangolli, Vilnis Ozols, and Roger Richardson, have been most generous with their time and have made many improvements in the text. Elsewhere, Professors A. Borel, C. Chevalley, J.G.M. Fell, R. Goodman, Harish-Chandra, S. Helgason, R. Jewitt, R.P. Langlands, H. Leptin, J.G.M. Mars, S. Rallis, L. Robertson, P. Sally, G. Schiffmann, R. Steinberg, and N. Wallach have all been helpful; in particular the numerous and detailed conversations which I had with Professor Harish-Chandra in the academic year 1969–70 at the Institute for Advanced Study were decisive in the later stages of the preparation of this work. I am also happy to express here my gratitude to the authorities of the National Science Foundation for providing me with support over the past few years. Finally I want to thank my indomitable typists Susan Lowrey, Susan Seeds, and Helen Warner for the great job which they did.

Prologue

We shall give here a brief overview of the subject matter which forms the content of the present volume and its immediate successor. A more technical survey of the field can be found in Harish-Chandra [34] (see also Schiffmann's Séminaire Bourbaki article (number 323)).

Let G be a locally compact unimodular group satisfying the second axiom of countability and which, moreover, is postliminaire; let $\hat{G}$ be the set of unitary equivalence classes of irreducible unitary representations of G equipped with the hull-kernel topology; fix a Haar measure d_G on G – then, according to a well-known theorem of I.E. Segal, there exists a unique positive measure μ on $\hat{G}$ such that

$$\int_G |f(x)|^2 d_G(x) = \int_{\hat{G}} \operatorname{tr}(\hat{U}(f)\hat{U}(f)^*) d\mu(\hat{U})$$

for all $f \in L^1(G) \cap L^2(G)$. The measure μ is called the Plancherel measure for $\hat{G}$ (associated with the given Haar measure on G). This being so, the basic problem in 'L^2 Harmonic Analysis' on G is the 'explicit' determination of μ. [To illustrate, suppose that G is compact; normalize d_G by the requirement $\int_G d_G = 1$ – then the abstract theory tells us that μ assigns to each point $\hat{U}$ in $\hat{G}$ mass $\dim(\hat{U})$. In many concrete situations one can actually compute $\dim(\hat{U})$ in terms of the group structure on G and thereby obtain an 'explicit' description of μ; this will be the case, for instance, when G is a compact connected semi-simple Lie group (Weyl's dimension formula . . .).]

Let us suppose that G is a Lie group. For an arbitrary G, very little is known about $\hat{G}$ and no attempt has yet been made to attack the above problem; on the other hand, in the two extreme cases of a nilpotent G (or even a solvable G) and a semi-simple G, great progress has been made on this question. Indeed, the work of Dixmier, Kirillov, and Pukanszky gives the explicit Plancherel formula in the nilpotent case whereas the labors of Harish-Chandra give the Plancherel formula in the semi-simple case. In the present work we shall discuss in detail the methods which lead to Harish-Chandra's Plancherel Theorem for a semi-simple G.

To be more precise, let us now assume that G is a connected semi-simple Lie group with finite center; fix a maximal compact subgroup K of G – then the first results in the Harmonic Analysis on G serve to relate the representation theory of K (which is known . . .) with the representation theory of G. Thus let $\hat{K}$ be the set of unitary equivalence classes of irreducible unitary representations of K; fix a class δ in $\hat{K}$ – then it can be shown that the number of times that δ occurs in the restriction to K of an arbitrary irreducible unitary representation of G is $\leqslant \dim(\delta)$ (the dimension of δ). This fact has a number of important consequences. In the first place an easy argument then leads at once to the conclusion that G is postliminaire (hence is type I). A second consequence is this: Let U be an irreducible unitary representation of G – then, for every $f \in C_c^\infty(G)$, the operator $U(f)$ $(= \int_G f(x)U(x)d_G(x))$ is of the trace class and, moreover, the map T_U defined by the rule

$$f \mapsto \operatorname{tr}(U(f)) \qquad (f \in C_c^\infty(G))$$

is a distribution on G (in the sense of Laurent Schwartz). The distribution T_U is called the character of U; it is not difficult to prove that T_U determines U to within unitary equivalence. Furthermore, the general theory of central eigendistributions of $\mathfrak{Z}$ on G ($\mathfrak{Z}$ the algebra of differential operators on G which commute with both left and right translation), when applied to a given T_U, serves to imply that there exists a locally summable function F_U (say) which is analytic on the set of regular elements in G and with the property that

$$T_U(f) = \int_G f(x)F_U(x)d_G(x)$$

for all $f \in C_c^\infty(G)$; otherwise said, in the sense of distribution theory, the character of a given irreducible unitary representation of G is a function.

In order to compute the Plancherel measure μ, one must have at hand a large supply of irreducible unitary representations of G. [For the purpose of determining μ, it is not necessary, however, to find 'all' the irreducible unitary representations of G; for, due to the existence of 'complementary series', the support $\hat{G}_r$ of μ in $\hat{G}$, i.e. the reduced dual of G, will in general be a proper subset of $\hat{G}$.] To begin with, one has to isolate the discrete series for G (since irreducible square integrable unitary representations always have non-zero Plancherel measure). In his well-known Acta paper Harish-Chandra [30] completely solves this deep and formidable problem; it turns out that the discrete series for G is not empty iff $\operatorname{rank}(G) = \operatorname{rank}(K)$ and, when that is so, the (formal) dimension of a given irreducible square integrable unitary representation can be explicitly computed in terms of parameters involving a compact Cartan subgroup of G (the formula for the dimension ($=$ formal degree) being strikingly analogous to that of Weyl's in the compact case). The other representations of G which enter into the Plancherel Theorem are constructed via the mechanism of 'induced representation'. Let P be a parabolic subgroup of G which is, moreover, cuspidal; let $P = \mathsf{M}AN$ be a Langlands decomposition for P 'compatible' with K (thus the split component A of P is the 'vector' part of a Cartan subgroup of G, P being cuspidal); let σ be a representation in the discrete series for M, ν a unitary character of A – then the representation $U^{\sigma,\nu}$ unitarily induced from P to G by $\sigma \otimes \nu$ ($\sigma \otimes \nu(mhn) = h^\nu\sigma(m)$, $mhn \in \mathsf{M}AN$) is irreducible 'in general' (those $U^{\sigma,\nu}$ which are not irreducible split into finitely many irreducible constituents and will not 'contribute' to μ). [One cannot, of course, carry out the above construction 'explicitly' unless the discrete series for M is known. Now M is, in general, neither connected nor semi-simple, but is always reductive; this is one of the reasons why it is necessary to develop the theory for reductive groups rather than just semi-simple groups.] To get the representations in the 'continuous series', one proceeds as follows. Let $\mathsf{J}^1, \ldots, \mathsf{J}^r$ be a maximal collection of mutually non-conjugate θ-stable Cartan subgroups of G (θ the Cartan involution per the pair (G,K)); let A_i be the vector part of J^i, $P_i = \mathsf{M}_iA_iN_i$ a (cuspidal) parabolic subgroup of G with split component A_i – then, in an obvious notation, the irreducible unitary representations of G (apart from the discrete series, when it exists) which figure in the Plancherel Theorem are the irreducible U^{σ_i,ν_i}. Because the discrete series representations σ_i are in a close correspondence with the characters of the unitary dual of the compact part of J^i, it follows that $\hat{G}_r$ (the support of μ in $\hat{G}$) is in a close correspondence with the disjoint union $\hat{\mathsf{J}}^1 \cup \ldots \cup \hat{\mathsf{J}}^r$ in a sense which can be made precise.

The proof of the Plancherel Theorem itself is based on the 'philosophy of cusp forms': Eisenstein integral, **c**-function, wave packet, Mass-Selberg relations etc. These topics will be dealt with systematically in a later volume. [For an expository account of these developments consult Harish-Chandra [36].] In the present volume and its immediate successor we shall go far enough into the theory to get the discrete series for G (cf. Chapter 10). For an example, the reader may wish to turn to the Epilogue where the Plancherel Theorem is established via the classical method of 'integration by parts' in the special case when rank $(G/K) = 1$ (the general case is best approached, however, via the philosophy alluded to above).

Contents

Chapter 1

The Structure of Real Semi-Simple Lie Groups

Chapter 2

The Universal Enveloping Algebra of a Semi-Simple Lie Algebra

Chapter 3

Finite Dimensional Representations of a Semi-Simple Lie Group

Chapter 4

Infinite Dimensional Group Representation Theory

Chapter 5

Induced Representations

Appendices

Contents Volume II

Chapter 1

The Structure of Real Semi-Simple Lie Groups

Introduction

In order to formulate and prove the deeper results in the representation theory of semi-simple Lie groups, it is necessary to have at hand detailed information on the structure of these groups. The purpose of this chapter is to set down the salient facts.

Section 1.1 of the present chapter is essentially preliminary in character. Thus, for the convenience of the reader, we start off by reviewing the standard generalities as regards the structure of complex semi-simple Lie algebras (since this material is assumed to be known, no proofs are given). Number's 1.1.2 and 1.1.3 provide a contemporary account (with complete proofs) of the elements of the theory of 'root systems' (cf. Bourbaki [5]...); here the reader will find the notions of Weyl group, Weyl chamber, parabolic subset, etc. in their most natural and general setting. As an application, let $\mathfrak{g}$ be a non-compact semi-simple Lie algebra over **R**, $\mathfrak{g} = \mathfrak{k} + \mathfrak{p}$ a Cartan decomposition of $\mathfrak{g}$ with Cartan involution θ; fix a maximal abelian subspace $\mathfrak{a}_{\mathfrak{p}}$ of $\mathfrak{p}$ – then, relative to the Euclidean structure (on $\mathfrak{g}$) canonically associated with θ, the elements of the set $\{\mathrm{ad}(H) : H \in \mathfrak{a}_{\mathfrak{p}}\}$ constitute a commutative, self-adjoint family of endomorphisms of $\mathfrak{g}$ and, as such, admit a simultaneous diagonalization. The corresponding collection Σ of non-zero eigenvalues, when viewed as a set of linear forms on $\mathfrak{a}_{\mathfrak{p}}$, is a root system in the dual of $\mathfrak{a}_{\mathfrak{p}}$ (cf. Proposition 1.1.3.1); on the other hand extend $\mathfrak{a}_{\mathfrak{p}}$ to a maximal abelian subalgebra $\mathfrak{a}$ of $\mathfrak{g}$ – then, as is well-known, the complexification $\mathfrak{a}_c$ of $\mathfrak{a}$ is a Cartan subalgebra of $\mathfrak{g}_c$. Let Φ denote the set of (non-zero) roots of the pair $(\mathfrak{g}_c, \mathfrak{a}_c)$, W_Φ the corresponding Weyl group; let W_Σ denote the Weyl group associated with the set of (non-zero) roots Σ of the pair $(\mathfrak{g}, \mathfrak{a}_{\mathfrak{p}})$ – then, roughly speaking, W_Σ may be viewed as the subgroup of W_Φ consisting of those elements which stabilize $\mathfrak{a}_{\mathfrak{p}}$ (cf. Proposition 1.1.3.3). Now fix a Weyl chamber $\mathscr{C}$ in $\mathfrak{a}_{\mathfrak{p}}$; such a choice of $\mathscr{C}$ determines an ordering on Σ and an Iwasawa decomposition $\mathfrak{g} = \mathfrak{k} + \mathfrak{a}_{\mathfrak{p}} + \mathfrak{n}^+$ for $\mathfrak{g}$; let G be a connected Lie group with Lie algebra $\mathfrak{g}$, $G = KA_{\mathfrak{p}}N^+$ the corresponding 'global' Iwasawa decomposition for G – we shall assume that G has finite center (so that K is compact). The objective of 1.1.4 is to study the nilpotent group N^+; among other things, it will be seen that to each $w \in W_\Sigma$ one may assign in a natural way closed subgroups N_w^+ and U_w^+ of N^+ such that $N^+ = N_w^+ U_w^+$ with $N_w^+ \cap U_w^+ = \{1\}$ – this result is of particular importance in the theory of parabolic subgroups of G (and in the theory of spherical functions...). The section concludes with a rapid survey of a few standard facts from the theory of reductive Lie algebras and groups, facts which will be used on and off in the sequel.

In 1.2 we take up the theory of parabolic subgroups of G. Let M denote the centralizer of $A_{\mathfrak{p}}$ in K, M^* the normalizer of $A_{\mathfrak{p}}$ in K (whence $W_\Sigma \sim \mathsf{M}^*/\mathsf{M}$...); let $P = \mathsf{M}A_{\mathfrak{p}}N^+$ – then P is a closed subgroup of G. Furthermore, following Furstenberg and Moore, it will be shown that P is a minimal boundary subgroup of G (so that the homogeneous space G/P is a maximal boundary of G); such a P will be termed a minimal parabolic subgroup of G. The importance of P resides in the fact that the double coset space $P\backslash G/P$ is finite, indeed is in a one-to-one correspondence with the Weyl group W_Σ of the pair $(\mathfrak{g}, \mathfrak{a}_\mathfrak{p})$ – this is the famous Bruhat Lemma (cf. Theorem 1.2.3.1). On the other hand, it can be shown that the triple (G, P, M^*) constitutes a Tits system; therefore, on the basis of a remarkable general theorem due to J. Tits (cf. 1.2.1), it follows that there are exactly 2^l abstract subgroups of G containing P ($l = \dim(A_\mathfrak{p})$ = split-rank of G) – such subgroups of G are termed parabolic. On the other hand, it is also possible to give a geometric description of the parabolic subgroups of G; thus it will turn out that they 'split' relative to the given Iwasawa decomposition of G in a manner entirely analogous to that of P itself (such a splitting will be abstracted in the notion of a Langlands decomposition . . .). A notable consequence of this result is the fact that the parabolic subgroups of G are necessarily closed.

Cartan subalgebras are the principal theme in 1.3. Let $\mathfrak{g}$ be a semi-simple Lie algebra over **R**, $\mathfrak{j}$ a maximal abelian subalgebra of $\mathfrak{g}$ such that the endomorphism $\mathrm{ad}(H)(=\mathrm{ad}_\mathfrak{g}(H))$ is semi-simple for every $H \in \mathfrak{j}$ – then $\mathfrak{j}$ is said to be a Cartan subalgebra of $\mathfrak{g}$. It is presupposed that the reader is familiar with the elements of the theory of Cartan subalgebras (in particular with the fact that Cartan subalgebras actually *exist* (consider the centralizer of a 'regular' element in $\mathfrak{g}$ . . .)); on the other hand, no a priori knowledge of the conjugacy problem will be assumed – thus consider the question: How many conjugacy classes (modulo the adjoint group of $\mathfrak{g}$) of Cartan subalgebras of $\mathfrak{g}$ are there? Let us recall that if $\mathfrak{g}$ is either compact or complex, then the answer is one; in general we are simply able to assert that there exists but a finite number (this is due to the fact that the set $\mathfrak{g}'$ of regular elements need not be connected in the real case) – however it is possible to describe these conjugacy classes precisely (cf. Theorem 1.3.1.10). The reader is also expected to be familiar with the basic facts centering around the Jacobson-Morosow Theorem (nilpotent elements, semi-simple elements, Jordan decomposition etc.); these notions are taken up in 1.3.5. The adjoint group G of $\mathfrak{g}$ operates in the natural fashion on $\mathfrak{g}$ itself; given an element X in $\mathfrak{g}$, it will be shown that the orbit $G\cdot X$ is closed iff X is semi-simple (i.e. iff $\mathrm{ad}(X)$ is semi-simple). [The fact that $G\cdot X$ closed $\Rightarrow X$ semi-simple is a ready consequence of the Jacobson-Morosow Theorem. As for the converse statement, its proof rests ultimately on some tools which will not be developed here; thus, using real algebraic geometry, one is led at once to the complex case – the proof is then completed by appealing to the Closed Orbit Lemma from the theory of algebraic groups.]

Returning to our semi-simple group G, by definition a Cartan subgroup J of G is the centralizer in G of a Cartan subalgebra $\mathfrak{j}$ of $\mathfrak{g}$. [Examples show that J need not be abelian or connected.] The Cartan subgroups of G play a role somewhat analogous to that played by the maximal tori in the compact case (cf. 1.4). For instance, owing to the result on the conjugacy of Cartan subalgebras alluded to above, it is possible to choose a maximal set $J^1, \ldots, J^r$ of mutually non-conjugate Cartan subgroups of G such that

$$G' = \bigcup_{i=1}^{r} \bigcup_{x \in G} x(J^i)'x^{-1}$$

where $(J^i)' = J^i \cap G'$ (G' the set of regular elements in G). An important role in the theory is played by the analog of the Weyl group in the compact case, namely by the (necessarily finite) group $W(G, J) = J^*/J$ (J^* the normalizer in G of the given Cartan subgroup J of G); see 1.4.2. The section concludes with a discussion of semi-simple and unipotent elements in G and the characterization of closed conjugacy classes in terms of semi-simple elements.

1.1 Preliminaries

1.1.1 The Structure of Complex Semi-Simple Lie Algebras

The purpose of this number is to write down some of what the reader is expected to be familiar with as far as the theory of semi-simple Lie algebras goes. Those proofs which are not given may be found in Helgason [2] or Jacobson [3].

Let $\mathfrak{g}_c$ be a semi-simple Lie algebra over the field **C** of complex numbers. Let $\mathfrak{j}_c \subset \mathfrak{g}_c$ be a *Cartan subalgebra* of $\mathfrak{g}_c$, that is, $\mathfrak{j}_c$ is maximal abelian and for each $H \in \mathfrak{j}_c$, the endomorphism ad (H) of $\mathfrak{g}_c$ is semi-simple. Let α be a linear function on the complex vector space $\mathfrak{j}_c$ and let $\mathfrak{g}_c^\alpha$ denote the linear subspace of $\mathfrak{g}_c$ given by $\mathfrak{g}_c^\alpha = \{X \in \mathfrak{g}_c : [H, X] = \alpha(H)X$ for all $H \in \mathfrak{j}_c\}$. The linear function α is called a *root* (of $\mathfrak{g}_c$ with respect to $\mathfrak{j}_c$) if $\mathfrak{g}_c^\alpha \neq \{0\}$ and in that case $\mathfrak{g}_c^\alpha$ is called a *root subspace.* By Φ we shall understand the set of non-zero roots of the pair $(\mathfrak{g}_c, \mathfrak{j}_c)$. [Abusing the language, we shall often refer to Φ as the 'set of roots' of the pair $(\mathfrak{g}_c, \mathfrak{j}_c)$.] Let B denote the Killing form on $\mathfrak{g}_c \times \mathfrak{g}_c$. Since $\mathfrak{g}_c$ is semi-simple, B is non-degenerate on $\mathfrak{g}_c \times \mathfrak{g}_c$. The restriction of B to $\mathfrak{j}_c \times \mathfrak{j}_c$ is also non-degenerate and thus for each **C**-linear function α there is a unique element $H_\alpha \in \mathfrak{j}_c$ such that $B(H, H_\alpha) = \alpha(H)$ for all $H \in \mathfrak{j}_c$. Letting $(\alpha, \beta) = B(H_\alpha, H_\beta)$ for all $\alpha, \beta \in \mathfrak{j}_c^\vee$ (the *dual* of $\mathfrak{j}_c$), it follows that (α, β) is a non-degenerate symmetric bilinear form on $\mathfrak{j}_c^\vee$.

Theorem 1.1.1.1 Let $\mathfrak{g}_c$ be a semi-simple Lie algebra over **C**, $\mathfrak{j}_c$ a Cartan subalgebra of $\mathfrak{g}_c$.

(i) The Lie algebra $\mathfrak{g}_c$ admits the root space decomposition $\mathfrak{g}_c = \mathfrak{j}_c + \sum_{\alpha\in\Phi} \mathfrak{g}_c^\alpha$ (direct sum).

(ii) The root spaces $\mathfrak{g}_c^\alpha$ $(\alpha \in \Phi)$ are one dimensional.

(iii) If α and β are any two roots (including 0) and $\beta \neq -\alpha$, then $\mathfrak{g}_c^\alpha$ and $\mathfrak{g}_c^\beta$ are orthogonal relative to B.

(iv) If α is a non-zero root, then the only multiples $c\alpha$ of α which are roots are α, 0, and $-\alpha (c \in \mathbf{C})$.

(v) For each $\alpha \in \Phi$ a vector $X_\alpha \in \mathfrak{g}_c^\alpha$ can be chosen so that for all $\alpha, \beta \in \Phi$, we have

(1) $[X_\alpha, X_{-\alpha}] = H_\alpha$, $[H, X_\alpha] = \alpha(H)X_\alpha$ (all $H \in \mathfrak{j}_c$);

(2) $[X_\alpha, X_\beta] = 0$ if $\alpha + \beta \neq 0$ and $\alpha + \beta \notin \Phi$;

(3) $[X_\alpha, X_\beta] = N_{\alpha,\beta}X_{\alpha+\beta}$ if $\alpha + \beta \in \Phi$ where the real constants $N_{\alpha,\beta}$ verify $N_{\alpha,\beta} = -N_{-\alpha,-\beta} = -N_{\beta,\alpha}$ and $N_{-\alpha,\alpha+\beta} = N_{\alpha+\beta,-\beta} = N_{-\beta,-\alpha}$.

(vi) Let $\beta \in \Phi$ and consider the set of all roots of the form $\beta + n\alpha$ (n an integer, α a non-zero root). The set $\beta + n\alpha$ is an uninterrupted progression and if $p \leqslant n \leqslant q$, then

$$-2\,\frac{\beta(H_\alpha)}{\alpha(H_\alpha)} = p + q$$

where $N_{\alpha,\beta}^2 = (q(1-p)/2)\alpha(H_\alpha)$.

The elements $\{X_\alpha : \alpha \in \Phi\}$ per Theorem 1.1.1.1(v) will be called a *Weyl basis* of $\mathfrak{g}_c$ mod $\mathfrak{j}_c$.

Theorem 1.1.1.2 Let $\mathfrak{g}_c$ be a semi-simple Lie algebra over $\mathbf{C}$, $\mathfrak{j}_c$ a Cartan subalgebra of $\mathfrak{g}_c$; put $\mathfrak{j}^* = \sum_{\alpha \in \Phi} \mathbf{R}H_\alpha$.

(i) The restriction of the Killing form B of $\mathfrak{g}_c$ to $\mathfrak{j}_c$ is real and strictly positive definite on $\mathfrak{j}^* \times \mathfrak{j}^*$.

(ii) The Cartan subalgebra $\mathfrak{j}_c$ splits: $\mathfrak{j}_c = \mathfrak{j}^* + \sqrt{-1}\,\mathfrak{j}^*$ (direct sum).

Let E be a finite dimensional vector space over the field $\mathbf{R}$ of real numbers, $E^\vee$ the dual space of E. The space E is said to be *ordered* by a relation $>$ if: (1) E is totally ordered by $>$; (2) $X > 0$, $Y > 0$ implies $X + Y > 0$ ($X, Y \in E$); (3) If $X > 0$ and r is a positive real number, then $rX > 0$ ($X \in E$). An element X in E is called *positive* if $X > 0$. There are two standard ways to introduce an order in $E^\vee$. Let $X_1, \ldots, X_n$ be a basis for E; define a function $\lambda \in E^\vee$ to be positive if $\lambda \neq 0$ and $\lambda(X_i) > 0$ where i is the least index such that $\lambda(X_i) \neq 0$ – this is called the lexicographic order in $E^\vee$ corresponding to the basis $\{X_1, \ldots, X_n\}$. On the other hand, let $\{\lambda_1, \ldots, \lambda_n\}$ be a basis for $E^\vee$. Since every $\lambda \in E^\vee$ can be written $\lambda = \sum_{i=1}^n r_i\lambda_i (r_i \in \mathbf{R})$, we may define λ to be positive if $\lambda \neq 0$ and $r_i > 0$ where i is the least index such that $r_i \neq 0$. The order so defined in $E^\vee$ is called the lexicographic order in $E^\vee$ corresponding to the basis $\{\lambda_1, \ldots, \lambda_n\}$. Let F be a subspace of E. Suppose that the duals $E^\vee$ and $F^\vee$ have been turned into ordered vector spaces. The orderings are said to be *compatible* if $\lambda \in E^\vee$ is positive whenever its restriction to F is positive. For example, let $\{X_1, \ldots, X_n\}$ be a basis for E such that $\{X_1, \ldots, X_m\}$ is a basis for F. The lexicographic orderings on $F^\vee$ and $E^\vee$ with respect to these bases are compatible.

Returning to the pair $(\mathfrak{g}_c, \mathfrak{j}_c)$, by Theorem 1.1.1.2 Φ may be regarded as a subset of the dual space of $\mathfrak{j}^*$. Evidently the dual space of $\mathfrak{j}^*$ may be identified with the set of all linear functions on $\mathfrak{j}_c$ whose restrictions to $\mathfrak{j}^*$ are real valued. Hence an order in the dual of $\mathfrak{j}^*$ induces an order on Φ.

Theorem 1.1.1.3 Let Φ^+ be the set of all positive roots of the pair $(\mathfrak{g}_c, \mathfrak{j}_c)$ under any given ordering of Φ – then there exists a system $\{\alpha_1, \ldots, \alpha_l\}$ of roots with the following properties.

(i) Each $\alpha_i \in \Phi^+$ and no α_i can be written as the sum of two other positive roots ($i = 1, \ldots, l$).

(ii) The α_i form a basis for $\mathfrak{j}_c^\vee$.

(iii) Every root α can be written in the form $\alpha = \sum_{i=1}^l n_i\alpha_i$ where the n_i are integers which are either all non-positive or all non-negative.

(iv) The system $\{\alpha_1, \ldots, \alpha_l\}$ is unique apart from a permutation. Moreover, Φ^+ is also the set of all positive roots in the lexicographic order corresponding to $\{\alpha_1, \ldots, \alpha_l\}$.

A system $\{\alpha_1, \ldots, \alpha_l\}$ of roots with the properties listed in Theorem 1.1.1.3 is called a *fundamental system* of roots. The matrix (A_{ij}), $A_{ij} = 2(\alpha_i, \alpha_j)/(\alpha_i, \alpha_i)$, is called a *Cartan matrix* for $\mathfrak{g}_c$ (relative to $\mathfrak{j}_c$). The Cartan matrix is non-singular. The off diagonal entries of this matrix are $A_{ij} = 0, -1, -2,$ or -3. Choose $X_{\alpha_i} \in \mathfrak{g}_c^{\alpha_i}$, $X_{-\alpha_i} \in \mathfrak{g}_c^{-\alpha_i}$ so that $[X_{\alpha_i}, X_{-\alpha_i}] = H_{\alpha_i}$. Let $X_i = X_{\alpha_i}$, $Y_i = 2X_{-\alpha_i}/(\alpha_i, \alpha_i)$, $H_i = 2H_{\alpha_i}/(\alpha_i, \alpha_i)$. The X_i, Y_i, H_i have the multiplication table for the three dimensional simple Lie algebra $\mathfrak{sl}(2, \mathbf{C})$ since: $[H_i, X_i] = 2X_i$, $[H_i, Y_i] = -2Y_i$, $[X_i, Y_i] = H_i$. The following relations hold among the X_i, Y_i, H_i:
(1) $[H_i, H_j] = 0$; (2) $[X_i, Y_j] = \delta_{ij}H_j$; (3) $[H_i, X_j] = A_{ij}X_j$;
(4) $[H_i, Y_j] = -A_{ij}Y_j$.

Theorem 1.1.1.4 Let $\{\alpha_1, \ldots, \alpha_l\}$ be a fundamental system of roots of $\mathfrak{g}_c$ relative to $\mathfrak{j}_c$ and let X_i, Y_i, H_i $(i = 1, \ldots, l)$ be as above – then the $3l$ elements X_i, Y_i, H_i generate $\mathfrak{g}_c$. For each positive root α we can select a representation of $\alpha = \alpha_{i_1} + \ldots + \alpha_{i_k}$ so that $\alpha_{i_1} + \ldots + \alpha_{i_m}$ is a root for every $m \leqslant k$. The elements H_i, $[\ldots[X_{i_1}, X_{i_2}]\ldots X_{i_k}]$, $[\ldots[Y_{i_1}, Y_{i_2}]\ldots Y_{i_k}]$ determined by the positive roots α form a basis for $\mathfrak{g}_c$. The multiplication table for this basis has rational coefficients which are determined by the Cartan matrix (A_{ij}).

Let E be a finite dimensional vector space over the field $\mathbf{R}$. A *complex structure* on E is an $\mathbf{R}$-linear endomorphism J of E such that $J^2 = -I$, where I is the identity mapping of E. A vector space E over $\mathbf{R}$ with a complex structure J can be turned into a vector space E^C over $\mathbf{C}$ by putting

$$(r + \sqrt{-1}s)X = rX + sJX \; (X \in E;\, r, s \in \mathbf{R}).$$

We call E^C the complex vector space associated with E.

If E is a finite dimensional vector space over $\mathbf{C}$, by restricting scalar multiplication to $\mathbf{R}$ we get a vector space E^R over $\mathbf{R}$ – then E^R has a complex structure J, namely multiplication by $\sqrt{-1}$. Evidently $(E^R)^C = E$.

A Lie algebra $\mathfrak{g}$ over $\mathbf{R}$ is said to have a complex structure J if: (1) J is a complex structure of the vector space $\mathfrak{g}$; (2) $\operatorname{ad}(X) \circ J = J \circ \operatorname{ad}(X)$ for all $X \in \mathfrak{g}$. The latter condition implies that $[JX, JY] = -[X, Y]$. Moreover $\mathfrak{g}^C$ is a Lie algebra over $\mathbf{C}$. On the other hand, if $\mathfrak{g}_c$ is a Lie algebra over $\mathbf{C}$, then the complex structure J on $\mathfrak{g}_c^R$ corresponding to multiplication by $\sqrt{-1}$ on $\mathfrak{g}_c$ turns $\mathfrak{g}_c^R$ into a Lie algebra over $\mathbf{R}$ with complex structure J.

Let $\mathfrak{g}_c$ be a Lie algebra over $\mathbf{C}$. A *real form* of $\mathfrak{g}_c$ is a subalgebra $\mathfrak{g}$ of the real Lie algebra $\mathfrak{g}_c^R$ such that $\mathfrak{g}_c^R = \mathfrak{g} + \sqrt{-1}\mathfrak{g}$ (vector space direct sum). Hence each $Z \in \mathfrak{g}_c$ can be written uniquely

$$Z = X + \sqrt{-1}Y \quad (X, Y \in \mathfrak{g}).$$

The map σ of $\mathfrak{g}_c$ onto itself given by $X + \sqrt{-1}\,Y \mapsto X - \sqrt{-1}\,Y$ is an automorphism of $\mathfrak{g}_c^R$.

Theorem 1.1.1.5 Every semi-simple Lie algebra $\mathfrak{g}_c$ over $\mathbf{C}$ has a compact real form $\mathfrak{u}$.

Thus, in case $\mathfrak{g}_c$ is a semi-simple Lie algebra over $\mathbf{C}$, Theorem 1.1.1.5 guarantees the existence of a semi-simple Lie algebra $\mathfrak{u} \subset \mathfrak{g}_c^R$ whose Killing form is strictly negative definite and whose complexification is $\mathfrak{g}_c$. If the vectors X_α are selected as in Theorem 1.1.1.1(v), then

$$\mathfrak{u} = \sum_{\alpha \in \Phi} \mathbf{R}(\sqrt{-1}\, H_\alpha) + \sum_{\alpha \in \Phi} \mathbf{R}(X_\alpha - X_{-\alpha}) + \sum_{\alpha \in \Phi} \mathbf{R}(\sqrt{-1}\,(X_\alpha + X_{-\alpha}))$$

is a compact real form for $\mathfrak{g}_c$.

Let $\mathfrak{g}$ be a semi-simple Lie algebra over $\mathbf{R}$, $\mathfrak{g}_c$ its complexification, σ the conjugation of $\mathfrak{g}_c$ with respect to $\mathfrak{g}$. A direct decomposition $\mathfrak{g} = \mathfrak{k} + \mathfrak{p}$ into a subalgebra $\mathfrak{k}$ and a vector subspace $\mathfrak{p}$ is called a *Cartan decomposition* if there exists a compact real form $\mathfrak{u}$ such that:

(1) $\sigma(\mathfrak{u}) \subset \mathfrak{u}$; (2) $\mathfrak{u} \cap \mathfrak{g} = \mathfrak{k}$; (3) $\sqrt{-1}\,\mathfrak{u} \cap \mathfrak{g} = \mathfrak{p}$.

Theorem 1.1.1.6 Every semi-simple Lie algebra $\mathfrak{g}$ over $\mathbf{R}$ admits a Cartan decomposition. If

$$\mathfrak{g} = \mathfrak{k}_1 + \mathfrak{p}_1, \ \mathfrak{g} = \mathfrak{k}_2 + \mathfrak{p}_2$$

are two Cartan decompositions of $\mathfrak{g}$, then there exists an element x in the adjoint group $\mathrm{Int}\,(\mathfrak{g})$ of $\mathfrak{g}$ such that $x(\mathfrak{k}_1) = \mathfrak{k}_2$, $x(\mathfrak{p}_1) = \mathfrak{p}_2$.

Theorem 1.1.1.7 Let $\mathfrak{g}$ be a semi-simple Lie algebra over $\mathbf{R}$ which is the direct sum $\mathfrak{g} = \mathfrak{k} + \mathfrak{p}$ where $\mathfrak{k}$ is a subalgebra and $\mathfrak{p}$ a vector subspace. The following conditions are equivalent:

(i) $\mathfrak{g} = \mathfrak{k} + \mathfrak{p}$ is a Cartan decomposition of $\mathfrak{g}$;
(ii) $B(Y, Y) < 0$ for $Y \neq 0$ in $\mathfrak{k}$,
$B(Z, Z) > 0$ for $Z \neq 0$ in $\mathfrak{p}$

and the map

$$\theta,\ Y + Z \mapsto Y - Z\ (Y \in \mathfrak{k},\ Z \in \mathfrak{p}),$$

is an automorphism of $\mathfrak{g}$.

Corollary 1.1.1.8 Let $\mathfrak{g}_c$ be a semi-simple Lie algebra over $\mathbf{C}$ and let $\mathfrak{u}$ be a compact real form of $\mathfrak{g}_c$. Let J denote the complex structure of $\mathfrak{g}_c^R$ which corresponds to multiplication by $\sqrt{-1}$ on $\mathfrak{g}_c$ – then $\mathfrak{g}_c^R = \mathfrak{u} + J\mathfrak{u}$ is a Cartan decomposition of $\mathfrak{g}_c^R$.

Let $\mathfrak{g}_c$ be a semi-simple Lie algebra over $\mathbf{C}$. A real form $\mathfrak{g}$ of $\mathfrak{g}_c$ is called *normal* if in each Cartan decomposition $\mathfrak{g} = \mathfrak{k} + \mathfrak{p}$ the space $\mathfrak{p}$ contains a maximal abelian subalgebra of $\mathfrak{g}$.

Theorem 1.1.1.9 Each semi-simple Lie algebra over $\mathbf{C}$ has a normal real form which is unique up to isomorphism.

If the vectors $X_\alpha(\alpha \in \Phi)$ are selected as in Theorem 1.1.1.1(v), then

$$\mathfrak{g} = \sum_{\alpha \in \Phi} \mathbf{R}H_\alpha + \sum_{\alpha \in \Phi} \mathbf{R}X_\alpha$$

is a normal real form for $\mathfrak{g}_c$.

We shall now review Iwasawa's structure theorem for semi-simple Lie algebras over $\mathbf{R}$ (cf. Helgason [2, p. 224]). Thus let $\mathfrak{g}$ be a semi-simple Lie algebra over $\mathbf{R}$, $\mathfrak{g} = \mathfrak{k} + \mathfrak{p}$ a Cartan decomposition of $\mathfrak{g}$, $\mathfrak{u} = \mathfrak{k} + \sqrt{-1}\,\mathfrak{p}$ the associated compact real form for $\mathfrak{g}_c$ ($\mathfrak{g}_c$ the complexification of $\mathfrak{g}$), and let σ and τ denote the conjugations of $\mathfrak{g}_c$ with respect to $\mathfrak{g}$ and $\mathfrak{u}$, respectively. The automorphism $\sigma\tau$ of $\mathfrak{g}_c$ will be denoted by θ. Let $\mathfrak{a}_\mathfrak{p}$ be a maximal abelian subspace of $\mathfrak{p}$. Let $\mathfrak{a}$ be any maximal abelian subalgebra of $\mathfrak{g}$ containing $\mathfrak{a}_\mathfrak{p}$. Put $\mathfrak{a}_\mathfrak{k} = \mathfrak{a} \cap \mathfrak{k}$, $\mathfrak{a}_\mathfrak{p} = \mathfrak{a} \cap \mathfrak{p}$ – then $\mathfrak{a} = \mathfrak{a}_\mathfrak{k} + \mathfrak{a}_\mathfrak{p}$ (direct sum of vector spaces). Let $\mathfrak{a}_c$, $\mathfrak{a}_{\mathfrak{p}_c}$, $\mathfrak{a}_{\mathfrak{k}_c}$, $\mathfrak{k}_c$ and $\mathfrak{p}_c$ denote the subspaces of $\mathfrak{g}_c$ generated by $\mathfrak{a}$, $\mathfrak{a}_\mathfrak{p}$, $\mathfrak{a}_\mathfrak{k}$, $\mathfrak{k}$, and $\mathfrak{p}$, respectively. The algebra $\mathfrak{a}_c$ is a Cartan subalgebra of $\mathfrak{g}_c$ and $\mathfrak{a}^* = \mathfrak{a}_\mathfrak{p} + \sqrt{-1}\,\mathfrak{a}_\mathfrak{k}$. Select compatible orderings in the dual spaces of $\mathfrak{a}_\mathfrak{p}$ and $\mathfrak{a}^*$, respectively. Since each root $\alpha \in \Phi$ is real valued on $\mathfrak{a}^*$ we get in this way an ordering of Φ. Let Φ^+ denote the set of positive roots. Divide the positive roots Φ^+ into two subclasses P_+ and P_- where P_+ consists of those positive roots which do not vanish identically on $\mathfrak{a}_\mathfrak{p}$ and P_- consists of those that do. Let $\mathfrak{n}_c^+ = \sum_{\alpha \in P_+} \mathfrak{g}_c^\alpha$, $\mathfrak{n}^+ = \mathfrak{g} \cap \mathfrak{n}_c^+$, $\mathfrak{s} = \mathfrak{a}_\mathfrak{p} + \mathfrak{n}^+$ – then $\mathfrak{n}_c^+$ and $\mathfrak{n}^+$ are nilpotent Lie algebras, $\mathfrak{s}$ is a solvable Lie algebra, and $\mathfrak{g} = \mathfrak{k} + \mathfrak{a}_\mathfrak{p} + \mathfrak{n}^+$ (vector space direct sum). This decomposition of $\mathfrak{g}$ is the *Iwasawa decomposition* of $\mathfrak{g}$; it depends on the choice of $\mathfrak{a}_\mathfrak{p}$ and the ordering of the roots. An alternative description of the Iwasawa decomposition will be given in 1.1.3.

Let $\mathfrak{m}$ denote the centralizer of $\mathfrak{a}_\mathfrak{p}$ in $\mathfrak{k}$ – then

$$\mathfrak{m}_c = \mathfrak{a}_{\mathfrak{k}_c} + \sum_{\alpha \in P_-} (\mathfrak{g}_c^\alpha + \mathfrak{g}_c^{-\alpha}).$$

Let $\mathfrak{g}$ be a semi-simple Lie algebra over $\mathbf{R}$, $\mathfrak{g} = \mathfrak{k} + \mathfrak{p}$ a Cartan decomposition of $\mathfrak{g}$. Retaining the above notations, let Φ denote the set of non-zero roots of the pair $(\mathfrak{g}_c, \mathfrak{a}_c)$.

Proposition 1.1.1.10 Let Γ be a subset of Φ satisfying the following conditions: If $\alpha \in \Gamma$, then $-\alpha \in \Gamma$; if $\alpha, \beta \in \Gamma$ and $\alpha + \beta \in \Phi$, then $\alpha + \beta \in \Gamma$. Let $\mathfrak{g}_c^\Gamma$ be the subspace of $\mathfrak{g}_c$ generated by the H_α and the X_α for $\alpha \in \Gamma$.

(i) The subspace $\mathfrak{g}_c^\Gamma$ is a semi-simple subalgebra of $\mathfrak{g}_c$ and $\mathfrak{g}_c^\Gamma \cap \mathfrak{a}_c$ is a Cartan subalgebra of $\mathfrak{g}_c^\Gamma$.

(ii) If $\mathfrak{g}_c^\Gamma$ is σ-stable, then $\mathfrak{g}^\Gamma = \mathfrak{g}_c^\Gamma \cap \mathfrak{g}$ is a real form of $\mathfrak{g}_c^\Gamma$.

(iii) If $\mathfrak{g}_c^\Gamma$ is σ-stable and τ-stable, then $\mathfrak{g}^\Gamma \cap \mathfrak{k} + \mathfrak{g}^\Gamma \cap \mathfrak{p}$ is a Cartan decomposition for $\mathfrak{g}^\Gamma$.

[Modulo (i), statements (ii) and (iii) are clear; in turn, the validity of (i) can be established, for instance, by verifying that the Killing form of $\mathfrak{g}_c^\Gamma$ (which is surely a subalgebra) is non-degenerate.]

1.1.2 Root Systems I – Basic Properties

Let E be a finite dimensional vector space over $\mathbf{R}$; let γ be a non-zero element in E – then by a *reflection with respect to* γ one understands any automorphism w_γ of E such that:

(i) $w_\gamma(\gamma) = -\gamma$;

(ii) The set $\mathscr{H}$ of elements in E which are invariant under w_γ is a hyperplane in E.

If w_γ is a reflection with respect to a non-zero vector γ in E, then it is clear that the hyperplane $\mathscr{H}$ is a supplement to the line $\mathbf{R}\gamma$ and that w_γ is of order 2.

Definition A subset Φ of E is said to be a *root system* in E if:

(i) Φ is finite, generates E, and does not contain 0;

(ii) For every $\alpha \in \Phi$, there exists a reflection w_α with respect to α which stabilizes Φ;

(iii) For every $\alpha, \beta \in \Phi$, the vector $w_\alpha(\beta) - \beta$ is an integral multiple of α.

Note The reflection $w_\alpha(= w$, say) in condition (ii) is unique. Thus suppose that w_1 and w_2 are two reflections with the property in question; let $\phi = w_1 w_2$ – then the automorphism ϕ leaves Φ stable, fixes α, and induces the identity on $E/\mathbf{R}\alpha$. The last two properties imply that the eigenvalues of ϕ are equal to 1; on the other hand, since ϕ stabilizes Φ, the finiteness of Φ, together with the fact that Φ spans E, ensures the existence of an integer $n \geqslant 1$ such that $\phi^n = 1$ – thus $\phi = 1$ and so $w_1 = w_2$.

Remark Most of what we shall have to say in the following is true if E is a finite dimensional vector space over $\mathbf{Q}$ (an observation which admits the obvious implication as far as the theory of semi-simple Lie algebras over an algebraically closed field $\mathbf{k}$ of characteristic zero is concerned . . .).

Let Φ be a root system in E – then the dimension of E is called the *rank* of Φ and the elements of Φ are called the *roots* of E relative to Φ. If $\alpha \in \Phi$, then the *co-root* associated with α is, by definition, that element $\alpha^\vee$ in the dual $E^\vee$ of E which vanishes on $\mathscr{H}$ (the hyperplane in E which is left invariant by the reflection w_α) and takes the value 2 at α (hence $w_\alpha(\gamma) = \gamma - \langle \gamma, \alpha^\vee \rangle \alpha$ (all $\gamma \in E$)); condition (iii) supra is thus equivalent to the requirement that $\langle \beta, \alpha^\vee \rangle \in \mathbf{Z}$ (all $\alpha, \beta \in \Phi$). It is not difficult to show that the collection $\Phi^\vee = \{\alpha^\vee : \alpha \in \Phi\}$ is, in its own right, a root system in $E^\vee$; we shall refer to $\Phi^\vee$ as the system of *co-roots* (associated with Φ).

Definition Let Φ be a root system in E. The group $W(= W_\Phi)$ generated by the reflections $\{w_\alpha\}$ is called the *Weyl group* of Φ.

Observe that W is finite since it may be viewed as a subgroup of the group of permutations of Φ.

Here is a useful device. Let Φ be a root system in E – then there exists a positive non-degenerate symmetric bilinear form $(.\,,.)$ on E which is invariant under the Weyl group W of Φ (this is obvious, W being finite). Such a form $(.\,,.)$ (which we assume to be fixed once and for all) equips E with the structure of a Euclidean space relative to which the elements of W are orthogonal transformations; in particular a reflection $w_\alpha(\alpha \in \Phi)$ may now be regarded as the reflection in the hyperplane perpendicular to α (of course we have that $w_\alpha(\gamma) = \gamma - 2(\gamma, \alpha)/(\alpha, \alpha)\alpha$ (all $\gamma \in E$)). The form $(.\,,.)$ may be used to identify E with its dual $E^\vee$; under such an identification the co-root $\alpha^\vee$ associated with a given α in Φ corresponds to the element $2\alpha/(\alpha, \alpha)$ in E; consequently condition (iii) above then becomes the crystallographic requirement: $2(\alpha, \beta)/(\alpha, \alpha) \in \mathbf{Z}$ (all $\alpha, \beta \in \Phi$).

Remark It is clear that by identifying E and its dual $E^\vee$ via the form $(.\,,.)$, one may speak of a root system dual to Φ; such identifications will usually be made without comment. [The root system dual to Φ is, however, not to be confused with the system of co-roots associated with Φ.]

A root system Φ in E is the *direct sum* of $\Phi_1 \subset E^1$ and $\Phi_2 \subset E^2$ if $E = E^1 + E^2$ (direct sum) and $\Phi = \Phi_1 \cup \Phi_2$. The root system Φ is said to be *irreducible* if it is not the direct sum of two sub-systems. It is evident that every root system is the direct sum of irreducible root systems.

Definition A system of roots Φ is said to be *reduced* if α and $-\alpha$ are the only roots in Φ proportional to α (all $\alpha \in \Phi$).

If Φ is not reduced, then it contains two proportional roots α and $t\alpha$ with $0 < t < 1$. An application of condition (iii) above shows that $2t \in \mathbf{Z}$, hence $t = \frac{1}{2}$ and thus the roots proportional to α are: $-\alpha, -\alpha/2, \alpha/2, \alpha$. On the other hand Φ always contains two reduced systems, viz. $\Phi_1 = \{\alpha : 2\alpha \text{ is not in } \Phi\}$, $\Phi_2 = \{\alpha : \alpha/2 \text{ is not in } \Phi\}$.

Example If $\mathfrak{g}_c$ is a semi-simple Lie algebra over $\mathbf{C}$, $\mathfrak{j}_c$ a Cartan subalgebra, Φ the set of non-zero roots of the pair $(\mathfrak{g}_c, \mathfrak{j}_c)$, $\mathfrak{j}^* = \sum_{\alpha \in \Phi} \mathbf{R}H_\alpha$, then taking for E the dual of $\mathfrak{j}^*$ (equipped with the Euclidean structure derived from the transport of the Killing form $B|\mathfrak{j}^*$) and Φ, we obtain an example of a reduced system of roots.

Important examples of non-reduced systems of roots will be exhibited presently.

Let Φ be a root system in E. If α, β are two roots, define the *Cartan integer* $A_{\alpha,\beta}$ by

$$A_{\alpha,\beta} = 2\frac{(\alpha, \beta)}{(\alpha, \alpha)} \in \mathbf{Z}.$$

Let $|\alpha| = (\alpha, \alpha)^{1/2}$ denote the length of α and let ϕ denote the angle between α and β (relative to the given Euclidean structure $(.\,,.)$); thus $(\alpha, \beta) = |\alpha||\beta| \cos(\phi)$ and so

$$A_{\alpha,\beta} = 2\frac{|\beta|}{|\alpha|}\cos(\phi).$$

It follows that $A_{\alpha,\beta}A_{\beta,\alpha} = 4\cos^2\phi$; since $A_{\alpha,\beta}$ is an integer $4\cos^2\phi$ can only take on the values 0, 1, 2, 3, or 4, the last case occurring only if α and β are proportional. Restricting ourselves to non-proportional α and β, one has the following possibilities (up to a switch of α and β).

(1) $A_{\beta,\alpha} = 0$, $A_{\alpha,\beta} = 0$, $\phi = \pi/2$.
(2) $A_{\beta,\alpha} = 1$, $A_{\alpha,\beta} = 1$, $\phi = \pi/3$, $|\beta| = |\alpha|$.
(3) $A_{\beta,\alpha} = -1$, $A_{\alpha,\beta} = -1$ $\phi = 2\pi/3$, $|\beta| = |\alpha|$.
(4) $A_{\beta,\alpha} = 1$, $A_{\alpha,\beta} = 2$, $\phi = \pi/4$, $|\beta| = \sqrt{2}|\alpha|$.
(5) $A_{\beta,\alpha} = -1$, $A_{\alpha,\beta} = -2$, $\phi = 3\pi/4$, $|\beta| = \sqrt{2}|\alpha|$.
(6) $A_{\beta,\alpha} = 1$, $A_{\alpha,\beta} = 3$, $\phi = \pi/6$, $|\beta| = \sqrt{3}|\alpha|$.
(7) $A_{\beta,\alpha} = -1$, $A_{\alpha,\beta} = -3$, $\phi = 5\pi/6$, $|\beta| = \sqrt{3}|\alpha|$.

Proposition 1.1.2.1 Let α, β be two non-proportional roots. If $A_{\alpha,\beta} > 0$, then $\alpha - \beta$ is a root.

Proof The preceding list shows that either $A_{\alpha,\beta} = 1$ or $A_{\beta,\alpha} = 1$. In the first case, $\alpha - \beta = -(\beta - A_{\alpha,\beta}\alpha) = -w_\alpha(\beta) \Rightarrow \alpha - \beta \in \Phi$. In the second case $\alpha - \beta = w_\beta(\alpha) \in \Phi$. ☐

Definition Let Φ be a root system in E. A subset Ψ of Φ is called a *fundamental system* of roots for Φ if:

(i) Ψ is a vector space basis for E;
(ii) Every root can be written as a linear combination

$$\sum_{\alpha\in\Psi} n_\alpha \alpha$$

where n_α are all integers of the same sign.

The elements in a fundamental system of roots are said to be *simple*.

[Of course in the case of the Example above this definition of a fundamental system of roots tallies with the usual one (Theorem 1.1.1.3).]

A subset Φ^+ of Φ is said to be a *positive system* if it consists of those roots which are positive relative to some ordering of E. [Of course this amounts to the specification of a subset of E which is closed under addition and under multiplication by positive scalars and satisfies trichotomy.] Evidently every fundamental system of roots for Φ is contained in a unique positive system; the converse is also true.

Proposition 1.1.2.2 Let Φ be a root system in E, Φ^+ a system of positive roots – then Φ^+ contains a unique fundamental system Ψ.

Proof Let Ψ be a subset of Φ^+ which generates Φ^+ under positive integral linear combinations and is minimal relative to this property – then $\alpha, \beta \in \Psi$, $\alpha \neq \beta \Rightarrow (\alpha, \beta) \leqslant 0$. [Suppose the contrary were true so that $A_{\alpha,\beta} > 0$; in view of Proposition 1.1.2.1, either $\alpha - \beta$ or $\beta - \alpha \in \Phi^+$; for sake of argument, assume that $\alpha - \beta \in \Phi^+$ – then $\alpha - \beta = \sum n_\gamma \gamma$ $(\gamma \in \Psi,\ n_\gamma \in \mathbf{N})$. If $n_\alpha = 0$ the last equation, written suitably, expresses α as a positive integral linear combination of the other elements of Ψ, a contradiction to the minimality of Ψ; if $n_\alpha \geqslant 1$, the equation in question expresses zero as a positive combination of elements of Ψ, hence of Φ^+, a contradiction once again.] This being so, it now follows that the elements of Ψ are linearly independent. [A linear relation on Ψ may be written $\sum r_\alpha \alpha = \sum r_\beta \beta$ with the two sums over disjoint subsets of Ψ and $r_\alpha, r_\beta \geqslant 0$; thus $(\sum r_\alpha \alpha, \sum r_\alpha \alpha) = (\sum r_\alpha \alpha, \sum r_\beta \beta) \leqslant 0$ and so $r_\alpha = 0$ (all α), the α's being positive; analogously we find that $r_\beta = 0$ (all β) . . . .] The very definition of 'fundamental system' implies that any fundamental system contained in Φ^+ consists of those elements in Φ^+ which are not positive integral linear combinations of others, hence is uniquely determined by Φ^+. □

Let Φ be a root system in E, $\Psi = \{\alpha_1, \ldots, \alpha_l\}$ a fundamental system of roots for $\Phi(l(= l_\Phi) = \dim(E))$, Φ^+ the corresponding set of positive roots.

Proposition 1.1.2.3 The Weyl group W of Φ is generated by the reflections w_{α_i} $(i = 1, \ldots, l)$. Moreover for any $\alpha \in \Phi$ there exists a $w \in W$ and an index i $(1 \leqslant i \leqslant l)$ such that α is proportional to $w(\alpha_i)$.

Proof It suffices to work with positive α. Write $\alpha = \sum_i n_i \alpha_i$ where the n_i are non-negative integers; put $N(\alpha) = \sum_i n_i$. If W^* is the subgroup of W generated by $\{w_{\alpha_1}, \ldots, w_{\alpha_l}\}$, then we are going to first prove by induction on $N(\alpha)$ that $\alpha = rw(\alpha_i)$ for some $w \in W^*$, some i $(1 \leqslant i \leqslant l)$ and a suitable positive number r. This is obvious if $N(\alpha) = 1$. So now suppose that $N(\alpha) \geqslant 2$ and α is not proportional to any $\alpha_i (1 \leqslant i \leqslant l)$. Because (α, α) is positive, (α, α_j) is positive for some j. In addition, since α is not proportional to α_j, $n_i > 0$ for some $i \neq j$. But $w_{\alpha_j}(\alpha) = \alpha - 2\{(\alpha, \alpha_j)/(\alpha_j, \alpha_j)\}\alpha_j$ and $-w_{\alpha_j}(\alpha)$ cannot be positive (for $n_i > 0$); therefore $w_{\alpha_j}(\alpha)$ is positive and $N(\alpha) > N(w_{\alpha_j}(\alpha))$. Our assertion now follows by applying the induction hypothesis to $w_{\alpha_j}(\alpha)$. If $\alpha = rw(\alpha_i)$, then clearly $w_\alpha = ww_{\alpha_i}w^{-1} \in W^*$. As the $w_\alpha (\alpha > 0)$ generate W, this proves that $W^* = W$. □

Corollary 1.1.2.4 Suppose that $\alpha > 0$ – then there exists an $i (1 \leqslant i \leqslant l)$ and a $w \in W$ such that $w_\alpha = ww_{\alpha_i}w^{-1}$.

Proposition 1.1.2.5 The reflection $w_{\alpha_i} (i = 1, \ldots, l)$ permutes the elements of Φ^+ different from α_i, $2\alpha_i$ amongst themselves and sends α_i, $2\alpha_i$ to $-\alpha_i$, $-2\alpha_i$, respectively (of course $2\alpha_i$ may not be present).

Proof Let $\alpha \in \Phi^+$ and write $\alpha = \sum_j n_j \alpha_j (n_j \in \mathbf{N})$; plainly $w_{\alpha_i}(\alpha) = \sum_{j \neq i} n_j \alpha_j + (n_i - n)\alpha_i$ where $n = 2(\alpha, \alpha_i)/(\alpha_i, \alpha_i)$. If $\alpha \neq \alpha_i, 2\alpha_i$, then some $n_j > 0$ $(j \neq i)$ – hence $w_{\alpha_i}(\alpha) > 0$. □

Remark Let G be a group, $\{x_i : i \in I\}$ a set of generators for G; a set $\{X_r = 1\}$, each X_r a word in the x_i's, is said to be a set of *defining relations* for G if every relation in G can be derived from the equations $X_r = 1$. This being so, let us mention, in connection with Proposition 1.1.2.3 above, the following interesting fact. If m_{ij} is the (necessarily finite) order of the element $w_i w_j$ of W, then

$$w_i^2 = (w_i w_j)^{m_{ij}} = 1 \qquad (w_i = w_{\alpha_i}, w_j = w_{\alpha_j}; 1 \leqslant i, j \leqslant l)$$

is a complete set of defining relations for W. [For a proof, see Séminaire Chevalley, Classification des Groupes de Lie Algébriques, 1956–1958, École Normale Superieure, Exposé 14.]

Theorem 1.1.2.6 Let Φ be a root system in E, W its Weyl group – then any two fundamental (or positive) systems for Φ are conjugate under a unique element from W.

Proof Earlier discussion tells us that we need only consider positive systems.

(1) Let Φ^+ and $\tilde{\Phi}^+$ be two positive systems for Φ – then we claim that there is at least one $w \in W$ with the property that $w\Phi^+ = \tilde{\Phi}^+$. To see this, we shall proceed by induction on the cardinality n of the set $\Phi^+ \cap (-\tilde{\Phi}^+)$. If $n = 0$, then it is clear that $\Phi^+ = \tilde{\Phi}^+$. If $n > 0$, then there is a root $\alpha \in \Psi$ (Ψ the unique fundamental system of roots for Φ which is contained in Φ^+ (cf. Proposition 1.1.2.2)) such that $\alpha \in \Phi^+ \cap (-\tilde{\Phi}^+)$; in view of Proposition 1.1.2.5, the cardinality of the set $w_\alpha \Phi^+ \cap (-\tilde{\Phi}^+)$ cannot exceed $n - 1$, and so the same must be true of the set $\Phi^+ \cap (-w_\alpha \tilde{\Phi}^+)$; by the induction hypothesis, $w_\alpha \tilde{\Phi}^+$ is conjugate to Φ^+ – therefore $\tilde{\Phi}^+$ is too.

(2) Let Φ^+ and $\tilde{\Phi}^+$ be two positive systems for Φ – then we claim that there is at most one $w \in W$ with the property that $w\Phi^+ = \tilde{\Phi}^+$. For this purpose it will clearly be sufficient to prove that $w\Phi^+ = \Phi^+$ $(w \in W) \Rightarrow w = 1$. We may assume that Φ is reduced. This being so, let us denote by $|w|$ the number of positive roots whose transforms under w are negative – then clearly $w\Phi^+ = \Phi^+$ iff $|w| = 0$. Let $w = w_1 w_2 \ldots w_n$ be a minimal expression for w as a product of reflections corresponding to elements in the fundamental system Ψ associated with Φ^+ (when $k = 1, \ldots, n$, we write $w_k = w_{\alpha_{i(k)}}$, $1 \leqslant i(k) \leqslant l$); it is easy to see that $n = |w|$; consequently $|w| = 0 \Rightarrow w = 1$. [The fact that $|w| \leqslant n$ follows at once from Proposition 1.1.2.5. Moreover this proposition tells us that if $|w| < n$, then of necessity $w_1 \ldots w_\nu(\alpha_{i(\nu+1)}) < 0$ for some $\nu \leqslant n - 1$; since $\alpha_{i(\nu+1)} > 0$, we have $w_\mu(w_{\mu+1} \ldots w_\nu(\alpha_{i(\nu+1)})) < 0$ and $w_{\mu+1} \ldots w_\nu(\alpha_{i(\nu+1)}) > 0$ for some $\mu \leqslant \nu$, whence $w_{\mu+1} \ldots w_\nu(\alpha_{i(\nu+1)}) = \alpha_{i(\mu)}$ (Proposition 1.1.2.5 again . . .) – thus $w_{\mu+1} w_{\mu+2} \ldots w_{\nu+1} = w_\mu w_{\mu+1} \ldots w_\nu$ and so

$w = w_1 \ldots \hat{w}_\mu \ldots \hat{w}_{\nu+1} \ldots w_n$ with w_μ and $w_{\nu+1}$ missing, a contradiction to the minimality of the expression for w.]

The proof of the theorem is now complete. □

Remark Let Φ be a reduced system of roots in E, Φ^+ a positive system for Φ, Ψ the associated fundamental system. Fix an element w in W and let $w = w_1 \ldots w_n$ be a minimal expression for w as a product of reflections corresponding to elements of $\Psi(w_k = w_{\alpha_{i(k)}},\ 1 \leqslant i(k) \leqslant l)$; let $\beta_k = w_n \ldots w_{k+1}(\alpha_{i(k)})$, $k = 1, \ldots, n$ – then the β_k are positive, pairwise distinct with $w(\beta_k) < 0$ and every root $\alpha > 0$ such that $w(\alpha) < 0$ is equal to one of the β_k. [Let $\Delta(w) = \{\alpha \in \Phi^+ : w(\alpha) < 0\}$ – then, in view of what was said in part (2) of the proof of the preceding theorem, $[\Delta(w)] = |w|$. But surely $\Delta(w)$ is contained in $\{\beta_k\} \ldots$.]

Let Φ be a root system in E; per Φ, we make the following agreements.

A hyperplane in E is said to be *singular* if it is orthogonal to a root $\alpha \in \Phi$. A *Weyl chamber* $\mathscr{C}$ is a connected component of the complement of the union of the singular hyperplanes; an element of E is said to be *regular* (or *Φ-regular*) if it lies in some Weyl chamber. The preceding theorem tells us that the Weyl group W of Φ acts *simply transitively* on the Weyl chambers (i.e., there is exactly one element of the Weyl group mapping a given Weyl chamber onto another one). [Let us note that with each Weyl chamber $\mathscr{C}$ there is associated an ordering of the roots defined by: $\alpha > 0$ if $(\alpha, \gamma) > 0$ for every γ in $\mathscr{C}$. Therefore our assertion is equivalent to the following statement: Given two fundamental systems of roots Ψ_1 and Ψ_2 for Φ there exists one and only one element $w \in W$ such that $w\Psi_1 = \Psi_2$.] In particular, given a Weyl chamber $\mathscr{C}$, there exists a unique element $w^\dagger$ in W which maps $\mathscr{C}$ into its negative $-\mathscr{C}$; if Φ^+ denotes the positive system of roots determined by $\mathscr{C}$, then clearly $\iota = -w^\dagger$ is an involutory permutation of Φ^+ We shall call ι the *opposition involution* (associated with $\mathscr{C}$).

Let Φ be a root system in E, W the Weyl group of Φ; let $\Psi = \{\alpha_1, \ldots, \alpha_l\}$ be a fundamental system of roots for Φ, $\mathscr{C}$ the Weyl chamber in E associated with Ψ.

Theorem 1.1.2.7 The closure of $\mathscr{C}$ (in E) is a fundamental domain for the action of W on E, i.e. the closure of $\mathscr{C}$ meets each W-orbit exactly once.

Proof (1) Every W-orbit meets $\mathscr{C}^{cl}$ at least once. To see this, let $\check{\mathscr{C}}^{cl}$ be the dual cone of $\mathscr{C}^{cl}$ (evidently $\mathscr{C}^{cl}$ is a simplicial cone with vertex at the origin) – then we may equip E with a partial ordering by requiring that $\gamma_1 \geqslant \gamma_2$ iff $\gamma_1 - \gamma_2 \in \check{\mathscr{C}}^{cl}$. Now fix an element γ in E; among the conjugates $\tilde{\gamma}$ of γ under W having the property that $\tilde{\gamma} \geqslant \gamma$, select a maximal one $\tilde{\gamma}_0$ (relative to the partial ordering at hand) – then, for $\alpha_i \in \Psi$, we have $w_{\alpha_i}(\tilde{\gamma}_0) \not\geqslant \tilde{\gamma}_0$ which implies that $(\alpha_i, \tilde{\gamma}_0) \geqslant 0$, whence $\tilde{\gamma}_0 \in \mathscr{C}^{cl}$. Of course this serves to establish the aforestated contention.

$$\check{\mathscr{C}}^{cl} = \{x \in E \mid (x, c) \geq 0 \ \forall c \in \mathscr{C}^{cl}\}$$

(2) Every W-orbit meets $\mathscr{C}^{cl}$ exactly once. Thus suppose that we have two elements γ and $\tilde{\gamma}$ in $\mathscr{C}^{cl}$ such that $w(\gamma) = \tilde{\gamma}$ for some $w \in W$ – we are to show that $\gamma = \tilde{\gamma}$. For this purpose, it will clearly be enough to prove that w is a product of fundamental reflections all of which fix γ. We shall argue by induction on $|w|$. [In general, for any element of W, $|.|$ will denote the number of positive roots in Φ whose transforms (under the element in question) are negative.] If $|w| = 0$, then $w = 1$ and there is no problem (let us bear in mind that W acts simply transitively on the Weyl chambers). Assuming now that $|w| > 0$, choose $\alpha_i \in \Psi$ such that $w(\alpha_i) < 0$ – then $0 \geqslant (\tilde{\gamma}, w(\alpha_i)) = (\gamma, \alpha_i) \geqslant 0 \Rightarrow (\gamma, \alpha_i) = 0 \Rightarrow w_{\alpha_i}(\gamma) = \gamma$. But $ww_{\alpha_i}(\gamma) = \tilde{\gamma}$ and $|ww_{\alpha_i}| < |w|$; therefore the induction assumption applied to ww_{α_i} yields our assertion.

This proves the theorem. □

Theorem 1.1.2.8 (Chevalley) Let F be a subset of E, W_F the subgroup of W which leaves F pointwise fixed – then W_F is generated by the reflections $w_{\alpha_i}(\alpha_i \in \Psi)$ which leave F pointwise fixed.

Proof (Steinberg) To begin with, it is clear that F may be taken linearly independent; by induction it may be assumed in addition that F consists of a single element γ (say); Theorem 1.1.2.7 then tells us that there is no loss of generality in assuming that γ lies in the closed Weyl chamber $\mathscr{C}^{cl}$; the present theorem is therefore seen to be a consequence of part (2) of the proof of Theorem 1.1.2.7. □

Let Φ be a root system in E, Φ^+ a system of positive roots. The set $\mathfrak{W}(= \mathfrak{W}_\Phi)$ comprised of those elements γ in E such that $\langle \gamma, \alpha^\vee \rangle$ is an integer for all $\alpha \in \Phi$ is a group, the *weight group* of Φ; the set $\mathfrak{W}^+(=\mathfrak{W}_\Phi^+)$ of *dominant weights* consists, by definition, of those γ in $\mathfrak{W}$ such that $\langle \gamma, \alpha^\vee \rangle \geqslant 0$ for all positive α (identifying the co-root $\alpha^\vee$ via the form $(.,.)$ with $2\alpha/(\alpha, \alpha)$, it becomes clear that a γ in $\mathfrak{W}$ is dominant iff $(\gamma, \alpha) \geqslant 0$ for all $\alpha \in \Phi^+$). The subgroup $\mathfrak{w}(= \mathfrak{w}_\Phi)$ of $\mathfrak{W}$ generated by the roots $\alpha \in \Phi$ is called the *root group* of Φ. The factor group $\mathfrak{W}/\mathfrak{w}$ is finite; its order is called the *connection index* of Φ.

Let Φ be a root system in E, Ξ_1 and Ξ_2 subsets of Φ. We shall say that Ξ_1 *normalizes* Ξ_2(in Φ) if $\alpha_1 \in \Xi_1, \alpha_2 \in \Xi_2, \alpha_1 + \alpha_2 \in \Phi \Rightarrow \alpha_1 + \alpha_2 \in \Xi_2$. If moreover $\Xi_1 \supset \Xi_2$ ($\Xi_1 = \Xi_2$), then we shall say that Ξ_2 is an *ideal* in Ξ_1 (is *closed* in Φ). A subset Ξ of Φ is said to be *symmetric* if $\Xi = -\Xi$.

Let Ξ be a subset of Φ; put $\Xi_s = \Xi \cap (-\Xi)$, $\Xi_u = \Xi \cap \complement(-\Xi)$; these are disjoint subsets of Ξ whose union is Ξ. Now suppose that Ξ is closed (in Φ) – then Ξ_s and Ξ_u are closed and Ξ_u is even an ideal in Ξ. [It is plain that Ξ_s is closed. If Ξ_u were not an ideal in Ξ, there would exist $\alpha \in \Xi$, $\beta \in \Xi_u$ such that $\alpha + \beta \in \Xi_s$ – but then $-\beta = -(\alpha + \beta) + \alpha \in \Xi$, which implies that $\beta \in \Xi_s$.]

Let $\Psi = \{\alpha_1, \ldots, \alpha_l\}$ be a fundamental system of roots for Φ, Φ^+ the corresponding set of positive roots.

Proposition 1.1.2.9 Let Ξ be a closed subset of Φ with the property that $\Phi = \Xi \cup (-\Xi)$ – then there exists a $w \in W$ such that $\Phi^+ \subset w\Xi$.

Proof (1) If Ξ verifies the hypotheses of our proposition, then so does $w\Xi$ for every $w \in W$. Therefore it suffices to prove that if $\Phi^+ \not\subset \Xi$ and if $\Phi^+ \cap \Xi$ possesses r elements, then there exists a reflection $w_i = w_{\alpha_i}$ $(1 \leqslant i \leqslant l)$ such that $\Phi^+ \cap w_i\Xi$ has at least $r + 1$ elements.

(2) Because Ξ is closed, there exists an $\alpha_i \in \Psi$ such that $\alpha_i \notin \Xi$. The reflection $w_i = w_{\alpha_i}$ permutes the elements of Φ^+ different from α_i, $2\alpha_i$ amongst themselves and sends α_i, $2\alpha_i$ to $-\alpha_i$, $-2\alpha_i$, respectively (of course $2\alpha_i$ may not be present); cf. Proposition 1.1.2.5. We shall now distinguish two cases.

(a) $2\alpha_i \notin \Xi$. Let $\beta \in \Phi^+ \cap \Xi$ – then $\beta \neq \alpha_i$, $2\alpha_i$ and so, in view of what was said above, $w_i(\beta) \in \Phi^+ \cap w_i\Xi$. Since $\alpha_i \notin \Xi$ and since $\Phi = \Xi \cup (-\Xi)$, it must be the case that $-\alpha_i \in \Xi$; therefore $\alpha_i \in \Phi^+ \cap w_i\Xi$. However $\alpha_i \neq w_i(\beta)$ $(\beta \in \Phi^+ \cap \Xi)$ – hence $\Phi^+ \cap w_i\Xi$ has at least $r + 1$ elements.

(b) $2\alpha_i \in \Xi$. In this situation $2\alpha_i \in \Phi^+ \cap \Xi$. If we consider those $\beta \in \Phi^+ \cap \Xi$ which are not equal to $2\alpha_i$, then we find, as in (a), that $\Phi^+ \cap w_i\Xi$ contains at least $r - 1$ elements. Now $-\alpha_i \in \Xi$; since Ξ is closed, $-\alpha_i - \alpha_i = -2\alpha_i \in \Xi$ too. But then α_i, $2\alpha_i \in \Phi^+ \cap w_i\Xi$ – hence $\Phi^+ \cap w_i\Xi$ has at least $r + 1$ elements in this case also.

The proof of the proposition is therefore complete. □

Corollary 1.1.2.10 In order that a subset Ξ of Φ be a set of positive roots for Φ, it is necessary and sufficient that: (1) Ξ be closed in Φ; (2) $\Phi = \Xi \cup (-\Xi)$; (3) Ξ contains but one element from each pair $(\alpha, -\alpha)$ $(\alpha \in \Phi)$.

Proof The stated conditions are certainly necessary. In order to establish the sufficiency of condition (3), we can, according to the preceding proposition, assume that $\Phi^+ \subset \Xi$. If $\Phi^+ \neq \Xi$, then there would exist an $\alpha \in \Xi$, $\alpha < 0$; therefore $-\alpha \in \Phi^+ \subset \Xi$, a contradiction. □

Let Θ be a subset of Ψ; we shall denote by $\langle\Theta\rangle$ the set of roots which are linear combinations of elements of Θ.

Corollary 1.1.2.11 Let Ξ be a subset of Φ. The following three conditions on Ξ are equivalent:

(i) Ξ is closed and $\Phi = \Xi \cup (-\Xi)$;

(ii) There exists a subset Θ of Ψ such that Ξ is conjugate under W to $\Phi^+ \cup \langle\Theta\rangle$;

(iii) There exists a *closed* half-space $\mathscr{S}$ of E such that $\Xi = \Phi \cap \mathscr{S}$.

Proof (i) $\Rightarrow$ (ii). Since Ξ is closed and $\Phi = \Xi \cup (-\Xi)$, there exists a $w \in W$ such that $\Phi^+ \subset w\Xi$ (cf. Proposition 1.1.2.9); put $\Theta = \Psi \cap (-w\Xi)$ – then we claim that $w\Xi = \Phi^+ \cup \langle\Theta\rangle$. To verify the claim, it will clearly be enough to establish that $w\Xi \cap \Phi^- = \langle\Theta\rangle \cap \Phi^-$. So suppose that $-\alpha \in \langle\Theta\rangle \cap \Phi^-$ $(\alpha > 0)$; α can be expressed as the sum of n elements in Θ; let us show by induction on n that $-\alpha \in w\Xi$. This is evident if $n = 1$. If $n > 1$, then one can write $\alpha = \beta + \gamma$ with $\beta \in \Phi$, $\gamma \in \Theta$ and β a sum of $n - 1$ elements in Θ. [In fact if $\{\alpha_i : 1 \leqslant i \leqslant p\}$ is a sequence of positive roots such that $\alpha_1 + \ldots + \alpha_p$ is a root, then there exists a permutation $\sigma \in \mathfrak{S}_p$ such that for every $i \in (1, 2, \ldots p)$, $\alpha_{\sigma(1)} + \ldots + \alpha_{\sigma(i)}$ is a root (use induction on p).] In view of the induction hypothesis, we have $-\beta \in w\Xi$; because $-\gamma \in w\Xi$ and $w\Xi$ is closed, it must therefore be the case that $-\alpha \in w\Xi$ – hence $\langle\Theta\rangle \cap \Phi^- \subset w\Xi \cap \Phi^-$. Conversely suppose that $-\alpha \in w\Xi \cap \Phi^-$ – then α can be written as the sum of n elements in Ψ and it will be shown by induction on n that $-\alpha \in \langle\Theta\rangle \cap \Phi^-$. This is plain if $n = 1$. If $n > 1$, we can write, according to what was said above, $\alpha = \beta + \gamma$ with $\beta \in \Phi$, $\gamma \in \Psi$ and β a sum of $n - 1$ elements in Ψ. Since $-\gamma = \beta + (-\alpha)$ and $w\Xi$ is closed, $-\gamma \in w\Xi$, whence $\gamma \in \Theta$. On the other hand, $-\beta = \gamma + (-\alpha)$ – thus $-\beta \in w\Xi$ and so by the induction hypothesis, $-\beta \in \langle\Theta\rangle \cap \Phi^-$. This implies that $-\alpha = -\beta - \gamma \in \langle\Theta\rangle \cap \Phi^-$, hence $w\Xi \cap \Phi^- \subset \langle\Theta\rangle \cap \Phi^-$.

(ii) $\Rightarrow$ (i). By assumption there exists a $w \in W$ such that $w\Xi = \Phi^+ \cup \langle\Theta\rangle$; since the essential point is to show that Ξ is closed, we may suppose that $w = 1$. Let $\alpha, \beta \in \Xi$ and suppose that $\alpha + \beta \in \Phi$ – then we are to prove that $\alpha + \beta \in \Xi$. This is obvious if $\alpha + \beta \in \Phi^+$. On the other hand, if $\alpha + \beta$ is negative, then we can write

$$\alpha + \beta = \textstyle\sum_\gamma n_\gamma \gamma \; (\gamma \in \Psi)$$

where the n_γ are non-positive integers. Because the coefficient of every element γ in $\Psi - \Theta$ relative to α or β is non-negative, it must therefore be the case that $n_\gamma = 0$ if $\gamma \in \Psi - \Theta$, whence $\alpha + \beta \in \Xi$ as desired.

(ii) $\Rightarrow$ (iii), (iii) $\Rightarrow$ (i). We shall leave it to the reader to convince himself of the validity of these implications.

This completes the proof of the corollary. □

Let Ξ be a subset of Φ. We shall say that Ξ is a *parabolic* subset of Φ if it verifies any of the three conditions enumerated in Corollary 1.1.2.11.

Proposition 1.1.2.12 Let Ξ_1 and Ξ_2 be W-conjugate parabolic subsets of Φ. If there exists a system Φ^+ of positive roots for Φ such that $\Phi^+ \subset \Xi_1$ and $\Phi^+ \subset \Xi_2$, then $\Xi_1 = \Xi_2$.

Proof To see this, note first that the argument given during the proof of '(i) $\Rightarrow$ (ii)' in Corollary 1.1.2.11 above guarantees the existence of subsets Θ_1 and Θ_2 in Ψ (the fundamental system of roots associated with Φ^+) such that $\Xi_1 = \Phi^+ \cup \langle\Theta_1\rangle$ and $\Xi_2 = \Phi^+ \cup \langle\Theta_2\rangle$. Owing to the fact that the only element in W which maps a Weyl chamber into itself is the identity, our assertion is thus seen to be a consequence of the following statement. Let Ψ_i be a fundamental system of roots for Φ, Θ_i a subset of Ψ_i, $\Xi_i = \Phi_i^+ \cup \langle\Theta_i\rangle$ $(i = 1, 2)$: If there exists an element of W which transforms Ξ_1 into Ξ_2, then there exists an element of W which transforms Ψ_1 into Ψ_2 and Θ_1 into Θ_2. [There is no loss of generality in assuming that $\Xi_1 = \Xi_2 = \Xi$ (say); this being so, let E_s be the subspace of E generated by $\Xi_s = \Xi \cap (-\Xi)$ – then Θ_1 and Θ_2 are two fundamental systems of roots for the root system Ξ_s in E_s. Let $W_s(= W_{\Xi_s})$ denote the subgroup of W generated by the Weyl reflections $w_\alpha(\alpha \in \Xi_s)$ – then there exists a $w_s \in W_s$ such that $w_s(\Theta_1) = \Theta_2$. Evidently w_s can be written as a product of reflections w_α with $\alpha \in \Theta_1$; because $\Xi - \Xi_s$ is stable for the $w_\alpha(\alpha \in \Theta_1)$, it must therefore be the case that $w_s(\Xi) = \Xi$. Consequently it can now be supposed that $\Xi_1 = \Xi_2 = \Xi$ (say), $\Theta_1 = \Theta_2 = \Theta$ (say) – but then it follows immediately that $\Psi_1 = \Psi_2 = \Psi$(say).] □

Remark Let Ξ be a closed subset of Φ such that $\Xi \cap (-\Xi) = \emptyset$ – then there exists a system Φ^+ of positive roots for Φ such that $\Xi \subset \Phi^+$. [The reader can find the details in Bourbaki [5, p.163].]

By a *horocylic* subset of Φ we shall understand a subset of the form Ξ_u, Ξ a parabolic subset of Φ. If Ξ is a parabolic subset of Φ, then so is $-\Xi$; therefore the horocylic subsets of Φ are the complements of the parabolic subsets. In view of Corollary 1.1.2.11 (iii), one may also characterize the horocylic subsets Γ of Φ by the condition: There exists an *open* half-space $\mathscr{S}$ in E such that $\Gamma = \Phi \cap \mathscr{S}$.

Let Θ be a subset of Ψ – then every transform of $\langle\Theta\rangle$ by W is closed and symmetric. It is not true that every closed and symmetric subset of Φ is of this form; on the other hand let Ω be a closed and symmetric subset of Φ – then the following four conditions on Ω are equivalent:

(i) If $E(\Omega)$ is the subspace of E generated by Ω, then $\Omega = \Phi \cap E(\Omega)$;
(ii) There exists a hyperplane $\mathscr{H}$ in E such that $\Omega = \Phi \cap \mathscr{H}$;
(iii) There exists $\Theta \subset \Psi$ and $w \in W$ such that $\Omega = w\langle\Theta\rangle$;
(iv) There exists a parabolic subset Ξ of Φ such that $\Omega = \Xi_s$.

[Evidently we have (i) $\Leftrightarrow$ (ii), (iii) $\Leftrightarrow$ (iv), (iii) $\Rightarrow$ (i). If (ii) is verified and if $\mathscr{S}$ is a closed half-space bounded by $\mathscr{H}$, then condition (iv) is met with $\Xi = \Phi \cap \mathscr{S}$.]

Let Θ be a subset of Ψ, $\Xi = \Phi^+ \cup \langle\Theta\rangle$ the corresponding parabolic subset of Φ – then Ξ induces, in a natural fashion, a decomposition of the Weyl group W of Φ. Thus write $\Xi = \Xi_s \cup \Xi_u$; let $W_s(= W_{\Xi_s})$ denote the subgroup of W generated by the Weyl reflections $w_\alpha(\alpha \in \Xi_s)$; let $W_u(= W_{\Xi_u})$ denote the set of all $w \in W$ such that $w\Phi^- \cap \Phi^+ \subset \Xi_u$. With these notations we have the following useful fact.

Proposition 1.1.2.13 Let Θ be a subset of Ψ, $\Xi = \Phi^+ \cup \langle\Theta\rangle$ the corresponding parabolic subset of Φ – then every element $w \in W$ can be written in one and only one way in the form

$$w = w_s w_u (w_s \in W_s,\ w_u \in W_u).$$

The proof of this result requires a couple of lemmas.

Given $w \in W$, let us agree to write Φ_w^+ for $w\Phi^- \cap \Phi^+$.

Lemma 1.1.2.14 The map which assigns to each $w \in W$ the set Φ_w^+ is a bijection between W and the family of subsets Γ of Φ^+ which have the following property: Γ and its complement $\Phi^+ - \Gamma$ in Φ^+ are closed.

Proof It is clear, in any event, that for every $w \in W$, the sets $w\Phi^- \cap \Phi^+$ and $w\Phi^+ \cap \Phi^+$ are closed, disjoint, and with union Φ^+. This being so, let Γ be a closed set of roots in Φ^+ with the property that its complement in Φ^+ is also closed; let Σ_Γ denote the set of all roots α such that either $\alpha \in \Phi^+ - \Gamma$ or $-\alpha \in \Gamma$. We intend to prove that Σ_Γ verifies the hypotheses of Corollary 1.1.2.10; once this is done, our lemma will follow without difficulty.

(1) Evidently $\Gamma \cap -\Gamma = \varnothing$, $(\Phi^+ - \Gamma) \cap (-(\Phi^+ - \Gamma)) = \varnothing$; on the other hand, one has $(\Phi^+ - \Gamma) \cap (-(-\Gamma)) = (\Phi^+ - \Gamma) \cap \Gamma = \varnothing$ – hence $\Sigma_\Gamma \cap -\Sigma_\Gamma = \varnothing$. Because $\Sigma_\Gamma \cup -\Sigma_\Gamma$ is a symmetric set of roots which contains Φ^+, it coincides with Φ.

(2) The set Σ_Γ is closed. Indeed, since Γ and $\Phi^+ - \Gamma$ are closed, it will be enough to show that if $\alpha \in \Phi^+ - \Gamma$ and $\beta \in -\Gamma$ are such that $\alpha + \beta \in \Phi$, then $\alpha + \beta \in \Sigma_\Gamma$. To this end, let us suppose first that $\alpha + \beta$ is positive; put $\mu = \alpha + \beta$ – then either $\mu \in \Gamma$ or $\mu \in \Phi^+ - \Gamma$. If $\mu \in \Gamma$, then the root $\alpha = \mu - \beta$ would belong to Γ since $-\beta \in \Gamma$ and Γ is closed, which contradicts the definition of α – hence $\mu \in \Phi^+ - \Gamma \subset \Sigma_\Gamma$. So suppose now that $\alpha + \beta$ is negative; put $-\mu = \alpha + \beta$ – then either $\mu \in \Gamma$ or $\mu \in \Phi^+ - \Gamma$. If $\mu \in \Phi^+ - \Gamma$, then the root $-\beta = \alpha + \mu$ would belong to $\Phi^+ - \Gamma$ (since $\Phi^+ - \Gamma$ is closed), a contradiction – hence $-\mu \in -\Gamma \subset \Sigma_\Gamma$. Therefore the set Σ_Γ is closed.

(3) It is clear that Σ_Γ contains but one element from each pair $(\alpha, -\alpha)$ $(\alpha \in \Phi)$; in view of (1), (2), and Corollary 1.1.2.10, there exists, therefore, a unique $w \in W$ such that $w\Phi^+ = \Sigma_\Gamma$. One then has that $w\Phi^- = -w\Phi^+ = -\Sigma_\Gamma$ and $-\Sigma_\Gamma \cap \Phi^+ = ((-(\Phi^+ - \Gamma)) \cup \Gamma) \cap \Phi^+ = \Gamma$.

This completes the proof of the lemma. □

Let us equip the set Ξ_s with the ordering which it inherits from Φ; the corresponding set of positive roots will be denoted by Ξ_s^+, the dominant weights in the weight group $\mathfrak{W}_s$ of Ξ_s by $\mathfrak{W}_s^+$.

Since $\Xi_u \subset \Phi^+$ is an ideal in Ξ, it is clear that a root in Ξ_s^+ is simple iff it is simple in Φ^+.

Lemma 1.1.2.15 Let $w \in W$ – then the following conditions are equivalent:

(i) $\Phi_w^+ = w\Phi^- \cap \Phi^+ \subset \Xi_u$;

(ii) $w^{-1}\Xi_s^+ \subset \Phi^+$;
(iii) $w\mathfrak{W}^+ \subset \mathfrak{W}_s^+$.

Proof (i) $\Rightarrow$ (ii). Since $w\Phi^- \cap \Phi^+ \subset \Xi_u$ and $(w\Phi^- \cap \Phi^+) \cup (w\Phi^+ \cap \Phi^+)$ is a partition of Φ^+, the subset Ξ_s^+ of Φ^+ must be contained in $w\Phi^+ \cap \Phi^+$ – therefore $w^{-1}\Xi_s^+ \subset \Phi^+$.

(ii) $\Rightarrow$ (i). Suppose to the contrary that there exists an $\alpha \in \Phi^-$ such that $w(\alpha) \in \Xi_s^+$ – then $\alpha \in w^{-1}\Xi_s^+ \subset \Phi^+$, a contradiction.

(ii) $\Leftrightarrow$ (iii). Assuming (ii), fix a $\gamma \in \mathfrak{W}^+$ – then $(\gamma, \alpha) \geqslant 0$ for all $\alpha \in \Phi^+$. In particular, if $\alpha \in \Xi_s^+$, then $(\gamma, w^{-1}(\alpha)) = (w(\gamma), \alpha) \geqslant 0$ and so $w(\gamma) \in \mathfrak{W}_s^+$. The converse is equally clear.

Hence the lemma. □

Corollary 1.1.2.16 Let w be an element in W verifying any one of the three equivalent conditions of the preceding lemma. If γ is an element of $\mathfrak{W}^+$ which assumes strictly positive (integral) values at the co-roots associated with the simple roots in Φ^+, then $w(\gamma)$ assumes strictly positive (integral) values at the co-roots associated with the simple roots in Ξ_s^+.

Proof of Proposition 1.1.2.13 (1) Every $w \in W$ can be written in at least one way in the form $w = w_s w_u (w_s \in W_s, w_u \in W_u)$. To this end, consider the sets $w\Phi^- \cap \Xi_s^+$, $w\Phi^+ \cap \Xi_s^+$; it is clear that they are complementary to one another in Ξ_s^+ and are closed; an application of Lemma 1.1.2.14 to the root system Ξ_s provides us with a $w_s \in W_s$ such that $w\Phi^- \cap \Xi_s^+ = w_s\Xi_s^- \cap \Xi_s^+$ – write $w_u = w_s^{-1}w$. Our objective then is to show that $w_u \in W_u$. Because $w\Phi^- \cap \Xi_s^+ = w_s\Xi_s^- \cap \Xi_s^+$, it is clear that $w_u\Phi^- \cap w_s^{-1}\Xi_s^+ = \Xi_s^- \cap w_s^{-1}\Xi_s^+$, whence $w_u\Phi^- \cap (w_s^{-1}\Xi_s^+ \cap \Xi_s^+) \subset \Xi_s^+ \cap \Xi_s^- = \varnothing$; since $(w_s^{-1}\Xi_s^+ \cap \Xi_s^+) \cup (w_s^{-1}\Xi_s^- \cap \Xi_s^+)$ is a partition of Ξ_s^+, one has that $w_u\Phi^- \cap \Xi_s^+ = w_u\Phi^- \cap (w_s^{-1}\Xi_s^- \cap \Xi_s^+)$; but $w_s^{-1}\Xi_s^- \cap \Xi_s^+ = -w_s^{-1}(w_s\Xi_s^- \cap \Xi_s^+) = -w_s^{-1}(w\Phi^- \cap \Xi_s^+)$, which implies that

$$w_u\Phi^- \cap \Xi_s^+ = w_u\Phi^- \cap (-w_s^{-1}(w\Phi^- \cap \Xi_s^+)) \subset w_u\Phi^- \cap (-w_u\Phi^-) = \varnothing.$$

Therefore $w_u\Phi^- \cap \Phi^+$ does not intersect Ξ_s^+ and so $w_u \in W_u$, as desired.

(2) Every $w \in W$ can be written in at most one way in the form $w = w_s w_u (w_s \in W_s, w_u \in W_u)$. To establish this it will clearly be enough to show if that $w_s w_u = \tilde{w}_u (w_s \in W_s;\ w_u, \tilde{w}_u \in W_u)$, then $w_u = \tilde{w}_u$. Thus, on the one hand, we have that $\tilde{w}_u w_u^{-1} = w_s$ which implies that $\tilde{w}_u w_u^{-1}(\Xi_s^+) \subset \Xi_s$; on the other hand $w_u^{-1}(\Xi_s^+)$ is contained in Φ^+ (Lemma 1.1.2.15) and so $\tilde{w}_u w_u^{-1}(\Xi_s^+) \subset \tilde{w}_u\Phi^+$. Plainly $\tilde{w}_u\Phi^+ = (\tilde{w}_u\Phi^+ \cap \Phi^+) \cup (\tilde{w}_u\Phi^+ \cap \Phi^-)$; since $\tilde{w}_u \in W_u$, it must be the case that $\tilde{w}_u\Phi^+ \cap \Phi^- = -(\tilde{w}_u\Phi^- \cap \Phi^+) \subset -\Xi_u$ – therefore $(\tilde{w}_u\Phi^+ \cap \Phi^-) \cap \Xi_s = \varnothing$, whence $w_s\Xi_s^+ \subset \tilde{w}_u w_u^{-1}(\Xi_s^+) \subset \Phi^+$. The last relation, in conjunction with the fact that $w_s\Xi_s^+ \subset \Xi_s$, leads to the conclusion that $w_s\Xi_s^+ \subset \Xi_s \cap \Phi^+ = \Xi_s^+$; because W_s permutes the Weyl chambers relative to Ξ_s in a simply transitive way, it follows that $w_s = 1$ – otherwise said, $w_u = \tilde{w}_u$.

This completes the proof of the proposition. □

For the theory of reductive groups, it will be necessary to extend slightly the notion of root system. Let E be a finite dimensional vector space over $\mathbf{R}$, F a subspace of E; let Φ be a finite subset of E, and E_Φ the subspace of E generated by Φ – assume that E is the direct sum of E_Φ and F. We shall then say that Φ is a *root system in* (E, F) provided Φ is a root system in the above sense for E_Φ. The Weyl group W of Φ is understood to act trivially on F; the Weyl chambers are the products of F with the Weyl chambers in E_Φ. The direct sum decomposition $E = E_\Phi + F$ permits one to realize $E_\Phi^\vee$ and $F^\vee$ as subspaces of $E^\vee$; the system $\Phi^\vee$ of co-roots associated with Φ can then be interpreted as a root system in $(E^\vee, F^\vee)$. The set comprised of those elements γ in E such that $\langle\gamma, \alpha^\vee\rangle$ is an integer for all $\alpha \in \Phi$ is a group, the *weight group* of Φ (thus all elements in F are weights); relative to some total ordering of Φ, those weights such that $\langle\gamma, \alpha^\vee\rangle \geqslant 0$ for all positive α are said to be *dominant* (in particular, then, every element of F is a dominant weight).

Appendix The only reduced irreducible root systems are the usual ones: $\mathbf{A}_l(l \geqslant 1)$, $\mathbf{B}_l(l \geqslant 2)$, $\mathbf{C}_l(l \geqslant 3)$, $\mathbf{D}_l(l \geqslant 4)$, $\mathbf{G}_2$, $\mathbf{F}_4$, $\mathbf{E}_6$, $\mathbf{E}_7$, and $\mathbf{E}_8$. They may all be realized in a suitable Euclidean space $\mathbf{R}^n$; thus let $\{e_1, \ldots, e_n\}$ be the standard basis in $\mathbf{R}^n$ – then the four 'classical' structures can be constructed as follows.

Construction of $\mathbf{A}_l$ The root system $\mathbf{A}_l$ may be realized in the hyperplane of $\mathbf{R}^{l+1}$ which is orthogonal to $e_1 + \ldots + e_{l+1}$; the elements of $\mathbf{A}_l$ are the $e_i - e_j (i \neq j)$; the $e_i - e_{i+1} (1 \leqslant i \leqslant l)$ comprise a fundamental system for $\mathbf{A}_l$; the Weyl group W of $\mathbf{A}_l$ may be identified with the group of permutations of the $e_1, \ldots, e_{l+1}$ (hence is of order $(l+1)!$).

Construction of $\mathbf{B}_l$ The root system $\mathbf{B}_l$ may be realized in $\mathbf{R}^l$. Thus $\mathbf{B}_l$ is comprised of the $\pm e_i \pm e_j (i \neq j)$ and the $\pm e_i$; the set

$$\{e_1 - e_2, e_2 - e_3, \ldots, e_{l-1} - e_l, e_l\}$$

may be taken as a fundamental system for $\mathbf{B}_l$ while the Weyl group in this case consists of the permutations and changes of sign of the e_i (hence is of order $2^l l!$) (for $l = 1$, $\mathbf{B}_1 = \mathbf{A}_1$).

Construction of $\mathbf{C}_l$ The root system $\mathbf{C}_l$ may be realized in $\mathbf{R}^l$. Thus $\mathbf{C}_l$ is comprised of the $\pm e_i \pm e_j (i \neq j)$ and the $\pm 2e_i$; the set

$$\{e_1 - e_2, e_2 - e_3, \ldots, e_{l-1} - e_l, 2e_l\}$$

is a fundamental system for $\mathbf{C}_l$ and the Weyl group is the same as that of $\mathbf{B}_l$ (for $l = 1$, $\mathbf{A}_1 = \mathbf{B}_1 = \mathbf{C}_1$; for $l = 2$, $\mathbf{C}_2 = \mathbf{B}_2$). [Evidently $\mathbf{C}_l$ may be identified with the system of co-roots associated with $\mathbf{B}_l$.]

Construction of $\mathbf{D}_l$ The root system $\mathbf{D}_l$ may be viewed as consisting of the $\pm e_i \pm e_j (i \neq j;\ 1 \leqslant i, j \leqslant l)$; the set

$$\{e_1 - e_2, e_2 - e_3, \ldots, e_{l-1} - e_l, e_{l-1} + e_l\}$$

is a fundamental system for $\mathbf{D}_l$; the Weyl group can be regarded as the permutations and even number of sign changes of the e_i (hence is of order

$2^{l-1}l!$) (for $l = 2$, this system is isomorphic to $\mathbf{A}_1 \times \mathbf{A}_1$; for $l = 3$ it is isomorphic to $\mathbf{A}_2$).

We shall not stop to give an explicit description of the five 'exceptional' structures (see Bourbaki [5] or Jacobson [3] or Serre [4]).

In conclusion let us mention that it can be shown that for every integer $l \geqslant 1$ there exists exactly one (to within isomorphism) irreducible non-reduced system – this is the system $\mathbf{BC}_l$ obtained by taking the union of $\mathbf{B}_l$ with $\mathbf{C}_l$ (the Weyl group of $\mathbf{BC}_l$ is that of $\mathbf{B}_l$).

1.1.3 Root Systems II — σ-Systems

Let E be a finite dimensional vector space over $\mathbf{R}$, Φ a root system in E – in what follows it will be assumed that Φ is *reduced.* Fix, once and for all, a W-invariant scalar product $(.\,,.)$ on E ($W(= W_\Phi)$) the Weyl group of Φ). Let us suppose now that we are given a linear involutive isometry $\sigma: E \longrightarrow E$ such that $\sigma\Phi = \Phi$ (with $\sigma \neq \pm 1$) – under these set of circumstances, we shall say that the pair (Φ, σ) (or simply Φ) is a *σ-system of roots.*

Definition A σ-system of roots Φ is said to be *normal* if, for all $\alpha \in \Phi$, $\alpha^\sigma - \alpha \notin \Phi$.

Let Φ be a normal σ-system of roots in E; let $E = E_+ + E_-$ be the direct decomposition of E into eigenspaces for σ (thus $E_\pm = \{\gamma \in E: \gamma^\sigma = \pm\gamma\}$); let $\Phi_+ = \{\alpha \in \Phi: \alpha \neq -\alpha^\sigma\}$, $\Phi_- = \{\alpha \in \Phi: \alpha = -\alpha^\sigma\}$; let Σ denote the collection $\{\tilde{\alpha}\}$ of vectors $\tilde{\alpha}$ obtained by projecting the elements of Φ_+ onto E_+ (relative to E_-) – then we intend to prove that Σ is a (in general non-reduced) system of roots in E_+.

Let W_Σ denote the group generated by the reflections $\{w_{\tilde{\alpha}}: \alpha \in \Phi_+\}$, $w_{\tilde{\alpha}}$ being defined by

$$\gamma \mapsto \gamma - 2\frac{(\gamma, \tilde{\alpha})}{(\tilde{\alpha}, \tilde{\alpha})}\tilde{\alpha} \quad (\gamma \in E_+);$$

here we are equipping E_+ with the Euclidean structure which it inherits from E – thus W_Σ is the Weyl group of the set Σ.

It is clear that Σ is finite, generates E_+, and does not contain zero (for we are assuming that $\sigma \neq \pm 1$). Given $\lambda \in \Sigma$, let

$$\Sigma(\lambda) = \{\alpha \in \Phi_+: \tilde{\alpha} = \lambda\}.$$

If $\alpha \in \Sigma(\lambda)$, then one has three possibilities:

(1) $\alpha = \alpha^\sigma$ – here $(\alpha, \alpha) = (\lambda, \lambda)$;

(2) $\alpha \neq \alpha^\sigma$ and $(\alpha, \alpha^\sigma) = 0$ – in this case $\lambda = (\tilde{\alpha} + \tilde{\alpha}^\sigma)/2$ and so $(\lambda, \lambda) = (\alpha, \alpha)/2 = (\alpha^\sigma, \alpha^\sigma)/2$;

(3) $\alpha \neq \alpha^\sigma$ and $(\alpha, \alpha^\sigma) < 0$ – again $\lambda = (\tilde{\alpha} + \tilde{\alpha}^\sigma)/2$. Since $|\alpha| = |\alpha^\sigma|$ and $A_{\alpha,\alpha^\sigma} < 0$, reference to our list of possibilities for A_{α,α^σ} shows that $A_{\alpha,\alpha^\sigma} = -1$ (cf. the table in 1.1.2), whence $(\alpha, \alpha) = -2\,(\alpha, \alpha^\sigma)$ and so

$$(\lambda, \lambda) = (\alpha, \alpha)/4 = (\alpha^\sigma, \alpha^\sigma)/4.$$

Observe that the case $\alpha \neq \alpha^\sigma$ and $(\alpha, \alpha^\sigma) > 0$ is ruled out by Proposition 1.1.2.1 and the normality assumption on Φ.

We shall now verify that Σ is a root system in E_+. Fix $\lambda \in \Sigma$, $\alpha \in \Sigma(\lambda)$; fix $\mu \in \Sigma, \beta \in \Sigma(\mu)$ – assume that λ and μ are not proportional. We shall show simultaneously that $A_{\lambda,\mu} \in \mathbf{Z}$ and $w_\lambda(\mu) \in \Sigma$ by considering each of the three cases above per the element λ.

(1)
$$A_{\lambda,\mu} = 2\frac{(\lambda, \mu)}{(\lambda, \lambda)} = 2\frac{(\alpha, \beta)}{(\alpha, \alpha)} \in \mathbf{Z}$$

and since

$$w_\alpha(\beta) = \beta - A_{\alpha,\beta}\alpha \in \Phi_+, \ (\beta - A_{\alpha,\beta}\alpha)^\sim = \mu - A_{\lambda,\mu}\lambda = w_\lambda(\mu) \in \Sigma.$$

(2)
$$\begin{aligned} A_{\lambda,\mu} &= 2\frac{(\lambda, \mu)}{(\lambda, \lambda)} = \frac{(\alpha + \alpha^\sigma, \beta)}{(\lambda, \lambda)} \\ &= 2\frac{(\alpha, \beta)}{(\alpha, \alpha)} + 2\frac{(\alpha^\sigma, \beta)}{(\alpha^\sigma, \alpha^\sigma)} \\ &= A_{\alpha,\beta} + A_{\alpha^\sigma,\beta} \in \mathbf{Z}. \end{aligned}$$

Let $\gamma = \beta - A_{\alpha,\beta}\alpha$ – then

$$A_{\alpha^\sigma,\gamma} = 2\frac{(\alpha^\sigma, \beta - A_{\alpha,\beta}\alpha)}{(\alpha^\sigma, \alpha^\sigma)} = A_{\alpha^\sigma,\beta}.$$

Let $\delta = \gamma - A_{\alpha^\sigma,\gamma}\alpha^\sigma \in \Phi_+$ – then

$$\tilde{\delta} = \mu - (A_{\alpha,\beta} + A_{\alpha^\sigma,\beta})\lambda = w_\lambda(\mu) \in \Sigma.$$

(3)
$$\begin{aligned} A_{\lambda,\mu} &= 2\frac{(\lambda, \mu)}{(\lambda, \lambda)} = 2\left\{2\frac{(\alpha, \beta)}{(\alpha, \alpha)} + 2\frac{(\alpha^\sigma, \beta)}{(\alpha^\sigma, \alpha^\sigma)}\right\} \\ &= 2(A_{\alpha,\beta} + A_{\alpha^\sigma,\beta}) \in \mathbf{Z}. \end{aligned}$$

Recall from above that $A_{\alpha,\alpha^\sigma} = -1$ in this case. Let $\gamma = \beta - A_{\alpha,\beta}\alpha$ – then

$$A_{\alpha^\sigma,\gamma} = 2\frac{(\alpha^\sigma, \beta - A_{\alpha,\beta}\alpha)}{(\alpha^\sigma, \alpha^\sigma)} = A_{\alpha^\sigma,\beta} + A_{\alpha,\beta}.$$

Let $\delta = \gamma - A_{\alpha^\sigma,\gamma}\alpha^\sigma$ – then

$$\begin{aligned} A_{\alpha,\delta} &= 2\frac{(\gamma - A_{\alpha^\sigma,\gamma}\alpha^\sigma, \alpha)}{(\alpha, \alpha)} = A_{\alpha,\gamma} + A_{\alpha^\sigma,\gamma} \\ &= A_{\alpha,\beta} - 2A_{\alpha,\beta} + A_{\alpha^\sigma,\beta} + A_{\alpha,\beta} = A_{\alpha^\sigma,\beta}. \end{aligned}$$

Let $\epsilon = \delta - A_{\alpha,\delta}\alpha \in \Phi_+$ – then

$$\begin{aligned} \tilde{\epsilon} &= \tilde{\beta} - A_{\alpha,\beta}\tilde{\alpha} - (A_{\alpha^\sigma,\beta} + A_{\alpha,\beta})\tilde{\alpha}^\sigma - A_{\alpha^\sigma,\beta}\tilde{\alpha} \\ &= \mu - 2(A_{\alpha,\beta} + A_{\alpha^\sigma,\beta})\lambda = \mu - A_{\lambda,\mu}\lambda = w_\lambda(\mu) \in \Sigma. \end{aligned}$$

[The reader will note that the following fact is contained in the preceding discussion. Let $\alpha \in \Phi_+$ – then $(\alpha, \alpha) = m(\tilde{\alpha}, \tilde{\alpha})$ $(m = 1, 2, 4)$.]

The above considerations can be summarized as follows.

Proposition 1.1.3.1 (Araki) Let Φ be a normal σ-system of roots in E – then the collection Σ is a root system in E_+.

In general the root systems Σ will furnish examples of non-reduced systems of roots; further remarks appear below.

Let Φ be a normal σ-system of roots in E. An ordering of Φ which satisfies the condition

$$\alpha \in \Phi - \Phi_-, \alpha > 0 \Rightarrow \alpha^\sigma > 0$$

is called a *σ-order.* A typical way in which to obtain a σ-order is to take the lexicographic order relative to a basis for E, the first l_+ members of which comprise a basis for E_+ ($l_+ = \dim(E_+)$). A fundamental system for Φ with respect to a σ-order is called a *σ-fundamental system*; if Ψ is a σ-fundamental system for Φ, then $\Psi \cap \Phi_-$ is a fundamental system for Φ_-.

Let $l(= l_\Phi)$ denote the rank of Φ, $l_-(= l_{\Phi_-})$ the rank of Φ_-. Select a σ-fundamental system $\{\alpha_1, \ldots, \alpha_{l-l_-}, \alpha_{l-l_-+1}, \ldots, \alpha_l\}$ of roots for Φ such that the set $\{\alpha_{l-l_-+1}, \ldots, \alpha_l\}$ constitutes a fundamental system for Φ_- (hence $\{\alpha_1, \ldots, \alpha_{l-l_-}\} \subset \Phi_+$).

Lemma 1.1.3.2 There exists a permutation $i \mapsto i'$ of order 2 of the set $(1, \ldots, l - l_-)$ such that

$$\alpha_i^\sigma = \alpha_{i'} + \sum_{j=l-l_-+1}^{l} n_j^i \alpha_j \quad (n_j^i \in \mathbf{N}, 1 \leqslant i \leqslant l - l_-).$$

Proof Write

$$\alpha_i^\sigma = \sum_{j=1}^{l} n_j^i \alpha_j \quad (n_j^i \in \mathbf{N}, 1 \leqslant i \leqslant l - l_-).$$

Since $\alpha_i^\sigma \in \Phi_+$, $n_{i'}^i > 0$ for at least one integer i' lying between 1 and $l - l_-$. Applying σ again, we get

$$\alpha_i = \sum_{j=1}^{l-l_-} \sum_{k=1}^{l} n_j^i n_k^j \alpha_k - \sum_{j=l-l_-+1}^{l} n_j^i \alpha_j$$

and so it follows that $n_{i'}^i = 1$, $n_j^i = 0$ $(1 \leqslant j \leqslant l - l_-, j \neq i')$. The assignment $i \mapsto i'$ $(1 \leqslant i \leqslant l - l_-)$ is obviously a permutation of order 2. Hence the lemma. □

Let $l - l_- = l_1 + 2l_2$, $l_1 + l_2 = l_+$, where, using the preceding lemma, we have put

$$i' = \begin{cases} i & \text{for } 1 \leqslant i \leqslant l_1, \\ i + l_2 & \text{for } l_1 + 1 \leqslant i \leqslant l_1 + l_2, \\ i - l_2 & \text{for } l_1 + l_2 + 1 \leqslant i \leqslant l_1 + 2l_2. \end{cases}$$

It is easy to see that the set $\Upsilon = \{\tilde{\alpha}_1, \ldots, \tilde{\alpha}_{l_+}\}$ is linearly independent and spans E_+ (so that $l_+ = \dim(E_+)$); moreover it is obvious that every element in Σ can be written as an integral linear combination of the

$\tilde{\alpha}_i (1 \leqslant i \leqslant l_+)$ where the integers involved are either all non-negative or non-positive – otherwise said, the set $\{\tilde{\alpha}_1, \ldots, \tilde{\alpha}_{l_+}\}$ is a fundamental system of roots for Σ.

Our next objective is to exhibit the connection that exists between the Weyl group W_Φ of Φ and the Weyl group W_Σ of Σ.

Let W_- denote the Weyl group of the root system Φ_-; $W_\sigma = \{w \in W_\Phi : \sigma w = w\sigma\}$ – then W_σ is the subgroup of W_Φ consisting precisely of those elements which stabilize E_+. Evidently W_- is a normal subgroup of W_σ.

Proposition 1.1.3.3 (Satake) The map $w \mapsto w | E_+$ is a surjective homomorphism of W_σ onto W_Σ with kernel W_-.

Roughly speaking, then, W_Σ consists of those elements in W_Φ which stabilize E_+.

The proof of Proposition 1.1.3.3 rests on the following lemmas.

Fix a σ-fundamental system $\{\alpha_1, \ldots, \alpha_{l-l_-+1}, \ldots, \alpha_l\}$ of roots for Φ such that $\{\alpha_{l-l_-+1}, \ldots, \alpha_l\}$ is a fundamental system for Φ_-.

Lemma 1.1.3.4 Let $w_\sigma \in W_\sigma$. In order that $w_\sigma \in W_-$ it is necessary and sufficient that $w_\sigma | E_+ = 1$. In fact it is sufficient that

$$\{w_\sigma(\tilde{\alpha}_1), \ldots, w_\sigma(\tilde{\alpha}_{l_+})\} = \{\tilde{\alpha}_1, \ldots, \tilde{\alpha}_{l_+}\}.$$

Proof If $w_- \in W_-$, then plainly $w_- | E_+ = 1$. Let us prove the sufficiency of the last condition. Since $w_\sigma \in W_\sigma$, $w_\sigma \Phi_- = \Phi_-$ and so $\{w_\sigma(\alpha_{l-l_-+1}), \ldots, w_\sigma(\alpha_l)\}$ is a fundamental system of roots for Φ_-. Therefore there exists $w_- \in W_-$ such that $\{w_\sigma(\alpha_{l-l_-+1}), \ldots, w_\sigma(\alpha_l)\} = \{w_-(\alpha_{l-l_-+1}), \ldots, w_-(\alpha_l)\}$. Hence, replacing w_σ by $w_-^{-1} w_\sigma = w$, we may assume in addition that $\{w(\alpha_{l-l_-+1}), \ldots, w(\alpha_l)\} = \{\alpha_{l-l_-+1}, \ldots, \alpha_l\}$. Now fix any index i between 1 and $l - l_-$ – then $w(\alpha_i) > 0$ (cf. Lemma 1.1.3.2). This means that the set of positive roots corresponding to the fundamental system $\{w(\alpha_1), \ldots, w(\alpha_l)\}$ is contained in that corresponding to $\{\alpha_1, \ldots, \alpha_l\}$. Therefore $w = 1$ and so $w_\sigma = w_-$, i.e. $w_\sigma \in W_-$. □

Lemma 1.1.3.5 For any $i (1 \leqslant i \leqslant l_+)$ the reflection $w_{\tilde{\alpha}_i}$ coincides with the restriction to E_+ of some element in W_σ.

Proof We consider two cases. If $\alpha_i + \alpha_i^\sigma \in \Phi$, then $w_{\tilde{\alpha}_i} = w_{\alpha_i + \alpha_i^\sigma} | E_+$. If $\alpha_i + \alpha_i^\sigma \notin \Phi$, then, since $\alpha_i - \alpha_i^\sigma \notin \Phi$ (Φ is normal), α_i is either orthogonal to α_i^σ, whence $w_{\alpha_i} w_{\alpha_i \sigma} | E_+ = w_{\tilde{\alpha}_i}$, or $\alpha_i = \alpha_i^\sigma$, whence $w_{\alpha_i} | E_+ = w_{\tilde{\alpha}_i}$. □

Proof of Proposition 1.1.3.3 As we have seen, the set Σ is a root system in E_+ while the set $\{\tilde{\alpha}_1, \ldots, \tilde{\alpha}_{l_+}\}$ is a fundamental system of roots for Σ. Let w_σ be any element in W_σ – then $w_\sigma \Sigma = \Sigma$ and so $\{w_\sigma(\tilde{\alpha}_1), \ldots, w_\sigma(\tilde{\alpha}_{l_+})\}$ is also a fundamental system of roots for Σ. But the group W_Σ acts (simply) transitively on the set of all fundamental systems of roots for

Σ; in particular there exists $w_\Sigma \in W_\Sigma$ such that $\{w_\sigma(\tilde{\alpha}_1), \ldots, w_\sigma(\tilde{\alpha}_{l_*})\} = \{w_\Sigma(\tilde{\alpha}_1), \ldots, w_\Sigma(\tilde{\alpha}_{l_*})\}$. Therefore $w_\Sigma^{-1} w_\sigma$ permutes the $\tilde{\alpha}_i$ amongst themselves. Proposition 1.1.2.3 and Lemma 1.1.3.5 imply that $W_\sigma | E_+ \supset W_\Sigma$; hence by Lemma 1.1.3.4 $w_\sigma | E_+ = w_\Sigma$ and the proposition follows. □

With these generalities out of the way, we shall now proceed to show how the normal σ-systems of roots arise in practice.

Let us agree to use the notations introduced near the end of 1.1.1 in the description of the Iwasawa decomposition of a semi-simple Lie algebra $\mathfrak{g}$ over $\mathbf{R}$. Put $\Phi_+ = P_+ \cup (-P_+)$, $\Phi_- = P_- \cup (-P_-)$. For a root $\alpha \in \Phi$, let α^σ denote the mapping $H \mapsto \overline{\alpha(\sigma(H))}$ ($H \in \mathfrak{a}_c$, the bar denoting complex conjugation) – then Φ_+ is stable under this action of σ while $\Phi_- = \{\alpha \in \Phi : \alpha^\sigma = -\alpha\}$ (cf. Helgason [2, p.222]). Evidently Φ_- may be viewed as a root system in its own right relative to the vector subspace in the dual of $\mathfrak{a}^*$ spanned by the roots α in Φ_-.

Select a Weyl basis $\{X_\alpha : \alpha \in \Phi\}$ for $\mathfrak{g}_c$ mod $\mathfrak{a}_c$ with respect to the compact real form $\mathfrak{u} = \mathfrak{k} + \sqrt{-1}\mathfrak{p}$ (so that in addition to the usual properties one has $\tau X_\alpha = -X_{-\alpha}$ for all $\alpha \in \Phi$); put $\sigma X_\alpha = k_\alpha X_{\alpha^\sigma}$ ($\alpha \in \Phi$) – then, since $\theta = \sigma\tau = \tau\sigma$, we have $\overline{k_\alpha} = k_{-\alpha}$ (all $\alpha \in \Phi$) and, moreover, $k_\alpha = -1$ if $\alpha \in \Phi_-$ (in fact $\alpha \in \Phi_- \Rightarrow X_\alpha \in \mathfrak{k}_c \Rightarrow \sigma X_\alpha = \sigma(\theta X_\alpha) = \tau X_\alpha = -X_{-\alpha} \Rightarrow k_\alpha = -1$).

Lemma 1.1.3.6 For all $\alpha \in \Phi$, $\alpha^\sigma - \alpha \notin \Phi$.

Proof If $\alpha \in \Phi_-$, then $\alpha^\sigma - \alpha = -2\alpha \notin \Phi$ (Φ is reduced). This being so, assume now that $\alpha \in \Phi_+$. If $\alpha^\sigma - \alpha \in \Phi$, then plainly $\alpha^\sigma - \alpha \in \Phi_-$ – hence $\sigma X_{\alpha^\sigma - \alpha} = -X_{\alpha - \alpha^\sigma}$. It is clear that

$$[X_\alpha, \sigma X_{-\alpha}] = [X_\alpha, k_{-\alpha} X_{-\alpha^\sigma}] = k_{-\alpha} N_{\alpha, -\alpha^\sigma} X_{\alpha - \alpha^\sigma}.$$

Applying σ to this identity gives

$$k_\alpha N_{\alpha^\sigma, -\alpha} X_{\alpha^\sigma - \alpha} = -\overline{k_{-\alpha}} N_{\alpha, -\alpha^\sigma} X_{\alpha^\sigma - \alpha}.$$

But (Theorem 1.1.1.1(v)) $N_{\alpha, -\alpha^\sigma} = -N_{-\alpha, \alpha^\sigma} = N_{\alpha^\sigma, -\alpha}$ which implies that $\bar{k}_\alpha = -k_{-\alpha} = k_{-\alpha}$. This is a contradiction, since $\alpha \neq 0 \Rightarrow \alpha^\sigma \neq 0$. □

The preceding lemma ensures us that Φ is a normal σ-system of roots – therefore the theory developed earlier in this number is applicable in the present situation. Thus, given $\alpha \in \Phi_+$, let $\tilde{\alpha}$ denote the restriction of α to $\mathfrak{a}_\mathfrak{p}$ – then the set Σ of restrictions to $\mathfrak{a}_\mathfrak{p}$ of the elements in Φ_+ is a root system in $\mathfrak{a}_\mathfrak{p}^\vee$ (where $\mathfrak{a}_\mathfrak{p}^\vee$ is equipped with the Euclidean structure obtained by transport of the restriction of the Killing form B to $\mathfrak{a}_\mathfrak{p}$). For each $\lambda \in \Sigma$ the number $m(\lambda)$ of $\alpha \in \Phi_+$ such that $\lambda = \tilde{\alpha}$ is called the *multiplicity* of λ. If $\lambda \in \Sigma$, then Q_λ will denote the vector in $\mathfrak{a}_\mathfrak{p}$ such that $\lambda(H) = B(H, Q_\lambda)$ for all $H \in \mathfrak{a}_\mathfrak{p}$. Observe that if $\alpha \in \Phi_+$, then $P_\alpha = H_\alpha - Q_{\tilde{\alpha}} \in \sqrt{-1}\mathfrak{a}_\mathfrak{k}$. Let W_Σ denote the group generated by the reflections $\{w_{\tilde{\alpha}} : \alpha \in \Phi_+\}$, $w_{\tilde{\alpha}}$ being defined by the rule $\gamma \mapsto \gamma - 2(\gamma, \tilde{\alpha})/(\tilde{\alpha}, \tilde{\alpha})\tilde{\alpha}$

($\gamma \in \mathfrak{a}_{\mathfrak{p}}^{\vee}$) (here of course $(\tilde{\alpha}, \tilde{\alpha}) = \tilde{\alpha}(Q_\alpha)$) – thus W_Σ is the Weyl group of Σ.

In the sequel we shall frequently write λ_i for $\tilde{\alpha}_i (i = 1, \ldots, l_\Sigma; l_\Sigma = l_+)$.

Example Consider the real form **AII** of the classical simple Lie algebra $\mathbf{A}_l$ (in the usual notations) – thus $\mathbf{A}_l$ is realized as the set $\mathfrak{sl}(l+1, \mathbf{C})$ of all $l+1$ by $l+1$ matrices with entries in $\mathbf{C}$ and of zero trace. The real form $\mathfrak{g} = \mathbf{AII}$ exists only if $l+1 = 2m$ is an even integer and may be described as follows. Let $J = \begin{pmatrix} 0 & I_m \\ -I_m & 0 \end{pmatrix}$ – then

$$\begin{aligned}\mathfrak{g} &= \{X \in \mathfrak{sl}(l+1, \mathbf{C}) : X = J\bar{X}J^{-1}\} \\ &= \left\{\begin{pmatrix} A & B \\ -\bar{B} & \bar{A} \end{pmatrix} : A, B \in \mathfrak{gl}(m, \mathbf{C}), \operatorname{tr}(A + \bar{A}) = 0\right\}.\end{aligned}$$

If we set

$$\begin{aligned}\mathfrak{k} &= \left\{\begin{pmatrix} A & B \\ -\bar{B} & \bar{A} \end{pmatrix} : A \in \mathfrak{u}(m), B \in \mathbf{S}(m, \mathbf{C})\right\}, \\ \mathfrak{p} &= \left\{\begin{pmatrix} A & B \\ -\bar{B} & \bar{A} \end{pmatrix} : A \in \mathbf{H}(m) \text{ with } \operatorname{tr}(A) = 0, B \in \mathfrak{o}(m, \mathbf{C})\right\},\end{aligned}$$

then $\mathfrak{g} = \mathfrak{k} + \mathfrak{p}$ is a Cartan decomposition of $\mathfrak{g}$. A maximal abelian subspace $\mathfrak{a}_\mathfrak{p}$ of $\mathfrak{p}$ is given by

$$\mathfrak{a}_\mathfrak{p} = \left\{D(a_1, \ldots, a_m, a_1, \ldots, a_m) : a_i \in \mathbf{R}, \sum_{i=1}^m a_i = 0\right\}$$

and a maximal abelian subalgebra $\mathfrak{a}$ of $\mathfrak{g}$ containing $\mathfrak{a}_\mathfrak{p}$ is

$$\mathfrak{a} = \left\{D(a_1, \ldots, a_m, \bar{a}_1, \ldots, \bar{a}_m) : a_i \in \mathbf{C}, \mathscr{R}\left(\sum_{i=1}^m a_i\right) = 0\right\}.$$

Plainly

$$\mathfrak{a}_c = \left\{D(a_1, \ldots, a_{2m}) : a_i \in \mathbf{C}, \sum_{i=1}^{2m} a_i = 0\right\}.$$

Let e_i be the linear form on $\mathfrak{a}_c$ defined by

$$e_i(A) = a_i \; (A = D(a_1, \ldots, a_{2m}) \in \mathfrak{a}_c).$$

The root system Φ of the pair $(\mathfrak{g}_c, \mathfrak{a}_c)$ is

$$\Phi = \{\pm(e_i - e_j) : 1 \leqslant i < j \leqslant 2m\}.$$

The Weyl group W_Φ of the pair $(\mathfrak{g}_c, \mathfrak{a}_c)$ consists of all permutations of the e_i's. With respect to an obvious selection of compatible orderings, the set

$$\{e_1 - e_{m+2}, \ldots, e_{m-1} - e_{2m}, e_{m+1} - e_1, \ldots, e_{2m} - e_m\}$$

is a fundamental system of roots for Φ such that $\{e_{m+1} - e_1, \ldots, e_{2m} - e_m\}$ is a fundamental system for Φ_-. Put $\lambda_i = \tilde{e}_i - \tilde{e}_{m+(i+1)}$ $(i = 1, \ldots, m-1)$ – then the λ_i constitute a fundamental system for Σ. Thus the rank l_Σ of Σ is: $l_\Sigma = m - 1$. Evidently in this case the multiplicity $m(\lambda_i)$ of λ_i is four while $2\lambda_i$ is not present. Put $f_i = \tilde{e}_i = \tilde{e}_{i+m}$ $(i = 1, \ldots, m)$ – then W_Σ is the group of all permutations of the f_i.

The reader who is interested in the classification of simple Lie algebras over $\mathbf{R}$ may wish to consult Araki [1] (see also Murakami [1]). In Araki [1] a completely new proof of the classification is given – the proof is in the same spirit as the one usually given for the corresponding classification problem of simple Lie algebras over **C** although, of course, more involved. In particular, a fundamental system $\{\lambda_i : 1 \leqslant i \leqslant l_\Sigma\}$ of roots for Σ is explicitly determined for each simple Lie algebra $\mathfrak{g}$ over **R** and the multiplicities $m(\lambda_i)$, $m(2\lambda_i)$ are computed.

Let $\mathfrak{g}$ be a non-compact semi-simple Lie algebra over **R**, $\mathfrak{g} = \mathfrak{k} + \mathfrak{p}$ a fixed Cartan decomposition of $\mathfrak{g}$ with corresponding Cartan involution θ. Let $\mathfrak{a}_\mathfrak{p}$ be a maximal abelian subspace in $\mathfrak{p}$ and extend $\mathfrak{a}_\mathfrak{p}$ to a maximal abelian subalgebra $\mathfrak{a}$ of $\mathfrak{g}$ containing $\mathfrak{a}_\mathfrak{p} : \mathfrak{a} = \mathfrak{a}_\mathfrak{k} + \mathfrak{a}_\mathfrak{p}$. It will prove to be convenient to re-interpret the Iwasawa decomposition of $\mathfrak{g}$ with respect to $\mathfrak{a}_\mathfrak{p}$.

Put $(X, Y)_\theta = -B(X, \theta Y)$ $(X, Y \in \mathfrak{g}$; B the Killing form of $\mathfrak{g})$ – then $(. , .)_\theta$ equips $\mathfrak{g}$ with the structure of a real Hilbert space. A simple calculation shows that the Hilbert space adjoint of $\mathrm{ad}(X)$ is $-\mathrm{ad}(\theta X)$ $(X \in \mathfrak{g})$.

For each $\lambda \in \mathfrak{a}_\mathfrak{p}^\vee$, let $\mathfrak{g}^\lambda = \{X \in \mathfrak{g} : [H, X] = \lambda(H)X \text{ for all } H \in \mathfrak{a}_\mathfrak{p}\}$. Since $\mathrm{ad}(\mathfrak{a}_\mathfrak{p})$ is a commuting family of self-adjoint operators in $\mathfrak{g}$, $\mathfrak{g}$ is the orthogonal (relative to $(. , .)_\theta$) direct sum of the subspaces $\mathfrak{g}^\lambda$. If $\lambda \in \mathfrak{a}_\mathfrak{p}^\vee$ and $\mathfrak{g}^\lambda \neq \{0\}$, call λ a *root* of $\mathfrak{g}$ with respect to $\mathfrak{a}_\mathfrak{p}$ – of course $\mathfrak{g}^0 = \mathfrak{m} + \mathfrak{a}_\mathfrak{p}$, $\mathfrak{m}$ the centralizer of $\mathfrak{a}_\mathfrak{p}$ in $\mathfrak{k}$. Fix a basis in $\mathfrak{a}_\mathfrak{p}^\vee$ and introduce an order into the set $\Sigma = \{\lambda \neq 0 : \mathfrak{g}^\lambda \neq \{0\}\}$; put $\Sigma^+ = \{\lambda \in \Sigma : \lambda > 0\}$. On the other hand this ordering in $\mathfrak{a}_\mathfrak{p}^\vee$ extends to a compatible order of $\mathfrak{a}_\mathfrak{p}^\vee$ with $(\mathfrak{a}^*)^\vee$ and it is obvious that Σ is precisely the set of restrictions to $\mathfrak{a}_\mathfrak{p}$ of the elements in $\Phi_+ = P_+ \cup (-P_+)$, Σ^+ the restrictions to $\mathfrak{a}_\mathfrak{p}$ of the elements in P_+. In particular Σ is a root system in $\mathfrak{a}_\mathfrak{p}^\vee$ (Proposition 1.1.3.1) and $\{\lambda_i : 1 \leqslant i \leqslant l_\Sigma)$ is a fundamental system of roots for Σ (recall that we have set $\lambda_i = \bar{\alpha}_i$, $i = 1, \ldots, l_\Sigma$). We emphasize that Σ is not reduced and that $m(\lambda) = \dim(\mathfrak{g}^\lambda) > 1$, in general. [By abuse of the language, we shall often refer to Σ as the 'set of roots' of the pair $(\mathfrak{g}, \mathfrak{a}_\mathfrak{p})$.]

If we write $\mathfrak{n}^+ = \Sigma_{\lambda > 0}\, \mathfrak{g}^\lambda$, then we have the orthogonal (relative to $(. , .)_\theta$) direct decomposition

$$\mathfrak{g} = (\mathfrak{m} + \mathfrak{a}_\mathfrak{p}) + \mathfrak{n}^+ + \mathfrak{n}^-, \; \mathfrak{n}^- = \theta(\mathfrak{n}^+) = \sum_{\lambda < 0} \mathfrak{g}^\lambda.$$

Since $[\mathfrak{g}^\lambda, \mathfrak{g}^\mu] \subset \mathfrak{g}^{\lambda + \mu}$ for all $\lambda, \mu \in \Sigma$, $\mathfrak{n}^+$ and $\mathfrak{n}^-$ are nilpotent subalgebras of $\mathfrak{g}$ – of course $\mathfrak{n}^+ = (\Sigma_{\alpha \in P_+} \mathfrak{g}_c^\alpha) \cap \mathfrak{g}$.

Keeping to the above notations, let G denote a connected semi-simple Lie group corresponding to the Lie algebra $\mathfrak{g}$; we shall agree to denote the extension to G of the Cartan involution θ by the same symbol. Let K, $A_\mathfrak{p}$, N^+, and N^- denote the analytic subgroups of G corresponding to $\mathfrak{k}$, $\mathfrak{a}_\mathfrak{p}$, $\mathfrak{n}^+$, and $\mathfrak{n}^-$, respectively – then K, $A_\mathfrak{p}$, N^+, N^-, and $A_\mathfrak{p}N^+$ are all closed subgroups of G with $A_\mathfrak{p}$, N^+ and N^- simply connected. As is well-known, the map $(k, h, n) \mapsto khn$ is an analytic diffeomorphism of the product manifold $K \times A_\mathfrak{p} \times N^+$ onto G. The rank of the associated Riemannian symmetric space G/K (i.e. the maximal dimension of a flat, totally geodesic submanifold in G/K) is called the *split-rank* of G; thus the split-rank of G is equal to the dimension of $A_\mathfrak{p}$ and, in view of standard conjugacy theorems, is independent of our choice of an Iwasawa decomposition for G. The root system Σ in $\mathfrak{a}_{\mathfrak{p}}^{\vee}$ (or by duality in $\mathfrak{a}_{\mathfrak{p}}$) is sometimes referred to as the set of roots of the symmetric space G/K.

Example Suppose that $G = \mathrm{Int}(\mathfrak{g})$ with $\mathfrak{g}$ simple. If G/K is a bounded symmetric domain, then there are two possibilities for the root system Σ:

G/K is a tube domain iff Σ is of type $\mathbf{C}_l$;

G/K is not a tube domain iff Σ is of type $\mathbf{BC}_l$ $\quad (l = l_\Sigma)$.

For definitions and details see Moore [2] and Korányi and Wolf [1].

The following result is frequently useful.

Lemma 1.1.3.7 Let $\mathfrak{g}$ be a semi-simple Lie algebra over $\mathbf{R}$, $\mathfrak{g} = \mathfrak{k} + \mathfrak{p}$ a Cartan decomposition of $\mathfrak{g}$ with Cartan involution θ; let G be a connected Lie group with Lie algebra $\mathfrak{g}$, K the analytic subgroup of G corresponding to $\mathfrak{k}$; let $\tilde{Y} = \mathrm{Ad}\,(k \exp X) Y$ where $Y, \tilde{Y} \in \mathfrak{g}$, $k \in K$ and $X \in \mathfrak{p}$ – then, if both Y and $\tilde{Y}$ are eigenvectors of θ, we have $[X, Y] = 0$.

Proof As above, we shall agree to denote the canonical extension of θ to G by the same symbol – then, since $\theta(\tilde{Y}) = \mathrm{Ad}\,(k \exp(-X))\theta(Y)$, it is clear that $e^{2\mathrm{ad}(X)} Y = \epsilon Y$ where $\epsilon = \pm 1$. Because X lies in $\mathfrak{p}$, the endomorphism $\mathrm{ad}\,(X)$ is semi-simple and all its eigenvalues are real – therefore it is obvious that $\epsilon = 1$ and $[X, Y] = 0$. ☐

Keeping to the above notations, let us bear in mind that G has finite center iff the analytic subgroup K is compact – it will be convenient to make this assumption in what follows.

We shall now recall another realization of the Weyl group W of the pair $(\mathfrak{g}, \mathfrak{a}_{\mathfrak{p}})$ (the subscript 'Σ' being dropped when no confusion can arise).

Let M denote the centralizer of $\mathfrak{a}_{\mathfrak{p}}$ in K and M^* the normalizer of $\mathfrak{a}_{\mathfrak{p}}$ in K – then W may be identified with the quotient M^*/M (cf. Helgason [2, p.244]). More precisely the map which assigns to each coset m^*M $(m^* \in M^*)$ the restriction of the inverse transpose of $\mathrm{Ad}\,(m^*)$ $(= \mathrm{Ad}_G(m^*))$ to $\mathfrak{a}_{\mathfrak{p}}^{\vee}$ is an isomorphism of M^*/M onto W; any element in M^* which represents a given element w of W under this isomorphism is said to be *associated* with w.

Convention Whenever it is convenient (and it frequently will be), the group W may be viewed as acting in $\mathfrak{a}_{\mathfrak{p}}$ by duality.

The following observation concerning the nature of the connected components of M is important. Put $Z(A_{\mathfrak{p}}) = \mathrm{Ad}_G^{-1}(\mathrm{Ad}_G(K) \cap \exp \sqrt{-1}\mathfrak{a}_{\mathfrak{p}})$ (exp being considered as a map into the adjoint group $\mathrm{Int}\,(\mathfrak{g}_c)$ of $\mathfrak{g}_c$); since K is compaet, $Z(A_{\mathfrak{p}})$ is a finite group.

Lemma 1.1.3.8 Let G be a connected semi-simple Lie group with finite center, $G = KA_{\mathfrak{p}}N^+$ an Iwasawa decomposition for G; let M denote the centralizer of $A_{\mathfrak{p}}$ in K – then $\mathsf{M} = \mathsf{Z}(A_{\mathfrak{p}})M$ (M the identity component of M).

Proof It is clear that we may limit ourselves to the case when G is the adjoint group of $\mathfrak{g}$; this being so, we may then consider G as a subgroup of the adjoint group Int $(\mathfrak{g}_c)$ of $\mathfrak{g}_c$. It is clear that $\mathsf{M} \supset \mathsf{Z}(A_{\mathfrak{p}})M$; to prove the reverse inclusion, first observe that M does not depend on the choice of $\mathfrak{a}_{\mathfrak{k}}$ but rather only on $\mathfrak{a}_{\mathfrak{p}}$ – hence $\mathfrak{a}_{\mathfrak{k}}$ is at our disposal. So fix an $m \in \mathsf{M}$; in the real analytic subgroup U of Int $(\mathfrak{g}_c)$ corresponding to $\mathfrak{u}$, there exists a maximal torus containing both m and $\exp(\sqrt{-1}\mathfrak{a}_{\mathfrak{p}})$. [Here we have to keep in mind the following fact: Let S be a compact connected Lie group, T a toral subgroup of S; suppose that s is an element of S which commutes with every element of T – then there exists a torus $\tilde{T} \subset S$ such that $T \subset \tilde{T}$ and $s \in \tilde{T}$ (cf. Helgason [2, p.247]).] This means that we can choose $\mathfrak{a}_{\mathfrak{k}}$ in such a way that $m \in \exp(\mathfrak{a}_{\mathfrak{k}} + \sqrt{-1}\mathfrak{a}_{\mathfrak{p}})$ – therefore $m \in \exp(\mathfrak{a}_{\mathfrak{k}} + \sqrt{-1}\,\mathfrak{a}_{\mathfrak{p}}) \cap K \subset \mathsf{Z}(A_{\mathfrak{p}})M$, as desired. □

The preceding result thus provides us with the structure of the group of components of M; it is to be noted however that $\mathsf{Z}(A_{\mathfrak{p}})$ need *not* be abelian (a counterexample may be found in the universal covering group of **SL**(3, **R**) – see 1.4.1).

As above, let Σ denote the set of roots of the pair $(\mathfrak{g}, \mathfrak{a}_{\mathfrak{p}})$; suppose that $w = w_\lambda$, $\lambda \in \Sigma$ – in this case one may choose an $m_w \in \mathsf{M}^*$ associated with w in a special way. Thus let X be any non-zero element in $\mathfrak{g}^\lambda$. There exist unique elements $Y \in \mathfrak{k}$, $Z \in \mathfrak{p}$ such that $X = Y + Z$; evidently neither Y nor Z is zero.

Lemma 1.1.3.9 If X is normalized so that $-B(Y, Y) = B(Q_\lambda, Q_\lambda)^{-1}$, then $[Y, Z] = B(Q_\lambda, Q_\lambda)^{-2}Q_\lambda$ and $m_w = \exp(\pi Y)$ is an element of M^* with the property that

$$\gamma^w = w_\lambda(\gamma) = \gamma - 2\frac{(\gamma, \lambda)}{(\lambda, \lambda)}\lambda \qquad (\gamma \in \mathfrak{a}_{\mathfrak{p}}^{\vee}).$$

Proof The computation in Helgason [2, p.246] shows that $[Y, Z] = B(Q_\lambda, Q_\lambda)^{-2}Q_\lambda$ and that $\mathrm{Ad}(\exp(tY))Q_\lambda = \cos(t)Q_\lambda + \sin(t)Z$ for all real t; hence $\mathrm{Ad}(m_w)Q_\lambda = -Q_\lambda$. On the other hand it is clear that $\mathrm{Ad}(m_w)H = H$ whenever $\lambda(H) = 0$ – therefore $m_w \in \mathsf{M}^*$ and, for any $H \in \mathfrak{a}_{\mathfrak{p}}$, we have

$$\mathrm{Ad}(m_w)H = H - 2\frac{\lambda(H)}{\lambda(Q_\lambda)}Q_\lambda.$$

The fact that

$$\gamma^w = w_\lambda(\gamma) = \gamma - 2\frac{(\gamma, \lambda)}{(\lambda, \lambda)}\lambda \qquad (\gamma \in \mathfrak{a}_{\mathfrak{p}}^{\vee})$$

now follows by duality. □

Appendix 1 Here, for the convenience of the reader, we shall give a table which describes the root systems for the simple Lie algebras over **R** (cf. Araki [1]). This description is accomplished by means of the *Satake diagram*: Consider the Dynkin diagram of the given σ-fundamental system $\Psi \subset \Phi$; every root in

$\Psi \cap \Phi_-$ is denoted by a black circle ● and every root in $\Psi - (\Psi \cap \Phi_-)$ by a white circle ○; if $i' = j$ ($i \neq j$, $1 \leqslant i, j \leqslant l - l_-$ – cf. Lemma 1.1.3.2), then α_i and α_j are connected by a curved arrow ↶↷. All other unexplained notations are as in the text proper (in particular $l = l_\Phi$, $l_+ = l_\Sigma$).

Type	$\Phi \longleftrightarrow \Psi$	$\Sigma \longleftrightarrow \Upsilon$	$m(\lambda_i)$	$m(2\lambda_i)$
AI	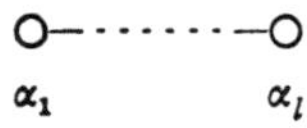	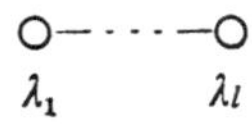	1	0
AII		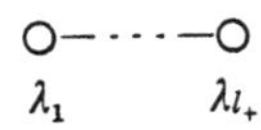 $(l = 2l_+ + 1, l_+ \geqslant 1)$	4	0
AIII	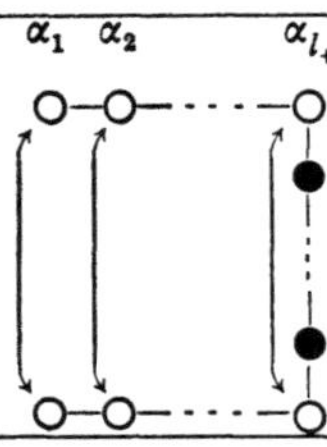	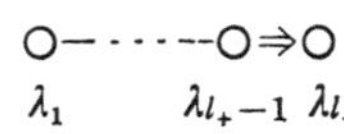 $\left(2 \leqslant l_+ \leqslant \frac{l}{2}\right)$	$2(i < l_+)$ $2(l - 2l_+ + 1)$ $(i = l_+)$	0 1
	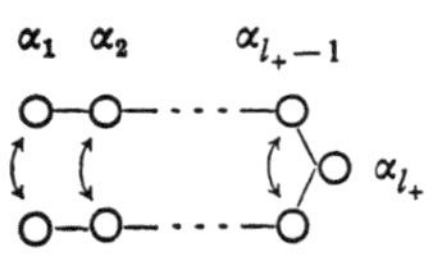	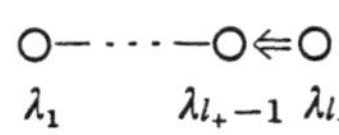 $(l = 2l_+ - 1, l_+ \geqslant 2)$	$2(i \leqslant l_+ - 1)$ $1(i = l_+)$	0 0
AIV	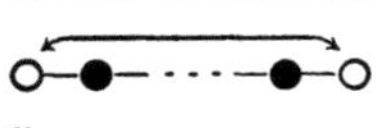		$2(l - 1)$	1
BI	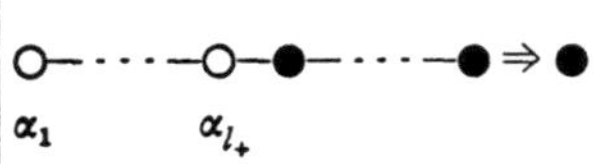	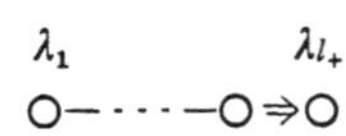 $(l \geqslant 2, 2 \leqslant l_+ \leqslant l)$	$1(i < l_+)$ $2(l - l_+) + 1$ $(i = l_+)$	0 0
BII	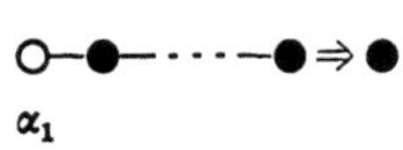		$2l - 1$	0
CI	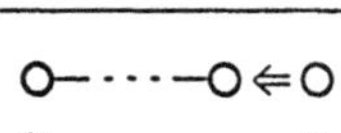	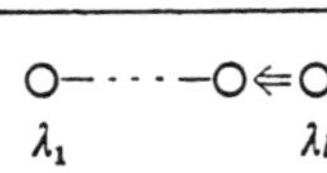	1	0

Type	$\Phi \longleftrightarrow \Psi$	$\Sigma \longleftrightarrow \Upsilon$	$m(\lambda_i)$	$m(2\lambda_i)$
CII	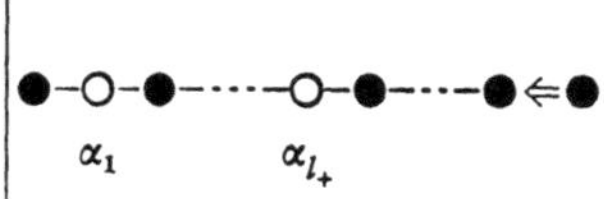	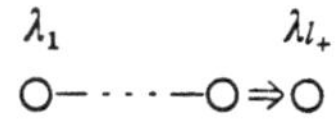	$4(i < l_+)$	0
		$\left(l \geqslant 3, 1 \leqslant l_+ \leqslant \frac{l-1}{2}\right)$	$4(l - 2l_+)$ $(i = l_+)$	3
	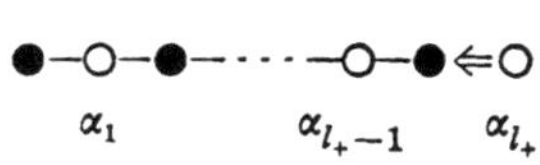	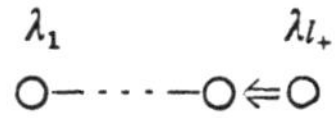	$4(i < l_+)$	0
		$(l = 2l_+, l_+ \geqslant 2)$	$3(i = l_+)$	0
DI	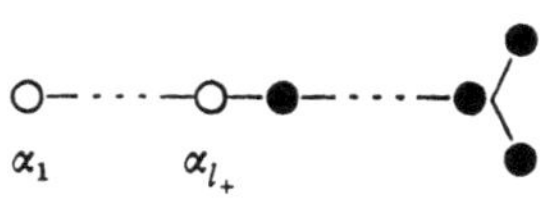	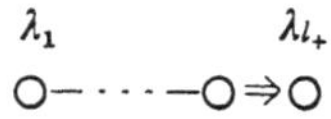	$1(i < l_+)$	0
		$(l \geqslant 4, 2 \leqslant l_+ \leqslant l-2)$	$2(l - l_+)$ $(i = l_+)$	0
	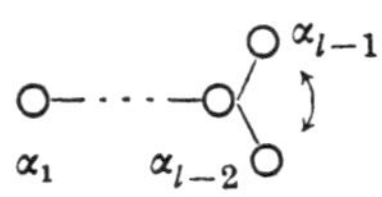	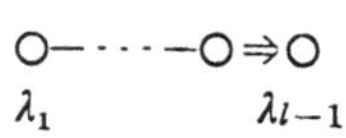	$1(i < l - 1)$	0
			$2(i = l - 1)$	0
	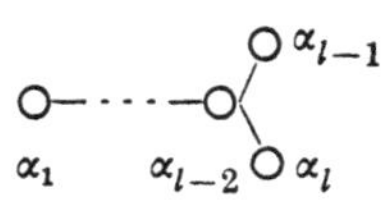	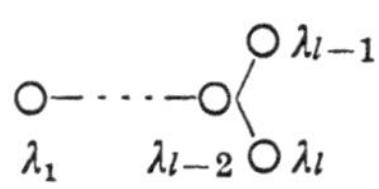	1	0
DII	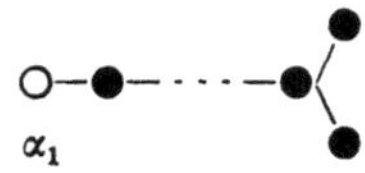	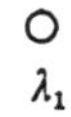	$2(l - 1)$	0
DIII	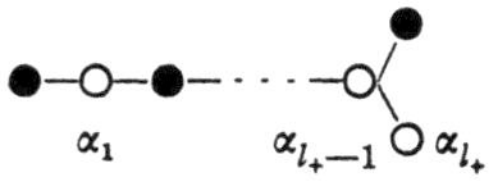	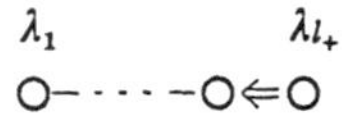	$4(i < l_+)$	0
		$(l = 2l_+, l_+ \geqslant 2)$	$1(i = l_+)$	0
	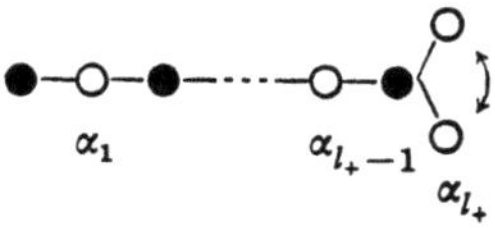	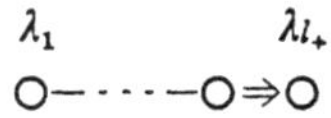	$4(i < l_+)$	0
		$(l = 2l_+ + 1, l_+ \geqslant 2)$	$4(i = l_+)$	1
EI	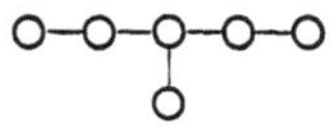	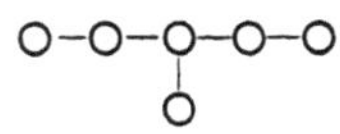	1	0

Type	$\Phi \longleftrightarrow \Psi$	$\Sigma \longleftrightarrow \Upsilon$	$m(\lambda_i)$	$m(2\lambda_i)$
EII	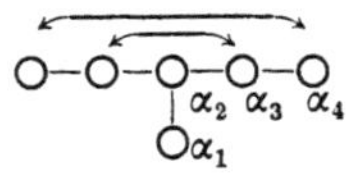	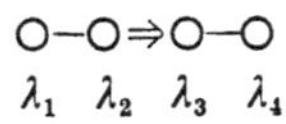	$1(i = 1, 2)$ $2(i = 3, 4)$	0 0
EIII	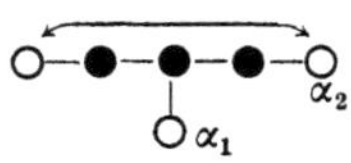	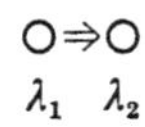	$6(i = 1)$ $8(i = 2)$	0 1
EIV	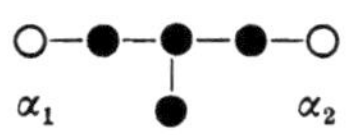	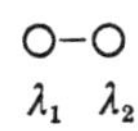	8	0
EV	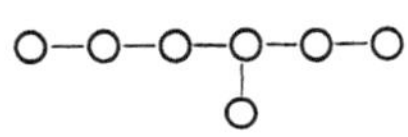	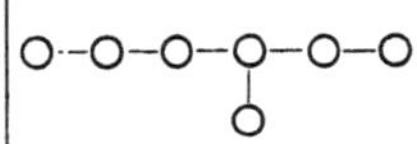	1	0
EVI	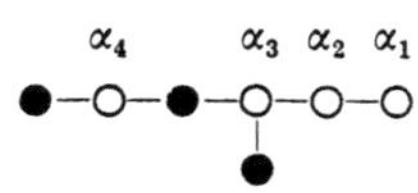	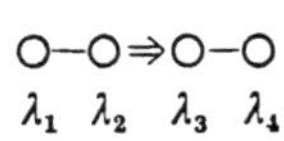	$1(i = 1, 2)$ $4(i = 3, 4)$	0 0
EVII	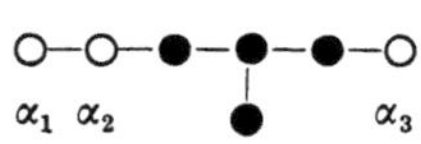	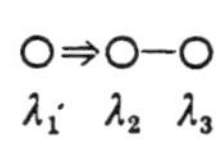	$1(i = 1)$ $8(i = 2, 3)$	0 0
EVIII	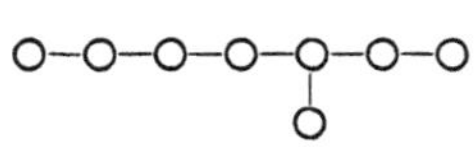	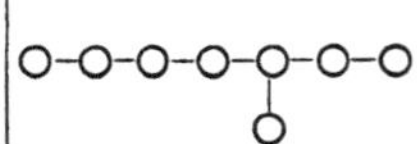	1	0
EIX	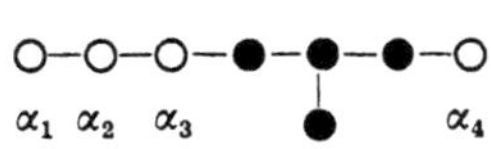	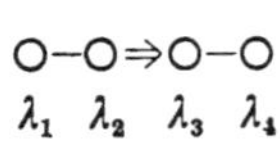	$1(i = 1, 2)$ $8(i = 3, 4)$	0 0
FI	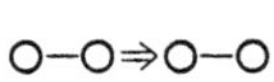	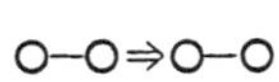	1	0
FII	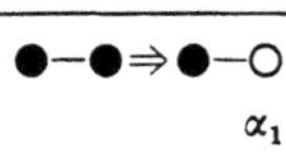		8	7
G	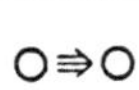	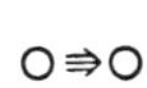	1	0

Appendix 2 Let Φ be a normal σ-system of roots in E; write, as usual, $E = E_+ + E_-$ and let Σ denote the root system in E_+ obtained by projecting the elements of Φ_+ onto E_+ – we shall give here a number of technical lemmas concerning the multiplicities $m(\lambda)$ of the elements $\lambda \in \Sigma$. [We recall that, by definition, the multiplicity $m(\lambda)$ of $\lambda \in \Sigma$ is the number of $\alpha \in \Phi_+$ such that $\tilde{\alpha} = \lambda$, i.e. $m(\lambda)$ is the cardinality of the set $\Sigma(\lambda) = \{\alpha \in \Phi_+ : \tilde{\alpha} = \lambda\}$.] In what follows it will be convenient to say that λ *is in* $\Sigma(\lambda)$ provided there exists an $\alpha \in \Sigma(\lambda)$ such that $\alpha = \alpha^\sigma$.

Lemma 1 Let $\lambda \in \Sigma$ – then $\lambda \in \Sigma(\lambda)$ iff $m(\lambda)$ is odd.

Proof If $\alpha \in \Sigma(\lambda)$, then so does α^σ. If $\alpha \in \Sigma(\lambda) - \{\lambda\}$, then $\alpha \neq \alpha^\sigma$ and hence $\Sigma(\lambda) - \{\lambda\}$ is a disjoint union of pairs (α, α^σ) – thus $\Sigma(\lambda) - \{\lambda\}$ has an even number of elements, which proves the lemma. □

Lemma 2 Let $\lambda \in \Sigma$ – we have:
(i) If $m(\lambda)$ is odd, then $2\lambda \notin \Sigma$;
(ii) If $m(\lambda)$ is even, and if $2\lambda \in \Sigma$, then $m(2\lambda)$ is odd.

Proof To prove (i), suppose to the contrary that $2\lambda \in \Sigma$ and let $\alpha \in \Phi_+$ be such that $\tilde{\alpha} = 2\lambda$ – then $(\alpha, \lambda) = 2(\lambda, \lambda) > 0$ as well as $|\alpha| \geqslant |\lambda|$. Bearing in mind that Φ is a reduced system of roots, inspection of the list of possibilities in 1.1.2 for the Cartan integer

$$A_{\alpha,\lambda} = 2\frac{(\alpha, \lambda)}{(\alpha, \alpha)}$$

shows that $|\alpha|^2|\lambda|^{-2} = 1, 2$, or 3. On the other hand $|\alpha|^2 \geqslant |\tilde{\alpha}|^2 = 4|\lambda|^2$, a contradiction. Let us now prove (ii). Since $m(\lambda)$ is even, there exists an $\alpha \in \Sigma(\lambda)$ such that $\alpha \neq \alpha^\sigma$. If $m(2\lambda)$ were even, there would exist a $\beta \in \Sigma(2\lambda)$ such that $\beta \neq \beta^\sigma$; because Φ is normal, it must be the case that $(\beta, \beta^\sigma) \leqslant 0$ (cf. Proposition 1.1.2.1); if $(\beta, \beta^\sigma) < 0$, then $\beta + \beta^\sigma \in \Phi$, hence $4\lambda \in \Sigma$, contradicting the fact that Σ is a root system in E_+ – thus $(\beta, \beta^\sigma) = 0$ and so $(\beta, \beta) = 8(\lambda, \lambda)$. The assumption that $m(2\lambda)$ is even implies that $2\lambda \notin \Phi$, hence $\alpha + \alpha^\sigma \notin \Phi$ – therefore $(\alpha, \alpha^\sigma) = 0$, whence $(\alpha, \alpha) = 2(\lambda, \lambda)$. Consequently $(\beta, \beta) = 4(\alpha, \alpha)$, a contradiction to the fact that Φ is reduced. □

Lemma 3 Let $\lambda \in \Sigma$. If $m(\lambda)$ is even, then:
(i) $2\lambda \in \Sigma$ iff there exists an element $\alpha \in \Sigma(\lambda)$ such that $(\alpha, \alpha^\sigma) < 0$ and in this case $(\beta, \beta^\sigma) < 0$ for all $\beta \in \Sigma(\lambda)$;
(ii) $2\lambda \notin \Sigma$ iff there exists an element $\alpha \in \Sigma(\lambda)$ such that $(\alpha, \alpha^\sigma) = 0$ and in this case $(\beta, \beta^\sigma) = 0$ for all $\beta \in \Sigma(\lambda)$.
[On the basis of what has been said above, this is immediate.]

Lemma 4 If $m(\lambda)$ is odd ($\lambda \in \Sigma$), then $(\alpha, \alpha^\sigma) = 0$ for all $\alpha \in \Sigma(\lambda) - \{\lambda\}$.

Proof If $(\alpha, \alpha^\sigma) < 0$ ($\alpha \in \Sigma(\lambda) - \{\lambda\}$), then $\alpha + \alpha^\sigma \in \Phi$ which implies that $2\lambda \in \Sigma$, contradicting Lemma 2(i). □

Lemma 5 (i) Let $\lambda \in \Sigma$ and suppose that $2\lambda \notin \Sigma$. If α, β are elements of $\Sigma(\lambda)$ such that $\alpha \neq \beta$, $\alpha \neq \beta^\sigma$, then $(\alpha, \beta) > 0$.
(ii) Let $\lambda \in \Sigma$ and suppose that $2\lambda \in \Sigma$. If α, β are elements of $\Sigma(\lambda)$ such that $\alpha \neq \beta$, $\alpha \neq \beta^\sigma$, then $(\alpha, \beta) \geqslant 0$ and $(\alpha, \beta) > 0$ iff $(\alpha, \beta^\sigma) = 0$.

Proof Let us prove (i). We may assume without loss of generality that $\beta \neq \beta^\sigma$ – then $(\alpha, \beta) + (\alpha, \beta^\sigma) = 2(\lambda, \lambda) > 0$. On the other hand, in view of Lemmas 3 and 4, $(\beta, \beta^\sigma) = 0$ – hence $(\beta, \beta) = 2(\lambda, \lambda)$ which implies that $A_{\beta,\alpha} + A_{\beta^\sigma,\alpha} = 2$. There are two possibilities for α: $\alpha \neq \alpha^\sigma$ or $\alpha = \alpha^\sigma$. If $\alpha \neq \alpha^\sigma$, then $(\alpha, \alpha^\sigma) = 0$ and $(\alpha, \alpha) = 2(\lambda, \lambda) = (\beta, \beta)$. Inspection of the list of possibilities for $A_{\alpha,\beta}$ in 1.1.2 shows that we must have $A_{\beta,\alpha} = A_{\beta^\sigma,\alpha} = 1 \Rightarrow (\alpha, \beta) > 0$. The case $\alpha = \alpha^\sigma \Rightarrow \alpha = \lambda \Rightarrow$

$$A_{\beta,\alpha} = 2(\lambda, \beta)/2(\lambda, \lambda) = 2(\lambda, \lambda)/2(\lambda, \lambda) = 1 \Rightarrow (\alpha, \beta) > 0,$$

which completes the proof of (i). To prove (ii) note first that $(\alpha, \beta + \beta^\sigma) = 2(\lambda, \lambda) > 0$. By Lemma 3(i), we see that $(\alpha, \alpha^\sigma) < 0$ and $(\beta, \beta^\sigma) < 0$, whence $(\alpha, \alpha) = (\beta, \beta) = 4(\lambda, \lambda)$. Thus $A_{\beta,\alpha} + A_{\beta^\sigma,\alpha} = 1$ which implies that $(\alpha, \beta) \geqslant 0$ and $(\alpha, \beta) > 0$ iff $(\alpha, \beta^\sigma) = 0$. □

1.1.4 Structure of the Nilpotent Constituent in an Iwasawa Decomposition

Let G be a connected semi-simple Lie group; write, after Iwasawa, $G = KA_\mathfrak{p}N^+$ – then the main task of the present number is to exhibit the connection that exists between the structure of N^+ (as well as N^-) and the Weyl group W of the pair $(\mathfrak{g}, \mathfrak{a}_\mathfrak{p})$. The notations of 1.1.3 will be used without comment.

We shall first establish two useful general lemmas on nilpotent groups.

Lemma 1.1.4.1 Let N be a connected, simply connected, real or complex nilpotent Lie group with Lie algebra $\mathfrak{n}$. Let $\mathfrak{n}_i (0 \leqslant i \leqslant r + 1)$ be a sequence of ideals in $\mathfrak{n}$ such that:

(1) $\mathfrak{n}_0 = \mathfrak{n}$, $\mathfrak{n}_{r+1} = \{0\}$;

(2) $\mathfrak{n}_i \supset \mathfrak{n}_{i+1}$ (strict inclusion) and $[\mathfrak{n}, \mathfrak{n}_i] \subset \mathfrak{n}_{i+1}$ $(0 \leqslant i \leqslant r)$.

If $\mathfrak{x}$ and $\mathfrak{y}$ are two mutually complementary subspaces of $\mathfrak{n}$ with the property that $\mathfrak{n}_i = \mathfrak{x} \cap \mathfrak{n}_i + \mathfrak{y} \cap \mathfrak{n}_i$ (direct sum of vector spaces) $(i = 0, \ldots)$, then the map $(X, Y) \mapsto \exp X \exp Y$ $(X \in \mathfrak{x}, \; Y \in \mathfrak{y})$ is an analytic diffeomorphism of $\mathfrak{x} \times \mathfrak{y}$ onto N.

Proof First note that $\mathfrak{n}_r$ is central and generates a closed simply connected central subgroup N_r of N such that N/N_r is simply connected (by hypothesis we have that $\mathfrak{n}_r \neq \{0\}$). Let $\tilde{\mathfrak{x}}, \tilde{\mathfrak{y}}$ be complementary subspaces to $\mathfrak{x} \cap \mathfrak{n}_r$ and $\mathfrak{y} \cap \mathfrak{n}_r$ in $\mathfrak{x}$ and $\mathfrak{y}$, respectively. By induction on r, the lemma may be assumed to be true for $\mathfrak{x}/(\mathfrak{x} \cap \mathfrak{n}_r)$ and $\mathfrak{y}/(\mathfrak{y} \cap \mathfrak{n}_r)$ in $\mathfrak{n}/\mathfrak{n}_r$. This implies that the map $(\tilde{X}, \tilde{Y}) \mapsto \exp \tilde{X} \exp \tilde{Y}$ is an analytic diffeomorphism of $\tilde{\mathfrak{x}} \times \tilde{\mathfrak{y}}$ onto $M = \exp(\tilde{\mathfrak{x}}) \exp(\tilde{\mathfrak{y}})$ and that $(m, n_r) \mapsto mn_r$ is an analytic diffeomorphism of $M \times N_r$ onto N. The statement of our lemma is therefore a consequence of: (1) The exponential map is an analytic diffeomorphism of $\mathfrak{n}$ onto N; (2) $\exp(X + Y) = \exp X \exp Y$ if $[X, Y] = 0$; (3) $\mathfrak{n}_r$ is central. □

Lemma 1.1.4.2 Let N be a connected, simply connected, real or complex nilpotent Lie group with Lie algebra $\mathfrak{n}$. Let $\mathfrak{n}_i$ $(0 \leqslant i \leqslant r + 1)$ be a sequence of ideals in $\mathfrak{n}$ such that:

(1) $\mathfrak{n}_0 = \mathfrak{n}$, $\mathfrak{n}_{r+1} = \{0\}$;

(2) $\mathfrak{n}_i \supset \mathfrak{n}_{i+1}$ (strict inclusion) and $[\mathfrak{n}, \mathfrak{n}_i] \subset \mathfrak{n}_{i+1}$ $(0 \leqslant i \leqslant r)$.

Let A_j $(1 \leqslant j \leqslant s)$ be endomorphisms of $\mathfrak{n}$ such that $A = \sum_{j=1}^{s} A_j$ is non-singular and $A_j\mathfrak{n}_i \subset \mathfrak{n}_i$ $(1 \leqslant j \leqslant s,\ 0 \leqslant i \leqslant r+1)$ – then the map

$$\phi,\ X \mapsto \log(\exp A_1X \exp A_2X \ldots \exp A_sX),$$

is a bijective polynomial map of $\mathfrak{n}$ onto itself whose inverse is also a polynomial map.

Proof In the first place it is clear that in any event ϕ is a polynomial map. Now if $r = 0$, then $\mathfrak{n}$ is abelian and $\phi(X) = AX$ and so all the statements of our lemma are obvious in this case. Thus we may assume that $r \geqslant 1$ and use induction on r. Let $X \mapsto X^*$ $(X \in \mathfrak{n})$ denote the projection of $\mathfrak{n}$ onto $\mathfrak{n}^* = \mathfrak{n}/\mathfrak{n}_r$, A_j^* the linear transformation in $\mathfrak{n}^*$ corresponding to A_j. Since $\mathfrak{n}_r^* = \{0\}$, the induction hypothesis is applicable to the triple $(\mathfrak{n}^*, \mathfrak{n}_i^*, A_j^*)$. Let $\mathfrak{m}$ be a subspace of $\mathfrak{n}$ complementary to $\mathfrak{n}_r$ and P the projection of $\mathfrak{n}$ onto $\mathfrak{m}$ corresponding to the decomposition $\mathfrak{n} = \mathfrak{m} + \mathfrak{n}_r$ – then we may choose a polynomial map π of $\mathfrak{m}$ into itself such that $\mathsf{P}X = \pi(\mathsf{P}\phi(X))$ for all $X \in \mathfrak{n}$. Since $A\mathfrak{n}_r \subset \mathfrak{n}_r$ and since A is non-singular, $A\mathfrak{n}_r = \mathfrak{n}_r$. It is clear that $\mathfrak{n}_r$ is central; thus $\phi(Y+Z) = \phi(Y) + AZ$ $(Y \in \mathfrak{n}, Z \in \mathfrak{n}_r)$ and so if $X = Y + Z$ $(Y \in \mathfrak{m}, Z \in \mathfrak{n}_r)$, it must then be the case that $AZ = \phi(X) - \phi(Y)$. On the other hand $Y = \mathsf{P}X = \pi(\mathsf{P}\phi(X))$, whence

$$\begin{aligned} X = Y + Z &= Y + A^{-1}(\phi(X) - \phi(Y)) \\ &= \pi(\mathsf{P}\phi(X)) + A^{-1}\phi(X) - A^{-1}\phi(\pi(\mathsf{P}\phi(X))). \end{aligned}$$

Put $\psi(X) = \pi(\mathsf{P}X) + A^{-1}X - A^{-1}\phi(\pi(\mathsf{P}X))$ $(X \in \mathfrak{n})$ – then ψ is a polynomial map and $X = \psi(\phi(X))(X \in \mathfrak{n})$, hence ϕ is injective. Finally, given $X \in \mathfrak{n}$, by the induction hypothesis we can choose $Y \in \mathfrak{m}$ such that $\phi(Y) \equiv X \bmod \mathfrak{n}_r$. Let $Z = A^{-1}(X - \phi(Y))$ – then $Z \in \mathfrak{n}_r$ and $\phi(Y+Z) = \phi(Y) + AZ = X$ which implies that ϕ is surjective. The proof of the lemma is therefore complete. □

For later use, let us note the following variant of Lemmas 1.1.4.1 and 1.1.4.2.

Lemma 1.1.4.3 Let N be a connected, simply connected, real or complex nilpotent Lie group with Lie algebra $\mathfrak{n}$. Let $\mathfrak{n}_i$ $(0 \leqslant i \leqslant r+1)$ be a sequence of ideals in $\mathfrak{n}$ such that:

(1) $\mathfrak{n}_0 = \mathfrak{n}$, $\mathfrak{n}_{r+1} = \{0\}$;

(2) $\mathfrak{n}_i \supset \mathfrak{n}_{i+1}$ (strict inclusion) and $[\mathfrak{n}, \mathfrak{n}_i] \subset \mathfrak{n}_{i+1}$ $(0 \leqslant i \leqslant r)$.

Suppose that $\mathfrak{x}$ and $\mathfrak{y}$ are two mutually complementary subspaces of $\mathfrak{n}$ such that $\mathfrak{n}_i = \mathfrak{x} \cap \mathfrak{n}_i + \mathfrak{y} \cap \mathfrak{n}_i$ (direct sum of vector spaces) $(i = 0, \ldots)$. Set $\mathfrak{x}_i = \mathfrak{x} \cap \mathfrak{n}_i$, $\mathfrak{y}_i = \mathfrak{y} \cap \mathfrak{n}_i$ and choose $\tilde{\mathfrak{x}}_k$, $\tilde{\mathfrak{y}}_k$ such that $\mathfrak{x}_i = \sum_{k=i}^{r} \tilde{\mathfrak{x}}_k$, $\mathfrak{y}_i = \sum_{k=i}^{r} \tilde{\mathfrak{y}}_k$ – then every element of N admits a unique representation in the form

$$n = \exp \tilde{X}_0 \ldots \exp \tilde{X}_r \exp \tilde{Y}_r \ldots \exp \tilde{Y}_0$$

where $\tilde{X}_k \in \tilde{\mathfrak{x}}_k$, $\tilde{Y}_k \in \tilde{\mathfrak{y}}_k$. Moreover, if $X \mapsto \phi(X)$ is an automorphism of $\mathfrak{n}$ leaving each $\mathfrak{n}_i$ invariant, if $\tilde{X}_i + \tilde{Y}_i \mapsto \tilde{X}_i' + \tilde{Y}_i'$ is the induced transformation on $\mathfrak{n}_i/\mathfrak{n}_{i+1} \sim \tilde{\mathfrak{x}}_i + \tilde{\mathfrak{y}}_i$, and if $\phi: N \to N$ is the induced automorphism of N with

$$\phi(n) = \exp \tilde{X}_0'' \ldots \exp \tilde{X}_r'' \exp \tilde{Y}_r'' \ldots \exp \tilde{Y}_0'',$$

then

$$\tilde{X}_i'' = \tilde{X}_i' + p_i(\tilde{X}_0, \ldots, \tilde{X}_{i-1}, \tilde{Y}_0, \ldots, \tilde{Y}_{i-1}),$$
$$\tilde{Y}_i'' = \tilde{Y}_i' + q_i(\tilde{X}_0, \ldots, \tilde{X}_{i-1}, \tilde{Y}_0, \ldots, \tilde{Y}_{i-1})$$

for certain polynomial functions p_i and q_i.

Let G be a connected semi-simple Lie group with finite center, $G = KA_{\mathfrak{p}}N^+$ an Iwasawa decomposition for G; as in 1.1.3 introduce the groups M, M^*, $W(= M^*/M)$, N^+, N^- etc.

Let $\mathfrak{a}_{\mathfrak{p}}$ denote the Lie algebra of $A_{\mathfrak{p}}$, Σ the set of roots of the pair $(\mathfrak{g}, \mathfrak{a}_{\mathfrak{p}})$. As usual, an element $H \in \mathfrak{a}_{\mathfrak{p}}$ is said to be *regular* in $\mathfrak{a}_{\mathfrak{p}}$ (or Σ-*regular*) if $\lambda(H) \neq 0$ for all $\lambda \in \Sigma$. The set of regular elements in $\mathfrak{a}_{\mathfrak{p}}$ will be denoted by $\mathfrak{a}_{\mathfrak{p}}'$; it is clear that an element H in $\mathfrak{a}_{\mathfrak{p}}$ belongs to $\mathfrak{a}_{\mathfrak{p}}'$ iff it lies in some Weyl chamber $\mathscr{C} \subset \mathfrak{a}_{\mathfrak{p}}$.

Let us recall the following theorem due to Harish-Chandra [8] (the proof may be found in Helgason [2, p.231].)

Theorem 1.1.4.4 Let G be a connected semi-simple Lie group with finite center, $G = KA_{\mathfrak{p}}N^+$ an Iwasawa decomposition for G; let H be an element in $\mathfrak{a}$ such that $\Pi_{\alpha \in P_+} \alpha(H) \neq 0$, $h = \exp H$. Then:

(i) The mapping ϕ, $n \mapsto \mathrm{Ad}\,(n)H - H$, is an analytic diffeomorphism of N^+ onto $\mathfrak{n}^+$;

(ii) The mapping ψ, $X \mapsto \exp(-H)\exp(H + X)$, is an analytic diffeomorphism of $\mathfrak{n}^+$ onto N^+;

(iii) The mapping ξ, $n \mapsto h^{-1} nhn^{-1}$, is an analytic diffeomorphism of N^+ onto itself.

[In particular this theorem is true for Σ-regular elements in $\mathfrak{a}_{\mathfrak{p}}$; of course there is an analogous result for N^- . . . .]

Fix an element w in the Weyl group W of the pair $(\mathfrak{g}, \mathfrak{a}_{\mathfrak{p}})$; choose an element m_w in M^* associated with w; given $\gamma \in \mathfrak{a}_{\mathfrak{p}}^\vee$, let γ^w denote the image of γ under the transpose of $\mathrm{Ad}\,(m_w^{-1})$ restricted to $\mathfrak{a}_{\mathfrak{p}}$. If λ belongs to Σ, then so does λ^w and $\mathrm{Ad}\,(m_w)\mathfrak{g}^\lambda = \mathfrak{g}^{\lambda^w}$ – therefore m_w (or rather ${}^t\mathrm{Ad}\,(m_w^{-1})$) permutes the elements of Σ. Define now subalgebras $\mathfrak{n}_w^+$ and $\mathfrak{u}_w^+$ of $\mathfrak{n}^+$ by

$$\mathfrak{n}_w^+ = \mathfrak{n}^+ \cap \mathrm{Ad}\,(m_w)\mathfrak{n}^+, \quad \mathfrak{u}_w^+ = \mathfrak{n}^+ \cap \mathrm{Ad}\,(m_w)\mathfrak{n}^-.$$

As the notations imply, $\mathfrak{n}_w^+$ and $\mathfrak{u}_w^+$ depend only on the residue class of m_w mod M; furthermore we have $\mathfrak{n}^+ = \mathfrak{n}_w^+ + \mathfrak{u}_w^+$ (direct sum of vector spaces). [In fact the Lie algebra $\mathfrak{n}_w^+$ (respectively $\mathfrak{u}_w^+$) is spanned by the $\mathfrak{g}^\lambda$ $(\lambda \in \Sigma^+)$ where $\lambda^{w^{-1}} > 0$ (respectively $\lambda^{w^{-1}} < 0$).]

Proposition 1.1.4.5 Let G be a connected semi-simple Lie group with finite center, $G = KA_{\mathfrak{p}}N^+$ an Iwasawa decomposition for G; fix an element w in the Weyl group W of the pair $(\mathfrak{g}, \mathfrak{a}_{\mathfrak{p}})$; set $N_w^+ = \exp(\mathfrak{n}_w^+)$, $U_w^+ = \exp(\mathfrak{u}_w^+)$ – then N_w^+ and U_w^+ are closed simply connected subgroups of N^+, $N_w^+ \cap U_w^+ = \{1\}$, and $N^+ = N_w^+ U_w^+ (= U_w^+ N_w^+)$.

Proof Let $\lambda_0 < \lambda_1 < \ldots < \lambda_r$ be the positive roots arranged in increasing order – then

$$\mathfrak{n}_i = \sum_{j \geqslant i} \mathfrak{g}^{\lambda_j},\ \mathfrak{x} = \mathfrak{n}_w^+,\ \mathfrak{y} = \mathfrak{u}_w^+\ (0 \leqslant i \leqslant r,\ \mathfrak{n}_{r+1} = \{0\})$$

satisfy the conditions of Lemma 1.1.4.1 from which we can read off our assertion. □

In the sequel, we shall find it convenient to write U_w^- for $m_w^{-1} U_w^+ m_w$; evidently $U_w^- = N^- \cap m_w^{-1} N^+ m_w$ and the Lie algebra $\mathfrak{u}_w^-$ of U_w^- is spanned by those $\mathfrak{g}^{-\lambda} (\lambda \in \Sigma^+)$ such that $\lambda^w < 0$.

A root λ of the pair $(\mathfrak{g}, \mathfrak{a}_{\mathfrak{p}})$ is said to be *indivisible* if $\lambda/2$ is not a root; the subset Σ_0 of Σ comprised of indivisible roots is a reduced root system. Given an element w in W, we shall agree to denote by $\Delta_0(w)$ the positive indivisible roots λ such that $w(\lambda) < 0$. Let $w = w_1 \ldots w_n$ be a minimal expression for w as a product of reflections with respect to the simple roots λ_i in the fundamental system Υ associated with Σ^+ ($w_k = w_{\lambda_{i(k)}}$, $1 \leqslant k \leqslant n$) – then $n = L(w)$, the *length* of w. Let us recall that the elements in $\Delta_0(w)$ admit a simple description; in fact they are precisely the roots $\mu_k = w_n \ldots w_{k+1}(\lambda_{i(k)})$, $k = 1, \ldots, n = L(w)$ (cf. the Remark following Theorem 1.1.2.6). In passing we observe that if w, w', and w'' are three elements in W such that $w = w'w''$ with $L(w) = L(w') + L(w'')$, then $\Delta_0(w) = \Delta_0(w'') \cup w''^{-1}\Delta_0(w')$; in particular $w''^{-1}\Delta_0(w')$ is a set of positive roots.

Proposition 1.1.4.6 (Schiffmann) Let G be a connected semi-simple Lie group with finite center, $G = KA_{\mathfrak{p}}N^+$ an Iwasawa decomposition for G; suppose that w, w', and w'' are three elements in the Weyl group W such that $w = w'w''$ with $L(w) = L(w') + L(w'')$ – then the mapping $(u', u'') \mapsto m_{w''}^{-1} u' m_{w''} u''$ of $U_{w'}^- \times U_{w''}^-$ into U_w^- is an isomorphism of analytic varieties.

Proof To begin with, observe that $\mathfrak{u}_w^- = \mathrm{Ad}(m_{w''}^{-1})\mathfrak{u}_{w'}^- + \mathfrak{u}_{w''}^-$. This being so, let us suppose that $L(w') = 1$ and that λ is a simple root such that $w' = w_\lambda$ – then one has $[\mathrm{Ad}(m_{w''}^{-1})\mathfrak{u}_{w_\lambda}^-, \mathfrak{u}_{w''}^-] \subset \mathfrak{u}_{w''}^-$. [In this case we have $\Delta_0(w_\lambda) = \lambda$; thus our problem is to show that if $\lambda'' \in \Delta_0(w'')$ and if $w''^{-1}(\lambda) + \lambda''$ is a root, then $w''^{-1}(\lambda) + \lambda'' \in \Delta_0(w'')$. It is clear that $\Delta_0(w)$ is a closed system of roots; therefore, if $w''^{-1}(\lambda) + \lambda''$ is a root but does not belong to $\Delta_0(w'')$, then we must necessarily have $w''^{-1}(\lambda) + \lambda'' \in w''^{-1}\Delta_0(w_\lambda)$ (cf. supra), that is we must have $\lambda + w''(\lambda'') = \lambda$, an impossibility. . . .] Therefore the nilpotent Lie algebra $\mathfrak{u}_w^-$ is the direct sum of two subalgebras, the one being normalized by the other; because U_w^- is simply connected, our proposition, in this situation, reduces to Lemma 1.1.4.1 in the theory of nilpotent Lie groups. . . . Turning to the general case, let us proceed by induction on $L(w')$; accordingly we may assume that the proposition has been established for all w' with $L(w') \leqslant n' - 1$ and that $L(w') = n'$. Choose a simple root λ such that $w' = w_\lambda \tilde{w}'$ with $L(\tilde{w}') = n' - 1$; set $\tilde{w}'' = \tilde{w}'w''$ – then $w = w'w'' = w_\lambda \tilde{w}'w'' = w_\lambda \tilde{w}''$ with $L(w) = 1 + L(\tilde{w}'')$, $L(\tilde{w}'w'') = L(\tilde{w}') + L(w'')$, and $L(w_\lambda \tilde{w}') = 1 + L(\tilde{w}')$. Consider now the commutative diagram

$$\begin{array}{ccc} U_{w_\lambda}^- \times U_{\tilde{w}'}^- \times U_{w''}^- & \longrightarrow & U_{w_\lambda}^- \times U_{\tilde{w}''}^- \\ \downarrow & & \downarrow \\ U_{w'}^- \times U_{w''}^- & \longrightarrow & U_w^- \end{array}$$

where the arrows are defined by means of the decompositions $w' = w_\lambda \tilde{w}'$, $\tilde{w}'' = \tilde{w}'w''$, $w = w_\lambda \tilde{w}''$, and $w = w'w''$. According to the induction hypothesis, three of these arrows are isomorphisms of analytic varieties; it is therefore the same with the fourth. □

Appendix This Appendix is devoted to giving the proof of the proposition whose statement appears below; the proposition will turn out to be of some importance in the theory of zonal spherical functions. So let G be a connected semi-simple Lie group with finite center, $G = KA_\mathfrak{p}N^+$ an Iwasawa decomposition for G; we shall assume that G is of split-rank 1, whence $\dim(A_\mathfrak{p}) = 1$. Choose an element λ in Σ^+ such that 2λ is the only other possible element in Σ^+; let $m(\lambda)$ (respectively $m(2\lambda)$) denote the multiplicity of λ (respectively 2λ); let H be the element in $\mathfrak{a}_\mathfrak{p}$ such that $\lambda(H) = 2$ – then $\mathfrak{a}_\mathfrak{p} = \mathbf{R}H$ and the complex dual of $\mathfrak{a}_\mathfrak{p}$ may be identified with $\mathbf{C}$ by assigning to each complex number c the linear form $tH \mapsto ct$ (hence, in particular, $\rho = 2^{-1}(m(\lambda)\lambda + m(2\lambda)2\lambda) \sim m(\lambda) + 2m(2\lambda)$). The quadratic form Q on $\mathfrak{g}$ defined by the rule $Q(X) = 4(X, X)_\theta/(H, H)_\theta (X \in \mathfrak{g})$ is positive and non-degenerate (here, of course, θ is the Cartan involution associated with the pair $(\mathfrak{g}, \mathfrak{k})$); evidently Q is K-invariant.

Proposition (Helgason-Schiffmann) Let us keep to the above notations; fix an element $X_{-\lambda}$(respectively $X_{-2\lambda}$) in $\mathfrak{g}^{-\lambda}$ (respectively $\mathfrak{g}^{-2\lambda}$); put

$$n = \exp(X_{-\lambda} + X_{-2\lambda}) \in N^-.$$

We have:

(i) If $n \in \kappa(n)\eta(n)N^+$ where $\kappa(n) \in K$ and $\eta(n) \in A_\mathfrak{p}$, then $\eta(n) = \exp(tH/2)$ with $e^{2t} = (1 + 2^{-1}Q(X_{-\lambda}))^2 + 2Q(X_{-2\lambda})$;

(ii) If $n \neq 1$, then there exists one and only one element h in $A_\mathfrak{p}$ such that $m_w n \in N^- \mathrm{M} h N^+$ – in fact $h = \exp(tH/2)$ with $e^{2t} = Q(X_{-\lambda})^2/4 + 2Q(X_{-2\lambda})$.

[Here m_w denotes an element in M^* associated with the unique non-trivial element w of the Weyl group W of the pair $(\mathfrak{g}, \mathfrak{a}_\mathfrak{p})$.]

As regards assertion (ii), the Bruhat decomposition for G tells us that $G = N^-\mathrm{M}A_\mathfrak{p}N^+ \cup m_w \mathrm{M}A_\mathfrak{p}N^+$ with $N^- \cap \mathrm{M}A_\mathfrak{p}N^+ = \{1\}$ (see 1. 2. 3); the existence and uniqueness of h are therefore clear. The a priori proof of the remaining assertions of our proposition is due to R. Godement; we shall proceed via a series of lemmas.

Lemma 1 Let $\mu \in \Sigma$; let $X \in \mathfrak{g}^{\mu}$ – then $[X, \theta(X)] = t(X)H$ with $4t(X) = -\mu(H)Q(X)$.

Proof The relation $[X, \theta(X)] = -\theta([X, \theta(X)])$ implies that $[X, \theta(X)]$ lies in $\mathfrak{p}$; on the other hand $[X, \theta(X)] \in [\mathfrak{g}^{\mu}, \mathfrak{g}^{-\mu}] \subset \mathfrak{g}^0(= \mathfrak{m} + \mathfrak{a}_{\mathfrak{p}})$ – therefore $[X, \theta(X)] \in \mathfrak{a}_{\mathfrak{p}}$. Because $\dim(\mathfrak{a}_{\mathfrak{p}}) = 1$, we can assume that $[X, \theta(X)] = t(X)H$; since $B([X, H], \theta(X)) + B(H, [X, \theta(X)]) = 0$, we have $-\mu(H)B(X, \theta(X)) = t(X)B(H, \theta(H))$, as desired. □

Lemma 2 Let $\mu = \pm\lambda$; let $X_{\mu} \in \mathfrak{g}^{\mu}$ and $X_{2\mu} \in \mathfrak{g}^{2\mu}$ – then $\mathrm{ad}(X_{\mu})^2(\theta(X_{2\mu})) \in \mathfrak{m}$.

Proof Since $\theta(\mathfrak{g}^{2\mu}) = \mathfrak{g}^{-2\mu}$, it follows that $\mathrm{ad}(X_{\mu})^2(\theta(X_{2\mu})) \in \mathfrak{g}^0$; but, as a simple calculation quickly shows, $B(H, \mathrm{ad}(X_{\mu})^2(\theta(X_{2\mu}))) = 0$ – therefore $\mathrm{ad}(X_{\mu})^2(\theta(X_{2\mu}))$ actually lies in $\mathfrak{m}$. □

Lemma 3 Let $X_{-\lambda} \in \mathfrak{g}^{-\lambda}$ and $X_{-2\lambda} \in \mathfrak{g}^{-2\lambda}$; put $X_{\lambda} = \theta(X_{-\lambda})$, $X_{2\lambda} = \theta(X_{-2\lambda})$ – then

(i) $\mathrm{ad}(X_{\lambda})\,\mathrm{ad}(X_{-\lambda})X_{2\lambda} = -4t(X_{-\lambda})X_{2\lambda}$;
(ii) $\mathrm{ad}(X_{\lambda})^2\,\mathrm{ad}(X_{-\lambda})^2X_{2\lambda} = 24t(X_{-\lambda})^2X_{2\lambda}$;
(iii) $\mathrm{ad}(X_{-\lambda})^4X_{2\lambda} = 24t(X_{-\lambda})^2X_{-2\lambda}$;
(iv) $\mathrm{ad}(X_{-\lambda})^3X_{2\lambda} = -6t(X_{-\lambda})[X_{\lambda}, X_{-2\lambda}]$;
(v) $Q(\mathrm{ad}(X_{-\lambda})^2X_{2\lambda}) = 24t(X_{-\lambda})^2Q(X_{-2\lambda})$;
(vi) $Q(\mathrm{ad}(X_{-\lambda})^4X_{2\lambda}) = (24)^2t(X_{-\lambda})^4Q(X_{-2\lambda})$;
(vii) $Q(\mathrm{ad}(X_{-\lambda})X_{2\lambda}) = 4t(X_{-\lambda})Q(X_{-2\lambda})$.

Proof Let us prove (i). Thus we have

$$\begin{aligned}\mathrm{ad}(X_{\lambda})\,\mathrm{ad}(X_{-\lambda})X_{2\lambda} &= \mathrm{ad}(X_{\lambda})[X_{-\lambda}, X_{2\lambda}]\\ &= [[X_{\lambda}, X_{-\lambda}], X_{2\lambda}] + [X_{-\lambda}, [X_{\lambda}, X_{2\lambda}]]\end{aligned}$$

and so, since $[X_{\lambda}, X_{2\lambda}] \in \mathfrak{g}^{3\lambda} = \{0\}$, it then follows from Lemma 1 that

$$\mathrm{ad}(X_{\lambda})\,\mathrm{ad}(X_{-\lambda})X_{2\lambda} = -t(X_{-\lambda})[H, X_{2\lambda}] = -4t(X_{-\lambda})X_{2\lambda}.$$

Let us prove (ii). Thus, bearing in mind Lemma 1 and relation (i), we find that

$$\begin{aligned}\mathrm{ad}(X_{\lambda})^2\,\mathrm{ad}(X_{-\lambda})^2X_{2\lambda} &= \mathrm{ad}(X_{\lambda})\{[\mathrm{ad}(X_{\lambda})X_{-\lambda}, \mathrm{ad}(X_{-\lambda})X_{2\lambda}]\\ &\quad + [X_{-\lambda}, \mathrm{ad}(X_{\lambda})\,\mathrm{ad}(X_{-\lambda})X_{2\lambda}]\}\\ &= \mathrm{ad}(X_{\lambda})\{-t(X_{-\lambda})[H, \mathrm{ad}(X_{-\lambda})X_{2\lambda}]\\ &\quad - 4t(X_{-\lambda})[X_{-\lambda}, X_{2\lambda}]\}.\end{aligned}$$

But $\mathrm{ad}(X_{-\lambda})X_{2\lambda} \in \mathfrak{g}^{\lambda}$, whence $[H, \mathrm{ad}(X_{-\lambda})X_{2\lambda}] = 2[X_{-\lambda}, X_{2\lambda}]$ – therefore

$$\mathrm{ad}(X_{\lambda})^2\,\mathrm{ad}(X_{-\lambda})^2X_{2\lambda} = -6t(X_{-\lambda})\,\mathrm{ad}(X_{\lambda})\,\mathrm{ad}(X_{-\lambda})X_{2\lambda} = 24t(X_{-\lambda})^2X_{2\lambda}.$$

Let us prove (iii). An application of the Cartan involution θ to the equality in (ii) gives

$$\mathrm{ad}(X_{-\lambda})^2\,\mathrm{ad}(X_{\lambda})^2X_{-2\lambda} = 24t(X_{-\lambda})^2X_{-2\lambda}.$$

Therefore

$$\mathrm{ad}(X_{-\lambda})^4X_{2\lambda} = \mathrm{ad}(X_{-\lambda})^2\,\mathrm{ad}(X_{\lambda})^2X_{-2\lambda} = 24t(X_{-\lambda})^2X_{-2\lambda}.$$

Let us prove (iv). Utilizing Lemma 1 and the transform of relation (i) under θ, we find that

$$\begin{aligned}\mathrm{ad}\,(X_{-\lambda})^3X_{2\lambda} &= \mathrm{ad}\,(X_{-\lambda})\,\mathrm{ad}\,(X_{-\lambda})^2X_{2\lambda} = \mathrm{ad}\,(X_{-\lambda})\,\mathrm{ad}\,(X_\lambda)^2X_{-2\lambda}\\ &= [\mathrm{ad}\,(X_{-\lambda})X_\lambda, \mathrm{ad}\,(X_\lambda)X_{-2\lambda}] + [X_\lambda, \mathrm{ad}\,(X_{-\lambda})\,\mathrm{ad}\,(X_\lambda)X_{-2\lambda}]\\ &= t(X_{-\lambda})[H, [X_\lambda, X_{-2\lambda}]] - 4t(X_{-\lambda})[X_\lambda, X_{-2\lambda}]\\ &= -6t(X_{-\lambda})[X_\lambda, X_{-2\lambda}].\end{aligned}$$

Let us prove (v). Thus, bearing in mind Lemma 2 and relation (iii), we have

$$\begin{aligned}B(\mathrm{ad}\,(X_{-\lambda})^2X_{2\lambda}, \theta(\mathrm{ad}\,(X_{-\lambda})^2X_{2\lambda})) &= B(\mathrm{ad}\,(X_{-\lambda})^2X_{2\lambda}, \mathrm{ad}\,(X_{-\lambda})^2X_{2\lambda})\\ &= B(\mathrm{ad}(X_{-\lambda})^4X_{2\lambda}, X_{2\lambda})\\ &= 24t(X_{-\lambda})^2B(X_{-2\lambda}, X_{2\lambda})\end{aligned}$$

which, according to the definition of Q, is equivalent to (v). Relation (vi) is a trivial consequence of (iii) while (vii) follows without difficulty from (i). Hence the lemma. □

Let us turn now to the proof of our proposition; without loss of generality it can be supposed that G is the adjoint group of $\mathfrak{g}$.

As for the first assertion, we shall assume to begin with that 2λ is a root of the pair $(\mathfrak{g}, \mathfrak{a}_\mathfrak{p})$. If $n = \exp(X_{-\lambda} + X_{-2\lambda}) \in N^-$ and if $X_{-2\lambda} \neq 0$, then $X_{2\lambda}$ is a dominant vector for the adjoint representation and one has $n(X_{2\lambda}) = \eta(n)^{2\lambda}\kappa(n)(X_{2\lambda})$, whence $Q(n(X_{2\lambda})) = \eta(n)^{4\lambda}Q(X_{2\lambda})$ (Q is K-invariant). The spaces $\mathfrak{g}^\mu$ ($\mu = 0, \pm\lambda, \pm 2\lambda$) are mutually orthogonal relative to Q; the components of $n(X_{2\lambda})$ are

$$\begin{aligned}&\mathfrak{g}^{2\lambda}:\ X_{2\lambda},\\ &\mathfrak{g}^{\lambda}:\ \mathrm{ad}\,(X_{-\lambda})X_{2\lambda},\\ &\mathfrak{g}^{0}:\ 2^{-1}\,\mathrm{ad}\,(X_{-\lambda})^2X_{2\lambda} + \mathrm{ad}\,(X_{-2\lambda})X_{2\lambda} = 2^{-1}\,\mathrm{ad}\,(X_{-\lambda})^2X_{2\lambda} + t(X_{-2\lambda})H,\\ &\mathfrak{g}^{-\lambda}:\ 6^{-1}\,\mathrm{ad}\,(X_{-\lambda})^3X_{2\lambda} + \mathrm{ad}\,(X_{-\lambda})\,\mathrm{ad}\,(X_{-2\lambda})X_{2\lambda}\\ &\qquad = -t(X_{-\lambda})\,\mathrm{ad}\,(X_\lambda)X_{-2\lambda} + 2t(X_{-2\lambda})X_{-\lambda},\\ &\mathfrak{g}^{-2\lambda}:\ 24^{-1}\,\mathrm{ad}\,(X_{-\lambda})^4X_{2\lambda} + 2^{-1}\,\mathrm{ad}\,(X_{-\lambda})^2\,\mathrm{ad}\,(X_{-2\lambda})X_{2\lambda} + 2^{-1}\,\mathrm{ad}\,(X_{-2\lambda})^2X_{2\lambda}\\ &\qquad = t(X_{-\lambda})^2X_{-2\lambda} + 2t(X_{-2\lambda})X_{-2\lambda}.\end{aligned}$$

Therefore

$$\begin{aligned}Q(n(X_{2\lambda})) &= Q(X_{2\lambda}) + Q(\mathrm{ad}\,(X_{-\lambda})X_{2\lambda}) + Q(2^{-1}\,\mathrm{ad}\,(X_{-\lambda})^2X_{2\lambda} + t(X_{-2\lambda})H)\\ &\quad + Q(2t(X_{-2\lambda})X_{-\lambda} - t(X_{-\lambda})\,\mathrm{ad}\,(X_\lambda)X_{-2\lambda}) + (t(X_{-\lambda})^2 + 2t(X_{-2\lambda}))^2Q(X_{-2\lambda}).\end{aligned}$$

Let us evaluate the different terms in this sum. According to Lemma 3 (vii), we have $Q(\mathrm{ad}\,(X_{-\lambda})X_{2\lambda}) = 4t(X_{-\lambda})Q(X_{-2\lambda})$; since $\mathfrak{a}_\mathfrak{p}$ and $\mathfrak{m}$ are mutually orthogonal, Lemma 2, in conjunction with Lemma 3 (v), tells us that

$$\begin{aligned}Q(2^{-1}\,\mathrm{ad}\,(X_{-\lambda})^2X_{2\lambda} + t(X_{-2\lambda})H) &= 4^{-1}Q(\mathrm{ad}\,(X_{-\lambda})^2X_{2\lambda}) + t(X_{-2\lambda})^2Q(H)\\ &= 6t(X_{-\lambda})^2Q(X_{-2\lambda}) + 4t(X_{-2\lambda})^2.\end{aligned}$$

Furthermore

$$\begin{aligned}&Q(2t(X_{-2\lambda})X_{-\lambda} - t(X_{-\lambda})\,\mathrm{ad}\,(X_\lambda)X_{-2\lambda}) = 4t(X_{-2\lambda})^2Q(X_{-\lambda})\\ &\quad + t(X_{-\lambda})^2Q(\mathrm{ad}\,(X_\lambda)X_{-2\lambda}) + 8B(2t(X_{-2\lambda})X_{-\lambda}, -t(X_{-\lambda})\,\mathrm{ad}\,(X_{-\lambda})X_{2\lambda})/B(H, H);\end{aligned}$$

the last term in this sum is zero since it is proportional to $B(X_{-\lambda}, \mathrm{ad}\,(X_{-\lambda})X_{2\lambda}) = -B(\mathrm{ad}\,(X_{-\lambda})X_{-\lambda}, X_{2\lambda}) = 0$; utilizing the transform of Lemma 3 (vii) by θ then provides us with the relation

$$Q(2t(X_{-2\lambda})X_{-\lambda} - t(X_{-\lambda})\,\mathrm{ad}\,(X_\lambda)X_{-2\lambda}) = 4t(X_{-2\lambda})^2Q(X_{-\lambda}) + 4t(X_{-\lambda})^3Q(X_{-2\lambda}).$$

We therefore have

$$Q(n(X_{2\lambda}))/Q(X_{2\lambda}) = [(1 + 2^{-1}Q(X_{-\lambda}))^2 + 2Q(X_{-2\lambda})]^2,$$

whence

$$e^{4t} = \eta(n)^{4\lambda} = Q(n(X_{2\lambda}))/Q(X_{2\lambda}) = [(1 + 2^{-1}Q(X_{-\lambda}))^2 + 2Q(X_{-2\lambda})]^2$$

which is indeed the sought for result. The preceding calculations were made under the assumption that $X_{-2\lambda}$ is not zero; however, in view of the fact that the mapping $n \mapsto \eta(n)$ of N^- into $A_\mathfrak{p}$ is continuous, the formula is also true if $X_{-2\lambda} = 0$. Finally, if 2λ is not a root of the pair $(\mathfrak{g}, \mathfrak{a}_\mathfrak{p})$, then $n = \exp(X_{-\lambda}) \in N^-$ and if $X_{-\lambda} \neq 0$, then X_λ is a dominant vector for the adjoint representation, whence $Q(n(X_\lambda))/Q(X_\lambda) = \eta(n)^{2\lambda}$. Since

$$\begin{aligned} n(X_\lambda) &= X_\lambda + [X_{-\lambda}, X_\lambda] + 2^{-1}[X_{-\lambda}, [X_{-\lambda}, X_\lambda]] \\ &= X_\lambda + t(X_{-\lambda})H + t(X_{-\lambda})X_{-\lambda}, \end{aligned}$$

we have

$$\begin{aligned} Q(n(X_\lambda)) &= Q(X_\lambda)(1 + 4t(X_{-\lambda})^2/Q(X_{-\lambda}) + t(X_{-\lambda})^2) \\ &= Q(X_\lambda)(1 + 2^{-1}Q(X_{-\lambda}))^2 \end{aligned}$$

from which the desired result then follows at once.

We come now to the second assertion. To begin with, let us suppose that 2λ is a root of the pair $(\mathfrak{g}, \mathfrak{a}_\mathfrak{p})$ and that $X_{2\lambda} \neq 0$; since $m_w n \in \tilde{n}mhN^+$ ($\tilde{n} \in N^-$; $mh \in MA_\mathfrak{p}$), we have $m_w n(X_{2\lambda}) = h^{2\lambda}\tilde{n}m(X_{2\lambda})$. The component of $m_w n(X_{2\lambda})$ according to $\mathfrak{g}^{-2\lambda}$ is

$$m_w[24^{-1}\,\mathrm{ad}\,(X_{-\lambda})^4 X_{2\lambda} + 2^{-1}\,\mathrm{ad}\,(X_{-\lambda})^2\,\mathrm{ad}\,(X_{-2\lambda})X_{2\lambda} + 2^{-1}\,\mathrm{ad}\,(X_{-2\lambda})^2 X_{2\lambda}]$$

while that of $h^{2\lambda}\tilde{n}m(X_{2\lambda})$ is $h^{2\lambda}m(X_{2\lambda})$; equating 'norms' of these components leads to the relation

$$(t(X_{-\lambda})^2 + 2t(X_{-2\lambda}))(Q(X_{-2\lambda}))^{1/2} = h^{2\lambda}Q(X_{-2\lambda})^{1/2},$$

from which we deduce that

$$e^{2t} = h^{2\lambda} = 4^{-1}Q(X_{-\lambda})^2 + 2Q(X_{-2\lambda}),$$

as we wished to prove. [By continuity the formula remains true even if $X_{2\lambda} = 0$.] The case when 2λ is not a root is treated in the same way; the details may be left to the reader.

1.1.5 Reductive Lie Algebras and Groups

For various reasons, which will be explained fully in due course, the natural setting for our theory is not the class of pairs (G, K) with G connected semi-simple and of finite center, K maximal compact in G. Instead it will be necessary for us to consider pairs (G, K), satisfying certain convenient axioms, where the Lie algebra of G is only assumed to be reductive. Accordingly we shall first recall a few standard facts concerning reductive Lie algebras; having done this, we shall then set down, once and for all, the conditions which are to be placed on our pairs (G, K).

Let $\mathbf{k}$ be a field of characteristic zero and $\mathfrak{g}$ a Lie algebra over $\mathbf{k}$. Let $\mathfrak{h}$ be a subalgebra of $\mathfrak{g}$ – then $\mathfrak{h}$ is said to be *reductive in* $\mathfrak{g}$ if the representation $X \mapsto \mathrm{ad}_{\mathfrak{g}}(X)$ of $\mathfrak{h}$ on $\mathfrak{g}$ is semi-simple. [We use the standard module terminology – see Chapter 2.] The Lie algebra $\mathfrak{g}$ is called *reductive* if it is reductive in itself. Plainly, if $\mathfrak{h}$ is reductive in $\mathfrak{g}$, then $\mathfrak{h}$ is reductive.

Let $[\mathfrak{g}, \mathfrak{g}]$ denote the derived algebra of $\mathfrak{g}$ (i.e. the linear subspace of $\mathfrak{g}$ spanned by the $[X, Y]$ $(X, Y \in \mathfrak{g})$), $\mathfrak{c}$ the center of $\mathfrak{g}$.

For the proofs of the following propositions see Bourbaki [4].

Proposition 1.1.5.1 The following conditions on the Lie algebra $\mathfrak{g}$ are equivalent:

(i) $\mathfrak{g}$ is reductive;

(ii) $\mathfrak{g}$ admits a faithful finite dimensional semi-simple representation;

(iii) $\mathfrak{g} = \mathfrak{c} + [\mathfrak{g}, \mathfrak{g}]$ with $\mathfrak{c} \cap [\mathfrak{g}, \mathfrak{g}] = \{0\}$ and $[\mathfrak{g}, \mathfrak{g}]$ a semi-simple ideal;

(iv) $\mathfrak{g}$ admits a finite dimensional representation such that the associated bilinear form (see 2.1.1) is non-degenerate.

Proposition 1.1.5.2 (i) Let $\mathfrak{g}$ be reductive; let U be a finite dimensional representation of $\mathfrak{g}$ – then U is semi-simple iff $U(\mathfrak{c})$ is comprised of semi-simple endomorphisms.

(ii) Let $\mathfrak{g}$ be reductive, $\mathfrak{h}$ a subalgebra of $\mathfrak{g}$ reductive in $\mathfrak{g}$; let U be a finite dimensional semi-simple representation of $\mathfrak{g}$ – then the restriction of U to $\mathfrak{h}$ is a semi-simple representation of $\mathfrak{h}$.

Examples (1) Let $\mathfrak{g}$ be a semi-simple Lie algebra over **R**, $\mathfrak{g} = \mathfrak{k} + \mathfrak{a}_{\mathfrak{p}} + \mathfrak{n}^+$ an Iwasawa decomposition of $\mathfrak{g}$. Claim: $\mathfrak{k}_c$ is reductive in $\mathfrak{g}_c$. For if K is the analytic subgroup of the adjoint group Int $(\mathfrak{g})$ of $\mathfrak{g}$ corresponding to $\mathfrak{k}$, then K is compact – hence $\mathfrak{k}$ is reductive in $\mathfrak{g} \Rightarrow \mathfrak{k}_c$ is reductive in $\mathfrak{g}_c$.

(2) In the notations of (1), let $\mathfrak{m}$ denote the centralizer of $\mathfrak{a}_{\mathfrak{p}}$ in $\mathfrak{k}$ – then $\mathfrak{m}_c$ is reductive in both $\mathfrak{k}_c$ and $\mathfrak{g}_c$.

Let $\mathfrak{g}$ be a semi-simple Lie algebra over **R**, $\mathfrak{g} = \mathfrak{k} + \mathfrak{p}$ a Cartan decomposition of $\mathfrak{g}$ with Cartan involution θ (cf. infra), $\mathfrak{u} = \mathfrak{k} + \sqrt{-1}\,\mathfrak{p}$ the corresponding compact real form of $\mathfrak{g}_c$, τ the conjugation of $\mathfrak{g}_c$ with respect to $\mathfrak{u}$.

Proposition 1.1.5.3 Let $\mathfrak{h}_c$ be a subalgebra of $\mathfrak{g}_c$ such that $\tau(\mathfrak{h}_c) = \mathfrak{h}_c$ – then $\mathfrak{h}_c$ is reductive in $\mathfrak{g}_c$.

Proof Evidently $\mathfrak{h}_c$ is the complexification of $\mathfrak{h}_c \cap \mathfrak{u}$; thus, since $\mathfrak{u}$ is compact, $\mathfrak{h}_c \cap \mathfrak{u}$ is reductive in $\mathfrak{u}$ and so $\mathfrak{h}_c$ is reductive in $\mathfrak{g}_c$. □

Corollary 1.1.5.4 Let $\mathfrak{h}$ be a subalgebra of $\mathfrak{g}$ such that $\theta(\mathfrak{h}) = \mathfrak{h}$ – then $\mathfrak{h}$ is reductive in $\mathfrak{g}$.

Let $\mathfrak{g}_c$ be a reductive Lie algebra over **C** – then $\mathfrak{g}_c$ is the direct sum of its center $\mathfrak{c}_c$ and the derived algebra $[\mathfrak{g}_c, \mathfrak{g}_c]$ which is semi-simple. Let $\mathfrak{j}_c$

be a Cartan subalgebra of $\mathfrak{g}_c$ – then $\mathfrak{j}_c = \mathfrak{c}_c + \mathfrak{j}_c \cap [\mathfrak{g}_c, \mathfrak{g}_c]$ where $\mathfrak{j}_c \cap [\mathfrak{g}_c, \mathfrak{g}_c]$ is a Cartan subalgebra of $[\mathfrak{g}_c, \mathfrak{g}_c]$. As has been pointed out already in 1.1.2 it is easy to extend to $\mathfrak{g}_c$ the notion of a root space decomposition with respect to the Cartan subalgebra $\mathfrak{j}_c$. Thus by a root α of $\mathfrak{g}_c$ (with respect to $\mathfrak{j}_c$) we mean a root α of $[\mathfrak{g}_c, \mathfrak{g}_c]$ (with respect to $\mathfrak{j}_c \cap [\mathfrak{g}_c, \mathfrak{g}_c]$). We extend α to a linear function on $\mathfrak{g}_c$ by setting $\alpha(C) = 0$ ($C \in \mathfrak{c}_c$). Let W be the Weyl group of $[\mathfrak{g}_c, \mathfrak{g}_c]$ with respect to $\mathfrak{j}_c \cap [\mathfrak{g}_c, \mathfrak{g}_c]$. If $w \in W$, then we extend w to a linear transformation on $\mathfrak{j}_c$ by writing $wC = C$ ($C \in \mathfrak{c}_c$). These conventions will be used whenever necessary without comment.

Let $\mathfrak{g}$ be a semi-simple Lie algebra over **R**. A *Cartan involution* θ of $\mathfrak{g}$ is an isomorphism of $\mathfrak{g}$ such that $\theta^2 = 1$ and the quadratic form $B(X, \theta(X)) = \mathrm{tr}\,(\mathrm{ad}\,(X)\,\mathrm{ad}\,(\theta X))$ ($X \in \mathfrak{g}$) is negative definite on $\mathfrak{g}$. There is, in the obvious way, a one-to-one correspondence between Cartan involutions of $\mathfrak{g}$ and Cartan decompositions of $\mathfrak{g}$.

Concerning Cartan involutions, we mention the following fact due to Mostow [1, p. 53].

Lemma 1.1.5.5 Let $\mathfrak{g}$ be a semi-simple Lie algebra over **R**, $\tilde{\mathfrak{g}}$ a semi-simple subalgebra of $\mathfrak{g}$. Let $\tilde{\theta}$ be a Cartan involution of $\tilde{\mathfrak{g}}$ – then $\tilde{\theta}$ can be extended to a Cartan involution θ of $\mathfrak{g}$.

Let us turn now to the basic assumptions and notations which will serve us for the remainder of the book.

Let $\mathfrak{g}$ be a reductive Lie algebra over **R** – thus $\mathfrak{g} = \mathfrak{c} + [\mathfrak{g}, \mathfrak{g}]$ where $\mathfrak{c}$ is the center of $\mathfrak{g}$, $[\mathfrak{g}, \mathfrak{g}]$ the derived algebra. Let θ be an automorphism of $\mathfrak{g}$ such that $\theta^2 = 1$ and let $\mathfrak{k}$ and $\mathfrak{p}$ be the subspaces of $\mathfrak{g}$ corresponding to the eigenvalues 1 and -1 respectively of θ. It will be supposed that the quadratic form $-\mathrm{tr}\,(\mathrm{ad}\,(X)\,\mathrm{ad}\,(\theta X))$ ($X \in \mathfrak{g}$) is positive definite on $[\mathfrak{g}, \mathfrak{g}]$ – otherwise said $\theta \mid [\mathfrak{g}, \mathfrak{g}]$ is a Cartan involution of $[\mathfrak{g}, \mathfrak{g}]$.

Let G be a connected Lie group with Lie algebra $\mathfrak{g}$ and K the analytic subgroup of G corresponding to $\mathfrak{k}$. Assume:

(1) K is compact;

(2) The map $(k, X) \mapsto k \exp X$ ($k \in K$, $X \in \mathfrak{p}$) defines an analytic diffeomorphism of $K \times \mathfrak{p}$ onto G.

Extend θ in the usual way to an automorphism of G.

A temporary, somewhat imprecise explanation for the origin of these assumptions is as follows. Suppose in addition that $\mathfrak{g}$ is semi-simple – thus, in this case, G is a connected semi-simple Lie group with finite center. Many of the proofs in the subsequent theory proceed by induction on $\dim(G)$. Such proofs are usually carried out via some procedure of reduction to a lower dimensional subgroup of G; however, it will often happen that this subgroup is no longer semi-simple but rather only reductive – therefore, in order to cite the induction hypothesis, it will be necessary to know the result holds already in the reductive case which in turn entails the more general assumptions made above.

Example The identity component G of $\mathbf{GL}(n, \mathbf{R})$ is not semi-simple but only reductive; however, the pair (G, K) (with the 'obvious' choice of K...) meets the assumptions laid down above.

Notation For brevity we shall write $\bar{\mathfrak{g}}$ for the derived algebra $[\mathfrak{g}, \mathfrak{g}]$; the analytic subgroup of G with Lie algebra $\bar{\mathfrak{g}}$ will be denoted by $\bar{G}$. We shall write $\mathfrak{c}_{\mathfrak{k}}$ for $\mathfrak{c} \cap \mathfrak{k}$, $\mathfrak{c}_{\mathfrak{p}}$ for $\mathfrak{c} \cap \mathfrak{p}$ (thus $\mathfrak{c} = \mathfrak{c}_{\mathfrak{k}} + \mathfrak{c}_{\mathfrak{p}}$ (direct sum)); the respective analytic subgroups of G will be denoted by $C_{\mathfrak{k}}$, $C_{\mathfrak{p}}$. Set $\hat{\mathfrak{g}} = \mathfrak{c}_{\mathfrak{k}} + \bar{\mathfrak{g}}$ so that $\mathfrak{g} = \hat{\mathfrak{g}} + \mathfrak{c}_{\mathfrak{p}}$ (direct sum); if $\hat{G}$ denotes the analytic subgroup of G with Lie algebra $\hat{\mathfrak{g}}$, then we have a direct product decomposition $G = \hat{G} \times C_{\mathfrak{p}}$. Note too that $\hat{G} = \bar{G}C_{\mathfrak{k}}$ (of course this product need not be direct).

Most of the standard structural results which hold for connected semi-simple Lie groups are valid with little, if any, change in the present, modestly more general situation. Take the Iwasawa decomposition for instance; here it may be formulated in the following way. Let $\mathfrak{a}_{\mathfrak{p}}$ be a maximal abelian subspace of $\mathfrak{p}$. Introduce an order into the space of real linear functions λ on $\mathfrak{a}_{\mathfrak{p}}$ and, for any such λ, let $\mathfrak{g}^{\lambda}$ denote the subspace consisting of those $X \in \mathfrak{g}$ for which $[H, X] = \lambda(H)X$ for all $H \in \mathfrak{a}_{\mathfrak{p}}$. Let Σ^+ be the set of positive roots for the pair $(\mathfrak{g}, \mathfrak{a}_{\mathfrak{p}})$; put $\mathfrak{n}^+ = \Sigma_{\lambda}\, \mathfrak{g}^{\lambda}$, $\lambda \in \Sigma^+$ – then the Iwasawa decomposition $\mathfrak{g} = \mathfrak{k} + \mathfrak{a}_{\mathfrak{p}} + \mathfrak{n}^+$ (direct sum of vector spaces) obtains. Of course one also has a global result: The map $(k, h, n) \mapsto khn$ of $K \times A_{\mathfrak{p}} \times N^+$ into G is a surjective analytic diffeomorphism. [All this follows at once from the usual Iwasawa decomposition for $\bar{G}$. Needless to say, the maximal abelian subspace $\mathfrak{a}_{\mathfrak{p}}$ above is the direct sum of $\mathfrak{c}_{\mathfrak{p}}$ with a maximal abelian subspace of $\bar{\mathfrak{p}} = \bar{\mathfrak{g}} \cap \mathfrak{p}$, the various root spaces $\mathfrak{g}^{\lambda}$ ($\lambda \neq 0$) are all contained in $\bar{\mathfrak{g}}$ etc.]

In actual practice, the direct factor $C_{\mathfrak{p}}$ will cause no difficulty; therefore our main interest is in the case when $\mathfrak{c}_{\mathfrak{p}} = \{0\}$. In this situation notice that the various possible θ's which might occur in our hypotheses above are all conjugate within the group of inner automorphisms of $\mathfrak{g}$; extending the usual terminology a bit, let us agree to call them *Cartan involutions* of $\mathfrak{g}$.

We shall conclude this number with a technical observation.

Lemma 1.1.5.6 Retain the above notations – then $\bar{G}$ is a closed subgroup of G.

To get at this, some preparation is needed. Let $i_c : \mathfrak{g} \to \mathfrak{g}_c$ denote the injection of $\mathfrak{g}$ into $\mathfrak{g}_c$ and G_c a complex analytic group corresponding to $\mathfrak{g}_c$; G_c is said to be a *complexification* of G if i_c can be extended to a homomorphism of G into G_c. [Note: It is definitely not requested that the extension i_c be injective.] Since $\bar{G} \cap C$ is finite, it is clear that G always admits a complexification (C the analytic subgroup of G corresponding to $\mathfrak{c}$). [To see that $\bar{G} \cap C$ is finite, it suffices to verify that $\bar{G} \cap C_{\mathfrak{k}}$ is finite. The group $C_{\mathfrak{t}} = \exp(\mathfrak{c}_{\mathfrak{t}})$ is a closed subgroup of K, hence compact. If $\hat{Z}$ denotes the center of $\hat{G}$, $\bar{Z}$ the center of $\bar{G}$, then $\hat{Z} = \bar{Z}C_{\mathfrak{t}}$; of course $\hat{Z}$ is compact while $\bar{Z}$ is discrete. Now $\bar{G} \cap C_{\mathfrak{t}} = \bar{Z} \cap C_{\mathfrak{t}}$; hence $\bar{G} \cap C_{\mathfrak{t}}$ is finite because $\bar{Z} \cap C_{\mathfrak{t}}$ is both compact and discrete. Note too that $\bar{Z}$ is finite (for $\hat{Z}/C_{\mathfrak{t}} = \bar{Z}/(\bar{Z} \cap C_{\mathfrak{t}})$ is compact).]

Returning to Lemma 1.1.5.6, first observe that since $G = \hat{G} \times C_{\mathfrak{p}}$ and since $\hat{G}$ is closed in G, it will be sufficient to verify that $\bar{G}$ is closed in $\hat{G}$ – hence we may assume that $C_{\mathfrak{p}} = \{0\}$. Therefore, Lemma 1.1.5.6 is a consequence of the following statement.

Lemma 1.1.5.7 Suppose that $\mathfrak{c}_{\mathfrak{p}} = \{0\}$; let Ξ be a semi-simple analytic subgroup of G whose Lie algebra is contained in $\bar{\mathfrak{g}}$ – then Ξ is closed in G and its center is finite.

Proof (1) The following well-known fact will be used: Let G be a semi-simple real analytic subgroup of $\mathbf{GL}(n, \mathbf{C})$ – then G is closed in $\mathbf{GL}(n, \mathbf{C})$. [Theorem of Goto – cf. Helgason [2, p.128].]

(2) Let G_c be a complexification of G, $i_c : G \to G_c$ the corresponding homomorphism. Let $\bar{G}_c$ be the complex analytic subgroup of G_c corresponding to $\bar{\mathfrak{g}}_c$ – then $\bar{G}_c$ is closed in G_c. [Thus let $G_c^!$ be the universal covering group of G_c with covering map ϕ. The group $G_c^!$ is the direct product of $\bar{G}_c^!$ and $C_c^!$, $\bar{G}_c^!$ the analytic subgroup of $G_c^!$ with Lie algebra $\bar{\mathfrak{g}}_c$, $C_c^!$ the analytic subgroup of $G_c^!$ with Lie algebra $\mathfrak{c}_c$. Let K_ϕ be the kernel of ϕ. Since the projection of K_ϕ into $\bar{G}_c^!$ is contained in the center of $\bar{G}_c^!$ and the center of $\bar{G}_c^!$ is finite, the projection of K_ϕ into $C_c^!$ is closed in $C_c^!$. There exists a natural continuous mapping of $G_c = G_c^!/K_\phi$ onto $G_c^!$ factored through the direct product of the projections of K_ϕ into $\bar{G}_c^!$ and $C_c^!$; because $\bar{G}_c$ is the identity component of the inverse image of $\bar{G}_c^!$ factored through the projection of K_ϕ into $\bar{G}_c^!$ under this mapping, $\bar{G}_c$ is closed in G_c.]

(3) The group $\bar{G}_c$, being a connected complex semi-simple Lie group, has a faithful finite dimensional representation U (say) (cf. Theorem 3.1.1.3). Owing to (1), then, $i_c(\Xi)$ is closed in $\bar{G}_c$ and so, in view of (2), is also closed in G_c; as the kernel of i_c, in the case at hand, is finite, it follows that Ξ is closed in G. Finally, since $U(i_c(\Xi))$ is a semi-simple linear group, it has a finite center (see 3.1.1); hence the same holds for Ξ.

This completes the proof of the lemma. □

1.2 The Bruhat Decomposition Parabolic Subgroups

1.2.1 Tits Systems

In this number we shall give, following Tits [1], an 'axiomatic' approach to the theory of parabolic subgroups of 'any' group G. Having done this, we shall then verify, in later numbers, that these 'axioms' are satisfied by the pairs (G, K) of number 1.1.5 and hence, as a consequence, find all the parabolic subgroups of G.

Let G be a group and P, M subgroups of G. The triple (G, P, M) is called a *Tits system* if the following conditions (1) – (3) are satisfied.

(1) G is generated by P and M, $N = P \cap M$ is a normal subgroup of M and the group $W = M/N$ admits a (finite) system of generators consisting of involutive elements w_i. [We assume that $w_i \neq 1$ and that $w_i \neq w_j$ for $i \neq j$.]

(2) For any w_i, $w_i P w_i^{-1} \neq P$.

(3) For any w_i and all $w \in W$, $w_i P w \subset (PwP) \cup (Pw_i wP)$.

The group W is called the *Weyl group* of the Tits system (G, P, M).

Fix now a Tits system (G, P, M) with Weyl group W.

Definition The conjugates of P in G will be called the *minimal parabolic subgroups* of G.

Definition A *parabolic subgroup* is any subgroup of G containing a minimal parabolic subgroup.

Our goal is to give a complete description of the parabolic subgroups of G.

Let us suppose that the generators w_i of W are indexed by the set $\Psi = (1, \ldots, l)$ (say). For any subset Θ of Ψ, let W_Θ be the subgroup of W generated by the w_i $(i \in \Theta)$ (hence $W_\Psi = W$); put $P_\Theta = PW_\Theta P$ (we are abusing the notation in the obvious fashion) – then P_Θ is a parabolic subgroup of G. In fact condition (3) above tells us that $W_\Theta P W_\Theta \subset PW_\Theta P$ and hence $(PW_\Theta P)(PW_\Theta P) \subset PW_\Theta P$, $(PW_\Theta P)^{-1} = PW_\Theta P$ which establishes our contention. [Thus, employing an evident abuse of the language, P_Θ is the subgroup of G generated by P and the generators $w_i (i \in \Theta)$ (of course $P_\varnothing = P$, $P_\Psi = G$).] In passing, observe that if $\Theta = \Theta_1 \cup \Theta_2$, then P_Θ is generated by P_{Θ_1} and P_{Θ_2}.

Here is the main result of the present number.

Theorem 1.2.1.1 (Tits) The subgroups P_Θ are the only subgroups of G containing P. They are all distinct and no two are conjugate to one another. There are, therefore, 2^l conjugacy classes of parabolic subgroups of G. The ones containing a given minimal parabolic subgroup form a lattice isomorphic to the lattice of subsets of a set of l elements. The parabolic subgroups are all their own normalizers.

Before turning to the necessary technical preliminaries which lead to the proof of Theorem 1.2.1.1, let us first obtain the *Bruhat decomposition* for G.

Since we are assuming that G is generated by P and M, it follows at once from (3) that $G = PWP$. This means that each double coset PxP

$(x \in G)$ can be written in the form PwP for some $w \in W$. We now wish to show that distinct elements $w \in W$ give rise to distinct double cosets PwP (consequently the number of double cosets is equal to the order of the Weyl group W). In order to prove this, recall that each element w of W is a product of certain generators w_i; let $L(w)$ denote the smallest integer such that w is a product of $L(w)$ generators. Suppose then that $PwP = Pw'P$ and $L(w) \leqslant L(w')$. If $L(w) = 0$, $w = 1$ and so $PwP = P \Rightarrow Pw'P = P \Rightarrow Nw' \subset P \cap M = N \Rightarrow w' = 1$ too. We may therefore argue by induction on $L(w)$. Suppose that $L(w) = k$ and $w = w_i w''$ where w_i is a generator and $L(w'') = k - 1$. Since $Pw_i w''P = Pw'P$, condition (3) tells us that $w''P \subset w_i Pw'P \subset (Pw'P) \cup (Pw_i w'P)$ – hence $Pw''P = Pw'P$ or $Pw''P = Pw_i w'P$. By the induction hypothesis we have that either $w'' = w'$ or $w'' = w_i w'$; but $L(w'') < L(w')$ – thus $w'' = w_i w'$ and so $w = w_i w'' = w'$.

The preceding discussion can be summarized as follows.

Theorem 1.2.1.2 (Bruhat-Tits) Let (G, P, M) be a Tits system with Weyl group W – then the mapping $w \mapsto PwP$ $(w \in W)$ is a one-to-one mapping of W onto the set of double cosets PxP $(x \in G)$ (hence $G = \bigcup_{w \in W} PwP$ (disjoint union)).

Retain the preceding notations and assumptions.

Lemma 1.2.1.3 Let $w \in W$ – then, for any w_i, wPw_i is contained in the union of PwP and Pww_iP.

Proof Since $w_i^2 = 1$, $wPw_i = (w_i P w^{-1})^{-1}$ – hence, using (3), we find that $wPw_i \subset (Pw^{-1}P)^{-1} \cup (Pw_i w^{-1}P)^{-1} = PwP \cup Pww_iP$. □

Lemma 1.2.1.4 Let $w \in W$ – then, for any w_i, $(w_i PwP) \cup (w_i Pw_i wP) = (PwP) \cup (Pw_i wP)$.

Proof It is obvious that the left hand side is contained in the right; to get the reverse inclusion one need only multiply through by w_i. □

If A, B are two subgroups of a group, then $\{A, B\}$ will denote the subgroup generated by A and B.

Lemma 1.2.1.5 If $L(w_i w) \geqslant L(w)$, then $w_i Pw$ is contained in $Pw_i wP$. If $L(w_i w) \leqslant L(w)$, then $w_i Pw$ is not contained in $Pw_i wP$; in the latter case w_i is contained in the subgroup generated by P and wPw^{-1}.

Proof (1) Assume that $L(w_i w) \geqslant L(w)$ – we shall use induction on $L(w)$. Let $L(w) = k$ and $w = w'w_j$ with $L(w') = k - 1$. If, to the contrary, $w_i Pw$ were not contained in $Pw_i wP$, then, in view of condition (3), $w_i Pw$ has a non-empty intersection with PwP – hence $w_i Pw'$ has a non-empty intersection with $PwPw_j$. Since $L(w_iw') \geqslant L(w')$, $w_iPw' \subset Pw_iw'P$ by induction. On the other hand, thanks to Lemma 1.2.1.3, $PwPw_j \subset (PwP) \cup (Pww_jP)$. Therefore $Pw_iw'P$ has a non-empty intersection with $(PwP) \cup (Pww_jP) \Rightarrow Pw_iw'P = PwP$ or $Pw_iw'P = Pww_jP$ and thus either $w_iw' = w$ or $w_iw' = ww_j = w'$ (Theorem 1.2.1.2). The latter equation is impossible, so $w_iw' = w$. This implies that $w_iw = w'$, whence $L(w_iw) < L(w)$, a contradiction.

(2) Now suppose that $L(w_iw) \leqslant L(w)$ – then $L(w_iw_iw) \geqslant L(w_iw)$ and we may apply part (1) to w_iw (thus $w_iPw_iw \subset PwP$). Note that $w_iPw_i \subset P \cup (Pw_iP)$ (condition (3) supra); on the other hand w_iPw_i is not contained in P (condition (2) supra). Therefore w_iPw_i has a non-empty intersection with Pw_iP; hence w_iPw has a non-empty intersection with Pw_iPw_iw, so also with PwP – thus w_iPw is not contained in Pw_iwP.

(3) By part (2), if $L(w_iw) \leqslant L(w)$, then w_iP has a non-empty intersection with $PwPw^{-1}$. Let x be a common element – then $w_iP = xP \subset PwPw^{-1}P \subset \{P, wPw^{-1}\}$, hence $w_i \in \{P, wPw^{-1}\}$.

The proof of the lemma is now complete. □

Corollary 1.2.1.6 Retain the above notations – then $L(w_iw) \neq L(w)$.

Lemma 1.2.1.7 For all $w \in W$, we have $w \in \{P, wPw^{-1}\}$. If $L(w) = k$ and $w = w_{i_1} \ldots w_{i_k}$, then

$$\{P, w\} = \{P, wPw^{-1}\} = P_\Theta$$

where $\Theta = (i_1, \ldots, i_k) \subset \Psi$.

Proof Plainly: $L(w_{i_1}w) < L(w) \Rightarrow w_{i_1} \in \{P, wPw^{-1}\}$ (Lemma 1.2.1.5); $L(w_{i_2}w_{i_1}w) < L(w_{i_1}w) \Rightarrow w_{i_2} \in \{P, w_{i_1}wPw^{-1}w_{i_1}^{-1}\} \subset \{P, wPw^{-1}\}$, etc. Conclusion: $w \in \{P, wPw^{-1}\}$. On the other hand $P_\Theta \subset \{P, w\}$ and since $w \in P_\Theta$, $P_\Theta = \{P, w\}$. □

Corollary 1.2.1.8 The system of generators $\mathscr{S} = \{w_1, \ldots, w_l\}$ for W is minimal.

Proof Suppose that $\mathscr{S}_i = \mathscr{S} - \{w_i\}$ generates W and let $w_i = w_{j_1} \ldots w_{j_k}$ be a word of minimal length expressing w_i as a product of elements of $\mathscr{S}_i$. The preceding reasonings apply to (G, P, M) and $\mathscr{S}_i$; in particular Lemma 1.2.1.7 tells us that the group $\{P, w_i\} = P \cup (Pw_iP)$ must contain $Pw_{j_1}P, \ldots, Pw_{j_k}P$ which is plainly false. □

Remark Let W be a finite group – then W is said to be a *group generated by reflections* if: (1) W admits a system of generators consisting of involutive elements $w_i (i \in I)$; (2) There exists a set of positive integers $\{m_{ij}\}$ with $m_{ij} = m_{ji} \geqslant 2$ for $i \neq j$ such that

$$w_i^2 = (w_iw_j)^{m_{ij}} = 1 \qquad (\text{all } i, j)$$

is a complete set of defining relations for W. Now suppose that (G, P, M) is a Tits system with Weyl group W – then it is a fact (due to Tits) that W is a group generated by reflections (cf. Bourbaki [5, p.25]). [On the other hand, not all finite groups generated by reflections are Weyl groups (this follows from a result of Feit and Higman).]

Proof of Theorem 1.2.1.1 (1) Let H be any subgroup of G containing P – then H is the union of a finite number of double cosets PwP, w running through a certain finite subset of W. With each $\{P, w\}$ we may, as in Lemma 1.2.1.7, associate a parabolic subgroup P_Θ of G. The union of these various subsets Θ gives a subset $\Theta_H \subset \Psi$ and $H = P_{\Theta_H}$.

(2) Let $P_{\Theta_1}, P_{\Theta_2}$ be parabolic subgroups of G containing P and suppose that $P_{\Theta_1} = xP_{\Theta_2}x^{-1}$ $(x \in G)$. Let $x \in PwP$ – then $P_{\Theta_1} = wP_{\Theta_2}w^{-1}$. But $w \in \{P, wPw^{-1}\} \subset P_{\Theta_1}$(Lemma 1.2.1.7) which implies that $P_{\Theta_1} = P_{\Theta_2}$ – thus distinct subgroups containing P can never be conjugate.

(3) If $\Theta_1 \neq \Theta_2$, then $P_{\Theta_1} \neq P_{\Theta_2}$. To see this, first observe that $P_{\Theta_1 \cap \Theta_2} = P_{\Theta_1} \cap P_{\Theta_2}$ – therefore if $P_{\Theta_1} = P_{\Theta_2}$ (with $\Theta_1 \neq \Theta_2$), then $P_{\Theta_1} = P_{\Theta_1 \cap \Theta_2}$ and so $W_{\Theta_1} = W_{\Theta_1 \cap \Theta_2}$. This means that at least one of the generators w_i is in the group generated by the rest – this is impossible (Corollary 1.2.1.8). Hence the 2^l groups P_Θ are distinct and since

$$\{P_{\Theta_1}, P_{\Theta_2}\} = P_{\Theta_1 \cup \Theta_2},\ P_{\Theta_1 \cap \Theta_2} = P_{\Theta_1} \cap P_{\Theta_2},$$

the subgroups P_Θ form a lattice isomorphic to the lattice of subsets of $(1, \ldots, l)$.

(4) Suppose that $xP_\Theta x^{-1} = P_\Theta$. Let $x \in PwP$ – then $wP_\Theta w^{-1} = P_\Theta$, whence $w \in \{P, wPw^{-1}\} \subset P_\Theta$. Thus $x \in P_\Theta$ and so P_Θ is self-normalizing.

The proof of Theorem 1.2.1.1 is complete. □

Corollary 1.2.1.9 Retain the above notations – then $W_{\Theta_1} \cap W_{\Theta_2} = W_{\Theta_1 \cap \Theta_2}$.

Let P_{Θ_1} and P_{Θ_2} be parabolic subgroups of G containing our fixed minimal parabolic subgroup P – then there are but a finite number of double cosets in G mod $P_{\Theta_1} \times P_{\Theta_2}$.

Proposition 1.2.1.10 There exists between the double cosets $P_{\Theta_1} x P_{\Theta_2}$ $(x \in G)$ and the double cosets $W_{\Theta_1} w W_{\Theta_2}$ $(w \in W)$ a one-to-one correspondence characterized by the relation

$$P_{\Theta_1} x P_{\Theta_2} = P W_{\Theta_1} w W_{\Theta_2} P.$$

Proof Let $x \in G$ – then $x \in PwP$ for some $w \in W$. Using Lemma 1.2.1.4 we get

$$P_{\Theta_1} x P_{\Theta_2} = P W_{\Theta_1} PwP W_{\Theta_2} P \subset P W_{\Theta_1} w W_{\Theta_2} P \subset P_{\Theta_1} w P_{\Theta_2} = P_{\Theta_1} x P_{\Theta_2}.$$

Since G is the disjoint union of the PwP $(w \in W)$, the proposition follows. □

Thus, Proposition 1.2.1.10 says that

$$P_{\Theta_1} \backslash G / P_{\Theta_2} = W_{\Theta_1} \backslash W / W_{\Theta_2}.$$

The following observation, which results immediately from Theorem 1.2.1.1, will be occasionally useful.

Lemma 1.2.1.11 Let P_1 and P_2 be parabolic subgroups of G. If P_1 and P_2 are conjugate and if $P_1 \cap P_2$ is parabolic, then $P_1 = P_2$.

Proposition 1.2.1.12 Let P_1 and P_2 be parabolic subgroups of G with $P_1 \subset P_2$; let $T(P_1, P_2) = \{x \in G: {}^xP_1 \subset P_2\}$ – then $T(P_1, P_2) = P_2$.

Proof In fact, if $x \in G$ has the property that ${}^xP_1 \subset P_2$, then $P_2 \cap {}^xP_2 \supset {}^xP_1$ – therefore, in view of the preceding lemma, $P_2 = {}^xP_2$ and so $x \in P_2$ (P_2 is self-normalizing). □

Proposition 1.2.1.13 Let $\mathscr{P}$, $\mathscr{Q}$ be two conjugacy classes of parabolic subgroups of G; suppose $P_1, P_2 \in \mathscr{P}$ and $Q_1, Q_2 \in \mathscr{Q}$ are such that $P_1 \cap Q_1$, $P_2 \cap Q_2$ are parabolic – then:

(i) Each element of $\mathscr{P}$ is contained in but one element of $\mathscr{Q}$ if $P_1 \subset Q_1$;

(ii) The groups $P_1 \cap Q_1$ and $P_2 \cap Q_2$ are conjugate;

(iii) The parabolic subgroups R_1, R_2 generated by P_1, Q_1 and P_2, Q_2, respectively, are conjugate.

Proof The assumption that $P_1 \subset Q_1$ guarantees us that each element of $\mathscr{P}$ is contained in at least one element of $\mathscr{Q}$; on the other hand Lemma 1.2.1.11 implies that such an element of $\mathscr{Q}$ is necessarily unique – hence (i). Adopt now the notations of Theorem 1.2.1.1. Choose $x_1, x_2 \in G$ such that ${}^{x_1}(P_1 \cap Q_1) \supset P$, ${}^{x_2}(P_2 \cap Q_2) \supset P$; there exist $\Theta_1, \Theta_2 \subset \Psi$ with the property that ${}^{x_1}P_1 = {}^{x_2}P_2 = P_{\Theta_1}$, ${}^{x_1}Q_1 = {}^{x_2}Q_2 = P_{\Theta_2}$. But then ${}^{x_1}(P_1 \cap Q_1) = {}^{x_2}(P_2 \cap Q_2) = P_{\Theta_1 \cap \Theta_2}$, ${}^{x_1}R_1 = {}^{x_2}R_2 = P_{\Theta_1 \cup \Theta_2}$. Assertions (ii) and (iii) of the present proposition are now clear. □

Let $\mathscr{P}$, $\mathscr{Q}$ be two conjugacy classes of parabolic subgroups of G. We shall write $\mathscr{P} \prec \mathscr{Q}$ provided each element of $\mathscr{P}$ is contained in an element of $\mathscr{Q}$; for this it is enough, in view of the preceding proposition, that there exists an element of $\mathscr{P}$ which is contained in an element of $\mathscr{Q}$. Note too that the parabolic subgroups of the form $P \cap Q$ ($P \in \mathscr{P}$, $Q \in \mathscr{Q}$) (resp. generated by two subgroups $P \in \mathscr{P}$, $Q \in \mathscr{Q}$ such that $P \cap Q$ is parabolic) form a single conjugacy class which will be denoted by $\mathscr{P} \curlywedge \mathscr{Q}$ (resp. $\mathscr{P} \curlyvee \mathscr{Q}$).

Proposition 1.2.1.14 Let P_1 and P_2 be parabolic subgroups of G – then $P_1 \cap P_2$ contains a conjugate of N.

Proof Conjugating P_1 and P_2 if necessary by an inner automorphism of G, it can be assumed that $P \subset P_1$. Let $x \in G$ be such that $xPx^{-1} \subset P_2$; owing to Theorem 1.2.1.2, there exist elements p and $\tilde{p}$ in P and a $w \in W$ such that $x = pw\tilde{p}$ (the usual abuse of notation being in force here); thus $P_2 \supset xPx^{-1} = pwPw^{-1}p^{-1} \supset pwNw^{-1}p^{-1} = pNp^{-1}$ (N is normal in M). Because $P_1 \supset P \supset pNp^{-1}$, the contention of the proposition is now clear. □

Appendix Let W be a finite group of order $[W]$ which is generated by reflections $\{w_i : i \in I\}$ (cf. the Remark following Corollary 1.2.1.8). If there exist non-empty sets $I_1, I_2 \subset I$ such that $I = I_1 \cup I_2$, $I_1 \cap I_2 = \varnothing$ and with the property that $(w_{i_1} w_{i_2})^2 = 1$ (all $i_1 \in I_1, i_2 \in I_2$), then W admits the obvious direct product decomposition and is said to be *reducible*; otherwise W is said to be *irreducible*. The irreducible finite groups which are generated by reflections were enumerated by Coxeter (cf. Bourbaki [5]): If $[I]$ (= cardinality of I) $= n$, then we represent W by a graph with n nodes, the ith and j^{th} nodes being joined by $m_{ij} - 2$ lines where $(w_i w_j)^{m_{ij}} = 1$. Schematically:

Name	Graph	$[W]$
$\mathbf{A}_l (l \geqslant 1)$		$(l+1)!$
$\mathbf{B}_l (l \geqslant 2)$		$2^l l!$
$\mathbf{D}_l (l \geqslant 4)$		$2^{l-1} l!$
$\mathbf{E}_6$		$72 \cdot 6!$
$\mathbf{E}_7$		$8 \cdot 9!$
$\mathbf{E}_8$		$192 \cdot 10!$
$\mathbf{F}_4$		1152
$\mathbf{G}_2$		12
$\mathbf{H}_3$		120
$\mathbf{H}_4$		$(120)^2$
$\mathbf{I}_2(l)$	(l-2 lines; $l = 5$ or $l \geqslant 7$) 	$2l$

It can be shown that the types $\mathbf{A}_l, \mathbf{B}_l, \ldots, \mathbf{G}_2$ correspond to the Weyl groups of the irreducible root systems. [In this connection let us recall the method by means of which the order $[W]$ of the Weyl group W of an irreducible reduced root system Φ may be calculated. Let $\Psi = \{\alpha_1, \ldots, \alpha_l\}$ be a fundamental system of roots for Φ, $\alpha_{\max}$ the greatest root for Φ relative to Ψ (so that if $\alpha_{\max} = \sum_{i=1}^{l} n_i \alpha_i$, then for every root $\sum_{i=1}^{l} m_i \alpha_i$ one has $n_i \geqslant m_i$ with $n_i \geqslant 1$ $(i = 1, \ldots, l)$), f the connection index of Φ – then the order of W is equal to $(l!) n_1 \ldots n_l f$ (cf. Bourbaki [5, p.178]).]

1.2.2 The Complex Case

In this number we shall record, for the most part without proof, certain standard structural facts as regards connected complex semi-simple Lie groups. The Bruhat decomposition for such groups (and its consequences) will turn out to be corollaries of the more general considerations presented in 1.2.3 infra.

Let $\mathfrak{g}_c$ be a complex semi-simple Lie algebra, $\mathfrak{j}_c$ a Cartan subalgebra of $\mathfrak{g}_c$. Let G_c be a connected Lie group with Lie algebra $\mathfrak{g}_c$ – then G_c is a complex Lie group (that is, G_c carries the structure of a complex analytic manifold with respect to which the map $(x, y) \mapsto xy^{-1}$ of $G_c \times G_c$ into G_c is holomorphic). Let J_c denote the complex analytic subgroup of G_c corresponding to $\mathfrak{j}_c$ – then J_c is a group of multiplicative type (i.e. isomorphic to a product of groups $\mathbf{GL}(1, \mathbf{C})$) which is, moreover, the Cartan subgroup of G_c associated with $\mathfrak{j}_c$. [Of course, in the case at hand, all Cartan subgroups are connected, abelian and mutually conjugate; cf. 1.4.1.]

Let Φ be the system of roots of the pair $(\mathfrak{g}_c, \mathfrak{j}_c)$, $\mathfrak{w}$ the root group of Φ, $\mathfrak{W}$ the weight group of Φ; as is known, both $\mathfrak{w}$ and $\mathfrak{W}$ are W-stable lattices in the dual of $\mathfrak{j}^*$ (W the Weyl group of Φ) while the quotient $\mathfrak{W}/\mathfrak{w}$ is a finite abelian group. Let $\hat{\mathfrak{w}}$ (respectively $\hat{\mathfrak{W}}$) denote the lattice in $\mathfrak{j}^*$ dual to $\mathfrak{w}$ (respectively $\mathfrak{W}$) (thus, in the language of co-roots, $\hat{\mathfrak{w}}$ is the weight group for $\Phi^\vee$ while $\hat{\mathfrak{W}}$ is the root group for $\Phi^\vee$); of course $\hat{\mathfrak{w}} \supset \hat{\mathfrak{W}}$.

Let e: $\mathfrak{j}_c \rightarrow J_c$ denote the map defined by the rule $H \mapsto \exp(2\pi\sqrt{-1}H)$ $(H \in \mathfrak{j}_c;\ \pi = 3.14\ldots)$ – then e is a homomorphism, $\mathfrak{j}_c$ being abelian.

The following results are proved in Helgason [2, Chapter 7].

Proposition 1.2.2.1 Retain the above notations – then:

(i) The homomorphism e: $\mathfrak{j}_c \rightarrow J_c$ is surjective and its kernel $\mathfrak{L}(G_c)$ lies between $\hat{\mathfrak{w}}$ and $\hat{\mathfrak{W}}$;

(ii) The homomorphism e: $\mathfrak{j}_c \rightarrow J_c$ defines an isomorphism of $\hat{\mathfrak{w}}/\mathfrak{L}(G_c)$ onto the center of G_c;

(iii) The canonical map $\pi_1(J_c) \rightarrow \pi_1(G_c)$ is surjective and defines, by passage to the quotient, an isomorphism of $\mathfrak{L}(G_c)/\hat{\mathfrak{W}}$ onto $\pi_1(G_c)$.

[One identifies $\pi_1(J_c)$ and $\hat{\mathfrak{L}}(G_c)$ via e; as the notation suggests, $\hat{\mathfrak{L}}(G_c)$ may be regarded as the dual of a group $\mathsf{L}(G_c)$ which will be defined later (cf. 3.1.1).]

Proposition 1.2.2.2 For every subgroup $\mathfrak{L}$ of $\mathfrak{j}_c$ such that $\hat{\mathfrak{w}} \supset \mathfrak{L} \supset \hat{\mathfrak{W}}$ there exists a connected complex semi-simple Lie group G_c with Lie algebra $\mathfrak{g}_c$ – unique to within isomorphism – such that $\mathfrak{L}(G_c) = \mathfrak{L}$.

[Therefore $G_c = \mathrm{Int}(\mathfrak{g}_c)$ iff $\mathfrak{L}(G_c) = \hat{\mathfrak{w}}$ while G_c is simply connected iff $\mathfrak{L}(G_c) = \hat{\mathfrak{W}}$.]

Because the group $\hat{\mathfrak{w}}/\hat{\mathfrak{W}} \sim \mathfrak{W}/\mathfrak{w}$ is finite, there exists but a finite number of connected complex semi-simple Lie groups with Lie algebra $\mathfrak{g}_c$ (to within isomorphism) – they all are obtained by taking the coverings of $\mathrm{Int}(\mathfrak{g}_c)$.

Example The following table describes the structure of $\mathfrak{W}/\mathfrak{w}$ for the various types of simple Lie algebras:

Type	$\mathfrak{W}/\mathfrak{w}$
$\mathbf{A}_l$ $(l \geqslant 1)$	$\mathbf{Z}_{l+1}$
$\mathbf{B}_l$ $(l \geqslant 2)$	$\mathbf{Z}_2$
$\mathbf{C}_l$ $(l \geqslant 3)$	$\mathbf{Z}_2$
$\mathbf{D}_{2l}$ $(l \geqslant 2)$	$\mathbf{Z}_2 \times \mathbf{Z}_2$
$\mathbf{D}_{2l+1}$ $(l \geqslant 2)$	$\mathbf{Z}_4$
$\mathbf{E}_6$	$\mathbf{Z}_3$
$\mathbf{E}_7$	$\mathbf{Z}_2$
$\mathbf{E}_8$	$\mathbf{Z}_1$
$\mathbf{F}_4$	$\mathbf{Z}_1$
$\mathbf{G}_2$	$\mathbf{Z}_1$

[One can, of course, also read off the connection index of the root system in question from this table.]

Let us keep to the above notations and assumptions. Fix a set Φ^+ of positive roots for Φ; put $\mathfrak{n}_c^- = \sum_{\alpha<0} \mathfrak{g}_c^\alpha$, $\mathfrak{n}_c^+ = \sum_{\alpha>0} \mathfrak{g}_c^\alpha$, $\mathfrak{b}_c = \mathfrak{j}_c + \mathfrak{n}_c^+$ – then $\mathfrak{g}_c = \mathfrak{n}_c^- + \mathfrak{b}_c$ (direct sum). The (solvable) subalgebra $\mathfrak{b}_c$ is called a *Borel subalgebra* of $\mathfrak{g}_c$; it can be shown that every solvable subalgebra of $\mathfrak{g}_c$ is conjugate, under an inner automorphism of $\mathfrak{g}_c$, to a subalgebra of $\mathfrak{b}_c$ – in particular $\mathfrak{b}_c$ is a maximal solvable subalgebra of $\mathfrak{g}_c$ (cf. Borel [4]).

Let B_c denote the analytic subgroup of G_c corresponding to $\mathfrak{b}_c$ – then B_c is closed. Evidently B_c is the semi-direct product of J_c and N_c^+ ($N_c^+ = \exp(\mathfrak{n}_c^+)$); furthermore B_c is a maximal connected closed solvable subgroup of G_c and all such are conjugate to B_c. The conjugates of B_c in G_c are called the *Borel subgroups* of G_c.

The quotient of the normalizer of J_c (in G_c) by J_c itself is a finite group which can be identified with the Weyl group W of the pair $(\mathfrak{g}_c, \mathfrak{j}_c)$. It will be shown in 1.2.3 that the triple comprised of G_c, B_c and the normalizer of J_c (in G_c) constitutes a Tits system (thus the minimal parabolic subgroups of G_c are the Borel subgroups). Consequently, if $l = \text{rank}(\Phi)$, then, on the basis of the generalities in 1.2.1, it follows that there are exactly 2^l subgroups of G_c containing B_c; any such parabolic subgroup P_c is a closed complex analytic Lie subgroup of G_c – in fact P_c is connected (because G_c is complex). Every Levi decomposition of the Lie algebra $\mathfrak{p}_c$ of P_c (cf. the Appendix to this number) gives rise to a corresponding Levi decomposition of P_c; thus if L_c is a Levi subgroup of P_c (hence L_c is a connected reductive complex analytic Lie group), then P_c can be written as the semi-direct product of L_c and its unipotent radical $R_u(P_c)$.

Let us suppose that G_c is simply connected. Fix a parabolic subgroup P_c of G_c – then the following facts are known about the complex manifolds $M = G_c/P_c$. The space G_c/P_c is a compact complex Kähler manifold with positive (ordinary) Euler characteristic and one obtains, up to a biholomorphic map, all such complex homogeneous spaces of G_c in this way. Moreover G_c/P_c is algebraic (i.e. admits a holomorphic imbedding into some complex projective space); varying $\mathfrak{g}_c$, to within a biholomorphic map, all simply connected algebraic homogeneous spaces arise in this manner. [For details see Wang [1].] Now fix a maximal compact subgroup U of G_c (thus the Lie algebra $\mathfrak{u}$ of U is a compact real form for $\mathfrak{g}_c$) – then the homogeneous space G_c/P_c also admits a 'compact' description. Indeed the maximal compact subgroup U of G_c acts transitively on M and so M can be identified with the quotient space U/L ($L = U \cap P_c$). [Since the real span of $\mathfrak{u}$ and $\mathfrak{p}_c$ is all of $\mathfrak{g}_c$, the U-orbit of the 'origin' in M must be open; on the other hand this orbit is also closed, U being compact.] The isotropy group $L = U \cap P_c$ is connected (M is simply connected); it is a maximal torus in the special case when our parabolic subgroup is minimal (i.e., when P_c is a Borel subgroup of G_c) and, in any event, is always the centralizer of a torus in U. There is a commutative diagram

$$\begin{array}{ccc} L & \longrightarrow & P_c \\ \downarrow & & \downarrow \\ U & & G_c \\ \downarrow & & \downarrow \\ U/L & \sim & G_c/P_c \end{array}$$

of principal fiber bundles where the principal bundle $P_c \to G_c \to G_c/P_c$ is holomorphic.

Remark The complex structure on $M = G_c/P_c = U/L$ can be described without reference to G_c. To see this, let f be a holomorphic function on some open subset $\mathcal{O} \subset M$; let $\tilde{f}$ be the pullback of f to G_c – then, since $\tilde{f}$ is P_c-invariant on the right, every $X \in \mathfrak{p}_c$, considered as a left invariant *real* tangent vector field on G_c, annihilates $\tilde{f}$. Because $\tilde{f}$ satisfies the Cauchy-Riemann equations on G_c, the restriction of $\tilde{f}$ to U, or, equivalently, the pullback of f to U, will be annihilated by every $X \in \mathfrak{p}_c$ when X is regarded as a left invariant *complex* tangent vector field on U. This being so, let $\pi\colon U \to M = U/L$ be the quotient map – then, according to what has just been said, the induced mapping π_* from the complexified tangent space of U at u (any $u \in U$), identified with $\mathfrak{g}_c$ by left translation, carries $\mathfrak{p}_c$ into the space of anti-holomorphic tangent vectors at $\pi(u)$. The kernel of π_* is the complexification of the Lie algebra of L; considerations of dimension then imply that $\pi_*(\mathfrak{n}_c^+) = \pi_*(\mathfrak{p}_c)$ ($\mathfrak{n}_c^+$ the nilpotent radical of $\mathfrak{p}_c$; cf. the Appendix to this number) is the full anti-holomorphic tangent space. Now let h be a C^∞ function on $\pi^{-1}(\mathcal{O}) \subset U$ with the property that $Xh = 0$ for every $X \in \mathfrak{p}_c$ (extended to a left invariant complex vector field); such a function must then be constant on each L-coset and hence drops to a C^∞ function f (say) on $\mathcal{O}$; in view of the characterization of the space of (0, 1)-tangent vectors which was derived above, it must be the case that f satisfies the Cauchy-Riemann equations. All this serves to show that

$$\pi^*(\mathfrak{H}(\mathcal{O})) = \{h \in C^\infty(\pi^{-1}(\mathcal{O}))\colon Xh = 0 \text{ (all } X \in \mathfrak{p}_c)\}.$$

[Here $\mathfrak{H}(\mathcal{O})$ is the ring of holomorphic functions on $\mathcal{O} \subset M$.]

Appendix Let $\mathfrak{g}$ be a reductive Lie algebra over an algebraically closed field $\mathbf{k}$ of characteristic zero. Let $\mathfrak{j}$ be a Cartan subalgebra of $\mathfrak{g}$, Φ the set of roots of the pair $(\mathfrak{g}, \mathfrak{j})$, W the Weyl group of Φ; fix a Borel subalgebra $\mathfrak{b}$ of $\mathfrak{g}$ containing $\mathfrak{j}$ (i.e. fix a maximal solvable subalgebra $\mathfrak{b}$ of $\mathfrak{g}$ containing $\mathfrak{j}$); let Φ^+ denote the system of positive roots singled out by this choice of $\mathfrak{b}$ (whence $\mathfrak{b} = \mathfrak{j} + \sum_{\alpha>0} \mathfrak{g}^\alpha$), $\Psi = \{\alpha_1, \ldots, \alpha_l\}$ the corresponding fundamental system. [As has been pointed out in 1.1.5, there is no difficulty in extending the traditional terminology to reductive Lie algebras . . . .] By a *parabolic subalgebra* of $\mathfrak{g}$ we shall understand any subalgebra of $\mathfrak{g}$ containing a Borel subalgebra of $\mathfrak{g}$; a $\mathfrak{b}$-parabolic subalgebra of $\mathfrak{g}$ is then, by definition, any subalgebra of $\mathfrak{g}$ which contains our fixed Borel subalgebra $\mathfrak{b}$.

Let $\mathfrak{p}$ be a $\mathfrak{b}$-parabolic subalgebra of $\mathfrak{g}$; then, since $\mathfrak{p}$ contains $\mathfrak{j}$, it is clear that there exists a subset Ξ of Φ such that

$$\mathfrak{p} = \mathfrak{j} + \sum_{\alpha \in \Xi} \mathfrak{g}^\alpha.$$

Obviously Ξ is closed and $\Phi = \Xi \cup (-\Xi)$ – otherwise said, Ξ is a parabolic subset of Φ (cf. Corollary 1.1.2.11). In the usual way, write $\Xi = \Xi_s \cup \Xi_u$ (thus $\Xi_s = \Xi \cap (-\Xi)$, $\Xi_u = \Xi \cap \complement(-\Xi)$) and put

$$\mathfrak{l} = \mathfrak{j} + \sum_{\alpha \in \Xi_s} \mathfrak{g}^\alpha, \; \mathfrak{n}^+ = \sum_{\alpha \in \Xi_u} \mathfrak{g}^\alpha.$$

Lemma Let $\mathfrak{p}$ be a $\mathfrak{b}$-parabolic subalgebra of $\mathfrak{g}$; define Ξ, $\mathfrak{l}$ and $\mathfrak{n}^+$ as above – then $\mathfrak{l}$ is a reductive subalgebra of $\mathfrak{g}$, $\mathfrak{n}^+$ is the nilpotent radical of $\mathfrak{p}$ and $\mathfrak{p}$ is the direct sum of $\mathfrak{l}$ and $\mathfrak{n}^+$.

Proof (1) Because Ξ_s is closed, $\mathfrak{l}$ is a subalgebra of $\mathfrak{g}$; to verify that $\mathfrak{l}$ is reductive in $\mathfrak{g}$, it will be enough to show that $\mathfrak{l}$ is reductive. [For the center of $\mathfrak{l}$ is contained in $\mathfrak{j}$ and the elements of $\mathfrak{j}$ are semi-simple in $\mathfrak{g}$; now quote Proposition 1.1.5.2(i).] This being so, choose a finite dimensional representation U of $\mathfrak{g}$ such that the associated bilinear form B (say) is non-degenerate. [Since $\mathfrak{g}$ is reductive, such a U exists – cf. Proposition 1.1.5.1(iv).] The fact that $\mathfrak{l}$ is reductive will follow, therefore, when it is shown that the restriction of B to $\mathfrak{l}$ is non-degenerate (Proposition 1.1.5.1(iv) again). Now the usual computation shows that B places $\mathfrak{j}$ in duality with itself and $\mathfrak{g}^\alpha$ in duality with $\mathfrak{g}^{-\alpha}$ ($\alpha \in \Phi^+$).

Furthermore $\mathfrak{j}$ and $\mathfrak{g}^\alpha$ are orthogonal (all $\alpha \in \Phi$); likewise $\mathfrak{g}^\alpha$ and $\mathfrak{g}^\beta$ are orthogonal if $\alpha + \beta \neq 0$. Because $\mathfrak{j} \subset \mathfrak{l}$ and Ξ_s is a symmetric set of roots, the restriction of B to $\mathfrak{l}$ is indeed non-degenerate.

(2) It is not difficult to see that $\mathfrak{n}^+$ is an ideal in $\mathfrak{p}$. Since $\mathfrak{n}^+$ is a subalgebra of the derived algebra of $\mathfrak{b}$ (which is itself nilpotent), $\mathfrak{n}^+$ is solvable and thus is contained in the radical $\mathfrak{r}$ of $\mathfrak{p}$; since $\mathfrak{n}^+$ is obviously contained in the derived algebra $[\mathfrak{p}, \mathfrak{p}]$ of $\mathfrak{p}$, we have $\mathfrak{n}^+ \subset \mathfrak{r} \cap [\mathfrak{p}, \mathfrak{p}]$. By definition, the nilpotent radical of $\mathfrak{p}$ is the intersection of the kernels of the finite dimensional simple $\mathfrak{p}$-modules and, as is well-known (Bourbaki [4, p.65]), is equal to $\mathfrak{r} \cap [\mathfrak{p}, \mathfrak{p}]$. The ideal $\mathfrak{n}^+$ certainly contains the nilpotent radical of $\mathfrak{p}$ (for $\mathfrak{p}/\mathfrak{n}^+$ is isomorphic to $\mathfrak{l}$ and $\mathfrak{l}$ is reductive), whence, in view of what was said above, is equal to the nilpotent radical of $\mathfrak{p}$.

The proof of the lemma is therefore complete (it being clear that $\mathfrak{p}$ is the direct sum of $\mathfrak{l}$ and $\mathfrak{n}^+$). □

The subalgebra $\mathfrak{l}$ will be called a *Levi subalgebra* of $\mathfrak{p}$ – thus $\mathfrak{l}$ is a reductive subalgebra of $\mathfrak{g}$ which is, moreover, of the same rank as $\mathfrak{g}$ ($\mathfrak{j}$ is a Cartan subalgebra of $\mathfrak{l}$). [This terminology is in accordance with that of Borel and Tits [1]; it does not, of course, strictly agree with the traditional language; cf. too Lemma 1.2.4.13.]

Remark Let $\mathfrak{p}$ be a $\mathfrak{b}$-parabolic subalgebra of $\mathfrak{g}$ – then we may associate with $\mathfrak{p}$ a subset Θ_s of Ψ, namely $\Theta_s = \Psi \cap (-\Xi_s)$ and, relative to the ordering inherited from Φ, Θ_s is the fundamental system for the positive roots Ξ_s^+ of the pair $(\mathfrak{l}, \mathfrak{j})$. On the other hand, with every subset of Ψ there may be associated, in a natural way, a $\mathfrak{b}$-parabolic subalgebra of $\mathfrak{g}$ (obvious construction). Because distinct subsets of Ψ lead to distinct $\mathfrak{b}$-parabolic subalgebras of $\mathfrak{g}$, it follows that there are precisely 2^l parabolic subalgebras containing a given Borel subalgebra of $\mathfrak{g}$.

1.2.3 Boundary Subgroups and Parabolic Subgroups of a Real Semi-Simple Lie Group

Let G be a connected semi-simple Lie group with finite center; the objective of the present number is to study a certain class of closed subgroups of G, the so-called minimal parabolic subgroups P – they may be characterized by any of the following equivalent conditions: (1) P is a minimal boundary subgroup of G; (2) P is maximal among the closed subgroups of G which are the extension of a compact group by a connected solvable group; (3) P is minimal among the closed subgroups of G which are equal to the normalizer in G of their Lie algebra, the complexification of the Lie algebra containing a Borel subalgebra of $\mathfrak{g}_c$ (cf. 1.2.2). This being so, fix one such P – then, as the terminology suggests, all other minimal parabolic subgroups of G are conjugate to P and, moreover, it will turn out that there is associated with the pair (G, P) a Tits system (hence the results obtained in 1.2.1 become available). We shall begin by establishing the Bruhat decomposition for G; after that we shall follow Furstenberg and introduce the notion of a boundary subgroup of G – in the course of this discussion the various equivalences listed above will become apparent.

Note All unexplained notations and conventions are as in 1.1.

Fix an Iwasawa decomposition $G = KA_{\mathfrak{p}}N^+$ for G; let M be the centralizer of $A_{\mathfrak{p}}$ in K; set $P = \mathsf{M}A_{\mathfrak{p}}N^+$ – then P is a closed subgroup of G (this is clear since both $A_{\mathfrak{p}}$ and M normalize N^+). [In passing let us note that once P is given, it may very well be the case that P can be written in the form $P = \mathsf{M}A_{\mathfrak{p}}N^+$ in many different ways (depending on our choice of an Iwasawa decomposition for G); any such decomposition of P will be called a *Langlands decomposition* (a notion with which we shall deal systematically in 1.2.4).]

We shall consider double cosets of the form PxP $(x \in G)$; since $\mathsf{M} \subset P$, we shall, by abuse of notation, write PwP for the double coset Pm_wP, m_w any element of M^* associated with $w \in W$ (= the Weyl group of the pair $(\mathfrak{g}, \mathfrak{a}_{\mathfrak{p}})$). [Analogous abuses of the notation will be used without comment in the sequel.] The following theorem, which gives the *Bruhat decomposition* for G (and is known in the literature as the Bruhat Lemma), is one of the basic structural facts on which the representation theory of semi-simple Lie groups rests.

Theorem 1.2.3.1 (Bruhat-Harish-Chandra) Let G be a connected semi-simple Lie group with finite center, $G = KA_{\mathfrak{p}}N^+$ an Iwasawa decomposition for G; set $P = \mathsf{M}A_{\mathfrak{p}}N^+$ – then the mapping $w \mapsto PwP$ $(w \in W)$ is a one-to-one mapping of W onto the set of double cosets PxP $(x \in G)$ (hence $G = \bigcup_{w \in W} PwP$ (disjoint union)).

For the applications, it will be convenient to have Theorem 1.2.3.1 stated in a slightly different form. Let us consider the orbits in G/P under the canonical action of P. Fix $w \in W$; let m_w be any element in M^* associated with w – then the subgroup U_w^+ 'parameterizes' the orbit Pm_wP of the coset $m_wP = wP$ in G/P. More precisely the map $u \mapsto um_wP$ is a one-to-one regular mapping of U_w^+ onto Pm_wP. [Thus, abusing the notation in the evident manner, we have $Pm_wP = N^+A_{\mathfrak{p}}\mathsf{M}m_wP = U_w^+N_w^+m_wP$ (cf. Proposition 1.1.4.5) $= U_w^+m_wm_w^{-1}N_w^+m_wP = U_w^+m_wP$ (because $m_w^{-1}N_w^+m_w \subset N^+$); hence our mapping is surjective. Next suppose that $u_1m_w = u_2m_wp(u_1, u_2 \in U_w^+; p \in P)$ – then $m_w^{-1}u_2^{-1}u_1m_w \in P \cap N^- = \{1\}$ (cf. Proposition 1.2.3.5 below), whence $u_1 = u_2$. Finally, since the regularity of our mapping is easy to check, the contention is clear.] Consequently, Theorem 1.2.3.1 now reads as follows.

Theorem 1.2.3.1 (bis) Let G be a connected semi-simple Lie group with finite center, $G = KA_{\mathfrak{p}}N^+$ an Iwasawa decomposition for G; set $P = \mathsf{M}A_{\mathfrak{p}}N^+$ – then $G = \bigcup_{w \in W} N^+wP = \bigcup_{w \in W} U_w^+wP$ (disjoint union).

We shall preface the proof of Theorem 1.2.3.1 with two technical lemmas.

The Lie algebra $\mathfrak{p}$ of $P(= \mathsf{M}A_{\mathfrak{p}}N^+)$ is $\mathfrak{p} = \mathfrak{m} + \mathfrak{a}_{\mathfrak{p}} + \mathfrak{n}^+$; of course $\mathfrak{m} + \mathfrak{a}_{\mathfrak{p}}$ is the centralizer of $\mathfrak{a}_{\mathfrak{p}}$ in $\mathfrak{g}$. [No confusion will arise from our double usage of the symbol '$\mathfrak{p}$'.] Given x in G, set $\mathfrak{p}_x = \mathfrak{p} \cap \mathrm{Ad}(x)\mathfrak{p}$.

Lemma 1.2.3.2 Retain the above notations – then, for every x in G, we have $\mathfrak{p} = \mathfrak{p}_x + \mathfrak{n}^+$

Proof Since $\mathfrak{p}_x + \mathfrak{n}^+ \subset \mathfrak{p}$, it suffices to show that $\dim(\mathfrak{p}_x + \mathfrak{n}^+) = \dim(\mathfrak{p})$. For this purpose write $x = khn\,(k \in K, h \in A_{\mathfrak{p}}, n \in N^+)$ – then $\mathrm{Ad}(hn)\mathfrak{p} = \mathfrak{p}$ and so we may suppose that $x = k \in K$. Relative to the Euclidean structure $(.\,,.)_\theta$ (cf. 1.1.3), $\mathrm{Ad}(k)$ is an orthogonal transformation which commutes with θ. The orthogonal complement $\mathfrak{p}^\perp$ of $\mathfrak{p}$ in $\mathfrak{g}$ is $\theta(\mathfrak{n}^+)$ and so $(\mathfrak{p} + \mathrm{Ad}(k)\mathfrak{p})^\perp = \theta(\mathfrak{n}^+ \cap \mathrm{Ad}(k)\mathfrak{n}^+) = \theta(\mathfrak{n}^+ \cap \mathfrak{p}_k)$; therefore, since

$$\dim(\mathfrak{p} + \mathrm{Ad}(k)\mathfrak{p}) + \dim(\mathfrak{p} + \mathrm{Ad}(k)\mathfrak{p})^\perp = \dim(\mathfrak{g}),$$

we have

$$\begin{aligned}\dim(\mathfrak{p}_k + \mathfrak{n}^+) &= \dim(\mathfrak{p}_k) + \dim(\mathfrak{n}^+) - \dim(\mathfrak{p}_k \cap \mathfrak{n}^+)\\ &= \dim(\mathfrak{n}^+) + 2\dim(\mathfrak{p}) - \dim(\mathfrak{g}) = \dim(\mathfrak{p}). \quad \square\end{aligned}$$

Lemma 1.2.3.3 Retain the above notations; fix $X \in \mathfrak{p}$ – then $X \in \mathfrak{a}_{\mathfrak{p}} + \mathfrak{n}^+$ iff the eigenvalues of $\mathrm{ad}(X)$ in $\mathfrak{g}_c$ are real.

Proof Write $X = X_1 + X_2 (X_1 \in \mathfrak{m}, X_2 \in \mathfrak{a}_{\mathfrak{p}} + \mathfrak{n}^+)$; since M is compact there exists an $m \in M$ such that $\mathrm{Ad}(m)X_1 \in \mathfrak{a}_{\mathfrak{k}}$ – therefore $\mathrm{Ad}(m)X = H + Y$ where $H \in \mathfrak{a}$ and $Y \in \mathfrak{n}^+$. The eigenvalues of $\mathrm{ad}(X)$ are therefore zero and $\alpha(H)(\alpha \in \Phi)$. The roots take real values on $\mathfrak{a}_{\mathfrak{p}}$ and pure imaginary values on $\mathfrak{a}_{\mathfrak{k}}$ – hence $\mathrm{ad}(X)$ admits real eigenvalues iff $H \in \mathfrak{a}_{\mathfrak{p}}$, hence iff $X = \mathrm{Ad}(m^{-1})(H + Y) \in \mathfrak{a}_{\mathfrak{p}} + \mathfrak{n}^+$. $\square$

Proof of Theorem 1.2.3.1 (1) Fix an $x \in G$ – we shall first prove that there exists an $m^* \in M^*$ such that $PxP = Pm^*P$. For this purpose, select a Σ-regular element H in $\mathfrak{a}_{\mathfrak{p}}$ and choose, according to Lemma 1.2.3.2, an $X \in \mathfrak{n}^+$ such that $H + X \in \mathfrak{p}_x$. Owing to Theorem 1.1.4.4, there exists an $n \in N^+$ such that $\mathrm{Ad}(n)H = H + X$, hence such that $X_x = \mathrm{Ad}(x^{-1}n)H \in \mathfrak{p}$. The eigenvalues of $\mathrm{Ad}(X_x)$ are the same as those of $\mathrm{ad}(H)$, thus are real; this being so, use Lemma 1.2.3.3 to write $X_x = \tilde{H} + \tilde{X}\,(\tilde{H} \in \mathfrak{a}_{\mathfrak{p}}, \tilde{X} \in \mathfrak{n}^+)$ – then the eigenvalues of $\mathrm{ad}(\tilde{H})$ are the same as those of $\mathrm{ad}(H)$, in particular $\tilde{H}$ also is Σ-regular. There exists, according to Theorem 1.1.4.4 again, $\tilde{n} \in N^+$ such that $\tilde{H} = \mathrm{Ad}(\tilde{n})X_x = \mathrm{Ad}(\tilde{n}x^{-1}n)H$; evidently $\mathrm{Ad}(n^{-1}x\tilde{n}^{-1})$ stabilizes $\mathfrak{m} + \mathfrak{a}_{\mathfrak{p}}$ since it maps the centralizer of $\tilde{H}$ into that of H. But $\mathfrak{a}_{\mathfrak{p}}$ consists precisely of those elements in $\mathfrak{m} + \mathfrak{a}_{\mathfrak{p}}$ such that $\mathrm{ad}(.)$ has real eigenvalues – therefore $\mathrm{Ad}(n^{-1}x\tilde{n}^{-1})$ stabilizes $\mathfrak{a}_{\mathfrak{p}}$ too, whence $m^* = n^{-1}x\tilde{n}^{-1} \in M^*$ and so $PxP = Pm^*P$, as desired.

(2) If $m_1^* \equiv m_2^* \pmod{M}$, then clearly $Pm_1^*P = Pm_2^*P(m_1^*, m_2^* \in M^*)$; it remains to show, therefore, that the equality $Pm_1^*P = Pm_2^*P$ $(m_1^*, m_2^* \in M^*)$ necessarily entails the congruence $m_1^* \equiv m_2^* \pmod{M}$. Let p_1, p_2 be elements in P such that $p_1m_1^* = m_2^*p_2$ $(m_1^*, m_2^* \in M^*)$; since $P = MA_{\mathfrak{p}}N^+ = N^+A_{\mathfrak{p}}M$, we have, for each $H \in \mathfrak{a}_{\mathfrak{p}}$, the relations

$$\operatorname{Ad}(p_1 m_1^*)H = \operatorname{Ad}(m_1^*)H + X_1,$$
$$\operatorname{Ad}(m_2^* p_2)H = \operatorname{Ad}(m_2^*)(H + X_2) = \operatorname{Ad}(m_2^*)H + X_3 + Y_3,$$

where $X_1, X_2, X_3 \in \mathfrak{n}^+$, $Y_3 \in \mathfrak{n}^-$. But $(\mathfrak{n}^+ + \mathfrak{n}^-) \cap \mathfrak{a}_\mathfrak{p} = \{0\}$ – thus $\operatorname{Ad}(m_1^*)H = \operatorname{Ad}(m_2^*)H$ and so $m_1^* \equiv m_2^* \pmod{M}$.

The proof of the theorem is now complete. □

Example Take $G = \mathbf{SL}(2, \mathbf{R})$; if $K = \mathbf{SO}(2)$, $A_\mathfrak{p} = \{\left(\begin{smallmatrix} h & 0 \\ 0 & h^{-1} \end{smallmatrix}\right) : h \in \mathbf{R}^+\}$ and $N^+ = \{\left(\begin{smallmatrix} 1 & n \\ 0 & 1 \end{smallmatrix}\right) : n \in \mathbf{R}\}$, then $G = KA_\mathfrak{p}N^+$ is an Iwasawa decomposition for G – here

$$M = \left\{\pm\begin{pmatrix} 1 & 0 \\ 0 & 1 \end{pmatrix}\right\}, \quad M^* = \left\{\pm\begin{pmatrix} 1 & 0 \\ 0 & 1 \end{pmatrix}, \pm\begin{pmatrix} 0 & 1 \\ -1 & 0 \end{pmatrix}\right\}$$

and so Bruhat's decomposition is given by $G = MA_\mathfrak{p}N^+ \cup N^+wMA_\mathfrak{p}N^+$ where $w = \left(\begin{smallmatrix} 0 & -1 \\ 1 & 0 \end{smallmatrix}\right)$.

Proposition 1.2.3.4 (Moore) Let G be a connected semi-simple Lie group with finite center, $G = KA_\mathfrak{p}N^+$ an Iwasawa decomposition for G; set $P = MA_\mathfrak{p}N^+$ – then P is the normalizer of N^+ in G. Moreover, there exists a unique double coset of P in G whose dimension is equal to that of G.

Proof (1) Obviously P is contained in the normalizer of N^+ in G; on the other hand, if x belongs to this normalizer, write $x = p_1 m_w p_2$ ($p_1, p_2 \in P$; $m_w \in M^*$) and then note that m_w normalizes N^+, hence permutes the elements of Σ^+ amongst themselves, hence represents 1 in W; therefore $m_w \in M$ and so $x \in P$. This establishes the first assertion of the proposition.

(2) Let $w_1, \ldots, w_{[W]}$ be an enumeration of the elements of W. There are but a finite number of P-orbits in G/P, thus at least one has dimension equal to that of G/P. But $\dim(G/P) = \dim(N^-) = \dim(N^+)$; therefore if $\dim(Pw_iP) = \dim(G)$, then $\dim(U_{w_i}^+) = \dim(N^+)$, thus $\mathfrak{u}_{w_i}^+ = \mathfrak{n}^+$ and so $\operatorname{Ad}(m_{w_i})\mathfrak{n}^- = \mathfrak{n}^+$ ($m_{w_i} \in M^*$ associated with w_i). Finally suppose that w_i and w_j represent two orbits of maximal dimension – then $\operatorname{Ad}(m_{w_i}m_{w_j}^{-1})(\mathfrak{n}^+) = \mathfrak{n}^+ \Rightarrow m_{w_i}m_{w_j}^{-1} \in P$ (by (1)) $\Rightarrow w_i = w_j$. Thus there is only one orbit of maximal dimension.

The proof of the proposition is therefore complete. □

Remark Let $w^\dagger$ denote that element in W which takes the positive Weyl chamber $\mathscr{C} \subset \mathfrak{a}_\mathfrak{p}$ to $(-\mathscr{C})$; part (2) of the proof of the preceding proposition shows that the double coset of P in G of maximal dimension is $Pw^\dagger P$. The mapping $-w^\dagger$, which clearly induces a permutation of Σ^+, is called the *opposition involution*; two subsets of Σ^+ which can be mapped into each other by $-w^\dagger$ are said to be *opposed.*

Proposition 1.2.3.5 Let G be a connected semi-simple Lie group with finite center, $G = KA_\mathfrak{p}N^+$ an Iwasawa decomposition for G; set $P = MA_\mathfrak{p}N^+$ – then N^-P is a dense open submanifold of G whose complement has zero Haar measure.

Proof First note that $N^- \cap P = \{1\}$. [Thus if $x \in N^- \cap P$, write $x = mh \exp X = \exp Y$ with $X \in \mathfrak{n}^+$, $Y \in \mathfrak{n}^-$; given a Σ-regular element H in $\mathfrak{a}_\mathfrak{p}$, we may, by Theorem 1.1.4.4, find $X_1 \in \mathfrak{n}^+$, $Y_1 \in \mathfrak{n}^-$ such that $\mathrm{Ad}(\exp Y)H = H + Y_1 = \mathrm{Ad}(mh)(H + X_1) = H + X_2$, $X_2 \in \mathfrak{n}^+$ – hence $Y_1 = 0 \Rightarrow x = 1$.] Therefore, since $\mathfrak{g} = \mathfrak{n}^- + \mathfrak{p}$ (direct sum of vector spaces), the map $(n, p) \mapsto np$ of $N^- \times P$ into G is open, regular and injective. Choose $m_{w^\dagger} \in M^*$ associated with $w^\dagger$ ($w^\dagger$ as defined in the Remark supra) – then $\dim(Pm_{w^\dagger}P) = \dim(G)$, $U^+_{w^\dagger} = N^+$ and $m^{-1}_{w^\dagger} U^+_{w^\dagger} m_{w^\dagger} = N^-$. Write $G = N^+ m_{w^\dagger} P \cup (\cup\, U^+_{w_i} m_{w_i} P)$ (disjoint union) with $\dim(U^+_{w_i} m_{w_i} P) < \dim(G)$; now multiply through by $m^{-1}_{w^\dagger}$ – plainly, then, G is the union of N^-P and a submanifold of lower dimension. Hence the result. □

As above, let $G = KA_\mathfrak{p}N^+$ be an Iwasawa decomposition for G; consider the group $M^*A_\mathfrak{p}$ – then $M^*A_\mathfrak{p}$ is the normalizer of $A_\mathfrak{p}$ in G. [Thus suppose that x normalizes $A_\mathfrak{p}(x \in G)$; write $x = k \exp X$ ($k \in K$, $X \in \mathfrak{p}$) – we are to show that $X \in \mathfrak{a}_\mathfrak{p}$. This, however, is clear in view of Lemma 1.1.3.7.] Here is a useful complement to the above discussion.

Proposition 1.2.3.6 Consider the double cosets $N^+m^*hN^+$ ($m^* \in M^*$, $h \in A_\mathfrak{p}$) – then $N^+m_1^*h_1N^+ = N^+m_2^*h_2N^+$ iff $m_1^* = m_2^*$, $h_1 = h_2$ ($m_i^* \in M^*$, $h_i \in A_\mathfrak{p}$; $i = 1, 2$).

Proof It is enough to show that $M^*A_\mathfrak{p} \cap (N^+m^*hN^+) = \{m^*h\}(m^* \in M^*, h \in A_\mathfrak{p})$. Let w denote the element of W which m^* represents – then $N^+m^*hN^+ = U_w^+N_w^+m^*hN^+ = U_w^+m^*hN^+$. Our assertion is therefore equivalent to the statement $M^*A_\mathfrak{p} \cap ((m^*h)^{-1}U_w^+m^*hN^+) = \{1\}$ which in turn will follow if we can establish that $M^*A_\mathfrak{p} \cap N^-N^+ = \{1\}$. So assume that we have an $x \in M^*A_\mathfrak{p}$ which can be written in the form $x = n_-n_+$ ($n_- \in N^-$, $n_+ \in N^+$) – we shall prove that $n_- = n_+ = 1$. For any $h \in A_\mathfrak{p}$, we have

$$(n_-^{-1}{}^xhn_-({}^xh)^{-1})({}^xhh^{-1})(hn_+h^{-1}n_+^{-1}) = 1.$$

The factors in the parentheses belong, respectively, to N^-, $A_\mathfrak{p}$, N^+ – hence each such factor must be equal to 1 (for $N^- \cap A_\mathfrak{p}N^+ = \{1\}$ (cf. the proof of Proposition 1.2.3.5); then quote Iwasawa). This means that $n_-, n_+ \in MA_\mathfrak{p}$ (= the centralizer of $A_\mathfrak{p}$ in G); consequently $n_- = n_+ = 1 \Rightarrow x = 1$. □

The construction of the boundary subgroups of a semi-simple Lie group is measure theoretic in character; accordingly we shall feel free to use the standard terminology which may be found in Appendices 1 and 2.

Definition Let G be a Lie group, H a closed subgroup and $B = G/H$ a compact homogeneous space of G – then B is said to be a *boundary* of G and H a *boundary subgroup* if for every probability measure μ on B

there is a sequence $\{x_n\}$ in G such that the sequence $\{x_n * \mu\}$ converges to a point measure on B.

[Since G operates on B, we have a C^∞ map from $G \times B$ into B; the symbol $x * \mu$ $(x \in G)$ then stands for the image of $\delta_x \times \mu$ under this map (cf. A2.3; for sake for brevity, we shall agree to write $x * \mu$ in place of $\delta_x * \mu$ $(x \in G)$).]

[In general, by a *probability measure* μ on a topological space X one understands a non-negative, bounded, regular, Borel measure μ such that $\mu(X) = 1$; convergence of measures on a compact space refers to weak convergence (that is, $\mu_n \longrightarrow \mu$ means that $\mu_n(f) \longrightarrow \mu(f)$ for every continuous function f on the space); needless to say, the set of probability measures on a separable compact space is sequentially compact with respect to weak convergence.]

Proposition 1.2.3.7 Let G be a Lie group, H a closed subgroup, and $B = G/H$ a compact homogeneous space of G – then the following conditions are equivalent:

(i) For every probability measure μ on B there is a sequence $\{x_n\}$ in G such that the sequence $\{x_n * \mu\}$ converges to a point measure on B;

(ii) For every probability measure μ on B there is a sequence $\{\mu_n\}$ of probability measures on G such that the sequence $\{\mu_n * \mu\}$ converges to a point measure on B;

(iii) There exists a quasi-invariant probability measure μ_0 on B and a sequence $\{\mu_n\}$ of probability measures on G such that the sequence $\{\mu_n * \mu_0\}$ converges to a point measure on B.

The proof depends upon the following lemma.

Lemma 1.2.3.8 Let G be a Lie group, B a compact homogeneous space of G; let μ be a positive measure on B and $\{\mu_n\}$ a sequence of probability measures on G such that the sequence $\{\mu_n * \mu\}$ converges to a point measure on B – then there exists a sequence $\{x_n\}$ in G such that the sequence $\{x_n * \mu\}$ is convergent to a point measure on B.

Proof Suppose that $\mu_n * \mu \longrightarrow \delta_b$ $(b \in B)$; let $\{\mathcal{O}_m : m = 1, 2, \ldots\}$ be a basis for the neighborhoods at b – then, since $\mu_n * \mu(\mathcal{O}_m) = \int_G x * \mu(\mathcal{O}_m)\, d\mu_n(x) > 1 - m^{-1}$ for sufficiently large n, it follows that there exists x_m in G with $x_m * \mu(\mathcal{O}_m) > 1 - m^{-1}$. It is therefore clear that $x_m * \mu \longrightarrow \delta_b$. □

Proof of Proposition 1.2.3.7 (i) ⇒ (ii) – this is clear.

(ii) ⇒ (i) – this is a consequence of Lemma 1.2.3.8.

(ii) ⇒ (iii) – this is clear.

(iii) ⇒ (ii). Let μ be any probability measure on B; fix a quasi-invariant measure π on G – then $\pi * \mu$ is quasi-invariant on B, whence $\pi * \mu$ is absolutely continuous with respect to μ_0. Since $\mu_n * \mu_0 \longrightarrow \delta_b$ for some b in B, it follows that $\mu_n * (\pi * \mu) \longrightarrow \delta_b$; but then, in view of the preceding lemma, there exists a sequence $\{x_n\}$ in G such that $x_n * \mu \longrightarrow \delta_b$.

The proof of the proposition is now complete. □

Examples (1) Let us suppose that G is abelian; in this case every subgroup H is normal and so G/H is a group with an invariant Haar measure; it is easy to see that G preserves this measure – therefore G/H cannot be a boundary unless $H = G$.

(2) Let us suppose that G is compact – then it is clear that only point measures can be sent into point measures and so G/H must consist of a single point if it is to be a boundary.

If K is a compact Lie group, then μ_K will denote normalized Haar measure on K.

Lemma 1.2.3.9 Let μ be a probability measure on a compact Lie group K whose support is not contained in a proper closed subgroup. If ν is a probability measure on K with $\mu * \nu = \nu$, then $\nu = \mu_K$.

Proof Suppose first that ν is absolutely continuous with respect to μ_K and that the Radon-Nikodym derivative $d\nu/d\mu_K$ is a continuous function f (say) on K. The equation $\mu * \nu = \nu$ implies that $f(k) = \int_K f(\tilde{k}^{-1}k)\, d\mu(\tilde{k})$ (all $k \in K$); let A be the subset of K on which f attains its maximum and let B be the subset of K comprised of those k in K such that $k^{-1}A \subset A$ – then B is a compact semi-group of K, whence is a closed subgroup of K. In view of the functional equation which f satisfies, it follows that $\mu(B) = 1$; therefore, by hypothesis, $B = K$ and so $A = K$, hence f is a constant and $\nu = \mu_K$. In general, if $\tilde{\nu}$ is absolutely continuous with respect to μ_K and if the Radon-Nikodym derivative $d\tilde{\nu}/d\mu_K$ is continuous, then the same is true of $\nu * \tilde{\nu}$ and $\mu * (\nu * \tilde{\nu}) = \nu * \tilde{\nu}$ – thus $\nu * \tilde{\nu} = \mu_K$ and, as this is the case for all $\tilde{\nu}$ of the kind described, it follows that $\nu = \mu_K$. □

A topological group G is said to have the *fixed-point property* if whenever it acts continuously on a locally convex, Hausdorff, topological vector space by linear transformations leaving a non-empty compact convex set $\mathscr{C}$ invariant it has a fixed point in the set. It is well-known that connected solvable Lie groups have the fixed point property (for a proof, see the Appendix to the present number); the same is true for compact groups ....

Example Let us suppose that G has the fixed point property – then the only boundary of G is the trivial one consisting of a single point. [Let B be a compact homogeneous space of G; because the set of probability measures on B forms a compact convex set in the weak topology on the space of measures on B, it follows that there is a probability measure μ on B which is G-stable; since G operates transitively on B, the measure μ will not be a point measure unless B is trivial ....]

The boundaries of a connected semi-simple Lie group G (with finite center) are all obtained from one of them, the so-called 'maximal boundary'; in turn its construction goes as follows. Write $G = KA_\mathfrak{p}N^+$, $S = A_\mathfrak{p}N^+$ (thus S is a closed connected solvable subgroup of G); let $\sigma: G \longrightarrow G/S$ be the natural map. Fix a quasi-invariant measure μ_0 on G/S (e.g. $\mu_0 = \mu_K * \delta_p$, $p \in G/S$) and let $\mathscr{C}$ be the compact convex set of measures on G/S which are limits of measures of the form $\mu * \mu_0$ where μ ranges over the probability measures on G.

Lemma 1.2.3.10 (Furstenberg) Retain the above notations – then there is a compact subgroup K_0 of K and a point $p_0 \in G/S$ such that the measure $\pi_0 = \mu_{K_0} * \delta_{p_0}$ is in $\mathscr{C}$ and is left fixed by S.

Proof Because $x*(\mu*\mu_0) = (x*\mu)*\mu_0$ (all $x \in G$), the group G operates on the compact convex set $\mathscr{C}$; therefore, since S is a connected solvable Lie group, there is a measure π in $\mathscr{C}$ which is stabilized by S; in view of the fact that K is homeomorphic to G/S, there exists a unique measure $\tilde{\pi}$ on K such that $\sigma(\tilde{\pi}) = \pi$ – we shall agree to regard $\tilde{\pi}$ as a measure on G. Let $s \in S$ – then it is clear that $\sigma(s*\tilde{\pi}) = s*\sigma(\tilde{\pi}) = s*\pi = \pi = \sigma(\tilde{\pi})$, thus, as a simple calculation quickly shows, $(s*\tilde{\pi})*\pi = s*(\tilde{\pi}*\pi) = \tilde{\pi}*\pi$ (all $s \in S$) and so $\tilde{\pi}*\pi$ is S-stable. This being the case, let $\pi_1 = \tilde{\pi}*\pi$; the same argument shows that $\pi_2 = \tilde{\pi}*\pi_1$ is invariant under S; in this way, therefore, we obtain a sequence $\{\pi_n\}$ of measures on G/S with $\pi_{n+1} = \tilde{\pi}*\pi_n$ and with each π_n invariant under S. Let $\tilde{\pi}_n$ be the measure on K such that $\sigma(\tilde{\pi}_n) = \pi_n$; clearly $\tilde{\pi}_n = \tilde{\pi}^n$, the n-fold convolution of $\tilde{\pi}$. Let K_0 be the closed subgroup of K generated by the support of $\tilde{\pi}$; the measures $\mu_n = n^{-1} \sum_{i=1}^n \tilde{\pi}^i$ have their support on K_0 and if a subsequence of $\{\mu_n\}$ converges to a measure μ, then $\tilde{\pi}*\mu = \mu$; by Lemma 1.2.3.9, $\mu = \mu_{K_0}$ so that $\mu_n \longrightarrow \mu_{K_0}$. Now $\sigma(\mu_n)(= n^{-1} \sum_{i=1}^n \pi_i)$ is S-stable, therefore the same is true of $\sigma(\mu_{K_0})$; set $\pi_0 = \sigma(\mu_{K_0})$ – then $\pi_0 = \mu_{K_0}*\sigma(\delta_1)$, is invariant under S and lies in $\mathscr{C}$. Hence the lemma. □

Keeping to the above notations, let $H(G)$ be the subgroup of G consisting of those elements which leave the measure π_0 of the preceding lemma invariant – then $H(G) = K_0S$. [It is clear that $K_0S \subset H(G)$. On the other hand, $H(G) = (H(G) \cap K)S$; if $k \in H(G) \cap K$, then $k*\pi_0 = \pi_0 \Rightarrow k*\sigma(\mu_{K_0}) = \sigma(\mu_{K_0}) \Rightarrow k*\mu_{K_0} = \mu_{K_0}$ ($\sigma: K \longrightarrow G/S$ is one-to-one); comparing the supports of $k*\mu_{K_0}$ and μ_{K_0} leads at once to the conclusion that $k \in K_0$ – therefore $H(G) = K_0S$, as contended.] Furthermore $H(G)$ is a boundary subgroup of G, i.e. $B(G) = G/H(G)$ is a boundary of G. [Thus, in the first place, we have a G-equivariant map τ: $G/S \longrightarrow G/H(G) = B(G)$. Because the measure π_0 lies in $\mathscr{C}$ there is a sequence $\{\mu_n\}$ of probability measures on G such that $\mu_n*\mu_0 \longrightarrow \pi_0$; therefore, in view of the G-equivariance of τ, we have $\mu_n*\tau(\mu_0) = \tau(\mu_n*\mu_0) \longrightarrow \tau(\pi_0)$; but $\tau(K_0p_0)$ is a point on $B(G)$, and so $\tau(\pi_0)$ is a point measure on $B(G)$ – the fact that $B(G)$ is a boundary of G now follows upon remarking that $\tau(\mu_0)$ is a quasi-invariant measure on $B(G)$]. Finally let us observe that any maximal compact subgroup of G operates transitively on $B(G)$. [It is clear that K is transitive on $B(G)$ (for K acts transitively on G/S . . .); on the other hand, every maximal compact subgroup of G is conjugate to K.]

A boundary B of G is said to be a *maximal boundary* if, given any other boundary $\tilde{B}$ of G, there exists a (continuous) G-equivariant map from B onto $\tilde{B}$; a closed subgroup H of G is said to be a *minimal boundary subgroup* provided $B = G/H$ is a maximal boundary (of G).

The following result tells us that the space $B(G)$ is a maximal boundary (hence that the subgroup $H(G)$ is a minimal boundary subgroup).

Proposition 1.2.3.11 Let G be a connected semi-simple Lie group with finite center; let B be a G-homogeneous space – then B is a boundary of G iff there exists a G-equivariant map carrying $B(G)$ onto B.

Proof Let us suppose that B is a boundary of G. Since $H(G) = K_0 S$ and preserves a measure on the homogeneous space $H(G)/S$, it follows without difficulty that $H(G)$ has the fixed point property (use the evident 'averaging' argument); consequently $H(G)$ leaves invariant some probability measure μ on the compact G-space B. Because B is a boundary, there exists a sequence $\{x_n\}$ in G such that the sequence $\{x_n * \mu\}$ converges to a point measure on B; however S leaves μ fixed, whence the G-orbit of μ is the same as the K-orbit of μ and so any limit of the sequence $\{x_n * \mu\}$ must be of the form $k * \mu$ (some $k \in K$) – therefore μ itself is necessarily a point measure. This means that $H(G)$ leaves fixed a point on the boundary B, hence $H(G)$ is contained in the stability group of this point; the existence of a G-equivariant map from $B(G)$ onto B is therefore obvious. The converse assertion is trivial. □

It follows, in particular, that the space $B(G)$ constructed above is essentially independent of the particular Iwasawa decomposition taken; that is, if two such spaces are constructed, then there will be an equivariant isomorphism between the two.

Corollary 1.2.3.12 Let G be a connected semi-simple Lie group with finite center – then the minimal boundary subgroups of G are all conjugate.

Proof Suppose that H is a boundary subgroup of G; according to what was said above, $H(G)$ is contained in the stability group of some point of G/H; this means that $H(G) \subset xHx^{-1}$ for some $x \in G$, i.e. that $H \supset x^{-1}H(G)x$; but the latter group is a boundary subgroup which is contained in H – thus, if H is minimal, then we must have $H(G) = xHx^{-1}$. □

Our next task will be to identify the subgroup K_0 which appears above.

Proposition 1.2.3.13 Let G be a connected semi-simple Lie group with finite center, $G = KA_{\mathfrak{p}}N^+$ an Iwasawa decomposition for G; let $H(G) = K_0A_{\mathfrak{p}}N^+$ – then actually $H(G) = \mathsf{M}A_{\mathfrak{p}}N^+$, hence $K_0 = \mathsf{M}$.

Proof We know that $P = \mathsf{M}A_{\mathfrak{p}}N^+$ is the normalizer of $S = A_{\mathfrak{p}}N^+$ in G (cf. Proposition 1.2.3.4); this being so, we shall first show that the normalizer of S in G is contained in the minimal boundary subgroup $H(G)$; the proof will then be completed by showing that G/P is a boundary.

(1) (Furstenberg) We shall agree to retain the earlier notations. Without loss of generality, it can be assumed that μ_0 is chosen as the unique K-invariant probability measure on G/S (μ_0 is unique because K is transitive on G/S); this being so, we then recall that K_0 has the property that $\sigma(\mu_{K_0})$ is a limit of measures $\mu_n * \mu_0$ where μ_n is a probability measure on G. Now let x_0 belong to the normalizer of S; the map $\bar{x}_0$ defined by the rule $xS \mapsto xx_0S = xSx_0$ $(x \in G)$ is obviously well-defined on G/S. Consider the measure $\bar{x}_0(\mu_0)$; we claim that $\bar{x}_0(\mu_0) = \mu_0$ – in fact $k*(\bar{x}_0(\mu_0)) = \bar{x}_0(k*\mu_0) = \bar{x}_0(\mu_0)$ (all $k \in K$), whence, in view of our supposition on μ_0, $\bar{x}_0(\mu_0) = \mu_0$. But then $\bar{x}_0(\sigma(\mu_{K_0})) = \lim_n \bar{x}_0(\mu_n*\mu_0) = \lim_n \mu_n*\bar{x}_0(\mu_0) = \lim_n \mu_n * \mu_0 = \sigma(\mu_{K_0})$; comparing the supports of the two sides, we find that $K_0Sx_0 = K_0S$, i.e. that $x_0 \in K_0S = H(G)$. Therefore the normalizer of S in G is contained in $H(G)$, as was to be shown.

(2) (Moore) Let us prove that G/P is a boundary. For this purpose set $Q = N^-P$ – then, as we know, Q is the one-one image in G/P of the subgroup N^- of G. We shall agree to identify Q with $\mathfrak{n}^-$ via the map $X \mapsto \exp(X)P$ $(X \in \mathfrak{n}^-)$; the relation $h\cdot\exp(X)P = \exp(\mathrm{Ad}(h)X)P$ $(X \in \mathfrak{n}^-)$, valid for all h in $A_{\mathfrak{p}}$, shows that $A_{\mathfrak{p}}$ leaves Q invariant and operates on Q just as it does on $\mathfrak{n}^-$; introduce coordinates x_i on $\mathfrak{n}^-$ (and corresponding coordinates on Q) which serve to diagonalize $\mathrm{Ad}_G(A_{\mathfrak{p}})\,|\,\mathfrak{n}^-$. Now fix an element H in the positive Weyl chamber and set $h_t = \exp(tH)(t \in \mathbf{R})$; let μ be an arbitrary quasi-invariant probability measure on G/P – then we intend to prove that $h_t*\mu$ tends to a point measure on G/P at P as $t \longrightarrow \infty$. Let f be a continuous function on G/P with compact support such that $0 \leqslant f \leqslant 1$ with $f = 1$ on $\mathscr{S}_r$ where $\mathscr{S}_r = \{q \in Q: |x_i(q)| \leqslant r\}$ – then, since the complement of Q in G/P has quasi-invariant measure zero, $\int_{G/P} f d(h_t*\mu) = \int_Q f d(h_t*\mu)$, whence

$$\mu(h_t^{-1}\cdot\mathscr{S}_r) \leqslant \int_Q f(h_t\cdot q)\, d\mu(q) \leqslant 1.$$

Choose $s > 0$ such that $\lambda(H) > s$ (all $\lambda \in \Sigma^+$) – then the eigenvalues of $\mathrm{Ad}(h_t^{-1})$ on $\mathfrak{n}^-$ all exceed $\exp(st)$ and so $\mathscr{S}_{r\exp(st)} \subset h_t^{-1}\cdot\mathscr{S}_r$ $(t > 0)$. Thus, as $t \longrightarrow \infty$,

$$1 \geqslant \mu(h_t^{-1}\cdot\mathscr{S}_r) \geqslant \mu(\mathscr{S}_{r\exp(st)}) \longrightarrow \mu(Q) = 1$$

and $\int_{G/P} f d(h_t*\mu) \longrightarrow 1$. Owing to the weak compactness of the set of probability measures on G/P, we can choose a sequence $\{t_n\}$ such that $h_{t_n} * \mu$ converges to a limit ν (say); it follows that $\int_{G/P} f\, d\nu = 1$. Since we can choose f to vanish outside of $\mathscr{S}_{2r}$, it is clear that ν is a point mass at the origin P in G/P – therefore $h_t * \mu$ tends to a point measure at P as $t \longrightarrow \infty$.

This completes the proof of the proposition. □

Example Let us consider the group G of holomorphic automorphisms of the Cartan domain $\mathscr{D}$ defined by

$$\mathscr{D} = \{Z \in \mathfrak{gl}(n, \mathbf{C}) : (2\sqrt{-1})^{-1}(Z - Z^*) > 0\}$$

where the '$*$' denotes conjugate transpose (positive means Hermitian positive). [Via the Cayley transform $Z \mapsto (Z - \sqrt{-1}I_n)(Z + \sqrt{-1}I_n)^{-1}$, $\mathscr{D}$ may be identified with the classical Cartan domain $\{Z : Z^*Z < I_n\}$.] The elements x of G can be represented by matrices in $\mathbf{SL}(2n, \mathbf{C})$

$$x = \begin{pmatrix} A & B \\ C & D \end{pmatrix}$$

where $xZ = (AZ + B)(CZ + D)^{-1}$; the requirement $x(\mathscr{D}) \subset \mathscr{D}$ forces the relations $A^*C = C^*A$, $D^*B = B^*D$, $A^*D - C^*B = I_n$. A maximal compact subgroup K of G is given by the stability group of $\sqrt{-1}I_n \in \mathscr{D}$ – thus

$$K = \left\{ \begin{pmatrix} \dfrac{U + V}{2} & \dfrac{U - V}{2} \\ \dfrac{V - U}{2} & \dfrac{U + V}{2} \end{pmatrix} : U, V \text{ unitary} \right\}.$$

The subgroup S comprised of those matrices of the form $\left(\begin{smallmatrix} A & B \\ 0 & A^{*-1} \end{smallmatrix}\right)$ where $AB^* = BA^*$ and A is upper triangular with positive entries on the diagonal is a complementary solvable subgroup to K (per Iwasawa); the normalizer P of S in G has the same form as S without restrictions on the diagonal coefficients (of A) – hence G/P is a maximal boundary of G and $\dim(G/P) = 2n^2 - n$. On the other hand it is of some significance to note that the maximal boundary of G can *not* in general be identified with the Bergman-Šilov boundary of $\mathscr{D}$; in fact if we take $\mathscr{D} = \{Z : Z^*Z < I_n\}$, then the Bergman-Šilov boundary corresponds to the set $\{Z : Z^*Z = I_n\}$, a space of dimension n^2. . . . It can be shown, however, that the Bergman-Šilov boundary is a G-equivariant image of the maximal boundary, hence is a boundary (cf. Furstenberg [1]).

Remarks (1) It is not difficult to show that the minimal boundary subgroups of our semi-simple group G are maximal subgroups with respect to the fixed point property; on the other hand the maximal compact subgroups of G are also maximal with respect to the fixed point property so that this condition does not characterize the minimal boundary subgroups (cf. Furstenberg [1]).

(2) The notion of boundary as introduced above can be used to study the compactifications of the Riemannian symmetric space G/K (cf. Furstenberg [1] and Moore [2]). Thus let P be a minimal boundary subgroup of G; let $m^+(G/P)$ be the space of probability measures on G/P (upon which G acts as a transformation group); let μ be the unique element in $m^+(G/P)$ which is K-invariant (K operates transitively on G/P) – then the assignment $x \mapsto x * \mu$ maps G into $m^+(G/P)$ and, since μ is K-invariant, this becomes an equivariant mapping ϕ of G/K into $m^+(G/P)$ which can be shown to be an injective homeomorphism. The closure $[\phi(G/K)]^{cl}$ may be regarded as a compactification of

G/K. Similarly, if B is any boundary of G, then one can define in the same way an equivariant mapping ϕ_B of G/K into $m^+(B)$; in general, however, ϕ_B need not be one-to-one. For further details and the relations of these compactifications with those of Satake [1], see Moore [2].

Example One can define the maximal boundary of our Riemannian symmetric space G/K in a geometric way, viz. by stipulating that it is to be the set B of all Weyl chambers in $\mathfrak{p}$; since the points of B are in a one-to-one correspondence with the points of the orbit space $K/M(\sim G/P \ldots)$, this is a legitimate agreement. To motivate such a 'geometric' notion of boundary, consider the non-Euclidean disk D; here we have a natural notion of boundary, namely the unit circle ∂D; this boundary notion refers, however, to the position of D in $\mathbf{R}^2$; accordingly, one is led to define the boundary of D as the set of all rays (= half-lines) issuing from the origin in D. . . .

Definition Let G be a connected semi-simple Lie group with finite center – then, by a *minimal parabolic subgroup* of G, we shall understand any minimal boundary subgroup of G.

In view of what has been said above, the minimal parabolic subgroups P of G are all conjugate to one another (cf. Corollary 1.2.3.12); this being so, fix one such P; relative to a suitable Iwasawa decomposition of G, write $P = MA_{\mathfrak{p}}N^+$ – then we intend to show that the triple (G, P, M^*) constitutes a Tits system which will thus serve to justify the adopted terminology (and hence make available the results obtained in 1.2.1). To get at this we shall need some preparation which will also be useful in 1.2.4.

Let $\mathfrak{g}$ be a semi-simple Lie algebra over $\mathbf{R}$, $\mathfrak{g} = \mathfrak{k} + \mathfrak{a}_{\mathfrak{p}} + \mathfrak{n}^+$ an Iwasawa decomposition for $\mathfrak{g}$; let Σ denote the set of roots of the pair $(\mathfrak{g}, \mathfrak{a}_{\mathfrak{p}})$, $\Upsilon = \{\lambda_1, \ldots, \lambda_l\}$ the fundamental system of positive roots implicitly associated with the given Iwasawa decomposition ($l = \dim(\mathfrak{a}_{\mathfrak{p}})$ = split-rank of $\mathfrak{g}$). For any subset Θ of Υ let $\langle\Theta\rangle$ denote the set comprised of those λ in Σ which are linear combinations of the elements of Θ; we shall agree to write $\langle\Theta\rangle^\pm$ for $\Sigma^\pm \cap \langle\Theta\rangle$. Put $\mathfrak{n}^\pm(\Theta) = \sum_\lambda \mathfrak{g}^\lambda$, $\lambda \in \langle\Theta\rangle^\pm$, and let $\mathfrak{g}(\Theta)$ be the subalgebra of $\mathfrak{g}$ generated by $\mathfrak{n}^+(\Theta) + \mathfrak{n}^-(\Theta)$ – then $\mathfrak{g}(\Theta)$ is θ-stable (θ the Cartan involution determined by the pair $(\mathfrak{g}, \mathfrak{k})$) and so $\mathfrak{g}(\Theta) = \mathfrak{k}(\Theta) + \mathfrak{p}(\Theta)$ where $\mathfrak{k}(\Theta) = \mathfrak{g}(\Theta) \cap \mathfrak{k}$, $\mathfrak{p}(\Theta) = \mathfrak{g}(\Theta) \cap \mathfrak{p}$.

Lemma 1.2.3.14 The Lie algebra $\mathfrak{g}(\Theta)$ is semi-simple.

Proof First note that $\mathfrak{g}(\Theta)$ is reductive in $\mathfrak{g}$; in fact the orthogonal complement in $\mathfrak{g}$ of any $\mathrm{ad}(\mathfrak{g}(\Theta))$-stable subspace is likewise $\mathrm{ad}(\mathfrak{g}(\Theta))$-stable, the $\mathrm{ad}(\mathfrak{g}(\Theta))$ being a self-adjoint family of linear transformations (relative to the Euclidean structure $(\cdot\,,\cdot)_\theta$) – therefore the derived algebra $[\mathfrak{g}(\Theta), \mathfrak{g}(\Theta)]$ is semi-simple. Fix $\lambda \in \langle\Theta\rangle^+$ and $X \in \mathfrak{g}^\lambda$ with $-B(X, \theta(X)) = 1$ – then $[\theta(X), X] = Q_\lambda \in \mathfrak{g}(\Theta)$, whence $[Q_\lambda, X] = \lambda(Q_\lambda)X \in [\mathfrak{g}(\Theta), \mathfrak{g}(\Theta)]$. It therefore follows that $\mathfrak{g}^\lambda \subset [\mathfrak{g}(\Theta), \mathfrak{g}(\Theta)]$ and so $\mathfrak{g}(\Theta) = [\mathfrak{g}(\Theta), \mathfrak{g}(\Theta)]$. □

As a consequence, $\mathfrak{g}(\Theta) = \mathfrak{k}(\Theta) + \mathfrak{p}(\Theta)$ is a Cartan decomposition of $\mathfrak{g}(\Theta)$.

For any $\lambda \in \Sigma$, let P_λ denote the orthogonal projection of $\mathfrak{g}$ onto $\mathfrak{g}^\lambda$; let $\mathsf{P}(\Theta)$ denote the orthogonal projection of $\mathfrak{g}$ onto $\mathfrak{g}(\Theta)$. Put $\mathfrak{a}_\mathfrak{p}(\Theta) = \sum_\lambda \mathbf{R}Q_\lambda$, $\lambda \in \langle\Theta\rangle^+$ – then, as was observed in the proof of Lemma 1.2.3.14, $\mathfrak{a}_\mathfrak{p}(\Theta) \subset \mathfrak{g}(\Theta)$. Let $\mathfrak{a}_\Theta$ denote the orthogonal complement (relative to $(\cdot\,,\cdot)_\theta$) of $\mathfrak{a}_\mathfrak{p}(\Theta)$ in $\mathfrak{a}_\mathfrak{p}$.

Lemma 1.2.3.15 Let us keep to the above notations – then $\mathfrak{a}_\mathfrak{p}(\Theta) = \mathfrak{g}(\Theta) \cap \mathfrak{a}_\mathfrak{p}$. Moreover $\mathfrak{a}_\mathfrak{p}(\Theta)$ is maximal abelian in $\mathfrak{p}(\Theta)$ and $\mathsf{P}(\Theta)(\mathfrak{a}_\mathfrak{p}) = \mathfrak{a}_\mathfrak{p}(\Theta)$.

Proof To see that $\mathfrak{a}_\mathfrak{p}(\Theta) = \mathfrak{g}(\Theta) \cap \mathfrak{a}_\mathfrak{p}$, fix an H in the orthogonal complement of $\mathfrak{a}_\mathfrak{p}(\Theta)$ in $\mathfrak{a}_\mathfrak{p}$; thus $\lambda_i(H) = (Q_{\lambda_i}, H)_\theta = 0$ for all $\lambda_i \in \Theta$ and so $\lambda(H) = 0$ (all $\lambda \in \langle\Theta\rangle$); this means that H commutes with $\mathfrak{g}(\Theta)$, whence $\mathfrak{g}(\Theta) \cap \mathfrak{a}_\Theta = \{0\}$ ($\mathfrak{g}(\Theta)$ being semi-simple) – therefore $\mathfrak{a}_\mathfrak{p}(\Theta) = \mathfrak{g}(\Theta) \cap \mathfrak{a}_\mathfrak{p}$. We claim next that $\mathfrak{a}_\mathfrak{p}(\Theta)$ is maximal abelian in $\mathfrak{p}(\Theta)$. Plainly $\mathfrak{g}^0 + \mathfrak{n}^+(\Theta) + \mathfrak{n}^-(\Theta)$ is a subalgebra of $\mathfrak{g}$, hence contains $\mathfrak{g}(\Theta)$; consequently

$$X = \mathsf{P}_0 X + \sum_\lambda \mathsf{P}_\lambda X + \sum_\lambda \mathsf{P}_{-\lambda} X, \quad \lambda \in \langle\Theta\rangle^+, \quad (\text{all } X \in \mathfrak{g}(\Theta)).$$

Suppose that $X \in \mathfrak{p}(\Theta)$ and commutes with $\mathfrak{a}_\mathfrak{p}(\Theta)$ – then

$$0 = [H, X] = \sum_\lambda \lambda(H)\mathsf{P}_\lambda X - \sum_\lambda \lambda(H)\mathsf{P}_{-\lambda} X, \quad \lambda \in \langle\Theta\rangle^+, \qquad (\text{all } H \in \mathfrak{a}_\mathfrak{p}(\Theta))$$

which implies that $\lambda(H)\mathsf{P}_{\pm\lambda}X = 0$ for $H \in \mathfrak{a}_\mathfrak{p}(\Theta)$ and $\lambda \in \langle\Theta\rangle^+$. Because $Q_\lambda \in \mathfrak{a}_\mathfrak{p}(\Theta)$ (all $\lambda \in \langle\Theta\rangle^+$), it therefore follows that $\mathsf{P}_{\pm\lambda}X = 0 \Rightarrow X = \mathsf{P}_0 X \Rightarrow X \in \mathfrak{g}(\Theta) \cap \mathfrak{a}_\mathfrak{p} \Rightarrow X \in \mathfrak{a}_\mathfrak{p}(\Theta)$ – hence $\mathfrak{a}_\mathfrak{p}(\Theta)$ is maximal abelian in $\mathfrak{p}(\Theta)$. The proof of the last assertion of our lemma may be left to the reader. □

Let us return now to the problem of showing that the triple (G, P, M^*) constitutes a Tits system; the burden of this verification rests on Lemma 1.2.3.16 below.

Fix $\lambda_i \in \{\lambda_1, \ldots, \lambda_l\}$ and set $w_i = w_{\lambda_i}$; evidently $\mathfrak{u}_{w_i}^+ = \mathfrak{g}^{\lambda_i} + \mathfrak{g}^{2\lambda_i}$ (of course $2\lambda_i$ could be zero in which case $\mathfrak{g}^{2\lambda_i}$ would not be present). Employing the notations of Lemma 1.2.3.14, take $\Theta = \{\lambda_i\}$, $\mathfrak{g}(\Theta) = \mathfrak{g}(\lambda_i)$ (say); thus $\mathfrak{g}(\lambda_i)$ is a semi-simple subalgebra of $\mathfrak{g}$ generated by $\mathfrak{g}^{\lambda_i} + \mathfrak{g}^{2\lambda_i} + \mathfrak{g}^{-\lambda_i} + \mathfrak{g}^{-2\lambda_i}$ admitting the Cartan decomposition $\mathfrak{g}(\lambda_i) = \mathfrak{g}(\lambda_i) \cap \mathfrak{k} + \mathfrak{g}(\lambda_i) \cap \mathfrak{p}$ and the subspace $\mathfrak{a}_\mathfrak{p}(\lambda_i) = \mathfrak{p}(\lambda_i) \cap \mathfrak{a}_\mathfrak{p}$ is maximal abelian in $\mathfrak{p}(\lambda_i)$ ($\mathfrak{p}(\lambda_i) = \mathfrak{g}(\lambda_i) \cap \mathfrak{p}$) – in fact $\mathfrak{a}_\mathfrak{p}(\lambda_i) = \mathbf{R}Q_{\lambda_i}$. Let $G(\lambda_i)$ denote the analytic subgroup of G corresponding to $\mathfrak{g}(\lambda_i)$ – then $G(\lambda_i) = K(\lambda_i)A_\mathfrak{p}(\lambda_i)N^+(\lambda_i)$ is an Iwasawa decomposition of $G(\lambda_i)$ where, of course, $A_\mathfrak{p}(\lambda_i)$ and $N^+(\lambda_i)$ denote the analytic subgroups of $G(\lambda_i)$ corresponding to $\mathfrak{a}_\mathfrak{p}(\lambda_i)$ and $\mathfrak{n}^+(\lambda_i) = \mathfrak{u}_{w_i}^+$, respectively. Owing to Lemma 1.1.3.9 it is possible to select an $m_{w_i} \in M^*$ associated with w_i such that $m_{w_i} \in K(\lambda_i)$ – during the ensuing discussion we shall assume that this has been done. Let $M(\lambda_i)$ denote the centralizer of $\mathfrak{a}_\mathfrak{p}(\lambda_i)$ in $K(\lambda_i)$ and set $P(\lambda_i) = M(\lambda_i)A_\mathfrak{p}(\lambda_i)N^+(\lambda_i)$; note that $M(\lambda_i) \subset M$, $P(\lambda_i) \subset P$. Theorem 1.2.3.1 (bis) implies that $G(\lambda_i) = P(\lambda_i) \cup U_{w_i}^+ m_{w_i} P(\lambda_i)$ (disjoint union) and, since $U_{w_i}^- (= m_{w_i}^{-1} U_{w_i}^+ m_{w_i})$, we have $G(\lambda_i) = m_{w_i}^{-1} P(\lambda_i) \cup U_{w_i}^- P(\lambda_i)$ (disjoint union) – hence $U_{w_i}^- P(\lambda_i)$ is a dense open submanifold of $G(\lambda_i)$ whose complement is a set of measure zero with respect to Haar measure on $G(\lambda_i)$.

Lemma 1.2.3.16 If $u \in U_{w_i}^-$ and $u \neq 1$, then $m_{w_i}^{-1} u \in U_{w_i}^- P(\lambda_i)$.

Proof By the above, either $m_{w_i}^{-1} u \in m_{w_i}^{-1} P(\lambda_i)$ or $m_{w_i}^{-1} u \in U_{w_i}^- P(\lambda_i)$. But

$$m_{w_i}^{-1} u \in m_{w_i}^{-1} P(\lambda_i) \Rightarrow u \in P(\lambda_i) \cap U_{w_i}^- = \{1\};$$

thus $m_{w_i}^{-1} u \in U_{w_i}^- P(\lambda_i)$. □

The fact that the triple (G, P, M^*) is a Tits system can now be seen as follows. In view of Theorem 1.2.3.1, it is clear that G is generated by P and M^*; of course $M = P \cap M^*$ is a normal subgroup of M^* and, moreover, the group $W = M^*/M$ admits a system of generators consisting of involutive elements $w_i = w_{\lambda_i}$ $(i = 1, \ldots, l;\ l = \dim(\mathfrak{a}_\mathfrak{p}))$ – otherwise said, all the assumptions of condition (1) (in 1.2.1) are met here. As for (2), observe that $w_i U_{w_i}^+ w_i^{-1} \subset N^-$ – therefore $w_i P w_i^{-1}$ is not contained in P. The fact that condition (3) is also satisfied is the gist of the following result.

Proposition 1.2.3.17 Fix w_i, $w \in W$ – then $w_i P w$ is contained in the union of PwP and $Pw_i wP$.

Proof It will actually be more convenient to show that wPw_i is contained in the union of PwP and Pww_iP; in view of the proof of Lemma 1.2.1.3, this is sufficient. Since $w_i = w_i^{-1}$, we have $wPw_i = wN^+A_\mathfrak{p}Mw_i = wN^+w_iA_\mathfrak{p}M = wU_{w_i}^+N_{w_i}^+w_iA_\mathfrak{p}M = wU_{w_i}^+w_iN_{w_i}^+A_\mathfrak{p}M \subset wU_{w_i}^+w_iP$; hence it will be enough to prove that $wU_{w_i}^+w_iP \subset (PwP) \cup (Pww_iP)$. Consider $wU_{w_i}^+w^{-1}$ – thus $wU_{w_i}^+w^{-1} = \exp(\mathfrak{g}^{\lambda_i^w} + \mathfrak{g}^{2\lambda_i^w})$ and so if $\lambda_i^w > 0$, then $wU_{w_i}^+w^{-1} \subset N^+$, whence $wU_{w_i}^+w_iP \subset Pww_iP$. We may therefore assume that $\lambda_i^w < 0$. Write $wU_{w_i}^+w_iP = ww_i^{-1}U_{w_i}^-P$; a typical element in the set $ww_i^{-1}U_{w_i}^-P$ has the form $ww_i^{-1}up$ $(u \in U_{w_i}^-,\ p \in P)$. If $u = 1$ we are in business; if $u \neq 1$, then $w_i^{-1}u \in U_{w_i}^-P$ (cf. Lemma 1.2.3.16 – we are employing the usual abuse of notation) and hence $ww_i^{-1}up \in wU_{w_i}^-P$. But the Lie algebra of $U_{w_i}^-$ is $\mathfrak{g}^{-\lambda_i} + \mathfrak{g}^{-2\lambda_i}$; since the assumption of the moment is that $\lambda_i^w < 0$, we have $wU_{w_i}^-w^{-1} \subset N^+$ and so it follows that $wU_{w_i}^-P \subset PwP$. The proof of our proposition is now complete. □

[The reader will notice that the above proof of the fact that the triple (G, P, M^*) is a Tits system utilizes already the Bruhat decomposition for G. On the other hand, if it were possible to show a priori that the triple (G, P, M^*) verifies conditions (1) – (3) supra, then the Bruhat decomposition would follow as a corollary – cf. Theorem 1.2.1.2.]

Let P be a minimal parabolic subgroup of G. The 2^l (l = split-rank of G) subgroups P_Θ containing P which are given to us by Theorem 1.2.1.1 are called the *standard* parabolic subgroups of G (with respect to the choice of P and $A_\mathfrak{p}$) – of course $P_\varnothing = P$, $P_\Upsilon = G$. Every parabolic subgroup of G is conjugate to one and only one standard parabolic subgroup. [Needless to say, our discussion could also be formulated in the language of boundary subgroups of G . . . .]

Example A *flag* $\mathfrak{F}$ in $\mathbf{C}^{l+1}$ is a nested sequence of subspaces

$$\{0\} \subset \mathscr{F}_1 \subset \mathscr{F}_2 \subset \cdots \subset \mathscr{F}_n \subset \mathscr{F}_{n+1} = \mathbf{C}^{l+1}.$$

The sequence $\{d_i\}$ ($d_i = \dim(\mathscr{F}_i)$, $i = 1, \ldots, n$) describes the *type* of the flag; if $d_i = i$ and $n = l$, then the flag is said to be *full*. The group $G = \mathbf{SL}(l+1, \mathbf{C})$ operates on $\mathbf{C}^{l+1}$ in the usual way. The parabolic subgroups of G are the stability groups of flags. To see this, let us determine a convenient set of standard parabolic subgroups for G. An Iwasawa decomposition of G is obtained by taking $K = \mathbf{SU}(l+1)$, $A_\mathfrak{p}$ = the group of positive diagonal matrices in $\mathbf{SL}(l+1, \mathbf{C})$, N^+ = the group of upper triangular matrices with 1 on the main diagonal. [The Lie algebra $\mathfrak{a}_\mathfrak{p}$ of $A_\mathfrak{p}$ consists of real diagonal matrices $D(a_1, \ldots, a_{l+1})$ ($a_i \in \mathbf{R}$) of trace zero; the roots Σ of the pair $(\mathfrak{g}, \mathfrak{a}_\mathfrak{p})$ are the forms $\pm(e_i - e_j)(1 \leqslant i < j \leqslant l+1)(e_i(D(a_1, \ldots, a_{l+1})) = a_i)$; the fundamental system of roots for Σ which determines our choice of N^+ is the set

$$\{(e_1 - e_2), \ldots, (e_l - e_{l+1})\},$$

hence $l_\Sigma = l$ here.] The centralizer M of $A_\mathfrak{p}$ in K is then seen to be the group of all matrices in $\mathbf{SL}(l+1, \mathbf{C})$ which are zero off of the main diagonal and of absolute value one on it; the minimal parabolic subgroup $P = \mathsf{M}A_\mathfrak{p}N^+$ is the group of upper triangular matrices in $\mathbf{SL}(l+1, \mathbf{C})$. Now let $e_1, \ldots, e_{l+1}$ be the canonical basis in $\mathbf{C}^{l+1}$ (these e_i have nothing to do with those above). The *canonical flag* in $\mathbf{C}^{l+1}$ is that flag with the property that $\mathscr{F}_i$ is generated by the first i vectors of the canonical basis; P is the stability group in G of the canonical flag in $\mathbf{C}^{l+1}$. [Notice too that G operates transitively on the set of all full flags in $\mathbf{C}^{l+1}$; hence G/P is the manifold of flags of type $(1, \ldots, l)$ and the minimal parabolic subgroups of G are the stability groups of full flags.] In general, one is able to associate with each increasing sequence of integers $1 \leqslant d_1 < \cdots < d_n \leqslant l$ the canonical flag of type $(d_1, \ldots, d_n)$ $(=(1, \ldots, l) - \Theta$, say):

$$(e_1, \ldots, e_{d_1}) \subset \cdots \subset (e_1, \ldots, e_{d_n}) \quad \text{(obvious notation)}.$$

The stability group of this flag is the standard parabolic subgroup $P_\Theta \supset P$. [Thus G/P_Θ is the manifold of flags of type $(d_1, \ldots, d_n)$; the conjugates of P_Θ in G are the stability groups of flags of type $(d_1, \ldots, d_n)$.]

Appendix We intend to establish here the following result.

Lemma Connected solvable Lie groups G have the fixed point property.

Proof Let E be a locally convex, Hausdorff, topological vector space on which G operates by linear transformations; let $\mathscr{C}$ be a non-empty, G-invariant compact convex subset of E – then we must prove that G has a fixed point in $\mathscr{C}$.

(1) Let us first suppose that G is any abelian group of linear transformations of E. Employing the evident notational conventions, for each $x \in G$ let $x_n = n^{-1}(1 + x + \cdots + x^{n-1})$; let $\tilde{G}$ denote the set of all products $x_{n_1} \dots x_{n_k}$ – then the elements of $\tilde{G}$ commute and, by convexity, $\mathscr{C}$ is $\tilde{G}$-stable. Let $\tilde{x}_1, \dots, \tilde{x}_r \in \tilde{G}$ – then, for each i, $1 \leqslant i \leqslant r$,

$$\tilde{x}_1 \dots \tilde{x}_r\mathscr{C} = \tilde{x}_i\tilde{x}_1 \dots \tilde{x}_{i-1}\tilde{x}_{i+1} \dots \tilde{x}_r\mathscr{C} \subset \tilde{x}_i\mathscr{C}$$

whence $\tilde{x}_1 \dots \tilde{x}_r\mathscr{C} \subset \bigcap_{i=1}^r \tilde{x}_i\mathscr{C}$ so that this intersection is non-empty. Owing to the compactness of $\mathscr{C}$ (expressed by the finite intersection property), we have $\bigcap_{\tilde{x} \in \tilde{G}} \tilde{x}\mathscr{C} \neq \varnothing$. Let a be an element in this latter intersection and let $x \in G$ – then $a \in x_n\mathscr{C}$ and so

$$a = n^{-1}(b + xb + \cdots + x^{n-1}b)$$

for a suitable element $b \in \mathscr{C}$. Therefore $xa - a = n^{-1}(x^nb - b) \in n^{-1}(\mathscr{C} + (-\mathscr{C}))$ (all $n \in \mathbf{N}^+$); using the compactness of $\mathscr{C}$, we then deduce that $xa = a$.

(2) Turning to the general case of a solvable G operating linearly on E, let $G = G_0 \supset G_1 \supset \cdots \supset G_n = \{1\}$ be the derived series of analytic subgroups of G. Suppose that our lemma holds for all connected solvable Lie groups whose derived series has length $< n$. Let $\mathscr{C}_1$ denote the subset of $\mathscr{C}$ consisting of those points which are left fixed under all $x_1 \in G_1$; by the induction assumption, $\mathscr{C}_1$ is not empty and is, of course, convex and compact; moreover $\mathscr{C}_1$ is G-stable, G_1 being normal in G. The closed subspace E_1 of E generated by $\mathscr{C}_1$ is locally convex and since G_1 operates trivially on it, G itself acts on E_1 as an abelian group; but then, by the first part of the proof, there exists a point in $\mathscr{C}_1$ which is left fixed by all the elements of G.

Hence the lemma. □

1.2.4 Levi Subgroups of a Parabolic Subgroup The Langlands Decomposition

Let G be a connected semi-simple Lie group with finite center. The objective of the present number is to continue the discussion initiated in 1.2.3; in particular we shall set down in a rather systematic fashion the basic properties which the parabolic subgroups of G possess. Thus let P be a minimal parabolic subgroup of G; let $G = KA_{\mathfrak{p}}N^+$ be an Iwasawa decomposition for G relative to which P splits: $P = MA_{\mathfrak{p}}N^+$ – then, to each subset $\Theta \subset \Upsilon$ $(= \{\lambda_1, \dots, \lambda_l\} \dots)$, we may assign a standard parabolic $(=$ boundary$)$ subgroup P_Θ of G which contains P. The first thing that we shall do is show that P_Θ splits relative to the given Iwasawa decomposition of G in a manner entirely analogous to that of P itself (cf. Theorem 1.2.4.8); we shall then derive an important generalization of the Iwasawa decomposition theorem which is due to Harish-Chandra; after this we shall introduce the notion of a Levi subgroup of an arbitrary parabolic subgroup of G and explore its consequences.

Let us agree to use in what follows the notations revolving around Lemmas 1.2.3.14 and 1.2.3.15.

Fix a subset Θ of Υ and consider the spectral resolution of the operators $\{\mathrm{ad}(H) : H \in \mathfrak{a}_\Theta\}$ on $\mathfrak{g}$ relative to the Euclidean structure $(.\,,.)_\theta$. If, for any element $\sigma \in \mathfrak{a}_\Theta^\vee$, we set

$$\mathfrak{g}^\sigma = \{X \in \mathfrak{g}: [H, X] = \sigma(H)X,\ H \in \mathfrak{a}_\Theta\},$$

then it is clear that $\mathfrak{g} = \mathfrak{g}^0 + \sum_\sigma \mathfrak{g}^\sigma$ (orthogonal direct sum) – here $\mathfrak{g}^0$ denotes the centralizer of $\mathfrak{a}_\Theta$ in $\mathfrak{g}$ while the sum is to be taken over those $\sigma \neq 0$ such that $\mathfrak{g}^\sigma \neq \{0\}$. Put $\mathfrak{n}_\Theta^+ = \sum_\lambda \mathfrak{g}^\lambda$, $\lambda \in \Sigma^+ - \langle\Theta\rangle^+$; $\mathfrak{n}_\Theta^- = \theta(\mathfrak{n}_\Theta^+)$.

Given $\lambda \in \mathfrak{a}_\mathfrak{p}^\vee$, let us agree to denote by $\tilde{\lambda}$ the restriction of λ to $\mathfrak{a}_\Theta$.

Lemma 1.2.4.1 Retain the preceding notations and conventions.

(i) Let $\mathscr{C}_\Theta = \{H \in \mathfrak{a}_\Theta: \lambda_i(H) > 0,\ \lambda_i \in \Upsilon - \Theta\}$ – then a root λ of the pair $(\mathfrak{g}, \mathfrak{a}_\mathfrak{p})$ belongs to $\Sigma^+ - \langle\Theta\rangle^+$ iff λ is positive on $\mathscr{C}_\Theta$.

(ii) If $\mathfrak{g}^\sigma \neq \{0\} (\sigma \neq 0)$, then there exists $\lambda \in \Sigma^+ - \langle\Theta\rangle^+$ such that $\tilde{\lambda} = \pm\sigma$.

(iii) We have $\mathfrak{n}_\Theta^+ = \sum_\lambda \mathfrak{g}^{\tilde{\lambda}}$, $\lambda \in \Sigma^+ - \langle\Theta\rangle^+$.

Proof It may be assumed that $\Theta \neq \Upsilon$. Given $\lambda \in \Sigma^+$, write $\lambda = \sum_{\lambda_i \in \Theta} n_i\lambda_i + \sum_{\lambda_i \in \Upsilon - \Theta} n_i\lambda_i$ with the n_i non-negative integers – then $\tilde{\lambda} = \sum_{\lambda_i \in \Upsilon - \Theta} n_i\tilde{\lambda}_i$. Since $\tilde{\lambda}_i \neq 0$ $(\lambda_i \in \Upsilon - \Theta)$, statement (i) is now clear. As for (ii), first observe that $\mathfrak{g} = \mathfrak{g}^0 + \mathfrak{n}_\Theta^+ + \mathfrak{n}_\Theta^-$ (direct sum of vector spaces); hence $\dim(\sum_\sigma \mathfrak{g}^\sigma) = \dim(\mathfrak{n}_\Theta^+ + \mathfrak{n}_\Theta^-)$. But obviously $\mathfrak{n}_\Theta^+ + \mathfrak{n}_\Theta^- \subset \sum_\sigma \mathfrak{g}^\sigma$ – therefore $\sum_\sigma \mathfrak{g}^\sigma = \mathfrak{n}_\Theta^+ + \mathfrak{n}_\Theta^-$. Now suppose that $X \in \mathfrak{g}^\sigma \neq \{0\} (\sigma \neq 0)$; select $X_{\pm\lambda} \in \mathfrak{g}^{\pm\lambda}$ $(\lambda \in \Sigma^+ - \langle\Theta\rangle^+)$ such that $X = \Sigma X_\lambda + \Sigma X_{-\lambda}$ – then, for all $H \in \mathfrak{a}_\Theta$, we have $[H, X] = \sigma(H)X$ which implies that

$$(*)\quad \sum_\lambda \lambda(H)X_\lambda - \sum_\lambda \lambda(H)X_{-\lambda} = \sum_\lambda \sigma(H)X_\lambda + \sum_\lambda \sigma(H)X_{-\lambda},\quad \lambda \in \Sigma^+ - \langle\Theta\rangle^+.$$

Equation $(*)$ gives (ii). Let $\lambda \in \Sigma^+ - \langle\Theta\rangle^+$ – then $(*)$ shows that $\mathfrak{g}^{\tilde{\lambda}} \subset \mathfrak{n}_\Theta^+$. Since $\mathfrak{n}_\Theta^+ \subset \sum_\lambda \mathfrak{g}^{\tilde{\lambda}}$, $\lambda \in \Sigma^+ - \langle\Theta\rangle^+$, (iii) follows. □

An element $\sigma \in \mathfrak{a}_\Theta^\vee$ is said to be a *root* of the pair $(\mathfrak{g}, \mathfrak{a}_\Theta)$ provided $\mathfrak{g}^\sigma \neq \{0\}$. The preceding result then tells us that the non-zero roots Σ_Θ of the pair $(\mathfrak{g}, \mathfrak{a}_\Theta)$ are precisely the distinct restrictions to $\mathfrak{a}_\Theta$ of the elements of $\Sigma - \langle\Theta\rangle$. [By abuse of the language we shall often refer to Σ_Θ as the 'set of roots' of the pair $(\mathfrak{g}, \mathfrak{a}_\Theta)$.] However our terminology is, to a certain degree, inaccurate as the following example shows.

Example It is natural to ask: Is the set $\Sigma_\Theta = \{\sigma\}$ of roots of the pair $(\mathfrak{g}, \mathfrak{a}_\Theta)$ a root system in $\mathfrak{a}_\Theta^\vee$? Abstractly our question may be formulated like this. Suppose that Φ is a root system in a finite dimensional Euclidean space E equipped with a Euclidean structure $(.\,,.)$; let $\Psi = \{\alpha_1, \ldots, \alpha_l\}$ be a fundamental system of roots for Φ; let Θ be a subset of Ψ, $E(\Theta)$ the space spanned by the roots which are integral linear combinations of the elements of Θ, E_Θ the orthogonal complement of $E(\Theta)$ in E, Φ_Θ the subset of E_Θ consisting of the orthogonal projections onto E_Θ of those roots which do not lie in $E(\Theta)$. Is it true that Φ_Θ is a root system in E_Θ? The answer, in general, is 'no' (consider, for instance, $G_2 \ldots$).

Although the set Σ_Θ need not be a root system for $\mathfrak{a}_\Theta^\vee$, nevertheless it is true that there exist l_Θ linearly independent roots $\sigma_1, \ldots, \sigma_{l_\Theta}$ such that every root $\sigma \in \Sigma_\Theta$ can be expressed uniquely as an integral linear combination of the σ_i, the integers being of the same parity. [Here $l_\Theta = \dim(\mathfrak{a}_\Theta^\vee)$; of course $l_\varnothing = \dim(\mathfrak{a}_\mathfrak{p}^\vee) = l$.] Thus consider the distinct members of the set $\{\tilde{\lambda}_i : \lambda_i \in \Upsilon - \Theta\}$; it is obvious that they span $\mathfrak{a}_\Theta^\vee$ and so our assertion will follow when it is established that they are linearly independent. Fix $\tilde{\lambda}_j$ (say); choose an $H \in \mathfrak{a}_\mathfrak{p}$ with the property that $\lambda_i(H) = 0$ $(\lambda_i \in \Theta)$, $\lambda_i(H) = 0$ if $\lambda_i \in \Upsilon - \Theta$ and $\tilde{\lambda}_i \neq \tilde{\lambda}_j$, $\lambda_i(H) = 1$ if $\lambda_i \in \Upsilon - \Theta$ and $\tilde{\lambda}_i = \tilde{\lambda}_j$. The statement in question is now clear since, of necessity, $H \in \mathfrak{a}_\Theta$.

Set $\mathfrak{p}_\Theta = \mathfrak{p} + \mathfrak{n}^-(\Theta)$ ($\mathfrak{p}$ the Lie algebra of P $(= MA_\mathfrak{p}N^+)$) – then clearly

$$\mathfrak{p}_\Theta = (\mathfrak{m} + \mathfrak{n}^+(\Theta) + \mathfrak{n}^-(\Theta) + \mathfrak{a}_\mathfrak{p}(\Theta)) + \mathfrak{a}_\Theta + \mathfrak{n}_\Theta^+.$$

Lemma 1.2.4.2 The space $\mathfrak{p}_\Theta$ is the normalizer of $\mathfrak{n}_\Theta^+$ in $\mathfrak{g}$ (hence $\mathfrak{p}_\Theta$ is a Lie algebra); in addition $\mathfrak{p}_\Theta$ is self-normalizing.

Proof To see that $\mathfrak{p}_\Theta$ is the normalizer of $\mathfrak{n}_\Theta^+$ in $\mathfrak{g}$, let us first verify that $\mathfrak{p}_\Theta$ actually does normalize $\mathfrak{n}_\Theta^+$. Certainly $[\mathfrak{m}, \mathfrak{n}_\Theta^+] \subset \mathfrak{n}_\Theta^+$, $[\mathfrak{a}_\mathfrak{p}, \mathfrak{n}_\Theta^+] \subset \mathfrak{n}_\Theta^+$, $[\mathfrak{n}^+(\Theta), \mathfrak{n}_\Theta^+] \subset \mathfrak{n}_\Theta^+$ ($\mathfrak{n}_\Theta^+$ is an ideal in $\mathfrak{n}^+$) and thus we need only check that $[\mathfrak{n}^-(\Theta), \mathfrak{n}_\Theta^+] \subset \mathfrak{n}_\Theta^+$. Choose $\lambda \in \langle\Theta\rangle^+$, $\mu \in \Sigma^+ - \langle\Theta\rangle^+$ – then $[\mathfrak{g}^{-\lambda}, \mathfrak{g}^\mu] \subset \mathfrak{g}^{-\lambda+\mu}$; because $\lambda \neq \mu$ and because $\lambda | \mathfrak{a}_\Theta = 0$, we see that $[\mathfrak{g}^{-\lambda}, \mathfrak{g}^\mu] \subset \mathfrak{n}_\Theta^+$, as desired. On the other hand $\mathfrak{g} = \mathfrak{p}_\Theta + \mathfrak{n}_\Theta^-$ (direct sum) – hence it suffices to show that no non-zero element $X \in \mathfrak{n}_\Theta^-$ normalizes $\mathfrak{n}_\Theta^+$. Write $X = \sum_\lambda \theta X_\lambda$, $\lambda \in \Sigma^+ - \langle\Theta\rangle^+$ with $X_{\lambda_0} \neq 0$ (say) – then $[X_{\lambda_0}, X] \notin \mathfrak{n}_\Theta^+$ and so X does not normalize $\mathfrak{n}_\Theta^+$. In a similar way we check that $\mathfrak{p}_\Theta$ is self-normalizing. □

[As the notation implies, it will turn out that $\mathfrak{p}_\Theta$ is the Lie algebra of P_Θ.]

Let $\mathfrak{l}_\Theta$ denote the centralizer of $\mathfrak{a}_\Theta$ in $\mathfrak{g}$.

Lemma 1.2.4.3 We have $\mathfrak{l}_\Theta = (\mathfrak{m} + \mathfrak{n}^+(\Theta) + \mathfrak{n}^-(\Theta) + \mathfrak{a}_\mathfrak{p}(\Theta)) + \mathfrak{a}_\Theta$.

Proof Obviously $\mathfrak{l}_\Theta$ contains the right hand side. To establish equality, recall that $\mathfrak{g} = \mathfrak{l}_\Theta + \mathfrak{n}_\Theta^+ + \mathfrak{n}_\Theta^-$ (direct sum); our lemma now follows by considerations of dimension. □

Put $\mathfrak{m}_\Theta = \mathfrak{l}_\Theta \cap \mathfrak{k} + [\mathfrak{l}_\Theta, \mathfrak{l}_\Theta] \cap \mathfrak{p}$ (where $\mathfrak{g} = \mathfrak{k} + \mathfrak{p} \ldots$).

Lemma 1.2.4.4 (i) We have $\mathfrak{m}_\Theta = \mathfrak{m} + \mathfrak{n}^+(\Theta) + \mathfrak{n}^-(\Theta) + \mathfrak{a}_\mathfrak{p}(\Theta)$.

(ii) The Lie algebra $\mathfrak{m}_\Theta$ is reductive in $\mathfrak{g}$; the center of $\mathfrak{m}_\Theta$ is contained in its intersection with $\mathfrak{k}$.

Proof Plainly $\mathfrak{m}$ (which is also equal to $\mathfrak{m}_\emptyset$!) is contained in $\mathfrak{l}_\Theta \cap \mathfrak{k}$; since $\mathfrak{n}^+(\Theta) + \mathfrak{n}^-(\Theta) + \mathfrak{a}_\mathfrak{p}(\Theta) \subset \mathfrak{g}(\Theta) \subset [\mathfrak{l}_\Theta, \mathfrak{l}_\Theta]$, it is evident that $\mathfrak{m}_\Theta \supset \mathfrak{m} + \mathfrak{n}^+(\Theta) + \mathfrak{n}^-(\Theta) + \mathfrak{a}_\mathfrak{p}(\Theta) = \tilde{\mathfrak{m}}_\Theta$ (say). As for the reverse inclusion, let's first check that $\mathfrak{l}_\Theta \cap \mathfrak{k} \subset \tilde{\mathfrak{m}}_\Theta$. Given $X \in \mathfrak{l}_\Theta \cap \mathfrak{k}$, write $X = Y + X^+ + X^- + H$ ($Y \in \mathfrak{m}$, $X^+ \in \mathfrak{n}^+(\Theta)$, $X^- \in \mathfrak{n}^-(\Theta)$, $H \in \mathfrak{a}_\mathfrak{p}$) – then $X = \theta X = Y + \theta X^+ + \theta X^- - H \Rightarrow X = 2^{-1}(2Y + (X^+ + \theta X^-) + (X^- + \theta X^+)) \in \mathfrak{m} + \mathfrak{n}^+(\Theta) + \mathfrak{n}^-(\Theta) + \mathfrak{a}_\mathfrak{p}(\Theta) \Rightarrow \mathfrak{l}_\Theta \cap \mathfrak{k} \subset \tilde{\mathfrak{m}}_\Theta$. In a similar way we find that $[\mathfrak{l}_\Theta, \mathfrak{l}_\Theta] \cap \mathfrak{p} \subset \tilde{\mathfrak{m}}_\Theta$ which proves (i). Turning to (ii), note that $\mathfrak{m}_\Theta$ is θ-stable – therefore $\mathfrak{m}_\Theta$ is reductive in $\mathfrak{g}$ (cf. Corollary 1.1.4.5). The last assertion may be seen as follows. Because $\mathfrak{a}_\mathfrak{k} + \mathfrak{a}_\mathfrak{p}(\Theta)$ is a maximal abelian subalgebra of $\mathfrak{m}_\Theta$, the center of $\mathfrak{m}_\Theta$ must be contained in $\mathfrak{a}_\mathfrak{k} + \mathfrak{a}_\mathfrak{p}(\Theta)$. Set $P_+(\Theta) = \{\alpha \in P_+ : \tilde{\alpha} \in \langle\Theta\rangle^+\}$; if $H = H_1 + H_2$ ($H_1 \in \mathfrak{a}_\mathfrak{k}$, $H_2 \in \mathfrak{a}_\mathfrak{p}(\Theta)$) is in this center, then $[H, X_\alpha] = \alpha(H)X_\alpha = 0$ for all $\alpha \in P_+(\Theta)$ – hence, for $\alpha \in P_+(\Theta)$,

$$\begin{aligned} 0 &= B(H_1, H_\alpha) + B(H_2, H_\alpha) \\ &= B(H_1, Q_\alpha) + B(H_1, P_\alpha) + B(H_2, Q_\alpha) + B(H_2, P_\alpha) \\ &= B(H_1, P_\alpha) + B(H_2, Q_\alpha) \end{aligned}$$

where, as usual (1.1.3), $H_\alpha = Q_\alpha + P_\alpha (Q_\alpha \in \mathfrak{a}_\mathfrak{p}, P_\alpha \in \sqrt{-1}\mathfrak{a}_\mathfrak{k})$. Since $\alpha \in P_+(\Theta)$, $\lambda = \tilde{\alpha} \in \langle\Theta\rangle^+$ and $Q_\alpha = Q_\lambda \in \mathfrak{a}_\mathfrak{p}(\Theta) = \sum_\lambda \mathbf{R} Q_\lambda$, $\lambda \in \langle\Theta\rangle^+$. On the other hand $B(H_1, P_\alpha)$ is pure imaginary – thus $B(H_1, P_\alpha) = B(H_2, Q_\alpha) = 0$ and so, in particular, $\tilde{\alpha}(H_2) = \lambda(H_2) = 0$ for all $\lambda \in \langle\Theta\rangle^+$, whence $H_2 = 0$. Therefore the center of $\mathfrak{m}_\Theta$ is contained in $\mathfrak{m}_\Theta \cap \mathfrak{k}$. □

Remark It is easy to explicate the last statement in (ii) above. Claim: The center of $\mathfrak{m}_\Theta$ is comprised precisely of those elements $H \in \mathfrak{a}_\mathfrak{k}$ which are annihilated by the $\alpha \in P_- \cup P_+(\Theta)$. In fact this is obvious provided we keep in mind that $\mathfrak{m}_c = \mathfrak{a}_{\mathfrak{k}_c} + \sum_{\alpha \in P_-} (\mathfrak{g}_c^\alpha + \mathfrak{g}_c^{-\alpha})$.

Having taken care of the necessary preparation in the algebra, let us turn now to the situation on the group G itself.

Let $A_\mathfrak{p}(\Theta)$, A_Θ, $N^\pm(\Theta)$, $N_\Theta^\pm$ denote the analytic subgroups of G corresponding, respectively, to $\mathfrak{a}_\mathfrak{p}(\Theta)$, $\mathfrak{a}_\Theta$, $\mathfrak{n}^\pm(\Theta)$, $\mathfrak{n}_\Theta^\pm$ – then we have a direct decomposition $A_\mathfrak{p} = A_\mathfrak{p}(\Theta)A_\Theta$ and a semi-direct decomposition $N^\pm = N^\pm(\Theta)N_\Theta^\pm$ (keep in mind that $\mathfrak{n}_\Theta^\pm$ is an ideal in $\mathfrak{n}^\pm$). Since $[\mathfrak{a}_\Theta, \mathfrak{n}_\Theta^+] \subset \mathfrak{n}_\Theta^+$, $A_\Theta N_\Theta^+$ is a group; in fact $A_\Theta N_\Theta^+$ is a closed solvable subgroup of G. Let M_Θ denote the analytic subgroup of G with Lie algebra $\mathfrak{m}_\Theta$ – then $M_\Theta A_\Theta$ (respectively $M_\Theta A_\Theta N_\Theta^+$) is a closed subgroup of G as it is the identity component of the centralizer of $\mathfrak{a}_\Theta$ (respectively of the normalizer of $\mathfrak{n}_\Theta^+$) in G (cf. the lemmas supra). In addition M_Θ itself is a closed subgroup of G. Let L_Θ denote the centralizer of $\mathfrak{a}_\Theta$ in G, $\mathsf{M}_\Theta(K) = \mathsf{L}_\Theta \cap K$ (thus $\mathsf{M}_\Theta(K)$ is the centralizer of $\mathfrak{a}_\Theta$ in K); put $\mathsf{M}_\Theta = \mathsf{M}_\Theta(K)M_\Theta$ (the identity component of M_Θ is, of course, M_Θ) – then $\mathsf{L}_\Theta = \mathsf{M}_\Theta A_\Theta$ (cf. infra). The identity component L_Θ of L_Θ is the analytic subgroup of G with Lie algebra $\mathfrak{l}_\Theta$. Agreeing to write $M_\Theta(K)$ for the identity component of $\mathsf{M}_\Theta(K)$, one clearly has $M_\Theta(K) = L_\Theta \cap K = M_\Theta \cap K$.

Lemma 1.2.4.5 The groups M_Θ and $\mathsf{M}_\Theta A_\Theta N_\Theta^+$ are closed subgroups of G. We have the Cartan decompositions $\mathsf{M}_\Theta = \mathsf{M}_\Theta(K)\exp(\mathfrak{m}_\Theta \cap \mathfrak{p})$, $M_\Theta = M_\Theta(K)\exp(\mathfrak{m}_\Theta \cap \mathfrak{p})$; furthermore $\mathsf{M}_\Theta(K) = \mathsf{Z}(A_\mathfrak{p})M_\Theta(K)$.

Proof The fact that M_Θ is a group is clear since $\mathfrak{m}_\Theta$ is stable under $\mathrm{Ad}(\mathsf{M}_\Theta(K))$ (bear in mind that $\mathsf{M}_\Theta(K) \subset K$). To see that $\mathsf{M}_\Theta A_\Theta N_\Theta^+$ is a group, note first that $\mathsf{M}_\Theta A_\Theta = A_\Theta \mathsf{M}_\Theta$, $hN_\Theta^+h^{-1} \subset N_\Theta^+$ ($h \in A_\Theta$); hence, to be ensured that $\mathsf{M}_\Theta A_\Theta N_\Theta^+$ is a group, we need only verify that $mN_\Theta^+m^{-1} \subset N_\Theta^+(m \in \mathsf{M}_\Theta)$ – this, however, is certainly true since $\mathrm{Ad}(m)\mathfrak{g}^\sigma \subset \mathfrak{g}^\sigma$ (all $m \in \mathsf{M}_\Theta$). [Recall that $\mathfrak{n}_\Theta^+ = \sum_\sigma \mathfrak{g}^\sigma$ ($\sigma = \tilde{\lambda}$, $\lambda \in \Sigma^+ - \langle\Theta\rangle^+$).] The groups $\mathsf{M}_\Theta = \mathsf{M}_\Theta(K)M_\Theta = \mathsf{M}_\Theta(K)A_\mathfrak{p}(\Theta)N^+(\Theta)$, $\mathsf{M}_\Theta A_\Theta N_\Theta^+ = \mathsf{M}_\Theta(K)A_\mathfrak{p}N^+$ are closed in G because they contain the center Z of G and their images in G/Z are closed being the product of a closed subgroup with the compact subgroup $\mathsf{M}_\Theta(K)/Z$. To see that the (reductive) groups M_Θ and M_Θ (= identity component of M_Θ) admit the asserted Cartan decompositions, one need only cite Lemma 1.1.3.7; as for the last result (which gives the structure of the group of components of $\mathsf{M}_\Theta(K)$ – cf. Lemma 1.1.3.8), we shall defer its proof to the first Appendix of this number. □

Example Let $\mathfrak{g} = \mathfrak{sl}(3, \mathbf{R})$, $\mathfrak{k} = \{X \in \mathfrak{g} : {}^tX = -X\}$, $\mathfrak{p} = \{X \in \mathfrak{g} : {}^tX = X\}$ – then $\mathfrak{g} = \mathfrak{k} + \mathfrak{p}$ is a Cartan decomposition of $\mathfrak{g}$ ('t' means transpose). Take $\mathfrak{a}_\mathfrak{p} = \{D(a_1, a_2, a_3) : a_i \in \mathbf{R},\ a_1 + a_2 + a_3 = 0\}$ ($D(a_1, a_2, a_3)$ a diagonal matrix with entries a_1, a_2, a_3) – then $\mathfrak{a}_\mathfrak{p}$ is a maximal abelian subalgebra of $\mathfrak{p}$. Let e_i be the linear form on $\mathfrak{a}_\mathfrak{p}$ given by the rule $e_i(A) = a_i$ for $A = D(a_1, a_2, a_3)$. The roots Σ of the pair $(\mathfrak{g}, \mathfrak{a}_\mathfrak{p})$ are the forms $\pm(e_i - e_j)(1 \leqslant i < j \leqslant 3)$; we may take $\Sigma^+ = \{(e_1 - e_2), (e_2 - e_3), (e_1 - e_3)\}$ – the corresponding fundamental system Υ is the two element set $\{(e_1 - e_2), (e_2 - e_3)\}$. Now let $\Theta = \{(e_1 - e_2)\}$; because the Killing form on $\mathfrak{g}$ is, up to a constant, $\mathrm{tr}(XY)$ (X, $Y \in \mathfrak{g}$), we see that

$$\mathfrak{a}_\Theta = \left\{\begin{pmatrix} a & 0 & 0 \\ 0 & a & 0 \\ 0 & 0 & -2a \end{pmatrix} : a \in \mathbf{R}\right\}.$$

This being the case, take $G = \mathbf{SL}(3, \mathbf{R})$ and let $G = KA_\mathfrak{p}N^+$ be the corresponding Iwasawa decomposition ($K = \mathbf{SO}(3)$) – then M_Θ may be identified with the group of real 2-by-2 matrices of determinant ± 1. In fact, by an elementary matrix computation, we quickly see that

$$\mathsf{M}_\Theta(K) = \begin{pmatrix} \mathbf{SO}(2) & 0 \\ 0 & 1 \end{pmatrix} \cup \epsilon\begin{pmatrix} \mathbf{SO}(2) & 0 \\ 0 & 1 \end{pmatrix} \quad \text{where } \epsilon = \begin{pmatrix} 1 & 0 & 0 \\ 0 & -1 & 0 \\ 0 & 0 & -1 \end{pmatrix}$$

from which our assertion follows at once. It is important to note, therefore, that M_Θ is in general *not* generated by its center and its identity component. On the other hand, in the present case,

$$\mathsf{M} = \mathsf{Z}(A_\mathfrak{p}) = \left\{ \begin{pmatrix} 1 & 0 & 0 \\ 0 & 1 & 0 \\ 0 & 0 & 1 \end{pmatrix}, \begin{pmatrix} -1 & 0 & 0 \\ 0 & -1 & 0 \\ 0 & 0 & 1 \end{pmatrix}, \begin{pmatrix} 1 & 0 & 0 \\ 0 & -1 & 0 \\ 0 & 0 & -1 \end{pmatrix}, \begin{pmatrix} -1 & 0 & 0 \\ 0 & 1 & 0 \\ 0 & 0 & -1 \end{pmatrix} \right\}$$

so that indeed $\mathsf{M}_\Theta(K) = \mathsf{Z}(A_\mathfrak{p})M_\Theta(K)$, $\mathsf{M}_\Theta = \mathsf{Z}(A_\mathfrak{p})M_\Theta$.

Remark Fix $\sigma \in \Sigma_\Theta$ – then, as can be verified immediately, the map $m \mapsto \mathrm{Ad}\,(m)|\mathfrak{g}^\sigma$ is a representation of M_Θ on $\mathfrak{g}^\sigma$. Claim: The absolute value of the determinant of the restriction of Ad (m) to $\mathfrak{g}^\sigma$ is 1 $(m \in \mathsf{M}_\Theta)$. This is obvious if $m \in \mathsf{M}_\Theta(K)$ (for $\mathsf{M}_\Theta(K)$ is compact). So suppose that $m \in M_\Theta$; write $m = k \exp X$ $(k \in M_\Theta(K),\ X \in \mathfrak{m}_\Theta \cap \mathfrak{p})$; since $M_\Theta(K)$ is compact and $\mathfrak{m}_\Theta \cap \mathfrak{p} \subset [\mathfrak{m}_\Theta, \mathfrak{m}_\Theta]$, it follows that

$$|\det(\mathrm{Ad}\,(k)|\mathfrak{g}^\sigma)| = |\det(\mathrm{Ad}\,(\exp X)|\mathfrak{g}^\sigma)| = 1$$

– therefore $|\det(\mathrm{Ad}\,(m)|\mathfrak{g}^\sigma)| = 1$ $(m \in M_\Theta)$. Hence the claim. Incidentally our considerations show that M_Θ can be characterized as the subgroup of L_Θ consisting of those x in L_Θ such that the restriction of Ad (x) to $\mathfrak{g}^\sigma(\sigma \in \Sigma_\Theta)$ has determinant ± 1 for all $\sigma \in \Sigma_\Theta$.

Let W denote the Weyl group of the pair $(\mathfrak{g}, \mathfrak{a}_\mathfrak{p})$, W_Θ the subgroup of W generated by the $w_{\lambda_i}(\lambda_i \in \Theta)$.

Lemma 1.2.4.6 The subgroup of W centralizing $\mathfrak{a}_\Theta$ is W_Θ.

Proof By a well-known principle, the centralizer of $\mathfrak{a}_\Theta$ in W is generated by those reflections $w_{\lambda_i}(\lambda_i \in \Upsilon)$ such that $w_{\lambda_i}(H) = H$ for all $H \in \mathfrak{a}_\Theta$, that is, for which $\lambda_i(H) = 0$ for all $H \in \mathfrak{a}_\Theta$ (cf. Theorem 1.1.2.8). This means that $Q_{\lambda_i} \in \mathfrak{a}_\mathfrak{p}(\Theta)$, whence $\lambda_i \in \Theta$. Therefore the subgroup of W centralizing $\mathfrak{a}_\Theta$ is indeed W_Θ. $\square$

Corollary 1.2.4.7 Retain the above notations – then

$$\mathsf{M}_\Theta(K) \cap \mathsf{M}^*/\mathsf{M} = W_\Theta.$$

Here is the basic structure theorem for parabolic subgroups of G.

Theorem 1.2.4.8 (Bruhat-Moore) Let G be a connected semi-simple Lie group with finite center, $G = KA_\mathfrak{p}N^+$ an Iwasawa decomposition for G; set $P = \mathsf{M}A_\mathfrak{p}N^+$ – then, for each subset $\Theta \subset \Upsilon$, $P_\Theta = \mathsf{M}_\Theta(K)A_\mathfrak{p}N^+ = \mathsf{M}_\Theta A_\Theta N_\Theta^+$ (hence the parabolic (= boundary) subgroups of G are closed in G). Furthermore the normalizer of $\mathfrak{n}_\Theta^+$ (or N_Θ^+) in G is P_Θ.

Proof Let Q_Θ be the normalizer of $\mathfrak{n}_\Theta^+$ in G – we shall prove that

$$P_\Theta \subset Q_\Theta \subset \mathsf{M}_\Theta(K)A_\mathfrak{p}N^+ \subset P_\Theta$$

from which the theorem will follow.

(1) Since $\mathfrak{n}_\Theta^+$ is an ideal in $\mathfrak{n}^+$ and since $\mathrm{Ad}\,(m)\mathfrak{g}^\lambda = \mathfrak{g}^\lambda$ $(m \in \mathsf{M}, \lambda \in \Sigma)$, it is clear that P normalizes $\mathfrak{n}_\Theta^+$. Choose $m_{w_i} \in \mathsf{M}^*$ associated with $w_{\lambda_i} = w_i$ $(\lambda_i \in \Theta)$. If λ belongs to $\Sigma^+ - \langle\Theta\rangle^+$, then so does λ^{w_i}. [Use the criterion of Lemma 1.2.4.1(i); thus if $\lambda^{w_i} \notin \Sigma^+ - \langle\Theta\rangle^+$, then, of

necessity, $\lambda^{w_i}(H) \leqslant 0$ for some $H \in \mathscr{C}_\Theta$, i.e. $\langle H, \lambda^{w_i}\rangle = \langle \mathrm{Ad}\,(m_{w_i})H, \lambda\rangle = \langle H, \lambda\rangle \leqslant 0$, a contradiction.] Therefore $P_\Theta = PW_\Theta P$ normalizes $\mathfrak{n}_\Theta^+$.

(2) We shall show next that $Q_\Theta \subset M_\Theta(K)A_\mathfrak{p}N^+$. So let $x \in Q_\Theta$ and write, via the Bruhat Lemma, $x = p_1 m^* p_2$ where $p_1, p_2 \in P$, $m^* \in M^* \cap Q_\Theta$ – then $m^* \in M_\Theta(K)$. [In fact one has the following criterion: If $m^* \in M^*$ and if Ad (m^*) normalizes $\mathfrak{n}_\Theta^+$, then $m^* \in M_\Theta(K)$. This being evident if $\Theta = \Upsilon$, assume that $\Theta \neq \Upsilon$ and let, as above, $\mathscr{C}_\Theta = \{H \in \mathfrak{a}_\Theta : \lambda_i(H) > 0, \lambda_i \in \Upsilon - \Theta\}$. Since Ad (m^*) normalizes $\mathfrak{n}_\Theta^+$, Ad (m^*) permutes the roots in $\Sigma^+ - \langle\Theta\rangle^+$ amongst themselves; consequently Ad $(m^*)(\mathscr{C}_\Theta) = \mathscr{C}_\Theta$ (cf. Lemma 1.2.4.1(i)). The 'Weyl chamber' $\mathscr{C}_\Theta$ is convex and so, by a rather standard argument, we see that Ad (m^*) has a fixed point in $\mathscr{C}_\Theta$, hence centralizes $\mathfrak{a}_\Theta \ldots$.] Since $M_\Theta(K)A_\mathfrak{p}N^+$ is a group containing P, it then follows that $x \in M_\Theta(K)A_\mathfrak{p}N^+$, as desired.

(3) It remains to prove that $M_\Theta(K)A_\mathfrak{p}N^+ \subset P_\Theta$ and for this it is enough to establish that $M_\Theta(K) \subset P_\Theta$. Thus let $x \in M_\Theta(K)$ and write $x = p_1 m^* p_2$ ($p_1, p_2 \in P$; $m^* \in M^*$). Since $M_\Theta(K)$ normalizes $\mathfrak{n}_\Theta^+$, so does m^*. But then, in view of what was said in (2) above, $m^* \in M_\Theta(K) \cap M^*$; on the basis of Corollary 1.2.4.7 we conclude that m^* represents an element of W_Θ – hence $M_\Theta(K)A_\mathfrak{p}N^+ \subset P_\Theta$.

The proof of the theorem is now complete. □

[In passing let us observe that the triple $(P_\Theta, P, M_\Theta(K) \cap M^*)$ constitutes a Tits system ....]

Our next task will be to obtain a 'generalized' Bruhat decomposition for G.

Fix a subset Θ of Υ; let $\Xi = \Sigma^+ \cup \langle\Theta\rangle$ be the corresponding parabolic subset of Σ and introduce the subsets $W_s(= W_\Theta)$ and W_u of W per Proposition 1.1.2.13 (W the Weyl group of the pair $(\mathfrak{g}, \mathfrak{a}_\mathfrak{p})$).

Proposition 1.2.4.9 (Borel-Kostant) Let G be a connected semi-simple Lie group with finite center, $G = KA_\mathfrak{p}N^+$ an Iwasawa decomposition for G; set $P = MA_\mathfrak{p}N^+$; fix a subset Θ of Υ and let P_Θ be the corresponding standard parabolic subgroup of G – then $G = \bigcup_{w_u \in W_u^{-1}} N^+ w_u P_\Theta = \bigcup_{w_u \in W_u^{-1}} U_{w_u}^+ w_u P_\Theta$ (disjoint union).

Proof Since $P_\Theta = \bigcup_{w_s \in W_s} N^+(\Theta) w_s^{-1} P$ and $w_u^{-1}\Xi_s^+ \subset \Sigma^+$ for all $w_u \in W_u$ (cf. Lemma 1.1.2.15(ii)), we have $N^+ w_u^{-1} P_\Theta = \bigcup_{w_s \in W_s} N^+ w_u^{-1} N^+(\Theta) w_s^{-1} P = \bigcup_{w_s \in W_s} N^+ w_u^{-1} w_s^{-1} P$ (all $w_u \in W_u$) – therefore $\bigcup_{w_u \in W_u^{-1}} U_{w_u}^+ w_u P_\Theta = \bigcup_{w_u \in W_u} N^+ w_u^{-1} P_\Theta = \bigcup_{w_u \in W_u} \bigcup_{w_s \in W_s} N^+(w_s w_u)^{-1} P = \bigcup_{w \in W} N^+ w P = G$, as we wished to prove. □

If P is a minimal parabolic subgroup of G, then the usual Bruhat decomposition for G can be regarded as giving the decomposition of the space G/P into N^+ orbits ($= P$ orbits); the 'generalized' Bruhat decomposition for G, as formulated above, may be interpreted as giving the decomposition of G/P_Θ into N^+ orbits. Thus fix $w_u \in W_u$; let m_{w_u} be any element in M^* associated with w_u – then the subgroup $U^+_{w_u^{-1}}$ 'parameterizes' the orbit $N^+w_u^{-1}P_\Theta$ of the coset $m_{w_u}^{-1}P_\Theta = w_u^{-1}P_\Theta$ in G/P_Θ. More precisely the map $u \mapsto um_{w_u}^{-1}P_\Theta$ is a one-to-one regular mapping of $U^+_{w_u^{-1}}$ onto $N^+w_u^{-1}P_\Theta$. [There are two things to be observed here, namely that $w_u U^+_{w_u^{-1}} w_u^{-1} \subset N_\Theta^-$ and that $N_\Theta^- \cap P_\Theta = \{1\}$ (cf. infra). The first point may be seen as follows. Given any $w \in W$, let us agree to write Σ_w^+ for $w\,\Sigma^- \cap \Sigma^+$ – then the elements of W_u are characterized by the fact that $\Sigma^+_{w_u} \subset \Xi_u$ $(= \Sigma^+ - \langle\Theta\rangle^+)$. The Lie algebra of $U^+_{w_u^{-1}}$ is spanned by those $\mathfrak{g}^\lambda$ $(\lambda \in \Sigma^+)$ such that $\lambda^{w_u} < 0$, i.e. is spanned by the $\mathfrak{g}^\lambda$, $\lambda \in \Sigma^+_{w_u^{-1}}$; the Lie algebra of $w_u U^+_{w_u^{-1}} w_u^{-1}$ is therefore spanned by the $\mathfrak{g}^\lambda(\lambda \in \Sigma^-)$ with λ in $w_u\Sigma^+_{w_u^{-1}} = -\Sigma^+_{w_u}$ – but $\Sigma^+_{w_u}$ is contained in Ξ_u $(w_u \in W_u)$ . . . .] Furthermore there is a unique N^+ orbit of maximal dimension; thus let $w^\dagger$ be the element in W which takes the positive Weyl chamber to its negative; write $w^\dagger = w_s^\dagger w_u^\dagger$ – then the orbit in question is $N^+(w_u^\dagger)^{-1}P_\Theta$ (since $\Sigma^+_{w_u^\dagger} = \Xi_u$).

The N^+ orbits in G/P_Θ are sometimes called *Schubert cells*.

Note The elements of W_u serve to 'parameterize' the action of N^+ on G/P_Θ; on the other hand the points of the double coset space $W_s\backslash W/W_s$ $(= W_\Theta\backslash W/W_\Theta)$ identify the orbits in G/P_Θ under the canonical action of P_Θ (the orbit spaces coinciding when $\Theta = \varnothing$).

Proposition 1.2.4.10 Let G be a connected semi-simple Lie group with finite center, $G = KA_\mathfrak{p}N^+$ an Iwasawa decomposition for G; set $P = MA_\mathfrak{p}N^+$ and fix a subset Θ of Υ – then $N_\Theta^- P_\Theta$ is a dense open submanifold of G whose complement has zero Haar measure.

Proof Assume that $\Theta \neq \Upsilon$ (otherwise our assertions are trivial) – then we claim that $N_\Theta^- \cap P_\Theta = \{1\}$. [To verify this we shall use the following elementary fact: Suppose that H is an element of $\mathfrak{l}_\Theta$ such that the determinant of the restriction of ad (H) to $\mathfrak{n}_\Theta^+$ is not zero – then the mapping $X \mapsto \exp(\mathrm{ad}\,(X))H - H$ is an analytic diffeomorphism of $\mathfrak{n}_\Theta^+$ onto itself (cf. Helgason [2, p. 231]). This being the case, fix an $x \in N_\Theta^- \cap P_\Theta$; write $x = mh\exp X = \exp Y (m \in M_\Theta,\ h \in A_\Theta,\ X \in \mathfrak{n}_\Theta^+,\ Y \in \mathfrak{n}_\Theta^-)$. Fix an element H in $\mathfrak{a}_\Theta$ with the property that $\sigma(H) \neq 0$ for every $\sigma \in \Sigma_\Theta$. The comment above then implies the existence of elements $X_1 \in \mathfrak{n}_\Theta^+$, $Y_1 \in \mathfrak{n}_\Theta^-$ such that $\mathrm{Ad}\,(\exp Y)H = H + Y_1 = \mathrm{Ad}\,(mh)(H + X_1) = H + X_2$, $X_2 \in \mathfrak{n}_\Theta^+$ (P_Θ normalizes $\mathfrak{n}_\Theta^+$) – therefore $X_1 = Y_1 = 0 \Rightarrow x = 1$.] Owing to the claim and the fact that $\mathfrak{g} = \mathfrak{n}_\Theta^- + \mathfrak{p}_\Theta$ (direct sum of vector spaces), we conclude that the map $(n, p) \mapsto np$ of $N_\Theta^- \times P_\Theta$ into G is open, regular, and injective. Write $G = U^+_{(w_u^\dagger)^{-1}}(w_u^\dagger)^{-1}P_\Theta \cup (\cup\, N^+w_u^{-1}P_\Theta)$ (disjoint union) with $\dim(N^+w_u^{-1}P_\Theta) < \dim(G)$ when $w_u \neq w_u^\dagger$; now multiply through by $w_u^\dagger$ – plainly, then, G is the union of $N_\Theta^- P_\Theta$ and a submanifold of lower dimension. Hence the result. □

Example Here is one way in which parabolic subgroups arise in practice. Let G be a connected semi-simple Lie group with finite center, $G = KA_{\mathfrak{p}}N^+$ an Iwasawa decomposition for G; let Σ denote the set of roots of the pair $(\mathfrak{g}, \mathfrak{a}_{\mathfrak{p}})$, $\mathscr{C}$ the open positive Weyl chamber in $\mathfrak{a}_{\mathfrak{p}}$ which defines the ordering of Σ. Fix an element H_0 in $\mathscr{C}^{cl}$ (thus $\lambda(H_0) \geqslant 0$ for $\lambda \in \Sigma^+$); let $\Theta(H_0)$ denote the subset of Υ comprised of those λ_i such that $\lambda_i(H_0) = 0$ – thus $\Theta(H_0)$ is empty iff $H_0 \in \mathscr{C}$ while, at the opposite extreme, $\Theta(H_0) = \Upsilon$ iff $H_0 = 0$. Since $H_0 \in \mathscr{C}^{cl}$, it is clear that $\lambda \in \Sigma$ vanishes at H_0 iff $\lambda \in \langle\Theta(H_0)\rangle$; hence $\mathfrak{l}_{\Theta(H_0)}$ is the centralizer of H_0 in $\mathfrak{g}$. Therefore, with each point $H_0 \in \mathscr{C}^{cl}$, there is associated in a natural way a parabolic subgroup $P_{\Theta(H_0)}$ of G.

The following important decomposition theorem generalizes the well-known result of Iwasawa.

Theorem 1.2.4.11 (Harish-Chandra) Let G be a connected semi-simple Lie group with finite center, $G = KA_{\mathfrak{p}}N^+$ an Iwasawa decomposition for G; set $P = MA_{\mathfrak{p}}N^+$ and fix a subset Θ of Υ – then the mapping ϕ defined by the prescription

$$(k, X, h, n) \mapsto k(\exp X)hn \quad (k \in K, X \in \mathfrak{m}_\Theta \cap \mathfrak{p}, h \in A_\Theta, n \in N_\Theta^+)$$

is an analytic diffeomorphism of the product manifold $K \times \mathfrak{m}_\Theta \cap \mathfrak{p} \times A_\Theta \times N_\Theta^+$ onto G.

Corollary 1.2.4.12 Retain the above notations – then the map $(k, X, n, h) \mapsto k(\exp X)nh$ is an analytic diffeomorphism of the product manifold $K \times \mathfrak{m}_\Theta \cap \mathfrak{p} \times N_\Theta^+ \times A_\Theta$ onto G.

Proof of Theorem 1.2.4.11 (1) Since $G = KA_{\mathfrak{p}}N^+ = KM_\Theta A_\Theta N_\Theta^+$ and $M_\Theta = (M_\Theta \cap K) \exp(\mathfrak{m}_\Theta \cap \mathfrak{p})$ it is evident that ϕ is surjective.

(2) We shall now prove that ϕ is injective.

(a) First we claim that $M_\Theta \cap A_\Theta N_\Theta^+ = \{1\}$. Thus let $m = h \exp X$ $(m \in M_\Theta,\ h \in A_\Theta,\ X \in \mathfrak{n}_\Theta^+)$; if $a \in A_\Theta$, then $m = ama^{-1} = h \exp(\mathrm{Ad}(a)X) \Rightarrow X = \mathrm{Ad}(a)X$ (all $a \in A_\Theta$) $\Rightarrow [H, X] = 0$ (all $H \in \mathfrak{a}_\Theta$) $\Rightarrow X \in \mathfrak{n}_\Theta^+ \cap \mathfrak{l}_\Theta = \{0\}$ – hence $m = h$. Write $m = k \exp X$ $(k \in K, X \in \mathfrak{p})$; owing to the uniqueness of this decomposition and Lemma 1.2.4.5, we see that $k \in M_\Theta \cap K$, $X \in \mathfrak{m}_\Theta \cap \mathfrak{p}$ – therefore $k = 1$, $X \in \mathfrak{a}_\Theta \cap (\mathfrak{m}_\Theta \cap \mathfrak{p}) = \{0\}$, whence $m = 1$ which serves to establish the claim.

(b) Second we claim that $K \cap (M_\Theta A_\Theta N_\Theta^+) = M_\Theta \cap K$. Thus let $k \in K \cap (M_\Theta A_\Theta N_\Theta^+)$ and write $k = mhn$ $(m \in M_\Theta, h \in A_\Theta, n \in N_\Theta^+)$ – then $\mathrm{Ad}(k)$ normalizes $\mathfrak{n}_\Theta^+$. Since $\mathrm{Ad}(k) = \theta \circ \mathrm{Ad}(k) \circ \theta^{-1}$, it follows that $\mathrm{Ad}(k)$ also normalizes $\mathfrak{n}_\Theta^-$; but then $\mathrm{Ad}(n) = \mathrm{Ad}(h^{-1}m^{-1}k)$ normalizes $\mathfrak{n}_\Theta^-$ and so $n = 1$ (for $M_\Theta A_\Theta N_\Theta^- \cap N_\Theta^+ = \{1\} \ldots$) – hence $k = mh$. Write $m = k_m \exp X$ $(k_m \in M_\Theta \cap K,\ X \in \mathfrak{m}_\Theta \cap \mathfrak{p})$ – then $k_m^{-1}k = (\exp X)h \in K \cap \exp \mathfrak{p} = \{1\}$ ($\mathfrak{m}_\Theta \cap \mathfrak{p}$ and $\mathfrak{a}_\Theta$ commute) $\Rightarrow h = \exp(-X) = 1 \Rightarrow k = k_m \in M_\Theta \cap K \Rightarrow K \cap (M_\Theta A_\Theta N_\Theta^+) = M_\Theta \cap K$, as desired.

Now assume that $k_1 m_1 h_1 n_1 = k_2 m_2 h_2 n_2$ $(k_i \in K,\ m_i \in \exp(\mathfrak{m}_\Theta \cap \mathfrak{p}),\ h_i \in A_\Theta,\ n_i \in N_\Theta^+;\ i = 1, 2)$ – then $k_1^{-1}k_2 \in K \cap (M_\Theta A_\Theta N_\Theta^+) =$

$M_\Theta \cap K$ ((b) supra) $\Rightarrow k_2 = k_1 m$ $(m \in M_\Theta \cap K)$, hence $m_1^{-1}(mm_2) \in M_\Theta \cap A_\Theta N_\Theta^+ = \{1\}$ ((a) supra) $\Rightarrow m_1 = mm_2 \Rightarrow m = 1 \Rightarrow m_1 = m_2 \Rightarrow k_1 = k_2$. By the standard Iwasawa decomposition theorem, $h_1 = h_2$, $n_1 = n_2$. Therefore ϕ is injective.

(3) It remains to establish the regularity of ϕ. Fix a point (k_0, X_0, h_0, n_0) $(k_0 \in K, X_0 \in \mathfrak{m}_\Theta \cap \mathfrak{p}, h_0 \in A_\Theta, n_0 \in N_\Theta^+)$; in order to check the regularity of ϕ at (k_0, X_0, h_0, n_0) we must verify that the following endomorphism f_0 of $\mathfrak{g}$ is non-singular:

$$f_0(X) = \mathrm{Ad}\,(n_0^{-1}h_0^{-1})\{(1 - e^{-\mathrm{ad}\,(X_0)})/\mathrm{ad}\,(X_0)\}(X) \quad (X \in \mathfrak{m}_\Theta \cap \mathfrak{p})$$
$$f_0(Y) = \mathrm{Ad}\,(n_0^{-1}h_0^{-1})e^{-\mathrm{ad}\,(X_0)}(Y) \quad (Y \in \mathfrak{k})$$
$$f_0(Z) = Z\ (Z \in \mathfrak{n}_\Theta^+)$$
$$f_0(H) = \mathrm{Ad}\,(n_0^{-1})H \quad (H \in \mathfrak{a}_\Theta).$$

So put $f_1 = e^{\mathrm{ad}\,(X_0)}\,\mathrm{Ad}\,(h_0 n_0) f_0$ – then $\det(f_0) = \det(f_1)$. If f_2 is the endomorphism of $\mathfrak{g}/(\mathfrak{a}_\Theta + \mathfrak{n}_\Theta^+)$ determined by f_1, then it is evident that $\det(f_0) = \det(\mathrm{Ad}_{\mathfrak{n}_\Theta^+}(h_0))\det(f_2)$ where $\mathrm{Ad}_{\mathfrak{n}_\Theta^+}(h_0)$ is the restriction of $\mathrm{Ad}\,(h_0)$ to $\mathfrak{n}_\Theta^+$. The endomorphism f_2 leaves every element in $\mathfrak{k}$ mod $(\mathfrak{a}_\Theta + \mathfrak{n}_\Theta^+)$ fixed, hence determines an endomorphism f_3 of $\mathfrak{g}/(\mathfrak{k} + \mathfrak{a}_\Theta + \mathfrak{n}_\Theta^+)$ and this latter space may be identified with $\mathfrak{m}_\Theta \cap \mathfrak{p}$ (the sum $\mathfrak{g} = \mathfrak{k} + \mathfrak{m}_\Theta \cap \mathfrak{p} + \mathfrak{a}_\Theta + \mathfrak{n}_\Theta^+$ being direct). If $X \in \mathfrak{m}_\Theta \cap \mathfrak{p}$, then

$$f_1(X) \equiv \{\sinh(\mathrm{ad}\,(X_0))/\mathrm{ad}\,(X_0)\}(X) \qquad \mathrm{mod}\ \mathfrak{k}.$$

Hence

$$\det(f_0) = \det(\sinh(\mathrm{ad}\,(X_0))/\mathrm{ad}\,(X_0))_{\mathfrak{m}_\Theta \cap \mathfrak{p}} \det(\mathrm{Ad}_{\mathfrak{n}_\Theta^+}(h_0)),$$

the suffix $\mathfrak{m}_\Theta \cap \mathfrak{p}$ denoting restriction to $\mathfrak{m}_\Theta \cap \mathfrak{p}$. Since $\mathrm{Ad}\,(h_0)$ and $\sinh(\mathrm{ad}\,(X_0))/\mathrm{ad}\,(X_0)$ are positive definite self-adjoint operators on $\mathfrak{g}$, it follows that $\det(f_0) > 0$. Therefore ϕ is regular.

The proof of our theorem is now complete. □

At this juncture it will be convenient to recall a classical lemma which is due to Mostow [3].

In general, if $\mathfrak{g}$ is a Lie algebra over a field $\mathbf{k}$ of characteristic zero, $\mathfrak{h}$ a subalgebra of $\mathfrak{g}$, then, by an inner automorphism from $\mathfrak{h}$, we shall understand any product of automorphisms of the form $\exp(\mathrm{ad}_{\mathfrak{g}}(X))$ with $X \in \mathfrak{h}$ and $\mathrm{ad}_{\mathfrak{g}}(X)$ nilpotent.

Lemma 1.2.4.13 (Mostow) Let $\mathfrak{g}$ be a Lie algebra over a field $\mathbf{k}$ of characteristic zero – then any two subalgebras of $\mathfrak{g}$ which are maximal with respect to the property of being reductive in $\mathfrak{g}$ are conjugate under an inner automorphism from the radical of the commutator subalgebra.

As above, let G be a connected semi-simple Lie group with finite center. Let P be a (not necessarily minimal) parabolic subgroup of G – then, by the *unipotent radical* $R_u(P)$ of P we shall understand, as is customary, the greatest connected normal subgroup of P whose elements are unipotent (cf. 1.4.3). [If we write $P = P_\Theta = \mathsf{L}_\Theta N_\Theta^+$ relative to some Iwasawa decomposition of G, then $N_\Theta^+ = R_u(P)$ (thus N_Θ^+ is actually an 'intrinsic' entity); one should also note here that $\mathfrak{n}_\Theta^+$ is precisely the nilpotent radical of $\mathfrak{p}$ (cf. the Appendix to 1.2.2).]

Definition Let P be a parabolic subgroup of G – then by a *Levi subgroup* L of P we mean a closed reductive subgroup L such that the product mapping $(x, u) \mapsto xu$ $(x \in \mathsf{L}, u \in R_u(P))$ defines an isomorphism of the analytic varieties $\mathsf{L} \times R_u(P)$ and P.

[This terminology is in accordance with that of Borel and Tits [1]. The definition does not request that a Levi subgroup be connected . . . .]

Thus if L is a Levi subgroup of P, then P can be written as the semi-direct product of L and its unipotent radical $R_u(P)$; any such presentation of P is said to be a *Levi decomposition* (of P).

Proposition 1.2.4.14 Let G be a connected semi-simple Lie group with finite center, P a parabolic subgroup of G – then P possesses a Levi subgroup. In addition any two such are conjugate by an element from $R_u(P)$.

Proof To avoid trivialities, we may assume that $P \neq G$. On the basis of Theorem 1.2.4.8, we may write, relative to a suitable Iwasawa decomposition of G, $P = P_\Theta = \mathsf{M}_\Theta A_\Theta N_\Theta^+ = \mathsf{L}_\Theta R_u(P)$ $(\mathsf{L}_\Theta = \mathsf{M}_\Theta A_\Theta)$ and so it is clear that P possesses a Levi subgroup, viz. L_Θ. This being the case, let L be an arbitrary Levi subgroup of P – then the conjugacy question will be resolved when it is shown that there exists an element $u \in R_u(P)$ such that $\mathsf{L} = {}^u\mathsf{L}_\Theta$. It is clear that the Lie algebra $\mathfrak{l}$ of L (= identity component of L) is maximal with respect to the property of being reductive in the Lie algebra of P (here we have to remind ourselves that the radical of a reductive Lie algebra is equal to its center – cf. Bourbaki [4, p.79]). Owing to Lemma 1.2.4.13, there exists an element p in the identity component of P which maps $\mathfrak{l}_\Theta$ to $\mathfrak{l}$; since $P = \mathsf{M}_\Theta A_\Theta N_\Theta^+$ and since $\mathsf{M}_\Theta A_\Theta$ normalizes $\mathfrak{l}_\Theta$, we can assume that $p = u \in R_u(P)$. Now the group ${}^u\mathsf{L}_\Theta$ is the centralizer of ${}^uA_\Theta$ in G and, moreover, L (= identity component of L) is contained in ${}^u\mathsf{L}_\Theta$; because $P = {}^u\mathsf{L}_\Theta R_u(P) = \mathsf{L}R_u(P)$, the fact that $\mathsf{L} = {}^u\mathsf{L}_\Theta$ will follow when it is established that L centralizes ${}^uA_\Theta$. So let x be an element of L which does not lie in L; write $x = \tilde{x}n$ $(\tilde{x} \in {}^u\mathsf{L}_\Theta, n \in R_u(P))$; fix an arbitrary element H in $\mathrm{Ad}\,(u)\mathfrak{a}_\Theta$ – then we claim that $\mathrm{Ad}\,(x)H = H$. To see this, first observe that $\mathrm{Ad}\,(n)H - H \in \mathfrak{n}_\Theta^+$; therefore, since P normalizes $\mathfrak{n}_\Theta^+$, $\mathrm{Ad}\,(\tilde{x})(\mathrm{Ad}\,(n)H - H) = \mathrm{Ad}\,(x)H - H \in \mathfrak{n}_\Theta^+$; but both H and $\mathrm{Ad}\,(x)H$ also belong to $\mathfrak{l}$, thus $\mathrm{Ad}\,(x)H - H \in \mathfrak{l} \cap \mathfrak{n}_\Theta^+ = \{0\}$ which implies that $\mathrm{Ad}\,(x)H = H$. Hence the claim. This completes the proof of our proposition. □

An abelian subgroup A of G is said to be *split* if, for every $h \in A$, $\mathrm{Ad}_G\,(h)$ can be diagonalized over **R**.

Let P be a parabolic subgroup of G, L a Levi subgroup of P; let A be a maximal connected split abelian subgroup lying in the center of L – then A is obviously unique and L is the centralizer of A in G (cf. the proof of the preceding proposition). Let $\mathsf{X}(\mathsf{L})$ denote the group of all continuous homomorphisms of L into the multiplicative group of real numbers; put

$$\mathsf{M} = \bigcap_{\chi \in \mathsf{X}(\mathsf{L})} \operatorname{Ker}(|\chi|)$$

where $|\chi|(x) = |\chi(x)|\,(x \in \mathsf{L})$ – then it is easy to see that $\mathsf{L} = \mathsf{M}A$ with $\mathsf{M} \cap A = \{1\}$ (bear in mind the Remark following Lemma 1.2.4.5 and the proof of Proposition 1.2.4.14). [The group M is, in general, neither connected nor semi-simple – however it has but finitely many connected components and is always reductive . . . .] The corresponding mapping of the product manifold $\mathsf{M} \times A \times R_u(P)$ onto P is an analytic diffeomorphism – the ensuing decomposition of P is said to be a *Langlands decomposition* of P and A a *split component* of P. Since any two Levi subgroups of P are conjugate in P by an element from the unipotent radical $R_u(P)$, the same must be true of any two split components of P; such an element from $R_u(P)$ is in fact unique . . . . By the *parabolic rank* of P (written prk (P)), we shall understand the dimension of any of its split components; if P is minimal, then the parabolic rank of P is equal to the split-rank of G; the parabolic subgroups of rank 1 are called *maximal parabolic subgroups* of G.

Notation Let P be a parabolic subgroup of G, L a Levi subgroup of P – then, on certain occasions, it will be necessary to employ a more precise notation and write L_P, M_P, and A_P for L, M, and A, respectively.

Remark Let P be a parabolic subgroup of G, $G = KA_{\mathfrak{p}}N^+$ an Iwasawa decomposition of G relative to which P splits: $P = P_\Theta = \mathsf{M}_\Theta A_\Theta N_\Theta^+$ – then, of course, the Levi subgroups (respectively split components of P) are simply the groups ${}^n\mathsf{L}_\Theta$ (respectively ${}^nA_\Theta$) $(n \in N_\Theta^+)$. Note too that if K is any maximal compact subgroup of G, then $G = KP$.

Let P be a parabolic subgroup of G, A a split component of P with Lie algebra $\mathfrak{a}$; for any linear function σ on $\mathfrak{a}$, let $\mathfrak{g}^\sigma$ denote the set of those X in the Lie algebra of $R_u(P)$ such that $[H, X] = \sigma(H)X$ (all $H \in \mathfrak{a}$) – then σ is said to be a (*positive*) *root* of the pair $(\mathfrak{g}, \mathfrak{a})$ if $\mathfrak{g}^\sigma \neq \{0\}$. The set of all (positive) roots of the pair $(\mathfrak{g}, \mathfrak{a})$ will be denoted by $\Sigma^+(\mathfrak{g}, \mathfrak{a})$; evidently the Lie algebra of $R_u(P)$ is spanned by the $\mathfrak{g}^\sigma(\sigma \in \Sigma^+(\mathfrak{g}, \mathfrak{a}))$. If $l\,(= l_P) = \dim(\mathfrak{a})$, then there exist l linearly independent positive roots $\sigma_1, \ldots, \sigma_l$ such that every other element in $\Sigma^+(\mathfrak{g}, \mathfrak{a})$ can be expressed as a positive integral linear combination of the σ_i (cf. the discussion preceding Lemma 1.2.4.2); the set $\{\sigma_1, \ldots, \sigma_l\}$ will be referred to as the set of *simple roots* of the pair $(\mathfrak{g}, \mathfrak{a})$ and will be denoted by $\Upsilon(\mathfrak{g}, \mathfrak{a})$. One can then imitate without difficulty the earlier constructions. Thus fix a subset Θ of $\Upsilon(\mathfrak{g}, \mathfrak{a})$; let $\mathfrak{a}_\Theta$ be the subspace of $\mathfrak{a}$ comprised of those elements H in $\mathfrak{a}$ such that

$\sigma_i(H) = 0$ for all $\sigma_i \in \Theta$; put $\mathfrak{n}_\Theta^+ = \sum_\sigma \mathfrak{g}^\sigma$, the sum being taken over those σ in $\Sigma^+(\mathfrak{g}, \mathfrak{a})$ which do not vanish identically on $\mathfrak{a}_\Theta$; let $\mathfrak{l}_\Theta$ denote the centralizer of $\mathfrak{a}_\Theta$ in $\mathfrak{g}$ and set $\mathfrak{p}_\Theta = \mathfrak{l}_\Theta + \mathfrak{n}_\Theta^+$; finally let P_Θ be the normalizer of $\mathfrak{p}_\Theta$ in G and N_Θ^+ (respectively A_Θ) the analytic subgroup of G corresponding to $\mathfrak{n}_\Theta^+$ (respectively $\mathfrak{a}_\Theta$) – then P_Θ is a parabolic subgroup of G containing P, N_Θ^+ is the unipotent radical of P_Θ and A_Θ is a split component of P_Θ. It is clear that $A_\Theta \subset A$ and $R_u(P_\Theta) \subset R_u(P)$; furthermore the (positive) roots of the pair $(\mathfrak{g}, \mathfrak{a}_\Theta)$ are the restrictions to $\mathfrak{a}_\Theta$ of those σ in $\Sigma^+(\mathfrak{g}, \mathfrak{a})$ which do not vanish identically on $\mathfrak{a}_\Theta$ while the simple roots are made up of the restrictions to $\mathfrak{a}_\Theta$ of the elements in the complement of Θ in $\Upsilon(\mathfrak{g}, \mathfrak{a})$.

Definition Let P_i be parabolic subgroups of G with split components A_i $(i = 1, 2)$ – then the pair (P_1, A_1) is said to *dominate* the pair (P_2, A_2), written $(P_1, A_1) \succ (P_2, A_2)$, if $P_1 \supset P_2$ and $A_1 \subset A_2$.

So, in the above situation, the pair (P_Θ, A_Θ) dominates (P, A); moreover, if Q is any parabolic subgroup of G containing P, then there exists a unique subset Θ of $\Upsilon(\mathfrak{g}, \mathfrak{a})$ such that $Q = P_\Theta$. On the other hand, if the pair (P_0, A_0) is minimal with respect to the above partial order, then, given any other pair (P, A), there exists, as we know, an element x in G and a unique subset Θ of $\Upsilon(\mathfrak{g}, \mathfrak{a}_0)$ such that $({}^xP, {}^xA) = ((P_0)_\Theta, (A_0)_\Theta)$.

Let (P_i, A_i) $(i = 1, 2)$ be two parabolic pairs of G; let $W(A_1, A_2)$ denote the set of all bijections $w: A_1 \to A_2$ such that $w(h_1) = xh_1x^{-1}$ (all $h_1 \in A_1$) for some $x \in G$.

Definition Two parabolic subgroups P_1 and P_2 of G are said to be *associated* if there exist split components $A_1 \subset P_1$ and $A_2 \subset P_2$ such that the set $W(A_1, A_2)$ is not empty.

[If the condition in question is fulfilled for one choice of A_1, A_2, then it is fulfilled for any choice of A_1, A_2; this follows from the fact that any two split components of a given parabolic subgroup P are conjugate within $R_u(P)$.]

It is easy to see that parabolic subgroups P_1 and P_2 are associated iff there exists an x in G such that P_1 and ${}^xP_2$ have a common Levi subgroup; it is also clear that conjugate parabolic subgroups of G are associated but the converse is not true. [Take $G = \mathbf{SL}(3, \mathbf{R})$; let P_1 be the subgroup of matrices which are upper triangular in the partition $3 = 2 + 1$ and take for P_2 the analogous subgroup relative to the partition $3 = 1 + 2$ – then P_1 and P_2 are associated but not conjugate . . . .] However if P_1 and P_2 are minimal, then P_1 and P_2 are associated iff they are conjugate. The relation of association is an equivalence relation, hence breaks the class $\mathfrak{P}$ of all parabolic subgroups of G into equivalence classes $\mathfrak{C}_\mathfrak{P}$; fix one such class $\mathfrak{C}_\mathfrak{P}$ and a minimal parabolic pair (P_0, A_0); choose a complete set of representatives $P_1, \ldots, P_r$ for $G\backslash\mathfrak{C}_\mathfrak{P}$ where $P_k = P_{\Theta_k} \supset P_0$ (all k) (let us keep in mind that, modulo conjugation by G, $\mathfrak{P}$ has but 2^{l_0} elements); let $\mathfrak{C}_k$ denote the class of all parabolic subgroups of G which are conjugate to P_k – then it is clear that $\mathfrak{C}_{\mathfrak{P}}$ is the disjoint union of the sets $\mathfrak{C}_k$.

Let (P_1, A_1), (P_2, A_2) be parabolic pairs of G – then the set $W(A_1, A_2)$ is finite. [Fix a minimal pair (P_0, A_0); it can be supposed that $(P_i, A_i) = (P_{\Theta_i}, A_{\Theta_i}) \succ (P_0, A_0)$ $(i = 1, 2)$. Fix a $w \in W(A_1, A_2)$; let x_w be a representative of w in G – then, since x_w maps A_1 onto A_2, it maps L_{Θ_1} onto L_{Θ_2} and thus takes A_0 onto a conjugate of A_0 in L_{Θ_2}, whence we may assume that x_w lies in M^* (= normalizer of A_0 in K). But then it is clear that the map $w \mapsto W_{\Theta_2} x_w W_{\Theta_1}$ of $W(A_1, A_2)$ into the double coset space $W_{\Theta_2} \backslash W / W_{\Theta_1}$ $(\sim P_{\Theta_2} \backslash G / P_{\Theta_1} \ldots)$ is injective.] In particular if $P_1 = P_2 = P$ (say), $A_1 = A_2 = A$ (say), then $W(A, A) = W(A)$ is a finite group which is canonically isomorphic to L^*/L (L^* the normalizer of A in G); it can very well be the case that the group $W(A)$ is trivial (cf. infra).

Fix a class $\mathfrak{C}_{\mathfrak{P}}$ of associated parabolic subgroups in $\mathfrak{P}$; let A be a split component of some element of $\mathfrak{C}_{\mathfrak{P}}$ and let $\Sigma(\mathfrak{g}, \mathfrak{a})$ be the non-zero roots of the pair $(\mathfrak{g}, \mathfrak{a})$. By definition, an element $H \in \mathfrak{a}$ is said to be *regular* in $\mathfrak{a}$ if $\sigma(H) \neq 0$ for every $\sigma \in \Sigma(\mathfrak{g}, \mathfrak{a})$; the connected components of the set of all regular elements in $\mathfrak{a}$ are called *Weyl chambers*; each Weyl chamber $\mathscr{C}$ singles out a set $\Sigma_{\mathscr{C}}^+(\mathfrak{g}, \mathfrak{a})$ of 'positive' roots of the pair $(\mathfrak{g}, \mathfrak{a})$; given such a $\mathscr{C}$, put $\mathfrak{n}_{\mathscr{C}}^+ = \sum_\sigma \mathfrak{g}^\sigma$ $(\sigma \in \Sigma_{\mathscr{C}}^+(\mathfrak{g}, \mathfrak{a}))$, $N_{\mathscr{C}}^+ = \exp(n_{\mathscr{C}}^+)$, and $P_{\mathscr{C}} = \mathsf{M}AN_{\mathscr{C}}^+$ – then it is easy to see that the mapping $\mathscr{C} \mapsto P_{\mathscr{C}}$ is a bijective mapping from the set $\{\mathscr{C}\}$ of all Weyl chambers $\mathscr{C}$ in $\mathfrak{a}$ onto the set $\mathfrak{P}_A$ of all parabolic subgroups of G which have A as a split component, and is, moreover, compatible with the action of $W(A)$. If $w \in W(A)$, $w \neq 1$, then $w\mathscr{C} \neq \mathscr{C}$ for any chamber $\mathscr{C}$, whence ${}^wP \neq P$ for any $P \in \mathfrak{P}_A$; on the other hand, two elements of $\mathfrak{P}_A$ are conjugate under G iff they are conjugate under $W(A)$ and each element of $\mathfrak{C}_{\mathfrak{P}}$ is G-conjugate to some element of $\mathfrak{P}_A$ – therefore we have bijections $G\backslash\mathfrak{C}_{\mathfrak{P}} \leftrightarrow W(A)\backslash\mathfrak{P}_A \leftrightarrow W(A)\backslash\{\mathscr{C}\}$. In particular suppose that A is the split component of a minimal parabolic subgroup of G – then the (necessarily minimal) parabolic subgroups of G which contain A as a split component are in a one-to-one correspondence with the Weyl chambers $\mathscr{C}$ in $\mathfrak{a}$ and, in addition, the Weyl group $W(A)$ (= W ...) permutes the elements of $\mathfrak{P}_A$ in a *simply transitive* way (a fact which is not true in general – cf. infra).

Examples (1) Fix a class $\mathfrak{C}_{\mathfrak{P}}$ of associated parabolic subgroups in $\mathfrak{P}$; let us suppose that prk $(\mathfrak{C}_{\mathfrak{P}}) = 1$, i.e. that the groups in the class $\mathfrak{C}_{\mathfrak{P}}$ are maximal parabolic subgroups of G. If $P \in \mathfrak{C}_{\mathfrak{P}}$ and $P = \mathsf{M}AR_u(P)$, then any group in $\mathfrak{C}_{\mathfrak{P}}$ is conjugate to P or to the 'opposed' group $Q = \mathsf{M}AR_u(Q)$; furthermore P and Q are conjugate iff $-1 \in W(A)$ (evidently, in the present case, either $W(A) = \{\pm 1\}$ or $W(A) = \{1\}$). To illustrate, take $G = \mathbf{SL}(3, \mathbf{R})$; agreeing to employ the notations which were used in the Example following Lemma 1.2.4.5, let

$$A = A_\Theta = \left\{\begin{pmatrix} h & 0 & 0 \\ 0 & h & 0 \\ 0 & 0 & h^{-2} \end{pmatrix} : h \in \mathbf{R}^+\right\}.$$

One may then check that any element in $K(= \mathbf{SO}(3))$ which normalizes A necessarily centralizes A, i.e. $W(A) = \{1\}$ (whence the Weyl chambers in $\mathfrak{a}$ are not permuted transitively by $W(A)$); an easy computation also shows that the order of the double coset space $W_\Theta \backslash W / W_\Theta$ is 2 (thus the injection of $W(A_1, A_2)$ into $W_{\Theta_2} \backslash W / W_{\Theta_1}$ is, in general, proper . . .). On the other hand, if rank $(G/K) =$ split-rank of $G = 1$, then $W(A)$ contains -1 and the maximal and minimal parabolic subgroups of G coincide.

(2) Let $\mathfrak{g} = \mathfrak{sp}(2, \mathbf{R})$ $(= \mathfrak{sp}(2, \mathbf{C}) \cap \mathfrak{gl}(2, \mathbf{R}))$ – then $\mathfrak{g}$ admits the Cartan decomposition $\mathfrak{g} = \mathfrak{k} + \mathfrak{p}$ where

$$\mathfrak{k} = \left\{\begin{pmatrix} A & B \\ -B & A \end{pmatrix} : A \in \mathfrak{o}(2), B \in \mathbf{S}(2, \mathbf{R})\right\}, \ \mathfrak{p} = \left\{\begin{pmatrix} A & B \\ B & -A \end{pmatrix} : A, B \in \mathbf{S}(2, \mathbf{R})\right\}.$$

Evidently $\mathfrak{a}_\mathfrak{p} = \{D(a_1, a_2, -a_1, -a_2) : a_i \in \mathbf{R}\}$ is a maximal abelian subalgebra of $\mathfrak{p}$ (in fact of $\mathfrak{g}$ . . .). Let e_i be the linear form on $\mathfrak{a}_\mathfrak{p}$ given by the rule $e_i(A) = a_i$ for $A = D(a_1, a_2, -a_1, -a_2)$ $(1 \leqslant i \leqslant 4)$; the roots Σ of the pair $(\mathfrak{g}, \mathfrak{a}_\mathfrak{p})$ are the forms $\pm(e_1 - e_2)$, $\pm(e_1 + e_2)$, $\pm 2e_1$, $\pm 2e_2$; the Weyl group W of the pair $(\mathfrak{g}, \mathfrak{a}_\mathfrak{p})$ consists of all permutations and changes of sign of the e_i's (hence is isomorphic to the dihedral group of order 8); a fundamental system Υ for Σ is the two element set $\{2e_2, (e_1 - e_2)\}$. Now take $G = \mathbf{Sp}(2, \mathbf{R})$ – then G admits 4 conjugacy classes of parabolic subgroups. Besides G itself and the 'obvious' minimal parabolic subgroup P having $A_\mathfrak{p}$ as split component, there are two maximal parabolic subgroups P_1 and P_2 containing P with respective split components A_1 and A_2 where

$$A_1 = \{D(h, 1, h^{-1}, 1) : h \in \mathbf{R}^+\}, A_2 = \{D(h, h, h^{-1}, h^{-1}) : h \in \mathbf{R}^+\}.$$

In the present case, both the groups $W(A_1)$ and $W(A_2)$ have order 2 – therefore P_1 and P_2 are not associated. The reader will have no trouble in explicating L_1 (or L_2) . . . .

(3) The number of association classes of parabolic subgroups of $G = \mathbf{SL}(n, \mathbf{C})$ is the number $p(n)$ of partitions of n. [A sequence $n_1, \ldots, n_r$ of integers such that $n_1 \geqslant n_2 \geqslant \cdots \geqslant n_r > 0$ is said to be a partition of $n_1 + n_2 + \cdots + n_r$, the n_i its parts.]

Appendix 1 Let G be a connected semi-simple Lie group with finite center, $G = KA_\mathfrak{p}N^+$ an Iwasawa decomposition for G; set $P = \mathsf{M}A_\mathfrak{p}N^+$ and fix a subset Θ of Υ; introduce $\mathsf{M}_\Theta(K)$, $M_\Theta(K)$, P_Θ etc. as above – then we have the following result.

Lemma Let $\mathsf{Z}(A_\mathfrak{p}) = \mathrm{Ad}_G^{-1}(\mathrm{Ad}_G(K) \cap \exp \sqrt{-1}\mathfrak{a}_\mathfrak{p})$ – then $\mathsf{M}_\Theta(K) = \mathsf{Z}(A_\mathfrak{p})M_\Theta(K)$.

The idea lying behind the proof of this lemma (which gives the structure of the group of components for $\mathsf{M}_\Theta(K)$) is as follows. The orbit $\mathcal{Q}_\Theta$ of P_Θ through the origin in G/P contains a Euclidean 'cell' which may be identified with $\mathfrak{n}^-(\Theta)$ via the map $X \mapsto \exp(X)P$ $(X \in \mathfrak{n}^-(\Theta))$; let μ_Θ be the unique $\mathsf{M}_\Theta(K)$-invariant probability measure on $\mathcal{Q}_\Theta$ and μ the unique K-invariant probability measure on G/P; fix an H in the Weyl chamber $\mathscr{C}_\Theta$ and set $h_t = \exp(tH)(t \in \mathbf{R})$ – then we shall prove below that $h_t * \mu$ converges to μ_Θ as $t \to \infty$. It will be shown that μ_Θ lives on $N^-(\Theta)P$ and vanishes on no non-empty open subset of this set; because $N^-(\Theta)P$ is contained in the orbit of $M_\Theta A_\Theta N_\Theta^+$ (= identity component of P_Θ) through the origin in G/P and this orbit is clearly closed, it follows that μ_Θ is supported on $(M_\Theta A_\Theta N_\Theta^+)P$ in G/P and the support of μ_Θ contains $N^-(\Theta)P$; since μ_Θ is $\mathsf{M}_\Theta(K)$-invariant, we deduce that $\mathsf{M}_\Theta(K) \subset (M_\Theta A_\Theta N_\Theta^+)P$, whence $\mathsf{M}_\Theta(K) = \mathsf{Z}(A_\mathfrak{p})M_\Theta(K)$ (cf. Lemma 1.1.3.8), as desired.

To prove that $h_t * \mu \to \mu_\Theta$, we shall show that every subsequence of $h_t * \mu$ which converges at all must converge to μ_Θ. So suppose that $h_n * \mu$ $(h_n =$

$\exp(t_n H))$ tends to a limit ν(say) – then we claim that $\nu = \mu_\Theta$. Let $k \in M_\Theta(K)$ (the centralizer of A_Θ in K); evidently $k * \nu = k * \lim(h_n * \mu) = \lim(h_n * k * \mu) = \nu$ – therefore ν is a probability measure on G/P invariant under $M_\Theta(K)$. Let us now recall that $N^-(\Theta)P$ is a Euclidean cell in the orbit $\mathscr{Q}_\Theta$; we intend to prove that the ν-measure of this cell is 1 – it will then follow easily that $\nu = \mu_\Theta$. To this end introduce coordinates x_i on $\mathfrak{n}^- \sim N^-P$ which serve to diagonalize $\mathrm{Ad}_G(A_\mathfrak{p})|\mathfrak{n}^-$; it can be supposed that a certain subset $\{\tilde{x}_i\}$ (say) of the x_i give coordinates on $\mathfrak{n}^-(\Theta)$. Let R be a closed subset of $N^-(\Theta)P$; let R_1 and R_2 be two neighborhoods of R such that $R \subset R_1 \subset R_1^{cl} \subset R_2$; for $\epsilon > 0$, let $\mathscr{S}_1(\epsilon)$ (respectively $\mathscr{S}_2(\epsilon)$) denote the set of points of the form $r_1 + s$ (respectively $r_2 + s$) such that $r_1 \in R_1$ (respectively $r_2 \in R_2$) with $|x_i(s)| < \epsilon$ and $|\tilde{x}_i(s)| = 0$; choose a continuous function f on N^-P such that $0 \leqslant f \leqslant 1$ with $f = 1$ on $\mathscr{S}_1(\epsilon)$ and such that the support of f is contained in $\mathscr{S}_2(2\epsilon)$ – then $\int f d\nu = \lim \int f(h_n \cdot q) d\mu(q)$. Since h_n acts trivially on $N^-(\Theta)P$ and since we can find a sequence λ_n tending to ∞ such that the eigenvalues of $\mathrm{Ad}(h_n^{-1})|\mathfrak{n}_\Theta^-$ exceed λ_n, we have $f(h_n \cdot q) = 1$ on $\mathscr{S}_1(\lambda_n \epsilon)$ and $f(h_n \cdot q)$ is supported on $\mathscr{S}_2(\infty)$ – thus

$$\mu(\mathscr{S}_1(\lambda_n \epsilon)) \leqslant \int f(h_n \cdot q) d\mu(q) \leqslant \mu(\mathscr{S}_2(\infty))$$

and so

$$\mu(\mathscr{S}_1(\infty)) \leqslant \int f d\nu \leqslant \mu(\mathscr{S}_2(\infty))$$

since $\mu(\mathscr{S}_1(\lambda_n \epsilon)) \uparrow \mu(\mathscr{S}_1(\infty))$ when $n \to \infty$. As $R_2 \downarrow R$, $\mu(\mathscr{S}_2(\infty)) \downarrow \mu(\mathscr{S}(\infty))$ (because μ is a finite measure). [Here we assign to the symbol $\mathscr{S}(\infty)$ the obvious meaning (relative to R).] But as $R_2 \downarrow R$ and $\epsilon \to 0$, $\int f d\nu \to \nu(R)$ – therefore $\nu(R) = \mu(\mathscr{S}(\infty))$. Now let $R \uparrow N^-(\Theta)P$ – then $\mu(\mathscr{S}(\infty)) \uparrow \mu(N^-P) = \mu(G/P) = 1$, whence $\nu(N^-(\Theta)P) = 1$. Since ν is also a probability measure, ν must live on $N^-(\Theta)P$; therefore the support of ν is contained in the closure $\mathscr{Q}_\Theta$ of $N^-(\Theta)P$; however $M_\Theta(K)$ is transitive on $\mathscr{Q}_\Theta$ and ν is $M_\Theta(K)$-invariant – therefore $\nu = \mu_\Theta$, as we wished to prove.

Appendix 2 Let G be a connected semi-simple Lie group with finite center; let P be a parabolic subgroup of G; fix a split component A of P and write $P = MAR_u(P)$ – then, as will be seen below, there exists a simple relation between the parabolic subgroups of M and the parabolic subgroups of G which are contained in P. However the group M is, in general, only reductive, and, more seriously, need not be connected (although the number of its components is finite) – therefore our theory is not immediately applicable to M. To get around this, first observe that the theory of parabolic subgroups, as developed above for G, is applicable without any change to those reductive pairs (G, K) of 1.1.5 for which $\mathfrak{c}_\mathfrak{p} = \{0\}$ (the general case is not really needed); therefore, in view of Lemma 1.2.4.4 (ii), we in any event have a theory of parabolic subgroups for the identity component M of M; furthermore, and this is the crucial point, the centralizer in M of any maximal connected split abelian subgroup of M meets every component of M (see the first Appendix to this number) – one may then extend without essential difficulty the theory of parabolic subgroups so as to be applicable to M (cf. infra).

Notation Let P be a parabolic subgroup of G, $P = MAR_u(P)$ a Langlands decomposition for P – then $MR_u(P)/R_u(P) \sim M$ and so we have a canonical homomorphism $\pi_{P|M}: MR_u(P) \to M$.

We shall now indicate how it is possible to extend the theory of parabolic subgroups, as developed in the earlier numbers of this section, to a certain class of non-connected reductive Lie groups.

Let $\mathfrak{g}$ be a reductive Lie algebra over $\mathbf{R}$, θ an involutive automorphism of $\mathfrak{g}$ whose restriction to $\bar{\mathfrak{g}}$ is a Cartan involution of $\bar{\mathfrak{g}}$; in all that follows it will be assumed that $\mathfrak{c}_{\mathfrak{p}} = \{0\}$. Let G be a Lie group with Lie algebra $\mathfrak{g}$. It will be supposed that G has but a finite number of connected components; we shall also assume that the identity component G of G verifies the general conditions laid down in 1.1.5 – in particular, then, the map $(k, X) \mapsto k \exp X$ is an analytic diffeomorphism of $K \times \mathfrak{p}$ onto G.

Concerning such a group G, one has the following structural result due to Mostow.

Theorem Put $E = \exp(\mathfrak{p})$ – then G has a compact subgroup $\mathsf{K} \supset K$ with the following properties:

(i) $kEk^{-1} = E(k \in \mathsf{K})$;

(ii) The multiplication $\mathsf{K} \times E \rightarrow \mathsf{G}$ is an isomorphism of disjoint unions of analytic manifolds;

(iii) For every compact subgroup $\tilde{\mathsf{K}}$ of G, there is an element e in E such that $e\tilde{\mathsf{K}}e^{-1} \subset \mathsf{K}$.

[The reader may wish to consult Hochschild [1, p. 180] for an exposition of Mostow's proof.]

It is to be observed that the compact subgroup K of G which figures in the preceding theorem is necessarily a *maximal* compact subgroup of G. [In fact if $\tilde{\mathsf{K}}$ is a compact subgroup of G containing K, then there is an element $x \in \mathsf{G}$ such that $x\tilde{\mathsf{K}}x^{-1} \subset \mathsf{K}$ – hence $x\mathsf{K}x^{-1} \subset x\tilde{\mathsf{K}}x^{-1} \subset \mathsf{K}$. By comparing dimensions and counting components, one verifies that these inclusions are actually equalities.] Needless to say, K has only a finite number of connected components; of course $\mathsf{K} \cap G = K$, $\mathsf{G} = \mathsf{K}G$. One should also note that the requirement $\mathsf{K} \supset K$ determines K uniquely, i.e. if $\tilde{\mathsf{K}} \supset K$ and verifies the other conditions, then $\tilde{\mathsf{K}} = \mathsf{K}$. [Owing to (iii) supra, we can find an $e \in E$ such that $e\tilde{\mathsf{K}}e^{-1} = \mathsf{K}$; but then $e(\tilde{\mathsf{K}} \cap G)e^{-1} = eKe^{-1} = K$, that is, e normalizes K. Conclusion: $e \in K \cap \exp(\mathfrak{p}) \Rightarrow e = 1$.]

Consider an Iwasawa decomposition $G = KA_{\mathfrak{p}}N^{+}$ for G – then it is certainly true that $\mathsf{G} = \mathsf{K}A_{\mathfrak{p}}N^{+}$. In addition, the usual arguments show that the map $(k, h, n) \mapsto khn$ is an analytic diffeomorphism of $\mathsf{K} \times A_{\mathfrak{p}} \times N^{+}$ onto G. Otherwise said, the Iwasawa decomposition theorem extends to the present situation without difficulty.

In order to make further progress, it will be necessary to make the following assumption.

Assumption Let $\mathfrak{h}$ be maximal among the abelian subalgebras of $\bar{\mathfrak{g}}$ whose image in $\mathrm{ad}(\mathfrak{g})$ is diagonalizable over $\mathbf{R}$ – then we shall assume that the centralizer of $\mathfrak{h}$ in G meets each component of G.

[In passing, note that if the assumption is met by one $\mathfrak{h}$, then it is met by every $\mathfrak{h}$ (since any two such $\mathfrak{h}$'s are conjugate under $\mathrm{Ad}(G)$, this is clear).]

Let $G = KA_{\mathfrak{p}}N^{+}$ be an Iwasawa decomposition for G, $\mathsf{G} = \mathsf{K}A_{\mathfrak{p}}N^{+}$ the Iwasawa decomposition for G; let $\mathbf{M}$ (respectively $\mathbf{M}^{*}$) denote the centralizer (respectively the normalizer) of $A_{\mathfrak{p}}$ in K; let $\mathbf{L}$ (respectively $\mathbf{L}^{*}$) denote the centralizer (respectively the normalizer) of $A_{\mathfrak{p}}$ in G – then, employing a simple variant of Lemma 1.1.3.7, we see without difficulty that $\mathbf{L} = \mathbf{M}A_{\mathfrak{p}}$, $\mathbf{L}^{*} = \mathbf{M}^{*}A_{\mathfrak{p}}$. Now, in view of our basic assumption, we have $\mathsf{G} = \mathbf{L}G$; but, as has just been pointed out, $\mathbf{L} = \mathbf{M}A_{\mathfrak{p}}$, whence $\mathsf{G} = \mathbf{M}G$. This being so, consider the group $\mathbf{M}^{*}/\mathbf{M}$; it is clearly finite and, moreover, is isomorphic to W (= the Weyl group of the pair $(\mathfrak{g}, \mathfrak{a}_{\mathfrak{p}})$). [Thus, since $\mathsf{G} = \mathbf{M}G$, it must be the case that $\mathbf{M}^{*} = M^{*}\mathbf{M}$ (M^{*} the normalizer of $A_{\mathfrak{p}}$ in K) – therefore $\mathbf{M}^{*}/\mathbf{M} = M^{*}\mathbf{M}/\mathbf{M} \sim M^{*}/M^{*} \cap \mathbf{M} = M^{*}/M = W$, as contended.]

Example Let G be a connected semi-simple Lie group with finite center, $G = KA_{\mathfrak{p}}N^+$ an Iwasawa decomposition for G; employing the usual notations, fix a subset $\Theta \subset \Upsilon$ and put $P_\Theta = \mathsf{M}_\Theta A_\Theta N_\Theta^+$ – then P_Θ is a parabolic subgroup of G. Since M_Θ has a compact center (cf. Lemma 1.2.4.4 (ii)) and since $\mathsf{M}_\Theta = \mathsf{Z}(A_{\mathfrak{p}})M_\Theta$ (cf. Appendix 1 supra), it follows that the pair $(\mathsf{M}_\Theta, \mathsf{M}_\Theta(K))$ can play the role of the pair (G, K) in the discussion above. [Evidently M_Θ admits the 'generalized' Iwasawa decomposition $\mathsf{M}_\Theta = \mathsf{M}_\Theta(K)A_{\mathfrak{p}}(\Theta)N^+(\Theta)$; the Weyl group of the pair $(\mathfrak{m}_\Theta, \mathfrak{a}_{\mathfrak{p}}(\Theta))$ may be identified with W_Θ.]

Definition By a *minimal parabolic subgroup* P of G we shall understand the normalizer in G of a minimal parabolic subgroup of G.

Let P be a minimal parabolic subgroup of G – then, by definition, there exists a minimal parabolic subgroup P (say) of G such that P is the normalizer of P in G. Because P is self-normalizing (in G), it follows that $\mathsf{P} \cap G = P$, whence P is unique. Evidently the minimal parabolic subgroups of G are all conjugate (by elements of G); moreover, they are self-normalizing (in G).

Let P be a minimal parabolic subgroup of G; put $P = \mathsf{P} \cap G$ (so that P is a minimal parabolic subgroup of G) and let $P = \mathsf{M}AR_u(P)$ be a Langlands decomposition for P; let $\mathbf{L}$ be the centralizer of A in G – then, in view of our basic assumption, we have $\mathsf{G} = \mathbf{L}G$, whence $\mathsf{P} = \mathbf{L}R_u(P)$ (thus P meets each component of G).

Let $G = KA_{\mathfrak{p}}N^+$ be an Iwasawa decomposition for G, $\mathsf{G} = \mathsf{K}A_{\mathfrak{p}}N^+$ the corresponding Iwasawa decomposition for G; set $P = \mathsf{M}A_{\mathfrak{p}}N^+$ and let P be the normalizer of P in G – then, by definition, P is a minimal parabolic subgroup of G and it is a simple matter to check that the triple $(\mathsf{G}, \mathsf{P}, \mathbf{M}^*)$ constitutes a Tits system. [Here, of course, we have to observe that G can be written as the disjoint union of the double cosets $\mathsf{P}w\mathsf{P}$ $(w \in W)$.] Consequently Theorem 1.2.1.1 is now available; in particular, therefore, there exist 2^l subgroups P_Θ containing P (l = split-rank of G), the so-called *standard* parabolic subgroups of G (with respect to our choice of P and $A_{\mathfrak{p}}$). A given standard parabolic subgroup P_Θ of G admits the decomposition $\mathsf{P}_\Theta = \mathbf{M}_\Theta(\mathsf{K})A_{\mathfrak{p}}N^+$ ($\mathbf{M}_\Theta(\mathsf{K})$ the centralizer of A_Θ in K); therefore the parabolic subgroups of G are closed in G. Every parabolic subgroup of G is conjugate to one and only one standard parabolic subgroup and meets each component of G; a given parabolic subgroup of G is the normalizer in G of a necessarily unique parabolic subgroup of G.

Let P be a parabolic subgroup of G; relative to P it is clear what one is to understand by the terms unipotent radical, Levi subgroup, split component, Langlands decomposition, parabolic rank. If P is a parabolic subgroup of G, $\mathsf{P} = \mathbf{M}\mathsf{A}R_u(\mathsf{P})$ a Langlands decomposition for P (such exist...), then $\mathbf{M}$ satisfies the same general hypotheses as G (including the basic assumption supra). [Therefore all of the preceding discussion on parabolic subgroups is applicable to $\mathbf{M}$....]

Let G be a connected semi-simple Lie group with finite center, P_0 a parabolic subgroup of G; let $P_0 = \mathsf{M}_0 A_0 R_u(P_0)$ be a Langlands decomposition for P_0 – then, in view of what has been said above, it makes sense to consider the parabolic subgroups of M_0. This being so, fix a parabolic subgroup P (say) of M_0 with Langlands decomposition $\mathsf{P} = \mathbf{M}\mathsf{A}R_u(\mathsf{P})$; put $\mathsf{M} = \mathbf{M}$, $A = \mathsf{A}A_0$, $R_u(P) = R_u(\mathsf{P})R_u(P_0)$, $P = \mathsf{M}AR_u(P)$ – then P is a parabolic subgroup of G with Langlands decomposition $P = \mathsf{M}AR_u(P)$ and such that $P \subset P_0$. Conversely, suppose that P is a parabolic subgroup of G which is contained in P_0 – then $\pi_{P_0 \mid \mathsf{M}_0}(P \cap \mathsf{M}_0 R_u(P_0))$ is a parabolic subgroup P (say) of M_0 whose corresponding 'P' is P itself. Otherwise said, the assignment $\mathsf{P} \mapsto P$ establishes a bijection between the set of parabolic subgroups of M_0 and the set of those parabolic subgroups of G which are contained in P_0.

1.3 Cartan Subalgebras

1.3.1 Conjugacy of Cartan Subalgebras in a Real Reductive Lie Algebra

Let $\mathfrak{g}$ be a semi-simple Lie algebra over the real numbers **R**. Let $\mathfrak{g} = \mathfrak{k} + \mathfrak{p}$ be a fixed Cartan decomposition of $\mathfrak{g}$ with corresponding Cartan involution θ. Let $\mathfrak{a}_\mathfrak{p}$ be a maximal abelian subspace in $\mathfrak{p}$; extend $\mathfrak{a}_\mathfrak{p}$ to a Cartan subalgebra $\mathfrak{a}$ of $\mathfrak{g}$, $\mathfrak{a} = \mathfrak{a}_\mathfrak{k} + \mathfrak{a}_\mathfrak{p}$, and by complexification to a Cartan subalgebra $\mathfrak{a}_c$ of $\mathfrak{g}_c$. Let Φ denote the set of roots of the pair $(\mathfrak{g}_c, \mathfrak{a}_c)$, W_Φ the corresponding Weyl group; let Σ denote the set of roots of the pair $(\mathfrak{g}, \mathfrak{a}_\mathfrak{p})$, W_Σ the corresponding Weyl group. These entities will remain fixed throughout the ensuing discussion; unexplained notations are as in 1.1 and 1.2.

Definition A *Cartan subalgebra* $\mathfrak{j}$ of $\mathfrak{g}$ is a subalgebra of $\mathfrak{g}$ such that:

(i) $\mathfrak{j}$ is maximal abelian;

(ii) For each $H \in \mathfrak{j}$, ad (H) is a semi-simple endomorphism of $\mathfrak{g}$.

The reader will note that a maximal abelian subalgebra of $\mathfrak{g}$ is not necessarily a Cartan subalgebra, i.e. (ii) is not a consequence of (i). For example, in $\mathfrak{sl}(2, \mathbf{R})$, consider $\mathfrak{j} = \{(\begin{smallmatrix} 0 & t \\ 0 & 0 \end{smallmatrix}): t \in \mathbf{R}\}$. It should also be pointed out that the above definition of Cartan subalgebra is tailored for the study of semi-simple or reductive Lie algebras only. [Let us recall that if $\mathfrak{g}$ is a Lie algebra over a field **k** of characteristic zero, then a subalgebra $\mathfrak{j}$ of $\mathfrak{g}$ is said to be a Cartan subalgebra of $\mathfrak{g}$ provided $\mathfrak{j}$ is nilpotent and self-normalizing. It is well-known that, in the case at hand, this general definition is equivalent to the one adopted above.]

The existence of Cartan subalgebras is established by the usual methods (cf. Serre [4, Chapter III]). [The notion of a regular element will be dealt with systematically in 1.3.4.]

Definition Let $\mathfrak{g}$ be a semi-simple Lie algebra over **R**. The *rank* of $\mathfrak{g}$ is the dimension of any Cartan subalgebra $\mathfrak{j}$ of $\mathfrak{g}$.

Evidently the rank is well-defined . . . .

Let $\mathfrak{j}$ be a Cartan subalgebra of $\mathfrak{g}$ – then the subalgebra $\mathfrak{j}_I$ of $\mathfrak{j}$ consisting of those X in $\mathfrak{j}$ such that ad (X) (viewed as an endomorphism of $\mathfrak{g}_c$) has pure imaginary eigenvalues will be called the *toroidal part* of $\mathfrak{j}$ while the subalgebra $\mathfrak{j}_R$ consisting of those X in $\mathfrak{j}$ such that ad (X) has real eigenvalues will be called the *vector part* of $\mathfrak{j}$. Plainly $\mathfrak{j}$ is the direct sum of its vector part and toroidal part.

Let $\mathfrak{j}^1$, $\mathfrak{j}^2$ be Cartan subalgebras of $\mathfrak{g}$ – then $\mathfrak{j}^1$ is said to be *conjugate* to $\mathfrak{j}^2$ if there exists $x \in$ Int $(\mathfrak{g})$ (Int $(\mathfrak{g})$ the adjoint group of $\mathfrak{g}$) such that

$x(\mathfrak{j}^1) = \mathfrak{j}^2$. The main goal of this number is to prove: There exist finitely many conjugacy classes of Cartan subalgebras for $\mathfrak{g}$. [Recall that if $\mathfrak{g}$ is compact or carries a complex structure, then there is only one conjugacy class of Cartan subalgebras for $\mathfrak{g}$ (cf. infra).] This will be accomplished through a series of reductions, the end result of which is not only the finiteness of the number of conjugacy classes but also the determination of an invariant by means of which the conjugacy classes can actually be computed in practice.

Example Let us take $\mathfrak{g} = \mathfrak{sl}(2, \mathbf{R})$ – then $\mathfrak{g}$ has two Cartan subalgebras, namely

$$\mathfrak{j}^1 = \left\{\begin{pmatrix} 0 & -t \\ t & 0 \end{pmatrix} : t \in \mathbf{R}\right\} \quad \text{and} \quad \mathfrak{j}^2 = \left\{\begin{pmatrix} t & 0 \\ 0 & -t \end{pmatrix} : t \in \mathbf{R}\right\},$$

and they are not conjugate under any inner automorphism from $\mathrm{Int}(\mathfrak{g})$. . . .

It is possible to establish the finiteness of the number of conjugacy classes of Cartan subalgebras in a real semi-simple Lie algebra by arguments significantly shorter than the one presented below (for instance via real algebraic geometry) – however such proofs are qualitative in nature and provide little, if any, information on how to go about selecting, in actual practice, a representative Cartan subalgebra from each conjugacy class. [To illustrate, suppose that $p_1, \ldots, p_l$ ($l = \mathrm{rank}\ (\mathfrak{g}_c)$) are a set of algebraically independent homogeneous generators for the algebra $I(\mathfrak{g}_c)$ of invariant polynomials on $\mathfrak{g}_c$ (see 2.1.5); let $\Omega: \mathfrak{g}_c \to \mathbf{C}^l$ be the map defined by the rule $\Omega(X) = (p_1(X), \ldots, p_l(X))$ $(X \in \mathfrak{g}_c)$; let $\mathscr{R}$ be the set of regular elements in $\mathfrak{g}_c$ – then the number of connected components of the set $\Omega(\mathfrak{g} \cap \mathscr{R})$ is the number of distinct conjugacy classes of Cartan subalgebras of $\mathfrak{g}$ (cf. Rothschild [1]).]

The following result tells us that, for the purposes of the present investigation, we need only consider θ-stable Cartan subalgebras.

Proposition 1.3.1.1 Let $\tilde{\mathfrak{j}}$ be a Cartan subalgebra of $\mathfrak{g}$ – then there exists an $x \in \mathrm{Int}(\mathfrak{g})$ such that $\mathfrak{j} = x(\tilde{\mathfrak{j}})$ is θ-stable (i.e. $\mathfrak{j} = \mathfrak{j} \cap \mathfrak{k} + \mathfrak{j} \cap \mathfrak{p} = \mathfrak{j}_{\mathfrak{k}} + \mathfrak{j}_{\mathfrak{p}}$) with $\mathfrak{j}_{\mathfrak{p}} \subset \mathfrak{a}_{\mathfrak{p}}$.

Proof It is not difficult to see that there exists a Cartan decomposition $\mathfrak{g} = \tilde{\mathfrak{k}} + \tilde{\mathfrak{p}}$ such that $\tilde{\mathfrak{j}} = \tilde{\mathfrak{j}} \cap \tilde{\mathfrak{k}} + \tilde{\mathfrak{j}} \cap \tilde{\mathfrak{p}}$. [There exists a compact real form $\tilde{\mathfrak{u}}$ of $\mathfrak{g}_c$ such that: (1) $\sigma(\tilde{\mathfrak{u}}) = \tilde{\mathfrak{u}}$; (2) $\tilde{\mathfrak{j}}_c \cap \tilde{\mathfrak{u}}$ is a Cartan subalgebra of $\tilde{\mathfrak{u}}$ (σ the conjugation of $\mathfrak{g}_c$ with respect to $\mathfrak{g}$). Thus construct the compact real form $\tilde{\mathfrak{g}}_{\mathfrak{k}}$ relative to $\tilde{\mathfrak{j}}_c$ per the discussion following Theorem 1.1.1.5. If $\tilde{\sigma}$ denotes the conjugation of $\mathfrak{g}_c$ with respect to $\tilde{\mathfrak{g}}_{\mathfrak{k}}$, then both σ and $\tilde{\sigma}$ leave $\tilde{\mathfrak{j}}_c$ invariant; put $\phi = (\sigma\tilde{\sigma})^{1/2}$ (cf. Helgason [2, pp.156–157]) – one may take $\tilde{\mathfrak{u}} = \phi(\tilde{\mathfrak{g}}_{\mathfrak{k}})$. This being so, put $\tilde{\mathfrak{k}} = \tilde{\mathfrak{u}} \cap \mathfrak{g}$, $\tilde{\mathfrak{p}} = (\sqrt{-1}\tilde{\mathfrak{u}}) \cap \mathfrak{g}$ – then $\mathfrak{g} = \tilde{\mathfrak{k}} + \tilde{\mathfrak{p}}$ is a Cartan decomposition of $\mathfrak{g}$ relative to which $\tilde{\mathfrak{j}}$ splits.] Theorem 1.1.1.6 gives $y \in \mathrm{Int}(\mathfrak{g})$ such that $y(\tilde{\mathfrak{k}}) = \mathfrak{k}$, $y(\tilde{\mathfrak{p}}) = \mathfrak{p}$. If $\tilde{\mathfrak{a}}_{\mathfrak{p}}$ is a maximal abelian subspace of $\mathfrak{p}$ containing $y(\tilde{\mathfrak{j}} \cap \tilde{\mathfrak{p}})$, then we may find a $k \in K$ such that $k(\tilde{\mathfrak{a}}_{\mathfrak{p}}) = \mathfrak{a}_{\mathfrak{p}}$. [Here K denotes the analytic subgroup of $\mathrm{Int}(\mathfrak{g})$ corresponding to $\mathfrak{k}$.] The element $x = ky$ will satisfy the requirements of the proposition. □

By a toral subgroup (or torus) of a Lie group we shall understand a compact, abelian, connected subgroup.

Proposition 1.3.1.2 (Maximum Principle) Let $\mathfrak{t}_1, \mathfrak{t}_2$ be subalgebras of $\mathfrak{g}$ generating tori T_1, T_2 in Int ($\mathfrak{g}$). If there exists a compact subgroup L with Lie algebra $\mathfrak{l}$ in Int ($\mathfrak{g}$) such that

$$[l(\mathfrak{t}_1), \mathfrak{t}_2] \subset \mathfrak{l}$$

for all $l \in L$, then there exists an element $l_0 \in L$ such that:
(i) Every element in $l_0 T_1 l_0^{-1}$ commutes with T_2;
(ii) $[l_0(\mathfrak{t}_1), \mathfrak{t}_2] = \{0\}$.

Proof Let B denote the Killing form for $\mathfrak{g}$. Given $X_1 \in \mathfrak{t}_1$, $X_2 \in \mathfrak{t}_2$, there exists an element $l_0 \in L$ such that $[l_0 X_1, X_2] = 0$. To see this, first note that $f(l) = B(lX_1, X_2)$ $(l \in L)$, being a continuous function, assumes a maximum at some point $l_0 \in L$ – hence

$$g_X(r) = B(e^{r\,\mathrm{ad}\,(X)}(l_0 X_1), X_2) \qquad (X \in \mathfrak{l}, r \in \mathbf{R})$$

as a function of r takes a maximum at $r = 0$. Thus

$$0 = g'_X(0) = B(X, [l_0 X_1, X_2])$$

and so, X being arbitrary and B negative definite on $\mathfrak{l}$ (cf. Helgason [2, p.123]), $[l_0 X_1, X_2] = 0$. Taking now X_i so that $e^{r\,\mathrm{ad}\,(X_i)}$ is dense in T_i $(i = 1, 2)$, the result follows. □

Remark It follows directly from the Maximum Principle that any two Cartan subalgebras in a compact or complex semi-simple Lie algebra $\mathfrak{g}$ are conjugate under the action of the adjoint group. This is of course immediate if $\mathfrak{g}$ is compact (since in that case the notion of Cartan subalgebra coincides with that of maximal abelian subalgebra). Turning to the case when $\mathfrak{g}$ is complex, let $\mathfrak{j}^1$ and $\mathfrak{j}^2$ be two Cartan subalgebras of $\mathfrak{g}$; let $\mathfrak{u}_i$ be a compact real form for $\mathfrak{g}$ containing $\mathfrak{j}_l^i$ $(i = 1, 2)$ – then, in view of Corollary 1.1.1.8 and Theorem 1.1.1.6, there exists an element x in Int ($\mathfrak{g}$) such that $x(\mathfrak{u}_1) = \mathfrak{u}_2$. Since $x(\mathfrak{j}_l^1)$ is a Cartan subalgebra of the compact semi-simple Lie algebra $\mathfrak{u}_2$, $x(\mathfrak{j}_l^1)$ is conjugate to $\mathfrak{j}_l^2$ — but then the same must be true of $\mathfrak{j}^1$ and $\mathfrak{j}^2$.

In the remainder of this number K will denote the analytic subgroup of Int ($\mathfrak{g}$) corresponding to $\mathfrak{k}$.

Proposition 1.3.1.3 Let $\mathfrak{j}^1$, $\mathfrak{j}^2$ be θ-stable Cartan subalgebras of $\mathfrak{g}$ such that $\mathfrak{j}_{\mathfrak{p}}^2 \subset \mathfrak{j}_{\mathfrak{p}}^1$ – then there exists an element $k \in K$ such that $kH = H$ for all $H \in \mathfrak{j}_{\mathfrak{p}}^2$ and $k(\mathfrak{j}_{\mathfrak{k}}^1) \subset \mathfrak{j}_{\mathfrak{k}}^2$. Moreover if $\mathfrak{j}_{\mathfrak{p}}^1 = \mathfrak{j}_{\mathfrak{p}}^2$, then $k(\mathfrak{j}_{\mathfrak{k}}^1) = \mathfrak{j}_{\mathfrak{k}}^2$.

Proof Let L_K^2 denote the centralizer of $\mathfrak{j}_{\mathfrak{p}}^2$ in K – then L_K^2 is a compact subgroup of Int ($\mathfrak{g}$). Let J_K^i be the analytic subgroup of K corresponding to $\mathfrak{j}_{\mathfrak{k}}^i$ $(i = 1, 2)$ – then J_K^i is a torus in Int ($\mathfrak{g}$). In order to apply the Maximum Principle to $\mathfrak{j}_{\mathfrak{k}}^1$, $\mathfrak{j}_{\mathfrak{k}}^2$ and L_K^2 we must check that $[l(\mathfrak{j}_{\mathfrak{k}}^1), \mathfrak{j}_{\mathfrak{k}}^2] \subset \mathfrak{l}^2$ for all $l \in \mathsf{L}_K^2$ ($\mathfrak{l}^2$ the Lie algebra of L_K^2). But

$$[\mathfrak{j}_\mathfrak{p}^2, [l(\mathfrak{j}_\mathfrak{k}^1), \mathfrak{j}_\mathfrak{k}^2]] = [l(\mathfrak{j}_\mathfrak{k}^1), [\mathfrak{j}_\mathfrak{p}^2, \mathfrak{j}_\mathfrak{k}^2]] + [\mathfrak{j}_\mathfrak{k}^2, [l(\mathfrak{j}_\mathfrak{k}^1), \mathfrak{j}_\mathfrak{p}^2]] = \{0\}$$

since $\mathfrak{j}^2$ is abelian and $[l(\mathfrak{j}_\mathfrak{k}^1), \mathfrak{j}_\mathfrak{p}^2] \subset l[\mathfrak{j}_\mathfrak{k}^1, \mathfrak{j}_\mathfrak{p}^1] = \{0\}$. By the Maximum Principle there exists an $l_0 \in \mathrm{L}_K^2$ such that $[l_0(\mathfrak{j}_\mathfrak{k}^1), \mathfrak{j}_\mathfrak{k}^2] = \{0\}$. On the other hand $\mathfrak{j}_\mathfrak{p}^2 \subset \mathfrak{j}_\mathfrak{p}^1$ implies that $[l_0(\mathfrak{j}_\mathfrak{k}^1), \mathfrak{j}_\mathfrak{p}^2] = \{0\}$ and so $l_0(\mathfrak{j}_\mathfrak{k}^1) \subset \mathfrak{j}_\mathfrak{k}^2$. □

Definition A θ-stable Cartan subalgebra $\mathfrak{j}$ of $\mathfrak{g}$ is said to be *standard* (with respect to the triple $(\mathfrak{k}, \mathfrak{p}, \mathfrak{a})$) if $\mathfrak{a}_\mathfrak{k} \subset \mathfrak{j}_\mathfrak{k}$ and $\mathfrak{j}_\mathfrak{p} \subset \mathfrak{a}_\mathfrak{p}$.

It is clear, on the basis of Proposition 1.3.1.3, that every conjugacy class of Cartan subalgebras of $\mathfrak{g}$ contains a standard member.

As usual let $\mathfrak{u}$ denote the compact real form $\mathfrak{u} = \mathfrak{k} + \sqrt{-1}\mathfrak{p}$ of the complexification $\mathfrak{g}_c$ of $\mathfrak{g}$. Whenever it proves to be convenient, we may identify K with the analytic subgroup of $\mathrm{Int}(\mathfrak{u})$ with Lie algebra $\mathfrak{k}$. Recall (1.1.3) that W_Σ is the Weyl group of the symmetric space $\mathrm{Int}(\mathfrak{g})/K$.

Proposition 1.3.1.4 Let $\mathfrak{j}^1, \mathfrak{j}^2$ be standard Cartan subalgebras of $\mathfrak{g}$ – then $\mathfrak{j}^1, \mathfrak{j}^2$ are conjugate iff $\mathfrak{j}_\mathfrak{p}^1$ and $\mathfrak{j}_\mathfrak{p}^2$ are conjugate under W_Σ.

Proof The sufficiency is plain in view of Proposition 1.3.1.3. As for the necessity, suppose that $x(\mathfrak{j}^1) = \mathfrak{j}^2$ for some $x \in \mathrm{Int}(\mathfrak{g})$ and write $x = ke^{\mathrm{ad}(X)}$ $(k \in K, X \in \mathfrak{p})$. Observing that $x(\mathfrak{j}_\mathfrak{p}^1) = \mathfrak{j}_\mathfrak{p}^2$, we see that $e^{\mathrm{ad}(X)}(\mathfrak{j}_\mathfrak{p}^1) \subset \mathfrak{p}$ and so Lemma 1.1.3.7 implies that $e^{\mathrm{ad}(X)}(H) = H$ for all $H \in \mathfrak{j}_\mathfrak{p}^1$ – hence $k(\mathfrak{j}_\mathfrak{p}^1) = \mathfrak{j}_\mathfrak{p}^2$. Let us apply the Maximum Principle to the subalgebras $\mathfrak{t}_1 = k(\sqrt{-1}\mathfrak{a}_\mathfrak{p})$, $\mathfrak{t}_2 = \sqrt{-1}\mathfrak{a}_\mathfrak{p}$ of $\mathfrak{u} = \mathfrak{k} + \sqrt{-1}\mathfrak{p}$ and to the centralizer L_K^2 of $\mathfrak{j}_\mathfrak{p}^2$ in $K \subset \mathrm{Int}(\mathfrak{u})$. It is not difficult to see that $\mathfrak{t}_1$ and $\mathfrak{t}_2$ generate tori T_1 and T_2 in $\mathrm{Int}(\mathfrak{u})$ (cf. Helgason [2, p.211]) and thus it remains to check that

$$[l(\mathfrak{t}_1), \mathfrak{t}_2] \subset \mathfrak{l}^2 \qquad (l \in \mathrm{L}_K^2).$$

But this is easily done using the Jacobi identity and the fact that $\mathfrak{j}^1, \mathfrak{j}^2$ are standard. Therefore there exists an $l_0 \in \mathrm{L}_K^2$ such that $[l_0 k(\mathfrak{a}_\mathfrak{p}), \mathfrak{a}_\mathfrak{p}] = \{0\}$, whence $l_0 k(\mathfrak{a}_\mathfrak{p}) = \mathfrak{a}_\mathfrak{p}$, $\mathfrak{a}_\mathfrak{p}$ being maximal abelian in $\mathfrak{p}$. The element $l_0 k$ is in W_Σ and takes $\mathfrak{j}_\mathfrak{p}^1$ to $\mathfrak{j}_\mathfrak{p}^2$. □

Corollary 1.3.1.5 All Cartan subalgebras of $\mathfrak{g}$ whose vector parts have maximal dimension are conjugate.

Proposition 1.3.1.6 Let $\mathfrak{j}^1, \mathfrak{j}^2$ be standard Cartan subalgebras of $\mathfrak{g}$. Let

$$\mathfrak{r}^i = \{X \in \mathfrak{a}_\mathfrak{p} : B(X, \mathfrak{j}_\mathfrak{p}^i) = 0\} \qquad (i = 1, 2).$$

Then $\mathfrak{j}^1, \mathfrak{j}^2$ are conjugate iff $\mathfrak{r}^1, \mathfrak{r}^2$ are conjugate under W_Φ.

Proof If $\mathfrak{j}^1, \mathfrak{j}^2$ are conjugate, then we know by Proposition 1.3.1.4 that $\mathfrak{j}_\mathfrak{p}^1$ is conjugate to $\mathfrak{j}_\mathfrak{p}^2$ under W_Σ. Since W_Σ may be taken as the subgroup of W_Φ whose elements stabilize $\mathfrak{a}_\mathfrak{p}$ (Proposition 1.1.3.3) and since these transformations are orthogonal relative to B the necessity is clear. Turning to the sufficiency, suppose $w \in W_\Phi$ is such that $w(\mathfrak{r}^1) = \mathfrak{r}^2$. The action of w on $\mathfrak{a}_\mathfrak{k} + \sqrt{-1}\mathfrak{a}_\mathfrak{p}$ is given by an element u of $\operatorname{Int}(\mathfrak{u})$ and extends by complexification to $\mathfrak{a}_c$; write $u = ke^{\sqrt{-1}\,\mathrm{ad}\,(X)}(k \in K,\ X \in \mathfrak{p})$; since $u(\mathfrak{r}^1) = \mathfrak{r}^2 \subset \mathfrak{p}$ it follows that $[X, Y] = 0$ for all $Y \in \mathfrak{r}^1$ (here we are using an obvious variant of Lemma 1.1.3.7) – therefore $k(\mathfrak{r}^1) = \mathfrak{r}^2$. An application of the Maximum Principle to $\mathfrak{t}_1 = k(\sqrt{-1}\mathfrak{a}_\mathfrak{p})$, $\mathfrak{t}_2 = \sqrt{-1}\mathfrak{a}_\mathfrak{p}$ and the centralizer L_K^2 of $\mathfrak{r}^2$ in $K \subset \operatorname{Int}(\mathfrak{u})$ gives an element $l_0 \in L_K^2$ with the property that $l_0k(\mathfrak{a}_\mathfrak{p}) = \mathfrak{a}_\mathfrak{p}$. But then l_0k represents an element of W_Σ and maps $\mathfrak{j}_\mathfrak{p}^1$ to $\mathfrak{j}_\mathfrak{p}^2$ – hence (Proposition 1.3.1.4) $\mathfrak{j}^1$ and $\mathfrak{j}^2$ are indeed conjugate. □

Our next task will be to determine when a subspace of $\mathfrak{a}_\mathfrak{p}$ is the vector part of a standard Cartan subalgebra. However, before proceeding to the precise formulation of the main result, some additional preparation is needed.

Let Φ denote the set of non-zero roots of the pair $(\mathfrak{g}_c, \mathfrak{a}_c)$. For any subspace $\mathfrak{r}$ of $\mathfrak{a}_\mathfrak{p}$ put

$$\Phi(\mathfrak{r}) = \{\alpha \in \Phi : H_\alpha \in \mathfrak{r}\}.$$

Let $\mathfrak{h} = \{X \in \mathfrak{a} : B(X, \mathfrak{r}) = 0\}$; let $\mathfrak{z}_c(\mathfrak{h})$ be the centralizer of $\mathfrak{h}$ in $\mathfrak{g}_c$, $\mathfrak{z}(\mathfrak{h}) = \mathfrak{z}_c(\mathfrak{h}) \cap \mathfrak{g}$ – plainly $\mathfrak{z}_c(\mathfrak{h}) = \mathfrak{a}_c + \sum_\alpha \mathfrak{g}_c^\alpha$, $\alpha \in \Phi(\mathfrak{r})$.

Definition A subspace $\mathfrak{r}$ of $\mathfrak{a}_\mathfrak{p}$ is called a *root space* (in $\mathfrak{a}_\mathfrak{p}$) if $\mathfrak{r} = \sum_\alpha \mathbf{R}H_\alpha$, $\alpha \in \Phi(\mathfrak{r})$.

Proposition 1.3.1.7 Let $\mathfrak{j}$ be a standard Cartan subalgebra of $\mathfrak{g}$ – then $\mathfrak{r} = \{X \in \mathfrak{a}_\mathfrak{p} : B(X, \mathfrak{j}_\mathfrak{p}) = 0\}$ is a root space.

Proof In the present case, we have $\mathfrak{h} = \mathfrak{a}_\mathfrak{k} + \mathfrak{j}_\mathfrak{p}$ and $\mathfrak{z}_c(\mathfrak{h}) \supset \mathfrak{j}_c$. If $\mathfrak{r}$ were not a root space, then there would exist a non-zero element $H \in \mathfrak{r}$ such that $\alpha(H) = 0$ for all $\alpha \in \Phi(\mathfrak{r})$. But then

$$[H, \mathfrak{j}_c] \subset [H, \mathfrak{z}_c(\mathfrak{h})] = \{0\}$$

since $X_\alpha \in \mathfrak{z}_c(\mathfrak{h})$ for all $\alpha \in \Phi(\mathfrak{r})$. This implies that $H \in \mathfrak{j}_\mathfrak{p}$, a contradiction. □

Let $\mathfrak{r}$ be a root space. Let $\mathfrak{g}_c^{\Phi(\mathfrak{r})}$ be defined as in Proposition 1.1.1.10 relative to the set $\Gamma = \Phi(\mathfrak{r})$. Observe that $\mathfrak{z}_c(\mathfrak{h}) = \mathfrak{h}_c + \mathfrak{g}_c^{\Phi(\mathfrak{r})}$.

Proposition 1.3.1.8 Let $\mathfrak{r}$ be a root space – then:

(i) $\mathfrak{g}_c^{\Phi(\mathfrak{r})}$ is a semi-simple Lie algebra over $\mathbf{C}$ with Cartan subalgebra $\mathfrak{r}_c$;

(ii) $\mathfrak{g}^{\Phi(\mathfrak{r})} = \mathfrak{g}_c^{\Phi(\mathfrak{r})} \cap \mathfrak{g}$ is a real form for $\mathfrak{g}_c^{\Phi(\mathfrak{r})}$;

(iii) $\mathfrak{g}^{\Phi(\mathfrak{r})} = \mathfrak{g}^{\Phi(\mathfrak{r})} \cap \mathfrak{k} + \mathfrak{g}^{\Phi(\mathfrak{r})} \cap \mathfrak{p}$ is a Cartan decomposition for $\mathfrak{g}^{\Phi(\mathfrak{r})}$;

(iv) $\mathfrak{z}(\mathfrak{h}) \cap \mathfrak{k} = \mathfrak{a}_{\mathfrak{k}} + (\mathfrak{g}^{\Phi(\mathfrak{r})} \cap \mathfrak{k})$ (direct sum).

Proof Statements (i), (ii), (iii) follow from Proposition 1.1.1.10. To prove (iv) note first that $\mathfrak{z}(\mathfrak{h}) = \mathfrak{h} + \mathfrak{g}^{\Phi(\mathfrak{r})}$ (for $\mathfrak{z}_c(\mathfrak{h})$, $\mathfrak{h}_c$, and $\mathfrak{g}_c^{\Phi(\mathfrak{r})}$ are each σ-stable). Now $\mathfrak{h} = \mathfrak{a}_{\mathfrak{k}} + (\mathfrak{h} \cap \mathfrak{p})$ (direct vector space sum) and so $\mathfrak{a}_{\mathfrak{k}} = \mathfrak{h} \cap \mathfrak{k}$. Since τ (the conjugation of $\mathfrak{g}_c$ with respect to $\mathfrak{u}$) stabilizes $\mathfrak{h}_c$ and $\mathfrak{g}_c^{\Phi(\mathfrak{r})}$, we have

$$\mathfrak{z}(\mathfrak{h}) \cap \mathfrak{k} = (\mathfrak{h} \cap \mathfrak{k}) + (\mathfrak{g}^{\Phi(\mathfrak{r})} \cap \mathfrak{k}) = \mathfrak{a}_{\mathfrak{k}} + (\mathfrak{g}^{\Phi(\mathfrak{r})} \cap \mathfrak{k}). \quad \square$$

We explicitly observe that if $\mathfrak{r}$ is a root space, then $\Phi(\mathfrak{r})$ is the root system of the semi-simple Lie algebra $\mathfrak{g}_c^{\Phi(\mathfrak{r})}$ with respect to the Cartan subalgebra $\mathfrak{r}_c$; furthermore $\mathfrak{g}^{\Phi(\mathfrak{r})}$ itself is *normal* (cf. infra).

Definition A semi-simple Lie algebra $\mathfrak{g}$ over $\mathbf{R}$ is said to be *normal* if $\mathfrak{g}$ contains a Cartan subalgebra whose toroidal part is zero.

Definition Let $\mathfrak{g}_c$ be a semi-simple Lie algebra over $\mathbf{C}$, $\mathfrak{j}_c$ a Cartan subalgebra of $\mathfrak{g}_c$, $\Phi(\Phi^+)$ the set of non-zero roots (positive roots) of the pair $(\mathfrak{g}_c, \mathfrak{j}_c)$. A subset $\Gamma = \{\alpha_1, \ldots, \alpha_m\}$ of distinct positive roots is said to be a *strongly orthogonal system* (of positive roots) if $\alpha_i \pm \alpha_j \notin \Phi$, $1 \leqslant i, j \leqslant m$.

Two strongly orthogonal systems Γ_1 and Γ_2 will be called *equivalent* if $\sum_{\alpha\in\Gamma_1} \mathbf{R}H_\alpha = \sum_{\alpha\in\Gamma_2} \mathbf{R}H_\alpha$. Two strongly orthogonal systems Γ_1 and Γ_2 will be called *conjugate* if there exists an element w in the Weyl group W of the pair $(\mathfrak{g}_c, \mathfrak{j}_c)$ such that $w\Gamma_1$ is equivalent to Γ_2. Evidently these relations are equivalence relations.

These general notions will be applied to the problem at hand shortly; the instrument of application is the following result.

Lemma 1.3.1.9 Let $\mathfrak{g}$ be a normal semi-simple Lie algebra over $\mathbf{R}$, $\mathfrak{g} = \mathfrak{k} + \mathfrak{p}$ a Cartan decomposition for $\mathfrak{g}$; let $\mathfrak{j}$ be a Cartan subalgebra of $\mathfrak{g}$ whose toroidal part is zero. The following conditions are equivalent:

(i) The ranks of $\mathfrak{k}$ and $\mathfrak{g}$ are equal;

(ii) There exists a strongly orthogonal system $\{\alpha_1, \ldots, \alpha_l\}$ of positive roots for the pair $(\mathfrak{g}_c, \mathfrak{j}_c)$ ($l = \text{rank}(\mathfrak{g})$).

Proof It suffices to work with a Cartan decomposition $\mathfrak{g} = \mathfrak{k} + \mathfrak{p}$ relative to which $\mathfrak{j}$ is θ-stable, i.e. $\mathfrak{j}_{\mathfrak{k}} = \{0\}$, $\mathfrak{j}_{\mathfrak{p}} = \mathfrak{j}$. Select a Weyl basis for $\mathfrak{g}_c$ mod $\mathfrak{j}_c$ with respect to $\mathfrak{u}$ – then the X_α may be taken in $\mathfrak{g}$ here and, moreover, we have that $\mathfrak{k} = \mathfrak{g} \cap \mathfrak{u} = \{\sum_{\alpha>0} r_\alpha(X_\alpha - X_{-\alpha}) : r_\alpha \in \mathbf{R}\}$ where $\mathfrak{u} = \mathfrak{k} + \sqrt{-1}\mathfrak{p}$. [For in this case all the roots α of the pair $(\mathfrak{g}_c, \mathfrak{j}_c)$ are real (i.e. $\alpha = \alpha^\sigma$, σ the conjugation of $\mathfrak{g}_c$ with respect to $\mathfrak{g}$ – see 1.3.2). Hence $\sigma(X_\alpha) \in \mathfrak{g}_c^\alpha \Rightarrow \sigma(X_\alpha) = c^2 X_\alpha$, c a unimodular complex number; replace X_α by cX_α, $X_{-\alpha}$ by $\bar{c}X_{-\alpha}$ – then we still have that $\tau(X_\alpha) = -X_{-\alpha}$

(τ the conjugation of $\mathfrak{g}_c$ with respect to $\mathfrak{u}$), $[X_\alpha, X_{-\alpha}] = H_\alpha$, $[X_\alpha, X_\beta] = N_{\alpha,\beta}X_{\alpha+\beta}$ where $N_{\alpha,\beta} = -N_{-\alpha,-\beta}$ etc.]

(ii) $\Rightarrow$ (i). Set $Y_\alpha = X_\alpha - X_{-\alpha}(\alpha > 0)$ – then for all α, $\beta > 0$

$$[Y_\alpha, Y_\beta] = N_{\alpha,\beta}Y_{\alpha+\beta} - N_{\alpha,-\beta}Y_{\alpha-\beta}.$$

Hence, if there exist l positive roots such that $\alpha_i \pm \alpha_j \notin \Phi$ (Φ the non-zero roots of the pair $(\mathfrak{g}_c, \mathfrak{j}_c)$), it follows that $\operatorname{rank}(\mathfrak{k}) = \operatorname{rank}(\mathfrak{g})$, as claimed.

(i) $\Rightarrow$ (ii). For each positive integer m define a proposition (P_m) by: (P_m) – The pair $(\mathfrak{g}_c, \mathfrak{j}_c)$ admits a strongly orthogonal system $\{\alpha_1, \ldots, \alpha_m\}$ of positive roots. We shall prove: $\operatorname{rank}(\mathfrak{k}) \geqslant m \Rightarrow (P_m)$ is valid for $(\mathfrak{g}_c, \mathfrak{j}_c)$. (P_1) is clearly valid. This being so, assume that (P_{m-1}) is and let $\mathfrak{t}$ be a Cartan subalgebra of $\mathfrak{k}$ containing $\sum_{i=1}^{m-1} \mathbf{R}Y_{\alpha_i}$. An arbitrary element $X \in \mathfrak{t}$ may be written in the form $X = \sum_{\beta>0} r_\beta Y_\beta$ ($r_\beta \in \mathbf{R}$). For each i $(1 \leqslant i \leqslant m-1)$ we have

$$0 = [Y_{\alpha_i}, X] = \sum_{\beta>0} r_\beta(N_{\alpha_i,\beta}Y_{\alpha_i+\beta} - N_{\alpha_i,-\beta}Y_{\alpha_i-\beta}).$$

Hence, if $\operatorname{rank}(\mathfrak{k}) = \dim(\mathfrak{t}) \geqslant m$, then there exists a non-zero $X \in \mathfrak{t}$ such that $X \notin \sum_{i=1}^{m-1} \mathbf{R}Y_{\alpha_i}$. Thus, for some positive root β different from the $\alpha_i(1 \leqslant i \leqslant m-1)$, r_β is not zero and so it must be the case that $N_{\alpha_i,\beta} = N_{\alpha_i,-\beta} = 0$. Calling this β, α_m, say, we find that (P_m) is valid. □

Henceforth reference to a strongly orthogonal system of roots will be relative to Φ^+, Φ^+ the set of positive roots of our fixed pair $(\mathfrak{g}_c, \mathfrak{a}_c)$. Now we are in a position to obtain the main result. Consider the question: When is a subspace of $\mathfrak{a}_\mathfrak{p}$ the vector part of a standard Cartan subalgebra $\mathfrak{j}$ of $\mathfrak{g}$? Thus let $\mathfrak{j}$ be a standard Cartan subalgebra of $\mathfrak{g}$. Form $\mathfrak{r} = \{X \in \mathfrak{a}_\mathfrak{p} : B(X, \mathfrak{j}_\mathfrak{p}) = 0\}$ which is a root space according to Proposition 1.3.1.7. Corresponding to $\mathfrak{r}$ we consider $\mathfrak{h} = \mathfrak{a}_\mathfrak{k} + \mathfrak{j}_\mathfrak{p}$ and $\mathfrak{z}_c(\mathfrak{h}) \supset \mathfrak{j}_c$. Since here $\mathfrak{z}(\mathfrak{h}) \cap \mathfrak{k} = \mathfrak{a}_\mathfrak{k} + \mathfrak{g}^{\Phi(\mathfrak{r})} \cap \mathfrak{k}$ (direct sum) and $\mathfrak{j}_\mathfrak{k}$ is maximal abelian in $\mathfrak{z}(\mathfrak{h}) \cap \mathfrak{k}$, we see that $\dim(\mathfrak{j}_\mathfrak{k}) = \dim(\mathfrak{j}) - \dim(\mathfrak{j}_\mathfrak{p}) = \dim(\mathfrak{a}_\mathfrak{k}) + \dim(\mathfrak{a}_\mathfrak{p}) - \dim(\mathfrak{j}_\mathfrak{p}) = \dim(\mathfrak{a}_\mathfrak{k}) + \dim(\mathfrak{r})$ and so we conclude that $\operatorname{rank}(\mathfrak{g}^{\Phi(\mathfrak{r})} \cap \mathfrak{k}) = \dim(\mathfrak{r}) = \operatorname{rank}(\mathfrak{g}^{\Phi(\mathfrak{r})})$. Ordering $\Phi(\mathfrak{r})$ in the obvious way, Lemma 1.3.1.9 implies that there exists a strongly orthogonal system $\{\alpha_1, \ldots, \alpha_r\}$ ($r = \dim(\mathfrak{r})$) of positive roots such that $\mathfrak{r} = \sum_{i=1}^r \mathbf{R}H_{\alpha_i}$. [In reality, Lemma 1.3.1.9 only guarantees us that $\alpha_i \pm \alpha_j \notin \Phi(\mathfrak{r})$; but it is then clear that $\alpha_i \pm \alpha_j \notin \Phi$ too.] Hence to each conjugacy class of Cartan subalgebras of $\mathfrak{g}$ we have assigned a strongly orthogonal system of positive roots such that the real span of the corresponding H_{α_i} is contained in $\mathfrak{a}_\mathfrak{p}$. On the other hand, let $\{\alpha_1, \ldots, \alpha_r\}$ be a strongly orthogonal system of positive roots such that $\mathfrak{r} = \sum_{i=1}^r \mathbf{R}H_{\alpha_i} \subset \mathfrak{a}_\mathfrak{p}$. Evidently the above argument can be read backwards to get a standard Cartan subalgebra $\mathfrak{j}$ such that $\mathfrak{a}_\mathfrak{p} = \mathfrak{j}_\mathfrak{p} + \mathfrak{r}$ with $\mathfrak{j}_\mathfrak{p}$ and $\mathfrak{r}$ orthogonal to each other relative to B. In particular, we have proved: *The number of conjugacy classes of Cartan subalgebras of a real semi-simple Lie algebra is finite.* Using the notion of conjugate strongly orthogonal systems of positive roots along with Propositions 1.3.1.4, 1.3.1.6 we obtain the following theorem.

Theorem 1.3.1.10 (Kostant-Sugiura) There is a one-to-one correspondence between conjugacy classes of Cartan subalgebras in a real semi-simple Lie algebra $\mathfrak{g}$ and the conjugacy classes of strongly orthogonal systems of positive roots in $\Phi(\mathfrak{a}_\mathfrak{p})$.

[In the terminology of 1.3.2, $\Phi(\mathfrak{a}_\mathfrak{p})$ is the set of real roots of the pair $(\mathfrak{g}_c, \mathfrak{a}_c)$.]

Corollary 1.3.1.11 There exists a finite number of Cartan subalgebras $\mathfrak{j}^1, \ldots, \mathfrak{j}^r$ of $\mathfrak{g}$ with the properties: (1) $\theta(\mathfrak{j}^i) = \mathfrak{j}^i$ $(i = 1, \ldots, r)$; (2) If $\mathfrak{j}$ is any Cartan subalgebra of $\mathfrak{g}$, then $\mathfrak{j}$ is conjugate to $\mathfrak{j}^i$ for exactly one i.

Suppose that $\mathfrak{g} = \mathfrak{g}_1 + \cdots + \mathfrak{g}_r$ is the decomposition of our semi-simple Lie algebra $\mathfrak{g}$ into simple ideals $\mathfrak{g}_i$. It is evident that every Cartan subalgebra $\mathfrak{j}$ of $\mathfrak{g}$ is the direct sum of Cartan subalgebras $\mathfrak{j}_i$ of $\mathfrak{g}_i$. So, if $\mathfrak{j}^1 = \mathfrak{j}^1_1 + \cdots + \mathfrak{j}^1_r$, $\mathfrak{j}^2 = \mathfrak{j}^2_1 + \cdots + \mathfrak{j}^2_r$ are two Cartan subalgebras of $\mathfrak{g}$ (with $\mathfrak{j}^1_i$, $\mathfrak{j}^2_i$ Cartan subalgebras of $\mathfrak{g}_i$), then $\mathfrak{j}^1$ and $\mathfrak{j}^2$ are conjugate under Int $(\mathfrak{g})$ iff $\mathfrak{j}^1_i$ and $\mathfrak{j}^2_i$ are conjugate under Int $(\mathfrak{g}_i)$. Consequently the number $N(\mathfrak{g})$ of conjugacy classes of Cartan subalgebras in $\mathfrak{g}$ is equal to the product of that of the $\mathfrak{g}_i$, i.e. $N(\mathfrak{g}) = \prod_i N(\mathfrak{g}_i)$ where $N(\mathfrak{g}_i)$ is the number of conjugacy classes of Cartan subalgebras in the simple Lie algebra $\mathfrak{g}_i$.

One may pose the problem: Given a semi-simple Lie algebra $\mathfrak{g}$ over $\mathbf{R}$ determine the number of conjugacy classes of Cartan subalgebras and a representative member in each class. In view of the above discussion, it is obviously enough to consider the case when $\mathfrak{g}$ is simple. In Sugiura [1] a complete solution to this problem may be found.

Examples (1) Consider the real simple Lie algebra of the Type (**AI**), i.e. $\mathfrak{g} = \mathfrak{sl}(l + 1, \mathbf{R}) = \{X \in \mathfrak{gl}(l + 1, \mathbf{R}) : \operatorname{tr}(X) = 0\}$. If $\mathfrak{k} = \{X \in \mathfrak{g} : {}^tX = -X\}$, $\mathfrak{p} = \{X \in \mathfrak{g} : {}^tX = X\}$, then $\mathfrak{g} = \mathfrak{k} + \mathfrak{p}$ is a Cartan decomposition of $\mathfrak{g}$ ('t' means transpose). Take $\mathfrak{a}_\mathfrak{p} = \{D(a_1, \ldots, a_{l+1}) : a_i \in \mathbf{R}, \sum_1^{l+1} a_i = 0\}$ ($D(a_1, \ldots, a_{l+1})$ a diagonal matrix with entries $a_1, \ldots, a_{l+1}$) – then $\mathfrak{a} = \mathfrak{a}_\mathfrak{p}$ is a Cartan subalgebra of $\mathfrak{g}$ contained in $\mathfrak{p}$. Hence $\mathfrak{g}$ is a normal real form for $\mathfrak{g}_c = \mathfrak{sl}(l + 1, \mathbf{C})$. Let e_i be the linear form on $\mathfrak{a}_c$ given by $e_i(A) = a_i$ if $A = D(a_1, \ldots, a_{l+1})$. The roots Φ of the pair $(\mathfrak{g}_c, \mathfrak{a}_c)$ are the forms $\pm(e_i - e_j)$ $(1 \leqslant i < j \leqslant l + 1)$ and the Weyl group W of the pair $(\mathfrak{g}_c, \mathfrak{a}_c)$ is the set of permutations of the e_i's. Any strongly orthogonal system is of the type $\{(e_{i_1} - e_{i_2}), \ldots, (e_{i_{2k-1}} - e_{i_{2k}}) : i_1 < i_2 < \cdots < i_{2k}\}$ where the $i_1, \ldots, i_{2k}$ are $2k$ different integers in the set $(1, 2, \ldots, l + 1)$. Plainly the toroidal part of the corresponding Cartan subalgebra is k dimensional and, in this case, since W consists of all permutations of the e_i's the integer k completely determines the conjugacy class. Hence the number N of conjugacy classes is: $N = [(l + 1)/2] + 1$. A representative from each class with dim $(\mathfrak{j}_\mathfrak{t}) = k$ is given by

$$\mathfrak{j} = \left\{ \left[\begin{array}{ccc|ccc|ccc} h_1 & & & -h_{k+1} & & & & & \\ & \ddots & & & \ddots & & & 0 & \\ & & h_k & & & -h_{2k} & & & \\ \hline h_{k+1} & & & h_1 & & & & & \\ & \ddots & & & \ddots & & & 0 & \\ & & h_{2k} & & & h_k & & & \\ \hline & & & & & & h_{2k+1} & & \\ & 0 & & & 0 & & & \ddots & \\ & & & & & & & & h_{l+1} \end{array}\right] : h_i \in \mathbf{R}, \text{ trace} = 0 \right\}.$$

(2) Adopt the notations of the Example following the proof of Lemma 1.1.3.6. Thus $l + 1 = 2m$ and if E_{ij} denotes the matrix $(\delta_{ai}\delta_{bj})$ $(1 \leqslant a \leqslant 2m, 1 \leqslant b \leqslant 2m)$, then, since the Killing form B is given in this case by $B(X, Y) = 2(l + 1) \operatorname{tr}(XY)$, we have

$$H_{e_i - e_j} = \frac{1}{2(l+1)}(E_{ii} - E_{jj}).$$

Hence the real simple Lie algebra $\mathfrak{g}$ of Type (**AII**) furnishes an example of a real non-compact Lie algebra in which all the Cartan subalgebras are mutually conjugate.

(3) It is possible to exhibit a non-compact simple Lie algebra containing non-conjugate Cartan subalgebras whose toroidal parts have the same dimension (an example of classical type may be found in the real simple Lie algebra of Type (**BI**)).

(4) To each standard Cartan subalgebra $\mathfrak{j}$ we have assigned a normal semi-simple Lie algebra $\mathfrak{g}^{\Phi(\mathfrak{r})}$. If $\mathfrak{j}^1$, $\mathfrak{j}^2$ are conjugate standard Cartan subalgebras then $\mathfrak{g}^{\Phi(\mathfrak{r}^1)}$ is isomorphic to $\mathfrak{g}^{\Phi(\mathfrak{r}^2)}$. The converse is not true. [A counterexample can be obtained by taking for $\mathfrak{g}$ the normal real form of $\mathfrak{g}_c$ where $\mathfrak{g}_c$ is of Type ($\mathbf{G}_2$): in this case $N(\mathfrak{g}) = 4$ and $\mathfrak{g}$ admits two non-conjugate standard Cartan subalgebras $\mathfrak{j}^1$ and $\mathfrak{j}^2$ both of toroidal dimension 1 with $\mathfrak{g}^{\Phi(\mathfrak{r}^1)} = \mathfrak{g}^{\Phi(\mathfrak{r}^2)} = \mathfrak{sl}(2, \mathbf{R})$.]

Let $\mathfrak{g}$ be a reductive Lie algebra over $\mathbf{R}$ – thus $\mathfrak{g} = \mathfrak{c} + [\mathfrak{g}, \mathfrak{g}]$ where $\mathfrak{c}$ denotes the center of $\mathfrak{g}$, $[\mathfrak{g}, \mathfrak{g}]$ the derived algebra of $\mathfrak{g}$. The definition of a Cartan subalgebra for a semi-simple $\mathfrak{g}$ extends without change to the case of a reductive $\mathfrak{g}$ as does the notion of rank. Thus a Cartan subalgebra $\mathfrak{j}$ of $\mathfrak{g}$ may be written $\mathfrak{j} = \mathfrak{j} \cap [\mathfrak{g}, \mathfrak{g}] + \mathfrak{c}$. It is obvious that the main result of this number extends to the case of a reductive $\mathfrak{g}$, namely: There exist finitely many conjugacy classes of Cartan subalgebras of $\mathfrak{g}$.

1.3.2 Classification of Roots

Let $\mathfrak{g}$ be a semi-simple Lie algebra over $\mathbf{R}$, $\mathfrak{j}$ a Cartan subalgebra of $\mathfrak{g}$. Select a Cartan involution θ of $\mathfrak{g}$ such that $\theta(\mathfrak{j}) = \mathfrak{j}$, $\mathfrak{g} = \mathfrak{k} + \mathfrak{p}$ the corresponding Cartan decomposition (cf. the first lines of the proof of

Proposition 1.3.1.1). Let Φ denote the set of roots of the pair $(\mathfrak{g}_c, \mathfrak{j}_c)$, Φ^+ the positive roots – we shall now classify the elements of Φ.

A root α is said to be *real* (*imaginary*) if $\alpha(H)$ is real (pure imaginary) for every $H \in \mathfrak{j}$; if α is neither real nor imaginary, then α is said to be *complex*. Since the roots assume real values on $\mathfrak{j}_\mathfrak{p} + \sqrt{-1}\mathfrak{j}_\mathfrak{k}$, it is clear that α is real (imaginary) iff α vanishes identically on $\mathfrak{j}_\mathfrak{k}(\mathfrak{j}_\mathfrak{p})$.

Fix $\alpha \in \Phi$ and consider the subalgebra $\mathfrak{l}_c^\alpha = \mathbf{C}H_\alpha + \mathbf{C}X_\alpha + \mathbf{C}X_{-\alpha}$ (here X_α, $X_{-\alpha}$, H_α have their usual meanings). Set $\mathfrak{l}^\alpha = \mathfrak{l}_c^\alpha \cap \mathfrak{g}$ and recall that σ denotes the conjugation of $\mathfrak{g}_c$ with respect to $\mathfrak{g}$.

Lemma 1.3.2.1 The following statements are equivalent:
(i) $\mathfrak{l}_c^\alpha$ is σ-stable;
(ii) $\dim_{\mathbf{R}}(\mathfrak{l}^\alpha) = 3$;
(iii) α is either real or imaginary.
[We shall leave the proof of this elementary fact to the reader.]

A root α is said to be *singular* if $\dim_{\mathbf{R}}(\mathfrak{l}^\alpha) = 3$ and if $\mathfrak{l}^\alpha$ is not compact; a root α is said to be *compact* if $\dim_{\mathbf{R}}(\mathfrak{l}^\alpha) = 3$ and if $\mathfrak{l}^\alpha$ is compact. So, since there are only two different real semi-simple Lie algebras of dimension 3, these last definitions classify the possible roots which occur in Lemma 1.3.2.1 into two disjoint subsets. On the other hand every real root is singular – for $H_\alpha \in \mathfrak{l}^\alpha$ and the eigenvalues of ad (H_α) are real. Hence every compact root is imaginary.

Notation As above, let $\mathfrak{g}$ be a semi-simple Lie algebra over $\mathbf{R}$, $\mathfrak{j}$ a Cartan subalgebra of $\mathfrak{g}$, Φ the set of roots of the pair $(\mathfrak{g}_c, \mathfrak{j}_c)$ – then Φ_R^+ will denote the set of positive real roots, Φ_C^+ the set of positive complex roots, Φ_{CP}^+ the set of positive compact roots, Φ_{SI}^+ the set of positive singular imaginary roots, Φ_I^+ the set of positive imaginary roots (hence $\Phi_I^+ = \Phi_{CP}^+ \cup \Phi_{SI}^+$).

Let us suppose that the root α is imaginary (hence $\alpha \mid \mathfrak{j}_\mathfrak{p} = 0$) – then, for $X_\alpha \in \mathfrak{g}_c^\alpha$, it is true that either $X_\alpha \in \mathfrak{k}_c$ or $X_\alpha \in \mathfrak{p}_c$. [To see this fix an $H \in \mathfrak{j}$ and write $H = H_\mathfrak{k} + H_\mathfrak{p}$ $(H_\mathfrak{k} \in \mathfrak{k},\ H_\mathfrak{p} \in \mathfrak{p})$ – then the relations $[H, X_\alpha] = \alpha(H_\mathfrak{k})X_\alpha$, $[H, \theta(X_\alpha)] = \theta[\theta(H), X_\alpha] = \alpha(\theta(H))\theta(X_\alpha) = \alpha(H_\mathfrak{k})\theta(X_\alpha)$ imply that $X_\alpha = c\theta(X_\alpha)$, c a constant. Since it is clear that $c^2 = 1$, it follows that either $X_\alpha \in \mathfrak{k}_c$ or $X_\alpha \in \mathfrak{p}_c$, as contended.] This observation has the following consequence. Assume in addition that $\mathfrak{j}_\mathfrak{p}$ is maximal abelian in $\mathfrak{p}$ – then the pair $(\mathfrak{g}_c, \mathfrak{j}_c)$ admits no singular imaginary root. [Invoking the notations employed in this situation, we thus have that $P_+ = \Phi_R^+ \cup \Phi_C^+$, $P_- = \Phi_{CP}^+$.]

[In passing, let us observe that if $\mathfrak{g}$ is only assumed to be a reductive Lie algebra over $\mathbf{R}$, then the definitions and results obtained above are still applicable to $\mathfrak{g}$ via $\bar{\mathfrak{g}}$ $(= [\mathfrak{g}, \mathfrak{g}])$.]

Let τ denote the conjugation of $\mathfrak{g}_c$ with respect to the compact real form $\mathfrak{u} = \mathfrak{k} + \sqrt{-1}\mathfrak{p}$; we recall that $\theta = \sigma\tau = \tau\sigma$. For $X \in \mathfrak{g}_c$, put $\|X\|_\tau^2 = -B(X, \tau(X))$.

Lemma 1.3.2.2 Given $\alpha \in \Phi^+$ we can select $E_\alpha \in \mathfrak{g}_c^\alpha$, $E_{-\alpha} \in \mathfrak{g}_c^{-\alpha}$ in such a way that $\tau(E_\alpha) = -E_{-\alpha}$ and $[E_\alpha, E_{-\alpha}] = \{2/\alpha(H_\alpha)\}H_\alpha$. Moreover, if $\alpha \in \Phi_R^+$, then we can assume that $E_\alpha, E_{-\alpha} \in \mathfrak{g}$.

Proof Because $\overline{\alpha(H)} = -\alpha(\theta(H))$ $(H \in \mathfrak{j})$, it is clear that for $X_\alpha \in \mathfrak{g}_c^\alpha$, $[H, \tau(X_\alpha)] = -\alpha(H)\tau(X_\alpha)(H \in \mathfrak{j})$ – therefore we can assume that $E_{-\alpha} = -\tau(E_\alpha)$. Normalize E_α by the requirement $B(E_\alpha, E_{-\alpha}) = -B(E_\alpha, \tau(E_\alpha)) = \|E_\alpha\|_\tau^2 = 2/\alpha(H_\alpha)$ – then, for $H \in \mathfrak{j}_c$, we have $B([E_\alpha, E_{-\alpha}], H) = B(E_{-\alpha}, [H, E_\alpha]) = \alpha(H)B(E_{-\alpha}, E_\alpha) = \{2/\alpha(H_\alpha)\}\alpha(H) \Rightarrow [E_\alpha, E_{-\alpha}] = \{2/\alpha(H_\alpha)\}H_\alpha$. Now suppose that $\alpha \in \Phi_R^+$. Plainly $[H, \sigma(E_\alpha)] = \alpha(H)\sigma(E_\alpha)$ $(H \in \mathfrak{j})$; since $\|\sigma(X)\|_\tau = \|X\|_\tau$ for any $X \in \mathfrak{g}_c$, it follows that $\sigma(E_\alpha) = c^2E_\alpha$ where c is a unimodular complex number. Therefore, replacing E_α by cE_α, we can assume that $\sigma(E_\alpha) = E_\alpha$ without altering the value of $\|E_\alpha\|_\tau$. But then $E_\alpha \in \mathfrak{g}$ and $E_{-\alpha} = -\tau(E_\alpha) = -\theta(E_\alpha)$ is also in $\mathfrak{g}$. □

Retain the notations of the preceding lemma.

Lemma 1.3.2.3 Fix $\alpha \in \Phi^+$ – then

$$\mathfrak{l}^\alpha = \begin{cases} \mathbf{R}H_\alpha + \mathbf{R}E_\alpha + \mathbf{R}E_{-\alpha} & \text{if } \alpha \in \Phi_R^+, \\ \mathbf{R}\sqrt{-1}H_\alpha + \mathbf{R}\sqrt{-1}(E_\alpha - E_{-\alpha}) + \mathbf{R}(E_\alpha + E_{-\alpha}) & \text{if } \alpha \in \Phi_{SI}^+, \\ \mathbf{R}\sqrt{-1}H_\alpha + \mathbf{R}(E_\alpha - E_{-\alpha}) + \mathbf{R}\sqrt{-1}(E_\alpha + E_{-\alpha}) & \text{if } \alpha \in \Phi_{CP}^+. \end{cases}$$

Proof Our assertion is evident if $\alpha \in \Phi_R^+$. So let us suppose that $\alpha \in \Phi_{SI}^+$ – then, as has been noted above, $E_\alpha \in \mathfrak{p}_c$ and therefore $E_{-\alpha} = -\tau(E_\alpha) = \sigma(E_\alpha)$. Hence $E_\alpha + E_{-\alpha}$ and $\sqrt{-1}(E_\alpha - E_{-\alpha})$ are in $\mathfrak{l}^\alpha$ in this case. If $\alpha \in \Phi_{CP}^+$, we see, in a similar way, that $E_\alpha - E_{-\alpha}$ and $\sqrt{-1}(E_\alpha + E_{-\alpha})$ lie in $\mathfrak{l}^\alpha$. The statements of the lemma are now obvious: □

Let K denote the analytic subgroup of Int $(\mathfrak{g})$ corresponding to $\mathfrak{k}$.

Lemma 1.3.2.4 Suppose that $\alpha \in \Phi_{CP}^+ \cup \Phi_R^+$ – then there exists an element $k \in K$ such that $kH = w_\alpha(H)$ for all $H \in \mathfrak{j}_c$.

Proof Put $H^* = \{2/\alpha(H_\alpha)\}H_\alpha$, $X^* = E_\alpha$, $Y^* = E_{-\alpha}$– then $[H^*, X^*] = 2X^*$, $[H^*, Y^*] = -2Y^*$, $[X^*, Y^*] = H^*$. A simple calculation shows that $\exp\{\pi/2 \operatorname{ad}(X^* - Y^*)\}H^* = -H^*$; on the other hand, it is clear that $\exp\{\pi/2 \operatorname{ad}(X^* - Y^*)\}H = H$ if $\alpha(H) = 0 (H \in \mathfrak{j}_c)$ – therefore $\exp\{\pi/2 \operatorname{ad}(X^* - Y^*)\} = w_\alpha$. Our contention now follows from the fact that $X^* - Y^* \in \mathfrak{k}$ if $\alpha \in \Phi_{CP}^+ \cup \Phi_R^+$. □

1.3.3 Fundamental Cartan Subalgebras

In this number we shall study an important special class of Cartan subalgebras; in a sense, these Cartan subalgebras are just the opposite of those which arise from Iwasawa decompositions (cf. infra).

Let $\mathfrak{g}$ be a semi-simple Lie algebra over $\mathbf{R}$, $\mathfrak{g} = \mathfrak{k} + \mathfrak{p}$ some fixed Cartan decomposition of $\mathfrak{g}$, θ the corresponding Cartan involution. Let ω, $X \mapsto \operatorname{tr}(\operatorname{ad}(X))^2$ $(X \in \mathfrak{g}_c)$, denote the Casimir polynomial of $\mathfrak{g}_c$. For any Cartan subalgebra $\mathfrak{j}$ of $\mathfrak{g}$, set $l_-(\mathfrak{j}) = \sup(\dim(\mathfrak{j}_-))$ where $\mathfrak{j}_-$ runs over all subspaces of $\mathfrak{j}$ on which ω is negative definite; put $l_- = \sup_{\mathfrak{j}} l_-(\mathfrak{j})$ ($\mathfrak{j}$ running through the Cartan subalgebras of $\mathfrak{g}$).

Definition A Cartan subalgebra $\mathfrak{j}$ of $\mathfrak{g}$ is said to be *fundamental* if $l_- = l_-(\mathfrak{j})$.

Since $l_-(x(\mathfrak{j})) = l_-(\mathfrak{j})$ $(x \in \operatorname{Int}(\mathfrak{g}))$, it is clear that if $\mathfrak{j}$ is fundamental, then so is any conjugate of $\mathfrak{j}$.

Lemma 1.3.3.1 The rank of $\mathfrak{k}$ is equal to l_-.

Proof Let $\mathfrak{j}$ be a fundamental Cartan subalgebra of $\mathfrak{g}$; it may be supposed that $\mathfrak{j}$ is θ-stable. Plainly $l_- = l_-(\mathfrak{j}) = \dim(\mathfrak{j}_{\mathfrak{k}})$, ω being negative definite on $\mathfrak{k}$ and positive definite on $\mathfrak{p}$ – hence $l_- \leqslant \operatorname{rank}(\mathfrak{k})$. On the other hand, if $\mathfrak{t}$ is a Cartan subalgebra of $\mathfrak{k}$, extend it to a Cartan subalgebra $\mathfrak{j}$ of $\mathfrak{g}$ (cf. Lemma 1.3.3.2) – then $l_- \geqslant l_-(\mathfrak{j}) \geqslant \dim(\mathfrak{t}) = \operatorname{rank}(\mathfrak{k})$, whence $l_- = \operatorname{rank}(\mathfrak{k})$. □

On the basis of the preceding lemma, it is clear that a θ-stable Cartan subalgebra $\mathfrak{j}$ of $\mathfrak{g}$ is fundamental iff $\mathfrak{j}_{\mathfrak{k}}$ is maximal abelian in $\mathfrak{k}$.

Lemma 1.3.3.2 Let $\mathfrak{t}$ be a Cartan subalgebra of $\mathfrak{k}$ and $\mathfrak{z}(\mathfrak{t})$ the centralizer of $\mathfrak{t}$ in $\mathfrak{g}$ – then $\mathfrak{z}(\mathfrak{t})$ is a θ-stable Cartan subalgebra of $\mathfrak{g}$.

Proof It is clear that $\theta(\mathfrak{z}(\mathfrak{t})) = \mathfrak{z}(\mathfrak{t})$; moreover $\mathfrak{z}(\mathfrak{t})$ is abelian. [Thus let $X, Y \in \mathfrak{z}(\mathfrak{t})$ and set $Z = [X, Y] \in \mathfrak{t}$ – then $-B(Z, Z) = B(Y, [X, Z]) = 0 \Rightarrow Z = 0 \Rightarrow \mathfrak{z}(\mathfrak{t})$ is abelian.] This being the case, let $\mathfrak{j}$ be any Cartan subalgebra of $\mathfrak{g}$ containing $\mathfrak{t}$ – then it is obvious that $\mathfrak{j}$ is contained in $\mathfrak{z}(\mathfrak{t})$ and so $\mathfrak{j} = \mathfrak{z}(\mathfrak{t})$, $\mathfrak{j}$ being maximal abelian in $\mathfrak{g}$. □

Proposition 1.3.3.3 Any two fundamental Cartan subalgebras of $\mathfrak{g}$ are conjugate under $\operatorname{Int}(\mathfrak{g})$.

Proof Let $\mathfrak{j}^1, \mathfrak{j}^2$ be two fundamental Cartan subalgebras of $\mathfrak{g}$; we may assume, without loss of generality, that they are θ-stable. Owing to Lemma 1.3.3.1, both $\mathfrak{j}^1_{\mathfrak{k}}$ and $\mathfrak{j}^2_{\mathfrak{k}}$ are maximal abelian in $\mathfrak{k}$. Consequently, if K denotes the analytic subgroup of $\operatorname{Int}(\mathfrak{g})$ corresponding to $\mathfrak{k}$, then there exists a $k \in K$ such that $\mathfrak{j}^1_{\mathfrak{k}} = k(\mathfrak{j}^2_{\mathfrak{k}})$, whence $\mathfrak{j}^1 = k(\mathfrak{j}^2)$ (Lemma 1.3.3.2). □

The second condition of the following result provides us with a useful alternative characterization of fundamental Cartan subalgebras.

Proposition 1.3.3.4 Let $\mathfrak{j}$ be a Cartan subalgebra of $\mathfrak{g}$ – then the following two conditions on $\mathfrak{j}$ are equivalent:
(i) $\mathfrak{j}$ is fundamental;
(ii) The pair $(\mathfrak{g}_c, \mathfrak{j}_c)$ admits no real roots.

Proof We may assume without loss of generality that $\mathfrak{j}$ is θ-stable.
(i) $\Rightarrow$ (ii). Lemma 1.3.3.2 implies that $\mathfrak{j}_c$ is the centralizer of $\mathfrak{j}_{\mathfrak{k}_c}$ in $\mathfrak{g}_c$; thus the pair $(\mathfrak{g}_c, \mathfrak{j}_c)$ cannot possibly admit a real root (such a root necessarily vanishes identically on $\mathfrak{j}_{\mathfrak{k}_c}$).
(ii) $\Rightarrow$ (i). If (ii) holds, then no root of the pair $(\mathfrak{g}_c, \mathfrak{j}_c)$ can vanish identically on $\mathfrak{j}_{\mathfrak{k}_c}$ – hence $\mathfrak{j}_c$ is the centralizer of $\mathfrak{j}_{\mathfrak{k}_c}$ in $\mathfrak{g}_c$. This means that $\mathfrak{j}_{\mathfrak{k}}$ is maximal abelian in $\mathfrak{k}$ – therefore $\mathfrak{j}$ is fundamental (Lemma 1.3.3.1).
Hence the proposition. ☐

1.3.4 Regular and Semi-Regular Elements in a Reductive Lie Algebra

Let $\mathbf{k}$ be a field of characteristic zero, $\mathfrak{g}$ a Lie algebra over $\mathbf{k}$. If $n = \dim(\mathfrak{g})$, then

$$\det(t - \operatorname{ad}(X)) = \sum_0^n d_i(X)t^i \qquad (X \in \mathfrak{g}),$$

where t is an indeterminate and the d_i are polynomial functions on $\mathfrak{g}$. Note that $d_n = 1$. Let l be the least integer such that $d_l \neq 0$ – then l is called the *rank* of $\mathfrak{g}$. An element $X \in \mathfrak{g}$ is said to be *regular* (in $\mathfrak{g}$) if $d_l(X) \neq 0$ – otherwise X is said to be *singular*. The set of regular elements will be denoted by $\mathfrak{g}'$.

Let $\mathfrak{g}$ be a reductive Lie algebra over $\mathbf{R}$, $\mathfrak{j}$ a Cartan subalgebra of $\mathfrak{g}$. Order the roots Φ of the pair $(\mathfrak{g}_c, \mathfrak{j}_c)$ in some way and set $\pi(H) = \prod_{\alpha>0} \alpha(H) (H \in \mathfrak{j})$. If l is the rank of $\mathfrak{g}$ and $d_l(X)$ is the coefficient of t^l in $\det(t - \operatorname{ad}(X))$, then, in particular, $d_l(H) = (-1)^r \pi(H)^2$ $(H \in \mathfrak{j})$ where $r = \frac{1}{2}(\dim(\mathfrak{g}) - \operatorname{rank}(\mathfrak{g}))$ – thus $\mathfrak{j}' = \mathfrak{j} \cap \mathfrak{g}'$ has only finitely many connected components. We shall call these components the *Weyl chambers* of $\mathfrak{j}$; evidently any given Weyl chamber is open in $\mathfrak{j}$.

[It should be pointed out that the present definition of rank agrees with that introduced in 1.3.1. For it is well-known that the centralizer in $\mathfrak{g}$ of a regular element X is a Cartan subalgebra, the dimension of which is equal to the rank of $\mathfrak{g}$ (in the present sense); moreover all Cartan subalgebras of $\mathfrak{g}$ arise in this fashion – cf. Serre [4, Chapter III] and Proposition 1.3.4.1.]

Notation If $\mathfrak{j}$ is a Cartan subalgebra of $\mathfrak{g}$, put $\mathfrak{g}(\mathfrak{j}) = \bigcup_{x \in G} x(\mathfrak{j}')$ $(G = \operatorname{Int}(\mathfrak{g}))$ where $\mathfrak{j}' = \mathfrak{j} \cap \mathfrak{g}'$.

In what follows it will be useful to bear in mind the obvious fact that if $\bar{X} \in \bar{\mathfrak{g}} = [\mathfrak{g}, \mathfrak{g}]$, $C \in \mathfrak{c}$ then $\bar{X} + C$ is regular in $\mathfrak{g}$ iff $\bar{X}$ is regular in $\bar{\mathfrak{g}}$.

Select a maximal set $\mathfrak{j}^i$ $(1 \leqslant i \leqslant r)$ of Cartan subalgebras of $\mathfrak{g}$ no two of which are conjugate – this may be done by the results in 1.3.1. [Note: This r has no relation to the r above.] Set $\mathfrak{g}(\mathfrak{j}^i) = \bigcup_{x \in G} x(\mathfrak{j}^i)'$ where $G = \text{Int}(\mathfrak{g})$, $(\mathfrak{j}^i)' = \mathfrak{j}^i \cap \mathfrak{g}'$ $(1 \leqslant i \leqslant r)$.

Proposition 1.3.4.1 Retain the above notations – then $\mathfrak{g}'$ is the disjoint union of $\mathfrak{g}(\mathfrak{j}^1), \ldots, \mathfrak{g}(\mathfrak{j}^r)$.

Proof It is clearly enough to consider the case when $\mathfrak{g}$ is semi-simple. Plainly $\mathfrak{g}' \supset \bigcup_{i=1}^r \mathfrak{g}(\mathfrak{j}^i)$ and this latter union is disjoint. So let $X \in \mathfrak{g}'$ – then the centralizer $\mathfrak{g}_X$ of X in $\mathfrak{g}$ is a Cartan subalgebra of $\mathfrak{g}$, X being regular. Therefore there exists an $x \in \text{Int}(\mathfrak{g})$ and an i $(1 \leqslant i \leqslant r)$ such that $\mathfrak{g}_X = x(\mathfrak{j}^i)$, whence $X \in \mathfrak{g}_X \cap \mathfrak{g}' \subset \mathfrak{g}(\mathfrak{j}^i)$. This establishes our proposition. □

Corollary 1.3.4.2 Retain the above notations – then $\mathfrak{g}'$ has but a finite number of connected components.

Proof As was mentioned above, $(\mathfrak{j}^i)'$ has only a finite number of connected components $(1 \leqslant i \leqslant r)$; since $\text{Int}(\mathfrak{g})$ is connected, it is clear that the same holds for each $\mathfrak{g}(\mathfrak{j}^i)$ – hence the corollary. □

In passing, observe that $\mathfrak{g}'$ is open and everywhere dense in $\mathfrak{g}$.

Example Let $\mathfrak{l}_c$ be the semi-simple Lie algebra of dimension 3 spanned over **C** by the three elements H, X, Y which satisfy the relations $[H, X] = 2X$, $[H, Y] = -2Y$, $[X, Y] = H$ – thus $\mathfrak{l}_c$ is isomorphic to $\mathfrak{sl}(2, \mathbf{C})$ (cf. below). Let ω denote the Casimir polynomial of $\mathfrak{l}_c$, i.e. $\omega(Z) = \text{tr}(\text{ad}(Z))^2$ $(Z \in \mathfrak{l}_c)$. Then

$$\omega(tH + cX + dY) = 8(t^2 + cd) \qquad (t, c, d, \in \mathbf{C}).$$

Now $\mathfrak{j}_c = \mathbf{C}H$ is a Cartan subalgebra of $\mathfrak{l}_c$ and there are only two roots α and $-\alpha$ given by $\alpha(H) = 2$; take α as the positive root – then ω coincides with $2\alpha^2$ on $\mathfrak{j}_c$. Hence an element $Z \in \mathfrak{l}_c$ is regular iff $\omega(Z) \neq 0$. Consider $\mathfrak{l} = \mathbf{R}H + \mathbf{R}X + \mathbf{R}Y$. Since $\omega(H) = 8 > 0$, $\mathfrak{l}$ is a non-compact real form of $\mathfrak{l}_c$. The singular set of $\mathfrak{l}$ consists of all points $tH + rX + sY$ $(t, r, s, \in \mathbf{R})$ satisfying the equation $t^2 + rs = 0$. Therefore, in this case, $\mathfrak{l}'$ has three connected components $\mathfrak{l}_1, \mathfrak{l}_2, \mathfrak{l}_3$ given by:

$$\mathfrak{l}_1 : t^2 + rs > 0; \; \mathfrak{l}_2 : t^2 + rs < 0, \; r > s; \; \mathfrak{l}_3 : t^2 + rs < 0, \; s > r.$$

Plainly $H \in \mathfrak{l}_1$ and $X - Y \in \mathfrak{l}_2$. The mapping

$$(tH + cX + dY) \mapsto \begin{pmatrix} t & c \\ d & -t \end{pmatrix} \qquad (t, c, d, \in \mathbf{C})$$

is an isomorphism of $\mathfrak{l}_c$ onto $\mathfrak{sl}(2, \mathbf{C})$. With this identification, $\mathfrak{l}$ becomes the real subalgebra of $\mathfrak{l}_c$ consisting of real matrices. Note that $\omega(Z) = -8\det(Z)$ and the two eigenvalues of Z are $\pm(-\det(Z))^{1/2}$ $(Z \in \mathfrak{l}_c)$. Call an element $Z \in \mathfrak{l}'$ *hyperbolic* or *elliptic* according to whether these eigenvalues are real or pure imaginary – then $\mathfrak{l}_1$ consists of hyperbolic and $\mathfrak{l}_2 \cup \mathfrak{l}_3$ of elliptic elements. Let $L = \text{Int}(\mathfrak{l})$ – then it is easy to see that every hyperbolic element is conjugate under L to tH for some real $t \neq 0$; since $\exp(\pi/2 \, \text{ad}(X - Y))$ maps H

to $-H$, it may be assumed that $t > 0$. On the other hand, every element in $\mathfrak{l}_2$ and $\mathfrak{l}_3$, respectively, is conjugate to $r(X - Y)$ for some $r > 0$ $(r < 0)$. However there exists no $l \in L$ such that $l(X - Y) = -(X - Y)$.

For certain important questions, one must take into account not only the regular elements in $\mathfrak{g}$ but the singular ones as well; usually it is sufficient to consider just the 'best' among these, the so-called semi-regular elements.

Let l be the rank of $\mathfrak{g}$.

Definition An element $\Gamma \in \mathfrak{g}$ is said to be *semi-regular* if:
(i) ad (Γ) is a semi-simple endomorphism of $\mathfrak{g}$;
(ii) The centralizer $\mathfrak{g}_\Gamma$ of Γ in $\mathfrak{g}$ is of dimension $l + 2$.

Let $\Gamma \in \mathfrak{g}$ be semi-regular. Owing to condition (i), the Lie algebra $\mathfrak{g}_\Gamma$ is reductive in $\mathfrak{g}$ (see Proposition 1.3.5.4) – therefore $\mathfrak{l}_\Gamma = [\mathfrak{g}_\Gamma, \mathfrak{g}_\Gamma]$ is semi-simple and $\mathfrak{g}_\Gamma = \mathfrak{l}_\Gamma + \mathfrak{c}_\Gamma$ (direct sum) where $\mathfrak{c}_\Gamma$ is the center of $\mathfrak{g}_\Gamma$. The proof of Proposition 1.3.5.4 below shows that it is possible to select a Cartan subalgebra $\mathfrak{j}_\Gamma$ of $\mathfrak{g}$ containing Γ; thus $\mathfrak{c}_\Gamma \subset \mathfrak{j}_\Gamma \subset \mathfrak{g}_\Gamma$ and so $\mathfrak{j}_\Gamma = \mathfrak{j}_\Gamma \cap \mathfrak{l}_\Gamma + \mathfrak{c}_\Gamma$ where $\mathfrak{j}_\Gamma \cap \mathfrak{l}_\Gamma$ is a Cartan subalgebra of $\mathfrak{l}_\Gamma$. Now $\dim(\mathfrak{g}_\Gamma) = l + 2 = \dim(\mathfrak{j}_\Gamma) + 2$; hence $\dim(\mathfrak{l}_\Gamma) = 2 + \dim(\mathfrak{l}_\Gamma \cap \mathfrak{j}) = 2 + \operatorname{rank}(\mathfrak{l}_\Gamma)$. Since $\mathfrak{l}_\Gamma$ is semi-simple, we deduce that $\dim(\mathfrak{l}_\Gamma) = 3$. Up to isomorphism, there are only two real semi-simple Lie algebras of dimension 3 – one of them is compact while the other is not. This being the case, we shall say that Γ is a semi-regular element of *compact* or *non-compact type* according to whether $\mathfrak{l}_\Gamma$ is compact or not.

[Roughly speaking, the significance of the semi-regular elements is this. In various analytical situations, computations on $\mathfrak{g}$ around Γ can be reduced to similar computations on $\mathfrak{l}_\Gamma$ around zero – presumably it will then be possible to proceed in an elementary way on $\mathfrak{l}_\Gamma$.]

Fix a semi-regular element Γ in $\mathfrak{g}$ of non-compact type; as above, let $\mathfrak{g}_\Gamma$ denote the centralizer of Γ in $\mathfrak{g}$ – then $\mathfrak{g}_\Gamma = \mathfrak{l}_\Gamma + \mathfrak{c}_\Gamma$. Because $\mathfrak{l}_\Gamma$ is non-compact, we can choose a basis $\{H^*, X^*, Y^*\}$ for $\mathfrak{l}_\Gamma$ over $\mathbf{R}$ such that $[H^*, X^*] = 2X^*$, $[H^*, Y^*] = -2Y^*$, $[X^*, Y^*] = H^*$; of course $\mathbf{R}H^*$ and $\mathbf{R}(X^* - Y^*)$ are Cartan subalgebras of $\mathfrak{l}_\Gamma$. Put $\mathfrak{h}_\Gamma^+ = \mathbf{R}H^* + \mathfrak{c}_\Gamma$, $\mathfrak{h}_\Gamma^- = \mathbf{R}(X^* - Y^*) + \mathfrak{c}_\Gamma$ – then $\mathfrak{h}_\Gamma^+$ and $\mathfrak{h}_\Gamma^-$ are Cartan subalgebras of $\mathfrak{g}_\Gamma$ and therefore also of $\mathfrak{g}$. [Recall that $\mathfrak{g}_\Gamma$ is reductive in $\mathfrak{g}$ and of the same rank as $\mathfrak{g}$.] Plainly $\Gamma \in \mathfrak{c}_\Gamma = \mathfrak{h}_\Gamma^+ \cap \mathfrak{h}_\Gamma^-$.

Lemma 1.3.4.3 Let Γ be a semi-regular element in $\mathfrak{g}$ of non-compact type; keeping to the above notations, let L_Γ be the analytic subgroup of Int $(\mathfrak{g})$ corresponding to $\mathfrak{l}_\Gamma$ – then any Cartan subalgebra of $\mathfrak{g}$ containing Γ is conjugate to $\mathfrak{h}_\Gamma^+$ or $\mathfrak{h}_\Gamma^-$ under L_Γ. Moreover $\mathfrak{h}_\Gamma^+$ and $\mathfrak{h}_\Gamma^-$ are not conjugate under any automorphism of $\mathfrak{g}$.

Proof (1) Let $\mathfrak{j}$ be a Cartan subalgebra of $\mathfrak{g}$ containing Γ. Of necessity $\mathfrak{j} \subset \mathfrak{g}_\Gamma$; hence $\mathfrak{j} = \mathfrak{j} \cap \mathfrak{l}_\Gamma + \mathfrak{c}_\Gamma$ where $\mathfrak{j} \cap \mathfrak{l}_\Gamma$ is a Cartan subalgebra of

$\mathfrak{l}_\Gamma$. But then $\mathfrak{j} \cap \mathfrak{l}_\Gamma$ is conjugate under L_Γ to either $\mathbf{R}H^*$ or $\mathbf{R}(X^* - Y^*)$ (see the Example following Corollary 1.3.5.2). As L_Γ leaves $\mathfrak{c}_\Gamma$ pointwise fixed, the first assertion of the lemma is now obvious.

(2) For the proof of the second assertion of our lemma, it may be assumed that $\mathfrak{g}$ is semi-simple. Let ω denote the Casimir polynomial of $\mathfrak{g}$ (i.e. $\omega(X) = \operatorname{tr}(\operatorname{ad}(X))^2$, $X \in \mathfrak{g}$) – then the restriction of ω to $\mathfrak{h}_\Gamma^+$ or $\mathfrak{h}_\Gamma^-$ is non-degenerate. It is a simple matter to verify that $\mathfrak{l}_\Gamma$ and $\mathfrak{c}_\Gamma$ are orthogonal under ω (i.e. under the Killing form). The mapping $\{H^*, X^*, Y^*\} \mapsto \{-H^*, -Y^*, -X^*\}$ defines a Cartan involution θ of $\mathfrak{l}_\Gamma$ which can, moreover, be extended to a Cartan involution of $\mathfrak{g}$; call the extension θ also (cf. Lemma 1.1.5.5) – then we have $\omega(H^*) > 0$ and $\omega(X^* - Y^*) < 0$. Let $s^+(\omega)$, $s^-(\omega)$, $s^0(\omega)$ denote the signatures of ω restricted to $\mathfrak{h}_\Gamma^+$, $\mathfrak{h}_\Gamma^-$, $\mathfrak{c}_\Gamma$, respectively; obviously, then, $s^+(\omega) = 1 + s^0(\omega)$, $s^-(\omega) = -1 + s^0(\omega)$ – hence $s^+(\omega) \neq s^-(\omega)$. But ω is invariant under any automorphism of $\mathfrak{g}$ – thus the signature of the restriction of ω to any conjugate of $\mathfrak{h}_\Gamma^+$ in $\operatorname{Aut}(\mathfrak{g})$ must be $s^+(\omega)$ and so it follows that $\mathfrak{h}_\Gamma^+$ cannot possibly be conjugate to $\mathfrak{h}_\Gamma^-$ in $\operatorname{Aut}(\mathfrak{g})$.

This completes the proof of the lemma. □

Agreeing to retain the above notations and assumptions as regards our semi-regular element Γ, let us consider the automorphism $\nu = \exp\{-\sqrt{-1}(\pi/4)\operatorname{ad}(X^* + Y^*)\}$ of $\mathfrak{g}_c$ ($\pi = 3.14\ldots$). Evidently ν maps $\mathfrak{l}_{\Gamma,c}$ onto itself and leaves the complexification of $\mathfrak{c}_\Gamma$ pointwise fixed; because $\nu(H^*) = \sqrt{-1}(X^* - Y^*)$, it is then clear that $\nu(\mathfrak{h}_{\Gamma,c}^+) = \mathfrak{h}_{\Gamma,c}^-$. In the obvious way, ν maps the roots of the pair $(\mathfrak{g}_c, \mathfrak{h}_{\Gamma,c}^+)$ onto those of the pair $(\mathfrak{g}_c, \mathfrak{h}_{\Gamma,c}^-)$. The following result is an immediate consequence of these observations.

Lemma 1.3.4.4 There exists exactly one root α_+ of the pair $(\mathfrak{g}_c, \mathfrak{h}_{\Gamma,c}^+)$ such that $\alpha_+(H^*) = 2$ and $\alpha_+ | \mathfrak{c}_\Gamma = 0$; α_+ is real while $\alpha_- = \alpha_+^\nu$ is a singular imaginary root of the pair $(\mathfrak{g}_c, \mathfrak{h}_{\Gamma,c}^-)$ – we have $\alpha_-(X^* - Y^*) = -2\sqrt{-1}$, $\alpha_- | \mathfrak{c}_\Gamma = 0$. If σ_{α_+} is the hyperplane in $\mathfrak{h}_\Gamma^+$ given by the equation $\alpha_+ = 0$, σ_{α_-} the hyperplane in $\mathfrak{h}_\Gamma^-$ defined by $\alpha_- = 0$, then $\sigma_{\alpha_+} = \sigma_{\alpha_-} = \mathfrak{c}_\Gamma$.

The following remarks should help to illustrate the preceding discussion.

Let us suppose that $\mathfrak{g}$ is semi-simple; let θ be a Cartan involution of $\mathfrak{g}$, $\mathfrak{j}$ a θ-stable Cartan subalgebra of $\mathfrak{g}$ – we shall then agree to subscribe to the notations introduced in 1.3.2. Fix a semi-regular element Γ in $\mathfrak{j}$ – then there exists exactly one root α_Γ (say) in Φ^+ such that $\alpha_\Gamma(\Gamma) = 0$ and so it is clear that the centralizer $\mathfrak{g}_{\Gamma,c}$ of Γ in $\mathfrak{g}_c$ is given by $\mathfrak{g}_{\Gamma,c} = \mathfrak{j}_c + \mathfrak{g}_c^{\alpha_\Gamma} + \mathfrak{g}_c^{-\alpha_\Gamma}$. Evidently $\mathfrak{g}_\Gamma$ is θ-stable; in the case at hand $\mathfrak{c}_\Gamma$ consists of those H in $\mathfrak{j}$ where $\alpha_\Gamma(H) = 0$; moreover it is easy to see that α_Γ is either real or imaginary (but not complex). Choose E_{α_Γ}, $E_{-\alpha_\Gamma}$ in accordance with Lemma 1.3.2.3; if $\alpha_\Gamma \in \Phi_{CP}^+$, then $\mathfrak{l}_\Gamma = \mathbf{R}\sqrt{-1}H_{\alpha_\Gamma} + \mathbf{R}(E_{\alpha_\Gamma} - E_{-\alpha_\Gamma}) +$

$\mathbf{R}\sqrt{-1}(E_{\alpha_\Gamma} + E_{-\alpha_\Gamma})$ (hence Γ is of the compact type in this situation); if $\alpha_\Gamma \in \Phi_R^+ \cup \Phi_{SI}^+$, then

$$\mathfrak{l}_\Gamma = \begin{cases} \mathbf{R}H_{\alpha_\Gamma} + \mathbf{R}E_{\alpha_\Gamma} + \mathbf{R}E_{-\alpha_\Gamma}) & (\alpha_\Gamma \in \Phi_R^+) \\ \mathbf{R}\sqrt{-1}H_{\alpha_\Gamma} + \mathbf{R}\sqrt{-1}(E_{\alpha_\Gamma} - E_{-\alpha_\Gamma}) + \mathbf{R}(E_{\alpha_\Gamma} + E_{-\alpha_\Gamma}) & (\alpha_\Gamma \in \Phi_{SI}^+) \end{cases}$$

and thus here Γ is of the non-compact type. This being so, let us assume now that $\alpha_\Gamma \in \Phi_R^+ \cup \Phi_{SI}^+$; put $H^* = \{2/\alpha_\Gamma(H_{\alpha_\Gamma})\}H_{\alpha_\Gamma}$, $X^* = E_{\alpha_\Gamma}$, $Y^* = E_{-\alpha_\Gamma}$ if $\alpha_\Gamma \in \Phi_R^+$ and $H^* = \sqrt{-1}(E_{\alpha_\Gamma} - E_{-\alpha_\Gamma})$, $X^* = 2^{-1}\{E_{\alpha_\Gamma} + E_{-\alpha_\Gamma} + \sqrt{-1}\{2/\alpha_\Gamma(H_{\alpha_\Gamma})\}H_{\alpha_\Gamma}\}$, $Y^* = 2^{-1}\{E_{\alpha_\Gamma} + E_{-\alpha_\Gamma} - \sqrt{-1}\{2/\alpha_\Gamma(H_{\alpha_\Gamma})\}H_{\alpha_\Gamma}\}$ if $\alpha_\Gamma \in \Phi_{SI}^+$ – then, in both cases, $\mathfrak{l}_\Gamma = \mathbf{R}H^* + \mathbf{R}X^* + \mathbf{R}Y^*$ with $[H^*, X^*] = 2X^*$, $[H^*, Y^*] = -2Y^*$, $[X^*, Y^*] = H^*$ (so that $\mathfrak{l}_\Gamma \cap \mathfrak{k} = \mathbf{R}(X^* - Y^*)$). We may take $\mathfrak{h}_\Gamma^+ = \mathbf{R}H^* + \mathfrak{c}_\Gamma$, $\mathfrak{h}_\Gamma^- = \mathbf{R}(X^* - Y^*) + \mathfrak{c}_\Gamma$ – hence $\mathfrak{h}_\Gamma^\pm$ is a θ-stable Cartan subalgebra of $\mathfrak{g}$.

Remark Let $\mathfrak{g}'$ denote the set of those elements in $\mathfrak{g}$ which are either regular or semi-regular – then it can be shown that $\mathfrak{g}'$ is connected. On the other hand, a given semi-regular element Γ of $\mathfrak{g}$ can lie in the closure of at most three distinct connected component of $\mathfrak{g}'$. [As we shall have no real need of these facts, we omit the proofs; the details may be found in Harish-Chandra [16].]

1.3.5 Semi-Simple and Nilpotent Elements in a Reductive Lie Algebra

Let $\mathfrak{g}$ be a reductive Lie algebra over $\mathbf{R}$. We shall record, in this number, some well-known results concerning semi-simple and nilpotent elements in $\mathfrak{g}$ – those facts with which the reader is assumed to be familiar will be stated without proof.

An element $X \in \mathfrak{g}$ is called *semi-simple* if ad (X) is semi-simple (thus ad (X) is diagonalizable over $\mathbf{C}$); an element $X \in \bar{\mathfrak{g}} = [\mathfrak{g}, \mathfrak{g}]$ is said to be *nilpotent* if ad (X) is nilpotent. The set of semi-simple elements in $\mathfrak{g}$ will be denoted by $\mathscr{S}$, the set of nilpotent elements by $\mathscr{N}$ – plainly $\mathscr{S} \cap \mathscr{N} = \{0\}$.

For the proof of the following proposition see, for instance, Bourbaki [4, p.79].

Proposition 1.3.5.1 Every element X in $\mathfrak{g}$ can be written in one and only one way in the form $X = X_s + X_n$ with $X_s \in \mathscr{S}$, $X_n \in \mathscr{N}$ and $[X_s, X_n] = \{0\}$. Moreover every element $Y \in \mathfrak{g}$ which commutes with X also commutes with X_s and X_n.

Fix an element X in $\mathfrak{g}$ and write, per the preceding proposition, $X = X_s + X_n$ – then X_s is called the *semi-simple component* of X while X_n is referred to as the *nilpotent component* of X.

Our definitions give preference to the adjoint representation of $\mathfrak{g}$; accordingly it is important to bear in mind the following result (cf. Bourbaki [4, p.78])

Lemma 1.3.5.2 Let E be a finite dimensional semi-simple $\mathfrak{g}$ – module (over $\mathbf{R}$), $X \mapsto U(X)$ the corresponding representation ($X \in \mathfrak{g}$). If $X \in \mathfrak{g}$ is semi-simple (nilpotent), then it is the same with the endomorphism $U(X)$.

We shall refer the reader to Jacobson [3, p.100] for the proof of the following important fact.

Proposition 1.3.5.3 (Jacobson-Morosow) Let $\mathfrak{g}$ be a semi-simple Lie algebra over a field $\mathbf{k}$ of characteristic zero. If $X \neq 0$ is any nilpotent element in $\mathfrak{g}$ (in the obvious sense), then there exist elements $H, Y \in \mathfrak{g}$ such that

$$[H, X] = 2X, [H, Y] = -2Y, [X, Y] = H.$$

Let $\mathfrak{h}$ be a subalgebra of $\mathfrak{g}$ which is reductive in $\mathfrak{g}$. By definition the representation $X \mapsto \mathrm{ad}_{\mathfrak{g}}(X)$ of $\mathfrak{h}$ on $\mathfrak{g}$ is semi-simple; owing to Lemma 1.3.5.2, an element $X \in \mathfrak{h}$ is semi-simple (or nilpotent) in $\mathfrak{h}$ iff the same holds in $\mathfrak{g}$. Let $X \in \mathfrak{h}$ – then it is clear that the semi-simple (or nilpotent) component of X in $\mathfrak{h}$ is the same as in $\mathfrak{g}$.

Proposition 1.3.5.4 Fix a semi-simple element Γ in $\mathscr{S}$; let $\mathfrak{g}_\Gamma$ denote the centralizer of Γ in $\mathfrak{g}$, $\mathfrak{g}^\Gamma$ the space $[\Gamma, \mathfrak{g}]$ (hence $\mathfrak{g} = \mathfrak{g}_\Gamma + \mathfrak{g}^\Gamma$ (direct sum)) – then the Lie algebra $\mathfrak{g}_\Gamma$ is reductive in $\mathfrak{g}$ and rank $(\mathfrak{g}_\Gamma)$ = rank $(\mathfrak{g})$.

Proof It may be assumed that $\mathfrak{g}$ is semi-simple. We first claim that there exists a Cartan subalgebra of $\mathfrak{g}$ to which Γ belongs. To see this consider the map $\phi\colon \mathfrak{g}_\Gamma \times \mathfrak{g}^\Gamma \to \mathfrak{g}$ given by

$$\phi(X, Y) = e^{\mathrm{ad}(Y)}(X + \Gamma) \quad (X \in \mathfrak{g}_\Gamma,\ Y \in \mathfrak{g}^\Gamma).$$

An elementary computation shows that the differential of ϕ at (0, 0) is the bijective linear map $(X, Y) \mapsto X + [Y, \Gamma]$ – hence the intersection of $\mathfrak{g}_\Gamma$ with $\mathfrak{g}'$ is not empty and so, the centralizer of a regular element being a Cartan subalgebra, the claim follows. Let θ denote the Cartan involution stabilizing $\mathfrak{j}$, $\mathfrak{j}$ a Cartan subalgebra of $\mathfrak{g}$ to which Γ belongs. In order to show that $\mathfrak{g}_\Gamma$ is reductive in $\mathfrak{g}$, it suffices to show that $\mathfrak{g}_\Gamma$ is θ-stable (cf. Corollary 1.1.5.4). Clearly $\mathfrak{g}_{\Gamma,c} = \mathfrak{j}_c + \sum_{\alpha \in \Phi_\Gamma} \mathfrak{g}_c^\alpha$ where Φ_Γ denotes the subset of Φ (= the roots of the pair $(\mathfrak{g}_c, \mathfrak{j}_c)$) consisting of those α with the property that $\alpha(\Gamma) = 0$; since the map $\alpha \mapsto \alpha^\theta$ is a permutation of Φ ($\alpha^\theta(H) = \alpha(\theta H)$, $H \in \mathfrak{j}_c$), the θ-stability of $\mathfrak{g}_\Gamma$ will follow when it is shown that Φ_Γ is θ-stable. Write $\Gamma = \Gamma_I + \Gamma_R$ where $\Gamma_I \in \mathfrak{j}_I$ (= $\mathfrak{j}_\mathfrak{k}$) and $\Gamma_R \in \mathfrak{j}_R$ (= $\mathfrak{j}_\mathfrak{p}$); let $\alpha \in \Phi_\Gamma$ – then $\alpha(\Gamma) = 0$, whence $\alpha(\Gamma_I) = \alpha(\Gamma_R) = 0$ (since $\alpha \mid \mathfrak{j}_I$ is pure imaginary while $\alpha \mid \mathfrak{j}_R$ is real). It therefore follows that $\alpha^\theta(\Gamma) = 0$ too; thus Φ_Γ is θ-stable and so $\mathfrak{g}_\Gamma$ itself is θ-stable, hence reductive in $\mathfrak{g}$. □

Let G be a connected Lie group with Lie algebra $\mathfrak{g}$ – then G operates on $\mathfrak{g}$ via the adjoint representation. [Note too that Ad (G) = Int $(\mathfrak{g})$ =

Int ($\bar{\mathfrak{g}}$); thus if C denotes the analytic subgroup of G with Lie algebra $\mathfrak{c}$, Z the center of G, and $\bar{Z}$ the center of $\bar{G}$, then Ad $(G) = \text{Int}(\mathfrak{g}) = G/Z = (\bar{G}C/C)/(\bar{Z}C/C) = \bar{G}/\bar{Z} = \text{Int}(\bar{\mathfrak{g}})$.] Given an $X \in \mathfrak{g}$, it is natural to ask: When is the orbit $G \cdot X = \{x \cdot X : x \in G\}$ a closed subset of $\mathfrak{g}$? Here is the answer.

Proposition 1.3.5.5 (Borel-Harish-Chandra) Let $X \in \mathfrak{g}$ – then the orbit $G \cdot X$ is closed in $\mathfrak{g}$ iff X is semi-simple.

Proof Suppose that $\Gamma \in \mathfrak{g}$ is semi-simple; it will be sufficient to verify that the orbit $\{x \cdot \Gamma : x \in \text{Int}(\mathfrak{g}_c)\}$ is closed in $\mathfrak{g}_c$. [The proof of this fact depends on a standard result in the theory of algebraic groups, the precise formulation of which may be found in the Appendix to this number. One should keep in mind that $G_c = \text{Int}(\mathfrak{g}_c)$ is a complex algebraic group defined over **R**. The subset of G_c consisting of the set of all points in G_c which are rational over **R** is a real algebraic group whose connected component of the identity for the usual topology is Int ($\mathfrak{g}$).] Let P_Γ denote the set of all $X \in \mathfrak{g}_c$ such that ad (X) has the same characteristic polynomial as ad (Γ) and which, in addition, annihilates the minimal polynomial of ad (Γ). This is an algebraic subset of $\mathfrak{g}_c$, stable under G_c. The minimal polynomial of ad (X) $(X \in P_\Gamma)$ divides the minimal polynomial of ad (Γ), hence has only simple factors and so X is also semi-simple (i.e. ad (X) is semi-simple). Consider now the centralizer $\mathsf{G}_{X,c}$ of X in G_c; its Lie algebra is the centralizer of X in $\mathfrak{g}_c$ and so the dimension of $\mathsf{G}_{X,c}$ is equal to the multiplicity of the eigenvalue zero of ad (X). But ad (X) has the same eigenvalues as ad (Γ) – hence $\dim(\mathsf{G}_{X,c}) = \dim(\mathsf{G}_{\Gamma,c})$. The G_c-orbit of X, the dimension of which is equal to $\dim(G_c) - \dim(\mathsf{G}_{X,c})$, has, therefore, the same dimension as the G_c-orbit of Γ, so P_Γ is a disjoint union of orbits of the same dimension. However, in the present case, the boundary of an orbit is a union of orbits of strictly smaller dimension (cf. Borel [4, §15]) and so it follows that the G_c-orbit of X is closed for every $X \in P_\Gamma$, in particular for $X = \Gamma$. As for the converse assertion of our proposition, it is an immediate consequence of Proposition 1.3.5.7 below. □

Lemma 1.3.5.6 Let $\mathcal{O}$ be a neighborhood of zero in $\bar{\mathfrak{g}}$ and $X \in \mathcal{N}$ – then there exists an $x \in G$ such that $x \cdot X \in \mathcal{O}$.

Proof Assume that X is not zero; by Proposition 1.3.5.3 there exists an H in $\bar{\mathfrak{g}}$ such that $[H, X] = 2X$ – but then $\text{Ad}(\exp - tH)X = e^{-t\,\text{ad}(H)}X = e^{-2t}X$ $(t \in \mathbf{R})$ and so $\text{Ad}(\exp - tH)X \in \mathcal{O}$ for large positive t. □

Proposition 1.3.5.7 Let $X \in \mathfrak{g}$, $X = X_s + X_n$ the decomposition of X per Proposition 1.3.5.1 $(X_s \in \mathcal{S}, X_n \in \mathcal{N})$ – then $X_s \in (G \cdot X)^{cl}$ (the closure being taken in $\mathfrak{g}$).

Proof Let $\mathfrak{g}_{X_s}$ denote the centralizer of X_s in G, G_{X_s} the analytic subgroup of G corresponding to $\mathfrak{g}_{X_s}$ – then $\mathfrak{g}_{X_s}$ is reductive in $\mathfrak{g}$ (cf. Proposition 1.3.5.4). Moreover $X_n \in \mathfrak{g}_{X_s}$ and so, in view of earlier remarks, is also nilpotent in $\mathfrak{g}_{X_s}$. Now apply the preceding lemma to X_n and $\mathfrak{g}_{X_s}$ – thus we get

$$X_s \in X_s + (G_{X_s} \cdot X_n)^{cl} = (G_{X_s} \cdot X)^{cl} \subset (G \cdot X)^{cl}$$

as we wished to prove. ☐

Remark Let us suppose that $\mathfrak{g}$ is semi-simple. Consider the set $\mathcal{N}$ of nilpotent elements in $\mathfrak{g}$; it is obvious that $x \cdot \mathcal{N} = \mathcal{N} (x \in G)$ and so G operates on $\mathcal{N}$. It follows from the work of Kostant [2] and Borel-Harish-Chandra [1. p.495] that $\mathcal{N}$ is the union of a *finite* number of G-orbits. To get a better idea of what is going on here, pass to the complexification $\mathfrak{g}_c$. By a TDS in $\mathfrak{g}_c$ we shall understand a three dimensional simple subalgebra of $\mathfrak{g}_c$; of course every TDS is isomorphic to $\mathfrak{sl}(2, \mathbf{C})$. Under the action of Int $(\mathfrak{g}_c)$, the set of all TDS's in $\mathfrak{g}_c$ break up into conjugacy classes. Concerning these conjugacy classes, Kostant [2, p.988] proved: The conjugacy classes of TDS's in $\mathfrak{g}_c$ are in a natural one-to-one correspondence with the orbits (under Int $(\mathfrak{g}_c)$) of non-zero nilpotent elements in $\mathfrak{g}_c$ – this correspondence is implemented by associating with the conjugacy class of $\mathfrak{l}$, $\mathfrak{l}$ a TDS in $\mathfrak{g}_c$, the orbit of any non-zero nilpotent element in $\mathfrak{l}$ (i.e. two TDS's $\mathfrak{l}_1$ and $\mathfrak{l}_2$ are conjugate iff X_1 and X_2 are conjugate where $X_i \in \mathfrak{l}_i$ $(i = 1, 2)$ are non-zero nilpotent elements). [For a short qualitative proof of the fact that, in the complex case, there are but finitely many conjugacy classes of nilpotent elements, see Richardson [1].] Turning back to the real case, to establish that $\mathcal{N}$ is the union of a finite number of G-orbits, one must appeal to the principle in algebraic groups alluded to already during the proof of Proposition 1.3.5.5 (see the Appendix infra). [At the time of this writing, no natural 'parameterization' of the G-orbits in $\mathcal{N}$ appears to be known.]

Appendix Part of the proof of Proposition 1.3.5.5 depends on a certain result from the theory of algebraic groups; we shall give here its precise statement after agreeing upon some terminology.

A *complex algebraic (matrix) group* is a subgroup $\mathbf{G}_c$ of $\mathbf{GL}(n, \mathbf{C})$ which consists of all invertible matrices $x = (x_{ij})$ whose coefficients annihilate some set of polynomials $\{p_\alpha(X_{11}, \ldots, X_{nn})\}$ with complex coefficients. The group $\mathbf{G}_c$ is said to be defined over a subfield $\mathbf{k}$ of $\mathbf{C}$ if the p_α may be chosen so as to have coefficients in $\mathbf{k}$; if $\mathbf{A}$ is a subring of $\mathbf{C}$, then $\mathbf{G}_\mathbf{A}$ will denote the subgroup of elements of $\mathbf{G}_c$ whose coefficients are in $\mathbf{A}$ and whose determinant is a unit of $\mathbf{A}$ – note, in particular, that $\mathbf{G}_\mathbf{R}$ is a real Lie group.

Let $\mathbf{G}_c$ be a complex algebraic group, E a finite dimensional vector space over $\mathbf{C}$. A *rational representation* U of $\mathbf{G}_c$ on E is a morphism $U: \mathbf{G}_c \to \mathbf{GL}(E)$; if $\mathbf{G}_c$ and $\mathbf{GL}(E)$ are defined over $\mathbf{k}$, then U is said to be defined over $\mathbf{k}$ provided U is a $\mathbf{k}$-morphism.

For the proof of the following lemma (which depends ultimately upon the results of Whitney [1]), the reader is referred to Borel and Harish-Chandra [1, p.495].

Lemma Let $\mathbf{G}_c$ be a connected complex algebraic group defined over $\mathbf{R}$, $U: \mathbf{G}_c \to \mathbf{GL}(E)$ a rational representation of $\mathbf{G}_c$ defined over $\mathbf{R}$, $Q = \mathbf{G}_c \cdot a$ $(a \in E)$ a $\mathbf{G}_c$-orbit in E – then $Q_\mathbf{R} = Q \cap E_\mathbf{R}$ is the union of a finite number of $G_\mathbf{R}$-orbits ($G_\mathbf{R}$ the topological component of the identity of $\mathbf{G}_\mathbf{R}$) which are closed if Q is so.

[The fact that U is defined over $\mathbf{R}$ presupposes that E admits an $\mathbf{R}$-structure, i.e. E contains a vector $\mathbf{R}$-subspace $E_{\mathbf{R}}$ (say) such that $E = E_{\mathbf{R}} \otimes_{\mathbf{R}} \mathbf{C}$.]

1.4 Cartan Subgroups

1.4.1 Structure Theorems

Let (G, K) be a pair verifying our general assumptions (1.1.5). In this section an important class of subgroups of G will be studied, the so-called Cartan subgroups. The present number will be devoted to establishing their basic properties.

Definition Let $\mathfrak{j}$ be a Cartan subalgebra of $\mathfrak{g}$ – then the centralizer J of $\mathfrak{j}$ in G is called the *Cartan subgroup* of G associated with $\mathfrak{j}$.

Let $\mathfrak{j}$ be a Cartan subalgebra of $\mathfrak{g}$, J the Cartan subgroup of G associated with $\mathfrak{j}$ – then J is closed in G and the identity component J of J is the analytic subgroup of G corresponding to $\mathfrak{j}$.

Examples (1) Let $G = \mathbf{SL}(2, \mathbf{R})$, $\mathfrak{g} = \mathfrak{sl}(2, \mathbf{R})$; pick a basis X, Y, H for $\mathfrak{sl}(2, \mathbf{R})$ such that $[X, Y] = H$, $[H, X] = 2X$, $[H, Y] = -2Y$ – thus take

$$X = \begin{pmatrix} 0 & 1 \\ 0 & 0 \end{pmatrix}, \quad Y = \begin{pmatrix} 0 & 0 \\ 1 & 0 \end{pmatrix}, \quad H = \begin{pmatrix} 1 & 0 \\ 0 & -1 \end{pmatrix}.$$

Let $\mathfrak{j}^1 = \{tH: t \in \mathbf{R}\}$; let J^1 be the Cartan subgroup of G associated with $\mathfrak{j}^1$ – then we have

$$\mathrm{J}^1 = \left\{\begin{pmatrix} e^t & 0 \\ 0 & e^{-t} \end{pmatrix}, \begin{pmatrix} -e^t & 0 \\ 0 & -e^{-t} \end{pmatrix} : t \in \mathbf{R}\right\}.$$

Hence J^1 is *not* connected. On the other hand, let $\mathfrak{j}^2 = \{t(X - Y): t \in \mathbf{R}\}$ – then the Cartan subgroup J^2 of G associated with $\mathfrak{j}^2$ is $\mathbf{SO}(2)$ which is connected (cf. Proposition 1.4.1.4 infra).

(2) (Langlands) Cartan subgroups need *not* be abelian. Thus take $\mathfrak{g} = \mathfrak{sl}(3, \mathbf{R})$,

$$\mathfrak{j} = \left\{\begin{pmatrix} X & 0 & 0 \\ 0 & Y & 0 \\ 0 & 0 & Z \end{pmatrix} : X + Y + Z = 0\right\},$$

and consider the centralizer J of $\mathfrak{j}$ in G, G the universal covering group of $\mathbf{SL}(3, \mathbf{R})$. The group $\mathbf{SU}(2)$ is a two-fold covering group of $\mathbf{SO}(3)$, the covering map being obtained by sending an element $u \in \mathbf{SU}(2)$ to the rotation $\phi(u)$ which sends (x_1, x_2, x_3) to (x_1', x_2', x_3') where

$$\begin{pmatrix} x_1' & x_2' + \sqrt{-1}x_3' \\ x_2' - \sqrt{-1}x_3' & -x_1' \end{pmatrix} = u \begin{pmatrix} x_1 & x_2 + \sqrt{-1}x_3 \\ x_2 - \sqrt{-1}x_3 & -x_1 \end{pmatrix} u^*$$

(u^* the adjoint of u).

The group $\mathbf{SO}(3)$ contains four diagonal elements with ± 1 on the diagonal and they all centralize $\mathfrak{j}$. The inverse image of these four elements in G consists of the eight elements

$$\left\{\pm\begin{pmatrix} 1 & 0 \\ 0 & 1 \end{pmatrix}, \pm\begin{pmatrix} \sqrt{-1} & 0 \\ 0 & -\sqrt{-1} \end{pmatrix}, \pm\begin{pmatrix} 0 & 1 \\ -1 & 0 \end{pmatrix}, \pm\begin{pmatrix} 0 & \sqrt{-1} \\ \sqrt{-1} & 0 \end{pmatrix}\right\}$$

and this is not abelian. . . . Note, too, that in the present case the toroidal part of our Cartan subalgebra $\mathfrak{j}$ is trivial, that is, $\mathfrak{j} = \mathfrak{a}_\mathfrak{p}$ is maximal abelian in $\mathfrak{p}$; furthermore, the non-abelian group above coincides here with $\mathsf{Z}(A_\mathfrak{p})$ (cf. Lemma 1.1.3.8).

In passing let us mention the following result which we shall not stop to establish since it won't be used in the sequel.

Proposition 1.4.1.1 The following conditions on a subgroup J of G are equivalent:

(i) J is a Cartan subgroup of G;

(ii) J is a maximal nilpotent subgroup with the property that every subgroup of finite index in J has finite index in its normalizer (in G).

[The latter condition on J is known as Chevalley's Criterion.]

Let $\mathfrak{j}$ be a θ-stable Cartan subalgebra of $\mathfrak{g}$, J the Cartan subgroup of G associated with $\mathfrak{j}$; put $\mathsf{J}_K = \mathsf{J} \cap K$, $J_\mathfrak{p} = \exp(\mathfrak{j} \cap \mathfrak{p}) = \exp(\mathfrak{j}_\mathfrak{p})$.

Proposition 1.4.1.2 Retain the above notations – then $\mathsf{J} = \mathsf{J}_K J_\mathfrak{p}$.

[The statement of our proposition is an immediate consequence of Lemma 1.1.3.7.]

Let $\mathfrak{j}$ be a θ-stable Cartan subalgebra of $\mathfrak{g}$, J the Cartan subgroup of G associated with $\mathfrak{j}$; let J_K denote the analytic subgroup of G corresponding to $\mathfrak{j}_\mathfrak{k}$ – then $J = J_K J_\mathfrak{p}$ (J the identity component of J). Most of the technical difficulties which arise in the study of Cartan subgroups may be traced directly to the fact that $J_K \neq \mathsf{J}_K$ in general, i.e. to the fact that J_K fails to be connected in general. On the other hand, one can, at least, write down the structure of the group of components of J. Thus define $\mathsf{Z}(J_\mathfrak{p})$ to be the set of all $j \in \mathsf{J}_K$ such that $i_c(j) \in \exp(\sqrt{-1}\mathfrak{j}_\mathfrak{p})$ for every complexification G_c of G. [Since $\bar{G} \cap C$ is finite (C the analytic subgroup of G with Lie algebra $\mathfrak{c}$), G always admits a complexification; cf. 1.1.5] Evidently $\mathsf{Z}(J_\mathfrak{p})$ is a finite subgroup of J_K.

Proposition 1.4.1.3 (Osborne-Rader) Retain the above notations – then $\mathsf{J}_K = \mathsf{Z}(J_\mathfrak{p}) J_K$.

Proof Let $i_c\colon G \to G_c$ be a complexification of G; let $\mathsf{Z}_{i_c}(J_\mathfrak{p})$ denote the set of those $j \in \mathsf{J}_K$ such that $i_c(j) \in \exp(\sqrt{-1}\mathfrak{j}_\mathfrak{p})$; evidently $\mathsf{Z}_{i_c}(J_\mathfrak{p})$ is a finite subgroup of J_K – therefore there exists a finite number $i_c^1, \ldots, i_c^r$ of complexifications of G such that $\mathsf{Z}(J_\mathfrak{p}) = \bigcap_{m=1}^r \mathsf{Z}_{i_c^m}(J_\mathfrak{p})$. Now it is obvious that $i_c(J_K) i_c(\mathsf{Z}_{i_c}(J_\mathfrak{p})) = i_c(\mathsf{J}_K)$ for any complexification i_c of G – thus our proposition will follow if we can produce a complexification i_c of G with the property that $\mathsf{Z}_{i_c}(J_\mathfrak{p}) = \mathsf{Z}(J_\mathfrak{p})$. So let $G_c^1, \ldots, G_c^r$ be the groups corresponding to $i_c^1, \ldots, i_c^r$ and consider the product $\prod_1^r G_c^m$. The map $X \mapsto (X, \ldots, X)$ (m times) identifies $\mathfrak{g}_c$ with a subalgebra of $\mathfrak{g}_c \times \cdots \times \mathfrak{g}_c$ (m times); denoting the corresponding analytic subgroup of $\prod_1^r G_c^m$ by G_c, we see without difficulty that the range of the map i_c, $x \mapsto (i_c^1(x), \ldots, i_c^m(x))(x \in G)$, is contained in G_c – but then it is clear that $\mathsf{Z}_{i_c}(J_\mathfrak{p}) \subset \bigcap_{m=1}^r \mathsf{Z}_{i_c^m}(J_\mathfrak{p})$, whence $\mathsf{Z}_{i_c}(J_\mathfrak{p}) = \mathsf{Z}(J_\mathfrak{p})$. □

The preceding proposition tells us, of course, that $\mathsf{J} = \mathsf{Z}(J_{\mathfrak{p}})J$.

Remark Suppose that G is imbedded in G_c with injection i_c and G_c is simply connected – then clearly $\mathsf{Z}(J_{\mathfrak{p}}) = K \cap \exp(\sqrt{-1}\,\mathfrak{j}_{\mathfrak{p}})$.

Example Take $G = \mathbf{SL}(2, \mathbf{R})$, $K = \mathbf{SO}(2)$, $G_c = \mathbf{SL}(2, \mathbf{C})$, $\mathfrak{j} = \{t\left(\begin{smallmatrix} 1 & 0 \\ 0 & -1 \end{smallmatrix}\right): t \in \mathbf{R}\}$ – then $G \subset G_c$ and G_c is simply connected. The Cartan subgroup of G associated with $\mathfrak{j}$ is given by

$$\mathsf{J} = \left\{\begin{pmatrix} e^t & 0 \\ 0 & e^{-t} \end{pmatrix}, \begin{pmatrix} -e^t & 0 \\ 0 & -e^{-t} \end{pmatrix} : t \in \mathbf{R}\right\}$$

and so $\mathsf{J}_K = \mathsf{J} \cap K$ is discrete and consists of two elements: $\left(\begin{smallmatrix} 1 & 0 \\ 0 & 1 \end{smallmatrix}\right)$, $\left(\begin{smallmatrix} -1 & 0 \\ 0 & -1 \end{smallmatrix}\right)$. Moreover

$$\begin{pmatrix} -1 & 0 \\ 0 & -1 \end{pmatrix} = \begin{pmatrix} e^{\pi\sqrt{-1}} & 0 \\ 0 & e^{-\pi\sqrt{-1}} \end{pmatrix} \in K \cap \exp(\sqrt{-1}\,\mathfrak{j}_{\mathfrak{p}}).$$

Hence $\mathsf{Z}(J_{\mathfrak{p}}) = \mathsf{J}_K$ in this case.

As has been noted above, the Cartan subgroups of G need not be connected; accordingly the following result is of interest.

Proposition 1.4.1.4 Let $\mathfrak{j}$ be a θ-stable Cartan subalgebra of $\mathfrak{g}$, J the Cartan subgroup of G associated with $\mathfrak{j}$; suppose that $\mathfrak{j}$ is fundamental – then J is connected and abelian.

Proof To say that $\mathfrak{j}$ is fundamental, means, as we know, that the pair $(\mathfrak{g}_c, \mathfrak{j}_c)$ admits no real root, i.e. that the pair $(\bar{\mathfrak{g}}_c, \bar{\mathfrak{j}}_c)$ admits no real root; in turn this entails that $\bar{\mathfrak{j}}_{\mathfrak{k}}$ is maximal abelian in $\bar{\mathfrak{k}}$ and so $\mathfrak{j}_{\mathfrak{k}} = \bar{\mathfrak{j}}_{\mathfrak{k}} + \mathfrak{c}_{\mathfrak{k}}$ is maximal abelian in $\mathfrak{k}$ – therefore J_K is a maximal torus in K. Because the elements of J_K all commute with the elements of J_K, it follows for familiar reasons that $\mathsf{J}_K = J_K$. In view of Proposition 1.4.1.2, the assertion of the present proposition is now clear. □

Let $\mathfrak{g}$ be a Lie algebra over $\mathbf{R}$, G a connected Lie group with Lie algebra $\mathfrak{g}$. Consider the polynomial

$$\det((t+1) - \operatorname{Ad}(x)) = \sum_0^n D_i(x)t^i \qquad (x \in G),$$

where t is an indeterminate and $n = \dim(G)$. The D_i are real analytic functions on G and $D_n = 1$. Let l be the least integer such that $D_l \not\equiv 0$ – then l is called the *rank* of G. Plainly: $l = \operatorname{rank}(\mathfrak{g})$. An element $x \in G$ is said to be *singular* or *regular* according to whether $D_l(x) = 0$ or not. The set of regular elements in G will be denoted by G' – thus G' is an open, dense subset of G while its complement, the set of singular elements, is a set of Haar measure zero (right or left). It is clear that G' is inner automorphism invariant and stable under multiplication by elements from the center of G.

One of the main results on Cartan subgroups of a reductive Lie group is the following theorem.

Theorem 1.4.1.5 (Harish-Chandra) Let J be a Cartan subgroup of G with Lie algebra $\mathfrak{j}$, J the identity component of J and Z ($\subset \mathsf{J}$) the center of G – then ZJ is an open subgroup of J lying in the center J_0 of J and J/ZJ is finite. Moreover J/Z is abelian and, in case G admits a faithful finite dimensional representation, J itself is abelian. Let $H_0 \in \mathfrak{j} \cap \mathfrak{g}'$ – then J is the centralizer of H_0 in G.

Before giving the proof of this theorem, let us make a few preliminary reductions. In the first place we may assume, without loss of generality, that $\mathfrak{j}$ is θ-stable (Proposition 1.3.1.1). It is perfectly clear that ZJ is an open subgroup of J lying in the center J_0 of J; in view of the identifications $\mathsf{J}/ZJ = \mathsf{J}_K J_{\mathfrak{p}}/ZJ_K J_{\mathfrak{p}} = \mathsf{J}_K/\hat{Z}J_K$, we see that J/ZJ is finite ($\hat{Z}$ the center of $\hat{G}$). To establish the other assertions of the theorem, we claim that it will be sufficient just to consider the case when G is semi-simple. Thus, employing the obvious notations, we have $G = \hat{G} \times C_{\mathfrak{p}}$, $\mathsf{J} = \hat{\mathsf{J}} \times C_{\mathfrak{p}}$, $Z = \hat{Z} \times C_{\mathfrak{p}}$, $\hat{G} = \bar{G}C_{\mathfrak{t}}$, $\hat{\mathsf{J}} = \bar{\mathsf{J}}C_{\mathfrak{t}}$. To verify, then, that J/Z is abelian, we note: $\mathsf{J}/Z = \hat{\mathsf{J}} \times C_{\mathfrak{p}}/\hat{Z} \times C_{\mathfrak{p}} = (\bar{\mathsf{J}}C_{\mathfrak{t}}/C_{\mathfrak{t}})/(\bar{Z}C_{\mathfrak{t}}/C_{\mathfrak{t}}) = (\bar{\mathsf{J}}/\bar{\mathsf{J}} \cap C_{\mathfrak{t}})/(\bar{Z}/\bar{Z} \cap C_{\mathfrak{t}}) = \bar{\mathsf{J}}/\bar{Z}$ (because $\bar{\mathsf{J}} \cap C_{\mathfrak{t}} = \bar{Z} \cap C_{\mathfrak{t}}$). Therefore, if we know that J/Z is abelian in the semi-simple case, then it is the same in general. Entirely similar comments apply to the remaining contentions. This being so, let us preface the proof of the theorem with a lemma.

Lemma 1.4.1.6 Let G_c be a connected semi-simple complex analytic group with Lie algebra $\mathfrak{g}_c$, $\mathfrak{j}_c$ a Cartan subalgebra of $\mathfrak{g}_c$, J_c the complex analytic subgroup of G_c with Lie algebra $\mathfrak{j}_c$ – then G_c' is connected and $G_c' = \bigcup_{x \in G_c} xJ_c'x^{-1}$ where $J_c' = J_c \cap G_c'$. In addition, if H_0 is regular in $\mathfrak{j}_c$ (i.e. if $H_0 \in \mathfrak{j}_c \cap \mathfrak{g}_c'$), then the centralizer of H_0 in G_c is J_c.

Proof Adopting the notations of 1.1, choose a compact real form $\mathfrak{u}$ of $\mathfrak{g}_c$ such that $\sqrt{-1}\mathfrak{j}^*$ is a Cartan subalgebra of $\mathfrak{u}$; let U be the compact real analytic subgroup of G_c corresponding to $\mathfrak{u}$. Ordering the roots of the pair $(\mathfrak{g}_c, \mathfrak{j}_c)$ in some way, set $\mathfrak{n}_c^+ = \sum_{\alpha>0} \mathfrak{g}_c^\alpha$. If N_c^+ is the complex analytic subgroup of G_c corresponding to $\mathfrak{n}_c^+$, then $G_c = UJ_cN_c^+$ with $J_cN_c^+$ closed in G_c.

(1) If l is the rank of G_c, then the singular set is the set of zeros of the *holomorphic* function D_l, hence is a complex analytic subvariety of G_c of one complex dimension (therefore of two real dimensions) less than the dimension of G_c – therefore G_c' is connected.

(2) Plainly $\mathscr{R} = \bigcup_{x \in G_c} xJ_c'x^{-1} \subset G_c'$; to establish equality it suffices to prove that $\mathscr{R}$ is both open and closed in G_c' (clearly $\mathscr{R}$ is not empty). Define a map ϕ from $G_c/J_c \times J_c'$ into G_c by the rule $\phi(xJ_c, j) = xjx^{-1}$ ($x \in G_c$, $j \in J_c'$); plainly ϕ is well-defined since J_c is abelian. It is not difficult to show that ϕ is regular (see Proposition 1.4.2.3) – hence $\mathscr{R} =$

$\phi(G_c/J_c \times J'_c)$ is open in G'_c. To see that $\mathscr{R}$ is closed in G'_c, consider the closed set $\bigcup_{u\in U} uJ_cN_c^+u^{-1}$; the intersection of this set with G'_c is

$$\bigcup_{u\in U} u(J_cN_c^+)'u^{-1}$$

where $(J_cN_c^+)' = (J_cN_c^+) \cap G'_c$. Now, in view of Theorem 1.1.4.4, we have

$$(J_cN_c^+)' = \bigcup_{n\in N_c^+} nJ'_cn^{-1}$$

and therefore, since $G_c = UJ_cN_c^+ = UN_c^+J_c$, it follows that

$$\bigcup_{u\in U} u(J_cN_c^+)'u^{-1} = \mathscr{R}.$$

Thus $\mathscr{R}$ is closed in G'_c.

(3) Finally let H_0 be a regular element in $\mathfrak{j}_c$ and suppose that $\mathrm{Ad}\,(x)H_0 = H_0$ $(x \in G_c)$ – we are to show that $x \in J_c$. Write $x = unj$ $(u \in U,$ $n \in N_c^+$, $j \in J_c)$ – then $\mathrm{Ad}\,(n)H_0 = \mathrm{Ad}\,(u^{-1})H_0$. Since H_0 is regular, $\mathfrak{j}_c$ is the centralizer of H_0 in $\mathfrak{g}_c$ – therefore $\mathrm{Ad}\,(n)\mathfrak{j}_c = \mathrm{Ad}\,(u^{-1})\mathfrak{j}_c$. We claim that $\mathrm{Ad}\,(n)\mathfrak{j}_c = \mathfrak{j}_c$. To see this, note that $\mathrm{Ad}\,(n)\mathfrak{j}_c \subset \mathfrak{j}_c + \mathfrak{n}_c^+$, $\mathrm{Ad}\,(u^{-1})$ $(\sqrt{-1}\mathfrak{j}^*) \subset \mathfrak{u}$ and $[\sqrt{-1}\mathfrak{j}^*, (\mathfrak{j}_c + \mathfrak{n}_c^+) \cap \mathfrak{u}] \subset \mathfrak{n}_c^+ \cap \mathfrak{u} = \{0\}$; hence $(\mathfrak{j}_c + \mathfrak{n}_c^+) \cap \mathfrak{u} = \sqrt{-1}\mathfrak{j}^* \Rightarrow \mathrm{Ad}\,(u^{-1})(\sqrt{-1}\mathfrak{j}^*) = \sqrt{-1}\mathfrak{j}^* \Rightarrow \mathrm{Ad}\,(n)\mathfrak{j}_c = \mathfrak{j}_c$ – therefore $n = 1$ (cf. Theorem 1.1.4.4). Thus we are left with the relation $\mathrm{Ad}\,(u)H_0 = H_0$ from which we wish to prove that $u \in J_c$. Write $H_0 = H_1 + \sqrt{-1}H_2$ where $H_1, H_2 \in \sqrt{-1}\mathfrak{j}^*$ – then $\mathrm{Ad}\,(u)H_i = H_i$ $(i = 1, 2)$ and we may choose a real number $r \in \mathbf{R}$ such that $H = H_1 + rH_2$ is a regular element in $\sqrt{-1}\mathfrak{j}^*$. Since $\mathrm{Ad}\,(u)H = H$ and since H is regular in $\sqrt{-1}\mathfrak{j}^*$, it follows that $\mathrm{Ad}\,(u)$ centralizes every element of $\sqrt{-1}\mathfrak{j}^*$. Let J^* be the analytic subgroup of U corresponding to $\sqrt{-1}\mathfrak{j}^*$ – then J^* is a maximal torus in U, and so $u \in J^* \subset J_c$, as desired.

This completes the proof of our lemma. □

[In passing, observe that J_c is the Cartan subgroup of G_c associated with $\mathfrak{j}_c$ (see Proposition 1.4.1.4).]

Proof of Theorem 1.4.1.5 As has been noted above, it may be supposed that $\mathfrak{g}$ is semi-simple. The fact that J/Z is abelian will follow if we can show that J is abelian whenever G admits a faithful finite dimensional representation. So assume that this is the case – then G can be imbedded in a complex analytic group G_c with Lie algebra $\mathfrak{g}_c$ such that the real analytic subgroup of G_c corresponding to $\mathfrak{g}$ is G. Let J_c denote, as in Lemma 1.4.1.6, the Cartan subgroup of G_c associated with $\mathfrak{j}_c$ – then $\mathsf{J} = J_c \cap G$, hence J is abelian. Let $H_0 \in \mathfrak{j}$ be regular. Using the obvious fact that an element of $\mathfrak{g}$ is regular in $\mathfrak{g}$ iff it is regular in $\mathfrak{g}_c$, we see that the centralizer of H_0 in G_c is J_c, therefore that the centralizer of H_0 in G is J. □

Two Cartan subgroups J^1 and J^2 of G are said to be *conjugate* (in G) if $\mathsf{J}^2 = x\mathsf{J}^1x^{-1}$ for some $x \in G$.

Select Cartan subalgebras $\mathfrak{j}^1, \ldots, \mathfrak{j}^r$ of $\mathfrak{g}$ per Corollary 1.3.1.11; let J^i denote the Cartan subgroup of G associated with $\mathfrak{j}^i$ $(i = 1, \ldots, r)$ – then it is evident that every Cartan subgroup J of G is conjugate to a J^i for exactly one i.

Basic for subsequent developments is the following theorem.

Theorem 1.4.1.7 (Harish-Chandra) Let $J^1, \ldots, J^r$ be a maximal set of mutually non-conjugate Cartan subgroups of G – then

$$G' = \bigcup_{i=1}^{r} \bigcup_{x \in G} x(J^i)' x^{-1}$$

where $(J^i)' = J^i \cap G'$.

Proof It is enough to consider the case when G is semi-simple and in fact it is permissible here to replace G by G/Z (Z the center of G) – thus let G_c be a complex analytic group with Lie algebra $\mathfrak{g}_c$ in which G is imbedded as the real analytic subgroup with Lie algebra $\mathfrak{g}$. Let $x \in G'$; if $\mathfrak{g}_{x,c}$ denotes the centralizer of x in $\mathfrak{g}_c$ (i.e. $\mathfrak{g}_{x,c} = \{X \in \mathfrak{g}_c : \mathrm{Ad}(x)X = X\}$), then $\mathfrak{g}_{x,c} \cap \mathfrak{g}$ is a Cartan subalgebra of $\mathfrak{g}$ (this follows from Lemma 1.4.1.6) – therefore, for some i $(1 \leqslant i \leqslant r)$ and some $x_i \in G$, it must be the case that $\mathfrak{g}_{x,c} \cap \mathfrak{g} = x_i(\mathfrak{j}^i)$. This being so, let J_x be the Cartan subgroup of G associated with $\mathfrak{g}_{x,c} \cap \mathfrak{g}$ – then $J_x = x_i J^i x_i^{-1}$, whence $x \in x_i(J^i)' x_i^{-1}$. As the reverse inclusion is clear, the proof is complete. □

Corollary 1.4.1.8 Let x be an element of G, $\mathfrak{g}_x$ the centralizer of x in $\mathfrak{g}$ – then $x \in G'$ iff $\mathfrak{g}_x$ is a Cartan subalgebra of $\mathfrak{g}$.

Notation If J is a Cartan subgroup of G, put $G(J) = \bigcup_{x \in G} xJ'x^{-1}$ where $J' = J \cap G'$.

Proposition 1.4.1.9 Let $\mathfrak{j}$ be a θ-stable Cartan subalgebra of $\mathfrak{g}$, J the Cartan subgroup of G associated with $\mathfrak{j}$; set $J' = J \cap G'$ – then J' has but a finite number of connected components.

Proof The basic idea here is to show that no component of J contains more than a finite number of components of J'; since J has only a finite number of components (cf. Proposition 1.4.1.3), it follows that the same must be true of J'.

(1) Let Φ denote the set of roots of the pair $(\mathfrak{g}_c, \mathfrak{j}_c)$, Φ^+ the positive roots in some ordering of the dual of $\mathfrak{j}^*$. For each root α we can define a character (not necessarily unitary) ξ_α of J by the rule $\mathrm{Ad}(j)X_\alpha = \xi_\alpha(j)X_\alpha (j \in J,\ X_\alpha \in \mathfrak{g}_c^\alpha)$; since $[X_\alpha, X_{-\alpha}] \in \mathfrak{j}_c$, it is clear that $\xi_{-\alpha} = \xi_\alpha^{-1}$.

(2) Fix a component of J; such a component is of the form $j_0 J (j_0 \in J_K)$. Choose complex numbers c_α such that $\xi_\alpha(j_0) = e^{c_\alpha} (\alpha \in \Phi^+)$; because J_K is compact, the c_α are all pure imaginary. Denote by $\sigma_{\alpha,n}$ the hyperplane in $\mathfrak{j}$ consisting of those elements $H \in \mathfrak{j}$ such that $\alpha(H) = -c_\alpha + 2\pi\sqrt{-1}n$ $(\alpha \in \Phi^+, n \in \mathbf{Z})$; obviously a given compact subset of $\mathfrak{j}$ meets at most a

finite number of these hyperplanes. Let $\mathfrak{j}^+$ be the complement in $\mathfrak{j}$ of the union of all these hyperplanes – then every point in $\mathfrak{j}$ admits a neighborhood $\mathcal{O}$ such that $\mathcal{O} \cap \mathfrak{j}^+$ decomposes into a finite number of components. Consequently, if ω is a compact subset of $\mathfrak{j}$, then there exists a neighborhood $\mathcal{O}$ of ω in $\mathfrak{j}$ for which $\mathcal{O} \cap \mathfrak{j}^+$ has but a finite number of components.

(3) Because J_K is compact, we can find a compact subset ω in $\mathfrak{j}_{\mathfrak{k}}$ such that $J_K \subset \exp(\omega)$. Put $\omega^+ = \mathfrak{j}^+ \cap (\mathfrak{j}_{\mathfrak{p}} + \omega)$; let $H^+ \in \omega^+$ – then $H^+ = H + H_0$ where $H \in \mathfrak{j}_{\mathfrak{p}}$, $H_0 \in \omega$. Consider $H_t^+ = tH + H_0$ $(t > 0)$ – then we claim that $H_t \in \omega^+$. For if not, then $\alpha(H_t^+) = -c_\alpha + 2\pi\sqrt{-1}n$ (some $\alpha \in \Phi^+$, $n \in \mathbf{Z}$); but $\alpha(H_t) = t\alpha(H) + \alpha(H_0)$ and $\alpha(H)$ is real while $\alpha(H_0)$ is pure imaginary; since c_α is pure imaginary, of necessity one would have $\alpha(H) = 0$, an impossibility as $H^+ \in \mathfrak{j}^+$.

(4) Let ω be as in (3). Owing to (2), there exists a neighborhood $\mathcal{O}$ of ω in $\mathfrak{j}$ such that $\mathcal{O} \cap \mathfrak{j}^+$ has a finite number of components. The discussion in (3) shows that every point of ω^+ can be joined to a point in $\mathcal{O}$ by a line lying entirely in ω^+ – hence ω^+ has a finite number of components too. But the image of ω^+ under the mapping $H \mapsto j_0 \exp H$ $(H \in \omega^+)$ is $\mathrm{J}' \cap (j_0 J)$ – therefore $j_0 J$ contains no more than a finite number of components of J'.

The proof of the proposition is now complete. □

Corollary 1.4.1.10 The set G' of regular elements in G has but a finite number of connected components.

Proof In view of Theorem 1.4.1.7 and the preceding proposition, this is clear. □

1.4.2 The Groups $W(G, \mathrm{J})$ and $W(G, \mathrm{J}_0)$

Throughout the present number, (G, K) will denote a reductive pair fulfilling our general assumptions (1.1.5).

Let $\mathfrak{j}$ be a θ-stable Cartan subalgebra of $\mathfrak{g}$, J the Cartan subgroup of G associated with $\mathfrak{j}$ – then $\mathrm{J} = \mathrm{J}_K J_{\mathfrak{p}}$ (cf. Proposition 1.4.1.2). Let J^* denote the normalizer of $\mathfrak{j}$ in G (or, equivalently, the normalizer of J in G).

Proposition 1.4.2.1 Retain the above notations – then $\mathrm{J}^* = \mathrm{J}_K^* J_{\mathfrak{p}}$ where $\mathrm{J}_K^* = \mathrm{J}^* \cap K$.

Proof Since in general $\theta \circ \mathrm{Ad}(x) \circ \theta = \mathrm{Ad}(\theta(x))$ for $x \in G$, it follows that $\theta(\mathrm{J}^*) = \mathrm{J}^*$. Given $j \in \mathrm{J}^*$, write $j = k \exp X$ $(k \in K, X \in \mathfrak{p})$. We have to show that $X \in \mathfrak{j}_{\mathfrak{p}}$. So take $H \in \mathfrak{j}_{\mathfrak{p}}$ and let $\tilde{H} = \mathrm{Ad}(k \exp X)H$; a consideration of eigenvalues tells us that $\tilde{H} \in \mathfrak{j}_{\mathfrak{p}}$ – therefore, by Lemma 1.1.3.7, $[X, H]$ is zero. Similar remarks apply if $H \in \mathfrak{j}_{\mathfrak{k}}$. Hence $X \in \mathfrak{j}_{\mathfrak{p}}$, as we wished to prove. □

Definition Let $\mathfrak{j}$ be a θ-stable Cartan subalgebra of $\mathfrak{g}$, J the Cartan subgroup of G associated with $\mathfrak{j}$ – then the group $W(G, J)$ is defined by

$$W(G, J) = J^*/J = J_K^*/J_K.$$

[Needless to say, the θ-stability assumption on $\mathfrak{j}$ is not really necessary for our definition of $W(G, J)$ . . . .]

Fix a θ-stable Cartan subalgebra $\mathfrak{j}$ of $\mathfrak{g}$; let J be the Cartan subgroup of G associated with $\mathfrak{j}$ – then it is evident that the group $W(G, J)$ is both compact and discrete, hence finite. In fact $W(G, J)$ may be viewed as a subgroup of the Weyl group $W = W(\mathfrak{g}_c, \mathfrak{j}_c)$ of the pair $(\mathfrak{g}_c, \mathfrak{j}_c)$ – thus $W(G, J)$ is precisely the subgroup of W consisting of those $w \in W$ for which there exists an $x \in G$ such that $\mathrm{Ad}(x)H = wH$ for all $H \in \mathfrak{j}$ (hence it is clear that $W(G, J)$ depends only on the pair $(\mathfrak{g}, \mathfrak{j})$, i.e. is independent of the choice of G). Consider the manifold $G/J \times \mathfrak{j}'$ – then $W(G, J)$ operates on $G/J \times \mathfrak{j}'$ in the following way (for notational agreements, see A2.4). Let $x \mapsto \dot{x} = xJ$ $(x \in G)$ denote the projection of G onto G/J. Fix a w in $W(G, J)$ and choose $j_w \in J^*$ lying in the coset w; given $\dot{x} \in G/J$ $(x \in G)$, let us agree to write $w \cdot \dot{x} = xj_w^{-1}J$; on the other hand, if $H \in \mathfrak{j}'$, then we set $w \cdot H = \mathrm{Ad}(j_w)H$; finally we shall write $w \cdot (\dot{x}, H) = (w \cdot \dot{x}, w \cdot H)$ $(x \in G/J, H \in \mathfrak{j}')$.

Proposition 1.4.2.2 Retain the above notations – then the mapping ϕ, $(\dot{x}, H) \mapsto \dot{x}H = xH$, of $G/J \times \mathfrak{j}'$ onto $\mathfrak{g}(\mathfrak{j})$ is an everywhere regular, $[W(G, J)]$-to-one map.

Proof We recall that $\mathfrak{g}(\mathfrak{j}) = \bigcup_{x \in G} x(\mathfrak{j}')$ (cf. 1.3.4, Notation) – hence ϕ does map $G/J \times \mathfrak{j}'$ onto $\mathfrak{g}(\mathfrak{j})$. It is clear that the complete inverse image under ϕ of a point $\dot{x}H$ $(\dot{x} \in G/J, H \in \mathfrak{j}')$ consists of the $[W(G, J)]$ points $(w \cdot \dot{x}, w \cdot H)$ $(w \in W(G, J))$ which are all distinct – hence ϕ is a $[W(G, J)]$-to-one map. It remains to verify the regularity of ϕ. Introduce an ordering into the set of roots of the pair $(\mathfrak{g}_c, \mathfrak{j}_c)$ and let $\pi = \prod_{\alpha > 0} \alpha$ – then π is a polynomial function on $\mathfrak{j}_c$. Fix a point $p_0 = (x_0J, H_0) \in G/J \times \mathfrak{j}'$; we intend to compute the action of the differential $(d\phi)_{p_0}$ on vectors in the tangent space to $G/J \times \mathfrak{j}'$ at p_0. Identify in the usual way the tangent space to G/J at x_0J with $\mathfrak{g}/\mathfrak{j}$; let $X \in \mathfrak{g}/\mathfrak{j}$; consider the curve $\gamma_X(t) = (x_0 \exp(tH)J, H_0)$ $(t \in \mathbf{R})$ – then

$$\phi \circ \gamma_X(t) = \mathrm{Ad}(\exp(\mathrm{Ad}(x_0)tX)x_0)H_0$$

and the tangent to $\phi \circ \gamma_X$ at $t = 0$ is evidently $\{-\mathrm{Ad}(x_0)\,\mathrm{ad}(H_0)(X)\}$. On the other hand let $H \in \mathfrak{j}$ – then the tangent to the curve $\phi \circ \gamma_H$ at $t = 0$, where $\gamma_H(t) = (x_0J, H_0 + tH)(t \in \mathbf{R})$, is $\mathrm{Ad}(x_0)H$. Let $\{X_1, \ldots, X_r\}$ be a basis for $\mathfrak{g}/\mathfrak{j}$, f_0 the endomorphism of $\mathfrak{g}$ given by

$$f_0X_i = -\mathrm{ad}(H_0)X_i, f_0H = H \quad (1 \leqslant i \leqslant r, H \in \mathfrak{j}).$$

Then f_0 passes to the quotient to define an endomorphism $\bar{f}_0$ of $\mathfrak{g}/\mathfrak{j}$. Now $|\det(\mathrm{Ad}(x_0))| = 1$, G being reductive and hence unimodular – consequently

$$|\det(\mathrm{Ad}(x_0)f_0)| = |\det(f_0)| = |\det(\tilde{f}_0)|$$
$$= \prod_{\alpha>0} |\alpha(H_0)|^2 = |\pi(H_0)|^2.$$

Since H_0 is regular, $\pi(H_0)$ is not zero – therefore ϕ is regular. □

Remark Since $\mathrm{J}/Z\mathrm{J}$ is finite (cf. Theorem 1.4.1.5) and $Z\mathrm{J}$ is abelian, the homogeneous space G/J admits a G-invariant measure $d_{G/\mathrm{J}}$ where of course $d_{G/\mathrm{J}}$ is determined to within a positive multiplicative constant. The action of the group $W(G, \mathrm{J})$ preserves this measure since the map $\dot{x} \mapsto w\cdot\dot{x}$ ($\dot{x} \in G/\mathrm{J}$, $w \in W(G, \mathrm{J})$) is the identity if raised to a sufficiently high integral power.

We shall now formulate the analogue of the preceding results for the group G.

Let $\mathfrak{j}$ be a θ-stable Cartan subalgebra of $\mathfrak{g}$, J the Cartan subgroup of G associated with $\mathfrak{j}$; let J_0 denote the center of J – then it is clear that $\mathrm{J}_0 = (\mathrm{J}_0 \cap K)J_{\mathfrak{p}}$.

Definition Let $\mathfrak{j}$ be a θ-stable Cartan subalgebra of $\mathfrak{g}$, J the Cartan subgroup of G associated with $\mathfrak{j}$ – then the group $W(G, \mathrm{J}_0)$ is defined by

$$W(G, \mathrm{J}_0) = \mathrm{J}^*/\mathrm{J}_0 = \mathrm{J}_K^*/\mathrm{J}_0 \cap K.$$

[Of course the θ-stability assumption on $\mathfrak{j}$ is not really necessary for our definition of $W(G, \mathrm{J}_0)$ . . . .]

Fix a θ-stable Cartan subalgebra $\mathfrak{j}$ of $\mathfrak{g}$; let J be the Cartan subgroup of G associated with $\mathfrak{j}$ – then it is evident that the group $W(G, \mathrm{J}_0)$ is both compact and discrete, hence finite. [Although one must distinguish $W(G, \mathrm{J}_0)$ from $W(G, \mathrm{J})$ in general, note that $W(G, \mathrm{J}_0) = W(G, \mathrm{J})$ if either G admits a faithful finite dimensional representation (cf. Theorem 1.4.1.5) or if $\mathfrak{j}$ is fundamental (cf. Proposition 1.4.1.4).] Consider the manifold $G/\mathrm{J}_0 \times \mathrm{J}'$ – then $W(G, \mathrm{J}_0)$ operates on $G/\mathrm{J}_0 \times \mathrm{J}'$ in the following way. Let $x \mapsto \dot{x} = x\mathrm{J}_0$ ($x \in G$) denote the projection of G onto G/J_0. Fix a w in $W(G, \mathrm{J}_0)$ and choose $j_w \in \mathrm{J}^*$ lying in the coset w; given $\dot{x} \in G/\mathrm{J}_0$ ($x \in G$), let us agree to write $w\cdot\dot{x} = xj_w^{-1}\mathrm{J}_0$; on the other hand, if $j \in \mathrm{J}'$, then we set $w\cdot j = j_w j j_w^{-1}$; finally we shall write $w\cdot(\dot{x}, j) = (w\cdot\dot{x}, w\cdot j)$ ($\dot{x} \in G/\mathrm{J}_0, j \in \mathrm{J}'$). [In passing observe that the entity $w\cdot j (j \in \mathrm{J}')$ is well-defined; this is why, since J may not be abelian, one considers J_0 . . . .]

Proposition 1.4.2.3 Retain the above notations – then the mapping ϕ, $(\dot{x}, j) \mapsto {}^x j = xjx^{-1}$, of $G/\mathrm{J}_0 \times \mathrm{J}'$ onto $G(\mathrm{J})$ $(= \bigcup_{x\in G} x\mathrm{J}'x^{-1})$ is an everywhere regular, $[W(G, \mathrm{J}_0)]$-to-one map.

Proof Only the regularity of ϕ needs to be proved. So fix a point $p_0 = (x_0\mathrm{J}_0, j_0)$ in $G/\mathrm{J}_0 \times \mathrm{J}'$ and make the usual identifications as regards the tangent space at p_0. Let $X \in \mathfrak{g}/\mathfrak{j}$; consider the curve $\gamma_X(t) = (x_0 \exp(tX)\mathrm{J}_0, j_0)$ ($t \in \mathbf{R}$) – then

$$\phi\circ\gamma_X(t) = x_0 j_0 x_0^{-1} \exp(t\,\mathrm{Ad}(x_0 j_0^{-1})X)\exp(-t\,\mathrm{Ad}(x_0)X)$$

and so the tangent to $\phi \circ \gamma_X$ at $t = 0$ may be taken to be $\mathrm{Ad}(x_0)[\mathrm{Ad}(j_0^{-1}) - 1]X$. Let $H \in \mathfrak{j}$ – then the tangent to the curve $\phi \circ \gamma_H$ at $t = 0$, where $\gamma_H(t) = (x_0 \mathsf{J}_0, j_0 \exp(tH))$ $(t \in \mathbf{R})$, may be identified with $\mathrm{Ad}(x_0)H$. It follows, as in Proposition 1.4.2.2, that

$$|\det(d\phi)_{p_0}| = |\det(\mathrm{Ad}(j_0^{-1}) - 1)_{\mathfrak{g}/\mathfrak{j}}|$$

where $(\mathrm{Ad}(j_0^{-1}) - 1)_{\mathfrak{g}/\mathfrak{j}}$ denotes the endomorphism viewed as a map on $\mathfrak{g}/\mathfrak{j}$. Since $j_0 \in \mathsf{J}'$, $|\det(d\phi)_{p_0}| \neq 0$ – therefore ϕ is regular. □

Remark The homogeneous space G/J_0 carries a G-invariant measure d_{G/J_0} – this is clear.

At this point we shall insert the following result which will be needed from time to time in the sequel.

Proposition 1.4.2.4 Let $\mathfrak{h}$ be a subalgebra of $\mathfrak{g}$, H the analytic subgroup of G corresponding to $\mathfrak{h}$, H^* the normalizer of $\mathfrak{h}$ in G; suppose that $\mathfrak{h}$ is reductive in $\mathfrak{g}$ with $\mathrm{rank}(\mathfrak{h}) = \mathrm{rank}(\mathfrak{g})$ – then the space H^*/HZ is finite (Z the center of G).

Proof Since the Lie algebra $\mathfrak{h}$ is reductive, there exists a finite number of Cartan subalgebras $\mathfrak{j} = \mathfrak{j}^1, \ldots, \mathfrak{j}^r$ of $\mathfrak{h}$ such that every Cartan subalgebra of $\mathfrak{h}$ is of the form $\mathrm{Ad}(\xi)(\mathfrak{j}^i) = \xi(\mathfrak{j}^i)$ for some $\xi \in H$ and some i $(1 \leqslant i \leqslant r)$; it will be assumed that $\mathfrak{j}$ is conjugate to $\mathfrak{j}^i$ $(1 \leqslant i \leqslant i_0)$ under H^* but not to $\mathfrak{j}^i$ $(i_0 < i \leqslant r)$; choose, accordingly, $\xi_i^* \in \mathsf{H}^*$ such that $\xi_i^*(\mathfrak{j}) = \mathfrak{j}^i$ $(1 \leqslant i \leqslant i_0, \xi_1^* = 1)$. Consider now an arbitrary element ξ^* in H^* – then $\xi^*(\mathfrak{j}) = \xi(\mathfrak{j}^i)$ for some $\xi \in H$ and some i $(1 \leqslant i \leqslant i_0)$, whence $\mathfrak{j} = \xi^{*-1}\xi\xi_i^*(\mathfrak{j})$. Let J be the Cartan subgroup of G associated with $\mathfrak{j}$, J^* the normalizer of $\mathfrak{j}$ in G – then, since J^*/JZ is finite, the same must be true of $\mathsf{H}^* \cap \mathsf{J}^*/JZ$. This being so, select elements $j_1, \ldots, j_s$ in $\mathsf{H}^* \cap \mathsf{J}^*$ such that $\mathsf{H}^* \cap \mathsf{J}^* = \bigcup_{k=1}^{s} j_k JZ$; plainly $\xi^{*-1}\xi\xi_i^* \in j_k JZ$ for some k $(1 \leqslant k \leqslant s)$; but H is normal in H^* – therefore we have $\xi^* \in \xi\xi_i^* JZ j_k^{-1} \subset \xi_i^* j_k^{-1} HZ$, from which we conclude that the order of H^*/HZ can not exceed rs. □

Remark Let G be a connected semi-simple Lie group with finite center, K a maximal compact subgroup of G with corresponding Cartan involution θ; let P be a parabolic subgroup of G; put $\mathsf{L} = P \cap \theta(P)$ and let A be a maximal connected split abelian subgroup lying in the center of L – then L is a Levi subgroup of P, while A is a split component of P. Write, in the usual way, $\mathsf{L} = \mathsf{M}A$ – then the parabolic subgroup P is said to be *cuspidal* if M admits a compact Cartan subgroup (the definition being, of course, independent of the given Levi decomposition for P). It is not difficult to prove that $P = \mathsf{M}AR_u(P)$ is cuspidal iff there exists a θ-stable Cartan subalgebra $\mathfrak{j}$ of $\mathfrak{g}$ such that $\mathfrak{a}$ (= Lie algebra of A) $= \mathfrak{j}_\mathfrak{p}$. This being the case, let us suppose that P is cuspidal and that $\mathfrak{j} = \mathfrak{j}_\mathfrak{k} + \mathfrak{j}_\mathfrak{p}$ is a θ-stable Cartan subalgebra of $\mathfrak{g}$ whose vector part $\mathfrak{j}_\mathfrak{p}$ is equal to $\mathfrak{a}$; let J be the Cartan subgroup of G associated with $\mathfrak{j}$ – then the group $W(A)$ (1.2.4) can be identified with the set of restrictions to $\mathfrak{a}$ of the elements in $W(G, \mathsf{J})$. In fact a consideration of eigenvalues tells us that a given element in $W(G, \mathsf{J})$ normalizes both $\mathfrak{j}_\mathfrak{k}$ and $\mathfrak{j}_\mathfrak{p}(=\mathfrak{a})$; on the other hand, let there be given an element $w \in W(A)$ – then we claim that there exists a $j_w \in \mathsf{J}_K^*$ such that $\mathrm{Ad}(j_w)H = wH$ (all $H \in \mathfrak{a}$). Thus let k_w be an element in K representing w; the space $k_w(\mathfrak{j}_\mathfrak{k})$ is a Cartan subalgebra of the compact Lie algebra $\mathfrak{l}_\mathfrak{k}$, whence there exists an $l \in L_K$ (the identity component of $\mathsf{L} \cap K$) such that $lk_w(\mathfrak{j}_\mathfrak{k}) = \mathfrak{j}_\mathfrak{k}$ – but $j_w = lk_w$ normalizes $\mathfrak{j}_\mathfrak{p}$ too, hence lies in J_K^*, and moreover $\mathrm{Ad}(j_w)|\mathfrak{a} = w$.

Example (Hirai) Let p and q be integers $\geqslant 1$ with $p \geqslant q$ and put $n = p + q$; let $G = \mathbf{U}(p, q)$ be the subgroup of $\mathbf{GL}(p + q, \mathbf{C})$ comprised of those matrices x which leave invariant the Hermitian form

$$z_1\bar{z}_1 + \cdots + z_p\bar{z}_p - z_{p+1}\bar{z}_{p+1} - \cdots - z_n\bar{z}_n,$$

i.e. such that $x^*I_{p,q}x = I_{p,q}$ (x^* = conjugate transpose of x) where $I_{p,q} = \left(\begin{smallmatrix} I_p & 0 \\ 0 & -I_q \end{smallmatrix}\right)$ – then G is a connected reductive Lie group. [Needless to say, there exists a canonical map from $\mathbf{T} \times \mathbf{SU}(p, q)$ onto $\mathbf{U}(p, q)$. . . .] Let K be the maximal compact subgroup of G consisting of all matrices of the form $\left(\begin{smallmatrix} k_p & 0 \\ 0 & k_q \end{smallmatrix}\right)$ where $k_p \in \mathbf{U}(p)$ and $k_q \in \mathbf{U}(q)$ ($\mathbf{U}(.)$ the unitary group of order .) – then our theory is applicable to the pair (G, K). The group G possesses $q + 1$ conjugacy classes of Cartan subgroups (we shall leave it to the reader to explicate Theorem 1.3.1.10 in the Lie algebra $\mathfrak{g}$ of G . . .); a representative from each class can be obtained as follows: For any i lying between 0 and q, write $\mathsf{J}^i = \mathsf{J}_K^i J_\mathfrak{p}^i$ where the elements of J_K^i and $J_\mathfrak{p}^i$ are given by the matrices

$$\mathsf{J}_K^i: D(e^{\sqrt{-1}\phi_1}, \ldots, e^{\sqrt{-1}\phi_{p-i}}, e^{\sqrt{-1}\theta_i}, \ldots, e^{\sqrt{-1}\theta_1}, e^{\sqrt{-1}\theta_1}, \ldots$$
$$e^{\sqrt{-1}\theta_i}, e^{\sqrt{-1}\psi_{q-i}}, \ldots, e^{\sqrt{-1}\psi_1}).$$

$$\mathsf{J}_\mathfrak{p}^i: \begin{bmatrix} I_{p-i} & & & & & & & & 0 \\ & \cosh(t_i) & 0 & & & & 0 & \sinh(t_i) & \\ & 0 & \cosh(t_{i-1}) & & & & \sinh(t_{i-1}) & 0 & \\ & & & \ddots & & ⋰ & & & \\ & & & & \cosh(t_1)\ \sinh(t_1) & & & & \\ & & & & \sinh(t_1)\ \cosh(t_1) & & & & \\ & & & ⋰ & & \ddots & & & \\ & 0 & \sinh(t_{i-1}) & & & & \cosh(t_{i-1}) & 0 & \\ & \sinh(t_i) & 0 & & & & 0 & \cosh(t_i) & \\ 0 & & & & & & & & I_{q-i} \end{bmatrix}$$

the variables ϕ, θ, ψ, and t all being real. Evidently, in the present case, each of the Cartan subgroups J^i $(i = 0, \ldots, q)$ is connected and abelian. Let us compute the groups $W(G, \mathsf{J}^i)$ $(i = 0, \ldots, q)$. For this purpose, it will be convenient to introduce some additional notation. Fix a j in J^i and then express it as the product of a matrix from J_K^i with a matrix in $J_\mathfrak{p}^i$; utilizing the notations introduced above, we find that the eigenvalues of j are the numbers $e^{\Phi_1}, \ldots, e^{\Phi_{p-i}}, e^{z_i}, \ldots, e^{z_1}, e^{z_{-1}}, \ldots, e^{z_{-i}}, e^{\Psi_{q-i}}, \ldots, e^{\Psi_1}$ where $\Phi_k = \sqrt{-1}\phi_k$ $(1 \leqslant k \leqslant p - i)$, $\Psi_k = \sqrt{-1}\psi_k$ $(1 \leqslant k \leqslant q - i)$, $z_k = t_k + \Theta_k$, $z_{-k} = -\bar{z}_k = -t_k + \Theta_k$ and $\Theta_k = \sqrt{-1}\theta_k$ $(1 \leqslant k \leqslant i)$. The group $W(G, \mathsf{J}^i)$ contains: (1) All permutations of the components $\Phi_1, \ldots, \Phi_{p-i}$; (2) All permutations of the components $\Psi_1, \ldots, \Psi_{q-i}$; (3) All permutations of the pairs $(z_1, z_{-1}), \ldots, (z_i, z_{-i})$; (4) The permutation of z_k and z_{-k}, that is, the transformation induced by changing the sign of $t_k (1 \leqslant k \leqslant i)$. In fact $W(G, J^i)$ is generated by the transformations appearing in (1)-(4), hence is of order $(p-i)!(q-i)!\, i!\, 2^i$ $(0 \leqslant i \leqslant q)$. [Given any element w in $W(G, J^i)$, it is clear that in the present situation the assignment $j \mapsto w \cdot j (j \in J^i)$ induces a permutation of the eigenvalues of j; this permutation must take place independently in the sets $\{e^{\Phi_1}, \ldots, e^{\Phi_{p-i}}, e^{\Psi_1}, \ldots, e^{\Psi_{q-i}}\}$, $\{e^{z_1}, \ldots, e^{z_i}, e^{z_{-1}}, \ldots, e^{z_{-i}}\}$ – therefore it is enough to consider the two extreme cases, namely when $i = 0$ or $i = p = q$. The details in both of these cases are easy and will be left to the reader to work out.]

1.4.3 Semi-Simple and Unipotent Elements in a Reductive Lie Group

Let (G, K) be a pair verifying our general assumptions (1.1.5). In this number we shall formulate the analogs of the results of 1.3.5 for the group G.

Let $x \in G$ – then, in the present number, $G \cdot x$ will denote the orbit $\{yxy^{-1} : y \in G\}$.

An element $x \in G$ is said to be *semi-simple* if the automorphism $\mathrm{Ad}(x)$ of $\mathfrak{g}$ is semi-simple. Note: If J is a Cartan subgroup of G, then every element in J is semi-simple – hence every regular element in G is semi-simple (cf. Theorem 1.4.1.7).

Let γ be a semi-simple element in G, $\mathfrak{g}_\gamma$ the centralizer of γ in $\mathfrak{g}$; let G_γ be the analytic subgroup of G corresponding to $\mathfrak{g}_\gamma$ – then G_γ is the identity component of G_γ, the centralizer of γ in G. If γ is central, put $\delta_\gamma = 1$; if γ is non-central, put

$$\delta_\gamma(x_\gamma) = \det(\mathrm{Ad}(\gamma x_\gamma)^{-1} - 1)_{\mathfrak{g}/\mathfrak{g}_\gamma} \qquad (x_\gamma \in G_\gamma).$$

It is clear that δ_γ is an analytic function on G_γ and, moreover, $\delta_\gamma(1) \neq 0$. Let $'G_\gamma$ be the set of all points $x_\gamma \in G_\gamma$ where $\delta_\gamma(x_\gamma) \neq 0$ – then $'G_\gamma$ is an open neighborhood of 1 in G_γ.

Lemma 1.4.3.1 Fix a semi-simple element γ in G; let ϕ denote the mapping $(x, x_\gamma) \mapsto x(\gamma x_\gamma)x^{-1}$ $(x \in G, x_\gamma \in G_\gamma)$ of $G \times G_\gamma$ into G – then ϕ is everywhere of rank n on $G \times {}'G_\gamma$ $(n = \dim(G))$.

Proof Because $\mathrm{Ad}(\gamma)$ is a semi-simple endomorphism of $\mathfrak{g}$, the minimal polynomial m_γ of $\mathrm{Ad}(\gamma)$ is a product of distinct primes, say $m_\gamma = p_1 \ldots p_r$. Suppose that $p_1(t) = t - 1$ (t an indeterminate) – then $\mathfrak{g}_\gamma = \{X \in \mathfrak{g} : p_1(\mathrm{Ad}(\gamma))X = 0\}$. Let $P_i = \{X \in \mathfrak{g} : p_i(\mathrm{Ad}(\gamma))X = 0\}$ $(i = 2, \ldots, r)$ – then each P_i is $\mathrm{Ad}(\gamma)$-stable and $\mathfrak{g} = \mathfrak{g}_\gamma + P_2 + \cdots + P_r$ (direct sum). Put $\mathfrak{g}^\gamma = P_2 + \cdots + P_r$; because $\mathrm{Ad}(x_\gamma)$ $(x_\gamma \in G_\gamma)$ commutes with $\mathrm{Ad}(\gamma)$, it is clear that $\mathfrak{g}^\gamma$ is $\mathrm{Ad}(x_\gamma)$-stable – hence $\mathfrak{g}^\gamma \sim \mathfrak{g}/\mathfrak{g}_\gamma$. Identify the tangent space to $G \times {}'G_\gamma$ at a point (x, x_γ) with $\mathfrak{g} \times \mathfrak{g}_\gamma$ – a simple calculation then shows that

$$d\phi_{(x, x_\gamma)}(X, X_\gamma) = \mathrm{Ad}(x)(X_\gamma + (\mathrm{Ad}(\gamma x_\gamma)^{-1} - 1)X) \qquad (X \in \mathfrak{g}, X_\gamma \in \mathfrak{g}_\gamma).$$

But if $x_\gamma \in {}'G_\gamma$, then $\mathfrak{g}_\gamma + (\mathrm{Ad}(\gamma x_\gamma)^{-1} - 1)\mathfrak{g}^\gamma = \mathfrak{g}_\gamma + \mathfrak{g}^\gamma = \mathfrak{g}$. Hence the lemma. □

Proposition 1.4.3.2 Fix a semi-simple element γ in G – then the centralizer $\mathfrak{g}_\gamma$ of γ in $\mathfrak{g}$ is reductive in $\mathfrak{g}$ with rank $(\mathfrak{g}_\gamma)$ = rank $(\mathfrak{g})$.

Proof We first claim that $\mathfrak{g}_\gamma$ necessarily contains a Cartan subalgebra $\mathfrak{j}$ of $\mathfrak{g}$. In fact, thanks to the preceding lemma, there exists $x \in G$, $x_\gamma \in {}'G_\gamma$ such that $x(\gamma x_\gamma)x^{-1} \in G'$ (the lemma says that ϕ is an open mapping) – thus $\gamma x_\gamma \in G'$ and so $\mathfrak{g}_{\gamma x_\gamma}$ is a Cartan subalgebra $\mathfrak{j}$ (say) of $\mathfrak{g}$ (cf. Corollary 1.4.1.8). Since $\delta_\gamma(x_\gamma) \neq 0$, it then follows that $\mathfrak{j} = \mathfrak{g}_{\gamma x_\gamma} \subset \mathfrak{g}_\gamma$. [Put $A_\gamma = \mathrm{Ad}(\gamma x_\gamma)^{-1} - 1$; because $\delta_\gamma(x_\gamma) \neq 0$, $\mathfrak{g}^\gamma \cap \mathrm{Ker}(A_\gamma) = \{0\}$ ($\mathfrak{g}^\gamma$ as in Lemma 1.4.3.1). If $X \in \mathrm{Ker}(A_\gamma)$ and if $X = X_\gamma + X^\gamma$ ($X_\gamma \in \mathfrak{g}_\gamma$, $X^\gamma \in \mathfrak{g}^\gamma$), then $0 = A_\gamma X = A_\gamma X_\gamma + A_\gamma X^\gamma \Rightarrow 0 = A_\gamma X_\gamma = A_\gamma X^\gamma \Rightarrow X^\gamma = 0$ ($\mathfrak{g}_\gamma$ and $\mathfrak{g}^\gamma$ being A_γ-stable).] Let J denote the Cartan subgroup of G associated with $\mathfrak{j}$, θ a 'Cartan involution' of $\mathfrak{g}$ stabilizing $\mathfrak{j}$; to verify that $\mathfrak{g}_\gamma$ is reductive in $\mathfrak{g}$, it will be enough to show that $\mathfrak{g}_\gamma$ is θ-stable .... Let Φ denote the set of roots of the pair $(\mathfrak{g}_c, \mathfrak{j}_c)$; each $\alpha \in \Phi$ determines a character ξ_α of J via the requirement $\mathrm{Ad}(j)X_\alpha = \xi_\alpha(j)X_\alpha$ ($j \in$ J, $X_\alpha \in \mathfrak{g}_c^\alpha$) – hence it is clear that $\mathfrak{g}_{\gamma,c} = \mathfrak{j}_c + \sum_{\alpha\in\Phi_\gamma} \mathfrak{g}_c^\alpha$ where Φ_γ denotes the subset of Φ consisting of those α which commute with γ, i.e. for which $\xi_\alpha(\gamma) = 1$. Since the map $\alpha \mapsto \alpha^\theta$ ($\alpha \in \Phi$) permutes the elements of Φ, the θ-stability of $\mathfrak{g}_\gamma$ will follow when it is shown that Φ_γ is θ-stable; but $\xi_{\alpha^\theta}(\gamma) = \overline{\xi_\alpha(\gamma^{-1})} = 1$ .... □

As a corollary, every semi-simple element in G is contained in a Cartan subgroup of G – hence the union of all Cartan subgroups of G is precisely the set of semi-simple elements in G.

Let $\mathcal{N}$ denote the set of all nilpotent elements in $\mathfrak{g}$; put $\mathcal{N}_G = \exp(\mathcal{N}) \subset G$ – then the elements of $\mathcal{N}_G$ are called the *unipotent* elements of G. [Let $u \in \mathcal{N}_G$ – then the endomorphism Ad (u) is unipotent, i.e. the only eigenvalue of Ad (u) is 1 (equivalently Ad (u) – 1 is nilpotent). In this connection let us bear in mind that the map $X \mapsto e^{\mathrm{ad}(X)}$ ($X \in \mathcal{N}$) is one-to-one on $\mathcal{N}$ ....]

Here is the group theoretic version of Proposition 1.3.5.3.

Proposition 1.4.3.3 (Borel) Every $x \in G$ can be written uniquely in the form $x = x_s x_u$ where x_s is a semi-simple element of G, x_u is a unipotent element of G and $x_s x_u = x_u x_s$. Let G_x denote the centralizer of x in G – then x_s and x_u lie in the center of G_x.

Proof A simple argument shows that it is sufficient to consider the case when $\mathfrak{g}$ is semi-simple and $G = \mathrm{Int}(\mathfrak{g})$, the adjoint group of G. Let $\mathrm{G}_{\mathbf{R}}$ denote the subgroup of $G_c = \mathrm{Int}(\mathfrak{g}_c)$ consisting of those elements in G_c which are rational over $\mathbf{R}$ – then G is the connected component of 1 (for the usual topology) in $\mathrm{G}_{\mathbf{R}}$. Our proposition is therefore a consequence of a standard result in the theory of algebraic groups (see Borel [4, p.47]). [The result which we have in mind here is, of course, the 'Jordan decomposition' of an element of an algebraic group.] □

Fix $x \in G$ – then the elements x_s and x_u per the preceding proposition are called the *semi-simple* and *unipotent components* of x, respectively.

Proposition 1.4.3.4 (Borel-Harish-Chandra) Let $x \in G$ – then the orbit $G \cdot x$ of x in G is closed iff x is semi-simple.

Proof Suppose that $\gamma \in G$ is semi-simple – then the fact that the orbit $G \cdot \gamma$ of γ in G is closed will follow provided we can show that the orbit of Ad (γ) in Int$(\mathfrak{g})$ is closed (in Int$(\mathfrak{g})$). [Since $G = \widehat{G} \times C_{\mathfrak{p}}$, we may write $\gamma = \hat{\gamma}c$ $(\hat{\gamma} \in \widehat{G},\ c \in C_{\mathfrak{p}})$; plainly $G \cdot \gamma = (\widehat{G} \cdot \hat{\gamma})c$ – it is therefore enough to consider the case when $\mathfrak{c}_{\mathfrak{p}} = \{0\}$. If $\mathfrak{c}_{\mathfrak{p}} = \{0\}$, then $G = \bar{G}C_{\mathfrak{t}}$ and so we may write $\gamma = \bar{\gamma}c\,(\bar{\gamma} \in \bar{G},\ c \in C_{\mathfrak{t}})$; plainly $G \cdot \gamma = (\bar{G} \cdot \bar{\gamma})c$ – this shows that it suffices to consider only the case when G is semi-simple and of finite center. Suppose, then, that $z_1, \ldots, z_n$ is an enumeration of the elements of Z (Z the center of G). It is clear that $G \cdot \gamma \subset \bigcup_{i=1}^{n} G \cdot (\gamma z_i)$; this latter union is closed if the Int $(\mathfrak{g})$-orbit of Ad (γ) is closed – our assertion now follows from the fact that, in the case at hand, the boundary of an orbit is a union of orbits of strictly smaller dimension.] So assume that $G = \text{Int}\,(\mathfrak{g})$ and $\gamma \in G$ is semi-simple; to verify that $G \cdot \gamma$ is closed in G it will be sufficient to show that the orbit $G_c \cdot \gamma$ is closed in $G_c = \text{Int}\,(\mathfrak{g}_c)$. [To see this, appeal to the lemma which is given in the Appendix to 1.3.5; thus take $E = \text{Hom}_{\mathbf{C}}\,(\mathfrak{g}_c, \mathfrak{g}_c)$ and put, for $a \in E$, $U(x)a = xax^{-1} (x \in G_c)$ – then U: $G_c \to \mathbf{GL}(E)$ is a rational representation of G_c which, moreover, is defined over $\mathbf{R}$.] This being so, let P_γ denote the set of all $x \in G_c$ such that Ad (x) has the same characteristic polynomial as Ad (γ) and which, in addition, annihilates the minimal polynomial of Ad (γ). Evidently P_γ is an algebraic subset of G_c, stable under G_c. Reasoning just as in the proof of Proposition 1.3.5.5, we see that P_γ consists of orbits of the same dimension – therefore $G_c \cdot \gamma$ is closed. As for the converse assertion of our proposition, it is an immediate consequence of Proposition 1.4.3.5 below. □

Proposition 1.4.3.5 Let $x \in G$; write, per Proposition 1.4.3.3, $x = x_s x_u$ – then $x_s \in (G \cdot x)^{cl}$ (the closure of $G \cdot x$ in G).

Proof Choose $X_u \in \mathscr{N}$ such that $x_u = \exp X_u$; assume that X_u is not zero – then $X_u \in \mathfrak{g}_{x_s}$ (the centralizer of x_s in $\mathfrak{g}$). Owing to Proposition 1.3.5.3, there exists an $H \in \mathfrak{g}_{x_s}$ such that $[H, X_u] = 2X_u$; put $h_t = \exp(-tH)$ $(t \in \mathbf{R})$ – then $h_t x h_{-t} = x_s \exp(e^{-2t}X_u) \to x_s$ as $t \to \infty$. □

Remark Let x be an element of G, J a Cartan subgroup of G – then the set $G \cdot x \cap$ J is finite. [Decomposing x into its semi-simple and unipotent components and using the fact that every element of J is semi-simple, we see that x may be taken semi-simple. Agreeing now to write γ for x, let $\mathfrak{g}_\gamma$ be the centralizer of γ in $\mathfrak{g}$, G_γ the analytic subgroup of G corresponding to $\mathfrak{g}_\gamma$. Select a maximal set $\mathfrak{j}^i (1 \leqslant i \leqslant r)$ of Cartan subalgebras of $\mathfrak{g}_\gamma$ such that: (1) No two of them are conjugate under G_γ; (2) For each i $(1 \leqslant i \leqslant r)$, there exists an element x_i in G such that $x_i(\mathfrak{j}) = \mathfrak{j}^i$ ($\mathfrak{j}$ the Lie algebra of J). Fix $j \in G\cdot\gamma \cap \mathsf{J}$ – then $j = x\gamma x^{-1}$ for some $x \in G$, whence $x^{-1}(\mathfrak{j}) \subset \mathfrak{g}_\gamma$. This shows that there exists an index i and an element $x_\gamma \in G_\gamma$ such that $x^{-1}(\mathfrak{j}) = x_\gamma(\mathfrak{j}^i)$ – therefore $y = xx_\gamma x_i \in \mathsf{J}^*$. Let w be the element of the group $W(G, \mathsf{J}_0)$ which y represents; plainly $y^{-1}jy = x_i^{-1}\gamma x_i \in \mathsf{J}$ – thus $j = w\cdot(x_i^{-1}\gamma x_i)$ and so $G\cdot\gamma \cap \mathsf{J}$ is indeed finite.]

Chapter 2

The Universal Enveloping Algebra of a Semi-Simple Lie Algebra

Introduction

Let $\mathfrak{g}$ be a reductive Lie algebra over **R**, G a connected Lie group with Lie algebra $\mathfrak{g}$. In the hands of Harish-Chandra (and others) the universal enveloping algebra $\mathfrak{G}$ of $\mathfrak{g}_c$ (the complexification of $\mathfrak{g}$) plays a prominent role in the representation theory of G. This will become apparent in Chapter 4 where we shall see, for instance, how a representation of G on a Banach space E, say, gives rise in a natural manner to various linear representations of $\mathfrak{G}$ on subspaces of E. Moreover the study of these representations of $\mathfrak{G}$ yields important information about the given representation of G. This chapter, then, which is algebraic in character, deals on the one hand with the structure of $\mathfrak{G}$ and, on the other, with various aspects of its representation theory. [Actually it will be just as easy to carry out most of the discussion in the context of a reductive Lie algebra $\mathfrak{g}$ over an algebraically closed field **k** of characteristic zero – unless specifically stated to the contrary, this will be the underlying assumption in what follows.]

We begin, in 2.1, with a review of $\mathfrak{g}$-modules and the associated customarily employed terminology. If E is a $\mathfrak{g}$-module and a is a non-zero element in E such that $X_E a = 0$ for all $X \in \mathfrak{g}$, then a is said to be a $\mathfrak{g}$-invariant (or simply an invariant). The theory of invariants dates to the classical algebraists; it has turned out that certain points from these older considerations along with developments of a more recent nature admit useful application to the representation theory of $\mathfrak{G}$. The purpose of 2.1, therefore, is to set up the machinery from invariant theory which will be needed in the sequel. Of particular importance here are the results of Chevalley (2.1.3) which, in one way or another, seem to lie at the heart of many of the central questions which face us.

In 2.2 we confine ourselves to the case $\mathbf{k} = \mathbf{C}$ and then proceed to establish a number of important 'finiteness' theorems due to Harish-Chandra (the proofs of which all resting ultimately on the results of Chevalley alluded to above) – these theorems will be applied later to the Harmonic Analysis on G. For example Theorem 2.2.1.1 (which is the *keystone* for all subsequent developments in the Harmonic Analysis on G) can be used to prove that the dimension of certain spaces of automorphic forms on G is finite or to solve various equations of convolution.

The discussion in 2.3 is devoted to a rapid recapitulation of various standard facts concerning the structure of $\mathfrak{G}$.

We turn, in 2.4, to the representation theory of $\mathfrak{g}$. As it was felt that the reader would already be familiar with the (classical) classification of the finite dimensional simple $\mathfrak{g}$-modules, we have limited ourselves here to merely outlining the relevant facts (cf. 2.4.1). [Let us recall that the main result along these lines, which is due to É. Cartan, establishes a one-to-one correspondence between such modules and the so-called 'dominant integral' linear functions on a Cartan subalgebra $\mathfrak{j}$ of $\mathfrak{g}$.] A similar attitude has been adopted towards Weyl's celebrated 'character formula'; thus in 2.4.2 we present a contemporary interpretation of this important result and then go on to prove an extension of it (due to Kostant) which will play an important role in the cohomology of $\mathfrak{g}$ (2.5). The exposition in 2.4.3 begins with a discussion of Weyl's character formula in a (perhaps) more familar setting and then moves to a discussion of the characters of $\mathfrak{g}$. A character of $\mathfrak{g}$ is a linear form $\kappa: \mathfrak{G} \to \mathbf{k}$ verifying:

(i) $\kappa(D_1 D_2) = \kappa(D_2 D_1)$, $\kappa(1) = 1$;

(ii) $\kappa(D_1^0 D_2) = \kappa(D_1)\kappa(D_2)$ $(D_1, D_2 \in \mathfrak{G})$.

[Here '^{0}' is the 'natural' projection of $\mathfrak{G}$ onto its center.] In particular to every finite dimensional simple $\mathfrak{g}$-module E_Λ (Λ a dominant integral linear function on $\mathfrak{j}$) there corresponds a character κ_Λ of $\mathfrak{g}$, viz. $\kappa_\Lambda(D) = \mathrm{tr}\,(U_\Lambda(D))/\dim\,(E_\Lambda)$ $(D \in \mathfrak{G}$, U_Λ the representation of $\mathfrak{g}$ on $E_\Lambda)$. One may pose the problem: Find all the characters of $\mathfrak{g}$. Harish-Chandra's solution to this question as well as an explicit formula for any character κ of $\mathfrak{g}$ is the gist of Theorem 2.4.3.2.

The present chapter terminates in 2.5 with a discussion of the algebraic foundations for the Borel-Weil Theorem. Thus let $\mathfrak{p}$ be a parabolic subalgebra of $\mathfrak{g}$, $\mathfrak{l}$ a Levi subalgebra of $\mathfrak{p}$, $\mathfrak{n}^+$ its nilpotent radical; let E_Λ be a finite dimensional simple $\mathfrak{g}$-module with highest weight Λ – then the cohomology spaces $H^q(\mathfrak{n}^+, E_\Lambda)$ can be regarded as (semi-simple) $\mathfrak{l}$-modules in a natural way and the main problem is to decompose them into simple $\mathfrak{l}$-modules. This is done by first computing the Euler-Poincaré characteristic of the cohomology space $H(\mathfrak{n}^+, E_\Lambda)$ in the representation ring of $\mathfrak{l}$ and then applying Kostant's generalization of Weyl's character formula.

2.1 Invariant Theory I – Generalities

2.1.1 Modules

We begin by recalling some basic module terminology together with a few well-known facts.

Let $\mathfrak{g}$ be a finite dimensional Lie algebra over a field $\mathbf{k}$ of characteristic zero (for our purposes, $\mathbf{k}$ will either be $\mathbf{R}$ or $\mathbf{C}$). A left $\mathfrak{g}$-module is a vector space E over $\mathbf{k}$ (possibly infinite dimensional) together with a $\mathbf{k}$-bilinear map $\mathfrak{g} \times E \to E$, denoted by $(X, a) \mapsto X_E a$, which satisfies the condition $[X, Y]_E a = (X_E Y_E - Y_E X_E)a$ for all $X, Y \in \mathfrak{g}$, $a \in E$. The corresponding Lie homomorphism U: $\mathfrak{g} \to \mathrm{Hom}_{\mathbf{k}}(E, E)$ is called a *linear representation* of $\mathfrak{g}$ and E is called the *representation space* (of U). We shall often write $U(X)$ instead of $X_E (X \in \mathfrak{g})$. Right $\mathfrak{g}$-modules are defined similarly. For brevity, a left $\mathfrak{g}$-module will be referred to simply as a $\mathfrak{g}$-module. If E is a finite dimensional $\mathfrak{g}$-module, then the bilinear form $(X, Y) \mapsto \mathrm{tr}\,(X_E Y_E)$ $(X, Y \in \mathfrak{g})$ is called the *bilinear form associated with* E (or with the corresponding representation of $\mathfrak{g}$).

The notions of direct sum of $\mathfrak{g}$-modules and isomorphism between $\mathfrak{g}$-modules are defined as usual, as are the concepts of simple and semi-simple $\mathfrak{g}$-modules. Let δ stand for an equivalence class of simple $\mathfrak{g}$-modules. If E is a $\mathfrak{g}$-module, let $E(\delta)$ be the sum of all simple submodules of E which are in the class δ. The various $E(\delta)$ so obtained as δ ranges over the set of all equivalence classes of simple $\mathfrak{g}$-modules are called the *isotypic submodules* of E. Evidently each $E(\delta)$ is a semi-simple $\mathfrak{g}$-module, being a sum of simple $\mathfrak{g}$-modules. In addition E itself is semi-simple iff $E = \sum_\delta E(\delta)$. Notice that this last sum is direct.

Let $\hat{\mathfrak{g}}$ denote the set of equivalence classes of *finite dimensional* simple $\mathfrak{g}$-modules. A semi-simple $\mathfrak{g}$-module E is said to be *locally finite* provided that $E = \sum_{\delta \in \hat{\mathfrak{g}}} E(\delta)$ (if E is a finite dimensional semi-simple $\mathfrak{g}$-module, the phrase 'locally finite' will be dropped).

Let E be a $\mathfrak{g}$-module, U the corresponding representation of $\mathfrak{g}$. Denoting by $\mathfrak{G}$ the universal enveloping algebra of $\mathfrak{g}$, E may of course also be viewed as a $\mathfrak{G}$-module (cf. 2.3.1).

Suppose that $\mathbf{k}$ is algebraically closed. Let $\delta \in \hat{\mathfrak{g}}$; fix once and for all a representation U_δ in the class δ together with a non-zero vector a_δ^* in the representation space E_δ of U_δ. Let $\mathfrak{A}_\delta$ denote the annihilator of a_δ^* in $\mathfrak{G}$, i.e.

$$\mathfrak{A}_\delta = \{D \in \mathfrak{G}: U_\delta(D)a_\delta^* = 0\}.$$

It is clear that $\mathfrak{G}/\mathfrak{A}_\delta$ and E_δ are equivalent as $\mathfrak{G}$-modules. Fix an element $D_\delta \in \mathfrak{G}$ such that $U_\delta(D_\delta)$ is a projection onto the one dimensional subspace determined by a_δ^* – this is possible in view of Burnside's Theorem.

Let E be a $\mathfrak{g}$-module, U the corresponding representation of $\mathfrak{g}$ on E. Let δ_0 be a class in $\hat{\mathfrak{g}}$ which occurs in E (i.e. $E(\delta_0) \neq \{0\}$). Put $E[\delta_0] = \{a \in E: U(\mathfrak{A}_{\delta_0})a = 0\}$ – clearly $E[\delta_0] \subset E(\delta_0)$.

Lemma 2.1.1.1 Retain the immediately preceding notations and assumptions.

(i) Let $\delta_1, \ldots, \delta_n$ be distinct elements in $\hat{\mathfrak{g}}$ – then there exists a $D \in \mathfrak{G}$ such that $U(D)$ is 1 in $E(\delta_i)$ and 0 in $E(\delta_j)$ $(1 \leqslant i, j \leqslant n)$.

(ii) Let $a_1, \ldots, a_n$ be elements of $\sum_{\delta \in \hat{\mathfrak{g}}} E(\delta)$; let $b_1, \ldots, b_n$ be elements of $E[\delta_0]$ – then there exists a $D \in \mathfrak{G}$ such that $U(D)a_i \in E[\delta_0]$, $U(D)b_i = b_i$ $(i = 1, \ldots, n)$.

Proof Statement (i) is a standard algebraic fact, namely the so-called Extended Burnside Theorem (see the Appendix to the present number). As for (ii), a simple argument shows that the restriction of $U(D_{\delta_0})$ to $E(\delta_0)$ is a projection of $E(\delta_0)$ onto $E[\delta_0]$. In addition, it may be supposed without loss of generality that $a_i \in E(\delta_i)$; this being so, choose, using (i), a $D_0 \in \mathfrak{G}$ such that $U(D_0)$ is 1 in $E(\delta_0)$ and 0 in those $E(\delta_i)$ such that $E(\delta_i) \neq E(\delta_0)$. The element $D = D_{\delta_0} D_0$ then fulfills the requirements in (ii). □

Let us turn now to the various ways in which one can form new modules from given ones. Let E_1 and E_2 be $\mathfrak{g}$-modules. The tensor product $E = E_1 \otimes_{\mathbf{k}} E_2$ can be made into a $\mathfrak{g}$-module in a unique way such that the rule

$$X_E(a_1 \otimes a_2) = (X_{E_1}a_1) \otimes a_2 + a_1 \otimes X_{E_2}a_2 \qquad (a_1 \in E_1, a_2 \in E_2)$$

holds. The corresponding representation U is called the *tensor product* of the representations U_1 and U_2 and is written $U = U_1 \otimes U_2$. The space $F = \mathrm{Hom}_{\mathbf{k}}(E_1, E_2)$ acquires the structure of a $\mathfrak{g}$-module if we put

$$X_F f = X_{E_2} \circ f - f \circ X_{E_1}$$

for all $f \in F$, $X \in \mathfrak{g}$. If E is a $\mathfrak{g}$-module, then the dual space $E^\vee$ of

E can be given the structure of a $\mathfrak{g}$-module in such a way that the corresponding representation $U^\vee$ satisfies the condition $U^\vee(X) = -{}^tU(X)$ for all $X \in \mathfrak{g}$ ('t' denotes transpose). $E^\vee$ is called the *dual* (or *contragredient*) module to E and $U^\vee$ is referred to as the *dual* (or *contragredient*) *representation* of the representation U. Observe that if E is a finite dimensional $\mathfrak{g}$-module, then E is simple (semi-simple) iff the $\mathfrak{g}$-module $E^\vee$ is simple (semi-simple). If E_1, E_2 are both finite dimensional $\mathfrak{g}$-modules, then $E_2 \otimes_{\mathbf{k}} E_1^\vee$ is $\mathfrak{g}$-isomorphic to $\mathrm{Hom}_{\mathbf{k}}(E_1, E_2)$ via the map which assigns to $\sum_i a_2^i \otimes \mathbf{z}_1^i (a_2^i \in E_2, \mathbf{z}_1^i \in E_1^\vee)$ the linear transformation

$$a_1 \mapsto \sum_i < a_1, \mathbf{z}_1^i > a_2^i (a_1 \in E_1).$$

Suppose that $\mathfrak{g}$ is, moreover, semi-simple – then we recall that the fundamental fact concerning $\mathfrak{g}$-modules is: Every finite dimensional $\mathfrak{g}$-module E is semi-simple. [This statement is false in general if E is infinite dimensional; cf. the second Example in 5.5.1.] The standard algebraic proof of this theorem may be found in Jacobson [3, p.79] (for the simple proof via Weyl's unitary trick see 4.3.2, Example 5 ($\mathbf{k} = \mathbf{C}$)). If $\mathfrak{g}$ is only assumed to be reductive, then a finite dimensional $\mathfrak{g}$-module E is semi-simple iff E is semi-simple when viewed as a $\mathfrak{c}$-module, $\mathfrak{c}$ the center of $\mathfrak{g}$ (Proposition 1.1.5.2(i)).

Returning to the general situation, let E_1 and E_2 be two finite dimensional semi-simple $\mathfrak{g}$-modules – then $E_1 \otimes_{\mathbf{k}} E_2$ is likewise semi-simple (see, for instance, Jacobson [3, p.83]). There are three immediate consequences of this remark, viz.:

(1) If E_1, E_2 are two finite dimensional semi-simple $\mathfrak{g}$-modules, then so is $\mathrm{Hom}_{\mathbf{k}}(E_1, E_2)$;

(2) Let E be a finite dimensional semi-simple $\mathfrak{g}$-module, $T = T(E)$ the tensor algebra over E, $S = S(E)$ the symmetric algebra over E, $\mathbf{\Lambda} = \mathbf{\Lambda}(E)$ the exterior algebra over E – then, with respect to the obvious $\mathfrak{g}$-structures, T and S are semi-simple (locally finite) $\mathfrak{g}$-modules, while $\mathbf{\Lambda}$ is a finite dimensional semi-simple $\mathfrak{g}$-module.

(3) Let E be a finite dimensional semi-simple $\mathfrak{g}$-module and suppose that in addition E is a Lie algebra: If the X_E $(X \in \mathfrak{g})$ are derivations, then the universal enveloping algebra $\mathfrak{E}$ of E is a semi-simple (locally finite) $\mathfrak{g}$-module (in the canonical $\mathfrak{g}$-structure inherited from E).

We shall conclude this number with a brief discussion of certain standard categorical notions.

As above, let $\mathfrak{g}$ be a finite dimensional Lie algebra over a field $\mathbf{k}$ of characteristic zero. Let $F(\mathfrak{g})$ be the free abelian group generated by the isomorphism classes of finite dimensional $\mathfrak{g}$-modules – thus an element of $F(\mathfrak{g})$ is a finite linear combination $\sum n_i E_i$ $(n_i \in \mathbf{Z}$, E_i a finite dimensional $\mathfrak{g}$-module). Let $R(\mathfrak{g})$ be the subgroup of $F(\mathfrak{g})$ generated by all elements $E - E' - E''$ where $0 \to E' \to E \to E'' \to 0$ is an exact sequence of $\mathfrak{g}$-modules. The *Grothendieck group* $K(\mathfrak{g})$ of the abelian category $\mathsf{C}(\mathfrak{g})$ of finite dimensional $\mathfrak{g}$-modules is, by definition, the quotient group $F(\mathfrak{g})/R(\mathfrak{g})$.

Since every finite dimensional $\mathfrak{g}$-module over $\mathbf{k}$ is of finite length, Jordan-Hölder's Theorem implies that $K(\mathfrak{g})$ is the free abelian group generated by the isomorphism classes of finite dimensional simple $\mathfrak{g}$-modules.

Definition By the $\mathfrak{g}$-*character* of a finite dimensional $\mathfrak{g}$-module E, we shall understand the class $\mathrm{ch}_{\mathfrak{g}}(E)$ of E in $K(\mathfrak{g})$.

It is clear that if E and F are finite dimensional semi-simple $\mathfrak{g}$-modules with the same $\mathfrak{g}$-character, then E and F are in fact isomorphic.

The group $K(\mathfrak{g})$ is a contravariant functor in $\mathfrak{g}$. Thus let $\mathfrak{g}$ and $\mathfrak{h}$ be Lie algebras; let p be a homomorphism of $\mathfrak{h}$ into $\mathfrak{g}$. If E is a finite dimensional $\mathfrak{g}$-module, then we may equip E with the structure of an $\mathfrak{h}$-module by writing $X_E a = p(X)_E a$ ($X \in \mathfrak{h}$, $a \in E$); let us denote the $\mathfrak{h}$-module so obtained by E_p. The operation which assigns to each finite dimensional $\mathfrak{g}$-module E, the $\mathfrak{h}$-module E_p is exact, hence leads to a homomorphism $p_!$ of $K(\mathfrak{g})$ into $K(\mathfrak{h})$; it is clear that $i_! = i$ (i the identity map in ...) – moreover if $\mathfrak{l}$ is a third Lie algebra and q is a homomorphism of $\mathfrak{l}$ into $\mathfrak{h}$, then one has $(p \circ q)_! = q_! \circ p_!$. In the event that p is the canonical injection of a subalgebra $\mathfrak{h}$ of $\mathfrak{g}$ into $\mathfrak{g}$, we shall refer to $p_!$ as the *homomorphism of restriction* and write $\mathrm{Res}(\mathfrak{g}, \mathfrak{h})$ instead of $p_!$.

If $0 \to E' \to E \to E'' \to 0$ is an exact sequence of $\mathfrak{g}$-modules and if F is a $\mathfrak{g}$-module, then the sequence $0 \to E' \otimes_{\mathbf{k}} F \to E \otimes_{\mathbf{k}} F \to E'' \otimes_{\mathbf{k}} F \to 0$ is still exact; the operation on $\mathsf{C}(\mathfrak{g}) \times \mathsf{C}(\mathfrak{g})$, with values in $\mathsf{C}(\mathfrak{g})$, which associates with each pair (E, F) the $\mathfrak{g}$-module $E \otimes_{\mathbf{k}} F$ is 'biadditive' and, evidently, defines a bilinear map from $K(\mathfrak{g}) \times K(\mathfrak{g})$ into $K(\mathfrak{g})$ – this map may be viewed as a multiplication relative to which $K(\mathfrak{g})$ acquires the structure of a commutative unitary ring, the unit element being the class of the trivial module.

The ring thus defined is called the *ring of representations* of $\mathfrak{g}$ and is denoted by $\mathfrak{R}(\mathfrak{g})$. We shall now proceed to establish an important fact concerning the structure of $\mathfrak{R}(\mathfrak{g})$.

Definition Let A be a commutative unitary ring. By a *λ-structure* on A we shall mean a homomorphism λ_t from the additive group of A into the multiplicative group $1 + A[[t]]^+$ (of all formal power series $1 + x_1 t + x_2 t^2 + \ldots$ with coefficients in A) verifying

$$\lambda_t(x) = 1 + xt + \ldots \qquad (x \in A).$$

A pair (A, λ_t) consisting of a commutative unitary ring A and a λ-structure on A is called a *λ-ring*. If (A, λ_t) is a λ-ring, we shall let $\lambda^i(x)$ denote the coefficient of t^i in $\lambda_t(x)$; thus $\lambda^0(x) = 1$ and $\lambda^1(x) = x (x \in A)$.

Example The ring of integers $\mathbf{Z}$ acquires the structure of a λ-ring in a natural fashion upon writing $\lambda_t(1) = 1 + t$, $\lambda_t(n) = (1 + t)^n$ (so that $\lambda^i(n) = \binom{n}{i}$, $n > 0$) – this λ-structure is called the *canonical* λ-structure (on $\mathbf{Z}$).

The ring of representations $\mathfrak{R}(\mathfrak{g})$ of a Lie algebra $\mathfrak{g}$ may be equipped with the structure of a λ-ring in the following way. Let E and F be finite dimensional $\mathfrak{g}$-modules; the equality $\mathbf{\Lambda}^i(E \otimes_{\mathbf{k}} F) = \sum_{r+s=i} (\mathbf{\Lambda}^r E) \otimes_{\mathbf{k}} (\mathbf{\Lambda}^s F)$, which is to be interpreted as an isomorphism of $\mathfrak{g}$-modules, implies that

$$\mathrm{ch}_{\mathfrak{g}}\,(\mathbf{\Lambda}^i(E \otimes_{\mathbf{k}} F)) = \sum_{r+s=i} \mathrm{ch}_{\mathfrak{g}}\,(\mathbf{\Lambda}^r E)\,\mathrm{ch}_{\mathfrak{g}}\,(\mathbf{\Lambda}^s F).$$

The operation which assigns to E the element

$$\lambda_t(E) = 1 + \mathrm{ch}_{\mathfrak{g}}(E)t + \ldots + \mathrm{ch}_{\mathfrak{g}}(\mathbf{\Lambda}^i(E))t^i + \ldots$$

in the multiplicative group $1 + \mathfrak{R}(\mathfrak{g})[[t]]^+$ is an additive operation in the category of (finite dimensional) semi-simple $\mathfrak{g}$-modules, which passes therefore to $\mathfrak{R}(\mathfrak{g})$ (the canonical injection from the category of (finite dimensional) semi-simple $\mathfrak{g}$-modules into $\mathsf{C}(\mathfrak{g})$ clearly induces an isomorphism between the corresponding Grothendieck groups). The λ-structure so obtained will be called the *canonical* λ-structure (on $\mathfrak{R}(\mathfrak{g})$).

Definition Let (A, λ_t) and (B, μ_t) be λ-rings. A ring morphism f: $A \to B$ is said to be a *homomorphism of λ-rings* if the diagram

$$\begin{array}{ccc} A & \xrightarrow{\lambda_t} & 1 + A[[t]]^+ \\ {\scriptstyle f}\downarrow & & \downarrow{\scriptstyle f_t} \\ B & \xrightarrow[\mu_t]{} & 1 + B[[t]]^+ \end{array}$$

is commutative (f_t the canonical extension of f to $A[[t]]$).

When (B, μ_t) is the λ-ring $\mathbf{Z}$ equipped with its canonical λ-structure, the homomorphism f is called an *augmentation* of (A, λ_t).

Example The rule which assigns to every finite dimensional $\mathfrak{g}$-module E its dimension is an augmentation of the λ-ring $\mathfrak{R}(\mathfrak{g})$ – this is clear since

$$\dim_{\mathbf{k}}\,(\mathbf{\Lambda}^i(E)) = \binom{\dim_{\mathbf{k}}(E)}{i}.$$

Example Let G be an abelian group. The group algebra $\mathbf{Z}[G]$ of G acquires the structure of a λ-ring, termed *canonical*, by writing, for the element e^x of the basis of $\mathbf{Z}[G]$ which corresponds to the element $x \in G$,

$$\lambda_t(e^x) = 1 + e^x t,\ \mathrm{Aug}\,(e^x) = 1.$$

The preceding notions are conveniently illustrated in the following situation. Consider a solvable Lie algebra $\mathfrak{b}$ over an algebraically closed field $\mathbf{k}$ of characteristic zero. Owing to Lie's Theorem, the ring of representations $\mathfrak{R}(\mathfrak{b})$ of $\mathfrak{b}$ is isomorphic to the ring of representations of the abelian Lie algebra $\mathfrak{j} = \mathfrak{b}/[\mathfrak{b}, \mathfrak{b}]$; since $\mathbf{k}$ is algebraically closed, the simple representations of $\mathfrak{j}$ all are of dimension 1 and conversely, that is, the simple representations of $\mathfrak{j}$ are characterized in the evident manner by linear functions on $\mathfrak{j}$ – therefore $K(\mathfrak{b}) \sim K(\mathfrak{j})$ is isomorphic to the free abelian group $\mathbf{Z}[\mathfrak{j}^\vee]$ generated by the elements in the $\mathbf{k}$-dual $\mathfrak{j}^\vee$ of $\mathfrak{j}$. But even more than this is true; for a simple argument tells us that the ring structure on $K(\mathfrak{j})$ coincides with the ring structure on the group algebra $\mathbf{Z}[\mathfrak{j}^\vee]$ of $\mathfrak{j}^\vee$. To summarize: If $\mathfrak{b}$ is a solvable Lie algebra over an algebraically closed field $\mathbf{k}$ of characteristic zero, then the representation ring $\mathfrak{R}(\mathfrak{b})$ (equipped with its canonical λ-structure) is isomorphic to the group algebra of $(\mathfrak{b}/[\mathfrak{b}, \mathfrak{b}])^\vee$ (equipped with its canonical λ-structure), the isomorphism being defined by attaching to each (finite dimensional) simple $\mathfrak{b}$-module its 'weight' (i.e. the linear function on $\mathfrak{b}$ which vanishes on $[\mathfrak{b}, \mathfrak{b}]$).

[These remarks may be applied, in particular, to a Borel subalgebra $\mathfrak{b}$ of a reductive Lie algebra $\mathfrak{g}$ over an algebraically closed field $\mathbf{k}$ of characteristic zero; in this case, $\mathfrak{b}/[\mathfrak{b}, \mathfrak{b}]$ will be a Cartan subalgebra of $\mathfrak{g}$.]

Appendix For completeness, let us give a proof of the so-called *Extended Burnside Theorem.* Let $\mathfrak{g}$ be a Lie algebra over an algebraically closed field $\mathbf{k}$ of characteristic zero, $\mathfrak{G}$ its universal enveloping algebra; let U be a representation of $\mathfrak{g}$ (equivalently $\mathfrak{G}$ ...) on a space E with the property that there exist finitely many distinct classes $\delta_1, \ldots, \delta_n$ in $\hat{\mathfrak{g}}$ such that $E = \sum_{i=1}^n E(\delta_i)$ – then there exist elements $D_1, \ldots, D_n$ in $\mathfrak{G}$ such that $U(D_i)a = a$ or 0 according as $a \in E(\delta_i)$ or $a \in E(\delta_j)$ $(i \neq j)$ $(1 \leqslant i, j \leqslant n)$.

Proof Let U_i be the representation of $\mathfrak{G}$ on $E(\delta_i)$ obtained by restricting U to $E(\delta_i)$; let $\mathfrak{K}_i$ be the kernel of U_i in $\mathfrak{G}$. Because the class δ_i is irreducible, $\mathfrak{K}_i$ is a maximal two-sided ideal in $\mathfrak{G}$ and $\mathfrak{K}_i \neq \mathfrak{K}_j$ if $i \neq j$ $(1 \leqslant i, j \leqslant n)$. Set $\mathfrak{K} = \bigcap_i \mathfrak{K}_i$ – then $\mathfrak{K}$ is the kernel of U and since $\delta_1, \ldots, \delta_n$ are all the distinct classes which occur in U, it is clear that $\mathfrak{K}_1, \ldots, \mathfrak{K}_n$ are the only maximal ideals in $\mathfrak{G}$ which contain $\mathfrak{K}$. Owing to Burnside's Theorem, the algebra $\bar{\mathfrak{G}} = \mathfrak{G}/\mathfrak{K}$ is finite dimensional and semi-simple; every maximal ideal in $\bar{\mathfrak{G}}$ is of the form $\bar{\mathfrak{K}}_i$ $(= \mathfrak{K}_i/\mathfrak{K})$ $(1 \leqslant i \leqslant n)$ and, in view of our hypotheses, these are all distinct; let $\bar{\mathfrak{L}}_i$ be the ideal complementary to $\bar{\mathfrak{K}}_i$ $(1 \leqslant i \leqslant n)$ – then the set $\{\bar{\mathfrak{L}}_1, \ldots, \bar{\mathfrak{L}}_n\}$ is a complete collection of distinct minimal ideals in $\bar{\mathfrak{G}}$. Because $\mathfrak{K}$ is the kernel of U, we may regard U as a faithful representation of $\bar{\mathfrak{G}}$ on E; evidently $U(\bar{\mathfrak{L}}_i)E \subset E(\delta_i)$ (for $\bar{\mathfrak{K}}_i\bar{\mathfrak{L}}_i = \{0\}$, $U(\bar{\mathfrak{K}}_i)U(\bar{\mathfrak{L}}_i)E = \{0\}\ldots$); moreover $E = U(\bar{\mathfrak{G}})E = \sum_{i=1}^n U(\bar{\mathfrak{L}}_i)E$ and so we see that $U(\bar{\mathfrak{L}}_i)E = E(\delta_i)$ $(1 \leqslant i \leqslant n)$. Let $\bar{D}_i$ be the unit element of the simple algebra $\bar{\mathfrak{L}}_i$ – then $\bar{D}_i\bar{D}_j = 0$ if $i \neq j$ $(1 \leqslant i, j \leqslant n)$. Select now an element D_i in $\mathfrak{G}$ such that its residue class mod $\mathfrak{K}$ is $\bar{D}_i$ – then $U(D_i)E = U(\bar{D}_i)E = E(\delta_i)$ $(1 \leqslant i \leqslant n)$. Consequently, if $a \in E(\delta_j)$, then $a = U(D_j)b$ $(b \in E)$, whence $U(D_i)a = U(\bar{D}_i\bar{D}_j)b = \delta_{ij}a$ $(1 \leqslant i, j \leqslant n)$. This proves the theorem. □

2.1.2 The Fundamental Theorem of Invariant Theory

Before proceeding to invariant theory proper, it will be well to first lay down some notational conventions.

Let E be a finite dimensional vector space over a field $\mathbf{k}$ of characteristic zero, $E^\vee$ its dual and $P(E)$ the algebra of $\mathbf{k}$-valued polynomial functions on E. Let $S = S(E)$ denote the symmetric algebra of E – whenever it proves to be convenient we shall identify $S(E)$ with $P(E^\vee)$. Let A be an automorphism of E – then A can be extended uniquely to an automorphism $u \mapsto u^A$ of $S(E)$. Moreover, for any $p \in P(E)$, define p^A to be the function $a \mapsto p(A^{-1}a)(a \in E)$ – then the map $p \mapsto p^A$ is an automorphism of $P(E)$. Denote the homogeneous elements of degree m in $S(E)$ (respectively $P(E)$) by $S^m(E)$ (respectively $P^m(E)$).

Suppose that E is the direct sum of two subspaces E_1, E_2. Let π_i denote the projection of E onto E_i $(i = 1, 2)$ corresponding to the decomposition $E = E_1 + E_2$; it is clear that π_i can be extended uniquely to a homomorphism of $S(E)$ onto $S(E_i)$ – call π_i the projection of $S(E)$ on $S(E_i)$. On the other hand, in the obvious way, $S(E_i) \subset S(E)$ and $\pi_i u = u$ for $u \in S(E_i)$.

Let $\mathfrak{g}$ be a finite dimensional Lie algebra over a field $\mathbf{k}$ of characteristic zero, E a $\mathfrak{g}$-module.

Definition An element $a \in E$ is said to be *invariant* (for the $\mathfrak{g}$-module structure on E or for the corresponding representation U of $\mathfrak{g}$ on E) if $X_E a = 0$ for all $X \in \mathfrak{g}$.

The motivation for this terminology arises from the fact that if G is a connected Lie group with Lie algebra $\mathfrak{g}$ and if U is a continuous representation of G on a finite dimensional vector space E whose differential is dU, then the element a is invariant for dU iff $a = U(x)a$ for all $x \in G$. [Of course the definition given above does not request that E be necessarily finite dimensional.]

As an example a $\mathbf{k}$-linear map $f: E_1 \to E_2$ is an invariant for the action of $\mathfrak{g}$ on $F = \mathrm{Hom}_{\mathbf{k}}(E_1, E_2)$ iff $f \circ X_{E_1} = X_{E_2} \circ f$ (all $X \in \mathfrak{g}$), i.e. iff f is a homomorphism of $\mathfrak{g}$-modules. For another example, suppose that E is a finite dimensional $\mathfrak{g}$-module with basis $\{a_i\}$ $(1 \leqslant i \leqslant n)$ and suppose that $\{\mathbf{z}_i\}$ $(1 \leqslant i \leqslant n)$ is the corresponding dual basis – then the identity map $I \in \mathrm{Hom}_{\mathbf{k}}(E, E)$ is evidently an invariant which, under the $\mathfrak{g}$-isomorphism between $\mathrm{Hom}_{\mathbf{k}}(E, E)$ and $E \otimes_{\mathbf{k}} E^\vee$, corresponds to the element $\sum_i a_i \otimes \mathbf{z}_i$ (which is therefore an invariant for the $\mathfrak{g}$-module $E \otimes_{\mathbf{k}} E^\vee$).

Notice that the subspace E_I of invariants in E is exactly the isotypic submodule $E(\delta^0)$, where δ^0 denotes the class of the zero representation of $\mathfrak{g}$ on a one dimensional space.

Lemma 2.1.2.1 Let E be a semi-simple $\mathfrak{g}$-module – then the submodule E_I of invariants in E admits one and only one complementary submodule E_R. E_R is spanned by the $X_E a$ $(X \in \mathfrak{g}, a \in E)$.

Proof Let $\tilde{E}_I$ be a complementary submodule of E_I. Given $a \in E$, write $a = a_I + \tilde{a}_I$ $(a_I \in E_I, \tilde{a}_I \in \tilde{E}_I)$ – then $X_E a = X_E \tilde{a}_I$ which implies that $E_R \subset \tilde{E}_I$. Let $\tilde{E}_R$ be a complementary submodule of E_R in $\tilde{E}_I$. For every $\tilde{a}_R \in \tilde{E}_R$, $X_E \tilde{a}_R \in E_R \cap \tilde{E}_R = \{0\}$; this being so for all $X \in \mathfrak{g}$, it must be the case that $\tilde{a}_R = 0$ – hence $E_R = \tilde{E}_I$. $\square$

Lemma 2.1.2.2 Let E and F be $\mathfrak{g}$-modules, F being assumed finite dimensional; let $b_1, \ldots, b_n$ be a $\mathbf{k}$-basis for F; suppose that $a_1, \ldots, a_n$ are elements of E, not all of which are zero, such that $\sum_{i=1}^{n} a_i \otimes b_i$ is an invariant for $E \otimes_{\mathbf{k}} F$ – then the subspace $\tilde{E}$ of E spanned by the a_i is a $\mathfrak{g}$-submodule of E. In addition, if F is simple, then the representation of $\mathfrak{g}$ on $\tilde{E}$ is contragredient to the representation of $\mathfrak{g}$ on F.

Proof Let $X_F b_i = \sum_j k_{ji} b_j$ ($k_{ji} \in \mathbf{k}$); since $\sum_i a_i \otimes b_i$ is an invariant for $E \otimes_{\mathbf{k}} F$, we see that $\sum_j (X_E a_j) \otimes b_j = -\sum_j (\sum_i k_{ji} a_i) \otimes b_j$ and so $X_E a_i = -\sum_j k_{ij} a_j$ which implies that $\tilde{E}$ is $\mathfrak{g}$-stable. Suppose now that F is simple. The space $\tilde{E} \otimes_{\mathbf{k}} F$ is $\mathfrak{g}$-isomorphic in the natural way to $\mathrm{Hom}_{\mathbf{k}}(F^\vee, \tilde{E})$ and under this isomorphism, the invariant element $\sum_i a_i \otimes b_i$ corresponds to the map $\mathbf{z} \mapsto \sum_i \langle b_i, \mathbf{z} \rangle a_i$ of $F^\vee$ into $\tilde{E}$. In turn this map is an invariant for the $\mathfrak{g}$-module structure on $\mathrm{Hom}_{\mathbf{k}}(F^\vee, \tilde{E})$, thus is a non-zero $\mathfrak{g}$-module homomorphism, and, $F^\vee$ being simple, is an isomorphism. Our assertion is now clear. □

We shall now take up the 'Fundamental Theorem of Invariant Theory'.

Theorem 2.1.2.3 (Hilbert-Godement) Let $\mathfrak{g}$ be a finite dimensional Lie algebra over $\mathbf{k}$, E a finite dimensional semi-simple $\mathfrak{g}$-module, $S = S(E)$ the symmetric algebra over E, and X_S the derivation of S which extends X_E (thus $X \mapsto X_S$ is a representation of $\mathfrak{g}$ on S) – then:

(i) The algebra S_I of invariants in S is a finitely generated algebra;

(ii) For every equivalence class $\delta \in \hat{\mathfrak{g}}$, $S(\delta)$ is a finite S_I-module.

The proof of the theorem will be prefaced with two simple lemmas.

Lemma 2.1.2.4 Let V_1, V_2, V_3 be representations of $\mathfrak{g}$ on the vector spaces F_1, F_2, F_3 respectively. Let there be given a $\mathbf{k}$-bilinear map $(b_1, b_2) \mapsto b_1 b_2$ of $F_1 \times F_2$ into F_3 such that

$$(V_1(X)b_1)b_2 + b_1(V_2(X)b_2) = V_3(X)(b_1 b_2)$$

for all $b_1 \in F_1$, $b_2 \in F_2$, $X \in \mathfrak{g}$.

(i) If $b_1 \in (F_1)_I$, then the map $b_2 \mapsto b_1 b_2$ is a $\mathfrak{g}$-module homomorphism.

(ii) If $b_2 \in F_2(\delta)$, then $b_1 b_2 \in F_3(\delta)$ ($b_1 \in (F_1)_I$).

The proof of this lemma is obvious. Consequence: Suppose that V is a representation of $\mathfrak{g}$ by derivations on a (not necessarily associative) algebra A – then the set of invariants A_I is a subalgebra of A and each isotypic component $A(\delta)$ is a right and left A_I-module. [In fact we may apply Lemma 2.1.2.4 in the case $A = F_1 = F_2 = F_3$, $V = V_1 = V_2 = V_3$].

Lemma 2.1.2.5 Under the conditions of Lemma 2.1.2.4, suppose in addition that F_2 and F_3 are finite dimensional and semi-simple. For every

$b_2 \in F_2$ (respectively $b_3 \in F_3$), let b_2^0 (respectively b_3^0) be its component in $(F_2)_I$ (respectively $(F_3)_I$). Let $b_1 \in (F_1)_I$ – then, for every $b_2 \in F_2$, $(b_1 b_2)^0 = b_1 b_2^0$.

Proof Since F_2 and F_3 are semi-simple, $F_2 = \Sigma F_2(\delta)$, $F_3 = \Sigma F_3(\delta)$. Let $b_2 \in F_2$; by linearity it may be assumed that $b_2 \in F_2(\delta)$ for some δ. If $\delta \neq \delta^0$, then $b_2^0 = 0$ and, as $b_1 b_2 \in F_3(\delta)$ (Lemma 2.1.2.4(ii)), we have $(b_1 b_2)^0 = 0 = b_1 b_2^0$. If $\delta = \delta^0$, then $b_2 = b_2^0$ and, since $b_1 b_2 \in (F_3)_I$, it follows that $(b_1 b_2)^0 = b_1 b_2 = b_1 b_2^0$. □

Proof of Theorem 2.1.2.3 Since E is a finite dimensional semi-simple $\mathfrak{g}$-module, S is a semi-simple locally finite $\mathfrak{g}$-module – this was noted above. Write $S = \sum_0^\infty S^m$, S^m the homogeneous elements of degree m in S. Because the $X_S (X \in \mathfrak{g})$ preserve degree, each S^m is $\mathfrak{g}$-stable and so $S^m = \sum_\delta (S^m \cap S(\delta))$; likewise $S(\delta) = \sum_m (S^m \cap S(\delta))$. Plainly the machinery developed in Lemmas 2.1.2.4, 2.1.2.5 is applicable here.

(1) Let us first prove that the algebra S_I of invariants is finitely generated. Let S^+ denote the ideal in S consisting of elements without a constant term, S_I^+ the ideal in S generated by $S_I \cap S^+$. By the Hilbert Basis Theorem, S_I^+ has a finite number of *ideal* generators, say $u_1, \ldots, u_n$, which, without loss of generality, may be taken homogeneous and in $S_I \cap S^+$. [This is clear since the X_S preserve degree.] Let $\tilde{S}_I$ be the subalgebra of S_I generated by 1 and the u_i. Claim: $\tilde{S}_I = S_I$. To establish the claim, it is sufficient to show that every homogeneous element $u \in S_I$ is in $\tilde{S}_I$ and in turn the proof of this will proceed by induction on the degree m of u. The case $m = 0$ being obvious, assume $m > 0$ and that the claim is valid for all homogeneous $u \in S_I$ of degree $<m$. Let u be of degree m; write $u = \sum_{i=1}^n u_i v_i$ where the $v_i \in S$. It may be supposed that the v_i are homogeneous and $\deg(v_i) = \deg(u) - \deg(u_i) < m$. Applying Lemma 2.1.2.5 and the notations thereof gives

$$u = u^0 = \sum_{i=1}^n (u_i v_i)^0 = \sum_{i=1}^n u_i v_i^0.$$

The elements v_i^0 are homogeneous elements in S_I of degree $<m$, hence they are in $\tilde{S}_I$ – thus $u \in \tilde{S}_I$ and so $\tilde{S}_I = S_I$, as claimed.

(2) Turning to the proof of (ii), Lemma 2.1.2.4 implies that each $S(\delta)$ is in fact an S_I-module ($\delta \in \hat{\mathfrak{g}}$). Let E_δ be a $\mathfrak{g}$-module in the class δ – then $L_\delta = \mathrm{Hom}_{\mathbf{k}}(E_\delta, S)$ is canonically a $\mathfrak{g}$-module. Since each S^m is a semi-simple $\mathfrak{g}$-module, so is $\mathrm{Hom}_{\mathbf{k}}(E_\delta, S^m)$; consequently L_δ is a semi-simple $\mathfrak{g}$-module. On the other hand, L_δ may be equipped in a natural way with the structure of an S-module – thus put $(ul)(a) = ul(a)$ for $u \in S$, $l \in L_\delta$, $a \in E_\delta$. Now E_δ is finite dimensional over $\mathbf{k}$ and so it is clear that L_δ is a finite S-module. But S is a Noetherian ring, hence L_δ is a Noetherian S-module (i.e. satisfies the ascending chain condition for S-submodules or, equivalently, every S-submodule has a finite S-basis). Given $u \in S$, $l \in L_\delta$, $X \in \mathfrak{g}$ it is easy to check that

$$(*)\quad (X_S u)l + u(X_{L_\delta} l) = X_{L_\delta}(ul).$$

Therefore Lemmas 2.1.2.4 and 2.1.2.5 are applicable per the pairing $S \times L_\delta \to L_\delta$. Consider $E_\delta \otimes_{\mathbf{k}} L_\delta$ – then $E_\delta \otimes_{\mathbf{k}} L_\delta$ acquires the structure of an S-module in the obvious way (thus write $u(a \otimes l) = a \otimes ul$ for $u \in S$, $a \in E_\delta$, $l \in L_\delta$). The canonical map ϕ, $a \otimes l \mapsto l(a)$, of $E_\delta \otimes_{\mathbf{k}} L_\delta$ onto S is evidently a homomorphism of S-modules. Moreover ϕ is a homomorphism of $\mathfrak{g}$-modules – thus, setting $M_\delta = E_\delta \otimes_{\mathbf{k}} L_\delta$, we have

$$\begin{aligned}\phi(X_{M_\delta}(a \otimes l)) &= \phi((X_{E_\delta}a) \otimes l + a \otimes X_{L_\delta} l) = l(X_{E_\delta}a) + (X_{L_\delta} l)(a)\\ &= l(X_{E_\delta}a) + X_S(l(a)) - l(X_{E_\delta}a) = X_S\phi(a \otimes l).\end{aligned}$$

Hence, since $(L_\delta)_I$ consists precisely of the $\mathfrak{g}$-homomorphisms of E_δ into S, it follows that $\phi(E_\delta \otimes_{\mathbf{k}} (L_\delta)_I) = S(\delta)$. Therefore, in order to show that $S(\delta)$ is a finite S_I-module, it suffices to show that $(L_\delta)_I$ is a finite S_I-module. Let $\tilde{L}_\delta$ be the S-submodule generated by $(L_\delta)_I$; since L_δ is a Noetherian S-module, there exists a finite number of elements $l_1, \ldots, l_n$ in $(L_\delta)_I$ which generate $\tilde{L}_\delta$. Let $(\tilde{L}_\delta)_I$ be the S_I-submodule generated by the l_i; the formula derived in $(*)$ implies that $(\tilde{L}_\delta)_I \subset (L_\delta)_I$ and the claim is that $(\tilde{L}_\delta)_I = (L_\delta)_I$. So let $l \in (L_\delta)_I$ and write $l = \sum_{i=1}^n u_i l_i$ $(u_i \in S)$; an application of Lemma 2.1.2.5 yields:

$$l = l^0 = \sum_{i=1}^n (u_i l_i)^0 = \sum_{i=1}^n u_i^0 l_i \in (\tilde{L}_\delta)_I.$$

Here $(\tilde{L}_\delta)_I = (L_\delta)_I$ which implies that $(L_\delta)_I$ is a finite S_I-module.

This completes the proof of the theorem. □

Let $\mathfrak{g}$ be a finite dimensional Lie algebra over $\mathbf{k}$, E a finite dimensional semi-simple $\mathfrak{g}$-module, $S = S(E)$ the symmetric algebra over E and X_S the derivation of S which extends $X_E (X \in \mathfrak{g})$. Let F be a finite dimensional semi-simple $\mathfrak{g}$-module. Put $L = S \otimes_{\mathbf{k}} F$ and define on L the structure of an S-module in the obvious way; call an element $l \in L$ homogeneous of degree m if $l \in S^m \otimes_{\mathbf{k}} F$.

The proof of the following proposition is contained in the proof of Theorem 2.1.2.3.

Proposition 2.1.2.6 In the above notations, the set of invariants L_I is a finite S_I-module – moreover the generators may be taken homogeneous.

Let E be a finite dimensional vector space over $\mathbf{k}$, G a group of automorphisms of E – then G operates on $S = S(E)$ and an element $u \in S(E)$ is said to be *G-invariant* if $u^x = u$ for all $x \in G$. The fundamental problem of invariant theory may be taken to be: Is the set S_I of G-invariants in S a finitely generated algebra over $\mathbf{k}$? This question, which is a special case of Hilbert's 14th problem, was answered in the negative by Nagata [1]. On the positive side of the ledger, consider the following examples.

Examples (1) If $\mathbf{k} = \mathbf{R}$ or $\mathbf{C}$ and if G is compact, then S_I is a finitely generated algebra over $\mathbf{k}$ – see Helgason [2, p.421].

(2) Let $\mathfrak{g}$ be a semi-simple Lie algebra over $\mathbf{R}$, G a connected Lie group with Lie algebra $\mathfrak{g}$. Extend the adjoint representation $x \mapsto \mathrm{Ad}(x)$ of G on $\mathfrak{g}$ to $S(\mathfrak{g}_c)$ and denote the corresponding set of G-invariants by $I(\mathfrak{g}_c)$ – then $I(\mathfrak{g}_c)$ is a finitely generated algebra over $\mathbf{C}$ in view of Theorem 2.1.2.3 (further remarks appear in 2.1.5).

(3) Let $\mathfrak{g}_c$ be a complex semi-simple Lie algebra, G_c the simply connected complex analytic group corresponding to $\mathfrak{g}_c$; let N_c be a maximal unipotent subgroup of G_c – then it can be shown that the ring $R(G_c/N_c)$ of rational functions on the quotient space G_c/N_c is a finitely generated algebra. Now let U be a finite dimensional holomorphic representation of G_c on a space E (say) – then G_c operates on the symmetric algebra $S(E)$, the corresponding set of G_c-invariants being a finitely generated algebra (cf. supra). Remarkably enough, the N_c-invariants in $S(E)$ also constitute a finitely generated algebra; for, as can be shown, the algebra of N_c-invariants in $S(E)$ is isomorphic to the G_c-invariants in the tensor product $R(G_c/N_c) \otimes_{\mathbf{C}} S(E)$ (cf. Hadziev [1]).

2.1.3 Invariants of Finite Groups Generated by Reflections

In this number we shall establish some results of a general nature concerning finite reflection groups; these results will be applied primarily to the Weyl groups which arise from the root systems which appear in the theory of semi-simple Lie algebras.

An invertible linear transformation of a finite dimensional vector space E over a field $\mathbf{k}$ of characteristic zero is said to be a *reflection* if it is of order two and leaves a hyperplane pointwise fixed. A group G of linear transformations in E is a *finite reflection group* if it is a finite group generated by reflections.

The following result is basic for subsequent developments.

Theorem 2.1.3.1 (Chevalley) If G is a finite reflection group operating in E, then the $\mathbf{k}$-algebra P_I of G-invariants in $P = P(E)$ is generated by n algebraically independent homogeneous elements and 1 ($n = \dim(E)$) – thus P_I is a polynomial ring.

Proof We shall let P^+ denote the ideal in P consisting of elements without a constant term, P_I^+ the ideal in P generated by $P_I \cap P^+$; given $p \in P$, we shall denote by $M(p)$ its average over G (thus

$$M(p) = [G]^{-1} \sum_{x \in G} p^x,$$

$[G]$ the order of G).

(1) Let $p_0, p_1, \ldots$ be elements of P_I such that p_0 is not in the ideal in P_I generated by the others and let $q_0, q_1, \ldots$ be homogenous elements in P such that $\sum_{i>0} p_i q_i = 0$ – then $q_0 \in P_I^+$. Thus first note that p_0 actually does not belong to the ideal in P generated by $p_1, \ldots$. [If the contrary were true, then we would necessarily have that $p_0 = \sum_{i>1} \tilde{p}_i p_i$ for certain $\tilde{p}_i \in P$; in turn this would imply that $p_0 = M(p_0) = \sum_{i>1} M(\tilde{p}_i) p_i$, a contradiction. . . .] This being so, let us proceed by induction on the degree $d = \deg(q_0)$ of q_0. If $d = 0$, then, in view of the immediately preceding

remarks, $q_0 = 0 \in P_I^+$. Assuming now that $d > 0$, let x be a reflection in G leaving pointwise fixed a hyperplane with equation $\mathscr{H}_x = 0$ – then, for each i, $\mathscr{H}_x$ divides $q_i - q_i^x$, whence $\sum_{i>0} p_i((q_i - q_i^x)/\mathscr{H}_x)) = 0$ and so, by the induction hypothesis $q_0 - q_0^x \in P_I^+$, that is $q_0 \equiv q_0^x \pmod{P_I^+}$. Because G is generated by reflections, a relation of this type must hold for all $x \in G$ – hence $q_0 \equiv M(q_0)(\mathrm{mod}\, P_I^+)$, which implies of course that $q_0 \in P_I^+$, as desired.

(2) Let $p_1, \ldots, p_\nu$ be a minimal set of homogeneous invariants which form an ideal basis for P_I^+ (such a basis exists by Hilbert's classical theorem) – we intend to prove that the p_i's are algebraically independent. If this is not the case, then let $r(p_1, \ldots, p_\nu) = 0$ be a non-trivial relation with all monomials in the p_i's of the same minimal degree in the underlying coordinates $x_1, \ldots, x_n$. Let $r_i = \partial r(p_1, \ldots, p_\nu)/\partial p_i$; by the choice of r, not all the r_i are zero. We may arrange the notation in such a way that $\{r_1, \ldots, r_\mu\}$ $(\mu \leqslant \nu)$ but no subset of it generates the ideal in P_I generated by all the r_i. Let $r_j = \sum_{i=1}^{\mu} q_{ij} r_i$ for $j = \mu + 1, \ldots, \nu$ where $q_{ij} \in P_I$ and all terms in the equation are homogeneous of the same degree – then, for $k = 1, 2, \ldots, n$, we have that

$$0 = \frac{\partial r}{\partial x_k} = \sum_{i=1}^{\nu} r_i \frac{\partial p_i}{\partial x_k} = \sum_{i=1}^{\mu} r_i \left(\frac{\partial p_i}{\partial x_k} + \sum_{j=\mu+1}^{\nu} q_{ij} \frac{\partial p_j}{\partial x_k} \right).$$

Owing to the observations in (1) supra, we see that

$$\frac{\partial p_1}{\partial x_k} + \sum_{j=\mu+1}^{\nu} q_{1j} \frac{\partial p_j}{\partial x_k} \in P_I^+.$$

Multiplying this relation through by x_k, summing over k, using Euler's formula, and writing $d_j = \deg(p_j)$, we find that

$$d_1 p_1 + \sum_{j=\mu+1}^{\nu} q_{1j}(d_j p_j) = \sum_{i=1}^{\nu} p_i q_i$$

where q_i belongs to the ideal in P generated by the x_k. Considerations of homogeneity imply that $q_1 = 0$; thus p_1 is in the ideal generated by $p_2, \ldots, p_\nu$, a contradiction.

(3) Retain the notations introduced in (2) – then we claim that the p_i's generate P_I as an algebra. To see this, fix a homogeneous element p (say) in P_I of positive degree; we may write $p = \sum_i p_i q_i$ for certain $q_i \in P$; by averaging, it can be assumed that each $q_i \in P_I$; since the degree of each q_i is less than the degree of p, it follows by induction that p is a polynomial in the p_i's.

(4) The discussion in (2) implies that $\nu \leqslant n$. On the other hand, if $Q(P)$ and $Q(P_I)$ denote the quotient fields of P and P_I respectively, then $Q(P)$ is a Galois extension of $Q(P_I)$ with Galois group G, hence $Q(P_I)$ has transcendence degree n over $\mathbf{k}$, whence $\nu \geqslant n$. [Let us note that the finiteness of G implies that every invariant in $Q(P)$ is the quotient of two invariant polynomials.]

The proof of the theorem is now complete. $\square$

Example Consider the four 'classical' reduced irreducible root systems $\mathbf{A}_l(l \geqslant 1)$, $\mathbf{B}_l(l \geqslant 2)$, $\mathbf{C}_l(l \geqslant 3)$, and $\mathbf{D}_l(l \geqslant 4)$ (see the Appendix to 1.1.2 in Chapter 1). Let $(.\,,.)$ be the usual inner product in $\mathbf{R}^n$, $\{e_1, \ldots, e_n\}$ the standard basis. The Weyl group W for $\mathbf{A}_l$ may be identified with the group of permutations of the $e_1, \ldots, e_{l+1}$ – here, then, the elementary symmetric polynomials $\sigma_2, \ldots, \sigma_{l+1}$ in the e_i are W-invariant and generate all other polynomials which are invariant under W. Turning to $\mathbf{B}_l$ and $\mathbf{C}_l$, we see that since the Weyl group W consists, in these two cases, of permutations and sign changes of the e_i, the elementary symmetric polynomials in $e_1^2, \ldots, e_l^2$ are W-invariant and generate all other polynomials which are invariant under W. As for $\mathbf{D}_l$, its Weyl group can be regarded as consisting of the permutations and *even* number of sign changes of the e_i – hence the generators for the W-invariants are the same as those for $\mathbf{B}_l$, $\mathbf{C}_l$ with the exception that $e_1^2 \ldots e_l^2$ has to be replaced by $e_1 \ldots e_l$.

Proposition 2.1.3.2 (Todd-Shephard) Let G be a finite reflection group in E; let $p_1, \ldots, p_n$ ($n = \dim(E)$) be algebraically independent homogeneous generators for the $\mathbf{k}$-algebra P_I of G-invariants in $P = P(E)$; let d_i denote the degree of p_i ($i = 1, \ldots, n$) – then the d_i are uniquely determined. In addition the d_i verify the relations $\prod_i d_i = [G]$, $\sum_i (d_i - 1) = N_G$ (N_G the number of reflections in G).

The proof of this result rests on the following lemma.

Lemma 2.1.3.3 If the p_i and the d_i are as in the statement of the preceding proposition, then

$$\prod_{i=1}^{n} (1 - t^{d_i})^{-1} = [G]^{-1} \sum_{x \in G} (\det(1 - xt)^{-1})$$

as a formal identity in t.

Proof Let $\epsilon_1, \ldots, \epsilon_n$ be the eigenvalues of x ($x \in G$) and $x_1, \ldots, x_n$ the corresponding eigenfunctions – then $\det(1 - xt)^{-1} = \prod_i (1 + \epsilon_i t + \epsilon_i^2 t^2 + \ldots)$. The coefficient of t^d is $\sum_{r_1 + r_2 + \cdots = d} \epsilon_1^{r_1} \epsilon_2^{r_2} \ldots$, i.e. the trace of x acting on the space of homogeneous polynomials in $x_1, \ldots, x_n$ of degree d (the monomials $x_1^{r_1} x_2^{r_2} \ldots$ form a basis for this space). By averaging we obtain the dimension of the space of invariant homogeneous polynomials of degree d; this dimension is the number of monomials $p_1^{r_1} p_2^{r_2} \ldots$ of degree d, i.e. the number of solutions to the equation $\sum_i d_i r_i = d$, i.e. the coefficient of t^d in $\prod_{i=1}^n (1 - t^{d_i})^{-1}$. □

Proof of Proposition 2.1.3.2 Let $x \in G$; consider $\det(1 - xt)$ – then it is clear that $\det(1 - xt) = (1 - t)^n$ if $x = 1$, that $\det(1 - xt) = (1 - t)^{n-1}(1 + t)$ if x is a reflection, that $\det(1 - xt)$ is a polynomial not divisible by $(1 - t)^{n-1}$ otherwise. Substituting this information into the identity derived in Lemma 2.1.3.3 and multiplying by $(1 - t)^n$ leads to the relation

$$(*) \quad \prod_i (1 + t + \ldots + t^{d_i - 1})^{-1} = [G]^{-1}\left[1 + \frac{N_G(1 - t)}{1 + t} + (1 - t)^2 P(t)\right]$$

where $P(t)$ is regular at $t = 1$. Setting $t = 1$, we get $\prod_i d_i^{-1} = [G]^{-1}$; differentiating and setting $t = 1$, we find that $(\prod_i d_i^{-1}) \sum_i (-(d_i - 1)/2) = [G]^{-1}(-N_G/2)$ – hence $\sum_i (d_i - 1) = N_G$. It remains to verify the uniqueness of the degrees d_i. So suppose that $\tilde{p}_1, \ldots, \tilde{p}_n$ is another system of algebraically independent homogeneous generators for P_I; let $\tilde{d}_i$ be the degree of $\tilde{p}_i$ $(i = 1, \ldots, n)$. The p_i can be expressed as polynomials in the $\tilde{p}_i$ with the determinant of the corresponding Jacobian not identically zero; after a rearrangement of the p_i, we have $\partial p_i/\partial \tilde{p}_i \neq 0$ (all i) – hence $d_i \geqslant \tilde{d}_i$. But $\sum_i (d_i - 1) = N_G = \sum_i (\tilde{d}_i - 1)$ – therefore $d_i = \tilde{d}_i$. □

Here is an interesting consequence of the preceding considerations.

Corollary 2.1.3.4 Let G be a finite group of automorphisms in the n dimensional vector space E; let P_I denote the **k**-algebra of G-invariants in P. If P_I is generated by n algebraically independent homogeneous elements (and 1), then G is generated by reflections.

Proof Let $\tilde{G}$ be the subgroup of G generated by the reflections in G; let $\tilde{p}_i$, $\tilde{d}_i$, and $N_{\tilde{G}}$ refer to $\tilde{G}$; let p_i, d_i, and N_G refer to G. The argument given near the end of the proof of Proposition 2.1.3.2 implies that $d_i = \tilde{d}_i$ (since $N_G = N_{\tilde{G}}$); but then, according to this proposition (applied to G in the evident manner), we have that $[G] = \prod_i d_i = \prod_i \tilde{d}_i = [\tilde{G}]$ – therefore $G = \tilde{G}$. □

Example Suppose that $n = 2$; let E have coordinates x_1, x_2; let $G = \{\pm 1\}$ (thus G is not a reflection group) – then P_I is generated by x_1^2, x_1x_2, x_2^2 and no smaller number will suffice.

As above, let E be a finite dimensional vector space over **k**; let G be a finite reflection group operating in E. The ideal P_I^+ in P generated by $P_I \cap P^+$ is G-stable; consequently G operates in a natural fashion on the quotient space P/P_I^+.

Proposition 2.1.3.5 The quotient space P/P_I^+ is a finite dimensional G-module of dimension $[G]$ ($[G]$ the order of G); furthermore the natural representation of G on P/P_I^+ is equivalent to the regular representation (in the usual sense for finite groups).

Proof On the basis of formula $(*)$ supra, one sees without essential difficulty that $\dim (P/P_I^+) = [G]$ (the Poincaré series attached to the space P/P_I^+ is the reciprocal of $(*)$); this being so, let $q_1, \ldots, q_r$ be homogeneous polynomials whose residue classes (mod P_I^+) form a basis for P/P_I^+ $(r = [G])$ – then, for $x \in G$, we have

$$q_j^x = \sum_i a_{ij}(x) q_i \qquad (j = 1, \ldots, r)$$

where the $a_{ij}(x)$ are homogeneous elements of P_I (in fact $a_{ij}(x) \in \mathbf{k}$).

The matrices $(a_{ij}(x))$ describe the natural representation of G on $Q(P)$, considered as a vector space over $Q(P_I)$. If we reduce the coefficients mod P_I^+ we get the natural representation of G on P/P_I^+, regarded as a vector space over $\mathbf{k}$; since this reduction does not affect the diagonal coefficients, both representations have the same character and thus are equivalent. But it was pointed out during the proof of Theorem 2.1.3.1 that $Q(P)$ is a Galois extension of $Q(P_I)$ with Galois group G; consequently the representation which the matrices $(a_{ij}(x))$ describe is equivalent to the regular representation of G. The second assertion of our proposition is now established. □

Let G be a finite reflection group operating in E, H a subgroup of G which is generated by reflections, $r = [G:H]$ the index of H in G; let $P_{I,G}$ denote the set of G-invariants in P, $P_{I,H}$ the H-invariants in P; let $P_{I,G}^*$ denote the space of elements in $P_{I,G}$ of strictly positive degree and similarly for $P_{I,H}^*$; let $P_{I,G}^+$ be the ideal in P generated by $P_{I,G}^*$ and similarly for $P_{I,H}^+$ (cf. supra). Finally, by $Q(P_{I,G})$ (respectively $Q(P_{I,H})$) we shall understand the quotient field of $P_{I,G}$ (respectively $P_{I,H}$) in the quotient field $Q(P)$ of P.

If $\mathbf{k}_1$ and $\mathbf{k}_2$ are fields, $\mathbf{k}_1 \supset \mathbf{k}_2$ a finite extension of $\mathbf{k}_2$, then $[\mathbf{k}_1 : \mathbf{k}_2]$ will denote the degree of the extension $\mathbf{k}_1/\mathbf{k}_2$.

Theorem 2.1.3.6 (Harish-Chandra) Let $r = [G:H]$ – then there exist r homogeneous elements $q_1 = 1, q_2, \ldots, q_r$ in $P_{I,H}$ such that $P_{I,H} = \sum_{i=1}^r P_{I,G} q_i$. Moreover the q_i are linearly independent over the quotient field $Q(P_{I,G})$ of $P_{I,G}$.

Proof As has been stated above, $Q(P)$ is a Galois extension of $Q(P_{I,G})$ with Galois group G; likewise the Galois group of $Q(P)/Q(P_{I,H})$ is H – hence $[Q(P_{I,H}) : Q(P_{I,G})] = [G:H] = r$.

(1) We claim that $P_{I,H} \cap P_{I,G}^+ = P_{I,H} P_{I,G}^*$. Thus if $p = \sum_i p_i p_i^* \in P_{I,H}$ ($p_i \in P$, $p_i^* \in P_{I,G}^*$), then $p = p^x = \sum_i p_i^x p_i^*$ ($x \in H$); this implies that

$$p = \sum_i p_{i,H} p_i^* \in P_{I,H} P_{I,G}^*$$

where $p_{i,H} = [H]^{-1} \sum_{x \in H} p_i^x$ – hence the claim. Consequently

$$\frac{P}{P_{I,G}^+} \supset \frac{(P_{I,H} + P_{I,G}^+)}{P_{I,G}^+} \sim \frac{P_{I,H}}{P_{I,H} \cap P_{I,G}^+} = \frac{P_{I,H}}{P_{I,H} P_{I,G}^*}.$$

(2) According to Proposition 2.1.3.5, $P/P_{I,G}^+$ is a finite dimensional G-module – in fact the natural representation of G on $P/P_{I,G}^+$ is equivalent to the regular representation of G. Therefore, by the representation theory of finite groups, the restriction to H of this representation contains the trivial representation of H exactly $r = [G:H]$ times; consequently $\dim(P_H/P_{I,G}^+) = r$, P_H denoting the set of those elements $p \in P$ whose residue class mod $P_{I,G}^+$ is left fixed by H. It is permissible to identify

$P_H/P_{I,G}^+$ with $P_{I,H}/P_{I,H}P_{I,G}^*$. [To see this, observe that $p - p^x \in P_{I,G}^+$ for any $p \in P_H$ and $x \in H$ and hence $p \equiv [H]^{-1} \sum_{x \in H} p^x \bmod P_{I,G}^+$ – thus $P_H = P_{I,H} + P_{I,G}^+$ and so our contention follows from the remarks in (1).]

(3) The considerations in (2) imply that $\dim(P_{I,H}/P_{I,H}P_{I,G}^*) = [G:H] = r$; this being so, select r homogeneous elements $q_1 = 1, q_2, \ldots, q_r$ in $P_{I,H}$ which have the property that their residue classes mod $P_{I,H}P_{I,G}^*$ form a basis for $P_{I,H}/P_{I,H}P_{I,G}^*$ over $\mathbf{k}$. Let $Q_H = \sum_{i=1}^r \mathbf{k}q_i$ – then we claim that $P_{I,H} = P_{I,G}Q_H$. To verify the claim, it clearly suffices to show that $P_{I,H} \cap P^m \subset P_{I,G}Q_H$ for all $m \geqslant 0$; proceeding by induction on m and observing that the case when $m = 0$ is obvious (since $P_{I,H} = P_{I,H}P_{I,G}^* + Q_H$) let us assume the result to be valid for all $P_{I,H} \cap P^i$, $i \leqslant m$ – then $P_{I,H} \cap P^{m+1} \subset P_{I,G}^* \sum_{i=0}^m P_{I,H} \cap P^i + Q_H$, which serves to establish the assertion for $P_{I,H} \cap P^{m+1}$.

(4) Because $Q(P_{I,H})/Q(P_{I,G})$ is a finite algebraic extension, $Q(P_{I,H}) = Q(P_{I,G})P_{I,H} = Q(P_{I,G})Q_H$ (cf. (3) supra). We noted earlier that $[Q(P_{I,H}) : Q(P_{I,G})] = r$ – therefore $q_1, \ldots, q_r$ are indeed linearly independent over $Q(P_{I,G})$.

This completes the proof of the theorem. □

Taking $H = 1$, $P_{I,G} = P_I$, $r = [G]$, yields the following corollary.

Corollary 2.1.3.7 There exist homogeneous elements $q_1 = 1, q_2, \ldots, q_r$ in P such that

$$P = \sum_{i=1}^{r} P_I q_i.$$

In addition $q_1, \ldots, q_r$ are linearly independent over $Q(P_I)$.

Otherwise said, P is a free abelian module over P_I (with respect to multiplication) of rank $[G]$.

The group G operates on the homomorphisms f of P into $\mathbf{k}$ – thus for $x \in G$, define f^x to be the homomorphism $p \mapsto f(p^{x^{-1}})$ $(p \in P)$.

Proposition 2.1.3.8 Retain the above notations but assume in addition that $\mathbf{k}$ is algebraically closed.

(i) Every homomorphism of P_I into $\mathbf{k}$ can be extended to a homomorphism of P into $\mathbf{k}$.

(ii) Two homomorphisms of P into $\mathbf{k}$ coincide on P_I iff they are conjugate under G.

Proof Statement (i) is proved in Helgason [2, p.420]. Before proving (ii), a preliminary observation is needed, namely: A finite set h_i of distinct homomorphisms of P into $\mathbf{k}$ are linearly independent. [For there exists a $p \in P$ such that the scalars $h_i(p)$ are distinct; putting

$$p_j = \prod_{i \neq j} \frac{p - h_i(p)}{h_j(p) - h_i(p)},$$

one has $h_i(p_j) = \delta_{ij}$ which implies that the h_i are linearly independent.] Turning now to (ii), it is clear that if f_1, f_2 are homomorphisms of P into $\mathbf{k}$ and if $f_1 = f_2^x$ for some $x \in G$, then f_1 and f_2 coincide on P_I. Conversely, if f_1, f_2 are homomorphisms of P into $\mathbf{k}$ whose restrictions to P_I are the same, then for all $p \in P$,

$$f_1(\sum_{x\in G} p^x) = f_2(\sum_{x\in G} p^x) \Rightarrow \sum_{x\in G} f_1^x = \sum_{x\in G} f_2^x \Rightarrow$$
$$n_1 \sum_{\alpha} f_{1,\alpha} = n_2 \sum_{\beta} f_{2,\beta}$$

where n_i is the order of the subgroup H_i of G stabilizing f_i ($i = 1, 2$) and all the terms in each of the partial sums are distinct. Owing to the earlier observation, one of the $f_{1,\alpha}$ must be equal to one of the $f_{2,\beta}$ – therefore f_1 and f_2 are conjugate under G. □

Remark The assumption that $\mathbf{k}$ is algebraically closed is not needed in the proof of part (ii) of the preceding proposition.

2.1.4 Symmetric Algebras and Formal Power Series

At this juncture some frankly technical material will be presented – the results however will be used rather frequently in the sequel.

Let E be a finite dimensional vector space over $\mathbf{R}$, E_c its complexification and $S(E_c)$ the symmetric algebra over E_c. Given $a_0 \in E$, let $\partial(a_0)$ denote the differential operator on E defined by

$$f(a; \partial(a_0)) = \left\{\frac{d}{dt} f(a + ta_0)\right\}_{t=0} \quad (a \in E, f \in C^\infty(E), t \in \mathbf{R}).$$

[For notational conventions, see A2.1.] Plainly the mapping $a \mapsto \partial(a)$ ($a \in E$) can be uniquely extended to an algebraic homomorphism of $S(E_c)$ (over $\mathbf{C}$) into the algebra $\mathfrak{D}(E)$ of all differential operators on E. In this way, for any $u \in S(E_c)$, we get a differential operator $\partial(u)$ on E. Let $\mathfrak{P}(E)$ denote the subalgebra of $\mathfrak{D}(E)$ generated by $P(E_c) \cup \partial(S(E_c))$ – we shall call $\mathfrak{P}(E)$ the algebra of *polynomial differential operators* on E. If A is an automorphism of E, then A extends in the usual way to an automorphism of $\mathfrak{P}(E)$ (A2.4); in particular observe that $(\partial(a))^A = \partial(Aa)$ ($a \in E_c$).

There exists a canonical pairing $\langle .\,, . \rangle$ of $P(E_c)$ with $S(E_c)$. Thus let $p \in P(E_c)$, $u \in S(E_c)$ and set $\langle p, u\rangle = p(0; \partial(u))$ – then the bilinear form $(p, u) \mapsto \langle p, u\rangle$ on $P(E_c) \times S(E_c)$ is non-degenerate. To see this, let $\{a_1, \ldots, a_n\}$ be a basis for E_c, $\{\mathbf{z}_1, \ldots, \mathbf{z}_n\}$ the corresponding dual basis in $E_c^\vee$ and observe that

$$\left\langle \prod_{i=1}^n \mathbf{z}_i^{m_i}, \prod_{i=1}^n a_i^{m_i'} \right\rangle = 0$$

unless $m_i = m_i'$ (m_i, m_i' non-negative integers, $1 \leqslant i \leqslant n$) in which case

$$\left\langle \prod_{i=1}^n \mathbf{z}_i^{m_i}, \prod_{i=1}^n a_i^{m_i} \right\rangle = \prod_{i=1}^n (m_i!).$$

It follows that $\langle . , . \rangle$ is indeed non-degenerate. Two simple consequences of this observation are:

(1) The homomorphism $\partial: S(E_c) \to \mathfrak{D}(E)$ is injective;

(2) The linear map ϕ of $P(E_c) \otimes_{\mathbf{C}} S(E_c)$ into $\mathfrak{P}(E)$, defined by the rule $\phi(p \otimes u) = p\partial(u)$, is bijective.

Often it is convenient to identify $S(E_c)$ with $P(E_c)$. Such a situation arises naturally when there is given a real non-degenerate symmetric bilinear form B on E. Extend this form by linearity (over $\mathbf{C}$) to E_c and then use it to identify E_c with its dual. Under this identification an element $a \in E_c$ corresponds to the linear function $B_a = B(. , a)$ and the map $a \mapsto B_a$ extends uniquely to an isomorphism ψ of $S(E_c)$ with $P(E_c)$. With this agreement the canonical pairing of $P(E_c)$ with $S(E_c)$ may be viewed as a pairing of $P(E_c)$ with itself and thus takes the form

$$\langle p, q \rangle = p(0; \partial(\psi^{-1}q)) \qquad (p, q \in P(E_c)).$$

From the symmetry of B it follows that $\langle p, q \rangle = \langle q, p \rangle$ $(p, q \in P(E_c))$.

Let E and F be two finite dimensional vector spaces over a field $\mathbf{k}$ of characteristic zero, $S(E)$ and $S(F)$ their respective symmetric algebras, $S^m(E)(S^m(F))$ the homogeneous elements of degree m in $S(E)(S(F))$. Suppose that E and F are in duality by a non-degenerate bilinear form $\langle . , . \rangle$. Define a bilinear from on $S^m(E) \times S^m(F)$ by the rule

$$\langle a_1 \dots a_m, b_1 \dots b_m \rangle = \sum_{\sigma} \prod_{i=1}^{m} \langle a_i, b_{\sigma(i)} \rangle \, (a_i \in E, b_j \in F),$$

the summation being taken over all permutations σ of the set $(1, \dots, m)$. It is clear that one has $\langle a_1 \dots a_m, b^m \rangle = m! \prod_{i=1}^m \langle a_i, b \rangle$, $\langle a^m, b^m \rangle = m! \langle a, b \rangle^m$. Picking dual bases $\{s_i\}$, $\{t_j\}$ in E and F, a computation analogous to the one above shows that this form is non-degenerate on $S^m(E) \times S^m(F)$. Therefore $S(E)$ and $S(F)$ may be placed in duality in such a way that $S^m(E)$ is orthogonal to $S^n(F)$ if $m \neq n$.

Since $S(E)$ is the direct sum of the $S^m(E)$, the dual of $S(E)$ may be identified with $S^*(F)$, $S^*(F)$ denoting the *direct product* of the $S^m(F)$. It is useful to bear in mind that $S(E)$ may be identified with the algebra of polynomials in the s_i while $S^*(F)$ may be taken as the algebra of formal power series in the t_j.

Given $b \in F$, consider the element $\mathbf{e}^b = \sum_{n \geq 0} (n!)^{-1} b^n \in S^*(F)$ – thus $\langle a_1 \dots a_m, \mathbf{e}^b \rangle = \prod_{i=1}^m \langle a_i, b \rangle$ and so the map f_b

$$u \mapsto \langle u, \mathbf{e}^b \rangle$$

is a homomorphism of $S(E)$ into $\mathbf{k}$ which maps $1 \mapsto 1$, $a \mapsto \langle a, b \rangle$. Notice that these 'exponentials' multiply by the usual rule of addition of exponents: $\mathbf{e}^{b_1 + b_2} = \mathbf{e}^{b_1}\mathbf{e}^{b_2} (b_1, b_2 \in F)$.

Consider the subalgebra $\mathfrak{F}$ of $S^*(F)$ generated by the $\mathbf{e}^{t_j}$ and 1. We shall leave it to the reader to prove the following simple lemma.

Lemma 2.1.4.1 For every $g \in S^*(F)$ and for every integer $m \geqslant 0$, there exists an $f \in \mathfrak{F}$ such that the components of $g - f$ of degree $\leqslant m$ are all zero.

2.1.5 Weyl Group Invariants

The objective of the present number is to draw a few simple consequences from the results of the preceding numbers in the context of a reductive Lie algebra $\mathfrak{g}$ over $\mathbf{R}$ (cf. too 2.2); for simplicity of exposition, we shall here, for the most part, deal explicitly only with the semi-simple case – the (modest) extensions to the reductive situation will be left to the reader.

Let $\mathfrak{g}$ be a semi-simple Lie algebra over $\mathbf{R}$, $\mathfrak{g} = \mathfrak{k} + \mathfrak{p}$ a fixed Cartan decomposition of $\mathfrak{g}$, $\mathfrak{a}_\mathfrak{p}$ a maximal abelian subspace of $\mathfrak{p}$; extend $\mathfrak{a}_\mathfrak{p}$ to a Cartan subalgebra $\mathfrak{a}_c$ of $\mathfrak{g}_c$ in the usual way. Let Φ denote the set of roots of the pair $(\mathfrak{g}_c, \mathfrak{a}_c)$, W_Φ the corresponding Weyl group; let Σ denote the set of roots of the pair $(\mathfrak{g}, \mathfrak{a}_\mathfrak{p})$, W_Σ the corresponding Weyl group.

The Killing form B of $\mathfrak{g}$ being non-degenerate, we may agree to subscribe to the conventions introduced in 2.1.4 and identify the symmetric algebra $S(\mathfrak{g}_c)$ with the algebra of all polynomial functions on $\mathfrak{g}_c$. Notice in particular that under this isomorphism the ideal $S(\mathfrak{g}_c)\mathfrak{k}_c$ goes into the ideal of polynomial functions on $\mathfrak{g}_c$ which vanish identically on $\mathfrak{p}_c$ whereas the subalgebra $S(\mathfrak{p}_c) \subset S(\mathfrak{g}_c)$ goes into an algebra of functions on $\mathfrak{g}_c$ which, under the restriction from $\mathfrak{g}_c$ to $\mathfrak{p}_c$, is isomorphic to $S(\mathfrak{p}_c)$. [The restriction of B to $\mathfrak{a}$ or $\mathfrak{a}_\mathfrak{p}$ is likewise non-degenerate and so the same conventions apply there too – observe that these identifications are compatible with the action of $\operatorname{Int}(\mathfrak{g})$, W_Φ, or W_Σ, B being invariant under these operations.]

[Consider a pair $(\mathfrak{g}, \theta)$ with $\mathfrak{g}$ a reductive Lie algebra over $\mathbf{R}$, θ an automorphism of $\mathfrak{g}$ such that $\theta^2 = 1$ and whose restriction to the derived algebra $[\mathfrak{g}, \mathfrak{g}]$ of $\mathfrak{g}$ is a Cartan involution (of $[\mathfrak{g}, \mathfrak{g}]$) (cf. 1.1.5); instead of using the Killing form to carry out the above identifications (which may not be non-degenerate . . .), one may use any real, symmetric, non-degenerate bilinear form B on $\mathfrak{g}$ such that: (1) $B(X, Y) = \operatorname{tr}(\operatorname{ad}(X)\operatorname{ad}(Y))$ $(X, Y \in \bar{\mathfrak{g}})$; (2) $B(X, Y) = 0$ $(X \in \mathfrak{c}, Y \in \bar{\mathfrak{g}})$; (3) $B(X, Y) = 0 (X \in \mathfrak{c}_\mathfrak{k}, Y \in \mathfrak{c}_\mathfrak{p})$.]

For each $X \in \mathfrak{g}_c$, let $d(X)$ $(= d_{\mathfrak{g}_c}(X))$ denote the unique derivation of $S(\mathfrak{g}_c)$ which coincides with $\operatorname{ad}(X)$ on $\mathfrak{g}_c$ – then, of course, the assignment $X \mapsto d(X)$ is a semi-simple (locally finite) representation of $\mathfrak{g}_c$ on $S(\mathfrak{g}_c)$. Observe that the restriction of this representation to $\mathfrak{k}_c$ equips $S(\mathfrak{g}_c)$ with the structure of a semi-simple, locally finite, $\mathfrak{k}_c$-module. [Each $S^m(\mathfrak{g}_c)$ is $\mathfrak{g}_c$-stable and semi-simple, hence $S^m(\mathfrak{g}_c)$ is a semi-simple $\mathfrak{k}_c$-module, $\mathfrak{k}_c$ being reductive in $\mathfrak{g}_c$ (Proposition 1.1.5.2(ii)). Since $S(\mathfrak{g}_c) = \sum_{m=0}^{\infty} S^m(\mathfrak{g}_c)$, $S(\mathfrak{g}_c)$ is clearly a semi-simple, locally finite, $\mathfrak{k}_c$-module.] Evidently, relative to this $\mathfrak{k}_c$-structure, $S(\mathfrak{p}_c)$ is a semi-simple, locally finite, $\mathfrak{k}_c$-module too.

Denote by $I(\mathfrak{g}_c)$ the set of invariants in $S(\mathfrak{g}_c)$ for the representation $X \mapsto d(X)$ of $\mathfrak{g}_c$, by $I(\mathfrak{p}_c)$ the set of invariants in $S(\mathfrak{p}_c)$ for the restriction of this representation to $\mathfrak{k}_c$; denote by $I(\mathfrak{a}_c)$ the set of invariants in $S(\mathfrak{a}_c)$ for the Weyl group W_Φ of the pair $(\mathfrak{g}_c, \mathfrak{a}_c)$, by $I(\mathfrak{a}_{\mathfrak{p}_c})$ the set of invariants in $S(\mathfrak{a}_{\mathfrak{p}_c})$ for the Weyl group W_Σ of the pair $(\mathfrak{g}, \mathfrak{a}_\mathfrak{p})$.

[If G is a connected Lie group with Lie algebra $\mathfrak{g}$, K the analytic subgroup of G corresponding to $\mathfrak{k}$, then it is clear that $I(\mathfrak{g}_c)$ is precisely the set of $\mathrm{Ad}_G(G)$ invariants in $S(\mathfrak{g}_c)$ while $I(\mathfrak{p}_c)$ is the set of $\mathrm{Ad}_G(K)$ invariants in $S(\mathfrak{p}_c)$.]

Fundamental for many questions is Chevalley's Restriction Theorem.

Theorem 2.1.5.1 (Chevalley) (i) The operation of restriction from $\mathfrak{g}_c$ to $\mathfrak{a}_c$ induces an isomorphism of $I(\mathfrak{g}_c)$ onto $I(\mathfrak{a}_c)$;

(ii) The operation of restriction from $\mathfrak{p}_c$ to $\mathfrak{a}_{\mathfrak{p}_c}$ induces an isomorphism of $I(\mathfrak{p}_c)$ onto $I(\mathfrak{a}_{\mathfrak{p}_c})$.

The proof of part (ii) of this theorem may be found in Helgason [2, p. 433]; the proof of part (i) makes free use of the notations and results of 2.4.1; accordingly the reader may wish to put off a detailed study until later on.

Proof of Theorem 2.1.5.1(i) (Steinberg) (1) The symmetric algebra $S(\mathfrak{a}_c)$ of $\mathfrak{a}_c$ is generated over $\mathbf{C}$ by $\{\Lambda^k\colon \Lambda \in \mathfrak{W}_\Phi^+, k = 0, 1, \ldots\}$. [Let $\tilde{S}(\mathfrak{a}_c)$ be the space so generated; if $\Omega_1, \Omega_2 \in \mathfrak{W}_\Phi^+$, then $(\Omega_1 + m\Omega_2)^k \in \tilde{S}(\mathfrak{a}_c)$ for $m = 0, 1, \ldots, k-1$ and since the matrix (van der Monde) of coefficients is invertible, each $\Omega_1^i\Omega_2^{k-i} \in \tilde{S}(\mathfrak{a}_c)$. By induction $\Omega_1^{i_1}\ldots\Omega_l^{i_l} \in \tilde{S}(\mathfrak{a}_c)$ $(\Omega_j \in \mathfrak{W}_\Phi^+;\ i_j \geqslant 0,\ j = 1, \ldots, l)$; this applies in particular to $\Omega_1 = \Lambda_1, \ldots, \Omega_l = \Lambda_l$ which yields the assertion in question.]

(2) For $\Lambda \in \mathfrak{W}_\Phi^+$ write $\mathrm{Sym}(\Lambda^k)$ for $\sum_{w \in W_\Phi} (\Lambda^k)^w$ $(k = 0, 1, \ldots)$ – thus $\mathrm{Sym}(\Lambda^k) \in I(\mathfrak{a}_c)$. Plainly, then, $I(\mathfrak{a}_c)$ is generated over $\mathbf{C}$ by $\{\mathrm{Sym}(\Lambda^k)\colon \Lambda \in \mathfrak{W}_\Phi^+, k = 0, 1, \ldots\}$.

(3) To complete the proof of the theorem, it is enough to show that the restriction map from $I(\mathfrak{g}_c)$ to $I(\mathfrak{a}_c)$ is surjective, or, equivalently, that each $\mathrm{Sym}(\Lambda^k)$ $(\Lambda \in \mathfrak{W}_\Phi^+)$ can be extended to all of $\mathfrak{g}_c$ in such a way as to lie in $I(\mathfrak{g}_c)$. [The operation of restriction is obviously injective . . . .] This will be done by using induction on the partial ordering: $\Lambda > \Omega$ if $\Lambda - \Omega$ is a sum of positive roots. Fix $\Lambda \in \mathfrak{W}_\Phi^+$; let E_Λ be a finite dimensional simple $\mathfrak{g}_c$-module with highest weight Λ, U_Λ the corresponding representation. The weights of U_Λ are Λ and its transforms under W_Φ and various $\Omega < \Lambda (\Omega \in \mathfrak{W}_\Phi^+)$ and their transforms under W_Φ. Summing on the kth powers gives the equation

$$\sum \lambda^k = \mathrm{Sym}(\Lambda^k) + \sum_{\Omega<\Lambda} c_\Omega \,\mathrm{Sym}(\Omega^k) \qquad (\Omega \in \mathfrak{W}_\Phi^+),$$

the sum on the left hand side being taken over the weights λ of U_Λ with multiplicities. Now observe that on $\mathfrak{a}_c$ the left hand side is the sum of the kth powers of the eigenvalues and hence can be extended to $\mathfrak{g}_c$ so as to lie in $I(\mathfrak{g}_c)$ by expressing it in terms of the elementary symmetric functions, these latter in terms of the coordinates relative to a basis of E_Λ. [If t is an indeterminate, then $\det(t - U_\Lambda(X)) = \sum_i p_i(X)t^i$ where the $p_i \in I(\mathfrak{g}_c)$ ($X \in \mathfrak{g}_c$) – on the other hand, for $H \in \mathfrak{a}_c$, $\det(t - U_\Lambda(H)) = \prod_\lambda (t - \lambda(H))$ etc.] But the second term on the right hand side can be extended by the induction hypothesis – hence so can $\operatorname{Sym}(\Lambda^k)$.

The proof of the theorem is now complete. □

Remark Part (i) of the above theorem together with Theorem 2.1.3.1 imply that $I(\mathfrak{g}_c)$ can actually be generated by l algebraically independent homogeneous polynomials and 1 ($l = \operatorname{rank}(\mathfrak{g}_c)$). [The Fundamental Theorem of Invariant Theory only guarantees us that $I(\mathfrak{g}_c)$ is finitely generated.] A remarkable result due to Chevalley says that it is possible to select these generators in a special way. Namely, there exist homogeneous generators $p_1, \ldots, p_l$ for $I(\mathfrak{g}_c)$ such that if we write $d_i = \deg(p_i) = m_i + 1$, then the integers m_i, called the *exponents* of $\mathfrak{g}_c$, are those integers such that

$$\prod_{i=1}^{l} (1 + t^{2m_i+1}) \, (m_1 \leqslant m_2 \leqslant \cdots \leqslant m_l)$$

is the Poincaré polynomial of $\mathfrak{g}_c$ (i.e. of the adjoint group of $\mathfrak{g}_c$). Call the p_i ($i = 1, \ldots, l$) *primitive* invariants (they are not necessarily unique although the corresponding collection of degrees is (Proposition 2.1.3.2)). Suppose now that $\mathfrak{g}_c$ is simple. An element $w \in W_\Phi$ is said to be a *Coxeter-Killing transformation* if it can be put in the form $w = w_{\alpha_1} \ldots w_{\alpha_l}$ where $\{\alpha_1, \ldots, \alpha_l\}$ is some fundamental system of roots for the pair $(\mathfrak{g}_c, \mathfrak{a}_c)$. Let $s = 1 + m_l$ – then $m_i + m_{l+1-i} = s$ ($i = 1, \ldots, l$) and the order of a Coxeter-Killing transformation is s. Moreover the eigenvalues of such a w are the numbers $e^{2\pi\sqrt{-1}m_i/s}$ ($i = 1, \ldots, l$). Finally, $ls = 2r$, r the number of positive roots. For details, see Bourbaki [5], Chevalley [2], Borel and Chevalley [1], Kostant [2, 4], Varadarajan [1] and the references given there. Here is a table which gives the d_i's for the irreducible Weyl groups:

W	d_i's
$\mathbf{A}_l$	$2, 3, \ldots, l+1$
$\mathbf{B}_l, \mathbf{C}_l$	$2, 4, \ldots, 2l$
$\mathbf{D}_l$	$2, 4, \ldots, 2l-2, l$
$\mathbf{E}_6$	2, 5, 6, 8, 9, 12
$\mathbf{E}_7$	2, 6, 8, 10, 12, 14, 18
$\mathbf{E}_8$	2, 8, 12, 14, 18, 20, 24, 30
$\mathbf{F}_4$	2, 6, 8, 12
$\mathbf{G}_2$	2, 6

For the classical structures, the entries in this table were confirmed by direct computation in 2.1.3 (cf. the Example preceding Proposition 2.1.3.2). Turning to the exceptional structures, let us determine, as an example, the d_i for $\mathbf{E}_8$; thus in this case $l = 8$, $r = 120$ and so $s = 30$; if $w = w_1 \ldots w_8$, then the numbers in the set $\{e^{2\pi\sqrt{-1}n/30} : (n, 30) = 1\}$ are eigenvalues for w (w acts rationally) and since $\phi(30) = 8 = l$ there are no others (ϕ being Euler's ϕ-function) – hence the numbers 2, 8, 12, 14, 18, 20, 24, and 30.

We shall now introduce some more notation and make a series of simple observations. Let $I(\mathfrak{g}_c; \mathfrak{p}_c)$ denote the algebra of restrictions to $\mathfrak{p}_c$ of the elements in $I(\mathfrak{g}_c)$ – thus $I(\mathfrak{g}_c; \mathfrak{p}_c)$ is a subalgebra of $I(\mathfrak{p}_c)$. Let

$I(\mathfrak{g}_c;\, \mathfrak{p}_c;\, \mathfrak{a}_{\mathfrak{p}_c})$ denote the algebra of restrictions to $\mathfrak{a}_{\mathfrak{p}_c}$ of the elements in $I(\mathfrak{g}_c;\, \mathfrak{p}_c)$. On the other hand, let $I(\mathfrak{g}_c;\, \mathfrak{a}_c)$ denote the algebra of restrictions to $\mathfrak{a}_c$ of the elements of $I(\mathfrak{g}_c)$ and $I(\mathfrak{g}_c;\, \mathfrak{a}_c;\, \mathfrak{a}_{\mathfrak{p}_c})$ the algebra of restrictions to $\mathfrak{a}_{\mathfrak{p}_c}$ of the elements of $I(\mathfrak{g}_c;\, \mathfrak{a}_c)$ – then $I(\mathfrak{g}_c;\, \mathfrak{p}_c;\, \mathfrak{a}_{\mathfrak{p}_c}) = I(\mathfrak{g}_c;\, \mathfrak{a}_c;\, \mathfrak{a}_{\mathfrak{p}_c})$.

Diagramatically we have

$$\begin{array}{ccc} I(\mathfrak{g}_c) & \xrightarrow{\;1\;} & I(\mathfrak{p}_c) \\ {\scriptstyle 3}\downarrow & & \downarrow{\scriptstyle 4} \\ I(\mathfrak{a}_c) & \xrightarrow[\;2\;]{} & I(\mathfrak{a}_{\mathfrak{p}_c}) \end{array}$$

where the maps 1, 2, 3, and 4 are restriction maps; 1 is onto iff 2 is onto (of course 3 and 4 are algebra isomorphisms).

Lemma 2.1.5.2 The ring $I(\mathfrak{g}_c;\, \mathfrak{a}_c)$ is a Noetherian ring.

Proof The Restriction Theorem of Chevalley stated above implies that $I(\mathfrak{g}_c;\, \mathfrak{a}_c) = I(\mathfrak{a}_c)$ and, since the Weyl group W_Φ is a finite group of linear transformations in $\mathfrak{a}_c$ which is generated by reflections, it follows from Theorem 2.1.3.1 that the algebra of polynomials $I(\mathfrak{a}_c)$ is generated by l algebraically independent homogeneous elements and the identity ($l = \dim(\mathfrak{a}_c) = \operatorname{rank}(\mathfrak{g}_c)$) which implies that $I(\mathfrak{a}_c)$ is Noetherian. $\square$

Lemma 2.1.5.3 The symmetric algebra $S(\mathfrak{a}_{\mathfrak{p}_c})$ is integral over $I(\mathfrak{g}_c;\, \mathfrak{a}_c;\, \mathfrak{a}_{\mathfrak{p}_c})$.

Proof Let t be an indeterminate, $p \in S(\mathfrak{a}_c)$ – then p satisfies the equation $\prod_{w \in W_\Phi} (t - p^w) = 0$ with leading coefficient 1 and coefficients in $I(\mathfrak{a}_c)$. Therefore the ring $S(\mathfrak{a}_c)$ is integral over the ring $I(\mathfrak{a}_c)$. The operation of restriction from $\mathfrak{a}_c$ to $\mathfrak{a}_{\mathfrak{p}_c}$ yields a homomorphism of $S(\mathfrak{a}_c)$ onto $S(\mathfrak{a}_{\mathfrak{p}_c})$ which maps $I(\mathfrak{a}_c)$ onto $I(\mathfrak{g}_c;\, \mathfrak{a}_c;\, \mathfrak{a}_{\mathfrak{p}_c})$. Hence the lemma. $\square$

Lemma 2.1.5.4 The symmetric algebra $S(\mathfrak{a}_{\mathfrak{p}_c})$ is a finite $I(\mathfrak{g}_c;\, \mathfrak{a}_c;\, \mathfrak{a}_{\mathfrak{p}_c})$-module.

Proof In fact, if $\tilde{\alpha}$ denotes the restriction of a root $\alpha \in P_+$ to $\mathfrak{a}_{\mathfrak{p}_c}$, then the preceding lemma tells us that $\tilde{\alpha}$ is integral over $I(\mathfrak{g}_c;\, \mathfrak{a}_c;\, \mathfrak{a}_{\mathfrak{p}_c})$. Since every element in $\mathfrak{a}_{\mathfrak{p}}^{\vee}$ is a linear combination of certain of the $\tilde{\alpha}$, the lemma follows. $\square$

In the sequel, write $I(\mathfrak{a}_c;\, \mathfrak{a}_{\mathfrak{p}_c})$ for $I(\mathfrak{g}_c;\, \mathfrak{a}_c;\, \mathfrak{a}_{\mathfrak{p}_c})$ – this is a legitimate agreement.

As will be seen, the following proposition and its corollary constitute an important step in the proof of a major theorem of Harish-Chandra (Theorem 2.2.1.1).

Proposition 2.1.5.5 The ring $I(\mathfrak{a}_{\mathfrak{p}_c})$ is a finite $I(\mathfrak{a}_c; \mathfrak{a}_{\mathfrak{p}_c})$-module.

Proof We have the inclusions: $I(\mathfrak{a}_c; \mathfrak{a}_{\mathfrak{p}_c}) \subset I(\mathfrak{a}_{\mathfrak{p}_c}) \subset S(\mathfrak{a}_{\mathfrak{p}_c})$. The ring $I(\mathfrak{a}_c; \mathfrak{a}_{\mathfrak{p}_c})$ is Noetherian while $S(\mathfrak{a}_{\mathfrak{p}_c})$ is a finite $I(\mathfrak{a}_c; \mathfrak{a}_{\mathfrak{p}_c})$-module (cf. supra). Thus, by standard algebra, $I(\mathfrak{a}_{\mathfrak{p}_c})$ is a finite $I(\mathfrak{a}_c; \mathfrak{a}_{\mathfrak{p}_c})$-module. ☐

The second part of Chevalley's Restriction Theorem immediately gives the following corollary.

Corollary 2.1.5.6 The ring $I(\mathfrak{p}_c)$ is a finite $I(\mathfrak{g}_c; \mathfrak{p}_c)$-module.

Remarks In view of Proposition 2.1.5.5, it is natural to ask: Under the restriction map from $\mathfrak{a}_c$ to $\mathfrak{a}_{\mathfrak{p}_c}$ does $I(\mathfrak{a}_c)$ map onto $I(\mathfrak{a}_{\mathfrak{p}_c})$? Helgason [5, p.591] has answered this question in the negative, namely:

Theorem 2.1.5.7 (Helgason) (i) Suppose that $\mathfrak{g}$ is a classical semi-simple Lie algebra (with real or complex structure) – then the restriction from $\mathfrak{a}_c$ to $\mathfrak{a}_{\mathfrak{p}_c}$ maps $I(\mathfrak{a}_c)$ onto $I(\mathfrak{a}_{\mathfrak{p}_c})$.

(ii) Part (i) does not hold in general for the real forms of the exceptional Lie algebras $\mathbf{E}_6, \mathbf{E}_7, \mathbf{E}_8$.

The proof of this theorem rests upon Cartan's classification and a case by case verification. During the course of the argument, it appears that it might be true that in any event the elements in $I(\mathfrak{a}_{\mathfrak{p}_c})$ are all obtained from rational invariants on $\mathfrak{a}_c$ by restriction – in fact (Helgason [5, p.597]):

Theorem 2.1.5.8 (Harish-Chandra-Helgason). Let $Q(I(\mathfrak{a}_c))$ and $Q(I(\mathfrak{a}_{\mathfrak{p}_c}))$ denote the quotient fields of $I(\mathfrak{a}_c)$ and $I(\mathfrak{a}_{\mathfrak{p}_c})$ respectively – then the restriction from $\mathfrak{a}_c$ to $\mathfrak{a}_{\mathfrak{p}_c}$ induces a mapping of $Q(I(\mathfrak{a}_c))$ onto $Q(I(\mathfrak{a}_{\mathfrak{p}_c}))$.

Example In the classical classification of É. Cartan consider 'the' real simple Lie algebra $\mathfrak{g}$ of type **DI**; **DI** may be realized as the Lie algebra of the orthogonal group for the quadratic form corresponding to the matrix

$$J_{m,n} = \begin{pmatrix} 0 & I_m & 0 \\ I_m & 0 & 0 \\ 0 & 0 & -I_{n-m} \end{pmatrix} \quad (n \geqslant m,\ n+m = 2l).$$

Thus $\mathfrak{g} = \{X \in \mathfrak{gl}(m+n, \mathbf{R}): {}^tXJ_{m,n} + J_{m,n}X = 0\}$ and a Cartan involution θ of $\mathfrak{g}$ is given by the map $X \mapsto J_{m,n}XJ_{m,n}$, which yields a corresponding Cartan decomposition $\mathfrak{g} = \mathfrak{k} + \mathfrak{p}$. Take, in the usual notations,

$$\mathfrak{a}_{\mathfrak{p}_c} = \{D(a_1, \ldots, a_m, -a_1, \ldots, -a_m, 0, \ldots, 0): a_i \in \mathbf{C}\}$$

and if $r = [\frac{1}{2}(n-m)]$ = greatest integer in $\frac{1}{2}(n-m)$,

$$\mathfrak{a}_c = \left\{H = \sum_1^m a_i(E_{ii} - E_{i+m,i+m}) + \sum_1^r c_i(E_{2m+r+i,2m+i} - E_{2m+i,2m+r+i}): a_i, c_i \in \mathbf{C}\right\}.$$

Let e_i denote the linear form on $\mathfrak{a}_c$ defined by

$$e_i(H) = \begin{cases} a_i & 1 \leqslant i \leqslant m, \\ \sqrt{-1}c_{i-m} & m+1 \leqslant i \leqslant [\frac{1}{2}(m+n)]. \end{cases}$$

Then the root system Φ of the pair $(\mathfrak{g}_c, \mathfrak{a}_c)$ is the set $\{\pm(e_i \pm e_j)\colon 1 \leqslant i < j \leqslant [\frac{1}{2}(m+n)]\}$ while the Weyl group W_Φ consists of all maps $e_i \mapsto \epsilon_i e_{\sigma(i)}$, where σ is any permutation of $(1, \ldots, l)$, $\epsilon_i = \pm 1$, and $\Pi_1^l \epsilon_i = 1$. Hence $I(\mathfrak{a}_c)$ is generated by $e_1 \ldots e_l$ and the symmetric polynomials in $e_1^2, \ldots, e_l^2$. Let us now show that $I(\mathfrak{a}_c; \mathfrak{a}_{\mathfrak{p}_c}) = I(\mathfrak{a}_{\mathfrak{p}_c})$. We may assume that $m < n$ so that $m + 1 \leqslant l$ (if $m = n$, $\mathfrak{g}$ is the normal real form for $\mathfrak{g}_c$ and the assertion is trivial). An element $w \in W_\Phi$ leaves $\mathfrak{a}_{\mathfrak{p}_c}$ invariant iff it maps $\sum_{m+1} \mathbf{C}e_i$, and therefore also $\sum_1^m \mathbf{C}e_i$, into itself. If $\tilde{e}_i$ denotes the restriction of e_i to $\mathfrak{a}_{\mathfrak{p}_c}$, it follows that W_Σ consists of all maps $\tilde{e}_i \mapsto \epsilon_i \tilde{e}_{\sigma(i)}$ where σ is any permutation of $(1, \ldots, m)$ and $\epsilon_i = \pm 1$ – hence $I(\mathfrak{a}_c; \mathfrak{a}_{\mathfrak{p}_c}) = I(\mathfrak{a}_{\mathfrak{p}_c})$.

Example Consider the real simple Lie algebra $\mathfrak{g}$ of type **EIV**; here $I(\mathfrak{a}_c)$ is generated by algebraically independent homogeneous generators of degrees 2, 5, 6, 8, 9, 12 (cf. the table supra); on the other hand, $I(\mathfrak{a}_{\mathfrak{p}_c})$ has generators of degrees 2 and 3, respectively (this fact can be read off from the diagrams given at the end of 1.1.3 in Chapter 1) – but $I(\mathfrak{a}_c)$ cannot contain any polynomial of degree 3 and so $I(\mathfrak{a}_c; \mathfrak{a}_{\mathfrak{p}_c}) \neq I(\mathfrak{a}_{\mathfrak{p}_c})$.

Let us agree to retain the above conventions and notations.

The groups W_Φ and W_Σ are finite reflection groups operating in $\mathfrak{a}_c$ and $\mathfrak{a}_{\mathfrak{p}_c}$ respectively – Corollary 2.1.3.7 then implies the following result.

Proposition 2.1.5.9 (i) The symmetric algebra $S(\mathfrak{a}_c)$ is a free abelian module over $I(\mathfrak{a}_c)$ of rank $[W]$ $(W = W_\Phi)$;

(ii) The symmetric algebra $S(\mathfrak{a}_{\mathfrak{p}_c})$ is a free abelian module over $I(\mathfrak{a}_{\mathfrak{p}_c})$ of rank $[W]$ $(W = W_\Sigma)$.

A useful extension of Proposition 2.1.5.9 will now be noted; for completeness we shall work directly with the reductive case.

So suppose that $\mathfrak{g}$ is a reductive Lie algebra over $\mathbf{R}$; write $\mathfrak{g} = \mathfrak{c} + [\mathfrak{g}, \mathfrak{g}]$ in the usual notations. Let G denote a connected Lie group with Lie algebra $\mathfrak{g}$; let $\mathfrak{h}$ be a subalgebra of $\mathfrak{g}$ such that $\mathfrak{h}$ is reductive in $\mathfrak{g}$ and rank $(\mathfrak{h})$ = rank $(\mathfrak{g})$; let H denote the analytic subgroup of G corresponding to $\mathfrak{h}$. Fix a Cartan subalgebra $\mathfrak{j}_c$ of $\mathfrak{h}_c$ and let $W(\mathfrak{g}_c, \mathfrak{j}_c)$ and $W(\mathfrak{h}_c, \mathfrak{j}_c)$ denote the Weyl groups of the pairs $(\mathfrak{g}_c, \mathfrak{j}_c)$ and $(\mathfrak{h}_c, \mathfrak{j}_c)$, respectively – then the group $W(\mathfrak{h}_c, \mathfrak{j}_c)$ may be regarded, in the obvious way, as a subgroup of $W(\mathfrak{g}_c, \mathfrak{j}_c)$. Choose a real, symmetric, non-degenerate bilinear form B on $\mathfrak{g}$ such that: (1) $B(X, Y) = \operatorname{tr}(\operatorname{ad}(X)\operatorname{ad}(Y))$ for $X, Y \in [\mathfrak{g}, \mathfrak{g}]$; (2) $B(X, Y) = 0$ for $X \in \mathfrak{c}$, $Y \in [\mathfrak{g}, \mathfrak{g}]$ – then B is invariant under G. Extend this form to $\mathfrak{g}_c$ by linearity over $\mathbf{C}$ and identify $\mathfrak{g}_c$ with its dual under B – thus $S(\mathfrak{g}_c) = P(\mathfrak{g}_c)$. In view of our assumptions on $\mathfrak{h}$, it is clear that the restriction $B_\mathfrak{h}$ of B to $\mathfrak{h}_c$ is non-degenerate; accordingly it is permissible to identify $\mathfrak{h}_c$ with its dual under $B_\mathfrak{h}$. If $p \mapsto p_\mathfrak{h}$ $(p \in S(\mathfrak{g}_c))$ is the natural projection of $S(\mathfrak{g}_c)$ onto $S(\mathfrak{h}_c)$ with respect to the decomposition $\mathfrak{g}_c = \mathfrak{h}_c + \mathfrak{g}_c/\mathfrak{h}_c$, then it is plain that $p_\mathfrak{h}$ is simply the restriction of the polynomial function p to $\mathfrak{h}_c$. Relative to the adjoint action of G on $\mathfrak{g}_c$, the set $I(\mathfrak{g}_c)$ of G-invariants in $S(\mathfrak{g}_c)$ consists, of course, of those $p \in S(\mathfrak{g}_c)$ such that $d(X)p = 0$ for all $X \in \mathfrak{g}_c$, $d(X)$ the canonical extension of ad (X) to $S(\mathfrak{g}_c)$. It is clear that if $p \in I(\mathfrak{g}_c)$, then $p_{\mathfrak{h}} \in I(\mathfrak{h}_c)$, $I(\mathfrak{h}_c)$ the set of H-invariants relative to the adjoint action of H on $\mathfrak{h}_c$.

Let $I(\mathfrak{g}_c; \mathfrak{h}_c)$ denote the image of $I(\mathfrak{g}_c)$ under the restriction map $p \mapsto p_{\mathfrak{h}}$ $(p \in I(\mathfrak{g}_c))$.

Proposition 2.1.5.10 Retain the above notations and assumptions – then the ring $I(\mathfrak{h}_c)$ is a free abelian module over $I(\mathfrak{g}_c; \mathfrak{h}_c)$ of rank $[W(\mathfrak{g}_c, \mathfrak{j}_c): W(\mathfrak{h}_c, \mathfrak{j}_c)]$.

Proof Using the evident notations, Chevalley's Restriction Theorem implies that $I(\mathfrak{h}_c)$ is isomorphic to the algebra $I(\mathfrak{h}_c; \mathfrak{j}_c)$ of $W(\mathfrak{h}_c, \mathfrak{j}_c)$-invariants in $S(\mathfrak{j}_c)$ while $I(\mathfrak{g}_c)$ is isomorphic to the algebra $I(\mathfrak{g}_c; \mathfrak{j}_c)$ of $W(\mathfrak{g}_c, \mathfrak{j}_c)$-invariants in $S(\mathfrak{j}_c)$. It is enough to show, therefore, that $I(\mathfrak{h}_c; \mathfrak{j}_c)$ is a free abelian module over $I(\mathfrak{g}_c; \mathfrak{j}_c)$ of rank $[W(\mathfrak{g}_c, \mathfrak{j}_c): W(\mathfrak{h}_c, \mathfrak{j}_c)]$ – this however is implicit in parts (3), (4) of the proof of Theorem 2.1.3.6. □

Remark Retain the immediately preceding notations and identifications; denote by $I^*(\mathfrak{g}_c)$ the space of elements in $I(\mathfrak{g}_c)$ having zero constant term, by $I^+(\mathfrak{g}_c)$ the ideal in $S(\mathfrak{g}_c)$ generated by $I^*(\mathfrak{g}_c)$. An element $q \in S(\mathfrak{g}_c)$ is said to be *G-harmonic* if $\partial(p)q = 0$ for all $p \in I^*(\mathfrak{g}_c)$. Plainly the space $H(\mathfrak{g}_c)$ of all G-harmonic polynomials in $S(\mathfrak{g}_c)$ is a G-submodule of $S(\mathfrak{g}_c)$ which is graded in the obvious way. Kostant [4] (see also Helgason [3]) has established the following points: (1) The space $H(\mathfrak{g}_c)$ of G-harmonic polynomials is precisely the space spanned by all powers X^m, $m = 0, 1, \ldots, X$ a nilpotent element in $\mathfrak{g}_c$; (2) $S(\mathfrak{g}_c) = I^+(\mathfrak{g}_c) + H(\mathfrak{g}_c)$ is a G-module direct sum so that $S(\mathfrak{g}_c) = I(\mathfrak{g}_c)H(\mathfrak{g}_c)$; (3) $S(\mathfrak{g}_c)$ is free as an $I(\mathfrak{g}_c)$-module (under multiplication) and the map of $I(\mathfrak{g}_c) \otimes_{\mathbf{C}} H(\mathfrak{g}_c) \rightarrow S(\mathfrak{g}_c)$ given by $p \otimes h \mapsto ph$ $(p \in I(\mathfrak{g}_c), h \in H(\mathfrak{g}_c))$ is a G-module isomorphism. Note that (3) reduces the study of the G-module structure of $S(\mathfrak{g}_c)$ to that of $H(\mathfrak{g}_c)$ – in turn the structure of $H(\mathfrak{g}_c)$ as a G-module has been completely determined (Kostant [4]).

We shall conclude with some elementary considerations of divisibility; all earlier notations and conventions are in force here.

Fix an order in the set Φ of roots of the pair $(\mathfrak{g}_c, \mathfrak{a}_c)$ and put $\pi_\Phi = \prod_{\alpha>0} \alpha$; similarly fix an order in the set Σ of roots of the pair $(\mathfrak{g}, \mathfrak{a}_\mathfrak{p})$ and let Σ_0^+ denote the subset of Σ^+ consisting of those elements in Σ^+ which are not integral multiples of other members of Σ^+ – write $\pi_\Sigma = \prod_{\lambda \in \Sigma_0^+} \lambda$. Plainly, for $w \in W_\Phi$, $\pi_\Phi^w = \det(w)\pi_\Phi$ where $\det(w) = \pm 1$; analogous remarks apply to π_Σ and W_Σ (cf. Proposition 1.1.2.5 and Corollary 1.1.2.4).

Definition (i) An element $p \in S(\mathfrak{a}_c)$ is said to be *skew* if $p^w = \det(w)p$ for $w \in W_\Phi$.

(ii) An element $p \in S(\mathfrak{a}_{\mathfrak{p}_c})$ is said to be *skew* if $p^w = \det(w)p$ for $w \in W_\Sigma$.

Proposition 2.1.5.11 (i) Let $p \in S(\mathfrak{a}_c)$ and suppose that p is skew – then $p = \pi_\Phi p_0$ for some $p_0 \in I(\mathfrak{a}_{\mathfrak{p}_c})$.

(ii) Let $p \in S(\mathfrak{a}_{\mathfrak{p}_c})$ and suppose that p is skew – then $p = \pi_\Sigma p_0$ for some $p_0 \in I(\mathfrak{a}_{\mathfrak{p}_c})$.

Proof We shall prove (i), the proof of (ii) being similar. For any root α let σ_α denote the hyperplane in $\mathfrak{a}_c$ defined by the equation $\alpha = 0$. If q is an arbitrary element in $S(\mathfrak{a}_c)$, then $q^{w_\alpha} - q$ vanishes on σ_α; therefore α divides $q^{w_\alpha} - q$ in the polynomial ring $S(\mathfrak{a}_c)$–hence $p = -1/2(p^{w_\alpha} - p)$ is divisible by α. But $S(\mathfrak{a}_c)$ is a unique factorization domain and the σ_α, corresponding to the distinct positive roots α, are likewise distinct – thus π_Φ divides p and so $p = \pi_\Phi p_0$ ($p_0 \in S(\mathfrak{a}_c)$). Since p and π_Φ are skew, $p_0 \in I(\mathfrak{a}_c)$. □

Corollary 2.1.5.12 (i) If p is a homogeneous element in $I(\mathfrak{a}_c)$ of positive degree, then $\partial(p)\pi_\Phi = 0$.

(ii) If p is a homogeneous element in $I(\mathfrak{a}_{\mathfrak{p}_c})$ of positive degree, then $\partial(p)\pi_\Sigma = 0$.

2.2 Invariant Theory II – Applications to Reductive Lie Algebras

2.2.1 A Theorem of Harish-Chandra

In this number (and the next) we shall establish theorems which, as will become apparent in due course, are the keystones on which the Harmonic Analysis on semi-simple Lie groups rests. To illustrate, suppose that G is a connected semi-simple Lie group with finite center, K a maximal compact subgroup of G – then the following result is a ready consequence of Theorem 2.2.1.1 infra (cf. Theorem 4.5.2.9): Every topologically completely irreducible Banach representation of G is K-finite.

Note In what follows, we shall use freely the notations and results of 2.3.1 infra.

Let $\mathfrak{g}$ be a reductive Lie algebra over **R**, $\theta: \mathfrak{g} \to \mathfrak{g}$ an automorphism of $\mathfrak{g}$ of order 2 whose restriction to the derived algebra $[\mathfrak{g}, \mathfrak{g}]$ of $\mathfrak{g}$ is a Cartan involution (of $[\mathfrak{g}, \mathfrak{g}]$); in the usual way, write $\mathfrak{g} = \mathfrak{k} + \mathfrak{p}$ and let $\mathfrak{G}$ (respectively $\mathfrak{K}$) denote the universal enveloping algebra of the complexification of $\mathfrak{g}$ (respectively of $\mathfrak{k}$). If λ denotes the canonical vector space isomorphism between the symmetric algebra $S(\mathfrak{g}_c)$ of $\mathfrak{g}_c$ and $\mathfrak{G}$, then it is clear that $\lambda(S(\mathfrak{k}_c)) = \mathfrak{K}(S(\mathfrak{k}_c)$ the symmetric algebra of $\mathfrak{k}_c$).

Let $\hat{\mathfrak{k}}_c$ denote the set $\{\delta\}$ of equivalence classes of finite dimensional simple $\mathfrak{k}_c$-modules.

Here is the main result of the present number.

Theorem 2.2.1.1 (Harish-Chandra) Let $\mathfrak{J}$ be a left ideal in $\mathfrak{K}$; assume that $\mathfrak{K}/\mathfrak{J}$ is finite dimensional and that the natural representation of $\mathfrak{k}_c$ on $\mathfrak{K}/\mathfrak{J}$ is semi-simple – then:

(i) The natural representation of $\mathfrak{k}_c$ on $\bar{\mathfrak{G}} = \mathfrak{G}/\mathfrak{G}\mathfrak{J}$ defines on $\bar{\mathfrak{G}}$ the structure of a semi-simple (locally finite) $\mathfrak{k}_c$-module;

(ii) Each isotypic $\mathfrak{k}_c$-submodule $\bar{\mathfrak{G}}(\delta)(\delta \in \hat{\mathfrak{k}}_c)$ is a finite module over the center $\mathfrak{Z}$ of $\mathfrak{G}$.

[There is no difficulty in seeing that each $\mathfrak{G}(\delta)(\delta \in \hat{\mathfrak{k}}_c)$ actually is a $\mathfrak{Z}$-module.]

The proof of this theorem will be broken up into a number of lemmas.

Let $S(\mathfrak{p}_c)$ denote the symmetric algebra over $\mathfrak{p}_c$; we recall that relative to the representation $X \mapsto d(X)$ $(X \in \mathfrak{k}_c)$, $S(\mathfrak{p}_c)$ is a semi-simple (locally finite) $\mathfrak{k}_c$-module (cf. 2.1.5).

Lemma 2.2.1.2 As $\mathfrak{k}_c$-modules, $S(\mathfrak{p}_c) \otimes_{\mathbf{C}} (\mathfrak{K}/\mathfrak{J})$ and $\bar{\mathfrak{G}}$ are isomorphic.

Proof It follows from the results outlined in 2.3.1 that the linear map of $E = S(\mathfrak{p}_c) \otimes_{\mathbf{C}} (\mathfrak{K}/\mathfrak{J})$ into $\bar{\mathfrak{G}}$ defined by the rule

$$u \otimes (D + \mathfrak{J}) \mapsto \lambda(u)(D + \mathfrak{G}\mathfrak{J}) \qquad (u \in S(\mathfrak{p}_c),\ D \in \mathfrak{K})$$

is an isomorphism of vector spaces; on the other hand, if $X \in \mathfrak{k}_c$, then

$$\begin{aligned} X_E(u \otimes (D + \mathfrak{J})) &= d(X)u \otimes (D + \mathfrak{J}) + u \otimes (XD + \mathfrak{J}) \\ &\mapsto \lambda(d(X)u)(D + \mathfrak{G}\mathfrak{J}) + \lambda(u)(XD + \mathfrak{G}\mathfrak{J}) \\ &= [X, \lambda(u)](D + \mathfrak{G}\mathfrak{J}) + \lambda(u)(XD + \mathfrak{G}\mathfrak{J}) \\ &= X\lambda(u)(D + \mathfrak{G}\mathfrak{J}) \end{aligned}$$

and so $S(\mathfrak{p}_c) \otimes_{\mathbf{C}} (\mathfrak{K}/\mathfrak{J})$ and $\bar{\mathfrak{G}}$ are indeed isomorphic as $\mathfrak{k}_c$-modules. □

Since $S(\mathfrak{p}_c)$ is a semi-simple (locally finite) $\mathfrak{k}_c$-module, $S(\mathfrak{p}_c) \otimes_{\mathbf{C}} (\mathfrak{K}/\mathfrak{J})$ is too (2.1.1); thus, in view of Lemma 2.2.1.2, $\bar{\mathfrak{G}}$ is a semi-simple (locally finite) $\mathfrak{k}_c$-module which establishes the first part of Theorem 2.2.1.1. However, the proof of part (ii) is more difficult – it depends upon a series of reductions which will now be undertaken.

Let $D \mapsto \bar{D}$ denote the natural mapping of $\mathfrak{G}$ onto $\bar{\mathfrak{G}}$.

Fix a class $\delta \in \hat{\mathfrak{k}}_c$; let E_δ be a simple $\mathfrak{k}_c$-module in the class δ. Let M_I denote the set of $\mathfrak{k}_c$-invariants for the $\mathfrak{k}_c$-module $M = \bar{\mathfrak{G}} \otimes_{\mathbf{C}} E_\delta$; observe that M may also be viewed as a $\mathfrak{G}$-module via the rule $D_1(\bar{D}_2 \otimes a) = \overline{(D_1 D_2)} \otimes a(D_1, D_2 \in \mathfrak{G};\ a \in E_\delta)$ – with this agreement it is clear that M_I is a $\mathfrak{Z}$-module ($\mathfrak{Z}$ the center of $\mathfrak{G}$).

Lemma 2.2.1.3 If M_I is a finite $\mathfrak{Z}$-module, then so is $\bar{\mathfrak{G}}(\check{\delta})$ ($\check{\delta}$ the class in $\hat{\mathfrak{k}}_c$ which is contragredient to δ).

Proof Let $\{a_1, \ldots, a_m\}$ be a $\mathbf{C}$-basis for E_δ; select elements $\bar{D}_{ij}$ in $\bar{\mathfrak{G}}$ such that the $\sum_{i=1}^m \bar{D}_{ij} \otimes a_i$ $(j = 1, \ldots, n)$ generate M_I as a $\mathfrak{Z}$-module – then Lemma 2.1.2.2 tells us that $\bar{\mathfrak{G}}(\check{\delta}) = \sum_{i,j} \mathfrak{Z}\bar{D}_{ij}$. □

Thanks to Lemma 2.2.1.3, Theorem 2.2.1.1(ii) will follow when it is shown that M_I is a finite $\mathfrak{Z}$-module.

The mapping ϕ defined by the prescription

$$u \otimes (D + \mathfrak{J}) \otimes a \mapsto \lambda(u)(D + \mathfrak{G}\mathfrak{J}) \otimes a \quad (u \in S(\mathfrak{p}_c),\ D \in \mathfrak{K},\ a \in E_\delta)$$

identifies the $\mathfrak{k}_c$-modules $L = S(\mathfrak{p}_c) \otimes_{\mathbf{C}} (\mathfrak{K}/\mathfrak{J}) \otimes_{\mathbf{C}} E_\delta$ and $M = \bar{\mathfrak{G}} \otimes_{\mathbf{C}} E_\delta$; of course $\phi(L_I) = M_I$, L_I the set of $\mathfrak{k}_c$-invariants in L. Put $E = (\mathfrak{K}/\mathfrak{J}) \otimes_{\mathbf{C}} E_\delta$ – then $S(\mathfrak{p}_c) \otimes_{\mathbf{C}} E$ may be regarded as an $S(\mathfrak{p}_c)$-module if we write $u_1(u_2 \otimes a) = u_1u_2 \otimes a$ $(u_1, u_2 \in S(\mathfrak{p}_c);\ a \in E)$. In the notations of 2.1.5, it is clear that L_I is an $I(\mathfrak{p}_c)$-module.

Lemma 2.2.1.4 L_I is a finite $I(\mathfrak{g}_c; \mathfrak{p}_c)$-module.

Proof For $I(\mathfrak{p}_c)$ is a finite $I(\mathfrak{g}_c; \mathfrak{p}_c)$-module (Corollary 2.1.5.6) and so it is enough to prove that L_I is a finite $I(\mathfrak{p}_c)$-module; in turn this is a consequence of Proposition 2.1.2.6. □

In view of what has been said above, the following lemma serves to complete the proof of Theorem 2.2.1.1(ii).

Lemma 2.2.1.5 M_I is a finite $\mathfrak{Z}$-module.

Proof As usual an element in $S(\mathfrak{p}_c) \otimes_{\mathbf{C}} E$ will be called homogeneous of degree m if it lies in $S^m(\mathfrak{p}_c) \otimes_{\mathbf{C}} E$; note that each $S^m(\mathfrak{p}_c) \otimes_{\mathbf{C}} E$ is $\mathfrak{k}_c$-stable. Owing to Lemma 2.2.1.4 we can find a finite number of homogeneous elements $l_i \in L_I$ $(1 \leqslant i \leqslant r)$, $l_i \in S^{m_i}(\mathfrak{p}_c) \otimes_{\mathbf{C}} E$ (say), such that $L_I = \sum_{i=1}^r I(\mathfrak{g}_c; \mathfrak{p}_c)l_i$. The space $\tilde{M}_I = \sum_{i=1}^r \mathfrak{Z}\phi(l_i)$ is contained in M_I – our lemma will follow when it is shown that $\tilde{M}_I = M_I$. Put $(L_I)_m = L_I \cap (\sum_{e=0}^m S^e(\mathfrak{p}_c) \otimes_{\mathbf{C}} E)$, $(M_I)_m = \phi((L_I)_m)$. The set of those l_i for which $m_i = 0$ generate $L_I \cap E$ – hence $(M_I)_0 \subset \tilde{M}_I$. By induction on m, it suffices, therefore, to prove that $(M_I)_{m+1} \subset \tilde{M}_I + (M_I)_m$ for all $m \geqslant 0$. Let $l \in L_I \cap (S^{m+1}(\mathfrak{p}_c) \otimes_{\mathbf{C}} E)$ and write $l = \sum_{i=1}^r u_il_i$, $u_i \in I(\mathfrak{g}_c; \mathfrak{p}_c)$. It may be supposed that $u_i \in S^{m+1-m_i}(\mathfrak{p}_c)$ $(S^e(\mathfrak{p}_c) = \{0\}$ if $e < 0)$. Let $\{a_1, \ldots, a_s\}$ be a $\mathbf{C}$-basis for E and let $l_i = \sum_{j=1}^s v_{ji} \otimes a_j$ with $v_{ji} \in S^{m_i}(\mathfrak{p}_c)$ – then, abusing the notation somewhat, we have

$$\phi(l) = \sum_{i=1}^r \sum_{j=1}^s \lambda(u_iv_{ji})a_j.$$

Each u_i is the restriction to $\mathfrak{p}_c$ of an element $p_i \in I(\mathfrak{g}_c)$; we may assume that $p_i \in S^{m+1-m_i}(\mathfrak{g}_c)$ – thus $u_i - p_i \in \sum_{e<m+1-m_i} S^e(\mathfrak{p}_c)S(\mathfrak{k}_c)$ and so

$$u_iv_{ji} - p_iv_{ji} \in \sum_{e<m} S^e(\mathfrak{p}_c)S(\mathfrak{k}_c)$$

$$\Rightarrow$$

$$\phi(l) - \sum_{i=1}^r \sum_{j=1}^s \lambda(p_iv_{ji})a_j \in \sum_{e<m} (\mathfrak{P}^e\mathfrak{K})E$$

where $\mathfrak{P}^e = \lambda(S^e(\mathfrak{p}_c))$ (Proposition 2.3.1.3). Now

$$\lambda(p_iv_{ji}) \equiv \lambda(p_i)\lambda(v_{ji}) \bmod \sum_{e<m} \lambda(S^e(\mathfrak{g}_c))$$

and since $\lambda(S^e(\mathfrak{g}_c)) \subset \sum_{f<e} \mathfrak{P}^f\mathfrak{K}$, we have

$$\phi(l) - \sum_{i=1}^{r} \sum_{j=1}^{s} \lambda(p_i)\lambda(v_{ji})a_j \in \sum_{e<m} (\mathfrak{P}^e\mathfrak{K})E.$$

Plainly $\mathfrak{K}E \subset E$ – therefore

$$\sum_{e<m} (\mathfrak{P}^e\mathfrak{K})E \subset \phi((\sum_{e<m} S^e(\mathfrak{p}_c)) \otimes_{\mathbf{C}} E).$$

But

$$\sum_{i=1}^{r} \sum_{j=1}^{s} \lambda(p_i)\lambda(v_{ji})a_j = \sum_{i=1}^{r} \lambda(p_i)\phi(l_i)$$

and, as the $\lambda(p_i) \in \mathfrak{Z}$, $\sum_{i=1}^{r} \lambda(p_i)\phi(l_i) \in \tilde{M}_I$ which implies that

$$\phi(l) - \sum_{i=1}^{r} \lambda(p_i)\phi(l_i) \in (M_I)_m.$$

Hence $\phi(l) \in \tilde{M}_I + (M_I)_m \Rightarrow (M_I)_{m+1} \subset \tilde{M}_I + (M_I)_m$.

The proof of the lemma is now complete. □

At the end, we see that the second assertion of Theorem 2.2.1.1 rests ultimately on the results obtained in 2.1.3, the main tool of reduction being Chevalley's Restriction Theorem (via Lemma 2.2.1.4).

2.2.2 Theorems of Finitude

Let $\mathfrak{g}$ be a reductive Lie algebra over $\mathbf{R}$, $\theta: \mathfrak{g} \to \mathfrak{g}$ an automorphism of $\mathfrak{g}$ of order 2 whose restriction to the derived algebra $[\mathfrak{g}, \mathfrak{g}]$ of $\mathfrak{g}$ is a Cartan involution (of $[\mathfrak{g}, \mathfrak{g}]$); employing the customary notations, write $\mathfrak{g} = \mathfrak{k} + \mathfrak{p}$ and let $\mathfrak{G}$ (respectively $\mathfrak{K}$) denote the universal enveloping algebra of the complexification of $\mathfrak{g}$ (respectively of $\mathfrak{k}$); let $\mathfrak{Z}$ denote the center of $\mathfrak{G}$.

Let $\hat{\mathfrak{k}}_c$ denote the set $\{\delta\}$ of equivalence classes of finite dimensional simple $\mathfrak{k}_c$-modules.

Let E be a (not necessarily finite dimensional) vector space over $\mathbf{C}$.

Definition A representation U of $\mathfrak{G}$ on E is said to be *quasi-simple* if there exists a homomorphism $\kappa: \mathfrak{Z} \to \mathbf{C}$ such that $U(Z) = \kappa(Z)U(1)$ for all $Z \in \mathfrak{Z}$.

In other words a quasi-simple representation U of $\mathfrak{G}$ is one which admits a character (see 2.4.3).

Theorem 2.2.2.1 (Harish-Chandra) Let κ be a character of $\mathfrak{G}$, δ_0 some fixed element in $\hat{\mathfrak{k}}_c$ – then, to within equivalence, there exist but a finite number of algebraically irreducible quasi-simple representations U of $\mathfrak{G}$ admitting the character κ and in which δ_0 occurs (i.e. in which $E(\delta_0) \neq \{0\}$, E the representation space of U). Furthermore the restriction of any such U to $\mathfrak{k}_c$ defines on the representation space of U the structure of a semi-simple, locally finite, $\mathfrak{k}_c$-module in which a given $\delta \in \hat{\mathfrak{k}}_c$ occurs (at most) a finite number of times.

The proof of this theorem depends upon two lemmas from the theory of associative algebras.

Lemma 2.2.2.2 Let A be an algebra over $\mathbf{C}$ with a unit 1, B a subalgebra of A (cf. 2.3.1), I a left ideal in A. Suppose that the following condition is satisfied: For all $x_1, x_2 \in A$, there exists $x \equiv 1 \bmod I$ such that $xx_1 \in B$, $xx_2 \in B$ – then, if M is a maximal left ideal in A containing I, $M \cap B$ is a maximal left ideal in B and the map $M \mapsto M \cap B$ is injective.

Proof (1) Let J be a left ideal in B such that $M \cap B \subset J$ and $J \neq M \cap B$. Choose $z \in J$, $z \notin M \cap B$; since M is maximal there exists $y \in A$ with $yz \equiv 1 \bmod M$; select $x \equiv 1 \bmod I$ such that $xy \in B$ – then $xyz \in J$, $xyz \equiv 1 \bmod M \Rightarrow xyz \equiv 1 \bmod J \Rightarrow J = B$ and so $M \cap B$ is a maximal left ideal in B.

(2) Let M_1, M_2 be two maximal left ideals in A containing I and such that $M_1 \cap B = M_2 \cap B$ – the claim is that $M_1 = M_2$. Suppose not – then we may write $1 = m_1 + m_2$ where $m_1 \in M_1$, $m_2 \in M_2$. Choose $x \equiv 1 \bmod I$ such that $xm_1 \in B$, $xm_2 \in B$ – then $x = xm_1 + xm_2 \in M_1 \cap B$ which implies that $1 \in M_1$, a contradiction.

Hence the lemma. □

Let A be an algebra over $\mathbf{C}$ with a unit 1, I a left ideal in A. Let A_I denote the set of $x \in A$ such that $Ix \subset I$ – then A_I is a subalgebra of A containing I and the center Z of A. Evidently I is a two-sided ideal in A_I. Let $x \mapsto \bar{x}$ denote the canonical map of A onto the space $\bar{A} = A/I$; its restriction to A_I is a homomorphism of A_I onto the algebra $\bar{A}_I = A_I/I$. Let J be a two-sided ideal in A and put $H = I + J$ – thus H is a left ideal in A and $H \cap A_I$ is a two-sided ideal in A_I. Let Ω denote the set of maximal left ideals in A which contain H.

Lemma 2.2.2.3 Suppose that the triple (A, A_I, I) verifies the hypotheses of Lemma 2.2.2.2 and, moreover, that $A_I/H \cap A_I$ is finite dimensional – then the set $\{U_M\}$ of irreducible representations of A defined canonically by the left ideals $M \in \Omega$ contains, apart from equivalence, but a finite number of distinct elements.

Proof Let $x \mapsto \tilde{x}$ denote the canonical homomorphism of A_I onto $\tilde{A}_I = A_I/H \cap A_I$ $(x \in A_I)$. The $M \cap A_I$, $M \in \Omega$, are maximal left ideals in $\tilde{A}_I$ containing $H \cap A_I$ – thus the $(M \cap A_I)^\sim$ are maximal left ideals in $\tilde{A}_I$. Let $\tilde{U}_M$ denote the irreducible representation of $\tilde{A}_I$ defined canonically by $(M \cap A_I)^\sim$ $(M \in \Omega)$. Owing to our assumptions, $\tilde{A}_I$ is finite dimensional and so there exists a finite subset Ω_F of Ω with the property that every $\tilde{U}_M$, $M \in \Omega$, is equivalent to a $\tilde{U}_N$, $N \in \Omega_F$. We now intend to show that every U_M, $M \in \Omega$, is equivalent to a U_N, $N \in \Omega_F$. For this purpose fix an $M \in \Omega$ – then there exists an $N \in \Omega_F$

such that $\tilde{U}_M$ is equivalent to $\tilde{U}_N$, whence there exists $x_0 \in A_I$, $x_0 \notin M$, such that

$$x \in A_I \text{ and } \tilde{x}\tilde{x}_0 \in (M \cap A_I)^{\sim} \Leftrightarrow x \in A_I \text{ and } \tilde{x} \in (N \cap A_I)^{\sim}.$$

Let $M_{x_0} = \{x \in A: xx_0 \in M\}$; evidently $M_{x_0} \in \Omega$ and U_M is equivalent to $U_{M_{x_0}}$ – we have:

$$x \in M_{x_0} \cap A_I \Leftrightarrow x \in A_I \text{ and } xx_0 \in M \Leftrightarrow x \in A_I \text{ and}$$
$$(xx_0)^{\sim} \in (M \cap A_I)^{\sim} \Leftrightarrow x \in A_I \text{ and } \tilde{x} \in (N \cap A_I)^{\sim} \Leftrightarrow x \in N \cap A_I.$$

These considerations show that $M_{x_0} \cap A_I = N \cap A_I$; from this we deduce that $M_{x_0} = N$ (Lemma 2.2.2.2) – therefore U_M is equivalent to U_N. □

We shall agree to retain the above notations.

Proof of Theorem 2.2.2.1 Let U be a representation of $\mathfrak{G}$ admitting the character κ and in which δ_0 occurs; let E be the representation space for U. Let $\mathfrak{J}_\kappa$ be the two-sided ideal in $\mathfrak{G}$ generated by the $Z - \kappa(Z)$ $(Z \in \mathfrak{Z})$. Fix a non-zero vector a_{δ_0} in $E[\delta_0]$ – then, in the notations of Lemma 2.1.1.1 (with the present $\mathfrak{k}_c$ playing the role of the $\mathfrak{g}$ there), $U(\mathfrak{A}_{\delta_0})a_{\delta_0} = 0$ and $\mathfrak{A}_{\delta_0}$ is a maximal left ideal in $\mathfrak{K}$. Let $\mathfrak{J}$ denote the left ideal in $\mathfrak{G}$ generated by $\mathfrak{A}_{\delta_0}$. If $\mathfrak{M}$ denotes the annihilator of a_{δ_0} in $\mathfrak{G}$, then $\mathfrak{M}$ is a maximal left ideal in $\mathfrak{G}$ containing $\mathfrak{H} = \mathfrak{J} + \mathfrak{J}_\kappa$ and, moreover, U is equivalent to $U_{\mathfrak{M}}$.

(1) Let $D \mapsto \bar{D}$, $D \mapsto \tilde{D}$, $D \mapsto D^*$ denote the canonical maps of $\mathfrak{G}$ onto the spaces $\bar{\mathfrak{G}} = \mathfrak{G}/\mathfrak{J}$, $\tilde{\mathfrak{G}} = \mathfrak{G}/\mathfrak{H}$, $\mathfrak{G}^* = \mathfrak{G}/\mathfrak{M}$, respectively – then $U_{\mathfrak{H}}$ is a quotient of $U_{\mathfrak{J}}$, $U_{\mathfrak{M}}$ is a quotient of $U_{\mathfrak{H}}$. Since $\mathfrak{K}/\mathfrak{A}_{\delta_0}$ is a finite dimensional simple $\mathfrak{k}_c$-module (in fact an element of the class δ_0), we may use Theorem 2.2.1.1 to conclude that $\bar{\mathfrak{G}} = \sum_{\delta \in \hat{\mathfrak{k}}_c} \bar{\mathfrak{G}}(\delta)$ with each $\bar{\mathfrak{G}}(\delta)$ a *finite* module over $\mathfrak{Z}$. Passing to the quotient and noting that every element of $\mathfrak{Z}$ is congruent mod $\mathfrak{J}_\kappa$ to a scalar, we see that $\tilde{\mathfrak{G}} = \sum_{\delta \in \hat{\mathfrak{k}}_c} \tilde{\mathfrak{G}}(\delta)$ where each $\tilde{\mathfrak{G}}(\delta)$ is *finite dimensional*. Finally, then, $\mathfrak{G}^* = \sum_{\delta \in \hat{\mathfrak{k}}_c} \mathfrak{G}^*(\delta)$ and, as each $\mathfrak{G}^*(\delta)$ is still finite dimensional, it follows that $\dim(E(\delta)) < \infty$ for all $\delta \in \hat{\mathfrak{k}}_c$ too. Therefore U is indeed $\mathfrak{k}$ – finite (cf. infra).

(2) To complete the proof of the theorem, we need only show that the triple $(\mathfrak{G}, \mathfrak{G}_{\mathfrak{J}}, \mathfrak{J})$ verifies the assumptions of Lemma 2.2.2.3 (as above, $\mathfrak{G}_{\mathfrak{J}} = \{D \in \mathfrak{G}: \mathfrak{J}D \subset \mathfrak{J}\}$). For this purpose, we shall first prove that $\bar{\mathfrak{G}}_{\mathfrak{J}} = \bar{\mathfrak{G}}[\delta_0]$ (cf. 2.1.1; we are, of course, viewing $\bar{\mathfrak{G}}$ as a $\mathfrak{k}_c$-module in the obvious way). Let $D \in \mathfrak{G}$ – then

$$\bar{D} \in \bar{\mathfrak{G}}_{\mathfrak{J}} \Leftrightarrow D \in \mathfrak{G}_{\mathfrak{J}} \Leftrightarrow \mathfrak{J}D \subset \mathfrak{J} \Leftrightarrow (\overline{\mathfrak{J}D}) = 0$$
$$\Leftrightarrow U_{\mathfrak{J}}(\mathfrak{J})\bar{D} = 0 \Leftrightarrow U_{\mathfrak{J}}(\mathfrak{A}_{\delta_0})\bar{D} = 0 \Leftrightarrow \bar{D} \in \bar{\mathfrak{G}}[\delta_0],$$

that is $\bar{\mathfrak{G}}_{\mathfrak{J}} = \bar{\mathfrak{G}}[\delta_0]$, from which it follows that $\mathfrak{G}_{\mathfrak{J}}/\mathfrak{H} \cap \mathfrak{G}_{\mathfrak{J}}$ is finite dimensional (since $\tilde{\mathfrak{G}}(\delta_0)$ is). It remains to establish that if $D_1, D_2 \in \mathfrak{G}$, then there exists $D \equiv 1 \bmod \mathfrak{J}$ such that $DD_1 \in \mathfrak{G}_{\mathfrak{J}}$, $DD_2 \in \mathfrak{G}_{\mathfrak{J}}$. Because

$\bar{\mathfrak{G}} = \sum_{\delta \in \hat{\mathfrak{k}}_c} \bar{\mathfrak{G}}(\delta)$, Lemma 2.1.1.1(ii) provides us with a $D \in \mathfrak{G}$ such that

$$U_{\mathfrak{J}}(D)\bar{D}_1 \in \bar{\mathfrak{G}}[\delta_0],\ U_{\mathfrak{J}}(D)\bar{D}_2 \in \bar{\mathfrak{G}}[\delta_0],\ U_{\mathfrak{J}}(D)\bar{1} = \bar{1},$$

i.e. such that $DD_1, DD_2 \in \mathfrak{G}_{\mathfrak{J}}$, $D \equiv 1 \bmod \mathfrak{J}$.

The proof of the theorem is now complete. □

Definition A representation of $\mathfrak{G}$ on E is said to be $\mathfrak{k}$-*finite* if:
(i) $E = \sum_{\delta \in \hat{\mathfrak{k}}_c} E(\delta)$;
(ii) $\dim(E(\delta)) < \infty$ for all $\delta \in \hat{\mathfrak{k}}_c$.

Proposition 2.2.2.4 Let U be an algebraically irreducible representation of $\mathfrak{G}$ on E which has the property that some class $\delta_0 (\delta_0 \in \hat{\mathfrak{k}}_c)$ occurs in the restriction of U to $\mathfrak{k}_c$ – then the following conditions are equivalent:
(i) U is algebraically completely irreducible (4.2.1);
(ii) U is quasi-simple;
(iii) U is $\mathfrak{k}$-finite.

Proof (i) ⇒ (ii) – this is clear; (ii) ⇒ (iii) – this follows from the preceding theorem; (iii) ⇒ (i) – for the proof, see Theorem 4.5.5.4. □

Remark The formidable problem of *classifying* the algebraic equivalence classes of algebraically irreducible $\mathfrak{k}$-finite representations of $\mathfrak{G}$ is one of the central questions in the Harmonic Analysis on semi-simple (or reductive) Lie groups – this will be discussed more throughly later on.

We shall now take a closer look at the algebraically completely irreducible representations of $\mathfrak{G}$.

Theorem 2.2.2.5 (Harish-Chandra) Let U be an algebraically completely irreducible representation of $\mathfrak{G}$ on E with character κ which has the property that some class $\delta_0 (\delta_0 \in \hat{\mathfrak{k}}_c)$ occurs in the restriction of U to $\mathfrak{k}_c$ – then there exists an integer $m_U \geqslant 1$ (depending, in general, on U) such that

$$\dim(E(\delta)) \leqslant m_U d(\delta)^2$$

for all $\delta \in \hat{\mathfrak{k}}_c$ (here $d(\delta)$ denotes the degree of δ).

In view of Theorem 2.2.2.1, we know that $E = \sum_{\delta \in \hat{\mathfrak{k}}_c} E(\delta)$ with $\dim(E(\delta)) < \infty$ (all $\delta \in \hat{\mathfrak{k}}_c$); on the other hand, Theorem 2.2.2.5 represents a substantial improvement on this result since it implies that there exists an a priori limitation on the dimensions of the various isotypic components $E(\delta)$ ($\delta \in \hat{\mathfrak{k}}_c$).

[Let G be a semi-simple Lie group with finite center (say) corresponding to $\mathfrak{g}$; it will turn out that Theorem 2.2.2.5 is very important for Harmonic Analysis on G – indeed, the fact that the characters of the irreducible unitary representations of G are actually distributions on G (in the sense of Laurent Schwartz) rests essentially on the present theorem (cf. Theorem 4.5.8.5).]

The proof of Theorem 2.2.2.5 is based on a result of Chevalley which may be formulated in the following manner. Fix a class $\delta \in \hat{\mathfrak{k}}_c$; let E_δ be a simple $\mathfrak{k}_c$-module in the class δ; consider the tensor product $L = S(\mathfrak{p}_c) \otimes_{\mathbf{C}} E_\delta$ – then L is a $\mathfrak{k}_c$-module in the natural way. On the other hand, L may also be viewed as an $S(\mathfrak{p}_c)$-module; consequently, the set L_I of $\mathfrak{k}_c$-invariants in L is a module over the set $I(\mathfrak{p}_c)$ of $\mathfrak{k}_c$-invariants in $S(\mathfrak{p}_c)$.

Theorem 2.2.2.6 (Chevalley) Retain the above notations – then there exist $d(\delta)$ homogeneous elements $l_1^*, \ldots, l_{d(\delta)}^*$ in L_I such that $L_I = \sum_{k=1}^{d(\delta)} I(\mathfrak{p}_c) l_k^*$.

Note The preceding theorem is obviously equally valid for finite dimensional semi-simple $\mathfrak{k}_c$-modules – this fact will be used without comment in the sequel.

We shall give the proof of this theorem in the Appendix to the present number.

Let $\mathfrak{P} = \lambda(S(\mathfrak{p}_c))$, $\mathfrak{P}^m = \lambda(S^m(\mathfrak{p}_c))$ $(m \geqslant 0)$ (λ the canonical vector space isomorphism of the symmetric algebra $S(\mathfrak{g}_c)$ of $\mathfrak{g}_c$ with $\mathfrak{G}$).

Lemma 2.2.2.7 We have:

(i) $\lambda(S^m(\mathfrak{p}_c)S(\mathfrak{k}_c)) \subset \sum_{e \leqslant m} \mathfrak{P}^e \mathfrak{K} (m \geqslant 0)$;

(ii) $\lambda(S^{m_1}(\mathfrak{p}_c)S(\mathfrak{k}_c))\lambda(S^{m_2}(\mathfrak{p}_c)S(\mathfrak{k}_c)) \subset \sum_{e \leqslant m_1+m_2} \mathfrak{P}^e \mathfrak{K}\ (m_1, m_2 \geqslant 0)$.

Proof The first assertion is an immediate consequence of the definition of λ and the fact that $[\mathfrak{k}_c, \mathfrak{p}_c] \subset \mathfrak{p}_c$. As for the second, first observe that it suffices to show that

$$(\mathfrak{P}^{m_1}\mathfrak{K})(\mathfrak{P}^{m_2}\mathfrak{K}) \subset \sum_{e \leqslant m_1+m_2} \mathfrak{P}^e \mathfrak{K}.$$

To this end, let $\mathfrak{p}_c^m$ denote the space spanned by the products of m elements in $\mathfrak{p}_c$ – then $\mathfrak{K}\mathfrak{p}_c^m \subset \mathfrak{p}_c^m \mathfrak{K}$. [This is clear for $m = 0$; proceeding by induction on m, let us suppose that our contention is true for some $m > 0$. Noting that $\mathfrak{K}\mathfrak{p}_c \subset \mathfrak{p}_c\mathfrak{K}$, we then have $\mathfrak{K}\mathfrak{p}_c^{m+1} = (\mathfrak{K}\mathfrak{p}_c^m)\mathfrak{p}_c \subset \mathfrak{p}_c^m \mathfrak{K}\mathfrak{p}_c \subset \mathfrak{p}_c^{m+1}\mathfrak{K}$, as desired.] But $\mathfrak{P}^m \subset \mathfrak{p}_c^m$ and so

$$(\mathfrak{P}^{m_1}\mathfrak{K})(\mathfrak{P}^{m_2}\mathfrak{K}) \subset \mathfrak{p}_c^{m_1}\mathfrak{K}\mathfrak{p}_c^{m_2}\mathfrak{K} \subset \mathfrak{p}_c^{m_1+m_2}\mathfrak{K}.$$

Therefore it will be enough to prove that $\mathfrak{p}_c^m \mathfrak{K} \subset \sum_{e \leqslant m} \mathfrak{P}^e \mathfrak{K}$; this is true for $m = 0, 1$; assume that it is true for m – then $\mathfrak{p}_c^{m+1}\mathfrak{K} \subset \sum_{e \leqslant m} \mathfrak{p}_c \mathfrak{P}^e \mathfrak{K}$. Since $\mathfrak{p}_c \mathfrak{P}^e \subset \lambda(S^1(\mathfrak{g}_c))\lambda(S^e(\mathfrak{g}_c)) \subset \sum_{f \leqslant e+1} \lambda(S^f(\mathfrak{g}_c))$ (Proposition 2.3.1.3(i)) $\subset \sum_{f \leqslant e+1} \mathfrak{P}^f \mathfrak{K}$ (Proposition 2.3.1.3(ii)), the result is likewise true for $m + 1$. □

Proof of Theorem 2.2.2.5 Fix a class δ_0 which occurs in $U|\mathfrak{k}_c$; let $\mathfrak{M}$ denote the maximal left ideal in $\mathfrak{G}$ as defined during the proof of Theorem

2.2.2.1 – then the representation U is equivalent to the natural representation $U_{\mathfrak{M}}$ of $\mathfrak{G}$ on $\mathfrak{G}/\mathfrak{M} = \mathfrak{G}^* = \sum_{\delta \in \hat{\mathfrak{k}}_c} \mathfrak{G}^*(\delta)$. Fix a representative E_{δ_0} of the class δ_0 in $\mathfrak{G}^*(\delta_0)$; in addition, for any class $\delta \in \hat{\mathfrak{k}}_c$, let $E_{\check{\delta}}$ be a simple $\mathfrak{k}_c$-module contragredient to the modules in the class δ. Consider the $\mathfrak{k}_c$-module $L = S(\mathfrak{p}_c) \otimes_{\mathbf{C}} E_{\delta_0} \otimes_{\mathbf{C}} E_{\check{\delta}}$ – then according to Chevalley's theorem (cf. supra), there exist $d(\delta_0)d(\delta)$ homogeneous elements $l_k^*(1 \leqslant k \leqslant d(\delta_0)d(\delta))$ in L_I such that $L_I = \sum_k I(\mathfrak{p}_c)l_k^*$. Since $I(\mathfrak{p}_c)$ is a finite $I(\mathfrak{g}_c; \mathfrak{p}_c)$ module (cf. Corollary 2.1.5.6), we may select a finite number of homogeneous elements $v_1, \ldots, v_r$ in $I(\mathfrak{p}_c)$ such that $I(\mathfrak{p}_c) = \sum_{e=1}^r I(\mathfrak{g}_c; \mathfrak{p}_c)v_e$ – therefore

$$L_I = \sum_k I(\mathfrak{p}_c)l_k^* = \sum_k \sum_e I(\mathfrak{g}_c; \mathfrak{p}_c)v_e l_k^*.$$

Put $m_U = rd(\delta_0)$; arrange the elements $v_e l_k^*$ in a sequence $l_i(1 \leqslant i \leqslant m_U d(\delta))$ – the l_i are homogeneous elements in L_I and $L_I = \sum_i I(\mathfrak{g}_c; \mathfrak{p}_c)l_i$. Let $\check{a}_1, \ldots, \check{a}_{d(\delta)}$ be a C-basis for $E_{\check{\delta}}$ and write

$$l_i = \sum_{j=1}^{d(\delta)} l_{ji} \otimes \check{a}_j \qquad (1 \leqslant i \leqslant m_U d(\delta))$$

where $l_{ji} \in S(\mathfrak{p}_c) \otimes_{\mathbf{C}} E_{\delta_0}$. The linear mapping ϕ of $S(\mathfrak{p}_c) \otimes_{\mathbf{C}} E_{\delta_0}$ into $\mathfrak{G}^*$ defined by the rule

$$\phi(u \otimes a) = U_{\mathfrak{M}}(\lambda(u))a \qquad (u \in S(\mathfrak{p}_c),\ a \in E_{\delta_0})$$

is a homomorphism of $\mathfrak{k}_c$-modules (cf. Lemma 2.2.1.2). Our intent now is to show that the $m_U d(\delta)^2$ elements $\phi(l_{ji})$ span $\mathfrak{G}^*(\delta)$; since m_U is independent of δ this would prove Theorem 2.2.2.5. [That the $\phi(l_{ji})$ really are in $\mathfrak{G}^*(\delta)$ is clear in view of Lemma 2.1.2.2.]

(1) Put $\mathfrak{G}_m^* = \phi(S^m(\mathfrak{p}_c) \otimes_{\mathbf{C}} E_{\delta_0})$ $(m \geqslant 0)$; since $\mathfrak{G} = \mathfrak{P}\mathfrak{K}$, $\mathfrak{G}^* = U_{\mathfrak{M}}(\mathfrak{P})E_{\delta_0} = \phi(S(\mathfrak{p}_c) \otimes_{\mathbf{C}} E_{\delta_0})$, whence $\mathfrak{G}^* = \sum_{m \geqslant 0} \mathfrak{G}_m^*$. Evidently each $\mathfrak{G}_m^*$ is $\mathfrak{k}_c$-stable; because $\mathfrak{G}^* = \sum_{\delta \in \hat{\mathfrak{k}}_c} \mathfrak{G}^*(\delta)$, it is then clear that $\mathfrak{G}^*(\delta) = \sum_{m \geqslant 0} (\mathfrak{G}^*(\delta) \cap \mathfrak{G}_m^*)$ and so, if $\mathscr{S}$ denotes the space spanned over $\mathbf{C}$ by the $\phi(l_{ji})$, it would be sufficient to prove that $\mathfrak{G}^*(\delta) \cap \mathfrak{G}_m^* \subset \mathscr{S}$ for all $m \geqslant 0$. Set $\mathfrak{G}_{-1}^* = \{0\}$; as the preceding statement is trivially true for $m = -1$, we may use induction on m and assume that $m \geqslant 0$. Fix a non-zero element D^* in $\mathfrak{G}^*(\delta) \cap \mathfrak{G}_m^*$ – our problem, therefore, is to show that $D^* \in \mathscr{S}$.

(2) The $\mathfrak{K}$-orbit in $\mathfrak{G}^*$ of D^* is a semi-simple $\mathfrak{k}_c$-module, all of whose simple components are in the class δ; for any such component E_δ (say), it will be shown that, necessarily, $E_\delta \subset \mathscr{S}$ – this, of course, will suffice to prove the theorem.

(3) Choose a basis $a_1, \ldots, a_{d(\delta)}$ for E_δ such that $\sum_j a_j \otimes \check{a}_j$ is an invariant for the $\mathfrak{k}_c$-module $E_\delta \otimes_{\mathbf{C}} E_{\check{\delta}}$. Since $E_\delta \subset \mathfrak{G}^*(\delta) \cap \mathfrak{G}_m^*$ and since $S^m(\mathfrak{p}_c) \otimes_{\mathbf{C}} E_{\delta_0}$ is a semi-simple $\mathfrak{k}_c$-module, we can select elements $\hat{a}_j \in S^m(\mathfrak{p}_c) \otimes_{\mathbf{C}} E_{\delta_0}$ such that $\phi(\hat{a}_j) = a_j$ and such that the space spanned over $\mathbf{C}$ by the $\hat{a}_j$ $(1 \leqslant j \leqslant d(\delta))$ is a simple $\mathfrak{k}_c$-module in the class δ. But then $\sum_j \hat{a}_j \otimes \check{a}_j \in L_I$ and hence

$$\sum_j \hat{a}_j \otimes \check{a}_j = \sum_i u_i l_i \qquad (u_i \in I(\mathfrak{g}_c; \mathfrak{p}_c)).$$

Let us suppose that $l_i \neq 0$ for $1 \leqslant i \leqslant i_0$ and $l_i = 0$ for $i_0 < i \leqslant m_U d(\delta)$ $(1 \leqslant i_0 \leqslant m_U d(\delta))$; that $u_i \neq 0$ if $1 \leqslant i \leqslant i_1$ and $u_i = 0$ if $i_1 < i \leqslant i_0$ $(1 \leqslant i_1 \leqslant i_0)$. Each u_i is the restriction to $\mathfrak{p}_c$ of an element $p_i \in I(\mathfrak{g}_c)$; we may assume that p_i is homogeneous of degree $m_i = m - n_i$ $(1 \leqslant i \leqslant i_1)$, n_i the degree of l_i – hence $p_i - u_i \in \sum_{e<m_i} S^e(\mathfrak{p}_c)S(\mathfrak{k}_c)$. Put

$$b_j = \sum_{1 \leqslant i \leqslant i_1} U_{\mathfrak{M}}(\lambda(p_i))\phi(l_{ji}).$$

Since $p_i \in I(\mathfrak{g}_c)$, $\lambda(p_i) \in \mathfrak{Z}$ and therefore

$$b_j = \sum_{1 \leqslant i \leqslant i_1} \kappa(\lambda(\mathfrak{p}_i))\phi(l_{ji}) \in \mathscr{S}$$

which implies that $a_j - b_j \in \mathfrak{G}^*(\delta)$.

(4) We are now going to prove that $a_j - b_j \in \mathscr{S}$; this will imply that $a_j \in \mathscr{S}$ and, in view of our earlier remarks, thus serve to complete the proof of the theorem. Write

$$a_j - b_j = \sum_{1 \leqslant i \leqslant i_1} \{\phi(u_i l_{ji}) - U_{\mathfrak{M}}(\lambda(p_i))\phi(l_{ji})\},$$

$$\begin{aligned}\phi(u_i l_{ji}) &- U_{\mathfrak{M}}(\lambda(p_i))\phi(l_{ji}) \\ &= \{\phi(u_i l_{ji}) - U_{\mathfrak{M}}(\lambda(u_i))\phi(l_{ji})\} - U_{\mathfrak{M}}(\lambda(p_i - u_i))\phi(l_{ji}).\end{aligned}$$

Since $p_i - u_i \in \sum_{e<m_i} S^e(\mathfrak{p}_c)S(\mathfrak{k}_c)$ $(1 \leqslant i \leqslant i_1)$ and $\phi(l_{ji}) \in \mathfrak{G}^*_{n_i}$, it follows that

$$\begin{aligned}U_{\mathfrak{M}}(\lambda(p_i - u_i))\phi(l_{ji}) &\in \sum_{e<m_i} U_{\mathfrak{M}}(\lambda(S^e(\mathfrak{p}_c)))\mathfrak{G}^*_{n_i} \\ &= \sum_{e<m_i} U_{\mathfrak{M}}(\lambda(S^e(\mathfrak{p}_c))\lambda(S^{n_i}(\mathfrak{p}_c)))E_{\delta_0} \subset \sum_{e<m} \mathfrak{G}^*_e,\end{aligned}$$

the last step being a consequence of the relation

$$\lambda(S^e(\mathfrak{p}_c))\lambda(S^{n_i}(\mathfrak{p}_c)) \subset \sum_{f \leqslant e+n_i} \lambda(S^f(\mathfrak{p}_c))\mathfrak{K} \quad \text{(Lemma 2.2.2.7)}.$$

On the other hand

$$\phi(u_i l_{ji}) - U_{\mathfrak{M}}(\lambda(u_i))\phi(l_{ji}) \in \sum_{e<m} U_{\mathfrak{M}}(\lambda(S^e(\mathfrak{g}_c)))E_{\delta_0} \subset \sum_{e<m} \mathfrak{G}^*_e,$$

as $\lambda(S^e(\mathfrak{g}_c)) \subset \sum_{f \leqslant e} \lambda(S^f(\mathfrak{p}_c))\mathfrak{K}$. Thanks to the induction hypothesis, then, we conclude that

$$a_j - b_j \in \mathfrak{G}^*(\delta) \cap (\sum_{e<m} \mathfrak{G}^*_e) = \sum_{e<m} (\mathfrak{G}^*(\delta) \cap \mathfrak{G}^*_e) \subset \mathscr{S}.$$

Hence the theorem. □

Appendix We shall give here a proof of Theorem 2.2.2.6 which is due to Stephen Rallis; Chevalley's original proof (which has never been published) is based on a different circle of ideas.

Let $\mathfrak{g}$ be a reductive Lie algebra over **R**, $\theta: \mathfrak{g} \to \mathfrak{g}$ an automorphism of $\mathfrak{g}$ of order 2 whose restriction to the derived algebra $[\mathfrak{g}, \mathfrak{g}]$ of $\mathfrak{g}$ is a Cartan involution (of $[\mathfrak{g}, \mathfrak{g}]$); write, in the usual way, $\mathfrak{g} = \mathfrak{k} + \mathfrak{p}$. Let $\mathfrak{g}_c$ be the complexification of $\mathfrak{g}$ (whence $\mathfrak{g}_c = \mathfrak{k}_c + \mathfrak{p}_c$ (direct sum)); let G_c be the adjoint group of $\mathfrak{g}_c$, K_c the analytic subgroup of G_c corresponding to ad $(\mathfrak{k}_c)$ – then K_c operates in the natural fashion on $\mathfrak{p}_c$ and hence on $S(\mathfrak{p}_c)$. We shall agree to denote by $\hat{K}_c$ the set of finite dimensional simple K_c-modules; it is clear that $\hat{K}_c$ can be regarded as a subset of $\hat{\mathfrak{k}}_c$.

Let $I(\mathfrak{p}_c)$ be the space of K_c-invariants in $S(\mathfrak{p}_c)$; let $I^*(\mathfrak{p}_c)$ be the subspace of $I(\mathfrak{p}_c)$ comprised of those elements having zero constant term, $I^+(\mathfrak{p}_c)$ the ideal in $S(\mathfrak{p}_c)$ generated by $I^*(\mathfrak{p}_c)$. An element $q \in S(\mathfrak{p}_c)$ is said to be *K_c-harmonic* if $\partial(p)q = 0$ for all $p \in I^*(\mathfrak{p}_c)$; it is clear that the space $H(\mathfrak{p}_c)$ of all K_c-harmonic polynomials in $S(\mathfrak{p}_c)$ is a K_c-submodule of $S(\mathfrak{p}_c)$ which is graded in the obvious way.

The proof of the following result may be found in Rallis [1].

Lemma 1 The symmetric algebra $S(\mathfrak{p}_c)$ is free as an $I(\mathfrak{p}_c)$-module (under multiplication) and the map of $I(\mathfrak{p}_c) \otimes_{\mathbf{C}} H(\mathfrak{p}_c) \to S(\mathfrak{p}_c)$ given by $p \otimes h \mapsto ph$ ($p \in I(\mathfrak{p}_c)$, $h \in H(\mathfrak{p}_c)$) is a K_c-module isomorphism.

Needless to say, Lemma 1 reduces the determination of the K_c-module structure on $S(\mathfrak{p}_c)$ to the analogous problem for $H(\mathfrak{p}_c)$; in turn the structure of $H(\mathfrak{p}_c)$ is explicated via the following lemma (cf. Rallis [1]).

Lemma 2 The space $H(\mathfrak{p}_c)$ of K_c-harmonic polynomials on $\mathfrak{p}_c$ is a semi-simple, locally finite, K_c-module – hence

$$H(\mathfrak{p}_c) = \sum_{\delta \in \hat{K}_c} H(\mathfrak{p}_c)(\delta).$$

Fix a maximal abelian subspace $\mathfrak{a}_{\mathfrak{p}_c}$ in $\mathfrak{p}_c$; let M_c denote the centralizer of $\mathfrak{a}_{\mathfrak{p}_c}$ in K_c – then the number of times that a given $\delta \in \hat{K}_c$ occurs in $H(\mathfrak{p}_c)$ (i.e. the length of $H(\mathfrak{p}_c)(\delta) \ldots$) is equal to the number of times that the trivial representation of M_c occurs in the restriction of δ to M_c.

[If we write $S(\mathfrak{p}_c) = \sum_{\delta \in \hat{K}_c} S(\mathfrak{p}_c)(\delta)$, then it follows from Lemmas 1 and 2 that $S(\mathfrak{p}_c)(\delta)$ is K_c-equivalent to $I(\mathfrak{p}_c) \otimes_{\mathbf{C}} H(\mathfrak{p}_c)(\delta)$ ($\delta \in \hat{K}_c$).]

Theorem (Chevalley) Fix a class $\delta \in \hat{\mathfrak{k}}_c$; let E_δ be a simple $\mathfrak{k}_c$-module in the class δ and put $L = S(\mathfrak{p}_c) \otimes_{\mathbf{C}} E_\delta$; let L_I be the set of $\mathfrak{k}_c$-invariants in L – then there exist $d(\delta)$ homogeneous elements $l_1^*, \ldots, l_{d(\delta)}^*$ in L_I such that $L_I = \sum_{k=1}^{d(\delta)} I(\mathfrak{p}_c) l_k^*$.

Proof (Rallis) Since $L = S(\mathfrak{p}_c) \otimes_{\mathbf{C}} E_\delta = I(\mathfrak{p}_c) \otimes_{\mathbf{C}} (H(\mathfrak{p}_c) \otimes_{\mathbf{C}} E_\delta)$, it is clear that $L_I = I(\mathfrak{p}_c) \otimes_{\mathbf{C}} (H(\mathfrak{p}_c) \otimes_{\mathbf{C}} E_\delta)_I$ where $(H(\mathfrak{p}_c) \otimes_{\mathbf{C}} E_\delta)_I$ is the set of $\mathfrak{k}_c$-invariants in $H(\mathfrak{p}_c) \otimes_{\mathbf{C}} E_\delta$; in view of Schur's Lemma and Lemma 2 supra, we see that actually

$$L_I = I(\mathfrak{p}_c) \otimes_{\mathbf{C}} (H(\mathfrak{p}_c)(\check{\delta}) \otimes_{\mathbf{C}} E_\delta)_I.$$

But Lemma 2 also tells us that the length of $H(\mathfrak{p}_c)(\check{\delta})$ cannot exceed $d(\check{\delta}) = d(\delta)$ – therefore $\dim((H(\mathfrak{p}_c)(\check{\delta}) \otimes_{\mathbf{C}} E_\delta)_I) \leqslant d(\delta)$, from which the assertion of the theorem then follows at once. □

[Let us note that the above proof implies that $L_I = \{0\}$ if the representations in the class δ do not 'drop' to $K_c \ldots$.]

2.3 On the Structure of the Universal Enveloping Algebra

2.3.1 Generalities

Let $\mathbf{k}$ be a field of characteristic zero.

In general by the term *algebra* over $\mathbf{k}$ *with a unit* we shall mean an associative algebra over $\mathbf{k}$ with an identity 1, by the term *subalgebra* we shall mean a subalgebra in the usual sense containing 1, and by the term *homomorphism of algebras* we shall mean a homomorphism in the usual sense mapping 1 into 1. In particular the notion of a **k**-*representation* of an algebra is to be given the usual meaning (cf. 4.2.1). If A is an algebra, let A_L denote the Lie algebra obtained from A per the bracket $[x, y] = xy - yx$ $(x, y \in A)$.

Let $\mathfrak{g}$ be a finite dimensional Lie algebra over a field **k** of characteristic zero.

We shall begin by reviewing some well-known properties of the universal enveloping algebra of $\mathfrak{g}$, properties with which the reader is assumed to be familiar (for details, see Bourbaki [4] or Jacobson [3]).

By definition a *universal enveloping algebra* of $\mathfrak{g}$ is a pair $(\mathfrak{G}, i)$ where $\mathfrak{G}$ is an algebra over **k** with a unit and i is a homomorphism of $\mathfrak{g}$ into $\mathfrak{G}_L$ with the following universal property: If A is any algebra over **k** with a unit and ϕ is any homomorphism of $\mathfrak{g}$ into A_L, then there exists a unique homomorphism ψ of $\mathfrak{G}$ into A such that $\phi = \psi \circ i$. In the usual way, one shows that $\mathfrak{G}$ exists and is essentially unique so that we shall speak of the universal enveloping algebra of $\mathfrak{g}$ and identify $\mathfrak{g}$ with its isomorphic replica in $\mathfrak{G}_L$.

Of course we hardly need to mention that the importance of the universal enveloping algebra resides in the fact that there exists a one-to-one correspondence between representations of $\mathfrak{g}$ and those of $\mathfrak{G}$. More precisely, every representation U of $\mathfrak{g}$ can be extended to a unique representation U of $\mathfrak{G}$ and conversely the restriction to $\mathfrak{g}$ of any representation of $\mathfrak{G}$ defines a representation of $\mathfrak{g}$.

Here are some of the basic properties of $\mathfrak{G}$.

Proposition 2.3.1.1 Let $\mathfrak{g}$ be a finite dimensional Lie algebra over **k**, $\mathfrak{G}$ its universal enveloping algebra.

(i) $\mathfrak{G}$ is generated by $\mathfrak{g}$ and 1.

(ii) Let $\mathfrak{G}^+$ be the ideal in $\mathfrak{G}$ generated by $\mathfrak{g}$ – then $\mathfrak{G} = \mathbf{k}1 + \mathfrak{G}^+$ (direct sum).

(iii) $\mathfrak{G}$ admits a unique anti-automorphism '+' such that $X^+ = -X$ for $X \in \mathfrak{g}$ (the *principal anti-automorphism* of $\mathfrak{G}$).

(iv) There exists an injective homomorphism ∇ of $\mathfrak{G}$ into $\mathfrak{G} \otimes_{\mathbf{k}} \mathfrak{G}$ such that $\nabla(X) = X \otimes 1 + 1 \otimes X$ for $X \in \mathfrak{g}$; ∇ is called the *diagonal map*.

(v) Any derivation of $\mathfrak{g}$ admits a unique extension to a derivation of $\mathfrak{G}$.

(vi) $\mathfrak{G}$ has no zero divisors; $\mathfrak{G}$ satisfies the ascending chain condition for left or right ideals.

(vii) If $X_1, \ldots, X_n$ is a basis for $\mathfrak{g}$, then the elements $X_1^{e_1} \ldots X_n^{e_n}$, $e_i = 0, 1, 2, \ldots$ $(X_i^0 = 1)$ form a basis for $\mathfrak{G}$ (Poincaré-Birkhoff-Witt).

Proposition 2.3.1.2 Let $\mathfrak{g}$ be a finite dimensional Lie algebra over **k**, $\mathfrak{G}$ its universal enveloping algebra.

(i) If $\mathfrak{h}$ is a subalgebra of $\mathfrak{g}$, then the subalgebra of $\mathfrak{G}$ generated by $\mathfrak{h}$ and 1 may be taken as the universal enveloping algebra of $\mathfrak{h}$.

(ii) Let $\mathfrak{h}_1, \mathfrak{h}_2$ be subalgebras of $\mathfrak{g}$ such that $\mathfrak{g} = \mathfrak{h}_1 + \mathfrak{h}_2$, $\mathfrak{h}_1 \cap \mathfrak{h}_2 = \{0\}$. If $\mathfrak{H}_1, \mathfrak{H}_2$ are the universal enveloping algebras of $\mathfrak{h}_1$ and $\mathfrak{h}_2$, then the bilinear map $(H_1, H_2) \mapsto H_1H_2$ of $\mathfrak{H}_1 \times \mathfrak{H}_2$ into $\mathfrak{G}$ defines a vector space bijection ϕ of $\mathfrak{H}_1 \otimes_{\mathbf{k}} \mathfrak{H}_2$ with $\mathfrak{G}$ such that $\phi(H_1 \otimes H_2) = H_1H_2$. Let $\mathfrak{J}_2$ be a left ideal in $\mathfrak{H}_2$, $\mathfrak{J}$ the left ideal in $\mathfrak{G}$ generated by $\mathfrak{J}_2$ – then $\phi(\mathfrak{H}_1 \otimes_{\mathbf{k}} \mathfrak{J}_2) = \mathfrak{J}$.

Remark As is shown in Jacobson [3, p.166], the fact that $\mathfrak{G}$ satisfies the ascending chain condition for left ideals implies that $\mathfrak{G}$ admits a *left quotient division ring* $\mathfrak{D}(\mathfrak{G})$, i.e.: (1) $\mathfrak{D}(\mathfrak{G})$ is a division ring; (2) $\mathfrak{G}$ is a subring of $\mathfrak{D}(\mathfrak{G})$; (3) Every element of $\mathfrak{D}(\mathfrak{G})$ has the form $D_1^{-1}D_2$ $(D_1, D_2 \in \mathfrak{G})$. For some consequences of this property of $\mathfrak{G}$ see Dixmier [19], Gelfand and Kirillov [1], Solomon and Verma [1] and the references given there.

Let $S = S(\mathfrak{g})$ denote the symmetric algebra of $\mathfrak{g}$ – then the connection that exists between S and $\mathfrak{G}$ may be described as follows. If X and Y are elements in $\mathfrak{g}$ denote their product in S by XY and their product in $\mathfrak{G}$ by $X \cdot Y$ (this convention will be adhered to only when it is necessary to avoid confusion). Let λ denote the canonical linear bijection of S with $\mathfrak{G}$ – thus if $\{X_1, \ldots, X_n\}$ is a basis for $\mathfrak{g}$, then

$$\lambda(X_1^{e_1} \ldots X_n^{e_n}) = \frac{1}{m!} \Sigma X_{i_{\sigma(1)}} \cdot \ldots \cdot X_{i_{\sigma(m)}},$$

where $m = \sum_i e_i$, $(i_1, \ldots, i_m)$ is a set of m integers such that exactly e_j of them are equal to j $(1 \leqslant j \leqslant n)$ and the sum is over all permutations σ of the m integers $(1, 2, \ldots, m)$. In Helgason [2, p.392], the map λ is called *symmetrization*; it is to be emphasized that λ does not, in general, establish an algebra isomorphism between S and $\mathfrak{G}$.

Let m be a non-negative integer; put $\mathfrak{G}^m = \lambda(S^m)$, S^m the homogeneous elements of degree m in S – then $\mathfrak{G} = \sum_{m \geqslant 0} \mathfrak{G}^m$ (direct sum of vector spaces). Note that $\mathfrak{G}^0 = \mathbf{k}$, $\mathfrak{G}^1 = \mathfrak{g}$ – however $\mathfrak{G}$ is *not* necessarily graded by the $\mathfrak{G}^m$. On the other hand, set $\mathfrak{G}_m = \mathfrak{G}^0 + \ldots + \mathfrak{G}^m$ – then $\mathfrak{G}_{m_1} \cdot \mathfrak{G}_{m_2} \subset \mathfrak{G}_{m_1+m_2}$ and thus $\mathfrak{G}$ is filtered by the $\mathfrak{G}_m$. With respect to this filtration of $\mathfrak{G}$ (the *standard filtration*), the associated graded algebra is S – of course $\mathfrak{G}_m/\mathfrak{G}_{m-1}$ is isomorphic as a space to S^m.

Proposition 2.3.1.3 Let $\mathfrak{g}$ be a finite dimensional Lie algebra over $\mathbf{k}$, $S = S(\mathfrak{g})$ its symmetric algebra, $\mathfrak{G}$ its universal enveloping algebra.

(i) If $u \in S^{m_1}$, $v \in S^{m_2}$, then

$$\lambda(uv) - \lambda(u)\lambda(v) \in \sum_{e < m_1 + m_2} \lambda(S^e).$$

(ii) Suppose that $\mathfrak{g}$ is the direct sum of two subspaces $\mathfrak{h}_1$ and $\mathfrak{h}_2$ – then the map $(u_1, u_2) \mapsto \lambda(u_1)\lambda(u_2)$ sets up a vector space isomorphism between $S(\mathfrak{h}_1) \otimes_{\mathbf{k}} S(\mathfrak{h}_2)$ and $\mathfrak{G}$. Furthermore, if m is a non-negative integer, then

$$\sum_{e\leqslant m}\lambda(S^e)=\sum_{e_1+e_2\leqslant m}\lambda(S^{e_1}(\mathfrak{h}_1))\lambda(S^{e_2}(\mathfrak{h}_2)).$$

Let $X\mapsto X_S$ be the usual representation of $\mathfrak{g}$ on S by derivations, S_I the corresponding set of invariants; let $X\mapsto X_{\mathfrak{G}}$ be the representation of $\mathfrak{g}$ on $\mathfrak{G}$ by derivations which extends the adjoint representation of $\mathfrak{g}$ – then

$$\lambda\circ X_S=X_{\mathfrak{G}}\circ\lambda \qquad (X\in\mathfrak{g}).$$

In addition $\lambda(S_I)=\mathfrak{Z}$, $\mathfrak{Z}$ the center of $\mathfrak{G}$ – thus $\mathfrak{Z}$ is the set $\mathfrak{G}_I$ of $\mathfrak{g}$-invariants in $\mathfrak{G}$ for the representation $X\mapsto X_{\mathfrak{G}}$ of $\mathfrak{g}$ on $\mathfrak{G}$. More generally any derivation of $\mathfrak{g}$ admits a unique extension to S and λ transforms this extension into the corresponding extension of the derivation to $\mathfrak{G}$.

Let A be an algebra filtered by subspaces $\{A_i\}$ $(i=0,1,\ldots)$, $A^\dagger$ the associated graded algebra. Let $x\in A$ and suppose that $x\in A_i$, $\notin A_{i-1}$ (A_{-1} being, by definition, $\{0\}$) – the element $x^\dagger=x+A_{i-1}$ is a homogeneous element of degree i in $A^\dagger$ and is called the *leading term* of x.

We shall now formulate an extension of the Fundamental Theorem of Invariant Theory. Thus let E be a finite dimensional semi-simple $\mathfrak{g}$-module – assume that E is a Lie algebra and that the representing endomorphisms $X_E(X\in\mathfrak{g})$ are derivations of E. Let $X\mapsto X_{\mathfrak{E}}$ be the representation of $\mathfrak{g}$ on the universal enveloping algebra $\mathfrak{E}$ of E which extends the representation $X\mapsto X_E$, $\mathfrak{E}_I$ the corresponding set of invariants; evidently $\mathfrak{E}=\sum_\delta\mathfrak{E}(\delta)$ and each $\mathfrak{E}(\delta)$ is a module over $\mathfrak{E}_I$. In the standard filtration $\{\mathfrak{E}_m\}$ of $\mathfrak{E}$ each $\mathfrak{E}_m$ is a semi-simple $\mathfrak{g}$-module – hence each $\mathfrak{E}(\delta)$ is a filtered $\mathfrak{E}_I$-module whose associated graded module is $S(\delta)$, $\delta\in\hat{\mathfrak{g}}(S=S(E))$.

Theorem 2.3.1.4 The algebra $\mathfrak{E}_I$ is finitely generated; in addition each $\mathfrak{E}(\delta)$ is a finite $\mathfrak{E}_I$-module $(\delta\in\hat{\mathfrak{g}})$.

Proof If E is abelian, this is just Theorem 2.1.2.3. Turning to the general case, let $u_1,\ldots,u_n$ be generators for the algebra S_I; we may assume that each u_i is the leading term of some $D_i\in\mathfrak{E}_I$. Claim: The D_i generate $\mathfrak{E}_I$, i.e. every $D\in\mathfrak{E}_I$ has the form $\sum_i\tilde{D}_iD_i$, $\tilde{D}_i\in\tilde{\mathfrak{E}}_I$ ($\tilde{\mathfrak{E}}_I$ the algebra generated by the D_i). It may be assumed that the leading term $D^\dagger$ of D is of degree m and we may suppose that the result holds for elements with leading terms of degree less than m – then $D^\dagger=\sum_i\tilde{u}_iu_i$, $\tilde{u}_i\in S_I$. There is no loss of generality in assuming that $\tilde{u}_i$ is homogeneous and that $\tilde{u}_iu_i$ is homogeneous of degree m – then $\tilde{u}_i$ is the leading term of an element $\tilde{D}_i\in\tilde{\mathfrak{E}}_I$ and the leading term of $\sum_i\tilde{D}_iD_i$ is $D^\dagger$. Therefore $D-\sum_i\tilde{D}_iD_i=\sum_i\tilde{D}'_iD_i$ $(\tilde{D}'_i\in\tilde{\mathfrak{E}}_I)$ (by the induction assumption). Hence the claim. In the same way one proves that each $\mathfrak{E}(\delta)$ is a finite $\mathfrak{E}_I$-module. □

Example Let $\mathfrak{g}$ be a semi-simple Lie algebra over $\mathbf{R}$, $\mathfrak{g}=\mathfrak{k}+\mathfrak{p}$ a Cartan decomposition of $\mathfrak{g}$ – then $\mathfrak{k}_c$ is reductive in $\mathfrak{g}_c$ and so $\mathfrak{g}_c$ is a semi-simple $\mathfrak{k}_c$-

module in the obvious way. In the notations of Theorem 2.3.1.4, E plays the role of the present $\mathfrak{g}_c$ while the role of $\mathfrak{E}_I$ is played by $\mathfrak{X}$, $\mathfrak{X}$ the centralizer of $\mathfrak{k}_c$ in $\mathfrak{G}$ ($\mathfrak{G}$ the universal enveloping algebra of $\mathfrak{g}_c$). Conclusions: (1) $\mathfrak{X}$ is a finitely generated algebra; (2) Each isotypic component $\mathfrak{G}(\delta)$ ($\delta \in \hat{\mathfrak{k}}_c$) is a finite $\mathfrak{X}$-module.

2.3.2 Existence of Sufficiently Many Finite Dimensional Representations

Retain the notations introduced in 2.3.1 – thus $\mathfrak{g}$ is a finite dimensional Lie algebra over **k**, **k** being a field of characteristic zero. The primary objective of this number is to prove: Given $D \neq 0$ in $\mathfrak{G}$ ($\mathfrak{G}$ the universal enveloping of $\mathfrak{g}$), there exists a finite dimensional representation U of $\mathfrak{G}$ such that $U(D) \neq 0$. We shall begin with a computation whose significance will become apparent presently.

Let E be a finite dimensional vector space over **k**. Let $\mathfrak{h} = \mathfrak{sl}(E)$, the endomorphisms of E with zero trace; let $\mathfrak{H}$ denote the universal enveloping algebra of $\mathfrak{h}$, $S(\mathfrak{h})$ the symmetric algebra of $\mathfrak{h}$. Let $X_1, \ldots, X_m$ be a basis for $\mathfrak{h}$ with the property that $1, X_1, \ldots, X_m$ is a basis for $L = \mathrm{Hom}_{\mathbf{k}}(E, E)$. Put $E_s = \overset{s}{\otimes} E$ – then the identity representation of $\mathfrak{h}$ on E extends in the usual way to a representation U_s of $\mathfrak{h}$ on E_s which in turn determines a homomorphism U_s of $\mathfrak{H}$ into $L_s = \mathrm{Hom}_{\mathbf{k}}(E_s, E_s)$. Let $\tilde{L}_s$ be the subspace of L_s generated by the elements $Y_1 \otimes \cdots \otimes Y_s$ where $Y_i \in L$ and $Y_i = 1$ for at least one i. It is clear that L_s may be identified with $\overset{s}{\otimes} L$ and so, for $X \in \mathfrak{h}$, the endomorphism $U_s(X)$ of E_s is given by

$$U_s(X) = \sum_{j=1}^{s} X^{(s,j)}$$

where $X^{(s,j)} = 1 \otimes \cdots \otimes 1 \otimes X \otimes 1 \otimes \cdots \otimes 1$, all the factors being equal to 1 except the jth which is X. Let $D \in \mathfrak{H}_s$; we propose to compute $U_s(D) \bmod \tilde{L}_s$. In the first place D may be written in the form

$$D = \sum_{r=0}^{s} \sum_{1 \leq i_1, \ldots, i_r \leq m} k^{i_1 \ldots i_r} X_{i_1} \cdots X_{i_r} \qquad (k^{i_1 \ldots i_r} \in \mathbf{k})$$

where $k^{i_1 \ldots i_r}$ are symmetric with respect to permutations of the indices; furthermore we have that

$$U_s(X_{i_1} \cdots X_{i_r}) = \sum_{1 \leq j_1, \ldots, j_r \leq s} X_{i_1}^{(s,j_1)} \cdots X_{i_r}^{(s,j_r)}.$$

Hence if $r < s$, then there is at least one integer in the set $(1, \ldots, s)$ which is not present among the j_i ($i = 1, \ldots, r$); so, if $r < s$, we have

$$U_s(X_{i_1} \cdots X_{i_r}) \equiv 0 \bmod \tilde{L}_s.$$

It remains to consider the case when $r = s$; it may be supposed that all the $j_1, \ldots, j_r$ are distinct – then

$$U_s(X_{i_1} \cdots X_{i_s}) \equiv \sum X_{i_1}^{(s,j_1)} \cdots X_{i_s}^{(s,j_s)} \bmod \tilde{L}_s$$

where the summation is taken over all permutations $j_1, \ldots, j_s$ of the set

$(1, \ldots, s)$. The right hand side of this expression is equal to

$$\sum_\sigma X_{\sigma(i_1)} \otimes \cdots \otimes X_{\sigma(i_s)}$$

the summation being extended over all permutations σ of the set $(i_1, \ldots, i_s)$. Therefore

$$U_s(D) \equiv s! \sum_{1 \leqslant i_1, \ldots, i_s \leqslant m} k^{i_1 \ldots i_s} X_{i_1} \otimes \cdots \otimes X_{i_s} \qquad \text{mod } \tilde{L}_s$$

where we have taken into account the fact that the $k^{i_1 \ldots i_s}$ are symmetric.

An important consequence of this computation will now be derived. Let $\mathfrak{K}_s$ denote the kernel of the representation U_s and put $\mathfrak{K} = \bigcap_1^\infty \mathfrak{K}_s$ – then we claim that $\mathfrak{K} = \{0\}$. Thus let $D \in \mathfrak{K}$; if t is the largest integer such that for some $(i_1, \ldots, i_t)$, $k^{i_1 \ldots i_t} \neq 0$ in the decomposition of D above, then, since $D \in \mathfrak{K}_t$, we have

$$\sum k^{i_1 \ldots i_t} X_{i_1} \otimes \cdots \otimes X_{i_t} \equiv 0 \qquad \text{mod } \tilde{L}_t.$$

Recalling that $1, X_1, \ldots, X_m$ is a basis for $L = \text{Hom}_{\mathbf{k}}(E, E)$, we see that the various $X_{i_1} \otimes \ldots \otimes X_{i_t}$ are linearly independent mod $\tilde{L}_t$ and so it must be the case that $k^{i_1 \ldots i_t} = 0$ for all $(i_1, \ldots, i_t)$, a contradiction. Therefore $\mathfrak{K} = \{0\}$ as claimed. All this implies: Given $D \neq 0$, $D \in \mathfrak{H}$, there exists a finite dimensional representation U of $\mathfrak{H}$ such that $U(D) \neq 0$. [For there exists an integer s such that $D \notin \mathfrak{K}_s$ – hence $U_s(D) \neq 0$.]

Let us now extend these conclusions to any finite dimensional Lie algebra $\mathfrak{g}$ over $\mathbf{k}$.

Lemma 2.3.2.1 Let E and $\tilde{E}$ be vector spaces over $\mathbf{k}$ of dimensions n and $n + 1$, respectively – then $\text{Hom}_{\mathbf{k}}(E, E)$ is isomorphic to a subalgebra of $\mathfrak{sl}(\tilde{E})$.

Proof We may assume that E is a hyperplane in $\tilde{E}$. Let $\tilde{a} \in \tilde{E}$, $\tilde{a} \notin E$; given $f \in \text{Hom}_{\mathbf{k}}(E, E)$, let $\tilde{f}$ denote the extension of f to $\tilde{E}$ with the property that $\tilde{f}(\tilde{a}) = -\text{tr}(f)\tilde{a}$ – then the map $f \mapsto \tilde{f}$ establishes the isomorphism in question. □

Corollary 2.3.2.2 Let $\mathfrak{g}$ be a finite dimensional Lie algebra over $\mathbf{k}$ – then $\mathfrak{g}$ admits a faithful finite dimensional representation by linear transformations of trace zero.

Proof By Ado's Theorem (Bourbaki [4, p.100]), $\mathfrak{g}$ has a faithful finite dimensional representation – now apply the preceding lemma. □

Theorem 2.3.2.3 (Harish-Chandra) Let $\mathfrak{g}$ be a finite dimensional Lie algebra over $\mathbf{k}$, $\mathfrak{G}$ its universal enveloping algebra. Given $D \neq 0$, $D \in \mathfrak{G}$, there exists a finite dimensional representation U of $\mathfrak{G}$ such that $U(D) \neq 0$.

Proof In view of Corollary 2.3.2.2, we may assume that $\mathfrak{g}$ is a subalgebra of $\mathfrak{sl}(E)$ for a certain finite dimensional vector space E – the theorem then follows from our earlier discussion. □

One usually expresses the conclusion of Theorem 2.3.2.3 by: $\mathfrak{G}$ admits sufficiently many finite dimensional representations.

Example (Kostant) Let $\mathfrak{g}$ be a semi-simple Lie algebra over **C**. The preceding theorem implies: If $\mathscr{S}$ is any one dimensional subspace of $\mathfrak{G}$, then there exists a finite dimensional irreducible representation U of $\mathfrak{G}$ which is faithful on $\mathscr{S}$. This is not true for higher dimensional subspaces of $\mathfrak{G}$. Thus let $Z \in \mathfrak{Z}$ ($\mathfrak{Z}$ the center of $\mathfrak{G}$), $D \in \mathfrak{G}$, where $D \neq 0$ and $Z \notin \mathfrak{G}_0$ – then D and ZD span a two dimensional space in $\mathfrak{G}$ but its image under any finite dimensional irreducible U is at most a one dimensional space (Schur's Lemma). However, if $\mathscr{S}$ is any finite dimensional subspace of $\mathscr{N}(\mathfrak{G})$ (see 2.3.3), then there does exist a finite dimensional irreducible U which is faithful on $\mathscr{S}$ (Kostant [4, p.403]).

2.3.3 The Reductive Case

Let $\mathfrak{g}$ be a reductive Lie algebra over a field **k** of characteristic zero, $\mathfrak{G}$ its universal enveloping algebra; let $X \mapsto X_{\mathfrak{G}}$ denote the representation of $\mathfrak{g}$ on $\mathfrak{G}$ which extends the adjoint representation – then, since $\mathfrak{g}$ is reductive, $\mathfrak{G}$ is a semi-simple, locally finite, $\mathfrak{g}$-module. The set of invariants $\mathfrak{G}_I$ in $\mathfrak{G}$ for this representation is $\mathfrak{Z}$, $\mathfrak{Z}$ the center of $\mathfrak{G}$ (cf. 2.3.1); consequently $\mathfrak{G} = \mathfrak{Z} + \mathfrak{G}_R$ (direct sum) where $\mathfrak{G}_R$ denotes the subspace of $\mathfrak{G}$ spanned by the $[X, D]$ ($X \in \mathfrak{g}$, $D \in \mathfrak{G}$) (cf. Lemma 2.1.2.1) – in fact $\mathfrak{G}_R = [\mathfrak{G}, \mathfrak{G}]$, $[\mathfrak{G}, \mathfrak{G}]$ the subspace of $\mathfrak{G}$ spanned by the elements $[D_1, D_2]$ ($D_1, D_2 \in \mathfrak{G}$). [It is clearly sufficient to verify that $[\mathfrak{G}, \mathfrak{G}] \subset \mathfrak{G}_R$; but if $D_1, D_2 \in \mathfrak{G}$ and if $D_1 = X_1 \ldots X_m$ (say) ($X_i \in \mathfrak{g}$), then

$$[D_1, D_2] = \sum_{i=1}^{m} [X_i, X_{i+1}X_{i+2} \ldots X_m D_2 X_1 \ldots X_{i-1}] \in \mathfrak{G}_R,$$

whence $[\mathfrak{G}, \mathfrak{G}] \subset \mathfrak{G}_R$, as desired.]

Let $D \mapsto D^0$ denote the projection of $\mathfrak{G}$ onto $\mathfrak{Z}$ (relative to $[\mathfrak{G}, \mathfrak{G}]$) – then this projection clearly has the following properties: (1) $(D_1 D_2)^0 = (D_2 D_1)^0$ ($D_1, D_2 \in \mathfrak{G}$); (2) $(D^0)^0 = D^0 (D \in \mathfrak{G})$; (3) $(D_1^0 D_2)^0 = D_1^0 D_2^0$ ($D_1, D_2 \in \mathfrak{G}$).

Proposition 2.3.3.1 Let $\mathfrak{g}$ be a reductive Lie algebra over **k**, $\mathfrak{G}$ its universal enveloping algebra; let $D \mapsto D^0$ denote the projection of $\mathfrak{G}$ onto its center $\mathfrak{Z}$ (relative to $[\mathfrak{G}, \mathfrak{G}]$); let $f: \mathfrak{G} \to \mathbf{k}$ be a linear function on $\mathfrak{G}$: In order that $f(D_1 D_2) = f(D_2 D_1)$ (all $D_1, D_2 \in \mathfrak{G}$), it is necessary and sufficient that $f(D) = f(D^0)$ for all $D \in \mathfrak{G}$.

Proof If f is a linear function on $\mathfrak{G}$ with the property in question, then f vanishes on $[\mathfrak{G}, \mathfrak{G}]$ and so $f(D) = f(D^0)$ (all $D \in \mathfrak{G}$); the converse statement is equally obvious. □

Let $\mathfrak{g}$ be a reductive Lie algebra over a field **k** of characteristic zero,

$\mathfrak{G}$ its universal enveloping algebra; in what follows it will be assumed that $\mathbf{k}$ is *algebraically closed.*

We shall now present some structural results which will play an important role in 2.4.3.

Let $\mathfrak{j}$ be a Cartan subalgebra of $\mathfrak{g}$, Φ the set of roots of the pair $(\mathfrak{g}, \mathfrak{j})$; let $\mathfrak{J}$ denote the universal enveloping algebra of $\mathfrak{j}$ – then $\mathfrak{J}$ may be identified with the symmetric algebra of $\mathfrak{j}$, $\mathfrak{j}$ being abelian.

Proposition 2.3.3.2 The smallest subspace of $\mathfrak{G}$ containing $\mathfrak{J}$ and stable for the adjoint representation of $\mathfrak{g}$ on $\mathfrak{G}$ is $\mathfrak{G}$ itself – thus $\mathfrak{G}$ is the sum of $\mathfrak{J}$ and $\mathfrak{G}_R = [\mathfrak{G}, \mathfrak{G}]$ and so every element $D \in \mathfrak{G}$ can be written in the form

$$D = H + \sum_i [X_i, D_i] \; (H \in \mathfrak{J}, X_i \in \mathfrak{g}, D_i \in \mathfrak{G}).$$

Proof Write $\mathfrak{g} = \mathfrak{c} + \bar{\mathfrak{g}}$ ($\mathfrak{c}$ the center of $\mathfrak{g}$, $\bar{\mathfrak{g}}$ the derived algebra of $\mathfrak{g}$); let $\mathfrak{C}$ and $\bar{\mathfrak{G}}$ denote the universal enveloping algebras of $\mathfrak{c}$ and $\bar{\mathfrak{g}}$, respectively – then $\mathfrak{G} = \mathfrak{C}\bar{\mathfrak{G}}$, which shows that it will be sufficient just to consider the case when $\mathfrak{g}$ is semi-simple. This being so, let $\mathscr{S} \subset \mathfrak{G}$ be the smallest subspace of $\mathfrak{G}$ stable for the adjoint representation of $\mathfrak{g}$ on $\mathfrak{G}$ and containing $\mathfrak{J}$. If $\lambda: S \rightarrow \mathfrak{G}$ denotes, as before, the canonical vector space bijection between S and $\mathfrak{G}$ (S the symmetric algebra of $\mathfrak{g}$), then $\lambda^{-1}(\mathscr{S}) \supset \mathfrak{J}$ and is stable for the usual representation of $\mathfrak{g}$ on S by derivations. It is enough to show, therefore, that $\lambda^{-1}(\mathscr{S}) = S$. Select elements $X_\alpha \in \mathfrak{g}^\alpha$ per usual; by induction on r, we find that

$$\begin{aligned} \operatorname{ad}(X_{\alpha_1})(X_{\alpha_2} \ldots X_{\alpha_r} H^{m-r+1}) &= \sum_{i=2}^{r} N_{\alpha_1, \alpha_i} X_{\alpha_2} \ldots X_{\alpha_1 + \alpha_i} \ldots X_{\alpha_r} H^{m-r+1} \\ &\quad + (m - r + 1)\alpha_1(H) X_{\alpha_1} \ldots X_{\alpha_r} H^{m-r} \qquad (H \in \mathfrak{j}), \end{aligned}$$

and so $\lambda^{-1}(\mathscr{S})$ contains all elements of the form $X_{\alpha_1} \ldots X_{\alpha_r} H^{m-r}$ ($X_{\alpha_i} \in \mathfrak{g}^{\alpha_i}$; $H \in \mathfrak{j}$ with $\alpha(H) \neq 0$ (all $\alpha \in \Phi$)). But $\mathfrak{J}$ is spanned by elements of the form H^m, H a regular element in $\mathfrak{j}$. Hence $S = \lambda^{-1}(\mathscr{S})$. □

Let $\mathfrak{b}$ be a Borel subalgebra of $\mathfrak{g}$ containing $\mathfrak{j}$, Φ^+ the set of positive roots in Φ singled out by this choice of $\mathfrak{b}$; write $\mathfrak{n}^- = \sum_{\alpha<0} \mathfrak{g}^\alpha$, $\mathfrak{n}^+ = \sum_{\alpha>0} \mathfrak{g}^\alpha$ (so that $\mathfrak{b} = \mathfrak{j} + \mathfrak{n}^+$); let $\mathfrak{N}^-$ (respectively $\mathfrak{N}^+$) denote the universal enveloping algebra of $\mathfrak{n}^-$ (respectively $\mathfrak{n}^+$) – then $\mathfrak{G}$ is isomorphic, as a vector space, to $\mathfrak{N}^- \otimes_{\mathbf{k}} \mathfrak{J} \otimes_{\mathbf{k}} \mathfrak{N}^+$. Let $\epsilon^-: \mathfrak{N}^- \rightarrow \mathbf{k}$ denote the projection of $\mathfrak{N}^-$ onto $\mathbf{k}$ determined by the splitting of $\mathfrak{N}^-$ into the direct sum of $\mathbf{k}$ and the ideal in $\mathfrak{N}^-$ generated by $\mathfrak{n}^-$ (cf. Proposition 2.3.1.1(ii)); define $\epsilon^+: \mathfrak{N}^+ \rightarrow \mathbf{k}$ in a similar way.

Let ϵ denote the map $\epsilon^- \otimes 1 \otimes \epsilon^+$ of $\mathfrak{G}$ onto $\mathfrak{J}$, γ^* the map $D \mapsto \epsilon(D^0)$ ($D \in \mathfrak{G}$).

Proposition 2.3.3.3 The restriction of γ^* to the center $\mathfrak{Z}$ of $\mathfrak{G}$ is an (algebra) isomorphism from $\mathfrak{Z}$ into $\mathfrak{J}$.

The proof of this result requires a couple of preliminary lemmas. Let $\mathfrak{Q}^+$ denote the left ideal $\mathfrak{G}\mathfrak{n}^+$ (in $\mathfrak{G}$).

Lemma 2.3.3.4 Every element $Z \in \mathfrak{Z}$ can be written uniquely in the form $Z = H + Q$ ($H \in \mathfrak{J}$, $Q \in \mathfrak{Q}^+$).

Proof Every element in $\mathfrak{G}$ can be written as a linear combination of elements of the form $X_{-\alpha_1} \ldots X_{-\alpha_r} H X_{\beta_1} \ldots X_{\beta_s}$ where the α_i and β_j are positive roots, $H \in \mathfrak{J}$. Such an element is a weight vector for the adjoint representation of $\mathfrak{g}$ on $\mathfrak{G}$, the corresponding weight being $\sum_{i=1}^r \alpha_i - \sum_{j=1}^s \beta_j$ (see 2.4.1). Hence an element of the center of $\mathfrak{G}$ can be written uniquely as a linear combination of such elements with $\sum \alpha_i = \sum \beta_j$. The assertion of the lemma is now clear. □

Lemma 2.3.3.5 Retain the preceding notations – then $\mathfrak{Q}^+ \cap \mathfrak{Z} = \{0\}$.

Proof It will be sufficient just to consider the case when $\mathfrak{g}$ is semi-simple (cf. the first lines of the proof of Proposition 2.3.3.2 and bear in mind that $\mathfrak{Z} = \mathfrak{C}\bar{\mathfrak{Z}}$, $\bar{\mathfrak{Z}}$ the center of $\bar{\mathfrak{G}}$). This being so, let us suppose that our assertion is false; fix a non-zero Z in $\mathfrak{Q}^+ \cap \mathfrak{Z}$. Owing to Theorem 2.3.2.3 there exists a finite dimensional irreducible representation U of $\mathfrak{G}$ on a vector space E with $U(Z) \neq 0$. Let Λ be the highest weight for U, a_Λ a $\mathfrak{b}$-primitive element in E (see 2.4.1). By Schur's Lemma there exists a non-zero constant $k \in \mathbf{k}$ such that $U(Z) = kU(1)$. On the other hand, $Z \in \mathfrak{Q}^+$ and $U(\mathfrak{Q}^+)a_\Lambda = \{0\}$ – therefore $k = 0$, a contradiction. □

Proof of Proposition 2.3.3.3 Using Lemma 2.3.3.4, given $Z \in \mathfrak{Z}$, write $Z = H + Q$ ($H \in \mathfrak{J}$, $Q \in \mathfrak{Q}^+$) – then $\gamma^*(Z) = \epsilon(Z) = \epsilon(H) + \epsilon(Q)$. But ϵ vanishes on $\mathfrak{Q}^+$ while it reduces to 1 on $\mathfrak{J}$, whence $Z - \gamma^*(Z) \in \mathfrak{Q}^+$. For $Z_1, Z_2 \in \mathfrak{Z}$, we can write

$$Z_1Z_2 - \gamma^*(Z_1)\gamma^*(Z_2) = (Z_1 - \gamma^*(Z_1))Z_2 + \gamma^*(Z_1)(Z_2 - \gamma^*(Z_2)) \in \mathfrak{Q}^+.$$

On the other hand, $Z_1Z_2 - \gamma^*(Z_1Z_2) \in \mathfrak{Q}^+$; in view of the uniqueness assertion in Lemma 2.3.3.4, it follows that $\gamma^*(Z_1Z_2) = \gamma^*(Z_1)\gamma^*(Z_2)$ and so γ^* is a homomorphism. To show that γ^* is injective it suffices to prove that $\mathfrak{Q}^+ \cap \mathfrak{Z} = \{0\}$ which was done in Lemma 2.3.3.5. □

Let W denote the Weyl group of the pair $(\mathfrak{g}, \mathfrak{j})$, $I(\mathfrak{j})$ the set of Weyl group invariants in $S(\mathfrak{j})$ ($\sim \mathfrak{J}$); as it stands, $\gamma^*(\mathfrak{Z}) \neq I(\mathfrak{j})$ – this situation will now be remedied.

Let B denote the symmetric $\mathfrak{g}$-invariant non-degenerate bilinear form obtained by taking the product of a non-degenerate bilinear form on the center of $\mathfrak{g}$ with the Killing form on the derived algebra of $\mathfrak{g}$. The restriction of B to $\mathfrak{j}$ is non-degenerate and may be used to identify $\mathfrak{j}$ with its dual $\mathfrak{j}^\vee$; we shall let H_α denote the element in $\mathfrak{j}$ which corresponds to α under this isomorphism ($\alpha \in \Phi$) (hence $\langle H, \alpha\rangle = B(H, H_\alpha)$ (all $H \in \mathfrak{j}$)) and write $(.\,,.)$ for the transport of $B|\mathfrak{j}$ to $\mathfrak{j}^\vee$.

(1) *The map* γ_1. We shall let γ_1 denote the automorphism of $S(\mathfrak{j})$ which maps $H \in \mathfrak{j}$ to $H + \rho(H)(\rho = \frac{1}{2}\sum_{\alpha>0}\alpha)$.

(2) *The map* γ_2. Define the endomorphism γ_2 of $S(\mathfrak{j})$ by the rule

$$\gamma_2(H) = \sum_{w\in W} \det(w)H^w/\prod_{\alpha>0} H_\alpha \qquad (H \in S(\mathfrak{j})).$$

It is easy to see that γ_2 maps $S(\mathfrak{j})$ onto $I(\mathfrak{j})$ (cf. Proposition 2.1.5.11(i)).

(3) *The map* γ_3. We shall let γ_3 denote the transpose of the operation of multiplication by $\sum_{w\in W}\det(w)e^{w(\rho)}/\prod_{\alpha>0}(\rho, \alpha)$ in $S^*(\mathfrak{j}^\vee)$ (for notation see 2.1.4). Since $S^*(\mathfrak{j}^\vee)$ may be viewed as a ring of formal power series and thus possesses no divisors of zero, it follows that γ_3 is surjective.

We shall defer until 2.4.3 the proof of the following result.

Proposition 2.3.3.6 In the above notations, we have $\gamma^*\circ\gamma_3 = \gamma_1\circ\gamma_2$.

Evidently, then, $\gamma^*(\mathfrak{Z}) = \gamma^*(\mathfrak{I}) = \gamma^*(\gamma_3(S(\mathfrak{j}))) = \gamma_1(I(\mathfrak{j}))$. Therefore the map $\gamma = \gamma_1^{-1}\circ\gamma^*$ is an algebra isomorphism of $\mathfrak{Z}$ onto $I(\mathfrak{j})$ – the *canonical isomorphism.* [In order to justify the term 'canonical' one should show that the definition of γ is actually independent of the choice for the ordering of the roots of the pair $(\mathfrak{g}, \mathfrak{j})$; the not entirely trivial details of this verification may be left to the reader to work out.] Let us note in passing that if $\mathfrak{C}$ is the subalgebra of $\mathfrak{G}$ generated by 1 and $\mathfrak{c}$, $\mathfrak{c}$ the center of $\mathfrak{g}$, then $\gamma|\mathfrak{C}$ is the identity.

Examples (1) Let $\mathfrak{g}$ be a semi-simple Lie algebra over **R**, B the Killing form for $\mathfrak{g}$. Let $X_1, \ldots, X_n$ be a basis for $\mathfrak{g}_c$ over **C**; put $g_{ij} = B(X_i, X_j)$ and let g^{ij} be the elements of the inverse matrix – the element

$$\omega = \sum_{i,j} g^{ij}X_iX_j$$

is called the *Casimir operator* of $\mathfrak{g}_c$. It is well-known and easy to prove that ω is independent of the choice of the basis and lies in $\mathfrak{Z}$. Suppose that $\mathfrak{g} = \mathfrak{k} + \mathfrak{p}$ is a Cartan decomposition for $\mathfrak{g}$ and $\mathfrak{j}$ is a θ-stable Cartan subalgebra of $\mathfrak{g}$. Let $H_1, \ldots, H_k, \ldots, H_l$ be a basis for $\mathfrak{j}$ such that $H_1, \ldots, H_k$ is a basis for $\mathfrak{j}_\mathfrak{k}$ and $H_{k+1}, \ldots, H_l$ is a basis for $\mathfrak{j}_\mathfrak{p}$ over **R** (these H_i have nothing to do with the co-roots H_i corresponding to the elements of a fundamental system of roots $\{\alpha_1, \ldots, \alpha_l\}$ for the pair $(\mathfrak{g}_c, \mathfrak{j}_c)$). Suppose further that $B(H_i, H_j) = -\delta_{ij}$ for $1 \leqslant i, j \leqslant k$ and $B(H_i, H_j) = \delta_{ij}$ for $k+1 \leqslant i, j \leqslant l$. If $\mathfrak{g}_c = \mathfrak{j}_c + \sum_{\alpha>0}(\mathfrak{g}_c^\alpha + \mathfrak{g}_c^{-\alpha})$, take $X_\alpha, X_{-\alpha}$ so that $B(X_\alpha, X_{-\alpha}) = 1$ – then

$$\begin{aligned}\omega &= -\Big(\sum_1^k H_i^2\Big) + \sum_{k+1}^l H_i^2 + \sum_{\alpha>0}(X_\alpha X_{-\alpha} + X_{-\alpha}X_\alpha)\\ &\equiv -\Big(\sum_1^k H_i^2\Big) + \sum_{k+1}^l H_i^2 + \sum_{\alpha>0} H_\alpha \qquad \Big(\text{mod} \sum_{\alpha>0}\mathfrak{G}X_\alpha\Big)\end{aligned}$$

since $X_\alpha X_{-\alpha} = X_{-\alpha}X_\alpha + H_\alpha$. Let $H_\rho = \frac{1}{2}\sum_{\alpha>0}H_\alpha$, $\rho = \frac{1}{2}\sum_{\alpha>0}\alpha$. If we apply the automorphism γ_1^{-1} to the above congruence we find that

$$\gamma(\omega) = \omega_\mathfrak{j} - (H_\rho, H_\rho) = \omega_\mathfrak{j} - \rho(H_\rho) \in I(\mathfrak{j}_c)$$

where, of course, $B(H_\rho, H) = \rho(H)$ $(H \in \mathfrak{j}_c)$ – here

$$\omega_\mathfrak{j} = -\Big(\sum_1^k H_i^2\Big) + \sum_{k+1}^l H_i^2.$$

More generally if $\mathfrak{g}$ is only assumed to be reductive, say $\mathfrak{g} = \mathfrak{c} + \bar{\mathfrak{g}}$, then $\mathfrak{Z} = \mathfrak{C}\bar{\mathfrak{Z}}$ (obvious notation). Let $C_1, \ldots, C_r$ be a basis for $\mathfrak{c}$ over $\mathbf{R}$. If $\omega = \bar{\omega} - \sum_i C_i^2$ ($\bar{\omega}$ = Casimir element of $\bar{\mathfrak{g}}$), then

$$\gamma(\omega) = \bar{\omega}_{\bar{\mathfrak{j}}} - \rho(H_\rho) - \sum_i C_i^2$$

where we have set $\bar{\mathfrak{j}}_c = \mathfrak{j}_c \cap \bar{\mathfrak{g}}_c$.

(2) Although the basis used in (1) for the definition of ω was useful in the computation of $\gamma(\omega)$, different choices are sometimes convenient in other applications. Thus choose a basis $\{X_1, \ldots, X_n\}$ for $\mathfrak{g}$ over $\mathbf{R}$ with the property that $\{X_1, \ldots, X_k\}$ is a basis for $\mathfrak{k}$, $X_{k+1}, \ldots, X_n$ is a basis for $\mathfrak{p}$ and assume moreover that $B(X_i, X_j) = -\delta_{ij}$ $(1 \leqslant i, j \leqslant k)$, $B(X_i, X_j) = \delta_{ij}(k+1 \leqslant i, j \leqslant n)$. Put $\omega_- = \sum_1^k X_i^2$, $\omega_+ = \sum_{k+1}^n X_i^2$ – then $\omega = -\omega_- + \omega_+$. Plainly ω_- is an element of the center of $\mathfrak{K}$, $\mathfrak{K}$ the universal enveloping algebra of $\mathfrak{k}_c$.

(3) Let us agree to adhere to the assumptions and notations immediately preceding Proposition 2.1.5.10. If $\mathfrak{Z}_\mathfrak{h}$ denotes the center of $\mathfrak{H}$, $\mathfrak{H}$ the universal enveloping algebra of $\mathfrak{h}_c$, then we are going to define an algebra isomorphism $\mu_{\mathfrak{g}/\mathfrak{h}}$ of $\mathfrak{Z}$ (the center of $\mathfrak{G}$) into $\mathfrak{Z}_\mathfrak{h}$. Thus fix a Cartan subalgebra $\mathfrak{j}_c$ of $\mathfrak{h}_c$ – then $\mathfrak{j}_c$ is a Cartan subalgebra of $\mathfrak{g}_c$ and one has the canonical isomorphisms

$$\gamma: \mathfrak{Z} \to I(\mathfrak{j}_c), \gamma_\mathfrak{h}: \mathfrak{Z}_\mathfrak{h} \to I(\mathfrak{h}_c; \mathfrak{j}_c).$$

Put $\mu_{\mathfrak{g}/\mathfrak{h}} = \gamma_\mathfrak{h}^{-1} \circ \gamma$ – then $\mu_{\mathfrak{g}/\mathfrak{h}}$ is an algebra isomorphism of $\mathfrak{Z}$ into $\mathfrak{Z}_\mathfrak{h}$ and, since any two Cartan subalgebras in $\mathfrak{h}_c$ are conjugate, it follows that $\mu_{\mathfrak{g}/\mathfrak{h}}$ is independent of the choice of $\mathfrak{j}_c$. In addition $\mathfrak{Z}_\mathfrak{h}$ is a free abelian module over $\mu_{\mathfrak{g}/\mathfrak{h}}(\mathfrak{Z})$ of rank $[W(\mathfrak{g}_c, \mathfrak{j}_c): W(\mathfrak{h}_c, \mathfrak{j}_c)]$ – this is clear from the proof of Proposition 2.1.5.10.

Adopt the notations of the Remark following Proposition 2.1.5.10 (which will be used without comment in what follows). As an application of the results outlined in that Remark, Kostant [4] obtained an important theorem on the structure of $\mathfrak{G}$.

Let $\mathcal{N}(\mathfrak{G})$ denote the subspace of $\mathfrak{G}$ spanned by all elements of the form X^k, $k = 0, 1, \ldots$, where $X \in \mathfrak{g}_c$ is nilpotent. Plainly $\mathcal{N}(\mathfrak{G})$ is a G-submodule of $\mathfrak{G}$ and the standard filtration on $\mathfrak{G}$ induces a filtration on $\mathcal{N}(\mathfrak{G})$ ($\mathcal{N}_m(\mathfrak{G}) = \mathcal{N}(\mathfrak{G}) \cap \mathfrak{G}_m$).

Lemma 2.3.3.7 Let $\lambda: S(\mathfrak{g}_c) \to \mathfrak{G}$ be the canonical linear bijection – then $\lambda(H(\mathfrak{g}_c)) = \mathcal{N}(\mathfrak{G})$, $H(\mathfrak{g}_c)$ the space of G-harmonic polynomials on $\mathfrak{g}_c$.

Proof This is evident since $\lambda(X^k) = \lambda(X)^k$ for any integer $k \geqslant 0$ $(X \in \mathfrak{g}_c)$. □

Let us agree to regard $\mathfrak{G}$ as a $\mathfrak{Z}$-module with respect to multiplication.

Theorem 2.3.3.8 (Kostant) Let $\mathfrak{g}_c$ be a reductive Lie algebra over $\mathbf{C}$ – then $\mathfrak{G}$ is free as a module over $\mathfrak{Z}$ and the mapping ϕ

$$\mathfrak{Z} \otimes_\mathbf{C} \mathcal{N}(\mathfrak{G}) \to \mathfrak{G}$$

given by $Z \otimes N \mapsto ZN$ is a G-module isomorphism $(Z \in \mathfrak{Z}, N \in \mathcal{N}(\mathfrak{G}))$.

Proof Recall that $\lambda(I(\mathfrak{g}_c)) = \mathfrak{Z}$.

(1) ϕ is surjective. To see this we shall proceed inductively on the degee m of the filtration $\{\mathfrak{G}_m\}$. Plainly $\mathfrak{G}_0 \subset \mathcal{N}(\mathfrak{G})$; hence assume that $\mathfrak{G}_m$ is a subset of the range of ϕ and let $D \in \mathfrak{G}_{m+1}$ – then $D = \lambda(u)$ where $u = \Sigma p_i h_i$ ($p_i \in$

$I(\mathfrak{g}_c)$, $h_i \in H(\mathfrak{g}_c)$) and $\deg(p_i) + \deg(h_i) \leqslant m + 1$. Consequently

$$D \equiv \phi(\sum \lambda(p_i) \otimes \lambda(h_i)) \bmod \mathfrak{G}_m$$

which shows that ϕ is surjective.

(2) ϕ is injective. Thus let D be a non-zero element in $\mathfrak{Z} \otimes_{\mathbf{C}} \mathcal{N}(\mathfrak{G})$ – then D is the image of an element $p \in I(\mathfrak{g}_c) \otimes_{\mathbf{C}} H(\mathfrak{g}_c)$ under the isomorphism $I(\mathfrak{g}_c) \otimes_{\mathbf{C}} H(\mathfrak{g}_c) \rightarrow \mathfrak{Z} \otimes_{\mathbf{C}} \mathcal{N}(\mathfrak{G})$ induced by λ and we may assume that $p = \Sigma p_i \otimes h_i$ where the p_i are non-zero elements of $I(\mathfrak{g}_c)$ and the h_i are homogeneous and linearly independent in $H(\mathfrak{g}_c)$. Let $m = \max(\deg(p_i) + \deg(h_i))$ – then $\Sigma p_i h_i \neq 0$ and $\deg(\Sigma p_i h_i) = m$, whence $\lambda(\Sigma p_i h_i) \not\equiv 0 \bmod \mathfrak{G}_{m-1}$. Now $D = \Sigma \lambda(p_i) \otimes \lambda(h_i)$; thus $\phi(D) = \Sigma \lambda(p_i)\lambda(h_i)$ and so

$$\lambda(\Sigma p_i h_i) \equiv \phi(D) \bmod \mathfrak{G}_{m-1} \Rightarrow \phi(D) \neq 0.$$

Since ϕ is evidently a G-module map, the theorem follows. □

2.4 Representations of a Reductive Lie Algebra

2.4.1 Simple Modules – The Theorem of Highest Weight

Let $\mathfrak{g}$ be a reductive Lie algebra over an algebraically closed field $\mathbf{k}$ of characteristic zero. The objective of this number (and the next) is to present, for the most part without proof, the well-known, standard facts as regards the finite dimensional simple $\mathfrak{g}$-modules (complete details may be found in Jacobson [3] or Serre [4]).

Let $\mathfrak{j}$ be a Cartan subalgebra of $\mathfrak{g}$, Φ the set of roots of the pair $(\mathfrak{g}, \mathfrak{j})$, W the Weyl group of Φ; fix a Borel subalgebra $\mathfrak{b}$ of $\mathfrak{g}$ containing $\mathfrak{j}$; let Φ^+ denote the system of positive roots singled out by this choice of $\mathfrak{b}$, $\Psi = \{\alpha_1, \ldots, \alpha_l\}$ the corresponding fundamental system.

[Write $\mathfrak{g} = \mathfrak{c} + [\mathfrak{g}, \mathfrak{g}] = \mathfrak{c} + \bar{\mathfrak{g}}$; put $\dot{\mathfrak{j}} = \mathfrak{j} \cap \bar{\mathfrak{g}}$ so that $\mathfrak{j} = \mathfrak{c} + \dot{\mathfrak{j}}$ (direct sum) – then, with an evident extension of the terminology of 1.1.2, Φ is a root system in $(\mathfrak{j}^\vee, \mathfrak{c}^\vee)$.]

Let E be a $\mathfrak{g}$-module (not necessarily finite dimensional), U the corresponding representation of $\mathfrak{g}$. A linear function λ on $\mathfrak{j}$ is said to be a *weight* for U (or E) if the set

$$E^\lambda = \{a \in E: U(H)a = \lambda(H)a \ (H \in \mathfrak{j})\}$$

does not consist of zero alone. The dimension of E^λ is called the *multiplicity* of λ.

Proposition 2.4.1.1 Let E be a $\mathfrak{g}$-module, U the corresponding representation of $\mathfrak{g}$.

(i) The sum $\sum_\lambda E^\lambda$, λ a weight, is direct; moreover one has $U(\mathfrak{g}^\alpha)E^\lambda \subset E^{\alpha+\lambda}$ (all $\alpha \in \Phi$) and so $\sum_\lambda E^\lambda$ is $\mathfrak{g}$-stable.

(ii) If $\tilde{E}$ is any $\mathfrak{g}$-stable subspace of E, then $\tilde{E} \cap (\sum_\lambda E^\lambda) = \sum_\lambda \tilde{E} \cap E^\lambda$

Theorem 2.4.1.2 (Weyl) Let E be a finite dimensional semi-simple $\mathfrak{g}$-module, U the corresponding representation – then the $U(H)$ $(H \in \mathfrak{j})$

form a semi-simple commutative family of endomorphisms of E and hence there exists a basis for E relative to which all the $U(H)$ ($H \in \mathfrak{j}$) are given by diagonal matrices. Let λ be a weight for U in E, E^λ the corresponding weight space. Consider the set of weights of the form $\lambda + n\alpha$ (n an integer, α a root of $\mathfrak{g}$ with respect to $\mathfrak{j}$) – then this set is an uninterrupted progression and if $p \leqslant n \leqslant q$ is this string, then the value of λ at the co-root in $\mathfrak{j}$ associated with α is $-(p+q)/2$. Furthermore $w_\alpha(\lambda)$ is a weight and $\dim(E^\lambda) = \dim(E^{w_\alpha(\lambda)})$ (all $\alpha \in \Phi$).

Let E be a $\mathfrak{g}$-module (not necessarily finite dimensional), U the corresponding representation of $\mathfrak{g}$; let Λ be a linear function on $\mathfrak{b}$ which vanishes on the derived algebra $[\mathfrak{b}, \mathfrak{b}]$ of $\mathfrak{b}$ – then a non-zero vector a_Λ in E is said to be $\mathfrak{b}$-*primitive of weight* Λ if $U(X)a_\Lambda = \Lambda(X)a_\Lambda$ (all $X \in \mathfrak{b}$).

[In passing let us note that since $\mathfrak{b}/[\mathfrak{b}, \mathfrak{b}]$ is isomorphic to $\mathfrak{j}$, the linear functions on $\mathfrak{b}$ which vanish on $[\mathfrak{b}, \mathfrak{b}]$ can be identified with the linear functions on $\mathfrak{j}$.]

Theorem 2.4.1.3 (Harish-Chandra) Let E be a simple $\mathfrak{g}$-module with a $\mathfrak{b}$-primitive element a_Λ of weight Λ – then:

(i) The vector a_Λ is unique to within a scalar factor; its weight Λ is called the *highest weight* of E;

(ii) The weights λ for E are all of the form $\lambda = \Lambda - \Sigma\, n_\alpha \alpha$, n_α a non-negative integer ($\alpha \in \Phi^+$); every weight appears with finite multiplicity (Λ itself being of multiplicity one); furthermore $E = \sum_\lambda E^\lambda$, λ a weight.

Theorem 2.4.1.4 (Harish-Chandra) Let Λ be a linear function on $\mathfrak{j}$ – then there exists a simple $\mathfrak{g}$-module E with highest weight Λ. Two simple $\mathfrak{g}$-modules E_1 and E_2 with highest weights Λ_1 and Λ_2, respectively, are isomorphic iff $\Lambda_1 = \Lambda_2$.

Remark Theorem 2.4.1.4 sets up a bijection between the dual $\mathfrak{j}^\vee$ of $\mathfrak{j}$ and the isomorphism classes of those simple $\mathfrak{g}$-modules which admit a highest weight. However there exist simple $\mathfrak{g}$-modules which are not $\mathfrak{b}$-primitive for any $\Lambda \in \mathfrak{j}^\vee$ (cf. Bouwer [1], Romm [2]) – such modules are necessarily infinite dimensional according to Theorem 2.4.1.5 below.

Let $\mathfrak{W}$ ($= \mathfrak{W}_\Phi$) denote the weight group of Φ, $\mathfrak{W}^+$ ($= \mathfrak{W}_\Phi^+$) the corresponding set of dominant weights. [Thus, by definition (compare 1.1.2), the elements of $\mathfrak{W}$ (respectively $\mathfrak{W}^+$) are linear functions on $\mathfrak{j}$ which take on integral values (respectively non-negative integral values) at the co-roots $H_i \in \mathfrak{j}$ corresponding to the α_i ($i = 1, \ldots, l$).]

The following fundamental result establishes a one-to-one correspondence between the isomorphism classes of finite dimensional simple $\mathfrak{g}$-modules and the dominant integral linear functions (= dominant weights).

Theorem 2.4.1.5 (É. Cartan-Harish-Chandra) (i) Every finite dimensional simple $\mathfrak{g}$-module E is $\mathfrak{b}$-primitive with highest weight dominant integral.

(ii) Every simple $\mathfrak{g}$-module E with highest weight dominant integral is finite dimensional.

A famous theorem due to Weyl enables one to compute the dimension of a finite dimensional simple $\mathfrak{g}$-module from the Dynkin diagram for the pair $(\mathfrak{g}, \mathfrak{j})$ (cf. Jacobson [3, p.257]).

Theorem 2.4.1.6 (Weyl) Let E_Λ be a finite dimensional simple $\mathfrak{g}$-module with highest weight Λ – then

$$\dim(E_\Lambda) = \frac{\prod_{\alpha>0}(\Lambda + \rho, \alpha)}{\prod_{\alpha>0}(\rho, \alpha)} \qquad \left(\rho = \frac{1}{2}\sum_{\alpha>0}\alpha\right).$$

[Here $(.\,,.)$ denotes a W-invariant non-degenerate symmetric bilinear form on the dual $\mathfrak{j}^\vee$ with the evident necessary properties; of course one could just as well write out the formula in terms of co-roots. . . .]

Let us suppose that $\mathfrak{g}$ is semi-simple – then, even though Theorem 2.4.1.5 enables one to 'parameterize' the finite dimensional simple $\mathfrak{g}$-modules by l-tuples of non-negative integers it still does not provide a satisfactory description of these modules in 'concrete' terms, i.e. one would like to have a 'model' for each simple $\mathfrak{g}$-module. We shall make a few general remarks along these lines. Let E_i denote the finite dimensional simple $\mathfrak{g}$-module corresponding to the dominant integral linear function Λ_i given by $\Lambda_i(H_j) = \delta_{ij}$, $i, j = 1, \ldots, l$; let a_i be a primitive element in E_i of weight Λ_i. If Λ is any dominant integral linear function on $\mathfrak{j}$, then $\Lambda = \sum_i m_i\Lambda_i$ where the m_i are non-negative integers. Let E be the module which is the tensor product of m_1 copies of E_1, m_2 copies of $E_2, \ldots, m_l$ copies of E_l. Put

$$a = \overbrace{a_1 \otimes \cdots \otimes a_1}^{m_1} \otimes \overbrace{a_2 \otimes \cdots \otimes a_2}^{m_2} \otimes \cdots \otimes \overbrace{a_l \otimes \cdots \otimes a_l}^{m_l}.$$

[Make the convention that if all the $m_i = 0$ then $E = \mathbf{k}$ and $\mathfrak{g}$ acts trivially on $\mathbf{k}$.] It is not difficult to see that the submodule of E generated by a is the simple module corresponding to Λ. Therefore the problem of explicitly determining the simple $\mathfrak{g}$-modules is reduced to that of identifying the modules E_i, the so-called *basic irreducible* modules for $\mathfrak{g}$. It is clearly enough to do this for the various simple Lie algebras and the solution here is, for the most part, classical – for details see Jacobson [3, Chapter 7] and Weyl [1].

Remark Suppose that $\mathbf{k} = \mathbf{R}$ (say) – then it is quite reasonable to consider real representations of a real semi-simple Lie algebra $\mathfrak{g}$. For the sorting out of the finite dimensional real simple $\mathfrak{g}$-modules see Iwahori [1]; the paper of Fell [5] is pertinent in this connection too.

Appendix 1 Let $\mathbf{k}$ be an algebraically closed field of characteristic zero; let us recall how the representation theory of $\mathfrak{sl}(2, \mathbf{k})$ goes (for details see Jacobson [3, p.83]). A basis for $\mathfrak{sl}(2, \mathbf{k})$ is given by $X = \left(\begin{smallmatrix} 0 & 1 \\ 0 & 0 \end{smallmatrix}\right)$, $Y = \left(\begin{smallmatrix} 0 & 0 \\ 1 & 0 \end{smallmatrix}\right)$, $H = \left(\begin{smallmatrix} 1 & 0 \\ 0 & -1 \end{smallmatrix}\right)$ and the multiplication table is $[X, Y] = H$, $[H, X] = 2X$, $[H, Y] = -2Y$. The basic fact concerning the representation theory of $\mathfrak{sl}(2, \mathbf{k})$ is the existence of a one-to-one correspondence between the isomorphism classes of finite dimensional simple $\mathfrak{sl}(2, \mathbf{k})$-modules and the set of non-negative integers. Precisely: For each integer $m = 0, 1, 2, \ldots$ there exists one and only one, up to isomorphism, simple $\mathfrak{sl}(2, \mathbf{k})$-module E_m of dimension $m + 1$; E_m has a basis $a_0, \ldots, a_m$ such that if U_m is the corresponding representation then

$$\begin{aligned} U_m(H)a_i &= (m - 2i)a_i, \\ U_m(Y)a_i &= (i + 1)a_{i+1},\ U_m(Y)a_m = 0, \\ U_m(X)a_i &= (m - i + 1)a_{i-1},\ U_m(X)a_0 = 0, i = 0, \ldots, m. \end{aligned}$$

Appendix 2 We intend to record here two elementary facts which are frequently useful – the notations and conventions will be those which were laid down at the beginning of the present number.

Lemma 1 Set $\rho = \frac{1}{2}\sum_{\alpha>0}\alpha$; if H_i is the co-root in $\mathfrak{j}$ associated with α_i $(i = 1, \ldots, l)$, then $\rho(H_i) = 1$ $(i = 1, \ldots, l)$.
[This statement is a ready consequence of Proposition 1.1.2.5.]

Lemma 2 Suppose that λ is an element in the space spanned over $\mathbf{Q}$ by the roots α in Φ – then λ dominates its transforms under the Weyl group W iff $\lambda(H_i) > 0$ $(i = 1, \ldots, l$; as in Lemma 1, H_i is the co-root in $\mathfrak{j}$ associated with α_i).
[Use the evident induction argument.]

2.4.2 The Formula of H. Weyl and B. Kostant

Let us agree to retain the notations and conventions introduced in 2.4.1; in particular, then, $\mathfrak{g}$ will denote a reductive Lie algebra over an algebraically closed field $\mathbf{k}$ of characteristic zero, $\mathfrak{j}$ a Cartan subalgebra of $\mathfrak{g}$, $\mathfrak{b}$ a Borel subalgebra of $\mathfrak{g}$ containing $\mathfrak{j}$, Φ^+ the set of positive roots of the pair $(\mathfrak{g}, \mathfrak{j})$ singled out by this choice of $\mathfrak{b}$, Ψ the fundamental system of roots for Φ contained in Φ^+.

We shall start off by reviewing Weyl's well-known 'character formula'.

Let $\mathfrak{R}(\mathfrak{g})$ (respectively $\mathfrak{R}(\mathfrak{b})$) denote the representation ring of $\mathfrak{g}$ (respectively $\mathfrak{b}$) (see 2.1.1); the representation theory of $\mathfrak{g}$ then tells us that the homomorphism of restriction $\operatorname{Res}(\mathfrak{g}, \mathfrak{b})$ from $\mathfrak{R}(\mathfrak{g})$ to $\mathfrak{R}(\mathfrak{b})$ is injective. On the other hand, as was pointed out in 2.1.1, one can identify $\mathfrak{R}(\mathfrak{b})$, in a natural way, with the group algebra $\mathbf{Z}[\mathfrak{j}^\vee]$; the weight group $\mathfrak{W}$ for the set of roots Φ of the pair $(\mathfrak{g}, \mathfrak{j})$ is stable under the action of the Weyl group W; the action of W on $\mathfrak{W}$ extends in the canonical fashion to the group algebra $\mathbf{Z}[\mathfrak{W}]$ of $\mathfrak{W}$ – then $\mathbf{Z}[\mathfrak{W}]$ is a subring of $\mathbf{Z}[\mathfrak{j}^\vee]$ and the image of $\operatorname{Res}(\mathfrak{g}, \mathfrak{b})$ $(\mathfrak{R}(\mathfrak{g}))$ is the subalgebra of $\mathbf{Z}[\mathfrak{W}]$ consisting of those elements in $\mathbf{Z}[\mathfrak{W}]$ which are W-invariant.

The formula of Weyl permits one to explicate $\operatorname{Res}(\mathfrak{g}, \mathfrak{b})$ in that it gives the images of the elements $\operatorname{ch}_{\mathfrak{g}}(E_\Lambda)$ $(\Lambda \in \mathfrak{W}^+)$ which comprise the canonical basis for $\mathfrak{R}(\mathfrak{g})$ (here E_Λ stands for an isomorphism class of simple $\mathfrak{g}$-modules with highest weight Λ dominant integral). Thus, employing the customary exponential notation for the basis elements of $\mathbf{Z}[\mathfrak{j}^\vee]$ which correspond to elements of $\mathfrak{j}^\vee$, Weyl's formula reads:

$$\operatorname{Res}(\mathfrak{g}, \mathfrak{b})(\operatorname{ch}_{\mathfrak{g}}(E_\Lambda)) = \frac{\sum_{w \in W} \det(w)\, e^{w(\Lambda+\rho)}}{\sum_{w \in W} \det(w) e^{w(\rho)}} \qquad \left(\rho = \frac{1}{2} \sum_{\alpha > 0} \alpha\right).$$

[Additional comments on Weyl's formula are given in 2.4.3]

Remark In passing, we remind the reader of the following fact. Consider the element $\Delta = \prod_{\alpha>0} (e^{\alpha/2} - e^{-\alpha/2})$, the product being calculated in the algebra $\mathbf{Z}[\mathfrak{W}/2]$ – then it is actually true that $\Delta \in \mathbf{Z}[\mathfrak{W}]$ since one can prove that $\Delta = \sum_{w \in W} \det(w) e^{w(\rho)}$.

We come now to the primary objective of the present number. Because the map $\operatorname{Res}(\mathfrak{g}, \mathfrak{b})$ is injective, the same must be true of all the homomorphisms of restriction $\operatorname{Res}(\mathfrak{g}, \mathfrak{p})$, where $\mathfrak{p}$ is any subalgebra of $\mathfrak{g}$ which contains $\mathfrak{b}$; that is, the map $\operatorname{Res}(\mathfrak{g}, \mathfrak{p})$ is injective for every $\mathfrak{b}$-parabolic subalgebra $\mathfrak{p}$ of $\mathfrak{g}$ – our goal is to explicate the image of $\operatorname{Res}(\mathfrak{g}, \mathfrak{p})(\operatorname{ch}_{\mathfrak{g}}(E_\Lambda))$ in $\mathfrak{R}(\mathfrak{p})$ ($\Lambda \in \mathfrak{W}^+$).

Let $\mathfrak{p}$ be a $\mathfrak{b}$-parabolic subalgebra of $\mathfrak{g}$ which we assume to be fixed once and for all; let Θ be the subset of Ψ (Ψ the fundamental system of roots for Φ determined by $\mathfrak{b}$) associated with $\mathfrak{p}$ in the usual way, $\mathfrak{l}$ the corresponding Levi subalgebra of $\mathfrak{p}$ (cf. the Appendix to 1.2.2 of Chapter 1) – we recall that $\mathfrak{l}$ is a subalgebra of $\mathfrak{g}$ which is reductive in $\mathfrak{g}$ and of the same rank as $\mathfrak{g}$ ($\mathfrak{j}$ being a Cartan subalgebra of $\mathfrak{l}$). Let us agree to use the notations and conventions revolving around Proposition 1.1.2.13 and supporting lemmas; in particular, then, $\Xi = \Phi^+ \cup \langle\Theta\rangle$ is the parabolic subset of Φ determined by $\mathfrak{p}$, Ξ_s the set of roots of the pair $(\mathfrak{l}, \mathfrak{j})$, Ξ_s^+ the positive roots (relative to the ordering inherited from Φ), W_s the Weyl group of Ξ_s etc. Because $\mathfrak{l}$ is reductive, the isomorphism classes of simple $\mathfrak{l}$-modules are in a one-to-one correspondence with the elements of $\mathfrak{W}_s^+$.

Lemma 2.4.2.1 If Λ is a dominant weight for Φ, then, for every $w_u \in W_u$, the weight $w_u(\Lambda + \rho) - \rho$ is a dominant weight for Ξ_s.

Proof Since the weight $\rho = \frac{1}{2} \sum_{\alpha>0} \alpha$ assumes the value 1 at each of the co-roots associated with the elements of Ψ (see Lemma 1 in Appendix 2 to 2.4.1), it is clear that the weight $\Lambda + \rho$ takes on *strictly positive* values at the co-roots associated with the elements of Ψ; because $w_u \in W_u$, our lemma is then seen to be an immediate consequence of Corollary 1.1.2.16. □

Remark Call an element γ in the weight group $\mathfrak{W}$ per the roots Φ *regular* if $\gamma^w \neq \gamma$ when $w \neq 1$ ($w \in W$). We shall now isolate an important special subset of $\mathfrak{W}_s^+$; thus let $\mathfrak{W}_s^+(\rho) = \{\Lambda_s \in \mathfrak{W}_s^+ : \Lambda_s + \rho \text{ is regular}\}$ – then the map $(\Lambda, w_u) \mapsto w_u(\Lambda + \rho) - \rho$ establishes a bijection between $\mathfrak{W}^+ \times W_u$ and $\mathfrak{W}_s^+(\rho)$. [The preceding lemma implies that $w_u(\Lambda + \rho) - \rho \in \mathfrak{W}_s^+(\rho)$ (all $\Lambda \in \mathfrak{W}^+$, $w_u \in W_u$). If $w_u(\Lambda + \rho) - \rho = \tilde{w}_u(\Lambda + \rho) - \rho$ (w_u, $\tilde{w}_u \in W_u$), then $\tilde{w}_u^{-1} w_u(\Lambda + \rho) = \Lambda + \rho$, whence $w_u = \tilde{w}_u$ (compare, for example, Theorem 1.1.2.7). Finally, if $\Lambda_s \in \mathfrak{W}_s^+(\rho)$, then there is a (unique) $w \in W$ such that $w^{-1}(\Lambda_s + \rho) \in \mathfrak{W}^+$; this w belongs to W_u (cf. Lemma 1.1.2.15) and the element $\Lambda = w^{-1}(\Lambda_s + \rho) - \rho \in \mathfrak{W}^+$ is sent to Λ_s by our mapping.]

Evidently $\mathfrak{R}(\mathfrak{l})$ may be identified with $\mathfrak{R}(\mathfrak{p})$ (since $\mathfrak{p}$ modulo its nilpotent radical may be identified with $\mathfrak{l}$).

Theorem 2.4.2.2 (Kostant) Let $\mathfrak{g}$ be a reductive Lie algebra over $\mathbf{k}$, $\mathfrak{j}$ a Cartan subalgebra of $\mathfrak{g}$, $\mathfrak{b}$ a Borel subalgebra of $\mathfrak{g}$ containing $\mathfrak{j}$, $\mathfrak{p}$ a $\mathfrak{b}$-parabolic subalgebra of $\mathfrak{g}$ with Levi subalgebra $\mathfrak{l}$; let $\Lambda \in \mathfrak{W}^+$ – then, in the ring $\mathfrak{R}(\mathfrak{l}) \sim \mathfrak{R}(\mathfrak{p})$, we have that

$$\operatorname{Res}(\mathfrak{g}, \mathfrak{p})(\operatorname{ch}_{\mathfrak{g}}(E_\Lambda)) = \frac{\sum_{w_u \in W_u} \det(w_u)\, \operatorname{ch}_{\mathfrak{l}}(E_{w_u(\Lambda+\rho)-\rho})}{\sum_{w_u \in W_u} \det(w_u)\, \operatorname{ch}_{\mathfrak{l}}(E_{w_u(\rho)-\rho})} \qquad \left(\rho = \frac{1}{2} \sum_{\alpha > 0} \alpha\right).$$

[On the basis of Lemma 2.4.2.1, the notation is clear; thus $E_{w_u(\Lambda+\rho)-\rho}$ denotes an isomorphism class of simple $\mathfrak{l}$-modules with highest weight $w_u(\Lambda + \rho) - \rho$ etc.]

Proof of Theorem 2.4.2.2 (Aribaud) Since $\operatorname{Res}(\mathfrak{p}, \mathfrak{b})$ (which may be identified with $\operatorname{Res}(\mathfrak{l}, \mathfrak{j})$) is injective, it will be sufficient, for the proof of our theorem, to verify that both sides of the equality in question are actually equal under an application of $\operatorname{Res}(\mathfrak{p}, \mathfrak{b})$. Because $\operatorname{Res}(\mathfrak{p}, \mathfrak{b})$ $\operatorname{Res}(\mathfrak{g}, \mathfrak{p}) = \operatorname{Res}(\mathfrak{g}, \mathfrak{b})$, Weyl's formula tells us that

$$\operatorname{Res}(\mathfrak{p}, \mathfrak{b}) \operatorname{Res}(\mathfrak{g}, \mathfrak{p})(\operatorname{ch}_{\mathfrak{g}}(E_\Lambda)) = \frac{\sum_{w \in W} \det(w)\, e^{w(\Lambda+\rho)}}{\sum_{w \in W} \det(w)\, e^{w(\rho)}}.$$

Applied to the reductive Lie algebra $\mathfrak{l}$, the same formula of Weyl implies that

$$\operatorname{Res}(\mathfrak{p}, \mathfrak{b})(\operatorname{ch}_{\mathfrak{l}}(E_{w_u(\Lambda+\rho)-\rho})) = \frac{\sum_{w_s \in W_s} \det(w_s)\, e^{w_s(w_u(\Lambda+\rho)-\rho+\rho_s)}}{\sum_{w_s \in W_s} \det(w_s)\, e^{w_s(\rho_s)}}$$

where $\rho_s = \frac{1}{2} \sum_{\alpha>0} \alpha$ ($\alpha \in \Xi_s^+$). The element $\rho_u = \rho - \rho_s$ is, evidently, invariant under the operations of W_s; thus $w_s(w_u(\Lambda + \rho) - \rho + \rho_s) = w_s w_u(\Lambda + \rho) - \rho_u$ and so

$$\begin{aligned}\frac{\sum_{w_s \in W_s} \det(w_s)\, e^{w_s(w_u(\Lambda+\rho)-\rho+\rho_s)}}{\sum_{w_s \in W_s} \det(w_s)\, e^{w_s(\rho_s)}} &= \frac{\sum_{w_s \in W_s} \det(w_s)\, e^{w_s w_u(\Lambda+\rho)}}{\sum_{w_s \in W_s} \det(w_s)\, e^{w_s(\rho_s)}} e^{-\rho_u} \\ &= \frac{\sum_{w_s \in W_s} \det(w_s)\, e^{w_s w_u(\Lambda+\rho)}}{\sum_{w_s \in W_s} \det(w_s)\, e^{w_s(\rho)}};\end{aligned}$$

consequently

$$\sum_{w_u \in W_u} \det(w_u)\operatorname{Res}(\mathfrak{p}, \mathfrak{b})(\operatorname{ch}_{\mathfrak{l}}(E_{w_u(\Lambda+\rho)-\rho})) = \sum_{w_u \in W_u} \frac{\sum_{w_s \in W_s} \det(w_s w_u)\, e^{w_s w_u(\Lambda+\rho)}}{\sum_{w_s \in W_s} \det(w_s)\, e^{w_s(\rho)}}$$

$$= \frac{\sum_{w \in W} \det(w)\, e^{w(\Lambda+\rho)}}{\sum_{w_s \in W_s} \det(w_s)\, e^{w_s(\rho)}}$$

(Proposition 1.1.2.13).

In a similar way we find that

$$\sum_{w_u \in W_u} \det(w_u)\operatorname{Res}(\mathfrak{p}, \mathfrak{b})(\operatorname{ch}_{\mathfrak{l}}(E_{w_u(\rho)-\rho})) = \frac{\sum_{w \in W} \det(w)\, e^{w(\rho)}}{\sum_{w_s \in W_s} \det(w_s)\, e^{w_s(\rho)}}.$$

Therefore

$$\frac{\sum_{w_u \in W_u} \det(w_u)\operatorname{Res}(\mathfrak{p}, \mathfrak{b})(\operatorname{ch}_{\mathfrak{l}}(E_{w_u(\Lambda+\rho)-\rho}))}{\sum_{w_u \in W_u} \det(w_u)\operatorname{Res}(\mathfrak{p}, \mathfrak{b})(\operatorname{ch}_{\mathfrak{l}}(E_{w_u(\rho)-\rho}))} = \frac{\sum_{w \in W} \det(w)\, e^{w(\Lambda+\rho)}}{\sum_{w \in W} \det(w)\, e^{w(\rho)}}$$

$$= \operatorname{Res}(\mathfrak{p}, \mathfrak{b})\operatorname{Res}(\mathfrak{g}, \mathfrak{p})(\operatorname{ch}_{\mathfrak{g}}(E_\Lambda)),$$

which serves to complete the proof of the theorem. □

2.4.3 The Characters of a Reductive Lie Algebra

Let $\mathfrak{g}$ be a reductive Lie algebra over an algebraically closed field **k** of characteristic zero, $\mathfrak{j}$ a Cartan subalgebra of $\mathfrak{g}$, $\mathfrak{b}$ a Borel subalgebra of $\mathfrak{g}$ containing $\mathfrak{j}$, Φ^+ the set of positive roots of the pair $(\mathfrak{g}, \mathfrak{j})$ singled out by this choice of $\mathfrak{b}$, Ψ the fundamental system of roots for Φ contained in Φ^+, W the Weyl group of Φ.

Let $\mathfrak{G}$ denote the universal enveloping algebra of $\mathfrak{g}$, $\mathfrak{Z}$ the center of $\mathfrak{G}$; let $D \mapsto D^0$ denote the canonical projection of $\mathfrak{G}$ onto $\mathfrak{Z}$ (cf. 2.3.3).

Definition A linear form $\kappa : \mathfrak{G} \to \mathbf{k}$ is called a *character* of $\mathfrak{g}$ if:
(1) $\kappa(D_1 D_2) = \kappa(D_2 D_1)$, $\kappa(1) = 1$;
(2) $\kappa(D_1^0 D_2) = \kappa(D_1)\kappa(D_2)$ $\quad (D_1, D_2 \in \mathfrak{G})$.

Thus one has $\kappa(D) = \kappa(D^0)$ for all $D \in \mathfrak{G}$; moreover κ is then a homomorphism of $\mathfrak{Z}$ into **k**. Note too that κ is completely determined by its restriction to $\mathfrak{J}$ (the universal enveloping algebra of $\mathfrak{j}$) – cf. Proposition 2.3.3.2.

Our ultimate goal is to find all the characters of $\mathfrak{g}$.

Let E_Λ be a finite dimensional simple $\mathfrak{g}$-module, U_Λ the corresponding representation ($\Lambda \in \mathfrak{W}^+$); extend U_Λ in the usual way to $\mathfrak{G}$ and put $\kappa_\Lambda(D) = \operatorname{tr}(U_\Lambda(D))/\dim(E_\Lambda)$ $(D \in \mathfrak{G})$ – then κ_Λ is a character of $\mathfrak{g}$. In fact (1) is obvious from well-known properties of the trace while (2) follows from Schur's Lemma. Plainly equivalent simple finite dimensional $\mathfrak{g}$-modules have the same character.

Proposition 2.4.3.1 Let E_{Λ_i} be a finite dimensional simple $\mathfrak{g}$-module with highest weight Λ_i, U_{Λ_i} the corresponding representation $(i = 1, 2)$. If $\kappa_{\Lambda_1} = \kappa_{\Lambda_2}$, then E_{Λ_1} and E_{Λ_2} are equivalent.

Proof Let $\mathfrak{J}_1 = \{D_1 \in \mathfrak{G}: \kappa_{\Lambda_1}(D_1 D) = 0 \text{ (all } D \in \mathfrak{G})\}$ – then $\mathfrak{J}_1$ is a left ideal in $\mathfrak{G}$. If $\kappa_{\Lambda_1}(D_1 D) = 0$ (all $D \in \mathfrak{G}$), then $\operatorname{tr}(U_{\Lambda_1}(D_1)U_{\Lambda_1}(D)) = 0$ (all $D \in \mathfrak{G}$), whence $U_{\Lambda_1}(D_1) = 0$ (for Burnside's Theorem implies that $U_{\Lambda_1}(\mathfrak{G}) = \operatorname{Hom}_{\mathbf{k}}(E_{\Lambda_1}, E_{\Lambda_1})$). Therefore $\mathfrak{G}/\mathfrak{J}_1$ is isomorphic to $\operatorname{Hom}_{\mathbf{k}}(E_{\Lambda_1}, E_{\Lambda_1})$. Defining $\mathfrak{J}_2$ analogously, we find that $\mathfrak{G}/\mathfrak{J}_2$ is isomorphic to $\operatorname{Hom}_{\mathbf{k}}(E_{\Lambda_2}, E_{\Lambda_2})$ – but $\mathfrak{J}_1 = \mathfrak{J}_2$ and so $\operatorname{Hom}_{\mathbf{k}}(E_{\Lambda_1}, E_{\Lambda_1})$ is isomorphic to $\operatorname{Hom}_{\mathbf{k}}(E_{\Lambda_2}, E_{\Lambda_2})$, from which the equivalence of E_{Λ_1} with E_{Λ_2} follows (cf. Jacobson [2, p. 235]). □

Suppose that U is an irreducible representation of $\mathfrak{g}$ on an infinite dimensional space E and suppose that $U(Z)$ is a scalar multiple of the identity for all $Z \in \mathfrak{Z}$, say $U(Z) = \kappa(Z)1$. Extend κ to $\mathfrak{G}$ by writing $\kappa(D) = \kappa(D^0)$ – then κ is a character of $\mathfrak{g}$ but it is no longer true that κ determines U to within equivalence. This will be taken up shortly.

Let E_Λ be a finite dimensional simple $\mathfrak{g}$-module, U_Λ the corresponding representation $(\Lambda \in \mathfrak{W}^+)$; let $E_\Lambda = \sum_\lambda E^\lambda$ be the decomposition of E_Λ into weight spaces, $m(\lambda)$ the multiplicity of the weight λ; let κ_Λ denote the character of E_Λ – then Weyl's 'character formula' provides one with an explicit expression for κ_Λ. Thus, as has been noted above, it is sufficient to determine κ_Λ on $\mathfrak{J}$. So let $H \in \mathfrak{J}$ – then $U_\Lambda(H)$ is a scalar on E^λ, its value being $f_\lambda(H) = \langle H, \mathbf{e}^\lambda \rangle$. [The notations of 2.1.4 are in force here.] Hence $\operatorname{tr}_{E_\Lambda}(U_\Lambda(H)) = \sum_\lambda m(\lambda) f_\lambda(H) = \langle H, \sum_\lambda m(\lambda)\, \mathbf{e}^\lambda \rangle$, while $\dim(E_\Lambda) = \sum_\lambda m(\lambda)$. We are led, therefore, to the consideration of the element $\sum_\lambda m(\lambda)\, \mathbf{e}^\lambda$ in the space $S^*(\mathfrak{j}^\vee)$; in turn, the formula of Weyl evaluates $\sum_\lambda m(\lambda)\, \mathbf{e}^\lambda$ since, as is well-known,

$$\operatorname{tr}_{E_\Lambda}(U_\Lambda(H)) = \langle H, \sum_\lambda m(\lambda)\, \mathbf{e}^\lambda \rangle = \left\langle H, \frac{\sum_{w \in W} \det(w)\, \mathbf{e}^{w(\Lambda+\rho)}}{\sum_{w \in W} \det(w)\, \mathbf{e}^{w(\rho)}} \right\rangle$$

(all $H \in \mathfrak{J}$).

The program of determining κ_Λ, in the case at hand, is then complete since the dimension of E_Λ is known through Theorem 2.4.1.6.

Remark Let us recall that Kostant has computed the multiplicities $m(\lambda)$ – thus if E_Λ is a finite dimensional simple $\mathfrak{g}$-module with highest weight Λ, then the multiplicity $m(\lambda)$ of a weight λ is given by

$$m(\lambda) = \sum_{w \in W} \det(w)\mathbf{P}(w(\Lambda + \rho) - (\lambda + \rho)),$$

where $\mathbf{P}$ denotes the partition function (cf. Jacobson [3, p.261]).

It is actually possible to extend the preceding considerations to those simple $\mathfrak{g}$-modules E_Λ which admit a $\mathfrak{b}$-primitive element a_Λ of weight Λ (say) $(\Lambda \in \mathfrak{j}^\vee)$ although it is no longer true in general that the corresponding characters determine the isomorphism classes of the simple

$\mathfrak{g}$-modules in question; on the other hand it will turn out that there are no other characters of $\mathfrak{g}$ besides the ones constructed in this way.

We shall agree to use the notations introduced in 2.3.3.

Let E_Λ be a simple $\mathfrak{g}$-module with a $\mathfrak{b}$-primitive element a_Λ of weight Λ, U_Λ the corresponding representation – let us define the character κ_Λ of U_Λ. We have seen that $E_\Lambda = E^\Lambda + \sum_{\lambda\neq\Lambda} E^\lambda$ where $\dim(E^\Lambda) = 1$; plainly E^Λ is $\mathfrak{Z}$-stable and so we may define a linear form α_Λ on $\mathfrak{Z}$ by the requirement $U_\Lambda(Z)a_\Lambda = \alpha_\Lambda(Z)a_\Lambda$ $(Z \in \mathfrak{Z})$. If we put $\kappa_\Lambda(D) = \alpha_\Lambda(D^0)$ $(D \in \mathfrak{G})$, then we get a character κ_Λ of $\mathfrak{g}$. Now in general, for arbitrary $D \in \mathfrak{G}$, we have $U_\Lambda(D)a_\Lambda \equiv \beta_\Lambda(D)a_\Lambda \bmod \sum_{\lambda\neq\Lambda} E^\lambda$, β_Λ a linear form on $\mathfrak{G}$ which extends α_Λ. Observe that: (1) If $D \in \mathfrak{N}^+$, then $U_\Lambda(D)a_\Lambda = \epsilon^+(D)a_\Lambda$; (2) If f_Λ denotes the homomorphism of $\mathfrak{I}$ into $\mathbf{k}$ determined by the conditions $1 \mapsto 1$, $H \mapsto \Lambda(H)$ $(H \in \mathfrak{j})$, then, for $H \in \mathfrak{I}$, $U_\Lambda(H)a_\Lambda = f_\Lambda(H)a_\Lambda$; (3) If $D \in \mathfrak{N}^-$, then $U_\Lambda(D)a_\Lambda \equiv \epsilon^-(D)a_\Lambda \bmod \sum_{\lambda\neq\Lambda} E^\lambda$. Therefore, for $D \in \mathfrak{G}$, $U_\Lambda(D)a_\Lambda \equiv f_\Lambda(\epsilon(D))a_\Lambda \bmod \sum_{\lambda\neq\Lambda} E^\lambda \Rightarrow \kappa_\Lambda(D) = \beta_\Lambda(D^0) = f_\Lambda(\epsilon(D^0))$, i.e. $\kappa_\Lambda = f_\Lambda \circ \gamma^*$.

Here is the main result on the characters.

Theorem 2.4.3.2 (Harish-Chandra) Let E_Λ be a simple $\mathfrak{g}$-module which admits a $\mathfrak{b}$-primitive element of weight Λ (say) $(\Lambda \in \mathfrak{j}^\vee)$ – then the corresponding character κ_Λ is given by the formula

$$\kappa_\Lambda(H) = \left\langle H, \prod_{\alpha>0} \frac{(\rho, \alpha)}{(\Lambda + \rho, \alpha)} \frac{\sum_{w\in W} \det(w)\, e^{w(\Lambda+\rho)}}{\sum_{w\in W} \det(w)\, e^{w(\rho)}} \right\rangle \qquad (H \in \mathfrak{I}).$$

Every character κ of $\mathfrak{g}$ is of the form κ_Λ for some $\Lambda \in \mathfrak{j}^\vee$. Moreover $\kappa_{\Lambda_1} = \kappa_{\Lambda_2}$ iff $\Lambda_1 + \rho = w(\Lambda_2 + \rho)$ for some $w \in W$.

Proof As has been said before, it is enough to determine κ_Λ on $\mathfrak{I}$.

(1) If $H \in \mathfrak{I}$, then $\kappa_\Lambda(H) = f_\Lambda(\gamma^*(H)) = \langle \gamma^*(H), e^\Lambda \rangle = \langle H, {}^t\gamma^*(e^\Lambda) \rangle$, i.e. $\kappa_\Lambda(H) = \langle H, {}^t\gamma_3^{-1} \circ {}^t\gamma_2 \circ {}^t\gamma_1(e^\Lambda) \rangle$ (Proposition 2.3.3.6). It will be shown in the proof of Proposition 2.4.3.3 below that ${}^t\gamma_2 \circ {}^t\gamma_1(e^\Lambda) = \sum_w \det(w) \cdot e^{w(\Lambda+\rho)} / \prod_{\alpha>0} (\Lambda + \rho, \alpha)$ – hence, in view of the definition of γ_3,

$$\kappa_\Lambda(H) = \left\langle H, \prod_{\alpha>0} \frac{(\rho, \alpha)}{(\Lambda + \rho, \alpha)} \frac{\sum_{w\in W} \det(w)\, e^{w(\Lambda+\rho)}}{\sum_{w\in W} \det(w)\, e^{w(\rho)}} \right\rangle \qquad (H \in \mathfrak{I})$$

which is the desired formula.

(2) Let κ be an arbitrary character of $\mathfrak{g}$ – then the restriction of κ to $\mathfrak{Z}$ is a homomorphism of $\mathfrak{Z}$ into $\mathbf{k}$. Proposition 2.1.3.8(i) implies that the homomorphism $\kappa \circ (\gamma^*)^{-1}$ of $\gamma_1(I(\mathfrak{j}))$ can be extended to a homomorphism of $\mathfrak{I}$ into $\mathbf{k}$ and this extension is evidently of the form $H \mapsto \langle H, e^\Lambda \rangle$ $(H \in \mathfrak{I})$ for some $\Lambda \in \mathfrak{j}^\vee$. Thus $\kappa \circ (\gamma^*)^{-1} = e^\Lambda$ or in other words $\kappa = e^\Lambda \circ \gamma^* = {}^t\gamma^*(e^\Lambda) = \kappa_\Lambda$.

(3) It is clear that $\kappa_{\Lambda_1} = \kappa_{\Lambda_2}$ iff ${}^t\gamma_1(e^{\Lambda_1})$ and ${}^t\gamma_1(e^{\Lambda_2})$ agree on $I(\mathfrak{j})$, i.e. iff $e^{\Lambda_1+\rho}$ and $e^{\Lambda_2+\rho}$ agree on $I(\mathfrak{j})$ which is the case iff $\Lambda_1 + \rho$ is conjugate to $\Lambda_2 + \rho$ under the Weyl group W (Proposition 2.1.3.8(ii)).

This proves the theorem. $\square$

Consequently, in contradistinction to the finite dimensional case, if $\Lambda_1 \neq \Lambda_2$ and yet $\Lambda_1 + \rho = w(\Lambda_2 + \rho)$ for some $w \in W$, then $\kappa_{\Lambda_1} = \kappa_{\Lambda_2}$ while the corresponding simple $\mathfrak{g}$-modules are not isomorphic.

Proposition 2.4.3.3 Retain the above notations – then $\gamma^* \circ \gamma_3 = \gamma_1 \circ \gamma_2$.

Proof If $\Lambda \in \mathfrak{j}^\vee$, then the element ${}'\gamma_1(\mathrm{e}^\Lambda)$ maps $H \in \mathfrak{j}$ to $\Lambda(H) + \rho(H)$, whence ${}'\gamma_1(\mathrm{e}^\Lambda) = \mathrm{e}^{\Lambda+\rho}$; it is also clear that $\langle \gamma_2(H), \mathrm{e}^\Lambda \rangle = \langle \sum_w \det(w) H^w / \prod_{\alpha>0} H_\alpha, \mathrm{e}^\Lambda \rangle = \sum_w \det(w) \langle H, \mathrm{e}^{w(\Lambda)} \rangle / \prod_{\alpha>0} (\Lambda, \alpha)$ (all $H \in \mathfrak{I}$) – therefore ${}'\gamma_2 \circ {}'\gamma_1(\mathrm{e}^\Lambda) = \sum_w \det(w) \mathrm{e}^{w(\Lambda+\rho)} / \prod_{\alpha>0} (\Lambda + \rho, \alpha)$. On the other hand, if $\Lambda \in \mathfrak{W}^+$, then

$$'\gamma^*(\mathrm{e}^\Lambda) = \frac{\prod_{\alpha>0} (\rho, \alpha) \sum_w \det(w) \mathrm{e}^{w(\Lambda+\rho)}}{\prod_{\alpha>0} (\Lambda + \rho, \alpha) \sum_w \det(w) \mathrm{e}^{w(\rho)}}$$

(Weyl's character and dimension formulae) – hence ${}'\gamma_3 \circ {}'\gamma^*(\mathrm{e}^\Lambda) = {}'\gamma_2 \circ {}'\gamma_1(\mathrm{e}^\Lambda)$ (all $\Lambda \in \mathfrak{W}^+$). Thanks to Lemma 2.1.4.1, this equality persists for all the elements in $S^*(\mathfrak{j}^\vee)$, otherwise said, $\gamma^* \circ \gamma_3 = \gamma_1 \circ \gamma_2$. □

Proposition 2.4.3.4 Let us suppose that $\mathfrak{g}$ is semi-simple – then there exists an element $Z_0 \in \mathfrak{Z}$ with the following property: If E_Λ is any finite dimensional simple $\mathfrak{g}$-module with highest weight Λ, U_Λ the corresponding representation, $d_\Lambda = \dim(E_\Lambda)$, then $U_\Lambda(Z_0) = d_\Lambda^2 U_\Lambda(1)$.

Proof The element $H_0 = (\prod_{\alpha>0} H_\alpha)^2 / (\prod_{\alpha>0} (\rho, \alpha))^2 \in I(\mathfrak{j})$, hence $Z_0 = \gamma^{-1}(H_0) \in \mathfrak{Z}$ – but

$$\begin{aligned} U_\Lambda(Z_0) &= \kappa_\Lambda(Z_0) U_\Lambda(1) = f_\Lambda \circ \gamma^* \circ (\gamma^*)^{-1} \circ \gamma_1(H_0) U_\Lambda(1) \\ &= \langle H_0, \mathrm{e}^{\Lambda+\rho} \rangle U_\Lambda(1) = \left(\frac{\prod_{\alpha>0} (\Lambda + \rho, \alpha)}{\prod_{\alpha>0} (\rho, \alpha)} \right)^2 U_\Lambda(1) \\ &= d_\Lambda^2 U_\Lambda(1). \quad \square \end{aligned}$$

2.5 Representations on Cohomology Groups

2.5.1 The Riemann-Roch Theorem for Lie Algebras

We shall begin by reviewing certain well-known generalities concerning Lie algebra cohomology (cf. Bourbaki [4, p.115, Ex.12]).

Let **k** be a field of characteristic zero.

Let $\mathfrak{g}$ be a finite dimensional Lie algebra over **k**, E a finite dimensional $\mathfrak{g}$-module. For $q > 0$, denote by $C^q(\mathfrak{g}, E)$ the space of alternating multilinear maps of $\mathfrak{g} \times \ldots \times \mathfrak{g}$ (q times) into E; put $C^0(\mathfrak{g}, E) = E$, and, for $q < 0$, $C^q(\mathfrak{g}, E) = \{0\}$. Let $C(\mathfrak{g}, E)$ denote the direct sum of the $C^q(\mathfrak{g}, E)$; the elements of $C(\mathfrak{g}, E)$ are called the *cochains of* $\mathfrak{g}$ *with values in* E – those in $C^q(\mathfrak{g}, E)$ are said to be of *degree* q. The space $C(\mathfrak{g}, E)$ acquires the structure of a $\mathfrak{g}$-module in the following way. For $q = 0$, put $\theta(X)a =$

$X_E a$ ($a \in E$, $X \in \mathfrak{g}$); for $f \in C^q(\mathfrak{g}, E)$, $q > 0$, put

$$(\theta(X)f)(X_1, \ldots, X_q) = X_E(f(X_1, \ldots, X_q)) - \sum_j f(X_1, \ldots, [X, X_j], \ldots, X_q) \quad (X, X_1, \ldots, X_q \in \mathfrak{g}).$$

The map $X \mapsto \theta(X)$ $(= \theta_{\mathfrak{g}}(X))$ defines a representation of $\mathfrak{g}$ on $C(\mathfrak{g}, E)$ and each $C^q(\mathfrak{g}, E)$ is θ-stable. On the other hand, for each $X \in \mathfrak{g}$, denote by $i(X)$ $(= i_E(X))$ the endomorphism of $C(\mathfrak{g}, E)$ which maps each subspace $C^q(\mathfrak{g}, E)$ into $C^{q-1}(\mathfrak{g}, E)$ and which, for $q > 0$, is defined by the rule

$$(i(X)f)(X_1, \ldots, X_{q-1}) = f(X, X_1, \ldots, X_{q-1}).$$

Then $i(X)^2 = 0$ – moreover, for all $X, Y \in \mathfrak{g}$, one has

$$\theta(X)i(Y) - i(Y)\theta(X) = i([X, Y]).$$

These two operations determine a unique $\mathfrak{g}$-endomorphism $d(= d_{\mathfrak{g}})$ of $C(\mathfrak{g}, E)$, mapping $C^q(\mathfrak{g}, E)$ into $C^{q+1}(\mathfrak{g}, E)$, such that

$$di(X) + i(X)d = \theta(X) \qquad (X \in \mathfrak{g}).$$

Explicitly, d is given by

$$df(X_1, \ldots, X_{q+1}) = \sum_{j<k} (-1)^{j+k} f([X_j, X_k], X_1, \ldots, \hat{X}_j, \ldots, \hat{X}_k, \ldots, X_{q+1}) + \sum_j (-1)^{j+1} (X_j)_E f(X_1, \ldots, \hat{X}_j, \ldots, X_{q+1}).$$

[Here the circumflex ^ over a letter means that it is to be omitted.] It may be shown that $d^2 = 0$ – thus the space $C(\mathfrak{g}, E) = \operatorname{Hom}_{\mathbf{k}}(\Lambda\mathfrak{g}, \mathbf{k}) \otimes_{\mathbf{k}} E$, together with the *coboundary operator* d, is a cochain complex. The elements of the kernel $Z^q(\mathfrak{g}, E)$ of the restriction of d to $C^q(\mathfrak{g}, E)$ are called *cocycles* of degree q with values in E while the elements of the image $B^q(\mathfrak{g}, E)$ of the restriction of d to $C^{q-1}(\mathfrak{g}, E)$ are called *coboundaries* of degree q with values in E. Since $d^2 = 0$, $B^q(\mathfrak{g}, E) \subset Z^q(\mathfrak{g}, E)$. The quotient space $Z^q(\mathfrak{g}, E)/B^q(\mathfrak{g}, E) = H^q(\mathfrak{g}, E)$ is called the space of cohomology of $\mathfrak{g}$ of degree q with values in E. The direct sum of the $H^q(\mathfrak{g}, E)$ will be denoted by $H(\mathfrak{g}, E)$ and referred to as the *space of cohomology* of $\mathfrak{g}$ with values in E. Observe that $H^0(\mathfrak{g}, E)$ is the set of $\mathfrak{g}$-invariants in E.

Note The adjoint representation of $\mathfrak{g}$ defines in the canonical way a representation $\mathrm{ad}^\vee$ (say) on $\operatorname{Hom}_{\mathbf{k}}(\Lambda\mathfrak{g}, \mathbf{k})$. This being so, suppose that the space $C(\mathfrak{g}, E)$ is identified with the tensor product $\operatorname{Hom}_{\mathbf{k}}(\Lambda\mathfrak{g}, \mathbf{k}) \otimes_{\mathbf{k}} E$ (in which case we write $i_{\mathbf{k}} \otimes 1$ for i) – then it is clear that our representation θ becomes the tensor product of $\mathrm{ad}^\vee$ with the given representation of $\mathfrak{g}$ on E.

Let B be a symmetric $\mathfrak{g}$-invariant non-degenerate bilinear form on $\mathfrak{g}$; let ω be the Casimir element associated with B (so that ω is an element of the center of the universal enveloping algebra of $\mathfrak{g}$). [Thus let $\{X_i\}$ be a basis for $\mathfrak{g}$, $\{\check{X}_i\}$ the corresponding dual basis relative to B – then $\omega = \sum_i X_i\check{X}_i$ (of course ω depends only on B and not on the choice of the basis

$\{X_i\}$).] If E is a finite dimensional $\mathfrak{g}$-module, then the element ω defines an endomorphism ω_E of E (a 'Casimir endomorphism' of E) which is compatible with the associated $\mathfrak{g}$-structure; we shall let $\boldsymbol{\Omega}_E$ denote the endomorphism $1 \otimes \omega_E$ of the space of cochains $C(\mathfrak{g}, E)$. Let i_ω denote the endomorphism of $C(\mathfrak{g}, E)$ which maps each subspace $C^q(\mathfrak{g}, E)$ into $C^{q-1}(\mathfrak{g}, E)$ and which, for $q > 0$, is defined by

$$(i_\omega f)(X_1, \ldots, X_{q-1}) = \sum_i (X_i)_E f(\check{X}_i, X_1, \ldots, X_{q-1}).$$

It will be left to the reader to convince himself that $\boldsymbol{\Omega}_E = di_\omega + i_\omega d$ (and hence, as a corollary, that $\boldsymbol{\Omega}_E d = d\boldsymbol{\Omega}_E$).

Suppose that $\mathfrak{s}$ and $\mathfrak{t}$ are subalgebras of $\mathfrak{g}$ such that $[\mathfrak{s}, \mathfrak{t}] \subset \mathfrak{t}$; under these circumstances one has an evident representation of $\mathfrak{s}$ on $\mathfrak{t}$ – we shall refer to it as the adjoint representation of $\mathfrak{s}$ on $\mathfrak{t}$ and denote it by $\mathrm{ad}_\mathfrak{t}$. Given a finite dimensional representation U of $\mathfrak{g}$ on E, one can, upon considering the restriction of U to $\mathfrak{t}$, form the cochain complex $C(\mathfrak{t}, E)$; in turn the Lie algebra $\mathfrak{s}$ operates on $C(\mathfrak{t}, E) = \mathrm{Hom}_\mathbf{k}(\boldsymbol{\Lambda}\mathfrak{t}, E) = \mathrm{Hom}_\mathbf{k}(\boldsymbol{\Lambda}\mathfrak{t}, \mathbf{k}) \otimes_\mathbf{k} E$ via the representation $\theta_\mathfrak{t}$ obtained by taking the tensor product of $\mathrm{ad}_\mathfrak{t}^\vee$ with the restriction of U to $\mathfrak{s}$ ($\mathrm{ad}_\mathfrak{t}^\vee$ the canonical representation of $\mathfrak{s}$ on $\mathrm{Hom}_\mathbf{k}(\boldsymbol{\Lambda}\mathfrak{t}, \mathbf{k})$ deduced from $\mathrm{ad}_\mathfrak{t}$).

Lemma 2.5.1.1 Retain the immediately preceding notations – then the representation $\theta_\mathfrak{t}$ of $\mathfrak{s}$ on $\mathrm{Hom}_\mathbf{k}(\boldsymbol{\Lambda}\mathfrak{t}, E)$ commutes with the coboundary operator $d_\mathfrak{t}$ of the cochain complex $C(\mathfrak{t}, E)$.

Proof The injection i of $\mathfrak{t}$ into $\mathfrak{g}$, leads, in the natural way, to a surjective homomorphism i (say) of the $\mathfrak{s}$-modules $C(\mathfrak{g}, E)$ and $C(\mathfrak{t}, E)$; on the one hand we have, therefore, that $\theta_\mathfrak{t}(X)i = i\theta_\mathfrak{g}(X)$ (all $X \in \mathfrak{s}$) while on the other, $d_\mathfrak{t} i = id_\mathfrak{g}$ – thus $d_\mathfrak{t}\theta_\mathfrak{t}(X)i = d_\mathfrak{t} i\theta_\mathfrak{g}(X) = id_\mathfrak{g}\theta_\mathfrak{g}(X) = i\theta_\mathfrak{g}(X)d_\mathfrak{g} = \theta_\mathfrak{t}(X)id_\mathfrak{g} = \theta_\mathfrak{t}(X)d_\mathfrak{t} i$ (all $X \in \mathfrak{s}$) and so our assertion follows from the surjectivity of i. $\square$

Let E be a finite dimensional $\mathfrak{g}$-module; in view of the preceding lemma, the representation $\theta_\mathfrak{t}$ of $\mathfrak{s}$ on $C(\mathfrak{t}, E)$ actually induces a representation of $\mathfrak{s}$ on the cohomology spaces $H^q(\mathfrak{t}, E)$. Introducing

$$\mathbf{X}_\mathfrak{s}(\mathfrak{t}; E) = \sum_q (-1)^q \mathrm{ch}_\mathfrak{s}(H^q(\mathfrak{t}, E)),$$

one has, for every exact sequence $0 \to E' \to E \to E'' \to 0$ of finite dimensional $\mathfrak{g}$-modules,

$$\mathbf{X}_\mathfrak{s}(\mathfrak{t}; E) = \mathbf{X}_\mathfrak{s}(\mathfrak{t}; E') + \mathbf{X}_\mathfrak{s}(\mathfrak{t}; E''),$$

since $0 \to E' \to E \to E'' \to 0$ is also an exact sequence of $\mathfrak{t}$-modules, and $H(\mathfrak{t}, .)$ is a cohomological functor. Consequently there exists a homomorphism $\chi_\mathfrak{s}(\mathfrak{t}; .)$ of $\mathfrak{R}(\mathfrak{g})$ into $\mathfrak{R}(\mathfrak{s})$ such that for every finite dimensional $\mathfrak{g}$-module E

$$\mathbf{X}_\mathfrak{s}(\mathfrak{t}; E) = \chi_\mathfrak{s}(\mathfrak{t}; \mathrm{ch}_\mathfrak{g}(E)).$$

Theorem 2.5.1.2 (Riemann-Roch) Let $\mathfrak{g}$ be a (finite dimensional) Lie algebra over a field $\mathbf{k}$ of characteristic zero; let $\mathfrak{s}$ and $\mathfrak{t}$ be subalgebras of $\mathfrak{g}$ such that $[\mathfrak{s}, \mathfrak{t}] \subset \mathfrak{t}$ – then, for every element x in the λ-ring $\mathfrak{R}(\mathfrak{g})$, one has

$$\chi_{\mathfrak{s}}(\mathfrak{t}\,; x) = \lambda_{-1}(\mathrm{ch}_{\mathfrak{s}}(\mathfrak{t}^\vee))\mathrm{Res}(\mathfrak{g}, \mathfrak{s})(x).$$

Proof (Aribaud) Let E be a finite dimensional $\mathfrak{g}$-module; if $Z^q(\mathfrak{t}, E)$ (respectively $B^q(\mathfrak{t}, E)$) is the $\mathfrak{s}$-module of cocycles (respectively of coboundaries) in $C^q(\mathfrak{t}, E)$, then

$$\mathrm{ch}_{\mathfrak{s}}(H^q(\mathfrak{t}, E)) = \mathrm{ch}_{\mathfrak{s}}(Z^q(\mathfrak{t}, E)) - \mathrm{ch}_{\mathfrak{s}}(B^q(\mathfrak{t}, E))$$

and

$$\mathrm{ch}_{\mathfrak{s}}(C^q(\mathfrak{t}, E)) = \mathrm{ch}_{\mathfrak{s}}(Z^q(\mathfrak{t}, E)) + \mathrm{ch}_{\mathfrak{s}}(B^{q+1}(\mathfrak{t}, E)).$$

Hence

$$\begin{aligned}\mathbf{X}_{\mathfrak{s}}(\mathfrak{t}\,; E) &= \sum_q (-1)^q(\mathrm{ch}_{\mathfrak{s}}(Z^q(\mathfrak{t}, E)) - \mathrm{ch}_{\mathfrak{s}}(B^q(\mathfrak{t}, E)))\\ &= \sum_q (-1)^q\mathrm{ch}_{\mathfrak{s}}(Z^q(\mathfrak{t}, E)) - \sum_q (-1)^q\mathrm{ch}_{\mathfrak{s}}(B^q(\mathfrak{t}, E))\\ &= \sum_q (-1)^q(\mathrm{ch}_{\mathfrak{s}}(Z^q(\mathfrak{t}, E)) + \mathrm{ch}_{\mathfrak{s}}(B^{q+1}(\mathfrak{t}, E)))\\ &= \sum_q (-1)^q\mathrm{ch}_{\mathfrak{s}}(C^q(\mathfrak{t}, E)).\end{aligned}$$

But

$$\mathrm{ch}_{\mathfrak{s}}(C^q(\mathfrak{t}, E)) = \mathrm{ch}_{\mathfrak{s}}(\mathbf{\Lambda}^q\mathfrak{t}^\vee \otimes_{\mathbf{k}} E) = \mathrm{ch}_{\mathfrak{s}}(\mathbf{\Lambda}^q\mathfrak{t}^\vee)\mathrm{ch}_{\mathfrak{s}}(E)$$

and

$$\mathrm{ch}_{\mathfrak{s}}(E) = \mathrm{Res}(\mathfrak{g}, \mathfrak{s})(\mathrm{ch}_{\mathfrak{g}}(E)).$$

Therefore

$$\begin{aligned}\mathbf{X}_{\mathfrak{s}}(\mathfrak{t}\,; E) &= (\sum_q (-1)^q\mathrm{ch}_{\mathfrak{s}}(\mathbf{\Lambda}^q\mathfrak{t}^\vee))\mathrm{Res}(\mathfrak{g}, \mathfrak{s})(\mathrm{ch}_{\mathfrak{g}}(E))\\ &= \lambda_{-1}(\mathrm{ch}_{\mathfrak{s}}(\mathfrak{t}^\vee))\mathrm{Res}(\mathfrak{g}, \mathfrak{s})(\mathrm{ch}_{\mathfrak{g}}(E)). \quad \square\end{aligned}$$

We shall now take up an important application of the preceding generalities.

Let $\mathbf{k}$ be an algebraically closed field of characteristic zero.

Theorem 2.5.1.3 Let $\mathfrak{g}$ be a reductive Lie algebra over $\mathbf{k}$, $\mathfrak{j}$ a Cartan subalgebra of $\mathfrak{g}$, $\mathfrak{b}$ a Borel subalgebra of $\mathfrak{g}$ containing $\mathfrak{j}$, $\mathfrak{p}$ a $\mathfrak{b}$-parabolic subalgebra of $\mathfrak{g}$ with Levi subalgebra $\mathfrak{l}$ and nilpotent radical $\mathfrak{n}^+$ – then, for every dominant weight $\Lambda \in \mathfrak{W}^+$,

$$\sum_q (-1)^q\mathrm{ch}_{\mathfrak{l}}(H^q(\mathfrak{n}^+, E_\Lambda)) = \sum_{w_u \in W_u} \det(w_u)\mathrm{ch}_{\mathfrak{l}}(E_{w_u(\Lambda+\rho)-\rho})$$

$$(\rho = \tfrac{1}{2} \sum_{\alpha>0} \alpha).$$

Proof (Aribaud) We shall agree to use without comment the notations and conventions revolving around Theorem 2.4.2.2. With $\mathfrak{l}$ and $\mathfrak{n}^+$ playing the roles of $\mathfrak{s}$ and $\mathfrak{t}$ above, respectively, Theorem 2.5.1.2, in conjunction with Theorem 2.4.2.2, tells us that we need only show that

$$\lambda_{-1}(\mathrm{ch}_{\mathfrak{l}}((\mathfrak{n}^+)^\vee)) = \sum_{w_u \in W_u} \det(w_u)\mathrm{ch}_{\mathfrak{l}}(E_{w_u(\rho)-\rho}) \qquad \text{in } \mathfrak{R}(\mathfrak{l}).$$

Because Res $(\mathfrak{l}, \mathfrak{j})$ is an injective homomorphism of λ-rings, the preceding equality will follow as soon as it is established that

$$\lambda_{-1}(\mathrm{ch}_{\mathfrak{j}}((\mathfrak{n}^+)^\vee)) = \mathrm{Res}(\mathfrak{l}, \mathfrak{j})(\lambda_{-1}(\mathrm{ch}_{\mathfrak{l}}((\mathfrak{n}^+)^\vee)) = \sum_{w_u \in W_u} \det(w_u)\mathrm{Res}(\mathfrak{l}, \mathfrak{j})(\mathrm{ch}_{\mathfrak{l}}(E_{w_u(\rho)-\rho}))$$

in $\mathfrak{R}(\mathfrak{j})$. Any symmetric $\mathfrak{g}$-invariant non-degenerate bilinear form on $\mathfrak{g}$ sets up, in the obvious fashion, an isomorphism between the $\mathfrak{j}$-module $(\mathfrak{n}^+)^\vee$ and the $\mathfrak{j}$-module $\mathfrak{n}^- = \sum_\alpha \mathfrak{g}^{-\alpha}$, $\alpha \in \Xi_u$ – thus $\mathrm{ch}_{\mathfrak{j}}((\mathfrak{n}^+)^\vee) = \mathrm{ch}_{\mathfrak{j}}(\mathfrak{n}^-) = \sum_\alpha e^{-\alpha}$, $\alpha \in \Xi_u$, and so

$$\lambda_{-1}(\mathrm{ch}_{\mathfrak{j}}((\mathfrak{n}^+)^\vee)) = \prod_{\alpha \in \Xi_u} (1 - e^{-\alpha}) = \frac{e^\rho \prod_{\alpha \in \Phi^+} (1 - e^{-\alpha})}{e^{\rho_s} \prod_{\alpha \in \Xi_s^+} (1 - e^{-\alpha})} e^{-\rho_u}.$$

A Remark in 2.4.2 provides us with the relations $e^\rho \prod_{\alpha \in \Phi^+} (1 - e^{-\alpha}) = \sum_{w \in W} \det(w) e^{w(\rho)}$, $e^{\rho_s} \prod_{\alpha \in \Xi_s^+}(1 - e^{-\alpha}) = \sum_{w_s \in W_s} \det(w_s)\, e^{w_s(\rho_s)}$; in view of this and Proposition 1.1.2.13, we are led to the equality

$$\lambda_{-1}(\mathrm{ch}_{\mathfrak{j}}((\mathfrak{n}^+)^\vee)) = \sum_{w_u \in W_u} \det(w_u) \frac{\sum_{w_s \in W_s} \det(w_s)\, e^{w_s w_u(\rho) - \rho_u}}{\sum_{w_s \in W_s} \det(w_s) e^{w_s(\rho_s)}}.$$

Since $w_s(\rho_u) = \rho_u$ (all $w_s \in W_s$), we can write $w_s w_u(\rho) - \rho_u = w_s(w_u(\rho) - \rho_u) = w_s((w_u(\rho) - \rho) + \rho_s)$; therefore, thanks to Weyl's formula, we have

$$\mathrm{Res}(\mathfrak{l}, \mathfrak{j})(\mathrm{ch}_{\mathfrak{l}}(E_{w_u(\rho)-\rho})) = \frac{\sum_{w_s \in W_s} \det(w_s)\, e^{w_s(w_u(\rho) - \rho + \rho_s)}}{\sum_{w_s \in W_s} \det(w_s)\, e^{w_s(\rho_s)}}$$

and so we finally obtain the desired relation, namely that

$$\lambda_{-1}(\mathrm{ch}_{\mathfrak{j}}((\mathfrak{n}^+)^\vee)) = \sum_{w_u \in W_u} \det(w_u)\mathrm{Res}(\mathfrak{l}, \mathfrak{j})(\mathrm{ch}_{\mathfrak{l}}(E_{w_u(\rho)-\rho})). \quad \square$$

2.5.2 Theorems of Bott and Kostant

In this number we shall obtain an important theorem due to Kostant which contains, as a corollary, a theorem first proved by Bott. Besides being of interest in its own right, the result may be regarded as one of the main steps in the proof of the Borel-Weil Theorem (see Chapter 3).

Let $\mathfrak{g}$ be a reductive Lie algebra over an algebraically closed field $\mathbf{k}$ of characteristic zero, $\mathfrak{j}$ a Cartan subalgebra of $\mathfrak{g}$, $\mathfrak{b}$ a Borel subalgebra of $\mathfrak{g}$ containing $\mathfrak{j}$, Φ^+ the set of positive roots of the pair $(\mathfrak{g}, \mathfrak{j})$ singled out by this choice of $\mathfrak{b}$, Ψ the fundamental system of roots for Φ contained in Φ^+; let $\mathfrak{p}$ be a $\mathfrak{b}$-parabolic subalgebra of $\mathfrak{g}$, Θ the subset of Ψ associated with $\mathfrak{p}$ in the usual way, $\mathfrak{l}$ the corresponding Levi subalgebra of $\mathfrak{p}$, $\mathfrak{n}^+$ the nilpotent radical of $\mathfrak{p}$; assign to the symbols W, W_s, W_u, $\mathfrak{W}$, $\mathfrak{W}_s$, $\mathfrak{W}^+$, $\mathfrak{W}_s^+$, Ξ_s and Ξ_u the customary meanings; let B denote the symmetric $\mathfrak{g}$-invariant non-degenerate bilinear form obtained by taking the product of a symmetric non-degenerate bilinear form on the center of $\mathfrak{g}$ with the

Killing form on the derived algebra of $\mathfrak{g}$ – such a form, when restricted to $\mathfrak{j}$, is non-degenerate and, moreover, places $\mathfrak{g}^{\alpha}$ in duality with $\mathfrak{g}^{-\alpha}$ ($\alpha \in \Phi^+$).

Given $w \in W$, let Φ_w^+ denote the subset of Φ^+ defined by $\Phi_w^+ = w\Phi^- \cap \Phi^+$; set $|w| = [\Phi_w^+]$ and call $|w|$ the *index* of w; since $\Phi_{w^{-1}}^+ = -w^{-1}(\Phi_w^+)$, it is clear that $|w| = |w^{-1}|$ (all $w \in W$). For any non-negative integer q, put $W(q) = \{w \in W: |w| = q\}$, $W_u(q) = \{w_u \in W_u: |w_u| = q\}$.

Let $\Lambda \in \mathfrak{W}^+$ – then Theorem 2.5.1.3 enables one to evaluate the Euler-Poincaré characteristic of $H(\mathfrak{n}^+, E_\Lambda)$ in the representation ring of $\mathfrak{l}$ (E_Λ a simple $\mathfrak{g}$-module with highest weight Λ). Our goal now is to compute the $\mathfrak{l}$-characters of the *semi-simple* $\mathfrak{l}$-modules $H^q(\mathfrak{n}^+, E_\Lambda)$. [Since $\mathfrak{l}$ is reductive in $\mathfrak{g}$, it is clear that $\mathfrak{n}^+$ and E_Λ actually are semi-simple $\mathfrak{l}$-modules (cf. Proposition 1.1.5.2), hence $C(\mathfrak{n}^+, E_\Lambda)$ and $H(\mathfrak{n}^+, E_\Lambda)$ are too .... It is useful to keep in mind that the semi-simplicity of $H^q(\mathfrak{n}^+, E_\Lambda)$ implies that the character $\mathrm{ch}_{\mathfrak{l}}(H^q(\mathfrak{n}^+, E_\Lambda))$ of $H^q(\mathfrak{n}^+, E_\Lambda)$ in $\mathfrak{R}(\mathfrak{l})$ determines $H^q(\mathfrak{n}^+, E_\Lambda)$ to within an isomorphism of $\mathfrak{l}$-modules.]

The main result may be formulated as follows.

Theorem 2.5.2.1 (Kostant) Let $\mathfrak{g}$ be a reductive Lie algebra over $\mathbf{k}$, $\mathfrak{j}$ a Cartan subalgebra of $\mathfrak{g}$, $\mathfrak{b}$ a Borel subalgebra of $\mathfrak{g}$ containing $\mathfrak{j}$, $\mathfrak{p}$ a $\mathfrak{b}$-parabolic subalgebra of $\mathfrak{g}$ with Levi subalgebra $\mathfrak{l}$ and nilpotent radical $\mathfrak{n}^+$ – then, for every dominant weight $\Lambda \in \mathfrak{W}^+$,

$$\mathrm{ch}_{\mathfrak{l}}(H^q(\mathfrak{n}^+, E_\Lambda)) = \sum_{w_u \in W_u(q)} \mathrm{ch}_{\mathfrak{l}}(E_{w_u(\Lambda+\rho)-\rho}) \qquad (q \in \mathbf{N}).$$

Furthermore, if $\mathrm{Len}\,(H^q(\mathfrak{n}^+, E_\Lambda))$ is the length of the semi-simple $\mathfrak{l}$-module $H^q(\mathfrak{n}^+, E_\Lambda)$, then

$$\mathrm{Len}\,(H^q(\mathfrak{n}^+, E_\Lambda)) = [W_u(q)] \qquad (q \in \mathbf{N}).$$

Specialized to the case when $\mathfrak{p} = \mathfrak{b}$ (so that $\mathfrak{l} = \mathfrak{j}$), Theorem 2.5.2.1 yields the following result due to Bott.

Corollary 2.5.2.2 (Bott). Let $\mathfrak{g}$ be a reductive Lie algebra over $\mathbf{k}$, $\mathfrak{j}$ a Cartan subalgebra of $\mathfrak{g}$, $\mathfrak{b}$ a Borel subalgebra of $\mathfrak{g}$ containing $\mathfrak{j}$, $\mathfrak{n}^+$ the nilpotent radical of $\mathfrak{b}$ – then, as Λ varies over $\mathfrak{W}^+$, the dimension of $H^q(\mathfrak{n}^+, E_\Lambda)$ remains constant, in fact

$$\dim_{\mathbf{k}}(H^q(\mathfrak{n}^+, E_\Lambda)) = [W(q)] \qquad (q \in \mathbf{N}).$$

The proof of Theorem 2.5.2.1 requires some preparation toward which we shall now direct our attention.

The isomorphism classes of (finite dimensional) simple $\mathfrak{l}$-modules form a basis for the free abelian group $\mathfrak{R}(\mathfrak{l})$; this being so, let $\square$ denote the linear map of the $\mathbf{k}$-space $\mathfrak{R}(\mathfrak{l}) \otimes_{\mathbf{Z}} \mathbf{k}$ into itself defined by the rule

$$\square\,(\mathrm{ch}_{\mathfrak{l}}(E_{\Lambda_s})) = B(\Lambda_s + \rho, \Lambda_s + \rho)\mathrm{ch}_{\mathfrak{l}}(E_{\Lambda_s})\ (\Lambda_s \in \mathfrak{W}_s^+).$$

Lemma 2.5.2.3 Suppose that Λ is an element of $\mathfrak{W}^+$, E_Λ a simple $\mathfrak{g}$-module with highest weight Λ – then the element $\mathrm{ch}_{\mathfrak{l}}(H(\mathfrak{n}^+, E_\Lambda))$ in $\mathfrak{R}(\mathfrak{l})$

is an eigenvector for $\square$ with eigenvalue $B(\Lambda + \rho, \Lambda + \rho)$ (as always, $\rho = \frac{1}{2} \sum_{\alpha>0} \alpha$).

In order not to interrupt our exposition, we shall defer the proof of this lemma (which is rather lengthy) to the Appendix of the present number.

Let us fix a family of non-zero elements $X_\alpha \in \mathfrak{g}^\alpha$ ($\alpha \in \Phi$) normalized by the condition that $B(X_\alpha, X_{-\alpha}) = 1$; the elements $X(\Gamma) = X_{\alpha_{i_1}} \wedge \ldots \wedge X_{\alpha_{i_k}}$ (the α_{i_j} being arranged in increasing order), where $\Gamma = \{\alpha_{i_1}, \ldots, \alpha_{i_k}\} \subset \Phi^+$, thus form a basis for the exterior algebra over $\sum_{\alpha>0} \mathfrak{g}^\alpha$ and so the $X(-\Gamma)$ ($\Gamma \subset \Phi^+$) constitute a basis for the exterior algebra over $\sum_{\alpha<0} \mathfrak{g}^\alpha$. It is clear that $\Lambda(\sum_{\alpha<0} \mathfrak{g}^\alpha)$ may be regarded as a $\mathfrak{j}$-module in the evident manner; in particular, the element $X(-\Gamma)$ is a weight vector of weight $-(\rangle\Gamma\langle)$. [If F is any subset of Φ, then we shall often write $\rangle F\langle$ in place of $\sum_{\alpha\in F} \alpha$.]

Lemma 2.5.2.4 In order that λ be a weight in the $\mathfrak{j}$-module $\Lambda(\sum_{\alpha<0} \mathfrak{g}^\alpha)$, it is necessary and sufficient that $\lambda + \rho$ be a weight in the $\mathfrak{j}$-module E_ρ; the multiplicity of λ in $\Lambda(\sum_{\alpha<0} \mathfrak{g}^\alpha)$ is equal to the multiplicity of $\lambda + \rho$ in E_ρ. If $w \in W$, then $w(\rho) - \rho$ is a weight of multiplicity 1 in $\Lambda(\sum_{\alpha<0} \mathfrak{g}_\alpha)$ and $X(-\Phi_w^+)$, where $\Phi_w^+ = w\Phi^- \cap \Phi^+$, is a weight vector for $w(\rho) - \rho$.

Proof According to Weyl's formula, we have

$$\mathrm{ch}_{\mathfrak{j}}(E_\rho) = \frac{\sum_{w\in W} \det(w)\, e^{2w(\rho)}}{\sum_{w\in W} \det(w)\, e^{w(\rho)}}.$$

Employing the well-known procedure, write $\sum_{w\in W} \det(w)\, e^{w(\rho)} = e^\rho \prod_{\alpha>0} (1 - e^{-\alpha})$, $\sum_{w\in W} \det(w)\, e^{2w(\rho)} = e^{2\rho} \prod_{\alpha>0} (1 - e^{-2\alpha})$; since $(1 - e^{-2\alpha}) = (1 - e^{-\alpha})(1 + e^{-\alpha})$, we deduce that $\mathrm{ch}_{\mathfrak{j}}(E_\rho) = e^\rho \prod_{\alpha>0} (1 + e^{-\alpha})$. The first assertion of our lemma now follows upon observing that $\mathrm{ch}_{\mathfrak{j}}(\Lambda(\sum_{\alpha<0} \mathfrak{g}^\alpha)) = \lambda_1(\mathrm{ch}_{\mathfrak{j}}(\sum_{\alpha<0} \mathfrak{g}^\alpha)) = \lambda_1(\sum_{\alpha>0} e^{-\alpha}) = \prod_{\alpha>0} (1 + e^{-\alpha})$. Turning to the second assertion, let us begin by noting that the multiplicity of $w(\rho) - \rho$ in $\Lambda(\sum_{\alpha<0} \mathfrak{g}^\alpha)$ is equal to the multiplicity of $w(\rho)$ in E_ρ, that is to the multiplicity of ρ in E_ρ, i.e. to 1; moreover $\rho - w(\rho) = 1/2(\rangle\Phi^+\langle) - 1/2(\rangle w\Phi^+\langle) = 1/2(\rangle w\Phi^- \cap \Phi^+\langle) + 1/2(\rangle w\Phi^+ \cap \Phi^+\langle) - 1/2(\rangle w\Phi^+ \cap \Phi^+\langle) - 1/2(\rangle w\Phi^+ \cap \Phi^-\langle) = 1/2(\rangle w\Phi^- \cap \Phi^+\langle) + 1/2(\rangle w\Phi^- \cap \Phi^+\langle) = \rangle\Phi_w^+\langle$ – the element $X(-\Phi_w^+)$ in $\Lambda(\sum_{\alpha<0} \mathfrak{g}^\alpha)$ is therefore a weight vector for the weight $w(\rho) - \rho$ (hence is unique to within a scalar factor, $w(\rho) - \rho$ being of multiplicity 1). $\square$

Because $\mathfrak{j}$ is contained in $\mathfrak{l}$, the $\mathfrak{l}$-module $C^q(\mathfrak{n}^+, E_\Lambda)$ can be regarded as a $\mathfrak{j}$-module in the natural way ($\Lambda \in \mathfrak{W}^+$).

Lemma 2.5.2.5 The weights λ in the $\mathfrak{j}$-module $C^q(\mathfrak{n}^+, E_\Lambda)$ which verify the condition

$$B(\lambda + \rho, \lambda + \rho) = B(\Lambda + \rho, \Lambda + \rho)$$

are of the form $\lambda = w_u(\Lambda + \rho) - \rho$, where $w_u \in W_u$ and $|w_u| = q$. The multiplicity of such a λ is 1.

Proof The bilinear form B places $\mathfrak{n}^+ = \sum_{\alpha \in \Xi_u} \mathfrak{g}^\alpha$ in duality with $\mathfrak{n}^- = \sum_{\alpha \in \Xi_u} \mathfrak{g}^{-\alpha}$; the representation of $\mathfrak{j}$ on $C^q(\mathfrak{n}^+, E_\Lambda)$ is therefore isomorphic to a subrepresentation of the representation of $\mathfrak{j}$ on $\Lambda^q(\sum_{\alpha<0} \mathfrak{g}^\alpha) \otimes_{\mathbf{k}} E_\Lambda$. According to Lemma 2.5.2.4, if λ is a weight in the $\mathfrak{j}$-module $\Lambda^q(\sum_{\alpha<0} \mathfrak{g}^\alpha) \otimes_{\mathbf{k}} E_\Lambda$, then $\lambda + \rho$ is a weight in the $\mathfrak{j}$-module $E_\Lambda \otimes_{\mathbf{k}} E_\rho$, the multiplicity of λ in $\Lambda^q(\sum_{\alpha<0} \mathfrak{g}^\alpha) \otimes_{\mathbf{k}} E_\Lambda$ being equal to that of $\lambda + \rho$ in $E_\Lambda \otimes_{\mathbf{k}} E_\rho$. This being so, choose a $w \in W$ such that $w(\lambda + \rho) \in \mathfrak{W}^+$; it follows, for familiar reasons, that the weight $w(\lambda + \rho)$ is of the form $\Lambda + \rho - \sum_{\alpha>0} n_\alpha \alpha (n_\alpha \in \mathbf{N})$; since B is W-invariant, we then have that

$$B(\lambda + \rho, \lambda + \rho) = B(w(\lambda + \rho), w(\lambda + \rho)) = B(\Lambda + \rho, \Lambda + \rho) - B(\Sigma n_\alpha \alpha, \Lambda + \rho) - B(\Lambda + \rho - \Sigma n_\alpha \alpha, \Sigma n_\alpha \alpha).$$

The expression $B(\Lambda + \rho - \Sigma n_\alpha \alpha, \Sigma n_\alpha \alpha)$ is a non-negative rational number, while $B(\Sigma n_\alpha \alpha, \Lambda + \rho)$ is a strictly positive rational number if $\Sigma n_\alpha \alpha \neq 0$; consequently the equality $B(\lambda + \rho, \lambda + \rho) = B(\Lambda + \rho, \Lambda + \rho)$ can obtain only if $w(\lambda + \rho) = \Lambda + \rho$, i.e. only if $\lambda = \tilde{w}(\Lambda + \rho) - \rho$ $(\tilde{w} = w^{-1})$. The multiplicity of the weight $\Lambda + \rho$ in $E_\Lambda \otimes_{\mathbf{k}} E_\rho$ is equal to 1; because $\lambda + \rho$ is W-conjugate to $\Lambda + \rho$, the same must be true of $\lambda + \rho$ – hence, in view of what has been said above, the multiplicity of λ in $\Lambda^q(\sum_{\alpha<0} \mathfrak{g}^\alpha) \otimes_{\mathbf{k}} E_\Lambda$ is likewise 1. If $a_{\tilde{w}(\Lambda)}$ is a weight vector for the weight $\tilde{w}(\Lambda)$ in E_Λ, then $X(-\Phi_{\tilde{w}}^+) \otimes a_{\tilde{w}(\Lambda)}$ is a weight vector for the weight $\lambda = \tilde{w}(\Lambda + \rho) - \rho$ in $\Lambda^q(\sum_{\alpha<0} \mathfrak{g}^\alpha) \otimes_{\mathbf{k}} E_\Lambda$; in particular, λ is a weight for the $\mathfrak{j}$-module $\Lambda^q \mathfrak{n}^- \otimes_{\mathbf{k}} E_\Lambda$ iff $X(-\Phi_{\tilde{w}}^+) \otimes a_{\tilde{w}(\Lambda)} \in \Lambda^q \mathfrak{n}^- \otimes_{\mathbf{k}} E_\Lambda$ – but $X(-\Phi_{\tilde{w}}^+) \otimes a_{\tilde{w}(\Lambda)} \in \Lambda^q \mathfrak{n}^- \otimes_{\mathbf{k}} E_\Lambda$ only if $X(-\Phi_{\tilde{w}}^+) \in \Lambda^q \mathfrak{n}^-$ and this condition is equivalent, by the very definition of $X(-\Phi_{\tilde{w}}^+)$, to the requirement that $\Phi_{\tilde{w}}^+ \subset \Xi_u$, whence $\tilde{w} \in W_u$ with $|\tilde{w}| = q$. □

Lemma 2.5.2.6 The highest weights $\Lambda_s (\Lambda_s \in \mathfrak{W}_s^+)$ for the simple $\mathfrak{l}$-submodules which occur in $H^q(\mathfrak{n}^+, E_\Lambda)(\Lambda \in \mathfrak{W}^+)$ are of the form $\Lambda_s = w_u(\Lambda + \rho) - \rho$, where $w_u \in W_u$ and $|w_u| = q$; the number of times the simple $\mathfrak{l}$-module $E_{w_u(\Lambda+\rho)-\rho}$ occurs in $H^q(\mathfrak{n}^+, E_\Lambda)$ cannot exceed unity $(w_u \in W_u$ with $|w_u| = q)$.

Proof It is clear that the number of times that E_{Λ_s} $(\Lambda_s \in \mathfrak{W}_s^+)$ occurs in $H^q(\mathfrak{n}^+, E_\Lambda)$ is less than or equal to the number of times that E_{Λ_s} occurs in $C^q(\mathfrak{n}^+, E_\Lambda)$, which itself cannot exceed the multiplicity of the weight Λ_s in $C^q(\mathfrak{n}^+, E_\Lambda)$. According to Lemma 2.5.2.3, those highest weights Λ_s which have the property that E_{Λ_s} actually does occur in $H^q(\mathfrak{n}^+, E_\Lambda)$ necessarily verify the relation $B(\Lambda_s + \rho, \Lambda_s + \rho) = B(\Lambda + \rho, \Lambda + \rho)$; owing to Lemma 2.5.2.5, then, such a Λ_s can be written $\Lambda_s = w_u(\Lambda + \rho) - \rho$, where $w_u \in W_u$ and $|w_u| = q$. Our assertions are now plain. □

Proof of Theorem 2.5.2.1 (Aribaud) The elements $E_{w_u(\Lambda+\rho)-\rho}$, w_u running through W_u, are linearly independent (cf. the Remark following Lemma 2.4.2.1). If $\mathrm{ch}_{\mathfrak{l}}(H^q(\mathfrak{n}^+, E_\Lambda)) = \sum_i n_i(q)\mathrm{ch}_{\mathfrak{l}}(E_{\Lambda_{s,i}(q)})(\Lambda_{s,i}(q) \in \mathfrak{W}_s^+)$, then we have

$$\sum_q (-1)^q \mathrm{ch}_{\mathfrak{l}}(H^q(\mathfrak{n}^+, E_\Lambda)) = \sum_{i,q} (-1)^q n_i(q)\mathrm{ch}_{\mathfrak{l}}(E_{\Lambda_{s,i}(q)})$$
$$= \sum_{\Lambda_s} [\sum_{\Lambda_{s,i}(q)=\Lambda_s} (-1)^q n_i(q)]\mathrm{ch}_{\mathfrak{l}}(E_{\Lambda_s}).$$

On the other hand, Theorem 2.5.1.3 provides us with the relation

$$\sum_q (-1)^q \mathrm{ch}_{\mathfrak{l}}(H^q(\mathfrak{n}^+, E_\Lambda)) = \sum_{w_u \in W_u} \det(w_u)\mathrm{ch}_{\mathfrak{l}}(E_{w_u(\Lambda+\rho)-\rho}).$$

The above mentioned linear independence of the $E_{w_u(\Lambda+\rho)-\rho}(w_u \in W_u)$ then enables one to deduce that $\Sigma(-1)^q n_i(q) \neq 0$, the sum being taken over those $\Lambda_{s,i}(q)$ such that $\Lambda_{s,i}(q) = w_u(\Lambda + \rho) - \rho$. It now follows that each $E_{w_u(\Lambda+\rho)-\rho}(w_u \in W_u)$ occurs at least once in $H(\mathfrak{n}^+, E_\Lambda)$; consequently each $E_{w_u(\Lambda+\rho)-\rho}(w_u \in W_u(q))$ occurs at least once in $H^q(\mathfrak{n}^+, E_\Lambda)$, whence, thanks to Lemma 2.5.2.6, exactly once. On the other hand, if Λ_s is not of the form $\Lambda_s = w_u(\Lambda + \rho) - \rho(w_u \in W_u(q))$, then E_{Λ_s} cannot occur in $H^q(\mathfrak{n}^+, E_\Lambda)$ (cf. Lemma 2.5.2.6). All the contentions of the theorem are now clear. □

Remark It is easy to compute the dimension of a given $H^q(\mathfrak{n}^+, E_\Lambda)$; in fact, agreeing to denote the restriction of B to $\mathfrak{j}$ by $(\, , \,)$, an application of Weyl's formula (Theorem 2.4.1.6) leads at once to

$$\dim_{\mathbf{k}}(H^q(\mathfrak{n}^+, E_\Lambda)) = \sum_{w_u \in W_u(q)} \frac{\prod_{\alpha \in \Xi_s^+} (w_u(\Lambda + \rho), \alpha)}{\prod_{\alpha \in \Xi_s^+} (\rho, \alpha)}.$$

[Here we have to bear in mind that ρ_u is orthogonal to every $\alpha \in \Xi_s$ (where, as usual, $\rho = \rho_s + \rho_u$).] Kostant's theorem tells us, among other things, that the number of simple components in $H^q(\mathfrak{n}^+, E_\Lambda)$ under the action of $\mathfrak{l}$ remains constant as Λ varies through $\mathfrak{W}^+$ (indeed, is equal to $[W_u(q)]$); on the other hand, the above formula for the dimension of $H^q(\mathfrak{n}^+, E_\Lambda)$ shows that, in general, $\dim_{\mathbf{k}} (H^q(\mathfrak{n}^+, E_\Lambda))$ can be made arbitrarily large by a suitable choice of Λ (cf. with Corollary 2.5.2.2).

Appendix Our objective here will be to give a proof of Lemma 2.5.2.3 supra – that is, we shall establish the following result.

Lemma Suppose that Λ is an element of $\mathfrak{W}^+$, E_Λ a simple $\mathfrak{g}$-module with highest weight Λ – then the element $\mathrm{ch}_{\mathfrak{l}}(H(\mathfrak{n}^+, E_\Lambda))$ in $\mathfrak{R}(\mathfrak{l})$ is an eigenvector for $\square$ with eigenvalue $B(\Lambda + \rho, \Lambda + \rho)$ $(\rho = \frac{1}{2} \sum_{\alpha>0} \alpha)$.

The proof will be divided up into several steps.

Since $C(\mathfrak{g}, E_\Lambda)$ (respectively $C(\mathfrak{n}^+, E_\Lambda)$) is a semi-simple $\mathfrak{l}$-module, there exists an endomorphism $\Gamma_{\mathfrak{g}}$ (respectively $\Gamma_{\mathfrak{n}^+}$) of $C(\mathfrak{g}, E_\Lambda)$ (respectively of $C(\mathfrak{n}^+, E_\Lambda)$) which reduces, on each simple $\mathfrak{l}$-component with dominant weight Λ_s (say), to scalar multiplication by $|\Lambda + \rho|^2 - |\Lambda_s + \rho|^2$. [If $\gamma_1, \gamma_2, \gamma \in \mathfrak{j}^\vee$, then we shall agree to write (γ_1, γ_2) for $B(\gamma_1, \gamma_2)$, $|\gamma|^2$ for $B(\gamma, \gamma)$.]

(1) If the endomorphism $\Gamma_{\mathfrak{n}^+}$ is homotopic to zero, then the element $\mathrm{ch}_{\mathfrak{l}}(H(\mathfrak{n}^+, E_\Lambda))$ in $\mathfrak{R}(\mathfrak{l})$ is an eigenvector for $\Box$ with eigenvalue $|\Lambda + \rho|^2$. [In fact the endomorphism $\Gamma_{\mathfrak{n}^+}$ of $C(\mathfrak{n}^+, E_\Lambda)$, which commutes with the coboundary operator $d_{\mathfrak{n}^+}$, then induces an endomorphism $\Gamma_{\mathfrak{n}^+}^*$ (say) of $H(\mathfrak{n}^+, E_\Lambda)$, the *null* endomorphism. But $\Gamma_{\mathfrak{n}^+}^*$ reduces, on each simple $\mathfrak{l}$-component E_{Λ_s} $(\Lambda_s \in \mathfrak{W}_s^+)$ of $H(\mathfrak{n}^+, E_\Lambda)$ to multiplication by the scalar $|\Lambda + \rho|^2 - |\Lambda_s + \rho|^2$ . . . .]

Our job now, then, is to prove that $\Gamma_{\mathfrak{n}^+}$ is homotopic to zero.

The injection of $\mathfrak{n}^+$ into $\mathfrak{g}$ induces a homomorphism p from the complex of $\mathfrak{l}$-modules $C(\mathfrak{g}, E_\Lambda)$ into the complex of $\mathfrak{l}$-modules $C(\mathfrak{n}^+, E_\Lambda)$; the diagram

$$\begin{array}{ccc} C(\mathfrak{g}, E_\Lambda) & \xrightarrow{p} & C(\mathfrak{n}^+, E_\Lambda) \\ {\scriptstyle \Gamma_{\mathfrak{g}}}\downarrow & & \downarrow{\scriptstyle \Gamma_{\mathfrak{n}^+}} \\ C(\mathfrak{g}, E_\Lambda) & \xrightarrow[p]{} & C(\mathfrak{n}^+, E_\Lambda) \end{array}$$

is commutative.

(2) If $\Gamma_{\mathfrak{g}}$ is homotopic to zero (so that $\Gamma_{\mathfrak{g}} = d_{\mathfrak{g}}h + hd_{\mathfrak{g}}$ for some endomorphism h of $C(\mathfrak{g}, E_\Lambda)$) and if there exists an endomorphism k of $C(\mathfrak{n}^+, E_\Lambda)$ such that $ph = kp$, then $\Gamma_{\mathfrak{n}^+}$ is homotopic to zero. [Thus one has $\Gamma_{\mathfrak{n}^+}p = p\Gamma_{\mathfrak{g}} = pd_{\mathfrak{g}}h + phd_{\mathfrak{g}} = d_{\mathfrak{n}^+}ph + phd_{\mathfrak{g}} = d_{\mathfrak{n}^+}kp + kpd_{\mathfrak{g}} = d_{\mathfrak{n}^+}kp + kd_{\mathfrak{n}^+}p = (d_{\mathfrak{n}^+}k + kd_{\mathfrak{n}^+})p$; because p is surjective, it follows that $\Gamma_{\mathfrak{n}^+} = d_{\mathfrak{n}^+}k + kd_{\mathfrak{n}^+}$, as desired.]

Choose a basis $\{H_i\}$ for $\mathfrak{j}$ and let $\{\check{H}_i\}$ denote the dual basis relative to the restriction of B to $\mathfrak{j}$; for each $\alpha \in \Phi$, choose a non-zero element $X_\alpha \in \mathfrak{g}^\alpha$ and normalize it by the condition $B(X_\alpha, X_{-\alpha}) = 1$ – the bracket $[X_\alpha, X_{-\alpha}]$ is then equal to the element H_α of $\mathfrak{j}$ which corresponds to α under the isomorphism between $\mathfrak{j}$ and its dual set up by B (in particular, $[H_\alpha, X_\alpha] = (\alpha, \alpha)X_\alpha$). Associated with B there is a Casimir element ω; as in 2.5.1, we shall denote by $\Omega_\Lambda(= \Omega_{E_\Lambda})$ the relevant endomorphism of $C(\mathfrak{g}, E_\Lambda)$ – it will be necessary to bear in mind that Ω_Λ is actually scalar multiplication by $|\Lambda + \rho|^2 - |\rho|^2$ (cf. Jacobson [3, p.254]).

For the moment, our discussion will be directed toward establishing the following fact: In $C(\mathfrak{g}, E_\Lambda)$, the endomorphism $\Gamma_{\mathfrak{g}}$ is homotopic to zero.

Let us start off by considering the 'Casimir endomorphism' $\omega_{\Lambda,s}$ of the $\mathfrak{l}$-module $C(\mathfrak{g}, E_\Lambda)$ associated with the restriction of $(.\,,.)$ to $\mathfrak{l}$ (which is non-degenerate). By definition, if θ is the canonical representation of $\mathfrak{g}$ on $C(\mathfrak{g}, E_\Lambda)$, then

$$\omega_{\Lambda,s} = \sum_{\alpha \in \Xi_s} \theta(X_{-\alpha})\theta(X_\alpha) + \sum_i \theta(H_i)\theta(\check{H}_i).$$

Let a_{Λ_s} be a primitive element of weight Λ_s in the $\mathfrak{l}$-module $C(\mathfrak{g}, E_\Lambda)$ for the Borel subalgebra of $\mathfrak{l}$ associated with Ξ_s^+ – then $\theta(X_\alpha)a_{\Lambda_s} = 0$ (all $\alpha \in \Xi_s^+$) and so $\sum_\alpha \theta(X_{-\alpha})\theta(X_\alpha)a_{\Lambda_s} = 0$, $\alpha \in \Xi_s^+$. This latter equality implies that

$$\begin{aligned} \sum_\alpha \theta(X_\alpha)\theta(X_{-\alpha})a_{\Lambda_s} &= \sum_\alpha \theta([X_\alpha, X_{-\alpha}])a_{\Lambda_s} + \sum_\alpha \theta(X_{-\alpha})\theta(X_\alpha)a_{\Lambda_s} \\ &= \sum_\alpha \theta(H_\alpha)a_{\Lambda_s} = \sum_\alpha \langle H_\alpha, \Lambda_s\rangle a_{\Lambda_s}, \end{aligned}$$

$\sum_\alpha$ denoting a sum over the α in Ξ_s^+ – hence

$$\sum_{\alpha \in \Xi_s} \theta(X_{-\alpha})\theta(X_\alpha)a_{\Lambda_s} = 2(\Lambda_s, \rho_s)a_{\Lambda_s}.$$

On the other hand, it is clear that

$$\sum_i \theta(H_i)\theta(\check{H}_i)a_{\Lambda_s} = \sum_i \langle H_i, \Lambda_s\rangle\langle \check{H}_i, \Lambda_s\rangle a_{\Lambda_s} = |\Lambda_s|^2 a_{\Lambda_s}.$$

Putting these observations together then leads to the conclusion that

$$\omega_{\Lambda,s}(a_{\Lambda_s}) = (|\Lambda_s|^2 + 2(\Lambda_s, \rho_s))a_{\Lambda_s} = (|\Lambda_s + \rho_s|^2 - |\rho_s|^2)a_{\Lambda_s};$$

Burnside's Theorem now tells us that on a given simple $\mathfrak{l}$-module

$$E_{\Lambda_s} \subset C(\mathfrak{g}, E_\Lambda),$$

the operator $\omega_{\Lambda,s}$ reduces to scalar multiplication by the number

$$|\Lambda_s + \rho_s|^2 - |\rho_s|^2.$$

As is easily seen, the endomorphism $\omega_{\Lambda,u} = \sum_{\alpha\in\Xi_u}\theta(H_\alpha)$ of $C(\mathfrak{g}, E_\Lambda)$ commutes with all the $\theta(X)(X \in \mathfrak{l})$, hence is an $\mathfrak{l}$-module endomorphism. If, as in the preceding paragraph, a_{Λ_s} is a primitive element of weight Λ_s in the $\mathfrak{l}$-module $C(\mathfrak{g}, E_\Lambda)$, then it is clear that we have $\omega_{\Lambda,u}(a_{\Lambda_s}) = \sum_{\alpha\in\Xi_u}\langle H_\alpha, \Lambda_s\rangle a_{\Lambda_s} = 2(\Lambda_s, \rho_u)a_{\Lambda_s}$; consequently (cf. supra) the operator $\omega_{\Lambda,u}$ reduces, on each simple $\mathfrak{l}$-module $E_{\Lambda_s} \subset C(\mathfrak{g}, E_\Lambda)$, to scalar multiplication by the number $2(\Lambda_s, \rho_u)$.

According to the results sketched out in 2.5.1, the relations

$$\theta(X) = (i_{\mathbf{k}}(X) \otimes 1)d_{\mathfrak{g}} + d_{\mathfrak{g}}(i_{\mathbf{k}}(X) \otimes 1),\ \theta(X)d_{\mathfrak{g}} = d_{\mathfrak{g}}\theta(X) \qquad (X \in \mathfrak{l})$$

obtain $(C(\mathfrak{g}, E_\Lambda) = \mathrm{Hom}_{\mathbf{k}}(\Lambda\mathfrak{g}, \mathbf{k}) \otimes_{\mathbf{k}} E_\Lambda \ldots)$ – hence

$$\begin{aligned}\omega_{\Lambda,s} = {} & [\sum_{\alpha\in\Xi_s}\theta(X_{-\alpha})(i_{\mathbf{k}}(X_\alpha) \otimes 1) + \sum_i \theta(H_i)(i_{\mathbf{k}}(\check{H}_i) \otimes 1)]\, d_{\mathfrak{g}} \\ & + d_{\mathfrak{g}}[\sum_{\alpha\in\Xi_s}\theta(X_{-\alpha})(i_{\mathbf{k}}(X_\alpha) \otimes 1) + \sum_i \theta(H_i)(i_{\mathbf{k}}(\check{H}_i) \otimes 1)].\end{aligned}$$

Let us agree to write X in place of $X_{E_\Lambda}(X \in \mathfrak{g})$; since $\theta(X) = \mathrm{ad}^\vee(X) \otimes 1 + 1 \otimes X$ (all $X \in \mathfrak{l}$), we have

$$\begin{aligned}&\sum_{\alpha\in\Xi_s}\theta(X_{-\alpha})(i_{\mathbf{k}}(X_\alpha) \otimes 1) + \sum_i \theta(H_i)(i_{\mathbf{k}}(\check{H}_i) \otimes 1) \\ &\quad = \sum_{\alpha\in\Xi_s}\mathrm{ad}^\vee(X_{-\alpha})i_{\mathbf{k}}(X_\alpha) \otimes 1 + \sum_{\alpha\in\Xi_s} i_{\mathbf{k}}(X_\alpha) \otimes X_{-\alpha} \\ &\quad\quad + \sum_i \mathrm{ad}^\vee(H_i)i_{\mathbf{k}}(\check{H}_i) \otimes 1 + \sum_i i_{\mathbf{k}}(\check{H}_i) \otimes H_i.\end{aligned}$$

Similar comments apply to the endomorphism $\omega_{\Lambda,u}$ – thus

$$\omega_{\Lambda,u} = \sum_{\alpha\in\Xi_u}\theta(X_\alpha)\theta(X_{-\alpha}) - \sum_{\alpha\in-\Xi_u}\theta(X_\alpha)\theta(X_{-\alpha}),$$

where

$$\begin{aligned}&\sum_{\alpha\in\Xi_u}\theta(X_\alpha)\theta(X_{-\alpha}) \\ &\quad = [\sum_{\alpha\in\Xi_u} i_{\mathbf{k}}(X_{-\alpha}) \otimes X_\alpha + \sum_{\alpha\in\Xi_u}\mathrm{ad}^\vee(X_\alpha)i_{\mathbf{k}}(X_{-\alpha}) \otimes 1]\, d_{\mathfrak{g}} \\ &\quad\quad + d_{\mathfrak{g}}[\sum_{\alpha\in\Xi_u} i_{\mathbf{k}}(X_{-\alpha}) \otimes X_\alpha + \sum_{\alpha\in\Xi_u}\mathrm{ad}^\vee(X_\alpha)i_{\mathbf{k}}(X_{-\alpha}) \otimes 1],\end{aligned}$$

and

$$\begin{aligned}&\sum_{\alpha\in\Xi_u}\theta(X_{-\alpha})\theta(X_\alpha) \\ &\quad = [\sum_{\alpha\in\Xi_u} i_{\mathbf{k}}(X_\alpha) \otimes X_{-\alpha} + \sum_{\alpha\in\Xi_u}\mathrm{ad}^\vee(X_{-\alpha})i_{\mathbf{k}}(X_\alpha) \otimes 1]\, d_{\mathfrak{g}} \\ &\quad\quad + d_{\mathfrak{g}}[\sum_{\alpha\in\Xi_u} i_{\mathbf{k}}(X_\alpha) \otimes X_{-\alpha} + \sum_{\alpha\in\Xi_u}\mathrm{ad}^\vee(X_{-\alpha})i_{\mathbf{k}}(X_\alpha) \otimes 1].\end{aligned}$$

(3) In $C(\mathfrak{g}, E_\Lambda)$, the endomorphism $\Gamma_\mathfrak{g}$ is homotopic to zero. [Retaining the preceding notations, let us put

$$\begin{aligned}
h'(\mathfrak{g}) &= \sum_{\alpha \in \Phi} i_{\mathbf{k}}(X_\alpha) \otimes X_{-\alpha} + \sum_i i_{\mathbf{k}}(\check{H}_i) \otimes H_i, \\
h'(\mathfrak{l}) &= \sum_{\alpha \in \Xi_s} i_{\mathbf{k}}(X_\alpha) \otimes X_{-\alpha} + \sum_i i_{\mathbf{k}}(\check{H}_i) \otimes H_i, \\
h''(\mathfrak{l}) &= \sum_{\alpha \in \Xi_s} \operatorname{ad}^\vee (X_{-\alpha}) i_{\mathbf{k}}(X_\alpha) + \sum_i \operatorname{ad}^\vee (H_i) i_{\mathbf{k}}(\check{H}_i), \\
h'_+ &= \sum_{\alpha \in \Xi_u} i_{\mathbf{k}}(X_{-\alpha}) \otimes X_\alpha, \\
h'_- &= \sum_{\alpha \in \Xi_u} i_{\mathbf{k}}(X_\alpha) \otimes X_{-\alpha}, \\
h''_+ &= \sum_{\alpha \in \Xi_u} \operatorname{ad}^\vee (X_\alpha) i_{\mathbf{k}}(X_{-\alpha}), \\
h''_- &= \sum_{\alpha \in \Xi_u} \operatorname{ad}^\vee (X_{-\alpha}) i_{\mathbf{k}}(X_\alpha).
\end{aligned}$$

Then

$$\begin{aligned}
\Omega_\Lambda &= h'(\mathfrak{g}) d_\mathfrak{g} + d_\mathfrak{g} h'(\mathfrak{g}), \\
\omega_{\Lambda, s} &= (h'(\mathfrak{l}) + h''(\mathfrak{l}) \otimes 1) d_\mathfrak{g} + d_\mathfrak{g}(h'(\mathfrak{l}) + h''(\mathfrak{l}) \otimes 1), \\
\omega_{\Lambda, u} &= (h'_+ - h'_- + (h''_+ - h''_-) \otimes 1) d_\mathfrak{g} + d_\mathfrak{g}(h'_+ - h'_- + (h''_+ - h''_-) \otimes 1).
\end{aligned}$$

The endomorphism $\Gamma_\mathfrak{g}$ admits the decomposition: $\Gamma_\mathfrak{g} = \Omega_\Lambda - \omega_{\Lambda, s} - \omega_{\Lambda, u}$. [To see this, write, as usual, $\rho = \rho_s + \rho_u$; it is easy to see that ρ_u is orthogonal to all the roots in Ξ_s – hence $|\rho|^2 = |\rho_s|^2 + |\rho_u|^2$. Similarly, if $\Lambda_s \in \mathfrak{W}_s^+$, then $|\Lambda_s + \rho|^2 = |(\Lambda_s + \rho_s) + \rho_u|^2 = |\Lambda_s + \rho_s|^2 + 2(\Lambda_s, \rho_u) + |\rho_u|^2$ – hence $|\Lambda + \rho|^2 - |\Lambda_s + \rho|^2 = |\Lambda + \rho|^2 - |\rho|^2 - (|\Lambda_s + \rho_s|^2 - |\rho_s|^2) - 2(\Lambda_s, \rho_u)$.] It follows, therefore, that

$$\Gamma_\mathfrak{g} = h d_\mathfrak{g} + d_\mathfrak{g} h$$

where $h = h' + h'' \otimes 1$, with $h' = h'(\mathfrak{g}) - h'(\mathfrak{l}) - h'_+ + h'_-$ and $h'' = h''_- - h''_+ - h''(\mathfrak{l})$.]

(4) There exists an endomorphism k' of $C(\mathfrak{n}^+, E_\Lambda)$ such that the diagram

$$\begin{array}{ccc}
C(\mathfrak{g}, E_\Lambda) & \xrightarrow{\ p\ } & C(\mathfrak{n}^+, E_\Lambda) \\
{\scriptstyle h'} \downarrow & & \downarrow {\scriptstyle k'} \\
C(\mathfrak{g}, E_\Lambda) & \xrightarrow[\ p\ ]{} & C(\mathfrak{n}^+, E_\Lambda)
\end{array}$$

is commutative. [This is obvious since $h' = 2h'_- = 2(\sum_{\alpha \in \Xi_u} i_{\mathbf{k}}(X_\alpha) \otimes X_{-\alpha})$.]

(5) There exists an endomorphism k'' of $C(\mathfrak{n}^+, \mathbf{k})$ such that the diagram

$$\begin{array}{ccc}
C(\mathfrak{g}, E_\Lambda) & \xrightarrow{\ p\ } & C(\mathfrak{n}^+, E_\Lambda) \\
{\scriptstyle h'' \otimes 1} \downarrow & & \downarrow {\scriptstyle k'' \otimes 1} \\
C(\mathfrak{g}, E_\Lambda) & \xrightarrow[\ p\ ]{} & C(\mathfrak{n}^+, E_\Lambda)
\end{array}$$

is commutative. [It suffices to just consider the case when $\Lambda = 0$ (so that $E_\Lambda = \mathbf{k}$); in this situation we have the identifications $C(\mathfrak{g}, \mathbf{k}) \sim \mathbf{\Lambda}\mathfrak{g}^\vee$, $C(\mathfrak{n}^+, \mathbf{k}) \sim \mathbf{\Lambda}(\mathfrak{n}^+)^\vee$. The bilinear form B sets up an isomorphism between $\mathfrak{g}$ and its dual $\mathfrak{g}^\vee$; thus, with each X in $\mathfrak{g}$, we may associate the linear form $B_X = B(X, .)$. Turning now to the proof of the assertion in question, let us note that it will be enough to verify that $h''(\operatorname{Ker}(p)) \subset \operatorname{Ker}(p)$ (the map $p: \mathbf{\Lambda}\mathfrak{g}^\vee \to \mathbf{\Lambda}(\mathfrak{n}^+)^\vee$ being surjective). The kernel of p is the two sided ideal $\mathfrak{a}$ in $\mathbf{\Lambda}\mathfrak{g}^\vee$ which is generated by $B_{\mathfrak{p}}$ (the orthogonal of $\mathfrak{n}^+$ in $\mathfrak{g}^\vee$ is $B_{\mathfrak{p}}$); the ideal $\mathfrak{a}$ is comprised of linear combinations of elements of the form $B_X \wedge B_{Y_1} \wedge \ldots \wedge B_{Y_n}$ ($X \in \mathfrak{p}$; $Y_1, \ldots, Y_n \in \mathfrak{g}$); since we want to show that p annihilates $h''(\mathfrak{a})$, let us begin by computing $i_{\mathbf{k}}(X_\alpha)(B_X \wedge B_{Y_1} \wedge \ldots \wedge B_{Y_n})$ ($\alpha \in \Phi$). [It will be useful to bear in mind that $h'' = h''_- - h''_+ - h''(\mathfrak{l}) = \sum_{\alpha \in \Xi_u} \mathrm{ad}^\vee (X_{-\alpha}) i_{\mathbf{k}}(X_\alpha) - \sum_{\alpha \notin \Xi_u} \mathrm{ad}^\vee (X_{-\alpha}) i_{\mathbf{k}}(X_\alpha) - \sum_i \mathrm{ad}^\vee (H_i) i_{\mathbf{k}}(\check{H}_i)$.] Thus we have

$$i_{\mathbf{k}}(X_\alpha)(B_X \wedge B_{Y_1} \wedge \ldots \wedge B_{Y_n}) = B(X, X_\alpha) B_{Y_1} \wedge \ldots \wedge B_{Y_n} \\ + \sum_k (-1)^k B(Y_k, X_\alpha) B_X \wedge B_{Y_1} \wedge \ldots \wedge \hat{B}_{Y_k} \wedge \ldots \wedge B_{Y_n}$$

and so

$$\sum_{\alpha \in \Xi_u} \mathrm{ad}^\vee (X_{-\alpha}) i_{\mathbf{k}}(X_\alpha)(B_X \wedge B_{Y_1} \ldots \wedge B_{Y_n}) \\ \underset{\mathfrak{a}}{\equiv} \sum_{\alpha \in \Xi_u; k} (-1)^k B(Y_k, X_\alpha) B_{[X_{-\alpha}, X]} \wedge B_{Y_1} \wedge \ldots \wedge \hat{B}_{Y_k} \wedge \ldots \wedge B_{Y_n},$$

$$\sum_{\alpha \notin \Xi_u} \mathrm{ad}^\vee (X_{-\alpha}) i_{\mathbf{k}}(X_\alpha)(B_X \wedge B_{Y_1} \wedge \ldots \wedge B_{Y_n}) \\ \underset{\mathfrak{a}}{\equiv} \sum_{\alpha \notin \Xi_u; k} B(X, X_\alpha) B_{Y_1} \wedge \ldots \wedge B_{[X_{-\alpha}, Y_k]} \wedge \ldots \wedge B_{Y_n} \\ \underset{\mathfrak{a}}{\equiv} \sum_{\alpha \notin \Xi_u; k} (-1)^{k-1} B(X, X_\alpha) B_{[X_{-\alpha}, Y_k]} \wedge B_{Y_1} \wedge \ldots \wedge \hat{B}_{Y_k} \wedge \ldots \wedge B_{Y_n}.$$

Because $\mathfrak{j} \subset \mathfrak{p}$, the preceding computation is applicable to $\sum_i \mathrm{ad}^\vee (H_i) i_{\mathbf{k}}(\check{H}_i)$; thus

$$\sum_i \mathrm{ad}^\vee (H_i) i_{\mathbf{k}}(\check{H}_i)(B_X \wedge B_{Y_1} \wedge \ldots \wedge B_{Y_n}) \\ \underset{\mathfrak{a}}{\equiv} \sum_{i,k} (-1)^{k-1} B(X, \check{H}_i) B_{[H_i, Y_k]} \wedge B_{Y_1} \wedge \ldots \wedge \hat{B}_{Y_k} \wedge \ldots \wedge B_{Y_n}.$$

Setting

$$Z(X, Y) = \sum_{\alpha \in \Xi_u} B(Y, X_\alpha)[X_{-\alpha}, X] + \sum_{\alpha \notin \Xi_u} B(X, X_\alpha)[X_{-\alpha}, Y] \\ + \sum_i B(X, \check{H}_i)[H_i, Y],$$

we see, therefore, that

$$h''(B_X \wedge B_{Y_1} \wedge \ldots \wedge B_{Y_n}) \\ \underset{\mathfrak{a}}{\equiv} \sum_k (-1)^k B_{Z(X, Y_k)} \wedge B_{Y_1} \wedge \ldots \wedge \hat{B}_{Y_k} \wedge \ldots \wedge B_{Y_n}.$$

In view of what has been established so far, then, the proof of our contention will be complete provided it is verified that $Z(X, Y)$ lies in $\mathfrak{p}$ whenever $X \in \mathfrak{p}$ and $Y \in \mathfrak{g}$. Because $Z(X, Y)$ depends linearly on X and Y, it may be assumed that $X = X_\gamma (\gamma \in \Xi_s \cup \Xi_u)$ or $X = H_i$ (resp. $Y = X_\alpha (\alpha \in \Phi)$ or $Y = H_i$). If $Y \in \mathfrak{p}$, then it is clear that $Z(X, Y) \in \mathfrak{p}$ (all $X \in \mathfrak{g}$); suppose next that $Y = X_{-\beta} (\beta \in \Xi_u)$, $\gamma \in \Xi_s \cup \Xi_u$ – then

$$Z(X_\gamma, X_{-\beta}) = \sum_{\alpha \in \Xi_u} B(X_{-\beta}, X_\alpha)[X_{-\alpha}, X_\gamma] + \sum_{\alpha \notin \Xi_u} B(X_\gamma, X_\alpha)[X_{-\alpha}, X_{-\beta}] \\ + \sum_i B(X_\gamma, \check{H}_i)[H_i, X_{-\beta}] = [X_{-\beta}, X_\gamma] + [X_\gamma, X_{-\beta}] = 0.$$

Entirely analogous considerations show that $Z(H_i, X_{-\beta}) = 0$ (all $\beta \in \Xi_u$). Hence (5).]

Putting together statements (1)–(5) now leads at once to the conclusion of our lemma.

Chapter 3

Finite Dimensional Representations of a Semi-Simple Lie Group

Introduction

Let G be a connected semi-simple Lie group with finite center, say – then it is not necessarily the case that G possesses enough finite dimensional representations to separate points (cf. 3.1.1). In fact G will, in general, admit no non-trivial finite dimensional unitary representations (cf. number 4.3.2). Nevertheless, despite these apparently discouraging facts, the finite dimensional representations of G are important – this will be elaborated on below.

We begin, in 3.1, with a discussion of the irreducible holomorphic representations of G_c, G_c a complex connected semi-simple Lie group. If $\mathfrak{u}$ is a compact real form for $\mathfrak{g}_c$, then an application of the Peter-Weyl Theorem to the real analytic subgroup U of G_c corresponding to $\mathfrak{u}$ yields without difficulty the important fact that G_c is a subgroup of $\mathbf{SL}(n, \mathbf{C})$ for some positive integer n, i.e. that G_c possesses a faithful linear holomorphic representation (cf. Theorem 3.1.1.3). Suppose in particular that G_c is simply connected – then the irreducible holomorphic representations U_Λ of G_c are parameterized by the dominant integral linear functions Λ on $\mathfrak{j}_c$ ($\mathfrak{j}_c$ a Cartan subalgebra of $\mathfrak{g}_c$). The algebraic version of Weyl's character formula then immediately gives a formula on J_c' for the character ξ_Λ of U_Λ (J_c the Cartan subgroup of G_c associated with $\mathfrak{j}_c$). The famous Borel-Weil Theorem is the subject of 3.1.2. This theorem says, essentially, that the irreducible holomorphic representations of G_c may be 'realized' on the holomorphic sections of certain holomorphic vector bundles or, more generally, on certain cohomology groups. The basic idea which underlies the general theory of induced representations (Chapter 5) may be found here already, namely represent the group on suitable sections of a vector bundle or even on suitable cohomology groups attached in some way or another to the bundle.

In 3.2 we focus our attention on the representation theory of a compact connected semi-simple Lie group. Our primary objective here is to derive the Plancherel Theorem for a compact G via the same method that Harish-Chandra uses to determine the contribution of the discrete series to the Plancherel formula in the case of a non-compact G; the heuristics in this situation are simple and instructive – moreover the invariant integral puts in a natural appearance both on the algebra and on the group.

The finite dimensional representations of a (real) connected semi-simple Lie group G (with finite center) are considered in 3.3. For the sake of simplicity assume that G is imbedded in its simply connected complexification G_c – then the finite dimensional representations of G may be identified with those of G_c. Fix a Cartan decomposition $\mathfrak{g} = \mathfrak{k} + \mathfrak{p}$ of $\mathfrak{g}$ and a maximal abelian subspace $\mathfrak{a}_\mathfrak{p} \subset \mathfrak{p}$; extend $\mathfrak{a}_\mathfrak{p}$ to a Cartan subalgebra $\mathfrak{a}$ of $\mathfrak{g}$ and by complexification to a Cartan subalgebra $\mathfrak{a}_c$ of $\mathfrak{g}_c$. By definition, a finite dimensional irreducible representation U_Λ of G on E_Λ (Λ dominant integral with respect to $\mathfrak{a}_c$) is said to be of class one relative to the pair (G, K) if the space E_Λ admits a K-invariant vector (i.e. if the trivial one dimensional representation of K occurs in $U_\Lambda | K$). A theorem due to É. Cartan, in the formulation of S. Helgason, characterizes those dominant integral Λ such that U_Λ is class one; using this, one may then establish without difficulty an important Criterion, due to Harish-Chandra, which serves to accomplish, in a remarkable way, the transcription of a geometrical condition (viz. that the scalar product of a vector H in $\mathfrak{a}_\mathfrak{p}$ with the elements of the positive Weyl chamber be non-negative) to a purely algebraic one (viz. that the highest weights of the class one representations of the pair (G, K) assume non-negative values at H). Now the powerful methods of representation theory can be employed to deduce various important inequalities (cf. 3.3.2), inequalities which play a fundamental role in the estimation of spherical functions, in the technique of descent, and in the theory of automorphic functions.

Throughout the present chapter we shall adhere to the following convention.

Convention Let G be a topological group – then by the term 'representation of G' we shall understand a *continuous* homomorphism of G into the group of automorphisms of a finite dimensional complex vector space.

3.1 Holomorphic Representations of a Complex Semi-Simple Lie Group

3.1.1 Generalities

Let $\mathfrak{g}$ be a Lie algebra over $\mathbf{R}$ (not necessarily semi-simple) and suppose that $\mathfrak{g}$ admits a complex structure J. Let G be a connected Lie group with Lie algebra $\mathfrak{g}$ – then G may be viewed as a complex Lie group.

A representation U of G on E is said to be *holomorphic* if, for all $a \in E$, $\mathbf{z} \in E^\vee$, the function $x \mapsto \langle U(x)a, \mathbf{z}\rangle$ is holomorphic on G. Thus U is holomorphic iff $dU \circ J(X) = \sqrt{-1} \circ dU(X)$ for all $X \in \mathfrak{g}$. [Recall that a necessary and sufficient condition that a differentiable map $f: G \rightarrow \mathbf{C}$ be holomorphic is that it be *almost complex*, i.e. $df \circ J = \sqrt{-1} \circ df$.] Let $\mathfrak{g}_c$ denote the complexification of $\mathfrak{g}$ and extend J by linearity to $\mathfrak{g}_c$; one has $J^2 = -1$, $[JX, Y] = J[X, Y]$ in $\mathfrak{g}$ and so these formulae persist in $\mathfrak{g}_c$. Let $\mathfrak{g}_c^+ = \{X \in \mathfrak{g}_c : JX = \sqrt{-1}X\}$, $\mathfrak{g}_c^- = \{X \in \mathfrak{g}_c : JX = -\sqrt{-1}X\}$ – then $\mathfrak{g}_c^+$, $\mathfrak{g}_c^-$ are ideals in $\mathfrak{g}_c$ and $\mathfrak{g}_c = \mathfrak{g}_c^+ + \mathfrak{g}_c^-$ (direct sum). Therefore the holomorphic representations of G lift to complex linear representations of $\mathfrak{g}_c$ which vanish on $\mathfrak{g}_c^-$ (for such representations are zero on the elements $JX - \sqrt{-1}X$ and these elements make up $\mathfrak{g}_c^-$). Likewise the anti-holomorphic representations of G (in the obvious sense) give rise to complex linear representations of $\mathfrak{g}_c$ which vanish on $\mathfrak{g}_c^+$.

Proposition 3.1.1.1 Every irreducible representation of a simply connected complex Lie group on a complex vector space is isomorphic to the tensor product of an irreducible holomorphic representation and an irreducible anti-holomorphic representation.

Let $\mathfrak{G}$, $\mathfrak{G}^+$, $\mathfrak{G}^-$ denote the universal enveloping algebras of $\mathfrak{g}_c$, $\mathfrak{g}_c^+$, $\mathfrak{g}_c^-$, respectively. In the present situation $\mathfrak{G}$ is isomorphic as an algebra to

$\mathfrak{G}^+ \otimes_{\mathbf{C}} \mathfrak{G}^-$ – therefore Proposition 3.1.1.1 is a consequence of the following lemma.

Lemma 3.1.1.2 Let A_1, A_2 be associative algebras over $\mathbf{C}$ (with a unit); set $A = A_1 \otimes_{\mathbf{C}} A_2$ – then every finite dimensional irreducible representation of A is the tensor product of an irreducible representation of A_1 with an irreducible representation of A_2.

[In view of Burnside's Theorem the converse is also true.]

Proof Let U be an irreducible representation of A on E; let E_1 be a minimal A_1-stable subspace of E, F the space of A_1-homomorphisms f of E_1 into E. Define a representation of A_2 on F by the rule $f \mapsto U(x_2) \circ f$ $(x_2 \in A_2, f \in F)$. Let F_2 be a minimal A_2-invariant subspace of F. The map $a \otimes f \mapsto f(a)$ of $E_1 \otimes_{\mathbf{C}} F_2$ into E is non-zero and commutes with the operations of A_1 and A_2 by construction. By Burnside's Theorem, the representation of $A_1 \otimes_{\mathbf{C}} A_2$ on $E_1 \otimes_{\mathbf{C}} F_2$ is irreducible; Schur's Lemma then implies that E is equivalent to $E_1 \otimes_{\mathbf{C}} F_2$ as an A-module. □

[In passing let us observe that the decomposition of a given simple A-module into the tensor product of a simple A_1-module with a simple A_2-module is essentially unique (the Jordan-Hölder Theorem . . .).]

Henceforth it will be assumed that $\mathfrak{g}$ is semi-simple (but carries a complex structure J, say). Let $\mathfrak{u}$ be a compact real form for the associated complex Lie algebra $\mathfrak{g}^C$ – then $\mathfrak{g} = \mathfrak{u} + J\mathfrak{u}$ is a Cartan decomposition of $\mathfrak{g}$, and $\mathfrak{u}_c$ is a semi-simple subalgebra of $\mathfrak{g}_c$ which is isomorphic to $\mathfrak{g}^C$. Define maps

$$i^+ : \mathfrak{u}_c \to \mathfrak{g}_c,\ i^- : \mathfrak{u}_c \to \mathfrak{g}_c$$

by $i^+(X) = \frac{1}{2}(X - \sqrt{-1}J(X))$, $i^-(X) = \frac{1}{2}(X + \sqrt{-1}J(X))$ $(X \in \mathfrak{u}_c)$ and put $\mathfrak{u}_c^+ = i^+(\mathfrak{u}_c)$, $\mathfrak{u}_c^- = i^-(\mathfrak{u}_c)$ – then i^+, i^- are isomorphisms of $\mathfrak{u}_c$ onto $\mathfrak{u}_c^+$, $\mathfrak{u}_c^-$, respectively, with $\mathfrak{u}_c^+ = \mathfrak{g}_c^+$, $\mathfrak{u}_c^- = \mathfrak{g}_c^-$.

Theorem 3.1.1.3 Every connected semi-simple complex Lie group admits a faithful linear holomorphic representation.

Proof Let G be a connected semi-simple (complex) Lie group with (real) Lie algebra $\mathfrak{g}$, U the real analytic subgroup of G corresponding to the compact real form $\mathfrak{u}$ of $\mathfrak{g}^C$. We are to produce a holomorphic representation $x \mapsto \mu(x)$, say, of G on a finite dimensional (complex) vector space E with the property that $x \neq 1 \Rightarrow \mu(x) \neq 1$ $(x \in G)$. Let $G^\prime$ denote the universal covering group of G, $U^\prime$ the real analytic subgroup of $G^\prime$ corresponding to $\mathfrak{u}$. Because $U^\prime$ is compact, it follows from the Peter-Weyl Theorem that $U^\prime$ admits a faithful (analytic) representation on a real finite dimensional vector space; by passing to the Lie algebra $\mathfrak{u}$, thence by complex linearity to $\mathfrak{u}_c \sim \mathfrak{g}^C$, and finally back to $G^\prime$ via simple connectivity, we obtain a holomorphic representation $\mu^{\natural}$, say, of $G^{\natural}$ on a finite dimensional complex vector space $E^{\natural}$, say, whose restriction to $U^{\natural}$ is still faithful. The center $Z^{\natural}$ of $G^{\natural}$ is contained in $U^{\natural}$; consequently, the direct sum of $\mu^{\natural}$ and the adjoint representation $\mathrm{Ad}^{\natural}$ gives a faithful holomorphic representation $\mu^{\natural} + \mathrm{Ad}^{\natural}$ of $G^{\natural}$ on $E^{\natural} + \mathfrak{g}^{C}$. Call $\phi: G^{\natural} \to G$ the covering map and K_{ϕ} the kernel of ϕ – then, of course, K_{ϕ} is a finite central subgroup of $G^{\natural}$. This being so, a well-known elementary lemma (cf. Hochschild [1, p. 199]) then ensures us that we may find a finite dimensional subspace E, say, of the tensor space $T(E^{\natural} + \mathfrak{g}^{C})$ over $E^{\natural} + \mathfrak{g}^{C}$ which is stable for the action of $\mu^{\natural} + \mathrm{Ad}^{\natural}$ and with the property that the corresponding representation μ, say, of $G^{\natural}$ on E has K_{ϕ} for its kernel. Since μ is holomorphic and $G^{\natural}/K_{\phi} \sim G$, the contention of our theorem is now clear. □

Remarks (1) Let G be a connected semi-simple Lie group with Lie algebra $\mathfrak{g}$ but drop for the moment the assumption that $\mathfrak{g}$ carries a complex structure – then the existence of a faithful finite dimensional representation of G is equivalent to the existence of sufficiently many finite dimensional representations. One direction of this assertion is clear. As for the other, let $N(G)$ denote the intersection of the kernels of all finite dimensional representations of G – then it was shown by Harish-Chandra [2, p.320] (see below) that G admits a representation with kernel $N(G)$. Hence, if G possesses sufficiently many finite dimensional representations, then G necessarily admits a faithful finite dimensional representation.

(2) Suppose that G is as in (1); then, if G admits a faithful finite dimensional representation U, the center of G is necessarily finite. Thus write $U = U_1 + \ldots + U_r$, the $U_i (1 \leqslant i \leqslant r)$ being irreducible. Let x_0 be an element in the center of G. By Schur's Lemma, $U_i(x_0) = c_i I_i$, $c_i \in \mathbf{C}$ (I_i the unit matrix of degree d_i, d_i the dimension of U_i). The mapping $x \mapsto \det(U_i(x))$ is an abelian representation of G – therefore $\det(U_i(x)) = 1$, G being connected and semi-simple. It follows that $c_i = \exp((m_i/d_i)2\pi\sqrt{-1})$ where $0 \leqslant m_i < d_i$ ($m_i \in \mathbf{Z}$) – this shows that there are only finitely many possibilities for x_0.

Example It is not true that a connected semi-simple Lie group with finite center necessarily admits a faithful finite dimensional representation (i.e. the converse to Remark 2 fails to hold in general). To see this, take G to be the universal covering group $\mathbf{SL}(3, \mathbf{R})^{\natural}$ of $\mathbf{SL}(3, \mathbf{R})$ – then the 'K' in $G = KA_{\mathfrak{p}}N^{+}$ may be identified with $\mathbf{Spin}(3)$ which is compact. But $\mathbf{SL}(3, \mathbf{R})^{\natural}$ contains a non-abelian Cartan subgroup (compare 1.4.1) and so cannot possibly admit a faithful finite dimensional representation (cf. Theorem 1.4.1.5).

[The following observation will help to clarify the preceding Example. Let $\mathfrak{g}$ be a semi-simple Lie algebra over $\mathbf{R}$, G a connected Lie group with Lie algebra $\mathfrak{g}$; let $G^{\natural}$ denote the simply connected Lie group corresponding to $\mathfrak{g}$, $G_c^{\natural}$ the simply connected Lie group corresponding to $\mathfrak{g}_c$ – the kernel in $G^{\natural}$ of the homomorphism $i_c: G^{\natural} \to G_c^{\natural}$ whose differential is the injection of $\mathfrak{g}$ into $\mathfrak{g}_c$ will be denoted by K_{i_c}. Let $\phi: G^{\natural} \to G$ be the covering map, K_{ϕ} the kernel of ϕ – then we claim that $N(G) = \phi(K_{i_c})$. For in the first place, it is obvious that $\phi(K_{i_c}) \subset N(G)$ – thus the claim will follow if we can produce a representation of G whose kernel is precisely $\phi(K_{i_c})$. To this end, first note that $G_c^{\natural}/i_c(K_{\phi})$ is a connected semi-simple complex Lie group ($i_c(K_{\phi})$ is a finite central subgroup), and so, by Theorem 3.1.1.3, admits a faithful linear holomorphic representation U; we may regard U as a holomorphic representation of $G_c^{\natural}$ with kernel $i_c(K_{\phi})$. But then the kernel of the representation $U \circ i_c$ of $G^{\natural}$ is $K_{\phi}K_{i_c}$ – therefore

$U \circ i_c$ induces a representation of G whose kernel is $\phi(K_{i_c})$. Hence the claim. This observation shows immediately that $\mathbf{SL}(3, \mathbf{R})^\natural$ does not possess a faithful finite dimensional representation; likewise it is also clear, for instance, that the finite covering groups of $\mathbf{SL}(2, \mathbf{R})$ are not faithfully representable even though they have finite center.]

Remark Let G be a compact Lie group, K a closed subgroup – it is classical that every irreducible representation of K is contained in some irreducible representation of G (see, for example, Chevalley [1, p.211]). Suppose now that G is a connected semi-simple Lie group, K a maximal compact subgroup of G. Later on we shall establish the following fact. Define $N(G)$ as above – then an irreducible representation $k \mapsto U(k)$ of K is contained in some irreducible (finite dimensional) representation of G iff $U(k) = 1$ for every $k \in K \cap N(G)$ (cf. Proposition 4.5.2.8).

We return now to our earlier assumptions and conventions. It will be, however, convenient to alter the notation slightly; thus in what follows $\mathfrak{g}_c$ will denote a complex semi-simple Lie algebra, G_c a connected complex analytic group with Lie algebra $\mathfrak{g}_c$ (so that $\mathfrak{g}_c$ is playing the role of $\mathfrak{g}^C$ above while G_c is playing the role of G).

As before, let $\mathfrak{u}$ be a compact real form for $\mathfrak{g}_c$, U the (real) analytic subgroup of G_c corresponding to $\mathfrak{u}$. Let H_c be a connected complex Lie group with Lie algebra $\mathfrak{h}_c$. Let $\mathrm{Hom}_{\mathbf{C}}(G_c, H_c)$ denote the set of holomorphic homomorphisms of G_c into H_c, $\mathrm{Hom}_{\mathbf{R}}(U, H_c)$ the set of real analytic homomorphisms of U into H_c – then, owing to the fact that $\pi_1(G_c) = \pi_1(U)$, the operation of restriction

$$\mathrm{Hom}_{\mathbf{C}}(G_c, H_c) \rightarrow \mathrm{Hom}_{\mathbf{R}}(U, H_c)$$

is bijective. This implies that G_c is determined to within isomorphism by U. [Conversely, it is true that given a compact connected semi-simple Lie group U there exists a connected semi-simple complex Lie group G_c which contains U as a maximal compact subgroup – cf. Chevalley [1, p.200] (of course G_c is essentially unique).]

Suppose in particular that G_c is simply connected. The well-known 'unitary trick' of Weyl may be expressed quite succinctly via the following diagram (cf. Serre [4, p.IV-10]):

$$\begin{array}{ccc} \mathrm{Hom}_{\mathbf{C}}(G_c, H_c) & \xrightarrow{a} & \mathrm{Hom}_{\mathbf{R}}(U, H_c) \\ {\scriptstyle b}\downarrow & & \downarrow{\scriptstyle c} \\ \mathrm{Hom}_{\mathbf{C}}(\mathfrak{g}_c, \mathfrak{h}_c) & \xrightarrow[d]{} & \mathrm{Hom}_{\mathbf{R}}(\mathfrak{u}, \mathfrak{h}_c). \end{array}$$

Here a and d are restriction maps while b and c have the obvious meaning – the point is that these canonical maps are *bijections*. In particular: The study of holomorphic representations of G_c is entirely equivalent to the study of representations of U on complex vector spaces.

Let $\mathfrak{t}$ be a maximal abelian subalgebra of $\mathfrak{u}$; extend $\mathfrak{t}$ to a Cartan subalgebra $\mathfrak{j}_c$ of $\mathfrak{g}_c$ and fix some order for the roots Φ of the pair $(\mathfrak{g}_c, \mathfrak{j}_c)$. Let J_c denote the subgroup (complex) of G_c corresponding to $\mathfrak{j}_c$.

By a *holomorphic character* of J_c we mean a holomorphic homomorphism $\xi : J_c \to \mathbf{C}^*$, $\mathbf{C}^*$ the multiplicative group of non-zero complex numbers. The group $\hat{J}_c$ of holomorphic characters of J_c is a free abelian group of rank equal to $\dim(J_c)$; observe that $J_c = \mathrm{Hom}(\hat{J}_c, \mathbf{C}^*)$. If $\xi \in \hat{J}_c$, then $d\xi \in \mathrm{Hom}_{\mathbf{C}}(\mathfrak{j}_c, \mathbf{C})$ – thus the map $\xi \mapsto d\xi$ is an injection of $\hat{J}_c$ into $\mathfrak{j}_c^\vee$ by means of which $\hat{J}_c$ will be identified with a subset of $\mathfrak{j}_c^\vee$.

Let $\mathfrak{w}$ denote the subgroup of $\mathfrak{j}_c^\vee$ generated by the roots, $\mathfrak{W}$ the integral linear functions on $\mathfrak{j}_c$. Noting that $\hat{\mathsf{L}}(G_c)$ may be regarded as the dual of $\hat{J}_c$, let us set $\mathsf{L}(G_c) = \hat{J}_c$ and then recast Propositions 1.2.2.1, 1.2.2.2 as follows:

(1) The group $\mathsf{L}(G_c)$ lies between $\mathfrak{w}$ and $\mathfrak{W}$;

(2) Conversely every subgroup of $\mathfrak{j}_c^\vee$ which lies between $\mathfrak{w}$ and $\mathfrak{W}$ is of the form $\mathsf{L}(G_c)$ for a convenient choice of G_c (G_c being, of course, unique to within isomorphism).

Proposition 3.1.1.4 Let U be a representation of $\mathfrak{g}_c$ on a finite dimensional vector space E. In order that U correspond to a (holomorphic) representation of G_c on E, it is both necessary and sufficient that the weights of U belong to $\mathsf{L}(G_c)$.

Proof The necessity is obvious. As for the sufficiency, let $G_c^!$ denote the universal covering group of G_c, $\phi : G_c^! \to G_c$ the covering map, K_ϕ the kernel of ϕ. Because U drops to a representation of $G_c^!$, it is enough to show that U passes to the quotient through K_ϕ; but K_ϕ may be identified with $\pi_1(G_c)$ (cf. Chevalley [1, p. 54]) and $\pi_1(G_c)$ may be identified with $\hat{\mathsf{L}}(G_c)/\hat{\mathfrak{W}}$ – the assertion is an immediate consequence of these facts. ☐

Remark If U is irreducible with highest weight Λ, then the conditions of Proposition 3.1.1.4 are met provided $\Lambda \in \mathsf{L}(G_c)$ – for the other weights are of the form $\Lambda - \sum_i n_i\alpha_i$, $\{\alpha_1, \ldots, \alpha_l\}$ the fundamental system of roots for Φ which determines the ordering ($n_i \in \mathbf{Z}$).

Suppose for simplicity that G_c is simply connected – then the irreducible (holomorphic) representations of G_c are parameterized by $\mathfrak{W}^+$, $\mathfrak{W}^+$ the dominant integral linear functions on $\mathfrak{j}_c$. So fix a $\Lambda \in \mathfrak{W}^+$; let U_Λ denote the corresponding irreducible holomorphic representation of G_c on E_Λ, say; put $\xi_\Lambda(x) = \mathrm{tr}(U_\Lambda(x))$ ($x \in G_c$) – then ξ_Λ will be called the *character* of U_Λ.

It is easy to give a formula for ξ_Λ on the regular set G_c'. Thus, since $G_c' = \bigcup_{x \in G_c} xJ_c'x^{-1}$ (cf. Theorem 1.4.1.7), it suffices to determine ξ_Λ on J_c'; for this latter purpose, let $j \in J_c'$ and write $j = \exp H$ ($H \in \mathfrak{j}_c$) – then Weyl's character formula leads at once to the relation

$$\xi_\Lambda(j) = \frac{\sum_{w \in W} \det(w)\, e^{\langle H, w(\Lambda+\rho)\rangle}}{\sum_{w \in W} \det(w)\, e^{\langle H, w(\rho)\rangle}} \qquad \left(\rho = \tfrac{1}{2}\sum_{\alpha>0}\alpha\right),$$

W the Weyl group of the pair $(\mathfrak{g}_c, \mathfrak{j}_c)$.

3.1.2 The Borel-Weil Theorem

Let $\mathfrak{g}_c$ be a complex semi-simple Lie algebra, G_c a connected Lie group with Lie algebra $\mathfrak{g}_c$ – then we shall agree to use without comment the notations and conventions which were introduced in the preceding number. Suppose now, for the sake of simplicity, that G_c is simply connected – then the irreducible holomorphic representations of G_c (or equivalently the irreducible unitary representations of U on complex vector spaces) are parameterized by the dominant integral linear functions $\mathfrak{W}^+$ on $\mathfrak{j}_c$ ($\mathfrak{j}_c$ being the complexification of a maximal abelian subalgebra $\mathfrak{t}$ of $\mathfrak{u}$). The problem that will be considered in the present number is this: Give a concrete geometric realization of these representations, i.e. exhibit model spaces for them – this is the content of the Borel-Weil Theorem.

It will be necessary to assume that the reader is familiar with the elements of sheaf-cohomology theory as well as the theory of holomorphic vector bundles – the book of Hirzebruch [3] is a good source for this material (see especially paragraphs 2 and 15).

The following notations will be employed.

Notations (1) Let M be a paracompact complex analytic manifold.

(a) The holomorphic vector bundle of covariant q-vectors of M will be denoted by $\mathbf{\Lambda}^q\mathbf{T}$ while $\overline{\mathbf{\Lambda}^q\mathbf{T}} = \mathbf{\Lambda}^q\bar{\mathbf{T}}$ will stand for the differentiable vector bundle over M which is conjugate to $\mathbf{\Lambda}^q\mathbf{T}$.

(b) Let $\mathfrak{S}$ be a sheaf over M – then the corresponding cohomology groups of M with coefficients in the sheaf $\mathfrak{S}$ will be denoted by $H^q(M, \mathfrak{S})$ $(q = 0, 1, \ldots)$.

(2) Let $\mathbf{E}$ be a holomorphic vector bundle over M.

(a) The cohomology groups of M with coefficients in the sheaf of germs of local holomorphic sections of $\mathbf{E}$ will be denoted, for brevity, by $H^q(M, \mathbf{E})$. More generally, write $H^{p,q}(M, \mathbf{E})$ for the cohomology groups $H^q(M, \mathbf{E} \otimes \mathbf{\Lambda}^p\mathbf{T})$ – thus $H^{0,q}(M, \mathbf{E}) = H^q(M, \mathbf{E})$ $(q = 0, 1, \ldots)$.

(b) Let $C^{p,q}(M, \mathbf{E}) = \mathbf{C}$-module of global (p, q)-forms over M with coefficients in $\mathbf{E}$, i.e. the global differentiable ($= C^\infty$) sections of the differentiable vector bundle $\mathbf{E} \otimes \mathbf{\Lambda}^p\mathbf{T} \otimes \overline{\mathbf{\Lambda}^q\mathbf{T}}$. A well-known theorem of Dolbeault says that

$$H^{p,q}(M, \mathbf{E}) \sim Z^{p,q}(M, \mathbf{E})/\bar{\partial}(C^{p,q-1}(M, \mathbf{E}))$$

where $Z^{p,q}(M, \mathbf{E})$ is the subspace of $C^{p,q}(M, \mathbf{E})$ on which $\bar{\partial}$ vanishes.

Fix a Borel subalgebra $\mathfrak{b}_c$ of $\mathfrak{g}_c$ containing the Cartan subalgebra $\mathfrak{j}_c$. Let Φ^+ be the set of positive roots of the pair $(\mathfrak{g}_c, \mathfrak{j}_c)$ singled out by this choice of $\mathfrak{b}_c$, Ψ the fundamental system of roots for Φ contained in Φ^+; let $\mathfrak{p}_c$ be a $\mathfrak{b}_c$-parabolic subalgebra of $\mathfrak{g}_c$, Θ the subset of Ψ associated with $\mathfrak{p}_c$ in the usual way, $\mathfrak{l}_c$ the corresponding Levi subalgebra of $\mathfrak{p}_c$, $\mathfrak{n}_c^+$

the nilpotent radical of $\mathfrak{p}_c$; assign to the symbols W, W_s, W_u, $\mathfrak{W}$, $\mathfrak{W}_s$, $\mathfrak{W}^+$, $\mathfrak{W}_s^+$, Ξ_s and Ξ_u the customary meanings and write W_s^+ for $\mathfrak{W}/W_s$.

Let P_c be the parabolic subgroup of G_c corresponding to $\mathfrak{p}_c$; consider the space $M = G_c/P_c$ – as was pointed out in number 1.2.2, G_c/P_c is a compact complex Kähler manifold. We also noticed there that the homogeneous space G_c/P_c admits a 'compact' description – thus the maximal compact subgroup U of G_c acts transitively on M and so M can be identified with the quotient space U/L ($L = U \cap P_c$). [We recall too that the Lie algebra of L is a real form for $\mathfrak{l}_c$.]

Let L_c (respectively N_c^+) denote the analytic subgroup of G_c corresponding to $\mathfrak{l}_c$ (respectively $\mathfrak{n}_c^+$) – then the parabolic subgroup P_c admits the Levi decomposition $P_c = L_c N_c^+$. The group L_c is, in general, not semi-simple; but it is a connected complex analytic reductive Lie group whose irreducible holomorphic representations are parameterized by the elements $\Lambda_s \in \mathsf{W}_s^+$ (cf. Proposition 3.1.1.4). Furthermore the irreducible holomorphic representations of P_c are parameterized by W_s^+. [Thus, if U_{Λ_s} is an irreducible holomorphic representation of L_c on E_{Λ_s} (say), then it may be extended to P_c by rendering it trivial on N_c^+; the converse is a simple consequence of the fact that N_c^+ is a normal, nilpotent subgroup of P_c with $\mathfrak{n}_c^+ \subset [\mathfrak{p}_c, \mathfrak{p}_c]$ (cf. Lemma 5.5.1.3).] Alternatively (but equivalently) the irreducible holomorphic representations of P_c are in a 'one-to-one' correspondence with the irreducible representations of L. [Thus a given irreducible representation of L determines a unique representation of $\mathfrak{l}_c$ which itself extends to a representation of $\mathfrak{p}_c$ by letting $\mathfrak{n}_c^+$ act trivially; this infinitesimal representation drops to an irreducible holomorphic representation of P_c (the fundamental group of P_c being equal to that of L); the resulting (irreducible) holomorphic extension of the given irreducible representation of L is unique . . . .]

Fix an irreducible holomorphic representation U_{Λ_s} of P_c on E_{Λ_s} ($\Lambda_s \in \mathsf{W}_s^+$). Viewing $M(= G_c/P_c)$ as the base space of the principal holomorphic fiber bundle with bundle space G_c and fiber P_c, form, in the usual way, the associated holomorphic vector bundle $\mathbf{E}_{\Lambda_s}$ with fiber E_{Λ_s} – thus the bundle space of $\mathbf{E}_{\Lambda_s}$ is $G_c \times E_{\Lambda_s}$ modulo the equivalence relation $(xp, a) \sim (x, U_{\Lambda_s}(p)a)$ ($x \in G_c$, $p \in P_c$; $a \in E_{\Lambda_s}$). This equivalence relation is preserved under the mapping $(y, a) \mapsto x\cdot(y, a) = (xy, a)$ ($x, y \in G_c$; $a \in E_{\Lambda_s}$); otherwise said, $\mathbf{E}_{\Lambda_s}$ is a G_c-homogeneous vector bundle. The space of holomorphic, respectively C^∞, sections of $\mathbf{E}_{\Lambda_s}$ may be identified with the space of those E_{Λ_s}-valued holomorphic, respectively C^∞, functions f on G_c which verify the relation $f(xp) = U_{\Lambda_s}(p^{-1})f(x)$ (all $x \in G_c$, $p \in P_c$). [This construction is, of course, nothing more than a variant of the procedure commonly encountered in the theory of induced representations (cf. number 5.1.1); since we are not interested in unitary representations here, there is no need to take into account the square root of the modular function.] On the other hand, as a C^∞ vector bundle, $\mathbf{E}_{\Lambda_s}$ is associated with the principal bundle $L \to U \to M$ via the restriction of U_{Λ_s} to L; the C^∞ sections of $\mathbf{E}_{\Lambda_s}$ may then be identified with the C^∞ functions $f: U \to E_{\Lambda_l}$ such that $f(ul) = U_{\Lambda_l}(l^{-1})f(u)$ for all $u \in U, l \in L$; the space of holomorphic sections of $\mathbf{E}_{\Lambda_l}$ is isomorphic to the space of E_{Λ_l}-valued C^∞ functions f on U which have the property that $Xf = -U_{\Lambda_l}(X)f$ (all $X \in \mathfrak{p}_c$) (this description is entirely analogous to that of the 'intrinsic' complex structure on M – cf. number 1.1.2).

In the natural way (by left translation) the action of G_c on $\mathbf{E}_{\Lambda_l}$ induces a G_c-module structure on the cohomology groups

$$H^q(M, \mathbf{E}_{\Lambda_l})\ (q = 0, 1, \ldots).$$

[Of course for $q > m$, m = complex dimension of M, $H^q(M, \mathbf{E}_{\Lambda_l}) = \{0\}$ – moreover, since M is compact, the vector spaces $H^q(M, \mathbf{E}_{\Lambda_l})$ are finite dimensional (cf. Hirzebruch [3, p.122]).] Let $U^q_{\Lambda_l}$ denote the corresponding representation (holomorphic) of G_c (or $\mathfrak{g}_c$) on $H^q(M, \mathbf{E}_{\Lambda_l})$; our goal now is to determine the decomposition of $U^q_{\Lambda_l}$ into irreducible constituents.

The analysis of $U^q_{\Lambda_l}$ $(q = 0, 1, \ldots)$ depends on Theorem 2.5.2.1 and Lemma 3.1.2.1 infra; thus, agreeing to employ the notations centering around Theorem 2.5.2.1 and supporting discussion, we have the following basic lemma.

Lemma 3.1.2.1 (Bott) Fix $\Lambda \in \mathfrak{W}^+$, $\Lambda_s \in W_s^+$ – then the multiplicity of E_Λ in $H^q(M, \mathbf{E}_{\Lambda_l})$ is the same as the multiplicity of $E^\vee_{\Lambda_l}$ in $H^q(\mathfrak{n}_c^+, E^\vee_\Lambda)$ $(q = 0, 1, \ldots)$.

We begin by collecting a series of miscellaneous remarks.

(1) Let L and R denote, respectively, the left and right regular representations of U on $L^2(U)$; thus $L(u)f(v) = f(u^{-1}v)$, $R(u)f(v) = f(vu)$ $(u, v \in U;\ f \in L^2(U))$. Employing an evident symbolism, the Peter-Weyl Theorem says that

$$C^\infty(U) \sim \sum_{\Lambda \in \mathfrak{W}^+} E_\Lambda \otimes_{\mathbf{C}} E^\vee_\Lambda$$

where $L | E_\Lambda \otimes_{\mathbf{C}} E^\vee_\Lambda = U_\Lambda \otimes 1$, $R | E_\Lambda \otimes_{\mathbf{C}} E^\vee_\Lambda = 1 \otimes U^\vee_\Lambda$ (U_Λ the irreducible unitary representation of U with highest weight Λ, $U^\vee_\Lambda$ the contragredient to U_Λ).

(2) If E is an L-module, then E_I will denote the corresponding set of L-invariants in E. We shall view $C^\infty(U)$ as an L-module relative to $R|L$; $\Lambda^q\mathfrak{n}_c^+$, $H^q(\mathfrak{n}_c^+, E_\Lambda)$ etc. are to be given the obvious L-module structures when applicable.

(3) The following fact is well-known. Let K be a compact Lie group; let U_1 and U_2 be finite dimensional representations of K with U_2 irreducible – then the number of times that U_2 is contained in U_1 is equal to the number of times that the trivial one dimensional unit representation is contained in $U_1 \otimes U^\vee_2$ (see Chevalley [1, p.183] or number 5.3.3).

Proof of Lemma 3.1.2.1 The group L acts on the right in the principal fibering $L \to U \to M = U/L$ and so, in view of (1) above, we have

$$C^\infty(U/L) \sim \sum_{\Lambda \in \mathfrak{W}^+} E_\Lambda \otimes_{\mathbf{C}} (E_\Lambda^\vee)_I.$$

As a C^∞ vector bundle, $\mathbf{E}_{\Lambda_i}$ is associated with the principal bundle $L \to U \to M = U/L$ via the restriction of U_{Λ_i} to L. This being so, consider the U-module $C^{0,q}(M, \mathbf{E}_{\Lambda_i})$ of $(0, q)$-forms on M with coefficients in $\mathbf{E}_{\Lambda_i}$; thus $C^{0,q}(M, \mathbf{E}_{\Lambda_i}) = \Gamma^\infty(M, \mathbf{E}_{\Lambda_i} \otimes \mathbf{\Lambda}^q\bar{\mathbf{T}})$, $\Gamma^\infty(M, \mathbf{E}_{\Lambda_i} \otimes \mathbf{\Lambda}^q\bar{\mathbf{T}})$ the C^∞ sections of the differentiable vector bundle $\mathbf{E}_{\Lambda_i} \otimes \mathbf{\Lambda}^q\bar{\mathbf{T}}$. Bearing in mind the relationship which exists between $\mathfrak{n}_c^+$ and the anti-holomorphic tangent vectors on M (cf. number 1.2.2), we see without difficulty that

$$\Gamma^\infty(U/L, \mathbf{E}_{\Lambda_i} \otimes \mathbf{\Lambda}^q\bar{\mathbf{T}}) \sim (C^\infty(U) \otimes_{\mathbf{C}} \mathbf{\Lambda}^q(\mathfrak{n}_c^+)^\vee \otimes_{\mathbf{C}} E_{\Lambda_i})_I,$$

whence

$$C^{0,q}(M, \mathbf{E}_{\Lambda_i}) \sim \sum_{\Lambda \in \mathfrak{W}^+} E_\Lambda \otimes_{\mathbf{C}} (C^q(\mathfrak{n}_c^+, E_\Lambda^\vee) \otimes_{\mathbf{C}} E_{\Lambda_i})_I$$

where $C^q(\mathfrak{n}_c^+, E_\Lambda^\vee)$ is the usual Lie algebra cochain group. One may then check that the diagram

$$\begin{array}{ccc} E_\Lambda \otimes_{\mathbf{C}} (C^q(\mathfrak{n}_c^+, E_\Lambda^\vee) \otimes_{\mathbf{C}} E_{\Lambda_i})_I & \xrightarrow{1 \otimes d \otimes 1} & E_\Lambda \otimes_{\mathbf{C}} (C^{q+1}(\mathfrak{n}_c^+, E_\Lambda^\vee) \otimes_{\mathbf{C}} E_{\Lambda_i})_I \\ \downarrow & & \downarrow \\ C^{0,q}(M, \mathbf{E}_{\Lambda_i}) & \xrightarrow[\bar\partial]{} & C^{0,q+1}(M, \mathbf{E}_{\Lambda_i}) \end{array}$$

is commutative ($d: C^q(\mathfrak{n}_c^+, E_\Lambda^\vee) \to C^{q+1}(\mathfrak{n}_c^+, E_\Lambda^\vee)$ the Lie algebra coboundary map). Dolbeault's Theorem, combined with the earlier observations, now leads to the formula

$$H^q(M, \mathbf{E}_{\Lambda_i}) = \sum_{\Lambda \in \mathfrak{W}^+} E_\Lambda \otimes_{\mathbf{C}} (H^q(\mathfrak{n}_c^+, E_\Lambda^\vee) \otimes_{\mathbf{C}} E_{\Lambda_i})_I.$$

From this relation we deduce that the number of times that E_Λ occurs in $H^q(M, \mathbf{E}_{\Lambda_i})$ is equal to the dimension of $(H^q(\mathfrak{n}_c^+, E_\Lambda^\vee) \otimes_{\mathbf{C}} E_{\Lambda_i})_I$ which in turn is equal to the number of times that $E_{\Lambda_i}^\vee$ occurs in $H^q(\mathfrak{n}_c^+, E_\Lambda^\vee)$ ((3) above). Hence the lemma. □

Let ι denote the opposition involution in Φ (cf. number 1.1.2).

Here is Bott's generalization of the Borel-Weil Theorem (as was conjectured by Borel and Hirzebruch).

Theorem 3.1.2.2 (Bott-Borel-Weil) Let $\mathfrak{g}_c$ be a semi-simple Lie algebra over $\mathbf{C}$, G_c the simply connected complex analytic Lie group corresponding to $\mathfrak{g}_c$; let $\mathfrak{j}_c$ be a Cartan subalgebra of $\mathfrak{g}_c$, $\mathfrak{b}_c$ a Borel subalgebra of $\mathfrak{g}_c$ containing $\mathfrak{j}_c$, $\mathfrak{p}_c$ a $\mathfrak{b}_c$-parabolic subalgebra of $\mathfrak{g}_c$; let P_c be the parabolic subgroup of G_c corresponding to $\mathfrak{p}_c$ and put $M = G_c/P_c$; fix $\Lambda_s \in W_s^+$ – we have:

(i) If $\Lambda_s^{\iota\prime} \notin \mathfrak{W}_s^+(\rho)$, then $H^q(M, \mathbf{E}_{\Lambda_s}) = \{0\}$ $(q = 0, 1, \ldots)$;

(ii) If $\Lambda_s^{\iota\prime} \in \mathfrak{W}_s^+(\rho)$, then, upon writing uniquely

$$\Lambda_s^{\iota\prime} = w_u(\Lambda' + \rho) - \rho \qquad (\Lambda \in \mathfrak{W}^+,\ w_u \in W_u),$$

one has $H^q(M, \mathbf{E}_{\Lambda_s}) = \{0\}$ for $q \neq |w_u|$ while for $q = |w_u|$

$$\dim (H^{|w_u|}(M, \mathbf{E}_{\Lambda_s})) = \dim (E_\Lambda)$$

and the holomorphic representation $U_{\Lambda_s}^{|w_u|}$ of G_c on $H^{|w_u|}(M, \mathbf{E}_{\Lambda_s})$ is equivalent to the irreducible holomorphic representation U_Λ of G_c on E_Λ.

[Here ι_s denotes the opposition involution in the root system Ξ_s (hence the highest weight of the simple $\mathfrak{l}_c$-module contragredient to E_{Λ_s} is $\Lambda_s^{\iota_s}(\Lambda_s \in \mathrm{W}_s^+)$); the other notations are as in the Remark following Lemma 2.4.2.1).]

Corollary 3.1.2.3 Retain the notations and assumptions of Theorem 3.1.2.2 but assume that $\mathfrak{p}_c = \mathfrak{b}_c$; let ξ be a holomorphic character of J_c – we have:

(i) If $-\xi \notin \mathfrak{W}^+(\rho)$, then $H^q(M, \mathbf{E}_\xi) = \{0\}$ $(q = 0, 1, \ldots)$;

(ii) If $-\xi \in \mathfrak{W}^+(\rho)$, then, upon writing uniquely

$$-\xi = w(\Lambda^\iota + \rho) - \rho \qquad (\Lambda \in \mathfrak{W}^+, w \in W),$$

one has $H^q(M, \mathbf{E}_\xi) = \{0\}$ for $q \neq |w|$ while for $q = |w|$

$$\dim (H^{|w|}(M, \mathbf{E}_\xi)) = \dim (E_\Lambda)$$

and the holomorphic representation $U_\xi^{|w|}$ of G_c on $H^{|w|}(M, \mathbf{E}_\xi)$ is equivalent to the irreducible holomorphic representation U_Λ of G_c on E_Λ.

Proof of Theorem 3.1.2.2 (Kostant) Assertions (i) and (ii) are both ready consequences of Theorem 2.5.2.1 and Lemma 3.1.2.1. Thus consider the first statement. If the space $H^q(M, \mathbf{E}_{\Lambda_s})$ were non-zero, then, of necessity, E_Λ would occur in $H^q(M, \mathbf{E}_{\Lambda_s})$ for some $\Lambda \in \mathfrak{W}^+$; but then, in view of Lemma 3.1.2.1, $E_{\Lambda_s}^\vee$ would occur in $H^q(\mathfrak{n}_c^+, E_\Lambda^\vee)$; however the structure of $H^q(\mathfrak{n}_c^+, E_\Lambda^\vee)$, as an $\mathfrak{l}_c$-module, is known through Theorem 2.5.2.1; in particular we deduce that the highest weight of $E_{\Lambda_s}^\vee$, i.e. $\Lambda_s^{\iota_s}$, must necessarily lie in $\mathfrak{W}_s^+(\rho)$, a contradiction to our original supposition on Λ_s. The second assertion is established in a similar fashion. ☐

It is now clear that the Bott-Borel-Weil Theorem provides an explicit, geometrical realization of the irreducible holomorphic representations of G_c.

Remark Borel and Weil considered the following special case of Corollary 3.1.2.3. Let ξ be a holomorphic character of J_c such that $-\xi$ is dominant, i.e. such that $-\xi \in \mathfrak{W}^+$. Owing to a well-known theorem of Kodaira on positive line bundles, it follows that $H^q(M, \mathbf{E}_\xi) = \{0\}$ for $q > 1$; the Riemann-Roch Theorem, in the formulation of Hirzebruch, can then be used to show that $\dim (H^0(M, \mathbf{E}_\xi)) = \dim (E_\Lambda)$ $(-\xi = \Lambda^\iota,\ \Lambda \in \mathfrak{W}^+)$ – of course this latter fact suggests that the representation U_ξ^0 is equivalent to U_Λ and, indeed, this is precisely what the original version of the Borel-Weil Theorem asserted (thus, among other things, Bott's generalization takes into account arbitrary holomorphic characters of J_c and not just the dominant ones). For an exposition of

Borel and Weil's original proof see J-P. Serre, 'Représentations linéaires et espaces homogènes kähleriens des groupes de Lie compacts', Séminaire Bourbaki, number 100, Paris (1954).

3.2 Unitary Representations of a Compact Semi-Simple Lie Group

3.2.1 The Invariant Integral on a Compact Semi-Simple Lie Algebra

In this number we shall collect a few simple facts as regards the invariant integral on a compact semi-simple Lie algebra. The results to be obtained should help to motivate later considerations (Vol. II, section 8.4); they will also be put to immediate use in 3.2.2 where, as we shall see, the Plancherel Theorem for a compact connected semi-simple Lie group is an immediate consequence of them.

So let $\mathfrak{g}$ be a compact semi-simple Lie algebra over $\mathbf{R}$, G a connected Lie group with Lie algebra $\mathfrak{g}$; let $\mathfrak{t}$ be a Cartan subalgebra of $\mathfrak{g}$, T the corresponding Cartan subgroup of G – then T is a maximal torus in G. Let $d_G(x)$ (respectively $d_T(t)$) denote the Haar measure on G (respectively T) normalized by the requirement $\int_G d_G(x) = 1$ (respectively $\int_T d_T(t) = 1$); the invariant measure $d_{G/T}(\dot{x})$ ($\dot{x} = xT$) on G/T can then be normalized in such a way that

$$\int_G f(x) d_G(x) = \int_{G/T} \left(\int_T f(xt) d_T(t) \right) d_{G/T}(\dot{x}) \qquad (f \in C(G)).$$

Relative to some ordering of the roots Φ of the pair $(\mathfrak{g}_c, \mathfrak{t}_c)$, set $\pi = \prod_{\alpha>0} \alpha$.

Definition Let $f \in C^\infty(\mathfrak{g})$ – then the *invariant integral* of f relative to the Cartan subalgebra $\mathfrak{t}$ is the function ϕ_f on $\mathfrak{t}$ defined by

$$\phi_f(H) = \pi(H) \int_G f(xH) d_G(x) \qquad (H \in \mathfrak{t}).$$

[In the compact case, therefore, one may form the invariant integral ϕ_f for any C^∞ function f on $\mathfrak{g}$; the singular elements play no role in the definition (cf. Vol. II, section 8.4).]

It is clear that $f \in C^\infty(\mathfrak{g}) \Rightarrow \phi_f \in C^\infty(\mathfrak{t})$ and that $f \in C_c^\infty(G) \Rightarrow \phi_f \in C_c^\infty(\mathfrak{t})$; moreover, if $f \in \mathscr{C}(\mathfrak{g})$, then $\phi_f \in \mathscr{C}(\mathfrak{t})$ (i.e. if f is rapidly decreasing on $\mathfrak{g}$, then ϕ_f is rapidly decreasing on $\mathfrak{t}$ (Vol. II, number 8.3.5)). [Thus first note that given $D \in \mathfrak{P}(\mathfrak{g})$ there exist a finite number of elements $D_1, \ldots, D_p \in \mathfrak{P}(\mathfrak{g})$ and analytic functions $a_1, \ldots, a_p$ on G such that $xD = \sum_{i=1}^p a_i(x) D_i$ ($x \in G$). Given $d \in \mathfrak{P}(\mathfrak{t})$, it is obvious that

$$\phi_f(H; d) = \int_G f(xH; xD) d_G(x) \qquad (H \in \mathfrak{t})$$

where $D = d \circ \pi$. Owing to the compactness of G, there exists a positive number M such that $|a_i(x)| \leqslant M$ $(1 \leqslant i \leqslant p, x \in G)$ – hence

$$|\phi_f(H; d)| \leqslant M \sum_{i=1}^{p} \sup_{X \in \mathfrak{g}} |f(X; D_i)|.$$

This proves that $\phi_f \in \mathscr{C}(\mathfrak{t})$; in fact that the mapping $f \mapsto \phi_f$ of $\mathscr{C}(\mathfrak{g})$ into $\mathscr{C}(\mathfrak{t})$ is continuous (cf. with Theorem 8.4.1.2).]

Proposition 3.2.1.1 Let $f \in C^\infty(\mathfrak{g})$ – then

$$\phi_{Df} = \tilde{\gamma}(D)\phi_f$$

for all $D \in \mathfrak{Z}(\mathfrak{g})$.

[For unexplained notations and conventions see Vol. II, number 8.2.2 (cf. in particular the discussion following the proof of Theorem 8.2.2.4).]

Proof Let $D \in \mathfrak{Z}(\mathfrak{g})$ – then, in view of the remarks following Proposition 8.2.2.1, there exist elements $S_i \in \mathfrak{S}^+$, $H_i \in S(\mathfrak{t}_c)$ and analytic functions a_i on $\mathfrak{t}'$ with the property that

$$f(xH;\, D) = f(x\colon H;\, \text{♀}(D)) + \sum_{i=1}^{p} a_i(H) f(x;\, S_i\colon H;\, \partial(H_i))$$

for all $H \in \mathfrak{t}'$. Since it is obvious that $\int_G f(x;\, d\colon H) d_G(x) = 0$ (any $d \in \mathfrak{G}\mathfrak{g}_c$), it follows from Theorem 8.2.2.4 that

$$\phi_{Df}(H) = \pi(H) \int_G f(x\colon H;\, \text{♀}(D)) d_G(x) = \phi_f(H;\, \tilde{\text{♀}}(D)) \qquad (H \in \mathfrak{t}').$$

But ϕ_{Df} and $\tilde{\text{♀}}(D)\phi_f$ are both in $C^\infty(\mathfrak{t})$; thus the last equality is actually valid on all of $\mathfrak{t}$, $\mathfrak{t}'$ being dense in $\mathfrak{t}$. ☐

Theorem 3.2.1.2 (Harish-Chandra) Let $f \in C^\infty(\mathfrak{g})$ – then $\phi_f(0;\, \partial(\pi)) = \langle \pi, \pi \rangle f(0)$ (all $f \in C^\infty(\mathfrak{g})$).

Proof Since π is a homogeneous polynomial, it is clear that the local expression of $\partial(\pi) \circ \pi$ at the origin is $\langle \pi, \pi \rangle$, $\langle \pi, \pi \rangle$ as defined in number 2.1.4. Therefore the value of $\partial(\pi)\phi_f$ at the origin is $\langle \pi, \pi \rangle f(0)$. ☐

[Evidently $\langle \pi, \pi \rangle = [W] \prod_{\alpha > 0} (\rho, \alpha)$, $[W]$ the order of the Weyl group W of the pair $(\mathfrak{g}_c, \mathfrak{t}_c)$, $\rho = \frac{1}{2} \sum_{\alpha > 0} \alpha$.]

Remark As will be seen in 3.2.2, the Plancherel formula for G is a ready consequence of Theorem 3.2.1.2; one of the main problems in the theory of the invariant integral relative to a non-compact G is to obtain a suitable analogue of Theorem 3.2.1.2.

Let B denote the Killing form on $\mathfrak{g}_c$.

Theorem 3.2.1.3 (Harish-Chandra) For all $H_1, H_2 \in \mathfrak{t}_c$,

$$\pi(H_1)\pi(H_2) \int_G \exp(B(xH_1, H_2)) d_G(x) = \prod_{\alpha > 0} (\rho, \alpha) \sum_{w \in W} \det(w) \exp(B(wH_1, H_2)).$$

Proof Choose an element $\tilde{H} \in \mathfrak{t}_c$ such that $\pi(\tilde{H}) \neq 0$ and put $f(X) = \exp(B(X, \tilde{H}))$ $(X \in \mathfrak{g})$ – then

$$\phi_f(H) = \pi(H) \int_G \exp(B(xH, \tilde{H})) d_G(x) \qquad (H \in \mathfrak{t})$$

and therefore, from Proposition 3.2.1.1, $\phi_{\partial(p)f} = \partial(\bar{p})\phi_f$ (all $p \in I(\mathfrak{g}_c)$; $\bar{p} = p|\mathfrak{t}_c$). But clearly $\partial(X)f = B(X, \tilde{H})f$ (all $X \in \mathfrak{g}_c$) – thus $\partial(q)f = q(\tilde{H})f$ (all $q \in S(\mathfrak{g}_c)$) and so $\partial(\bar{p})\phi_f = p(\tilde{H})\phi_f$ (all $p \in I(\mathfrak{g}_c)$). Altering the notation a bit, in view of Chevalley's Restriction Theorem (Theorem 2.1.5.1(i)) we may now conclude that $\partial(p)\phi_f = p(\tilde{H})\phi_f$ for all $p \in I(\mathfrak{t}_c)$. Corollary 8.2.2.9 then gives unique constants $c_w (w \in W)$ such that

$$\phi_f(H) = \sum_{w \in W} c_w \exp(B(wH, \tilde{H})) \qquad (H \in \mathfrak{t}).$$

On the other hand, since G is compact, we can select, for each $w \in W$, an element $x_w \in G$ with the property that $x_w H = wH$ $(H \in \mathfrak{t})$; therefore w maps $\mathfrak{t}$ into itself. From this we deduce that $\phi_f(wH) = \det(w)\phi_f(H)$ and so

$$\phi_f(H) = [W]^{-1} \sum_{w \in W} \det(w)\phi_f(wH) = [W]^{-1} c \sum_{w \in W} \det(w) \exp(B(wH, \tilde{H})) \qquad (H \in \mathfrak{t})$$

where $c = \sum_{w \in W} \det(w) c_w$. Put $\psi_w(H) = \exp(B(wH, \tilde{H}))$ $(H \in \mathfrak{t})$; obviously $\partial(q)\psi_w = q(w^{-1}\tilde{H})\psi_w$ (all $q \in S(\mathfrak{t}_c)$), whence $\partial(\pi)\psi_w = \det(w)\pi(\tilde{H})\psi_w$. Thus, using Theorem 3.2.1.2, we have

$$\langle \pi, \pi \rangle = \phi_f(0; \partial(\pi)) = [W]^{-1} c\pi(\tilde{H})[W] = c\pi(\tilde{H}),$$

from which it follows that

$$\pi(H)\pi(\tilde{H}) \int_G \exp(B(xH, \tilde{H})) d_G(x) = \prod_{\alpha > 0} (\rho, \alpha) \sum_{w \in W} \det(w) \exp(B(wH, \tilde{H}))$$

for $H \in \mathfrak{t}$, $\tilde{H} \in \mathfrak{t}_c'$. Both sides of this relation may be regarded as holomorphic functions on $\mathfrak{t}_c \times \mathfrak{t}_c$ – therefore equality persists for all $H_1, H_2 \in \mathfrak{t}_c$. ☐

Application Let $\mathfrak{g}$ be a real non-compact semi-simple Lie algebra, $\mathfrak{j}$ a Cartan subalgebra of $\mathfrak{g}$; let W denote the Weyl group of the pair $(\mathfrak{g}_c, \mathfrak{j}_c)$. Fix a linear function λ on $\mathfrak{j}_c$ – then there exists one and only one central analytic function f_λ on $\mathfrak{g}$ such that

$$\pi(H) f_\lambda(H) = \sum_{w \in W} \det(w) e^{\lambda(wH)} \qquad (H \in \mathfrak{j}).$$

[The uniqueness of f_λ is obvious. As for its existence, let U denote the real analytic subgroup of $G_c = \mathrm{Int}(\mathfrak{g}_c)$ corresponding to the compact real form $\mathfrak{u} = \mathfrak{k} + \sqrt{-1}\mathfrak{p}$ of $\mathfrak{g}_c$ (the usual notations being in force here); consider the function f defined by the rule

$$f(X{:}\ Y) = \int_U \exp(B(uX, Y)) du \qquad (X, Y \in \mathfrak{g}_c),$$

du being normalized Haar measure on U – then f is obviously holomorphic on $\mathfrak{g}_c \times \mathfrak{g}_c$ and, moreover, $f(xX{:}\ Y) = f(X{:}\ Y)$ (all $x \in G_c$). Let H_λ denote that element in $\mathfrak{j}_c$ such that $B(H, H_\lambda) = \lambda(H)$ (all $H \in \mathfrak{j}_c$) – then, in view of Theorem 3.2.1.3 supra, we have

$$\pi(H)\pi(H_\lambda) f(H{:}\ H_\lambda) = c \sum_{w \in W} \det(w) e^{\lambda(wH)} \qquad (H \in \mathfrak{j}_c),$$

where c is a non-zero number independent of H and λ. Now set $f_\lambda(X) = c^{-1}\pi(H_\lambda) f(X{:}\ H_\lambda)$ $(X \in \mathfrak{g})$ – then f_λ obviously meets all of our requirements.]

3.2.2 The Plancherel Theorem for a Compact Connected Semi-Simple Lie Group

As in 3.2.1, G will denote a compact connected semi-simple Lie group with Lie algebra $\mathfrak{g}$; for simplicity it will be assumed in what follows that G is simply connected. This being so, fix a maximal torus T in G with Lie algebra $\mathfrak{t}$ – then, as we know, the irreducible unitary representations of G are parameterized by the dominant integral linear functions Λ on $\mathfrak{t}_c$. [In fact the Borel-Weil Theorem even provides us with an explicit realization of these representations. . . .] Given a $\Lambda \in \mathfrak{W}^+$, we shall denote by U_Λ the corresponding irreducible unitary representation of G on E_Λ (say); the degree d_Λ of U_Λ (i.e. the dimension of E_Λ) is given, of course, by the formula

$$d_\Lambda = \frac{\prod_{\alpha>0} (\Lambda + \rho, \alpha)}{\prod_{\alpha>0} (\rho, \alpha)} \qquad (\rho = \tfrac{1}{2} \sum_{\alpha>0} \alpha).$$

We shall agree to write χ_Λ for the 'normalized' character

$$x \mapsto d_\Lambda \operatorname{tr}(U_\Lambda(x)) \; (x \in G)$$

on G.

In these notations, then, the Plancherel Theorem for G asserts, essentially, that

$$f(1) = \sum_{\Lambda \in \mathfrak{W}^+} \int_G f(x) \chi_\Lambda(x) d_G(x)$$

for all $f \in C^\infty(G)$. [The reader who is unfamiliar with the whys and wherefores behind the Plancherel Theorem may turn to Chapter 7 for amplification.] Before giving the proof, a few preliminary remarks are needed.

Thus first note that since G is simply connected, the unitary character group $\hat{T}$ of T may be identified with the lattice $\mathfrak{W}$ of all integral linear functions on $\mathfrak{t}_c$ (cf. Proposition 3.1.1.4); in fact any $\tau \in \mathfrak{W}$ gives rise to a unitary character ξ_τ of T by the prescription $\xi_\tau(\exp H) = e^{\tau(H)}$ $(H \in \mathfrak{t})$ and all such arise in this way. Next, owing to Weyl's character formula, one has

$$\chi_\Lambda(t) \Delta_T(t) = d_\Lambda \sum_{w \in W} \det(w)\, \xi_{w(\Lambda+\rho)}(t)$$

for all $t \in T$ (as usual, $\Delta_T(t) = \xi_\rho(t) \prod_{\alpha>0} (1 - \xi_\alpha(t^{-1}))$, $t \in T$). Finally suppose given a regular element $\tau \in \hat{T} (= \mathfrak{W})$ – then there exists a unique pair $(\Lambda, w) \in \mathfrak{W}^+ \times W$ such that $\tau = w(\Lambda + \rho)$ (here we have to bear in mind the Remark which follows Lemma 2.4.2.1).

So fix an $f \in C^\infty(G)$; for $t \in T$, put $\Phi_f(t) = \Delta_T(t) \int_G f(xtx^{-1}) d_G(x)$ (thus $\Phi_f \in C^\infty(T) \ldots$) – then, in view of the normalizations agreed to in 3.2.1, Weyl's well-known integral formula tells us that

$$\int_G f(x) d_G(x) = [W]^{-1} \int_T \overline{\Delta_T(t)} \Phi_f(t) d_T(t).$$

Consequently

$$\sum_{\Lambda \in \mathfrak{W}^+} \int_G f(x) \chi_\Lambda(x) d_G(x) = [W]^{-1} \sum_{\Lambda \in \mathfrak{W}^+} \int_T \overline{\Delta_T(t)} \chi_\Lambda(t) \Phi_f(t) d_T(t).$$

Now $\overline{\Delta_T(t)} = (-1)^r \Delta_T(t)$ (r the number of positive roots of the pair $(\mathfrak{g}_c, \mathfrak{t}_c)$) and $\det(w) \prod_{\alpha>0} (\Lambda + \rho, \alpha) = \prod_{\alpha>0} (w(\Lambda + \rho), \alpha)$ $(w \in W)$ – therefore

$$\sum_{\Lambda \in \mathfrak{W}^*} \int_G f(x)\chi_\Lambda(x)d_G(x)$$
$$= \frac{(-1)^r}{[W]\prod_{\alpha>0}(\rho, \alpha)} \sum_{\Lambda, w} \int_T \det(w) \prod_{\alpha>0} (\Lambda + \rho, \alpha)\xi_{w(\Lambda+\rho)}(t)\, \Phi_f(t)\, d_T(t)$$
$$= \frac{(-1)^r}{[W]\prod_{\alpha>0}(\rho, \alpha)} \sum_{\Lambda, w} \int_T \prod_{\alpha>0} (w(\Lambda + \rho), \alpha)\xi_{w(\Lambda+\rho)}(t)\Phi_f(t)\, d_T(t)$$
$$= \frac{(-1)^r}{[W]\prod_{\alpha>0}(\rho, \alpha)} \sum_{\tau \in \mathfrak{W}} \int_T \prod_{\alpha>0} (\tau, \alpha)\xi_\tau(t)\, \Phi_f(t)\, d_T(t),$$

the $\tau \in \mathfrak{W}$ for which $\prod_{\alpha>0}(\tau, \alpha) = 0$ making no contribution. [The preceding manipulations can of course be justified on the obvious grounds. . . .]

On the other hand put $\check{f}(X) = f(\exp X)(X \in \mathfrak{g})$ – then $\check{f} \in C^\infty(\mathfrak{g})$. Furthermore, for $H \in \mathfrak{t}'$, we have

$$\Phi_f(\exp H) = \Delta_T(\exp H) \int_G f(\exp(xH))\, d_G(x)$$
$$= \prod_{\alpha>0} \left\{\frac{e^{\alpha(H)/2} - e^{-\alpha(H)/2}}{\alpha(H)}\right\} \phi_f(H)$$

and so, thanks to Theorem 3.2.1.2, it follows that

$$f(1) = \check{f}(0) = \frac{1}{[W]\prod_{\alpha>0}(\rho, \alpha)} \phi_f(0; \partial(\pi))$$
$$= \frac{1}{[W]\prod_{\alpha>0}(\rho, \alpha)} \lim_{H\to 0} \Phi_f\ (\exp H; \Pi) \qquad (\Pi = \prod_{\alpha>0} H_\alpha).$$

For $\tau \in \mathfrak{W}$, put $\hat{\Phi}_f(\tau) = \int_T \xi_\tau(t)\Phi_f(t)d_T(t)$ (thus $\hat{\Phi}_f$ is simply the Fourier transform of Φ_f); assuming that the Haar measure on $\hat{T}$ is normalized so that Fourier inversion is valid (thus each $\tau \in \mathfrak{W}$ is assigned mass 1), we find that

$$f(1) = \frac{1}{[W]\prod_{\alpha>0}(\rho, \alpha)} \lim_{H\to 0} \Phi_f(\exp H; \Pi)$$
$$= \frac{1}{[W]\prod_{\alpha>0}(\rho, \alpha)} \int_{\hat{T}} \lim_{H\to 0} \xi_{-\tau}(\exp H; \Pi)\hat{\Phi}_f(\tau)d_{\hat{T}}(\tau)$$
$$= \frac{(-1)^r}{[W]\prod_{\alpha>0}(\rho, \alpha)} \int_{\hat{T}} \prod_{\alpha>0} (\tau, \alpha)\left(\int_T \xi_\tau(t)\Phi_f(t)d_T(t)\right) d_{\hat{T}}(\tau)$$
$$= \frac{(-1)^r}{[W]\prod_{\alpha>0}(\rho, \alpha)} \sum_{\tau \in \mathfrak{W}} \int_T \prod_{\alpha>0} (\tau, \alpha)\xi_\tau(t)\Phi_f(t)d_T(t),$$

our manipulations again being justified for the obvious reasons. A comparison with the results of the preceding paragraph then gives the Plancherel formula.

Remark Let $f, g \in C^\infty(G)$; an application of the Plancherel Theorem to $f*g^*(g^*(x) = \overline{g(x^{-1})}, x \in G)$ provides us with the relation

$$f*g^*(1) = \int_G f(x)\overline{g(x)}\,d_G(x) = \sum_{\Lambda\in\mathfrak{W}^+} d_\Lambda \operatorname{tr}(U_\Lambda(f)U_\Lambda(g)^*),$$

the formula being actually valid for all $f, g \in L^2(G)$ (see Chapter 4 for an explanation of the notation). Define a measure $d\mu(\Lambda)$ on $\mathfrak{W}^+$ by attaching to each Λ the mass d_Λ – then one has

$$\int_G |f(x)|^2\,d_G(x) = \int_{\mathfrak{W}^+} \operatorname{tr}(U_\Lambda(f)U_\Lambda(f)^*)\,d\mu(\Lambda)$$

for all $f \in L^2(G)$. Let $\hat{G}$ denote the set of unitary equivalence classes of (continuous) irreducible unitary representations of G; in the case at hand $\hat{G}$ may be identified with $\mathfrak{W}^+$ while $d\mu(\Lambda)$ may be regarded as the *Plancherel measure* for $\hat{G}$.

With a view toward motivating subsequent developments (the theory of the discrete series on a non-compact G), let us draw here a few simple consequences from the above considerations.

Let τ be a regular element in $\mathfrak{W}$. As has been observed earlier, there exists a unique pair $(\Lambda, w) \in \mathfrak{W}^+ \times W$ such that $\tau = w(\Lambda + \rho)$; put $\epsilon(\tau) = \operatorname{sign}\prod_{\alpha>0}(\tau, \alpha)$ – then $\epsilon(\tau) = \det(w)$ (for $\prod_{\alpha>0}(\tau,\alpha) = \prod_{\alpha>0}(w(\Lambda+\rho),\alpha) = \det(w)\prod_{\alpha>0}(\Lambda+\rho,\alpha)$ and $\prod_{\alpha>0}(\Lambda+\rho,\alpha) > 0$). Consider now the distribution Θ_τ on G defined by the rule

$$\Theta_\tau(f) = \epsilon(\tau)\int_G f(x)\operatorname{tr}(U_\Lambda(x))\,d_G(x) \qquad (f \in C^\infty(G)).$$

Then, in the terminology of Vol. II, number 8.3.3, Θ_τ is a central eigendistribution (of $\mathfrak{Z}$) on G. [It is obvious that Θ_τ is central. Moreover, for any $a \in E_\Lambda$, $U_\Lambda(Z)a = \kappa_\Lambda(Z)a = \langle\gamma(Z), e^{\Lambda+\rho}\rangle a$ (all $Z \in \mathfrak{Z}$; cf. number 2.4.3) – thus $Z\chi_\Lambda = \langle\gamma(Z), e^{\Lambda+\rho}\rangle\chi_\Lambda$ (all $Z \in \mathfrak{Z}$) and so $Z\cdot\Theta_\tau = \langle\gamma(Z), e^{\Lambda+\rho}\rangle\Theta_\tau$ (all $Z \in \mathfrak{Z}$).]

The transformation $f \mapsto \Phi_f$ $(f \in C^\infty(G))$ enables one to relate the Harmonic Analysis on G to the Fourier Analysis on T; in fact, as a simple computation shows, for any regular τ in $\mathfrak{W}$, one has $\Theta_\tau(f) = (-1)^r\hat{\Phi}_f(\tau)$ (all $f \in C^\infty(G)$). This being so, let δ denote the Dirac measure concentrated at 1 – then, in view of what was said above, the Plancherel formula for G can be written in the form

$$\delta = \sum_{\Lambda\in\mathfrak{W}^+} d_\Lambda\Theta_\Lambda = \frac{1}{[W]\prod_{\alpha>0}(\rho,\alpha)}\sum_{\tau\in\mathfrak{W}}\prod_{\alpha>0}(\tau,\alpha)\Theta_\tau$$

where we make the convention that $\prod_{\alpha>0}(\tau,\alpha)\Theta_\tau = 0$ if τ is singular.

Later on, in the case of a non-compact connected semi-simple G (with finite center) we shall see that G admits a discrete series iff rank (G) = rank (K) and, in that event, it will be shown that the contribution of the discrete series to the Plancherel formula for G is given by a relation exactly the same as the one above. It should also be noted that these remarks add a degree of credibility to the heuristic statement: The Plancherel formula for G may be viewed as an 'expansion' of δ into central eigendistributions (of $\mathfrak{Z}$). [This will be taken up somewhat more precisely in due course.]

Summary Let G be a simply connected compact semi-simple Lie group, T a maximal torus in G; let $\mathfrak{W}$ $(=\hat{T})$ denote the set of integral linear functions on $\mathfrak{t}_c$, $\mathfrak{W}'$ the regular elements in $\mathfrak{W}$ – then, to each $\tau \in \mathfrak{W}'$, there corresponds a unique element $\hat{U}_\tau \in \hat{G}$ whose character is given by $\epsilon(\tau)\Theta_\tau$. The mapping $\tau \mapsto \hat{U}_\tau$ of $\mathfrak{W}'$ into $\hat{G}$ is surjective and the degree $d_{\hat{U}_\tau}$ of the class $\hat{U}_\tau$ is given by

$$d_{\hat{U}_\tau} = \frac{|\prod_{\alpha>0} (\tau, \alpha)|}{\prod_{\alpha>0} (\rho, \alpha)}$$

Finally $\hat{U}_{\tau_1} = \hat{U}_{\tau_2}$ $(\tau_1, \tau_2 \in \mathfrak{W}')$ iff τ_1 and τ_2 are conjugate under the Weyl group W of the pair $(\mathfrak{g}_c, \mathfrak{t}_c)$ – symbolically $\hat{G} \sim W\backslash\mathfrak{W}'$.

Example Let us take $G = \mathbf{SU}(2)$ – then the finite dimensional irreducible unitary representations of G are parameterized by the set $\{0, \frac{1}{2}, 1, \frac{3}{2}, \ldots\}$, the representation of 'weight' k being $(2k+1)$-dimensional. The Plancherel Theorem for G thus asserts that

$$\int_G |f(x)|^2 d_G(x) = \sum_k (2k+1) \operatorname{tr}(U_k(f)U_k(f)^*) \qquad (\text{all } f \in L^2(G)).$$

3.3 Finite Dimensional Class One Representations of a Real Semi-Simple Lie Group

3.3.1 The Theorem of É. Cartan and S. Helgason

Let $\mathfrak{g}$ be a semi-simple Lie algebra over $\mathbf{R}$, $\mathfrak{g} = \mathfrak{k} + \mathfrak{p}$ a fixed Cartan decomposition of $\mathfrak{g}$. Select a maximal abelian subspace $\mathfrak{a}_\mathfrak{p}$ of $\mathfrak{p}$ and extend it in the usual way to a Cartan subalgebra $\mathfrak{a}(=\mathfrak{a}_\mathfrak{k} + \mathfrak{a}_\mathfrak{p})$ of $\mathfrak{g}$; introduce compatible orderings into the dual spaces of $\mathfrak{a}_\mathfrak{p}$ and $\mathfrak{a}_\mathfrak{p} + \sqrt{-1}\mathfrak{a}_\mathfrak{k}$; let Σ^+ (respectively Φ^+) denote the corresponding set of positive roots of the pair $(\mathfrak{g}, \mathfrak{a}_\mathfrak{p})$ (respectively $(\mathfrak{g}_c, \mathfrak{a}_c)$).

Let G_c denote the simply connected complex Lie group corresponding to $\mathfrak{g}_c$, G the analytic subgroup of G_c corresponding to $\mathfrak{g}$ – then it is clear that the irreducible finite dimensional representations of G (on complex vector spaces) may be identified with the irreducible holomorphic representations of G_c or, equivalently, with the irreducible representations of $\mathfrak{g}_c$.

Let K denote the analytic subgroup of G corresponding to $\mathfrak{k}$ – then K is a maximal compact subgroup of G. . . .

Definition An irreducible representation U of G on a finite dimensional vector space E (over $\mathbf{C}$) is said to be of *class one* (relative to K) if there exists a non-zero vector $a \in E$ such that $U(k)a = a$ for all $k \in K$.

Remarks (1) If the representation U is of class one (relative to K) and if a is a non-zero vector in E such that $U(k)a = a$ for all $k \in K$, then a is unique to within a scalar factor (cf. Theorem 4.5.2.7).

(2) Every finite dimensional $\mathfrak{g}_c$-module determines, in the obvious way, a finite dimensional semi-simple $\mathfrak{k}_c$-module (cf. Proposition 1.1.5.2(ii)). Thus, from the infinitesimal point of view, a class one representation U of the pair (G, K) is one in which the trivial zero representation of $\mathfrak{k}_c$ occurs in the restriction of U to $\mathfrak{k}_c$.

The irreducible finite dimensional representations of G may be parameterized by the dominant integral linear functions Λ on $\mathfrak{a}_c$; what, then, is the necessary and sufficient condition on Λ to ensure that the corresponding representation of the pair (G, K) is of class one?

Theorem 3.3.1.1 (É. Cartan-S. Helgason) Retain the above notations and assumptions; fix a dominant integral linear function Λ in $\mathfrak{W}^+$ – then Λ determines a class one representation of the pair (G, K) iff:

(i) $\Lambda | \mathfrak{a}_{\mathfrak{k}}$ vanishes;

(ii) $(\Lambda, \lambda)/(\lambda, \lambda)$ is a non-negative integer (all $\lambda \in \Sigma^+$).

[We shall agree to extend the linear function $\lambda (\lambda \in \Sigma^+)$ to all of $\mathfrak{a}^* (= \mathfrak{a}_{\mathfrak{p}} + \sqrt{-1}\mathfrak{a}_{\mathfrak{k}})$ by rendering it trivial on $\sqrt{-1}\mathfrak{a}_{\mathfrak{k}}$.]

Proof *Necessity* Let U_Λ be a finite dimensional irreducible class one representation of the pair (G, K) on E_Λ with highest weight $\Lambda (\Lambda \in \mathfrak{W}^+)$ – then Λ, when restricted to $\mathfrak{a}_{\mathfrak{k}}$, vanishes identically. Thus fix a non-zero vector a_Λ in the weight space E^Λ determined by Λ; since $G = KA_{\mathfrak{p}}N^+$ and since U_Λ is irreducible, it must be the case that E_Λ is spanned by vectors of the form $U_\Lambda(k)a_\Lambda$ $(k \in K)$. Set $\mathsf{P} = \int_K U_\Lambda(k)dk$, dk being normalized Haar measure on K – then P is a projection on E_Λ with $\mathsf{P}(E_\Lambda) = \mathbf{C}\mathsf{P}(a_\Lambda)$. Now choose a non-zero K-invariant vector a_K in E_Λ; write $a_K = c\mathsf{P}(a_\Lambda)$ – then, in view of the fact that $U_\Lambda(X)\circ\mathsf{P} = \mathsf{P}\circ U_\Lambda(X) = 0$ (all $X \in \mathfrak{k}_c$), we have $\Lambda(H)\mathsf{P}(a_\Lambda) = \mathsf{P}\circ U_\Lambda(H)(a_\Lambda) = 0$ (all $H \in \mathfrak{a}_{\mathfrak{k}}$), whence $\Lambda | \mathfrak{a}_{\mathfrak{k}} = 0$, a_K being non-zero. As for the integrality condition, one may reason as follows. In the Appendix to the present number, it will be shown that the set $\{H \in \sqrt{-1}\mathfrak{a}_{\mathfrak{p}} : \exp H \in K\}$ consists of the integral linear combinations of the vectors $2\pi\sqrt{-1}Q_\lambda/(\lambda, \lambda)$ $(\lambda \in \Sigma)$; this being so, it then follows that

$$a_K = U_\Lambda\left(\exp\frac{2\pi\sqrt{-1}Q_\lambda}{(\lambda, \lambda)}\right) a_K$$

$$= \exp\left\{\frac{2\pi\sqrt{-1}\,(\Lambda, \lambda)}{(\lambda, \lambda)}\right\} a_K \qquad (\lambda \in \Sigma^+)$$

which serves to prove that $(\Lambda, \lambda)/(\lambda, \lambda)$ is a non-negative integer for all $\lambda \in \Sigma^+$.

Sufficiency Let U_Λ be a finite dimensional irreducible representation of G on E_Λ with highest weight $\Lambda (\Lambda \in \mathfrak{W}^+)$ which verifies conditions (i) and (ii) supra; let a_Λ be a non-zero vector in the weight space E^Λ determined by Λ – then we claim that a_Λ is $\mathsf{M}N^+$-invariant (M the centralizer of $A_\mathfrak{p}$ in K). Of course we need only verify the claim for $m \in \mathsf{M}$. So write $\mathsf{M} = K \cap \exp(\sqrt{-1}\mathfrak{a}_\mathfrak{p})\cdot M$ (M the identity component of M – cf. the proof of Lemma 1.1.3.8); because Λ vanishes on $\mathfrak{a}_{\mathfrak{k}_c}$, $\Lambda - \alpha(\alpha \in P_-)$ is not a weight of U_Λ (for $\Lambda - \alpha$ is the reflection of $\Lambda + \alpha$ in the hyperplane $\alpha = 0$ ($\alpha \in P_-$) . . .), whence $U_\Lambda(M)a_\Lambda = a_\Lambda$; on the other hand, in view of the explicit description of the set $K \cap \exp(\sqrt{-1}\mathfrak{a}_\mathfrak{p})$ which was mentioned above and the integrality condition on Λ, we infer that $U_\Lambda(K \cap \exp(\sqrt{-1}\mathfrak{a}_\mathfrak{p}))a_\Lambda = a_\Lambda$ – therefore $U_\Lambda(\mathsf{M})a_\Lambda = a_\Lambda$, as claimed. Now set $a_K = \int_K U_\Lambda(k)a_\Lambda dk$ – then a_K is obviously K-invariant. To see that it is non-zero, we proceed as follows. Owing to the compactness of K, it can be assumed that E_Λ carries the structure of a Hilbert space relative to which the Hilbert space adjoint of $U_\Lambda(x)$ is $U_\Lambda(\theta(x^{-1}))$ ($x \in G$). [Here θ is the Cartan involution associated with the pair $(\mathfrak{g}, \mathfrak{k})$.] Agreeing to use a standard integration formula (cf. Vol. II, number 8.1.3) and bearing in mind that the Haar measure on N^- is invariant under the transformations $n \mapsto mnm^{-1}$ ($m \in \mathsf{M}$), we find that

$$\begin{aligned}(a_K, a_\Lambda) &= \int_{\mathsf{M}\times N^-} (U_\Lambda(m\kappa(n))a_\Lambda, a_\Lambda)e^{-2\rho(H(n))}\, dm d_{N^-}(n)\\ &= \int_{N^-} (U_\Lambda(\kappa(n))a_\Lambda, a_\Lambda)e^{-2\rho(H(n))}\, d_{N^-}(n)\\ &= \int_{N^-} (U_\Lambda(n)a_\Lambda, a_\Lambda)e^{-(\Lambda+2\rho)(H(n))}\, d_{N^-}(n)\\ &= (a_\Lambda, a_\Lambda)\int_{N^-} e^{-(\Lambda+2\rho)(H(n))}\, d_{N^-}(n) \neq 0\end{aligned}$$

where, in passing to the last equality, we used the fact that $(U_\Lambda(n)a_\Lambda, a_\Lambda) = (a_\Lambda, U_\Lambda^*(n)a_\Lambda) = (a_\Lambda, U_\Lambda(\theta(n^{-1}))a_\Lambda) = (a_\Lambda, a_\Lambda)$ (all $n \in N^- = \theta(N^+)$ . . .). [The integral over N^- really does not vanish – cf. Vol. II, Chapter 9.]

The proof of the theorem is therefore complete. □

[Evidently the preceding theorem (or rather the proof thereof) serves to establish the following point: A finite dimensional irreducible representation of G admits a K-invariant vector iff it admits an $\mathsf{M}N^+$-invariant vector.]

Application (Harish-Chandra) Keeping to the above notations and assumptions, let L_0 denote the set of all (complex-valued) linear functions on $\mathfrak{a}_\mathfrak{p}$ with the property that there exists a class one representation of the pair (G, K) whose highest weight coincides on $\mathfrak{a}_\mathfrak{p}$ with the given linear function. Take now an arbitrary dominant integral Λ and let $\tilde{\Lambda}$ denote the restriction of Λ to $\mathfrak{a}_\mathfrak{p}$ – then $2\tilde{\Lambda} \in \mathsf{L}_0$. [Let Ω be the linear function on $\mathfrak{a}_\mathfrak{p} + \sqrt{-1}\mathfrak{a}_\mathfrak{k}$ obtained by rendering $2\tilde{\Lambda}$ trivial on $\sqrt{-1}\mathfrak{a}_\mathfrak{k}$; our contention will follow when it is shown that Ω is the highest weight of a class one representation of the pair (G, K). To this end, let us first show that

$(\Omega, \lambda)/(\lambda, \lambda)$ is a non-negative integer for all $\lambda \in \Sigma^+$. Thus fix a λ in Σ^+ along with an element $\alpha \in P_+$ such that $\lambda = \tilde{\alpha}$. If $\alpha = \alpha^\sigma$, then $(\alpha, \alpha) = (\lambda, \lambda)$, whence $(\Omega, \lambda)/(\lambda, \lambda) = 2(\Lambda, \alpha)/(\alpha, \alpha) \in \mathbf{N}$; if $\alpha \neq \alpha^\sigma$, then either $(\alpha, \alpha) = 2(\lambda, \lambda)$ or $(\alpha, \alpha) = 4(\lambda, \lambda)$ (cf. number 1.1.3) and $\lambda = (\tilde{\alpha} + \tilde{\alpha}^\sigma)/2$, whence

$$\begin{aligned}\frac{(\Omega, \lambda)}{(\lambda, \lambda)} &= \frac{2(\Lambda, \lambda)}{(\lambda, \lambda)} \\ &= \frac{2(\Lambda, 2^{-1}(\alpha + \alpha^\sigma))}{n(\alpha, \alpha)} = \frac{(\Lambda, \alpha)}{n(\alpha, \alpha)} + \frac{(\Lambda, \alpha^\sigma)}{n(\alpha^\sigma, \alpha^\sigma)} \in \mathbf{N}\end{aligned}$$

since $n = 2^{-1}$ or 4^{-1}. To see that Ω is dominant integral, write

$$2\frac{(\Omega, \alpha)}{(\alpha, \alpha)} = \begin{cases} \dfrac{2}{n}\dfrac{(\Omega, \tilde{\alpha})}{(\tilde{\alpha}, \tilde{\alpha})} & \text{if } 2\tilde{\alpha} \notin \Sigma, \\ \dfrac{4}{n}\dfrac{(\Omega, 2\tilde{\alpha})}{(2\tilde{\alpha}, 2\tilde{\alpha})} & \text{if } 2\tilde{\alpha} \in \Sigma, \end{cases} \qquad \alpha \in P_+$$

n being such that $(\alpha, \alpha) = n(\tilde{\alpha}, \tilde{\alpha})$. In the first case $n = 1, 2$ while in the second $n = 1, 2, 4$ (cf. Appendix 2 to number 1.1.3).]

Remark Agreeing to use the notations introduced during the proof of Theorem 3.3.1.1, let U_Λ be a class one representation of the pair (G, K) on a finite dimensional vector space E_Λ with highest weight $\Lambda(\Lambda \in \mathfrak{W}^+)$; fix a non-zero unit vector $a_K \in E_\Lambda$ which is invariant under K (it being tacitly supposed that E_Λ is supplied with the structure of a Hilbert space relative to which $U_\Lambda | K$ is unitary . . .) – then we have

$$(U_\Lambda(x)a_K, a_K) = \int_K e^{\Lambda(H(xk))}\, dk \qquad (\text{all } x \in G).$$

Thus, for $x \in G$, $(U_\Lambda(x)a_K, a_K) = |c|^2(\mathsf{P}U_\Lambda(x)\mathsf{P}(a_\Lambda), a_\Lambda)$ and

$$\mathsf{P}U_\Lambda(x)\mathsf{P}(a_\Lambda) = \int_K \mathsf{P}U_\Lambda(xk)a_\Lambda\, dk = \left(\int_K e^{\Lambda(H(xk))}\, dk\right)\mathsf{P}(a_\Lambda),$$

from which we can read off our assertion, a_K being a unit vector in E_Λ. Needless to say, the function

$$x \mapsto \int_K e^{\Lambda(H(xk))}\, dk \qquad (x \in G)$$

is a zonal spherical function on G . . . .

Example Consider the pair (G, K) with $G = \mathbf{SL}(2, \mathbf{R})$ and $K = \mathbf{SO}(2)$; since the finite dimensional irreducible representations of G are parameterized by the non-negative integers $0, 1, 2, \ldots$, it follows from Theorem 3.3.1.1 that the finite dimensional class one representations of the pair (G, K) correspond to the set of non-negative even integers $0, 2, 4, \ldots$.

Appendix Let G be a connected semi-simple Lie group with finite center imbedded in its simply connected 'complexification' G_c; write, after Iwasawa, $G = KA_\mathfrak{p}N^+$ and let Σ denote the set of roots of the pair $(\mathfrak{g}, \mathfrak{a}_\mathfrak{p})$ – then we have the following result.

Lemma (Helgason) Retain the above notations and assumptions – then the set $\mathfrak{L} = \{H \in \sqrt{-1}\mathfrak{a}_\mathfrak{p} : \exp H \in K\}$ consists of the integral linear combinations of the vectors $2\pi\sqrt{-1}Q_\lambda/(\lambda, \lambda)$ $(\lambda \in \Sigma)$.

Proof In the first place we have $\mathfrak{L} = \{H \in \sqrt{-1}\mathfrak{a}_\mathfrak{p} : \exp(2H) = 1\}$ (cf. Helgason [2, p.272]); this being so, let Γ be the group generated by W (the Weyl group of the pair $(\mathfrak{g}, \mathfrak{a}_\mathfrak{p})$) and $\mathfrak{L}$. For each $\lambda \in \Sigma$, let $w_{\lambda,n}$ be the reflection in the plane $\lambda = n\pi\sqrt{-1}(n \in \mathbf{Z})$; if H_0 lies in this plane, then $w_{\lambda,n}(H) = w_\lambda(H) - w_\lambda(H_0) + H_0$ $(H \in \sqrt{-1}\mathfrak{a}_\mathfrak{p})$; on the other hand,

$$\exp(2(w_\lambda(H_0) - H_0)) = \exp\left(\frac{4n\pi\sqrt{-1}Q_\lambda}{(\lambda, \lambda)}\right) = 1$$

(cf. Helgason [2, p. 258]) – therefore $w_{\lambda,n} \in \Gamma$ (all $\lambda \in \Sigma$). Arguing as in Helgason [2, p.279], one then sees that Γ is actually generated by the $w_{\lambda,n}$. Now let L be the lattice generated by the $2\pi\sqrt{-1}Q_\lambda/(\lambda, \lambda)$ $(\lambda \in \Sigma)$; in view of what was said above, it is clear that L is contained in $\mathfrak{L}$ and so our lemma will follow when it is shown that L is Γ-stable; this however follows from the explicit expression for the reflection $w_{\lambda,n}$ (cf. supra) and the fact that

$$\frac{2(\lambda, \mu)}{(\lambda, \lambda)} \in \mathbf{Z} \quad (\text{all } \lambda, \mu \in \Sigma). \qquad \square$$

3.3.2 Inequalities

In this number we shall derive a few inequalities of a geometrical nature using the methods of representation theory; these results turn out to be of particular importance in the theory of spherical functions.

As in 3.3.1, we shall let G be a connected semi-simple Lie group with finite center imbedded in its simply connected 'complexification' G_c; write, after Iwasawa, $G = KA_\mathfrak{p}N^+$ and let Σ denote the set of roots of the pair $(\mathfrak{g}, \mathfrak{a}_\mathfrak{p})$; let $\mathscr{C}$ be the positive Weyl chamber in $\mathfrak{a}_\mathfrak{p}$ (relative to the given Iwasawa decomposition of G...) and let $\mathfrak{a}_\mathfrak{p}^+$ denote the subset of $\mathfrak{a}_\mathfrak{p}$ comprised of those H such that $(H, \mathscr{C}) \geqslant 0$ – then we have the following Criterion.

Criterion (Harish-Chandra) Let L_0 denote the set of all linear functions on $\mathfrak{a}_\mathfrak{p}$ with the property that there exists a class one representation of the pair (G, K) whose highest weight coincides on $\mathfrak{a}_\mathfrak{p}$ with the given linear function – then an element H in $\mathfrak{a}_\mathfrak{p}$ lies in $\mathfrak{a}_\mathfrak{p}^+$ iff $\eta(H) \geqslant 0$ for all $\eta \in \mathsf{L}_0$.

[Let $\lambda_1, \ldots, \lambda_l$ be the fundamental system of roots for Σ determined by $\mathscr{C}$; choose elements $H_i \in \mathfrak{a}_\mathfrak{p}$ such that $\langle H_i, \lambda_j \rangle = \delta_{ij}$ $(1 \leqslant i, j \leqslant l)$ – then it is clear that $\mathscr{C}$ consists of all elements of the form $\sum_1^l t_i H_i$ where the t_i are positive real numbers. Moreover an element $H \in \mathfrak{a}_\mathfrak{p}$ obviously lies in $\mathfrak{a}_\mathfrak{p}^+$ iff $(H, H_i) \geqslant 0$ for every i. Consider the linear function μ_i on $\mathfrak{a}_c$ defined by the rule $\mu_i(H) = B(H_i, H)$ $(H \in \mathfrak{a}_c)$ – then we have $\mu_i(H_\alpha) = B(H_i, Q_\alpha + P_\alpha) = \tilde{\alpha}(H_i) \geqslant 0$ for all $\alpha \in \Phi^+$ (Φ^+ the positive roots of the pair $(\mathfrak{g}_c, \mathfrak{a}_c)$). Select a positive integer m such that $2m\mu_i(H_\alpha)/(\alpha, \alpha)$ is a non-negative integer for every $\alpha \in \Phi^+$ and every i $(1 \leqslant i \leqslant l)$; let η_i denote the restriction of $2m\mu_i$ to $\mathfrak{a}_\mathfrak{p}$ – then we claim that $\eta_i \in \mathsf{L}_0$. In fact $m\mu_i$ is dominant integral (by construction) and so our claim follows upon citing an observation which was made in 3.3.1. Given any linear function η on $\mathfrak{a}_\mathfrak{p}$, we may write $\eta = \sum_i c_i\eta_i$ where $c_i = (\eta, \lambda_i)/2m$; if in addition $\eta \in \mathsf{L}_0$, then $\eta \geqslant w_{\lambda_i}(\eta)$ (w_{λ_i} the Weyl reflection in W corresponding to λ_i) and so $2mc_i = (\eta, \lambda_i) \geqslant 0$. Now fix an $H \in \mathfrak{a}_\mathfrak{p}$ – then our remarks show that $(H, H_i) \geqslant 0$ iff $\eta(H) \geqslant 0$ $(i = 1, \ldots, l)$ which serves to establish the Criterion.]

Given a finite dimensional representation U of G on E (say), it will be tacitly assumed in what follows that E carries the structure of a Hilbert space relative to which the Hilbert space adjoint of $U(x)$ is $U(\theta(x^{-1}))$ $(x \in G)$. [Here θ is the Cartan involution associated with the pair (G, K).]

Lemma 3.3.2.1 Let W denote the Weyl group of the pair $(\mathfrak{g}, \mathfrak{a}_\mathfrak{p})$; let H_1, $H_2 \in \mathscr{C}$ – then $(H_1, H_2) > (H_1, wH_2)$ for all $w \neq 1$ in W.

Proof Fix an element $w \neq 1$ in W and select an $m_w \in M^*$ associated with w. Let U_Λ be a finite dimensional irreducible representation of G on E_Λ with highest weight Λ – then $U_\Lambda(wH) = U_\Lambda(m_w)U_\Lambda(H)U_\Lambda(m_w^{-1})$ (all $H \in \mathfrak{a}_\mathfrak{p}$). Suppose now that H_2 belongs to $\mathscr{C}$ – then, since $\Lambda(H_2)$ is the largest eigenvalue of $U_\Lambda(H_2)$, it follows that $\Lambda(wH_2) \leqslant \Lambda(H_2)$, whence $H_2 - wH_2 \in \mathfrak{a}_\mathfrak{p}^+$. On the other hand, $H_2 \neq wH_2$ and so it must be the case that $\eta_i(H_2 - wH_2) \neq 0$ for some i (η_i as above) – therefore, for $H_1 \in \mathscr{C}$, we have $(H_1, H_2 - wH_2) = (2m)^{-1} \sum_{i=1}^{l} \lambda_i(H_1)\eta_i(H_2 - wH_2) > 0$, as we wished to prove. □

Lemma 3.3.2.2 Retain the above notations and assumptions – then $\mathscr{C}$ is contained in $\mathfrak{a}_\mathfrak{p}^+$ and, moreover, $\log h - H(hk) \in \mathfrak{a}_\mathfrak{p}^+$ for all $h \in \exp(\mathscr{C}^{cl})$ and $k \in K$.

Proof As always, if $h \in A_\mathfrak{p}$, then $\log h$ denotes the unique element in $\mathfrak{a}_\mathfrak{p}$ for which $\exp(\log h) = h$. It will be convenient to prove the two assertions together and, in turn, for this purpose we shall apply the Criterion obtained above. Thus let U_Λ be an irreducible representation of G on a finite dimensional vector space E_Λ with highest weight Λ; let a_Λ be a unit vector in E_Λ belonging to Λ – then, since $U_\Lambda(H)$ $(H \in \mathfrak{a}_\mathfrak{p})$ is a self-adjoint operator of trace zero, it must be the case that $\Lambda(H) \geqslant 0$ (all $H \in \mathscr{C}$). On the other hand, if $h \in \exp(\mathscr{C}^{cl})$, then h^Λ is the largest eigenvalue of the self-adjoint operator $U_\Lambda(h)$ – therefore

$$\| U_\Lambda(hk)a_\Lambda \| \leqslant h^\Lambda \| U_\Lambda(k)a_\Lambda \| = h^\Lambda \qquad (\text{all } k \in K).$$

[Recall that if T is a self-adjoint operator in a finite dimensional vector space E over $\mathbf{C}$, then the operator norm $\|T\|$ of T cannot exceed the absolute value of the largest eigenvalue in its spectrum.] But it is obvious that $\| U_\Lambda(hk)a_\Lambda \| = e^{\Lambda(H(hk))}$, hence $\Lambda(\log h - H(hk)) \geqslant 0$ $(h \in \exp(\mathscr{C}^{cl}), k \in K)$. There can be no doubt, then, that $\eta(H) \geqslant 0$ $(H \in \mathscr{C})$ and $\eta(\log h - H(hk)) \geqslant 0$ $(h \in \exp(\mathscr{C}^{cl}), k \in K)$ for all $\eta \in \mathsf{L}_0$. □

Let W denote the Weyl group of the pair $(\mathfrak{g}, \mathfrak{a}_\mathfrak{p})$; let $w^\dagger$ denote that element in W which sends Σ^+ to Σ^-, i.e. that element in W which takes the positive Weyl chamber $\mathscr{C}$ to its negative; put $\imath = -w^\dagger$ – then $\imath$ is the opposition involution in Σ (cf. number 1.1.2).

Lemma 3.3.2.3 Retain the above notations and assumptions – then $(\log h)^\imath + H(hk) \in \mathfrak{a}_\mathfrak{p}^+$ for all $h \in \exp(\mathscr{C}^{cl})$ and $k \in K$.

Proof Fix an element $m_{w^\dagger}$ in M^* which is associated with $w^\dagger$ – then, for $h \in \exp(\mathscr{C}^{cl})$, we have $h^\imath = m_{w^\dagger}h^{-1}m_{w^\dagger}^{-1}$ (here we have set $h^\imath = \exp((\log h)^\imath)$). It therefore follows that $H(h^\imath m_{w^\dagger}\kappa(hk)) = H(h^{-1}\kappa(hk)) = -H(hk)$, whence

$$(\log h)^\imath + H(hk) = \log h^\imath - H(h^\imath m_{w^\dagger}\kappa(hk)) \in \mathfrak{a}_\mathfrak{p}^+,$$

as desired. □

Let us agree to write ρ in place of $2^{-1} \sum_{\lambda>0} m(\lambda)\lambda$.

Lemma 3.3.2.4 Retain the above notations and assumptions – then $\rho(\log h) \geqslant |\rho(H(hk))|$ for all $h \in \exp(\mathscr{C}^{cl})$ and $k \in K$.

Proof It is obvious that ρ lies in $\mathscr{C}$; furthermore $\rho(\log h^{\iota}) = \rho(\log h)$ (all $h \in \exp(\mathscr{C}^{cl})$) – therefore our lemma follows from the fact that both $\log h - H(hk)$ and $\log h^{\iota} + H(hk)$ are in $\mathfrak{a}_{\mathfrak{p}}^{+}$ ($h \in \exp(\mathscr{C}^{cl})$, $k \in K$). □

Let N^- denote the analytic subgroup of G corresponding to $\mathfrak{n}^-(= \theta(\mathfrak{n}^+)\ldots)$.

Lemma 3.3.2.5 Retain the above notations and assumptions; fix $h \in \exp(\mathscr{C}^{cl})$ and $n \in N^-$ – then both $H(n)$ and $H(n) - H(hnh^{-1})$ lie in $\mathfrak{a}_{\mathfrak{p}}^{+}$ (hence $\rho(H(n)) \geqslant 0$).

Proof Let U_Λ be a finite dimensional irreducible representation of G on E_Λ with highest weight Λ, a_Λ a unit vector in E_Λ belonging to Λ. It is clear that $\| U_\Lambda(n)a_\Lambda \| = e^{\Lambda(H(n))}$; in addition, since h^Λ is the largest eigenvalue of the self-adjoint operator $U_\Lambda(h)$ ($h \in \exp(\mathscr{C}^{cl})$), we have

$$e^{\Lambda(H(hnh^{-1}))} = \| U_\Lambda(hnh^{-1})a_\Lambda \| \leqslant h^\Lambda \| U_\Lambda(nh^{-1})a_\Lambda \| = \| U_\Lambda(n)a_\Lambda \| = e^{\Lambda(H(n))}.$$

This shows that $\Lambda(H(n) - H(hnh^{-1})) \geqslant 0$, whence $H(n) - H(hnh^{-1}) \in \mathfrak{a}_{\mathfrak{p}}^{+}$, Λ being an arbitrary element in $\mathfrak{W}^+$. Next, choose an $X \in \mathfrak{n}^-$ such that $n = \exp X$ and then note that $U_\Lambda(\exp X)a_\Lambda - a_\Lambda$ is a sum of vectors belonging to weights lower than Λ – thus a_Λ and $U_\Lambda(n)a_\Lambda - a_\Lambda$ are mutually orthogonal and so

$$\| U_\Lambda(n)a_\Lambda \|^2 = \| a_\Lambda \|^2 + \| U_\Lambda(n)a_\Lambda - a_\Lambda \|^2 \geqslant 1.$$

Conclusion: $\| U_\Lambda(n)a_\Lambda \| = e^{\Lambda(H(n))} \geqslant 1 \Rightarrow \Lambda(H(n)) \geqslant 0 \Rightarrow H(n) \in \mathfrak{a}_{\mathfrak{p}}^{+}$. The fact that $\rho(H(n)) \geqslant 0$ ($n \in N^-$) is clear since $\rho \in \mathscr{C}$. Hence the lemma. □

Recall É. Cartan's 'polar coordinate decomposition' of G/K: The map

$$(kM, h) \mapsto khK$$

is a differentiable map of $K/M \times A_{\mathfrak{p}}$ onto G/K and if we put $A_{\mathfrak{p}}' = \exp(\mathfrak{a}_{\mathfrak{p}}')$ ($\mathfrak{a}_{\mathfrak{p}}'$ the Σ-regular elements in $\mathfrak{a}_{\mathfrak{p}}$), then the restriction to $K/M \times A_{\mathfrak{p}}'$ is a covering of order $[W]$ (for the proof, see Helgason [2, p.381]). There is also a well-known description of the orbit space for the action of K on G/K, viz: $K\backslash G/K = W\backslash A_{\mathfrak{p}}$ – here $W\backslash A_{\mathfrak{p}}$ denotes the space obtained from $A_{\mathfrak{p}}$ by identifying the points which are conjugate under the canonical action of W on $A_{\mathfrak{p}}$. [The proof of this fact goes as follows. Let $x \in G$; write $x = k_1hk_2$ ($k_1, k_2 \in K$; $h \in A_{\mathfrak{p}}$). If $x = \tilde{k}_1\tilde{h}\tilde{k}_2$ is another such representation of x, then an application of the automorphism θ gives $kh^2k^{-1} = \tilde{h}^2$ ($k \in K$); this implies that h^2 and $\tilde{h}^2$ (and therefore h and $\tilde{h}$) are conjugate under W. ...] Evidently, then, $G = K\exp(\mathscr{C}^{cl})K$ while $K\exp(\mathscr{C})K$ is dense in G.

Lemma 3.3.2.6 Retain the above notations and assumptions; let h and n be two elements in $A_{\mathfrak{p}}$ and N^-, respectively; choose an $h^+ \in \exp(\mathscr{C}^{cl})$ such that $nh \in Kh^+K$ – then $\rho(\log h^+) \geqslant \rho(H(n)) + \rho(\log h)$.

Proof Fix an irreducible representation U_ρ of G on E_ρ whose highest weight with respect to $\mathfrak{a}_{\mathfrak{p}}$ is ρ; let a_ρ be a unit vector in E_ρ belonging to the weight ρ. Writing $nh = k_1 h^+ k_2$ ($k_1, k_2 \in K$), we find that $\|U_\rho(nh)a_\rho\| = \|U_\rho(h^+k_2)a_\rho\|$, whence $\rho(H(n)) + \rho(\log h) = \rho(H(h^+k_2))$; since h^+ lies in $\exp(\mathscr{C}^{cl})$, the present lemma now follows upon citing Lemma 3.3.2.4. □

Remark Retain the notations which were introduced during the proof of the preceding lemma – then we claim that $\rho(\log h^+) + \log\sqrt{d_\rho} \geqslant |\rho(\log h)|$ ($d_\rho = \dim(E_\rho)$). Thus, for $T \in \mathrm{Hom}_{\mathbf{C}}(E_\rho, E_\rho)$, let us agree to write $\|T\|_{HS}^2$ for $\mathrm{tr}(TT^*)$ (T^* the adjoint of T); of course $\|T\|_{HS}$ is nothing more than the Hilbert-Schmidt norm of T ($T \in \mathrm{Hom}_{\mathbf{C}}(E_\rho, E_\rho)$). Since it is clear that $\|U_\rho(h^+)\|_{HS} = \|U_\rho(nh)\|_{HS}$, we have

$$(h^+)^{2\rho} \leqslant \|U_\rho(h^+)\|_{HS}^2 \leqslant d_\rho(h^+)^{2\rho}.$$

Now fix a w in W and choose an element m_w in M^* which is associated with w; on the basis of the discussion in number 1.1.4, we can write $m_w n m_w^{-1} = n^- n^+$ where $n^- \in N^- \cap m_w N^- m_w^{-1}$ and $n^+ \in N^+ \cap m_w N^- m_w^{-1}$ – therefore

$$\begin{aligned}\|U_\rho(h^+)\|_{HS} &= \|U_\rho(m_w n m_w^{-1} \cdot m_w h m_w^{-1})\|_{HS}\\ &= \|U_\rho(n^- m_w h m_w^{-1} \tilde{n}^+)\|_{HS} \qquad (\tilde{n}^+ = (w\cdot h)^{-1} n^+ (w\cdot h) \in N^+),\end{aligned}$$

whence

$$\begin{aligned}&\|U_\rho(h^+)\|_{HS}^2 \geqslant \|U_\rho(n^- m_w h m_w^{-1})a_\rho\|^2 \geqslant (w\cdot h)^{2\rho}\\ \Rightarrow\;& \\ &\rho(\log h^+) + \log\sqrt{d_\rho} \geqslant \max_{w\in W} \rho(w\cdot h) \geqslant |\rho(\log h)|\end{aligned}$$

which establishes the claim.

Chapter 4

Infinite Dimensional Group Representation Theory

Introduction

Let G be a Lie group, E a complete locally convex linear space over $\mathbf{C}$. A continuous representation U of G on E is a homomorphism $x \mapsto U(x)$ of G into the group of topological automorphisms of E such that the map $(x, a) \mapsto U(x)a$ of $G \times E$ into E is continuous. Although our ultimate concern will be with continuous representations on a Banach space (or even a Hilbert space), necessity dictates the need to formulate the fundamental notions of representation theory for more general spaces E. For instance, one has canonical continuous representations of G on the spaces $C^\infty(G)$ and $C_c^\infty(G)$ (equipped with their usual topologies) or on the space of distributions on G. There are other reasons too. Thus, as will be seen in 4.4, the study of Banach representations leads one quickly to the consideration of continuous representations on a Fréchet space. On the other hand, 'analytic continuation' of Banach representations can sometimes lead to continuous representations which simply cannot be realized on a Banach space.

In 4.1 the basic concepts and definitions of representation theory are presented. Methods of forming new representations from old are described; in particular the contragredient representation and conjugate representation are considered. Also the procedure of 'lifting' the given representation of G to a representation of the algebra $M_c(G)$ (the set of compactly supported Radon measures on G) is examined.

Representations of an associative algebra A by continuous operators on a Banach space E are the principal theme in 4.2. The various definitions of irreducibility for such representations are investigated – an important role here is played by those representing operators which are of finite rank. The notion of Naimark equivalence is introduced and is shown to be an equivalence relation in a certain important special case. Finally these results are applied to Banach representations of G by considering their 'lift' to the associative algebra $M_c(G)$.

In 4.3 a few simple facts about unitary representations are collected. In particular it is shown that unitary equivalence coincides with Naimark equivalence in this case.

It is not until 4.4 that the assumption that G is a Lie group is really used. If U is a continuous representation of G on E, then the subspace E_∞ consisting of those $a \in E$ for which the map $x \mapsto U(x)a$ is an element of $C^\infty(G; E)$ constitutes a dense U-stable subspace of E – the space of differentiable vectors in E. The space E_∞ can be canonically equipped with a finer topology with which, if E is a Banach space, it acquires the structure of a Fréchet space. The corresponding continuous representation U_∞ of G on E_∞ is 'nicer' than U; the passage from U to U_∞ plays an important role in the sequel. On the other hand U may also be studied infinitesimally. Thus U gives rise, by differentiation, to a representation of the universal enveloping algebra $\mathfrak{G}$ of $\mathfrak{g}_c$, $\mathfrak{g}$ the Lie algebra of G. Although this approach has its benefits (especially in the unitary case –

see 4.4.4), the connection between U and this 'infinitesimal representation' of $\mathfrak{G}$ is not as close as one would hope. In 4.4.5, the link between Banach representations U of G and the corresponding representations of $\mathfrak{G}$ is restored. This necessitates the introduction of analytic vectors, a concept due to Harish-Chandra. A vector $a \in E$ is said to be analytic for U if the map $x \mapsto U(x)a$ of G into E is analytic. An important theorem due to Nelson says that the analytic vectors are dense in E. Their use permits one to establish a rather close relationship between Banach representations of G and representations of $\mathfrak{G}$.

In 4.5 we take up the study of an important class of Lie groups G, namely those which admit a large compact subgroup. A compact subgroup $K \subset G$ is said to be large (in G) if for each $\delta \in \hat{K}$ there exists an integer $m(\delta) \geqslant 1$ such that δ occurs no more than $m(\delta)$ times in every topologically completely irreducible Banach representation of G. Examples of groups which admit large compact subgroups include the connected semi-simple Lie groups with finite center and the 'motion' groups. The basic properties of largeness are then developed and the implications for the representation theory of G are explored. The notion of infinitesimal equivalence is introduced and is shown to coincide with Naimark equivalence in the cases which will arise in the applications. A basic consequence of the existence of a uniformly large compact subgroup K is that in any irreducible unitary representation U of G the operator $U(f)$ is of the trace class for all $f \in C_c^\infty(G)$; it then turns out that the mapping $f \mapsto \operatorname{tr}(U(f))$ $(f \in C_c^\infty(G))$ defines a distribution T_U on G – the character of U. It is shown in 4.5.8 that two irreducible unitary representations of G are unitarily equivalent iff they have the same character. The section concludes with a treatment of the square integrable representations of G, due to Harish-Chandra, which is self-contained and makes no use of the theory of von-Neumann algebras.

4.1 Representations on a Locally Convex Space

4.1.1 Basic Concepts

Although our primary concern is with continuous representations of a locally compact group G on a Banach space, nevertheless the more general situation of a continuous representation on a locally convex space will often occur. For example one has canonical continuous representations of G by translation on the spaces $C(G)$ and $C_c(G)$ equipped with their usual topologies. [If G is a Lie group, then G may also be represented continuously by translation on the spaces $C^\infty(G)$ and $C_c^\infty(G)$.] However, a systematic treatment of representation theory in this more general setting will not be undertaken – only the material directly pertinent for the sequel will be given and that in a concise form.

Let E be a locally convex, complete, Hausdorff, topological vector space over **C** whose topology is described by the family of continuous semi-norms $\{|\cdot|_\alpha : \alpha \in \mathfrak{A}\}$.

Let G be a locally compact group which is, furthermore, countable at infinity. In this number the symbol d_G will always stand for left Haar measure on G.

Definition A homomorphism $x \mapsto U(x)$ of G into the group of automorphisms of E is called a *representation* of G on E.

Definition A homomorphism $x \mapsto U(x)$ of G into the group of topological automorphisms of E is called a *continuous representation* of G on E if the map $(x, a) \mapsto U(x)a$ of $G \times E$ into E is continuous.

The continuity condition in this definition is equivalent to:

(a) For every $a \in E$ the map $x \mapsto U(x)a$ of G into E is continuous (that is the map $x \mapsto U(x)$ of G into $\mathfrak{L}_s(E; E)$ is continuous);

(b) For every compact subset $\omega \subset G$ the set of operators $\{U(x) : x \in \omega\}$ is equicontinuous (that is for any neighborhood $\mathscr{P}_1$ of zero in E there exists another neighborhood $\mathscr{P}_2$ of zero in E such that $U(\omega)(\mathscr{P}_2) \subset \mathscr{P}_1$).

[It is clear that a continuous representation $x \mapsto U(x)$ of G on E satisfies (a) and (b). Conversely, suppose (a) and (b) are met by some representation U of G – in view of (a), the continuity of U will follow when it is shown that $U(x_n)a_n \to 0$ in E whenever $x_n \to x$ in G and $a_n \to 0$ in E. So suppose $x_n \to x$ in G; pick a compact neighborhood ω of x in G such that $x_n \in \omega$ for all $n \geqslant n_0$, say – then the set of operators $\{U(x_n) : n \geqslant n_0\}$ is equicontinuous. Fix a neighborhood $\mathscr{P}_1$ of zero in E; choose a neighborhood $\mathscr{P}_2$ of zero in E such that $U(x_n)(\mathscr{P}_2) \subset \mathscr{P}_1$ (all $n \geqslant n_0$) and arrange the notation so that $n \geqslant n_0 \Rightarrow a_n \in \mathscr{P}_2$. Then $U(x_n)a_n \in \mathscr{P}_1$ for all $n \geqslant n_0$, which establishes the continuity of U.]

Observe that (b) follows from (a) if E is barrelled since the Banach-Steinhaus Theorem is then in force (and is directly applicable...) – in particular this is so if E is a Banach space, Fréchet space, or a Montel space. On the other hand, if a representation U satisfies (b), then in order to show that U is a continuous representation, it suffices to verify (a) for a running through a total subset $\mathscr{S}$ of E – this follows from familiar properties of equicontinuous sets. [Thus let $x \in G$, $a \in E$ be arbitrary and suppose $x_n \to x$ in G – we are to show that $U(x_n)a \to U(x)a$ in E. It may be assumed that $\{U(x_n)\}$ is equicontinuous. Given a convex neighborhood $\mathscr{P}_1$ of zero in E, choose a convex symmetric neighborhood $\mathscr{P}_2$ of zero in E such that $U(x)(\mathscr{P}_2) \subset (1/3)\mathscr{P}_1$ and $U(x_n)(\mathscr{P}_2) \subset (1/3)\mathscr{P}_1$ (all n). Next pick an element b in the span of $\mathscr{S}$ such that $a - b \in \mathscr{P}_2$ and then choose an index n_0 such that $n \geqslant n_0 \Rightarrow U(x_n)b - U(x)b \in (1/3)\mathscr{P}_1$. Then, for all $n \geqslant n_0$, we have

$$\begin{aligned} U(x_n)a - U(x)a &= U(x_n)(a - b) + (U(x_n)b - U(x)b) + U(x)(b - a) \\ &\in (1/3)\mathscr{P}_1 + (1/3)\mathscr{P}_1 + (1/3)\mathscr{P}_1 \subset \mathscr{P}_1 \end{aligned}$$

which shows that $U(x_n)a \to U(x)a$ in E, as desired.]

Example (Bourbaki) Let E be a Hilbert space with an orthonormal basis $\{a_y : y \in \mathbf{R}\}$ in a one-to-one correspondence with the real line $\mathbf{R}$. Define $U(x)a_y = a_{x+y}$ $(x \in \mathbf{R})$ – then the assignment $x \mapsto U(x)$ is a representation of $\mathbf{R}$ by isometries of E and in fact the set $\{U(x) : x \in \mathbf{R}\}$ is equicontinuous. Nevertheless U is not a continuous representation of $\mathbf{R}$ on E.

Note that condition (b) implies:

(c) Given a compact set $\omega \subset G$ and an $\alpha \in \mathfrak{A}$, one may choose a $\beta \in \mathfrak{A}$ such that for all $x \in \omega$, $a \in E$

$$| U(x)a |_\alpha \leqslant | a |_\beta.$$

On the other hand one has the following result.

Proposition 4.1.1.1 Let U be a homomorphism of G into the group of topological automorphisms of E. Suppose that:

(i) The map $x \mapsto U(x)a$ of G into E is continuous for every $a \in E$;

(ii) There exists a neighborhood $\mathcal{O}$ of 1 in G with the property: Given $\alpha \in \mathfrak{A}$ one may choose a $\beta \in \mathfrak{A}$ such that

$$| U(x)a |_\alpha \leqslant | a |_\beta$$

for all $x \in \mathcal{O}$, $a \in E$.

Then U is a continuous representation of G on E.

Proof Fix $x \in G$ and $a \in E$; write

$$U(xy)b - U(x)a = U(x)U(y)(b - a) + U(x)(U(y)a - a)$$

for $y \in \mathcal{O}$, $b \in E$. Plainly if $y \to 1$ and $b \to a$, then $U(xy)b \to U(x)a$. □

Example Suppose that G is a Lie group; consider the space $C^\infty(G; E)$ equipped with the usual topology (Appendix 2). [Let us recall that the topology of $C^\infty(G; E)$ can be described by the semi-norms $| f |_{D,\omega}^{\alpha} = \sup_{\omega} | Df(x) |_\alpha$, D a left invariant differential operator on G.] The *left regular* representation L of G (with respect to the canonical left action of G on itself) on $C^\infty(G; E)$ is defined by $L(x)f = f^x (x \in G)$ where $f^x(y) = f(x^{-1}y)$ $(y \in G)$. To verify that L is in fact a continuous representation of G, note that $| L(x)f |_{D,\omega}^{\alpha} = | f |_{D,x^{-1}\omega}^{\alpha}$ and, since $L(x)$ and D commute as endomorphisms of $C^\infty(G; E)$,

$$|L(x)f - f|_{D,\omega}^{\alpha} = \sup_{\omega} | L(x)Df - Df |_\alpha \to 0$$

as $x \to 1$ in G. Hence L is a continuous representation of G by Proposition 4.1.1.1. One may also consider the *right regular* representation R of G on $C^\infty(G; E)$ – thus $R(x)f = f^x$ $(x \in G)$ where $f^x(y) = f(yx)$ $(y \in G)$. We verify as above that R is a continuous representation of G on $C^\infty(G; E)$.

There are of course many other examples of a similar nature – we thus have 'left regular' representations of G on $C_c^\infty(G; E)$ (if G is a Lie group), $L^p(G)$ (with respect to left Haar measure, $p \geqslant 1$), $C_c(G; E)$ etc. Since the reader is expected to be more or less familiar with the details involved here we shall not give them.

Remark In the sequel we shall sometimes use the term 'continuous representation of G on E' even when E is not necessarily complete; no difficulty will arise from this practice.

Let $M_c(G)$ denote the algebra, with respect to convolution '$*$', of Radon measures on G with compact support, $M_\omega(G)$ the space of Radon measures on G with support contained in the compact set ω. If $\mu \in M_c(G)$ let $\|\mu\| = \int_G d|\mu|(x)$, $|\mu|$ the total variation of μ. Equip $M_c(G)$ with the inductive limit of the topologies defined by the norm $\|\mu\|$ on the spaces $M_\omega(G)$ (ω a compact subset of G) – in this fashion $M_c(G)$ acquires the structure of a locally convex, Hausdorff, topological vector space.

Let U be a continuous representation of G on E. Since E is complete, U lifts in the well-known way to a homomorphism $\mu \mapsto U(\mu)$ of the algebra $M_c(G)$ into $\mathfrak{L}(E; E)$, i.e. each $U(\mu)$ is a continuous endomorphism of E and $U(\mu * \nu) = U(\mu)U(\nu)$ for all μ, ν in $M_c(G)$. Here $U(\mu)a = \int_G U(x)a\,d\mu(x)$ for every $a \in E$; thus for all $\mathbf{z} \in E^*$ (E^* the topological dual of E), $\langle U(\mu)a, \mathbf{z}\rangle = \int_G \langle U(x)a, \mathbf{z}\rangle d\mu(x)$. [Fix an element $\mu \in M_c(G)$; since, for each $a \in E$, the map $x \mapsto U(x)a$ is continuous, we can form the E-valued Bochner integral (compare 5.1.1) $U(\mu)a = \int_G U(x)a\,d\mu(x)$ – plainly $U(\mu)a$ is linear in both μ and a. In addition $U(\mu)$ defines a continuous endomorphism of E. To see this, let ω denote the (compact) support of μ. If $a_n \to 0$ in E, then $U(x)a_n \to 0$ uniformly for x in ω and so $U(\mu)a_n \to 0$ in E, as desired. To verify that $U(\mu * \nu) = U(\mu)U(\nu)$ ($\mu, \nu \in M_c(G)$) fix $a \in E$, $\mathbf{z} \in E^*$ – then

$$\begin{aligned}\langle U(\mu*\nu)a, \mathbf{z}\rangle &= \int_G \langle U(x)a, \mathbf{z}\rangle d(\mu*\nu)(x) = \int_G\int_G \langle U(xy)a, \mathbf{z}\rangle\, d\mu(x)d\nu(y)\\ &= \int_G \langle U(\mu)U(y)a, \mathbf{z}\rangle\, d\nu(y) = \int_G \langle U(y)a, {}^tU(\mu)\mathbf{z}\rangle d\nu(y)\\ &= \langle U(\mu)U(\nu)a, \mathbf{z}\rangle \Rightarrow U(\mu*\nu) = U(\mu)U(\nu)\end{aligned}$$

('t' denotes transpose).] Let $f \in C_c(G)$ – then $U(f)$ will denote the operator obtained by identifying f with the measure fd_G (d_G denoting left Haar measure on G). If $\delta_x, f \mapsto f(x)$, is the Dirac measure at x, then of course $U(x) = U(\delta_x)$ ($x \in G$). Let us determine the continuity properties of the map $\mu \mapsto U(\mu)$ ($\mu \in M_c(G)$).

Proposition 4.1.1.2 The map $\mu \mapsto U(\mu)$ of $M_c(G)$ (equipped with the topology described above) into $\mathfrak{L}_b(E; E)$ is continuous.

Proof Let a run through a bounded set in E and let ω be a compact subset of G. Since $U(\omega)$ is equicontinuous, the set $U(\omega)a$ is still bounded in E, call it B; since E is locally convex, the closed balanced convex hull B' of B is bounded. Moreover, for each $\mu \in M_\omega(G)$, we have

$$U(\mu)a = \int_G U(x)a\,d\mu(x) \in \|\mu\|\, B'$$

by known generalities. Therefore $U(\mu)a \to 0$ uniformly in a if $\mu \to 0$ in $M_\omega(G)$ and hence whenever $\mu \to 0$ in $M_c(G)$ by the very definition of the topology in $M_c(G)$. □

Consider the restriction of the map $\mu \mapsto U(\mu)$ to the space $C_c(G)$. We shall view $C_c(G)$ as the inductive limit of the spaces $C_\omega(G)$, ω a compact subset of G, where $C_\omega(G)$ is equipped with the topology of *uniform convergence*. Clearly, then, the assignment $f \mapsto U(f)$ is a continuous homomorphism of the convolution algebra $C_c(G)$ into $\mathfrak{L}_b(E; E)$. [That $C_c(G)$ actually is a convolution algebra results from the fact that G is countable at infinity.] Thus we see that a continuous representation of G on E leads in a natural way to a representation of $C_c(G)$ on E (cf. 4.2.1). Before presenting the converse, let us note the following technical device. Fix an $x \in G$. The relatively compact neighborhoods $\mathcal{O}$ of x form a directed system under inclusion. For each such $\mathcal{O}$ select an element $g_{\mathcal{O}} \in C_c(G)$ such that: (i) spt $g_{\mathcal{O}} \subset \mathcal{O}$; (ii) $g_{\mathcal{O}} \geqslant 0$; (iii) $\int_G g_{\mathcal{O}}(x)d_G(x) = 1$. The net $\{g_{\mathcal{O}}\}$ in $C_c(G)$ has the following properties: (1) For every $f \in C_c(G)$, $g_{\mathcal{O}}*f \rightarrow \delta_x*f$ in $C_c(G)$; (2) For every $a \in E$, $U(g_{\mathcal{O}})a \rightarrow U(x)a$ in E.

Proposition 4.1.1.3 Let $f \mapsto U(f)$ be a continuous homomorphism of the algebra $C_c(G)$ into $\mathfrak{L}_b(E; E)$ – then U corresponds to a continuous representation of G on E iff:

(i) The set of vectors of the form $U(f)a (f \in C_c(G), a \in E)$ constitutes a total subset of E;

(ii) The set of operators $U(f)$ for f with support contained in a fixed compact subset of G and such that $\int_G |f| \, d_G(x) \leqslant 1$, is equicontinuous.

Proof *Necessity* Condition (i) follows from the approximate identity discussion above, since, taking $x = 1$, we have $U(g_{\mathcal{O}})a \rightarrow a$ for every $a \in E$. To see (ii), observe that the set of operators $U(\mu)$ with μ ranging through a bounded subset of $M_c(G)$ (hence a bounded subset of $M_\omega(G)$, ω some compact subset of G) is equicontinuous.

Sufficiency Fix $x \in G$ – then our problem is to define $U(x)$. This will be done by first defining $U(x)$ on vectors $\sum_{i=1}^n U(f_i)a_i$ ($f_i \in C_c(G)$, $a_i \in E$) and then extending by continuity. Thus put, by definition, $U(x)(\sum_{i=1}^n U(f_i)a_i) = \sum_{i=1}^n U(\delta_x*f_i)a_i$ – then $U(x)$ is well-defined. [Thus suppose that $\sum_{i=1}^n U(f_i)a_i = 0$ – then $\lim_{\mathcal{O}} U(g_{\mathcal{O}})(\sum_{i=1}^n U(f_i)a_i) = \lim_{\mathcal{O}} \sum_{i=1}^n U(g_{\mathcal{O}}*f_i)a_i = \sum_{i=1}^n U(\delta_x*f_i)a_i = 0. \ldots$] Moreover $U(x)$ is continuous on the set of vectors of the form $\sum_{i=1}^n U(f_i)a_i$. This is so because for $\mathcal{O}$ 'close to x' the set of operators $U(g_{\mathcal{O}})$ is equicontinuous in view of (ii). Since E is complete we may then extend $U(x)$ to a continuous endomorphism of E by means of the usual argument. One has to verify that U really is a continuous representation of G. The homomorphism property is clear. Since the map $x \mapsto \delta_x*f$ of G into $C_c(G)$ is continuous for each $f \in C_c(G)$, the continuity of the map $x \mapsto U(x)a$ ($a \in E$) follows at once. If ω is a compact subset of G, then the set $\{U(x) : x \in \omega\}$ is equicontinuous since it is contained in the closure of the set of operators $U(g_{\mathcal{O}})$, $\mathcal{O}$ in a relatively compact neighborhood of ω. [Let Ω be a relatively compact neighborhood of ω in G within which the $\mathcal{O}$'s arising from the x's in ω are contained – then the $U(g_{\mathcal{O}})$ ($\mathcal{O} \subset \Omega$) are equicontinuous (cf. (ii)). Moreover for any $x \in \omega$, $U(g_{\mathcal{O}}) \rightarrow U(x)$ in $\mathfrak{L}_s(E; E)$ (by construction). Because the closure in $\mathfrak{L}_s(E; E)$ of any equicontinuous set is again equicontinuous, our contention is clear.] Hence the assignment $x \mapsto U(x)$ defines a continuous representation of G on E. It remains to show that the lift $\tilde{U}$ of U to $C_c(G)$ gets us back where we started from. Recall that if f, $g \in C_c(G)$, then

$$f*g(y) = \int_G \delta_x*g(y)f(x)d_G(x) \qquad (y \in G).$$

Hence:

$$\begin{aligned}\tilde{U}(f)(U(g)a) &= \int_G f(x)U(\delta_x*g)ad_G(x) = U(f*g)a \\ &= U(f)(U(g)a) \qquad (g \in C_c(G), a \in E)\end{aligned}$$

where the formula for convolution is used along with Fubini's Theorem.

The proof of the proposition is now complete. □

Remark Condition (ii) of Proposition 4.1.1.3 implies that the map $f \mapsto U(f)$ of $C_\omega(G)$ into $\mathfrak{L}_b(E; E)$ is continuous for every compact subset ω of G, hence is a continuous map of $C_c(G)$ into $\mathfrak{L}_b(E; E)$.

4.1.2 Operations on Representations

Let E be a locally convex, complete, Hausdorff, topological vector space. Let E^* denote the topological dual of E; the symbols E_s^*, E_c^*, E_b^* stand, respectively, for E^* equipped with the topology of weak convergence, compact convergence, or bounded convergence.

Let G be a locally compact group countable at infinity; throughout the present number, the symbol d_G will stand for left Haar measure on G.

Let U be a continuous representation of G on E – then the map U^*, $x \mapsto {}^tU(x^{-1})$ $(x \in G)$, is a representation of G on E_b^* ('t' denoting transpose). Let us discuss the continuity properties of this representation. In the first place U^* satisfies condition (b) of 4.1.1 since the transpose of an equicontinuous set in $\mathfrak{L}(E; E)$ is equicontinuous in $\mathfrak{L}(E_b^*; E_b^*)$. But the assignment $x \mapsto {}^tU(x^{-1})\mathbf{z}$ is not, in general, a continuous map of G into E_b^* for each $\mathbf{z}$ in E^* (cf. the Example infra). Let, then, $E^\vee$ be the subspace of E^* consisting of those $\mathbf{z}$ for which the rule $x \mapsto {}^tU(x^{-1})\mathbf{z}$ defines a continuous map of G into E_b^*.

Proposition 4.1.2.1 The space $E^\vee$ is closed in E_b^*.

Proof Let $\mathbf{z}$ belong to the closure of $E^\vee$ in E_b^* and let $\{\mathbf{z}_n\}$ be a net in $E^\vee$ converging to $\mathbf{z}$. Let $\mathscr{P}_1$ be a neighborhood of zero in E_b^* – then we must find a neighborhood $\mathscr{O}_1$ of 1 in G such that $x \in \mathscr{O}_1 \Rightarrow {}^tU(x^{-1})\mathbf{z} - \mathbf{z} \in \mathscr{P}_1$. Let $\mathscr{P}_2$ be a convex neighborhood of zero in E_b^* such that $\mathscr{P}_2 + \mathscr{P}_2 \subset \mathscr{P}_1$. Select a relatively compact neighborhood $\mathscr{O}_2$ of 1 in G and choose a convex symmetric neighborhood $\mathscr{P}_3$ of zero in E_b^* such that ${}^tU(x^{-1})\mathscr{P}_3 \subset \mathscr{P}_2$ $(x \in \mathscr{O}_2)$ with $\mathscr{P}_3 + \mathscr{P}_3 \subset \mathscr{P}_2$. Choose an index n_0 so large that $\mathbf{z} - \mathbf{z}_{n_0} \in \mathscr{P}_3$. Write

$$\begin{aligned}{}^tU(x^{-1})\mathbf{z} - \mathbf{z} &= {}^tU(x^{-1})(\mathbf{z} - \mathbf{z}_{n_0}) + {}^tU(x^{-1})\mathbf{z}_{n_0} \\ &\quad - \mathbf{z}_{n_0} + (\mathbf{z}_{n_0} - \mathbf{z}) \qquad (x \in G).\end{aligned}$$

Shrinking $\mathcal{O}_2$, if necessary, to $\mathcal{O}_1$, say, we may assume that ${}^tU(x^{-1})\mathbf{z}_{n_0} - \mathbf{z}_{n_0} \in \mathscr{P}_3$ $(x \in \mathcal{O}_1)$ since $\mathbf{z}_{n_0} \in E^\vee$. Hence ${}^tU(x^{-1})\mathbf{z} - \mathbf{z} \in \mathscr{P}_1$ for all $x \in \mathcal{O}_1$, as desired. □

Remark $E^\vee$ is not necessarily complete (E_b^* need not be complete). [Let us recall that E_b^* is complete whenever E is bornological, in particular whenever E is an LF-space.]

Plainly $E^\vee$ is stable for the restriction of U^* to $E^\vee$.

Definition The continuous representation $x \mapsto {}^tU(x^{-1})$ of G on $E^\vee$ will be called the representation *contragredient* to U; we shall denote it by $U^\vee$.

As above, equip $C_c(G)$ with the inductive limit of the topologies of uniform convergence on the spaces $C_\omega(G)$ (ω a compact subset of G). It is easy to see that the map $f \mapsto {}^tU(f'')$ of $C_c(G)$ into $\mathfrak{L}(E_b^*; E_b^*)$ satisfies condition (ii) of Proposition 4.1.1.3. [As in Appendix 2,

$$f''(x) = \delta_G(x^{-1})f(x^{-1})\ (x \in G).]$$

On the other hand this condition implies that the map $f \mapsto {}^tU(f'')$ is continuous from $C_\omega(G)$ into $\mathfrak{L}_b(E_b^*; E_b^*)$ for every compact set $\omega \subset G$, hence of $C_c(G)$ into $\mathfrak{L}_b(E_b^*; E_b^*)$. Since the map $x \mapsto \delta_x * f$ of G into $C_c(G)$ is continuous, the map $x \mapsto {}^tU((\delta_x * f)'')\mathbf{z} = {}^tU(x^{-1}){}^tU(f'')\mathbf{z}$ of G into E_b^* is continuous for every $\mathbf{z} \in E^*$. Therefore $E^\vee$ contains all elements of the form ${}^tU(f)\mathbf{z}$ ($f \in C_c(G)$, $\mathbf{z} \in E^*$) – in fact $E^\vee$ is the closure in E_b^* of the space spanned by the elements ${}^tU(f)\mathbf{z}$ ($f \in C_c(G)$, $\mathbf{z} \in E^*$). [If $\{g_\mathcal{O}\}$ is the net considered in the preceding number ($\mathcal{O}$ running through the relatively compact neighborhoods of 1), then $\lim_\mathcal{O} {}^tU(g_\mathcal{O})\mathbf{z} = \mathbf{z}$ for each $\mathbf{z}$ in $E^\vee$.]

Proposition 4.1.2.2 The space $E^\vee$ is weakly dense in E^*.

Proof For if not then choose a non-zero element $a \in E$ such that $\langle a, E^\vee \rangle = 0$; select a net $\{g_\mathcal{O}\}$ ($g_\mathcal{O} \in C_c(G)$) such that $U(g_\mathcal{O})a \to a$ in E – then, for each $\mathbf{z} \in E^*$, we have $\langle U(g_\mathcal{O})a, \mathbf{z} \rangle = \langle a, {}^tU(g_\mathcal{O})\mathbf{z} \rangle = 0$, contradicting the fact that $a \neq 0$. □

Corollary 4.1.2.3 If E is semi-reflexive, then $E^\vee = E^*$.

Proof In this case a closed subset in E_b^* is also weakly closed (Kelley and Namioka [1, p. 190]) – therefore our contention follows from the preceding propositions. □

Example Take $E = L^1(G)$ (the set of all complex valued functions on G absolutely summable with respect to the left Haar measure d_G). Let U be the left regular representation of G on $L^1(G)$ – then E^* is the set of essentially bounded Haar measurable functions on G and $E^\vee$ is the subspace of E^* consisting of the bounded (right) uniformly continuous functions on G.

Remark It is not difficult to show that the assignment $x \mapsto {}^tU(x^{-1})$ is a continuous representation of G on E_c^*.

Let $i : E \to \bar{E}$ be a Hermitian linear isomorphism of E onto $\bar{E}$, i.e. i is an isomorphism of the real vector space structures while $i(ca) = \bar{c}i(a)$ $(c \in \mathbf{C})$. If U is a continuous representation of G on E, then by the representation *conjugate* to U (with respect to i) we mean the continuous representation $x \mapsto i \circ U(x) \circ i^{-1}$ of G on $\bar{E}$ – we shall denote it by $\bar{U}$. If U is a unitary representation of G on E (i.e. a continuous representation of G on a Hilbert space E such that the representing operators $U(x)$ $(x \in G)$ are unitary), then by taking for i the canonical map of E onto its dual we see that $\bar{U} = U^\vee$.

The notations of Appendix 2 will be used without comment. Let U^i be a continuous representation of the group G_i on a space E_i (E_i satisfying the assumption made in 4.1.1, $i = 1, 2$) – then for each pair $(x_1, x_2) \in G_1 \times G_2$ the operator $U^1(x_1) \hat{\otimes} U^2(x_2)$ is a continuous invertible endomorphism of the projective tensor product $E_1 \hat{\otimes} E_2$.

Proposition 4.1.2.4 The map $(x_1, x_2) \mapsto U^1(x_1) \hat{\otimes} U^2(x_2)$ is a continuous representation of $G_1 \times G_2$ on $E_1 \hat{\otimes} E_2$.

Proof (1) Let ω_i be a compact subset of $G_i (i = 1, 2)$; we shall first verify that the set $\{U^1(x_1) \hat{\otimes} U^2(x_2) : (x_1, x_2) \in \omega_1 \times \omega_2\}$ is equicontinuous. So let $\mathscr{P}$ be a neighborhood of zero in $E_1 \hat{\otimes} E_2$, $\mathscr{P}^\circ$ its polar in $(E_1 \hat{\otimes} E_2)^*$ – then $\mathscr{P}^\circ$ is an equicontinuous subset of $(E_1 \hat{\otimes} E_2)^*$, as follows from the definitions. Since $U^i(\omega_i)$ is an equicontinuous subset of E_i $(i = 1, 2)$ we easily check that the set of continuous bilinear forms $\mathbf{z}_{x_1, x_2}, (a_1, a_2) \mapsto \mathbf{z}(U^1(x_1)a_1, U^2(x_2)a_2)$, $(\mathbf{z} \in \mathscr{P}^\circ, (x_1, x_2) \in \omega_1 \times \omega_2)$, is an equicontinuous subset of $(E_1 \hat{\otimes} E_2)^*$, whence is the polar of a neighborhood, say $\mathscr{P}$, of zero in $E_1 \hat{\otimes} E_2$. But then $U^1(x_1) \hat{\otimes} U^2(x_2)(\mathscr{P}) \subset \mathscr{P}$ for all $(x_1, x_2) \in \omega_1 \times \omega_2$ which establishes the equicontinuity.

(2) The canonical image of $E_1 \times E_2$ in $E_1 \hat{\otimes} E_2$ is a total subset of $E_1 \hat{\otimes} E_2$. Therefore, since the rule $(x_1, x_2) \mapsto U^1(x_1) \hat{\otimes} U^2(x_2)$ already verifies condition (b) of 4.1.1, we need only show that the map $(x_1, x_2) \mapsto (U^1(x_1) \hat{\otimes} U^2(x_2)) (a_1 \otimes a_2)$ is continuous for each pair $(a_1, a_2) \in E_1 \times E_2$. However, since the map $(x_1, x_2) \mapsto (U^1(x_1)a_1, U^2(x_2)a_2)$ is continuous and since the canonical map $(a_1, a_2) \mapsto a_1 \otimes a_2$ of $E_1 \times E_2$ into $E_1 \hat{\otimes} E_2$ is continuous, the assertion is obvious.

The proof of the proposition is now complete. □

Naturally enough $U_1 \hat{\otimes} U_2$ will be referred to as the *projective tensor product* of U_1 with U_2.

Suppose that $G_1 = G_2 = G$; let us agree to identify G with the diagonal subgroup $\tilde{G}$ of $G \times G$ consisting of all elements of the form (x, x)

$(x \in G)$ – then, by abuse of the language, the restriction of the representation $U_1 \hat{\otimes} U_2$ of $G \times G$ to $\tilde{G}$ will be called the *projective product* of U_1 with U_2 and will be viewed as a continuous representation of G.

Remark Let E_1 and E_2 be Hilbert spaces. The 'classical' tensor product of E_1 with E_2 is formed as follows: Consider the set of all bounded linear operators S from E_2^* to E_1 such that trace $(SS^*) < +\infty$ (S^* the Hilbert space adjoint of S) – then this set acquires the structure of a Hilbert space if we define the inner product by $(S_1, S_2) = \text{trace}\,(S_1 S_2^*)$. The Hilbert space thus obtained is called the *Hilbert product* of E_1 and E_2 and is denoted by $E_1 \times E_2$; of course it can be identified with the class of Hilbert-Schmidt operators from E_2^* to E_1. On the other hand, $E_1 \hat{\otimes} E_2$ can be identified with the trace class operators from E_2^* to E_1 and so $E_1 \hat{\otimes} E_2$ is in general smaller than the Hilbert product of E_1 with E_2 (cf. Grothendieck [1, p. 50]). If U^i is a unitary representation of G_i on E_i $(i = 1, 2)$, then the map $S \mapsto U^1(x_1)S({}^tU^2(x_2))$ $(S \in E_1 \times E_2)$ is, for each pair $(x_1, x_2) \in G_1 \times G_2$, a unitary operator in $E_1 \times E_2$ – call it $U^1(x_1) \times U^2(x_2)$. Then the assignment $(x_1, x_2) \mapsto U^1(x_1) \times U^2(x_2)$ is a unitary representation of $G_1 \times G_2$ on $E_1 \times E_2$ which will be referred to as the *Hilbert tensor product* of U^1 with U^2 and written $U^1 \times U^2$. Suppose in particular that $G_1 = G_2 = G$ – then the restriction of the unitary representation $U_1 \times U_2$ to the diagonal subgroup $\tilde{G}$ will be called the *Hilbert product* of U^1 with U^2 and will be viewed as a unitary representation of G.

Let E be a Fréchet space; let U be a continuous representation of G on E; let $E_1 \subset E_2$ be closed U-stable subspaces of E – then the corresponding continuous representation of G on the Fréchet space E_2/E_1 is called a *subquotient* of U. A topologically irreducible subrepresentation of a subquotient of U is called an *irreducible component* of U. [Let us recall that, in general, a continuous representation U of G is said to be *topologically irreducible* if it possesses no non-trivial, closed, invariant subspaces.]

4.1.3 Intertwining Forms and Operators

Let E, F be spaces satisfying the assumptions made in 4.1.1; let U (respectively V) be a continuous representation of the locally compact group G on E (respectively F).

Definition The continuous representations U and V are said to be *equivalent* if there exists a linear bicontinuous bijection $Q : E \to F$ such that $QU(x) = V(x)Q$ for all $x \in G$.

Definition A continuous linear map $S: E \to F$ is said to be an *intertwining operator* for U and V if $SU(x) = V(x)S$ for all $x \in G$.

The set $\mathfrak{I}(U, V)$ of all intertwining operators for U and V forms a vector space the dimension of which will be denoted by $I(U, V)$.

In the theory of unitary representations, intertwining operators play an important part (cf. 4.3), one reason being that in this special case $I(U, V) = I(V, U)$. But this equality fails in general.

Example Take for G the real line $\mathbf{R}$, for U the left regular representation of G on $L^1(G)$, and for V the representation $x \mapsto \chi(x)$, χ a unitary character of G. An application of Fourier transforms gives $I(U, V) = 1$ and $I(V, U) = 0$.

In order to reestablish this symmetry for more general representations, Bruhat [1, p. 156] posed the following definition.

Definition Let E (respectively F) be locally convex, complete, Hausdorff, barrelled spaces; let U (respectively V) be a continuous representation of G on E (respectively F). A separately continuous bilinear form $\mathbf{B}$ on $E \times F$ will be called an *intertwining form* of U with V if for every $x \in G$ we have

$$\mathbf{B}(U(x)a, V(x)b) = \mathbf{B}(a, b) \qquad (a \in E, b \in F).$$

The dimension of the space $\mathrm{i}(U, V)$ of intertwining forms will be called the *intertwining number* of U with V and written $i(U, V)$ $(= i(V, U) \ldots)$.

Remarks (1) The point behind limiting the definition to barrelled spaces is that in this case $\mathbf{B}$ is hypo-continuous relative to bounded subsets of E and F – that is, if a runs through a bounded subset of E, then the set of linear forms $\{\mathbf{B}(a,.)\}$ is equicontinuous and if b runs through a bounded subset of F, then the set of linear forms $\{\mathbf{B}(.,b)\}$ is equicontinuous. To treat the general case, one must simply assume $\mathbf{B}$ possesses this property.

(2) If U and V are unitary representations, the notions of intertwining form and intertwining operator coalesce.

Proposition 4.1.3.1 Retain the above notations and assumptions – then $i(U, V) = I(U, V^\vee)$.

Proof Since E and F are barrelled, the map $T: E \rightarrow F_b^*$, $\langle b, Ta\rangle = \mathbf{B}(a, b)(a \in E, b \in F)$, is continuous; clearly $TU(x) = {}^tV(x^{-1})T$ for all $x \in G$ – thus $T(E) \subset F^\vee$ and so each intertwining form $\mathbf{B}$ of U with V gives rise to an intertwining operator for U and $V^\vee$. The converse is likewise clear. Hence $i(U, V) = I(U, V^\vee)$. □

Let $\tilde{U}$ (respectively $\tilde{V}$) be a continuous representation of G on $\tilde{E}$ (respectively $\tilde{F}$); let S (respectively T) be an intertwining operator for $\tilde{U}$ and U (respectively for $\tilde{V}$ and V) such that $S(\tilde{E})$ (respectively $T(\tilde{F})$) is dense in E (respectively F). If $\mathbf{B}$ is an intertwining form for U and V, the form $\tilde{\mathbf{B}}(\tilde{a}, \tilde{b}) = \mathbf{B}(S\tilde{a}, T\tilde{b})$ $(\tilde{a} \in \tilde{E}, \tilde{b} \in \tilde{F})$ is an intertwining form for $\tilde{U}$ and $\tilde{V}$; moreover, $\tilde{\mathbf{B}} = 0 \Rightarrow \mathbf{B} = 0$ – hence $i(U, V) \leqslant i(\tilde{U}, \tilde{V})$, an observation which will be used in due course.

4.2 Representations on a Banach Space

4.2.1 Banach Representations of Associative Algebras

Although we are interested in the representations of groups, it will be convenient to begin by discussing representations of associative algebras

A over the complex numbers $\mathbf{C}$. The main reason for this is, as we have already noted in 4.1, that a continuous representation of the group G on a space E lifts in a natural manner to a representation of the algebra $M_c(G)$ (or $C_c(G)$) on E. In turn knowledge of properties of this representation of $M_c(G)$ (or $C_c(G)$) will yield valuable information about the given representation of G.

Let A be an associative algebra over $\mathbf{C}$.

A *representation* U of A is a homomorphism $x \mapsto U(x)$ ($x \in A$) of A into the algebra of all linear endomorphisms of a complex linear space E (the representation space of U). By dim (U) we mean dim (E). Representations U and V of A on linear spaces E and F, respectively, are said to be *algebraically equivalent* if E and F are equivalent as A-modules in the usual sense. A representation U of A on E is said to be *algebraically irreducible* if E contains no non-trivial subspace invariant under the $\{U(x) : x \in A\}$.

Definition A representation U of A on E is said to be *algebraically completely irreducible* if, given any two ordered sets of vectors $\{a_1, \ldots, a_n\}$, $\{b_1, \ldots, b_n\}$ in E such that the a_i's are linearly independent, then there exists $x \in A$ such that $U(x)a_i = b_i$ ($i = 1, \ldots, n$).

Thus if U is an algebraically completely irreducible representation of A on E, then the set $\{U(x) : x \in A\}$ is a dense ring of endomorphisms of E in the finite topology (see Jacobson [2, p. 248]). In view of the Jacobson Density Theorem (Jacobson [2, p. 274]) a representation U of A on E is algebraically completely irreducible iff: (1) A is algebraically irreducible; (2) The only endomorphisms of E commuting with all of the $U(x)$'s ($x \in A$) are scalar multiples of the identity.

By a *Banach representation* U of A on a Banach space E we mean a homomorphism U of A into the algebra of all bounded linear operators on E.

Definition A Banach representation U of A is said to be *topologically irreducible* if E contains no non-trivial closed subspace invariant under the set $\{U(x) : x \in A\}$.

Definition A Banach representation U of A is said to be *topologically completely irreducible* (TCI) if, given any bounded linear operator S on E and elements $a_1, \ldots, a_n$ in E, then there exists, for every $\epsilon > 0$, an element $x \in A$ such that

$$\|(U(x) - S)a_i\| < \epsilon \qquad (i = 1, \ldots, n).$$

Thus a Banach representation $x \mapsto U(x)$ of A is TCI provided the collection $\{U(x) : x \in A\}$ is dense in the space $\mathfrak{L}(E; E)$ when equipped with the topology of simple convergence.

Remark Let E be a normed linear space (not necessarily complete); in the sequel it will be occasionally useful to consider representations of A by bounded linear transformations on such a space E. The two definitions given above are, of course, applicable in this situation; they (and others) will be employed whenever it is convenient without further comment.

The following statement is clear.

Lemma 4.2.1.1 Let U be a Banach representation of A on E – then U is TCI iff the following condition is met: Given any two ordered sets of vectors $\{a_1, \ldots, a_n\}$, $\{b_1, \ldots, b_n\}$ in E such that the a_i's are linearly independent, there exists a net $\{x_k\}$ of elements of A such that $U(x_k)a_i \rightarrow b_i$ $(i = 1, \ldots, n)$.

Let U be a Banach representation of A. We shall write Ker (U) for the kernel of U. If I is a two-sided ideal in A with $I \not\subset$ Ker (U) and if U is topologically irreducible, then certainly $U|I$ is topologically irreducible too. In addition we have the following result.

Proposition 4.2.1.2 Let U be a TCI Banach representation of A on E; suppose that I is a two-sided ideal in A with $I \not\subset$ Ker (U) – then $U|I$ is TCI.

Proof If U is finite dimensional, then our result is obvious in view of Burnside's Theorem – hence we may assume that U is infinite dimensional.

(1) Our first objective is to prove that the rank of $U(x)$ $(x \in I)$ (i.e. dim $(U(x)E)$) can be made arbitrarily large upon a suitable selection of the element $x \in I$. If the contrary were true, then it would be possible to find a positive integer n with the property that $n = \max\{\text{rank}\,(U(x)) : x \in I\} < \infty$; let us show that this is impossible. There exist linearly independent vectors $a_0, a_1, \ldots, a_n$ in E, linearly independent vectors $b_0, b_1, \ldots, b_n$ in E, and elements x, y in I such that $U(x)a_0 = b_0$, $U(y)a_i = b_i$ $(i = 1, \ldots, n)$. By the topological complete irreducibility of U there are nets $\{x_k\}$, $\{y_k\}$ of elements of A such that $U(x_k)a_0 \rightarrow a_0$, $U(y_k)a_0 \rightarrow 0$, $U(x_k)a_i \rightarrow 0$, $U(y_k)a_i \rightarrow a_i$ $(i = 1, \ldots, n)$. Therefore $U(xx_k + yy_k)a_i \rightarrow b_i$ for $i = 0, 1, \ldots, n$. This shows that for large k, $U(xx_k + yy_k)$ has rank at least $n + 1$; as $xx_k + yy_k \in I$, this contradicts our assumption about n.

(2) Let $\{a_1, \ldots, a_n\}$ be any finite set of linearly independent elements of E, $\{b_1, \ldots, b_n\}$ any other ordered set of vectors in E. In view of (1) and the fact that U is TCI, there are elements $x \in I$ and $y \in A$ such that the $U(xy)a_i$ are linearly independent $(i = 1, \ldots, n)$. Choose a net $\{x_k\}$ of elements of A such that $U(x_kxy)a_i \rightarrow b_i$ for each i; since $x_kxy \in I$, it follows that $U|I$ is indeed topologically completely irreducible.

Hence the proposition. □

Given a Banach representation U of A on E we have four kinds of irreducibility and so it is natural to ask for the relations which exist among them. In the first place, if dim $(E) < \infty$, then Burnside's Theorem shows that the various notions are all equivalent.

Proposition 4.2.1.3 Let U be a Banach representation of A on E. If U is topologically completely irreducible, then U is topologically irreducible.

Proof It suffices to prove that the closure of the orbit under A of every non-zero element in E is the whole space E. If to the contrary there exists an $a \in E$ for which this is not true, select a non-zero element $b \in E - \{U(x)a : x \in A\}^{cl}$ ('*cl*' denotes closure). Using the Hahn-Banach Theorem, choose $\mathbf{z} \in E^*$ such that $\mathbf{z}$ vanishes on $\{U(x)a : x \in A\}^{cl}$ and $\langle b, \mathbf{z}\rangle = 1$. By the topological complete irreducibility of U there exists a net $\{x_k\}$ in A such that $U(x_k)a \to b$. Hence in particular $0 \to 1$, a contradiction. ☐

Proposition 4.2.1.4 (Dieudonné) Let U be an algebraically irreducible Banach representation of A on E – then U is algebraically completely irreducible.

Proof In view of the Jacobson Density Theorem it suffices to show that the set of endomorphisms $\mathscr{S}$ of E commuting with all the $\{U(x) : x \in A\}$ consists of scalar multiplications alone. Now $\mathscr{S}$ can be given the structure of a normed algebra as follows. Choose a non-zero vector $a \in E$ and for $S \in \mathscr{S}$ define

$$\|S\| = \inf \|U(x)\|$$

where x runs through the elements in A for which $Sa = U(x)a$. It is easy to check that $\mathscr{S}$ is a normed algebra (it is not claimed that $\mathscr{S}$ is complete though). But $\mathscr{S}$ is a division algebra by Schur's Lemma and hence is the complex field by Gelfand-Mazur's Theorem. ☐

Remark Plainly Proposition 4.2.1.4 is valid if U is an algebraically irreducible representation of A by bounded linear transformations on a normed linear space (i.e. completeness of E is not needed).

Summing up, if U is a Banach representation of A on E, then we have: U is algebraically irreducible $\Leftrightarrow$ U is algebraically completely irreducible $\Rightarrow$ U is topologically completely irreducible $\Rightarrow$ U is topologically irreducible.

Remark Let U be a topologically irreducible Banach representation of A on E – then, in general, it is *unknown* whether U must necessarily be topologically completely irreducible or not.

It is very important to determine when topological irreducibility implies topological complete irreducibility. In turn an answer to this question rests on the behavior of the representation contragredient to the given Banach representation U of A.

Let $\tilde{A}$ denote the reverse algebra of A, i.e. $\tilde{A}$ has the same linear operations as A but multiplication is given by: $(xy)_{\tilde{A}} = (yx)_A$, $(x, y \in A)$.

Let U be a Banach representation of A on E.

Definition The representation U^* of A *contragredient* to U is the Banach representation $x \mapsto {}^tU(x)$ of $\tilde{A}$ where, of course, ${}^tU(x)$ is the usual Banach space adjoint of $U(x)$.

Recall that the *rank* of an endomorphism of a linear space is by definition the dimension of its range.

Let U be a Banach representation of A on E. Set $I(U) = \{x \in A : U(x)$ is of finite rank$\}$ – then $I(U)$ is a two-sided ideal in A. [In passing, observe that $U(x)$ is of finite rank iff ${}^tU(x)$ is $(x \in A)$.] Let $E_0(E_0^*)$ be the linear span of the ranges of the $U(x)$ (${}^tU(x)$) with $x \in I(U)$; clearly $E_0(E_0^*)$ is U-stable (U^*-stable). Let $U_0(U_0^*)$ denote the restriction of $U(U^*)$ to $E_0(E_0^*)$.

Definition Let U be a Banach representation of A on E – then U is said to be *finite-dimensionally spanned* (FDS) if:

(a) E_0 is dense in E; (b) E_0^* is dense in E_s^*.

[In passing observe that if U is topologically irreducible, then it will be FDS provided that $U(x)$ is of non-zero finite rank for at least one x in A.]

Proposition 4.2.1.5 If U is FDS, then the following five conditions are equivalent:

(i) U is topologically irreducible;
(ii) U is topologically completely irreducible;
(iii) U_0 is algebraically irreducible;
(iv) U_0^* is algebraically irreducible;
(v) U_0 and U_0^* are algebraically completely irreducible.

Proof (i) $\Rightarrow$ (iii). Let $\tilde{E}$ be a non-zero U_0-stable subspace of E_0 – then $(\tilde{E})^{cl} = E$. Thus, for any $x \in A$, $U(x)(\tilde{E})$ is dense in the range of $U(x)$. In particular, if $x \in I(U)$, then $U(x)(\tilde{E}) = U(x)E$. So, for all $x \in I(U)$, $U(x)E \subset \tilde{E} \Rightarrow E_0 \subset \tilde{E} \Rightarrow \tilde{E} = E_0$.

(iii) $\Rightarrow$ (i). Let $\tilde{E}$ be a non-zero U-stable closed subspace of E – then $\tilde{E} \cap E_0 \neq \{0\}$. [Because $U(x)\tilde{E} \subset \tilde{E} \cap E_0 (x \in I(U))$, one need only show that $U(x)\tilde{E} \neq \{0\}$ for some $x \in I(U)$. Now if $U(x)a = 0$ for all $x \in I(U)$ and $a \in \tilde{E}$, then for every $\mathbf{z} \in E^*$, $0 = \langle U(x)a, \mathbf{z}\rangle = \langle a, {}^tU(x)\mathbf{z}\rangle$, i.e. the annihilator $\tilde{E}^\perp$ of $\tilde{E}$ in E^* contains E_0^* and since E_0^* is weakly dense in E^*, $\tilde{E} = \{0\}$ – this is a contradiction.] Therefore $\tilde{E} \cap E_0$ is a non-zero U_0-stable subspace of E_0. From the algebraic irreducibility of U_0 we deduce that $\tilde{E} \supset E_0$ – hence $\tilde{E} = E$, E_0 being dense in E.

(i) $\Leftrightarrow$ (iv). U is topologically irreducible iff U^* is weakly topologically irreducible (i.e. E^* admits no non-trivial weakly closed U^*-stable subspaces); arguing as above, our assertion then follows immediately.

(i) $\Rightarrow$ (v). For (iii) and (iv) are both valid now – the assertion is therefore a consequence of the Remark following Proposition 4.2.1.4.

(v) $\Rightarrow$ (ii). Let $\{a_1, \ldots, a_n\}$ be linearly independent elements in E and select elements $\{\mathbf{z}_1, \ldots, \mathbf{z}_n\}$ in E_0^* such that the matrix $(\langle a_i, \mathbf{z}_j\rangle)$ is non-singular – this can be done since E_0^* is weakly dense in E^*. Because condition (v) is in force, there exists $x_0 \in I(U)$ such that ${}^tU(x_0)\mathbf{z}_j = \mathbf{z}_j$ $(j = 1, \ldots, n)$. In particular the matrix $(\langle U(x_0)a_i, \mathbf{z}_j\rangle) = (\langle a_i, \mathbf{z}_j\rangle)$ is non-singular and so the $U(x_0)a_i$ are linearly independent elements of E_0. Let $\{b_1, \ldots, b_n\}$ be any set of n elements in E. For each $\epsilon > 0$ choose elements b_i^ϵ $(i = 1, \ldots, n)$ in E_0 such that $\|b_i - b_i^\epsilon\| < \epsilon$ and then, using (v), choose $x_\epsilon \in A$ such that $U(x_\epsilon)U(x_0)a_i = b_i^\epsilon$. It is now clear that there exists a net $\{y_\epsilon\}$ $(y_\epsilon = x_\epsilon x_0)$ in A such that $U(y_\epsilon)a_i \to b_i$ in E which establishes the assertion.

Since we have (i) $\Rightarrow$ (iv) $\Rightarrow$ (i) $\Rightarrow$ (iii) $\Rightarrow$ (i) $\Rightarrow$ (v) $\Rightarrow$ (ii) $\Rightarrow$ (i), the proof of the present proposition is complete. □

Let us recall the notion of the closure of a linear operator. Let T be a linear operator from the Banach space E to the Banach space F with domain of definition Dom(T). Consider the graph of T, i.e. the subset of $E + F$ consisting of all pairs (a, Ta) with $a \in$ Dom (T). T is said to be *closed* if its graph is a closed subset of $E + F$. On the other hand, T is said to *admit closure* if the closure of the graph of T in $E + F$ is the graph of some linear operator, say $\bar{T}$. $\bar{T}$ (if it exists) is called the closure of T. Evidently T admits closure iff the relations $a_n \to 0$ $(a_n \in \text{Dom}(T))$, $Ta_n \to b \Rightarrow b = 0$.

Let U, V be Banach representations of A on E, F respectively.

Definition U and V are said to be *Naimark related* (*equivalent*) if there exists a closed one-to-one linear map T of a dense U-stable subspace $\tilde{E}$ of E onto a dense V-stable subspace $\tilde{F}$ of F such that $V(x)Ta = TU(x)a$ for all $a \in \tilde{E}$, $x \in A$.

It is to be emphasized that in general Naimark relatedness is not transitive, hence not an equivalence relation (cf. below, 4.2.2). On the other hand it is easy to see that Naimark relatedness is indeed an equivalence relation when confined to FDS Banach representations of A.

Theorem 4.2.1.6 (Fell) Two topologically irreducible FDS Banach representations U and V of A on E and F respectively are Naimark equivalent iff Ker (U) = Ker (V).

Proof It is clear that if U and V are Naimark equivalent, then Ker(U) = Ker(V) – hence our chief concern is with the converse assertion. Plainly Ker (U) = Ker (U_0) = Ker (U_0^*), Ker (V) = Ker (V_0) = Ker (V_0^*); as will be seen below, if Ker (U) = Ker (V), then there exist linear bijective maps $S : E_0 \to F_0$, $S^* : E_0^* \to F_0^*$ such that $\langle a, \mathbf{z}\rangle = \langle Sa, S^*\mathbf{z}\rangle$ for all

$a \in E_0$, $\mathbf{z} \in E_0^*$ and with the property that

$$SU_0(x)a = V_0(x)Sa, \; S^*U_0^*(x)\mathbf{z} = V_0^*(x)S^*\mathbf{z} \quad (x \in A; a \in E_0, \mathbf{z} \in E_0^*).$$

From this (see (1)), it will be deduced that U and V are Naimark equivalent.

(1) (Mackey) Suppose that S and S^* are as above – claim: U and V are Naimark equivalent. To see this, first observe that the linear operator S admits closure. Thus suppose that $a_n \to 0$ ($a_n \in E_0$), $Sa_n \to b$ – we are to show that $b = 0$. Let $\mathbf{z} \in E_0^*$ – then $0 = \lim \langle a_n, \mathbf{z}\rangle = \lim \langle Sa_n, S^*\mathbf{z}\rangle = \langle b, S^*\mathbf{z}\rangle \Rightarrow b = 0$, F_0^* being weakly dense in F^*. Therefore S admits closure. In the same way we see that $\bar{S}$ is one-to-one. Finally $\bar{S}U(x)a = V(x)\bar{S}a$ for all $a \in \operatorname{Dom}(\bar{S})$. Thus let $a \in \operatorname{Dom}(\bar{S})$; if $a_n \in E_0$ and $a_n \to a$, then $U(x)a_n \in E_0$ too and $U(x)a_n \to U(x)a \Rightarrow V(x)\bar{S}a = \lim V(x)\bar{S}a_n = \lim \bar{S}U(x)a_n = \bar{S}U(x)a$, as desired. Hence U and V are indeed Naimark equivalent.

(2) In view of (1), our main task is to produce the maps S and S^*. To get at this, we need some preparation which will be given here and in (3). Let $x \in A$ – claim: $U_0(x)$ is of rank 1 iff $\{y \in A : U_0(xyx) = 0\}$ is of codimension 1 in A. [A similar statement is, of course, valid for V_0.] Thus suppose that $U_0(x)$ is of rank 1 – then there is a non-zero linear functional λ on A such that $U_0(xyx) = \lambda(y)U_0(x)$ $(y \in A)$. Hence $\{y \in A : U_0(xyx) = 0\}$ has codimension 1. On the other hand, assume that the rank of $U_0(x)$ is greater than 1; let a_1', a_2' be elements of E_0 such that $a_1 = U_0(x)a_1'$, $a_2 = U_0(x)a_2'$ are linearly independent. Since U_0 is algebraically completely irreducible (Proposition 4.2.1.5 (v)), we may find elements $y_1, y_2 \in A$ such that $U_0(y_1)a_1 = a_1'$, $U_0(y_1)a_2 = 0$, $U_0(y_2)a_1 = 0$, $U_0(y_2)a_2 = a_2'$ – therefore

$$U_0(xy_1x)a_1' = a_1, \; U_0(xy_1x)a_2' = 0, \; U_0(xy_2x)a_1' = 0, \; U_0(xy_2x)a_2' = a_2$$

which implies that the operators $U_0(xy_1x)$, $U_0(xy_2x)$ are linearly independent. Consequently the kernel of the map $y \mapsto U_0(xyx)$ has codimension greater than one. Hence the claim.

(3) Let $\Phi(U_0)$ be the linear subspace of the algebraic dual of A spanned by the maps $x \mapsto \langle U_0(x)a, \mathbf{z}\rangle$ $(a \in E_0, \mathbf{z} \in E_0^*)$; assign to $\Phi(V_0)$ a similar meaning – we intend to show that if $\Phi(U_0) \cap \Phi(V_0) \neq \{0\}$, then there exist linear bijective maps $S : E_0 \to F_0$, $S^* : E_0^* \to F_0^*$ with the properties stated earlier.

(a) Let $\phi \in \Phi(U_0)$; assume that $\phi \neq 0$. Owing to Proposition 4.2.1.5 (v), ϕ can be written in the form

$$\phi(x) = \sum_{i=1}^{r} \langle U_0(x)a_i, \mathbf{z}_i\rangle \qquad (x \in A)$$

where the $a_1, \ldots, a_r$ are linearly independent in E_0, the $\mathbf{z}_1, \ldots, \mathbf{z}_r$ are linearly independent in E_0^* and r is unique – call r the rank of ϕ relative to U_0. Let

$$J_0 = \{x \in A : \phi(yx) = 0, \text{ all } y \in A\},$$
$$J_0^* = \{y \in A : \phi(yx) = 0, \text{ all } x \in A\}.$$

Evidently $J_0(J_0^*)$ is a left (right) ideal in A – moreover A/J_0 and A/J_0^* are in duality under the bilinear form

$$\langle x + J_0, y + J_0^* \rangle = \phi(yx) \qquad (x, y \in A).$$

Let π_0 (respectively π_0^*) denote the natural representation of A on A/J_0 (respectively of $\tilde{A}$ on A/J_0^*). Claim: There exist linear bijective maps $T: A/J_0 \to E_0 + \ldots + E_0$ (direct sum r times), $T^*: A/J_0^* \to E_0^* + \ldots + E_0^*$ (direct sum r times) such that $\langle x + J_0, y + J_0^* \rangle = \langle T(x + J_0), T^*(y + J_0^*) \rangle$ $(x, y \in A)$ and with the property that

$$T\pi_0(y)(x + J_0) = r \cdot U_0(y)T(x + J_0),$$
$$T^*\pi_0^*(x)(y + J_0^*) = r \cdot U_0^*(x)T^*(y + J_0^*)$$

for all $x, y \in A$. [Here $r \cdot U_0$ denotes the direct sum of r copies of U_0; similarly for $r \cdot U_0^*$.] In fact, since U_0 and U_0^* are algebraically completely irreducible, it is clear that

$$J_0 = \{x \in A : U_0(x)a_i = 0, \text{ all } i\}, \quad J_0^* = \{y \in A : U_0^*(y)\mathbf{z}_i = 0, \text{ all } i\}.$$

Define T and T^* by the prescription:

$$T(x + J_0) = (U_0(x)a_1, \ldots, U_0(x)a_r),$$
$$T^*(y + J_0^*) = (U_0^*(y)\mathbf{z}_1, \ldots, U_0^*(y)\mathbf{z}_r) \qquad (x, y \in A).$$

Plainly T and T^* have the desired properties.

(b) Suppose that $0 \neq \phi \in \Phi(U_0) \cap \Phi(V_0)$; express ϕ as in (a). Choosing $y \in A$ such that $U_0(y)a_1 = a_1$ and $U_0(y)a_i = 0$ for $i > 1$, we see that the functional $\psi(x) = \phi(xy)$ $(x \in A)$ belongs to $\Phi(U_0)$ and is of rank 1 relative to U_0. On the other hand $\psi \in \Phi(V_0)$ and has rank r (say) relative to V_0. But then, due to the irreducibility of U_0 and U_0^* and the discussion in (a), $r = 1$. The existence of the maps S and S^* is now clear.

(4) Let us wrap up the proof of the present theorem. Assuming that Ker $(U) =$ Ker (V), we are to show that U and V are Naimark equivalent; in view of (1) and (3), it is enough to prove that $\Phi(U_0) \cap \Phi(V_0)$ contains a non-zero element. Select an $x \in A$ such that $U_0(x)$ (and hence also $U_0^*(x)$) has rank 1. By (2) and the fact that Ker $(U_0) =$ Ker (V_0), we see that $V_0(x)$ and $V_0^*(x)$ likewise have rank 1. Let $a_0, \mathbf{z}_0, b_0, \mathbf{w}_0$ be non-zero vectors in the ranges of $U_0(x)$, $U_0^*(x)$, $V_0(x)$ and $V_0^*(x)$, respectively. Put $\phi(y) = \langle U_0(y)a_0, \mathbf{z}_0 \rangle$, $\psi(y) = \langle V_0(y)b_0, \mathbf{w}_0 \rangle$ $(y \in A)$. Let $y \in A$ – claim: $\phi(y) = 0$ iff $xyx \in$ Ker (U_0). Thus: $\phi(y) = 0$ iff $0 = \langle U_0(yx)a, U_0^*(x)\mathbf{z} \rangle = \langle U_0(xyx)a, \mathbf{z} \rangle$ (all $a \in E_0, \mathbf{z} \in E_0^*$) iff $U_0(xyx) = 0$. The same holds for ψ. Therefore ϕ and ψ are non-zero and have the same kernel; hence $\phi = c\psi$ for some non-zero complex number c. Conclusion: $\Phi(U_0) \cap \Phi(V_0)$ contains a non-zero element.

This proves Theorem 4.2.1.6. □

In order to formulate the next result we need to recall some ring-theoretic terminology. Let A be a ring. A two-sided ideal I in A is said to be *primitive* if it is the kernel of an algebraically irreducible representation of A. A itself is said to be primitive if A admits a faithful algebraically irreducible representation.

A is said to be *simple* iff: (i) $A^2 \neq \{0\}$; (ii) There are no proper ideals other than $\{0\}$.

The sum of all minimal left (right) ideals in A is called the *left* (*right*) *socle* of A. The left (right) socle exists iff A contains minimal left (right) ideals. If the left and right socles exist and are equal, then the resulting two-sided ideal is called the *socle* of A. The structure theorem for primitive rings with non-zero socles (proved in Jacobson [4, p. 75]) is:

Theorem 4.2.1.7 The following two conditions on a ring A are equivalent:
(i) A is a primitive ring with a non-zero socle;
(ii) A is isomorphic to a subring of the ring of linear transformations over a division ring Δ which is dense in the finite topology and contains non-zero linear transformations of finite rank.

If A is represented as in (ii) its socle is the set of linear transformations of finite rank contained in this ring. Moreover the socle of A is a simple ring which is contained in every non-zero two-sided ideal of A.

Proposition 4.2.1.8 Let A be an associative algebra over **C**. Let U be an FDS topologically irreducible Banach representation of A; let Ker (U) be the kernel of U – then Ker (U) is a primitive ideal of A and $I(U)/\text{Ker}\,(U)$ is the socle of $A/\text{Ker}\,(U)$. In fact $I(U)/\text{Ker}\,(U)$ is the smallest non-zero two-sided ideal of $A/\text{Ker}\,(U)$.

Proof Using the notations of Proposition 4.2.1.5, it is clear that Ker (U) is the kernel of U_0 and, since U_0 is algebraically irreducible, Ker (U_0) is primitive. But according to Proposition 4.2.1.5, U_0 is in fact algebraically completely irreducible. Applying Theorem 4.2.1.7 (ii) we get the assertion of the present proposition. □

Let A be an associative algebra over **C**.

Let U be a Banach representation of A on E; a Banach representation V of A is said to be a *subquotient* of U if there exist closed U-stable subspaces E_1, E_2 of E with $E_1 \subset E_2$ such that V is (homeomorphically) equivalent to the natural representation of A induced by U on the factor space E_2/E_1. [Thus a subquotient of U is a subrepresentation (acting on a closed subspace) of a quotient representation, or a quotient representation of a subrepresentation (on a closed subspace).] A subquotient of a subquotient of U is clearly a subquotient of U. A topologically irreducible subrepresentation of a subquotient of U is called an *irreducible component* of U.

Definition A Banach representation U of A on E is said to be *discrete* if every non-zero subquotient of U has a subquotient V such that either $V(x) = 0$ for all x in A or V is topologically irreducible (it being assumed, of course, that the representation space for V is non-zero).

It is clear that every finite dimensional representation of A is discrete.

Example If A is a *-algebra and U is a *-representation of A (4.3.1), then U is discrete iff it is unitarily equivalent to a direct sum of *-representations which are either topologically irreducible or zero.

Definition Let U be a Banach representation of A on E. By a *reducing series* for U we shall understand a nested family $\mathscr{E}$ of closed U-stable subspaces of E such that $\{0\}$, E belong to $\mathscr{E}$ and with the property that if $\mathscr{F} \subset \mathscr{E}$, then $\bigcap_{F \in \mathscr{F}} F$ and $(\bigcup_{F \in \mathscr{F}} F)^{cl}$ lie in $\mathscr{E}$.

[As used here, 'nested' means that if $E_1, E_2 \in \mathscr{E}$, then either $E_1 \subset E_2$ or $E_2 \subset E_1$.]

Any maximal nested family of closed U-stable subspaces of E is a reducing series for U.

Definition (Fell) Let U be a Banach representation of A on E – then a *Jordan-Hölder series* for U is a maximal reducing series for U, that is, a maximal nested family of closed U-stable subspaces of E.

In view of Zorn's Lemma, every nested family of closed U-stable subspaces is contained in a Jordan-Hölder series.

Let U be a Banach representation of A on E, $\mathscr{E}$ a reducing series for U. Two subspaces E_1 and E_2 in $\mathscr{E}$ with $E_1 \subsetneq E_2$ are said to be *adjacent* in $\mathscr{E}$ if there is no $\tilde{E}$ in $\mathscr{E}$ such that $E_1 \subsetneq \tilde{E} \subsetneq E_2$. In particular, this will be the case if either E_2/E_1 is one dimensional or the subquotient of U acting on E_2/E_1 is topologically irreducible – in either of these cases, E_1 and E_2 are called *strictly adjacent.*

Definition Let U be a Banach representation of A on E – then a reducing series for U is said to be *discrete* (respectively *strictly discrete*) if, whenever $E_1, E_2, \in \mathscr{E}$ and $E_1 \subsetneq E_2$, there are subspaces $\tilde{E}_1$, $\tilde{E}_2$ in $\mathscr{E}$ with $E_1 \subset \tilde{E}_1 \subsetneq \tilde{E}_2 \subset E_2$ such that $\tilde{E}_1$ and $\tilde{E}_2$ are adjacent in $\mathscr{E}$ (respectively strictly adjacent).

A discrete Jordan-Hölder series is certainly strictly discrete.

Proposition 4.2.1.9 Let U be a Banach representation of A on E; let $\mathscr{E}$ be a strictly discrete reducing series for U – then $\mathscr{E}$ is a Jordan-Hölder series for U.

Proof Suppose, if possible, that $\tilde{E}$ is a closed U-stable subspace of E not in $\mathscr{E}$ for which $\mathscr{E} \cup \{\tilde{E}\}$ is nested; let $\mathscr{E}_1 = \{E_1 \in \mathscr{E} : E_1 \subset \tilde{E}\}$, $\mathscr{E}_2 = \{E_2 \in \mathscr{E} : \tilde{E} \subset E_2\}$, $\tilde{E}_1 = (\bigcup \mathscr{E}_1)^{cl}$, $\tilde{E}_2 = \bigcap \mathscr{E}_2$; because $\tilde{E} \notin \mathscr{E}$, we have $\tilde{E}_1 \subsetneq \tilde{E} \subsetneq \tilde{E}_2$ – but $\tilde{E}_1$ and $\tilde{E}_2$ are clearly adjacent, hence strictly adjacent (in $\mathscr{E}$), which contradicts the inclusions $\tilde{E}_1 \subsetneq \tilde{E} \subsetneq \tilde{E}_2$. □

Proposition 4.2.1.10 (Fell) Let U be a Banach representation of A on E.

(i) If U is discrete, then some Jordan-Hölder series for U is discrete.

(ii) If every Jordan-Hölder series for U is discrete, then U is discrete.

Proof We shall prove (i) leaving the proof of (ii) for the reader. Let $\mathscr{P}$ be a maximal family of strictly adjacent pairs (F_1, F_2) (with $F_1 \subset F_2$) having the property that if (F_1, F_2) and $(\tilde{F}_1, \tilde{F}_2)$ are distinct pairs in $\mathscr{P}$, then either $F_2 \subset \tilde{F}_1$ or $\tilde{F}_2 \subset F_1$; let $\mathscr{F} = \bigcup \{\{F_1\} \cup \{F_2\} : (F_1, F_2) \in \mathscr{P}\}$ (thus $\mathscr{F}$ is a nested family of closed U-stable subspaces); let $\mathscr{E}$ be the smallest reducing series containing $\mathscr{F}$ (an intersection of reducing series is a reducing series) – then $\mathscr{E}$ is a discrete Jordan-Hölder series for U. □

Examples (Fell) (1) The converse of Proposition 4.2.1.10 (i) is false, i.e. there exists an associative algebra A and a Banach representation U of A such that U is non-discrete but admits a discrete Jordan-Hölder series. [Let A be the *-algebra of all continuous complex valued functions on $[0, 1]$; $E_1 = L^2([0, 1])$ (relative to Lebesgue measure), $E_2 = L^2([0, 1])$ (relative to counting measure) and put $E = E_1 + E_2$ (Hilbert space direct sum – evidently E is non-separable). We may represent A on E by bounded linear operators through the prescription $f \mapsto U(f)$, $U(f)(a_1 + a_2) = fa_1 + fa_2$, $(f \in A;\ a_i \in E_i,\ i = 1, 2)$ – it is clear that U is not discrete. Let $I = [0, 1] \times \{0, 1\}$; for $(r, s) \in I$, let $E_1(r, s)$ be the closed linear subspace of E_1 consisting of all functions vanishing outside of $[0, r]$ and let $E_2(r, s)$ be the closed linear subspace of E_2 comprised of those functions vanishing off of $[0, r)$ if $s = 0$ and outside $[0, r]$ if $s = 1$. Putting $E(r, s) = E_1(r, s) + E_2(r, s)$, we find that $\{E(r, s) : (r, s) \in I\}$ is a strictly discrete reducing series, hence (Proposition 4.2.1.9) a Jordan-Hölder series, for U.]

(2) The converse of Proposition 4.2.1.10 (ii) is false, i.e. there exists an associative algebra A and a Banach representation U of A such that U is discrete but admits a non-discrete Jordan-Hölder series. [Consider the representation of **C** by scalar operators on an infinite dimensional separable Hilbert space.]

4.2.2 Banach Representations of Groups

As usual let G be a locally compact group countable at infinity. By a *Banach representation* of G we shall mean a continuous representation of G on a Banach space E.

Proposition 4.2.2.1 Let $x \mapsto U(x)$ be a homomorphism of G into the group of topological automorphisms of a Banach space E. The following three conditions are equivalent:

(i) The map $(x, a) \mapsto U(x)a$ is a continuous map of $G \times E$ into E;

(ii) For each fixed $a \in E$, the map $x \mapsto U(x)a$ is a continuous map of G into E;

(iii) For each fixed $a \in E$, $\mathbf{z} \in E^*$, the map $x \mapsto \langle U(x)a, \mathbf{z}\rangle$ is a continuous map of G into **C**.

Proof Obviously (i) ⇒ (ii) ⇒ (iii); in addition (ii) ⇒ (i) (4.1.1). Hence it remains to verify that (iii) ⇒ (ii). Equipping E with the weak topology induced by E^*, (iii) says that the map $x \mapsto U(x)a$ is a weakly continuous map of G into E for each $a \in E$. We are to prove that it is strongly continuous, i.e. continuous in the norm topology of E. Let ω be a compact neighborhood of 1 in G – then for any $a \in E$ the set

$U(\omega)a = \{U(x)a : x \in \omega\}$ is a weakly compact subset of E. An application of the Uniform Boundedness Theorem gives a constant $M > 0$ such that $\|U(x)\| < M$ for all $x \in \omega$. Let $\mathcal{O}_\omega$ be a symmetric neighborhood of 1 in G with $\mathcal{O}_\omega^2 \subset \omega$; let $\{g_\mathcal{O}\}$ be the net used in 4.1.1 ($\mathcal{O}$ running through the relatively compact neighborhoods of 1 in G – we may assume that $\mathcal{O} \subset \mathcal{O}_\omega$). Fix $a \in E$; because the strongly closed convex hull of the weakly compact set $U(\omega)a$ is again weakly compact (Theorem of Krein and Šmulian), one can, by a well-known principle, define for each $x \in \mathcal{O}_\omega$ the weakly convergent left invariant Haar integral $\int_G g_\mathcal{O}(x^{-1}y)U(y)ad_G(y)$ as the unique element $a(x) \in E$ such that for all $\mathbf{z} \in E^*$

$$\langle a(x), \mathbf{z}\rangle = \int_G g_\mathcal{O}(x^{-1}y)\langle U(y)a, \mathbf{z}\rangle d_G(y) \quad (x \in \mathcal{O}_\omega).$$

Since

$$\left\| U(x)\left(\int_G g_\mathcal{O}(y)U(y)ad_G(y)\right) - \int_G g_\mathcal{O}(y)U(y)ad_G(y)\right\|$$

$$\leqslant M\|a\|\int_G |g_\mathcal{O}(x^{-1}y) - g_\mathcal{O}(y)|\, d_G(y) \qquad (x \in \mathcal{O}_\omega)$$

and this last expression tends to zero as $x \to 1$, we see that the map $x \mapsto U(x)(\int_G g_\mathcal{O}(y)U(y)ad_G(y))$ $(x \in G)$ is strongly continuous. The set F of all vectors b in E for which the map $x \mapsto U(x)b$ $(x \in G)$ is strongly continuous is plainly a strongly closed subspace of E and so is weakly closed. But for any $a \in E$, the net $\{\int_G g_\mathcal{O}(x)U(x)ad_G(x)\}$ is in F and so F, being weakly closed, must contain the weak limit of this net, i.e. a. Hence $E = F$. □

It is now time to tie up Banach representations of the group G with Banach representations of certain associative algebras so that the material in the preceding number becomes available. In what follows d_G will denote left Haar measure on G. Let us keep in mind that under the identification $f \mapsto fd_G$, the algebra $C_c(G)$ becomes a two-sided ideal in $M_c(G)$.

Let U be a Banach representation of G on E – then U lifts to a Banach representation of $M_c(G)$ on E which will again be denoted by U (thus $U(\mu * \nu) = U(\mu)U(\nu)$ for all $\mu, \nu \in M_c(G)$). For any $a \in E, \mathbf{z} \in E^*$, we have

$$\langle U(\mu)a, \mathbf{z}\rangle = \int_G \langle U(x)a, \mathbf{z}\rangle\, d\mu(x) \qquad (\mu \in M_c(G)).$$

We shall refer to the Banach representation U of $M_c(G)$ as the *integrated form* of the Banach representation U of G. The Banach representation U of G is said to be *topologically irreducible* (*topologically completely irreducible*) if its integrated form is so.

Consider the restriction $U | C_c(G)$ of U to the ideal $C_c(G)$. If the original Banach representation $x \mapsto U(x)$ of G is non-zero, then so is $U | C_c(G)$; in fact the restriction of U to $C_c(G)$ completely determines the integrated form. If the integrated form is topologically completely irreducible, then so is $U | C_c(G)$ (cf. Proposition 4.2.1.2).

There is of course another definition of irreducibility for Banach representations of G.

Definition Let U be a Banach representation of G – then U is said to be *topologically irreducible* if U admits no non-trivial closed G-invariant subspace.

Proposition 4.2.2.2 Let U be a Banach representation of G on E. The following conditions are equivalent:

(i) U is topologically irreducible per G;

(ii) U is topologically irreducible per $M_c(G)$.

Proof (i) $\Rightarrow$ (ii). If $\tilde{E} \neq \{0\}$ is a closed subspace of E stable for $U(M_c(G))$, then $\tilde{E}$ is, in particular, stable for the set $\{U(\delta_x) : x \in G\}$; but $U(\delta_x) = U(x)$ $(x \in G)$ and so $\tilde{E} = E$.

(ii) $\Rightarrow$ (i). If $\tilde{E} \neq \{0\}$ is a closed subspace of E stable for $U(G)$, then $\tilde{E}$ is a Banach space on which we have a Banach representation of G which lifts to a Banach representation of $M_c(G)$ on $\tilde{E}$. Because U is topologically irreducible per $M_c(G)$ and $\tilde{E}$ is U-stable, $\tilde{E} = E$.

Hence the proposition. □

In the sequel we shall fail to distinguish between the two concepts of topological irreducibility.

Let us note the following analogue of Schur's Lemma for Banach representations of G.

Proposition 4.2.2.3 Let U be a topologically completely irreducible Banach representation of G on E – then the algebra of intertwining operators for U consists of scalar multiples of the identity alone.

Proof Suppose that $S : E \rightarrow E$ is a bounded linear transformation such that $SU(x) = U(x)S$ for all $x \in G$ – then $SU(\mu) = U(\mu)S$ for all $\mu \in M_c(G)$. The claim is that S is a scalar multiple of the identity. To see this, first note that for any vector $a \in E$, a and Sa are linearly dependent. [Suppose, if possible, that for some non-zero $a \in E$, a and Sa are linearly independent – then, since U is TCI, there exists a net $\{\mu_k\}$ in $M_c(G)$ such that $U(\mu_k)a \rightarrow a$, $U(\mu_k)Sa \rightarrow a$, whence, for all $\mathbf{z} \in E^*$, $\langle a, \mathbf{z}\rangle = \lim \langle U(\mu_k)Sa, \mathbf{z}\rangle = \lim \langle SU(\mu_k)a, \mathbf{z}\rangle = \lim \langle U(\mu_k)a, {}^tS\mathbf{z}\rangle = \langle a, {}^tS\mathbf{z}\rangle = \langle Sa, \mathbf{z}\rangle \Rightarrow a = Sa \ldots$.] This being the case, it then follows immediately that for each $a \in E$ there exists a complex number $c(a)$ such that $Sa = c(a)a$. It remains to prove that the $c(a)$ are the same for all $a \in E$. So let a_1, a_2 be non-proportional elements in E and select a net $\{\mu_k\}$ in $M_c(G)$ such that $U(\mu_k)a_1 \rightarrow a_2$ in E. In particular, then, for every

$\mathbf{z} \in E^*$, we have

$$\langle U(\mu_k)a_1, {}^t S\mathbf{z}\rangle = c(a_1)\langle U(\mu_k)a_1, \mathbf{z}\rangle \to \langle a_2, {}^t S\mathbf{z}\rangle = c(a_2)\langle a_2, \mathbf{z}\rangle.$$

Hence $c(a_1) = c(a_2)$. □

Let Z denote the center of G. If U is a topologically completely irreducible Banach representation of G, then the preceding proposition provides us with a complex valued homomorphism ζ_U of Z such that $U(z) = \zeta_U(z)I$ for each z in Z (I the identity operator in the representation space of U) – we shall call ζ_U the *central character* of U.

Here is a result which seems to be a part of the folklore of the subject.

Proposition 4.2.2.4 Let G be a compact topological group, U a topologically irreducible Banach representation of G on a Banach space E – then E is finite dimensional.

[In comparison, recall the well-known fact that if U is a topologically irreducible unitary representation (4.3.1) of G on a Hilbert space E, then E is finite dimensional.]

Proof By the Uniform Boundedness Theorem, there exists $M > 0$ such that $\| U(x)\| < M$ for all $x \in G$. Fix a non-zero element $\mathbf{z} \in E^*$ and put

$$(a, b) = \int_G \langle U(x)a, \mathbf{z}\rangle\overline{\langle U(x)b, \mathbf{z}\rangle}\, d_G(x) \qquad (a, b \in E),$$

d_G denoting normalized Haar measure on G.

(1) $(\,.\,,\,.\,)$ is an inner product on E. [For if $(a, a) = 0$, then

$$\int_G |\langle U(x)a, \mathbf{z}\rangle|^2\, d_G(x) = 0$$

which implies that $\langle U(x)a, \mathbf{z}\rangle = 0$ for every $x \in G$. Let $\tilde{E} = \{a \in E : \langle U(x)a, \mathbf{z}\rangle = 0 \text{ for all } x \in G\}$ – then $\tilde{E}$ is a closed U-stable subspace of E, and so, by irreducibility, $\tilde{E} = \{0\}$, whence $a = 0$.]

(2) Let F denote the completion of E with respect to the norm $|||\,.\,|||$ induced on E by $(\,.\,,\,.\,)$. If $a_n \to a$ in $E (a_n, a \in E)$, then

$$|||a_n - a|||^2 \leqslant \|\mathbf{z}\|^2 M^2 \|a - a_n\|^2 \to 0.$$

Denote the image of E in F under the canonical injection $i : E \to F$ by $F(E)$.

(3) If $a \in F(E)$, then $|||U(x)a|||^2 = |||a|||^2$ $(x \in G)$ by invariance of the Haar measure. Thus $U(x)$ is, for each $x \in G$, an isometry. Extend it by continuity to all of F. Then (2) shows that the assignment $x \mapsto U(x)$ is a unitary representation of G on F. Since G is compact, we have an orthogonal direct sum decomposition

$$F = \Sigma F_j$$

into finite dimensional non-zero subspaces, all of which are irreducible for U. Let $\mathsf{P}_j : F \to F_j$ denote the orthogonal projection of F onto F_j.

(4) Select any j and consider the map $a \mapsto \mathsf{P}_j \circ i(a)$ of E into F_j; (5) below implies that this map is one-to-one; since $\dim(F_j) < \infty$, the finite dimensionality of E follows.

(5) Suppose that $\mathsf{P}_j \circ i(a) = 0$. Claim: $a = 0$. For let $F_j^{\perp}$ be the orthogonal complement of F_j in F. If $i(a) \neq 0$, then $F(E) \cap F_j^{\perp} \neq \{0\}$. Let $E_j = i^{-1}(F(E) \cap$

$F_j^\perp$) – then E_j is a non-zero closed subspace since (2) implies that i is continuous. However E_j is also U-stable – therefore $E_j = E$, whence $F(E) \cap F_j^\perp = F(E)$, contradicting the fact that $F(E)$ is dense in F.

The proof of the proposition is now complete. □

Example (Fell) It is reasonable to restrict our attention primarily to Banach representations since the ultimate application of the theory will be to semi-simple Lie groups. Nevertheless it is important to realize that Banach spaces may form too narrow a framework for certain classes of groups. For instance, let G be the Heisenberg group, that is, the three dimensional nilpotent Lie group comprised of all triples of real numbers, multiplication being given by $(x, y, z)(x', y', z') = (x + x', y + y', z + z' + xy')$. [Otherwise said, G is the group, under matrix multiplication, of all matrices of the form

$$\begin{pmatrix} 1 & x & z \\ 0 & 1 & y \\ 0 & 0 & 1 \end{pmatrix} \qquad (x, y, z \in \mathbf{R}).]$$

The infinite dimensional irreducible unitary representations of G are well-known (see number 5.4.2, Example 3). For each non-zero real number r there is a unique (infinite dimensional) irreducible unitary representation U^r of G with the property that, for each real z, U^r sends the central element $(0, 0, z)$ of G to the scalar operator $e^{\sqrt{-1}\, rz} \cdot I$. By a process of 'analytic continuation' one might expect that it would be possible to obtain non-unitary topologically irreducible representations U^r having the same property for *complex* r. The point here is that such a U^r can not be realized on a Banach space. Thus write $\phi_1(x) = (x, 0, 0)$, $\phi_2(y) = (0, y, 0)$, $\phi_3(z) = (0, 0, z)$ $(x, y, z$ real); let us suppose that U is a homomorphism of G into the group of topological automorphisms of some Banach space E such that $U(\phi_3(z)) = e^{\sqrt{-1}\, rz} \cdot I$ for all real z, r a non-real complex number – then, since $\phi_1(-1)\phi_2(-y)\phi_1(1) = \phi_3(y)\phi_2(-y)$, we have

$$U(\phi_1(-1))U(\phi_2(-y))U(\phi_1(1)) = e^{\sqrt{-1}\, ry}\, U(\phi_2(-y))$$

for all real y. Hence

$$|e^{\sqrt{-1}\, ry}| \cdot \| U(\phi_2(-y))\| \leqslant \| U(\phi_1(-1))\| \cdot \| U(\phi_2(-y))\| \cdot \| U(\phi_1(1))\|$$

or

$$|e^{\sqrt{-1}\, ry}| \leqslant \| U(\phi_1(-1))\| \cdot \| U(\phi_1(1))\|$$

for all real y, an impossibility because r is not real. This shows that such a U cannot exist. To handle phenomena like this, one can, for instance, pass to the more general notion of a linear system representation (cf. Fell [4]).

Definition A *semi-norm* ρ on G is a positive real valued lower semi-continuous function which is bounded on compact subsets of G and satisfies

$$\rho(xy) \leqslant \rho(x)\rho(y)$$

for all $x, y \in G$.

A typical example of a semi-norm is obtained by considering $\rho(x) = \| U(x)\|$ $(x \in G)$, U a Banach representation of G on E ($\|.\|$ is the usual operator norm). [Plainly ρ is bounded on compact subsets of G in view of the Uniform Boundedness Theorem; on the other hand, if B is the unit ball in E, then $\rho(x) = \sup_{a \in B} \| U(x)a \|$ and so, since the functions $x \mapsto \| U(x)a \|$ are continuous on G ($a \in E$), ρ is lower semi-continuous.]

If ρ is any semi-norm on G, then $M_c(G)$ acquires the structure of a normed algebra if we put

$$\| \mu \|_\rho = \int_G \rho(x) \, d|\mu|(x) \qquad (\mu \in M_c(G))$$

($|\mu|$ the total variation of μ). In particular if $\rho(x) = \| U(x) \|$ ($x \in G$), then the integrated form of U is continuous with respect to $\| . \|_\rho$; in fact $\| U(\mu) \| \leqslant \| \mu \|_\rho$ for $\mu \in M_c(G)$ (compare with Proposition 4.1.1.2).

Let ρ be a semi-norm on G. We shall let $C_\rho(G)$ denote the Banach algebra obtained by completing $C_c(G)$ with respect to the ρ-norm

$$\| f \|_\rho = \int_G | f(x) | \rho(x) d_G(x) \qquad (f \in C_c(G)).$$

If a semi-norm ρ arises from a Banach representation of G, then $\| U(f) \| \leqslant \| f \|_\rho$ ($f \in C_c(G)$) – hence the map $f \mapsto U(f)$ can be extended to all of $C_\rho(G)$. On the other hand, if $f \mapsto U(f)$ is a representation of $C_\rho(G)$ on a Banach space E with $\| U(f) \| \leqslant r \| f \|_\rho$ (ρ some semi-norm on G, r a positive constant independent of f) such that the set $\{U(f)a : f \in C_\rho(G),\ a \in E\}$ is total in E, then this representation generates a Banach representation $x \mapsto U(x)$ of G on E verifying $\| U(x) \| \leqslant s\rho(x)$ ($x \in G$, s a positive constant independent of x), the lift of which is the initial representation of $C_\rho(G)$.

Let us conclude this number by stating how the concepts of Naimark relatedness and FDS representation will be used in conjunction with Banach representations of G.

Two Banach representations U and V of G are said to be *Naimark related* if their integrated forms on $M_c(G)$ are. Since $C_c(G)$ is weakly dense in $M_c(G)$, it is easy to see that U and V are Naimark related iff their integrated forms cut down to $C_c(G)$ are Naimark related. A Banach representation U of G is said to be FDS if its integrated form is FDS, or equivalently, if its integrated form restricted to $C_c(G)$ is FDS.

Example (Tomita) Let us give an example to show that Naimark relatedness fails to be transitive in general. Take for G the additive group $\mathbf{Z}$ of integers. Let $\mathbf{T}$ denote the circle group: $\mathbf{T} = \{c \in \mathbf{C} : |c| = 1\}$. Let μ_1 and μ_2 be two positive Radon measures on $C(\mathbf{T})$ neither of which is absolutely continuous with respect to the other but whose closed supports are all of $\mathbf{T}$. Put $E_1 = L^2(\mu_1)$, $E_2 = L^2(\mu_2)$, $E_3 = C(\mathbf{T})$. Set, for $n \in G$,

$$U^i(n)f_i(t) = t^n f_i(t) \qquad (t \in \mathbf{T}),$$

$f_i \in E_i$ ($i = 1, 2, 3$). Plainly U^1 and U^2 are unitary representations of G on the Hilbert spaces E_1 and E_2, respectively. If $C(\mathbf{T})$ is equipped with the sup norm, then U^3 is a Banach representation of G on E_3. The canonical injection of E_3 into E_1 (respectively E_2) serves to establish the Naimark relatedness of U^3 and U^1 (respectively U^3 and U^2). However U^1 and U^2 are not Naimark related – to show this it suffices to prove that U^1 and U^2 are not unitarily equivalent (see Proposition 4.3.1.4). Suppose that $Q : L^2(\mu_1) \to L^2(\mu_2)$ is a surjective linear isometry such that $QU^1(n)f_1 = U^2(n)Qf_1$ for all $n \in G$, $f_1 \in L^2(\mu_1)$. Let

$g = Q(1)$ (1 the constant function 'one' on $\mathbf{T}$). If $\hat{\mu}_1$ denotes the Fourier-Stieltjes transform of μ_1, then

$$(U^1(n)1, 1) = \int_{\mathbf{T}} t^n d\mu_1(t) = \hat{\mu}_1(n)$$

and

$$(U^1(n)1, 1) = (U^2(n)g, g) = \int_{\mathbf{T}} t^n g(t)\overline{g(t)}\, d\mu_2(t) = \hat{\nu}(n) \qquad (n \in G)$$

where $d\nu(t) = |g(t)|^2\, d\mu_2(t)$. The uniqueness of Fourier-Stieltjes transforms implies that $d\mu_1(t) = |g(t)|^2\, d\mu_2(t)$, contradicting the hypothesis that μ_1 was not absolutely continuous with respect to μ_2. These considerations show, then, that Naimark relatedness may not be transitive. However it should be noted that the U^i ($i = 1, 2, 3$) above are not topologically irreducible Banach representations – it appears to be unknown whether Naimark relatedness is transitive or not when confined only to topologically completely irreducible Banach representations.

The behavior of Naimark relatedness, in general, can be rather pathological. For instance, it may very well happen that a topologically irreducible Banach representation of G is Naimark equivalent to one which is reducible.

Example (Berezin) Let us agree to use the notations of Example 1 in number 5.4.2 – there the reader will find written down the topologically irreducible unitary representations of the group G of affine transformations of the real line $\mathbf{R}$. Consider U^+ – then U^+ can be realized on $L_+^2 = L^2([0, \infty))$ via the prescription: $U^+(\alpha, \beta)f(t) = \sqrt{\alpha}\, e^{\sqrt{-1}\,\beta t} f(\alpha t)$ ($f \in L_+^2, 0 \leqslant t < \infty$). Let F denote the space of all square summable, continuous functions on $[0, \infty)$ which vanish at infinity – then F is a dense subspace of L_+^2. One may equip F with the structure of a Banach space in the following way: Put

$$\|f\| = \|f\|_2 + \|f\|_\infty \qquad (f \in F).$$

Evidently F is U^+-stable; moreover the assignment $(\alpha, \beta) \mapsto U^+(\alpha, \beta)|F = V^+(\alpha, \beta)$ is a Banach representation of G on F. The natural injection of F into L_+^2 serves to establish the Naimark equivalence of V^+ and U^+ - but V^+ is not topologically irreducible. In fact the subspace of F consisting of those f which vanish at the origin is closed and V^+-stable.

Fortunately, as regards the groups in which we are really interested (cf. 4.5), Naimark relatedness will behave in a satisfactory manner.

4.3 Representations on a Hilbert Space

4.3.1 Generalities

In this number we shall merely note a few useful general properties of unitary representations, it being presupposed that the reader is familiar with the elements of the theory (see, e.g. Dixmier [15, § 13] or Hewitt and Ross [1, § 21–22]).

As usual, let G be a locally compact group countable at infinity. By a *Hilbert representation* U of G we shall mean a continuous representation of G on a Hilbert space E; a Hilbert representation U of G is said to be a *unitary representation* provided that all the operators $U(x)$ $(x \in G)$ are unitary.

The Hilbert representations U and V of G on E and F, respectively, are said to be *unitarily equivalent* if there exists a linear surjective isometry $Q: E \to F$ such that $QU(x) = V(x)Q$ for all $x \in G$.

Notation The set of unitary equivalence classes of topologically irreducible unitary representations of G will be denoted by $\hat{G}$.

Let U be a unitary representation of G on a Hilbert space E; fix vectors a, b in E and let $\phi_{a,b}$ denote the function on G defined by the prescription $x \mapsto (U(x)a, b)$ $(x \in G)$ – then $\phi_{a,b}$ is said to be a *coefficient* of U. Evidently $\phi_{a,b}$ is a bounded, continuous function on G and, moreover,

$$(\phi) \quad \phi_{U(\mu)a,b} = \phi_{a,b}*\mu', \phi_{a,U(\mu)b} = \bar{\mu}*\phi_{a,b} \qquad (a, b \in E;\ \mu \in M_c(G)).$$

This being so, by the functions of positive type *associated with* U we shall understand the coefficients $\phi_{a,a}$ $(a \in E)$. [In passing observe that the coefficients of $\bar{U}$ (the contragredient of U) are the complex conjugates of the coefficients of U.]

Example Let $L^2(G)$ denote the space of complex valued functions on G which are square summable with respect to left Haar measure. For $x \in G$, let $L(x)$ denote the operator in $L^2(G)$ defined by $L(x)f(y) = f(x^{-1}y)(y \in G, f \in L^2(G))$ – then the assignment $x \mapsto L(x)$ is a unitary representation of G on $L^2(G)$, the *left regular representation*. On the other hand, for $x \in G$, let $R(x)$ denote the operator in $L^2(G)$ defined by $R(x)f(y) = \delta_G(x)^{1/2}f(yx)$ $(y \in G, f \in L^2(G))$ – then the assignment $x \mapsto R(x)$ is a unitary representation of G on $L^2(G)$, the *right regular representation*. Evidently L and R are injective. Given $f \in L^2(G)$, define $\check{f} \in L^2(G)$ by the rule $\check{f}(x) = \delta_G(x)^{-1/2}f(x^{-1})$ $(x \in G)$ – then the map $f \mapsto \check{f}$ is an isometric isomorphism of $L^2(G)$ onto itself which transforms L into R. Therefore L and R are unitarily equivalent.

Lemma 4.3.1.1 (Mackey) Let U, V be unitary representations of G on Hilbert spaces E, F respectively; let S be an intertwining operator for U and V; let $\tilde{E}$ denote the kernel of S, $\tilde{F}$ the closure of the range of S; let $\tilde{E}^\perp$ denote the orthogonal complement of $\tilde{E}$ in E – then $\tilde{E}, \tilde{F}$ are invariant subspaces for U, V, respectively, and the restriction of U to $\tilde{E}^\perp$ is unitarily equivalent to the restriction of V to $\tilde{F}$.

Proof (1) Let us first check the invariance of $\tilde{E}$ under U (from which, since U is unitary, follows the invariance of $\tilde{E}^\perp$). If $a \in \tilde{E}$, then $SU(x)a = V(x)Sa = 0 \Rightarrow U(x)a \in \tilde{E}$ (all $x \in G$); the invariance of $\tilde{F}$ under V is also clear.

(2) Let $S = S_1S_2$ denote the polar decomposition of S (thus S_1 is an isometry mapping $\tilde{E}^\perp$ onto $\tilde{F}$ while $S_2 = (S^*S)^{1/2}$ is a positive Hermitian operator taking E into $\tilde{E}^\perp$ (S^* the Hilbert space adjoint of S)). Let $\tilde{U}$ denote the restriction of U to $\tilde{E}^\perp$, $\tilde{V}$ the restriction of V to $\tilde{F}$. Claim:

$S_1\tilde{U}(x) = \tilde{V}(x)S_1$ (all $x \in G$). Thus first note that $S_2\tilde{U}(x) = \tilde{U}(x)S_2$ (all $x \in G$); in fact $S\tilde{U}(x) = \tilde{V}(x)S \Rightarrow \tilde{U}(x)^*S^* = S^*\tilde{V}(x)^*$ ($x \in G$) and so

$$S^*S\tilde{U}(x) = S^*\tilde{V}(x)S = S^*\tilde{V}(x^{-1})^*S = \tilde{U}(x^{-1})^*S^*S,$$

that is, $S_2^2\tilde{U}(x) = \tilde{U}(x)S_2^2$ ($x \in G$). Since S_2 is a strong limit of polynomials in S_2^2 (Riesz and Nagy [1, p. 265]), it follows that $S_2\tilde{U}(x) = \tilde{U}(x)S_2$ (all $x \in G$). Finally, $S_1S_2\tilde{U}(x) = S_1\tilde{U}(x)S_2 = \tilde{V}(x)S_1S_2$ ($x \in G$) – therefore $(S_1\tilde{U}(x) - \tilde{V}(x)S_1)S_2 = 0$, whence $S_1\tilde{U}(x) = \tilde{V}(x)S_1$ for all $x \in G$, the range of S_2 being dense in $\tilde{E}^\perp$.

This completes the proof of the lemma. □

Corollary 4.3.1.2 Let U, V be unitary representations of G on Hilbert spaces E, F respectively. If there exists an injective S intertwining U and V such that $S(E)$ is dense in F, then U and V are unitarily equivalent.

Corollary 4.3.1.3 Let U, V be topologically irreducible unitary representations of G on Hilbert spaces E, F. If there exists a non-zero operator S intertwining U and V, then U and V are unitarily equivalent.

Remarks (1) Lemma 4.3.1.1 is valid (in the obvious sense) for closed densely defined unbounded interwining operators (for the existence of a polar decomposition in this case see Dunford and Schwartz [2, p. 1249]).

(2) If $I(U, V)$ is the dimension of the space $\mathfrak{I}(U, V)$ of operators intertwining U and V (cf. 4.1.3), then Lemma 4.3.1.1 shows that $I(U, V) = 0$ iff U and V are *disjoint*, i.e. iff no subrepresentation of U is unitarily equivalent to a subrepresentation of V.

Proposition 4.3.1.4 Let U, V be unitary representations of G on Hilbert spaces E, F respectively. The following two conditions are equivalent:

(i) U and V are unitarily equivalent;

(ii) U and V are Naimark equivalent.

Proof (i) ⇒ (ii). We remind the reader that to say U and V are Naimark equivalent means, by definition, that their integrated forms are so. If $Q : E \to F$ is a surjective linear isometry such that $QU(x) = V(x)Q$ for all $x \in G$, then plainly $QU(\mu) = V(\mu)Q$ for all $\mu \in M_c(G)$. Hence U and V are Naimark related.

(ii) ⇒ (i). If U and V are Naimark related, then, by definition, there exists a closed one-to-one linear map T of a dense U-stable subspace of E onto a dense V-stable subspace of F such that $TU(\mu) = V(\mu)T$ for all $\mu \in M_c(G)$. In particular, taking $\mu = \delta_x$ ($x \in G$), we have $TU(x) = V(x)T$ for all $x \in G$. By Remark 1 above, and Corollary 4.3.1.2, U and V are unitarily equivalent.

It therefore follows that (i) ⇔ (ii), as contended. □

Let us now recall the definition of a von-Neumann algebra. A *von-*

Neumann algebra $\mathfrak{R}$ is an algebra of bounded operators in a complex Hilbert space E which contains the identity operator, is closed under the adjoint operation and is closed in the weak operator topology (i.e. $T_n \to T$ in the weak operator topology if for every pair of vectors $a, b \in E$, $(T_n a, b) \to (Ta, b)$). For instance if U is a unitary representation of G on E, then the algebra $\mathfrak{J}(U, U)$ of intertwining operators for U is a von-Neumann algebra. If $\mathfrak{R}$ is a von-Neumann algebra, define $\mathfrak{R}'$ to be the set of all bounded operators in E commuting with all the elements in $\mathfrak{R}$ – then $\mathfrak{R}'$ is also a von-Neumann algebra.

An important result, the proof of which may be found in Dunford and Schwartz [2, p. 885], is:

Theorem 4.3.1.5 (von-Neumann) Let $\mathfrak{R}$ be a von-Neumann algebra – then $\mathfrak{R} = (\mathfrak{R}')'$.

Before giving an application of this theorem, we recall the following well-known fact.

Proposition 4.3.1.6 Let U be a unitary representation of G on a Hilbert space E – then U is topologically irreducible iff the algebra $\mathfrak{J}(U, U)$ of intertwining operators for U consists of scalar multiples of the identity operator alone.

Proof If U is reducible, then there exists a non-trivial closed U-invariant subspace of E. The orthogonal projection of E onto this subspace intertwines U and this projection is not a scalar. Conversely, suppose that S is a non-zero bounded self-adjoint operator on E commuting with all the $U(x)$ $(x \in G)$. Let $\{\mathsf{P}_t : t \in \mathbf{R}\}$ be the family of projections on E for the operator S given to us by the Spectral Theorem – then each P_t commutes with all the $U(x)$ and so, by the irreducibility of U, $\mathsf{P}_t = 0$ or I (I the identity operator in E). Consequently there is a real number t_0 such that $\mathsf{P}_t = 0$ for $t < t_0$ and $\mathsf{P}_t = I$ for $t > t_0$ – hence $S = t_0 I$. Turning to the general case, if S is any bounded operator on E commuting with all the $U(x)$ $(x \in G)$, then S^* does the same. We see, therefore, upon writing $S = (S + S^*)/2 + \sqrt{-1}(S - S^*)/2\sqrt{-1}$, that S is a linear combination of self-adjoint operators which commute with all the $U(x)$ $(x \in G)$ – it now follows from what was said above that S is a scalar multiple of I. □

Remarks (1) It is in this proposition that the importance for unitary representations of the number $I(U, U)$ lies: U is topologically irreducible iff $I(U, U) = 1$. [In the general case consideration of $I(U, U)$ will not give information about the irreducibility of U but only on the possibility of decomposing U into a (topological) direct sum: $I(U, U) = 1 \Rightarrow U$ is indecomposable. However it is perfectly possible that U can be reducible without being decomposable as a direct sum, i.e. indecomposability does not in general imply irreducibility.]

(2) Let U and V be unitary representations of G. Since $V^\vee = \bar{V}$ in this case (cf. 4.1.2), Proposition 4.1.3.1 implies that $i(U, V) = I(U, \bar{V})$. Therefore U is

irreducible iff $i(U, \bar{U}) = 1$. If U is irreducible, then $i(U, \bar{V})$ is the number of times that V contains U as a discrete direct component.

Recall that a **-algebra* A is an associative algebra over $\mathbf{C}$ admitting an *adjoint* operation $x \mapsto x^*$ such that for all $x, y \in A$ and $c \in \mathbf{C}$:

(i) $(x + y)^* = x^* + y^*$;
(ii) $(cx)^* = \bar{c}x^*$;
(iii) $(xy)^* = y^*x^*$;
(iv) $(x^*)^* = x$.

If A is also a normed algebra and $\|x^*\| = \|x\|$ $(x \in A)$, then A is called a *normed *-algebra.* A representation U of a *-algebra A by bounded operators on a Hilbert space E is said to be a **-representation* of A if $U(x)^* = U(x^*)$ for all $x \in A$.

Plainly Proposition 4.3.1.6 is also applicable to *-representations of a *-algebra. In particular, if U is a unitary representation of G, then the integrated form of U is a *-representation of the normed *-algebra $M_c(G)$ $(\|\mu\| = \int_G d|\mu|(x), d\mu^*(x) = \overline{d\mu(x^{-1})})$ and if U is topologically irreducible, then the only bounded linear operators commuting with all the $U(\mu)$ $(\mu \in M_c(G))$ are the scalar multiples of the identity.

[We remind the reader that if U is a unitary representation of G, then U actually lifts to a *-representation of the Banach algebra of bounded measures on G; in particular U lifts to a *-representation of $L^1(G)$ $(f^* = \delta_G^{-1}\bar{f}', f \in L^1(G))$. Moreover if U is topologically irreducible, then so is the lift of U to $L^1(G)$ (and conversely) although $U(L^1(G))$ need not be algebraically irreducible.]

Proposition 4.3.1.7 Let U be a unitary representation of G on a Hilbert space E. The following statements are equivalent:

(i) U is topologically irreducible;
(ii) U is topologically completely irreducible.

Proof (i) $\Rightarrow$ (ii). Let $\mathfrak{A}_n = \{a_1, \ldots, a_n\}$ denote any set of n elements in E. For each n and each $\mathfrak{A}_n$ and each $S \in \mathfrak{L}(E; E)$, we must show that there exists a net $\{\mu_k\}$ in $M_c(G)$ such that

$$U(\mu_k)a_i \to Sa_i \qquad (i = 1, \ldots, n).$$

Let $\mathfrak{R}$ denote the set of all bounded linear operators S on E with the property that there exists a net $\{\mu_k\}$ in $M_c(G)$ such that for every $\mathfrak{A}_n$

$$U(\mu_k)a_i \to Sa_i \qquad (i = 1, \ldots, n).$$

Then $\mathfrak{R}$ is a von-Neumann algebra. Since $\mathfrak{R} \supset \{U(\mu) : \mu \in M_c(G)\}$, $\mathfrak{R}'$ consists of scalar multiples of the identity alone (Proposition 4.3.1.6). Hence $\mathfrak{R} = (\mathfrak{R}')'$ (Theorem 4.3.1.5) consists of all the bounded linear operators on E.

(ii) $\Rightarrow$ (i). This has been observed earlier (Proposition 4.2.1.3).

It therefore follows that (i) $\Leftrightarrow$ (ii), as contended. □

In view of this proposition, we need not distinguish, in the unitary case, between topologically irreducible and topologically completely irreducible representations of G (henceforth, such representations will frequently just be referred to as irreducible).

Remark Suppose that G satisfies the second axiom of countability – then the representation space for any irreducible unitary representation of G is evidently separable.

If $U_1, U_2, \ldots$ are unitary representations of G on Hilbert spaces $E_1, E_2, \ldots$, then we define, as usual, their (Hilbert space) direct sum $U = U_1 + U_2 + \ldots$ as follows. Let E be the direct sum of the E_i (in the sense of Hilbert space theory). Given $a = (a_1, a_2, \ldots)$ in $E(\sum_i \|a_i\|^2 < \infty)$, $x \in G$, put $U(x)a = (U_1(x)a_1, U_2(x)a_2, \ldots)$ – then U is a unitary representation of G on E, the *direct sum* of the U_i.

Let U be a unitary representation of G on E. We shall say that U is *discretely decomposable* if U is unitarily equivalent to a direct sum $U_1 + U_2 + \ldots$, where the U_i are topologically irreducible. If V is irreducible, then the dimension of $\mathfrak{J}(U, V)$ is equal to the number of indices i for which V is unitarily equivalent to U_i – call this number the *multiplicity* of V in $U = U_1 + U_2 + \ldots$. It is clear that U is uniquely determined by the non-negative integers (the *spectrum* of U) which are the dimensions of the spaces $\mathfrak{J}(U, V)$ whenever U is discretely decomposable. [Thus to know all irreducible unitary representations (to within unitary equivalence) is to know all discretely decomposable unitary representations (to within unitary equivalence).] As the reader is well aware, discretely decomposable unitary representations are rather special – nevertheless they will play a surprisingly important role at many points later on.

Proposition 4.3.1.8 Let U be a unitary representation of G on the separable Hilbert space E. Suppose that for any neighborhood $\mathcal{O}$ of the identity in G, there is a non-negative integrable function f on G with support in $\mathcal{O}$ such that $f = f^*$, $\int_G f(x)\, d_G(x) = 1$ with $U(f)$ completely continuous – then U is discretely decomposable. In fact a given topologically irreducible unitary representation of G occurs at most a finite number of times in the direct sum decomposition of U.

Proof (Langlands) Consider the families of closed mutually orthogonal subspaces of E which are invariant and irreducible under the action of G. If these families are ordered by inclusion there will be a maximal one. Let the (Hilbert space) direct sum of the subspaces in some maximal family be F – in order to prove the first assertion, it is necessary to show that $E = F$. Assuming that this is not the case, let $F^\perp$ be the orthogonal complement of F in E. Choose $a \in F^\perp$ with $\|a\| = 1$ and then choose $\mathcal{O}$ so that $\|a - U(x)a\| < 1/2$ if x is in $\mathcal{O}$. Pick f as in the statement of the proposition–then $\|U(f)a - a\| < 1/2$ and hence $U(f)a \neq 0$. The restriction of $U(f)$ to $F^\perp$ is self-adjoint and thus has a non-zero eigenvalue λ. Let

E^λ be the (finite dimensional) space of eigenvectors belonging to the eigenvalue λ. Select from the family of non-zero subspaces of E^λ obtained by intersecting E^λ with closed invariant subspaces of $F^\perp$ a minimal one $\mathscr{S}$. Take the intersection $\mathscr{T}$ of all closed invariant subspaces of $F^\perp$ containing $\mathscr{S}$. Since $\mathscr{T} \neq \{0\}$, a contradiction will result when it is shown that the restriction of U to $\mathscr{T}$ is irreducible. If this were not the case, then $\mathscr{T}$ would be the orthogonal direct sum of two closed invariant subspaces $\mathscr{T}_1$ and $\mathscr{T}_2$. Because $E^\lambda \cap \mathscr{T}_i$ is contained in $E^\lambda \cap \mathscr{T} = \mathscr{S}$ for $i = 1, 2$, the space $E^\lambda \cap \mathscr{T}_i$ is either $\{0\}$ or $\mathscr{S}$. However $U(f)\mathscr{T}_i \subset \mathscr{T}_i$ and so $\mathscr{S} = \mathscr{S} \cap \mathscr{T}_1 + \mathscr{S} \cap \mathscr{T}_2$ (orthogonal direct sum); consequently $\mathscr{S} \cap \mathscr{T}_i = \mathscr{S}$ for $i = 1$ or $i = 2$, an impossibility. As for the second assertion, one need only observe that if some topologically irreducible unitary representation of G occured with infinite multiplicity, then, for some f, $U(f)$ would have a non-zero eigenvalue of infinite multiplicity. □

Example Let G be a locally compact group satisfying the second axiom of countability; let Γ be a discrete subgroup of G such that G/Γ is compact – then the left regular representation $L_{G/\Gamma}$ of G on $L^2(G/\Gamma)$ is discretely decomposable, the spectrum of $L_{G/\Gamma}$ being comprised of *finite* non-negative integers. [It can be easily shown that $L_{G/\Gamma}(f)$ is completely continuous for every $f \in C_c(G)$; therefore the assertion follows from Proposition 4.3.1.8.]

Let U be a unitary representation of G on E – then it is natural to ask: In what sense can U be decomposed into irreducibles? Of course, as the reader probably knows already, the answer to this important question (and others) depends on the theory of direct integrals of Hilbert spaces. Since there are excellent accounts of these aspects of the theory readily available in the literature (cf. Dixmier [2], [15]), there is no point in reproducing the details of such results here.

Example It can very well happen that a unitary representation U of G on E contains no irreducible subrepresentation (i.e. E admits no closed U-stable subspace on which U defines an irreducible unitary representation of G). For instance, take $G = \mathbf{R}$, $E = L^2(G)$ and define U by the prescription $U(x)f(y) = e^{\sqrt{-1}\,xy} f(y)$ ($f \in L^2(G)$; $x, y \in G$) – then U is a unitary representation of G on E. If $\mathfrak{B}$ is a Borel subset of G, then the set $E_{\mathfrak{B}} = \{f : f \in L^2(G), f \mid (G - \mathfrak{B}) = 0\}$ is a closed U-invariant subspace of E and every closed U-invariant subspace of E is of this form. If $E_{\mathfrak{B}}$ is to be non-trivial, $\mathfrak{B}$ must be of positive measure; writing $\mathfrak{B} = \mathfrak{B}_1 \cup \mathfrak{B}_2$, where the $\mathfrak{B}_i$ ($i = 1, 2$) are disjoint Borel sets of positive measure, amounts to writing $E_{\mathfrak{B}}$ as the orthogonal direct sum of $E_{\mathfrak{B}_1}$ and $E_{\mathfrak{B}_2}$. Therefore no irreducible subrepresentation of U can exist. [Plainly these considerations admit obvious generalizations which may be left to the reader.] This example shows already the futility in attempting to decompose (in analogy with a compact G) an arbitrary unitary representation of a non-compact G as a direct sum of irreducibles; on the other hand, the phenomenon at hand admits a ready interpretation in terms of direct integrals . . . .

4.3.2 Examples

In this number we shall offer some complements to the generalities presented in 4.3.1.

Example 1 *Compact Lie Groups.* Let K be a compact Lie group; for convenience let us review some of the highlights in the representation theory of K (complete details may be found, for instance, in Dixmier [15, §15]). As is usual, let $\hat{K}$ denote the (necessarily countable) set of unitary equivalence classes of irreducible unitary representations of K – then every class $\delta \in \hat{K}$ is finite dimensional. Every unitary representation of K on a separable Hilbert space is discretely decomposable; in particular the left regular representation of K on $L^2(K)$ decomposes into a discrete direct sum of irreducibles, with every $\delta \in \hat{K}$ occurring in this decomposition exactly $d(\delta)$ times ($d(\delta)$ the degree of δ, i.e. the dimension of δ). Fix an irreducible unitary representation μ_δ (on E_δ, say) in the class δ – then, for every $f \in L^2(K)$ one has $\| f \|_2^2 = \int_K |f(k)|^2 dk = \sum_{\delta \in \hat{K}} d(\delta)\mathrm{tr}(\mu_\delta(f)\mu_\delta^*(f))$ (the Plancherel Theorem for K). [We shall not stop to write down explicitly the Schur Orthogonality Relations. . . .] A function $f \in C(K)$ is said to be *K-finite* if the span of its right and left translates under K is finite dimensional. In this connection it is useful to bear in mind that the following conditions on an element $f \in C(K)$ are equivalent: (1) f is a representative function (i.e. f is a coefficient in some finite dimensional (continuous) representation of K); (2) f is left K-finite (i.e. the left K-translates of f span a finite dimensional space); (3) f is right K-finite (i.e. the right K-translates of f span a finite dimensional space).

Example 2 *A Theorem of van der Waerden.* Let K be a connected semisimple compact Lie group. Let U be a representation of K by unitary operators on a finite dimensional Hilbert space E – then U is necessarily continuous, i.e. U is necessarily a unitary representation of K. For a proof of this remarkable fact see van der Waerden [1]. [Of course this result is patently false if the semisimplicity assumption on K is dropped.]

Example 3 *Bichteler's Theorem.* Let G be a locally compact topological group. Suppose that every topologically irreducible representation of G by unitary operators on a Hilbert space is continuous – then, of necessity, G is discrete (cf. Bichteler [1]). Consequently, in view of Example 2, a connected semisimple compact Lie group must admit at least one *non-continuous* infinite dimensional topologically irreducible representation by unitary operators on a Hilbert space. [Such representations have been exhibited, in a number of special cases, by L. Robertson (unpublished).]

Example 4 *A Theorem of Segal and von Neumann.* The well-known Gelfand-Raikov Theorem asserts that if G is a locally compact topological group, then, for every x in G different from 1, there exists a topologically irreducible unitary representation U of G such that $U(x) \neq I$ (I the identity operator in the representation space of U). Of course U may be infinite dimensional and for certain classes of groups it always is. Thus let G be a simple non-compact Lie group (that is the Lie algebra $\mathfrak{g}$ of G is simple, hence has no proper ideals). We shall prove that G admits no finite dimensional unitary representation other than the trivial representation $x \mapsto I_n$, n a finite cardinal. This will be done in a number of steps.

(a) Let $\tilde{G}$ be a connected Lie group with Lie algebra $\tilde{\mathfrak{g}}$, $\tilde{H}$ a normal analytic subgroup with Lie algebra $\tilde{\mathfrak{h}}$. Suppose that there exists a sequence of elements $\{X_n\}$ in $\tilde{\mathfrak{g}}$ such that $\mathrm{Ad}(\exp X_n)X \to 0$ for each $X \in \tilde{\mathfrak{h}}$ – then the restriction to $\tilde{H}$ of any finite dimensional unitary representation U of $\tilde{G}$ is the identity representation.

Proof Let $X \in \tilde{\mathfrak{h}}$ and consider $\exp(X_n)\exp(X)\exp(-X_n) = \exp(\mathrm{Ad}(\exp X_n)X)$ – thus $\lim_{n\to\infty} \exp(X_n)\exp(X)\exp(-X_n) = 1$, and so, $\tilde{H}$ being connected, we deduce that $\lim_{n\to\infty} \exp(X_n)\xi\exp(-X_n) = 1$ for every $\xi \in \tilde{H}$. From this it follows at once that $\mathrm{sp}(U(\xi)) = 1$ ($\xi \in \tilde{H}$) (sp denoting the trace divided by the dimension of U). It remains to show that if $\operatorname{sp}(U(x_0)) = 1$ for some $x_0 \in \tilde{G}$, then $U(x_0) = I$ (I the identity operator in the representation space E of U). So suppose that $\operatorname{sp}(U(x_0)) = 1$; put $R = U(x_0)$ and write $R = S + \sqrt{-1}T$ where S and T are self-adjoint – then $S = \frac{1}{2}(R + R^*)$ and $\|S\| \leqslant 1$. Moreover (and this is the key observation) $I - S$ is positive semi-definite. [Fix a non-zero vector $a \in E$ and put $\phi(x) = (a, U(x)a)$ $(x \in \tilde{G})$ – then ϕ is a positive definite function on $\tilde{G}$ and so, in particular, we have $|\phi(x)| \leqslant \phi(1) = (a, a)$ (all $x \in \tilde{G}$), whence $(a, (I - S)a) = (a, a) - \frac{1}{2}(a, U(x_0)a) - \frac{1}{2}(a, U(x_0^{-1})a) = \phi(1) - \mathscr{R}\phi(x_0) \geqslant 0$.] Therefore we have: $\operatorname{sp}(R) = 1 = \operatorname{sp}(S) + \sqrt{-1}\operatorname{sp}(T)$ $\Rightarrow \operatorname{sp}(S) = 1 \Rightarrow \operatorname{sp}(I - S) = 0 \Rightarrow I - S = 0 \Rightarrow I = S \Rightarrow R = I + \sqrt{-1}T$. But $RR^* = I = I + T^2 \Rightarrow T^2 = 0 \Rightarrow T = 0$, T being self-adjoint; consequently $R(= U(x_0)) = I \ldots .$ □

(b) Let $\tilde{G}$ denote a connected Lie group with a two dimensional Lie algebra $\tilde{\mathfrak{g}}$. Let $\{X, Y\}$ denote a basis for $\tilde{\mathfrak{g}}$ such that $[X, Y] = -X$ – then any finite dimensional unitary representation of $\tilde{G}$ is the identity representation on the one parameter subgroup generated by X.

Proof The one parameter subgroup generated by X is normal in $\tilde{G}$ since the algebra generated by X in $\tilde{\mathfrak{g}}$ is an ideal. For each positive integer n set $X_n = -nY$ – then $\operatorname{Ad}(\exp X_n)(-X) = e^{-n}(-X)$. Statement (b) now follows from (a). □

(c) Let $\mathfrak{g}$ be a simple non-compact Lie algebra over **R** – then $\mathfrak{g}$ contains a two dimensional subalgebra $\{X, Y\}$ where $[X, Y] = -X$.

Proof Let $\mathfrak{g} = \mathfrak{k} + \mathfrak{a}_{\mathfrak{p}} + \mathfrak{n}^+$ be an Iwasawa decomposition for $\mathfrak{g}$. Select a non-zero root α in P_+ (P_+ is not empty) and fix a non-zero $X_\alpha \in \mathfrak{g}_c^\alpha$. Writing $X_\alpha = X_1 + \sqrt{-1}X_2$ where $X_1, X_2 \in \mathfrak{g}$ and bearing in mind that α is real on $\mathfrak{a}_{\mathfrak{p}}$, we obtain $[H, X_i] = \alpha(H)X_i$ for $H \in \mathfrak{a}_{\mathfrak{p}}$, $i = 1, 2$. At least one of the vectors X_i is non-zero, say X_1. Choosing $H_0 \in \mathfrak{a}_{\mathfrak{p}}$ such that $\alpha(H_0) \neq 0$, set $X = X_1$, $Y = \alpha(H_0)^{-1}H_0$ – then $[X, Y] = -X$. □

Let us return to our simple non-compact connected Lie group G – then it is immediate, on the basis of statements (a), (b), and (c) above, that G admits no non-trivial finite dimensional unitary representation U. [Indeed, for any such U, the identity component of the set $\{x \in G : U(x) = I\}$ is a connected, normal subgroup of G of *positive* dimension, hence is equal to G itself.]

Various extensions of this result are possible. Let G be a non-compact semi-simple connected Lie group with Lie algebra $\mathfrak{g}$. Let $\mathfrak{g} = \mathfrak{g}_1 + \ldots + \mathfrak{g}_r$ be the direct decomposition of $\mathfrak{g}$ into simple ideals; the analytic subgroups G_i of G corresponding to the $\mathfrak{g}_i$ $(i = 1, \ldots, r)$ are called the *simple constituents* of G. Passing to the simply connected covering group of G we see: If G is a connected semi-simple Lie group, none of whose simple constituents is compact, then G admits no non-trivial finite dimensional unitary representation. In particular: If G is a connected semi-simple complex analytic Lie group, then G admits no non-trivial finite dimensional unitary representation. [Recall that a compact connected complex analytic Lie group is necessarily abelian.]

Example 5 *Weyl's Unitary Trick*. Let G be a connected non-compact semi-simple Lie group with Lie algebra $\mathfrak{g}$, U a continuous representation of G on a finite dimensional Banach space E. Claim: E is a semi-simple G-module, i.e. E may be written as the algebraic direct sum of U-stable, U-irreducible subspaces. To see this, first let $\mathfrak{g} = \mathfrak{k} + \mathfrak{p}$ be a Cartan decomposition for $\mathfrak{g}$, $\mathfrak{g}_c = \mathfrak{g} + \sqrt{-1}\,\mathfrak{g}$ the complexification of $\mathfrak{g}$, $\mathfrak{u} = \mathfrak{k} + \sqrt{-1}\mathfrak{p}$ the corresponding compact real form of $\mathfrak{g}_c$. Now the derivative of U gives rise to a representation of $\mathfrak{g}$ in the usual way and it is clear that E is a semi-simple G-module iff E is a semi-simple $\mathfrak{g}$-module iff E is a semi-simple $\mathfrak{g}_c$-module iff E is a semi-simple $\mathfrak{u}$-module. Let $G_\mathfrak{u}$ denote the simply connected Lie group with Lie algebra $\mathfrak{u}$ – then $G_\mathfrak{u}$ is compact and the representation of $\mathfrak{u}$ 'pulls down' to a continuous representation of $G_\mathfrak{u}$, call it $U_\mathfrak{u}$. Let $(.\,,.)$ denote any inner product in E and put $((a, b)) = \int_{G_\mathfrak{u}} (U_\mathfrak{u}(x)a, U_\mathfrak{u}(x)b)dx (a, b \in E)$ (dx is normalized Haar measure on $G_\mathfrak{u}$) – then $U_\mathfrak{u}$ is a unitary representation of $G_\mathfrak{u}$ when E is equipped with the Hilbert space structure $((.\,,.))$ – thus E is a semi-simple $G_\mathfrak{u}$-module and so is a semi-simple $\mathfrak{u}$-module.

4.4 Differentiable Vectors Analytic Vectors

4.4.1 Passage to U_∞

Let G be a Lie group countable at infinity, d_G left Haar measure on G. Let $\mathfrak{g}$ denote the Lie algebra of G, $\mathfrak{g}_c = \mathfrak{g} + \sqrt{-1}\,\mathfrak{g}$ the complexification of $\mathfrak{g}$, $\mathfrak{G}$ the universal enveloping algebra of $\mathfrak{g}_c$; unless specifically stated to the contrary, in what follows the elements of $\mathfrak{G}$ are to be viewed as right invariant differential operators on G.

Let E denote, as in 4.1, a locally convex, complete, Hausdorff, topological vector space whose topology is described by the family of continuous semi-norms $\{|\,.\,|_\alpha : \alpha \in \mathfrak{A}\}$.

Let $x \mapsto U(x)$ be a continuous representation of G on $E (x \in G)$.

Definition A vector $a \in E$ is said to be *differentiable* (for U) if the map $\tilde{a}$, $x \mapsto U(x)a$, is an element of $C^\infty(G; E)$.

Example Let U be the left regular representation of G on $L^2(G)$. If $f \in C_c^\infty(G)$, then f is differentiable for U. To see this it suffices to show that for each $X \in \mathfrak{g}$ (viewed as a right invariant differential operator on G) we have

$$\lim_{t\to 0} \int_G |t^{-1}[f(\exp(-tX)x) - f(x)] - Xf(x)|^2 d_G(x) = 0.$$

Let $M = \sup_{x\in G} |Xf(x)|^2$. If t is such that $0 < |t| \leqslant 1$, then

$$\begin{aligned} &|t^{-1}[f(\exp(-tX)x) - f(x)] - Xf(x)|^2 \\ &\leqslant 2|t^{-1}[f(\exp(-tX)x) - f(x)]|^2 + 2|Xf(x)|^2 \leqslant 4M. \end{aligned}$$

By the Bounded Convergence Theorem we may pass to the limit under the integral sign (everything in sight has compact support) and this gives the result.

Let U be a continuous representation of G on E; let E_∞ denote the set of differentiable vectors in E (for U) – clearly E_∞ is a U-stable linear subspace of E. Let $a \in E$; if $f \in C_c^\infty(G)$, then, as an easy computation shows, $U(f)a \in E_\infty$. [One may also reason as follows. Let $a \in E$ – then we may identify the map $x \mapsto U(x)a = \tilde{a}(x)$ $(x \in G)$ with an element of $\mathfrak{L}(C_c^\infty(G); E)$. To verify that $U(f)a \in E_\infty$, one must check that the map $x \mapsto U(x)U(f)a = \tilde{a}*f'(x)$ $(f'(x) = f(x^{-1}), x \in G)$ is an element of $C^\infty(G; E)$ – in view of Proposition A2.4.1 (Appendix 2), this is clear.]

Definition The set E_α of all finite linear combinations of vectors $U(f)a$ where $f \in C_c^\infty(G)$, $a \in E$, will be called the *Gårding subspace* of E.

Proposition 4.4.1.1 (Gårding) Let U be a continuous representation of G on E – then E_α is dense in E.

Proof As in 4.1.1 consider the net $\{g_\mathcal{O}\}$, $\mathcal{O}$ running through the relatively compact neighborhoods of 1. Since G is a Lie group, we may assume in addition that $g_\mathcal{O} \in C_c^\infty(G)$ – hence $\lim_\mathcal{O} U(g_\mathcal{O})a = a$ for every $a \in E$. □

Remark Let K be a compact subgroup of G. Observe that the elements $g_\mathcal{O}$ of the net used in the proof of Proposition 4.4.1.1 may be chosen so that $g_\mathcal{O}(kxk^{-1}) = g_\mathcal{O}(x)\ (k \in K, x \in G)$.

Let U be a continuous representation of G on E, E_∞ the space of differentiable vectors in E (for U) – then, as we have just seen, E_∞ is dense in E.

We shall now place a topology on E_∞, a topology which is, in general, finer than that inherited from E. Thus consider the map $a \mapsto \tilde{a}$ of E_∞ into $C^\infty(G; E)$; this map identifies E_∞ with a *closed* subspace of $C^\infty(G; E)$. [If $\tilde{a}_n$ converges to f in $C^\infty(G; E)$ $(a_n \in E_\infty)$, then, for all $x \in G$, one has $f(x) = U(x)f(1)$ (since $\tilde{a}_n(x) = U(x)\tilde{a}_n(1)$) and so we see that f is the function $\tilde{a}$ associated with the element $a = f(1)$ of E_∞.] Henceforth, unless specifically stated to the contrary, we shall consider E_∞ as equipped with the relative topology induced by $C^\infty(G; E)$ (via the identification $a \mapsto \tilde{a}$) – then E_∞ is complete for this topology and is a Fréchet space if E is a Fréchet space (in particular, if E is a Banach space).

Let $D_c(G)$ denote the algebra, under convolution, of distributions on G with compact support. A standard argument shows that our continuous representation U lifts to a representation of $D_c(G)$ on E_∞. [Fix T in $D_c(G)$ – then T determines a map from $C^\infty(G) \otimes_C E$ into E, whence, by continuity, a map from $C^\infty(G) \hat{\otimes} E$ into E. But (cf. A2.2, Example 2), $C^\infty(G) \hat{\otimes} E = C^\infty(G; E)$ and so, for $a \in E_\infty$, we can put $U(T)a = \int \tilde{a}(x)\, dT(x)$, a vector which actually does lie in E_∞ since:

$$(*) \quad U(x)U(T)a = \tilde{a} * T'(x) \qquad (a \in E_\infty, x \in G).$$

To see this, employ the conventions and notations of A2.4 – thus, for all $x \in G$, we have $\tilde{a} * T'(x) = \langle((\tilde{a} \circ L_x)\delta_G)', T'\rangle = \langle((\tilde{a} \circ L_x)\delta_G)\delta_G^{-1}, T\rangle = U(x)U(T)a$.] However the endomorphism $U(T)$ of E_∞ $(T \in D_c(G))$ is not, in general, continuous when E_∞ is equipped with the topology inherited from E. On the other hand, formula $(*)$, in conjunction with Proposition A2.4.1 (Appendix 2), shows that the operators $U(T)$ $(T \in D_c(G))$ are continuous provided we place on E_∞ the topology agreed to in the previous paragraph. In particular the operators $U(x)$ $(x \in G)$ restricted to E_∞ define a *continuous* representation of G on E_∞ which one may view as the restriction of the right regular representation of G on $C^\infty(G; E)$ to the closed subspace of functions $\tilde{a}(a \in E_\infty)$. Plainly the map $x \mapsto U(x)a$ is an element of $C^\infty(G; E_\infty)$ for each $a \in E_\infty$, i.e. $(E_\infty)_\infty = E_\infty$.

Definition The restriction U_∞ of U to E_∞ (equipped with the above topology) will be referred to as the *differentiable representation* associated with U.

The reader will note that the continuous representation U is topologically irreducible iff the associated differentiable representation U_∞ is. [Thus suppose that U is reducible; let $\tilde{E}$ be a non-trivial closed U-stable subspace of E – then $\tilde{E} \cap E_\infty$ is a U_∞-stable subspace of E_∞, closed for the topology induced on E_∞ by the space $C^\infty(G; E)$. If $f \in C_c^\infty(G)$, $a \in \tilde{E}$, then $U(f)a \in \tilde{E} \cap E_\infty$ – this means that $\tilde{E} \cap E_\infty$ is dense in $\tilde{E}$, therefore is non-zero and is, moreover, distinct from E_∞ since its closure in E, $\tilde{E}$, is distinct from E, and E is the closure of E_∞ in E. Consequently U_∞ is reducible. On the other hand, suppose that U is irreducible; let $\tilde{E}_\infty$ be a non-zero closed U_∞-stable subspace of E_∞ – to verify that $E_\infty = \tilde{E}_\infty$, it will be enough to show that $\tilde{E}_\infty$ contains the Gårding subspace, $E_\propto$. Because $\tilde{E}_\infty$ is dense in E, given $a \in E$ there exists a net $\{a_n\}$ in $\tilde{E}_\infty$ such that $a_n \to a$ in E. Since the functions $\tilde{a}_n$ then tend to $\tilde{a}$ uniformly on compact subsets of G, it must be the case that for $f \in C_c^\infty(G)$, $\tilde{a}_n * f' \to \tilde{a} * f'$ in $C^\infty(G; E)$ (Proposition A2.4.1 of Appendix 2), hence that $U(f)a_n \to U(f)a$ in E_∞ – this shows that $\tilde{E}_\infty \supset E_\propto$, $\tilde{E}_\infty$ being closed in E_∞.]

Let U be a continuous representation of G on E – then U lifts to a representation of $\mathfrak{G}$ on E_∞ in the following way.

For $X \in \mathfrak{g}$, $a \in E_\infty$ define

$$U_\infty(X)a = \lim_{t \to 0} \frac{U(\exp(tX))a - a}{t} \qquad (t \in \mathbf{R}),$$

where, of course, the limit is taken in the topology of E – then $U_\infty(X)a$ is again in E_∞ and in fact the assignment $X \mapsto U_\infty(X)$ is a representation of $\mathfrak{g}$ on the set of continuous endomorphisms of E_∞ (E_∞ being equipped with the topology inherited from $C^\infty(G; E)$). The extension of U_∞ to $\mathfrak{G}$ will be denoted by the same symbol.

Remark As is pointed out in A2.4, we may identify $\mathfrak{G}$ with the algebra of distributions on G whose support is $\{1\}$. If $D \mapsto T_D$ is the identification in question ($D \in \mathfrak{G}$), then $U_\infty(D) = U(T_D)$. [Thus let $X \in \mathfrak{g}$ – then, for $a \in E_\infty$, $\mathbf{z} \in E^*$, $\langle U(T_X)a, \mathbf{z}\rangle = X(\langle U(.)a, \mathbf{z}\rangle')(1) = d/dt \langle U(\exp(tX))a, \mathbf{z}\rangle|_{t=0} = \langle U_\infty(X)a, \mathbf{z}\rangle \Rightarrow U_\infty(X) = U(T_X)$ etc.] Of course this is true regardless of whether we regard $\mathfrak{G}$ as the algebra of right or left invariant differential operators on G (because if we view the elements of $\mathfrak{G}$ as left invariant differential operators on G, then the distribution T_D is defined by the rule $\langle f, T_D\rangle = Df(1)$ rather than $\langle f, T_D\rangle = Df'(1)$ ($f \in C_c^\infty(G)$)).

The verification of the following elementary fact may be left to the reader (proceed as in the Example supra).

Proposition 4.4.1.2 Let U be a continuous representation of G on E, E_∞ the Gårding subspace of E (for U) – then E_∞ is stable under $U_\infty(\mathfrak{G})$. In fact

$$U_\infty(D)U(f)a = U(Df)a \qquad (a \in E, f \in C_c^\infty(G),\ D \in \mathfrak{G}).$$

Remark The preceding proposition is also true if we regard the elements of $\mathfrak{G}$ as left invariant differential operators on G – however the formula is more complicated since one, in general, will have to differentiate the modular function δ_G.

Let U be a continuous representation of G on E, E_∞ the space of differentiable vectors in E (for U), U_∞ the representation of $\mathfrak{G}$ on E_∞ obtained by 'differentiating' U. One would hope that a study of U_∞ would yield useful information about U itself (at least when G is connected). For we have seen already in the study of finite dimensional representations of a connected Lie group G that the corresponding infinitesimal representation of the Lie algebra $\mathfrak{g}$ and its extension to $\mathfrak{G}$ play an important role. One difficulty that arises in the infinite dimensional case is that the operators $U_\infty(D)$ $(D \in \mathfrak{G})$ are in general not continuous when E_∞ is equipped with the relative topology inherited from E (in particular, if E is a Banach space, the $U_\infty(D)$ are in general unbounded). A more serious problem is this. If E is finite dimensional, then U induces a representation of $\mathfrak{G}$ on all of E and moreover the closure of any $\mathfrak{G}$-stable subspace of E is G-stable. This property is one of the main connections between finite dimensional representations of G and $\mathfrak{G}$. However, in the infinite dimensional case, as the following examples show, this property is not possessed in general by U_∞-stable subspaces of E_∞.

Examples (1) Let $G = \mathbf{R}$, $\mathbf{R}$ the real line. Let L be the left regular representation of G on $E = L^2(\mathbf{R})$ – then the space $C_c^\infty(0, 1)$ is contained in E_∞ (cf. the Example supra) and, in fact, is stable under differentiation. However it obviously is not invariant under translation nor is its E-closure.

(2) (L. Robertson) The Gårding space is dense and U_∞-stable (Proposition 4.4.1.2) so it is natural to wonder whether the phenomenon illustrated in Example 1 could perhaps be avoided if we just stuck to $U_\infty | E_\infty$ – the answer again is 'no'. Thus let $G = \mathbf{R}$, $\mathbf{R}$ the real line. Take for E the continuous functions on $\mathbf{R}$ which vanish at infinity, equipped with the sup norm topology. For $x, y \in \mathbf{R}, f \in E$, let $L(x)f = f \circ L_{-x}$ where $L_x(y) = x + y$. Evidently the assignment $x \mapsto L(x)$ is a Banach representation of G on E; here $E_\infty \supset \{f * g : f \in C_c^\infty(G), g \in E\}$ ('$*$' denoting convolution). Let $\epsilon > 0$ be a small positive number; by taking $g \in E$ $(g \geqslant 0)$ such that g vanishes on a large neighborhood of zero and $f \in C_c^\infty(G)$ such that f is supported on a small neighborhood of zero one sees that E_∞ contains non-zero elements which vanish in $(-\epsilon, \epsilon)$. Consider the subspace of E_∞ consisting of those elements which vanish in some open subinterval of $(-1, 1)$ containing zero. This subspace is stable under differentiation but its E-closure is certainly not invariant under translation.

Although these examples show that the relationship between the initial representation U of G and the corresponding infinitesimal representation U_∞ of $\mathfrak{G}$ on E_∞ is not as close as one would like, nevertheless the study of

U_∞ will prove to be profitable. On the other hand, as will be seen in 4.4.5, it is possible to find another subspace of E_∞, stable under $U_\infty(\mathfrak{G})$ and dense in E, in which the pathology of the preceding examples will not appear.

Remark In the case that our representation U is unitary, the Gårding space may be interpreted in the following way. Let $X \in \mathfrak{g}$; let $dU(X)$ denote the skew-adjoint (cf. 4.4.4 infra) infinitesimal generator of the (strongly continuous) one parameter unitary group $t \mapsto U(\exp(tX))$ in E. [The existence of $dU(X)$ is ensured by Stone's Theorem (Riesz and Nagy [1, p. 385]); of course $dU(X)$ is unbounded in general.] Let Dom $(dU(X))$ denote the domain of $dU(X)$. Consider $\bigcap \{\text{Dom}\,(dU(X)) : X \in \mathfrak{g}\}$ – then a priori this intersection may very well be $\{0\}$. [It is easy to find examples (say on $L^2(\mathbf{R})$) of two skew-adjoint operators, the intersection of whose domains is $\{0\}$.] However, as we know, $E_\alpha \subset \cap \{\text{Dom}\,(dU(X)) : X \in \mathfrak{g}\} \ldots.$

Let U be a continuous representation of G on E, E_α the Gårding subspace of E; let us agree to denote by E_α^* the linear subspace of E^* spanned by the ${}^tU(f)\mathbf{z}$ ($f \in C_c^\infty(G)$, $\mathbf{z} \in E^*$) – then it is clear that E_α^* is weakly dense in E^*.

Lemma 4.4.1.3 For every T in $D_c(G)$ there exists a unique endomorphism ${}^tU(T)\colon E_\alpha^* \to E_\alpha^*$ with the property that

$$\langle U(T)a, \mathbf{z}\rangle = \langle a, {}^tU(T)\mathbf{z}\rangle$$

for all $a \in E_\infty$, $z \in E_\alpha^*$; the assignment $T \mapsto {}^tU(T)$ is a representation of the reverse algebra $\tilde{D}_c(G)$ on E_α^*.

Proof The spaces E_α, E_α^* are in duality under the pairing $(a, \mathbf{z}) \mapsto \langle a, \mathbf{z}\rangle$; bearing in mind that the algebra $C_c^\infty(G)$ is a two-sided ideal in $D_c(G)$, we can then define the operator ${}^tU(T)$ ($T \in D_c(G)$) unambiguously on E_α^* via the prescription:

$${}^tU(T)\left[\sum_{i=1}^r {}^tU(f_i)\mathbf{z}_i\right] = \sum_{i=1}^r {}^tU(f_i * T)\mathbf{z}_i \qquad (f_i \in C_c^\infty(G),\ \mathbf{z}_i \in E^*).$$

For $T \in D_c(G), f, g \in C_c^\infty(G)$, $a \in E$, $\mathbf{z} \in E^*$, we have:

$$\begin{aligned}\langle U(T)U(f)a, {}^tU(g)\mathbf{z}\rangle &= \langle U(T*f)a, {}^tU(g)\mathbf{z}\rangle\\ = \langle a, {}^tU(g*T*f)\mathbf{z}\rangle &= \langle U(f)a, {}^tU(g*T)\mathbf{z}\rangle\\ &= \langle U(f)a, {}^tU(T){}^tU(g)\mathbf{z}\rangle\end{aligned}$$

which shows that ${}^tU(T)$ ($T \in D_c(G)$) has the required property. The last assertion of our lemma is now obvious. □

[In passing observe that

$$\langle U(T)a, \mathbf{z}\rangle = \langle a, {}^tU(T)\mathbf{z}\rangle \qquad (a \in E_\infty,\ \mathbf{z} \in E_\alpha^*)$$

for all $T \in D_c(G)$ (direct verification).]

At this juncture we shall make a couple of simple observations per the TCI Banach representations of G.

Proposition 4.4.1.4 Let U be a TCI Banach representation of G on E – then the center of $D_c(G)$ is represented by scalars on E_∞.

Proof Consider first the restriction to E_α of the lift of U to $D_c(G)$ – then, for any Z in the center of $D_c(G)$, we have $U(T)U(Z) = U(Z)U(T)$ and ${}^tU(T){}^tU(Z) = {}^tU(Z){}^tU(T)$ (all $T \in D_c(G)$). Arguing now just as in the proof of Proposition 4.2.2.3 (keeping in mind that E_α^* is weakly dense in E^*), we deduce that $U(Z)$ is a scalar operator, $\kappa_U(Z)$ (say), on E_α. To complete the proof of our proposition, we need only show that $U(Z) = \kappa_U(Z)$ on all of E_∞. So fix $a \in E_\infty$; suppose that $a_n \to a$ in E ($a_n \in E_\alpha$) – then, for any $\mathbf{z} \in E_\alpha^*$, we have

$$\begin{aligned}\kappa_U(Z)\langle a, \mathbf{z}\rangle &= \lim \langle U(Z)a_n, \mathbf{z}\rangle \\ &= \lim \langle a_n, {}^tU(Z)\mathbf{z}\rangle \\ &= \langle a, {}^tU(Z)\mathbf{z}\rangle = \langle U(Z)a, \mathbf{z}\rangle,\end{aligned}$$

whence $U(Z) = \kappa_U(Z)$ on E_∞ too. □

Example It can happen that the center of $D_c(G)$ is trivial (i.e. $\sim \mathbf{C}$). Thus take for G the '$\alpha t + \beta$' group (cf. 5.4.2 ,Example 1); G has two distinct infinite dimensional irreducible unitary representations, U^+ and U^- and, in the above notations, $\kappa_{U^+} = \kappa_{U^-}$ even though U^+ and U^- are distinguished by the hull-kernel topology (cf. Vol. II, 7.1.1). The left regular representation L of G is the Hilbert space direct sum of aleph-nought copies of U^+ and aleph-nought copies of U^-; consequently, L represents the center of $D_c(G)$ by scalars. But L is faithful on $D_c(G)$. . . .

It will now be convenient to suppose that, in addition, G is *connected*; because the center $\mathfrak{Z}$ of $\mathfrak{G}$ is then contained in the center of $D_c(G)$, we conclude: If U is a TCI Banach representation of a connected Lie group G on E, then there exists a homomorphism $Z \mapsto \kappa_U(Z)$ of $\mathfrak{Z}$ into $\mathbf{C}$ such that $U_\infty(Z) = \kappa_U(Z)$ on $E_\infty(Z \in \mathfrak{Z})$.

Part of the discussion in the preceding paragraph is isolated in the following definition.

Definition Let U be a Banach representation of a connected Lie group G on E. U is said to be *quasi-simple* if there exists a homomorphism κ_U of $\mathfrak{Z}$ ($\mathfrak{Z}$ the center of $\mathfrak{G}$) into $\mathbf{C}$ such that $U_\infty(Z)a = \kappa_U(Z)a$ for all $Z \in \mathfrak{Z}$ and $a \in E_\infty$; κ_U is then called the *infinitesimal character* of U.

For later reference, let us explicitly record:

Proposition 4.4.1.5 Let U be a TCI Banach representation of a connected Lie group G – then U is quasi-simple.

Corollary 4.4.1.6 (Segal) Let U be a topologically irreducible unitary representation of a connected Lie group G – then U is quasi-simple.

Proof According to Proposition 4.3.1.7, U is, in fact, TCI – hence our corollary follows upon citing Proposition 4.4.1.5. □

We shall conclude this number with some general considerations involving differentiable representations.

Let U be a continuous representation of G on E.

Proposition 4.4.1.7 (Bruhat) The following three conditions on U are equivalent:

(i) The map $a \mapsto \tilde{a}$ is a topological isomorphism of E into $C^\infty(G; E)$ (otherwise said, $E = E_\infty$ with coincidence of topologies);

(ii) The operators $U_\infty(D)$ $(D \in \mathfrak{G})$ are continuous in E_∞ when E_∞ is equipped with the relative topology inherited from E;

(iii) The map $x \mapsto U(x)$ of G into $\mathfrak{L}_s(E; E)$ is once differentiable.

Proof (i) $\Leftrightarrow$ (ii). We have seen above that (i) $\Rightarrow$ (ii). So assume (ii). Let $X \in \mathfrak{g}$; agreeing to interpret the elements of $\mathfrak{G}$ as left invariant differential operators on G during the present discussion, for each $a \in E_\infty$ one has

$$X\tilde{a}(x) = \frac{d}{dt}\tilde{a}(x \exp(tX))\bigg|_{t=0} = U(x)U_\infty(X)a$$

and hence for all $D \in \mathfrak{G}$,

$$(*) \quad D\tilde{a}(x) = U(x)U_\infty(D)a \qquad (a \in E_\infty).$$

Now suppose that $a_n \in E_\infty$ converges to b in E. Owing to the continuity of the representation U, the C^∞ functions $\tilde{a}_n$ converge uniformly on compact subsets of G to $\tilde{b}$. Since the operators $U_\infty(D)$ are assumed to be continuous, each derivative $D\tilde{a}_n$ converges uniformly on compact subsets of G to the continuous function $x \mapsto U(x)U_\infty(D)b$; thus $\tilde{b} \in C^\infty(G; E)$ and so $E = E_\infty$. In order to prove that the initial topology of E coincides with that inherited from $C^\infty(G; E)$ it suffices to prove that if $a_n \to 0$ in E, then $\tilde{a}_n \to 0$ in $C^\infty(G; E)$. But this is clear from $(*)$.

(i) $\Leftrightarrow$ (iii). Plainly (i) $\Rightarrow$ (iii). To prove that (iii) $\Rightarrow$ (i) it suffices to prove that (iii) $\Rightarrow$ (ii). The fact that (iii) is valid implies in particular that $U_\infty(X)$ $(X \in \mathfrak{g})$ is a continuous endomorphism of E_∞ when E_∞ is equipped with the relative topology inherited from E. Since every element D in $\mathfrak{G}$ is a sum of products of elements in $\mathfrak{g} + \sqrt{-1}\,\mathfrak{g}$, it is clear that (iii) $\Rightarrow$ (ii). This completes the proof of the proposition. □

Remarks (1) Observe that if E is a Fréchet space, then $E = E_\infty$ implies (i) since two comparable topologies on a Fréchet space are identical (Closed Graph Theorem).

(2) Let U be a continuous representation of G on E satisfying (i) – then the map $x \mapsto U(x)$ of G into $\mathfrak{L}_b(E; E)$ is continuous. [Let $a \in E$ describe a bounded subset of E – then it is enough to verify that as $x \to 1$ in G, $U(x)a \to a$ in E *uniformly* in a. But for every $D \in \mathfrak{G}$, the functions $x \mapsto D\tilde{a}(x) = U(x)U_\infty(D)a$ are uniformly bounded so long as x remains within a compact subset of G and a describes a bounded subset of E – hence the sought for uniform convergence follows from Taylor's Theorem (with remainder...).]

(3) The reader will note that the conditions of Proposition 4.4.1.7 are all equivalent to the following statement: The map $x \mapsto U(x)$ of G into $\mathfrak{L}_b(E; E)$

is C^∞. [For it is clear that the present statement implies condition (ii) and hence (i). On the other hand suppose that U verifies (i) – then, as we have seen, the map $x \mapsto U(x)$ of G into $\mathfrak{L}_s(E; E)$ is C^∞. Now, according to a classical lemma of Laurent Schwartz (see Schwartz, 'Un lemme sur la dérivation des fonctions vectorielles d'une variable réele', Ann. Inst. Fourier, vol. 2 (1950), pp. 17-18), a C^∞ map of G into $\mathfrak{L}_s(E; E)$ is actually a C^∞ map of G into $\mathfrak{L}_b(E; E)$ iff all its derivatives are continuous maps of G into $\mathfrak{L}_b(E; E)$. Taking this fact into account, together with the relation $DU(x) = U(x)U_\infty(D)$ ($D \in \mathfrak{G}$), we see that it suffices to prove (since $U_\infty(D)$ is a continuous operator) that the map $x \mapsto U(x)$ of G into $\mathfrak{L}_b(E; E)$ is continuous – but this is the point of Remark 2.]

Definition Let U be a continuous representation of G on E – then U is said to be *differentiable* if it satisfies any one of the above three equivalent conditions.

Observe that the left (right) regular representation of G on $C^\infty(G; E)$ or $C_c^\infty(G; E)$ is differentiable since condition (ii) is evidently satisfied. On the other hand, if $x \mapsto U(x)$ is any continuous representation of G on E, then the associated differentiable representation $x \mapsto U_\infty(x)$ of G on E_∞ is differentiable.

Proposition 4.4.1.8 Let U (respectively V) be a continuous representation of G on E (respectively F). Assume that E and F are barrelled – then

$$i(U, V) \leqslant i(U_\infty, V_\infty).$$

Proof This is a consequence of the remarks given at the end of 4.1.3. □

Let U be a continuous representation of G on E, $U^\vee$ the contragredient representation of G on $E^\vee$.

Proposition 4.4.1.9 If U is differentiable and if E_b^* is complete, then $E^\vee = E^*$ and $U^\vee$ is differentiable.

Proof Remark 2 implies that the map $x \mapsto U(x)$ of G into $\mathfrak{L}_b(E; E)$ is continuous; hence the map $x \mapsto U^\vee(x)$ of G into $\mathfrak{L}_s(E_b^*; E_b^*)$ is continuous (since transposition is a continuous map of $\mathfrak{L}_b(E; E)$ into $\mathfrak{L}_s(E_b^*; E_b^*)$) – therefore $E^\vee = E^*$. The operators $U_\infty^\vee(D)$ ($D \in \mathfrak{G}$), being the transposes of continuous operators, are continuous in E_b^* – thus $U^\vee$ is differentiable (Proposition 4.4.1.7 (ii)). □

Proposition 4.4.1.10 In the notations of Proposition 4.1.2.4, if in addition U^i is differentiable, then the projective tensor product representation $U^1 \hat{\otimes} U^2$ of $G_1 \times G_2$ on $E_1 \hat{\otimes} E_2$ is differentiable.

Proof Let $a_i \in E_i$, X_i an element of the Lie algebra of G_i, t a real number – then

$$\begin{aligned}&t^{-1}[U^1 \hat{\otimes} U^2(\exp t(X_1, X_2)) - 1](a_1 \otimes a_2)\\&= [t^{-1}(U^1(\exp(tX_1)) - 1)a_1] \otimes a_2\\&\quad + (U^1(\exp(tX_1)) \hat{\otimes} 1)(a_1 \otimes [t^{-1}(U^2(\exp(tX_2)) - 1)a_2]).\end{aligned}$$

Letting $t \to 0$, we see that the right hand side of this equality approaches $U_\infty^1(X_1)a_1 \otimes a_2 + a_1 \otimes U_\infty^2(X_2)a_2$ – thus every element in $E_1 \otimes_C E_2$ is once differentiable and its derivative is still in $E_1 \otimes_C E_2$, whence $E_1 \otimes_C E_2 \subset (E_1 \hat{\otimes} E_2)_\infty$. But the operators $U_\infty(X_1) \otimes 1 + 1 \otimes U_\infty(X_2)$ are continuous (in the projective topology); our proposition is therefore seen to be a consequence of Proposition 4.4.1.7 (ii) (or, more precisely, the argument thereof). □

Proposition 4.4.1.11 Let U be a continuous representation of G on a Fréchet space E, $\tilde{E}$ a closed U-stable subspace of E; let π denote the canonical projection of E onto $E/\tilde{E}$ – then the space $(E/\tilde{E})_\infty$ of differentiable vectors in $E/\tilde{E}$ for the quotient representation $\tilde{U}$ is the image $\pi(E_\infty)$ of E_∞ in $E/\tilde{E}$.

Proof It is clear that $\pi(E_\infty) \subset (E/\tilde{E})_\infty$. On the other hand suppose that $a \in (E/\tilde{E})_\infty$; since E is a Fréchet space, there exists a function $\tilde{f} \in C^\infty(G; E)$ such that $\pi(\tilde{f}(x)) = \tilde{U}(x^{-1})a$ for all $x \in G$. Let ϕ be an element of $C_c^\infty(G)$ such that $\int_G \phi(x)d_G(x) = 1$; define a function f on G by the rule $f(x) = \int_G U(y)\tilde{f}(xy)\phi(xy)d_G(y)$ $(x \in G)$ – then $f \in C^\infty(G; E)$ (cf. the proof of Lemma 5.1.1.4) and one has $f(x) = U(x^{-1})f(1)$ $(x \in G)$, which shows that $f(1) \in E_\infty$. The fact that a actually lies in $\pi(E_\infty)$ now results upon observing that

$$\begin{aligned}\pi(f(1)) &= \int_G \pi(U(x)\tilde{f}(x))\phi(x)d_G(x)\\&= \int_G \tilde{U}(x)\pi(\tilde{f}(x))\phi(x)d_G(x)\\&= a\int_G \phi(x)d_G(x) = a. \quad \square\end{aligned}$$

4.4.2 Absolute Convergence of the Fourier Series

In this number an important extension of a result familiar from Harmonic Analysis (mod 2π) will be proved.

Let E denote a locally convex, complete, Hausdorff, topological vector space whose topology is described by the family of continuous semi-norms $\{|\cdot|_\alpha : \alpha \in \mathfrak{A}\}$. Let $\{a_i : i \in I\}$ be an indexed family of elements in E. We shall say that the series $\sum_{i \in I} a_i$ *converges* if, given any neighborhood $\mathscr{P}$ of zero in E, there exists a finite subset $F_\mathscr{P} \subset I$ such that if F_1, F_2 are any two finite subsets of I containing $F_\mathscr{P}$, then $\sum_{i \in F_1} a_i - \sum_{i \in F_2} a_i \in \mathscr{P}$. Ordering the finite subsets of I by inclusion, the net $\{s_F = \sum_{i \in F} a_i : F$ a finite subset of $I\}$ must then converge to a limit s in E, E being complete.

We write $s = \sum_{i \in I} a_i$. Moreover the series is said to *converge absolutely* if $\sum_{i \in I} |a_i|_\alpha < \infty$ for every $\alpha \in \mathfrak{A}$. Plainly absolute convergence implies convergence.

Let K be a compact Lie group, U a continuous representation of K on E. Let $\hat{K}$ be the set of all equivalence classes of finite dimensional irreducible representations of K (this abuse of notation is justified on the obvious grounds). For each $\delta \in \hat{K}$, let ξ_δ denote the character of δ, $d(\delta)$ the degree of δ (i.e. dimension of δ) and $\chi_\delta = d(\delta)\xi_\delta$; put

$$\mathsf{P}(\delta) = U(\bar{\chi}_\delta) = d(\delta) \int_K \overline{\xi_\delta(k)} U(k)\, dk,$$

dk normalized Haar measure on K. Since $\bar{\chi}_\delta * \bar{\chi}_\delta = \bar{\chi}_\delta$ (Schur Orthogonality Relations), $\mathsf{P}(\delta)$ is a continuous projection of E onto $\mathsf{P}(\delta)E = E(\delta)$. Observe that $E(\delta)$ is the isotypic K-submodule of E of type δ, i.e. $E(\delta)$ consists of those vectors in E, the linear span of whose K-orbit is finite dimensional and splits into irreducible K-submodules of type δ.

Theorem 4.4.2.1 (Harish-Chandra) Let a be a differentiable vector in E (for U) – then the Fourier series

$$\sum_{\delta \in \hat{K}} \mathsf{P}(\delta)a$$

converges absolutely to a.

The proof of this theorem requires some lemmas.

In the first place, since K is compact, there exists a positive definite quadratic form Q on the Lie algebra $\mathfrak{k}$ of K which is invariant under the adjoint representation of K. Let $X_1, \ldots, X_n$ be a basis for $\mathfrak{k}$ orthonormal with respect to Q. Put $\Omega = 1 - \sum_1^n X_i^2$ – then Ω belongs to the center of $\mathfrak{K}$, $\mathfrak{K}$ the universal enveloping algebra of $\mathfrak{k}_c$. Fix a unitary representation $\mu_\delta \in \delta$. By Schur's Lemma, $\mu_\delta(\Omega) = c(\delta)\mu_\delta(1)$ where $c(\delta)$ is a certain complex number. But the $\mu_\delta(X_i)$ are skew-adjoint operators, hence $c(\delta)$ is real, in fact $c(\delta) \geqslant 1$; moreover,

$$\Omega\xi_\delta = c(\delta)\xi_\delta.$$

Thus we have $\mathsf{P}(\delta)U_\infty(\Omega)a = c(\delta)\mathsf{P}(\delta)a \ (a \in E_\infty)$.

Lemma 4.4.2.2 Fix an $\alpha \in \mathfrak{A}$ – then there exists an $\alpha_0 \in \mathfrak{A}$ such that

$$|\mathsf{P}(\delta)a|_\alpha \leqslant c(\delta)^{-m} d(\delta)^2 |U_\infty(\Omega^m)a|_{\alpha_0}$$

for $\delta \in \hat{K}$, any integer $m \geqslant 0$ and any differentiable vector $a \in E_\infty$.

Proof Because K is compact, the set $\{U(k) : k \in K\}$ is equicontinuous – therefore there exists an $\alpha_0 \in \mathfrak{A}$ such that $|U(k)b|_\alpha \leqslant |b|_{\alpha_0}$ for all $b \in E$, $k \in K$, whence

$$|\mathsf{P}(\delta)b|_\alpha = |U(\bar{\chi}_\delta)b|_\alpha \leqslant d(\delta)^2 |b|_{\alpha_0}$$

since $\sup|\chi_\delta| \leqslant d(\delta)^2$. We have just seen that if a is differentiable, then $\mathsf{P}(\delta)a = c(\delta)^{-m}\mathsf{P}(\delta)U_\infty(\Omega^m)a$ $(m \geqslant 0)$. Hence the lemma. □

Lemma 4.4.2.3 Retain the above assumptions and notations – then $\sum_{\delta\in\hat{K}} d(\delta)^2 c(\delta)^{-m} < \infty$ if m is large enough.

Proof (1) We first reduce the proof of the lemma to the case when K is connected. Let K_0 denote the identity component of K, $N = [K : K_0]$ the finite index of K_0 in K. Let $\hat{K}_0$ denote the set of all equivalence classes of finite dimensional irreducible representations of K_0. If $\delta \in \hat{K}$, $\delta_0 \in \hat{K}_0$, then let $[\delta : \delta_0]$ denote the number of times δ_0 occurs in the restriction of δ to K_0. Finally, for each $\delta_0 \in \hat{K}_0$, let $\hat{K}(\delta_0) = \{\delta \in \hat{K} : [\delta : \delta_0] \geqslant 1\}$. It follows from the Frobenius Reciprocity Theorem that

$$\sum_{\delta\in\hat{K}(\delta_0)} [\delta : \delta_0]d(\delta) = Nd(\delta_0)$$

(see Corollary 5.3.3.6). If ξ_{δ_0} denotes the character of a class $\delta_0 \in \hat{K}_0$, then $\Omega\xi_{\delta_0} = c(\delta_0)\xi_{\delta_0}$ where $c(\delta_0) = c(\delta)$ for every $\delta \in \hat{K}(\delta_0)$ – hence

$$\begin{aligned}\sum_{\delta\in\hat{K}} c(\delta)^{-m}d(\delta)^2 &\leqslant \sum_{\delta_0\in\hat{K}_0} c(\delta_0)^{-m} \sum_{\delta\in\hat{K}(\delta_0)} d(\delta)^2 \\ &\leqslant N^2 \sum_{\delta_0\in\hat{K}_0} c(\delta_0)^{-m}d(\delta_0)^2.\end{aligned}$$

This estimate shows that one may assume that K is connected.

(2) Fix a maximal torus T with Lie algebra $\mathfrak{t}$ in our compact connected Lie group K. Let L_T denote the lattice of all $\lambda \in \mathrm{Hom}(\sqrt{-1}\mathfrak{t}, \mathbf{R})$ such that $e^{\lambda(H)} = 1$ for those H which lie in the kernel of the exponential map of $\mathfrak{t}$ onto T. Introduce an order in $\mathrm{Hom}(\sqrt{-1}\mathfrak{t}, \mathbf{R})$ and put $\rho = \frac{1}{2}\sum_{\alpha>0}\alpha$ where α runs over the positive roots of the pair $(\mathfrak{k}_c, \mathfrak{t}_c)$. [Since K is compact, $\mathfrak{k}_c$ is reductive.] For any $\delta \in \hat{K}$, let $\Lambda(\delta)$ denote the highest weight of δ with respect to $\mathfrak{t}$ – then:

(i) $\Lambda(\delta) \in \mathsf{L}_T$;

(ii) There exists a polynomial function p on $\mathrm{Hom}(\sqrt{-1}\mathfrak{t}, \mathbf{R})$ such that $d(\delta) = p(\Lambda(\delta))$ for all $\delta \in \hat{K}$ (cf. Theorem 2.4.1.6);

(iii) There exists a positive definite quadratic form q on $\mathrm{Hom}(\sqrt{-1}\mathfrak{t}, \mathbf{R})$ such that $c(\delta) = 1 + q(\Lambda(\delta) + \rho) - q(\rho)$ $(\delta \in \hat{K})$.

This being so, choose a compact set ω in $\mathrm{Hom}(\sqrt{-1}\mathfrak{t}, \mathbf{R})$ such that $q(\lambda + \rho) - q(\rho) \geqslant q(\lambda)/2$ for all $\lambda \in \mathrm{Hom}(\sqrt{-1}\mathfrak{t}, \mathbf{R})$ outside of ω. Let $\hat{\omega} = \{\delta \in \hat{K} : \Lambda(\delta) \in \omega\}$ – then $\hat{\omega}$ is finite since $\mathsf{L}_T \cap \omega$ is finite. Let $\complement\hat{\omega}$ denote the complement of $\hat{\omega}$ in $\hat{K}$ – then

$$\sum_{\delta\in\complement\hat{\omega}} c(\delta)^{-m}d(\delta)^2 \leqslant 2^m \sum_{\lambda\in\mathsf{L}_T} (1 + q(\lambda))^{-m}\,|p(\lambda)^2| < \infty$$

if m is large enough.

Hence the lemma. □

Returning to the proof of Theorem 4.4.2.1, the preceding lemmas plainly imply that the series $\sum_{\delta\in\hat{K}} \mathsf{P}(\delta)a$ is absolutely convergent. Put $a_0 =$

$\sum_{\delta\in\hat{K}} \mathsf{P}(\delta)a$ – then for each $\delta_0 \in \hat{K}$ we have $\mathsf{P}(\delta_0)(a_0 - a) = \mathsf{P}(\delta_0)a - \mathsf{P}(\delta_0)a = 0$. Hence the proof of Theorem 4.4.2.1 will be complete as soon as the following result is established.

Lemma 4.4.2.4 Let $b \in E$ and suppose that $\mathsf{P}(\delta)b = 0$ for every $\delta \in \hat{K}$ – then $b = 0$.

Proof Fix an $\alpha \in \mathfrak{A}$ and choose an $\alpha_0 \in \mathfrak{A}$ such that $|U(k)b|_\alpha \leqslant |b|_{\alpha_0}$ for all $k \in K$. Fix an $\epsilon > 0$ – then there exists a neighborhood $\mathcal{O}_\epsilon$ of 1 in K with the property that $|U(k)b - b|_\alpha \leqslant \epsilon$ for all $k \in \mathcal{O}_\epsilon$. Choose $f \in C(K)$ such that $f \geqslant 0$, $\int_K f dk = 1$, and $f = 0$ outside $\mathcal{O}_\epsilon$ – then $|U(f)b - b|_\alpha \leqslant \epsilon$. Using the Peter-Weyl Theorem, select a K-finite function $h \in C(K)$ such that $\sup_{k\in K} |h(k) - f(k)| \leqslant \epsilon$ – then $|U(f)b - U(h)b|_\alpha \leqslant \epsilon |b|_{\alpha_0}$, whence $|U(h)b - b|_\alpha \leqslant \epsilon(|b|_{\alpha_0} + 1)$. On the other hand, since h is K-finite, we can find a finite subset F of $\hat{K}$ such that $h = \sum_{\delta\in F} \bar{\chi}_\delta * h$. Now $\bar{\chi}_\delta$ is in the center of $L^1(K)$; thus $\mathsf{P}(\delta)U(h)b = U(\bar{\chi}_\delta * h)b = U(h)\mathsf{P}(\delta)b = 0$ and so $U(h)b = \sum_{\delta\in F} \mathsf{P}(\delta)U(h)b = 0$ – hence $|b|_\alpha \leqslant \epsilon(|b|_{\alpha_0} + 1)$. This implies that $\alpha(b) = 0$, ϵ being arbitrary. Therefore $b = 0$, as we wished to prove. □

Remark The proof of Theorem 4.4.2.1 (especially Lemmas 4.4.2.2, 4.4.2.3) shows that there exists an element $D_\infty \in \mathfrak{K}$ with the property: Given $\alpha \in \mathfrak{A}$, there exists $\alpha_0 \in \mathfrak{A}$ such that for all $a \in E_\infty$

$$\sum_{\delta\in\hat{K}} |\mathsf{P}(\delta)a|_\alpha \leqslant |U_\infty(D_\infty)a|_{\alpha_0}.$$

In particular, if U is a Banach representation of K, then there exists a $D_\infty \in \mathfrak{K}$ such that for all $a \in E_\infty$

$$\sum_{\delta\in\hat{K}} \|\mathsf{P}(\delta)a\| \leqslant \|U_\infty(D_\infty)a\|.$$

4.4.3 A Density Theorem Fourier Series in Function Spaces

Let us agree to use the notations employed in 4.4.1 – thus, in particular, G denotes a Lie group countable at infinity, U a continuous representation of G on the (complete) locally convex space E, and E_∞ the space of differentiable vectors in E (for U). Fix a compact subgroup K of G. The restriction of U to K defines a continuous representation of K on E and so the notations and results of 4.4.2 are available.

Theorem 4.4.3.1 The space $\sum_{\delta\in\hat{K}} E_\infty \cap E(\delta)$ is dense in E.

A couple of preliminary remarks are needed. We have already observed in 4.4.1 that the regular representations of G on $C^\infty(G)$ and $C_c^\infty(G)$ are differentiable. If $x \mapsto L(x)$ is the left regular representation of G on $C^\infty(G)$ (or $C_c^\infty(G)$), if $x \mapsto R(x)$ is the right regular representation of G on $C^\infty(G)$ (or $C_c^\infty(G)$), then, identifying $\chi_\delta(\delta \in \hat{K})$ with an element of $M_c(G)$, we have

$$\mathsf{P}_L(\delta)f = \bar{\chi}_\delta * f, \ \mathsf{P}_R(\delta)f = f * \chi_\delta$$

where $\mathsf{P}_L(\delta) = L(\bar{\chi}_\delta)$, $\mathsf{P}_R(\delta) = R(\bar{\chi}_\delta)$ and f denotes an element in $C^\infty(G)$ (or $C_c^\infty(G)$). [Note that $\delta_G | K = 1$, δ_G the modular function on G.] Recalling that $\bar{\xi}_\delta$ is the character of the class contragredient to δ, Theorem 4.4.2.1 implies:

Proposition 4.4.3.2 (i) Fix $f \in C^\infty(G)$ – then the series

$$\sum_{\delta \in \hat{K}} \chi_\delta * f, \quad \sum_{\delta \in \hat{K}} f * \chi_\delta$$

converge absolutely to f in $C^\infty(G)$.

(ii) Fix $f \in C_c^\infty(G)$ – then the series

$$\sum_{\delta \in \hat{K}} \chi_\delta * f, \quad \sum_{\delta \in \hat{K}} f * \chi_\delta$$

converge absolutely to f in $C_c^\infty(G)$.

Proof of Theorem 4.4.3.1 Fix $\alpha \in \mathfrak{A}$, $\epsilon > 0$, $a \in E$ – then Proposition 4.4.1.1 gives us an element $f \in C_c^\infty(G)$ such that $| U(f)a - a|_\alpha \leqslant \epsilon$. Choose a compact set ω in G such that $K\omega = \omega$ and spt $(f) \subset \omega$. Let $\mu(g) = \int_G |g| \, d_G(x) (g \in C_c^\infty(G))$ – then μ is a continuous semi-norm on $C_c^\infty(G)$. For any finite subset F of $\hat{K}$ put $\bar{\chi}_F = \sum_{\delta \in F} \bar{\chi}_\delta$ – then spt $(f - \bar{\chi}_F * f) \subset \omega$, whence $| U(f - \bar{\chi}_F * f)a|_\alpha \leqslant c\mu(f - \bar{\chi}_F * f)$ where $c = \sup_{x \in \omega} | U(x)a|_\alpha < \infty$. Proposition 4.4.3.2 implies that we can choose F so that $\mu(f - \bar{\chi}_F * f) \leqslant \epsilon/c$ – therefore

$$| U(\bar{\chi}_F * f)a - a|_\alpha \leqslant | U(f - \bar{\chi}_F * f)a|_\alpha + | U(f)a - a|_\alpha \leqslant 2\epsilon.$$

Since $U(\bar{\chi}_F * f)a \in \sum_{\delta \in \hat{K}} E_\infty \cap E(\delta)$, the theorem follows. $\square$

Corollary 4.4.3.3 The closure $(E_\infty \cap E(\delta))^{cl}$ of $E_\infty \cap E(\delta)$ in E is $E(\delta)$ (cf. infra).

Proposition 4.4.3.4 Let $\tilde{E}$ be a subspace of E stable for the set $\{U(k) : k \in K\}$; put $\tilde{E}(\delta) = \tilde{E} \cap E(\delta)$ $(\delta \in \hat{K})$. If $\sum_{\delta \in \hat{K}} \tilde{E}(\delta)$ is dense in E, then $\tilde{E}(\delta)^{cl} = E(\delta)$.

Proof Plainly $\sum_{\delta \in \hat{K}} \tilde{E}(\delta)$ is K-stable. Let $a \in E(\delta)$ and suppose that $a_n \to a$ where $a_n \in \tilde{E}$. Since $\mathsf{P}(\delta)$ is a continuous map, $\mathsf{P}(\delta)a_n \to \mathsf{P}(\delta)a = a$; but the $\mathsf{P}(\delta)a_n \in \tilde{E}(\delta)$ – therefore $\tilde{E}(\delta)^{cl} = E(\delta)$, as contended. $\square$

Example Consider the right regular representation R of G on $E = C_c^\infty(G)$. The dual E^* of E is the space of distributions on G. Equip E^* with the topology of uniform convergence on bounded subsets of E (this is the usual topology on the space of distributions (Schwartz [1, p. 71]) – then E_b^* is complete. Since R is differentiable, the contragredient representation $R^\vee$ is too (cf. Proposition 4.4.1.9). Denote the class contragredient to an element $\delta \in \hat{K}$ by $\check{\delta}$. If

$$\mathsf{P}_{R^\vee}(\delta) = R^\vee(\bar{\chi}_\delta) = \int_K \bar{\chi}_\delta(k) R^\vee(k) dk \qquad (\delta \in \hat{K}),$$

then it is easy to see that $\mathrm{P}_{R^\vee}(\delta) = {}^t\mathrm{P}_R(\check{\delta})$. [Let $f \in E, T \in E^*$ – then

$$\begin{aligned}\langle f, \mathrm{P}_{R^\vee}(\delta)T\rangle &= \left\langle f, \int_K \bar{\chi}_\delta(k) R^\vee(k) T dk\right\rangle = \int_K \bar{\chi}_\delta(k)\langle f, {}^tR(k^{-1})T\rangle\, dk \\ &= \left\langle \int_K \chi_\delta(k) R(k) f dk, T\right\rangle = \langle \mathrm{P}_R(\check{\delta})f, T\rangle \\ &= \langle f, {}^t\mathrm{P}_R(\check{\delta})T\rangle \Rightarrow \mathrm{P}_{R^\vee}(\delta) = {}^t\mathrm{P}_R(\check{\delta}).]\end{aligned}$$

For $f \in E$ (respectively $T \in E^*$), set $\mathrm{P}_R(\delta)f = f_\delta = f * \chi_\delta$ (respectively $\mathrm{P}_{R^\vee}(\delta)T = T_\delta = T * \chi_\delta$ (convolution of distributions)) – then $T(f_\delta) = T_{\check{\delta}}(f)$. We have seen above that

$$\sum_{\delta \in \hat{K}} f_\delta \to f$$

in $E = C_c^\infty(G)$ – therefore $T(\sum_{\delta \in \hat{K}} f_\delta) = \sum_{\delta \in \hat{K}} T_\delta(f) = T(f)$ and of course

$$\sum_{\delta \in \hat{K}} T_\delta \to T$$

in E_b^*. The distribution T_δ will be called the δth*-Fourier component* of T.

The following observation will be useful later on. Define a representation U of $G \times G$ on $C^\infty(G)$ (or $C_c^\infty(G)$) by the prescription $U(x, y)f = L(x)R(y)f$ $(x, y \in G,\ f \in C^\infty(G)$ (or $C_c^\infty(G)$)). Since $U(x, y)f - f = L(x)(R(y)f - f) + (L(x)f - f)$, Proposition 4.1.1.1 implies that U is a continuous representation of $G \times G$ on $C^\infty(G)$ (or $C_c^\infty(G)$); in fact U is differentiable – hence:

Proposition 4.4.3.5 Let E denote either one of the spaces $C^\infty(G)$ or $C_c^\infty(G)$ – then, for each $f \in E$, the series

$$\sum_{\delta_1, \delta_2 \in \hat{K}} \chi_{\delta_1} * f * \chi_{\delta_2}$$

converges absolutely to f in E.

4.4.4 Elliptic Elements in the Enveloping Algebra

As in 4.4.1, let G be a Lie group countable at infinity, d_G left Haar measure on G; let $\mathfrak{G}$ denote the universal enveloping algebra of the complexification of the Lie algebra of G – we recall, according to our standing convention, that the elements of $\mathfrak{G}$ are to be viewed as right invariant differential operators on G (unless the contrary is specifically stated).

Let U be a unitary representation of G on a Hilbert space E equipped with an inner product $(.\,,.)$; let E_∞ denote the space of differentiable vectors in E (for U) – then the primary objective of the present number is to consider the operators $U_\infty(D)$ $(D \in \mathfrak{G})$ in more detail at least for certain D.

Before presenting the main results, for convenience let us review some of the basic terminology concerning unbounded linear operators on a Hilbert space; it is definitely presupposed, however, that the reader is familiar with this theory (see Riesz and Nagy [1, Chapter 8] or Dunford and Schwartz [2, Chapter

12]). Let T be a linear operator on a Hilbert space E with domain Dom (T) – then T is said to be *densely defined* if Dom (T) is dense in E. T is said to be *symmetric* if T is densely defined and $(Ta, b) = (a, Tb)$ for all $a, b \in$ Dom (T), *skew-symmetric* if T is densely defined and $(Ta, b) = -(a, Tb)$ for all $a, b \in$ Dom (T). The *adjoint* T^* of a densely defined linear operator T is defined on the set Dom (T^*) of all $b \in E$ for which there exists a vector $c \in E$ such that $(Ta, b) = (a, c)$ for all $a \in$ Dom (T); by definition $T^*b = c$. T is said to be *self-adjoint* in case $T = T^*$, *skew-adjoint* if $T = -T^*$, *essentially self-adjoint* if $\bar{T} = T^*$, *essentially skew-adjoint* if $\bar{T} = -T^*$ ($\bar{T}$ denoting, as usual, the closure of T, if it exists). Two unbounded self-adjoint operators are said to *commute* (or to be *permutable*) provided their spectral resolutions commute. [Along these lines one must exercise a modicum of caution. Thus, for instance, Nelson [1, p. 604] has given an example of two essentially self-adjoint operators defined on a common invariant domain and which commute on that domain but whose spectral resolutions do not commute. More precisely Nelson showed that there exist two symmetric operators T and S on a Hilbert space E having a common invariant dense domain $\mathfrak{D}$ such that for all real t and s, $tT + sS$ is essentially self-adjoint and such that for all $a \in \mathfrak{D}$, $TSa = STa$, but such that the spectral resolutions of $\bar{T}$ and $\bar{S}$ do not commute (cf. the Example preceding Theorem 4.4.6.6). On the other hand it is possible to prove the following (cf. Nelson [1, p. 603]): Let T and S be symmetric operators on a Hilbert space E and let $\mathfrak{D}$ be a dense linear subspace of E such that $\mathfrak{D}$ is contained in the domain of T, S, T^2, TS, ST, and S^2 and such that $TSa = STa$ for all $a \in \mathfrak{D}$ – then, if the restriction of $T^2 + S^2$ to $\mathfrak{D}$ is essentially self-adjoint, T and S are essentially self-adjoint and $\bar{T}$ and $\bar{S}$ commute.]

Let $D \mapsto D^*$ denote the involutory conjugate linear anti-automorphism of $\mathfrak{G}$ such that $X^* = -X$ if $X \in \mathfrak{g}$ – thus $(X + \sqrt{-1}\,Y)^* = -X + \sqrt{-1}\,Y (X, Y \in \mathfrak{g})$.

Remark If $D \in \mathfrak{G}$, then, as usual (Appendix 2) D^t will denote the adjoint of D with respect to the left invariant measure d_G, i.e. for all $f, g \in C_c^\infty(G)$

$$\int_G Df(x)g(x)d_G(x) = \int_G f(x)D^t g(x)d_G(x).$$

Since we are regarding the elements of $\mathfrak{G}$ as right invariant differential operators on G, the remarks in Appendix 2 show that $D^t = D^+$ where '+' is the principal anti-automorphism of $\mathfrak{G}$ over $\mathbf{C}$; observe that $(D^*)^t = (D^t)^*$ (all $D \in \mathfrak{G}$).

Definition An element $D \in \mathfrak{G}$ is said to be *Hermitian* if $D^* = D$.

Let U be a unitary representation of G on E. If $a, b \in E_\infty$, then for $X \in \mathfrak{g}$,

$$\begin{aligned}(U_\infty(X)a, b) &= \lim_{t \to 0} (t^{-1}[U(\exp(tX))a - a], b) \\ &= \lim_{t \to 0} (a, t^{-1}[U(\exp(-tX))b - b]) = -(a, U_\infty(X)b),\end{aligned}$$

i.e. the operators $U_\infty(X)$ $(X \in \mathfrak{g})$ in E_∞ are skew-symmetric. Moreover we clearly have

$$(U_\infty(D)a, b) = (a, U_\infty(D^*)b) \qquad (a, b \in E_\infty;\ D \in \mathfrak{G}).$$

Consequently, if D is a Hermitian element in $\mathfrak{G}$, then $U_\infty(D)$ is a symmetric operator. [In passing also observe that the above relation implies that $U_\infty(D)$ ($D \in \mathfrak{G}$) always admits closure.]

We shall be concerned with such questions as: If $D \in \mathfrak{G}$ is Hermitian, then does $U_\infty(D)$ admit a self-adjoint extension? In turn the results to be established will be used in the next number during the discussion of analytic vectors.

Recall that a complex valued continuous function f on G is said to be *positive definite* if

$$\sum_{i,j=1}^{n} c_i \bar{c}_j f(x_i^{-1} x_j) \geqslant 0$$

for all finite sets $x_1, \ldots, x_n$ of elements in G and any complex numbers $c_1, \ldots, c_n$.

Lemma 4.4.4.1 Let f be a C^∞ positive definite function on G – then $(D^*Df)(1) \geqslant 0$ for all $D \in \mathfrak{G}$.

Proof The algebra $D_c(G)$ is, in the natural way, a *-algebra. If $D \mapsto T_D$ is the identification of $\mathfrak{G}$ with the algebra of distributions on G whose support is $\{1\}$, then it is easy to verify that $T_{D^*} = (T_D)^*$ (all $D \in \mathfrak{G}$). On the basis of well-known approximation arguments (cf. Dixmier [15, pp. 256–257]), the assertion of the lemma then follows immediately. □

Lemma 4.4.4.2 Let $\mathfrak{D}$ be a dense linear subspace of a Hilbert space E. Let S and T be linear transformations whose domains are $\mathfrak{D}$ and whose ranges are contained in $\mathfrak{D}$ such that $T \subset S^*$. If TS is essentially self-adjoint, then $\bar{T} = S^*$.

Proof Plainly T admits closure since $T \subset S^*$. It suffices to prove that the graph of S^* contains no element orthogonal to the graph of $\bar{T}$. Suppose to the contrary that $\{a_0, b_0\} \in E + E$ is a non-zero element of the graph of S^* which is orthogonal to the graph of T – then $(a, a_0) + (Ta, b_0) = (a, a_0) + (Ta, S^*a_0) = 0$ for all $a \in \mathfrak{D}$. Now $S\mathfrak{D} \subset \mathfrak{D}$ by assumption; therefore $(Sa, a_0) + (TSa, b_0) = (a, b_0) + (TSa, b_0) = 0$ for all $a \in \mathfrak{D}$. But $I + TS$ has a dense range – hence $b_0 = 0$. This means that $(a, a_0) = 0$ for all $a \in \mathfrak{D}$ and so $a_0 = 0$, contradicting the fact that $\{a_0, b_0\}$ is non-zero. □

Let $D \in \mathfrak{G}$ be arbitrary – then the relation

$$(U_\infty(D)a, b) = (a, U_\infty(D^*)b) \qquad (a, b \in E_\infty),$$

shows that $U_\infty(D^*) \subset U_\infty(D)^*$. The main result in this number gives a sufficient condition on D in order to ensure that $\overline{U_\infty(D^*)} = U_\infty(D)^*$.

Definition Let $D \in \mathfrak{G}$ – then D is said to be *elliptic* if D is elliptic as a linear partial differential operator on G.

[Let $X_1, \ldots, X_n$ be a basis for $\mathfrak{g}$ and write $D = p(X_1, \ldots, X_n)$ where p is a non-commutative polynomial whose homogeneous part of highest degree is q, say – then D is said to be *elliptic* if $q(r_1, \ldots, r_n) \neq 0$ for all non-zero n-tuples $(r_1, \ldots, r_n)$ of real numbers. For instance $\Delta = X_1^2 + \ldots + X_n^2$ is elliptic; so is $(1 - \Delta)^m$ for any positive integer m.]

Theorem 4.4.4.3 (Nelson and Stinespring) Let U be a unitary representation of G on a Hilbert space E. If D is an elliptic element in $\mathfrak{G}$, then the closure of $U_\infty(D^*)$ is the adjoint of $U_\infty(D)$.

Proof Let us divide the proof into two steps.

(1) Suppose that it is possible to write $D = D_1^* D_1$ for some $D_1 \in \mathfrak{G}$ – then we have to prove that $S = U_\infty(D)$ is essentially self-adjoint. For this purpose it suffices to show: (i) $(Sa, a) \geqslant 0$ for all $a \in E_\infty$; (ii) $S + I$ has a dense range ($(S + I)^{-1}$ is then bounded and densely defined). (i) is clear since $(Sa, a) = (U_\infty(D_1)a, U_\infty(D_1)a) \geqslant 0$ for all $a \in E_\infty$. As for (ii), suppose that a_0 is an element of E orthogonal to the range of $S + I$ – then, in particular, $((S + I)U(f)a_0, a_0) = 0$ for all $f \in C_c^\infty(G)$, i.e. (cf. Proposition 4.4.1.2)

$$\int_G (D + 1)f(x)(U(x)a_0, a_0)d_G(x) = 0 \qquad (f \in C_c^\infty(G)).$$

This equation simply says that the function $x \mapsto (U(x)a_0, a_0)$ is a distribution solution to an elliptic equation; hence by the Regularity Theorem (see, e.g., L. Nirenberg, 'Remarks on strongly elliptic partial differential equations', Comm. Pure Appl. Math., vol. 8 (1955), pp. 649–675), the function $x \mapsto (U(x)a_0, a_0)$ is analytic on G, and, if $D_2 = D_1^\dagger$,

$$(D_2 D_2^* + 1)(U(x)a_0, a_0) = 0$$

in the ordinary pointwise sense. Evaluating at $x = 1$ gives

$$(D_2 D_2^*)(U(x)a_0, a_0)\Big|_{x=1} = -(a_0, a_0)$$

which implies that $a_0 = 0$ in view of Lemma 4.4.4.1. Hence the range of $S + I$ is dense.

(2) Let $D \in \mathfrak{G}$ be a general elliptic element. The discussion in (1) implies that $U_\infty(D^*D) = U_\infty(D^*)U_\infty(D)$ is essentially self-adjoint; hence the closure of $U_\infty(D^*)$ is $U_\infty(D)^*$ by Lemma 4.4.4.2.

The proof of the theorem is now complete. ☐

[It is in general false that Hermitian elements of the enveloping algebra $\mathfrak{G}$ are sent into operators essentially self-adjoint on E_∞. This is so even for second order Hermitian elements; for an example (with $G = \mathbf{SL}(2, \mathbf{R})$) see below.]

Proposition 4.4.4.4 Let U be a unitary representation of G on a Hilbert space E. Let D_0 be a (non-constant) Hermitian elliptic element of $\mathfrak{G}$. If D is any element of $\mathfrak{G}$ with the property that $U_\infty(D^*D)U_\infty(D_0)a = U_\infty(D_0)U_\infty(D^*D)a$ for all $a \in E_\infty$, then the closure of $U_\infty(D^*)$ is the adjoint of $U_\infty(D)$.

Proof (1) Let r be an integer greater than the order of D and set $S = U_\infty(D_0^{2r})$, $T = U_\infty(D^*D)$ and $R = S + T$ – then R, S and T are all symmetric operators. In addition, in view of Theorem 4.4.4.3, R and S are essentially self-adjoint since $D_0^{2r} + D^*D$ and D_0^{2r} are Hermitian elliptic elements of $\mathfrak{G}$. Our hypotheses tell us that $RSa = SRa$ for all $a \in E_\infty$ – we shall, however, need the stronger fact that their closures $\bar{R}$ and $\bar{S}$ commute, i.e. have commuting spectral resolutions. To see this first observe that the bounded operators $(I + S)^{-1}(I + R)^{-1}$ and $(I + R)^{-1}(I + S)^{-1}$ agree on their common domain, the range of $(I + S)(I + R) = (I + R)(I + S)$. Theorem 4.4.4.3 implies that $(I + S)(I + R)$ is essentially self-adjoint and, since $S + R + SR$ is positive, has a dense range. Hence the spectral resolutions of $\bar{R}$ and $\bar{S}$ commute.

(2) Let $\tilde{T} = \bar{R} - \bar{S}$ – then $\tilde{T}$ is essentially self-adjoint by the Spectral Theorem. In order to complete the proof of the present proposition, it is sufficient, thanks to Lemma 4.4.4.2, to show that $\tilde{T} \subset \bar{T}$. Let a belong to the domain of $\tilde{T}$ – then a is in the domain of $\bar{R}$ and thus we may choose $\{a_n\}$ in E_∞ such that $a_n \rightarrow a$ and $Ra_n \rightarrow \bar{R}a$. On the other hand, $\|Sb\| \leqslant \|Rb\|$ for all $b \in E_\infty$ since

$$(Rb, Rb) = (Sb, Sb) + (Tb, Tb) + 2(TU_\infty(D_0^r)b, U_\infty(D_0^r)b)$$

and all terms are positive – therefore $Sa_n \rightarrow \bar{S}a$ and so $Ta_n \rightarrow \tilde{T}a$, i.e. $\tilde{T} \subset \bar{T}$, as desired.

The proof of the proposition is now complete. $\square$

Examples Let U be a unitary representation of G on a Hilbert space E. Let $\mathfrak{Z}$ denote the center of $\mathfrak{G}$.

(1) If G is abelian or compact, then $\mathfrak{Z}$ contains a Hermitian elliptic element. Conclusion: For every $D \in \mathfrak{G}$, $\overline{U_\infty(D^*)} = U_\infty(D)^*$.

(2) Let G be arbitrary. Let $X_1, \ldots, X_n$ be a basis for $\mathfrak{g}$ – then $\Delta = X_1^2 + \ldots + X_n^2$ is a Hermitian elliptic element of $\mathfrak{G}$. For any $Z \in \mathfrak{Z}$, Z^*Z commutes with Δ. Conclusion: For every $Z \in \mathfrak{Z}$, $\overline{U_\infty(Z^*)} = U_\infty(Z)^*$.

(3) Suppose that G is (connected) semi-simple with finite center, K a maximal compact subgroup of G. We have seen that the Casimir element ω in $\mathfrak{G}$ may be written: $\omega = -\omega_- + \omega_+$ where ω_- is a certain elliptic element in the enveloping algebra of the complexification of the Lie algebra of K (cf. number 2.3.3, Example 2). The very construction of ω shows that ω, ω_-, ω_+ are all Hermitian; in addition $\omega_- + \omega_+$ is elliptic. Putting these observations together gives: If D is an element of $\mathfrak{G}$ commuting with ω_-, then $\overline{U_\infty(D^*)} = U_\infty(D)^*$.

(4) Let G be arbitrary. Let $X \in \mathfrak{g}$ – then $\sqrt{-1}\, U_\infty(X)$ is essentially self-adjoint on E_∞. The proof may be left to the reader. A corollary to this remark is the fact that the domain of $\overline{U_\infty(X)}$ consists precisely of those vectors $a \in E$ such that

$$\lim_{t \rightarrow 0} \frac{U(\exp(tX))a - a}{t}$$

exists.

Let U be a unitary representation of G on a Hilbert space E, E_∞ the space of differentiable vectors in E (for U). As above, let $X_1, \ldots, X_n$ be a basis for $\mathfrak{g}$; put $\Delta = \sum_i X_i^2$ – then we know (Theorem 4.4.4.3) that the closure of the operator $U_\infty(\Delta)$ is a self-adjoint operator on E. It turns out that the space E_∞ is definable in terms of $\overline{U_\infty(\Delta)}$ – in fact:

Theorem 4.4.4.5 (Nelson) $E_\infty = \bigcap_1^\infty \mathrm{Dom}\,(\overline{U_\infty(\Delta)^m})$.

We shall preface the proof with a few technical lemmas, some of which will also be used in other connections later on.

The proof of the following elementary combinatorial fact may be left to the reader.

Lemma 4.4.4.6 Let $X_1, \ldots, X_n, T$ be operators on a normed linear space E with a common invariant domain $\mathfrak{D}$ – then

$$X_n \ldots X_1 T = \sum_{k=0}^{n} \sum_{\sigma} \{\mathrm{ad}(X_{\sigma(k)}) \ldots \mathrm{ad}(X_{\sigma(1)})(T)\} X_{\sigma(n)} \ldots X_{\sigma(k+1)},$$

the summation $\sum_\sigma$ running over the $\binom{n}{k}$ permutations of $1, \ldots, n$ such that $\sigma(n) > \sigma(n-1) > \ldots > \sigma(k+1)$, $\sigma(k) > \sigma(k-1) > \ldots \sigma(1)$. [Here

$$\mathrm{ad}\,(X_i)(.) = [X_i, .],\ i = 1, \ldots, n.]$$

Lemma 4.4.4.6 (bis) Let $X_1, \ldots, X_n, T$ be operators on a normed linear space E with a common invariant domain $\mathfrak{D}$ – then

$$TX_n \ldots X_1 = X_n \ldots X_1 T$$
$$- \sum_{k=1}^{n} \sum_{\sigma} \{\mathrm{ad}(X_{\sigma(k)}) \ldots \mathrm{ad}(X_{\sigma(1)})(T)\} X_{\sigma(n)} \ldots X_{\sigma(k+1)},$$

the summation $\sum_\sigma$ running over the $\binom{n}{k}$ permutations of $1, \ldots, n$ such that $\sigma(n) > \sigma(n-1) > \ldots \sigma(k+1)$, $\sigma(k) > \sigma(k-1) > \ldots > \sigma(1)$.

[This is an easy variant of the preceding result.]

Lemma 4.4.4.7 Let $X_1, \ldots, X_n, T$ be symmetric operators on a Hilbert space E with a common invariant domain $\mathfrak{D}$; assume in addition that T is essentially self-adjoint. If there exist positive (finite) constants M and $M_m (m \geqslant 1)$ such that

$$\sum_i \|X_i a\| \leqslant M(\|a\| + \|Ta\|),\ \sum_{(m)} \|\mathrm{ad}\,(X_{i_1}) \ldots \mathrm{ad}\,(X_{i_m})(T)a\|$$
$$\leqslant M_m(\|a\| + \|Ta\|) \qquad (\text{all } a \in \mathfrak{D}),$$

then, for any finite sequence $i_1, \ldots, i_m$,

$$\mathrm{Dom}\,(\bar{T}^m) \subset \mathrm{Dom}\,(\bar{X}_{i_1} \ldots \bar{X}_{i_m}).$$

[Here the symbol $\sum_{(m)}$ denotes a sum over all i_j with $1 \leqslant i_j \leqslant n$ $(j = 1, \ldots, m)$.]

Proof We shall begin with the case $m = 1$. Let $a_0 \in \mathrm{Dom}\,(\bar{T})$ – then we can find $a_r \in \mathfrak{D}$ such that $a_r \to a_0$ and $Ta_r \to \bar{T}a_0$. Hence

$\|X_i(a_r - a_s)\| \leqslant M(\|a_r - a_s\| + \|T(a_r - a_s)\|) \to 0$ as $r, s \to \infty$, which shows that a_0 also lies in Dom $(\bar{X}_i)$ $(i = 1, \ldots, n)$. [Note too that the same argument proves that a_0 belongs to the domain of the closure of ad $(X_{i_1}) \ldots$ ad $(X_{i_m})(T)$ for any finite sequence $i_1, \ldots, i_m$.] To handle the case of an arbitrary m, it will be sufficient, in view of what we have seen when $m = 1$, to verify that

$$(*) \quad \mathrm{Dom}\,(\bar{T}^m) \subset \mathrm{Dom}\,(\bar{T}\bar{X}_{i_1} \ldots \bar{X}_{i_{m-1}}).$$

Proceeding inductively, then, assume that $(*)$ holds for m and let a_0 be in Dom $(\bar{T}^{m+1})$; because $\bar{T} = T^*$, we need only show that $\bar{X}_{i_1} \ldots \bar{X}_{i_m} a_0$ belongs to Dom (T^*), i.e. that the linear function $a \mapsto (X_{i_m} \ldots X_{i_1} Ta, a_0)$ is continuous on $\mathfrak{D}$. To this end, write $X_{i_m} \ldots X_{i_1} T = TX_{i_m} \ldots X_{i_1} + S$, where $TX_{i_m} \ldots X_{i_1}$ is the term $k = 0$ in Lemma 4.4.4.6 and S is the remainder; the induction hypothesis, in conjunction with Lemma 4.4.4.6, tells us that the function $a \mapsto (Sa, a_0)$ $(a \in \mathfrak{D})$ is continuous; since $a_0 \in$ Dom $(\bar{T}^{m+1})$, T^*a_0 is in Dom $(\bar{T}^m)$ and so $(TX_{i_m} \ldots X_{i_1}a, a_0) = (X_{i_m} \ldots X_{i_1}a, T^*a_0)$ is, by the induction hypothesis, also a continuous function of a $(a \in \mathfrak{D})$ – the validity of relation $(*)$ is now clear. □

In actual practice, the X_i above will usually be skew-symmetric; of course this does not alter the conclusion of the lemma in any way.

Let E be a Hilbert space, $\mathfrak{g}$ a (finite dimensional) Lie algebra of skew-symmetric operators on E with a common invariant domain $\mathfrak{D}$. Let $\mathfrak{G}$ be the associative algebra over the field of *real* numbers generated by $\mathfrak{g}$ and the identity operator I on $\mathfrak{D}$; let $\mathfrak{G}_m$ denote the elements in $\mathfrak{G}$ of order $\leqslant m (m = 0, 1, \ldots)$. Let $X_1, \ldots, X_n$ be a basis for $\mathfrak{g}$; put $\Delta = \sum_i X_i^2$.

Lemma 4.4.4.8 Let $T \in \mathfrak{G}_2$ – then there exists a positive (finite) constant M_T such that $\|Ta\| \leqslant M_T \|(\Delta - I)a\|$ for all $a \in \mathfrak{D}$.

Proof It is clear that I, the X_i, and the

$$X_{ij} = X_iX_j + X_jX_i \qquad (1 \leqslant i, j \leqslant n)$$

constitute a set of linear generators for $\mathfrak{G}_2$; obviously it will be enough to verify the assertion of the lemma for them.

(1) Fix $a \in \mathfrak{D}$ – then $\|a\| \leqslant \|\Delta a\| + \|(\Delta - I)a\| \leqslant 2\|(\Delta - I)a\|$ (since $\|\Delta a\|^2 \leqslant (\Delta^2 a, a) - 2(\Delta a, a) + (a, a) = ((\Delta - I)^2 a, a) = ((\Delta - I)a, (\Delta - I)a))$. To take care of the X_i, first observe that $\sum_i \|X_ia\|^2 = \sum_i (X_ia, X_ia) = (-\Delta a, a) \leqslant \|(\Delta - I)a\|^2/2$. The Schwarz inequality for finite sequences now implies that $\sum_i \|X_ia\| \leqslant \sqrt{n/2}\,\|(\Delta - I)a\|$, which establishes our contention in this case.

(2) It remains to consider the X_{ij}. Let $\mathfrak{P}$ be the set of elements in $\mathfrak{G}$ of the form $\sum_k D_k^+ D_k$ (here $D \mapsto D^+$ is the principal anti-automorphism of $\mathfrak{G}$ over $\mathbf{R}$ – thus $(X_{i_1} \ldots X_{i_m})^+ = (-1)^m X_{i_m} \ldots X_{i_1}$).

(a) If $T = \sum t_{ij} X_{ij}$ $(t_{ij} \in \mathbf{R})$, then there exists $t \geqslant 0$ such that

$-t\Delta + T \in \mathfrak{P}$. [This assertion is an immediate consequence of the following identities:

$$-\Delta + X_{ij} = (X_i - X_j)^+(X_i - X_j) + \sum_{k \neq i,j} X_k^+ X_k,\ i \neq j;$$
$$-2\Delta + X_{ii} = 2\sum_{k \neq i} X_k^+ X_k;$$
$$-\Delta - X_{ij} = (X_i + X_j)^+(X_i + X_j) + \sum_{k \neq i,j} X_k^+ X_k.]$$

(b) Write $4\Delta^2 - X_{ij}^2 = (2\Delta - X_{ij})(2\Delta + X_{ij}) + D_1(D_1 \in \mathfrak{G}_3)$; in view of the identities in (a) above, we have $-(2\Delta - X_{ij}) = \sum_k Y_k^+ Y_k$ with the $Y_k \in \mathfrak{g}$ and, similarly, $-(2\Delta + X_{ij}) = \sum_l Z_l^+ Z_l$ $(Z_l \in \mathfrak{g})$ – hence $4\Delta^2 - X_{ij}^2 = \sum_{k,l}(Y_k Z_l)^+(Y_k Z_l) + D_2(D_2 \in \mathfrak{G}_3)$. Now $\mathfrak{G}_3$ is spanned by the operators which span $\mathfrak{G}_2$ together with those of the form

$$X_{ijk} = X_i X_j X_k + X_i X_k X_j + X_j X_i X_k + X_j X_k X_i + X_k X_i X_j + X_k X_j X_i$$

– therefore we may write $D_2 = \Sigma t_{ij} X_{ij} + D_3(t_{ij} \in \mathbf{R})$, where D_3 is a real linear combination of the X_i and the X_{ijk}. It is clear that D_3 is symmetric; but the X_i and the X_{ijk} are skew-symmetric, so that D_3 is also skew-symmetric and hence equal to zero. Owing to (a) above, then, there exists $t \geqslant 0$ such that $-t\Delta + D_2$ is in $\mathfrak{P}$; consequently $4\Delta^2 - X_{ij}^2 - t\Delta \in \mathfrak{P}$. Completing the square, we have, for all $a \in \mathfrak{D}$, $\| X_{ij} a \|^2 \leqslant ((4\Delta^2 - t\Delta)a, a) \leqslant \|(2\Delta - (t/4))a\|^2$. Now let M be the maximum of $\sqrt{2}$ and $\sqrt{t}/2$ – then $\|X_{ij}a\| \leqslant M\|(\Delta - I)a\|$ $(1 \leqslant i, j \leqslant n)$, as we wished to show.

This completes the proof of the lemma. □

Lemma 4.4.4.9 Retain the assumptions and notations of Lemma 4.4.4.8 – then $\sum_i \|X_i a\| \leqslant \sqrt{n/2}\,\|(\Delta - I)a\|$ (all $a \in \mathfrak{D}$). Furthermore, there exists a positive (finite) constant M such that for $m \geqslant 1$,

$$\sum_{(m)} \|\mathrm{ad}(X_{i_1}) \ldots \mathrm{ad}(X_{i_m})(\Delta)a\| \leqslant M^m \|(\Delta - I)a\| \qquad (a \in \mathfrak{D}).$$

Proof The fact that $\sum_i \|X_i a\| \leqslant \sqrt{n/2}\,\|(\Delta - I)a\|$ $(a \in \mathfrak{D})$ was established in part (1) of the proof of the preceding lemma. Turning to the second assertion, first note that for any $T \in \mathfrak{G}_2$, there exists, according to Lemma 4.4.4.8, a positive constant M_T such that

$$\|Ta\| \leqslant M_T \|(\Delta - I)a\| \qquad (a \in \mathfrak{D});$$

this being so, set

$$|T| = \sup_{\|(\Delta - I)a\| = 1} \|Ta\| \qquad (a \in \mathfrak{D}).$$

Then $|\,.\,|$ is a semi-norm on $\mathfrak{G}_2$. The operators $\mathrm{ad}(X_i)$ carry $\mathfrak{G}_2$ into itself – therefore, since $\dim(\mathfrak{G}_2) < \infty$, there exists $M_i > 0$ such that

$$|\mathrm{ad}(X_i)(T)| \leqslant M_i |T| \qquad (T \in \mathfrak{G}_2).$$

Put $M = n \max_i M_i$ – then, for $a \in \mathfrak{D}$,

$$\sum_{(m)} \|\mathrm{ad}(X_{i_1}) \ldots \mathrm{ad}(X_{i_m})(\Delta)a\| \leqslant \sum_{(m)} M_{i_1} \ldots M_{i_m} |\Delta| \cdot \|(\Delta - I)a\|$$
$$\leqslant M^m \|(\Delta - I)a\|,$$

it being clear that $|\Delta| \leqslant 1$ (cf. part (1) of the proof of Lemma 4.4.4.8). □

Proof of Theorem 4.4.4.5 The operators $U_\infty(X)(X \in \mathfrak{g})$ are skew-symmetric and admit a common invariant domain, namely E_∞; the operator $U_\infty(\Delta)$ is symmetric and essentially self-adjoint; on the basis of Lemma 4.4.4.9, it is then clear that all the assumptions of Lemma 4.4.4.7 are met in the present situation. Consequently, every vector in $\bigcap_1^\infty \mathrm{Dom}\,(\overline{U_\infty(\Delta)^m})$ also belongs to $\mathrm{Dom}\,(\overline{U_\infty(X_{i_1})} \ldots \overline{U_\infty(X_{i_m})})$ for all finite sequences $i_1, \ldots, i_m$. Now, as pointed out in Example 4 supra, the domain of $\overline{U_\infty(X)}$ $(X \in \mathfrak{g})$ consists precisely of those vectors $a \in E$ such that

$$\lim_{t \to 0} \frac{U(\exp(tX))a - a}{t}$$

exists. Therefore, if $a \in \bigcap_1^\infty \mathrm{Dom}\,(\overline{U_\infty(\Delta)^m})$, then $\tilde{a}$ $(\tilde{a}(x) = U(x)a,\ x \in G)$ has partial derivatives of all orders at $x = 1$, that is, $\tilde{a} \in C^\infty(G; E)$ and so $\bigcap_1^\infty \mathrm{Dom}\,(\overline{U_\infty(\Delta)^m}) \subset E_\infty$. Since the reverse inclusion is obvious, the assertion of the theorem follows. □

Important as the conclusion of Theorem 4.4.4.5 is for theoretical purposes, nevertheless its applicability in a given concrete situation seems to be rather limited. On the other hand, it is possible to obtain another description of the space E_∞ which turns out to be very useful in the applications.

As above, then, let U be a unitary representation of G on a Hilbert space E, E_∞ the space of differentiable vectors in E (for U). Given $X \in \mathfrak{g}$, let $dU(X)$ denote the skew-adjoint infinitesimal generator of the (strongly continuous) one parameter unitary group $t \mapsto U(\exp(tX))$ in E – then $dU(X) \supset U_\infty(X)$ with, of course, $\mathrm{Dom}\,(dU(X))$ being the set of vectors $a \in E$ for which the function $t \mapsto \tilde{a}(\exp(tX))$ is once differentiable. Evidently if $a \in E_\infty$, then $a \in \mathrm{Dom}\,(dU(X)^m)$ $(m = 1, 2, \ldots)$ for all $X \in \mathfrak{g}$. On the other hand we have the following result.

Theorem 4.4.4.10 (Goodman) Let $X_1, \ldots, X_n$ be a basis for $\mathfrak{g}$; suppose that a vector $a_0 \in E$ lies in the domain of $dU(X_i)^m$ for $m = 1, 2, \ldots$ and $1 \leqslant i \leqslant n$ – then $a_0 \in E_\infty$.

Remark The theorem at hand admits the following interpretation in terms of the one parameter subgroups $G_i = \{\exp(tX_i) : -\infty < t < \infty\}$ $(i = 1, \ldots n)$: If a vector $a_0 \in E$ is a differentiable vector for $U|G_i$, $i = 1, \ldots, n$, then a_0 is a differentiable vector for U. [The space of differentiable vectors for $U|G_i$ is, of course, the set $\bigcap_1^\infty \mathrm{Dom}\,(dU(X_i)^m)$.]

Proof of Theorem 4.4.4.10 It will be convenient, for the proof of the present theorem, to let d_G denote right Haar measure on G and then to regard the elements of $\mathfrak{G}$ as left invariant differential operators on G; bear in mind that in this situation the adjoint $D^\natural$ of an element $D \in \mathfrak{G}$ (taken with respect to d_G) is D^+.

(1) Let a_0 be a vector in E per the statement of the theorem. As has been noted in the Remark supra, the function $t \mapsto U(\exp(tX_i))a_0$ is an E-valued C^∞ function on the line. Suppose now that $f \in C_c^\infty(G)$, $a \in E$ – then

$$\begin{aligned}
&\int_G f(x)(U(x)dU(X_i)^m a_0, a)d_G(x)\\
&\quad= \left(\frac{d}{dt}\right)^m \int_G f(x)(U(x\exp(tX_i))a_0, a)d_G(x)\Big|_{t=0}\\
&\quad= \left(\frac{d}{dt}\right)^m \int_G f(x\exp(-tX_i))(U(x)a_0, a)d_G(x)\Big|_{t=0}\\
&\quad= (-1)^m \int_G X_i^m f(x)(U(x)a_0, a)d_G(x).
\end{aligned}$$

(2) Let Δ_m be the differential operator $X_1^{2m} + \dots + X_n^{2m}$ – then the computation in (1) tells us that for each $a \in E$, the function ϕ, $\phi(x) = (U(x)a_0, a)$ $(x \in G)$, is a distribution solution to the equation $\Delta_m\phi = \Phi_m$ where $\Phi_m(x) = \sum_{i=1}^n (U(x)dU(X_i)^{2m}a_0, a)$ $(x \in G)$. But Φ_m is a continuous function and Δ_m is an elliptic operator of order $2m$, so by the Local Regularity Theorem for distribution solutions of elliptic equations, ϕ has locally L^2 derivatives of order $\leqslant 2m$. [See, for instance, Bers, John, and Schechter [1, p. 190]. Tacitly we are using the fact that Δ_m has C^∞ coefficients and Haar measure has C^∞ non-vanishing density relative to local coordinates on G.] Since this is true for all m, ϕ is C^∞ by the classical Lemma of Sobolev (cf. Sobolev [1, Chapter 1]) – therefore $a_0 \in E_\infty$ (recall that weak differentiability is the same as differentiability; cf. A 2.2).

This completes the proof of the theorem. □

In reading the examples below one should be aware of the following fact (often used implicitly . . .): Let R be the right regular representation of $\mathbf{R}$ on $L^2(\mathbf{R})$; if $f \in L^2(\mathbf{R})$ lies in Dom $(dR(d/dt))$, then f coincides almost everywhere with an (absolutely) continuous function. [Let $g = dR(d/dt)f$; define the function F on $\mathbf{R}$ by the rule $F(t) = \int_a^t f(s)ds$ $(t \in \mathbf{R})$ – then F is absolutely continuous with derivative f (almost everywhere). We shall assume that a has been chosen so that this relation persists there too. Let us now show that $f(t) = f(a) + \int_a^t g(s)ds$ $(t \in \mathbf{R})$. By definition we have $\lim_{t\to 0} \int_{\mathbf{R}} |g(s) - t^{-1}[f(t+s) - f(s)]|^2 ds = 0$ and so we see that for almost all fixed r,

$$\begin{aligned}
&\left|\int_a^r g(s)ds - t^{-1}[F(r+t) - F(r)] + t^{-1}[F(a+t) - F(a)]\right|\\
&= \left|\int_a^r g(s)ds - \int_a^r t^{-1}[f(t+s) - f(s)]ds\right|\\
&\leqslant |r-a|^{1/2}\left(\int_a^r |g(s) - t^{-1}[f(t+s) - f(s)]|^2 ds\right)^{1/2} \to 0
\end{aligned}$$

as $t \to 0$. Our contention now follows without difficulty.]

Examples (1) Consider the Heisenberg group G – thus G is the three dimensional nilpotent Lie group of all matrices of the form

$$(x, y, z) = \begin{pmatrix} 1 & x & z \\ 0 & 1 & y \\ 0 & 0 & 1 \end{pmatrix} \qquad (x, y, z \in \mathbf{R})$$

whose Lie algebra $\mathfrak{g}$ is spanned by elements X, Y, Z with relation $[X, Y] = Z$, Z central. The infinite dimensional irreducible unitary representations of G are worked out in number 5.4.2, Example 3; as is seen there, these representations may be parameterized by the line $\mathbf{R}$ (minus $\{0\}$) and realized on $L^2(\mathbf{R})$. Thus for each non-zero $r \in \mathbf{R}$, the irreducible unitary representation U^r is given by the formula

$$U^r((x, y, z))f(t) = \exp(\sqrt{-1}r(z + ty))f(t + x) \quad (f \in L^2(\mathbf{R}), -\infty < t < \infty).$$

Evidently, then, $dU^r(X) = d/dt$, $dU^r(Y) =$ multiplication by $\sqrt{-1}rt$, $dU^r(Z) =$ multiplication by $\sqrt{-1}r$ (with the usual maximal domains as skew-adjoint operators). We now ask: What is the space of differentiable vectors for the representations U^r? Theorem 4.4.4.10 provides the answer – indeed it tells us that the space in question is the Schwartz space $\mathscr{C}(\mathbf{R})$ of rapidly decreasing C^∞ functions on the line. As a matter of fact, Theorem 4.4.4.10 gives a superficially weaker condition: $f \in \mathscr{C}(\mathbf{R})$ iff all the derivatives of f belong to $L^2(\mathbf{R})$ and pf is in $L^2(\mathbf{R})$ for all polynomials p. [It follows from the theorem that if f satisfies the above condition, then $p(d^nf/dt^n) \in L^2(\mathbf{R})$.] Finally, under U^r_∞, $\mathfrak{G}$ maps onto all ordinary differential operators with polynomial coefficients. With Theorem 4.4.4.3 in mind, we remark that there are many among these which are symmetric but not essentially self-adjoint.

(2) Take for G the '$\alpha t + \beta$' group, i.e. the (non-abelian solvable) group of affine transformations of the real line $\mathbf{R}$. In number 5.4.2, Example 1, the irreducible unitary representations of G are determined; consider U^+; as is shown there U^+ can be realized on $L^2(\mathbf{R})$, the explicit formula being $U^+(\alpha, \beta)f(t) = \exp(\sqrt{-1}e^t\beta)f(t + \log \alpha)(f \in L^2(\mathbf{R}), -\infty < t < \infty)$. The Lie algebra $\mathfrak{g}$ of G has a basis $\{X, Y\}$ and relation $[X, Y] = Y$; one has $\exp(tX) = (e^t, 0)$, $\exp(tY) = (1, t)$ $(t \in \mathbf{R})$. In this case, $dU^+(X) = d/dt$, $dU^+(Y) =$ multiplication by $\sqrt{-1}e^t$ (with the usual maximal domains as skew-adjoint operators). According to Theorem 4.4.4.10, the space of differentiable vectors in $L^2(\mathbf{R})$ for U^+ consists of the C^∞ functions f with $f^{(n)}(t)$ and $e^{nt}f(t)$ in $L^2(\mathbf{R})$ for $n = 0, 1, 2, \ldots$. [Such a function then has the property that $e^{mt}f^{(n)}(t)$ belongs to $L^2(\mathbf{R})$ for $m, n = 0, 1, 2, \ldots$.] Evidently the second condition is only a restriction at $t = +\infty$. Consider now the operator $U^+_\infty(XY + YX) = 2\sqrt{-1}e^t(d/dt) + \sqrt{-1}e^t$; because $XY + YX$ is a Hermitian element in $\mathfrak{G}$, $U^+_\infty(XY + YX)$ is symmetric – but $U^+_\infty(XY + YX)$ has no self-adjoint extension. [It may be shown, for instance, that the operator in question has deficiency indices (0, 1) – this is not quite obvious however.] On the other hand, $U^+_\infty(X^2 + Y^2) = d^2/dt^2 - e^{2t}$; according to Theorem 4.4.4.3, the closure of this operator is a self-adjoint, negative operator on $L^2(\mathbf{R})$.

(3) (Nelson and Stinespring) Let G be the group of all 2×2 real unimodular matrices – our purpose here is to give still another example which points up the fact that some kind of condition (say ellipticity) is needed for the validity of Theorem 4.4.4.3. Let us agree to use the notations centering around the discussion which appears later on in number 5.5.1; as is shown there, G admits an irreducible unitary representation $U^{+,0}$ which can be realized on $L^2(\mathbf{T})$, $\mathbf{T}$ the real numbers modulo 2π – moreover it is easy to see that $\mathfrak{g}$ has a basis $\{X, Y_1, Y_2\}$ with the property that $dU(X) = -d/dt$, $dU(Y_1) = \sin t(d/dt) + \frac{1}{2}\cos t$, $dU(Y_2) = \cos t(d/dt) - \frac{1}{2}\sin t$ (we are writing U for

$U^{+,0}$). Let $D = XY_1 + Y_1X$ – then $U_\infty(D) = -2d/dt(\sin t)d/dt + \frac{1}{2}\sin t$. Let T be the operator with the same formula as $U_\infty(D)$ but with domain $C^\infty(\mathbf{T})$ – then T and $U_\infty(D)$ have the same closure but the operator T is not essentially self-adjoint. To see this consider the functions g_1 and g_2 defined by

$$g_1(t) = \begin{cases} \log|\sin t|, & \frac{-\pi}{2} \leqslant t \leqslant \frac{\pi}{2} \\ 0, & \text{elsewhere on } \mathbf{T} \end{cases}; \quad g_2(t) = \begin{cases} 1, & 0 \leqslant t \leqslant \pi \\ 0, & \text{, elsewhere on } \mathbf{T}. \end{cases}$$

The distribution derivative of g_1 is the map $f \mapsto$ P.V. $\int_{-\pi/2}^{\pi/2} f(t)(\cos t/\sin t)dt$, the symbol P.V. standing for the Cauchy principal value of the integral. Therefore the operator with the same formula as T applied to g_1 in the sense of distributions yields the function h_1 defined by

$$h_1(t) = \begin{cases} 2\sin t + \frac{1}{2}\sin t\, g_1(t), & \frac{-\pi}{2} \leqslant t \leqslant \frac{\pi}{2} \\ 0, & \text{elsewhere on } \mathbf{T}. \end{cases}$$

Since g_1 and h_1 are both in $L^2(\mathbf{T})$, this shows that $h_1 = T^*g_1$; in a similar way we find that $h_2 = T^*g_2$ where

$$h_2(t) = \begin{cases} \frac{1}{2}\sin t, & 0 \leqslant t \leqslant \pi \\ 0, & \text{elsewhere on } \mathbf{T}. \end{cases}$$

However $(T^*g_1, g_2) - (g_1, T^*g_2) = \int_0^{\pi/2} 2\sin t\, dt \neq 0$ and so T (and hence $U_\infty(D)$) is not essentially self-adjoint.

4.4.5 Density of Analytic Vectors – The Theorem of Nelson

Let G be a Lie group which is countable at infinity, $\mathfrak{g}$ the Lie algebra of G and $\mathfrak{G}$ the universal enveloping algebra of the complexification of $\mathfrak{g}$. Let E be a Banach space, U a continuous representation of G on E. As has been seen in 4.4.1 we may represent $\mathfrak{G}$ by differentiation on E_∞, the space of differentiable vectors in E (for U); there we noticed, though, that the connection between U and the corresponding infinitesimal representation U_∞ of $\mathfrak{G}$ was not as close as one might like – the main objective of the present number is to remedy this situation.

We shall begin by recalling some well-known properties of analytic maps.

Let M be a real analytic manifold, E a Banach space. A mapping $f: M \to E$ is said to be *analytic* at a point $p \in M$ if there exists a coordinate system $\{x_1, \ldots, x_m\}$ on a neighborhood $\mathcal{O}$ of p such that $x_1(p) = \ldots = x_m(p) = 0$ with

$$f(q) = \Sigma\, a_{i_1 \ldots i_m} x_1^{i_1}(q) \ldots x_m^{i_m}(q) = \Sigma\, a_I x^I(q) \qquad (q \in \mathcal{O})$$

where the coefficients $a_I = a_{i_1 \ldots i_m} \in E$ and the series converges absolutely. The map f is said to be *analytic on M* if f is analytic at every point of M.

Notation The space of analytic maps on M with values in E will be denoted by $C^\omega(M; E)$.

[For details on Banach space valued analytic maps see Dieudonné [2, Chapter 9].]

Call a map $f: M \to E$ *weakly analytic* if for every $\mathbf{z} \in E^*$ the function $p \mapsto \langle f(p), \mathbf{z}\rangle$ is analytic. A proof of the following lemma may be found in Browder [1, p. 669].

Lemma 4.4.5.1 Let f be a mapping of M into E – then f is analytic iff f is weakly analytic.

[In contradistinction to weak differentiability and weak holomorphy, this result is false if E is replaced by just any locally convex, complete, Hausdorff, topological vector space – for examples and a discussion see Neto [1].]

Let $f: M \to E$ and suppose that f is analytic at a point $p \in M$. Let X_p be an element in the tangent space to M at p. If $\{x_1, \ldots, x_m\}$ is a coordinate system around p with $x_j(p) = 0$ $(j = 1, \ldots, m)$ in which the expansion $f = \sum a_I x^I$ is valid, then put

$$X_p f = \sum_1^m a_j(X_p x_j)$$

where a_j is the value of a_I for $I = (0, \ldots, 0, 1, 0, \ldots, 0)$ (1 in the jth slot). It is easy to check that the value of $X_p f$ is independent of the choice of the coordinates $\{x_1, \ldots, x_m\}$ at p.

The following lemma is obvious.

Lemma 4.4.5.2 (i) Consider two mappings $g: M \to M$ and $f: M \to E$. If g is analytic at $p \in M$ and f is analytic at $g(p)$, then the composite map $f \circ g: M \to E$ is analytic at p.

(ii) Suppose that the map $f: M \to E$ is analytic at p. If $S: E \to F$ is a continuous linear map of E into a Banach space F, then the map $S \circ f$ is analytic at p.

(iii) Let M, N be real analytic manifolds, $h: M \to N$ a map analytic at $p \in M$. Let $f: N \to E$ and suppose that f is analytic at $q = h(p)$. If X_p is in the tangent space to M at p, then

$$X_p(f \circ h) = dh_p(X_p)(f)$$

where dh_p denotes the differential of h at p.

Lemma 4.4.5.3 Let $f: M \to E$ be analytic on M, X an analytic vector field on M.

(i) The map $p \mapsto X_p f$ (i.e. Xf) is analytic on M.

(ii) If Y is another analytic vector field on M and $Z = [X, Y] = XY - YX$, then

$$Zf = X(Yf) - Y(Xf).$$

Proof Only (ii) needs proof. Let $\mathbf{z} \in E^*$. If $g: M \to E$ is analytic, by Lemma 4.4.5.2 (ii), so is $\mathbf{z} \circ g$. If W is any analytic vector field on M, then $\mathbf{z} \circ Wg = W(\mathbf{z} \circ g)$ as is easily seen. Specialize now and take $g = Zf - (X(Yf) - Y(Xf))$. Since

$$\mathbf{z} \circ g(p) = Z_p(\mathbf{z} \circ f) - (X_p[Y(\mathbf{z} \circ f)] - Y_p[X(\mathbf{z} \circ f)]) = 0$$

for all $p \in M$, $\mathbf{z} \circ g = 0$ and thus, by the Hahn-Banach Theorem, $g = 0$. □

Let U be a Banach representation of G on E.

Definition A vector $a \in E$ is said to be *analytic* (for U) if the map $\tilde{a}$, $x \mapsto U(x)a$, is an element of $C^\omega(G; E)$.

Let E_ω denote the set of analytic vectors in E (for U); plainly E_ω is a linear subspace of E_∞. It will be shown in due course that analytic vectors exist in profusion in E; in fact are dense in E.

Consider the map $\tilde{a}$, $x \mapsto U(x)a$, $a \in E$. In order to check that $\tilde{a}$ is analytic, it suffices to verify analyticity at 1. For let $x \in G$ – then $\tilde{a} = U(x) \circ \tilde{a} \circ L_{x^{-1}}$ and so if $\tilde{a}$ is analytic at 1, then it is analytic at x (Lemma 4.4.5.2).

Similar reasoning shows that E_ω is U-stable. We next observe that E_ω is a $\mathfrak{G}$-submodule of E_∞, i.e. the restriction to E_ω of the assignment $D \mapsto U_\infty(D)$ $(D \in \mathfrak{G})$ defines a representation of $\mathfrak{G}$ on E_ω. In fact, if $X \in \mathfrak{g}$, $a \in E_\omega$, then, regarding X as a left invariant vector field on G, we have

$$X\tilde{a}(x) = U(x)U_\infty(X)a \qquad (x \in G).$$

Since the function $x \mapsto X\tilde{a}(x)$ is analytic (Lemma 4.4.5.3 (i))), $U_\infty(X)a \in E_\omega$ – hence E_ω is indeed U_∞-stable. The representation property is a consequence of Lemma 4.4.5.3(ii).

This representation of $\mathfrak{G}$ on the space of analytic vectors E_ω will be denoted by U_ω.

The justification for the term analytic vector is contained in:

Theorem 4.4.5.4 (Harish-Chandra) Let $a \in E_\omega$ – then there exists a neighborhood $\mathcal{O}$ of zero in $\mathfrak{g}$ such that the series

$$\sum_0^\infty \frac{1}{m!} U_\omega(X^m)a$$

converges to $U(\exp X)a$ for all $X \in \mathcal{O}$.

Proof Contrary to our standing convention, we shall view the elements of $\mathfrak{G}$, during the present proof, as left invariant differential operators on G. Let $f: G \to E$ be analytic at $x \in G$ – then there exists a starlike neighborhood $\mathcal{O}$ of zero in $\mathfrak{g}$ such that the Taylor expansion

$$f(x \exp X) = \sum_0^\infty \frac{1}{m!}(X^m f)(x)$$

holds for all $X \in \mathcal{O}$ (cf. Helgason [2, p. 95]). Applying this with $f = \tilde{a}$, $x = 1$, gives

$$\tilde{a}(\exp X) = \sum_0^\infty \frac{1}{m!}(X^m \tilde{a})(1)$$

and since $X^m \tilde{a}(1) = U_\omega(X^m)a$, the theorem follows. □

Let $a \in E_\infty$. Following a rather customary practice, the set $\mathfrak{G}a = \{U_\infty(D)a : D \in \mathfrak{G}\}$ will be referred to as the $\mathfrak{G}$-orbit of a.

Corollary 4.4.5.5 Suppose that G is connected. Let $a \in E_\omega$ – then the closure $(\mathfrak{G}a)^{cl}$ of the $\mathfrak{G}$-orbit of a is U-stable.

Proof Let $\mathbf{z}$ be any element in E^* which vanishes identically on $\mathfrak{G}a$. Let b be any element in $\mathfrak{G}a$ – then $b \in E_\omega$ and by the preceding theorem there exists a neighborhood $\mathcal{O}$ of zero in $\mathfrak{g}$ such that for all $X \in \mathcal{O}$

$$U(\exp X)b = \sum_0^\infty \frac{1}{m!} U_\omega(X^m)b.$$

Thus $\langle U(\exp X)b, \mathbf{z}\rangle = 0$, and so, G being connected, it must therefore be the case that $\langle U(x)b, \mathbf{z}\rangle = 0$ for all $x \in G$. The Hahn-Banach Theorem now implies that $U(x)b \in (\mathfrak{G}a)^{cl}$ $(x \in G)$; consequently $U(x)(\mathfrak{G}a) \subset (\mathfrak{G}a)^{cl}$, whence $U(x)(\mathfrak{G}a)^{cl} \subset (\mathfrak{G}a)^{cl}$ for all $x \in G$. □

As was pointed out in 4.4.1, a major defect of the representation U_∞ of $\mathfrak{G}$ on E_∞ was the fact that the closures of U_∞-stable subspaces of E_∞ were not necessarily U-stable. That this flaw is not present in the representation U_ω of $\mathfrak{G}$ on E_ω is the gist of the next result.

Proposition 4.4.5.6 Suppose that G is connected. Let E_ω^0 be a U_ω-stable subspace of E_ω – then the closure $(E_\omega^0)^{cl}$ of E_ω^0(in E) is U-stable.

Proof Let $b \in (E_\omega^0)^{cl}$. Fix an $x \in G$. Given $\epsilon > 0$, we must find an element $a \in E_\omega^0$ such that $\| U(x)b - a \| < \epsilon$. If $M = \| U(x) \|$, choose $a_0 \in E_\omega^0$ so that $\| b - a_0 \| < \epsilon/2M$. By the preceding corollary, $U(x)a_0 \in (\mathfrak{G}a_0)^{cl} \subset (E_\omega^0)^{cl}$ and so we may select an element $a \in \mathfrak{G}a_0 \subset E_\omega^0$ such that $\| a - U(x)a_0 \| < \epsilon/2$ – hence

$$\| U(x)b - a \| \leqslant \| U(x)(b - a_0) \| + \| U(x)a_0 - a \| < \epsilon,$$

as we wished to prove. □

Our next task will be to show that E_ω is dense in E; needless to say this will enhance the usefulness of Proposition 4.4.5.6.

Theorem 4.4.5.7 (Nelson) Let U be a Banach representation of G on E, E_ω the space of analytic vectors in E (for U) – then E_ω is dense in E.

[If E is finite dimensional, then $E_\omega = E$ – this is so because a continuous homomorphism from one Lie group into another is necessarily analytic].

This result was first proved by Harish-Chandra [5] for connected semi-simple G and then in full generality by Nelson [1]. The proof requires some preparation which we shall undertake in a moment – but first let's give a few words of motivation.

Heuristics Recall that the density of E_∞ in E was shown by considering the vectors

$$\int_G f(x)U(x)ad_G(x) \qquad (f \in C_c^\infty(G), a \in E).$$

The main idea behind establishing the density of E_ω in E is to replace the functions f by analytic functions p which, together with all their derivatives, decrease very rapidly at infinity – this is necessary in order to secure the convergence of the integral

$$\int_G p(x)U(x)a d_G(x) \qquad (a \in E).$$

For a suitable choice of p this integral will also give analytic vectors for U. This idea was already used by Gelfand [1] in 1939 in his investigation of Banach representations U of the real line $\mathbf{R}$. In order to show that such a representation possessed a dense set of analytic vectors, the integral

$$(*) \quad a_t = \frac{1}{\sqrt{4\pi t}} \int_{-\infty}^{\infty} e^{-x^2/4t} U(x) a dx \qquad (t > 0,\ a \in E)$$

was considered. Since $\| U(.) \|$ is at most of exponential increase (the exact definitions are given below), the integral defining a_t is absolutely convergent and $a_t \to a$ as $t \to 0$. Moreover it can be shown that the integral remains absolutely convergent when the kernel is extended to the complex domain and from this it follows that a_t is an analytic vector for U. The whole point here is that the kernel in $(*)$ is the fundamental solution of the heat equation $\partial u/\partial t = \partial^2 u/\partial x^2$ on the line. With this observation in mind, Nelson [1] employed the fundamental solution of the heat equation $\partial u/\partial t = \Delta u$ on G in order to extend Gelfand's argument to the general case. A slight variant of this approach was later used by Gårding [2] in order to obtain the same result – this is the method which we shall follow.

As usual d_G will denote a left invariant Haar measure on G. The elements $D \in \mathfrak{G}$ are to be viewed as right invariant differential operators on G.

Let $X_1, \ldots, X_n$ be a basis for the Lie algebra $\mathfrak{g}$ of G; set $\Delta = X_1^2 + \ldots + X_n^2$. If L is the left regular representation of G on $L^2(G)$, then the closure Λ of $L_\infty(\Delta)$ is a self-adjoint negative operator (Theorem 4.4.4.3). The functional calculus for self-adjoint operators on a Hilbert space (Dunford and Schwartz [2, pp. 1767–1771]) then tells us in particular: For each $f \in C_c^\infty(G)$ and each $t > 0$, the function

$$p_t(x) = e^{t\Lambda} f(x) \qquad (x \in G)$$

is in the domain of Λ for fixed t and satisfies the heat equation

$$\left(\frac{\partial}{\partial t} - \Delta\right) p_t(x) = 0.$$

Moreover $p_t \to f$ in $L^2(G)$ as $t \to 0$.

In the special case when $G = \mathbf{R}$, $\mathbf{R}$ the real line,

$$p_t(x) = \frac{1}{\sqrt{4\pi t}} \int_{-\infty}^{\infty} e^{-(x-y)^2/4t} f(y)\, dy \qquad (x \in G).$$

We shall next determine the regularity of these solutions p_t to the heat equation on G – this information is contained in the following classical theorem from the theory of partial differential equations, due to Avner Friedman. [See Friedman, 'On the regularity of the solutions of non-linear elliptic and parabolic systems of partial differential equations', J. Math. and Mech., vol. 7 (1958), pp. 51–92; for a short proof see Gårding [2, p. 87].]

Theorem 4.4.5.8 (Friedman) Let $\mathscr{I}$ be an open subinterval of the real line **R**.

(i) Let $u = u(t, x)$ be a locally square integrable (complex valued) function in $\mathscr{I} \times G$; suppose, in the sense of distribution theory, that $(\partial/\partial t - \Delta)\cdot u = 0$ – then $u \in C^\infty(\mathscr{I} \times G)$. Moreover, for each fixed $t \in \mathscr{I}$, u is analytic in x in some complex neighborhood G_c of G where G_c is independent of u.

(ii) Let $u \in C^\infty(\mathscr{I} \times G; E)$, E a Banach space; suppose that $(\partial/\partial t - \Delta)u = 0$ – then, for each fixed $t \in \mathscr{I}$, u is analytic in x in some complex neighborhood G_c of G where G_c is independent of u.

Remark The statement that (for fixed t) u is analytic in x in some complex neighborhood G_c of G is to be given the following interpretation. One may find a coordinate neighborhood $\mathcal{O}$ around 1 in which a system $\Phi = \{\phi_1, \ldots, \phi_n\}$ of analytic coordinates is valid with the property: There exists a neighborhood $\mathcal{O}_c$ of zero in $\mathbf{C}^n$ and a complex analytic function $\tilde{u}$ (depending on t) in $\mathcal{O}_c$ such that $\tilde{u} \,|\, \Phi(\mathcal{O}) = u \circ \Phi^{-1}$. Transporting $\mathcal{O}$ by left translation gives a set of local charts on G in each of which u admits a complex analytic continuation, whence the term u is analytic in some complex neighborhood G_c of G.

Returning to Theorem 4.4.5.7, fix a vector $a \in E$ – then the question of whether or not $\tilde{a} \in C^\omega(G; E)$ is a purely local one which may be settled in a neighborhood of 1. Accordingly we may assume, without loss of generality, that G is connected; with this in mind, we shall now proceed to develop the technical machinery needed for the proof of Nelson's Theorem.

Assume in what follows that G is *connected*. Equip G with the structure of a Riemannian manifold in which the left translations are all isometries. Since this structure then admits a transitive group of isometries, it is complete; therefore by a classical theorem of Hopf and Rinow (Helgason [2, p. 56]) balls in G are compact. Let $d(x, y)$ $(x, y \in G)$ denote the corresponding Riemannian metric on G. In particular we have:

$$d(1, x) = d(1, x^{-1}),$$

$$d(xy, 1) = d(y, x^{-1}) \leqslant d(1, x) + d(1, y) \qquad (x, y \in G).$$

Let $d(x) = d(1, x)$ $(x \in G)$ – then $d(xy) \leqslant d(x) + d(y)$ (all $x, y \in G$). Let $B_r = B_r(1) = \{x : d(x) \leqslant r\}$ denote the ball in G of center 1 and radius r.

Lemma 4.4.5.9 There exists an integer k such that B_{m+1} (m a non-negative integer) can be covered by at most k^m left translates of B_1 – here k is independent of m.

Proof Let $x_1, \ldots, x_k$ be elements of B_2 such that the $x_i B_1$ cover B_2. Denote by x_L any product of l of the elements x_i. We shall prove that the $x_M B_1$ are a covering of B_{m+1} – this will prove the lemma, there being at most k^m elements x_M. Let us proceed by induction on m. Fix $x \in B_{m+2}$ – then there exists a $y \in B_{m+1}$ such that $d(y) \leqslant m+1$, $d(y, x) = d(y^{-1}x) \leqslant 1$. By the induction hypothesis there exists x_M and $z_y \in B_1$ such that $y = x_M z_y$; since $z = y^{-1}x \in B_1$ while $x = x_M z_y z$, our lemma thus follows from the fact that $B_1 B_1 \subset B_2$. □

Let $v(B) = \int_B d_G(x)$ denote the volume of a ball B – then for any $x \in G$, $v(xB) = v(B)$ and Lemma 4.4.5.9 shows that for each non-negative integer m one has: $v(B_{m+1}) \leqslant k^m v(B_1)$.

Let E be a Banach space. A map $f: G \rightarrow E$ is said to be at most of *exponential increase* on G if there exists a constant $r \geqslant 0$ such that $\|f(x)\| \leqslant e^{r(1+d(x))}$ for all $x \in G$.

Example Let U be a continuous representation of G on a Banach space E – then the map $x \mapsto U(x)$ $(x \in G)$ is at most of exponential increase on G. [Put $\rho(x) = \|U(x)\|$, $\sigma(x) = \log(\rho(x))$ $(x \in G)$ – then $\sigma(xy) \leqslant \sigma(x) + \sigma(y)$ $(x, y \in G)$ and, moreover, $\sigma(x) \leqslant r(d(x)+1)$, $r = \sup\{|\sigma(x)| : d(x) \leqslant 1\}$ (which, in view of the properties of ρ (4.2.2), is finite, B_1 being compact). To see this, fix an $x \in B_{m+1} - B_m$ – then there exists $x_j \in B_j$ $(j = 1, \ldots, m)$ such that $d(x_j, x_{j+1}) = d(x_{j+1}^{-1}x_j) \leqslant 1$, $d(x, x_m) = d(x_m^{-1}x) \leqslant 1$, whence $\sigma(x) = \sigma(x_1(x_1^{-1}x_2)\ldots(x_m^{-1}x)) \leqslant \sigma(x_1) + \sigma(x_1^{-1}x_2) + \ldots + \sigma(x_m^{-1}x) \leqslant r(m+1) = rm + r \leqslant r(d(x)+1)$. Our assertion about ρ is now evident.]

Note Let $D \in \mathfrak{G}$ – then we may view D as either a right or left invariant differential operator on G. Temporarily let us agree to write D^R when D is to be regarded as a right invariant differential operator and D^L when D is to be regarded as a left invariant differential operator. Now fix D in $\mathfrak{G}$ – then there exist linearly independent elements $D_1, \ldots, D_r$ in $\mathfrak{G}$ and analytic functions $a_1, \ldots, a_r$ on G such that $D^R = \sum_{i=1}^{r} a_i(x) D_i^L$ $(x \in G)$. It is easy to see (cf. the Example supra) that the a_i $(i = 1, \ldots, r)$ are at most of exponential increase on G. [Of course, a similar representation of D^L in terms of right invariant differential operators is also valid].

Let $X_1, \ldots, X_n$ be a basis for the Lie algebra $\mathfrak{g}$ of G. By the *gradient* of a C^1 function f, we shall understand the vector valued function $\nabla f = \{X_1 f, \ldots, X_n f\}$; write $|\nabla f(x)|^2 = \sum_k |X_k f(x)|^2$ $(x \in G)$. It will be convenient to employ the same notation for functions which are not differentiable – thus put, by definition,

$$|Xf(x)| = \limsup_{t \to 0} |f(\exp(-tX)x) - f(x)|/|t| \qquad (x \in G,\ X \in \mathfrak{g}).$$

Lemma 4.4.5.10 The distance function $x \mapsto d(1, x) = d(x)$ has a bounded gradient; in fact $|\nabla d(x)| \leqslant |\nabla d(1)|$ $(x \in G)$.

Proof Let $x, y \in G$ – then $d(x) = d(yy^{-1}x) \leqslant d(y) + d(y^{-1}x)$, whence $|d(y^{-1}x) - d(x)| \leqslant d(y)$. Therefore, if $X \in \mathfrak{g}$, then $|Xd(x)| \leqslant$

$\limsup_{t\to 0} |d(\exp(tX))|/|t| = |Xd(1)|$ which is finite. Our contention is now clear. □

In what follows, it will be assumed that the Riemannian structure on G has been chosen in such a way as to ensure that $|\nabla d(1)| = 1$.

Let $(.\,,.)$ denote the L^2 inner product calculated over G relative to the left Haar measure on G. We shall write $(\nabla.,\nabla.)$ for $\sum_k (X_k.,X_k.)$ whenever the latter makes sense.

Lemma 4.4.5.11 Fix $f \in C_c^\infty(G)$; per f, put $u(t,x) = p_t(x)$ $(t > 0, x \in G)$. Let $h: G \to \mathbf{R}$ be a bounded C^1 function on G with a bounded gradient. The following identities then obtain:

(i) $(\partial u/\partial t, h^2u) + (h^2u, \partial u/\partial t) = \partial/\partial t(hu, hu)$;

(ii) $(\nabla u, \nabla(h^2u)) + (\nabla(h^2u), \nabla u) = 2(\nabla(hu), \nabla(hu)) - 2(u\nabla h, u\nabla h)$;

(iii) $\mathscr{R}\{(\partial u/\partial t, h^2u) + (\nabla u, \nabla(h^2u))\}$
$$= \tfrac{1}{2}\partial/\partial t(hu, hu) + (\nabla(hu), \nabla(hu)) - (u\nabla h, u\nabla h).$$

Proof To verify (i), one need only differentiate the right hand side with respect to t (in view of our hypotheses, it is legitimate to differentiate under the integral sign). The proof of the second identity is just as straightforward. The third can be obtained by adding (i) and (ii) and then dividing by 2. □

The next lemma is important as it will serve to establish the absolute convergence of certain integrals.

Lemma 4.4.5.12 (Gaffney) Let
$$f \in C_c^\infty(G), p_t(x) = e^{t\Delta}f(x)\ (t > 0, x \in G)$$
– then, for all real numbers s,
$$\int_G e^{sd(x)}p_t(x)\overline{p_t(x)}d_G(x) \leqslant e^{(1/2)s^2t}\int_G e^{sd(x)}f(x)\overline{f(x)}d_G(x).$$

Proof Fix $\delta > 0$ – then, as is well-known, there exists a C^∞ function d_δ on G such that both $|d - d_\delta| \leqslant \delta$ and $|\nabla d_\delta| \leqslant 1 + \delta$ on G (see, e.g. Gaffney [1]). On the other hand, if the distance function d is truncated from above at n $(n \in \mathbf{N}^+)$, then an analogous approximation obtains for $d_n = \min(d, n)$; moreover, on any compact set, the approximating function $d_{\delta,n}$ eventually equals d_δ as n increases. Now take $h = \exp(sd_{\delta,n}/2(1 + \delta))$ – then h is bounded and has a bounded gradient (in fact $|\nabla h| \leqslant |s|h/2$). Since $(\nabla u, \nabla(h^2u)) = -(\Delta u, h^2u)$, $(\partial/\partial t - \Delta)u = 0$ (in the case under consideration), the third identity of Lemma 4.4.5.11 tells us that $0 = \tfrac{1}{2}\partial/\partial t(hu, hu) + (\nabla(hu), \nabla(hu)) - (u\nabla h, u\nabla h)$ – thus $\partial/\partial t(hu, hu) \leqslant 2(u\nabla h, u\nabla h) \leqslant \tfrac{1}{2}s^2(hu, hu)$ and so
$$\frac{\partial}{\partial t}(e^{-(1/2)s^2t}(hu, hu)) \leqslant 0.$$

Now integrate with respect to t over this last inequality to get that

$$\int_\epsilon^t \frac{\partial}{\partial t}(e^{-(1/2)s^2t}(hu, hu))dt = e^{-(1/2)s^2t}(hu, hu) - e^{-(1/2)s^2\epsilon}(hp_\epsilon, hp_\epsilon)$$
$$\leqslant 0 \qquad (0 < \epsilon < t).$$

As $\epsilon \to 0$, $p_\epsilon \to f(L^2)$ – therefore

$$e^{-(1/2)s^2t}(hu, hu) \leqslant (hf, hf).$$

Using dominated convergence, our lemma then follows by first letting n tend to infinity and then letting δ approach zero. □

In the notations introduced above, let $q = q_t(x)$ denote either one of the functions

$$\left(\frac{\partial}{\partial t}\right)^n D^L p_t(x), \ \left(\frac{\partial}{\partial t}\right)^n D^R p_t(x) \qquad (D \in \mathfrak{G}),$$

n a non-negative integer.

Corollary 4.4.5.13 Let $\phi \in C(G)$ and suppose that ϕ is at most of exponential increase on G – then

$$\int_G \phi(x)q_t(x)\overline{q_t(x)}d_G(x)$$

is bounded for t bounded.

Proof Since Δ is a right invariant differential operator on G, the statement follows at once from the preceding lemma if $q = (\partial/\partial t)^n D^L p_t$. On the other hand, if $q = (\partial/\partial t)^n D^R p_t$, then we may write

$$q_t(x) = \sum_{i=1}^r a_i(x)\left(\frac{\partial}{\partial t}\right)^n D_i^L p_t(x) \qquad (x \in G).$$

Since the coefficients $a_i(x)$ $(x \in G)$ are at most of exponential increase on G, our assertion follows once again from Lemma 4.4.5.12. □

One more preliminary to the main theorem is needed; namely, Banach space valued solutions to the heat equation will be considered in:

Lemma 4.4.5.14 (Gårding) Let E be a Banach space. Let $h \in C(G; E)$ and suppose that h is at most of exponential increase on G. If $f \in C_c^\infty(G)$, then

$$P_t(x) = \int_G p_t(x^{-1}y)h(y)d_G(y) \qquad (x \in G)$$

is a solution to the heat equation, i.e.,

$$\left(\frac{\partial}{\partial t} - \Delta\right)P_t(x) = 0 \qquad (t > 0).$$

Moreover, for each fixed $t > 0$, $P_t(x)$ is an analytic function of x and

$$\lim_{t \to 0}\left\|P_t(1) - \int_G f(y)h(y)d_G(y)\right\| = 0.$$

Proof To be able to apply Theorem 4.4.5.8 (ii) to deduce the analyticity of $P_t(x)$ in x for fixed t, we must first prove that $P_t(x) \in C^\infty(\mathbf{R}^+ \times G; E)$. In turn this will be done by showing that all the distribution derivatives $(\partial/\partial t)^n DP_t(x)$ are in reality continuous functions on $\mathbf{R}^+ \times G$ (the convergence of the integrals involved here is established below). [Corollary 4.4.5.13 will ensure us that we need only consider right invariant differential operators D.]

(1) Let $q = q_t(x)$ denote the function $(\partial/\partial t)^n Dp_t(x)$ (n a non-negative integer, $p_t(x) = e^{t\Delta}f(x)$). Consider the integral

$$Q_t(x) = \int_G q_t(x^{-1}y)h(y)d_G(y) = \int_G q_t(y)g(x, y)d_G(y) \qquad (x \in G)$$

where $g(x, y) = h(xy)$ $(x, y \in G)$ – then $Q_t(x)$ is defined and continuous for $t > 0$. For note that

$$\left[\int_G |q_t(y)|\cdot\|g(x, y)\|d_G(y)\right]^2$$
$$\leqslant \left(\int_G |q_t(y)|^2 e^{\lambda d(y)} \|g(x, y)\|^2 d_G(y)\right)\left(\int_G e^{-\lambda d(y)} d_G(y)\right),$$

where λ is real. Taking λ large enough, we can be sure that $\int_G e^{-\lambda d(y)} d_G(y)$ is finite. [Let k be the integer of Lemma 4.4.5.9; if $\lambda > \log k$, then the integral $\int_G e^{-\lambda d(y)} d_G(y)$ is finite. In fact this integral is majorized by the sum

$$\sum_m e^{-\lambda m}(v(B_{m+1}) - v(B_m)) \leqslant \sum_m e^{-\lambda m} v(B_{m+1}) \leqslant v(B_1) \sum_m e^{-(\lambda - \log k)m} < \infty.]$$

In addition $\|g(x, y)\| \leqslant e^{r(1+d(y))}$ where r is *bounded* if x is bounded. Therefore, in view of Corollary 4.4.5.13, we may conclude that

$$\int_G |q_t(y)|\cdot\|g(x, y)\|d_G(y)$$

is bounded if x and t remain bounded. Let $G_N = \{x : d(x) \geqslant N\}$ – then the above considerations show that

$$\lim_{N\to\infty} \int_{G_N} |q_t(y)|\cdot\|g(x, y)\|d_G(y) = 0$$

uniformly for x and t bounded. Hence

$$Q_t(x) = \lim_{N\to\infty} \int_{G-G_N} q_t(y)g(x, y)d_G(y) \qquad (x \in G)$$

is continuous since it is the uniform limit of continuous functions.

(2) Let $w \in C_c^\infty(\mathbf{R}^+ \times G)$. Bearing in mind that D is a right invariant differential operator, we have

$$\int p_t(x^{-1}y)\left[\left(\frac{\partial}{\partial t}\right)^n D\right]^t w(t, x)dtd_G(x)$$
$$= \int q_t(x^{-1}y)w(t, x)dtd_G(x) \qquad (y \in G).$$

Multiplying by $h(y)d_G(y)$ and integrating yields

$$\int P_t(x)\left[\left(\frac{\partial}{\partial t}\right)^n D\right]^t w(t, x)dtd_G(x) = \int Q_t(x)w(t, x)dtd_G(x),$$

the interchange of the order of integration being valid by Fubini's Theorem. Conclusion: The derivatives $(\partial/\partial t)^n DP_t(x)$ (taken in the sense of distributions) are continuous functions, whence by a well-known theorem (Schwartz [2, p. 47]) $P_t(x) \in C^\infty(\mathbf{R}^+ \times G; E)$.

(3) Since the integrals $Q_t(x)$ converge unformly for t and x bounded, advanced calculus permits one to differentiate under the integral sign – in particular:

$$\left(\frac{\partial}{\partial t} - \Delta\right)P_t(x) = 0 \qquad (t > 0).$$

The analyticity of $P_t(x)$ in x for fixed t therefore follows from Theorem 4.4.5.8 (ii).

(4) It remains to prove that

$$\lim_{t\to 0} P_t(1) = \int_G f(y)h(y)d_G(y).$$

Let $\epsilon > 0$ be given. First choose $\lambda > 0$ so that $\int_G e^{-\lambda d(y)}d_G(y) < \infty$ (cf. (1)). Next choose $r > 0$ so that $e^{\lambda d(y)} \| h(y) \|^2 \leqslant e^{r(1+d(y))}$ for all $y \in G$. Put

$$M_1^2 = M \int_G e^{rd(y)} f(y)\overline{f(y)}d_G(y),\ M = \exp\left(r + \left(\frac{r^2}{2}\right)\right).$$

Select a compact subset $\omega \supset \operatorname{spt}(f)$ large enough to secure the inequality

$$\left(\int_{G-\omega} e^{-\lambda d(y)}d_G(y)\right)^{1/2} \leqslant \frac{\epsilon}{2M_1}.$$

If

$$M_2 = \sup_{y\in\omega} e^{\lambda d(y)} \| h(y) \|^2,\ M_3^2 = M_2\left(\int_G e^{-\lambda d(y)}d_G(y)\right),$$

then choose t, $0 < t < 1$ say, close enough to zero in order to ensure that

$$\left(\int_\omega |p_t(y) - f(y)|^2 d_G(y)\right)^{1/2} \leqslant \frac{\epsilon}{2M_3}.$$

This is possible since $p_t \to f$ in $L^2(G)$ as $t \to 0$. We then have

$$\left\| P_t(1) - \int_G f(y)h(y)d_G(y)\right\| \leqslant \int_G |p_t(y) - f(y)|\cdot\| h(y)\| \, d_G(y)$$

$$\leqslant \int_\omega |p_t(y) - f(y)|\cdot\| h(y)\| \, d_G(y) + \int_{G-\omega} |p_t(y)|\cdot\| h(y)\| \, d_G(y)$$

$$\leqslant \left(\int_\omega |p_t(y) - f(y)|^2 e^{\lambda d(y)} \| h(y)\|^2 \, d_G(y)\right)^{1/2}\left(\int_G e^{-\lambda d(y)} \, d_G(y)\right)^{1/2}$$

$$+ \left(\int_G |p_t(y)|^2 e^{r(1+d(y))} \, d_G(y)\right)^{1/2}\left(\int_{G-\omega} e^{-\lambda d(y)} \, d_G(y)\right)^{1/2}$$

$$\leqslant \frac{\epsilon}{2} + \left(M\int_G e^{rd(y)} f(y)\overline{f(y)} \, d_G(y)\right)^{1/2}\left(\int_{G-\omega} e^{-\lambda d(y)} \, d_G(y)\right)^{1/2}$$

$$\leqslant \frac{\epsilon}{2} + \frac{\epsilon}{2} = \epsilon,$$

the Gaffney estimate (Lemma 4.4.5.12) being used to establish that

$$\int_G |p_t(y)|^2 e^{rd(y)} d_G(y) \leqslant e^{r^2/2} \int_G e^{rd(y)} f(y)\overline{f(y)}\, d_G(y) \qquad (0 < t < 1).$$

The proof of Gårdings lemma is now quite complete. □

Proof of Theorem 4.4.5.7 As has been observed earlier, we may assume that G is connected; the machinery developed above is then applicable. We already know that the vectors of the form

$$\int_G f(y)U(y)a d_G(y) \qquad (f \in C_c^\infty(G), a \in E)$$

are dense in E. Replace $f(y)$ in the integrand by $p_t(y) = e^{t\Delta}f(y)$ and put

$$a_t = \int_G p_t(y)U(y)a d_G(y) \qquad (t > 0).$$

The proof of Lemma 4.4.5.14 shows that the integral defining a_t is absolutely convergent. [Earlier remarks tell us that the function $y \mapsto U(y)a$ is at most of exponential increase on G.] In addition it is clear that

$$U(x)a_t = \int_G p_t(x^{-1}y)U(y)a d_G(y) \qquad (x \in G).$$

Lemma 4.4.5.14 implies that $a_t \in E_\omega$ with

$$\lim_{t \to 0} a_t = \int_G f(y)U(y)a d_G(y).$$

Therefore E_ω is dense in E, as we wished to prove. □

Remarks Let U be a unitary representation of G on E.

(1) It is natural to ask: Given $a \in E_\infty$, when is $a \in E_\omega$? According to R. Goodman, the answer is: A vector $a \in E_\infty$ lies in E_ω iff there exists a constant $M > 0$ such that $\| U_\infty(\Delta)^m a \| \leqslant M^m(2m)!$ for all $m \in \mathbf{N}^+$ – we shall give the proof in 4.4.6. [Here $\Delta = X_1^2 + \ldots + X_n^2$ is the Laplacian relative to a basis $X_1, \ldots, X_n$ for $\mathfrak{g}$.]

(2) In view of Theorem 4.4.4.10, the study of the space E_∞ boils down to a study of the corresponding spaces relative to certain one parameter subgroups of G. The situation for E_ω is more delicate since separate real analyticity does not imply joint analyticity – hence the analogue of Theorem 4.4.4.10 for analytic vectors may not be true. It is possible, however, to prove the following. Suppose that G can be written as the semi-direct product of closed subgroups H_1 and H_2; let E_ω^i denote the space of analytic vectors in E for $U|H_i$ $(i = 1, 2)$ – then $E_\omega = E_\omega^1 \cap E_\omega^2$ (cf. Goodman [2]). [Consequently Theorem 4.4.4.10 admits a counterpart in the analytic case at least if G is solvable.]

Let G be a Lie group countable at infinity, U a Banach representation of G on E, E_ω the space of analytic vectors in E (for U). Fix a compact subgroup K of G – we now intend to establish an analogue of Theorem 4.4.3.1 with E_∞ replaced by E_ω.

The following elementary lemma is proved in Helgason [2, p. 399].

Lemma 4.4.5.15 Let $f \in C^\omega(G; E)$, $h \in C^\omega(K)$ – then the map

$$x \mapsto \int_K f(xk)h(k)dk \qquad (x \in G)$$

also lies in $C^\omega(G; E)$.

For $\delta \in \hat{K}$, set $E_\omega(\delta) = E_\omega \cap E(\delta)$ – then $E_\omega \cap (\sum_{\delta \in \hat{K}} E(\delta)) = \sum_{\delta \in \hat{K}} E_\omega(\delta)$, E_ω being K-stable.

Theorem 4.4.5.16 Let U be a Banach representation of G on E – then

$$\sum_{\delta \in \hat{K}} E_\omega(\delta)$$

is dense in E.

Proof Let $a \in E$ and $\epsilon > 0$ be given. Since E_ω is dense in E, there exists $b \in E_\omega$ such that $\|a - b\| < \epsilon/2$. But b is differentiable for $U|K$ – hence, by Theorem 4.4.2.1, we can find a finite subset F of $\hat{K}$ such that $\|\sum_{\delta \in F} \mathsf{P}(\delta)b - b\| < \epsilon/2$. Lemma 4.4.5.15 implies that $\mathsf{P}(\delta)b \in E_\omega$ $(\delta \in \hat{K})$ – therefore $\sum_{\delta \in F} \mathsf{P}(\delta)b \in \sum_{\delta \in \hat{K}} E_\omega(\delta)$. Since it is clear that

$$\|a - \textstyle\sum_{\delta \in F} \mathsf{P}(\delta)b\| < \epsilon,$$

the proof of the present theorem is now complete. □

Corollary 4.4.5.17 If for some $\delta \in \hat{K}$, $\dim(E(\delta)) < \infty$, then $E(\delta) \subset E_\omega$.

Proof In fact Proposition 4.4.3.4 implies that $E_\omega \cap E(\delta) = E(\delta)$. □

Let U be a Banach representation of G on E. We shall henceforth write: $E_K = \sum_{\delta \in \hat{K}} E_\omega(\delta)$.

Proposition 4.4.5.18 The space E_K is U_ω-stable.

Proof First note that a vector $a \in E$ lies in $\sum_{\delta \in \hat{K}} E(\delta)$ iff its K-orbit is finite dimensional. Consequently, if $a \in E_K$, $D \in \mathfrak{G}$, then all we need do is show that the K-orbit of $U_\omega(D)a$ is finite dimensional. But for $k \in K$, we have $U(k)U_\omega(D)a = U_\omega(\mathrm{Ad}(k)D)U(k)a$; since $\mathrm{Ad}(k)$ $(k \in K)$ does not increase degree, the elements $\mathrm{Ad}(k)D$ span a finite dimensional subspace of $\mathfrak{G}$ – however $a \in E_K$ and so its K-orbit is finite dimensional. It is now clear that the K-orbit of $U_\omega(D)a$ is finite dimensional too. □

To summarize, given a pair (G, K) and a Banach representation U of G on E, we may then represent $\mathfrak{G}$ by differentiation on E_∞, E_ω or E_K. It is natural to wonder which of these three representations of $\mathfrak{G}$ is the 'best' one to use in any given situation. Evidently, in making such a choice, one is guided primarily by the use which is to be made of the 'infinitesimal' representation at hand. Thus in the unitary case, for questions of operator theory, we have found that the representation of $\mathfrak{G}$ on E_∞ was important. On the other hand, one of our goals is to establish an intimate connection between the TCI Banach representations of G and certain algebraically completely irreducible representations of $\mathfrak{G}$; as will be seen in 4.5.5, this connection can be established, for the pairs (G, K) of interest to us, via the representation of $\mathfrak{G}$ on E_K.

Fix a pair (G, K) as above. We shall now indicate briefly how it is possible to establish the density of analytic vectors for certain topologically irreducible Banach representations of G without having to appeal to Nelson's Theorem. Let us assume, moreover, that G is connected.

Proposition 4.4.5.19 Let U be a topologically irreducible representation of G on a Banach space E. Suppose that $0 < \dim(E(\delta)) < \infty$ for some $\delta \in \hat{K}$ – then E_ω is dense in E.

Before giving a proof of this, we shall make a couple of general observations and establish a lemma. Given a distribution $T \in D_c(G)$, let T_K denote its average over K – thus, by definition, $\langle f, T_K \rangle = \langle f_K, T \rangle$, f_K denoting the map $x \mapsto \int_K f(kxk^{-1})dk$ $(f \in C_c^\infty(G))$. In particular, if $D \in \mathfrak{G}$, then, in view of the identification $D \mapsto T_D$ of $\mathfrak{G}$ with the distributions on G whose support is $\{1\}$ (A 2.4), it is obvious what one is to understand by the symbol D_K; evidently $D = D_K$ iff $\mathrm{Ad}(k)D = D$ for all $k \in K$.

Working in local coordinates and keeping in mind the classical theorem of S. Bernstein, it becomes clear that there exists a $D \in \mathfrak{G}$ with the following property: If f is a C^∞ function on G verifying the relation $Df = \lambda f + g$, where λ is a scalar and g is an analytic function, then f itself is analytic. [For definiteness, we may assume that D is represented around 1 by an operator of the form $r + \Sigma r_i \partial/\partial x_i + \Sigma r_{ij} \partial^2/\partial x_i \partial x_j$ where the coefficients r_{ij} constitute a strictly positive definite matrix.] Note that this implies that every C^∞ solution of an equation of the form $(D - \lambda)^n f = 0$ $(n \in \mathbf{N})$ is analytic. [To see this proceed by induction on n. Assuming the contention is true for $n - 1$, put $g = (D - \lambda)f$ – then g is analytic because we have $(D - \lambda)^{n-1} g = 0$ and so, since $Df = \lambda f + g$, f itself is analytic, as desired.]

An easy averaging argument now yields the following result.

Lemma 4.4.5.20 There exists a $D \in \mathfrak{G}$ such that $D = D_K$ and having the property that every C^∞ solution f of the equation $(D - \lambda)^n f = 0$ is, in fact, analytic $(\lambda \in \mathbf{C}, n \in \mathbf{N})$.

Proof of Proposition 4.4.5.19 Because E_ω is a U-stable subspace of E, in view of the topological irreducibility of U, it will be enough to show that non-trivial analytic vectors actually do exist. For this purpose we shall prove that every element of $E(\delta)$ is weakly analytic, hence strongly analytic (Lemma 4.4.5.1). To this latter end, first note that $E(\delta) \subset E_\infty$ (Corollary 4.4.3.3) – thus it makes sense to apply the endomorphisms $U(T)$ to elements of $E(\delta)$ $(T \in D_c(G))$. In particular, if we choose $D \in \mathfrak{G}$ per the preceding lemma, then $U(T_D) = U_\infty(D)$ commutes with the $U(k)$ $(k \in K)$ – hence $U_\infty(D)E(\delta) \subset E(\delta)$. Because $E(\delta)$ is finite dimensional, by linear algebra we can find a basis $\{a_i\}$ for $E(\delta)$ such that every a_i satisfies a relation of the form $(U_\infty(D) - \lambda_i)^n a_i = 0$ $(\lambda_i \in \mathbf{C})$. Fix $\mathbf{z} \in E^*$; let f_i denote the function $x \mapsto \langle U(x)a_i, \mathbf{z} \rangle$ $(x \in G)$ – then f_i is annihilated by $(D - \lambda_i)^n$, whence f_i is analytic. Obviously this serves to prove our proposition. □

4.4.6 Analytic Domination – Applications to Representation Theory

We shall keep to the notations and conventions introduced in 4.4.5.

Let U be a unitary representation of the Lie group G on a Hilbert space E, E_ω the space of analytic vectors in E (for U). Our first task will be to formulate a result which describes the space E_ω in rather precise terms (Theorem 4.4.6.1). This description is implemented via the following definition.

Definition Let T be an operator on the Banach space E. An *analytic vector for T* is an element $a \in E$ such that $T^m a$ is defined for all $m \in \mathbf{N}$ and having the property that $\sum_0^\infty \| T^m a \| t^m/m! < \infty$ for some $t > 0$ (that is, the power series expansion of $\exp(tT)a$ is defined and has a positive radius of absolute convergence).

Evidently if T is bounded then every vector in E is an analytic vector for T so that only unbounded operators are of interest. Note too that an analytic vector a for T of necessity satisfies an estimate $\| T^m a \| \leqslant M^m m!$ for some positive (finite) constant M $(m = 1, 2, \ldots)$.

Examples (1) If $E = C([0, 1])$ is the Banach space of continuous functions on the closed unit interval in the sup norm topology and if T is differentiation, then an analytic vector for T is simply an analytic function on $[0, 1]$.

(2) Let U be a Banach representation of G on E; let $X_1, \ldots, X_n$ be a basis for the Lie algebra $\mathfrak{g}$ of G – then a vector $a \in E_\infty$ is an analytic vector for U iff a is an 'analytic vector' for $\{U_\infty(X_i)\}$, that is, iff there exists $t > 0$ such that the series

$$\sum_0^\infty \frac{1}{m!} \sum_{(m)} \| U_\infty(X_{i_1}) \ldots U_\infty(X_{i_m})a \| t^m$$

is convergent, or, equivalently, iff there exists a positive (finite) constant M such that $\| U_\infty(X_{i_1}) \ldots U_\infty(X_{i_m})a \| \leqslant M^m m!$ for indices i_j, $1 \leqslant i_j \leqslant n$ $(j = 1, \ldots, m)$. [Here $\sum_{(m)}$ denotes a sum over all the i_j.]

(3) Let $\mathcal{O}$ be an open set in $\mathbf{R}^n$; let T be a linear elliptic partial differential operator on $\mathcal{O}$ with analytic coefficients (the domain of T being the set of all functions f in $E = L^2(\mathcal{O})$ such that the distribution $T \cdot f$ is also in $L^2(\mathcal{O})$); let f be an analytic vector for T (in $L^2(\mathcal{O})$) – then, of necessity, f is an analytic function on $\mathcal{O}$ (cf. Nelson [1]; see too Kotake and Narasimhan [1]). [In passing observe that f is in any event a C^∞ function on $\mathcal{O}$ (for by the Regularity Theorem any function which is in the domain of T^m (all $m \in \mathbf{N}$) is automatically $C^\infty \ldots$).] Needless to say this generalizes the well-known result which asserts that an L^2-solution f to the equation $T \cdot f = 0$ must actually be analytic. As an application suppose in addition that T is self-adjoint – then any element of a spectral subspace corresponding to a bounded interval is an analytic function (cf. the Appendix to the present number).

Returning to G, let $X_1, \ldots, X_n$ be a basis for $\mathfrak{g}$; put $\Delta = \sum_i X_i^2$ and consider $U_\infty(1 - \Delta)$ – then, as we know, the operator $U_\infty(1 - \Delta)$ is symmetric on E_∞ and its closure, which we denote by A, is a positive self-adjoint operator $\geqslant I$ (Theorem 4.4.4.3). Let $B = A^{1/2}$ – then:

Theorem 4.4.6.1 (Goodman) The space E_ω is precisely the set of analytic vectors for B.

The proof of this important fact depends on the notion of analytic domination and certain a priori norm estimates – cf. infra.

Definition Let E be a normed linear space. Suppose that A and $X_1, \ldots, X_n$ are (everywhere defined) linear operators on E – then the operator A is said to *analytically dominate* the family $\{X_i\}$ if every analytic vector a for A verifies an estimate $\| X_{i_1} \ldots X_{i_m} a \| \leqslant M^m m!$, M a positive (finite) constant independent of the indices i_j (but depending, in general, on a).

[In the case $m = 1$, one may paraphrase the definition as follows: A analytically dominates the operator X if every analytic vector for A is an analytic vector for X.]

Note Needless to say the notion of analytic vector makes sense in a normed linear space E (i.e. the completeness of E is not a necessary requirement).

Lemma 4.4.6.2 (Nelson) Let E be a normed linear space. Suppose that A and X are (everywhere defined) linear operators on E such that

$$\| Xa \| \leqslant \| Aa \|, \ \| \operatorname{ad}(X)^m(A)a \| \leqslant m! \, \| Aa \| \qquad (\text{all } a \in E)$$

– then A analytically dominates X.

Proof (Cartier-Goodman) Suppose that a is an analytic vector for A – then of necessity there exists a positive (finite) constant M such that $\| A^m a \| \leqslant M^m m!$ $(m = 1, 2, \ldots)$. This being so, set

$$C_{r,s} = \| AX^r A^s a \| \qquad (r, s \in \mathbf{N})$$

– then we claim that the $C_{r,s}$ verify the recursive inequalities

$$C_{0,s} \leqslant M^{s+1}(s+1)!, \ C_{r+1,s} \leqslant C_{r,s+1} + \sum_{k=0}^{r} (r+1)! \frac{C_{k,s}}{k!}.$$

The first relation is, of course, automatic (due to our assumption on the vector a). As for the second, we may proceed as follows. Bearing in mind the algebraic identity

$$X^{r+1}A^{s+1}a = AX^{r+1}A^s a + \sum_{k=0}^{r} \binom{r+1}{k} [\operatorname{ad}(X)^{r+1-k}(A)] X^k A^s a,$$

we derive that

$$\begin{aligned} C_{r+1,s} &= \| AX^{r+1}A^s a \| \\ &\leqslant \| X^{r+1}A^{s+1}a \| + \sum_{k=0}^{r} \binom{r+1}{k} \| [\operatorname{ad}(X)^{r+1-k}(A)] X^k A^s a \| \\ &\leqslant \| AX^r A^{s+1} a \| + \sum_{k=0}^{r} \frac{(r+1)!}{k!(r+1-k)!} \cdot (r+1-k)! \, \| AX^k A^s a \| \\ &= C_{r,s+1} + \sum_{k=0}^{r} (r+1)! \frac{C_{k,s}}{k!}, \end{aligned}$$

as desired. Now let $D_{r,s}$ be defined recursively by replacing the inequalities supra by equalities – then certainly $C_{r,s} \leqslant D_{r,s}$ $(r, s \in \mathbf{N})$. Introducing the generating function

$$\Phi(\alpha, \beta) = \sum_{r,s=0}^{\infty} \frac{D_{r,s}}{r!s!}\alpha^r\beta^s,$$

we obtain the equation

$$(\Phi) \quad \begin{cases} \Phi_\alpha(\alpha, \beta) = \Phi_\beta(\alpha, \beta) + \left[\dfrac{\alpha}{1-\alpha}\cdot\Phi(\alpha, \beta)\right]_\alpha, \\ \Phi(0, \beta) = M(1 - M\beta)^{-2}, \end{cases}$$

the subscripts denoting partial differentiation with respect to the indicated variable. Owing to the Cauchy-Kowalesky Theorem, equation (Φ) admits a unique analytic solution for α and β near zero; indeed the classical method of characteristics leads at once to the formula

$$\Phi(\alpha, \beta) = M(1-\alpha)[(1-2\alpha)(1 - M\beta - M\alpha/2 + M\log(1-2\alpha)^{1/4})^2]^{-1}.$$

Set $\beta = 0$ – then

$$\Phi(\alpha, 0) = M(1-\alpha)[(1-2\alpha)(1 - M\alpha/2 + M\log(1-2\alpha)^{1/4})^2]^{-1}$$

is analytic for α near zero, whence $D_{r,0} \leqslant R^r r!$ for some positive (finite) constant R which depends *only* on M. So, for r large enough,

$$\|X^{r+1}a\| \leqslant \|AX^r a\| \leqslant D_{r,0} \leqslant R^r r! \leqslant R^{r+1}(r+1)!$$

which is equivalent to the contention of our lemma. □

Lemma 4.4.6.2 (bis) Let E be a normed linear space. Suppose that A and $X_1, \ldots, X_n$ are (everywhere defined) linear operators on E such that

$$\|X_i a\| \leqslant \|Aa\|, \|\mathrm{ad}(X_{i_1}) \ldots \mathrm{ad}(X_{i_m})(A)a\| \leqslant m!\,\|Aa\| \quad (\text{all } a \in E)$$

– then A analytically dominates the family $\{X_i\}$.

[This is a straightforward generalization of the preceding result (here we have to keep in mind Lemma 4.4.4.6 (bis) . . .).]

Lemma 4.4.6.3 (Nelson) Let E be a normed linear space. Suppose that A and X are (everywhere defined) linear operators on E such that

$$\|Xa\|, \|X^2a\| \leqslant \|Aa\|, \|\mathrm{ad}(X)^m(A)a\| \leqslant m!\,\|Aa\| \qquad (\text{all } a \in E).$$

Let a be an element of E such that $\|A^m a\| \leqslant M^m(2m)!$ for some positive (finite) constant M – then a is an analytic vector for X.

Proof Let $\tilde{E} = E + E$ (direct sum) with the norm $\|\tilde{a}\| = \|a_1 + a_2\| = \|Xa_1\| + \|a_2\|$ $(\tilde{a} \in \tilde{E}; a_1, a_2 \in E)$; put

$$\tilde{A} = \begin{pmatrix} 0 & 1 \\ A & 0 \end{pmatrix}, \tilde{X} = \begin{pmatrix} X & 0 \\ 0 & X \end{pmatrix}.$$

Then we claim that $\tilde{A}$ analytically dominates $\tilde{X}$ on $\tilde{E}$. To see this, we shall apply the criterion which was derived in Lemma 4.4.6.2. Thus in the first place is clear that

$$\|\tilde{X}(a_1 + a_2)\| = \|X^2 a_1\| + \|X a_2\|$$
$$\leqslant \|X a_2\| + \|A a_1\| = \|\tilde{A}(a_1 + a_2)\| \qquad (a_1, a_2 \in E).$$

Secondly we have

$$\operatorname{ad}(\tilde{X})^m(\tilde{A}) = \begin{pmatrix} 0 & 0 \\ \operatorname{ad}(X)^m(A) & 0 \end{pmatrix}$$

so that

$$\|\operatorname{ad}(\tilde{X})^m(\tilde{A})(a_1 + a_2)\| = \|\operatorname{ad}(X)^m(A)a_1\| \leqslant m!\,\|Aa_1\|$$
$$\leqslant m!\,\|\tilde{A}(a_1 + a_2)\| \qquad (a_1, a_2 \in E).$$

It therefore follows that every analytic vector for $\tilde{A}$ is an analytic vector for $\tilde{X}$. Now

$$\tilde{A}^{2m} = \begin{pmatrix} A^m & 0 \\ 0 & A^m \end{pmatrix}, \quad \tilde{A}^{2m+1} = \begin{pmatrix} 0 & A^m \\ A^{m+1} & 0 \end{pmatrix}$$

and it is easy to see that if $\|A^m a\| \leqslant M^m(2m)!$, then $a + 0$ is necessarily an analytic vector for $\tilde{A}$, hence is an analytic vector for X. . . . □

Lemma 4.4.6.3 (bis) Let E be a normed linear space. Suppose that A and $X_1, \ldots, X_n$ are (everywhere defined) linear operators on E such that

$$\|X_i a\|,\ \|X_i X_j a + X_j X_i a\| \leqslant \|Aa\|,\ \|\operatorname{ad}(X_{i_1}) \ldots \operatorname{ad}(X_{i_m})(A)a\| \leqslant m!\,\|Aa\| \qquad (\text{all } a \in E).$$

Let a be an element of E such that $\|A^m a\| \leqslant M^m(2m)!$ for some positive (finite) constant M – then a is an analytic vector for the family $\{X_i\}$ (i.e. a verifies an estimate $\|X_{i_1} \ldots X_{i_m} a\| \leqslant M^m m!$ for some positive (finite) constant M which is independent of the indices i_j).

[We shall leave the proof for the reader to work out.]

As above, let $A = \overline{U_\infty(1 - \Delta)}$ – then A is a positive self-adjoint operator $\geqslant I$ on E. [In passing, let us recall that the space E_∞ of C^∞ vectors for E is definable in terms of A; thus, according to Theorem 4.4.4.5, we have $E_\infty = \bigcap_1^\infty \operatorname{Dom}(A^m)$.] If c is a complex number, then the operator A^c is defined via the operational calculus for self-adjoint operators; it is useful to bear in mind that $\operatorname{Dom}(A^{c_1}) \subset \operatorname{Dom}(A^{c_2})$ if $\mathscr{R}(c_1) \geqslant \mathscr{R}(c_2)$.

Proof of Theorem 4.4.6.1 Let a be an analytic vector for $B(= A^{1/2})$ – then $a \in E_\infty$ and, moreover, verifies an estimate $\|A^m a\| \leqslant M^m(2m)!$ for some positive (finite) constant M. We claim that a is an analytic vector for the family $U_\infty(X_1), \ldots, U_\infty(X_n)$ of skew-symmetric operators on E_∞ (in the sense of Lemma 4.4.6.3 (bis)); in fact, on the basis of the estimates provided by Lemmas 4.4.4.8 and 4.4.4.9, this is clear (modulo Lemma 4.4.6.3 (bis), of course). Now, according to Example 2 above, a vector

$a \in E_\infty$ lies in E_ω iff there exists a positive (finite) constant M such that $\| U_\infty(X_{i_1}) \ldots U_\infty(X_{i_m})a \| \leqslant M^m m!$ for all indices i_j; this being so, it then follows that every analytic vector for B belongs to E_ω. Turning to the converse, observe that $A^m a$ $(a \in E_\infty)$ is the sum of $(n+1)^m$ terms of the form $U_\infty(X_{i_1}) \ldots U_\infty(X_{i_k})a$ with $k \leqslant 2m$; therefore, if $a \in E_\omega$, then $\| A^m a \| \leqslant (n+1)^m M^{2m}(2m)!$ for some $M < \infty$, from which it follows that a is an analytic vector for B. Hence the theorem. □

Corollary 4.4.6.4 Let $a \in E_\infty$ – then $a \in E_\omega$ iff there exists a positive (finite) constant M such that $\| U_\infty(\Delta)^m a \| \leqslant M^m(2m)!$ $(m = 1, 2, \ldots)$.

Remark Let U be a unitary representation of G on the Hilbert space E, E_ω the space of analytic vectors in E (for U); as we have seen in 4.4.5, E_ω is dense in E. This fact can also be deduced from Theorem 4.4.6.1; thus B is a self-adjoint operator on E and so, thanks to the Spectral Theorem, has a dense set of analytic vectors (see the Appendix to this number) – but these vectors all lie in E_ω, whence E_ω is dense in E as contended.

Suppose that G is represented on a function space, with Δ acting as a differential operator – one may then pose the function theoretic question: Characterize, in terms of the given function space, the space of analytic vectors for $(1 - \Delta)^{1/2}$. Before presenting some examples along these lines, we shall first prove a useful little lemma.

Let T be a self-adjoint operator on a Hilbert space E – then, according to the measure theoretic version of the Spectral Theorem, T is unitarily equivalent to a multiplication operator on a certain L^2 space. This being so, the motivation lying behind the following result becomes clear.

Lemma 4.4.6.5 Let ϕ be a measurable, real valued function on the measure space $(X; \mu)$; let $f \in L^2(X; \mu)$ – then the following conditions are equivalent:

(i) $e^{z\phi} f \in L^2(X; \mu)$ for $|z| < r$ $(z \in \mathbf{C})$;
(ii) $\phi^m f \in L^2(X; \mu)$ for all m and $\sum_0^\infty \| \phi^m f \| t^m/m! < \infty$ for $0 \leqslant t < r$.

Proof (ii) $\Rightarrow$ (i). We have the pointwise equality $e^{z\phi} f = \sum_0^\infty (\phi^m f) z^m/m!$ on X; so, if $|z| < r$, then the series on the right converges absolutely in L^2 and hence the L^2 limit must equal $e^{z\phi} f$.

(i) $\Rightarrow$ (ii). On the disk $|z| < 2r$, the function F defined by the rule $z \mapsto \int e^{\sqrt{-1} z\phi} |f|^2 \, d\mu$ is holomorphic (the integral being absolutely convergent there). Let $u(t) = \mathscr{R}(F(t)) = \int \cos(t\phi) |f|^2 \, d\mu$; since u is even and C^∞ near 0, we have $u'(0) = 0$, whence $\lim_{t \to 0} t^{-2}[u(0) - u(t)]$ exists; thus, by Fatou's Lemma, $\int \phi^2 |f|^2 \, d\mu < \infty$. By dominated convergence, it then follows that

$$u''(t) = -\int \cos(\phi t) \phi^2 |f|^2 \, d\mu$$

and hence, by induction, that $\phi^m f \in L^2(X; \mu)$ for all m. This being so, another application of dominated convergence yields the equality $F^{(2m)}(0) = (-1)^m \| \phi^m f \|^2$. Agreeing to write $M(s)$ for $\sup_{|z|=s} |F(z)|$, one then has, by Cauchy's inequalities, $\| \phi^m f \|^2 \leqslant M(s)(2m)! \, s^{-2m} \leqslant M(s)(m!)^2(s/2)^{-2m}$ for any $s < 2r$, which implies (ii).

Hence the lemma. □

Examples (1) Consider the left regular representation L of $\mathbf{R}$ on $L^2(\mathbf{R})$. To determine the analytic vectors for L, it is enough, thanks to Theorem 4.4.6.1, to determine the analytic vectors for the operator $B = [1 - (d/dt)^2]^{1/2}$. If $f \in L^2(\mathbf{R})$ and if $\hat{f}$ denotes the Fourier transform of f, then, since B corresponds to multiplication by $(1 + t^2)^{1/2}$ under Fourier transformation, we deduce from Lemma 4.4.6.5 that the set of analytic vectors for L is comprised of those $f \in L^2(\mathbf{R})$ having the property that $e^{s|t|}\hat{f}(t) \in L^2(\mathbf{R})$ for some $s > 0$ (s depending on f). This space of functions was characterized long ago by Paley and Wiener [1, Theorems I and IV]: A function $f \in L^2(\mathbf{R})$ is an analytic vector for translation iff it is the restriction to the real axis of a function F which is holomorphic in some strip $|\mathscr{I}(z)| < r$, $r > 0$ and which satisfies

$$\sup_{|y|<r} \int_{\mathbf{R}} |F(x + \sqrt{-1}\,y)|^2\, dx < \infty.$$

[A similar result can be established for any connected, simply connected nilpotent Lie group; cf. Goodman [6].] In order to avoid any confusion it should be pointed out that there exist square summable analytic functions f on $\mathbf{R}$, all of whose derivatives lie in L^2, but such that f is not an analytic vector for the regular representation; consider, for instance, the function f on $\mathbf{R}$ defined by the rule

$$f(x) = \sum_{n=1}^{\infty} 2^{-n}[(x + n)^2 + n^{-1}]^{-1} \qquad (x \in \mathbf{R}).$$

(2) Let us determine the analytic vectors for the irreducible unitary representation U (say) corresponding to $r = 1$ of the Heisenberg group G (see Example 1 following Theorem 4.4.4.10). If $\Delta = X^2 + Y^2 + Z^2$, then the operator $U_\infty(\Delta)$ differs by a constant from the 'harmonic oscillator' Hamiltonian $d^2/dt^2 - t^2$ (acting on $\mathscr{C}(\mathbf{R})$). It is a classical fact that this latter operator has a discrete spectrum $\lambda_n = -2n - 1$ of multiplicity one, the corresponding normalized eigenfunctions ϕ_n being the Hermite functions:

$$\phi_n(t) = (\sqrt{\pi}\,2^n n!)^{-1/2}(-1)^n e^{t^2/2}\left(\frac{d}{dt}\right)^n(e^{-t^2}) \qquad (t \in \mathbf{R}).$$

[These results are established, for instance, in N. Wiener's famous book 'The Fourier integral and its applications' – see in particular §6.] This being, so we then see by an elementary computation that the set of analytic vectors for U consists of those L^2 functions f whose Hermite coefficients $c_n = \int_{\mathbf{R}} f(t)\phi_n(t)dt$ verify $\sum_0^\infty e^{s\sqrt{n}}|c_n|^2 < \infty$ for some $s > 0$ (or, equivalently, $c_n = O(e^{-s\sqrt{n}})$ for some $s > 0$). These functions were characterized in terms of their analytic continuations by Hille (cf. Hille, 'Contributions to the theory of Hermitian series II: The representation problem', Trans. A.M.S., vol. 47 (1940), pp. 80–94): A function f in $L^2(\mathbf{R})$ is an analytic vector for the irreducible unitary representation U of the Heisenberg group G iff f is the restriction to the real axis of a function F which is holomorphic in a strip $|\mathscr{I}(z)| < r$ and satisfies $|F(x + \sqrt{-1}\,y)| \leqslant Me^{-s|x|}(|y| < r)$ for positive r, s, M. [In passing observe that in view of the relation $U^r_\infty(X^m Y^n Z^k) = r^{n+k}U^1_\infty(X^m Y^n Z^k)$ $(r \neq 0)$ each U^r has the *same* space of analytic vectors . . . .]

(3) Consider the '$\alpha t + \beta$' group G (see Example 2 following Theorem 4.4.4.10). Viewing G as the group of 2×2 matrices $\left(\begin{smallmatrix}\alpha & \beta\\ 0 & 1\end{smallmatrix}\right)$ $(\alpha > 0, \beta \in \mathbf{R})$ under matrix multiplication, it is clear that G is the semi-direct product of the subgroups $H_1 = \{\left(\begin{smallmatrix}\alpha & 0\\ 0 & 1\end{smallmatrix}\right) : \alpha > 0\}$ and $H_2 = \{\left(\begin{smallmatrix}1 & \beta\\ 0 & 1\end{smallmatrix}\right) : \beta \in \mathbf{R}\}$; of course $H_1 = \{\exp(tX) : t \in \mathbf{R}\}$, $H_2 = \{\exp(tY) : t \in \mathbf{R}\}$ where $X = \left(\begin{smallmatrix}1 & 0\\ 0 & 0\end{smallmatrix}\right)$, $Y = \left(\begin{smallmatrix}0 & 1\\ 0 & 0\end{smallmatrix}\right)$ and $[X, Y] = Y$. We now ask: What is the space of analytic vectors for the irreducible unitary representation $U^\pm$ of G? The answer is: A function $f \in L^2(\mathbf{R})$ is an analytic vector for the irreducible unitary representation $U^\pm$ of the '$\alpha t + \beta$'

group G iff it is the restriction to the real axis of a function F which is holomorphic in some strip $|\mathscr{I}(z)| < r, r > 0$, where F has the property that

$$\sup_{|y|<r} \int_{\mathbf{R}} |F(x + \sqrt{-1}y)|^2 \, dx < \infty$$

and satisfies the growth condition $|F(x + \sqrt{-1}y)| \leqslant Me^{-sx^2}(|y| < r)$ for positive s, M. [The point here is that G is the semi-direct product of the subgroups H_1, H_2; one may then utilize Remark 2 following the proof of Theorem 4.4.5.7.]

Let G be a simply connected Lie group with Lie algebra $\mathfrak{g}$. There exists a well-known relation between finite dimensional Banach representations of G and finite dimensional representations of $\mathfrak{g}$. It is natural to ask to what extent this connection persists in the infinite dimensional case; we shall restrict ourselves to the unitary situation.

Thus let E be a Hilbert space, $\mathfrak{D}$ a dense subspace of E on which there is defined a representation $U_{\mathfrak{D}}$ of $\mathfrak{G}$ with the property that $U_{\mathfrak{D}}(\sqrt{-1}X)$ is symmetric for every $X \in \mathfrak{g}$. The question is: Under what conditions does there exist a unitary representation $x \mapsto U(x)$ of G on E such that $\mathfrak{D} \subset E_\infty$ and $U_\infty(D)|\mathfrak{D} = U_{\mathfrak{D}}(D)$ for all $D \in \mathfrak{G}$?

In view of Example 4 following Proposition 4.4.4.4 supra one might think that a sufficient condition would be: $U_{\mathfrak{D}}(\sqrt{-1}X)$ is essentially self-adjoint ($X \in \mathfrak{g}$). However Nelson [1, p. 603] gave a counter-example refuting this conjecture.

Example (Nelson) Let M be the C^∞ manifold obtained from the closed band lying between two concentric squares by making the following identifications (cf. the figure infra): (1) The points A, B, C, D are to be identified, the points A′, B′, C′, D′ are to be omitted; (2) The line segments with the same label are to be identified (via translation).

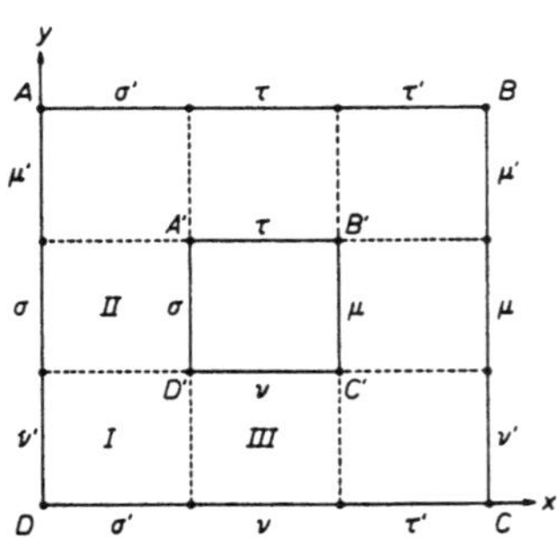

The Lebesgue measure in the plane defines a measure on M; let E be the associated L^2-space. Let $\mathfrak{g}$ be the (abelian) Lie algebra consisting of all operators of the form $x(\partial/\partial x) + y(\partial/\partial y)$ $(x, y \in \mathbf{R})$ with domain $\mathfrak{D} = C_c^\infty(M)$ – then a given $x(\partial/\partial x) + y(\partial/\partial y)$ is certainly skew-symmetric and is, moreover, essentially skew-adjoint .... But $\mathfrak{g}$ does not give rise to a group representation. Thus let X be the closure of $\partial/\partial x$, Y the closure of $\partial/\partial y$; suppose that f has its support in square I – then $e^X e^Y f$ has support in square II while $e^Y e^X f$ has support in square III, whence $e^X e^Y \neq e^Y e^X$ .... [In passing one should note that these considerations also serve to establish the following point: There exist two symmetric operators T and S on a Hilbert space E having a common invariant dense domain $\mathfrak{D}$ such that for all real t and s, $tT + sS$ is essentially self-adjoint and such that for all $a \in \mathfrak{D}$, $TSa = STa$ but such that the spectral resolutions of $\bar{T}$ and $\bar{S}$ do not commute.]

Let $X_1, \ldots, X_n$ be a basis for $\mathfrak{g}$; put $\Delta = X_1^2 + \ldots + X_n^2$. Agreeing to retain the above notations and assumptions, the positive result of Nelson [1] may be formulated as follows.

Theorem 4.4.6.6 (Nelson) If $U_{\mathfrak{D}}(\Delta)$ is essentially self-adjoint, then there is a unique unitary representation U of G on E such that $\mathfrak{D} \subset E_\infty$ and $U_\infty(D) | \mathfrak{D} = U_{\mathfrak{D}}(D)$ for all $D \in \mathfrak{G}$.

Before stating the main lemma on which the proof of Theorem 4.4.6.6 rests, we shall make a few general observations and introduce some notation. In the first place, let us agree to write X_i for $U_{\mathfrak{D}}(X_i)$, T for $U_{\mathfrak{D}}(\Delta)$ – then all the assumptions of Lemmas 4.4.4.8 and 4.4.4.9 are met here in the obvious way. The estimates of Lemma 4.4.4.9, in conjunction with the fact that T is essentially self-adjoint, imply that Lemma 4.4.4.7 is also available; this lemma tells us that the restrictions to $\bar{\mathfrak{D}} = \bigcap_1^\infty \mathrm{Dom}\,(\bar{T}^m)$ of the $\bar{X}_i$ leave $\bar{\mathfrak{D}}$ invariant (and hence provide an extension of the representation $U_{\mathfrak{D}}$ to $\bar{\mathfrak{D}}$, say $U_{\bar{\mathfrak{D}}}$) – moreover it is plain that the estimates on the X_i and T in $\mathfrak{D}$ hold with no change for the $\bar{X}_i | \bar{\mathfrak{D}}$ and $\bar{T} | \bar{\mathfrak{D}}$ in $\bar{\mathfrak{D}}$ (the M being $\sqrt{n/2}$, the M_m being the M^m of Lemma 4.4.4.9). [Thus, for instance, let $a \in \bar{\mathfrak{D}}$ – then $a \in \mathrm{Dom}\,(\bar{T})$ and so we can find $a_r \in \mathfrak{D}$ such that $a_r \to a$, $Ta_r \to \bar{T}a$, whence $\sum_i \| \bar{X}_i a \| = \sum_i \lim_r \| X_i a_r \| \leqslant \sqrt{n/2} \lim_r (\| a_r \| + \| Ta_r \|) = \sqrt{n/2}(\| a \| + \| \bar{T}a \|)$. Similar comments apply to the commutator estimates.]

Given $a \in \bar{\mathfrak{D}}$, put

$$| a |_m = \sum_{(m)} \| U_{\bar{\mathfrak{D}}}(X_{i_1}) \ldots U_{\bar{\mathfrak{D}}}(X_{i_m}) a \|$$

where $m = 0, 1, 2, \ldots$ ($| a |_0 = \| a \|$).

Lemma 4.4.6.7 Retain the assumptions of Theorem 4.4.6.6 – then there exists a subspace $\bar{\mathfrak{S}}$ of $\bar{\mathfrak{D}}$, which is dense in E, and a positive (finite) constant M such that

$$\sum_{m=0}^{\infty} | a |_m t^m / m! < \infty$$

for *all* $a \in \bar{\mathfrak{S}}$ so long as $| t | < M$.

Proof (Goodman) Let A be the closure of $U_{\mathfrak{D}}(1 - \Delta)$ – then, in particular, A is self-adjoint. Every analytic vector for A lies in $\bar{\mathfrak{D}}$; moreover, there exists a positive (finite) constant $\bar{M}$ such that

$$\| A^m a \| \leqslant m! \Rightarrow \| U_{\bar{\mathfrak{D}}}(X_{i_1}) \ldots U_{\bar{\mathfrak{D}}}(X_{i_m}) a \| \leqslant \bar{M}^m m! \qquad (m = 1, 2 \ldots),$$

the constant $\bar{M}$ being *independent* of a so long as a verifies the estimate on the left of the implication. [Here we have to bear in mind Lemma 4.4.6.2 (bis) together with the *proof* of Lemma 4.4.6.2 . . . .] This being so, let $\mathfrak{S}$ be the set of bounded vectors for A; by definition, then, a vector a lies in $\mathfrak{S}$ iff there exists a positive (finite) constant C_a such that $\| A^m a \| \leqslant C_a^m$ $(m = 0, 1, 2, \ldots)$ (it is tacitly assumed that $A^m a$ is defined for all m); of course, $\mathfrak{S}$ is contained in $\mathfrak{D}$ and, thanks to the Spectral Theorem, is dense in E. Given $a \in \mathfrak{S}$, let us agree to write Γ_a for $\sup_{m \geqslant 0} [C_a^m/m!]$ $(\leqslant e^{C_a})$ – then, for all $a \in \mathfrak{S}$ and all t, $0 \leqslant t < \infty$, we have

$$\sum_{m=0}^{\infty} |a|_m t^m/m! \leqslant \Gamma_a \sum_{m=0}^{\infty} \left|\frac{a}{\Gamma_a}\right|_m t^m/m!$$

$$\leqslant \Gamma_a \sum_{m=0}^{\infty} \bar{M}^m n^m t^m,$$

and so, $\bar{M}$ being independent of the a in $\mathfrak{S}$, the existence of the sought for constant M is now clear. □

Given $X \in \mathfrak{g}$, $X = \sum_i t_i X_i (t_i \in \mathbf{R})$, write $\| X \|_\infty = \sup_i |t_i|$; let $\| \mathrm{ad}\,(X) \|_\infty$ denote the norm of ad (X) as a linear transformation on the normed linear space $(\mathfrak{g}, \| \cdot \|_\infty)$.

Proof of Theorem 4.4.6.6 Let C be a positive constant $< 1/8$ such that $\theta(C) < M/4$ (here $\theta(t) = (e^{2t} - 1)/(2 - e^{2t})(t > 0)$; note that $\theta(t) > 2t$ and $\theta(t) \simeq 2t$ as $t \to 0$); we shall assume further that C is chosen sufficiently small so as to ensure that on the neighborhood $\mathcal{O}_C = \{X \in \mathfrak{g} : \| X \|_\infty < \theta(C)\}$ the exponential map $X \mapsto \exp X$ is regular and injective and, moreover, that the Campbell-Hausdorff formula holds there. With this agreement, for $X \in \mathcal{O}_C$, define $U(\exp X)$ to be the unitary operator $\exp \overline{(U_{\mathfrak{D}}(X))}$. [If X is any element of $\mathfrak{g}$ whatsoever, then, under the current hypotheses, $\sqrt{-1}\, \overline{U_{\mathfrak{D}}(X)}$ is self-adjoint; this follows at once from Lemma 4.4.6.7 and the lemma in the Appendix to this number.] The main step in the proof then consists in showing that if

$$\max (\| X \|_\infty, \| \mathrm{ad}\,(X) \|_\infty, \| Y \|_\infty, \| \mathrm{ad}\,(Y) \|_\infty) < C$$

and $a \in \mathfrak{S}$, then $U(\exp X)U(\exp Y)a = U(\exp X \exp Y)a$. Once this is established, the assertions of the theorem will follow quickly.

(1) The Campbell-Hausdorff formula is the formal identity $e^S e^T = e^{L(S,T)}$ where

$$L(S, T) = \sum_{n=1}^{\infty} \frac{(-1)^{n+1}}{n} \sum_{(p,q) \in \sigma_n} c(p, q) Z_{p,q}(S, T).$$

Here $\sigma_n = \{(p, q) : p, q \in \mathbf{N}^n, p_i + q_i > 0\}$, $c(p, q) = (p!\, q!\, |p + q|)^{-1}$, and

$$Z_{p,q}(S, T) = \begin{cases} \mathrm{ad}(S)^{p_1} \mathrm{ad}(T)^{q_1} \ldots \mathrm{ad}(S)^{p_n} \mathrm{ad}(T)^{q_n - 1}(T) & \text{if} \quad q_n > 0 \\ \mathrm{ad}(S)^{p_1} \mathrm{ad}(T)^{q_1} \ldots \mathrm{ad}(S)^{p_n - 1}(S) & \text{if} \quad q_n = 0. \end{cases}$$

[We are employing the usual notation: If $p \in \mathbf{N}^n$, then $p! = p_1! \ldots p_n!$,

$|p| = p_1 + \ldots + p_n$.] We shall now prove that the series $L(U_{\mathfrak{D}}(X), U_{\mathfrak{D}}(Y))a$ converges absolutely in E to $U_{\mathfrak{D}}(L(X, Y))a$ for all $a \in \bar{\mathfrak{S}}$ (X, Y as above). Thus, since $U_{\mathfrak{D}}$ is a representation, $Z_{p,q}(U_{\mathfrak{D}}(X), U_{\mathfrak{D}}(Y)) = U_{\mathfrak{D}}(Z_{p,q}(X, Y))$; because ad (.) is a bounded linear transformation on $\mathfrak{g}$, one has $\|Z_{p,q}(X, Y)\|_\infty < C^{|p|+|q|}$ and so, if $a \in \bar{\mathfrak{S}}$, then $\|Z_{p,q}(U_{\mathfrak{D}}(X), U_{\mathfrak{D}}(Y))a\| \leqslant C^{|p|+|q|}|a|_1$. But

$$\sum_{n=1}^{\infty} \frac{1}{n} \sum_{(p,q)\in\sigma_n} c(p, q)t^{|p|+|q|} \leqslant \sum_{n=1}^{\infty} (e^{2t} - 1)^n = \theta(t) \qquad (t > 0)$$

if $e^{2t} < 2$ – hence $\|L(X, Y)\|_\infty < \theta(C) < M/4$ and the series $L(U_{\mathfrak{D}}(X), U_{\mathfrak{D}}(Y))a$ converges absolutely in E to $U_{\mathfrak{D}}(L(X, Y))a$ ($a \in \bar{\mathfrak{S}}$), as contended. [Observe that, in general, if $\|Z\|_\infty \to 0$, then $\|U_{\mathfrak{D}}(Z)a\| \to 0$ ($a \in \bar{\mathfrak{D}}$).]

(2) Let X, Y be as above. It is clear that the series $\sum_0^\infty U_{\mathfrak{D}}(Y)^m a/m!$ is absolutely convergent for all $a \in \bar{\mathfrak{S}}$. [In fact

$$\sum_0^\infty \frac{\|U_{\mathfrak{D}}(Y)^m a\|}{m!} \leqslant \sum_0^\infty |a|_m \frac{\|Y\|_\infty^m}{m!} < \infty$$

since $\|Y\|_\infty < C < M$).] Therefore, in view of the Spectral Theorem, $U(\exp Y)a = \sum_0^\infty U_{\mathfrak{D}}(Y)^m a/m!$. Let $b = U(\exp Y)a$ ($a \in \bar{\mathfrak{S}}$) – then the series $\sum_0^\infty U_{\mathfrak{D}}(X)^m b/m!$ is absolutely convergent. [In fact

$$\begin{aligned}\textstyle\sum_0^\infty \|U_{\mathfrak{D}}(X)^m b\|/m! &\leqslant \textstyle\sum_0^\infty |b|_m \|X\|_\infty^m/m! \\ &\leqslant \textstyle\sum_{m=0}^\infty \sum_{k=0}^\infty |a|_{k+m} \|X\|_\infty^m \|Y\|_\infty^k/k!\,m! \\ &= \textstyle\sum_{m=0}^\infty \{\sum_{k=0}^m \binom{m}{k} \|X\|_\infty^{m-k} \|Y\|_\infty^k\} |a|_m/m! \\ &= \textstyle\sum_{m=0}^\infty |a|_m (\|X\|_\infty + \|Y\|_\infty)^m/m! < \infty\end{aligned}$$

since $\|X\|_\infty + \|Y\|_\infty < 2C < M$.] Therefore the series defining $U(\exp X)U(\exp Y)a$ ($a \in \bar{\mathfrak{S}}$) is absolutely convergent. On the other hand the series defining $\exp L(U_{\mathfrak{D}}(X), U_{\mathfrak{D}}(Y))a$ ($a \in \bar{\mathfrak{S}}$) is absolutely convergent by the estimates in (1). [In fact

$$\begin{aligned}\textstyle\sum_0^\infty \|L(U_{\mathfrak{D}}(X), U_{\mathfrak{D}}(Y))^m a\|/m! &\leqslant \textstyle\sum_0^\infty \|U_{\mathfrak{D}}(L(X, Y))^m a\|/m! \\ &\leqslant \textstyle\sum_0^\infty |a|_m \|L(X, Y)\|_\infty^m/m! < \infty\end{aligned}$$

since $\|L(X, Y)\|_\infty < \theta(C) < M$.] Owing to the Campbell-Hausdorff formula, then, we deduce that $U(\exp X)U(\exp Y)a = \exp L(U_{\mathfrak{D}}(X), U_{\mathfrak{D}}(Y))a = U(\exp L(X, Y))a$; but $\exp L(X, Y) = \exp X \exp Y$ – hence

$$U(\exp X)U(\exp Y)a = U(\exp X \exp Y)a \quad (a \in \bar{\mathfrak{S}}).$$

This shows that U defines a local unitary representation of G (notice that U is continuous since $\|\exp(\overline{U_{\mathfrak{D}}(X)})a - a\| \to 0$ as $X \to 0$ in $\mathfrak{g}$ ($a \in \bar{\mathfrak{S}}$) and $\bar{\mathfrak{S}}$ is dense in E); as G is simply connected, there is a unique extension of U to a unitary representation of G on E.

All the assertions of the present theorem are now obvious or immediate. □

The following corollary is clear provided we bear in mind Theorem 4.4.4.5.

Corollary 4.4.6.8 Let G be a simply connected Lie group with Lie algebra $\mathfrak{g}$; let $X_1, \ldots, X_n$ be a basis for $\mathfrak{g}$; let E be a Hilbert space, $\mathfrak{D}$ a dense subspace of E, and $U_{\mathfrak{D}}$ a representation of $\mathfrak{g}$ by skew-symmetric operators on $\mathfrak{D}$ – then there is a unitary representation U of G on E such that $\mathfrak{D}$ is the space of differentiable vectors for U with $U_\infty(X) = U_{\mathfrak{D}}(X)$ $(X \in \mathfrak{g})$ iff $U_{\mathfrak{D}}(\Delta) = \sum_i U_{\mathfrak{D}}(X_i^2)$ is essentially self-adjoint and $\mathfrak{D} = \bigcap_1^\infty \text{Dom}\,(\overline{U_{\mathfrak{D}}(\Delta)}^m)$.

Appendix The result to be established here relates the standard extension problem with the notion of analytic vector.

Lemma Let T be a closed symmetric operator on a Hilbert space E – then T is self-adjoint iff it has a dense set of analytic vectors.

[Of course the lemma remains true if symmetric is replaced by skew-symmetric, self-adjoint by skew-adjoint.]

Proof *Necessity* If T is self-adjoint and if $\mathsf{P}(.)$ is the resolution of the identity for T, then any vector in the range of the spectral projection $\mathsf{P}(\mathfrak{B})$, $\mathfrak{B}$ a bounded Borel subset of the plane, is an analytic vector for T; such vectors are dense in E by the Spectral Theorem.

Sufficiency For each $r > 0$, let E_r denote the set of vectors $a \in E$ such that $\sum_0^\infty \|T^m a\| r^m/m! < \infty$ – then E_r is a linear subspace of E. Put $E_0 = \bigcup_{r>0} E_r$ – then our assumption is that E_0 is dense in E.

(1) For $a \in E_r$, $|t| \leqslant r$, let $U(t)a = \sum_0^\infty (\sqrt{-1}T)^n a t^n/n!$; it is clear that this series converges absolutely in E. We shall now prove that if $a \in E_r$ and $|t_1| + |t_2| \leqslant r$, then $U(t_2)a \in E_{|t_1|}$ and $U(t_1)U(t_2)a = U(t_1 + t_2)a$. Thus let $b = U(t_2)a$, $b_i = \sum_{n=0}^i (\sqrt{-1}T)^n a t_2^n/n!$ – then b_i is in Dom (T) and $T(b_i - b_k) \to 0$ as $i, k \to \infty$ (since $a \in E_{|t_2|}$). Because T is a closed operator, this means that $b \in \text{Dom}\,(T)$ and $Tb_i \to Tb$; but $Tb_i \to U(t_2)Ta$, so that $TU(t_2)a = U(t_2)Ta$ – hence

$$\sum_{k=0}^{l} \frac{(\sqrt{-1}T)^k}{k!}\left(\sum_{n=0}^{\infty} \frac{(\sqrt{-1}T)^n}{n!} a t_2^n\right) t_1^k = \sum_{n=0}^{\infty} \sum_{k=0}^{l} \frac{(\sqrt{-1}T)^{k+n}}{k!\,n!} a t_1^k t_2^n$$

converges absolutely as $l \to \infty$ to

$$\sum_{n=0}^{\infty} \frac{(\sqrt{-1}T)^n}{n!} a(t_1 + t_2)^n = U(t_1 + t_2)a,$$

which is what we wanted to prove.

(2) The next step consists in showing that there is a one parameter unitary group $t \mapsto U(t)$ on E such that $U(t)a = \sum_0^\infty (\sqrt{-1}T)^n a t^n/n!$ so long as $a \in E_r$ and $|t| \leqslant r$. To this end, first observe that if $a \in E_r$ and $|t| < r$, then

$$\begin{aligned} \frac{d}{dt}(U(t)a, U(t)a) &= 2\mathscr{R} \lim_{h \to 0} (h^{-1}[U(t + h) - U(t)]a, U(t)a) \\ &= 2\mathscr{R}(\sqrt{-1}TU(t)a, U(t)a) = 0, \end{aligned}$$

T being symmetric – thus $\|U(t)a\| = \|a\|$ $(a \in E_r, |t| < r)$ and so

$$\sum_0^\infty \|T^m U(t)a\| r^m/m! = \sum_0^\infty \|U(t)T^m a\| r^m/m! = \sum_0^\infty \|T^m a\| r^m/m! < \infty,$$

provided a is in E_r and $|t| < r$. Conclusion: If $|t| < r$, then $U(t)E_r \subset E_r$.

Consider now an arbitrary real number t; let n_t be the integer such that $|t| < n_t r$; define $U(t)a$, for $a \in E_r$, by $U(t)a = (U(t/n_t))^{n_t}a$ – then $U(t)$ $(t \in \mathbf{R})$ is well-defined on E_r, takes E_r into itself, and verifies $U(t_1 + t_2)a = U(t_1)U(t_2)a$, $\|U(t)a\| = \|a\|$ $(a \in E_r)$. This may be done for each $r > 0$, in a consistent fashion, so that $U(t)$ is defined on E_0, takes E_0 into itself and verifies $U(t_1 + t_2)a = U(t_1)U(t_2)a$, $\|U(t)a\| = \|a\|$ $(a \in E_0)$. In view of the density of E_0 in E and the fact that $\|U(t)\| = 1$, the existence of the one parameter unitary group $t \mapsto U(t)$ referred to above is now clear.

(3) By construction, U is strongly continuous – hence, thanks to Stone's Theorem, there is a unique self-adjoint operator S on E such that $U(t) = \exp(\sqrt{-1}tS)$ $(t \in \mathbf{R})$. We intend to prove that $S = T$. Let T_0 be the restriction of T to E_0 – then $T_0 \subset S$. [Thus let $a \in E_0$ – then

$$\sqrt{-1}Sa = \lim_{h\to 0} h^{-1}[U(h) - 1]a = \sqrt{-1}Ta.]$$

Let S_0 be an arbitrary self-adjoint extension of T_0; put $U_0(t) = \exp(\sqrt{-1}tS_0)$ $(t \in \mathbf{R})$ – then, for $a \in E_r$, $|t| \leqslant r$, we have $U(t)a = \sum_0^\infty(\sqrt{-1}T)^n at^n/n! = \sum_0^\infty(\sqrt{-1}S_0)^n at^n/n!$ (since T and S_0 agree on E_r). But this series is absolutely convergent, hence is equal to $U_0(t)a$ (by the Spectral Theorem). Conclusion: If $|t| \leqslant r$, then $U(t)$ and $U_0(t)$ agree on E_r. Repeating the argument used in the definition of $U(t)$ on E_0 for all real t, we conclude that it is actually true that $U(t)$ and $U_0(t)$ agree on E_0, therefore on E itself. Consequently $S = S_0$ and so T_0 admits a unique self-adjoint extension; in view of well-known generalities, it now follows that the closure of T_0 is self-adjoint. Because T is a closed symmetric extension of T_0, T is self-adjoint (in fact, $S = S_0 = \bar{T}_0 = T$).

This establishes the sufficiency and, in so doing, completes the proof of the lemma. □

4.4.7 The Paley – Wiener Space

Let T be a self-adjoint operator on a Hilbert space E; let P be the projection valued measure on $\mathbf{R}$ associated with T by the Spectral Theorem (so that $T = \int_{\mathbf{R}} t d\mathsf{P}(t)$) – then, for $a \in E_n(T) = \mathsf{P}([-n, n])(E)$ $(n \in \mathbf{N})$, the function $z \mapsto e^{\sqrt{-1}zT}a$ is an entire function of z $(z \in \mathbf{C})$ with

$$(*) \quad |(e^{\sqrt{-1}zT}a, a)| \leqslant Me^{N|\mathscr{I}(z)|} \qquad (M, N \text{ positive constants}).$$

Definition The *Paley-Wiener space* for a self-adjoint operator T on a Hilbert space E is the space $E_{PW}(T) = \bigcup_1^\infty E_n(T)$.

Example If $T = 1/\sqrt{-1}\,(d/dt)$ in its usual self-adjoint realization on $E = L^2(\mathbf{R})$, then $E_n(T) = \{f : \hat{f}(t) = 0 \text{ for } |t| > n\}$ ($\hat{f}$ the Fourier transform of f) – otherwise said, $E_{PW}(T)$ is the classical Paley-Wiener space of functions on the line whose Fourier transforms have compact support.

Let T be a self-adjoint operator on E, $E_{PW}(T)$ the Paley-Wiener space for T – then it is clear that $E_{PW}(T)$ is a subspace of the space of analytic vectors for T (in general proper, as simple examples show). Given $a \in E$ it is natural to ask: When does $a \in E_{PW}(T)$? Here is a sufficient condition.

Proposition 4.4.7.1 Let $a \in E$; suppose that the function f, $t \mapsto (e^{\sqrt{-1}tT}a, a)$ $(t \in \mathbf{R})$, extends to an entire function verifying the estimate $(*)$ above – then $a \in E_{PW}(T)$

Proof By Bochner's Theorem the function f is the Fourier-Stieltjes transform of a positive finite Borel measure μ_f on $\mathbf{R}$; by the Hörmander-Schwartz generalization of the classical Paley-Wiener Theorem, μ_f is supported on $[-N, N]$ (cf. Hörmander [1]). Now T is unitarily equivalent to multiplication by a measurable function ϕ on a measure space $L^2(X; \mu)$ and under such a unitary equivalence our vector a becomes an element of $L^2(X; \mu)$ (ψ, say) with $\mu_f(\mathfrak{B}) = \int_{\phi^{-1}(\mathfrak{B})} |\psi(x)|^2 d\mu(x)$ for every Borel subset $\mathfrak{B} \subset \mathbf{R}$. Therefore ψ is supported on $\{x \in X : |\phi(x)| \leqslant N\}$ which implies that $a \in E_{PW}(T)$. □

Consider now a pair (G, K), K a compact subgroup of G – we shall assume that K is large in G (see 4.5.1). Let U be an irreducible unitary representation of G on E, E_ω the space of analytic vectors in E (for U), $E_K = \sum_{\delta \in \hat{K}} E_\omega(\delta)$ (cf. 4.4.5) – then, since K is large, $E_\omega(\delta) = E(\delta)$ for every $\delta \in \hat{K}$ and so $E_K = \sum_{\delta \in \hat{K}} E(\delta)$ (algebraic sum, of course). We know that the space E_ω coincides with the space of analytic vectors for a certain self-adjoint operator on E (Theorem 4.4.6.1). What can be said about the space E_K? Let us assume that the Laplacian Δ has been chosen so as to commute with K. [Since K is compact, there exists an $\mathrm{Ad}_G(K)$-invariant inner product on $\mathfrak{g}$; let $\{X_i\}$ be an orthonormal basis for $\mathfrak{g}$ with respect to this inner product – then $\Delta = \sum_i X_i^2$ commutes with K (cf. Lemma 4.4.5.20).] Denoting the closure of the operator $U_\infty(1 - \Delta)$ by A, we have:

Theorem 4.4.7.2 (Goodman) Let U be an irreducible unitary representation of G on E – then E_K is equal to the Paley-Wiener space for A.

Proof (1) Nelson and Stinespring have shown that A^{-1} is a completely continuous operator (see Lemma 2 in the Appendix to this number) – therefore A has a discrete spectrum $\{\lambda_n\}$, $\lambda_n \to \infty$, with finite multiplicity for each eigenvalue. Consequently the space $E_{PW}(A)$ consists of the algebraic linear span of the eigenvectors for A.

(2) The eigenvectors for A all lie in E_∞ (since $E_\infty = \bigcap_1^\infty \mathrm{Dom}\,(A^m)$; cf. Theorem 4.4.4.5). Because $U(k)U_\infty(\Delta) = U_\infty(\Delta)U(k)$ $(k \in K)$, the eigenspaces of A are all K-finite and so every $a \in E_{PW}(A)$ is K-finite. Conversely, if $a \in E$, let $a_n = \mathsf{P}_n a$ where P_n is the projection of E onto the λ_n-eigenspace – then $U(K)a_m$ is orthogonal to $U(K)a_n$ if $m \neq n$. If in addition a is K-finite, then $a_n = 0$ except for finitely many n, which implies that $a \in E_{PW}(A)$.

Hence the theorem. □

Under the preceding set of circumstances, we shall frequently refer to E_K as the *Paley-Wiener space* (for U).

Remark Suppose that G is a connected semi-simple Lie group with finite center, K a maximal compact subgroup (or, more generally, suppose that we have a pair (G, K) verifying the assumptions laid down in number 1.1.5) – then, in the notations of Example 3 following Proposition 4.4.4.4, $\Delta = \omega_- + \omega_+$ is a Laplacian which commutes with K. Because $U_\infty(-\omega_- + \omega_+)$

is a scalar in any irreducible unitary representation (Corollary 4.4.1.6), it follows that in this case we may use ω_- in place of $\Delta = \omega_- + \omega_+$ in Theorems 4.4.6.1, 4.4.7.2.

Example Suppose that $G = \mathbf{SU}(1, 1)$ ($\sim \mathbf{SL}(2, \mathbf{R})$) – thus the elements x of G are 2×2 complex matrices $\begin{pmatrix} \alpha & \beta \\ \bar{\beta} & \bar{\alpha} \end{pmatrix}$ with $|\alpha|^2 - |\beta|^2 = 1$. G acts on the unit disc $D = \{z \in \mathbf{C} : |z| \leqslant 1\}$ via the prescription $z \mapsto x \cdot z = (\alpha z + \beta)/(\bar{\beta} z + \bar{\alpha})$ leaving the boundary ∂D invariant. The subgroup K of G consisting of all matrices

$$k(\theta) = \begin{pmatrix} e^{\sqrt{-1}\theta/2} & 0 \\ 0 & e^{-\sqrt{-1}\theta/2} \end{pmatrix} \qquad (\theta \in \mathbf{R})$$

is a maximal compact subgroup; evidently $k(\theta)$ operates on D by the rotation $z \mapsto e^{\sqrt{-1}\theta} z$. Aside from the trivial one dimensional representation, the irreducible unitary representations of G, determined by Bargmann [1], fall into three distinct classes (with one exception in the principal P-series . . .).

(a) *The Principal P-Series.* The representations in this series can be realized on $L^2(\partial D)$; since ω_- is represented by $d^2/d\theta^2$ (∂D being parametrized as $\{e^{\sqrt{-1}\theta}\}$), the space of analytic vectors for such a representation consists of those functions f in $L^2(\partial D)$ whose Fourier series $\sum_n a_n e^{\sqrt{-1}n\theta}$ has the property that its coefficients verify the relation $a_n = O(e^{-\epsilon|n|})$ for some $\epsilon > 0$. [To put it a different way, the analytic vectors for a representation in the principal P-series constitute the space of functions on ∂D which are holomorphic on a neighborhood of ∂D.] On the other hand, the Paley-Wiener space is comprised of the trigonometric polynomials . . . .

(b) *The Complementary Series.* The representation space E^σ for a member of the complementary series consists of all distributions on ∂D whose Fourier coefficients a_n satisfy the condition $\sum_n \lambda_n(\sigma) |a_n|^2 < \infty$ where

$$\lambda_n(\sigma) = \frac{\Gamma(2^{-1} + \sigma)}{\Gamma(2^{-1} - \sigma)} \cdot \frac{\Gamma(|n| + 2^{-1} - \sigma)}{\Gamma(|n| + 2^{-1} + \sigma)} \qquad (0 < \sigma < 2^{-1}).$$

Since $\lambda_n(\sigma) \sim [\Gamma(2^{-1} + \sigma)/\Gamma(2^{-1} - \sigma)]\, |n|^{-2\sigma}$ (Stirling's formula . . .), the space of analytic vectors for a given representation in the complementary series is the same as that for the representations in the principal P-series (and likewise for the Paley-Wiener space).

(c) *The Discrete Series.* For a suitable choice of l, a given member of the discrete series can be realized on the space of all functions

$$f(z) = \sum_{n=0}^{\infty} a_n z^n$$

holomorphic in the interior of D such that

$$\sum_{n=0}^{\infty} \binom{l-1+n}{n}^{-1} |a_n|^2 < \infty \qquad (l \in \mathbf{N}^+).$$

In this realization, K acts by rotations. Therefore the space of analytic vectors for such a representation consists of the functions holomorphic on a neighborhood of the closed unit disc whereas the polynomials in z make up the Paley-Wiener space.

Appendix Our objective is to prove Lemma 2 below; in turn, its proof rests on the following result.

Let G be a Lie group countable at infinity.

Lemma 1 Let $X_1, \ldots, X_n$ be a basis for the Lie algebra $\mathfrak{g}$ of G; let $D = \sum_i X_i^2 + X$ $(X \in \mathfrak{g})$; let U be a bounded Banach representation of G on E – then the closure T of $U_\infty(D)$ is the infinitesimal generator (in the sense of Hille-Yosida) of a bounded strongly continuous one parameter semi-group of operators on E. Moreover there exists, for each $t > 0$, a function f_t in $L^1(G)$ such that the resolvent $R_t = (t - T)^{-1}$ is equal to $U(f_t)$ – in fact, if m is an integer $> n/4$, then there exists $f_{m,t} \in L^1(G) \cap L^2(G)$ such that $R_t^m = U(f_{m,t})$.

The proof of this lemma, taken in all its details, is lengthy and highly technical – we shall refer the reader to Nelson and Stinespring [1, pp. 552–553] for a discussion and the necessary additional references.

A Lie group G is said to be *liminaire* if, under every irreducible unitary representation of G, $L^1(G)$ is represented by completely continuous operators. Here is a criterion for G to be liminaire.

Lemma 2 Let U be a unitary representation of G on a Hilbert space E. If, for some element $D \in \mathfrak{G}$, $U_\infty(D)$ has an inverse which is contained in a completely continuous operator, then $U(f)$ is completely continuous for all $f \in L^1(G)$; conversely, if $U(f)$ is completely continuous for all $f \in L^1(G)$, then the closure of $U_\infty(1 - \Delta)$ $(\Delta = \sum_i X_i^2)$ possesses a completely continuous inverse.

Proof Let T be the completely continuous operator containing the inverse of $U_\infty(D)$; if $f \in C_c^\infty(G)$, then $U(f)$ is completely continuous since $U(f) = TU(Df)$ (Proposition 4.4.1.2); because $C_c^\infty(G)$ is dense in $L^1(G)$ and the completely continuous operators are closed in the uniform operator topology, the direct assertion of the lemma is now clear. As for the converse, it is an immediate consequence of Lemma 1. □

We shall see in 4.5.7 that if G admits a large compact subgroup K, then G is liminaire – hence the application of Lemma 2 in the proof of Theorem 4.4.7.2 is legitimate.

4.5 Large Compact Subgroups

4.5.1 The Algebras $C_{c,\delta}(G)$, $I_{c,\delta}(G)$

Let G be a locally compact unimodular group countable at infinity, K a compact subgroup of G. Let $\hat{K}$ denote the set of all equivalence classes of finite dimensional irreducible representations of K; for each $\delta \in \hat{K}$, let ξ_δ denote the character of δ, $d(\delta)$ the degree of δ and $\chi_\delta = d(\delta)\xi_\delta$.

Let E be a Banach space.

Definition A Banach representation U of G on E is said to be *K-finite* if $U(\bar{\chi}_\delta)$ is of finite rank for every $\delta \in \hat{K}$.

Thus U is K-finite iff each $\delta \in \hat{K}$ occurs with finite multiplicity in $U|K$.

A K-finite Banach representation U of G on E is FDS. [Of course this means that we must show that its integrated form is FDS. The fact that E_0 is dense in E is clear. To see that E_0^* is dense in E_s^*, fix $a \in E$ – then $\langle a, {}^t\mathrm{P}(\delta)\mathbf{z}\rangle = 0$ (all $\delta \in \hat{K}, \mathbf{z} \in E^*$) $\Rightarrow \mathrm{P}(\delta)a = 0$ (all $\delta \in \hat{K}$) $\Rightarrow a = 0$ (Lemma 4.4.2.4). [As usual (4.4.2), $\mathrm{P}(\delta) = U(\bar{\chi}_\delta)$, $E(\delta) = \mathrm{P}(\delta)E$ ($\delta \in \hat{K}$).] Hence $\sum_{\delta \in \hat{K}} {}^t\mathrm{P}(\delta)E^*$ is weakly dense in E^* which implies that E_0^* is too.] Therefore a K-finite topologically irreducible Banach representation of G is necessarily topologically completely irreducible (Proposition 4.2.1.5).

Example (Fell) There exists a locally compact group G, a compact subgroup K of G and an irreducible unitary representation U of G such that some elements of $\hat{K}$ occur with positive finite multiplicity and others with infinite multiplicity in $U|K$. [Thus let $\mathbf{T}$ be the circle group, $\mathbf{Z}$ the additive group of integers and $\mathbf{T}^{\mathbf{Z}} = \prod_{n \in \mathbf{Z}} \mathbf{T}$; let G be the semi-direct product of $\mathbf{T}^{\mathbf{Z}}$ and $\mathbf{Z}$, with $\mathbf{Z}$ operating on $\mathbf{T}^{\mathbf{Z}}$ via the prescription $(m, t) \mapsto m \cdot t$, where $(m \cdot t)_n = t_{n-m}$ ($m, n \in \mathbf{Z}$; $t \in \mathbf{T}^{\mathbf{Z}}$); let ξ_0 be the character $t \mapsto t_0$ of $\mathbf{T}^{\mathbf{Z}}$, $U = U^{\xi_0}$ the corresponding unitarily induced representation of G (which can be shown to be irreducible . . .); let K be the compact subgroup of $\mathbf{T}^{\mathbf{Z}}$ consisting of those t such that $t_m = t_n$ whenever $m \neq 0$, $n \neq 0$ – these entities possess the requisite properties.]

Remark Let U be a K-finite Banach representation of G, V a Banach representation of G Naimark equivalent to U – then V is, evidently, K-finite too. This being so, suppose in addition that U is topologically irreducible – then V is also topologically irreducible. [Thus, in the notations of 4.5.4 infra, the Naimark equivalence of U and V automatically entails the algebraic equivalence of U_K and V_K; because U is topologically irreducible, U_K is algebraically irreducible (Lemma 4.5.4.3) – hence V_K is algebraically irreducible and so V itself is topologically irreducible (Lemma 4.5.4.3 again).] Consequently the pathology illustrated in Berezin's Example (4.2.2) is not present here.

Definition (Mackey) The compact subgroup K is said to be *large* if, for each $\delta \in \hat{K}$, there exists an integer $m(\delta) \geqslant 1$ such that δ occurs no more than $m(\delta)$ times in every topologically completely irreducible (TCI) Banach representation of G.

Thus if K is large, then every TCI Banach representation U of G is K-finite – in fact the ranks of the various $U(\bar{\chi}_\delta)$ ($\delta \in \hat{K}$) are even *uniformly* bounded (by $m(\delta)d(\delta)$).

Examples of groups which admit large compact subgroups include the 'motion' groups and the connected semi-simple Lie groups (with finite center) – cf. 4.5.2. [In any event, it is clear that a compact group is large in itself; it is equally evident that any compact subgroup of an abelian Lie group is large.]

A large compact subgroup K of G is said to be *uniformly large* provided there exists a positive integer M such that $m(\delta)/d(\delta) \leqslant M$ (all $\delta \in \hat{K}$); the pairs (G, K) (with K large in G) which are of interest to us will satisfy this stronger requirement (cf. 4.5.2). [If K is uniformly large in G, then the ranks of the various $U(\bar{\chi}_\delta)$ are bounded by $Md(\delta)^2$ ($\delta \in \hat{K}$); the significance of this condition will become apparent when we take up the theory of characters on G.]

In order to give conditions which will ensure that a given compact subgroup of G is large, some algebraic preparation is needed.

The associative algebra $\mathfrak{gl}(n, \mathbf{C})$ of n by n complex matrices is of dimension n^2 – hence, if r is an integer $> n^2$, then one has

$$(*) \sum_{\sigma} (\operatorname{sgn} \sigma) X_{\sigma(1)} \cdots X_{\sigma(r)} = 0$$

for $X_1, \ldots, X_r \in \mathfrak{gl}(n, \mathbf{C})$ (σ ranging over all permutations of the set $(1, \ldots, r)$). Let $r(n)$ be the smallest integer r such that the identity $(*)$ prevails in $\mathfrak{gl}(n, \mathbf{C})$.

Lemma 4.5.1.1 (Kaplansky) Retain the above notations and assumptions – then $r(n) \geqslant r(n-1) + 2$ $(n = 1, 2, \ldots)$.

Proof Let $t = r(n-1) - 1$; let $X_1, \ldots, X_t \in \mathfrak{gl}(n-1, \mathbf{C})$ be such that $Y = \sum_{\sigma} (\operatorname{sgn} \sigma) X_{\sigma(1)} \cdots X_{\sigma(t)} \neq 0$. Put $Y = (y_{ij})$ – then it is clear that there exist k and l such that $y_{kl} \neq 0$. Let X'_i (respectively Y') be the matrix obtained from X_i (respectively Y) by adding a row and a column of zeros. For convenience in writing, let X'_{t+1}(respectively X'_{t+2}) denote the matrix in $\mathfrak{gl}(n, \mathbf{C})$ all of whose elements are zero except at the ln^{th} – spot (respectively nn^{th} – spot) which is to be 1. In the sum

$$\sum_{\tau} (\operatorname{sgn} \tau) X'_{\tau(1)} X'_{\tau(2)} \cdots X'_{\tau(t+2)}$$

all the terms in which X'_{t+2} is not in the last place are zero (since $X'_{t+2} X'_j = 0$ for $j < t+2$); those for which X'_{t+2} is in the last place are also zero provided that X'_{t+1} does not immediately precede X'_{t+2} (since $X'_j X'_{t+2} = 0$ for $j < t+1$). Hence we are left with $Y' X'_{t+1} X'_{t+2} = Y' X'_{t+1}$. Because $y_{kl} \neq 0$, this matrix is non-zero – therefore $r(n) > t + 2 = r(n-1) + 1$. □

Let A be an associative algebra over $\mathbf{C}$.

Definition A set $\mathfrak{R}$ of representations of A is said to be *complete* if for every non-zero x in A there exists $U \in \mathfrak{R}$ such that $U(x) \neq 0$.

Proposition 4.5.1.2 (Kaplansky-Godement) If there exists a complete set $\mathfrak{R}$ of representations of A whose dimensions are $\leqslant n$ (n a fixed integer), then every topologically completely irreducible Banach representation of A is of dimension $\leqslant n$.

Proof Let $r(n) = r$ and suppose that U is a TCI Banach representation of A on E of dimension $> n$. Let F be a subspace of E of dimension $n + 1$ – then Lemma 4.5.1.1 implies that $r(n+1) > r(n)$ and so there exist operators $T_1, \ldots, T_r$ in $\operatorname{Hom}_{\mathbf{C}}(F, F)$ such that $[T_1, \ldots, T_r] \neq 0$. Here we have put

$$[T_1, \ldots, T_r] = \sum_{\sigma} (\operatorname{sgn} \sigma) T_{\sigma(1)} \cdots T_{\sigma(r)} \qquad (T_i \in \operatorname{Hom}_{\mathbf{C}}(F, F)).$$

Extend the T_i to bounded linear transformations $\tilde{T}_i$ on E. Owing to the fact that U is TCI, there exists a net $\{x_k\}$ in A such that $U(x_k) \to \tilde{T}_1$ in $\mathfrak{L}_s(E; E)$ – thus $0 \neq [\tilde{T}_1, \ldots, \tilde{T}_r] = \lim [U(x_k), \tilde{T}_2, \ldots, \tilde{T}_r]$ and so we can find $x_1 \in A$ with the property that $[U(x_1), \tilde{T}_2, \ldots, \tilde{T}_r] \neq 0$. Iteration then leads to the conclusion that there exists $x_1, \ldots, x_r$ in A such that $[U(x_1), \ldots, U(x_r)] = U([x_1, \ldots, x_r]) \neq 0$, a contradiction (the image of $[x_1, \ldots, x_r]$ under any element in $\mathfrak{R}$ being zero). □

Let G be a locally compact unimodular group countable at infinity, K a compact subgroup of G. For $f \in C_c(G)$, $\delta \in \hat{K}$ put

$$_\delta f(x) = \int_K \chi_\delta(k) f(kx) dk, \; f_\delta(x) = \int_K \chi_\delta(k^{-1}) f(xk) dk \qquad (x \in G),$$

where dk is normalized Haar measure on K (hence $_\delta f = \bar{\chi}_\delta * f$, $f_\delta = f * \chi_\delta$). Let $^\delta C_c(G)$ (respectively $C_c^\delta(G)$) denote the subspace of $C_c(G)$ consisting of all functions of the form $_\delta f$ (respectively f_δ) ($f \in C_c(G)$); since $_\delta(f*g) = {_\delta f}*g$ ($f, g \in C_c(G)$), $^\delta C_c(G)$ is a right ideal in $C_c(G)$; likewise $C_c^\delta(G)$ is a left ideal in $C_c(G)$. Let $C_{c,\delta}(G) = \{f \in C_c(G) : f = f_\delta = f_{\check{\delta}}\}$ ($\check{\delta}$ the class in $\hat{K}$ contragredient to δ) – then $C_{c,\delta}(G)$ is a subalgebra of $C_c(G)$ and the map $f \mapsto \bar{\chi}_\delta * f * \bar{\chi}_\delta$ (convolution of measures in $M_c(G)$) is a projection of $C_c(G)$ onto $C_{c,\delta}(G)$.

Remarks (1) Equip $C_c(G)$ with the inductive limit topology; let L and R be the canonical continuous representations of G on $C_c(G)$ by left and right translation, respectively. For any $\delta \in \hat{K}$, define the continuous projections $\mathsf{P}_L(\delta)$ and $\mathsf{P}_R(\delta)$ as usual – thus, for $f \in C_c(G)$, $\mathsf{P}_L(\delta)f = \bar{\chi}_\delta * f$, $\mathsf{P}_R(\delta) = f * \chi_\delta$ and so the space $C_{c,\delta}(G)$ consists of those elements f in $C_c(G)$ which transform to the left under δ and to the right under $\check{\delta}$. Plainly, then, $C_{c,\delta}(G)$ is a closed subspace of $C_c(G)$ and the map $f \mapsto \bar{\chi}_\delta * f * \bar{\chi}_\delta$ is a continuous projection of $C_c(G)$ onto $C_{c,\delta}(G)$.

(2) Of course one need not confine attention to just $C_c(G)$ – other spaces can (and will) be considered as well (e.g. $C_c^\infty(G)$ if G is a Lie group).

Let U be a Banach representation of G on E – put, as usual, $\mathsf{P}(\delta) = U(\bar{\chi}_\delta)$, $E(\delta) = \mathsf{P}(\delta)E$ ($\delta \in \hat{K}$). Plainly

$$\mathsf{P}(\delta)U(f)\mathsf{P}(\delta) = U(g), \; g = \bar{\chi}_\delta * f * \bar{\chi}_\delta,$$

for all $f \in C_c(G)$; hence $E(\delta)$ is stable under the $U(f)$ ($f \in C_{c,\delta}(G)$) and so, letting $U_\delta(f)$ denote the restriction of $U(f)$ to $E(\delta)$, one obtains a representation $f \mapsto U_\delta(f)$ of $C_{c,\delta}(G)$ on $E(\delta)$.

Lemma 4.5.1.3 Let U be a TCI Banach representation of G on E – then, for every $\delta \in \hat{K}$, the representation $f \mapsto U_\delta(f)$ of $C_{c,\delta}(G)$ on $E(\delta)$ is TCI too.

Proof Let T be a bounded linear operator on $E(\delta)$; extend it to E by the condition that $\tilde{T}a = T\mathsf{P}(\delta)a$ for all $a \in E$. Now, since U is TCI, there exist functions $f_i \in C_c(G)$ such that the $U(f_i)$ converge simply on

E to $\tilde{T}$; setting $g_i = \bar{\chi}_\delta * f_i * \bar{\chi}_\delta$, it is then clear that $g_i \in C_{c,\delta}(G)$ and that $U_\delta(g_i) \to T$ simply on $E(\delta)$. □

Definition A set $\mathfrak{R}$ of Banach representations of G is said to be *complete* if the corresponding set of representations of the algebra $C_c(G)$ is complete.

The next result admits important applications – see 4.5.2.

Theorem 4.5.1.4 (Godement) Let $\mathfrak{R}$ be a complete set of Banach representations of G. For a given $\delta \in \hat{K}$ and a given finite integer m, assume that δ is contained at most m times in every member of $\mathfrak{R}$ – then δ is contained at most m times in every TCI Banach representation of G.

Proof For $f \in C_{c,\delta}(G)$ it is clear that $U(f) = 0$ iff $U_\delta(f) = 0$ – hence the representations $f \mapsto U_\delta(f)$ associated with the various members of $\mathfrak{R}$ form a complete set of representations of $C_{c,\delta}(G)$ whose dimensions are $\leqslant n = md(\delta)$. Proposition 4.5.1.2 then implies that every TCI Banach representation of $C_{c,\delta}(G)$ has dimension $\leqslant n$. Therefore the theorem follows from Lemma 4.5.1.3. □

Evidently Theorem 4.5.1.4 may be used to determine when a compact subgroup K of G is large – thus one need only exhibit a complete set $\mathfrak{R}$ of Banach representations of G in which $\delta \in \hat{K}$ occurs at most $m(\delta)$ times (say). For certain pairs (G, K) this is easy to do – cf. 4.5.2.

Proposition 4.5.1.5 Suppose that K is a compact subgroup of G. Let n be a positive integer – the following conditions are equivalent:

(i) Every TCI Banach representation of $C_{c,\delta}(G)$ is of finite dimension $\leqslant n$;

(ii) There are enough representations of $C_{c,\delta}(G)$ of finite dimension $\leqslant n$ to separate the points of $C_{c,\delta}(G)$;

(iii) If U is any TCI Banach representation of G on E, then $\dim(E(\delta)) \leqslant n$;

(iv) There are enough Banach representations of G satisfying $\dim(E(\delta)) \leqslant n$ to separate the points of $C_c(G)$.

Proof We have seen that (ii) ⇒ (i) (Proposition 4.5.1.2) and that (i) ⇒ (iii). Now (iii) ⇒ (iv) in virtue of the Gelfand-Raikov Theorem while the implication (iv) ⇒ (ii) is clear. □

Consider a pair (G, K), K a large compact subgroup of G – then, for each $\delta \in \hat{K}$, there exists an integer $m(\delta) \geqslant 1$ such that the rank of $U(\bar{\chi}_\delta)$ cannot exceed $m(\delta)d(\delta)$ for any TCI Banach representation U of G. As has been noted earlier, this implies, in particular, that every TCI Banach representation of G is FDS; consequently (4.2.1) Naimark relatedness is an equivalence relation when confined to the TCI Banach representations of G.

Notation In what follows $\bar{\bar{G}}$ will denote the set of Naimark equivalence classes of TCI Banach representations of G.

[Tacitly this agreement presupposes that G admits a large compact subgroup.]

One of the central problems in the Harmonic Analysis on G is to 'classify' $\bar{\bar{G}}$; in view of Lemma 4.5.1.3, it stands to reason that the study of this problem might well be facilitated by a systematic investigation of the TCI Banach representations of the various algebras $C_{c,\delta}(G)$ (one ignores, for the time being, any topology on $C_{c,\delta}(G)$) – here is some additional evidence along these lines.

Proposition 4.5.1.6 Let G be a locally compact unimodular group, K a large compact subgroup of G; let U^1 and U^2 be TCI Banach representations of G and suppose that δ occurs in both U^1 and U^2 – then U^1 is Naimark equivalent to U^2 iff U_δ^1 is algebraically equivalent to U_δ^2.

Proof Let Ker (U^1) denote the kernel of the lift of U^1 to $C_c(G)$; similarly for U^2 – then, since both U_δ^1 and U_δ^2 are finite dimensional algebraically irreducible representations of $C_{c,\delta}(G)$, it is enough, in view of Theorem 4.2.1.6, to show that Ker (U^1) = Ker (U^2) iff Ker (U_δ^1) = Ker (U_δ^2). Evidently

$$\mathrm{Ker}\,(U_\delta^1) = \mathrm{Ker}\,(U^1) \cap C_{c,\delta}(G),$$

$$\mathrm{Ker}\,(U^1) = \{f \in C_c(G): \bar{\chi}_\delta * g * f * h * \bar{\chi}_\delta \in \mathrm{Ker}\,(U_\delta^1), \text{ all } g, h \in C_c(G)\};$$

similarly for U^2 – our contention is now clear. ☐

Let $\bar{\bar{G}}(\delta)$ denote the subset of $\bar{\bar{G}}$ consisting of those classes in which a given $\delta \in \hat{K}$ occurs (hence $\bar{\bar{G}} = \bigcup_\delta \bar{\bar{G}}(\delta)$). In view of the preceding proposition, the map $\bar{\bar{U}} \mapsto \bar{\bar{U}}_\delta$ of $\bar{\bar{G}}(\delta)$ into the set $\bar{\bar{C}}_{c,\delta}(G)$ of Naimark equivalence classes of TCI Banach representations of $C_{c,\delta}(G)$ is *injective* (the range of this map will be considered in due course (Vol. II, number 6.1.1); notice that the elements of $\bar{\bar{C}}_{c,\delta}(G)$ are finite dimensional – in fact are uniformly bounded in dimension (Proposition 4.5.1.5 (i))).

Let K be a compact subgroup of G; we shall now examine the structure of the algebras $C_{c,\delta}(G)$ ($\delta \in \hat{K}$) in more detail.

In what follows we shall denote by $\mathfrak{I}_c(G)$ the set of functions $f \in C_c(G)$ which are K-central – thus $\mathfrak{I}_c(G)$ is a subalgebra of $C_c(G)$ and the operator $f \mapsto f_K, f_K(x) = \int_K f(kxk^{-1})dk$ $(x \in G)$, is a projection of $C_c(G)$ onto $\mathfrak{I}_c(G)$. One has various simple rules of computation:

(1) $(f_K * g)_K = f_K * g_K = (f * g_K)_K \qquad (f, g \in C_c(G));$

(2) $(\bar{\chi}_\delta * f)_K = \bar{\chi}_\delta * f_K, (f * \bar{\chi}_\delta)_K = f_K * \bar{\chi}_\delta \qquad (f \in C_c(G));$

(3) $\phi * f = f * \phi \qquad (f \in \mathfrak{Z}_c(G),\ \phi \in C(K))$.

Put $I_{c,\delta}(G) = \mathfrak{Z}_c(G) \cap C_{c,\delta}(G)$ ($\delta \in \hat{K}$) – then, according to what has just been said, $I_{c,\delta}(G)$ is a subalgebra of $C_c(G)$ and $f \mapsto f_K$ maps $C_{c,\delta}(G)$ onto $I_{c,\delta}(G)$. If $f \in \mathfrak{Z}_c(G)$ and if $\bar{\chi}_\delta * f = f$, then also $f = f * \bar{\chi}_\delta$; this means that the map $f \mapsto \bar{\chi}_\delta * f$ is a projection of $\mathfrak{Z}_c(G)$ onto $I_{c,\delta}(G)$; evidently $I_{c,\delta}(G)$ is a closed subspace (in the inductive limit topology) of $C_{c,\delta}(G)$. Note too, that if $f \mapsto f^*$ is the usual involution in $C_c(G)$ (thus $f^*(x) = \overline{f(x^{-1})}$, $x \in G$), then $(f_K)^* = (f^*)_K$ and so $I_{c,\delta}(G)$ is a *-subalgebra of $C_c(G)$.

Example Consider the subalgebra $I_{c,\delta}(G)$ corresponding to the class δ of the trivial one dimensional representation ($k \mapsto 1$) of K – then it is clear that the elements f of $I_{c,\delta}(G)$ are biinvariant under K, i.e. $f(k_1 x k_2) = f(x)$ for all $k_1, k_2 \in K$; $x \in G$. In this special case, the δ will be dropped from the notation; we shall write $I_c(G)$ for $I_{c,\delta}(G)$.

The various algebras $I_{c,\delta}(G)$ ($\delta \in \hat{K}$) will play an important role in the sequel – a first indication of their significance is the following result.

Proposition 4.5.1.7 Suppose that K is a large compact subgroup of G. Let U be a TCI Banach representation of G on E – then the set of operators $U_\delta(f)$, $f \in I_{c,\delta}(G)$, is the centralizer of the representation $k \mapsto U_\delta(k)$ of K on $E(\delta)$($U_\delta(k) = U(k) \mid E(\delta)$, $k \in K$).

Proof Plainly: $f \in I_{c,\delta}(G) \Rightarrow U_\delta(f) U_\delta(k) = U_\delta(k) U_\delta(f) \quad (k \in K)$. Conversely let $T \in \mathrm{Hom}_{\mathbf{C}}(E(\delta), E(\delta))$ and assume that T commutes with every $U_\delta(k)$ ($k \in K$). Since the representation $f \mapsto U_\delta(f)$ of $C_{c,\delta}(G)$ on $E(\delta)$ is finite dimensional and irreducible (Lemma 4.5.1.3), it must be the case, thanks to Burnside's Theorem, that $T = U_\delta(f)$ for some $f \in C_{c,\delta}(G)$. But

$$T = \int_K U_\delta(k) T U_\delta(k^{-1}) dk = \int_K U_\delta(k) U_\delta(f) U_\delta(k^{-1}) dk = U_\delta(f_K) \ldots . \quad \square$$

Remark Retain the notations and assumptions of Proposition 4.5.1.7 – then the representation $k \mapsto U_\delta(k)$ of K on $E(\delta)$ decomposes into m (say) equivalent irreducible representations and so it is clear that its centralizer is isomorphic with the algebra $\mathfrak{gl}(m, \mathbf{C})$ of $m \times m$ matrices; hence, by Proposition 4.5.1.7, there exists an isomorphism $U_\delta(f) \mapsto \mu_\delta(f)$ of the algebra of operators $\{U_\delta(f) : f \in I_{c,\delta}(G)\}$ onto $\mathfrak{gl}(m, \mathbf{C})$. Plainly the assignment $f \mapsto \mu_\delta(f)$ is an m-dimensional irreducible representation of $I_{c,\delta}(G)$ with $\mathrm{tr}(U_\delta(f)) = d(\delta)\mathrm{tr}(\mu_\delta(f))$ ($f \in I_{c,\delta}(G)$).

Let $I(\delta) = C(K) * \bar{\chi}_\delta$; of course $I(\delta)$ is a *-algebra of $d(\delta) \times d(\delta)$ matrices over $\mathbf{C}$($\delta \in \hat{K}$).

Proposition 4.5.1.8 (Dieudonné) Let K be a compact subgroup of G – then the bilinear map $(\phi, f) \mapsto \phi * f$ of $I(\delta) \times I_{c,\delta}(G)$ into $C_c(G)$, when lifted to a linear map of $I(\delta) \otimes_{\mathbf{C}} I_{c,\delta}(G)$ into $C_c(G)$, establishes a *-algebra isomorphism of the tensor product *-algebra $I(\delta) \otimes_{\mathbf{C}} I_{c,\delta}(G)$ with $C_{c,\delta}(G)$.

The proof of this important structural fact rests on the following classical lemma from the theory of rings (cf. Jacobson [4, p. 118]).

Lemma 4.5.1.9 Let $\mathfrak{A}$ be an associative algebra over $\mathbf{C}$ with a unit, $\mathfrak{S}$ a finite dimensional simple subalgebra with the same unit as $\mathfrak{A}$ and with center $\mathbf{C}$. If $\mathfrak{S}'$ is the centralizer of $\mathfrak{S}$ in $\mathfrak{A}$, then the canonical homomorphism $\mathfrak{S} \otimes_{\mathbf{C}} \mathfrak{S}' \to \mathfrak{A}$ is an isomorphism.

Proof of Proposition 4.5.1.8 Let $\mathfrak{A}$ denote the subspace $I(\delta) + C_{c,\delta}(G)$ of $M_c(G)$ – then $\mathfrak{A}$ is a subalgebra of $M_c(G)$ with unit $\bar{\chi}_\delta$. Suppose first that K is nowhere dense in G (which will be the case in the majority of our applications) so that the sum $I(\delta) + C_{c,\delta}(G)$ is direct; clearly the centralizer of $\mathfrak{S} = I(\delta)$ in $\mathfrak{A}$ is the subalgebra $\mathfrak{S}' = \mathbf{C}\bar{\chi}_\delta + I_{c,\delta}(G)$. Therefore, due to what was said above, the canonical homomorphism $\mathfrak{S} \otimes_{\mathbf{C}} \mathfrak{S}' \to \mathfrak{A}$ is an isomorphism – however the image of $I(\delta) \otimes_{\mathbf{C}} I_{c,\delta}(G)$ under this homomorphism is contained in $C_{c,\delta}(G)$ and hence, as the sum $I(\delta) + C_{c,\delta}(G)$ is direct, we deduce that $I(\delta) \otimes_{\mathbf{C}} I_{c,\delta}(G) \sim C_{c,\delta}(G)$. On the other hand if K is open in G, then the continuous functions on K can be identified with elements in $C_c(G)$ – in particular $I(\delta) \subset C_{c,\delta}(G)$. In this case the algebra $\mathfrak{A} = C_{c,\delta}(G)$ admits a unit, viz. $\bar{\chi}_\delta$; furthermore, $\bar{\chi}_\delta$ is a unit for the simple central subalgebra $\mathfrak{S} = I(\delta)$; as the centralizer of $\mathfrak{S}$ in $\mathfrak{A}$ is $I_{c,\delta}(G)$, we see once again that $I(\delta) \otimes_{\mathbf{C}} I_{c,\delta}(G) \sim C_{c,\delta}(G)$. □

There is a simple but important consequence which may be drawn from Proposition 4.5.1.8, namely: There exists a natural one-to-one correspondence between the equivalence classes of finite dimensional algebraically irreducible representations of $I_{c,\delta}(G)$ and those of $C_{c,\delta}(G)$ – thus if μ_δ is a finite dimensional algebraically irreducible representation of $I_{c,\delta}(G)$ and if $I_{d(\delta)}$ is the identity representation of the full matrix algebra $I(\delta) = C(K)*\bar{\chi}_\delta$ on $\mathbf{C}^{d(\delta)}$, then $I_{d(\delta)} \otimes \mu_\delta$ determines an algebraically irreducible representation of $I(\delta) \otimes_{\mathbf{C}} I_{c,\delta}(G) \sim C_{c,\delta}(G)$ and conversely.

Let us assume that K is a large compact subgroup of G – then there exists an integer $m(\delta) \geqslant 1$ such that δ occurs no more than $m(\delta)$ times in every TCI Banach representation U of $G (\delta \in \hat{K})$. According to Proposition 4.5.1.5, these assumptions entail that every TCI Banach representation of $C_{c,\delta}(G)$ is of finite dimension $\leqslant m(\delta)d(\delta)$ and so, in view of the above discussion, every finite dimensional algebraically irreducible representation of $I_{c,\delta}(G)$ is bounded in dimension by $m(\delta)$. [Notice, in particular, that if U is a TCI Banach representation of G on E in which δ occurs $m \leqslant m(\delta)$ times, U_δ the corresponding algebraically irreducible representation of $C_{c,\delta}(G)$ on $E(\delta)$ (Lemma 4.5.1.3), then the algebraically irreducible representation of the algebra $I_{c,\delta}(G)$ canonically associated with U_δ may be taken as the rule which assigns to each $U_\delta(f)$ ($f \in I_{c,\delta}(G)$) the m by m matrix $\mu_\delta(f)$ in the centralizer of the representation $k \mapsto U_\delta(k)$ of K on $E(\delta)$ (cf. the Remark supra).] On the other hand the discussion following Proposition 4.5.1.6 provides one with an injection of $\bar{G}(\delta)$

into $\bar{C}_{c,\delta}(G)$, i.e. with an injection of $\bar{G}(\delta)$ into $\bar{I}_{c,\delta}(G)$; with a view toward classifying the elements of $\bar{G}(\delta)$, it is evident that a characterization of the range of this last injection would be highly desirable – this question will be considered later on (Theorem 6.1.1.11).

Proposition 4.5.1.10 Suppose that K is a compact subgroup of G. Let m be a positive integer – the following conditions are equivalent:

(i) Every TCI Banach representation of $I_{c,\delta}(G)$ is of finite dimension $\leqslant m$;

(ii) There are enough representations of $I_{c,\delta}(G)$ of finite dimension $\leqslant m$ to separate the points of $I_{c,\delta}(G)$;

(iii) If U is any TCI Banach representation of G on E, then δ occurs no more than m times in $U|K$;

(iv) There are enough Banach representations of G in which δ occurs no more than m times to separate the points of $C_c(G)$.

Proof Bearing in mind the above discussion, the proof of the present proposition is seen to be the same as that of Proposition 4.5.1.5. □

Let ρ be a semi-norm on G (4.2.2); one defines, in the obvious way, subalgebras $C_{\rho,\delta}(G)$, $I_{\rho,\delta}(G)$ of the Banach algebra $C_\rho(G)$ ($\delta \in \hat{K}$).

We shall conclude this number with a result on the density of certain subalgebras which will be needed later during the discussion of spherical functions on G.

Lemma 4.5.1.11 Let K be a compact subgroup of G; let ρ be a semi-norm on G – then, for every $\delta \in \hat{K}$, the set

$$\{\delta_k * f : k \in K, f \in I_{\rho,\delta}(G)\}$$

is total in $C_{\rho,\delta}(G)$ (δ_k the unit mass at the point $k \in K$).

In order to prove this we need some notations and a lemma.

Notations (1) If $\mu \in M(G)$, define a measure $\mu_K \in M(G)$ by the rule: $\mu_K(f) = \mu(f_K)$ ($f \in C_c(G)$).

(2) Given a continuous function ϕ on G and a measure $\mu \in M_c(G)$, set

$$\phi(\mu) = \mu(\phi) = \int_G \phi(x)d\mu(x);$$

if $d\mu(x) = f(x)d_G(x)$ for some $f \in C_c(G)$, we write $\phi(f)$ or $f(\phi)$.

Lemma 4.5.1.12 Let ϕ be a function or a measure on G satisfying $\bar{\chi}_\delta * \phi = \phi * \bar{\chi}_\delta = \phi$ for some $\delta \in \hat{K}$ and such that $(\delta_{k_1} * \phi * \delta_{k_2})_K = 0$ (all $k_1, k_2 \in K$) – then $\phi = 0$.

Proof Consider the vector space $\mathscr{S}$ spanned by the measures $\mu * \phi * \nu$ with μ, ν arbitrary elements in $M(K)$ – then $\mathscr{S}$ is finite dimensional. [Since $\bar{\chi}_\delta * \phi = \phi * \bar{\chi}_\delta$, it is enough to consider those $\mu, \nu \in M(K)$ verifying $\mu * \bar{\chi}_\delta = \mu$ and $\bar{\chi}_\delta * \nu = \nu$, respectively.] Define a continuous representation U of $K \times K$

on $\mathscr{S}$ by the rule: $U(k_1, k_2)s = \delta_{k_1}*s*\delta_{k_2^{-1}}$ $(s \in \mathscr{S}; k_1, k_2 \in K)$. A simple calculation shows that

$$\int_K \int_K U(k_1, k_2)\bar{\chi}_\delta(k_1)\chi_\delta(k_2)dk_1dk_2 = I \qquad (I \text{ the identity operator on } \mathscr{S}).$$

On the other hand, in the well-known way, the multiplicity of the irreducible representation $\delta \times \check{\delta}$ of $K \times K$ in U is

$$d(\delta)^{-2} \int_K \int_K \operatorname{tr}(U(k_1, k_2))\bar{\chi}_\delta(k_1)\chi_\delta(k_2)dk_1dk_2 = \dim(\mathscr{S})d(\delta)^{-2}.$$

Therefore, if $\mathscr{S} \neq \{0\}$, then the representation $k \mapsto U(k, k)$ of K on $\mathscr{S}$ contains the trivial one dimensional representation of K; in turn this implies that $\mathscr{S}$ contains a non-vanishing element s such that $s_K = s$. But the fact that $(\delta_{k_1}*\phi*\delta_{k_2})_K = 0 \Rightarrow (\mu*\phi*\nu)_K = 0$ $(\mu, \nu \in M(K)) \Rightarrow s_K = 0$, a contradiction. □

Proof of Lemma 4.5.1.11 Let $f \in I_{p,\delta}(G)$, $k \in K$ – then $\delta_k*f \in C_{p,\delta}(G)$ (for $\bar{\chi}_\delta*\delta_k*f = \delta_k*\bar{\chi}_\delta*f = \delta_k*f = \delta_k*f*\bar{\chi}_\delta$). In order to prove the lemma it suffices to show: If ϕ is a continuous linear form on $C_{p,\delta}(G)$ such that $\phi(\delta_k*f) = 0$ for all $f \in I_{p,\delta}(G)$, $k \in K$, then $\phi = 0$. To show that such a ϕ necessarily vanishes, extend it to a continuous linear form on $C_p(G)$ (this is possible, $C_{p,\delta}(G)$ being a closed subspace of $C_p(G)$). Moreover, it may be supposed that such an extension (which still will be denoted by ϕ) verifies the relation

$$\phi(\bar{\chi}_\delta*f*\bar{\chi}_\delta) = \phi(f)$$

for all $f \in C_c(G)$. Now, noting that $(\bar{\chi}_\delta*f)_K = \bar{\chi}_\delta*f_K \in I_{p,\delta}(G)$ $(f \in C_c(G))$, we have $\phi(\delta_k*f_K) = \phi(\delta_k*\bar{\chi}_\delta*f_K*\bar{\chi}_\delta) = 0$; so, for all $f \in C_c(G)$, $\phi(\delta_{k_1}*f_K*\delta_{k_2}) = (\delta_{k_1^{-1}}*\phi*\delta_{k_2^{-1}})(f_K) = (\delta_{k_1^{-1}}*\phi*\delta_{k_2^{-1}})_K(f) = 0$ $(k_1, k_2 \in K)$ – hence $(\delta_{k_1}*\phi*\delta_{k_2})_K = 0$ (all $k_1, k_2 \in K$). Because $\chi_\delta*\phi = \phi*\chi_\delta = \phi$, it then follows from Lemma 4.5.1.12 (with δ replaced by $\check{\delta}$), that $\phi = 0$, as we wished to prove. □

4.5.2 Groups with Large Compact Subgroups

In this number we shall give some typical examples of pairs (G, K) which have the property that K is a large compact subgroup of G. The actual verification of largeness requires, however, some tools from the next chapter (viz., induced representations); accordingly we shall start off by presenting, without proof, two main classes of non-compact locally compact groups which admit large compact subgroups – the reader may then return to the details when convenient.

Examples (1) We shall say that G is a *motion group* provided that G is the semi-direct product of a closed normal abelian subgroup H and a compact subgroup K. [For instance take for G the Euclidean group of real n-space, i.e. take $H = \mathbf{R}^n$, $K = \mathbf{O}(n)$ and let G denote the group generated by the rotations in K and the translations on H.] According to Theorem 4.5.2.1 infra, the compact subgroup K of a motion group G is uniformly large in G – in fact a given $\delta \in \hat{K}$ occurs no more than $d(\delta)$ times in the restriction to K of any TCI Banach representation of G (so that we may take $m(\delta) = d(\delta)\ldots$).

(2) Let G be a connected semi-simple Lie group with finite center, K a maximal compact subgroup of G – then a given $\delta \in \hat{K}$ occurs no more than $d(\delta)$ times in the restriction to K of any TCI Banach representation of G

(Theorem 4.5.2.11 below) (hence K is uniformly large in G with $m(\delta) = d(\delta)$, $\delta \in \hat{K}$). [An analogous statement is valid for the reductive pairs (G, K) of 1.1.5.]•

Theorem 4.5.2.1 (Godement) Suppose that H and K are two closed subgroups of G with $K \cap H = \{1\}$ and such that $G = KH$, H being abelian and K compact – then K is a large compact subgroup of G. In fact a given $\delta \in \hat{K}$ occurs no more than $d(\delta)$ times in any TCI Banach representation of G.

To prove this we shall need a lemma. In what follows, by a *character* of H (or of any topological group) we shall understand a continuous homomorphism $\hat{\xi}$ of H into $\mathbf{C}^*$ (that is, we are not requesting that $\hat{\xi}$ be unitary). Consider now the family $\{U^{\hat{\xi}}\}$ of Banach representations of G which are continuously induced by the characters $\hat{\xi}$ of H (cf. number 5.3.1). [In the case at hand, it is clear that the representation space ${}^{\hat{\xi}}C(G)$ for $U^{\hat{\xi}}$ may be identified (by restriction) with the space $C(K)$; ${}^{\hat{\xi}}C(G)$ then acquires the structure of a Banach space upon writing $\|a\| = \sup_{k \in K} |a(k)|$ $(a \in {}^{\hat{\xi}}C(G))$.] We intend to prove that the collection $\{U^{\hat{\xi}}\}$ constitutes a complete set of representations for G.

Lemma 4.5.2.2 The Banach representations $U^{\hat{\xi}}$, $\hat{\xi}$ running through the characters of H, form a complete system of representations of G (i.e. of $C_c(G)$).

Proof Let $f \in C_c(G)$ be such that $U^{\hat{\xi}}(f) = 0$ for every character $\hat{\xi}$ of H – we are to show that $f = 0$. Normalize the Haar measures on G, K and H so that $d_G(x) = dk\, d_H(\xi)$ if $x = k\xi$ (dk denoting normalized Haar measure on K). Let a be an element in ${}^{\hat{\xi}}C(G)$ (thus a verifies the relation $a(x\xi) = \langle \xi^{-1}, \hat{\xi} \rangle a(x)$ $(\xi \in H, x \in G)$) – then, for $k \in K$, we have

$$\begin{aligned} U^{\hat{\xi}}(f)a(k) &= \int_G f(x)a(x^{-1}k)d_G(x) = \int_G f(kx^{-1})a(x)d_G(x) \\ &= \int_K \int_H f(k\xi\tilde{k}^{-1}) \langle \xi, \hat{\xi} \rangle\, a(\tilde{k})d_H(\xi)d\tilde{k} \\ &\Rightarrow \int_H f(k\xi\tilde{k}) \langle \xi, \hat{\xi} \rangle d_H(\xi) = 0 \qquad (\text{all } k, \tilde{k} \in K). \end{aligned}$$

But $\hat{\xi}$ is an arbitrary character of the abelian group H – therefore $f(k\xi\tilde{k}) = 0$ for all $k, \tilde{k} \in K, \xi \in H \Rightarrow f = 0$, as desired. □

Proof of Theorem 4.5.2.1 For any character $\hat{\xi}$ of H, the functions in the representation space ${}^{\hat{\xi}}C(G)$ for $U^{\hat{\xi}}$ may be identified (by restriction) with the space $C(K)$. Under this identification, the representation $k \mapsto U^{\hat{\xi}}(k)$ of K reduces to the left regular representation of K on $C(K)$, hence, by well-known facts about compact groups, contains δ exactly $d(\delta)$ times. The present theorem is then a consequence of Lemma 4.5.2.2 and Theorem 4.5.1.4. □

Remark Bearing in mind Lemma 5.5.1.4, the reader will have no difficulty in convincing himself that Theorem 4.5.2.1 is valid as it stands when the assumption that $K \bigcap H = \{1\}$ is dropped.

Example (Fell) A locally compact group G is said to be of *FDS type* provided that all of its TCI Banach representations are FDS; such a G need not possess a large compact subgroup. Thus, for instance, let us consider the group G comprised of all pairs $\langle (x, y), t \rangle$ ($(x, y) \in \mathbf{R}^2$, $t \in \mathbf{R}$) with multiplication given by

$$\langle (x, y), t \rangle \cdot \langle (\tilde{x}, \tilde{y}), \tilde{t} \rangle$$
$$= \langle (x, y) + (\tilde{x} \cos(2\pi t) - \tilde{y} \sin(2\pi t), \tilde{x} \sin(2\pi t) + \tilde{y} \cos(2\pi t)), t + \tilde{t} \rangle.$$

Otherwise said, G is the simply connected covering group of the proper Euclidean group G_0 of the plane; evidently $G_0 = G/Z$, Z the center of G (in passing notice that $Z = \{\langle 0, n \rangle : n \in \mathbf{Z}\}$). We claim that G is of FDS type. To see this, let U be an arbitrary TCI Banach representation of G on E (say) – then it will be enough to produce a $\mu \in M_c(G)$ such that $U(\mu)$ is of non-zero finite rank. Let ζ_U be the central character of U (so that, by definition, $U(z) = \zeta_U(z)I$ (all $z \in Z$)) – then there certainly is a continuous character ζ (say) of G which agrees with the central character of U on Z. This being so, let U_0 be the TCI Banach representation of G_0 defined by the prescription

$$U_0(\pi(x)) = \zeta(x^{-1})U(x) \qquad (x \in G),$$

$\pi : G \to G_0$ the canonical map. Because G_0 has a large compact subgroup (cf. supra), there is a $\mu_0 \in M_c(G_0)$ such that $U_0(\mu_0)$ is of non-zero finite rank; let ω be a compact subset of G such that $\pi(\omega)$ carries μ_0; choose a μ in $M_c(G)$ whose support is contained in ω and such that $\pi(\zeta d\mu) = \mu_0$ – then $U(\mu) = U_0(\mu_0)$, whence $U(\mu)$ is of non-zero finite rank. On the other hand, the only compact subgroup of G is $\langle (0, 0) \rangle$ . . . .

Let G be a connected semi-simple Lie group which, moreover, admits a *faithful finite dimensional representation* – then G has finite center (compare 3.1.1).

Proposition 4.5.2.3 Suppose that G is as above – then G admits a complete set of finite dimensional irreducible representations.

Proof Let $\mathscr{S}$ be the linear span of the set of all coefficients of finite dimensional (continuous) irreducible representations of G, i.e. the linear span of all functions on G of the form $x \mapsto \langle U(x)a, \mathbf{z} \rangle$ ($x \in G$) where U is a finite dimensional (continuous) irreducible representation of G, $a \in E$, $\mathbf{z} \in E^\vee$ (E the representation space of U). Since every finite dimensional representation of G is semi-simple, it is clear that $\mathscr{S}$ contains all coefficients of all finite dimensional representations of G. Furthermore, the hypotheses imply that $\mathscr{S}$ is closed under multiplication (tensor product of representations) and under complex conjugation (complex conjugate representations), and that $\mathscr{S}$ separates the points of G. Owing to the Stone-Weierstrass Theorem, then, $\mathscr{S}$ is uniformly on compacta dense in $C(G)$. Therefore, if f is a non-zero element in $C_c(G)$, there exists some $s \in \mathscr{S}$ such that $\int_G f(x)s(x)d_G(x) \neq 0$ or, in other words, there exists an irreducible finite dimensional representation U of G such that $U(f) \neq 0$. □

Theorem 4.5.2.4 Let G be a connected semi-simple Lie group with a faithful finite dimensional representation, K a maximal compact subgroup of G. For a given $\delta \in \hat{K}$ and a given finite integer m, assume that δ is contained at most m times in every finite dimensional (continuous) irreducible representation of G – then δ is contained at most m times in every TCI Banach representation of G.

Proof This is an immediate consequence of Proposition 4.5.2.3 and Theorem 4.5.1.4. □

Let $G = KA_{\mathfrak{p}}N^+$ be an Iwasawa decomposition for G; put $S = A_{\mathfrak{p}}N^+$ – then S is a closed connected solvable subgroup of G. Agreeing to use the notations introduced in number 5.5.1, let us consider the family of (Banach) representations U^{L^ν} of G continuously induced by the characters L^ν of S – we claim that the U^{L^ν} form a complete system of representations for $C_c(G)$. To get at this, we need a lemma.

Lemma 4.5.2.5 Given a finite dimensional irreducible representation U of G, we can find a character L^ν of S such that U is a subrepresentation of U^{L^ν}.

Proof Let E be the representation space for U. Consider the contragredient representation $U^\vee$ of G on the dual space $E^\vee$ of E. Since S is a connected solvable group, Lie's Theorem (see the Appendix to this number) tells us that there exists a non-zero vector $\mathbf{z} \in E^\vee$ and a (complex valued) linear function ν on $\mathfrak{a}_{\mathfrak{p}}$ such that $U^\vee(s)\mathbf{z} = L^{\nu+\rho}(s^{-1})\mathbf{z}$ for all $s \in S$. Assign to each $a \in E$ the continuous function ϕ_a defined by $\phi_a(x) = \langle U(x^{-1})a, \mathbf{z}\rangle$ $(x \in G)$; plainly $\phi_a(xs) = L^{\nu+\rho}(s^{-1})\phi_a(x)$ $(s \in S, x \in G)$ – therefore the mapping ϕ, $a \mapsto \phi_a$ $(a \in E)$, is a linear, one-to-one (U is irreducible) mapping of E onto a finite dimensional subspace of the representation space for the continuously induced representation U^{L^ν} of G. Since ϕ intertwines U and U^{L^ν}, the lemma follows. □

Remark Notice that S, itself, does not play a crucial role in Lemma 4.5.2.5. In fact it is obvious that the lemma remains valid when S is replaced by any closed connected solvable subgroup H of G whose product with K is all of G (of course one then utilizes the continuously induced (Banach) representations U^ξ (ξ a character of H)).

Proposition 4.5.2.6 (Godement) The Banach representations U^{L^ν} of G (continuously) induced by the characters L^ν of S form a complete system of representations of G (i.e. of $C_c(G)$).

Proof This is clear in view of Proposition 4.5.2.3 and the preceding lemma. □

Proposition 4.5.2.6 combined with the argument given during the proof of Theorem 4.5.2.1 yields:

Theorem 4.5.2.7 (Godement – Harish-Chandra) Let G be a connected semi-simple Lie group with a faithful finite dimensional representation, K a maximal compact subgroup of G – then each δ ($\delta \in \hat{K}$) occurs at most $d(\delta)$ times in every TCI Banach representation of G.

Suppose, in particular, that G is a connected complex semi-simple Lie group (recall (Theorem 3.1.1.3) that such a G admits a faithful finite dimensional representation) – then, due to the fact that the minimal parabolic subgroups (i.e. the Borel subgroups) of G are closed connected and solvable, it is possible to refine Theorem 4.5.2.7 to a certain extent. Thus let $P = MA_{\mathfrak{p}}N^+$ be a Borel subgroup of G; of course, in the case at hand, M is a maximal torus in K. Claim: If $m(\delta)$ ($\delta \in \hat{K}$) denotes the smallest integer such that every irreducible representation of M occurs at most $m(\delta)$ times in the restriction of δ to M, then δ occurs at most $m(\delta)$ times in every TCI Banach representation of G. Thus, in the first place, the (Banach) representations U^{ξ} of G continuously induced by the characters ξ of P form a complete system of representations of $C_c(G)$ (see the Remark following Lemma 4.5.2.5); furthermore, the Frobenius Reciprocity Theorem implies that the multiplicity of δ in $U^{\xi}|K$ is equal to the multiplicity of $\hat{\xi}|M$ in δ – the claim now follows from the completeness of the U^{ξ} and Theorem 4.5.1.4.

Example Let $G = \mathbf{SL}(2, \mathbf{C})$, $K = \mathbf{SU}(2)$ – then every $\delta \in \hat{K}$ occurs at most once in every TCI Banach representation of G.

Let G be a connected semi-simple Lie group, K a maximal compact subgroup of G; let $N(G)$ denote the intersection of the kernels of all finite dimensional continuous representations of G.

Proposition 4.5.2.8 (Harish-Chandra) Let U be an irreducible representation of K – then U is contained in some finite dimensional irreducible representation of G iff $U(k) = 1$ for every $k \in K \cap N(G)$.

Proof (Godement) By considering $G/N(G)$ we may assume that G has sufficiently many representations; as we know (compare 3.1.1), this is equivalent to the existence of a faithful finite dimensional representation for G. Now, if U were contained in no finite dimensional irreducible representation of G, then by Proposition 4.5.2.3 we would have $C_{c,\delta}(G) = \{0\}$ (δ the class of U). In turn this would imply that $\bar{\chi}_\delta * f * \bar{\chi}_\delta = 0$ for every $f \in C_c(G)$, hence also $\bar{\chi}_\delta * \mu * \bar{\chi}_\delta = 0$ for every $\mu \in M_c(G)$, hence $\bar{\chi}_\delta = 0$, an impossibility. □

Consider a pair (G, K), where G is a connected semi-simple Lie group with a faithful finite dimensional representation, K a maximal compact subgroup of G – then Theorem 4.5.2.7 tells us that K is (uniformly) large in G. Unfortunately, however, if G is an arbitrary connected semi-simple Lie group with finite center, then G need *not* admit a faithful finite dimensional representation (cf. number 3.1.1), and so we are not ensured, a priori, that a maximal compact subgroup K of G is actually large in G – we propose now to consider this question.

As usual, let $\mathfrak{G}$ denote the universal enveloping algebra of $\mathfrak{g}_c$ ($\mathfrak{g}$ the Lie algebra of G); let $\mathfrak{K}$ denote the universal enveloping algebra of $\mathfrak{k}_c$ ($\mathfrak{k}$ the Lie algebra of K).

Theorem 4.5.2.9 (Harish-Chandra) Let G be a connected semi-simple Lie group with finite center, K a maximal compact subgroup of G; let U be a topologically irreducible Banach representation of G on E – then U is quasi-simple iff U is topologically completely irreducible and, moreover, in either of these two equivalent events, U is K-finite.

Proof Suppose that U is quasi-simple – we shall prove that U is K-finite and hence (Proposition 4.2.1.5) that U is topologically completely irreducible. Fix a non-zero vector $a \in \sum_{\delta \in \hat{K}} E_\omega(\delta)$ and consider the $\mathfrak{G}$-orbit of a – thus the closure of this orbit is U-stable (Proposition 4.4.5.5) and hence, by the irreducibility of U, is all of E. We then claim that $\mathfrak{G}a = \sum_{\delta \in \hat{K}} E_\omega(\delta)$, where $\dim(E_\omega(\delta)) < \infty$ (all $\delta \in \hat{K}$). [Of course this claim has the consequence that $E_K = \sum_{\delta \in \hat{K}} E(\delta) \subset E_\omega$ (Proposition 4.4.3.4).] To verify the claim, first observe that $\mathfrak{G}a \subset \sum_{\delta \in \hat{K}} E_\omega(\delta)$ and hence $\mathfrak{G}a = \sum_{\delta \in \hat{K}} \mathfrak{G}a \cap E_\omega(\delta)$. Let $\mathfrak{J}$ denote the annihilator of a in $\mathfrak{K}$, $\mathfrak{K}$ the universal enveloping algebra of $\mathfrak{k}_c$ – then $\mathfrak{J}$ is a left ideal in $\mathfrak{K}$ and $\mathfrak{K}/\mathfrak{J} \sim \mathfrak{K}a$ (as $\mathfrak{K}$-modules). Therefore $\mathfrak{K}/\mathfrak{J}$ is finite dimensional and the natural representation of $\mathfrak{K}$ on $\mathfrak{K}/\mathfrak{J}$ is semi-simple. Adopt the notations of Theorem 2.2.1.1; identify the classes in $\hat{K}$ with classes in $\hat{\mathfrak{k}}_c$ (of course, since K may not be simply connected, not all classes in $\hat{\mathfrak{k}}_c$ need arise in this way). Write $\bar{\mathfrak{G}} = \mathfrak{G}/\mathfrak{G}\mathfrak{J} = \sum_{\delta \in \hat{\mathfrak{k}}_c} \bar{\mathfrak{G}}(\delta)$, each $\bar{\mathfrak{G}}(\delta)$ being a *finite* $\mathfrak{Z}$-module ($\mathfrak{Z}$ the center of $\mathfrak{G}$). The map which assigns to each coset $D + \mathfrak{G}\mathfrak{J}$ the element $U_\omega(D)a$ is a surjection of $\mathfrak{G}$-modules and so, thanks to the quasi-simplicity of U, it follows that $\dim(\mathfrak{G}a \cap E_\omega(\delta)) < \infty$ for every $\delta \in \hat{K}$. Since $\mathfrak{G}a$ is dense in E, the fact that $\dim(E_\omega(\delta)) < \infty$ ($\delta \in \hat{K}$) is now clear (Proposition 4.4.3.4). On the other hand, any topologically completely irreducible Banach representation of G is necessarily quasi-simple (Proposition 4.4.1.5) and thus, by the above, is K-finite. Hence the theorem. □

The above theorem implies that to each TCI Banach representation U of G on E, one may assign an algebraically completely irreducible representation of $\mathfrak{G}$ on $E_K = \sum_{\delta \in \hat{K}} E(\delta)$, namely U_K (cf. Theorem 4.5.5.4); moreover this representation of $\mathfrak{G}$ determines U to within infinitesimal equivalence and hence to within Naimark equivalence (cf. Theorem 4.5.5.2). On the basis of these observations and Theorem 2.2.2.5, we have:

Theorem 4.5.2.10 (Harish-Chandra) Let G be a connected semi-simple Lie group with finite center, K a maximal compact subgroup of G; let U be a TCI Banach representation of G on E – then there exists an integer $m_U \geqslant 1$ such that $\dim(E(\delta)) \leqslant m_U d(\delta)^2$ $(\delta \in \hat{K})$.

This theorem is, of course, a decided improvement on Theorem 4.5.2.9 – for it says that not only is every TCI Banach representation U of G K-finite but that one even has an a priori bound on the dimensions of the various K-isotypic submodules which occur in the restriction of U to K. On the other hand, the number m_U conceivably depends on U (so that Theorem 4.5.2.10 still does not say that K is large in G) – the following result should allay any doubts on this score.

Theorem 4.5.2.11 (Harish-Chandra) Let G be a connected semi-simple Lie group with finite center, K a maximal compact subgroup of G – then a given $\delta \in \hat{K}$ occurs no more than $d(\delta)$ times in the restriction to K of any TCI Banach representation of G.

We shall give a proof of this theorem in number 5.5.1 (cf. Corollary 5.5.1.8).

Note Let us agree to retain the above notations and assumptions – then the fact that K is uniformly large in G may be seen as follows. Write, after Iwasawa, $G = KA_{\mathfrak{p}}N^+$; extend $\mathfrak{a}_{\mathfrak{p}}$ to a Cartan subalgebra $\mathfrak{a}$ of $\mathfrak{g}$; let W_Φ denote the Weyl group of the pair $(\mathfrak{g}_c, \mathfrak{a}_c)$, W_- the subgroup of W_Φ generated by the w_α $(\alpha \in P_-)$ – then we intend to show that a given δ in $\hat{K}$ occurs no more than $\sqrt{[W_\Phi : W_-]}d(\delta)$ times in any TCI Banach representation U of G. [The point here is that this estimate on the multiplicities does not lie as deep as the one stated in Theorem 4.5.2.11.] Owing to Theorem 4.5.1.4, we need only work with irreducible unitary representations of G. So let U be an irreducible unitary representation of G on a Hilbert space E; fix a class δ in $\hat{K}$ which occurs in $U|K$; let $a_1, \ldots, a_n$ be an orthonormal basis for $E(\delta)$ $(n = \dim(E(\delta))$; for i, j lying between 1 and n, put $\phi_{ij}(x) = (U(x)a_i, a_j)$ $(x \in G)$ – then, in view of Lemma 4.5.1.3 (and Burnside's Theorem . . .), the ϕ_{ij} are linearly independent over **C**. It is easy to check that $\chi_\delta * \phi_{ij} = \phi_{ij} = \phi_{ij} * \chi_\delta$; thus, for example $\chi_\delta * \phi_{ij}(x) = \int_K \overline{\chi_\delta(k)}(U(kx)a_i, a_j)dk = (U(x)a_i, \mathrm{P}(\delta)a_j) = \phi_{ij}(x)$ (all $x \in G$). In addition we have $Z\phi_{ij} = \kappa_U(Z)\phi_{ij}$ (all $Z \in \mathfrak{Z}$ (= center of $\mathfrak{G}$)). [Here, of course, κ_U is the infinitesimal character of U.] This being the case, it then follows from a principle which is spelled out in Vol. II, number 9.1.2, that the dimension of the space spanned by the ϕ_{ij} is necessarily bounded by the number $[W_\Phi : W_-]d(\delta)^4$; otherwise said, $n^2 \leqslant [W_\Phi : W_-]d(\delta)^4$, whence $\dim(E(\delta)) \leqslant \sqrt{[W_\Phi : W_-]}d(\delta)^2$ . . . .

Problem Consider the pairs (G, K) envisioned by Harish-Chandra [34, §§ 14–15] (K per Bruhat-Tits . . .): Is K large in G?

Appendix Let G be a connected solvable topological group – then Godement's [7, p. 549] version of Lie's Theorem is:

Theorem Let U be a continuous representation of G on a finite dimensional vector space E (over $\mathbf{C}$) – then E contains a G-stable flag $\mathfrak{F}$.

Let H be a subgroup of G. We shall say that a non-zero vector $a \in E$ is an *H-eigenvector* if the line determined by a is H-stable. Obviously an H-eigenvector $a \in E$ determines a character $\hat{\xi}_a$ of H, viz.: $U(\xi)a = \langle \xi, \hat{\xi}_a \rangle a$ $(\xi \in H)$ – it is clear that the number of distinct characters $\hat{\xi}_a$ of H arising from eigenvectors $a \in E$ is finite (in fact is $\leqslant \dim(E)$).

Lemma Let a be an H-eigenvector in E. If H is normal in G, then $\langle x^{-1}\xi x, \hat{\xi}_a \rangle = \langle \xi, \hat{\xi}_a \rangle$ for all $x \in G$ and $\xi \in H$.

Proof It is clear that $\langle x^{-1}\xi x, \hat{\xi}_a \rangle = \langle \xi, \hat{\xi}_{U(x)a} \rangle$ $(x \in G)$; as there are only a finite number of characters of H of the form $\hat{\xi}_{U(x)a}$ for $x \in G$, the subgroup $G_a = \{x \in G : \hat{\xi}_{U(x)a} = \hat{\xi}_a\}$ is of finite index in G. However G_a is closed in G (being the set of common zeros of the continuous functions $x \mapsto \langle x^{-1}\xi x, \hat{\xi}_a \rangle - \langle \xi, \hat{\xi}_a \rangle$ as ξ ranges over H) and so, by the connectedness of G, it follows that $G_a = G$. □

Proof of the Theorem (1) Suppose in addition that U is irreducible; as will be seen below, this necessarily entails that $\dim(E) = 1$. But then by induction on $\dim(E)$, the statement of the present theorem becomes clear.

(2) Put $G^{(1)} = G$, $G^{(n)} = (G^{(n-1)}, G^{(n-1)})$ $(n > 1)$ – here $G^{(2)} = (G^{(1)}, G^{(1)})$ is the commutator subgroup of G. Of course the assumption of solvability implies that $G^{(n)} = \{1\}$ for some n. Assuming now that U is irreducible, let us prove that $\dim(E) = 1$. This will be done by induction on the smallest number n such that $G^{(n)} = \{1\}$. If $n = 1$, then $G = \{1\}$ and the theorem is trivial – hence we may assume that $n > 1$ so that $G^{(2)} \neq G^{(1)} = G$. By induction the theorem is true for $G^{(2)}$ because $G^{(2)}$ is connected. [Thus the set $\mathscr{S}$ of elements in G of the form $xyx^{-1}y^{-1}$ $(x, y \in G)$ is connected, $\mathscr{S}$ being the image of $G \times G$ under the continuous mapping $(x, y) \mapsto xyx^{-1}y^{-1}$. Let $\mathscr{S}^m$ denote the set of elements in G which can be expressed as the product of m elements in $\mathscr{S}$ – then $\mathscr{S}^m$ is connected too. Since $\mathscr{S}^m \subset \mathscr{S}^{m+1}$ and since $G^{(2)}$ is the union of the $\mathscr{S}^m$, we see that $G^{(2)}$ is likewise connected, the $\mathscr{S}^m$ having a common point, viz. 1]. Therefore the space E admits a $G^{(2)}$-eigenvector $\bar{a}$, say. The lemma supra implies that the set of all a in E which are $G^{(2)}$-eigenvectors relative to the character determined by $\bar{a}$ is G-stable; by the irreducibility of U, this set must actually be equal to E. Fix an $x \in G$; let H_x be the subgroup of G generated by x and $G^{(2)}$. Because $H_x \supset G^{(2)}$, H_x is normal. Let a_x be a non-zero eigenvector in E for the operator $U(x)$ – then, by the above, a_x is an H_x-eigenvector. Let $\hat{\xi}_x$ be the corresponding character. Owing to the lemma supra, the set of all $a \in E$ such that $U(\xi)a = \langle \xi, \hat{\xi}_x \rangle a$ for all $\xi \in H_x$ is G-stable and thus is all of E. In particular, $U(x)a \in \mathbf{C}a$ for each $a \in E$ which implies that $\dim(E) = 1$, x being arbitrary in G.

Hence the theorem. □

Remark Examples show that the assumption of connectedness cannot be removed in Lie's Theorem (consider any non-commutative solvable finite group).

4.5.3 Properties of Largeness

Let G be a locally compact unimodular group countable at infinity. If $K_1 \subset G$ is large and K_2 is a compact subgroup of G containing K_1, then K_2 is also large – this is clear. On the other hand:

Proposition 4.5.3.1 If K_2 is a large compact subgroup of G and $K_1 \subset K_2$ is a closed subgroup of G such that K_2/K_1 is finite, then K_1 is also a large compact subgroup of G.

Proof The Frobenius Reciprocity Theorem (cf. Corollary 5.3.3.6) tells us that there are only finitely many irreducible representations of K_2 whose restrictions to K_1 contain a given irreducible representation of K_1. □

Corollary 4.5.3.2 Suppose that G is a Lie group with a large compact subgroup K – then G admits a large connected compact subgroup.

Proof Since G is a Lie group, the identity component of K will fulfill the requirements of the corollary. □

Remark Let G be a unimodular Lie group, K a compact subgroup of G. Let U be a Banach representation of G on E; assume that there exists an integer $m_U \geqslant 1$ such that $\dim(E(\delta)) \leqslant m_U d(\delta)^2$ (all $\delta \in \hat{K}$) – then there exists an integer n_U such that $\dim(E(\delta_0)) \leqslant n_U d(\delta_0)^2$ (all $\delta_0 \in \hat{K}_0$) (K_0 the identity component of K). [Let us agree to use the notations of Lemma 4.4.2.3. Fix $\delta_0 \in \hat{K}_0$; if $\delta_1, \ldots, \delta_r$ are the elements of $\hat{K}$ such that $[\delta_i : \delta_0] \geqslant 1$ $(i = 1, \ldots, r)$, then δ_0 occurs at most $m_U d(\delta_i)[\delta_i : \delta_0]$ times in $E(\delta_i)$ – thus δ_0 occurs at most $m_U \sum_i d(\delta_i)[\delta_i : \delta_0] = m_U[K : K_0]d(\delta_0)$ times in E and so we may take $n_U = m_U[K : K_0]$.]

Proposition 4.5.3.3 If K is a large compact subgroup of G and H is a closed subgroup of G with $K \subset H$, then K is large in H.

Proof Owing to the largeness of K in G, there exists a family $\mathfrak{R}$ of Banach representations of G which separates the points of $C_c(G)$ and has the property that δ occurs no more than $m(\delta)$ times in any member of $\mathfrak{R}$ (Proposition 4.5.1.5 (iv)). Because $\mathfrak{R}$ distinguishes points of $C_c(G)$, it also distinguishes points of $M_c(G)$ and hence of $C_c(H)$ (since $C_c(H) \subset M_c(G)$). Therefore our assertion follows from Proposition 4.5.1.5 (iv). □

The preceding result admits a partial converse, namely:

Proposition 4.5.3.4 (Fell) Let K and H be closed subgroups of G with $K \subset H \subset G$. If K is large in H and if G/H is finite, then K is large in G.

The proof requires a couple of lemmas and some tools from the next chapter; accordingly the reader may want to put off the study of this result for a while.

Lemma 4.5.3.5 If H is a subgroup of G such that G/H is finite (i.e. such that H has finite index in G), then the normal subgroup $\bigcap_{x \in G} xHx^{-1}$ has finite index in G.

Proof The assertion is a consequence of the following statements.

(1) Let H_1 and H_2 be subgroups of G with $H_1 \subset H_2 \subset G$. If G/H_2 and H_2/H_1 are finite, then G/H_1 is finite.

(2) If H_1, H_2 are subgroups of G and G/H_1 is finite, then $H_2/(H_1 \cap H_2)$ is finite.

(3) If $H_1, \ldots, H_n$ are subgroups of G, each of finite index in G, then $H_1 \cap \ldots \cap H_n$ has finite index in G. [Use (1) and (2) along with induction on n.]

(4) Let H be as in the statement of our lemma. If F is a finite subset of G intersecting each coset xH in exactly one point, then $\bigcap_{x \in G} xHx^{-1} = \bigcap_{x \in F} xHx^{-1}$; by (3), $\bigcap_{x \in F} xHx^{-1}$ has finite index in G.

The proof of the lemma is now complete. □

Lemma 4.5.3.6 If K is a large compact subgroup of G, and H is a closed subgroup of G of finite index in G, then $K \cap H$ is large in H.

Proof By part (2) of the proof of Lemma 4.5.3.5, $K \cap H$ is of finite index in K; Proposition 4.5.3.1 then implies that $K \cap H$ is large in G – therefore our lemma follows from Proposition 4.5.3.3. □

Proof of Proposition 4.5.3.4 The (normal) subgroup $\bigcap_{x \in G} xHx^{-1}$ is of finite index in G (Lemma 4.5.3.5) and $K \cap (\bigcap_{x \in G} xHx^{-1})$ is large in $\bigcap_{x \in G} xHx^{-1}$ (Lemma 4.5.3.6) – therefore it is sufficient to establish Proposition 4.5.3.4 under the additional assumption that H is normal. This being so, consider now the collection $\mathfrak{R}$ of all unitarily induced representations U^L, L running through $\hat{H}$; since $\mathfrak{R}$ weakly contains the left regular representation of G (see the Appendix to this number) and since the left regular representation of G separates the points of $C_c(G)$, the same must be true of $\mathfrak{R}$; otherwise said, the set $\mathfrak{R} = \{U^L : L \in \hat{H}\}$ is complete (4.5.1). Choose points $x_1, \ldots, x_n$ $(n = [G : H])$ in G such that $G = \bigcup_i x_i H$ (disjoint union). If L is an irreducible unitary representation of H, then, thanks to an easy extension of Mackey's Subgroup Theorem (Theorem 5.3.4.1), $U^L | H$ is the (Hilbert space) direct sum of the L^{x_i}, $L^{x_i}(\xi) = L(x_i \xi x_i^{-1})$ $(\xi \in H; i = 1, \ldots, n)$. In turn this shows that if a given $\delta \in \hat{K}$ occurs at most $m(\delta)$ times in an arbitrary irreducible unitary representation of H, then it occurs at most $[G : H]m(\delta)$ times in an arbitrary U^L $(L \in \hat{H})$. Consequently K is large in G (Proposition 4.5.1.5 (iv)). □

Example Consider pairs (G, K), (G, K) per the second Appendix to number 1.2.4; thus K is large in G (cf. Theorem 4.5.2.11) and, of course, G/G is finite – therefore, in view of Proposition 4.5.3.4, we see that K is large in G (with $m(\delta) = [\mathsf{G} : G]d(\delta)$ $(\delta \in \hat{K})$, G being normal in G). [From this it follows, of course, that K is large in G.]

In conclusion let us mention the following result.

Proposition 4.5.3.7 If K_i is a large compact subgroup of G_i $(i = 1, 2)$, then $K_1 \times K_2$ is large in $G_1 \times G_2$.

Proof Since G_1 and G_2 are type I (Theorem 4.5.7.3 infra), every irreducible unitary representation of $G_1 \times G_2$ is of the form $U^1 \times U^2$, U^i an irreducible unitary representation of G_i $(i = 1, 2)$ (cf. Dixmier [15, p. 251]). If $\delta_i \in \hat{K}_i$ $(i = 1, 2)$, then the multiplicity of $\delta_1 \times \delta_2$ in $U^1 \times U^2$ is the product of the multiplicity of δ_1 in $U^1 | K_1$ with that of δ_2 in $U^2 | K_2$ – therefore our proposition follows from Proposition 4.5.1.5 (iv) (in view of the Gelfand-Raikov Theorem, the irreducible unitary representations of $G_1 \times G_2$ separate the points of $C_c(G_1 \times G_2)$). □

Appendix Let G be a locally compact group countable at infinity. We recall that a unitary representation U of G is said to be *weakly contained* in a collection $\mathfrak{U}$ of unitary representations of G provided that every function of positive type associated with U is the uniform limit on compacta of sums of functions of positive type associated with $\mathfrak{U}$. It is well-known that every unitary representation of G is weakly contained in $\hat{G}$ (cf. Dixmier [2, p. 262]).

For the proof of the following result, which does not fit conveniently within our frame of exposition, we shall refer the reader to Fell [2, p. 260].

Theorem (Fell) Let H be a closed subgroup of the group G; let L_0 be a unitary representation of H – then the unitarily induced representation U^{L_0} is weakly contained in the collection $\{U^L : L \in \hat{H}\}$.

Corollary Let H be a closed subgroup of the group G – then the left regular representation of G is weakly contained in the collection $\{U^L : L \in \hat{H}\}$.

Proof In the preceding theorem, take for L_0 the left regular representation of H – then U^{L_0} is unitarily equivalent to the left regular representation of G (Corollary 5.1.3.6). □

4.5.4 Naimark Equivalence

Let G be a locally compact unimodular group countable at infinity, K a compact subgroup of G. Let $\hat{K}$ denote the set of all equivalence classes of finite dimensional irreducible representations of K; for each finite subset F of $\hat{K}$, let $\bar{\chi}_F = \sum_{\delta \in F} \bar{\chi}_F$ (hence $\bar{\chi}_F$ is a central idempotent in $C(K)$) – we shall then let $C_{c,F}(G)$ denote the subalgebra of $C_c(G)$ consisting of those f which verify the relation $f = \bar{\chi}_F * f * \bar{\chi}_F$.

Let U be a K-finite Banach representation of G on E – then each $E(\delta)$ $(\delta \in \hat{K})$ is a finite dimensional space. If F is a finite subset of $\hat{K}$, then the (finite dimensional) subspace $E_F = \sum_{\delta \in F} E(\delta)$ is, evidently, stable under the $U(f)$ $(f \in C_{c,F}(G))$; hence, letting $U_F(f)$ denote the restriction of $U(f)$ to E_F $(f \in C_{c,F}(G))$, one obtains a representation $f \mapsto U_F(f)$ of $C_{c,F}(G)$ on E_F.

Lemma 4.5.4.1 Let U be a K-finite Banach representation of G on E – then U is topologically irreducible iff each U_F is algebraically irreducible (F a finite subset of $\hat{K}$).

The proof is straightforward and will be left to the reader.

[Let U be a K-finite Banach representation of G on E; if U is topologically irreducible, then each U_δ ($\delta \in \hat{K}$) is algebraically irreducible (cf. Lemma 4.5.1.3) – on the other hand, the converse of this is certainly not true . . . .]

Theorem 4.5.4.2 (Fell) Suppose that $\hat{K}$ is in addition countable. Let U^i be a K-finite Banach representation of G on E_i ($i = 1, 2$) – then U^1 and U^2 are Naimark equivalent iff U_F^1 and U_F^2 are algebraically equivalent for all finite subsets F of $\hat{K}$.

Before giving the proof, it will be convenient to present some additional generalities.

Let $C_*(G)$ denote the subalgebra of $C_c(G)$ consisting of the linear span of the $\bar{\chi}_{\delta_1} * C_c(G) * \bar{\chi}_{\delta_2}$ ($\delta_1, \delta_2 \in \hat{K}$) – then, relative to the inductive limit topology, $C_*(G)$ is dense in $C_c(G)$ (cf. 4.4.3–in particular Proposition 4.4.3.5).

Let U be a K-finite Banach representation of G on E; let $E_K = \sum_{\delta \in \hat{K}} E(\delta)$ and denote by U_K the (algebraic) representation of $C_*(G)$ on E_K defined by $U_K(f) = U(f) | E_K (f \in C_*(G))$. [It is to be observed that for each f in $C_*(G)$, $U(f)$ is of finite rank.]

The proof of the following-result proceeds along the same lines as that of Proposition 4.2.1.5.

Lemma 4.5.4.3 Let U be a K-finite Banach representation of G on E – then the following conditions are equivalent:

(i) U is topologically irreducible;

(ii) U_K is algebraically irreducible;

(iii) U_K is algebraically completely irreducible.

Proof of Theorem 4.5.4.2 It is clear that if U^1 and U^2 are Naimark equivalent, then U_F^1 and U_F^2 are algebraically equivalent for all finite subsets F of $\hat{K}$. Turning to the converse assertion, let us begin by arranging the elements of $\hat{K}$ in a sequence, $\hat{K} = \{\delta_1, \delta_2, \ldots\}$; agreeing to write U_n^i for $U_{F_n}^i$ where $F_n = \{\delta_1, \ldots, \delta_n\}$ ($i = 1, 2$), we shall assume that $U_n^1 = U_n^2$ ($n = 1, 2, \ldots$) and then prove below that U_K^1 and U_K^2 are algebraically equivalent – this will suffice to establish the Naimark equivalence of U^1 and U^2 (cf. Theorem 4.5.5.2 . . .).

(1) Let $E_n^i = \sum_{\delta \in F_n} E^i(\delta)$ – then E_n^i is a monotone increasing sequence of finite dimensional spaces whose union is E_K^i ($i = 1, 2$). We shall construct inductively a sequence $\{Q_n\}$ with the property that, for each n, Q_n is an equivalence of U_n^1 and U_n^2 and such that $Q_{n+1} | E_n^1 = Q_n$; $Q = \bigcup_n Q_n$ will then set up the required equivalence between U_K^1 and U_K^2.

(2) Let n be an integer $\geqslant 1$. Assuming that an equivalence $Q_n : E_n^1 \rightarrow$

E_n^2 of U_n^1 and U_n^2 has already been constructed with the property that for each $m > n$ there exists an equivalence $P_m : E_m^1 \to E_m^2$ of U_m^1 and U_m^2 such that $P_m | E_n^1 = Q_n$, we shall proceed in (3) below to construct an equivalence $Q_{n+1} : E_{n+1}^1 \to E_{n+1}^2$ of U_{n+1}^1 and U_{n+2}^2 having the property that for each $m > n + 1$ there exists an equivalence $P_m : E_m^1 \to E_m^2$ of U_m^1 and U_m^2 such that $P_m | E_{n+1}^1 = Q_{n+1}$ and with $Q_{n+1} | E_n^1 = Q_n$ $(n \geqslant 1)$ – this will serve, of course, to permit the inductive construction of the Q_n per (1) supra.

(3) Let $\mathfrak{J}_m$ be the linear space of all U_m^1, U_m^2 intertwining operators and $\mathfrak{E}_m$ the subset of $\mathfrak{J}_m$ consisting of all U_m^1, U_m^2 equivalences. If $m > n \geqslant 1$, let $\mathfrak{S}_m = \{S : S \in \mathfrak{J}_m, \ S | E_n^1 = Q_n\}$, $\mathfrak{T}_m = \mathfrak{S}_m \cap \mathfrak{E}_m$; if $m > n = 0$, let $\mathfrak{S}_m = \mathfrak{J}_m$, $\mathfrak{T}_m = \mathfrak{E}_m$; in all cases where $m > n$, let $\sigma_m(S) = S | E_{n+1}^1$ $(S \in \mathfrak{S}_m)$ – thus $\mathfrak{S}_m$ is an affine subspace of $\mathfrak{J}_m$ and $\mathfrak{T}_m$ consists of the points in $\mathfrak{S}_m$ where a certain polynomial function on $\mathfrak{J}_m$ does not vanish. By our assumption on Q_n, $\mathfrak{T}_m$ is a non-empty dense open subset of $\mathfrak{S}_m$; since σ_m is an affine map, it follows that $\sigma_m(\mathfrak{T}_m)$ is a dense open subset of $\sigma_m(\mathfrak{S}_m)$. As an affine subspace of $\mathfrak{J}_{n+1}$, it is clear that $\sigma_m(\mathfrak{S}_m)$ is monotone decreasing in m; let $\mathfrak{Z} = \bigcap_{m>n} \sigma_m(\mathfrak{S}_m)$ – then, for all $m \geqslant m_0$ (say), $\sigma_m(\mathfrak{S}_m) = \mathfrak{Z}$. Because $\sigma_m(\mathfrak{T}_m)$ is a dense open subset of $\mathfrak{Z}$ whenever $m \geqslant m_0$, the Category Theorem tells us that $\bigcap_{m \geqslant m_0} \sigma(\mathfrak{T}_m)$ is non-empty. Any element Q_{n+1} in this intersection will meet the requirements laid down in (2) above.

The proof of the theorem is now complete. □

Remark Let U^i be a K-finite Banach representation of G on E^i $(i = 1, 2)$; in connection with the preceding theorem, it is natural to conjecture that if U_δ^1 and U_δ^2 are algebraically equivalent for every $\delta \in \hat{K}$, then U^1 and U^2 are Naimark equivalent (cf. Proposition 4.5.1.6) – however, as can be shown by example, this is definitely *not* the case. [For instance, let $G = KH$ be the proper Euclidean group of the plane (thus $G = KH$ where $K = \mathbf{SO}(2)$ and $H = \mathbf{R}^2$); let U^1 be the K-finite Banach representation of G continuously induced by the trivial character of H; let U^2 be the K-finite Banach representation of G continuously induced by the character $(x_1, x_2) \mapsto \exp(\sqrt{-1}x_1 + x_2)$ $(x_1, x_2$ real) of H – then U^1 and U^2 are not Naimark equivalent but nevertheless $U_\delta^1 \sim U_\delta^2$ for all δ in $\hat{K}$.]

Problem Can the countability assumption on $\hat{K}$ be dropped in Theorem 4.5.4.2?

Proposition 4.5.4.4 Let G be a locally compact unimodular group, K a compact subgroup of G; let U and V be K-finite topologically irreducible Banach representations of G on E and F, respectively. If there exists a non-zero vector b_0 in F such that $U(f) = 0 \Rightarrow V(f)b_0 = 0$ $(f \in C_c(G))$, then U and V are Naimark equivalent.

Proof There exists a class $\delta_0 \in \hat{K}$ which occurs in both U and V. [Because $b_0 \neq 0$, there exists $\delta_0 \in \hat{K}$ such that $\mathbf{Q}(\delta_0)b_0 \neq 0$ $(\mathbf{Q}(\delta_0) = V(\bar{\chi}_{\delta_0}))$ (cf. Lemma 4.4.2.4); this being so, select an $f \in C_c(G)$ such that $V(f)\mathbf{Q}(\delta_0)b_0 = V(f * \bar{\chi}_{\delta_0})b_0 \neq 0$ – then, by hypothesis, $U(f * \bar{\chi}_{\delta_0}) =$

$U(f)\mathsf{P}(\delta_0) \neq 0 \Rightarrow \mathsf{P}(\delta_0) \neq 0 \ldots$.] Let U_{δ_0} and V_{δ_0} denote the corresponding (algebraically) irreducible representations of the algebra $C_{c,\delta_0}(G)$ on $E(\delta_0)$ and $F(\delta_0)$, respectively (see 4.5.1); to establish the Naimark equivalence of U and V it will be sufficient, in view of Proposition 4.5.1.6, to show that U_{δ_0} and V_{δ_0} are algebraically equivalent. [Let us bear in mind that U and V are actually TCI (see the discussion at the beginning of 4.5.1); in addition the K-finiteness of our representations guarantees that Proposition 4.5.1.6 is legitimately applicable here (the largeness of K not being essential there).] Since U_{δ_0} and V_{δ_0} are finite dimensional and irreducible, we need only prove that they have the same kernel. So suppose that $f \in \operatorname{Ker}(U_{\delta_0})$ – then $U_{\delta_0}(f) = 0 \Rightarrow U(f) = 0 \Rightarrow V(f)b_0 = 0 \Rightarrow V_{\delta_0}(f)\mathsf{Q}(\delta_0)b_0 = 0 \Rightarrow \operatorname{Ker}(U_{\delta_0}) = \operatorname{Ker}(V_{\delta_0})$ (the kernels being maximal ideals in $C_{c,\delta_0}(G)$). ☐

4.5.5 Infinitesimal Equivalence

Let G be a connected unimodular Lie group countable at infinity, K a compact connected subgroup of G. Let $\hat{K}$ denote the set of all equivalence classes of finite dimensional irreducible representations of K.

Lemma 4.5.5.1 Let U be a K-finite Banach representation of G on E – then $E_K = \sum_{\delta \in \hat{K}} E(\delta) \subset E_\omega$, E_ω the analytic vectors in E (for U).

Proof This is an immediate consequence of Corollary 4.4.5.17. ☐

Let $\mathfrak{G}$ be the universal enveloping algebra of the complexification of the Lie algebra $\mathfrak{g}$ of G; let U be a K-finite Banach representation of G on E – then, in view of Proposition 4.4.5.18 and Lemma 4.5.5.1, we may represent $\mathfrak{G}$, by differentiation, on the space $E_K = \sum_{\delta \in \hat{K}} E(\delta)$. We shall agree to denote this representation of $\mathfrak{G}$ by U_K (thus $U_K = U_\omega | E_K$).

Definition (Harish-Chandra) Let U and V be K-finite Banach representations of G on E and F, respectively – then U and V are said to be *infinitesimally equivalent* if the representations U_K and V_K of $\mathfrak{G}$ on E_K and F_K, respectively, are algebraically equivalent.

It is clear that if U and V are equivalent, then they are infinitesimally equivalent.

Theorem 4.5.5.2 (Godement) Let G be a connected unimodular Lie group, K a compact connected subgroup of G – then two K-finite Banach representations U and V of G on E and F, respectively, are infinitesimally equivalent iff they are Naimark equivalent.

Proof (A) Assume that U_K and V_K are equivalent under an isomorphism $Q: E_K \to F_K$ – then we have to prove that U and V are Naimark equivalent.

(1) Let $\mathfrak{Z}_{\mathfrak{k}}$ denote the center of $\mathfrak{K}$, $\mathfrak{K}$ the universal enveloping algebra of the complexification $\mathfrak{k}_c$ of the Lie algebra $\mathfrak{k}$ of K. Each $\delta \in \hat{K}$ gives rise to a homomorphism $\kappa_\delta : \mathfrak{Z}_{\mathfrak{k}} \rightarrow \mathbf{C}$ (the infinitesimal character of δ) with the property that $\kappa_{\delta_1} = \kappa_{\delta_2} \Rightarrow \delta_1 = \delta_2 (\delta_1, \delta_2 \in \hat{K})$ (K is connected). Hence a vector $a \in E_K$ belongs to $E(\delta)$ iff $U_K(Z)a = \kappa_\delta(Z)a$ for all $Z \in \mathfrak{Z}_{\mathfrak{k}}$. An analogous characterization obtains for the vectors in F_K. Since $Q \circ U_K(Z) = V_K(Z) \circ Q$ $(Z \in \mathfrak{Z}_{\mathfrak{k}})$, we see that $Q(E(\delta)) = F(\delta)$ for each $\delta \in \hat{K}$.

(2) Let us prove next that Q admits closure. Thus suppose that $a_n \rightarrow 0$ $(a_n \in E_K)$ and $Qa_n \rightarrow b$ $(b \in F)$ – we have to show that $b = 0$. Let $\mathsf{Q}(\delta) = V(\bar{\chi}_\delta)$ $(\delta \in \hat{K})$; since $\mathsf{P}(\delta)a_n \rightarrow 0$ and $E(\delta)$ is finite dimensional, $Q(\mathsf{P}(\delta)a_n) \rightarrow 0$; but, in view of (1), $Q\mathsf{P}(\delta) = \mathsf{Q}(\delta)Q \Rightarrow \mathsf{Q}(\delta)b = 0 \Rightarrow b = 0$, δ being arbitrary (cf. Lemma 4.4.2.4).

(3) Let $T = \bar{Q}$ – then, using an argument similar to the one in (2), we see that T is one-to-one. Therefore, in order to conclude that U and V are Naimark equivalent, we need only show that the domain and range of T are stable for the $U(f)$ and $V(f)$, respectively $(f \in C_c(G))$ and that $TU(f)a = V(f)Ta$ $(a \in \text{Dom}(T))$. To this end, let $\mathscr{S}$ denote the linear span of the ${}_\delta f$ $(\delta \in \hat{K}, f \in C_c(G)$ – we recall that ${}_\delta f = \bar{\chi}_\delta * f$ (see 4.5.1)); it will be shown in (4) below that $QU({}_\delta f)a = V({}_\delta f)Qa$ (all $a \in E_K$). This being so, fix an $a \in E_K$, $f \in C_c(G)$ and then choose a net $\{f_n\}$ in $\mathscr{S}$ such that

$$U(f_n)a \rightarrow U(f)a, \quad V(f_n)Qa \rightarrow V(f)Qa.$$

Since T is closed, we conclude that $U(f)a$ is in the domain of T – in fact that $TU(f)a = V(f)Ta$. Finally, applying the continuity of the $U(f)$ and $V(f)$ still again, we easily see that the domain and range of T are indeed stable for the $U(f)$ and $V(f)$, respectively $(f \in C_c(G))$ and that $TU(f)a = V(f)Ta$ $(a \in \text{Dom}(T))$.

(4) Fix $\delta \in \hat{K}$. Let $\mathbf{w}$ be an element in the dual of $F(\delta)$, $\tilde{\mathbf{w}}$ some continuous extension of $\mathbf{w}$ to F. Since Q maps $E(\delta)$ onto $F(\delta)$ ((1) above), $\mathbf{z} = {}^tQ\mathbf{w}$ belongs to the dual of $E(\delta)$; let $\tilde{\mathbf{z}}$ denote some continuous extension of $\mathbf{z}$ to E. Fix $a \in E_K$ – then a is an analytic vector in E (for U) (Lemma 4.5.5.1) and so, for all X close enough to zero in $\mathfrak{g}$, Theorem 4.4.5.4 tells us that

$$\begin{aligned}
\langle U(\exp X)a, {}^t\mathsf{P}(\delta)\tilde{\mathbf{z}}\rangle &= \sum_0^\infty \frac{1}{m!}\langle U_K(X^m)a, {}^t\mathsf{P}(\delta)\tilde{\mathbf{z}}\rangle \\
&= \sum_0^\infty \frac{1}{m!}\langle \mathsf{P}(\delta)U_K(X^m)a, \mathbf{z}\rangle \\
&= \sum_0^\infty \frac{1}{m!}\langle \mathsf{Q}(\delta)V_K(X^m)Qa, \mathbf{w}\rangle \\
&= \langle V(\exp X)Qa, {}^t\mathsf{Q}(\delta)\tilde{\mathbf{w}}\rangle.
\end{aligned}$$

The connectedness of G then implies that

$$\langle U(x)a, {}^t\mathsf{P}(\delta)\tilde{\mathbf{z}}\rangle = \langle V(x)Qa, {}^t\mathsf{Q}(\delta)\tilde{\mathbf{w}}\rangle$$

(all $x \in G$); using this fact, we can now show that for any $f \in C_c(G)$, $QU({}_\delta f)a = V({}_\delta f)Qa$ (all $a \in E_K$). In the first place, it is clear that $U({}_\delta f)a \in E(\delta)$, $V({}_\delta f)Qa \in F(\delta)$ – thus, since $Q(E(\delta)) = F(\delta)$, it will be sufficient to verify that $\langle QU({}_\delta f)a, \mathbf{w}\rangle = \langle V({}_\delta f)Qa, \mathbf{w}\rangle$ for each $\mathbf{w}$ in the dual of $F(\delta)$. As we have

$$\begin{aligned}\langle QU({}_\delta f)a, \mathbf{w}\rangle &= \langle U({}_\delta f)a, {}^t Q\mathbf{w}\rangle = \langle U({}_\delta f)a, {}^t\mathrm{P}(\delta)\tilde{\mathbf{z}}\rangle \\ &= \int_G {}_\delta f(x)\langle U(x), {}^t\mathrm{P}(\delta)\tilde{\mathbf{z}}\rangle d_G(x) \\ &= \int_G {}_\delta f(x)\langle V(x)Qa, {}^t\mathrm{Q}(\delta)\tilde{\mathbf{w}}\rangle d_G(x) \\ &= \langle V({}_\delta f)Qa, {}^t\mathrm{Q}(\delta)\tilde{\mathbf{w}}\rangle = \langle V({}_\delta f)Qa, \mathbf{w}\rangle,\end{aligned}$$

our contention follows.

(B) Assume that U and V are Naimark equivalent – then we have to prove that they are infinitesimally equivalent.

(1) The Naimark equivalence of U and V provides us with dense subspaces $\tilde{E} \subset E$, $\tilde{F} \subset F$ such that $\tilde{E}$ is stable for the $\{U(\mu) : \mu \in M_c(G)\}$, $\tilde{F}$ is stable for the $\{V(\mu) : \mu \in M_c(G)\}$ and a closed one-to-one linear map T: $\tilde{E} \to \tilde{F}$ with

$$TU(\mu)a = V(\mu)Ta$$

for all $a \in \tilde{E}$, $\mu \in M_c(G)$. This being the case, observe next that $E_0 \subset \tilde{E}$, $F_0 \subset \tilde{F}$. Thus, for any $\mu \in M_c(G)$, $U(\mu)(\tilde{E})$ is dense in the range of $U(\mu)$ while at the same time $U(\mu)(\tilde{E}) \subset \tilde{E}$; so in particular, if $U(\mu)$ is of finite rank, then $U(\mu)(\tilde{E}) = U(\mu)(E) \subset \tilde{E}$ which implies that $E_0 \subset \tilde{E}$. In a similar way we find that $F_0 \subset \tilde{F}$.

(2) It is clear that for each $\delta \in \hat{K}$, $T(E(\delta)) = F(\delta)$; put $Q = T|E_K$ – then $Q : E_K \to F_K$. We shall now show that Q sets up an algebraic equivalence between U_K and V_K – this means that we have to prove that $QU_K(D)a = V_K(D)Qa$ (all $a \in E_K$, $D \in \mathfrak{G}$). Let $X \in \mathfrak{g}$; for any real $t \neq 0$,

$$\frac{U(\exp(tX))a - a}{t} \in \tilde{E}$$

and

$$\lim_{t\to 0} \frac{U(\exp(tX))a - a}{t} = U_K(X)a \in E_K \subset \tilde{E} \qquad (a \in E_K).$$

Therefore, since T is *closed* and

$$T\left(\frac{U(\exp(tX))a - a}{t}\right) = \frac{V(\exp(tX))Qa - Qa}{t} \to V_K(X)Qa$$

as $t \to 0$ ($a \in E_K$), we must have $QU_K(X)a = V_K(X)Qa$ ($a \in E_K$); of course this proves that U_K and V_K are algebraically equivalent and hence, by definition, that U and V are infinitesimally equivalent.

The proof of the theorem is now complete. □

Corollary 4.5.5.3 (Harish-Chandra) Let G be a connected unimodular Lie group, K a compact connected subgroup of G; let U and V be K-finite unitary representations of G – then U and V are unitarily equivalent iff they are infinitesimally equivalent.

Proof This is an immediate consequence of Proposition 4.3.1.4 and the preceding theorem. □

The next theorem establishes an intimate relationship between representations of G and $\mathfrak{G}$.

Theorem 4.5.5.4 (Harish-Chandra) Let G be a connected unimodular Lie group, K a compact connected subgroup of G; let U be a K-finite Banach representation of G on E – then the following conditions are equivalent:

(i) U is topologically irreducible;
(ii) U_K is algebraically irreducible;
(iii) U_K is algebraically completely irreducible.

Proof (iii) $\Rightarrow$ (ii) – this is clear.

(ii) $\Rightarrow$ (i). Let $\tilde{E}$ be a closed non-zero U-stable subspace of E; let $\tilde{E}_K = \sum_{\delta \in \hat{K}} \mathrm{P}(\delta)\tilde{E}$ – then $\tilde{E}_K$ is a U_K-stable subspace of E_K. Owing to the algebraic irreducibility of U_K, it must be the case that either $\tilde{E}_K$ is $\{0\}$ or E_K itself; the former is impossible since $\tilde{E} \neq \{0\}$ – thus $\tilde{E}$ is dense in E and so $\tilde{E} = E$.

(i) $\Rightarrow$ (iii). We shall begin by verifying that U_K is algebraically irreducible. Thus let $\tilde{E}$ be a non-zero U_K-stable subspace of E_K – then $\tilde{E} = \sum_{\delta \in \hat{K}} \tilde{E} \cap E(\delta)$. [It is enough to show that, if $\delta_1, \ldots, \delta_n$ are distinct elements of $\hat{K}$, if $a_i \in E(\delta_i)$ for each i, and if $a = \sum_1^n a_i \in \tilde{E}$, then each $a_i \in \tilde{E}$. Using the notations employed in part (A) of the proof of Theorem 4.5.5.2, choose a $Z \in \mathfrak{Z}_{\mathfrak{k}}$ such that $\kappa_{\delta_1}(Z) = 1$ and $\kappa_{\delta_i}(Z) = 0$ for $i > 1$ – then $a_1 = \sum_i \kappa_{\delta_i}(Z)a_i = U_K(Z)a \in \tilde{E}$. In a similar way we find that $a_i \in \tilde{E}$ for $i > 1$.] Furthermore $\tilde{E}^{cl} = E$ – this is a consequence of Proposition 4.4.5.6 and the topological irreducibility of U. Now suppose that $\tilde{E} \neq E_K$ – then, for some $\delta_0 \in \hat{K}$, we would have $\tilde{E} \cap E(\delta_0)$ properly contained in $E(\delta_0)$. Select a non-zero element $\mathbf{z}$ in the dual of $E(\delta_0)$ such that $\mathbf{z}$ vanishes identically on $\tilde{E} \cap E(\delta_0)$; let $\tilde{\mathbf{z}}$ denote some continuous extension of $\mathbf{z}$ to E and consider ${}^t\mathrm{P}(\delta_0)\tilde{\mathbf{z}}$. If $\delta \neq \delta_0$, then

$$\langle \tilde{E} \cap E(\delta), {}^t\mathrm{P}(\delta_0)\tilde{\mathbf{z}} \rangle = \langle \mathrm{P}(\delta_0)(\tilde{E} \cap E(\delta)), \mathbf{z} \rangle = 0;$$

hence ${}^t\mathrm{P}(\delta_0)\tilde{\mathbf{z}}$ vanishes on $\tilde{E}$ and by continuity on $\tilde{E}^{cl} = E$, a contradiction. Therefore U_K is algebraically irreducible. It remains to show that U_K is algebraically completely irreducible; for this purpose, it suffices, thanks to the Jacobson Density Theorem, to prove that the division algebra $\mathscr{S}$ of endomorphisms of E_K commuting with all the $U_K(D)$ ($D \in \mathfrak{G}$) is one dimensional. So let $S \in \mathscr{S}$ – then each $E(\delta)$ ($\delta \in \hat{K}$) is stable under S.

[For let $a \in E(\delta_1)$ and write $Sa = a_1 + \ldots + a_n$ where $a_i \in E(\delta_i)$ and $\delta_1, \ldots, \delta_n$ are all distinct. Choose a $Z \in \mathfrak{Z}_{\mathfrak{k}}$ such that $\kappa_{\delta_1}(Z) = 1$ and $\kappa_{\delta_i}(Z) = 0$ for $i > 1$ – then

$$\begin{aligned} a_1 + \ldots + a_n = Sa = SU_K(Z)a &= U_K(Z)Sa \\ = a_1 \Rightarrow a_2 = \ldots &= a_n = 0 \Rightarrow Sa \in E(\delta_1).] \end{aligned}$$

This being the case, fix a non-zero vector a in $E(\delta)$ (say) – then the map $S \mapsto Sa$ ($S \in \mathscr{S}$) is linear and one-to-one (since $\mathscr{S}$ is a division algebra) and has a finite dimensional range (since $\dim(E(\delta)) < \infty$) – therefore $\mathscr{S}$ is a finite dimensional division algebra over **C**, hence is one dimensional. Consequently U_K is algebraically completely irreducible.

The proof of Theorem 4.5.5.4 is now complete. □

Naturally Theorem 4.5.5.4 assumes special significance when applied to pairs (G, K) with K large in G.

Example (Harish-Chandra-Dixmier) Let $\mathfrak{g} = \mathfrak{k} + \mathfrak{p}$ be a semi-simple Lie algebra over **R**; let G be the simply connected Lie group corresponding to $\mathfrak{g}$, K the analytic subgroup of G corresponding to $\mathfrak{k}$ – assume that G has finite center so that K is compact. Let U be a topologically irreducible unitary representation of G on a Hilbert space E – then U is, in particular, K-finite (4.5.2). Owing to Theorem 4.5.5.4, the representation U_K of $\mathfrak{G}$ on the dense subspace E_K (the Paley-Wiener space for U) is algebraically completely irreducible – that is to say, in the terminology of number 2.2.2, U_K is a quasi-simple representation of $\mathfrak{G}$. Moreover, for every $X \in \mathfrak{g}$, $U_K(\sqrt{-1}X)$ is symmetric. Summary: To each topologically irreducible unitary representation U of G one may assign a pre-Hilbert space E_K and an algebraically irreducible $\mathfrak{k}$-finite representation of $\mathfrak{G}$ on E_K such that $\sqrt{-1}X$ is represented by a symmetric operator for all $X \in \mathfrak{g}$ – moreover this representation of $\mathfrak{G}$ determines U to within unitary equivalence (Corollary 4.5.5.3). Conversely let $U_{\mathfrak{D}}$ be an algebraically irreducible $\mathfrak{k}$-finite representation of $\mathfrak{G}$ on a pre-Hilbert space $\mathfrak{D}$ such that $U_{\mathfrak{D}}(\sqrt{-1}X)$ is symmetric for all $X \in \mathfrak{g}$ (so $U_{\mathfrak{D}}$ is quasi-simple; cf. number 2.2.2). Let E denote the completion of $\mathfrak{D}$ – then there exists a unique topologically irreducible unitary representation U of G on E such that $E_K = \mathfrak{D}$ and $U_K = U_{\mathfrak{D}}$. Indeed, this is a more or less immediate consequence of Nelson's Theorem (Theorem 4.4.6.6). Thus define ω_-, ω_+ as in Example 3 following Proposition 4.4.4.4 – then $\Delta = 2\omega_- + \omega$. Since U is algebraically completely irreducible, $U_{\mathfrak{D}}(\omega)$ is a scalar; therefore, in order to verify that $U_{\mathfrak{D}}(\Delta)$ is essentially self-adjoint, it is enough to do it for $U_{\mathfrak{D}}(\omega_-)$, or, equivalently, for $\zeta_- = -\omega_-$. Now each subspace $\mathfrak{D}(\delta)$ ($\delta \in \hat{K}$) is stable under the restriction of $U_{\mathfrak{D}}$ to $\mathfrak{k}$; because K is simply connected, there is induced therein a representation of K which is even unitary since

$$(a, U_{\mathfrak{D}}(X)b) = -(U_{\mathfrak{D}}(X)a, b) \; (X \in \mathfrak{k};\, a, b \in \mathfrak{D}(\delta)).$$

From this it follows at once that if $\delta_1 \neq \delta_2$, then $\mathfrak{D}(\delta_1)$ and $\mathfrak{D}(\delta_2)$ are mutually orthogonal subspaces of $\mathfrak{D}$. Plainly $(U_{\mathfrak{D}}(\zeta_-)a, a) \geqslant 0$ for all $a \in \mathfrak{D}$; moreover $U_{\mathfrak{D}}(\zeta_-) \mid \mathfrak{D}(\delta)$ is a scalar $\geqslant 0$ (cf. the discussion following the statement of Theorem 4.4.2.1) – therefore $I + U_{\mathfrak{D}}(\zeta_-)$ has a dense range and so $U_{\mathfrak{D}}(\zeta_-)$ is essentially self-adjoint. Quoting the theorem of Nelson referred to above and bearing in mind Proposition 4.4.3.4, our contention is now clear. Conclusion: The problem of classifying the topologically irreducible unitary representations of G is equivalent to classifying the algebraically irreducible $\mathfrak{k}$-finite representations of $\mathfrak{G}$ which are *infinitesimally unitary* in the sense that the representation space carries the structure of a pre-Hilbert space with respect to which the operators representing the $\sqrt{-1}X (X \in \mathfrak{g})$ are symmetric.

Remark Retain the notations and assumptions of the preceding Example. Harish-Chandra [5, p. 233] proved the following result: Let U be a K-finite topologically irreducible Banach representation of G on E; if U is infinitesimally unitary (i.e. if E_K admits the structure of a pre-Hilbert space per the preceding discussion), then it is infinitesimally equivalent to a unique (apart from unitary equivalence) topologically irreducible unitary representation of G. It is not difficult to see that this result is in fact a consequence of the above considerations.

A few general remarks will help to put Theorem 4.5.5.4 into perspective. First let us recall some ring theoretic terminology.

Definition Let A be a ring. An ideal $I \subset A$ is called a *prime* ideal iff $I_1 I_2 \subset I$ for any two ideals I_1, I_2 in A implies that either $I_1 \subset I$ or $I_2 \subset I$.

Of course, if A is commutative, then this definition is equivalent to the element-wise condition: $xy \in I \Rightarrow x \in I$ or $y \in I \, (x, y \in A)$.

Lemma 4.5.5.5 If I is a primitive ideal, then I is prime.

Proof Let E be an irreducible A-module such that $I = \{x \in A : xE = \{0\}\}$. Let I_1, I_2 be ideals in A such that $I_1 \not\subset I, I_2 \not\subset I$ – we must show that $I_1 I_2 \not\subset I$. Since $I_1 \not\subset I$, there exists an $a_1 \neq 0$ such that $I_1 a_1 \neq \{0\}$ – hence, by irreducibility, $I_1 a_1 = E$. Likewise there exists $a_2 \neq 0$ in E such that $I_2 a_2 = E$ – therefore $I_1 I_2 a_2 \neq \{0\} \Rightarrow I_1 I_2 \not\subset I$. □

Lemma 4.5.5.6 An ideal I in A is prime iff $xAy \subset I \Rightarrow x \in I$ or $y \in I$.

Proof Assume that I is prime and that $xAy \subset I$ – then $(AxA)(AyA) \subset I \Rightarrow AxA \subset I$ or $AyA \subset I$. Let $\mathfrak{X}$ denote the principal ideal generated by x – then $\mathfrak{X}^3 \subset AxA$. Consequently, if $AxA \subset I$, then $\mathfrak{X}^3 \subset I \Rightarrow \mathfrak{X} \subset I \Rightarrow x \in I$; similarly $AyA \subset I \Rightarrow y \in I$. Conversely suppose that I is an ideal in A for which the condition of the lemma holds. Let I_1 and I_2 be ideals in A such that $I_1 \not\subset I$, $I_2 \not\subset I$; select $x_1 \in I_1, x_2 \in I_2$ such that $x_1 \notin I, x_2 \notin I$; since $x_1 A x_2 \subset I_1 I_2$, we must have $I_1 I_2 \not\subset I$. □

Now retain the assumption that G is *connected* but drop for the moment the supposition concerning a compact subgroup K of G. Let U be a topologically irreducible Banach representation of G on E. Consider the representation U_∞ of $\mathfrak{G}$ on the space E_∞ of differentiable vectors in E (for U); let I_∞ denote the kernel of U_∞. If $U^\vee$ is the contragredient representation of G on $E^\vee$, $E_\infty^\vee$ the subspace of differentiable vectors for $U^\vee$, then $E_\infty^\vee$ is weakly dense in E^* and for all $a \in E_\infty$, $\mathbf{z} \in E_\infty^\vee$, $D \in \mathfrak{G}$, we have

$$\langle U_\infty(D)a, \mathbf{z}\rangle = \langle a, U_\infty^\vee(D^+)\mathbf{z}\rangle$$

($D \mapsto D^+$ the principal anti-automorphism of $\mathfrak{G}$ over $\mathbf{C}$).

Note that $U_\infty(D) = 0 \Leftrightarrow U^\vee_\infty(D^+) = 0$ $(D \in \mathfrak{G})$; plainly $\operatorname{Ker}(U_\infty) = \operatorname{Ker}(U_\omega)$.

Remark In the event that U is unitary, we have

$$(U_\infty(D)a, b) = (a, U_\infty(D^*)b) \qquad (a, b \in E_\infty;\ D \in \mathfrak{G}).$$

Therefore in this case I_∞ is $*$-stable and $\mathfrak{G}/I_\infty$ is a $*$-algebra in the obvious way (for the definition of the map $D \mapsto D^*(D \in \mathfrak{G})$, see 4.4.4).

It is in general false that U_∞ is an algebraically irreducible representation of $\mathfrak{G}$ even if U is unitary (Dixmier [17]) although I_∞ will be a primitive ideal in certain important special cases (cf. infra). However it is always true that I_∞ is prime.

Proposition 4.5.5.7 (Dixmier) If U is a topologically irreducible Banach representation of G, then $I_\infty = \operatorname{Ker}(U_\infty)$ is a prime ideal in $\mathfrak{G}$.

Proof Let $D_1, D_2 \in \mathfrak{G}$ and suppose that $U_\infty(D_1\mathfrak{G}D_2) = 0$ – we have to show that either $U_\infty(D_1) = 0$ or $U_\infty(D_2) = 0$. If $U_\infty(D_2)(E_\omega) = \{0\}$, then $U^\vee_\infty(D_2^+) = 0 \Rightarrow U_\infty(D_2) = 0$; if $U_\infty(D_2)(E_\omega) \neq \{0\}$, select $a \in E_\omega$ such that $U_\infty(D_2)a \neq 0$ – then $U_\infty(D_2)a \in E_\omega$ and $\mathfrak{G}U_\infty(D_2)a$ is dense in E (Corollary 4.4.5.5), whence $U^\vee_\infty(D_1^+) = 0 \Rightarrow U_\infty(D_1) = 0$. □

Let $\hat{G}$ denote the set of unitary equivalence classes of topologically irreducible unitary representations of G, $\operatorname{Prim}(\mathfrak{G})$ the set of primitive ideals in $\mathfrak{G}$.

Remarks (1) Let K be a connected compact subgroup of G as above; let U be a K-finite topologically irreducible Banach representation of G – then $\operatorname{Ker}(U_\infty) = \operatorname{Ker}(U_\omega) = \operatorname{Ker}(U_K)$ and so $\operatorname{Ker}(U_\infty)$ is primitive (Theorem 4.5.5.4).

(2) Let U be a topologically irreducible unitary representation of G, $\hat{U}$ its unitary equivalence class in $\hat{G}$ – then $\operatorname{Ker}(U_\infty)$ is primitive if G is nilpotent (Kirillov [1], Dixmier [14]) or if G is solvable (Dixmier [17]). The map $\hat{U} \mapsto \operatorname{Ker}(\hat{U}_\infty)$ of $\hat{G}$ into $\operatorname{Prim}(\mathfrak{G})$ is a bijection between $\hat{G}$ and the $*$-stable elements in $\operatorname{Prim}(\mathfrak{G})$ if G is nilpotent (Dixmier [14]) but this may be false if G is solvable (Dixmier [16]).

4.5.6 Jordan-Hölder Series – Multiplicities

Let G be a locally compact unimodular group countable at infinity, K a *large* compact subgroup of G. Let $\hat{K}$ denote the set of all equivalence classes of finite dimensional irreducible representations of K.

Definition Let U be a Banach representation of G on E – then by a *reducing series* or *Jordan-Hölder series* for U we shall mean such a series for the integrated form of U. The series is *discrete* for U if it is discrete for the integrated form of U.

[Naturally one could just as well employ the lift of U to $C_c(G)$ instead of $M_c(G)$.]

Let U be a K-finite Banach representation of G on E; if $\tilde{U}$ is a subrepresentation of U acting on a closed subspace $\tilde{E}$ of E, then the K-finiteness of U clearly implies that of $\tilde{U}$ and that of the quotient representation $U/\tilde{U}$ on $E/\tilde{E}$.

Proposition 4.5.6.1 Let U be a K-finite Banach representation of G on E – then every reducing series $\mathscr{E}$ for U is discrete (in particular U itself is discrete).

Proof Because a subquotient of a K-finite Banach representation is also K-finite, it will be enough to prove that there is some pair of elements of $\mathscr{E}$ which are adjacent in $\mathscr{E}$. Let δ_0 be an element of $\hat{K}$ which occurs in $U|K$ – then it is clear that for any $\tilde{E} \in \mathscr{E}$, we have $\mathsf{P}(\delta_0)\tilde{E} = \tilde{E} \cap E(\delta_0)$. This being so, let

$$\mathscr{E}_1 = \{\tilde{E}_1 \in \mathscr{E}: \tilde{E}_1 \cap E(\delta_0) = \{0\}\}, \mathscr{E}_2 = \{\tilde{E}_2 \in \mathscr{E}: \tilde{E}_2 \cap E(\delta_0) \neq \{0\}\};$$

since $\mathsf{P}(\delta_0)$ is continuous, $E_1 = (\bigcup \mathscr{E}_1)^{cl}$ belongs to $\mathscr{E}_1$; since $\mathscr{E}_2$ is non-empty and nested and $E(\delta_0)$ is finite dimensional, $E_2 = \bigcap \mathscr{E}_2$ belongs to $\mathscr{E}_2$ – therefore $E_1 \subsetneq E_2$. Now, for any $\tilde{E} \in \mathscr{E}$, either $\tilde{E} \in \mathscr{E}_1$, in which case $\tilde{E} \subset E_1$, or else $\tilde{E} \in \mathscr{E}_2$, in which case $E_2 \subset \tilde{E}$ – hence E_1 and E_2 are adjacent in $\mathscr{E}$. The fact that U is discrete is then seen to be a consequence of Proposition 4.2.1.10(ii). □

As is well-known, if U is a finite dimensional representation of an algebra A, then any two Jordan-Hölder series for U have the same irreducible quotients, with the same multiplicities – we shall now show that this is also true for K-finite representations of G (which, in view of the preceding proposition, are automatically discrete).

Let $\bar{G}$ be the set of Naimark equivalence classes of TCI Banach representations of G (cf. 4.5.1 – we recall that, by assumption, K is large in G). Let U be a K-finite Banach representation of G on E; we shall say that a class $\bar{V} \in \bar{G}$ *occurs in* U if some irreducible component of U belongs to $\bar{V}$ (i.e. if some topologically irreducible subquotient of U belongs to $\bar{V}$). If $\mathscr{E}$ is a Jordan-Hölder series for U, then $\bar{V}(\bar{V} \in \bar{G})$ is said to *occur in* $\mathscr{E}$ if there are adjacent subspaces E_1 and E_2 in $\mathscr{E}$ (with $E_1 \subsetneq E_2$) such that the quotient representation on E_2/E_1 deduced from U is in the class $\bar{V}$. [To avoid undue discursiveness, we shall often just say that V occurs in U (or in $\mathscr{E}$).]

Theorem 4.5.6.2 (Fell) Let U be a K-finite Banach representation of G on E; let $\mathscr{E}$ be a Jordan-Hölder series for U; given $\bar{V}$ in $\bar{G}$, let $m(\mathscr{E}; \bar{V})$ denote the cardinality of the set of all pairs (E_1, E_2) (with $E_1, E_2 \in \mathscr{E}$, $E_1 \subsetneq E_2$) having the property that the subquotient of U acting on E_2/E_1 is topologically irreducible and in the class $\bar{V}$. If $\delta \in \hat{K}$ and if δ occurs in $V|K$, then $m(\mathscr{E}; \bar{V})$ equals the multiplicity of V_δ in U_δ (hence $m(\mathscr{E}; \bar{V})$ is finite and independent of $\mathscr{E}$).

[Consequently $\bar{V}$ occurs in U iff $m(\mathscr{E}; \bar{V}) \geqslant 1$.]

The number $m(\mathscr{E}; \bar{V})(\bar{V} \in \bar{G})$ (being independent of the choice of the Jordan-Hölder series $\mathscr{E}$) will be denoted by $m(U; \bar{V})$ (or even $m(U; V)$) and referred to as the *multiplicity* of $\bar{V}$ (or V) in U.

Remark Let U be a K-finite Banach representation of G on E and assume that there exists a linearly independent family $\{E_i : i \in I\}$ of closed U-stable subspaces of E whose linear span is dense in E; if $U_i = U | E_i$, then it is not difficult to see that, for each $\bar{V} \in \bar{G}$, $m(U; V) = \sum_i m(U_i; V)$ (the summands on the right being zero except for finitely many i) – thus a class $\bar{V}$ in $\bar{G}$ occurs in U iff it occurs in U_i for some i. [These observations are applicable, in particular, when U is a K-finite unitary representation of G on a separable Hilbert space E, say (such a representation is discretely decomposable – cf. Theorem 4.5.7.1 infra and Proposition 4.3.1.8).]

Theorem 4.5.6.2 above is an immediate consequence of the following lemmas.

Lemma 4.5.6.3 Let U be a K-finite Banach representation of G on E, $\tilde{E}$ a closed U-stable subspace of E; let $\tilde{U}$ and $U/\tilde{U}$ denote the subrepresentation and quotient representation deduced from U, acting on $\tilde{E}$ and $E/\tilde{E}$, respectively – then, for $\delta \in \hat{K}$, $\tilde{U}_\delta$ is the subrepresentation of U_δ acting on $\tilde{E} \cap E(\delta)$ whereas $(U/\tilde{U})_\delta$ is the quotient representation deduced from U_δ, acting on $E(\delta)/\tilde{E} \cap E(\delta)$.

[The proof of this lemma is straightforward.]

Lemma 4.5.6.4 Let U be a K-finite Banach representation of G on E; let $\mathscr{E}$ be a Jordan-Hölder series for U; let $V \in \bar{V} \in \bar{G}$; let δ be an element of $\hat{K}$ which occurs in $V | K$ – then the following conditions are equivalent:

(i) V occurs in U;

(ii) V_δ occurs (as a subquotient) in U_δ;

(iii) V occurs in $\mathscr{E}$.

Proof (i) $\Rightarrow$ (ii) – this follows from Lemma 4.5.6.3 above.

(iii) $\Rightarrow$ (i) – this is obvious.

(ii) $\Rightarrow$ (iii). For each $\tilde{E}$ in $\mathscr{E}$, let $\tilde{E}(\delta) = \tilde{E} \cap E(\delta)$ $(= \mathrm{P}(\delta)\tilde{E})$ – thus $\{\tilde{E}(\delta) : \tilde{E} \in \mathscr{E}\}$ is a nested family of U_δ-stable subspaces of $E(\delta)$. Choose $\tilde{E}_1(\delta)$, $\tilde{E}_2(\delta)$ in this family with $\tilde{E}_1(\delta) \subsetneq \tilde{E}_2(\delta)$ and having the property that for all $\tilde{E} \in \mathscr{E}$, either $\tilde{E}(\delta) \subset \tilde{E}_1(\delta)$ or $\tilde{E}_2(\delta) \subset \tilde{E}(\delta)$ (such a choice is possible, $E(\delta)$ being finite dimensional). Let E_1 be the closure of $\bigcup \{\tilde{E} \in \mathscr{E}: \tilde{E}(\delta) \subset \tilde{E}_1(\delta)\}$; let $E_2 = \bigcap \{\tilde{E} \in \mathscr{E}: \tilde{E}_2(\delta) \subset \tilde{E}(\delta)\}$ – then $E_i \in \mathscr{E}$, $E_i \cap E(\delta) = \tilde{E}_i(\delta)$ $(i = 1, 2)$. Furthermore E_1 and E_2 are adjacent in $\mathscr{E}$, hence strictly adjacent. [In fact, if $\tilde{E} \in \mathscr{E}$, then either $\tilde{E}(\delta) \subset \tilde{E}_1(\delta)$, in which case $\tilde{E} \subset E_1$, or else $\tilde{E}_2(\delta) \subset \tilde{E}(\delta)$, in which case $E_2 \subset \tilde{E}$.] Consequently, if W is the subquotient of U acting on E_2/E_1, then W is K-finite and topologically irreducible; Lemma 4.5.6.3 tells us that W_δ is the subquotient of U_δ acting on $\tilde{E}_2(\delta)/\tilde{E}_1(\delta)$ – in particular W_δ is algebraically irreducible. On the other hand $\{\tilde{E}(\delta) : \tilde{E} \in \mathscr{E}\}$ is a Jordan-Hölder series for U_δ; because V_δ occurs in U_δ, we may assume that $\tilde{E}_1(\delta)$ and $\tilde{E}_2(\delta)$

were chosen so that $V_\delta = W_\delta$ – but this implies that V and W are Naimark equivalent (Proposition 4.5.1.6), whence V occurs in $\mathscr{E}$.

The proof of the lemma is now complete. □

4.5.7 Theorems of Finitude

Let G be a locally compact unimodular group satisfying the second axiom of countability, K a compact subgroup of G. Let $\hat{K}$ denote the set of all equivalence classes of finite dimensional irreducible representations of K.

Theorem 4.5.7.1 Let U be a K-finite Banach representation of G on E; put $\rho(x) = \| U(x) \|$ $(x \in G)$ – then, for every $f \in C_\rho(G)$, the operator $U(f)$ is completely continuous.

Proof If $\bar{\chi}_\delta * f = f (\delta \in \hat{K})$, then we have $\mathsf{P}(\delta)U(f) = U(f)$, and so, since $\mathsf{P}(\delta)$ has finite rank, the theorem is obvious in this case. Now take an arbitrary $f \in C_\rho(G)$; in the decomposition of the left regular representation of K on $C_\rho(G)$ the corresponding projection operators $\mathsf{P}_L(\delta)$ are given by $f \mapsto \bar{\chi}_\delta * f$ – therefore our f is a strong limit of linear combinations of functions $\bar{\chi}_\delta * f$. Since $\| U(f) \| \leqslant \| f \|_\rho$, it follows that the operator $U(f)$ is a uniform limit of operators of finite rank; of course this suffices to prove the theorem. □

Corollary 4.5.7.2 If U is a K-finite unitary representation of G, then $U(f)$ is completely continuous for every f in $L^1(G)$.

Before we formulate our next result, it will be convenient to digress momentarily and recall some well-known terminology.

Definition Let G be a locally compact group which satisfies the second axiom of countability; let $C^*(G)$ be the group C^*-algebra of G (cf. Vol. II, number 7.1.1).

(i) G is said to be *liminaire* provided $C^*(G)$ is liminaire. [Thus, for every topologically irreducible $*$-representation U of $C^*(G)$, the operator $U(f)$ is completely continuous (all $f \in C^*(G)$).]

(ii) G is said to be *postliminaire* provided $C^*(G)$ is postliminaire. [Thus every non-zero quotient C^*-algebra of $C^*(G)$ possesses a non-zero closed two-sided ideal which is liminaire.]

[If G is liminaire, then G is postliminaire but the converse is false.]

We recall that for G to be liminaire, it is both necessary and sufficient that in any irreducible unitary representation of G the operator $U(f)$ $(f \in L^1(G))$ is completely continuous (all $f \in L^1(G)$).

Note The reader will find a complete discussion of the above generalities in Dixmier [15].

Example Suppose that G is a real linear algebraic group – then G is postliminaire (for the proof, see Dixmier [4]).

The importance of the following result should be evident.

Theorem 4.5.7.3 (Harish-Chandra) Let G be a locally compact unimodular group satisfying the second axiom of countability. If G admits a large compact subgroup K, then G is type I.

Proof The preceding corollary implies that G is liminaire – therefore our assertion is clear on the basis of well-known generalities (cf. Dixmier [15, p. 271]). □

We shall now confine our attention to (not necessarily unitary) representations U of G on a Hilbert space E (say); this being so, the reader will find in the Appendix to this number a rapid summary of the basic facts concerning Hilbert-Schmidt operators and the like – these results will be used without comment in what follows.

Theorem 4.5.7.4 (Harish-Chandra) Let U be a continuous representation of G on a Hilbert space E with the property that there exists an integer $m_U \geqslant 1$ such that $\dim(E(\delta)) \leqslant m_U d(\delta)^2$ $(\delta \in \hat{K})$; let f be a measurable and square integrable function on G which vanishes outside of a compact set – then the operator $\int_G f(x)U(x)d_G(x)$ is of the Hilbert-Schmidt class.

Proof Let us begin with some remarks of a general nature. First of all, notice, in view of the Schwarz inequality, that the operator $\int_G f(x)U(x)d_G(x)$ is a well-defined bounded linear operator. Next let $Q: E \rightarrow E$ be a linear bicontinuous bijection of E; put $\tilde{U}(x) = QU(x)Q^{-1}(x \in G)$ – then

$$\left\| \int_G f(x)U(x)d_G(x) \right\|_{HS} = \left\| Q^{-1} \int_G f(x)\tilde{U}(x)d_G(x)Q \right\|_{HS}$$
$$\leqslant \|Q^{-1}\| \cdot \left\| \int_G f(x)\tilde{U}(x)d_G(x) \right\|_{HS} \|Q\|$$

and so we see that it is enough to prove that the corresponding operator for an equivalent representation is of the HS-class. Here is how this observation will be used. Consider the restriction of U to K; since K is compact, there exists a linear bicontinuous bijection Q of E such that the representation $k \mapsto QU(k)Q^{-1}$ $(k \in K)$ is unitary. Set $\tilde{U}(x) = QU(x)Q^{-1}$ $(x \in G)$ – then $\tilde{U}$ is a (continuous) representation of G on E which is equivalent to U and, moreover, has the property that the various subspaces $\tilde{E}(\delta)$ consisting of those elements in E which transform under $\tilde{U}|K$ according to δ are mutually orthogonal. Consequently, without loss of generality, it may be supposed that our representation U has the property that $U|K$ is unitary and hence that the $E(\delta)$ $(\delta \in \hat{K})$ are mutually orthogonal.

(1) Let ω be a compact set outside of which f vanishes; let $\mathcal{O}_\omega$ be a compact neighborhood of ω in G; let $\phi \in C_c^+(G)$ be such that $\phi = 1$ on $K\mathcal{O}_\omega$. Choose a sequence $\{g_n\}$ of continuous functions on G which vanish outside of $\mathcal{O}_\omega$ and such that $\int_G |f(x) - g_n(x)|^2 d_G(x) \to 0$ as $n \to \infty$. Plainly then, $g_n \to f(L^1)$ and so it follows that as $n \to \infty$,

$$\left\| \int_G (f(x) - g_n(x))U(x)d_G(x) \right\| \to 0.$$

On the other hand, we shall see in (2) below that the sequence of operators $T_n = \int_G g_n(x)U(x)d_G(x)$ is a Cauchy sequence with respect to the Hilbert-Schmidt norm $\|\cdot\|_{HS}$. Therefore, since the space of HS-operators is complete with respect to this norm, there exists an operator T in this class such $\|T_n - T\|_{HS} \to 0$ (the discussion in (2) infra shows that each T_n is in the Hilbert-Schmidt class); but $\|T_n - T\| \leqslant \|T_n - T\|_{HS}$ – hence $\|T_n - T\| \to 0$. However we have just seen that $\|T_n - \int_G f(x)U(x)d_G(x)\| \to 0$ – thus $T = \int_G f(x)U(x)d_G(x)$ and so $\int_G f(x)U(x)d_G(x)$ is of the Hilbert-Schmidt class, as desired.

(2) It remains to prove that the sequence $\{T_n\}$ of operators is a Cauchy sequence relative to the HS-norm. For this purpose we shall establish the following estimate: Let $g \in C_c(G)$ be such that g vanishes outside of $\mathcal{O}_\omega$ – then there exists $N > 0$ (independent of g) such that $\|U(g)\|_{HS} \leqslant N\|g\|_2$. To get this, observe that

$$\int_G g(x)U(x)d_G(x) = \int_K \left(\int_G g(kx)U(kx)d_G(x) \right) dk$$

$\Rightarrow$

$$\left\| \int_G g(x)U(x)d_G(x) \right\|_{HS} \leqslant \int_G \left\| \int_K g(kx)U(kx)dk \right\|_{HS} d_G(x)$$

$$\leqslant \int_G \left\| \int_K g(kx)U(k)dk \right\|_{HS} \|U(x)\| d_G(x).$$

By hypothesis there exists an integer $m_U \geqslant 1$ such that no $\delta \in \hat{K}$ occurs more than $m_U d(\delta)$ times in $U|K$. On the other hand every $\delta \in \hat{K}$ occurs exactly $d(\delta)$ times in the left regular representation L of K on $L^2(K)$ and so, since the representation $k \mapsto U(k)$ is unitary, we may conclude that

$$\left\| \int_K g(kx)U(k)dk \right\|_{HS}^2 \leqslant m_U \left\| \int_K g(kx)L(k)dk \right\|_{HS}^2.$$

But the Peter-Weyl Theorem tells us that

$$\left\| \int_K g(kx)L(k)dk \right\|_{HS}^2 = \int_K |g(kx)|^2 dk;$$

therefore

$$\left\| \int_K g(kx)U(k)dk \right\|_{HS} \leqslant (m_U)^{1/2} \left(\int_K |g(kx)|^2 dk \right)^{1/2}.$$

Let $M = \sup_{K\mathcal{O}_\omega} \| U(x) \|$ – then

$$\begin{aligned}\left\| \int_G g(x)U(x)d_G(x) \right\|_{HS} &\leqslant M(m_U)^{1/2} \int_G \left(\int_K | g(kx) |^2 dk \right)^{1/2} d_G(x) \\ &= M(m_U)^{1/2} \int_G \phi(x) \left(\int_K | g(kx) |^2 dk \right)^{1/2} d_G(x) \\ &\leqslant N \left(\int_G | g(x) |^2 d_G(x) \right)^{1/2},\end{aligned}$$

where $N = M(m_U)^{1/2} \| \phi \|_2$. It is now a simple matter to show that the sequence $\{T_n\}$ is Cauchy with respect to the HS-norm – thus

$$\left\| \int_G (g_m(x) - g_n(x))U(x)d_G(x) \right\|_{HS} \leqslant N \| g_m - g_n \|_2$$

and so, as the sequence $\{g_n\}$ is L^2-convergent (to f), our contention is clear.

This completes the proof of the theorem. □

Corollary 4.5.7.5 Let G be a locally compact unimodular group satisfying the second axiom of countability, K a uniformly large compact subgroup of G; let U be a TCI Hilbert representation of G on E; let f be a measurable and square integrable function on G which vanishes outside of a compact set – then the operator $\int_G f(x)U(x)d_G(x)$ is of the Hilbert-Schmidt class.

Remark To say that G is liminaire means that under every irreducible unitary representation of G, $L^1(G)$ is represented by completely continuous operators; in this connection Baggett [1] proved: Let G be a locally compact group, U a unitary representation of G on a Hilbert space E; if $U(f)$ is a Hilbert-Schmidt operator for every $f \in L^1(G)$, then E is finite dimensional (cf., however, Dixmier [15, p. 85]). [Consequently, if U is an infinite dimensional (irreducible) unitary representation of G, then there must exist an element $f \in L^1(G)$ for which the operator $U(f)$ has infinite dimensional range.]

Continuing in the same vein as above, we shall now derive an important sharpening of Theorem 4.5.7.4 in the case when G is a Lie group (namely Theorem 4.5.7.6 infra). As will be seen in due course, this result plays a central role in the theory of characters on G.

Let G be a unimodular Lie group.

Theorem 4.5.7.6 (Harish-Chandra) Let U be a continuous representation of G on a Hilbert space E with the property that there exists an integer $m_U \geqslant 1$ such that $\dim (E(\delta)) \leqslant m_U d(\delta)^2 (\delta \in \hat{K})$; let $f \in C_c^\infty(G)$ – then the operator $\int_G f(x)U(x)d_G(x)$ is of the trace class.

Corollary 4.5.7.7 Let G be a unimodular Lie group, K a uniformly large compact subgroup of G; let U be a TCI Hilbert representation of

G on E; let $f \in C_c^\infty(G)$ – then the operator $\int_G f(x)U(x)d_G(x)$ is of the trace class.

Before giving the proof of Theorem 4.5.7.6 it will be convenient to first make a series of elementary observations. Let $\mathfrak{L}(E; E)$ denote the space of all bounded linear transformations on E equipped with the norm topology – then $\mathfrak{L}(E; E)$ is a Banach space. Let $\mathfrak{E}$ denote the closure in $\mathfrak{L}(E; E)$ of the space spanned by the operators $U(f)(f \in C_c^\infty(G))$. For $x \in G$ one has

$$U(x)U(f) = \int_G f(x^{-1}y)U(y)d_G(y), \; U(f)U(x^{-1}) = \int_G f(yx)U(y)d_G(y)$$

and therefore it is clear that if $A \in \mathfrak{E}$, then both $U(x)A$ and $AU(x^{-1})$ are also in $\mathfrak{E}$.

Lemma 4.5.7.8 (i) The representations l and r of G on $\mathfrak{E}$ defined by

$$l(x)A = U(x)A, \; r(x)A = AU(x^{-1}) \; (x \in G, A \in \mathfrak{E})$$

are Banach representations of G;

(ii) For every $f \in C_c^\infty(G)$, the operator $U(f)$ is differentiable under l and r;

(iii) For any class $\delta \in \hat{K}$, let $\mathsf{P}(\delta), \mathsf{P}_l(\delta), \mathsf{P}_r(\delta)$ denote the canonical projections corresponding to U, l and r, respectively – then $\mathsf{P}_l(\delta)A = \mathsf{P}(\delta)A$; $\mathsf{P}_r(\delta)A = A\mathsf{P}(\check{\delta})$ $(A \in \mathfrak{E})$;

(iv) Let L (respectively R) denote the left (respectively right) regular representation of G on $L^2(G)$ – then, for $D_1, D_2 \in \mathfrak{G}$, one has

$$l_\infty(D_1)r_\infty(D_2)U(f) = U(L_\infty(D_1)R_\infty(D_2)f).$$

[The proofs of the above assertions are all straightforward; thus, for example, in order to establish the continuity of l and r, it is sufficient to prove that

$$\lim_{x\to 1, y\to 1} \| U(x)AU(y^{-1}) - A \| = 0$$

for all A in $\mathfrak{E}$. So fix an $\epsilon > 0$; given $A \in \mathfrak{E}$, choose $f \in C_c^\infty(G)$ such that $\| A - U(f) \| \leqslant \epsilon$. Let $\omega = \omega^{-1}$ be a compact neighborhood of 1 in G; put $M = \sup_{z\in\omega} \| U(z) \|$ – then, for all $x, y \in \omega$, we have

$$\| U(x)AU(y^{-1}) - U(x)U(f)U(y^{-1}) \| \leqslant \epsilon M^2,$$

whence

$$\begin{aligned}\| U(x)AU(y^{-1}) - A \| &\leqslant \epsilon(M^2 + 1) + \| U(x)U(f)U(y^{-1}) - U(f) \| \\ &\leqslant \epsilon(M^2 + 1) + \int_G | f(x^{-1}zy) - f(z) | \cdot \| U(z) \| \, d_G(z).\end{aligned}$$

Fix a neighborhood Ω of 1 in G $(\Omega \subset \omega)$ such that $| f(x^{-1}zy) - f(z) | \leqslant \epsilon$ whenever $x, y \in \Omega$; fix a real, non-negative, compactly supported, continuous function F on G which is equal to 1 on spt (f) – then

$$\|U(x)AU(y^{-1})-A\|\leqslant\epsilon(M^2+1)+\epsilon N\cdot\int_G F(z)d_G(z)$$
$$(N=\sup_{z\in\omega\cdot\text{spt}(f)\cdot\omega}\|U(z)\|)$$

provided $x, y\in\Omega\ldots$.]

Define now a Banach representation $\mathfrak{U}$ of the group $G\times G$ on $\mathfrak{E}$ as follows:

$$\mathfrak{U}(x,y)A=l(x)r(y)A=U(x)AU(y^{-1})\qquad(x,y\in G;\ A\in\mathfrak{E}).$$

It is obvious that every vector A in $\mathfrak{E}$ of the form $A=U(f)$ $(f\in C_c^\infty(G))$ is differentiable under $\mathfrak{U}$. In addition, the canonical projections for the representation $\mathfrak{U}$ (with respect to the subgroup $K\times K$) are the operators $\mathsf{P}_l(\delta_1)\mathsf{P}_r(\delta_2)$ $(\delta_1,\delta_2\in\hat{K})$.

There is no loss of generality in assuming that K is connected (cf. the Remark following Corollary 4.5.3.2) – therefore, according to Proposition 2.4.3.4, there exists a *central* element $D_0\in\mathfrak{K}$ with the property that $\mu_\delta(D_0)=d(\delta)^2\mu_\delta(1)$ for every $\delta\in\hat{K}$ (μ_δ any representation in the class δ). [Here $\mathfrak{K}$ denotes the universal enveloping algebra of $\mathfrak{k}_c$, $\mathfrak{k}$ the derived algebra of the Lie algebra $\mathfrak{k}$ of K.] Applying Theorem 4.4.2.1 to the representation $\mathfrak{U}$ and the differentiable element $U(L_\infty(D_0)R_\infty(D_0)f)$, we see that

$$\sum_{\delta_1,\delta_2\in\hat{K}}\|\mathsf{P}_l(\delta_1)\mathsf{P}_r(\delta_2)U(L_\infty(D_0)R_\infty(D_0)f)\|<\infty.$$

However

$$\begin{aligned}\mathsf{P}_l(\delta_1)\mathsf{P}_r(\delta_2)U(L_\infty(D_0)R_\infty(D_0)f)&=\mathsf{P}_l(\delta_1)\mathsf{P}_r(\delta_2)l_\infty(D_0)r_\infty(D_0)U(f)\\&=d(\delta_1)^2d(\delta_2)^2\mathsf{P}_l(\delta_1)\mathsf{P}_r(\delta_2)U(f).\end{aligned}$$

Hence

$$(*)\quad\sum_{\delta_1,\delta_2\in\hat{K}}d(\delta_1)^2d(\delta_2)^2\|\mathsf{P}(\delta_1)U(f)\mathsf{P}(\delta_2)\|<\infty.$$

Proof of Theorem 4.5.7.6 (1) We shall first assume that the various subspaces $E(\delta)$ $(\delta\in\hat{K})$ are mutually orthogonal – we may then choose an orthonormal basis $\{a_i : i\in I\}$ for E by selecting an orthonormal basis $\{a_i : i\in I(\delta)\}$ for each $E(\delta)$ $(\delta\in\hat{K})$. Bearing in mind that $\dim(E(\delta))\leqslant m_U d(\delta)^2$ for all $\delta\in\hat{K}$, we have

$$\begin{aligned}\sum_{i,j\in I}|(U(f)a_i,a_j)|&=\sum_{\delta_1,\delta_2\in\hat{K}}\sum_{i\in I(\delta_1)}\sum_{j\in I(\delta_2)}|(U(f)a_i,a_j)|\\&\leqslant\sum_{\delta_1,\delta_2\in\hat{K}}\dim(E(\delta_1))\dim(E(\delta_2))\|\mathsf{P}(\delta_1)U(f)\mathsf{P}(\delta_2)\|\\&\leqslant m_U^2\sum_{\delta_1,\delta_2\in\hat{K}}d(\delta_1)^2d(\delta_2)^2\|\mathsf{P}(\delta_1)U(f)\mathsf{P}(\delta_2)\|<\infty\end{aligned}$$

((*) above).

This proves that $U(f)$ is of the trace class and even somewhat more (see the Appendix to this number, in particular Lemma 4).

(2) Now discard the assumption about the mutual orthogonality of

the spaces $E(\delta)$. As was noted during the proof of Theorem 4.5.7.4, one may always find a linear bicontinuous bijection Q of E such that, relative to the representation $x \mapsto QU(x)Q^{-1}$, the spaces $E(\delta)$ are, in fact, mutually orthogonal. Hence by (1) the operator $QU(f)Q^{-1}$ is summable – therefore $U(f)$ is of the trace class ($f \in C_c^\infty(G)$).

This completes the proof of the theorem. □

Remark It is natural to ask: To what extent is the hypothesis that f be C^∞ necessary in Theorem 4.5.7.6? In this connection let us mention the following result: If U is unitary and if the restriction of U to K is contained in the left regular representation L of K on $L^2(K)$ (i.e. if $m_U = 1$), then the operator $\int_G f(x)U(x)d_G(x)$ is of the trace class provided that $f \in C_c^{4p}(G)$ where p is an integer $> 1/4$ times the dimension of K. [Let $X_1, \ldots, X_k$ be a basis for the Lie algebra $\mathfrak{k}$ of K; put $\omega_- = \sum_i X_i^2$; according to Lemma 1 in the Appendix to 4.4.7, if p is an integer $> k/4$, then there exists a function $f_p \in L^1(K) \cap L^2(K)$ such that $L(f_p)$ is the closure of the inverse of $L_\infty((\omega_- - 1)^p)$; since K is compact, $L(f_p)$ is of Hilbert-Schmidt type and hence, a fortiori, $(U|K)(f_p)$ is too. Now suppose that f is a function on G with compact support; if f has $2p$ continuous derivatives, then $U(f) = (U|K)(f_p)U((\omega_- - 1)^p f)$ and so $U(f)$ is of Hilbert-Schmidt type; if f has $4p$ continuous derivatives, then $U(f) = [(U|K)(f_p)]^2 U((\omega_- - 1)^{2p} f)$ and so $U(f)$ is of the trace class.]

Appendix In this Appendix we shall collect some well-known facts concerning Hilbert-Schmidt operators and trace class operators – details may be found, for instance, in Dunford and Schwartz [2, Chapter XI].

Let E be a Hilbert space, $\{a_i : i \in I\}$ a complete orthonormal set in E. A bounded linear operator T on E is said to be a *Hilbert-Schmidt operator* in case the quantity $\|T\|_{HS} = (\sum_i \|Ta_i\|^2)^{1/2}$ is finite. The number $\|T\|_{HS}$ is called the *Hilbert-Schmidt norm* of T; its value is independent of the choice of the orthonormal basis $\{a_i : i \in I\}$. The set of Hilbert-Schmidt operators on E is a (two-sided) *-ideal in the Banach algebra $\mathfrak{L}(E; E)$ of all bounded linear operators on E; every Hilbert-Schmidt operator is completely continuous (but not conversely) and is the limit in the Hilbert-Schmidt norm of a sequence of operators with finite dimensional range – moreover the class of Hilbert-Schmidt operators is a Banach algebra under the norm $\|.\|_{HS}$ (in fact is an H^*-algebra ...).

Lemma 1 Let T be a Hilbert-Schmidt operator on E.

(i) The operator norm of T cannot exceed the Hilbert-Schmidt norm, i.e. $\|T\| \leqslant \|T\|_{HS}$;

(ii) If U is a unitary operator on E, then UTU^{-1} is a Hilbert-Schmidt operator and $\|T\|_{HS} = \|UTU^{-1}\|_{HS}$;

(iii) If A is a bounded linear operator on E, then $\|AT\|_{HS} \leqslant \|A\|\cdot\|T\|_{HS}$ and $\|TA\|_{HS} \leqslant \|T\|_{HS}\|A\|$.

A bounded linear operator A on E is said to be of the *trace class* if it is of the form $A = ST$, where S and T are Hilbert-Schmidt operators; if A is of the trace class and if $\{a_i : i \in I\}$ is an orthonormal basis for E, then the number $\sum_i (Aa_i, a_i)$ (which is independent of the choice of the a_i) is called the *trace of A* and is written tr (A) – the series $\sum_i (Aa_i, a_i)$ is absolutely convergent. The set of all trace class operators on E is a (two-sided) *-ideal in $\mathfrak{L}(E; E)$ containing the operators of finite rank; every trace class operator is a Hilbert-Schmidt operator (but not conversely); the norm $A \mapsto \|A\|_{TC} = \text{tr}((A^*A)^{1/2})$ equips the set of trace class operators with the structure of a Banach *-algebra in which the operators of finite rank are everywhere dense.

Lemma 2 The trace 'tr' is a linear functional on the set of trace class operators. If A is of the trace class and if T is an arbitrary bounded linear operator on E, then the products AT and TA are also in the trace class and one has $\operatorname{tr}(AT) = \operatorname{tr}(TA)$.

Lemma 3 Let A be a self-adjoint operator on E which is of the trace class – then A is completely continuous and its trace is equal to the sum of its eigenvalues (counted with multiplicities), the series being absolutely convergent.

Let A be a bounded linear operator on E – then A is said to be *summable* if there exists a linear homeomorphism Q of E and an orthonormal basis $\{a_i : i \in I\}$ for E such that

$$\sum_{i,j} |(QAQ^{-1}a_i, a_j)| < \infty.$$

Lemma 4 Let $\{a_i : i \in I\}$ be an orthonormal basis for E, A a bounded linear operator on E such that $\sum_{i,j} |(Aa_i, a_j))| < \infty$; if S and T are bounded linear operators on E, then SAT, TSA, ATS are all of the trace class and $\operatorname{tr}(SAT) = \operatorname{tr}(TSA) = \operatorname{tr}(ATS)$.

Corollary Let A be a bounded linear operator on E; if A is summable, then A is of the trace class.

4.5.8 Characters

Let G be a unimodular Lie group countable at infinity, K a uniformly large compact subgroup of G. Let $\hat{K}$ denote the set of all equivalence classes of finite dimensional irreducible representations of K. Let $\mathfrak{G}$ denote the universal enveloping algebra of $\mathfrak{g}_c$, $\mathfrak{g}_c$ the complexification of the Lie algebra $\mathfrak{g}$ of G; let $\mathfrak{Z}$ denote the center of $\mathfrak{G}$, Z the center of G.

In this number we shall lay the foundations for the theory of characters on G.

Let $C_*^\infty(G)$ denote the subalgebra of $C_c^\infty(G)$ consisting of the linear span of the $\bar{\chi}_{\delta_1} * C_c^\infty(G) * \bar{\chi}_{\delta_2}$ $(\delta_1, \delta_2 \in \hat{K})$ – then, relative to the inductive limit topology, $C_*^\infty(G)$ is dense in $C_c^\infty(G)$ (cf. Proposition 4.4.3.5).

Let U be a K-finite Banach representation of G on E. For each $f \in C_*^\infty(G)$, the operator $U(f)$ is of finite rank and so it makes sense to consider the linear functional T_U on $C_*^\infty(G)$ defined by the rule $f \mapsto \operatorname{tr}(U(f))$ $(f \in C_*^\infty(G))$ – we shall call T_U the *character* of U.

Remark Let U be a K-finite Banach representation of G on E, T_U its character. If T_U is continuous on $C_*^\infty(G)$ in the inductive limit topology, then T_U admits a unique extension to a distribution on G (of course this does not necessarily mean that $U(f)$ is of the 'trace class' with trace $T_U(f)$ for all f in $C_c^\infty(G)$).

Theorem 4.5.8.1 (Fell) Let U^1 and U^2 be K-finite Banach representations of G on E^1 and E^2, respectively – then $T_{U^1} = T_{U^2}$ iff $m(U^1; \bar{V}) = m(U^2; \bar{V})$ for all $\bar{V} \in \bar{G}$.

Proof (1) Let us assume that $T_{U^1} = T_{U^2}$; let $V \in \bar{V} \in \bar{G}$ and choose a $\delta \in \hat{K}$ which occurs in $V|K$ – then, by Theorem 4.5.6.2, $m(U^1; V)$ is the multiplicity of V_δ in U_δ^1 and similarly for $m(U^2; V)$. If $f \in C_{c,\delta}^\infty(G)$, then $\operatorname{tr}(U_\delta^1(f)) = T_{U^1}(f) = T_{U^2}(f) = \operatorname{tr}(U_\delta^2(f))$; in turn, this implies that $\operatorname{tr}(U_\delta^1(f)) = \operatorname{tr}(U_\delta^2(f))$ for all $f \in C_{c,\delta}(G)$. [Fix $f \in C_{c,\delta}(G)$; choose f_n in $C_{c,\delta}^\infty(G)$ such that $f_n \to f$ in $C_c(G)$ – then, in $\operatorname{Hom}_{\mathbf{C}}(E^i(\delta), E^i(\delta))$, $U_\delta^i(f_n) \to U_\delta^i(f)$ which implies that $\operatorname{tr}(U_\delta^i(f_n)) \to \operatorname{tr}(U_\delta^i(f))$ $(i = 1, 2)$. Because $\operatorname{tr}(U_\delta^1(f_n)) = \operatorname{tr}(U_\delta^2(f_n))$ (all n), it follows that $\operatorname{tr}(U_\delta^1(f)) = \operatorname{tr}(U_\delta^2(f))$.] Consequently, since characters determine multiplicities for finite dimensional representations of algebras, the multiplicities of V_δ in U_δ^1 and U_δ^2 are the same, and so $m(U^1; V) = m(U^2; V)$, as contended.

(2) Conversely, suppose that $m(U^1; \bar{V}) = m(U^2; \bar{V})$ for all $\bar{V} \in \bar{G}$ – then we are to prove that $T_{U^1} = T_{U^2}$. For this purpose, it will be enough to fix $\delta \in \hat{K}$ and show that $\operatorname{tr}(U^1(f)) = \operatorname{tr}(U^2(f))$ for all f in $C_{c,\delta}^\infty(G)$ (since $\operatorname{tr}(U^1(f)) = \operatorname{tr}(U^2(f)) = 0$ if $f \in \bar{\chi}_{\delta_1} * C_c^\infty(G) * \bar{\chi}_{\delta_2} (\delta_1 \neq \delta_2)$). The proof of Lemma 4.5.6.4 shows that each irreducible subquotient Q (say) of U_δ^1 or U_δ^2 is of the form V_δ ($V \in \bar{V} \in \bar{G}$, say). Therefore, from our hypothesis and Lemma 4.5.6.4, we deduce that $m(U_\delta^1; Q) = m(U^1; V) = m(U^2; V) = m(U_\delta^2; Q)$; because of the arbitrariness of Q and the fact that multiplicities determine characters for finite dimensional representations, we conclude, in particular, that $\operatorname{tr}(U^1(f)) = \operatorname{tr}(U_\delta^1(f)) = \operatorname{tr}(U_\delta^2(f)) = \operatorname{tr}(U^2(f))$ for all $f \in C_{c,\delta}^\infty(G)$.

This completes the proof of the theorem. □

Corollary 4.5.8.2 If U^1 and U^2 are K-finite Banach representations having the same character and if U^1 is topologically irreducible, then U^2 is also topologically irreducible and U^1 and U^2 are Naimark equivalent.

Proof On the basis of the preceding theorem, the topological irreducibility of U^2 is clear; the Naimark equivalence of U^1 and U^2 follows upon citing Proposition 4.5.1.6. □

There is no difficulty in verifying that if U^1 and U^2 are Naimark equivalent TCI Banach representations of G (and hence K-finite, K being large in G), then the characters T_{U^1} and T_{U^2} are equal; taking Corollary 4.5.8.2 into account, then, we see that it makes sense to consider the character $T_{\bar{U}}$ of the class $\bar{U} \in \bar{G}$.

It is also meaningful to speak of the character $T_{\hat{U}}$ of a class $\hat{U} \in \hat{G}$ – this is clear in view of the following result.

Corollary 4.5.8.3 (Harish-Chandra) Let U^i be a topologically irreducible unitary representation of G $(i = 1, 2)$ – then U^1 and U^2 are unitarily equivalent iff they have the same character.

Remark It should be observed that Theorem 4.5.8.1 implies that two K-finite *unitary* representations of G are determined to within Naimark equivalence by their characters; on the other hand, this is definitely false if it is a question of two arbitrary K-finite Banach representations of G (one may already find counterexamples when $G = \mathbf{SL}(2, \mathbf{R})$ or even when G is the proper Euclidean group of the plane (cf. the Remark in 4.5.4)).

Proposition 4.5.8.4 Let $U^1, \ldots, U^r$ be TCI Banach representations of G on $E^1, \ldots, E^r$; suppose that no two of them are Naimark equivalent – then the characters $T_{U^1}, \ldots, T_{U^r}$ are linearly independent.

Proof Consider a dependence relation of the form $\sum c_i T_{U^i} = 0$ ($c_i \in \mathbf{C}$) – we have to prove that all the c_i are zero. So suppose, to the contrary, that $c_1 \neq 0$; fix a class $\delta \in \hat{K}$ which occurs in $U^1 | K$; by rearranging the terms, if necessary, we can assume that δ occurs in $U^i | K$ for $i = 1, \ldots, s$ but not in $U^i | K$ for $i = s+1, \ldots, r$. If $f \in C_{c,\delta}^\infty(G)$, then

$$c_1 T_{U^1}(f) + \ldots + c_r T_{U^r}(f) = c_1 T_{U^1}(f) + \ldots + c_s T_{U^s}(f) = 0.$$

Now the representation U_δ^i of $C_{c,\delta}^\infty(G)$ on $E^i(\delta)$ is clearly algebraically irreducible ($C_{c,\delta}^\infty(G)$ is dense in $C_{c,\delta}(G)$) and, moreover, the U_δ^i are pairwise inequivalent (this follows from our assumption on the U^i and Proposition 4.5.1.6) ($i = 1, \ldots, s$); by standard algebra, then, the functions $f \mapsto \operatorname{tr}(U_\delta^i(f)) = T_{U^i}(f)$ ($f \in C_{c,\delta}^\infty(G)$, $i = 1, \ldots, s$) are linearly independent (cf. the Appendix infra), which contradicts our supposition on c_1. □

It is only natural to ask: When does a character of a given K-finite Banach representation of G extend to a distribution on G? The following important result gives an answer to this question.

Theorem 4.5.8.5 (Harish-Chandra) Let U be a continuous representation of G on a Hilbert space E with the property that there exists an integer $m_U \geqslant 1$ such that $\dim(E(\delta)) \leqslant m_U d(\delta)^2$ ($\delta \in \hat{K}$) – then, for every $f \in C_c^\infty(G)$, the operator $\int_G f(x)U(x)d_G(x)$ is of the trace class and the mapping T_U defined by the rule

$$f \mapsto \operatorname{tr}\left(\int_G f(x)U(x)d_G(x)\right) \qquad (f \in C_c^\infty(G))$$

is a distribution on G.

Proof Let us agree to adopt the notations which were used during the proof of Theorem 4.5.7.6 (there it was shown that the operator $\int_G f(x)U(x)d_G(x)$ is of the trace class for all $f \in C_c^\infty(G)$); in particular we may assume that K is connected and that the $E(\delta)$ ($\delta \in \hat{K}$) are mutually orthogonal.

(1) Let $f \in C_c^\infty(G)$ – then we claim that there exists an element $D_\infty \in \mathfrak{K}$, which is independent of f, with the property that $|T_U(f)| \leqslant m_U \| U(L_\infty(D_\infty D_0)f) \|$. Thus, to begin with, we have

$$|T_U(f)| \leqslant \sum_{\delta \in \hat{K}} |\operatorname{tr}(\mathsf{P}(\delta)U(f)\mathsf{P}(\delta))| \leqslant \sum_{\delta \in \hat{K}} \dim(E(\delta)) \, \| \mathsf{P}(\delta)U(f) \|$$
$$\leqslant m_U \sum_{\delta \in \hat{K}} d(\delta)^2 \, \| \mathsf{P}(\delta)U(f) \| = m_U \sum_{\delta \in \hat{K}} \| \mathsf{P}(\delta)U(L_\infty(D_0)f) \|.$$

The vector $U(L_\infty(D_0)f)$ is a differentiable vector for the Banach representation l of G on $\mathfrak{E}$ (cf. Lemma 4.5.7.8); the Remark following Lemma

4.4.2.4 then provides us with an element $D_\infty \in \mathfrak{K}$ (which does not depend on f) such that

$$\sum_{\delta \in \hat{K}} \| \mathsf{P}(\delta) U(L_\infty(D_0) f) \| = \sum_{\delta \in \hat{K}} \| \mathsf{P}_l(\delta) U(L_\infty(D_0) f) \| \quad \text{(Lemma 4.5.7.8 (iii))}$$

$$\leqslant \| l_\infty(D_\infty) U(L_\infty(D_0) f) \| = \| U(L_\infty(D_\infty D_0) f \| \quad \text{(Lemma 4.5.7.8 (iv)).}$$

Therefore $|T_U(f)| \leqslant m_U \| U(L_\infty(D_\infty D_0) f) \|$ (all $f \in C_c^\infty(G)$), as claimed.

(2) Now suppose that ω is a compact set in G and $\{f_n\}$ is a sequence of functions in $C_c^\infty(G)$ such that each f_n vanishes outside of ω and, for any $D \in \mathfrak{G}$, $L_\infty(D) f_n \to 0$ uniformly on ω – then

$$\| U(L_\infty(D) f_n) \| \leqslant \int_G | L_\infty(D) f_n(x) | \cdot \| U(x) \| \, d_G(x) \to 0$$

as $n \to \infty$ (all $D \in \mathfrak{G}$). Therefore, thanks to the estimate in (1), we see that

$$| T_U(f_n) | \leqslant m_U \| U(L_\infty(D_\infty D_0) f_n) \| \to 0$$

which shows that T_U is a distribution.

Hence the theorem. □

The reader will observe that the hypotheses of Theorem 4.5.8.5 are fulfilled by every TCI Hilbert representation of G (since K is *uniformly* large in G), in particular by every irreducible unitary representation of G (so that in these cases the character T_U, defined initially only on $C_*^\infty(G)$, admits a natural (unique) extension to all of $C_c^\infty(G)$ (which will, of course, be denoted by the same symbol and referred to as the character of U)).

Remark Let us suppose that U is an irreducible unitary representation of G on E, T_U its character – then T_U is of *positive type*, that is, $T_U(f * f^*) \geqslant 0$ for all $f \in C_c^\infty(G)$ (here $f \mapsto f^*$ is the usual involution in $C_c^\infty(G)$ – thus $f^*(x) = \overline{f(x^{-1})}$ $(x \in G)$). It is equally evident that in this case one has $T_U(f^*) = \overline{T_U(f)}$ $(f \in C_c^\infty(G))$.

A distribution T on G is said to be *central* if it is invariant under the canonical action of G on itself through inner automorphisms; a distribution T on G is said to be $\mathfrak{Z}$*-finite* if the space spanned by the $Z \cdot T$ $(Z \in \mathfrak{Z})$ is finite dimensional (in particular T is said to be an *eigendistribution* of $\mathfrak{Z}$ if there exists a homomorphism $\kappa : \mathfrak{Z} \to \mathbf{C}$ such that $Z \cdot T = \kappa(Z) T$ (all $Z \in \mathfrak{Z}$)); a distribution T on G is said to be an *eigendistribution* of Z (Z the center of G) if there exists a homomorphism $\zeta : Z \to \mathbf{C}^*$ with the property that $T^z = \zeta(z) T$ (all $z \in Z$ – here T^z denotes the 'right translate' of T by z).

Proposition 4.5.8.6 Let G be a connected unimodular Lie group, K a uniformly large compact connected subgroup of G; let U be a TCI

Hilbert representation of G on E, T_U the character of U; let ζ_U denote the central character of U, κ_U the infinitesimal character of U – then T_U is a central distribution on G verifying

$$T_U^z = \zeta_U(z)T_U \qquad (z \in Z),\ Z\cdot T_U = \kappa_U(Z)T_U \qquad (Z \in \mathfrak{Z}).$$

Proof The G-centrality of T_U is clear in view of standard properties of the function 'tr'. To see that T_U is an eigendistribution of the center Z of G, fix $f \in C_c^\infty(G)$ and observe that

$$\begin{aligned} T_U^z(f) = T_U(f^{z^{-1}}) = \operatorname{tr}(U(f^{z^{-1}})) &= \operatorname{tr}\Big(\int_G f^{z^{-1}}(x)U(x)d_G(x)\Big) \\ &= \operatorname{tr}\Big(\int_G f(xz^{-1})U(x)d_G(x)\Big) \\ &= \operatorname{tr}\Big(\int_G f(x)U(xz)d_G(x)\Big) = \zeta_U(z)T_U(f). \end{aligned}$$

Hence $T_U^z = \zeta_U(z)T_U$ (all $z \in Z$). As for the last assertion of our proposition one can, for instance, reason as follows. By passing to an equivalent representation, if necessary, we may assume that the $E(\delta)$ ($\delta \in \hat{K}$) are mutually orthogonal; this being so, for all $f \in C_c^\infty(G)$ we then have

$$\begin{aligned} Z\cdot T_U(f) = T_U(Z^t f) &= \sum_{\delta\in\hat{K}} \int_G Z^t f(x)\psi_\delta^U(x)d_G(x) \\ &= \sum_{\delta\in\hat{K}} \int_G f(x)Z\psi_\delta^U(x)d_G(x) = \kappa_U(Z) \sum_{\delta\in\hat{K}} \int_G f(x)\psi_\delta^U(x)d_G(x) \\ &= \kappa_U(Z)T_U(f), \end{aligned}$$

which implies that $Z\cdot T = \kappa_U(Z)T\,(Z \in \mathfrak{Z})$. [Here we have taken the liberty of introducing the spherical trace functions ψ_δ^U of Vol. II, number 6.1.1; we have also used the fact that $Z\psi_\delta^U = \kappa_U(Z)\psi_\delta^U$ (all $Z \in \mathfrak{Z}$) – cf. Proposition 6.1.2.1.] □

We shall now assume that G is a connected semi-simple Lie group with finite center Z – our goal will be to obtain some theorems of a general nature on the characters of G.

Note Everything that we shall have to say applies with no essential change to the reductive pairs (G, K) of number 1.1.5.

The theorems in question rest on two facts from the theory of distributions on G (see Vol. II, numbers 8.3.3 and 8.3.4).

Theorem 4.5.8.7 (Harish-Chandra) Let T be a central eigendistribution (of $\mathfrak{Z}$) on G – then, in the sense of distributions, T agrees with a locally summable function F_T (say) on G. Furthermore, the function F_T is analytic on the set G' of regular elements in G.

Theorem 4.5.8.8 (Harish-Chandra) Let T be a central eigendistribution (of $\mathfrak{Z}$) on G, F_T the locally summable function on G which agrees with T (in the sense of distributions); let $\mathfrak{j}$ be a Cartan subalgebra of $\mathfrak{g}$, J the Cartan subgroup of G associated with $\mathfrak{j}$ – then there exists a linear function Λ on $\mathfrak{j}_c$ with the following property. Given $j_0 \in J'$, we can find polynomial functions p_w ($w \in W$) on $\mathfrak{j}_c$ such that

$$F_T(j_0 \exp H) = |D_l(j_0 \exp H)|^{-1/2} \sum_{w \in W} p_w(H) e^{w(\Lambda+\rho)(H)}$$

for all H lying sufficiently near zero in $\mathfrak{j}$. The linear function $\Lambda + \rho$ is unique to within an operation of W – moreover if $[W_\Lambda]$ is the order of the subgroup W_Λ consisting of those $w \in W$ for which $w(\Lambda + \rho) = \Lambda + \rho$, then the degree of every p_w is necessarily less than $[W_\Lambda]$.

[The usual notations are in force here – thus W denotes the Weyl group of the pair $(\mathfrak{g}_c, \mathfrak{j}_c)$, ρ one-half the sum of the positive roots in some ordering etc. A comment on the linear function Λ is also in order. Let κ be the homomorphism of $\mathfrak{Z}$ into $\mathbf{C}$ with the property that $Z \cdot T = \kappa(Z)T$ ($Z \in \mathfrak{Z}$); Theorem 2.4.3.2 tells us that we can write $\kappa = \kappa_\Lambda$ for some linear function Λ on $\mathfrak{j}_c$ – this is the Λ which appears in the statement of Theorem 4.5.8.8.]

Let U be a TCI Hilbert representation of G, T_U its character – then, of course, Theorems 4.5.8.7 and 4.5.8.8 are applicable to the distribution T_U. In particular, note the resemblance of the 'local formula' for $F_{T_U} = F_U$ to those of Weyl for the irreducible characters of a compact connected semi-simple Lie group.

Remark Let G be a connected, simply connected nilpotent Lie group; let U be an irreducible unitary representation of G – then, for every C^∞ rapidly decreasing function f on G, the operator $U(f)$ is of the trace class and the mapping $f \mapsto \operatorname{tr}(U(f))$ defines a tempered distribution T_U (say) on G. In contradistinction with the semi-simple case, however, the character T_U is in general not defined by a locally summable function or even a measure (cf. Dixmier [7], [8], [9])).

Select Cartan subalgebras $\mathfrak{j}^1, \ldots, \mathfrak{j}^r$ of $\mathfrak{g}$ per Corollary 1.3.1.11; let J^i denote the Cartan subgroup of G associated with $\mathfrak{j}^i$ ($i = 1, \ldots, r$) – we recall that $G' = \bigcup_{i=1}^r \bigcup_{x \in G} x(J^i)'x^{-1}$ (Theorem 1.4.1.7). [The set G' of regular elements in G is, in general, not connected; on the other hand, it is always possible to find a finite number of connected components $G_1, \ldots, G_M$ of G' such that $ZG_i \cap ZG_j = \varnothing$ if $i \neq j$ and with the property that $G' = \bigcup_{k=1}^M ZG_k$ (cf. Corollary 1.4.1.10).]

Theorem 4.5.8.9 (Harish-Chandra) Let G denote a connected semi-simple Lie group with finite center Z; let $\mathfrak{Z}$ denote the center of $\mathfrak{G}$; let κ and ζ be non-zero homomorphisms of $\mathfrak{Z}$ and Z, respectively; let $\mathfrak{S}(\kappa, \zeta)$ denote the set of all TCI Hilbert representations U of G such that $\kappa_U = \kappa$, $\zeta_U = \zeta$ – then, apart from Naimark equivalence, $\mathfrak{S}(\kappa, \zeta)$ contains only a finite number of representations.

The theorem at hand depends on the following lemma.

Lemma 4.5.8.10 Retain the notations of Theorem 4.5.8.9; given $U \in \mathfrak{S}(\kappa, \zeta)$, let T_U denote its character, F_U the locally summable function on G which agrees with T_U in the sense of distribution theory – then the dimension of the linear space $\mathfrak{F}$ spanned over $\mathbf{C}$ by the functions F_U ($U \in \mathfrak{S}(\kappa, \zeta)$) cannot exceed $M[W]$ ($[W]$ the order of the Weyl group of $\mathfrak{g}_c$ with respect to any Cartan subalgebra $\mathfrak{j}_c$).

Proof Let us bear in mind that $G' = \bigcup_{k=1}^{M} ZG_k$. By assumption, if $U \in \mathfrak{S}(\kappa, \zeta)$, then we have

$$F_U(xz) = \zeta_U(z)F_U(x) = \zeta(z)F_U(x) \qquad (x \in G, z \in Z);$$

of course a similar relation holds for any $F \in \mathfrak{F}$; therefore a given $F \in \mathfrak{F}$ is completely determined by its restriction F_k to the components G_k $(k = 1, \ldots, M)$. Now the elements of $\mathfrak{F}$ are clearly G-central; furthermore, each component G_k is of the form $\bigcup_{x \in G} x(\mathrm{C}^{i_k})x^{-1}$ where C^{i_k} denotes a component of $(\mathrm{J}^{i_k})'$ $(1 \leqslant i_k \leqslant r)$ – therefore $F_k = F | G_k$ is known when it is known on C^{i_k} (or even on some non-empty open subset of C^{i_k}). Let F_k^c denote the restriction of F to C^{i_k} and let $\mathfrak{F}_k^c$ be the linear space consisting of all the $F_k^c (F \in \mathfrak{F})$ – then Theorem 4.5.8.8 tells us that $\dim(\mathfrak{F}_k^c) \leqslant [W]$. The linear map $F \mapsto (F_1^c, \ldots, F_M^c)$ of $\mathfrak{F}$ into the direct sum of the $\mathfrak{F}_k^c$ is, in view of what has been said above, injective. Hence $\dim(\mathfrak{F}) \leqslant M[W]$. □

Proof of Theorem 4.5.8.9 Let $F_{U^1}, \ldots, F_{U^n}$ be a basis for the linear space $\mathfrak{F}$ introduced in the preceding lemma; let U be an element of $\mathfrak{S}(\kappa, \zeta)$ – then there exist complex numbers c_i such that $F_U = \Sigma c_i F_{U^i}$, whence $T_U = \Sigma c_i T_{U^i}$. The validity of the theorem will be settled when it is shown that U is, of necessity, Naimark equivalent to one of the U^i. But if this were not the case, then, since the characters are linearly independent (Proposition 4.5.8.4), we would be led to an evident contradiction. Hence the theorem. □

The set G' has only a finite number of connected components (cf. Corollary 1.4.1.10). [Needless to say, the number of components of G' is, in general, strictly greater than M $(G' = \bigcup_{k=1}^{M} ZG_k)$.] With this in mind, it is clear that the proof of Theorem 4.5.8.9 yields the following result.

Theorem 4.5.8.11 (Harish-Chandra) Let G denote a connected semi-simple Lie group with finite center Z; let $\mathfrak{Z}$ denote the center of $\mathfrak{G}$; let κ be a homomorphism of $\mathfrak{Z}$ into $\mathbf{C}$ – then, apart from Naimark equivalence (= infinitesimal equivalence), there exists but a finite number of TCI Hilbert representations of G which have the infinitesimal character κ.

Remark It is instructive to compare Theorem 4.5.8.11 with the following weaker statement (which is an immediate consequence of Theorem 2.2.2.1). Let G denote a connected semi-simple Lie group with finite center Z; let $\mathfrak{Z}$ denote the center of $\mathfrak{G}$; let κ be a homomorphism of $\mathfrak{Z}$ into $\mathbf{C}$, δ a class in $\hat{K}$ – then, apart from Naimark equivalence (= infinitesimal equivalence), there exists but a finite number of TCI Hilbert representations of G which have the infinitesimal character κ and in which δ occurs.

Appendix The following lemma is frequently useful.

Lemma Let A be an associative algebra over an algebraically closed field $\mathbf{k}$ of characteristic zero; let $U_1, \ldots, U_r$ be non-trivial finite dimensional irreducible representations of A no two of which are equivalent – then the linear functions

$$x \mapsto \operatorname{tr}(U_i(x)) \qquad (x \in A; 1 \leqslant i \leqslant r)$$

are linearly independent over $\mathbf{k}$.

Proof There is clearly no loss of generality in assuming that A has a unit 1, say. This being the case, let U be the direct sum of the representations $U_1, \ldots, U_r$ – then $U(A)$ is a finite dimensional semi-simple algebra. Now the assignment $U(x) \mapsto U_i(x)$ $(x \in A)$ is, for each i $(1 \leqslant i \leqslant r)$, an irreducible representation of $U(A)$; thus $U(\operatorname{Ker}(U_i))$ is a maximal two-sided ideal in $U(A)$ and so, $U(A)$ being semi-simple, there exists a two-sided ideal I_i in $U(A)$ which is complementary to $U(\operatorname{Ker}(U_i))$ $(1 \leqslant i \leqslant r)$. Evidently each I_i is a simple ideal in $U(A)$; moreover, $I_i \neq I_j$ $(i \neq j)$. [In fact if $i \neq j$, then U_i is not equivalent to U_j, whence $\operatorname{Ker}(U_i) \neq \operatorname{Ker}(U_j) \ldots$.] Because $U(A)$ is the direct sum of its simple ideals, it follows that $U(\operatorname{Ker}(U_i)) \supset I_j$ $(i \neq j)$. Choose an element $e_i \in A$ such that $U(e_i)$ is the unit element of the simple algebra I_i – then, since $U(e_i) \in U(\operatorname{Ker}(U_j))$ $(i \neq j)$, we have $U_j(e_i) = 0$. One the other hand, $U(1 - e_i) \in U(\operatorname{Ker}(U_i))$ and so $U_i(1 - e_i) = 0$. Putting these facts together then leads to the conclusion that $\operatorname{tr}(U_j(e_i)) = 0$ if $i \neq j$ while $\operatorname{tr}(U_i(e_i)) = d_i$, d_i the degree of U_i. The assertion of the lemma is now obvious. □

4.5.9 Square Integrable Representations

We shall begin by formulating, in a general setting, certain results from the unitary representation theory of locally compact topological groups; after that, we shall specialize to the pairs (G, K) which have been the main object of study in the present section.

Let G be a locally compact unimodular group satisfying the second axiom of countability.

Definition An irreducible unitary representation U of G on E is said to be *square integrable* if there exists a non-zero vector a in E such that the coefficient $x \mapsto (U(x)a, a)$ is square integrable on G.

It turns out that if U is square integrable, then *every* coefficient of U lies in $L^2(G)$; furthermore there will then exist a unique number $d_U > 0$ (depending on the normalization of the Haar measure on G) such that

$$\int_G |(U(x)a, b)|^2 d_G(x) = d_U^{-1} \, \|a\|^2 \, \|b\|^2 \qquad (\text{all } a, b \in E).$$

The number d_U is called the *formal degree* (*dimension*) of U. [If G is compact (so that every irreducible unitary representation U of G is square integrable) and if d_G is normalized Haar measure on G, then d_U is the dimension of U in the usual sense.] The notion of square integrability, as well as that of formal degree, are invariant under unitary equivalence. A unitary equivalence class $\hat{U} \in \hat{G}$ is said to be *discrete* if every member of the class is square integrable; if $\hat{U}$ is discrete, then we write $d_{\hat{U}}$ for $d_U(U \in \hat{U})$. The set $\hat{G}_d$ of all discrete classes in $\hat{G}$ is called the *discrete series* (of irreducible unitary representations) for G.

Note If $\hat{G}_d$ is non-empty, then the center Z of G is necessarily compact (the converse assertion being false). Thus let U be an (irreducible) square integrable unitary representation of G; let a and b be two non-zero elements of E – then, in view of Proposition 4.3.1.6, the function $x \mapsto |(U(x)a, b)|$ $(x \in G)$ is biinvariant under Z and so

$$\int_G |(U(x)a, b)|^2 d_G(x) = \int_{G/Z} |(U(\dot{x})a, b)|^2 d_{G/Z}(\dot{x}) \int_Z d_Z(z) < \infty$$
$$(\dot{x} = xZ;\ x \in G)$$

which shows that the Haar measure of Z is finite . . . .

The (irreducible) square integrable unitary representations of G admit an alternative characterization – indeed, they are precisely those irreducible unitary representations of G which occur as (irreducible) subrepresentations of the left (equivalently right) regular representation of G on $L^2(G)$ (hence the term 'discrete series'). This being so, let U be an (irreducible) square integrable unitary representation of G on a Hilbert space E; let $\mathscr{E}_U$ be the smallest closed subspace of $L^2(G)$ containing the conjugates of the coefficients of U – then $\mathscr{E}_U$ depends only on the unitary equivalence class $\hat{U}$ of U. [In passing it should be pointed out that if V satisfies the same general conditions as U, then $\mathscr{E}_U$ and $\mathscr{E}_V$ are mutually orthogonal unless $\hat{U} = \hat{V}$; this follows from the Schur Orthogonality Relations (cf. infra).] Proceeding just as in the compact case, one shows without difficulty that the restriction to $\mathscr{E}_U$ of the left regular representation L of G on $L^2(G)$ is unitarily equivalent to dim $(U)U$ while the restriction to $\mathscr{E}_U$ of the right regular representation R of G on $L^2(G)$ is unitarily equivalent to dim $(\bar{U})\bar{U}$ ($\bar{U}$ the conjugate of U). Let $L^2(G)_d$ be the smallest closed subspace of $L^2(G)$ containing the $\mathscr{E}_{\hat{U}}$ $(\hat{U} \in \hat{G}_d)$ – then the restriction of L (or R) to $L^2(G)_d$ is certainly discretely decomposable and, moreover,

$$L^2(G)_d = \sum_{\hat{U} \in \hat{G}_d} \mathscr{E}_{\hat{U}},$$

the (orthogonal) sum being taken in the sense of Hilbert space theory.

These results, first noticed by Bargmann [1] for the case $G = \mathbf{SL}(2, \mathbf{R})$, were proved in complete generality by Godement [1, 2]; Godement's proofs depend on some theorems from the theory of von Neumann algebras (cf. the exposition of Dixmier [15, §14]) – on the other hand, as was observed by Harish-Chandra [11], it is possible to proceed somewhat more directly in the case of the pairs (G, K) of interest to us. [As far as Lie groups G are concerned, only certain reductive groups are known to have a non-empty discrete series; in fact, a celebrated theorem due to Harish-Chandra asserts that among the pairs (G, K) of number 1.1.5, only those with $\operatorname{rank}(G) = \operatorname{rank}(K)$ admit a non-empty discrete series – we shall deal with this question later on (cf. Vol. II, Chapter 10).]

Let G be a locally compact unimodular group satisfying the second axiom of countability, K a large compact subgroup of G. Let $\hat{K}$ denote the set of equivalence classes of finite dimensional irreducible representations of K.

Lemma 4.5.9.1 Let U be an irreducible unitary representation of G on a Hilbert space E; suppose that there exist two non-zero elements a_0, b_0 in E such that $\int_G |(U(x)a_0, b_0)|^2 d_G(x) < \infty$ – then there exists a (finite) positive real number d_U such that

$$\int_G |(U(x)a, b)|^2 d_G(x) = d_U^{-1} \|a\|^2 \|b\|^2$$

for all a, b in E.

Proof (1) We shall start off by showing that, without loss of generality, both a_0 and b_0 may be assumed to lie in $E(\delta_0)$, δ_0 a class in $\hat{K}$ which occurs in $U|K$. To this end, observe that the sets

$$\mathscr{S}(a_0) = \{U(f)a_0 : f \in C_c(G)\},\ \mathscr{S}(b_0) = \{U(f)b_0 : f \in C_c(G)\}$$

are dense in E, and hence contain $E_K = \sum_{\delta \in \hat{K}} E(\delta)$; this being so, fix a class δ_0 in $\hat{K}$ such that $E(\delta_0) \neq \{0\}$; let a, b be two non-zero elements in $E(\delta_0)$ – then the coefficient $x \mapsto (U(x)a, b)$ $(x \in G)$ is square integrable. To see this, it is obviously sufficient to investigate the square integrability of the function $x \mapsto (U(x)U(f)a_0, U(g)b_0)$ $(f, g \in C_c(G))$ – but

$$\begin{aligned}
&\int_G^* |(U(x)U(f)a_0, U(g)b_0)|^2 d_G(x)\\
&\leqslant \int_G \int_G \int_G \int_G \int_G^* (a_0, U(y^{-1}xz)b_0)\overline{(a_0, U(\bar{y}^{-1}x\bar{z})b_0)}\\
&\times f(y)\overline{f(\bar{y})g(z)}g(\bar{z})d_G(x)d_G(y)d_G(\bar{y})d_G(z)d_G(\bar{z})\\
&< \infty,
\end{aligned}$$

where we have used the Schwarz inequality, the invariance of Haar measure on G under both left and right translation, and the fact that f and g have compact support. [For use below, observe that we have actually shown that the function $(U(.)a, b)$ is square integrable for any $a \in \mathscr{S}(a_0)$, $b \in \mathscr{S}(b_0)$.]

(2) In view of (1), we may now assume that a_0 and b_0 belong to

$E(\delta_0)$. Let $b \neq 0$ be an element in $\mathscr{S}(b_0)$; let ϕ_b denote the function defined by the rule $x \mapsto (U(x)a_0, b)$ $(x \in G)$; let R denote the right regular representation of G on $L^2(G)$, F_b the smallest R-closed subspace of $L^2(G)$ containing ϕ_b; let V_b denote the unitary representation of G on F_b obtained by restricting R to F_b – then, according to a lemma which may be found in the Appendix to the present number, U and V_b are unitarily equivalent. Let $Q_b : E \to F_b$ be a unitary equivalence between U and V_b; let S_b be the linear mapping of $\mathscr{S}(a_0)$ into F_b defined by

$$S_b(U(f)a_0) = \phi_b * f' \qquad (f \in C_c(G));$$

put $T_b = Q_b^{-1}S_b$ – then T_b is a linear map of $\mathscr{S}(a_0)$ into E such that $T_bU(f) = U(f)T_b$ $(f \in C_c(G))$. Consequently T_b maps each $E(\delta)$ $(\delta \in \hat{K})$ into itself, thus leaves E_K invariant and so, in view of the aforementioned intertwining property, is a scalar multiple of the identity on E_K (cf. Lemma 4.5.4.3).

(3) Owing to the observations in (2), there exists a complex number t_b (say) such that for all $f \in C_*(G)$ (cf. 4.5.4), we have

$$\begin{aligned} \| S_b(U(f)a_0) \| &= \| Q_b^{-1}S_b(U(f)a_0) \| = \| T_b(U(f)a_0) \| \\ &= |t_b| \cdot \| U(f)a_0 \|; \end{aligned}$$

since the vectors $U(f)a_0$ $(f \in C_*(G))$ fill up E_K (Lemma 4.5.4.3) and since

$$\| S_b(U(f)a_0) \|^2 = \int_G |(U(x)(U(f)a_0), b)|^2 d_G(x),$$

it follows that

$$\int_G |(U(x)a, b)|^2 d_G(x) = \int_G |(a, U(x)b)|^2 d_G(x) = |t_b|^2 \, \| a \|^2$$

for all $a \in E_K$. [If $b = 0$, we put $t_b = 0$ so that the above relation continues to hold in that case as well.] Inverting the roles of a and b then leads to the conclusion that

$$|t_a|^2 \, \| b \|^2 = |t_b|^2 \, \| a \|^2 \qquad (\text{all } a, b \in E_K).$$

Therefore there exists a real number $d_U > 0$ with the property that

$$|t_a|^2 = d_U^{-1} \, \| a \|^2 \qquad (\text{all } a \in E_K),$$

whence

$$(*) \quad \int_G |(U(x)a, b)|^2 d_G(x) = d_U^{-1} \, \| a \|^2 \, \| b \|^2 \qquad (a, b \in E_K).$$

(4) It remains to show that relation $(*)$ continues to persist for all $a, b \in E$ – this, however, requires some preparation. Let $a \in E$, $b \in E_K$; choose a sequence $\{a_n\}$ in E_K such that $a_n \to a$; put

$$\Phi_n(x) = |(U(x)a_n, b)|^2, \; \Phi(x) = |(U(x)a, b)|^2 \qquad (x \in G).$$

Then the sequence $\{\Phi_n\}$ converges to Φ uniformly on G. [In fact, for all $x \in G$, we have

$$\begin{aligned}|\Phi_n(x) - \Phi(x)| &= |(U(x)a_n, b)\overline{(U(x)a_n, b)} - (U(x)a, b)\overline{(U(x)a, b)}| \\ &\leqslant |(U(x)a_n, b)|\cdot|\overline{(U(x)(a_n - a), b)}| \\ &\quad + |\overline{(U(x)a, b)}|\cdot|(U(x)(a_n - a), b)| \\ &\leqslant \|a_n - a\|\cdot\|a_n\|\cdot\|b\|^2 + \|a_n - a\|\cdot\|a\|\cdot\|b\|^2\end{aligned}$$

from which we can read off our assertion.] Owing to Fatou's Lemma, we have

$$\int_G \Phi(x)d_G(x) = \int_G \liminf_{n\to\infty} \Phi_n(x)d_G(x) \leqslant \liminf_{n\to\infty} \int_G \Phi_n(x)d_G(x) < \infty$$

and so it follows that Φ lies in $L^1(G)$. This being so, consider the (linear) mapping $a \mapsto (U(.)a, b)$ of E into $L^2(G)$; in view of what has just been said, this mapping is everywhere defined on E and, as is easily seen, closed; therefore, thanks to the Closed Graph Theorem, it is actually continuous. In particular, then, $\int_G \Phi_n \to \int_G \Phi$; but

$$\int_G \Phi_n = d_U^{-1}\|a_n\|^2\|b\|^2$$

and so

$$\int_G \Phi = d_U^{-1} \lim_{n\to\infty} \|a_n\|^2\|b\|^2 = d_U^{-1}\|a\|^2\|b\|^2;$$

otherwise said,

$$(*) \quad \int_G |(U(x)a, b)|^2 d_G(x) = d_U^{-1}\|a\|^2\|b\|^2$$

for all $a \in E$, $b \in E_K$. Finally, if we fix $a \in E$, $b \in E$ and approach b by elements b_n (say) in E_K, then we can argue as above to conclude that relation $(*)$ is valid for all $a, b \in E$.

The proof of the lemma is now quite complete. □

[As above, the number d_U will be referred to as the formal degree (or dimension) of U.]

Corollary 4.5.9.2 Let U be an irreducible unitary representation of G on a Hilbert space E – then the following conditions are equivalent:

(i) U is square integrable;

(ii) U is unitarily equivalent to a subrepresentation of the left (or right) regular representation of G on $L^2(G)$.

Proof That (i) ⇒ (ii) is contained, essentially, in part (2) of the proof of the preceding lemma; that (ii) ⇒ (i) may be seen as follows. Agreeing to identify E with a closed L-stable subspace of $L^2(G)$ (and hence U with $L|E$), let P denote the orthogonal projection of $L^2(G)$ onto E – then $\mathsf{P}(C_c(G))$ is dense in E. Fix a non-zero element f_0 in E; fix a non-zero element $g_0 \in \mathsf{P}(C_c(G))$ and let h_0 be an element in $C_c(G)$ such that $g_0 =$

$\mathsf{P}(h_0)$ – then certainly

$$\|(L(.)f_0, g_0)\|_2 \leqslant \|f_0\|_2 \, \|h_0\|_1 < \infty,$$

whence, on the basis of the lemma supra, U is indeed square integrable. □

Remark If G is not compact, then the left regular representation of G on $L^2(G)$ contains no non-zero finite dimensional subrepresentation (cf. Corollary 5.3.3.7).

Remarkably enough, the Schur Orthogonality Relations are valid for square integrable representations of G.

Theorem 4.5.9.3 (Godement) Let U and V be (irreducible) square integrable unitary representations of G on Hilbert spaces E and F, respectively; let $a_1, a_2 \in E$, $b_1, b_2 \in F$.

(i) If U and V are not unitarily equivalent, then

$$\int_G (U(x)a_1, a_2)\overline{(V(x)b_1, b_2)}d_G(x) = 0;$$

(ii) If U and V are unitarily equivalent under a unitary equivalence $Q : E \to F$, then

$$\int_G (U(x)a_1, a_2)\overline{(V(x)b_1, b_2)}d_G(x) = d_U^{-1}(Qa_1, b_1)\overline{(Qa_2, b_2)},$$

where $d_U(= d_V)$ is the formal degree of U.

Proof Thanks to Lemma 4.5.9.1, it is clear that we have

$$\int_G |(U(x)a_1, a_2)\overline{(V(x)b_1, b_2)}| \, d_G(x) \leqslant (d_U d_V)^{-1/2} \|a_1\| \cdot \|a_2\| \cdot \|b_1\| \cdot \|b_2\|.$$

This being so, fix $a_2 \in E$, $b_2 \in F$ and put

$$\mathbf{B}(\,.\,,\,.\,) = \int_G (U(x).\,, a_2)\overline{(V(x).\,, b_2)}d_G(x).$$

Then $\mathbf{B}$ is a bounded conjugate bilinear functional on $E \times F$ and so we can find a bounded linear operator $T : E \to F$ such that $\mathbf{B}(.\,,.) = (T.\,,.)$. The operator T intertwines U and V; in fact, for $x \in G$,

$$\begin{aligned}
(TU(x).\,,.) &= \int_G (U(y)U(x).\,, a_2)\overline{(V(y).\,, b_2)}d_G(x) \\
&= \int_G (U(y).\,, a_2)\overline{(V(y)V(x^{-1}).\,, b_2)}d_G(x) \\
&= (T.\,, V(x^{-1}).) = (V(x)T.\,,.) \\
\Rightarrow TU(x) &= V(x)T \qquad (\text{all } x \in G).
\end{aligned}$$

In view of Corollary 4.3.1.3, statement (i) of our theorem is now evident.

As for the second assertion, note first that it suffices just to consider the case when $U = V$ (for

$$\int_G (U(x)a_1, a_2)\overline{(V(x)b_1, b_2)}d_G(x)$$
$$= \int_G (U(x)a_1, a_2)\overline{(U(x)Q^{-1}b_1, Q^{-1}b_2)}d_G(x)$$
$$= d_U^{-1}(a_1, Q^{-1}b_1)\overline{(a_2, Q^{-1}b_2)}$$
$$= d_U^{-1}(Qa_1, b_1)\overline{(Qa_2, b_2)}).$$

Accordingly the operator T is now a bounded linear operator on E which commutes with all the $U(x)$ $(x \in G)$ and so, due to the irreducibility of U, is a scalar multiple of the identity (say t_{a_2, b_2}). The relation

$$\int_G (U(x)a_1, a_2)\overline{(U(x)b_1, b_2)}d_G(x)$$
$$= \int_G (U(x)b_2, b_1)\overline{(U(x)a_2, a_1)}d_G(x)$$

implies that

$$t_{a_2, b_2}(a_1, b_1) = t_{b_1, a_1}(b_2, a_2)$$

for all $a_1, a_2, b_1, b_2 \in E$. Therefore statement (ii) follows from Lemma 4.5.9.1. □

Corollary 4.5.9.4 Let U be an (irreducible) square integrable unitary representation of G on a Hilbert space E; put $\phi_{\cdot,\cdot}(x) = (U(x)\cdot, \cdot)$ $(x \in G)$ – then, for all $a_1, a_2, b_1, b_2 \in E$, we have

$$\phi_{a_1, a_2} * \phi_{b_1, b_2} = d_U^{-1}(a_1, b_2)\phi_{b_1, a_2}.$$

Proof Bearing in mind the second assertion of Theorem 4.5.9.3, we have

$$\phi_{a_1, a_2} * \phi_{b_1, b_2}(x) = \int_G (U(y)a_1, a_2)(U(y^{-1}x)b_1, b_2)d_G(y)$$
$$= \int_G (U(y)a_1, a_2)\overline{(U(y)b_2, U(x)b_1)}d_G(y)$$
$$= d_U^{-1}(a_1, b_2)\overline{(a_2, U(x)b_1)} = d_U^{-1}(a_1, b_2)\phi_{b_1, a_2}(x)$$
$$(x \in G). \quad \square$$

Proposition 4.5.9.5 (Harish-Chandra) Let U be an irreducible unitary representation of G on a Hilbert space E – then the following conditions are equivalent:

(i) There exists a non-zero vector a in E such that the function $x \mapsto (U(x)a, a)$ is integrable on G;

(ii) There exists a dense subspace $\tilde{E}$ of E such that for all $\tilde{a}, \tilde{b}$ in $\tilde{E}$, the function $x \mapsto (U(x)\tilde{a}, \tilde{b})$ is integrable on G.

Proof Let us suppose that U verifies condition (i); fix elements $f, g \in C_c(G)$ – then, for every $x \in G$, one has

$$(U(f)a, U(x)U(g)a) = \int_G \int_G f(y)\overline{g(z)}(U(y)a, U(x)U(z)a)d_G(y)d_G(z)$$
$$= \int_G \int_G f(y)\overline{g(z)}(a, U(y^{-1}xz)a)d_G(y)d_G(z).$$

Therefore

$$\int_G^* |(U(f)a, U(x)U(g)a)| \, d_G(x)$$
$$\leqslant \int_G \int_G \int_G^* |f(y)\overline{g(z)}(a, U(y^{-1}xz)a)| \, d_G(x)d_G(y)d_G(z)$$
$$= \int_G \int_G \int_G^* |f(y)g(z)(a, U(x)a)| \, d_G(x)d_G(y)d_G(z) < \infty.$$

Consequently the requirements of condition (ii) are met by the subspace $\tilde{E} = U(C_c(G))a$. □

Definition An irreducible unitary representation U of G is said to be *integrable* if it satisfies either one of the equivalent conditions of Proposition 4.5.9.5.

It is clear that an integrable unitary representation of G is square integrable; as will be seen by example later on, the converse to this latter statement is false.

Remark Let U be an integrable unitary representation of G on a Hilbert space E – then, for every $a, b \in E_K$, the function $x \mapsto (U(x)a, b)$ is integrable on G (cf. part (1) of the proof of Lemma 4.5.9.1).

Proposition 4.5.9.6 (Kunze-Stein) Let U be a unitary representation of G on a Hilbert space E; let $p, q \in [1, \infty]$ be such that $p^{-1} + q^{-1} = 1$ – then the following conditions are equivalent:

(i) There exists a finite constant $M > 0$ such that

$$\| U(f) \| \leqslant M \| f \|_p \qquad \text{for all } f \in L^1(G) \cap L^p(G);$$

(ii) The function $x \mapsto (U(x)a, b)$ lies in $L^q(G)$ (all $a, b \in E$).

Proof (i) ⇒ (ii). Let $f \in L^1(G) \cap L^p(G)$ – then, for $a, b \in E$, we have

$$\left| \int_G (U(x)a, b) f(x) d_G(x) \right| \leqslant |(U(f)a, b)| \leqslant M \| f \|_p \| a \| \cdot \| b \|.$$

Limiting ourselves to those f's for which $\|f\|_p = 1$ and taking the sup over the left hand side leads to the conclusion that

$$\| (U(\,.\,)a, b) \|_q \leqslant M \| a \| \cdot \| b \|.$$

Therefore the function $x \mapsto (U(x)a, b)$ lies in $L^q(G)$ (all $a, b \in E$), as desired.

(ii) $\Rightarrow$ (i). Fix $b \in E$ and consider the mapping $a \mapsto (U(.)a, b)$ of E into $L^q(G)$; in view of our hypotheses, this mapping is linear, everywhere defined on E, and, as is easily seen, closed – consequently there exists a constant $M_b > 0$ such that $\|(U(.)a, b)\|_q \leqslant M_b \|a\|$ (all $a \in E$) (Closed Graph Theorem). Similar considerations imply the existence of a constant $M_a > 0$ with the property that $\|(U(.)a, b)\|_q \leqslant M_a \|b\|$ (all $b \in E$). Now let f be any function in $L^p(G)$ – then the prescription $(U(f)a, b) = \int_G f(x)(U(x)a, b)d_G(x)$ $(a, b \in E)$ defines a bounded linear operator $U(f)$ (say) on E. [Thanks to Hölder's inequality, the integral in question is, at least, well-defined; in addition, the above estimates on the coefficients $(U(.)a, b)$ tell us that

$$|(U(f)a, b)| \leqslant M_a \|b\| \cdot \|f\|_p, \; |(U(f)a, b)| \leqslant M_b \|a\| \cdot \|f\|_p \qquad (a, b \in E).$$

Thus the vector $U(f)a$ is well-defined for every $a \in E$. A straightforward argument then shows that $U(f)$ is a closed operator and, hence, due to the Closed Graph Theorem, a bounded operator on E.] This being so, consider the mapping $f \mapsto U(f)$ from $L^p(G)$ to $\mathfrak{L}(E; E)$ (viewed as a Banach space under the usual operator norm). As we have just seen, this mapping is everywhere defined (on $L^p(G)$) and is, of course, linear – it is also closed. [In fact suppose that $f_n \to f$ in the L^p-norm and that $U(f_n) \to U_0$ (say) in the operator norm – then $(U(f_n)a, b) \to (U_0 a, b)$ for every $a, b \in E$. But

$$|(U(f_n)a, b)| \leqslant M_b \|a\| \cdot \|f_n\|_p \Rightarrow (U(f_n)a, b) \to (U(f)a, b)$$
$$\Rightarrow (U(f)a, b) = (U_0 a, b) \Rightarrow U(f) = U_0,$$

from which the closedness of the map $f \mapsto U(f)$ follows.] The Closed Graph Theorem then provides us with a constant $M > 0$ such that $\|U(f)\| \leqslant M \|f\|_p$ (all $f \in L^p(G)$). Hence (i).

This completes the proof of the proposition. □

In what follows, it will be assumed that G is in addition a (unimodular) Lie group with a *uniformly* large compact subgroup K.

Our final task in this number will be to obtain a useful expression for the character of an (irreducible) square integrable unitary representation of G.

Theorem 4.5.9.7 (Harish-Chandra) Let U be an (irreducible) square integrable unitary representation of G on a Hilbert space E; let T_U denote the character of U, d_U the formal degree of U – then, for any unit vector a in E, we have,

$$T_U(f) = d_U \int_G d_G(x)\Big(\int_G f({}^x y)(U(y)a, a) d_G(y)\Big) \qquad (\text{all } f \in C_c^\infty(G)).$$

Proof Fix $f \in C_c^\infty(G)$; as we know, there exists an orthonormal basis $\{a_i : i \in I\}$ for E such that $\sum_{i,j} |(U(f)a_i, a_j)| < \infty$ (see the proof of

Theorem 4.5.7.6). Since

$$\begin{aligned}(U(f)U(x)a, U(x)a) &= (U(f)U(x)a, \sum_i (U(x)a, a_i)a_i)\\ &= \sum_i \overline{(U(x)a, a_i)}(U(f)U(x)a, a_i)\\ &= \sum_i \overline{(U(x)a, a_i)}(U(x)a, U(f^*)a_i)\\ &= \sum_i \sum_j \overline{(U(x)a, a_i)}(U(x)a, a_j)(U(f)a_j, a_i),\end{aligned}$$

the Schwarz inequality and the Schur Orthogonality Relations (Theorem 4.5.9.3) imply that

$$\begin{aligned}&\int_G |(U(f)U(x)a, U(x)a)|\, d_G(x)\\ &\leqslant \sum_{i,j} |(U(f)a_i, a_j)| \int_G |(U(x)a, a_i)\overline{(U(x)a, a_j)}|\, d_G(x)\\ &\leqslant \sum_{i,j} |(U(f)a_i, a_j)| \left(\int_G |(U(x)a, a_i)|^2\, d_G(x)\right)^{1/2}\\ &\quad\times \left(\int_G |(U(x)a, a_j)|^2\, d_G(x)\right)^{1/2}\\ &= d_U^{-1} \sum_{i\,j} |(U(f)a_i, a_j)| < \infty.\end{aligned}$$

Owing to a theorem of Lebesgue, then, the series for $(U(f)U(x)a, U(x)a)$ may be integrated term by term – we thus have

$$\begin{aligned}&\int_G d_G(x)\left(\int_G f({}^x y)(U(y)a, a)d_G(y)\right)\\ &= \sum_i \sum_j (U(f)a_j, a_i) \int_G (U(x)a, a_j)\overline{(U(x)a, a_i)}d_G(x)\\ &= d_U^{-1} \sum_i \sum_j (U(f)a_j, a_i)(a_i, a_j)\\ &= d_U^{-1} \sum_i (U(f)a_i, a_i) = d_U^{-1} T_U(f),\end{aligned}$$

which serves to prove the theorem. □

In general, the double integral in the above theorem is not absolutely convergent and therefore the order of the two integrations cannot be interchanged (we shall leave it to the reader to determine what happens if G is compact . . .).

Theorem 4.5.9.7 (bis) Let U be an (irreducible) square integrable unitary representation of G on a Hilbert space E; let T_U denote the character of U, d_U the formal degree of U – then, for all $a, b \in E$, we have

$$(a, b)T_U(f) = d_U \int_G d_G(x)\left(\int_G f({}^x y)(U(y)a, b)\, d_G(y)\right) \qquad (\text{all } f \in C_c^\infty(G)).$$

[This is but a simple variant of the preceding considerations . . . .]

Appendix Let us agree to retain the above assumptions and notations.

Lemma Let U be an irreducible unitary representation of G on a Hilbert space E; fix a class δ_0 in $\hat{K}$ which occurs in $U|K$ and suppose that there exist non-zero elements $a_0 \in E(\delta_0)$, $b \in E$ such that the function ϕ_b defined by the prescription $x \mapsto (U(x)a_0, b)$ $(x \in G)$ lies in $L^2(G)$; let R denote the right regular representation of G on $L^2(G)$, F_b the smallest R-closed subspace of $L^2(G)$ containing ϕ_b, V_b the unitary representation of G on F_b obtained by restricting R to F_b – then U and V_b are unitarily equivalent.

Proof Granting for the moment that our unitary representation V_b is irreducible (and thus K-finite, K being large in G), it will be enough, in view of Proposition 4.5.4.4, to show that $U(f) = 0 \Rightarrow V_b(f)\phi_b = 0$ $(f \in C_c(G))$. [Here, of course, it is necessary to keep in mind Proposition 4.3.1.4.] So suppose that f is an element of $C_c(G)$ with the property that $U(f) = 0$ – then

$$\begin{aligned} V_b(f)\phi_b &= \phi_b * f' \Rightarrow V_b(f)\phi_b(x) \\ &= \int_G \phi_b(xy)f(y)d_G(y) = (U(x)U(f)a_0, b) \\ &= 0 \Rightarrow V_b(f)\phi_b = 0, \end{aligned}$$

as desired. [In passing, note that we have actually proved that $U(f)a_0 = 0 \Rightarrow V_b(f)\phi_b = 0$ $(f \in C_c(G))$.] It remains to verify that our unitary representation V_b is irreducible. By construction, $\phi_b \in F_b(\delta_0)$; since the elements of the form $V_b(f)\phi_b (f \in C_c(G))$ are dense in F_b, $V_b(C_{c,\delta_0}(G))\phi_b$ is dense in $F_b(\delta_0)$ – in fact $V_b(C_{c,\delta_0}(G))\phi_b = F_b(\delta_0)$. [To see this, observe that we can define a linear map Q of $E(\delta_0)$ onto $V_b(C_{c,\delta_0}(G))\phi_b$ by the requirement $Q(U(f)a_0) = V_b(f)\phi_b$ $(f \in C_{c,\delta_0}(G))$ – then $\dim(V_b(C_{c,\delta_0}(G))\phi_b) \leqslant \dim(E(\delta_0)) < \infty \Rightarrow V_b(C_{c,\delta_0}(G))\phi_b = F_b(\delta_0)$.] Now suppose that $\tilde{F}_b$ is a closed V_b-stable subspace of F_b; let $\tilde{F}_b^{\perp}$ denote the orthogonal complement of $\tilde{F}_b$ in F_b – then, since the canonical representation of $C_{c,\delta_0}(G)$ on $F_b(\delta_0)$ is algebraically irreducible (cf. supra), it must be the case that either $\tilde{F}_b$ or $\tilde{F}_b^{\perp}$ contains $F_b(\delta_0)$. If $\tilde{F}_b \supset F_b(\delta_0)$, then $\phi_b \in \tilde{F}_b$ which implies, in view of the definition of V_b, that $\tilde{F}_b = F_b$; similar comments apply to $\tilde{F}_b^{\perp}$. It follows that V_b is irreducible, as desired. Hence the lemma. □

Chapter 5

Induced Representations

Introduction

Let G be a locally compact group, $\hat{G}$ its unitary dual. A first problem in the Harmonic Analysis on G is to 'identify' $\hat{G}$, i.e. to find all possible irreducible unitary representations of G. [Of course, when G is abelian or compact, $\hat{G}$ is, in principle at least, 'known'.] Ideally, one would like to have at hand a systematic method for classifying the elements of $\hat{G}$; although this appears to be too much to hope for in general, nevertheless there is a basic construction by means of which it is possible to write down irreducible unitary representations for large classes of groups – this is the notion of (unitarily) induced representation. For certain classes of groups *all* the irreducible unitary representations can be obtained as unitarily induced representations, while for other classes this is the case for a large number of the irreducible unitary representations. It turns out that whenever G admits a closed normal subgroup H, there is a procedure (using induced representations) whereby one can classify the elements of $\hat{G}$ in terms of those of $\hat{H}$ and $(G/H)^{\wedge}$; naturally this technique for reducing the study of irreducible unitary representations of G to those of successively smaller groups will fail as soon as one reaches groups which are 'simple', i.e. have no non-trivial closed normal subgroups (nevertheless, even for 'simple' groups, the concept of induced representation is of fundamental importance).

In order to get a better understanding of what the theory of induced representations is all about, let us assume that G is a finite group (in which case the theory is classical, dating back to Frobenius [1]). Let H be a subgroup of G, L a unitary representation of H on E. [When dealing with finite groups, the term 'unitary representation' will mean 'unitary representation on a finite dimensional vector space over **C**'.] The unitary representation U^L of G *induced* from L is defined as follows. Let E^L denote the space of all E-valued functions f on G satisfying $f(x\xi) = L(\xi^{-1})f(x)$ $(\xi \in H, x \in G)$. The inner product

$$(f, g) = [H]^{-1} \sum_{x \in G} (f(x), g(x)) \qquad (f, g \in E^L)$$

equips E^L with the structure of a Hilbert space. Given $x \in G$, let $U^L(x)$ denote the linear operator on E^L defined by $U^L(x)f(y) = f(x^{-1}y)$ $(x, y \in G; f \in E^L)$ – then it is immediate that U^L is a unitary representation of G on E^L, the unitary representation of G *induced* by the unitary representation L of the subgroup H. A basic question now is this: When is U^L irreducible? [In passing note that U^L cannot be irreducible unless L itself is irreducible (for if L^1, L^2 are unitary representations of H, then $U^{L^1+L^2}$ is unitarily equivalent to $U^{L^1} + U^{L^2}$).] Of course, on general grounds, we know that U^L is irreducible iff $i(U^L, \overline{U^L}) = i(U^L, U^{\bar{L}}) = 1$ (i the intertwining number of U^L and $\overline{U^L}$) – this suggests, then, the following problem: Find a formula for $i(U^L, U^{\bar{L}})$. As will be seen below, such a formula is contained in the so-called Intertwining Number Theorem.

Let H_i be a subgroup of G, L^i a unitary representation of H_i on a Hilbert space E^i, U^{L^i} the corresponding unitary representation of G induced by L^i $(i = 1, 2)$.

Theorem (Intertwining Number Theorem) Let Q be an H_2, H_1 double coset in G; choose an element $x \in Q$ and let $i(L^1, L^2; Q)$ denote the intertwining number for the unitary representations $\zeta \mapsto L^1(\zeta)$, $\zeta \mapsto L^2(x\zeta x^{-1})$ of the subgroup $H_x = H_1 \cap (x^{-1}H_2x)$ ($i(L^1, L^2; Q)$ is independent of the choice of x in Q) – then

$$i(U^{L^1}, U^{L^2}) = \sum_Q i(L^1, L^2; Q),$$

the sum $\sum_Q$ running over all the H_2, H_1 double cosets Q in G.

Proof (1) There exists a canonical map π^i of $C(G; E^i)$ (the space of all E^i-valued functions on G) onto E^{L^i} $(i = 1, 2)$: π^i is defined by the prescription

$$\pi^i(f_i)(x_i) = \sum_{\xi_i \in H_i} L^i(\xi_i)f_i(x_i\xi_i) \qquad (x_i \in G, f_i \in C(G; E^i); i = 1, 2).$$

If $L_i(R_i)$ is the left (right) regular representation of G on $C(G; E^i)$ $(i = 1, 2)$, then, as is easily checked,

$$\pi^i \circ L_i(x) = U^{L^i}(x) \circ \pi^i,$$
$$\pi^i(R_i(\xi_i)f_i) = \pi^i(L^i(\xi_i^{-1})f_i) \qquad (x \in G, \xi_i \in H_i, f_i \in C(G; E^i); i = 1, 2).$$

(2) Let $\mathbf{B}$ be a U^{L^1}, U^{L^2} intertwining form and put

$$\mathbf{B}^\natural(f_1, f_2) = \mathbf{B}(\pi^1(f_1), \pi^2(f_2))$$

where $f_1 \in C(G; E^1)$, $f_2 \in C(G; E^2)$ – then $\mathbf{B}^\natural$ is a bilinear form on $C(G; E^1) \times C(G; E^2)$. Consequently there exists a unique function $T^\natural$ on $G \times G$, whose values are bilinear forms on $E^1 \times E^2$, such that $\mathbf{B}^\natural$ is given by the formula

$$\mathbf{B}^\natural(f_1, f_2) = \sum_{x_1, x_2 \in G} T^\natural_{x_1, x_2}(f_1(x_1), f_2(x_2)) \qquad (f_i \in C(G; E^i); i = 1, 2).$$

Moreover it follows from the intertwining property of $\mathbf{B}$ that

$$T^\natural_{zx_1, zx_2} = T^\natural_{x_1, x_2} \qquad (x_i \ (i = 1, 2), z \in G).$$

[Thus, if $f_i \in C(G; E^i)$ $(i = 1, 2)$, then

$$\begin{aligned}\sum_{x_1, x_2} T^\natural_{zx_1, zx_2}(f_1(x_1), f_2(x_2)) &= \mathbf{B}^\natural(L_1(z)f_1, L_2(z)f_2)\\ &= \mathbf{B}(\pi^1(L_1(z)f_1), \pi^2(L_2(z)f_2))\\ &= \mathbf{B}(U^{L^1}(z)\pi^1(f_1), U^{L^2}(z)\pi^2(f_2))\\ &= \mathbf{B}(\pi^1(f_1), \pi^2(f_2))\\ &= \sum_{x_1, x_2} T^\natural_{x_1, x_2}(f_1(x_1), f_2(x_2))\\ &\Rightarrow T^\natural_{zx_1, zx_2} = T^\natural_{x_1, x_2},\end{aligned}$$

f_1 and f_2 being arbitrary.]

(3) Put $T_x = T^\natural_{x,1}$ $(x \in G)$ – then $T_{x_2^{-1}x_1} = T^\natural_{x_1, x_2}$ (since $T_{x_2^{-1}x_1} = T^\natural_{x_2^{-1}x_1, 1} = T^\natural_{x_1, x_2}$ (by (2))) and, of course, T completely determines $\mathbf{B}$. We shall call T the *G-trace* of $\mathbf{B}$. The group $H_2 \times H_1$ operates to the left on G in the obvious fashion, namely

$$(\xi_2, \xi_1)\cdot x = \xi_2 x \xi_1^{-1} \qquad (x \in G; (\xi_2, \xi_1) \in H_2 \times H_1);$$

relative to this action of $H_2 \times H_1$, the function T transforms in a certain way – thus, for $f_i \in C(G; E^i)$ $(i = 1, 2)$,

$$\begin{aligned}\sum_{x_1, x_2} T_{\xi_2^{-1}x_2^{-1}x_1\xi_1}(f_1(x_1), f_2(x_2)) &= \sum_{x_1, x_2} T^{\xi}_{x_1\xi_1, x_2\xi_2}(f_1(x_1), f_2(x_2)) \\ = \sum_{x_1, x_2} T^{\xi}_{x_1, x_2}(f_1(x_1\xi_1^{-1}), f_2(x_2\xi_2^{-1})) &= \mathbf{B}^{\xi}(R_1(\xi_1^{-1})f_1, R_2(\xi_2^{-1})f_2) \\ = \mathbf{B}^{\xi}(L^1(\xi_1)f_1, L^2(\xi_2)f_2),\end{aligned}$$

from which it follows that

$$\text{(A)} \quad T_{\xi_2^{-1}x\xi_1}(a_1, a_2) = T_x(L^1(\xi_1)a_1, L^2(\xi_2)a_2) \qquad (a_1 \in E^1, a_2 \in E^2).$$

Conversely, if T is any function on G whose values are bilinear forms on $E^1 \times E^2$ and which verifies condition (A), then one checks without difficulty (reverse the argument) that there is a unique U^{L^1}, U^{L^2} intertwining form $\mathbf{B}$ on G whose G-trace is T. Therefore, letting $\mathbf{T}$ denote the space of all functions T on G whose values are bilinear forms on $E^1 \times E^2$ and which verify condition (A), we conclude that $i(U^{L^1}, U^{L^2}) = \dim(\mathbf{T})$. Clearly, then, our next task is to obtain an expression for $\dim(\mathbf{T})$.

(4) The group G is, of course, the union of the H_2, H_1 double cosets Q. Let $Q = H_2xH_1$ be any such double coset containing the group element x. Let $\mathbf{T}(Q)$ denote the space of all functions T on Q whose values are bilinear forms on $E^1 \times E^2$ and which verify condition (A) above. If $T \in \mathbf{T}$, then the restriction T_Q of T to Q belongs to $\mathbf{T}(Q)$; since the conditions imposed by (A) on the different double cosets Q are independent of one another, $\dim(\mathbf{T}) = \sum_Q \dim(\mathbf{T}(Q))$ – on the other hand, one checks (cf. the discussion following Proposition 5.3.2.2) that the dimension of $\mathbf{T}(Q)$ is equal to the intertwining number for the unitary representations $\zeta \mapsto L^1(\zeta)$, $\zeta \mapsto L^2(x\zeta x^{-1})$ of the subgroup $H_x = H_1 \cap (x^{-1}H_2x)$.

Since $i(U^{L^1}, U^{L^2}) = \dim(\mathbf{T}) = \sum_Q \dim(\mathbf{T}(Q))$, the proof of the Intertwining Number Theorem is complete. □

As we shall now see, a number of the fundamental facts about unitary representations of finite groups are ready consequences of the Intertwining Number Theorem.

Let us recall that unitary representations U and V of G are said to be disjoint if they have no common non-zero unitarily equivalent subrepresentations; for this to be the case, it is necessary and sufficient that $i(U, \bar{V}) = 0$.

Corollary 1 (Mackey-Shoda) Let H be a subgroup of G, L an irreducible unitary representation of H – then U^L is irreducible iff for every $x \in G - H$, the unitary representations $\zeta \mapsto L(\zeta)$, $\zeta \mapsto L(x\zeta x^{-1})$ of the subgroup $H_x = H \cap (x^{-1}Hx)$ are disjoint.

Proof Bearing in mind that $\overline{U^L} = U^{\bar{L}}$, we see that our corollary follows upon an application of the Intertwining Number Theorem in the case $H = H_1 = H_2$, $L^2 = \bar{L}$. □

Corollary 2 (Mackey-Shoda) Let H be a normal subgroup of G, L an irreducible unitary representation of H – then U^L is irreducible iff for every $x \in G - H$, the unitary representations $\zeta \mapsto L(\zeta)$, $\zeta \mapsto L(x\zeta x^{-1})$ of H are disjoint.

A disjointness criterion is equally easy to obtain.

Corollary 3 (Mackey-Shoda) Let H_i be a subgroup of G, L^i a unitary representation of H_i on a Hilbert space E^i, U^{L^i} the corresponding unitary representation of G induced by L^i $(i = 1, 2)$ – then U^{L^1} and U^{L^2} are disjoint iff for every x in G, the unitary representations $\zeta \mapsto L^1(\zeta)$, $\zeta \mapsto L^2(x\zeta x^{-1})$ of the subgroup $H_x = H_1 \cap (x^{-1}H_2x)$ are disjoint.

Corollary 4 (Mackey-Shoda) Let H_i be a subgroup of G, L^i a unitary representation of H_i on a Hilbert space E^i, U^{L^i} the corresponding unitary representation of G induced by L^i $(i = 1, 2)$; assume that U^{L^1} and U^{L^2} are irreducible – then U^{L^1} and U^{L^2} are unitarily equivalent iff there exists an x in G such that the unitary representations $\zeta \mapsto L^1(\zeta)$, $\zeta \mapsto L^2(x\zeta x^{-1})$ of the subgroup $H_x = H_1 \cap (x^{-1}H_2x)$ are not disjoint.

Corollary 5 (Frobenius Reciprocity Theorem) Let H be a subgroup of G, L an irreducible unitary representation of H; let M be an irreducible unitary representation of G – then U^L contains M exactly as many times as the restriction $M|H$ of M to H contains L.

Proof In the Intertwining Number Theorem, take $H_1 = H$, $H_2 = G$, $x = 1$ and replace M by $\bar{M}$; then there is only one H_2, H_1 double coset and we obtain $i(U^L, \bar{M}) = i(L, \bar{M}|H)$; but $i(U^L, \bar{M}) = i(M, \overline{U^L})$, $i(L, \bar{M}|H)$ are the multiplicities of M in U^L and of L in $M|H$, respectively. Hence the corollary. □

Let us return now to our locally compact group G. A primary goal of the present chapter then is this: Determine to what extent the Intertwining Number Theorem and its corollaries remain true for G. Unfortunately, as they stand, the assertions of these results are, in general, false (we shall have more to say about this later on); therefore, somewhat more accurately, our program will be to obtain weakened or modified versions of the Intertwining Number Theorem and its corollaries, at least for certain G.

In the first place, of course, one must have at hand a definition of (unitarily) induced representation. Actually there are several (equivalent) ways to formulate such a definition – the various possibilities are examined systematically in 5.1.1.

This being so, let us see what problems are encountered when one tries to generalize the Intertwining Number Theorem from finite groups to infinite groups; for this purpose it will be convenient to assume that G is a Lie group. Thus let H_i be a closed subgroup of G, L^i a unitary representation of H_i on a Hilbert space E^i, U^{L^i} the corresponding unitarily induced representation of G – assume for simplicity that L^i is one dimensional $(i = 1, 2)$. There is no difficulty in constructing the analogue of the map π^i – it will now be a continuous linear map of $C_c(G)$ onto a dense subspace of the representation space for U^{L^i} $(i = 1, 2)$ (see 5.1.1). Proceeding as in the proof of the Intertwining Number Theorem above for finite groups, let $\mathbf{B}$ be a U^{L^1}, U^{L^2} intertwining form; composing $\mathbf{B}$ with π^1 and π^2 leads to a bilinear form $\mathbf{B}^\natural$ on $C_c(G) \times C_c(G)$; one might then guess that the natural candidate for $T^\natural$ would be a measure on $G \times G$ (so that

$$\mathbf{B}^\natural(f_1, f_2) = \int_{G\times G} f_1(x_1)f_2(x_2)\, dT^\natural(x_1, x_2), \qquad f_i \in C_c(G); i = 1, 2)$$

which would mean that the 'G-trace' T of $\mathbf{B}$ would be a measure on G verifying

$$dT^\natural(x_1, x_2) = d_G(x_2)\, dT(x_2^{-1}x_1)$$

(d_G being left Haar measure on G). Unfortunately, however, the general bilinear form on $C_c(G) \times C_c(G)$ does not come from a measure on $G \times G$ (cf. Grothendieck [1, Chapter 1, §4, numbers 4, 5]). On the other hand this difficulty can be overcome by passing from $C_c(G)$ to $C_c^\infty(G)$; since $C_c^\infty(G)$ is (unlike $C_c(G)$) a nuclear space, Schwartz's Kernel Theorem tells us that (separately continuous) bilinear forms on $C_c^\infty(G) \times C_c^\infty(G)$ always come from distributions on $G \times G$; hence the $T^\natural$ above will be a distribution on $G \times G$, while the 'G-trace' T of $\mathbf{B}$ becomes a distribution on G (cf. Theorem 5.3.1.1 (which is the analogue of the Intertwining Number Theorem)). The next step in the estimation of the intertwining number $i(U^{L^1}, U^{L^2})$ is to obtain an expression for the dimension of the space $\mathbf{T}$ of all G-traces T arising from U^{L^1}, U^{L^2} intertwining forms $\mathbf{B}$ (the distributions T in question verify an analogue of condition (A) supra); here we encounter a second difficulty. Thus, in the case of finite groups, we restricted such T to each H_2, H_1 double coset Q and then added up, so to speak, the different contributions to $i(U^{L^1}, U^{L^2})$ from the various possible Q. But, as it stands, the double cosets Q need not be open in G (rather are submanifolds of lower dimension than G) – therefore, it is in general meaningless to speak of 'the restriction of T to Q'. To get around this, one makes a simplifying assumption, namely that there exist in G at most *countably* many H_2, H_1 double cosets Q. [Such a hypothesis is not at all unreasonable; consider, for instance, the case when G is a connected semi-simple Lie group with finite center and the H_i are parabolic subgroups of G.] It turns out that in this situation the double cosets Q have certain special properties which enable one to majorize $i(U^{L^1}, U^{L^2})$ by a sum $\sum_Q \dim(\mathbf{T}(Q))$, where roughly speaking, for each H_2, H_1 double coset Q, $\mathbf{T}(Q)$ is a space of distributions on G which verify the analogue of condition (A) and have their support contained in Q. It remains, therefore, to estimate $\dim(\mathbf{T}(Q))$ for each Q – as this is somewhat subtle, it seems best to omit the heuristics and wait until we come to the precise statements (5.2 contains the necessary technical preliminaries for this undertaking; Theorem 5.3.2.3 is then the main result along these lines). However, even after all this, the sought for analogue of Corollary 1 still lies off in the distance; only after considerably more preparation, which we shall not stop to describe here, will it be possible to obtain it (Theorem 5.3.2.15, a theorem of utmost importance for the representation theory of connected semi-simple Lie groups). An analogue of Corollary 2 does not lie quite so deep – cf. Proposition 5.3.2.21; finally, questions of reciprocity will be dealt with in 5.3.3.

In the theory of unitary representations of finite groups G, the fundamental notion of a system of imprimitivity is well-known and classical; the intimate relationship which exists between systems of imprimitivity and unitarily induced representations is made clear through the so-called Imprimitivity Theorem, a theorem which permits one to completely analyze $\hat{G}$ in terms of the unitary duals of its normal subgroups in many important special cases. The Imprimitivity Theorem was extended to infinite groups by Mackey; just as in the finite case, it frequently yields, in typical applications (e.g. 'regular' semi-direct products), a very complete and satisfying description of *all* the irreducible unitary representations of numerous non-compact, non-abelian infinite groups. As might be expected, the Imprimitivity Theorem for infinite groups requires a great deal of topological and measure theoretic preparation (spectral measures, transformation group algebras etc.); on the other hand its usefulness in the representation theory of semi-simple Lie groups appears to be limited (this should not be surprising, however, . . .) – accordingly we shall content ourselves with presenting, in outline form only, the main results in this important circle of ideas (5.4.1) (one may consult Fell [8] for a systematic exposition). By way of illustration, a number of examples are computed in 5.4.2.

The present chapter concludes in 5.5 with a first application of the theory of induced representations to semi-simple Lie groups (a topic with which we shall deal systematically in Volume III when we come to the theory of the 'continuous series'). Thus let G be a connected semi-simple Lie group with finite center, $G = KA_{\mathfrak{p}}N^+$ an Iwasawa decomposition for G – then, as we shall see, it is possible to associate with each (not necessarily unitary) character μ of $A_{\mathfrak{p}}N^+$ an 'induced' representation U^μ of G which may be realized on the Hilbert space $L^2(K)$ (in a not necessarily unitary way). This being so, an important theorem due to Harish-Chandra asserts that every TCI Banach representation of G is Naimark equivalent to some irreducible component of some U^μ – we shall refer to this result as the Subquotient Theorem. An immediate consequence of the Subquotient Theorem is the basic fact that the maximal compact subgroup K is uniformly large in G (so that we have a 'good' theory of characters on G). Let M be the centralizer of $A_{\mathfrak{p}}$ in K; put $P = MA_{\mathfrak{p}}N^+$ – then P is a minimal parabolic subgroup of G. By definition, the principal P-series for G is the set comprised of those unitary representations of G which are unitarily induced from the finite dimensional irreducible unitary representations of P. In 5.5.2, we utilize the generalities alluded to earlier and establish a theorem which serves to ensure that 'most' of the members of the principal P-series are actually irreducible. [In this connection, let us recall that the Bruhat Lemma implies that the double coset space $P\backslash G/P$ is finite. . . .] The characters of the principal P-series are computed in 5.5.3; the section then terminates with a variant of the classical Riemann-Lebesgue Lemma for the elements of the principal P-series. [Since these latter results make more or less free use of certain material from Vol. II, Chapter 8, the reader may wish to put off their detailed study for the time being.]

Acknowledgement The author would like to thank Professor J.M.G. Fell for being kind enough to allow him to use some of his results and remarks on induced representations. In particular the above discussion on finite groups is based on unpublished lecture notes of his; we should also like to point out that Fell's proof of Lemma 5.2.4.4 represents a far-reaching simplification over Bruhat's original argument.

5.1 Unitarily Induced Representations

5.1.1 The Definition

Let G be a Lie group (assumed, as always, to be countable at infinity), H a closed subgroup of G, L a unitary representation of H on a Hilbert space E. The objective of the present number is to show how one may associate with L a unitary representation U^L of G – the unitary representation of G induced by L. Taking our cue from the case of finite groups (mentioned in the Introduction), the representation space E^L for U^L will be defined as a certain space of E-valued (locally summable) functions on G (cf. infra); in all honesty, however, this definition (its utility notwithstanding) lacks motivation – accordingly, with a view toward a more geometric description, we shall indicate at the end an alternative approach, namely that of representing G on the square integrable cross sections of a Hilbert bundle $(\mathbf{E}, \pi_{\mathbf{E}})$ sitting over G/H.

Conventions In what follows the term 'measurable', applied to vector valued functions, will mean Bochner (strongly) measurable; all Haar measures are assumed to be *left* invariant (unless specifically stated to the contrary).

Let L be a unitary representation of H on a Hilbert space E.

Definition Let $\mathscr{E}^L$ be the linear space of all functions f from G to E such that:

(i) f is d_G-measurable;
(ii) $f(x\xi) = (\delta_H(\xi)/\delta_G(\xi))^{1/2}L(\xi^{-1})f(x)$ whenever $\xi \in H$, $x \in G$;
(iii) $\| f(\cdot)\|^2$ is locally summable on G.

It is clear that $x \mapsto \| f(x)\|^2$ is d_G-measurable ($f \in \mathscr{E}^L$); moreover, the polarization identity together with (iii) imply that $x \mapsto (f(x), g(x))$ is d_G-measurable and locally summable ($f, g \in \mathscr{E}^L$).

Given $\phi \in C_c(G)$, set $\dot{\phi}(\dot{x}) = \int_H \phi(x\xi)d_H(\xi)$ ($\dot{x} = xH$) – then the assignment $\phi \mapsto \dot{\phi}$ is a continuous surjection of $C_c(G)$ onto $C_c(G/H)$ (cf. infra).

Lemma 5.1.1.1 Let $f, g \in \mathscr{E}^L$ – then we may define a Radon measure $\mu_{f,g}$ on G/H by the rule:

$$\int_G (f(x), g(x))\phi(x)d_G(x) = \int_{G/H} \dot{\phi}(\dot{x})d\mu_{f,g}(\dot{x}) \qquad (\phi \in C_c(G)).$$

Proof To verify the lemma, one need only prove that $\dot{\phi} = 0 \Rightarrow \int_G (f(x), g(x))\phi(x)d_G(x) = 0$. For this purpose, fix an arbitrary $\psi \in C_c(G)$. Noting that $\dot{\phi} = 0 \Rightarrow$

$$\int_H \phi(x\xi)d_H(\xi) = \int_H \phi(x\xi^{-1})\delta_H(\xi^{-1})d_H(\xi) = 0,$$

we have

$$\begin{aligned} 0 &= \int_G \int_H (f(x), g(x))\psi(x)\phi(x\xi^{-1})\delta_H(\xi^{-1})d_H(\xi)d_G(x) \\ &= \int_H \int_G (f(x), g(x))\psi(x)\phi(x\xi^{-1})\delta_H(\xi^{-1})d_G(x)d_H(\xi) \\ &= \int_H \int_G (f(x\xi), g(x\xi))\psi(x\xi)\phi(x)\delta_G(\xi)\delta_H(\xi^{-1})d_G(x)d_H(\xi) \\ &= \int_G (f(x), g(x))\phi(x)\left[\int_H \psi(x\xi)d_H(\xi)\right]d_G(x). \end{aligned}$$

All that remains to do is choose ψ so that the function $x \mapsto \int_H \psi(x\xi)d_H(\xi)$ is 1 on the compact support of ϕ. □

Definition Let E^L denote the subset of $\mathscr{E}^L$ consisting of those f such that $\mu_{f,f}(G/H) < \infty$.

Due to the fact that

$$\| f(\cdot) + g(\cdot)\|^2 \leqslant 2\| f(\cdot)\|^2 + 2\| g(\cdot)\|^2 \qquad (f, g \in \mathscr{E}^L),$$

it is clear that E^L is a linear subspace of $\mathscr{E}^L$. Evidently $f, g \in E^L \Rightarrow \mu_{f,g}(G/H) < \infty$. Set $(f, g) = \mu_{f,g}(G/H)$ $(f, g \in E^L)$; plainly $(.,.)$ is a positive semi-definite Hermitian form on E^L. Agreeing to identify functions which are equal almost everywhere, we then see that the form $(.,.)$ equips E^L with the structure of a pre-Hilbert space. In order to show completeness, we need the following estimate. Write $\|f\|$ for $(f,f)^{1/2}(f \in E^L)$.

Lemma 5.1.1.2 For each compact subset ω of G, there exists a constant r_ω such that for all $f \in E^L$,

$$\int_\omega \|f(x)\| \, d_G(x) \leqslant r_\omega \|f\|.$$

Proof Choose $\phi \in C_c^+(G)$ such that $\phi = 1$ on ω – then

$$\int_\omega \|f(x)\|^2 d_G(x) \leqslant \int_G \phi(x) \|f(x)\|^2 d_G(x) = \mu_{f,f}(\dot{\phi}) \leqslant \|\dot{\phi}\|_\infty \|f\|^2.$$

Therefore, thanks to the Schwarz inequality, we may take

$$r_\omega = \left(\|\dot{\phi}\|_\infty \int_\omega d_G(x)\right)^{1/2}. \quad \square$$

Proposition 5.1.1.3 The space E^L is complete.

Proof Let $\{f_n\}$ be a Cauchy sequence in E^L; in order to show that $\{f_n\}$ has a limit in E^L we may, by passing to a subsequence if necessary, assume that $\|f_n - f_{n+1}\| < 2^{-n}$.

(1) We begin by proving that $\lim_{n\to\infty} f_n(x)$ exists in E for almost all $x \in G$. Thus let ω be a compact subset of G – then, owing to Lemma 5.1.1.2, we have

$$\int_\omega \|f_n(x) - f_{n+1}(x)\| \, d_G(x) < 2^{-n} r_\omega$$

$$\Rightarrow$$

$$\int_\omega \left(\sum_1^\infty \|f_n(x) - f_{n+1}(x)\|\right) d_G(x) < r_\omega.$$

Therefore, for almost all $x \in \omega$, $\{f_n(x)\}$ is Cauchy in E. Put $f(x) = \lim_{n\to\infty} f_n(x)$ or 0 according to whether the limit exists or not. Of course f is d_G-measurable and verifies, almost everywhere,

$$f(x\xi) = \left(\frac{\delta_H(\xi)}{\delta_G(\xi)}\right)^{1/2} L(\xi^{-1}) f(x) \qquad (\xi \in H, x \in G).$$

(2) We still have to show that $\|f(\cdot)\|^2$ is locally summable, that $\|f\| < \infty$, and that $\|f_n - f\| \to 0$. Let $\phi \in C_c^+(G)$ – then, iterating the parallelogram identity for E, we find that

$$\int_G \|f_n(x) - f_{n+m}(x)\|^2 \phi(x) d_G(x)$$
$$\leqslant \sum_{i=1}^\infty 2^i \int_G \|f_{n+i-1}(x) - f_{n+i}(x)\|^2 \phi(x) d_G(x)$$
$$\leqslant \sum_{i=1}^\infty 2^i \|f_{n+i-1} - f_{n+i}\|^2 \|\dot{\phi}\|_\infty < 2^{-2n+2} \|\dot{\phi}\|_\infty.$$

By Fatou's Lemma, $\int_G \|f_n(x) - f(x)\|^2 \phi(x) d_G(x) \leqslant 2^{-2n+2} \|\dot{\phi}\|_\infty$. Letting ω be compact in G and taking $\phi = 1$ on ω, we see that $\|f_n(\cdot) - f(\cdot)\|^2$ is summable on ω, whence $f_n - f \in \mathscr{E}^L \Rightarrow f \in \mathscr{E}^L$. On the other hand, as ϕ was arbitrary in $C_c^+(G)$, our computation also tells us that

$$\|f_n - f\| \leqslant 2^{-2n+2} \Rightarrow f \in E^L;$$

finally, it is now obvious that $f_n \to f$ in E^L.

The proposition is thereby established. □

A priori, our Hilbert space E^L might consist of zero alone; to show that this is not the case, we shall now discuss a procedure, introduced by Mackey and elaborated upon by Bruhat, which will serve to establish, among other things, that E^L contains plenty of functions.

Convention Let E and F be two Hausdorff topological vector spaces over $\mathbf{C}$, f a linear map of E onto F. We shall say that f is a *strict morphism* if the canonical bijection of $E/f^{-1}(0)$ onto F which is associated with f is an isomorphism (of topological vector spaces). For this, it is necessary and sufficient that f be continuous and open.

It will be convenient to make a temporary change in our hypotheses; thus, let E be a Fréchet space, L a differentiable representation of H on E – consider the space ${}^LC^\infty(G; E)$ of all functions f on G with values in E such that:

(a) $f(x\xi) = \rho_H(\xi)^{1/2} L(\xi^{-1}) f(x)$ for all $\xi \in H$ and $x \in G$;

(b) The canonical image in G/H of the support of f is compact (briefly f has compact support mod H);

(c) $f \in C^\infty(G; E)$.

[Here ρ_H is to be taken as in Appendix 1; in particular, then, $\rho_H(\xi) = (\delta_H(\xi)/\delta_G(\xi))$ $(\xi \in H)$.]

The space ${}^LC^\infty(G; E)$ may be topologized in the following way. Let ω be a compact subset of G; let ${}^LC_\omega^\infty(G; E)$ be the subspace of ${}^LC^\infty(G; E)$ comprised of those functions with support contained in ωH. Place on ${}^LC_\omega^\infty(G; E)$ the relative topology inherited from $C^\infty(G; E)$; in this topology ${}^LC_\omega^\infty(G; E)$ is a Fréchet space. Equip ${}^LC^\infty(G; E)$ with the strict inductive limit of the topologies of the ${}^LC_\omega^\infty(G; E)$ – then ${}^LC^\infty(G; E)$ is an LF-space.

Given $f \in C_c^\infty(G; E)$, put

$$f^L(x) = \int_H \rho_H(\xi)^{-1/2} L(\xi) f(x\xi) d_H(\xi)$$

$$(x \in G;\ d_H \text{ left Haar measure on } H).$$

[The integral appearing on the right here exists in the Bochner sense, since the integrand is continuous and has compact support.]

Lemma 5.1.1.4 (Bruhat) The map $\pi^L, f \mapsto f^L$, is a continuous surjection of $C_c^\infty(G; E)$ onto ${}^LC^\infty(G; E)$ – in fact, it is a strict morphism.

Proof (1) Plainly f^L satisfies (a) and (b) above. As for (c) observe that for each $X \in \mathfrak{g}$ (viewed as a right invariant differential operator on G),

$$Xf^L(x) = \lim_{t\to 0} \int_H \rho_H(\xi)^{-1/2} L(\xi)\left[\frac{f(\exp(-tX)x\xi) - f(x\xi)}{t}\right] d_H(\xi)(x \in G).$$

Since spt(f) is compact and since the operators $L(\xi)$ constitute an equicontinuous set when ξ runs through a compact subset of H, we see that the limit on the right hand side, as $t \to 0$, exists. Hence $f^L \in C^\infty(G; E)$ and for each right invariant differential operator $D \in \mathfrak{G}$ we have

$$Df^L(x) = \int_H \rho_H(\xi)^{-1/2} L(\xi) Df(x\xi) d_H(\xi) \quad (x \in G).$$

In addition note that $f \in C_\omega^\infty(G; E) \Rightarrow f^L \in {}^LC_\omega^\infty(G; E)$ (ω a compact subset of G).

(2) Let us now establish the continuity of the map $f \mapsto f^L$. It suffices to show that the map $f \mapsto f^L$ of $C_\omega^\infty(G; E)$ into ${}^LC_\omega^\infty(G; E)$ is continuous (ω a compact subset of G). So suppose $f_n \to 0$ in $C_\omega^\infty(G; E)$. We must prove that $Df_n^L \to 0$ on each compact subset $\omega_1 \subset G$ ($D \in \mathfrak{G}$). Let $\mathscr{P}_1$ be a closed convex balanced neighborhood of zero in E; per $\mathscr{P}_1$, select a neighborhood $\mathscr{P}_2$ of zero in E such that $a \in \mathscr{P}_2$, $\xi \in \omega_1^{-1}\omega \cap H \Rightarrow L(\xi)a \in \mathscr{P}_1$. [Evidently the function $\xi \mapsto f(x\xi)$ ($f \in C_\omega^\infty(G; E)$) has its support in $\omega_1^{-1}\omega \cap H$ if $x \in \omega_1$.] Now we may find an index n_0 so that $n \geqslant n_0 \Rightarrow Df_n(x) \in \mathscr{P}_2$ for all $x \in G$; therefore, if $n \geqslant n_0$, then

$$Df_n^L(x) \in \left(\int_{\omega_2} \rho_H(\xi)^{-1/2} d_H(\xi)\right)\mathscr{P}_1 \qquad (\omega_2 = \omega_1^{-1}\omega \cap H)$$

for all $x \in \omega_1$ and so the continuity of the map $f \mapsto f^L$ follows.

(3) We have yet to prove that $f \mapsto f^L$ is surjective; in so doing it will be seen that our mapping is actually a strict morphism. First of all observe that if $\alpha \in C_c^\infty(G/H)$ and $h \in {}^LC^\infty(G; E)$, then the function $x \mapsto \alpha(\dot{x})h(x)$ is still in ${}^LC^\infty(G; E)$ (the map $x \mapsto \pi(x) = \dot{x}$ denoting, as usual, the canonical projection of G onto G/H). Hence ${}^LC^\infty(G; E)$ is a module over $C_c^\infty(G/H)$. Viewing G in the usual way as a principal fiber bundle over G/H, select a covering $\{\mathcal{O}_i\}$ of G/H by open sets in each of which there exists a C^∞ section $\dot{x} \mapsto s_i(\dot{x})$ of $\pi^{-1}(\mathcal{O}_i)$ fibered by H. Let $\{\phi_i\}$ be a C^∞ partition of unity subordinate to this covering and let ψ be a function in $C_c^\infty(H)$ such that $\int_H \rho_H(\xi)^{-1/2}\psi(\xi)d_H(\xi) = 1$. For $h \in {}^LC^\infty(G; E)$, put

$${}_Lh(x) = \sum_i \phi_i(\dot{x})\psi(s_i(\dot{x})^{-1}x)L(x^{-1}s_i(\dot{x}))h(s_i(\dot{x})) \qquad (x \in G).$$

Then ${}_Lh \in C_c^\infty(G; E)$, $h \mapsto {}_Lh$ is a continuous map from ${}^LC^\infty(G; E)$ into $C_c^\infty(G; E)$ and ${}_Lh^L = h$, which shows that $h \mapsto {}_Lh$ is a continuous right inverse to $f \mapsto f^L$. Therefore the map $f \mapsto f^L$ of $C_c^\infty(G; E)$ into ${}^LC^\infty(G; E)$ is indeed a strict morphism.

Hence the lemma. □

Remarks (1) The proof of Lemma 5.1.1.4 shows that the map $f \mapsto f^L$ of $C_c^p(G; E)$ into the space ${}^LC_c^p(G; E)$ of p times $(0 \leqslant p \leqslant \infty)$ differentiable functions on G with values in E verifying (a) and (b) above, equipped with the evident topology, is a strict morphism. Take in particular for L the representation $\xi \mapsto \rho_H(\xi)^{1/2}(\dim(E) = 1)$ – then ${}^LC_c^p(G)$ is canonically identified with the space $C_c^p(G/H)$, and thus the map $f \mapsto f^L$ of $C_c^p(G)$ into $C_c^p(G/H)$ is a strict morphism, a well-known result due to Weil [1].

(2) Let us suppose that L is merely a continuous representation of H on our Fréchet space E; then, of course, the natural counterpart to the space ${}^LC^\infty(G; E)$ (in the differentiable case) is the space ${}^LC^0(G; E) = {}^LC(G; E)$ of continuous E-valued functions f on G, having compact support modulo H and verifying the relation

$$f(x\xi) = \rho_H(\xi)^{1/2}L(\xi^{-1})f(x) \qquad (\xi \in H, x \in G).$$

Equipping the spaces $C_c(G; E)$, ${}^LC(G; E)$ with the evident inductive limit topologies, one verifies without difficulty that the map

$$\pi^L, f \mapsto f^L, f^L(x) = \int_H \rho_H(\xi)^{-1/2}L(\xi)f(x\xi)d_H(\xi) \qquad (x \in G),$$

is a strict morphism of $C_c(G; E)$ onto ${}^LC(G; E)$ (cf. Remark 1).

We shall now return to the situation initially under consideration; thus, in particular, we are once again working with a unitary representation L of H on a Hilbert space E. As in number 4.4.1, let L_∞ be the differentiable representation of H on E_∞ canonically associated with L. As we know, E_∞ is a Fréchet space; hence it makes sense to consider the space ${}^{L_\infty}C^\infty(G; E_\infty)$ – plainly ${}^{L_\infty}C^\infty(G; E_\infty) \subset {}^LC(G; E) \subset E^L$. [Let us recall that the topology on E_∞ is not the relative topology induced by E but rather the (in general finer) topology which E_∞ inherits from $C^\infty(H; E)$ via the map $a \mapsto \tilde{a}, \tilde{a}(\xi) = L(\xi)a$ $(a \in E_\infty, \xi \in H)$.]

Lemma 5.1.1.5 The injection of ${}^{L_\infty}C^\infty(G; E_\infty)$ into E^L is continuous and, in fact, admits a continuous extension to ${}^LC(G; E)$ – furthermore, ${}^{L_\infty}C^\infty(G; E_\infty)$ is dense in E^L.

Proof (1) Let us begin by estimating $\|f^{L_\infty}\|(f \in C_c^\infty(G; E_\infty))$. Choose $\phi \in C_c^+(G)$ such that $\phi = 1$ on $\widehat{\operatorname{spt}(\dot{f})}$. Let $g \in E^L$. Since f^{L_∞} vanishes outside of $\omega H(\omega = \operatorname{spt}(f))$, the measure $\mu_{f^{L_\infty}, g}$ must be carried by $\widehat{\operatorname{spt}(\dot{f})}$. Hence, using Fubini's Theorem, we find that

$$(f^{L_\infty}, g) = \mu_{f^{L_\infty}, g}(G/H) = \int_G \phi(x)(f^{L_\infty}(x), g(x))d_G(x)$$

$$= \int_G \int_H \phi(x)\, \delta_G(\xi)^{1/2}\, \delta_H(\xi)^{-1/2}\, (f(x\xi), L(\xi^{-1})g(x))d_H(\xi)d_G(x)$$

$$= \int_G \int_H \phi(x)\, \delta_G(\xi)\, \delta_H(\xi)^{-1}\, (f(x\xi), g(x\xi))d_H(\xi)d_G(x)$$

$$= \int_G \int_H \phi(x)\, \delta_G(\xi^{-1})\, (f(x\xi^{-1}), g(x\xi^{-1}))d_H(\xi)d_G(x)$$

$$= \int_H \int_G \phi(x\xi)(f(x), g(x))d_G(x)d_H(\xi) = \int_G (f(x), g(x))d_G(x).$$

Conclusion:

$$|(f^{L_\infty}, g)| \leqslant r_{\mathrm{spt}(f)} \| f \|_\infty \| g \| \Rightarrow \| f^{L_\infty} \| \leqslant r_{\mathrm{spt}(f)} \| f \|_\infty$$

(cf. Lemma 5.1.1.2).

(2) The estimate in (1), together with Lemma 5.1.1.4, implies that the injection of ${}^{L_\infty}C^\infty(G; E_\infty)$ into E^L is continuous; the argument also shows that this injection admits a continuous extension to ${}^{L}C(G; E)$. [As was shown in part (3) of the proof of Lemma 5.1.1.4, the map $f \mapsto f^{L_\infty}$ of $C_c^\infty(G; E_\infty)$ onto ${}^{L_\infty}C^\infty(G; E_\infty)$ admits a continuous right inverse $h \mapsto {}_{L_\infty}h$ $(h \in {}^{L_\infty}C^\infty(G; E_\infty))$. So suppose, for instance, that $h_n \to 0$ in ${}^{L_\infty}C^\infty(G; E_\infty)$; to verify that $h_n \to 0$ in E^L, first note that ${}_{L_\infty}h_n \to 0$ in $C_c^\infty(G; E_\infty)$. This being the case, select a compact set $\omega \subset G$ such that $\mathrm{spt}\,({}_{L_\infty}h_n) \subset \omega$ (all n); in view of the preceding estimate, we then have that

$$\| {}_{L_\infty}h_n^{L_\infty} \| = \| h_n \| \leqslant r_\omega \| {}_{L_\infty}h_n \|_\infty \to 0$$

as $n \to \infty$.]

(3) Let $f \in C_c^\infty(G)$, $a \in E_\infty$ – then

$$(f \otimes a)^{L_\infty}(x) = \int_H \rho_H(\xi)^{-1/2} f(x\xi) L_\infty(\xi) a \, d_H(\xi) \qquad (x \in G).$$

To prove that ${}^{L_\infty}C^\infty(G; E_\infty)$ is dense in E^L, it suffices to show that the $(f \otimes a)^{L_\infty}$ $(f \in C_c^\infty(G), a \in E_\infty)$ span a dense subspace of E^L (since E^L is complete); in turn, for this, it will be enough to show that only zero is orthogonal to all the $(f \otimes a)^{L_\infty}$. If $((f \otimes a)^{L_\infty}, g) = 0$ for some $g \in E^L$ (all $f \in C_c^\infty(G)$, $a \in E_\infty$), then, by (1) above,

$$\int_G f(x)(a, g(x)) d_G(x) = 0 \qquad (f \in C_c^\infty(G), a \in E_\infty).$$

Hence $(a, g(x)) = 0$ almost everywhere on G (all $a \in E_\infty$). Therefore $g(x) = 0$ almost everywhere on G.

The proof of Lemma 5.1.1.5 is now complete. □

Corollary 5.1.1.6 Let $a \in E$, $\epsilon > 0$, $x \in G$ – then there is a differentiable function f in E^L such that $\| f(x) - a \| < \epsilon$.

Remark The discussion in part (3) of the proof of Lemma 5.1.1.5 establishes the following fact. Let $\mathscr{S}$ be a subset of E which is total in E – then the

$$(f \otimes a)^{L_\infty} \ (f \in C_c^\infty(G), a \in \mathscr{S})$$

are total in E^L.

With these preliminaries out of the way, let us proceed to the definition of *the unitary representation U^L of G induced by the unitary representation L of H on E*. The representation space for U^L will be the Hilbert space E^L; given $f \in E^L$, $U^L(x)f$ will simply be, by definition, $f \circ L_{x^{-1}}$, i.e. the left translate of f by x^{-1} $(x \in G)$. Of course it should be checked that U^L really is a unitary representation of G on E^L, i.e. that:

(1) $U^L(xy) = U^L(x)U^L(y) \quad (x, y \in G), U^L(1) = 1;$

(2) $\| U^L(x)f \| = \| f \| \quad$ for $x \in G, f \in E^L$;

(3) The map $x \mapsto (U^L(x)f, f)$ is continuous on G for $f \in E^L$.

Obviously (1) is true. As for (2), fix $x \in G, f \in E^L$ and set $g = U^L(x)f$ – then, for $\phi \in C_c(G)$,

$$\begin{aligned}\mu_{g,g}(\dot{\phi}) &= \int_G \phi(y) \| g(y) \|^2 \, d_G(y) = \int_G \phi(xy) \| f(y) \|^2 \, d_G(y)\\ &= \mu_{f,f}^x(\dot{\phi})\\ &\Rightarrow \mu_{g,g} = \mu_{f,f}^x \Rightarrow \mu_{g,g}(G/H) = \mu_{f,f}^x(G/H) < \infty \Rightarrow g \in E^L,\end{aligned}$$

in fact that $\| U^L(x)f \| = \| f \|$, as desired. Turning to (3), first note that it is sufficient to show that $x \mapsto (U^L(x)(f \otimes a)^{L_\infty}, (f \otimes a)^{L_\infty})$ is continuous on G (all $f \in C_c^\infty(G)$, $a \in E_\infty$) – however (cf. supra)

$$(U^L(x)(f \otimes a)^{L_\infty}, (f \otimes a)^{L_\infty}) = \int_G f(x^{-1}y)(a, (f \otimes a)^{L_\infty}(y)) d_G(y)$$

and the right hand side of this equation is plainly a continuous function of x by standard facts on integration. Therefore U^L is indeed a unitary representation of G on the Hilbert space E^L.

In the sequel, we shall frequently refer to U^L as the unitary representation of G *unitarily induced* by L.

Notation If there is any doubt as to which 'large' group G is under consideration, then the more precise notations ${}_GU^L$, ${}_GE^L$ will be employed.

Remarks (1) The formation of U^L is a function only of the unitary equivalence class of L, i.e. if L_1 is unitarily equivalent to L_2, then U^{L_1} is unitarily equivalent to U^{L_2}.

(2) Observe that the definition of the unitary representation U^L induced by L is fixed once Haar measures have been fixed on G and H. Moreover, for different choices of Haar measure we obtain the same E^L with the norm changed by a positive constant. Thus the present definition of U^L is, to a certain extent at least, 'intrinsic'.

(3) Let α be a homeomorphic automorphism of G. If L is a unitary representation of H, U^L the corresponding unitarily induced representation of G, then U^L is unitarily equivalent to U^{L^α}, L^α the unitary representation of $\alpha(H)$ defined by $L^\alpha(\xi) = L(\alpha^{-1}(\xi))$ ($\xi \in \alpha(H)$).

Examples (1) Take $H = \{1\}$, L the trivial one dimensional representation of H – then U^L is the left regular representation of G on $L^2(G)$.

(2) Let L be the trivial one dimensional representation of H, V some unitary representation of G – then $U^{(V|H)}$ is unitarily equivalent to $U^L \times V$ (here $U^L \times V$ denotes the Hilbert product of the unitary representations U^L and V – cf. the Remark following Proposition 4.1.2.4).

(3) For certain important classes of groups, *every* irreducible unitary representation is unitarily equivalent to one induced by a unitary representation of some subgroup. Here is a case in point. Suppose that G is a connected, simply

connected nilpotent Lie group with Lie algebra $\mathfrak{g}$. Fix an element $\mathbf{z} \in \mathfrak{g}^\vee$ and let $B_\mathbf{z}$ denote the bilinear form $(X_1, X_2) \mapsto \langle [X_1, X_2], \mathbf{z}\rangle$ $(X_1, X_2 \in \mathfrak{g})$. We shall say that a subalgebra $\mathfrak{h}$ of $\mathfrak{g}$ is *subordinate* to $\mathbf{z}$ if it is isotropic with respect to $B_\mathbf{z}$. If $\mathfrak{h}$ is subordinate to $\mathbf{z}$, if H is the analytic subgroup of G corresponding to $\mathfrak{h}$, then there exists a continuous unitary character $\chi_\mathbf{z}$ of H such that

$$\chi_\mathbf{z}(\exp X) = \exp(\sqrt{-1}\langle X, \mathbf{z}\rangle) \qquad (X \in \mathfrak{h}).$$

Put $V(\mathfrak{h}, \mathbf{z}) = U^{\chi_\mathbf{z}}$ – then:

(i) In order that the (unitary) representation $V(\mathfrak{h}, \mathbf{z})$ be irreducible, it is necessary and sufficient that $\mathfrak{h}$ be of maximal dimension in the set of subalgebras of $\mathfrak{g}$ subordinate to $\mathbf{z}$;

(ii) Two irreducible unitary representations of this type are unitarily equivalent iff the corresponding $\mathbf{z}$ lie on the same orbit under the coadjoint representation of G;

(iii) Every irreducible unitary representation of G is obtained in this way by a convenient choice of $\mathbf{z}$ and $\mathfrak{h}$.

[These results are due to Kirillov [1]; for an exposition see Pukanszky [2] – cf. also Dixmier [3], [5], [6], [7], [8], [9] (papers [5] and [7] contain a number of examples).]

(4) It is natural to ask: To what extent do the results in (3) extend to connected, simply connected solvable Lie groups? The situation here is substantially more complicated since such groups need not be of type I (a connected, simply connected nilpotent Lie group is of type I). [The classical example is due to Mautner. Fix an irrational real number r; let $\mathbf{R}$ operate on $\mathbf{C}^2$ by the rule

$$(z_1, z_2) \mapsto t\cdot(z_1, z_2) = (e^{2\pi\sqrt{-1}t}z_1, e^{2\pi\sqrt{-1}rt}z_2) \qquad (t \in \mathbf{R});$$

let M denote the semi-direct product of $\mathbf{R}$ and $\mathbf{C}^2$ relative to this action – then it can be shown that M is not of type I (cf. Auslander and Moore [1, p. 137]). Along these lines, let us mention the following facts. In order that G be liminaire (compare 4.5.7), it is necessary that G be of type R (i.e. that $\mathrm{Ad}(x)$, as a transformation on the Lie algebra of G, has eigenvalues of modulus one (all $x \in G$)); if G is of type R, then G is of type I iff it is liminaire (see Auslander and Moore [1, Chapter V]).] As a first step Bernat [1] proved that the Kirillov theory carries over, in its essential points, to those G for which the exponential map is surjective (cf. also Takenouchi [2]). [Statement (i) in (3) does not hold, in general, for such G.] Other investigations have been made by Auslander and Kostant [1], Auslander and Moore [1], Brezin [1], Dixmier [13], [16], [17], [18], and Pukanszky [3].

Remark Let L be a unitary representation of H on a Hilbert space E, U^L the unitarily induced representation of G on E^L; let E^L_∞ be the space of differentiable vectors in E^L (for U^L) – then one can characterize the space E^L_∞ as follows: E^L_∞ consists precisely of those functions f in $C^\infty(G; E)$ such that Df belongs to E^L for all right invariant differential operators D on G (including the constants...). [The sufficiency of the stated condition is a ready consequence of Theorem 4.4.4.10; the necessity follows from the inclusion

$${}_{L_\infty}C^\infty(G; E_\infty) \subset E^L_\infty$$

together with an appropriate variant of Sobolev's Lemma. Details may be found in Poulsen [1], [2]. Incidentally Poulsen also shows that if f is an analytic vector for U^L, then f is actually an analytic function from G to E. . . .]

Let L be a unitary representation of $H(\subset G)$ on a Hilbert space E, U^L 'the' corresponding unitarily induced representation of G. For certain purposes, it will be convenient to have at hand other 'realizations' of U^L (such alternative descriptions are especially desirable in the applications).

Thus let ρ_H be a rho-function on G per Appendix 1, μ_H the corresponding quasi-invariant measure on G/H. Let E^L denote the set of those E-valued functions f on G such that:

(i) For each $a \in E$, the function $x \mapsto (f(x), a)$ is a Borel function on G;

(ii) For all $\xi \in H$, $x \in G$, $f(x\xi) = \rho_H(\xi)^{1/2}L(\xi^{-1})f(x)$;

(iii) The integral

$$\int_{G/H} \rho_H(x)^{-1} \| f(x) \|^2 \, d\mu_H(\dot{x})$$

is finite.

[In connection with condition (iii), observe, in view of (ii), that the integrand is constant on left H cosets and hence defines a function on G/H.] If f and g are any two functions in E^L, then one proves without difficulty that the integral $\int_{G/H} \rho_H(x)^{-1}(f(x), g(x)) \, d\mu_H(\dot{x})$ is absolutely convergent; denote its value by (f, g) – then, after functions which agree almost everywhere are identified, E^L becomes a Hilbert space under the inner product $(.\,, .)$. Letting G operate on E^L by left translation leads, in view of standard properties of the measure μ_H, to a unitary representation U^L of G on E^L, a representation which is, moreover, unitarily equivalent to the U^L discussed earlier. Accordingly, whenever it is convenient to do so, we may take the U^L at hand as a realization of 'the' unitary representation of G induced by the unitary representation L of H.

Remark The reader will observe that the choice of a C^∞ rho-function above was not essential; any rho-function on G would do.

The unitary representation U^L of G induced by L is sometimes referred to as a 'multiplier' representation – here is why. Let E^L denote the Hilbert space consisting of those Borel functions f on G with values in E (i.e. such that $x \mapsto (f(x), a)$ is Borel for all $a \in E$) verifying the relation $f(x\xi) = L(\xi^{-1})f(x)$ ($\xi \in H, x \in G$) and having the property that

$$\int_{G/H} (f(x), f(x)) \, d\mu_H(\dot{x}) < \infty.$$

For $x \in G$, $f \in E^L$, put $U^L(x)f = s_H(x^{-1}, .)^{1/2} f \circ L_{x^{-1}}$ (s_H as defined in Appendix 1) – then $U^L(x)f \in E^L$ and $x \mapsto U^L(x)$ is a unitary representation of G on E^L which is clearly unitarily equivalent to the 'intrinsic' realization of U^L discussed above. The factor

$$s_H(x, .) = \rho_H(x.) = \frac{\rho_H(x.)}{\rho_H(\,.\,)} \qquad (x \in G)$$

is a multiplier (cf. Appendix 1); this accounts for the term 'multiplier' representation which is frequently applied to this realization of U^L.

Let π denote the canonical map $x \mapsto \dot{x} = xH$ of G onto G/H. By a *Borel section* s of G fibered by H we shall understand a Borel map $\dot{x} \mapsto s(\dot{x})$

of G/H into G such that $\pi \circ s$ is the identity map on G/H. Borel sections always exist. [We shall prove the following: There exists a Borel set $\mathfrak{B}$ in G such that $\mathfrak{B}$ intersects each left H coset in exactly one point. Thus, since G is countable at infinity, we can find compact sets $\omega_1 \subset \omega_2 \subset \cdots$ with the property that every compact subset of G is contained in some ω_i. By a classical result in measure theory, there exists for each i a Borel subset $\mathfrak{B}_i \subset \omega_i$ such that $\pi(\mathfrak{B}_i) = \pi(\omega_i)$ and such that π is one-to-one on $\mathfrak{B}_i$ (cf. Federer and Morse, 'Some properties of measurable functions B.A.M.S., vol. 49 (1943), Theorem 5.1). The $\mathfrak{B}_i$ may be chosen so as to ensure in addition that $\mathfrak{B}_i \subset \mathfrak{B}_{i+1}$ $(i = 1, 2, \ldots)$. In fact if $\mathfrak{B}_1, \ldots, \mathfrak{B}_i$ have been so chosen, then we may define $\mathfrak{B}_{i+1}$ to be $(\mathfrak{B}'_{i+1} - \pi^{-1}(\pi(\mathfrak{B}_i))) \cup \mathfrak{B}_i$ where $\mathfrak{B}'_{i+1}$ is any Borel subset of ω_{i+1} on which π is one-to-one and has range $\pi(\omega_{i+1})$. Because $\mathfrak{B}_i$ is a Borel set in a complete metric space and π is continuous and one-to-one on $\mathfrak{B}_i$, a famous theorem of Kuratowski tells us that $\pi(\mathfrak{B}_i)$ is a Borel set – hence $\mathfrak{B}_{i+1}$ is a Borel set too. The set $\mathfrak{B} = \cup \mathfrak{B}_i$ will then satisfy the requirement of our contention. [In passing, observe that $\mathfrak{B}$ also has the property that for any compact subset ω of G, the closure of $\pi^{-1}(\pi(\omega)) \cap \mathfrak{B}$ is compact.] To define our Borel section s, simply let $s(\dot{x})$ be the unique element in $\mathfrak{B}$ such that $\pi(s(\dot{x})) = \dot{x}$ (that s really is a Borel function is clear in view of the Kuratowski result referred to above).]

Let s be a Borel section of G, per the immediately preceding discussion, normalized so that $s(\dot{1}) = 1$ – then the map $x \mapsto (s(\dot{x})^{-1})x$ is a Borel function from G to H which will be denoted by t. This being so, set

$$\rho_H(x) = \frac{\delta_H(t(x))}{\delta_G(t(x))} \qquad (x \in G).$$

Evidently ρ_H is a Borel function on G; moreover, it is easy to see that ρ_H is also a rho-function (the boundedness of ρ_H on compact sets being a consequence of the fact that $(\pi^{-1}(\pi(\omega)) \cap s(G/H))^{cl}$ is compact for any compact subset ω of G – cf. supra). Calling μ_H the associated quasi-invariant measure on G/H, put $xs(\dot{y}) = s(\widehat{xy}^{\,\bullet})w(x, \dot{y})$ $(x, y \in G)$ (hence $w(x, \dot{y}) \in H$) – then we may realize U^L on $L^2_{\mu_H}(G/H; E)$ as follows: Given $x \in G, f \in L^2_{\mu_H}(G/H; E)$, define $U^L(x)f$ by the prescription

$$U^L(x)f(\dot{y}) = \rho_H(w(x^{-1}, \dot{y}))^{1/2}L(w(x^{-1}, \dot{y})^{-1})f(\widehat{x^{-1}y}^{\,\bullet}).$$

It is an easy matter to check that $x \mapsto U^L(x)$ is a unitary representation of G on $L^2_{\mu_H}(G/H; E)$, unitarily equivalent to the intrinsic version. [This description of the induced representation is important since, as will be seen later on, it often leads in the applications to simpler explicit formulas, the space G/H being smaller than G.]

Remark Up to now we have given preference to the space G/H of left H cosets and, accordingly, left Haar measures. In the theoretical considerations to follow we shall stick to this convention. However, in actual practice, it is sometimes convenient to work with the space $H\backslash G$ of cosets of opposite parity (and hence right Haar measures). Let us point out, for instance, the form that the first realization of U^L discussed above takes in this situation. Thus one considers the space E^L of all Borel functions f on G with values in E verifying the relation

$$f(\xi x) = \rho_H(\xi)^{1/2}L(\xi)f(x) \qquad (\xi \in H, x \in G)$$

and such that

$$\int_{H\backslash G} \rho_H(x)^{-1}\,\|f(x)\|^2\,d\mu_H(\dot{x}) < \infty.$$

[Here ρ_H denotes a rho-function on G constructed relative to *right* Haar measures on G and H.] Letting G operate on E^L by right translation leads to the desired unitary representation U^L of G. Of course, as above, one may also employ Borel sections to describe U^L. [Let s be a Borel section of G fibered by H (per earlier discussion), normalized so that $s(\dot{1}) = \dot{1}$; let t denote the (Borel) function

$$x \mapsto x(s(\dot{x})^{-1}) \qquad (\dot{x} = Hx, x \in G);$$

set

$$\rho_H(x) = \frac{\delta_H(t(x))}{\delta_G(t(x))} \qquad (x \in G)$$

(δ_H, δ_G denoting modular functions on H, G, respectively, relative to right Haar measure) – then ρ_H is a rho-function on G. Put $s(\dot{y})x = w(\dot{y}, x)s(\dot{\widehat{yx}})$ $(x, y \in G)$; denoting the quasi-invariant measure on $H\backslash G$ corresponding to ρ_H by μ_H, the realization of U^L on $L^2_{\mu_H}(H\backslash G; E)$ then takes the form

$$U^L(x)f(\dot{y}) = \rho_H(w(\dot{y}, x))^{1/2}L(w(\dot{y}, x))f(\dot{\widehat{yx}}) \qquad (f \in L^2_{\mu_H}(H\backslash G; E)).]$$

This becomes particularly important when G is the semi-direct product of a normal H and another closed subgroup – for that subgroup is then a natural choice for the Borel section s (or rather $s(H\backslash G)$). As an example, assume that $H\backslash G$ is isomorphic to $\mathbf{R}$, the additive group of real numbers. If L is a unitary character of H and π is the obvious homomorphism of $H\backslash G$ onto $\mathbf{R}$, then we may realize U^L on $L^2(\mathbf{R})$ by the rule

$$(\dagger) \qquad U^L(x)f(t) = A(x, t)f(t + \pi(x)) \qquad (f \in L^2(\mathbf{R}), t \in \mathbf{R}).$$

[If $x = \xi_x r_x (\xi_x \in H, r_x \in \mathbf{R})$ and $\pi(y_t) = t$, then $A(x, t) = L(y_t\xi_x y_t^{-1})$. Because H is a closed *normal* subgroup of G, it is, of course, true that the modular functions of G and H coincide on H – consequently the Lebesgue measure on $\mathbf{R}$ is invariant under the action of G.]

In the author's opinion, it is desirable to have at hand a geometric interpretation of the inducing mechanism; as will be seen below, a convenient vehicle for this is the notion of a Hilbert bundle. To set things straight from the beginning, however, it must be emphasized that the bundle-theoretic approach to the concept of unitarily induced representation (as discussed above) does not simplify, in any essential way, the underlying analysis. What, then, is the point of the procedure? It appears to be this: Once the present definition of induced representation has been formulated in the language of fiber bundles, important generalizations of the entire process immediately suggest themselves. Further comments along these lines will be given later on.

We shall now proceed to recall a few elementary facts about Hilbert bundles and to establish the requisite terminology – we do not, however, intend to write a treatise here on the theory of fibrations.

Let X be a fixed Hausdorff topological space. A *bundle over* X is a pair (B, π) where B is a Hausdorff topological space and π is a continuous open map of B onto X.

Let $\mathfrak{B} = (B, \pi)$ be a bundle over X. We shall call X the *base space*, B the *bundle space*, and π the *bundle projection* of $\mathfrak{B}$. For $x \in X$, the set $\pi^{-1}(x)$ is called the *fiber over* x – we shall often denote it by $\mathfrak{B}_x$. A *cross section* for $\mathfrak{B}$ is a map $s: X \to B$ such that $\pi \circ s$ is the identity map on X; if $b \in B$ and $s(\pi(b)) = b$, then we say that *s passes through b*. If to every $b \in B$ there is a *continuous cross section s* passing through b, then $\mathfrak{B}$ is said to have *enough cross sections*.

A *Hilbert bundle* $\mathfrak{B}$ over X is a bundle (B, π) over X together with operations and a norm making each fiber $\mathfrak{B}_x$ $(x \in X)$ into a complex Hilbert space and satisfying the following conditions:

(1) The map $b \mapsto \|b\|$ is continuous on B to $\mathbf{R}$;

(2) The operation of addition is continuous on $\{(b_1, b_2) \in B \times B: \pi(b_1) = \pi(b_2)\}$ to B;

(3) The operation of scalar multiplication is a continuous map of $\mathbf{C} \times B$ into B;

(4) If $x \in X$ and $\{b_n\}$ is a net in B such that $\|b_n\| \to 0$ and $\pi(b_n) \to x$, then $b_n \to 0_x$ in B.

[Let us clarify the notation. First of all we are writing $\|\,.\,\|$ for the norm on the fiber under consideration; second, 0_x denotes the zero element of the Hilbert space $\mathfrak{B}_x = \pi^{-1}(x)$ $(x \in X)$; finally we emphasize, in connection with condition (2), that addition is defined only if the elements in question lie on the same fiber over X.]

Let (B, π) be a Hilbert bundle over X – then our assumptions imply that the topology of B relativized to each fiber $\mathfrak{B}_x$ is the norm topology of $\mathfrak{B}_x$. [In fact suppose $b_n \to b$ in B where b_n, b lie on the fiber $\mathfrak{B}_x$. By condition (2), $\|b_n - b\| \to 0_x$; hence by (1), $\|b_n - b\| \to 0$ and so $b_n \to b$ in the norm topology. On the other hand, condition (4) tells us that the argument can be reversed.]

Example Let E be any Hilbert space. If we put $B = X \times E$ (equipped with the product topology) and set $\pi(x, a) = x(x \in X, a \in E)$, then each fiber $\{x\} \times E$ acquires the structure of a Hilbert space via the bijection $a \mapsto (x, a)$; plainly $\mathfrak{B} = (B, \pi)$ is a Hilbert bundle over X, the *trivial bundle with constant fiber E*. In the case at hand, the cross sections s for $\mathfrak{B}$ may be viewed as functions $s: X \to E$.

Let $\mathfrak{B} = (B, \pi)$ be a Hilbert bundle over X, $\mathfrak{S}(\mathfrak{B})$ the set of continuous cross sections for $\mathfrak{B}$ – then it is clear that $\mathfrak{S}(\mathfrak{B})$ is a complex linear space (under pointwise addition and scalar multiplication), whose zero element is the cross section $x \mapsto 0_x$ $(x \in X)$. [Note too that if $s \in \mathfrak{S}(\mathfrak{B})$ and σ is a continuous complex valued function on X, then $x \mapsto \sigma(x)s(x)$ $(x \in X)$ is in $\mathfrak{S}(\mathfrak{B})$.] Now assume that X is locally compact; define the *uniform on compacta* topology for $\mathfrak{S}(\mathfrak{B})$ as that in which a net $\{s_n\}$ in $\mathfrak{S}(\mathfrak{B})$ converges to a continuous cross section

s iff, for each compact subset $\omega \subset X$, $\sup_{x \in \omega} \| s_n(x) - s(x) \| \to 0$ as n becomes large.

Lemma 5.1.1.7 Let $\mathfrak{B} = (B, \pi)$ be a Hilbert bundle over a locally compact space X. Let $\mathscr{S}$ be a linear subspace of $\mathfrak{S}(\mathfrak{B})$ with the following properties:
(i) $\mathscr{S}$ is closed under multiplication by continuous functions on X;
(ii) For each $x \in X$, $\{s(x) : s \in \mathscr{S}\}$ is dense in $\mathfrak{B}_x$.
Then $\mathscr{S}$ is uniformly on compacta dense in $\mathfrak{S}(\mathfrak{B})$.

Proof Let ω be a compact subset of X, $t \in \mathfrak{S}(\mathfrak{B})$. Fix $\epsilon > 0$. We can choose (1) a finite open covering $\mathcal{O}_1, \ldots, \mathcal{O}_r$ of ω, (2) for each $i = 1, \ldots, r$ an element $s_i \in \mathscr{S}$ such that $\| s_i(x) - t(x) \| < \epsilon$ (all $x \in \mathcal{O}_i$), and (3) for each i a continuous non-negative function σ_i on X vanishing outside of $\mathcal{O}_i$ and having the property that $\sum_i \sigma_i(x) = 1$ ($x \in \omega$) – then $s = \sum_i \sigma_i s_i \in \mathscr{S}$ and $\sup_{x \in \omega} \| s(x) - t(x) \| < \epsilon$, which completes the proof. □

It will be necessary to have at hand a few simple facts concerning integration theory in Hilbert bundles $\mathfrak{B}$ over locally compact spaces X.

Let $\mathfrak{B} = (B, \pi)$ be a Hilbert bundle over a locally compact Hausdorff space X – *assume* that $\mathfrak{B}$ has enough cross sections.

Fix a (positive) regular Borel measure μ on X. We shall regard μ as being defined on the sigma-ring comprised of those Borel subsets of X which are sigma-bounded (i.e. contained in a countable union of compact sets) – thus any μ-summable scalar function on X automatically vanishes outside of some sigma-bounded set. If f is any non-negative function on X, then its *μ-upper integral* $\int_X^* f(x) d\mu(x)$ is defined in the usual way. Now fix a real number p with $1 \leqslant p < \infty$; let $\mathscr{L}^p(\mathfrak{B}; \mu)$ denote the set of those cross sections s for $\mathfrak{B}$ such that $\int_X^* \| s(x) \|^p d\mu(x) < \infty$ – then $\mathscr{L}^p(\mathfrak{B}; \mu)$ is a linear space under the obvious pointwise operations. Evidently the map $s \mapsto (\int_X^* \| s(x) \|^p d\mu(x))^{1/p}$ defines a semi-norm $\| \cdot \|_p$ on $\mathscr{L}^p(\mathfrak{B}; \mu)$; letting $\mathscr{N}^p(\mathfrak{B}; \mu) = \{s \in \mathscr{L}^p(\mathfrak{B}; \mu) : \| s \|_p = 0\}$, then $\mathscr{L}^p(\mathfrak{B}; \mu)/\mathscr{N}^p(\mathfrak{B}; \mu)$ becomes a Banach space under the norm induced by $\| \cdot \|_p$ (which we also call $\| \cdot \|_p$). All of this is rather familiar and well-known. As is the custom, we shall fail to distinguish notationally between $\mathscr{L}^p(\mathfrak{B}; \mu)$ and $\mathscr{L}^p(\mathfrak{B}; \mu)/\mathscr{N}^p(\mathfrak{B}; \mu)$.

Lemma 5.1.1.8 Let $\{s_n\}$ be a sequence of elements in $\mathscr{L}^p(\mathfrak{B}; \mu)$ converging in $\mathscr{L}^p(\mathfrak{B}; \mu)$ to s – then there exists a subsequence $\{s_m\}$ of $\{s_n\}$ such that $s_m(x) \to s(x)$ in $\mathfrak{B}_x$ for μ almost all x.

[The proof is quite canonical.]

Let $\mathfrak{S}_c(\mathfrak{B})$ denote the linear space of all continuous cross sections s for $\mathfrak{B}$ which have compact support – then $\mathfrak{S}_c(\mathfrak{B})$ is a linear subspace of $\mathscr{L}^p(\mathfrak{B}; \mu)$. [Note too that there are elements of $\mathfrak{S}_c(\mathfrak{B})$ which pass through any preassigned point of B.] Consequently the closure of $\mathfrak{S}_c(\mathfrak{B})$ in $\mathscr{L}^p(\mathfrak{B}; \mu)$ is a Banach space (under $\| \cdot \|_p$), which we shall call the *cross sectional L^p space of $\mathfrak{B}$ with respect to μ* and denote by $L^p(\mathfrak{B}; \mu)$ (or simply $L^p(\mathfrak{B})$ if μ is understood).

Evidently, if $s \in L^p(\mathfrak{B}; \mu)$, then $x \mapsto \| s(x) \|$ is μ-measurable (Lemma 5.1.1.8).

Remark Suppose that $\mathfrak{B}$ is the trivial bundle over X with constant fiber E. By the regularity of μ, $L^1(\mathfrak{B}; \mu)$ contains all functions $\psi_W a$ where $a \in E$ and W is a μ-measurable subset of X with $\mu(W) < \infty$ – here

$$\psi_W a(x) = \begin{cases} a & \text{if } x \in W \\ 0 & \text{if } x \notin W. \end{cases}$$

The linear span of the set of all such $\psi_W a$ is dense in $L^1(\mathfrak{B}; \mu)$; the continuous function I on this span defined by

$$I\Big(\sum_{i=1}^{r} \psi_{W_i} a_i\Big) = \sum_{i=1}^{r} \mu(W_i) a_i$$

admits a unique continuous linear extension to all of $L^1(\mathfrak{B}; \mu)$. If $s \in L^1(\mathfrak{B}; \mu)$, then $I(s)$ is called the *Bochner integral* of s with respect to μ; of course $I(s)$ is an element of E and will be denoted by $\int_X s(x)\, d\mu(x)$.

A cross section s for $\mathfrak{B}$ is said to be *locally μ-measurable* if, for each compact subset ω of X, there is a sequence $\{s_n\}$ of continuous cross sections for $\mathfrak{B}$ such that $s_n(x) \to s(x)$ for μ almost all x in ω. The set of locally μ-measurable cross sections for $\mathfrak{B}$ is a linear space which is, evidently, closed under multiplication by locally μ-measurable complex valued functions on X.

Lemma 5.1.1.9 Fix a real number p with $1 \leqslant p < \infty$ – then an element $s \in \mathscr{L}^p(\mathfrak{B}; \mu)$ belongs to $L^p(\mathfrak{B}; \mu)$ iff it is locally μ-measurable.

Proof The fact that elements of $L^p(\mathfrak{B}; \mu)$ are locally μ-measurable is, in view of Lemma 5.1.1.8, clear. Conversely, suppose that s is a locally μ-measurable function in $\mathscr{L}^p(\mathfrak{B}; \mu)$; let ω be a compact subset of X, r a positive real number; put

$$s_{\omega,r}(x) = \psi_\omega(x) \min(1, r\,\|s(x)\|^{-1}) s(x) \qquad (x \in X),$$

ψ_ω the characteristic function of ω – then $s_{\omega,r}$ is bounded in norm by r and has compact support. Owing to the Lebesgue Dominated Convergence Theorem, we shall be guaranteed that s belongs to $L^p(\mathfrak{B}; \mu)$ provided we can verify first that the same is true for the $s_{\omega,r}$ (all compact $\omega \subset X$, $r > 0$). So fix a compact subset ω of X and a positive real number r; choose a sequence $\{s_n\}$ of continuous cross sections for $\mathfrak{B}$ such that $s_n(x) \to s(x)$ for μ almost all x in ω; put

$$(s_n)_{\omega,r}(x) = \psi_\omega(x) \min(1, r\,\|s_n(x)\|^{-1}) s_n(x) \qquad (x \in X).$$

Then $(s_n)_{\omega,r} \to s_{\omega,r}$ pointwise μ almost everywhere on X. Hence, by standard arguments, for any $\epsilon > 0$ we can find a μ-measurable subset ω_ϵ of ω with $\mu(\omega_\epsilon) < \epsilon(2r)^{-p}$ and an n such that

$$\|(s_n)_{\omega,r}(x) - s_{\omega,r}(x)\|^p < \epsilon\mu(\omega)^{-1} \qquad (\text{all } x \in \omega - \omega_\epsilon).$$

But then

$$\begin{aligned}\|(s_n)_{\omega,r} - s_{\omega,r}\|_p^p &\leqslant \int_{\omega_\epsilon} \|(s_n)_{\omega,r}(x) - s_{\omega,r}(x)\|^p\, d\mu(x) \\ &\quad + \int_{\omega-\omega_\epsilon} \|(s_n)_{\omega,r}(x) - s_{\omega,r}(x)\|^p\, d\mu(x) \\ &\leqslant 2\epsilon,\end{aligned}$$

which shows that $s_{\omega,r}$ belongs to $L^p(\mathfrak{B}; \mu)$ since $(s_n)_{\omega,r}$ does (s_n is continuous and μ is regular). □

Lemma 5.1.1.10 Let $\mathscr{S}$ be a linear subspace of $\mathfrak{S}_c(\mathfrak{B})$ with the following properties:

(i) $\mathscr{S}$ is closed under multiplication by continuous functions on X;
(ii) For each $x \in X$, $\{s(x) : s \in \mathscr{S}\}$ is dense in $\mathfrak{B}_x$.
Then $\mathscr{S}$ is dense in $L^p(\mathfrak{B}; \mu)$.

Proof It is enough to approximate a given $s \in \mathfrak{S}_c(\mathfrak{B})$ uniformly on X by functions in $\mathscr{S}$ whose supports are all contained in some fixed compact set. Let σ be a function in $C_c(X)$ which assumes the value 1 on the compact support of s. Owing to Lemma 5.1.1.7 there is a net $\{s_n\}$ in $\mathscr{S}$ which approaches s uniformly on ω, $\omega = \text{spt}(\sigma)$. But then $\sigma s_n \to \sigma s = s$ uniformly on ω, hence uniformly on X. Since the σs_n all vanish off of ω, we are done. □

Let's take a closer look at the space $L^2(\mathfrak{B}; \mu)$. In the first place, the polarization identity together with the continuity of the norm shows that the inner product on the fibers of our Hilbert bundle $\mathfrak{B}$ is a continuous function on the set $\{(b_1, b_2) \in B \times B : \pi(b_1) = \pi(b_2)\}$ – therefore the norm $\|\,.\,\|_2$ on $\mathfrak{S}_c(\mathfrak{B})$ comes from an inner product $(s, t) = \int_X (s(x), t(x))d\mu(x)$ $(s, t \in \mathfrak{S}_c(\mathfrak{B}))$ and so it follows that $L^2(\mathfrak{B}; \mu)$ is a Hilbert space. Moreover, if $s, t \in L^2(\mathfrak{B}; \mu)$, then the polarization identity shows that the function $x \mapsto (s(x), t(x))$ is μ-measurable, hence by Schwarz's inequality μ-summable, with $|\int_X (s(x), t(x))d\mu(x)| \leqslant \|s\|_2 \|t\|_2$. Consequently, upon approximating s and t by elements of $\mathfrak{S}_c(\mathfrak{B})$, we find that the inner product (s, t) in $L^2(\mathfrak{B}; \mu)$ itself is still given by the formula

$$(s, t) = \int_X (s(x), t(x))\, d\mu(x).$$

Let G be a Lie group countable at infinity, H a closed subgroup of G, ρ_H a continuous rho-function on G (normalized by the condition $\rho_H(1) = 1$). Let $\xi \mapsto L(\xi)$ be a unitary representation of H on a Hilbert space E – then we may define a continuous right action of H on the product space $G \times E$ as follows:

$$(x, a)\cdot\xi = (x\xi, \rho_H(\xi)^{1/2}L(\xi^{-1})a) \qquad (x \in G, a \in E).$$

The orbit under H of a pair $(x, a) \in G \times E$ will be denoted by $(x, a)H$; the space of all such orbits equipped with the quotient topology will be called $\mathbf{E}$ (or $\mathbf{E}(G, H, L)$ if more precision is needed). Evidently the space $\mathbf{E}$ is Hausdorff and the projection $\pi_{\mathbf{E}}$, $(x, a)H \mapsto \dot{x} = xH$, ($x \mapsto \dot{x}$ the canonical map of G onto G/H), of $\mathbf{E}$ onto G/H is continuous and open. [Thus, in the product space $(G \times E) \times (G \times E)$, consider those pairs $((x_1, a_1), (x_2, a_2))$ having the property that (x_1, a_1) and (x_2, a_2) lie on the same H-orbit in $G \times E$; owing to the continuity of L, it is a simple matter to verify that this set of pairs is closed in $(G \times E) \times (G \times E)$ etc.] Therefore $(\mathbf{E}, \pi_{\mathbf{E}})$ is a bundle over G/H. Our objective now is to place on $(\mathbf{E}, \pi_{\mathbf{E}})$ the structure of a Hilbert bundle. For this purpose (and others) we shall use without comment the following elementary fact (which is a direct consequence of the openness of $\pi_{\mathbf{E}}$): If $x_n \to x$ in G and $(x_n, a_n)H \to (x, a)H$ in $\mathbf{E}$, then $a_n \to a$ in E. This being so, introduce linear operations on each fiber of $(\mathbf{E}, \pi_{\mathbf{E}})$ by the (legitimate) formula $c_1(x, a_1)H + c_2(x, a_2)H = (x, c_1a_1 + c_2a_2)H$ $(c_1, c_2 \in \mathbf{C})$ and then equip each fiber with the structure of a Hilbert space via the rule $((x, a_1)H, (x, a_2)H)_{\rho_H} = \rho_H(x)^{-1}(a_1, a_2)$ (since $\rho_H(x\xi) = \rho_H(x)\rho_H(\xi)$ $(x \in G, \xi \in H)$, our definition is meaningful). Let us verify that $(\mathbf{E}, \pi_{\mathbf{E}})$ is a Hilbert bundle over G/H. Denoting the entity in question by $\mathfrak{E}(\rho_H)$ (to emphasize the dependence of its norms on the choice of ρ_H), one may, for instance, prove that $\mathfrak{E}(\rho_H)$ satisfies condition (4) of the definition of a Hilbert bundle as follows.

Suppose that $\{\mathbf{a}_n\}$ is a net in $\mathbf{E}$ such that $\|\mathbf{a}_n\|_{\rho_H} \to 0$ and $\pi_{\mathbf{E}}(\mathbf{a}_n) \to \dot{x} = xH$ (say). To verify that $\mathbf{a}_n \to \mathbf{0}_{\dot{x}}$, it will be sufficient to find a subnet of $\{\mathbf{a}_n\}$ converging to $\mathbf{0}_{\dot{x}}$. Because the canonical map of G onto G/H is open, the hypothesis that $\pi_{\mathbf{E}}(\mathbf{a}_n) \to \dot{x}$ permits us to pass to a subnet and write $\mathbf{a}_n = (x_n, a_n)H$ ($x_n \in G, a_n \in E$), where $x_n \to x$ in G. Since $\rho_H(x_n) \to \rho_H(x) \neq 0$ and $\|\mathbf{a}_n\|_{\rho_H}^2 = \rho_H(x_n)^{-1}\|a_n\|^2 \to 0$, we have $a_n \to 0$ in E – therefore $\mathbf{a}_n = (x_n, a_n)H \to (x, 0)H = \mathbf{0}_{\dot{x}}$, which serves to show that condition (4) of the definition of a Hilbert bundle is met in the case at hand. The other conditions which $\mathfrak{E}(\rho_H)$ must satisfy in order to be a Hilbert bundle are proved similarly. We shall call $\mathfrak{E}(\rho_H)$ the *Hilbert bundle induced by L (and ρ_H)*. [In passing, let us note that our Hilbert bundle $\mathfrak{E}(\rho_H)$ is, in addition, 'G-homogeneous'; in fact G operates continuously to the left on $\mathbf{E}$ via the prescription $(y, a)H \mapsto x\cdot(y, a)H = (xy, a)H$, hence moves the fibers of $\mathfrak{E}(\rho_H)$ around by linear transformations.]

Our next task will be to represent the group G by unitary operators on the square integrable cross sections of $\mathfrak{E}(\rho_H)$; in turn, this will lead us once again to the unitary representation U^L of G induced by the unitary representation L of H. To this end, first observe that there exists a one-to-one correspondence between cross sections s for $\mathfrak{E}(\rho_H)$ and functions $f: G \to E$ having the property that $f(x\xi) = \rho_H(\xi)^{1/2}L(\zeta^{-1})f(x)$ ($\xi \in H$, $x \in G$) – the correspondence is given by $s(xH) = (x, f(x))H$. If f and s are connected in this way, then we shall say that f is the *unitary form* of s and s is the *bundle form* of f. Evidently a cross section s is continuous (on G/H) iff its unitary form f is continuous (on G); from this, it follows more or less at once that the Hilbert bundle $\mathfrak{E}(\rho_H)$ admits enough cross sections (Lemma 5.1.1.4 and subsequent discussion) – therefore it makes sense to consider the Hilbert space $L^2(\mathfrak{E}(\rho_H); \mu_H)$ (μ_H the quasi-invariant measure on G/H associated with the rho-function ρ_H). It is not difficult to see that the Hilbert space $L^2(\mathfrak{E}(\rho_H); \mu_H)$ is actually independent of the choice of ρ_H; accordingly, we shall henceforth write E^L in place of $L^2(\mathfrak{E}(\rho_H); \mu_H)$. [Thus let σ_H be another rho-function on G satisfying the same conditions as ρ_H, ν_H the associated quasi-invariant measure on G/H; let s be a cross section for $(\mathbf{E}, \pi_{\mathbf{E}})$. If $s(xH) = (x, a)H$, then

$$\|s(xH)\|_{\sigma_H}^2 = \sigma_H(x)^{-1}\|a\|^2 = \rho_H(x)\sigma_H(x)^{-1}\|s(xH)\|_{\rho_H}^2$$

and so

$$\int_{G/H}^{*} \|s(\dot{x})\|_{\sigma_H}^2 \, d\nu_H(\dot{x}) = \int_{G/H}^{*} \|s(\dot{x})\|_{\rho_H}^2 \, d\mu_H(\dot{x}),$$

from which it follows that the same functions belong to $L^2(\mathfrak{E}(\rho_H); \mu_H)$ as belong to $L^2(\mathfrak{E}(\sigma_H); \nu_H)$ and that their norms in the two spaces are the same. Consequently the Hilbert space E^L is indeed invariantly constructed.] Abusing the notation a little, we shall agree to denote the space of all unitary forms of elements in E^L by the same symbol; of course, under the correspondence $s \leftrightarrow f$ mentioned above, this space acquires the structure of a Hilbert space, the inner product being given by $(f, g) = \int_{G/H} \rho_H(x)^{-1}(f(x), g(x))\, d\mu_H(\dot{x})$.

The bundle-theoretic definition of the unitarily induced representation U^L is as follows: For $x \in G$, put

$$U^L(x)s(yH) = x \cdot s(x^{-1}yH) \qquad (s \in E^L).$$

Naturally, it is necessary to show that U^L really is a unitary representation of G. In the first place, it is clear that $U^L(x)s$ is at least a cross section for $(\mathbf{E}, \pi_{\mathbf{E}})$; if s is continuous, then certainly so is $U^L(x)s$ – therefore, since a general $s \in E^L$ is locally μ-measurable (Lemma 5.1.1.9), we see, by passing to sequential limits, that $U^L(x)s$ is too. In addition, we have

$$\int_{G/H}^{*} \| U^L(x)s(\dot{y}) \|_{\rho_H}^2 \, d\mu_H(\dot{y}) = \int_{G/H}^{*} \| x \cdot s(x^{-1} \cdot \dot{y}) \|_{\rho_H}^2 \, d\mu_H(\dot{y})$$

$$= \int_{G/H}^{*} \frac{\rho_H(xy)}{\rho_H(y)} \| x \cdot s(\dot{y}) \|_{\rho_H}^2 \, d\mu_H(\dot{y}) = \int_{G/H}^{*} \| s(\dot{y}) \|_{\rho_H}^2 \, d\mu_H(\dot{y}) < \infty.$$

Thus, on the basis of Lemma 5.1.1.9, we deduce that $U^L(x)s \in E^L$; in fact that $\| U^L(x)s \| = \| s \|$ $(s \in E^L)$. Now it is perfectly clear that $U^L(xy) = U^L(x)U^L(y)$ $(x, y \in G)$, $U^L(1) = 1$; as it is not difficult to verify the continuity of the map $x \mapsto U^L(x)$, we conclude that U^L is a unitary representation of G on E^L.

The transport of U^L to the space of all unitary forms of elements in E^L is given by $U^L(x)f = f \circ L_{x^{-1}}$ $(f \in E^L)$; keeping this in mind, the reader will easily convince himself that the present bundle-theoretic definition of 'unitarily induced representation' is (unitarily) equivalent to the one given at the beginning of this number.

The preceding bundle-theoretic construction makes no use of the local triviality of G/H (and is valid for general locally compact groups); moreover, the definition of Hilbert bundle is, strictly speaking, not in accord with the customary version – however, it has the advantage of allowing one to place Hilbert space structures on the fibers without having to worry about transition functions.

The philosophy lying behind all this can be summarized as follows. Given a unitary representation L of a closed subgroup H of a Lie group G, one may proceed to represent G on the compactly supported C^∞ sections of a certain vector bundle; upon equipping the space of all such sections with the pre-Hilbert space structure inherent in the situation, by continuity G can then be represented on the corresponding Hilbert space completion E^L (say) – the resulting unitary representation U^L (say) is 'the' unitary representation of G unitarily induced by L. [This rough description can, of course, be made precise.] In actual practice, it will often happen that the vector bundle under consideration has additional structure – for instance, it might be a holomorphic vector bundle. Under these circumstances, one could instead try to represent G on the holomorphic sections of the bundle (or even on cohomology groups attached, in some way or another, to the bundle) – this has been encountered already during the discussion of the Borel-Weil Theorem. We shall return to such questions later on. In any event, it should now be clear that the notion of 'unitarily induced representation' is, after all, but an important special case in a somewhat more general scheme of things.

5.1.2 Unitarily Induced Representations and Measures of Positive Type

Let us agree to retain the notations and conventions introduced in 5.1.1; the objective of the present number is to establish:

Theorem 5.1.2.1 (Blattner) Let α be a measure of positive type on the closed subgroup H of G; let β be the injection of $(\delta_G/\delta_H)^{1/2}\alpha$ into G – then β is a measure of positive type on G. Let A be the unitary representation of H generated by α, B the unitary representation of G generated by β – then the unitarily induced representation U^A of G is unitarily equivalent to B.

[The well-known connection between measures of positive type and unitary representations will be reviewed in a moment.]

Before starting on the proof of the theorem, it will be convenient to prove a preliminary lemma; for the purposes of this lemma we shall modify the terminology of Appendix 1 as follows: By a rho-function on G (relative to H) we shall understand any complex valued locally d_G-summable function ρ on G verifying $\rho(x\xi) = (\delta_H(\xi)/\delta_G(\xi))\rho(x)$ ($\xi \in H$, $x \in G$). Per a rho-function ρ, one may define a measure μ_ρ on G/H by the rule

$$\int_G f(x)\rho(x)d_G(x) = \int_{G/H} d\mu_\rho(\dot{x}) \int_H f(x\xi)d_H(\xi) \qquad (f \in C_c(G)).$$

Lemma 5.1.2.2 Let α be a measure on H, f an element of $C_c(G)$. For $x \in G$, let ϕ_x^f denote the function

$$\xi \mapsto \left(\frac{\delta_G(\xi)}{\delta_H(\xi)}\right)^{1/2} f(x\xi) \qquad (\xi \in H);$$

put

$$\rho_f(x) = \int_H ((\phi_x^f)^* * \phi_x^f)(\xi)d\alpha(\xi) \qquad (x \in G).$$

Then ρ_f is a continuous rho-function on G. Moreover the corresponding measure μ_{ρ_f} on G/H has compact support – in fact

$$\mu_{\rho_f}(G/H) = \int_H \left(\frac{\delta_G(\xi)}{\delta_H(\xi)}\right)^{1/2} (f^* * f)(\xi)d\alpha(\xi).$$

[The star operations (compare 4.3.1) on ϕ_x^f and f are, of course, to be taken in $C_c(H)$ and $C_c(G)$, respectively; similarly for the convolutions.]

Proof (1) Let f be any element in $C_c(G)$ whatsoever – then it is clear that the function $x \mapsto \int_H (\delta_G(\xi)/\delta_H(\xi))f(x\xi)d_H(\xi)$ $(x \in G)$ is a continuous rho-function on G. In addition, it is easy to verify that the corresponding measure on G/H has compact support, the total mass of G/H being $\int_G f(x)d_G(x)$.

(2) An elementary computation shows that

$$\int_H \left(\frac{\delta_G(\xi)}{\delta_H(\xi)}\right)^{1/2} (f^* * f)(\xi)d\alpha(\xi) = \int_G g(x)d_G(x)$$

where

$$g(x) = f(x)\int_H (\delta_H(\xi)\delta_G(\xi))^{-1/2}\overline{f(x\xi^{-1})}d\alpha(\xi) \qquad (x \in G).$$

Since it is clear that $g \in C_c(G)$, the discussion in (1) is applicable to g; letting ρ_f denote the rho-function on G obtained from g per that discussion, a routine calculation gives

$$\rho_f(x)\left(= \int_H \frac{\delta_G(\xi)}{\delta_H(\xi)} g(x\xi)d_H(\xi)\right) = \int_H ((\phi_x^f)^* * \phi_x^f)(\xi)d\alpha(\xi) \quad (x \in G).$$

Bearing in mind the remarks in (1), the assertions of the lemma are now evident. □

Corollary 5.1.2.3 Let α be a measure of positive type on H, β the injection of $(\delta_G/\delta_H)^{1/2}\alpha$ into G – then β is a measure of positive type on G.

Proof Because the measure α is of positive type, it is true, in particular, that $\alpha((\phi_x^f)^* * \phi_x^f) \geqslant 0$ (all $x \in G, f \in C_c(G)$); this means that the measure μ_{ρ_f} of Lemma 5.1.2.2 is, for each $f \in C_c(G)$, non-negative – hence $\mu_{\rho_f}(G/H) = \beta(f^* * f) \geqslant 0$ $(f \in C_c(G))$, as desired. □

Proof of Theorem 5.1.2.1 The first assertion of our theorem is contained in Corollary 5.1.2.3. Turning to the second, let us begin by recalling the construction of A and B. Thus let I_α and I_β be the null ideals of α and β in $C_c(H)$ and $C_c(G)$, respectively; let π_α and π_β be the canonical linear maps of $C_c(H)$ onto $C_c(H)/I_\alpha$ and of $C_c(G)$ onto $C_c(G)/I_\beta$, respectively – then the representation space E_α for A is the completion of $C_c(H)/I_\alpha$ with respect to the inner product

$$(\pi_\alpha(f_1), \pi_\alpha(f_2)) = \int_H (f_2^* * f_1)(\xi)d\alpha(\xi) \qquad (f_i \in C_c(H), i = 1, 2),$$

while the representation space E_β for B is the completion of $C_c(G)/I_\beta$ with respect to the inner product

$$(\pi_\beta(f_1), \pi_\beta(f_2)) = \int_H (f_2^* * f_1)(\xi)\left(\frac{\delta_G(\xi)}{\delta_H(\xi)}\right)^{1/2} d\alpha(\xi)$$
$$(f_i \in C_c(G), i = 1, 2).$$

(1) Let $f \in C_c(G)$. For $x \in G$, let ϕ_x^f denote the function

$$\xi \mapsto \left(\frac{\delta_G(\xi)}{\delta_H(\xi)}\right)^{1/2} f(x\xi) \qquad (\xi \in H);$$

of course $\phi_x^f \in C_c(H)$. Put

$$\Phi_f(x) = \pi_\alpha(\phi_x^f) \qquad (x \in G).$$

Then Φ_f is an E_α-valued function on G verifying

$$\Phi_f(x\xi) = \left(\frac{\delta_H(\xi)}{\delta_G(\xi)}\right)^{1/2} A(\xi^{-1})\Phi_f(x) \qquad (\xi \in H, x \in G).$$

As it is clear that Φ_f is continuous with compact support modulo H, we see that $\Phi_f \in E^A$ (the representation space for the unitarily induced representation U^A) – therefore the mapping $\Phi, f \mapsto \Phi_f$, is a linear map of $C_c(G)$ into E^A.

(2) Claim: If $f \in C_c(G)$, then $\|\Phi_f\|$ (in E^A) $= \|\pi_\beta(f)\|$ (in E_β). Thus, keeping in mind that for fixed f, Φ_f is an E_α-valued function on G, the very definition of $\|\Phi_f\|$ tells us that $\|\Phi_f\|^2 = \mu_{\rho_f}(G/H)$ where

$$\rho_f(x) = \|\pi_\alpha(\phi_x^f)\|^2 = \int_H ((\phi_x^f)^* * \phi_x^f)(\xi) d\alpha(\xi) \qquad (x \in G);$$

on the other hand, by Lemma 5.1.2.2, $\mu_{\rho_f}(G/H) = \|\pi_\beta(f)\|^2$. Hence the claim.

(3) In view of (2), we can define an isometry Q of E_β into E^A, satisfying $Q(\pi_\beta(f)) = \Phi_f$ $(f \in C_c(G))$. Since it is a simple matter to verify that Q intertwines B and U^A, the proof of the theorem will be complete when it is established that Q is a linear isometry of E_β *onto* E^A. For this, it will be enough to show that the range of Q is dense in E^A; in turn, the density of $Q(E_\beta)$ in E^A will follow provided we can prove that the range of Q contains all elements in E^A of the form $(f \otimes \pi_\alpha(g))^A$ $(f \in C_c(G), g \in C_c(H))$ (cf. the Remark following Corollary 5.1.1.6). So let $f \in C_c(G)$, $g \in C_c(H)$ – then the image under Φ of the function

$$x \mapsto \int_H \left(\frac{\delta_G(\xi)}{\delta_H(\xi)}\right)^{1/2} f(x\xi)g(\xi^{-1})d_H(\xi) \qquad (x \in G)$$

is $(f \otimes \pi_\alpha(g))^A$. Hence $Q(E_\beta)$ is dense in E^A, as desired.

The proof of the theorem is now complete. □

5.1.3 Elementary Properties of Unitarily Induced Representations

The purpose of this number is to present some of the identifications which are commonly used in the theory of unitarily induced representations.

As in 5.1.1, G will denote a Lie group countable at infinity.

Proposition 5.1.3.1 Let H be a closed subgroup of G, L a unitary representation of H on a Hilbert space E – then $\overline{U^L}$ is unitarily equivalent to $U^{\overline{L}}$.

[As is the custom for unitary representations, the 'bar' denotes the contragredient representation – cf. number 4.1.2.]

Proposition 5.1.3.2 Let H be a closed subgroup of G, L^i a unitary representation of H on a Hilbert space E^i $(i \in I)$ – then $U^{\Sigma L_i}$ is unitarily equivalent to $\sum U^{L^i}$ ($\sum$ denoting a sum in the sense of Hilbert space theory).

In the case when the relevant Hilbert spaces E^i are separable, Mackey [3, p. 123] has shown that Proposition 5.1.3.2 is valid for direct integrals too, i.e. that $U^{\int L^i d\mu(i)}$ is unitarily equivalent to $\int U^{L^i} d\mu(i)$.

The reader will observe that Proposition 5.1.3.2 implies, in particular, that a unitarily induced representation U^L of G cannot be irreducible if L is not irreducible, that is irreducibility of L is a necessary (but not sufficient!) condition for the irreducibility of U^L.

Proposition 5.1.3.3 Let H_i be a closed subgroup of the Lie group G_i, L^i a unitary representation of H_i on a Hilbert space E^i $(i = 1, 2)$ – then $U^{L^1 \times L^2}$ is unitarily equivalent to $U^{L^1} \times U^{L^2}$.

[The Hilbert tensor products $L^1 \times L^2$ and $U^{L^1} \times U^{L^2}$ being, of course, unitary representations of $H_1 \times H_2$ and $G_1 \times G_2$, respectively.]

Corollary 5.1.3.4 Suppose that G is the direct product $H \times \tilde{H}$ of closed subgroups H and $\tilde{H}$; let L be a unitary representation of H on a Hilbert space E – then ${}_GU^L$ is unitarily equivalent to the Hilbert tensor product of L with the left regular representation of $\tilde{H}$.

A word or two about the proofs of these results are now in order; Propositions 5.1.3.1 and 5.1.3.2 are immediate; as for Proposition 5.1.3.3, it seems to be easiest to proceed using the methods of Proposition 5.1.3.5 infra – we shall leave the details to the reader.

Our next task will be to consider the result on inducing a representation in 'stages'.

Proposition 5.1.3.5 Let H, $\tilde{H}$ be closed subgroups of G with $H \subset \tilde{H}$. Let L be a unitary representation of H on a Hilbert space E. If $\tilde{L} = {}_{\tilde{H}}U^L$, then ${}_GU^L$ is unitarily equivalent to ${}_GU^{\tilde{L}}$.

Proof (Fell) (1) We shall begin by establishing the contention of the proposition in the case when L is generated by a measure α of positive type on H (hence, in particular, whenever L is cyclic). Let $\tilde{\alpha}$ be the injection of $(\delta_{\tilde{H}}/\delta_H)^{1/2}\alpha$ into $\tilde{H}$, β the injection of $(\delta_G/\delta_H)^{1/2}\alpha$ into G – then $\tilde{\alpha}$ and β are measures of positive type on $\tilde{H}$ and G, respectively (Corollary 5.1.2.3). But β is the injection of $(\delta_G/\delta_{\tilde{H}})^{1/2}\tilde{\alpha}$ into G – therefore the unitary equivalence of ${}_G U^L$ and ${}_G U^{\tilde{L}}$ follows upon a threefold application of Theorem 5.1.2.1.

(2) An arbitrary L can be written $L = \sum L^i$, the L^i being cyclic. Bearing in mind that the operation of induction is interchangeable with the operation of taking (Hilbert space) direct sums (Proposition 5.1.3.2), an application of (1) to each L^i then leads at once to the desired result for L.

This completes the proof of the proposition. □

Corollary 5.1.3.6 Let H be a closed subgroup of G. If L is the left regular representation of H on $L^2(H)$, then U^L is unitarily equivalent to the left regular representation of G on $L^2(G)$.

5.2 Quasi-Invariant Distributions

5.2.1 The Global Situation

It is a very important problem to give conditions which will enable one to decide on the irreducibility of a unitarily induced representation. As might be expected these conditions revolve around the computation of intertwining numbers; in turn the intertwining numbers will be estimated in terms of the dimensions of certain canonically constructed spaces of distributions. It is in these later estimates that the so-called quasi-invariant distributions make their appearance. Our purpose here, then, is to develop the fundamental properties of quasi-invariant distributions with a view to their later application to induced representations. The results, although somewhat technical, are obtained by straightforward computations. Throughout, the notations and facts on quasi-invariant measures given in Appendix 1 will be used (for the most part without comment).

Let G be a Lie group operating as a differentiable Lie transformation group on a C^∞ manifold M. Let E be a Fréchet space.

Definition An *E-multiplier* on M relative to G is a map $A : G \times M \to \mathfrak{L}(E; E)$ such that:

(i) $A(1, p) = I$ (I the identity map in E);

(ii) $A(xy, p) = A(y, p)A(x, y \cdot p)$, $(x, y \in G; p \in M)$.

In addition, A is said to be *differentiable* if the map $(x, p) \mapsto A(x, p)$ of $G \times M$ into $\mathfrak{L}_s(E; E)$ is C^∞ and if the image of a compact subset of $G \times M$ by any derivative of A is equicontinuous in $\mathfrak{L}(E; E)$.

[Only Fréchet spaces will be encountered in the applications; however the definition and most of the theory is valid for locally convex, complete, Hausdorff, topological vector spaces too.]

Remarks (1) Let H be a closed subgroup of G, $M = G/H$, ρ_H a C^∞ rho-function on G relative to H. The multiplier s_H corresponding to ρ_H is a differentiable **R**-multiplier (Appendix 1).

(2) Let U be a differentiable representation of G on E and set $A(x, p) = U(x^{-1})$ $(x \in G, p \in M)$ – then A is a differentiable E-multiplier on M.

(3) If A is differentiable E-multiplier on M, then the map $\xi \mapsto A(\xi^{-1}, p)$ is, for each fixed $p \in M$, a differentiable representation of the stabilizer G_p of p in G.

Let A be an E-multiplier on M relative to G – then $A(x, p)$ has an inverse, namely $A(x^{-1}, x\cdot p)$ $(x \in G, p \in M)$. [Thus, in the definition above, replace x and y by x^{-1} and x respectively to get $I = A(1, p) = A(x, p)A(x^{-1}, x\cdot p)$; in this equation replace x and p by x^{-1} and $x\cdot p$ respectively to get $I = A(x^{-1}, x\cdot p)A(x, p)$ – our contention is now clear.]

The proof of the following elementary, technical lemma will be left to the reader.

Lemma 5.2.1.1 Let E be a locally convex, complete, Hausdorff topological vector space; let M be a C^∞ manifold; let $A: M \to \mathfrak{L}(E; E)$ be a C^∞ map of M into $\mathfrak{L}_s(E; E)$ with the property that the image of any compact subset of M by any derivative of A is equicontinuous in $\mathfrak{L}(E; E)$ – then the map which assigns to each $f \in C_c^\infty(M; E)$ the function $p \mapsto A(p)f(p)$ $(p \in M)$ is a continuous endomorphism of $C_c^\infty(M; E)$.

Let A be a differentiable E-multiplier on M. According to Lemma 5.2.1.1, the rule which assigns to each $f \in C_c^\infty(M; E)$ the function $p \mapsto A(x^{-1}, p)f(p)$ $(x \in G, p \in M)$ is a topological automorphism of $C_c^\infty(M; E)$ – this is exactly what is needed in order to justify the next definition.

Definition Let A be a differentiable E-multiplier on M. An E-distribution T on M is said to be *quasi-invariant* (with multiplier A) provided

$$\int_M f(x\cdot p)\, dT(p) = \int_M A(x^{-1}, p)f(p)\, dT(p) \qquad (x \in G, f \in C_c^\infty(M; E)).$$

[In other words, an E-distribution T on M is quasi-invariant with multiplier A if $T^x = {}^tA(x^{-1}, .)T$ or, symbolically, if

$$dT(x\cdot p) = {}^tA(x, p)\, dT(p) \qquad (x \in G).$$

Of course the notion of a quasi-invariant distribution is an extension of that of a quasi-invariant measure (Appendix I).]

Suppose first that we take $M = G$ and consider the canonical left action of G on itself, viz. $y \mapsto x\cdot y = xy$ $(x, y \in G)$. Taking $A = I$, we have:

Proposition 5.2.1.2 Let T be a left invariant E-distribution on G, i.e.,

$$\int_G f(x\cdot y)\, dT(y) = \int_G f(xy)\, dT(y) = \int_G f(y)\, dT(y)$$

for all $x \in G$ – then there exists a unique element $\mathbf{z} \in E^*$ such that

$$\langle f, T\rangle = \int_G \langle f(x), \mathbf{z}\rangle \, d_G(x) \qquad (f \in C_c^\infty(G;E)),$$

that is T is a 'multiple' of the left invariant Haar measure.

Proof The E-distribution T determines in a canonical way a left invariant distribution S on G with values in E_b^* (see A 2.3, Example 1). It suffices to prove, therefore, that there exists an element $\mathbf{z} \in E^*$ such that $\langle f, S\rangle = (\int_G f(x) d_G(x))\mathbf{z}$ for all $f \in C_c^\infty(G)$. Now for every $f \in C_c^\infty(G)$, $S * f$ is a left invariant element of $C_c^\infty(G; E_b^*)$, hence is a constant. But by the usual regularization argument (Schwartz [1, p. 75]), S can be approximated in $\mathfrak{L}_b(C_c^\infty(G); E_b^*)$ by suitable $S * f$. Thus S must be a constant, say $\mathbf{z}$, too. □

The proof of Proposition 5.2.1.2 suffices to establish the following corollary.

Corollary 5.2.1.3 Let T be a left invariant distribution on G with values in E – then there exists a unique vector $a \in E$ such that

$$\langle f, T\rangle = \left(\int_G f(x) d_G(x)\right) a$$

for all $f \in C_c^\infty(G)$.

Returning to the general situation, let us suppose that G acts transitively on M – then M may be written in the form G/H, H a closed subgroup of G. Of course G operates on $M = G/H$ via the prescription $\dot{y} = yH \mapsto x \cdot \dot{y} = xyH$ (where, as usual, $x \mapsto \dot{x}$ ($\dot{x} = xH$, $x \in G$) denotes the canonical projection of G onto G/H).

The following theorem will play a major role in subsequent developments.

Theorem 5.2.1.4 (Bruhat) Let T be a quasi-invariant E-distribution with differentiable E-multiplier A on G/H – thus

$$\int_{G/H} f(x \cdot \dot{y}) \, dT(\dot{y}) = \int_{G/H} A(x^{-1}, \dot{y}) f(\dot{y}) \, dT(\dot{y}) \qquad (x \in G, f \in C_c^\infty(G/H; E)).$$

Then there exists a unique $\mathbf{z} \in E^*$ such that:

(i) ${}^tA(\xi, \dot{1})\mathbf{z} = \rho_H(\xi)\mathbf{z} \quad (\xi \in H)$;

(ii) $\langle f, T\rangle = \int_{G/H} \langle f(\dot{x}), \rho_H(x)^{-1}\,{}^tA^{-1}(x^{-1}, \dot{x})\mathbf{z}\rangle \, d\mu_H(\dot{x}) \qquad (f \in C_c^\infty(G/H; E))$.

[To avoid any possibility of confusion, let us explicitly observe that

$$\rho_H(x\xi)^{-1}\,{}^tA^{-1}((x\xi)^{-1}, \dot{x})\mathbf{z} = \rho_H(x)^{-1}\,{}^tA^{-1}(x^{-1}, \dot{x})\mathbf{z} \text{ (all } \xi \in H)$$

(hence the right hand side in (ii) is meaningful). In fact

$$\rho_H(x\xi)^{-1} = \rho_H(x)^{-1}\rho_H(\xi)^{-1}$$

(Appendix 1) while

$$A((x\xi)^{-1}, \dot{x}) = A(\xi^{-1}x^{-1}, \dot{x}) = A(x^{-1}, \dot{x})A(\xi^{-1}, \dot{1})$$
$$\Rightarrow {}^tA^{-1}((x\xi)^{-1}, \dot{x})\mathbf{z} = {}^tA^{-1}(x^{-1}, \dot{x})^tA^{-1}(\xi^{-1}, \dot{1})\mathbf{z} = \rho_H(\xi)^tA^{-1}(x^{-1}, \dot{x})\mathbf{z} \text{ (by (i))}$$
$$\Rightarrow \rho_H(x\xi)^{-1t}A^{-1}((x\xi)^{-1}, \dot{x})\mathbf{z} = \rho_H(x)^{-1t}A^{-1}(x^{-1}, \dot{x})\mathbf{z}.]$$

Proof of Theorem 5.2.1.4 (1) As is well-known, the map $f \mapsto \dot{f}$, $\dot{f}(\dot{x}) = \int_H f(x\xi)\, d_H(\xi)$, is a continuous surjection of $C_c^\infty(G; E)$ onto $C_c^\infty(G/H; E)$; in addition, since A is a differentiable E-multiplier, the rule which assigns to each $f \in C_c^\infty(G; E)$ the function $x \mapsto A(x^{-1}, \dot{x})f(x)$ is a topological automorphism of $C_c^\infty(G; E)$ – therefore the map which assigns to each $f \in C_c^\infty(G; E)$ the function f^A, $\dot{x} \mapsto \int_H A((x\xi)^{-1}, \dot{x})f(x\xi)\, d_H(\xi)$, is a continuous surjection of $C_c^\infty(G; E)$ onto $C_c^\infty(G/H; E)$. Let T_A be the transpose of T with respect to this latter map – then T_A is an E-distribution on G. We shall first show that T_A is left invariant under the canonical left action of G on itself. Thus, let $f \in C_c^\infty(G; E)$ – then, for all $x \in G$, we have

$$\int_G f(x\cdot y)\, dT_A(y) = \int_G f(xy)\, dT_A(y)$$
$$= \int_{G/H} dT(\dot{y}) \int_H A((y\xi)^{-1}, \dot{y})f(xy\xi)\, d_H(\xi)$$
$$= \int_{G/H} dT(\dot{y}) \int_H A((x^{-1}xy\xi)^{-1}, \dot{y})f(xy\xi)\, d_H(\xi)$$
$$= \int_{G/H} dT(\dot{y})\, A^{-1}(x^{-1}, x\cdot\dot{y}) \int_H A((xy\xi)^{-1}, x\cdot\dot{y})f(xy\xi)\, d_H(\xi)$$
$$= \int_{G/H} dT(\dot{y}) \int_H A((y\xi)^{-1}, \dot{y})f(y\xi)\, d_H(\xi) = \int_G f(y)\, dT_A(y).$$

Hence T_A is indeed left invariant. Proposition 5.2.1.2 then implies the existence of a unique element $\mathbf{z} \in E^*$ such that $dT_A(x) = \mathbf{z}d_G(x)$. Now fix $\eta \in H$. Then, on the one hand we have,

$$\int_G f(x\eta)\, dT_A(x) = \int_G \langle f(x\eta), \mathbf{z}\rangle\, d_G(x)$$
$$= \delta_G(\eta)^{-1}\langle f, T_A\rangle,$$

while on the other,

$$\int_G f(x\eta)\, dT_A(x) = \int_{G/H} dT(\dot{x}) \int_H A((x\xi)^{-1}, \dot{x})f(x\xi\eta)\, d_H(\xi)$$
$$= \delta_H^{-1}(\eta) \int_{G/H} dT(\dot{x}) \int_H A((x\xi\eta^{-1})^{-1}, \dot{x})f(x\xi)\, d_H(\xi)$$
$$= \delta_H^{-1}(\eta) \int_{G/H} dT(\dot{x}) \int_H A((x\xi)^{-1}, \dot{x})A(\eta, \dot{1})f(x\xi)\, d_H(\xi)$$
$$= \delta_H^{-1}(\eta)\langle A(\eta, \dot{1})f, T_A\rangle.$$

This implies that ${}^tA(\eta, \dot{1})\mathbf{z} = \rho_H(\eta)\mathbf{z}$ (because $\rho_H(\eta) = \delta_H(\eta)/\delta_G(\eta)$ $(\eta \in H)$ (Appendix 1)).

(2) Define an E-distribution S on $C_c^\infty(G/H; E)$ by the rule

$$\langle f, S\rangle = \int_{G/H} \langle f(\dot{x}), \rho_H(x)^{-1\,t}A^{-1}(x^{-1}, \dot{x})\mathbf{z}\rangle \, d\mu_H(\dot{x}) \qquad (f \in C_c^\infty(G/H; E)).$$

Since the map $f \mapsto f^A$ is a continuous surjection, in order to prove that $S = T$ it suffices to prove that $S_A = T_A$ (S_A the transpose of S under the map $f \mapsto f^A$). So let $f \in C_c^\infty(G; E)$ – then

$$\begin{aligned}
&\langle f, S_A\rangle \\
&= \int_{G/H} \Big\langle \int_H A((x\xi)^{-1}, \dot{x}) f(x\xi)\, d_H(\xi), \rho_H(x)^{-1\,t}A^{-1}(x^{-1}, \dot{x})\mathbf{z}\Big\rangle \, d\mu_H(\dot{x}) \\
&= \int_{G/H} d\mu_H(\dot{x}) \int_H \rho_H(x)^{-1}\langle f(x\xi), {}^tA((x\xi)^{-1}, \dot{x})^t A^{-1}(x^{-1}, \dot{x})\mathbf{z}\rangle \, d_H(\xi) \\
&= \int_{G/H} d\mu_H(\dot{x}) \int_H \rho_H(x)^{-1}\langle f(x\xi), {}^tA(\xi^{-1}, \dot{1})\mathbf{z}\rangle \, d_H(\xi) \\
&= \int_{G/H} d\mu_H(\dot{x}) \int_H \rho_H(x\xi)^{-1}\langle f(x\xi), \mathbf{z}\rangle \, d_H(\xi) \\
&= \int_G \langle f(x), \mathbf{z}\rangle \, d_G(x) = \langle f, T_A\rangle.
\end{aligned}$$

The theorem is now completely proved. □

Remark Theorem 5.2.1.4(ii) shows that in reality, with reference to the quasi-invariant measure μ_H, our quasi-invariant distribution T is a function.

Let M be a C^∞ manifold. If H is a closed subgroup of G, then G operates on the product space $M \times G/H$ via the rule $(p, yH) \mapsto (p, xyH)$ ($p \in M; x, y \in G$). With this agreement, the following extension of Theorem 5.2.1.4 holds.

Proposition 5.2.1.5 Let T be a quasi-invariant E-distribution with differentiable E-multiplier A on $M \times G/H$ – then there exists on M a unique E-distribution S such that:

(i) $$\int_M A(\xi : p, \dot{1}) f(p)\, dS(p) = \rho_H(\xi) \int_M f(p)\, dS(p) \qquad (f \in C_c^\infty(M; E), \xi \in H);$$

(ii) $$\int_{M \times G/H} f(p, \dot{x})\, dT(p, \dot{x}) = \int_{G/H} d\mu_H(\dot{x}) \int_M \rho_H(x)^{-1} A^{-1}(x^{-1} : p, \dot{x}) f(p, \dot{x})\, dS(p) \qquad (f \in C_c^\infty(M \times G/H; E)).$$

Proof (1) Writing

$$\int_{M \times G} f(p, x)\, dT_A(p, x) = \int_{M \times G/H} dT(p, \dot{x}) \int_H A((x\xi)^{-1} : p, \dot{x}) f(p, x\xi)\, d_H(\xi) \qquad (f \in C_c^\infty(M \times G; E)),$$

one shows as in the proof of Theorem 5.2.1.4 that T_A is left invariant under the (obvious) action of G on $M \times G$. Let $g \in C_c^\infty(M; E)$, $h \in C_c^\infty(G)$. For fixed g, $h \mapsto T_A(gh)$ is a left invariant scalar distribution on G; therefore there exists a complex number c_g such that $T_A(gh) = c_g \int_G h(x)\, d_G(x)$ (all $h \in C_c^\infty(G)$). The map $g \mapsto c_g$ is an E-distribution, call it S, on M. Thus

$$\int_{M\times G} g(p)h(x)\, dT_A(p, x) = \left(\int_M g(p)\, dS(p)\right)\left(\int_G h(x)\, d_G(x)\right),$$

and so by continuity we see that

$$\int_{M\times G} f(p, x)\, dT_A(p, x) = \int_G d_G(x) \int_M f(p, x)\, dS(p)$$

for all $f \in C_c^\infty(M \times G; E)$.

(2) Let us now verify conditions (i) and (ii). Fix $\eta \in H$. On the one hand we have

$$\int_{M\times G} f(p, x\eta)\, dT_A(p, x) = \delta_G(\eta)^{-1}\langle f, T_A\rangle,$$

while on the other,

$$\begin{aligned}
&\int_{M\times G} f(p, x\eta)\, dT_A(p, x)\\
&= \int_{M\times G/H} dT(p, \dot{x}) \int_H A((x\xi)^{-1}:p, \dot{x})f(p, x\xi\eta)\, d_H(\xi)\\
&= \delta_H(\eta)^{-1} \int_{M\times G/H} dT(p, \dot{x}) \int_H A((x\xi\eta^{-1})^{-1}:p, \dot{x})f(p, x\xi)\, d_H(\xi)\\
&= \delta_H(\eta)^{-1} \int_{M\times G/H} dT(p, \dot{x}) \int_H A((x\xi)^{-1}:p, \dot{x})A(\eta:p, \dot{1})f(p, x\xi)\, d_H(\xi)\\
&= \delta_H(\eta)^{-1}\langle A(\eta:p, \dot{1})f, T_A\rangle.
\end{aligned}$$

This computation serves to establish the validity of condition (i); a similar calculation yields (ii).

The proof of the proposition is therefore complete. □

Corollary 5.2.1.6 Let B be a C^∞ manifold on which G acts (to the left) in such a way as to define on B the structure of a differentiable principal fiber bundle; let U be a differentiable representation of G on E, T an E-distribution on B, quasi-invariant with multiplier $A(x, b) = U(\dot{x}^{-1})$ ($x \in G, b \in B$) – then T vanishes on the subspace $\mathscr{S}$ of $C_c^\infty(B; E)$ consisting of those f such that for every $b \in B$

$$\int_G U(x^{-1})f(x\cdot b)\, d_G(x) = 0.$$

Proof An easy partition of unity argument shows that it is enough just to consider the case when B is a trivial bundle, i.e. when $B = M \times G$, M a C^∞ manifold – of course G now operates on B via the rule $(p, y) \mapsto$

(p, xy) $(p \in M; x, y \in G)$. Owing to Proposition 5.2.1.5, then, there exists on M an E-distribution S such that for all $f \in \mathscr{S}$

$$\begin{aligned}\langle f, T\rangle &= \int_M \left\{\int_G U(x^{-1})f(p, x)\, d_G(x)\right\} dS(p)\\ &= \int_M \left\{\int_G U(x^{-1})f(x\cdot(p, 1))\, d_G(x)\right\} dS(p) = 0. \quad \square\end{aligned}$$

5.2.2 The Local Situation

Let G be a Lie group, H a closed subgroup of G. In practice it often turns out that an E-distribution T will not be defined on all of $M = G/H$ but only locally, say in some neighborhood $\mathscr{P}$ of $\dot{1}$ in G/H. Let us see what changes are necessary in order to formulate an analogue of Theorem 5.2.1.4 in such a situation.

In the first place a local analogue of Proposition 5.2.1.2 is needed. Let E be a Fréchet space; let T be an E-distribution defined in some open neighborhood $\mathcal{O}$ of 1 in G – then the transform T^x of T under the map $y \mapsto xy$ $(x, y \in G)$ is an E-distribution defined on $x\cdot\mathcal{O}$. Assume now that T is left invariant, i.e. that $T = T^x$ in $\mathcal{O} \cap x\cdot\mathcal{O}$ $(x \in G)$. Putting the various distributions T^x together by the usual 'pasting' argument (Schwartz [1, p. 27]), one obtains an E-distribution on G which is plainly left invariant in the usual sense. Proposition 5.2.1.2 then gives us a unique element $\mathbf{z} \in E^*$ such that

$$\int_{\mathcal{O}} f(x)\, dT(x) = \int_{\mathcal{O}} \langle f(x), \mathbf{z}\rangle\, d_G(x)$$

for all $f \in C_c^\infty(\mathcal{O}; E)$.

Keeping to the above notations, fix a neighborhood $\mathscr{P}$ of $\dot{1}$ in G/H; relative to $\mathscr{P}$, let us make the following agreement.

Definition An *E-multiplier* on $\mathscr{P}$ relative to G is a map A with values in $\mathfrak{L}(E; E)$ such that:

(i) $A(x, \dot{y})$ is defined if $\dot{y} \in \mathscr{P} \cap x^{-1}\cdot\mathscr{P}$;

(ii) $A(xy, \dot{z}) = A(y, \dot{z})A(x, y\cdot\dot{z})$ whenever this relation makes sense;

(iii) $A(1, \dot{y}) = I$ for all $\dot{y} \in \mathscr{P}$.

If A satisfies a differentiability condition as before, then A will be called a *differentiable E-multiplier* on $\mathscr{P}$.

Henceforth let A be a differentiable E-multiplier on $\mathscr{P}$. Observe that $A(\xi^{-1}, \dot{1})$ is defined for all $\xi \in H$, whence $\xi \mapsto A(\xi^{-1}, \dot{1})$ is a differentiable representation of H on E.

Let $\pi : G \to G/H$ denote the canonical projection $x \mapsto \dot{x}$ of G onto G/H. Consider the expression $A(x^{-1}, \dot{x})$. It is defined if $\dot{x} \in \mathscr{P} \cap x\cdot\mathscr{P}$, i.e. if $x \in \pi^{-1}(\mathscr{P})$ – hence, for $\xi \in H$, $x \in \pi^{-1}(\mathscr{P})$, we have

$$A((x\xi)^{-1}, \dot{x}) = A(x^{-1}, \dot{x})A(\xi^{-1}, \dot{1}).$$

Definition Let A be a differentiable E-multiplier on $\mathscr{P}$. An E-distribution T on $\mathscr{P}$ is said to be *quasi-invariant* (with multiplier A) provided

$$\int_{\mathscr{P}} f(x\cdot\dot{y})\,dT(\dot{y}) = \int_{\mathscr{P}} A(x^{-1},\dot{y})f(\dot{y})\,dT(\dot{y})$$

for all $x \in G$ and all $f \in C_c^\infty(\mathscr{P}; E)$ such that $\operatorname{spt}(f) \subset \mathscr{P} \cap x\cdot\mathscr{P}$.

Here is the analogue of Theorem 5.2.1.4.

Theorem 5.2.2.1 Let T be a quasi-invariant E-distribution on $\mathscr{P}$ with differentiable E-multiplier A (on $\mathscr{P}$) – then there exists a unique element $\mathbf{z} \in E^*$ such that:

(i) ${}^tA(\xi, \dot{1})\mathbf{z} = \rho_H(\xi)\mathbf{z}$ for all $\xi \in H$;

(ii) $\langle f, T\rangle = \int_{\mathscr{P}} \langle f(\dot{x}), \rho_H(x)^{-1\,t}A^{-1}(x^{-1},\dot{x})\mathbf{z}\rangle\, d\mu_H(\dot{x})$ $(f \in C_c^\infty(\mathscr{P}; E))$.

Proof Since A is a differentiable E-multiplier on $\mathscr{P}$ relative to G, the map which assigns to each $f \in C_c^\infty(\pi^{-1}(\mathscr{P}); E)$ the function $x \mapsto A(x^{-1},\dot{x})f(x)\,(x \in \pi^{-1}(\mathscr{P}))$ is a topological automorphism of $C_c^\infty(\pi^{-1}(\mathscr{P}); E)$. Hence the map which assigns to each $f \in C_c^\infty(\pi^{-1}(\mathscr{P}); E)$ the function $f^A, \dot{x} \mapsto \int_H A((x\xi)^{-1},\dot{x})f(x\xi)\,d_H(\xi)$, is a continuous surjection of $C_c^\infty(\pi^{-1}(\mathscr{P}); E)$ onto $C_c^\infty(\mathscr{P}; E)$. Let T_A be the transpose of T relative to this latter map – then T_A is an E-distribution on $\pi^{-1}(\mathscr{P})$. This being the case, let us first verify that T_A is left invariant under the action of G. Fix $x \in G$ and let $f \in C_c^\infty(\pi^{-1}(\mathscr{P}); E)$ with $\operatorname{spt}(f) \subset \pi^{-1}(\mathscr{P}) \cap x\cdot\pi^{-1}(\mathscr{P})$ – we must show that

$$\int_{\mathcal{O}} f(xy)\,dT_A(y) = \int_{\mathcal{O}} f(y)\,dT_A(y) \qquad (\text{where } \mathcal{O} = \pi^{-1}(\mathscr{P})).$$

Plainly

$$\int_{\mathcal{O}} f(xy)\,dT_A(y) = \int_{\mathscr{P}} dT(\dot{y}) \int_H A((y\xi)^{-1},\dot{y})f(xy\xi)\,d_H(\xi).$$

By assumption, $\operatorname{spt}(f^A) \subset \mathscr{P} \cap x\cdot\mathscr{P}$; define a function h on $\mathscr{P} \cap x\cdot\mathscr{P}$ via the prescription

$$h(\dot{y}) = A(x, x^{-1}\cdot\dot{y})f^A(\dot{y}).$$

Since $A(x, x^{-1}\cdot\dot{y})$ is defined if $x^{-1}\cdot\dot{y} \in \mathscr{P} \cap x^{-1}\cdot\mathscr{P}$, i.e. if $\dot{y} \in \mathscr{P} \cap x\cdot\mathscr{P}$, this definition makes sense; of course $\operatorname{spt}(h) \subset \mathscr{P} \cap x\cdot\mathscr{P}$. Now

$$\begin{aligned} h(x\cdot\dot{y}) &= A(x,\dot{y})\int_H A((xy\xi)^{-1}, x\cdot\dot{y})f(xy\xi)\,d_H(\xi) \\ &= \int_H A((y\xi)^{-1},\dot{y})f(xy\xi)\,d_H(\xi) \end{aligned}$$

and so the following manipulation is meaningful:

$$\begin{aligned} &\int_{\mathscr{P}} dT(\dot{y}) \int_H A((y\xi)^{-1},\dot{y})f(xy\xi)\,d_H(\xi) \\ &= \int_{\mathscr{P}} dT(\dot{y}) \int_H A((x^{-1}xy\xi)^{-1},\dot{y})f(xy\xi)\,d_H(\xi) \end{aligned}$$

$$= \int_{\mathscr{P}} dT(\dot{y})A(x, \dot{y}) \int_H A((xy\xi)^{-1}, x\cdot\dot{y})f(xy\xi)\, d_H(\xi)$$

$$= \int_{\mathscr{P}} h(x\cdot\dot{y})\, dT(\dot{y}) = \int_{\mathscr{P}} A(x^{-1}, \dot{y})h(\dot{y})\, dT(\dot{y})$$

$$= \int_{\mathscr{P}} f^A(\dot{y})\, dT(\dot{y}) = \int_{\mathscr{O}} f(y)\, dT_A(y).$$

Thus T_A is left invariant. Earlier remarks give us a unique element $\mathbf{z} \in E^*$ such that $dT_A(x) = \mathbf{z}d_G(x)$. The rest of the proof is virtually identical to that of Theorem 5.2.1.4 and will be omitted. □

5.2.3 A Fundamental Estimate

Let M be a C^∞ manifold on which our Lie group G operates to the left. Let U be a differentiable representation of G on a Fréchet space E. We shall consider the space $\mathbf{T}(Q)$ of those E-distributions T on M such that:

(a) $T^x = {}^tU(x)T$ for all $x \in G$;

(b) The support of T is contained in a *closed* G-orbit Q.

Our problem is: Obtain a convenient majorization for the dimension of $\mathbf{T}(Q)$. This is given in Theorem 5.2.3.1 below, a theorem which plays an important role in the sequel. The proof requires some preparatory discussion and an application of Theorem 5.2.2.1 of the previous number.

So let $Q = G\cdot p_0$ be a closed G-orbit in M, $G_{p_0} = H$ the stabilizer of p_0 in G – then we may identify Q with G/H. [A priori, the submanifold topology on Q (A 2.4) is finer than the topology of M relativised to Q; but since G is countable at infinity and Q is closed, it follows from a well-known result due to Arens that the natural bijection $G/H \to Q$ is a homeomorphism of G/H, endowed with the quotient topology, onto Q, endowed with the relative topology.]

In a sufficiently small open relatively compact neighborhood $\mathscr{O}_M(p_0)$ of p_0 in M there exists a local coordinate system $\{u_1, \ldots, u_q, v_1, \ldots, v_r\}$ such that $\mathscr{P} = Q \cap \mathscr{O}_M(p_0)$ is defined by the equations $v_i = 0$. An E-distribution T in $\mathbf{T}(Q)$, restricted to $\mathscr{O}_M(p_0)$, is of finite order and has support contained in $\mathscr{P}$. By Proposition A2.1.2 we can write

$$T = \sum_{|I| \leq n} (-1)^{|I|} D^I T_I$$

where

$$D^I = D_1^{i_1} \cdots D_r^{i_r} = \partial^{|I|}/\partial v_1^{i_1} \cdots \partial v_r^{i_r} \quad \left(|I| = i_1 + \cdots + i_r, D_i = \frac{\partial}{\partial v_i}\right)$$

and the T_I are extensions to $\mathscr{O}_M(p_0)$ of E-distributions on $\mathscr{P}$. It is important to bear in mind that this decomposition is unique.

Here n is the transversal order of T on $\mathscr{O}_M(p_0)$ and is the same at all points of $Q = G\cdot p_0$ in view of condition (a) above. In the considerations which follow we shall use the symbol $\equiv$ to indicate that the relation in question is true modulo distributions whose transversal order is less than n.

We point out explicitly that if in addition the orbit Q were open in M, then $r = 0$ and one must take $n = 0$ in what follows.

Let $p \in \mathscr{P}$. If δ_p is the Dirac measure concentrated at p, then $D^I\delta_p$ will denote the distribution $f \mapsto (D^If)(p)$ (a temporary but useful convention). The group G operates linearly on the E-distributions on M (A2.4); in particular, since the transversal order is invariant under translation, we have

$$(D^I\delta_p)^x \equiv \sum_{|J|=n} \lambda_{JI}(x, p)D^J\delta_{x\cdot p}$$

for $|I| = n, p \in \mathscr{P} \cap x^{-1}\cdot\mathscr{P}$ (where we order the indices I for $|I| = n$ in the usual way). Here the functions $\lambda_{IJ}(x,p)$ are defined and differentiable for $p \in \mathscr{P} \cap x^{-1}\cdot\mathscr{P}$.

Fix $x \in G$. Take an $f \in C_c^\infty(M; E)$ such that $\operatorname{spt}(f) \cap Q \subset \mathscr{P} \cap x^{-1}\cdot\mathscr{P}$. If $p \in \mathscr{P} \cap x^{-1}\cdot\mathscr{P}$, then

$$\begin{aligned}(D^If^{x^{-1}})(p) &= (D^I\delta_p)(f^{x^{-1}}) = (D^I\delta_p)^x(f)\\ &\equiv \sum_{|J|=n} \lambda_{JI}(x, p)D^J\delta_{x\cdot p}(f) \equiv \sum_{|J|=n} \lambda_{JI}(x, p)D^Jf(x\cdot p).\end{aligned}$$

Hence

$$\begin{aligned}T^x(f) = T(f^{x^{-1}}) &= \sum_{|I|\leqslant n} (-1)^{|I|}D^IT_I(f^{x^{-1}})\\ &= \sum_{|I|\leqslant n} \int D^If^{x^{-1}}(p)\, dT_I(p)\\ &\equiv \sum_{|I|=n}\sum_{|J|=n} \int \lambda_{JI}(x, p)D^Jf(x\cdot p)\, dT_I(p).\end{aligned}$$

Now let us use condition (a) above and obtain an alternative formula for $T^x(f)$. Thus

$$\begin{aligned}T^x(f) &= T(U(x)\circ f)\\ &\equiv \sum_{|I|=n} \int D^If(p)\, d({}^tU(x)T_I)(p)\\ &\equiv \sum_{|I|=n} \int D^If(x\cdot p)\, d({}^tU(x)T_I)(x\cdot p).\end{aligned}$$

In view of the uniqueness of the decomposition of a distribution into the derivatives of transversal extensions we get

$$d({}^tU(x)T_I)(x\cdot p) = \sum_{|J|=n} \lambda_{IJ}(x, p)\, dT_J(p)$$

or

$$(*)\quad dT_I^x(p) = \sum_{|J|=n} \lambda_{IJ}(x^{-1}, p)^tU(x)\, dT_J(p)$$

where $p \in \mathscr{P} \cap x\cdot\mathscr{P}$.

In a moment we shall interpret (*) as a condition of quasi-invariance for a certain distribution.

There are $s = \binom{r+n-1}{n}$ of the distributions T_I for $|I| = n$. [If E is a finite dimensional vector space of dimension r over $\mathbf{C}$, say, $S(E)$ the symmetric algebra over E, $S^n(E)$ the subspace of homogeneous elements of degree n, then $\dim(S^n(E)) = \binom{r+n-1}{r-1} = \binom{r+n-1}{n}$.] We may consider their sum $S = \sum T_I$ as an $E \hat{\otimes} \mathbf{C}^s$-distribution on $\mathscr{P}$ (or as an F-distribution, F a direct sum of s copies of E, if preferred). Here $\mathbf{C}^s$ is the complex $\mathbf{C}$-space of dimension s. Let $f \in C_c^\infty(\mathscr{P}; E \hat{\otimes} \mathbf{C}^s)$ with $\operatorname{spt}(f) \subset \mathscr{P} \cap x \cdot \mathscr{P}$ – thus $f \leftrightarrow (f_I : |I| = n)$ and so

$$\begin{aligned} S^x(f) &= \sum_{|I|=n} T_I^x(f_I) \\ &= \sum_{|J|=n} T_J\Big(\sum_{|I|=n} \lambda_{IJ}(x^{-1}, p)U(x)\circ f_I\Big). \end{aligned}$$

If $\Lambda_n(x, p)$ denotes the operator whose matrix with respect to the canonical basis in $\mathbf{C}^s$ is $(\lambda_{IJ}(x, p))$, then we have

$$S^x(f) = S(U(x) \hat{\otimes} {}^t\Lambda_n(x^{-1}, p)(f)) \qquad (p \in \mathscr{P} \cap x \cdot \mathscr{P}).$$

Now identify the orbit $Q = G \cdot p_0$ with the homogeneous space G/H. Set $A(x, \dot{y}) = U(x^{-1}) \hat{\otimes} {}^t\Lambda_n(x, \dot{y})$ $(x, y \in G)$; because the matrix coefficients $\lambda_{IJ}(x, \dot{y})$ are defined if $\dot{y} \in \mathscr{P} \cap x^{-1} \cdot \mathscr{P}$, the same applies to $A(x, \dot{y})$. In addition, for all $f \in C_c^\infty(\mathscr{P}; E \hat{\otimes} \mathbf{C}^s)$ such that $\operatorname{spt}(f) \subset \mathscr{P} \cap x \cdot \mathscr{P}$, we have

$$\int_{\mathscr{P}} f(x \cdot \dot{y})\, dS(\dot{y}) = \int_{\mathscr{P}} A(x^{-1}, \dot{y}) f(\dot{y})\, dS(\dot{y}).$$

On the other hand A is a differentiable $E \hat{\otimes} \mathbf{C}^s$-multiplier on $\mathscr{P}$. [The only property that isn't obvious is the relation $A(xy, \dot{z}) = A(y, \dot{z})A(x, y \cdot \dot{z})$ $(x, y, z \in G)$. To see this, note that

$$\begin{aligned} (D^I\delta_{\dot{z}})^{xy} &= ((D^I\delta_{\dot{z}})^y)^x \\ &\equiv \Big(\sum_J \lambda_{JI}(y, \dot{z}) D^J\delta_{y\cdot\dot{z}}\Big)^x \\ &\equiv \sum_K \Big(\sum_J \lambda_{KJ}(x, y\cdot\dot{z})\lambda_{JI}(y, \dot{z})\Big) D^K\delta_{(xy)\cdot\dot{z}}; \end{aligned}$$

since

$$(D^I\delta_{\dot{z}})^{xy} \equiv \sum_K \lambda_{KI}(xy, \dot{z}) D^K\delta_{(xy)\cdot\dot{z}},$$

we have

$$\Lambda_n(xy, \dot{z}) = \Lambda_n(x, y\cdot\dot{z})\Lambda_n(y, \dot{z}).$$

Therefore

$$\begin{aligned} A(xy, \dot{z}) &= U((xy)^{-1}) \hat{\otimes} {}^t\Lambda_n(xy, \dot{z}) \\ &= (U(y^{-1}) \hat{\otimes} {}^t\Lambda_n(y, \dot{z}))\circ(U(x^{-1}) \hat{\otimes} {}^t\Lambda_n(x, y\cdot\dot{z})) \\ &= A(y, \dot{z})A(x, y\cdot\dot{z}).] \end{aligned}$$

Hence S is quasi-invariant and we are in the local situation discussed in the previous number – in particular Theorem 5.2.2.1 is available.

This theorem gives us a unique element $\mathbf{z} \in (E \hat{\otimes} \mathbf{C}^s)^*$ such that

$\rho_H(\xi)^{-1}{}^t(U(\xi^{-1}) \hat{\otimes} {}^t\Lambda_n(\xi, \dot{1}))\mathbf{z} = \mathbf{z}$ for all $\xi \in H$. But to the continuous linear form $\mathbf{z}$ there corresponds, in the canonical way, a continuous bilinear form $\mathbf{z}$ (say) on $E \times \mathbf{C}^s$; for all $a \in E$, $b \in \mathbf{C}^s$, $\xi \in H$, we then have

$$\begin{aligned}\mathbf{z}(U(\xi)a, \rho_H(\xi)^t\Lambda_n(\xi^{-1}, \dot{1})b) &= \langle U(\xi)a \otimes \rho_H(\xi)^t\Lambda_n(\xi^{-1}, \dot{1})b, \mathbf{z}\rangle \\ &= \langle a \otimes b, \mathbf{z}\rangle = \mathbf{z}(a, b).\end{aligned}$$

That is, $\mathbf{z}$ is an intertwining form for the representations $\xi \mapsto U(\xi)$ and $\xi \mapsto \rho_H(\xi)\Lambda_n^\vee(\xi, \dot{1})$ of H. Let $i(E; p_0, n)$ denote the intertwining number of these representations.

We have shown, therefore, that the map which assigns to each $T \in \mathbf{T}(Q)$ the intertwining form $\mathbf{z}$ is a linear map of the subspace $\mathbf{T}^n(Q)$ of those distributions T in $\mathbf{T}(Q)$ whose transversal order does not exceed n into the space of intertwining forms for the representations $\xi \mapsto U(\xi)$ and $\xi \mapsto \rho_H(\xi)\Lambda_n^\vee(\xi, \dot{1})$ $(\xi \in H)$ with kernel $\mathbf{T}^{n-1}(Q)$. Thus

$$\dim(\mathbf{T}^n(Q)/\mathbf{T}^{n-1}(Q)) \leqslant i(E; p_0, n)$$

and so

$$\dim(\mathbf{T}(Q)) \leqslant \sum_{n=0}^{\infty} i(E; p_0, n).$$

Plainly the number $i(E; p_0, n)$ depends only on the orbit $Q = G \cdot p_0$ and not on the point p_0 (since changing p_0 leads to a subgroup of G conjugate to H). Hence we shall write $i(E; Q, n)$ for $i(E; p_0, n)$.

The preceding discussion serves to prove:

Theorem 5.2.3.1 (Bruhat) The dimension of the space of distributions $\mathbf{T}(Q)$ is less than or equal to $\sum_{n=0}^{\infty} i(E; Q, n)$.

As we shall see, Theorem 5.2.3.1 will enable us to obtain very simply estimates on the dimension of the space of intertwining forms for certain types of induced representations of G.

5.2.4 The Case of Countably Many Orbits

Let us agree to retain the notations and assumptions of 5.2.3; in particular, then, M denotes a C^∞ manifold, G a Lie group operating to the left on M. In what follows, it will be supposed, in addition, that there exist in M *at most* countably many G-orbits.

Let U be a differentiable representation of G on the Fréchet space E. Let $\mathbf{T}$ denote the space of E-distributions T on M such that

$$T^x = {}^tU(x)T \qquad (\text{all } x \in G).$$

The objective of the present number is to estimate the dimension of $\mathbf{T}$ by means of Theorem 5.2.3.1 – in order to do this, however, we shall first have to digress and develop some technical lemmas.

Keep to the preceding notations and assumptions.

Lemma 5.2.4.1 (i) Each G-orbit $Q \subset M$ is open relative to its closure Q^{cl} in M;

(ii) Given a G-orbit $Q \subset M$, we may find a G-stable open subset Ω_Q of M containing Q as a closed submanifold;

(iii) Let $p \in M$ – then the natural bijection $G/G_p \to G \cdot p$ is a diffeomorphism.

Proof Let Q be any G-orbit in M – then Q is either open or nowhere dense. [Thus let $p \in M$; either there exists a compact subset $\omega \subset G$ such that $\omega \cdot p$ has a non-empty interior in M or there does not – in the former case $G \cdot p = \bigcup_{x \in G} x \cdot (\mathrm{Int}(\omega \cdot p))$ is open while in the latter case $G \cdot p$ is a countable union of nowhere dense closed sets, hence is of the first category. Conclusion: Every orbit is either open or of the first category. This implies that the union of all the non-open orbits (there are countably many) is of the first category and is closed (being the complement of the union of the open orbits). Hence, from the Baire Category Theorem, we see that this union is a closed nowhere dense set which shows, in particular, that every orbit is either open or nowhere dense.] The closure Q^{cl} of Q in M is G-stable and may be viewed as a G-transformation space in its own right; applying the preceding reasoning to the dense orbit Q in Q^{cl}, we find that Q is open in Q^{cl}, establishing (i). Turning to (ii), put $\Omega_Q = Q \cup (M - Q^{cl})$; according to (i), Ω_Q is open in M. Evidently Ω_Q is G-stable. The fact that Q is closed in Ω_Q follows upon remarking that $Q = \Omega_Q \cap Q^{cl}$. As for (iii), the argument proceeds along rather standard lines and will be omitted (cf. Helgason [2, Chapter 2, §3]). □

[In passing, notice that Lemma 5.2.4.1 (iii) implies, in the case at hand, that the submanifold topology on any given orbit Q (see A 2.4) coincides with the topology of M relativised to Q.]

To make further progress, it will be necessary to define a decreasing transfinite sequence $\{S_\alpha\}$ (α running over all ordinals) of closed G-stable subsets of M – thus:

(i) Put $S_0 = M$;

(ii) For each ordinal α, let $S_{\alpha+1}$ be the complement in S_α of the union of the family of all G-orbits which are contained in S_α and open in S_α;

(iii) If β is a limit ordinal, then set $S_\beta = \bigcap_{\alpha<\beta} S_\alpha$.

Lemma 5.2.4.2 $S_\alpha = \varnothing$ (the empty set) for some countable ordinal α.

Proof If S_α is non-empty, then, by Lemma 5.2.4.1 (i) (applied to the G-transformation space S_α), we have $S_{\alpha+1} \neq S_\alpha$. But $S_\alpha - S_{\alpha+1}$ is a union of (at most) countably many orbits. Hence the lemma. □

The following result is clear.

Lemma 5.2.4.3 Let Q be a G-orbit in M – then there is a unique ordinal α such that Q is contained in S_α and open in S_α.

Under the conditions of the preceding lemma, it will be convenient to say that Q and α are *associated.*

Given a G-orbit $Q \subset M$, put $\Omega_Q = Q \cup (M - S_\alpha)$ where α is the ordinal with which Q is associated – then Ω_Q is a G-stable open subset of M such that $\Omega_Q \cap S_\alpha = Q$ (hence Q is closed in Ω_Q).

We are now in a position to formulate the main lemma. Let $\mathbf{T}$ be some fixed linear space of E-distributions on M. For each ordinal α, let $\mathbf{T}_\alpha = \{T \in \mathbf{T}: \operatorname{spt}(T) \subset S_\alpha\}$ – thus $\mathbf{T}_0 = \mathbf{T}$ and the $\mathbf{T}_\alpha$ are linear subspaces of $\mathbf{T}$, decreasing in α and eventually equal to $\{0\}$ for some countable α (Lemma 5.2.4.2). Let $\mathbf{Q}_\alpha$ be the family of all orbits associated with a given ordinal α. For each $T \in \mathbf{T}_\alpha$ and each Q in $\mathbf{Q}_\alpha$, let T_Q be the restriction of T to Ω_Q – then the support of T_Q is contained in the Ω_Q-closed set Q. If Q is an orbit associated with the ordinal α, set $\mathbf{T}[Q] = \{T_Q | T \in \mathbf{T}_\alpha\}$; evidently $\mathbf{T}[Q]$ is a linear space of distributions on Ω_Q, the support of each member being contained in Q.

Lemma 5.2.4.4 (Bruhat) Retain the preceding notations and assumptions – then $\dim(\mathbf{T}) \leqslant \sum_Q \dim(\mathbf{T}[Q])$, the sum $\sum_Q$ ranging over the (at most) countably many G-orbits in M.

Proof (Fell) We may assume that $\sum_Q \dim(\mathbf{T}[Q]) < \infty$. This being so, we shall prove below (by transfinite induction) that for each ordinal α

$$(*) \quad \dim(\mathbf{T}) \leqslant \dim(\mathbf{T}_\alpha) + \sum_{\substack{Q \in \cup \mathbf{Q}_\beta \\ \beta < \alpha}} \dim(\mathbf{T}[Q]).$$

An application of this result to an α for which $\mathbf{T}_\alpha = \{0\}$ will then yield the required result.

(1) Relation $(*)$ is certainly true if $\alpha = 0$; the next thing that has to be done is to show that if $(*)$ holds for a given ordinal α, then it also holds for $\alpha + 1$. To this end, let us first prove that

$$\dim(\mathbf{T}_\alpha) \leqslant \dim(\mathbf{T}_{\alpha+1}) + \sum_{Q \in \mathbf{Q}_\alpha} \dim(\mathbf{T}[Q]) \qquad (\alpha \text{ any ordinal}).$$

Thus consider the map $f_\alpha : \mathbf{T}_\alpha \to \sum_{Q \in \mathbf{Q}_\alpha} \mathbf{T}[Q]$ given by $(f_\alpha(T))_Q = T_Q$ $(T \in \mathbf{T}_\alpha)$; obviously f_α is linear. Moreover $f_\alpha(T) = 0$ iff $\operatorname{spt}(T) \subset S_\alpha - \bigcup_{Q \in \mathbf{Q}_\alpha} \Omega_Q = S_\alpha - \bigcup_{Q \in \mathbf{Q}_\alpha} Q = S_{\alpha+1}$; i.e. $\operatorname{Ker}(f_\alpha) = \mathbf{T}_{\alpha+1}$. Therefore

$$\begin{aligned} \dim(\mathbf{T}_\alpha) &\leqslant \dim(\operatorname{Ker}(f_\alpha)) + \dim(\operatorname{Im}(f_\alpha)) \\ &\leqslant \dim(\mathbf{T}_{\alpha+1}) + \sum_{Q \in \mathbf{Q}_\alpha} \dim(\mathbf{T}[Q]), \end{aligned}$$

as desired. Now suppose that $(*)$ holds for a given ordinal α – then

$$\dim(\mathbf{T}) \leqslant \dim(\mathbf{T}_\alpha) + \sum_{\substack{Q \in \cup \mathbf{Q}_\beta \\ \beta < \alpha}} \dim(\mathbf{T}[Q])$$

$$\leqslant \dim(\mathbf{T}_{\alpha+1}) + \sum_{Q \in \mathbf{Q}_\alpha} \dim(\mathbf{T}[Q]) + \sum_{\substack{Q \in \cup \mathbf{Q}_\beta \\ \beta < \alpha}} \dim(\mathbf{T}[Q])$$
$$= \dim(\mathbf{T}_{\alpha+1}) + \sum_{\substack{Q \in \cup \mathbf{Q}_\beta \\ \beta < \alpha+1}} \dim(\mathbf{T}[Q]),$$

which shows that $(*)$ holds for $\alpha + 1$ too.

(2) Let β be a limit ordinal; suppose that $(*)$ is true for all $\alpha < \beta$ – then we must prove that it is also true for β. In view of our assumptions, $\mathbf{T}[Q] = \{0\}$ for all but finitely many Q; thus the map f_α in (1) is zero but for finitely many α and so $\mathbf{T}_{\alpha+1} = \operatorname{Ker}(f_\alpha) = \mathbf{T}_\alpha$ except for finitely many α. It follows that there exists an $\alpha_0 < \beta$ such that $\mathbf{T}_\alpha = \mathbf{T}_{\alpha+1}$ whenever $\alpha_0 \leqslant \alpha < \beta$. In (3) below we shall establish that $\mathbf{T}_{\alpha_0} = \mathbf{T}_\alpha$ for all α with $\alpha_0 \leqslant \alpha \leqslant \beta$. Accepting this momentarily, let us suppose that $(*)$ is true for all $\alpha < \beta$ – then, in particular, it is true for $\alpha = \alpha_0$ and we have

$$\dim(\mathbf{T}) \leqslant \dim(\mathbf{T}_{\alpha_0}) + \sum_{\substack{Q \in \cup \mathbf{Q}_\gamma \\ \gamma < \alpha_0}} \dim(\mathbf{T}[Q])$$
$$\leqslant \dim(\mathbf{T}_\beta) + \sum_{\substack{Q \in \cup \mathbf{Q}_\gamma \\ \gamma < \beta}} \dim(\mathbf{T}[Q]),$$

which shows that $(*)$ also holds for β.

(3) The fact that $\mathbf{T}_{\alpha_0} = \mathbf{T}_\alpha$ for all α with $\alpha_0 \leqslant \alpha \leqslant \beta$ will be proved by transfinite induction. Our contention is true if $\alpha = \alpha_0$; if it is true for α, then it is true for $\alpha + 1$ (since $\mathbf{T}_\alpha = \mathbf{T}_{\alpha+1}$ for those α such that $\alpha_0 \leqslant \alpha < \beta$ – cf. (2)); finally let γ be a limit ordinal with $\alpha_0 < \gamma \leqslant \beta$ and such that $\mathbf{T}_{\alpha_0} = \mathbf{T}_\alpha$ for those α verifying $\alpha_0 \leqslant \alpha < \gamma$ – then any T in $\mathbf{T}_{\alpha_0}$ has its support contained in S_α $(\alpha_0 \leqslant \alpha < \gamma)$, hence has its support contained in $\bigcap_{\alpha<\gamma} S_\alpha = S_\gamma$ which implies that $\mathbf{T}_{\alpha_0} = \mathbf{T}_\gamma$, finishing the induction.

The proof of Lemma 5.2.4.4 is now complete. □

Let us return to the problem which was posed at the beginning of the present number – thus we would like to obtain an estimate for the dimension of the space $\mathbf{T}$ of those E-distributions T on M which verify the relation $T^x = {}^tU(x)T$ (all $x \in G$). To this end, for each orbit Q in M, introduce the G-stable open subset Ω_Q of M in which Q is imbedded as a *closed* submanifold (see the discussion following Lemma 5.2.4.3 supra). Let $\mathbf{T}(Q)$ denote the space of E-distributions on Ω_Q such that:

(a) $T^x = {}^tU(x)T$ for all $x \in G$;

(b) The support of T is contained in Q.

Since Q is closed in Ω_Q, Theorem 5.2.3.1 is applicable and gives us a majorization for the dimension of $\mathbf{T}(Q)$ (the role of M in 5.2.3 is played here by Ω_Q). On the other hand, it is clear that the space $\mathbf{T}[Q]$ is contained in $\mathbf{T}(Q)$; furthermore Lemma 5.2.4.4 tells us that $\dim(\mathbf{T}) \leqslant \sum_Q \dim(\mathbf{T}[Q])$. Putting these facts together now leads to:

Theorem 5.2.4.5 (Bruhat) The dimension of $\mathbf{T}$ is less than or equal to the sum of the numbers $i(E; Q, n)$ as Q runs through the countably many G-orbits in M and n runs through the set of non-negative integers.

[We recall that if Q is an open orbit, then it is necessary to take $i(E; Q, n) = 0$ for $n > 0$.]

5.3 Irreducibility of Unitarily Induced Representations

5.3.1 On the Notion of Induced Representation

Let G be a Lie group countable at infinity, H a closed subgroup of G; let L be a unitary representation of H on a Hilbert space E – then, as we have seen in 5.1.1, L determines, in a rather natural way, a unitary representation U^L of G (the unitary representation of G unitarily induced by L). This being so, it is natural to wonder what form the definition of U^L should take when the unitary requirement on L is relaxed – the objective of the present number is to consider this question (although only briefly).

As above, let H be a closed subgroup of our Lie group G; let L be a differentiable representation of H on a Fréchet space E – the first thing that shall be done is to associate with L a differentiable representation U^L of G, the differentiable representation of G differentiably induced by L. To this end, consider the space ${}^LC^\infty(G; E)$ introduced already in the discussion centering around Lemma 5.1.1.4 (keep in mind that the map $\pi^L, f \mapsto f^L$, of $C_c^\infty(G; E)$ onto ${}^LC^\infty(G; E)$ is a strict morphism). It is evident that left translation defines a representation U^L of G on ${}^LC^\infty(G; E)$ – in fact:

Proposition 5.3.1.1 The representation U^L of G on ${}^LC^\infty(G; E)$ is differentiable.

Proof The left regular representation of G on $C_c^\infty(G; E)$ is, of course, differentiable; on the other hand, the map $f \mapsto f^L$ commutes with left translation – it is then immediate that U^L is equivalent to the quotient representation of G on $C_c^\infty(G; E)/\text{Ker}\,(f \mapsto f^L)$ (deduced from the left regular representation of G on $C_c^\infty(G; E)$), which proves that U^L is indeed differentiable. □

The above representation U^L of G on ${}^LC^\infty(G; E)$ will be called *the differentiable representation of* G *induced from the differentiable representation* L *of* H. [For brevity, we shall often simply say that U^L is *differentiably induced* by L.]

Let L be an arbitrary continuous representation of H on the Fréchet space E – we shall now take up the more subtle notion of what one is to understand by a continuous representation (of G) induced from L.

In the first place, in view of what has been said above, we may consider the differentiable representation U^{L_∞} of G on ${}^{L_\infty}C^\infty(G; E_\infty)$; on the other hand, we also have a natural continuous representation U^L of G by left translation on the space ${}^LC(G; E)$ (cf. Remark 2 following Lemma 5.1.1.4)

– we shall refer to U^L as *the continuous representation of G induced from the continuous representation L of H.* [For brevity, we shall frequently just say that U^L is *continuously induced* by L.] The point of view that will be taken now is this: Any candidate for the title 'continuous representation induced by L' must contain a subrepresentation equivalent to U^L. More precisely let us agree to the following definition.

Definition (Bruhat) Let L be a continuous representation of H on the Fréchet space E. A continuous representation U of G on a space $E(U)$ ($E(U)$ not necessarily Fréchet) is said to be *induced from L* if there exists a continuous linear injection i of ${}^{L_\infty}C^\infty(G; E_\infty)$ into $E(U)$ such that:

(i) $i({}^{L_\infty}C^\infty(G; E_\infty))$ is dense in $E(U)$;

(ii) i extends to a continuous linear injection of ${}^LC(G; E)$ into $E(U)$ such that $i\circ U^L(x) = U(x)\circ i$ (all $x \in G$).

As far as notation goes, we shall, in the sequel, usually denote continuous representations U of G which are induced from L (in the sense of the preceding definition) by U^L too; at first glance, this appears to be an unsatisfactory arrangement – however no confusion will ever arise since the symbol U^L will always be accompanied by a qualifying phrase such as 'the continuous representation of G continuously induced by L'. If the occasion warrants it, the more precise notation ${}_GU^L$ will be employed.

The formation of a continuous representation U^L of G induced by L is in general not canonical, i.e. may be carried out in possibly many different ways. Typically, one extends U^{L_∞} by changing the topology in the space and then completing – however the U^L so defined will, of course, depend on the choice of this topology.

Examples (1) Let L be a continuous representation of H on a Fréchet space E; let U^L denote the corresponding continuously induced representation of G on ${}^LC(G; E)$ – then U^L is induced by L.

(2) Let L be a unitary representation of H on a Hilbert space E; let U^L denote the corresponding unitarily induced representation of G on the Hilbert space E^L (5.1.1) – then U^L is induced by L (this is the gist of Lemma 5.1.1.5).

Remark It should be pointed out that certain G admit unitary representations which are induced, in the sense of Bruhat's general definition, by a *non-unitary* representation L and so are constructed by a procedure different from that given in 5.1.1. For instance, as will be discussed later, certain unitary representations of the complementary series for a connected semi-simple G with finite center will arise in this fashion.

5.3.2 Estimation of the Intertwining Number

In this number the material of 5.2 will be utilized in order to obtain criteria which will enable one to decide on the irreducibility of certain types of induced representations. In turn these results are part of the foundation on which the discussion of the representation theory of connected semi-simple Lie groups rests.

Let H_i be a closed subgroup of the Lie group G and let L_i be a continuous representation of H_i on a *Fréchet space* E^i ($i = 1, 2$). Let U^{L^1} (respectively U^{L^2}) be a (not 'the'!) continuous representation of G induced by L^1 (respectively L^2) (in the sense of 5.3.1) – our first job will be to obtain a majorization for the intertwining number $i(U^{L^1}, U^{L^2})$. In turn, in view of the definition of induced representation and the remarks made near the end of 4.1.3 in Chapter 4, it suffices to estimate $i(U^{L^1\infty}, U^{L^2\infty})$ ($U^{L^i\infty}$ the differentiable representation of G differentiably induced by L^i_∞, $i = 1, 2$).

[Recall the definition of the canonical strict morphism $\pi^{L^i} = \pi^i$, $f_i \mapsto f_i^{L^i\infty}$, of $C_c^\infty(G; E^i_\infty)$ onto ${}^{L^i\infty}C^\infty(G; E^i_\infty)$ – thus for $x_i \in G$ we have

$$f_i^{L^i\infty}(x_i) = \int_{H_i} \rho_i(\xi_i)^{-1/2} L^i_\infty(\xi_i) f_i(x_i\xi_i) d_i(\xi_i)$$
$$(f_i \in C_c^\infty(G; E^i_\infty), i = 1, 2).$$

Here we are writing for brevity $\rho_i = \rho_{H_i}$, $d_i = d_{H_i}$ etc. ($i = 1,2$). Observe, then, that this implies that ${}^{L^i\infty}C^\infty(G; E^i_\infty)$ is a barrelled space since it is a quotient of $C_c^\infty(G; E^i_\infty)$ and $C_c^\infty(G; E^i_\infty)$ is an *LF*-space ($i = 1, 2$) – hence it makes sense to consider $i(U^{L^1\infty}, U^{L^2\infty})$.]

Before proceeding further, it will be convenient to simplify our notations a little. Thus, in view of what has been said so far, it is apparent that we want to deal with the following situation. Let L^i be a differentiable representation of H_i on a Fréchet space E^i ($i = 1, 2$) – then our objective is to estimate the intertwining number $i(U^{L^1}, U^{L^2})$ of the two differentiably induced representations U^{L^1} and U^{L^2} of G. The first step along these lines is Theorem 5.3.2.1 below – this theorem reduces the problem of estimating the dimension of a space of intertwining forms to a seemingly more tractable one, namely that of estimating the dimension of a certain space of distributions. But this space of distributions is of the type studied in 5.2.4 – hence the machinery developed there is applicable and will lead us quickly to the sought for estimates.

The group $H_2 \times H_1$ operates to the left on G in the obvious way, namely $(\xi_2, \xi_1)\cdot x = \xi_2 x \xi_1^{-1}$ ($x \in G$; $(\xi_2, \xi_1) \in H_2 \times H_1$).

Theorem 5.3.2.1 (Bruhat) Let L^i be a differentiable representation of H_i on a Fréchet space E^i, U^{L^i} the corresponding differentiably induced representation of $G(i = 1, 2)$ – then the intertwining number $i(U^{L^1}, U^{L^2})$ is equal to the dimension of the vector space $\mathbf{T}$ of $E^1 \hat{\otimes} E^2$-distributions T on G verifying condition

$$\text{(A)} \quad T^{(\xi_2,\xi_1)} = {}^tU(\xi_2, \xi_1)T \text{ (all } (\xi_2, \xi_1) \in H_2 \times H_1).$$

Here U is the continuous representation of $H_2 \times H_1$ on $E^1 \hat{\otimes} E^2$ given by

$$U(\xi_2, \xi_1) = (\delta_G(\xi_1\xi_2^{-1})\delta_1(\xi_1)\delta_2(\xi_2))^{1/2}(L^1(\xi_1) \hat{\otimes} L^2(\xi_2)).$$

[In view of Proposition 4.4.1.10, U is actually a differentiable representation of $H_2 \times H_1$ on the Fréchet space $E^1 \hat{\otimes} E^2$ (Proposition A 2.2.2).]

Proof (1) Let $\mathbf{B}$ be a U^{L^1}, U^{L^2} intertwining form and put

$$\mathbf{B}^\natural(f_1, f_2) = \mathbf{B}(\pi^1(f_1), \pi^2(f_2))$$

where $f_1 \in C_c^\infty(G; E^1), f_2 \in C_c^\infty(G; E^2)$ – then $\mathbf{B}^\natural$ is a separately continuous bilinear form on $C_c^\infty(G; E^1) \times C_c^\infty(G; E^2)$. Now for $\xi_1 \in H_1, x_1 \in G$, we have

$$\begin{aligned}\pi^1(f_1(.\xi_1))(x_1) &= \int_{H_1} \rho_1(\eta_1)^{-1/2} L^1(\eta_1) f_1(x_1\eta_1\xi_1)\, d_1(\eta_1)\\ &= \rho_1(\xi_1)^{1/2}\delta_1(\xi_1^{-1}) \int_{H_1} \rho_1(\eta_1)^{-1/2}\\ &\quad \times L^1(\eta_1) L^1(\xi_1^{-1}) f_1(x_1\eta_1)\, d_1(\eta_1)\\ &= (\delta_G(\xi_1)\delta_1(\xi_1))^{-1/2}\pi^1(L^1(\xi_1^{-1})f_1)(x_1)\end{aligned}$$

and so we see that

$$\pi^1(f_1(.\xi_1)) = (\delta_G(\xi_1)\delta_1(\xi_1))^{-1/2}\pi^1(L^1(\xi_1^{-1})f_1).$$

In a similar way we find that

$$\pi^2(f_2(.\xi_2)) = (\delta_G(\xi_2)\delta_2(\xi_2))^{-1/2}\pi^2(L^2(\xi_2^{-1})f_2) \qquad (\xi_2 \in H_2).$$

Hence, if $z \in G$,

$$\begin{aligned}\mathbf{B}^\natural(f_1(z.\xi_1), f_2(z.\xi_2)) &= (\delta_G(\xi_1\xi_2)\delta_1(\xi_1)\delta_2(\xi_2))^{-1/2}\\ &\quad \times \mathbf{B}^\natural(L^1(\xi_1^{-1})f_1, L^2(\xi_2^{-1})f_2).\end{aligned}$$

(2) Example 8 of A 2.2 implies that

$$C_c^\infty(G; E^1) \bar{\otimes} C_c^\infty(G; E^2) = C_c^\infty(G \times G; E^1 \hat{\otimes} E^2).$$

By definition of the inductive tensor product $\bar{\otimes}$, the separately continuous bilinear form $\mathbf{B}^\natural$ may be identified with an element of the dual of $C_c^\infty(G; E^1) \bar{\otimes} C_c^\infty(G; E^2)$, that is, with an $E^1 \hat{\otimes} E^2$-distribution, say $T^\natural$, on $G \times G$. Thus, in particular, for $f_1 \in C_c^\infty(G; E^1)$, $f_2 \in C_c^\infty(G; E^2)$ we have

$$\int_{G\times G} f_1(x_1) \otimes f_2(x_2)\, dT^\natural(x_1, x_2) = \mathbf{B}^\natural(f_1, f_2)$$

and, in view of the relations developed in (1),

$$\begin{aligned}&\int_{G\times G} f_1(zx_1\xi_1) \otimes f_2(zx_2\xi_2)\, dT^\natural(x_1, x_2)\\ &= (\delta_G(\xi_1\xi_2)\delta_1(\xi_1)\delta_2(\xi_2))^{-1/2} T^\natural(L^1(\xi_1^{-1})f_1 \otimes L^2(\xi_2^{-1})f_2).\end{aligned}$$

(3) The group G acts on $G \times G$ via the operations $(x_1, x_2) \mapsto z\cdot(x_1, x_2) = (zx_1, zx_2)$ while $H_1 \times H_2$ acts on $G \times G$ by the rule $(x_1, x_2) \mapsto (\xi_1, \xi_2)\cdot(x_1, x_2) = (x_1\xi_1^{-1}, x_2\xi_2^{-1})$. Relative to these operations, what has

been established up to now is this: To each non-zero U^{L^1}, U^{L^2} intertwining form $\mathbf{B}$ there corresponds a non-zero $E^1 \hat{\otimes} E^2$-distribution T^t on $G \times G$ which is invariant with respect to the action of G on $G \times G$ and quasi-invariant with respect to the action of $H_1 \times H_2$ on $G \times G$. Symbolically:

$$\begin{aligned} &dT^t(z^{-1}x_1\xi_1, z^{-1}x_2\xi_2) \\ &= (\delta_G(\xi_1\xi_2)\delta_1(\xi_1)\delta_2(\xi_2))^{1/2}\,{}^t(L^1(\xi_1) \hat{\otimes} L^2(\xi_2))\, dT^t(x_1, x_2). \end{aligned}$$

(4) The group G also operates on $G \times G$ via the prescription $(x_1, x_2) \mapsto z\cdot(x_1, x_2) = (x_1, zx_2)$. Let ϕ denote the manifold automorphism of $G \times G$ given by $(x_1, x_2) \mapsto (x_2^{-1}x_1, x_2)$, S^t the image of T^t under ϕ (so that $S^t(h) = T^t(h\circ\phi)$ ($h \in C_c^\infty(G \times G; E^1 \hat{\otimes} E^2)$) – then S^t is invariant under the action of G at hand. [It suffices to verify that $T^t(h^z\circ\phi) = T^t(h\circ\phi)$ (all $z \in G$, $h \in C_c^\infty(G \times G; E^1 \hat{\otimes} E^2)$) – thus

$$\begin{aligned} T^t(h^z\circ\phi) &= \int h^z(x_2^{-1}x_1, x_2)\, dT^t(x_1, x_2) \\ &= \int h(x_2^{-1}x_1, z^{-1}x_2)\, dT^t(x_1, x_2) \\ &= \int h(x_2^{-1}zz^{-1}x_1, z^{-1}x_2)\, dT^t(x_1, x_2) \\ &= \int h((z^{-1}x_2)^{-1}z^{-1}x_1, z^{-1}x_2)\, dT^t(x_1, x_2) \\ &= \int h\circ\phi(z^{-1}x_1, z^{-1}x_2)\, dT^t(x_1, x_2) \\ &= \int h\circ\phi(x_1, x_2)\, dT^t(x_1, x_2) \\ &= T^t(h\circ\phi), \end{aligned}$$

as desired.] This being so, Proposition 5.2.1.5 (with $M = G$, $H = \{1\}$, A the trivial multiplier) gives us a unique $E^1 \hat{\otimes} E^2$-distribution T on G having the property that

$$S^t(h) = \int_G d_G(x_2) \int_G h(x_1, x_2)\, dT(x_1) \quad (h \in C_c^\infty(G \times G; E^1 \hat{\otimes} E^2))$$

or, in other words,

$$\int_{G\times G} h(x_1, x_2)\, dT^t(x_1, x_2) = \int_G d_G(x_2) \int_G h(x_2x_1, x_2)\, dT(x_1)$$
$$(h \in C_c^\infty(G \times G; E^1 \hat{\otimes} E^2)).$$

We shall call T the *G-trace* of $\mathbf{B}$.

(5) The connection between T^t and T may be given in the symbolic relationship: $dT^t(x_1, x_2) = d_G(x_2)\, dT(x_2^{-1}x_1)$. Now on the one hand we have

$$\begin{aligned} dT^t(x_1\xi_1, x_2\xi_2) &= d_G(x_2\xi_2)\, dT(\xi_2^{-1}x_2^{-1}x_1\xi_1) \\ &= \delta_G(\xi_2)\, d_G(x_2)\, dT(\xi_2^{-1}x_2^{-1}x_1\xi_1), \end{aligned}$$

while on the other

$$dT^{\mathfrak{t}}(x_1\xi_1, x_2\xi_2) =$$
$$(\delta_G(\xi_1\xi_2)\delta_1(\xi_1)\delta_2(\xi_2))^{1/2}\, d_G(x_2)\ {}^{t}(L^1(\xi_1)\,\hat{\otimes}\, L^2(\xi_2))\, dT(x_2^{-1}x_1);$$

therefore

$$dT(\xi_2^{-1}x_1\xi_1) = (\delta_G(\xi_1\xi_2^{-1})\delta_1(\xi_1)\delta_2(\xi_2))^{1/2}\, {}^{t}(L^1(\xi_1)\,\hat{\otimes}\, L^2(\xi_2))\, dT(x_1),$$

that is

$$\text{(A)}\quad T^{(\xi_2,\xi_1)} = (\delta_G(\xi_1\xi_2^{-1})\delta_1(\xi_1)\delta_2(\xi_2))^{1/2}\, {}^{t}(L^1(\xi_1)\,\hat{\otimes}\, L^2(\xi_2))T.$$

(6) Conversely let T be an $E^1 \hat{\otimes} E^2$-distribution on G verifying condition (A). In order to show that T is the G-trace of some U^{L^1}, U^{L^2} intertwining form, first backtrack and define an $E^1 \hat{\otimes} E^2$-distribution $T^{\mathfrak{t}}$ on $G \times G$ by the relation $dT^{\mathfrak{t}}(x_1, x_2) = d_G(x_2)\, dT(x_2^{-1}x_1)$; of course $T^{\mathfrak{t}}$ satisfies

$$dT^{\mathfrak{t}}(z^{-1}x_1\xi_1, z^{-1}x_2\xi_2)$$
$$= (\delta_G(\xi_1\xi_2)\delta_1(\xi_1)\delta_2(\xi_2))^{1/2}\, {}^{t}(L^1(\xi_1)\,\hat{\otimes}\, L^2(\xi_2))\, dT^{\mathfrak{t}}(x_1, x_2).$$

Now put

$$\mathbf{B}^{\mathfrak{t}}(f_1, f_2) = T^{\mathfrak{t}}(f_1 \otimes f_2) \qquad (f_1 \in C_c^\infty(G; E^1), f_2 \in C_c^\infty(G; E^2)).$$

Then $\mathbf{B}^{\mathfrak{t}}$ is a separately continuous bilinear form on $C_c^\infty(G; E^1) \times C_c^\infty(G; E^2)$ which satisfies the same relations as the $\mathbf{B}^{\mathfrak{t}}$ appearing in (1) above. Claim: $\mathbf{B}^{\mathfrak{t}}$ passes to the quotient to define a bilinear form $\mathbf{B}$ on ${}^{L^1}C^\infty(G; E^1) \times {}^{L^2}C^\infty(G; E^2)$. To see this, let us for the moment consider G as a principal H_2-bundle, the action of H_2 on G being left multiplication. Our distribution T is quasi-invariant with respect to this action of H_2 on G, the multiplier being $A(\xi_2, x) = (\delta_G(\xi_2)\delta_2(\xi_2^{-1}))^{1/2}\,(1 \,\hat{\otimes}\, L^2(\xi_2^{-1}))$. Thanks to Corollary 5.2.1.6, we know that T vanishes at those $h \in C_c^\infty(G; E^1 \hat{\otimes} E^2)$ which have the property that

$$\int_{H_2} (\delta_G(\xi_2)\delta_2(\xi_2^{-1}))^{1/2}(1 \,\hat{\otimes}\, L^2(\xi_2^{-1}))h(\xi_2 x)\,d_2(\xi_2) = 0$$

for all $x \in G$. In a similar way, if we regard G as a principal H_1-bundle under the left action $x \mapsto \xi_1 \cdot x = x\xi_1^{-1}$, then Corollary 5.2.1.6 tells us once again that T vanishes at those $h \in C_c^\infty(G; E^1 \hat{\otimes} E^2)$ verifying

$$\int_{H_1} (\delta_G(\xi_1)\delta_1(\xi_1))^{-1/2}(L^1(\xi_1^{-1}) \,\hat{\otimes}\, 1)h(x\xi_1^{-1})\, d_1(\xi_1) = 0 \qquad (\text{all } x \in G).$$

Let $f_i \in C_c^\infty(G; E^i)$ $(i = 1, 2)$ – then

$$\mathbf{B}^{\mathfrak{t}}(f_1, f_2) = \int_G d_G(x_2) \int_G f_1(x_2x_1) \otimes f_2(x_2)\, dT(x_1)$$

will vanish provided $\int_G f_1(x_2x_1) \otimes f_2(x_2)\, dT(x_1) = 0$ for all x_2 in G. In turn this will be true provided

$$(\pi^1(f_1))(x_2x_1) \otimes f_2(x_2)$$
$$= \int_{H_1} (\delta_G(\xi_1)\delta_1(\xi_1^{-1}))^{1/2}(L^1(\xi_1)f_1(x_2x_1\xi_1) \otimes f_2(x_2))\, d_1(\xi_1) = 0$$

for all $x_1, x_2 \in G$, hence provided $\pi^1(f_1) = 0$. Similarly

$$\mathbf{B}'(f_1, f_2) = \int_G d_G(x_2) \int_G f_1(x_2) \otimes \delta_G(x_1^{-1}) f_2(x_2 x_1^{-1})\, dT(x_1)$$

will vanish if

$$\int_G f_1(x_2) \otimes \delta_G(x_1^{-1}) f_2(x_2 x_1^{-1})\, dT(x_1) = 0$$

for all x_2 in G which will be the case if for all $x_1, x_2 \in G$

$$\delta_G(x_1^{-1})(f_1(x_2) \otimes \pi^2(f_2)(x_2x_1^{-1}))$$
$$= \int_{H_2} (\delta_G(\xi_2)\delta_2(\xi_2^{-1}))^{1/2}(1 \mathbin{\hat{\otimes}} L^2(\xi_2^{-1}))(f_1(x_2)$$
$$\otimes\, \delta_G(x_1^{-1}\xi_2^{-1}) f_2(x_2x_1^{-1}\xi_2^{-1})\, d_2(\xi_2)$$
$$= 0,$$

i.e., if $\pi^2(f_2) = 0$. Conclusion: $\mathbf{B}'(f_1, f_2) = 0$ whenever either $\pi^1(f_1) = 0$ or $\pi^2(f_2) = 0$. Hence $\mathbf{B}'$ does indeed pass to the quotient to define a bilinear form $\mathbf{B}$ on ${}^{L^1}C^\infty(G; E^1) \times {}^{L^2}C^\infty(G; E^2)$. Clearly $\mathbf{B}$ is an intertwining form for U^{L^1} and U^{L^2} and, of course, our original distribution T is the G-trace of $\mathbf{B}$.

To summarize, we have shown that the map $\mathbf{B} \mapsto T$ which assigns to each U^{L^1}, U^{L^2} intertwining form $\mathbf{B}$ its G-trace T is a linear bijection of all U^{L^1}, U^{L^2} intertwining forms onto the space $\mathbf{T}$ of all $E^1 \mathbin{\hat{\otimes}} E^2$-distributions T on G which verify condition (A) – in particular, $i(U^{L^1}, U^{L^2}) = \dim(\mathbf{T})$. All the assertions of the present theorem are now obvious. □

Remark The reader will observe that the Schwartz Kernel Theorem (A 2.2, Example 8) is the keystone on which the proof of Theorem 5.3.2.1 rests.

In order to be able to apply the results in 5.2.4, we shall make an assumption which will be in force throughout the remainder of this number.

Assumption There exist at most countably many orbits in G for the canonical action $x \mapsto (\xi_2, \xi_1)\cdot x = \xi_2 x \xi_1^{-1}$ of $H_2 \times H_1$ on G.

[The orbits in G under the action of the group $H_2 \times H_1$ are, of course, precisely the double cosets $H_2 x H_1$ $(x \in G)$.]

We shall now reinterpret Theorem 5.2.4.5 (as well as the discussion preceding it) in the present situation. Let $\mathbf{T}$ denote the space of $E^1 \mathbin{\hat{\otimes}} E^2$-distributions on G verifying

$$\text{(A)} \quad T^{(\xi_2, \xi_1)} = {}^tU(\xi_2, \xi_1)T \qquad \text{for all } (\xi_2, \xi_1) \in H_2 \times H_1.$$

Here

$$U(\xi_2, \xi_1) = (\delta_G(\xi_1\xi_2^{-1})\delta_1(\xi_1)\delta_2(\xi_2))^{1/2}(L^1(\xi_1) \hat{\otimes} L^2(\xi_2));$$

as mentioned above U is a differentiable representation of $H_2 \times H_1$ on the Fréchet space $E^1 \hat{\otimes} E^2$. For each orbit $Q = H_2 xH_1$ in G, let Ω_Q denote the $H_2 \times H_1$-stable open subset of G, defined in the discussion following Lemma 5.2.4.3, in which $H_2 xH_1$ is imbedded as a closed submanifold. [The M and G there being G and $H_2 \times H_1$, respectively, in the present context; $H_2 \times H_1$ operates on G per supra.] Let $\mathbf{T}(Q)$ denote the space of $E^1 \hat{\otimes} E^2$-distributions on Ω_Q such that:

(a) $T^{(\xi_2,\xi_1)} = {}^tU(\xi_2, \xi_1)T$ for all $(\xi_2, \xi_1) \in H_2 \times H_1$;

(b) The support of T is contained in Q.

As was noted in 5.2.4, $\dim(\mathbf{T}) \leqslant \sum_Q \dim(\mathbf{T}(Q))$. Hence:

Proposition 5.3.2.2 The intertwining number $i(U^{L^1}, U^{L^2})$ is less than or equal to $\sum_Q \dim(\mathbf{T}(Q))$, Q running through the countably many (at most) double cosets $H_2 xH_1$ $(x \in G)$.

Consider now an arbitrary double coset $Q = H_2 xH_1$ $(x \in G)$. If $S \in \mathbf{T}(Q)$, then the left translate T of S by x^{-1} is an $E^1 \hat{\otimes} E^2$-distribution on $\Omega_x = x^{-1}\Omega_Q$ with $\operatorname{spt}(T) \subset x^{-1}Q = Q_x$ (say). Let $f \in C_c^\infty(\Omega_x; E^1 \hat{\otimes} E^2)$ – then, for $\xi_1 \in H_1, \xi_2 \in x^{-1}H_2x$, we have

$$\begin{aligned} T^{(\xi_2,\xi_1)}(f) &= S^{x^{-1}}(f^{(\xi_2,\xi_1)^{-1}}) \\ &= \int_{\Omega_Q} f(\xi_2 x^{-1}y\xi_1^{-1})\, dS(y) \\ &= \int_{\Omega_Q} f(x^{-1}(x\xi_2 x^{-1})y\xi_1^{-1})\, dS(y) \\ &= (\delta_G(\xi_1\xi_2^{-1})\delta_1(\xi_1)\delta_2(x\xi_2x^{-1}))^{1/2} \\ &\quad \times \int_{\Omega_Q} L^1(\xi_1) \hat{\otimes} L^2(x\xi_2x^{-1}) f(x^{-1}y)\, dS(y), \end{aligned}$$

that is,

$$(\mathrm{A}_x) \quad T^{(\xi_2,\xi_1)} = {}^tV(\xi_2, \xi_1)T$$

where V is the differentiable representation of $(x^{-1}H_2x) \times H_1$ on $E^1 \hat{\otimes} E^2$ given by:

$$V(\xi_2, \xi_1) = (\delta_G(\xi_1\xi_2^{-1})\delta_1(\xi_1)\delta_2(x\xi_2x^{-1}))^{1/2} L^1(\xi_1) \hat{\otimes} L^2(x\xi_2x^{-1})$$
$$(\xi_1 \in H_1, \xi_2 \in x^{-1}H_2x).$$

Thus: The space $\mathbf{T}(Q)$ is isomorphic to the space $\mathbf{T}(Q_x)$ of $E^1 \hat{\otimes} E^2$-distributions on $\Omega_x = x^{-1}\Omega_Q$ with support contained in $Q_x = x^{-1}H_2xH_1$ and verifying condition (A_x).

Put $H_x = H_1 \cap (x^{-1}H_2x)$ and let $\tilde{H}_x$ denote the subgroup of $(x^{-1}H_2x) \times H_1$ consisting of the pairs (ζ, ζ) with $\zeta \in H_x$. Identify, in the obvious way, $(x^{-1}H_2x) \times H_1/\tilde{H}_x$ with Q_x (map $(\xi_2, \xi_1)\tilde{H}_x \mapsto \xi_2^{-1}\xi_1$).

Our next task is to estimate the dimension of the space $\mathbf{T}(Q)$ or, equivalently, of $\mathbf{T}(Q_x)$. This has essentially already been accomplished in 5.2.3 – thus, we should study the intertwining number of the differentiable representations $(\zeta, \zeta) \mapsto V(\zeta, \zeta)$ and $(\zeta, \zeta) \mapsto \rho_{\tilde{H}_x}(\zeta, \zeta)\Lambda_n^\vee((\zeta, \zeta), \dot{1})$ of $\tilde{H}_x$, or, what amounts to the same thing, of the differentiable representations $\zeta \mapsto V(\zeta, \zeta)$ and $\zeta \mapsto \rho_{\tilde{H}_x}(\zeta, \zeta)\Lambda_n^\vee((\zeta, \zeta), \dot{1})$ of H_x. Actually it proves to be more convenient to modify these considerations a bit.

First let us compute $\rho_{\tilde{H}_x}(\zeta, \zeta)$. Set for brevity $\delta_x(\zeta) = \delta_{\tilde{H}_x}(\zeta, \zeta)$ $(\zeta \in H_x)$ – then the left Haar measure on $(x^{-1}H_2x) \times H_1$ is $d_2(x\xi_2x^{-1})\, d_1(\xi_1)$, whence

$$\rho_{\tilde{H}_x}(\zeta, \zeta) = \frac{\delta_x(\zeta)}{\delta_1(\zeta)\delta_2(x\zeta x^{-1})} \qquad (\zeta \in H_x).$$

On the other hand

$$V(\zeta, \zeta) = (\delta_1(\zeta)\delta_2(x\zeta x^{-1}))^{1/2} L^1(\zeta) \hat{\otimes} L^2(x\zeta x^{-1}) \qquad (\zeta \in H_x).$$

So put, for each non-negative integer n,

$$A_n(\zeta) = (\delta_1(\zeta)\delta_2(x\zeta x^{-1}))^{1/2}\delta_x(\zeta^{-1})\Lambda_n((\zeta, \zeta), \dot{1}) \qquad (\zeta \in H_x)$$

and let $i(L^1, L^2; Q, n)$ denote the intertwining number for the differentiable representations $\zeta \mapsto A_n^\vee(\zeta)$ and $\zeta \mapsto L^1(\zeta) \hat{\otimes} L^2(x\zeta x^{-1})$ of H_x. [This is a legitimate agreement, i.e. the number in question does not depend on our choice of x (so long as x belongs to Q).] Theorem 5.2.3.1 immediately gives

$$\dim(\mathbf{T}(Q)) \leqslant \sum_{n=0}^{\infty} i(L^1, L^2; Q, n).$$

Therefore follows:

Theorem 5.3.2.3 (Bruhat) Let L^i be a differentiable representation of H_i on a Fréchet space E^i, U^{L^i} the corresponding differentiably induced representation of $G (i = 1, 2)$ – then the intertwining number $i(U^{L^1}, U^{L^2})$ is less than or equal to the sum of the numbers $i(L^1, L^2; Q, n)$ as Q runs through the countably many (at most) double cosets H_2xH_1 $(x \in G)$ and n runs through the set of non-negative integers.

[If Q is an open orbit, then it is necessary to take $i(L^1, L^2; Q, n) = 0$ for $n > 0$; in particular, if either H_1 or H_2 itself is open, then $i(U^{L^1}, U^{L^2}) \leqslant \Sigma_Q\, i(L_1, L_2; Q, 0)$.]

Remark It is actually possible to prove:

$$i(U^{L^1}, U^{L^2}) \leqslant \sum_{Q} \sum_{n=0}^{2\dim(G)+1} i(L^1, L^2; Q, n).$$

The proof of this fact rests on a closer study of the G-traces T arising from the U^{L^1}, U^{L^2} intertwining forms $\mathbf{B}$ – it can be shown that such a distribution is necessarily of finite order less than or equal to $2\dim(G) + 1$ (cf. Bruhat [1, p. 180]).

It is not difficult to interpret the representations $\zeta \mapsto A_n(\zeta)$ of $H_x = H_1 \cap (x^{-1}H_2x)$; for this, it is obviously enough to consider the representations $\zeta \mapsto \Lambda_n((\zeta, \zeta), \dot{1})$ $(n = 1, 2, \ldots)$. Thus, in the first place, for $|I| = |J| = n$, we can write $D^I = D_{k_1} \cdots D_{k_n}$ where $1 \leqslant k_1 \leqslant \cdots \leqslant k_n \leqslant r$, and $D^J = D_{l_1} \cdots D_{l_n}$ where $1 \leqslant l_1 \leqslant \cdots \leqslant l_n \leqslant r$ (the notations of 5.2.3 being in force here). Computation then shows that

$$\lambda_{JI} = \sum_{\sigma} \lambda_{\sigma(l_1)k_1} \cdots \lambda_{\sigma(l_n)k_n}$$

where the sum is taken over all distinct permutations σ of the integers $l_1, \ldots, l_n$. Hence the operator $\Lambda_n((\zeta, \zeta), \dot{1})$ is the restriction to the subspace of symmetric tensors in $\otimes^n \mathbf{C}^r$ of the tensor product $\otimes^n \Lambda_1((\zeta, \zeta), \dot{1})$. On the other hand it is easy to determine Λ_1. Thus, if $\mathfrak{g}$ denotes the Lie algebra of $\mathfrak{g}$, $\mathfrak{h}_i$ the Lie algebra of H_i $(i = 1, 2)$, then it is clear that $\mathfrak{h}_1 + \mathrm{Ad}(x^{-1})\mathfrak{h}_2$ is stable for $\mathrm{Ad}|H_x$ – moreover, Λ_1 is the representation of H_x in

$$\mathfrak{A}_x = \frac{\mathfrak{g}_c}{\mathfrak{h}_{1,c} + \mathrm{Ad}(x^{-1})\mathfrak{h}_{2,c}} \sim \mathbf{C}^r$$

obtained from the adjoint representation of H_x on $\mathfrak{g}_c$ by passage to the quotient, which implies, in view of what was said above, that Λ_n is the representation of H_x on $S^n(\mathfrak{A}_x)$ $(n > 1)$ canonically deduced from Λ_1 ($S^n(\mathfrak{A}_x)$ the subspace of homogeneous elements of degree n in $S(\mathfrak{A}_x)$, $S(\mathfrak{A}_x)$ the symmetric algebra of $\mathfrak{A}_x$).

Let us keep to the preceding assumptions and notations. Thus H_1 and H_2 are closed subgroups of G having the property that there exist in G at most countably many H_2, H_1 double cosets; if L^i is a differentiable representation of H_i on a Fréchet space E^i, U^{L^i} the corresponding differentiably induced representation of $G (i = 1, 2)$, then Theorem 5.3.2.3 provides us with an upper bound for the intertwining number $i(U^{L^1}, U^{L^2})$. Our next objective is to obtain certain lower bounds for $i(U^{L^1}, U^{L^2})$.

Actually, the result toward which we now want to move may be stated as follows.

Proposition 5.3.2.4 Let Φ be the family of those H_2, H_1 double cosets Q with the property that for some x in Q (and so for all x in Q) there exists a finite H_1-invariant measure ν_1 on $H_1/H_1 \cap (x^{-1}H_2x)$ and a finite H_2-invariant measure ν_2 on $H_2/(xH_1x^{-1}) \cap H_2$; let L^i be a unitary representation of H_i on a Hilbert space E^i, U^{L^i} the corresponding unitarily induced representation of $G (i = 1, 2)$ – then

$$i(U^{L^1}, U^{L^2}) \geqslant \sum_{Q \in \Phi} i(L^1, L^2; Q, 0).$$

[Here $i(L^1, L^2; Q, 0)$ denotes the intertwining number of the representations $\zeta \mapsto L^1(\zeta)$, $\zeta \mapsto L^2(x\zeta x^{-1})$ of $H_x = H_1 \cap (x^{-1}H_2x)$.]

The proof of this proposition requires some preparation which will also be useful in other connections as well.

So let us return to the situation described above. Fix an element $x \in G$ and let $Q = H_2xH_1$ be the H_2, H_1 double coset to which x belongs. As we know, Q is a manifold, stable under the canonical left action of $H_2 \times H_1$ on G; the stability subgroup of $H_2 \times H_1$ for the element $x \in Q$ is $\{(x\zeta x^{-1}, \zeta) : \zeta \in H_x\}$ where $H_x = H_1 \cap (x^{-1}H_2x)$.

Consider the multiplier A on $(H_2 \times H_1) \times G$ (relative to the action of $H_2 \times H_1$ on G) defined by

$$A((\xi_2, \xi_1), y) = (\delta_G(\xi_1\xi_2^{-1})\delta_1(\xi_1)\delta_2(\xi_2))^{-1/2}(L^1(\xi_1^{-1}) \hat{\otimes} L^2(\xi_2^{-1})) \qquad ((\xi_2, \xi_1) \in H_2 \times H_1; y \in G).$$

Restricted to $(H_2 \times H_1) \times Q$, A is, of course, a multiplier with respect to the (transitive) action of $H_2 \times H_1$ on Q. Owing to Theorem 5.2.1.4, then, the following result is now clear.

Proposition 5.3.2.5 There is a natural one-to-one correspondence between the space of all quasi-invariant $E^1 \hat{\otimes} E^2$-distributions S on Q with multiplier A (relative to the action of $H_2 \times H_1$ on Q) and the space of all intertwining forms $\mathbf{z}$ for the representations

$$\zeta \mapsto L^1(\zeta) \hat{\otimes} L^2(x\zeta x^{-1}) \text{ and } \zeta \mapsto \delta_x(\zeta)(\delta_1(\zeta)\delta_2(x\zeta x^{-1}))^{-1/2} \quad (\zeta \in H_x)$$

of H_x.

[Let us recall that the intertwining condition on $\mathbf{z}$ reads

$$\delta_x(\zeta)(\delta_1(\zeta)\delta_2(x\zeta x^{-1}))^{-1/2}\, {}^t(L^1(\zeta) \hat{\otimes} L^2(x\zeta x^{-1}))\mathbf{z} = \mathbf{z} \qquad (\zeta \in H_x).]$$

The orbit Q is open in its closure Q^{cl} (Lemma 5.2.4.1(i)) – hence $Q^{cl} - Q$ is closed in G. Each $E^1 \hat{\otimes} E^2$-distribution S on Q gives rise to an $E^1 \hat{\otimes} E^2$-distribution T on the (open) set $\mathcal{O} = G - (Q^{cl} - Q) = Q \cup (G - Q^{cl})$ if we require that $\langle f, T\rangle = \langle f | Q, S\rangle (f \in C_c^\infty(\mathcal{O}; E^1 \hat{\otimes} E^2))$. [In fact Q is closed in $\mathcal{O}$; hence if $f \in C_c^\infty(\mathcal{O}; E^1 \hat{\otimes} E^2)$, then $f | Q \in C_c^\infty(Q; E^1 \hat{\otimes} E^2)$.] Evidently $\mathcal{O}$ is $H_2 \times H_1$-stable; if S is quasi-invariant on Q with multiplier A, then surely T is also quasi-invariant on $\mathcal{O}$ with multiplier A (or, more precisely, $A | (H_2 \times H_1) \times \mathcal{O}$). Note too that the support of T is contained in Q – indeed, T has transversal order zero relative to Q. Consequently, if our orbit Q is *closed* in G (so that $\mathcal{O} = G$) and if S is quasi-invariant with multiplier A, then the distribution T belongs to the space $\mathbf{T}$ defined earlier. In this fashion, we may associate with each closed orbit Q a certain linear subspace of $\mathbf{T}$; furthermore, since it is a question of closed orbits, the distributions corresponding to different subspaces are linearly independent. Bearing in mind Theorem 5.3.2.1, we now see that we have found the following (admittedly unsatisfactory) lower bound for $i(U^{L^1}, U^{L^2})$.

Proposition 5.3.2.6 $i(U^{L^1}, U^{L^2}) \geqslant \sum_Q i(L^1, L^2; Q, 0)$, the sum $\sum_Q$ running over the closed H_2, H_1 double cosets.

Turning back to the general case of a not necessarily closed Q, our next objective is this: Given $\mathbf{z}$, S and T as above, obtain an expression for

$\langle f, T\rangle$ in terms of $\mathbf{z}$ (at least for certain $f \in C_c^\infty(\mathcal{O}; E^1 \hat{\otimes} E^2)$). As will become apparent in due course, this formula occupies a central position as regards the development of the remaining results in the present number.

We need a technical lemma.

Lemma 5.3.2.7 Let ω be any compact subset of G such that $\omega \cap (Q^{cl} - Q)x^{-1} = \varnothing$ – then there exists a compact subset $\tilde{\omega}$ of G such that

$$(\omega(xH_1x^{-1})) \cap H_2 \subset \tilde{\omega}((xH_1x^{-1}) \cap H_2).$$

Proof (1) Let π denote the natural map $G \rightarrow G/xH_1x^{-1}$. The group H_2 operates on G/xH_1x^{-1} by left multiplication. Claim: There are but countably many orbits in G/xH_1x^{-1} for this action of H_2. Thus, if $y \in G$, then

$$\pi^{-1}(H_2 \cdot \pi(y)) = \bigcup_{\xi_2 \in H_2} \xi_2 yxH_1x^{-1} = (H_2 yxH_1)x^{-1}$$

which shows that the π-inverse image of any orbit in G/xH_1x^{-1} under H_2 is the right translate by x^{-1} of some H_2, H_1 double coset. As there are only countably many of the latter, the claim follows.

(2) The stability subgroup for $\pi(1)$ under the action of H_2 on G/xH_1x^{-1} is $(xH_1x^{-1}) \cap H_2$. Owing to Lemma 5.2.4.1(iii), the natural map

$$\phi: H_2/(xH_1x^{-1}) \cap H_2 \rightarrow G/xH_1x^{-1}$$

given by $\xi_2((xH_1x^{-1}) \cap H_2) \mapsto \pi(\xi_2)$ is a homeomorphism; its range, $\pi(H_2)$, is that subset of G/xH_1x^{-1} such that $\pi^{-1}(\pi(H_2)) = Qx^{-1}$.

(3) Because ω is compact, $\pi(\omega)$ is too. Let us verify that $\pi(\omega) \cap \pi(H_2)$ is ccmpact. Thus, since $\omega \cap (Q^{cl}x^{-1} - Qx^{-1}) = \varnothing$, ω is a compact subset of

$$G - (Q^{cl}x^{-1} - Qx^{-1}) = Qx^{-1} \cup (G - Q^{cl}x^{-1}).$$

But, Q being closed in $Q \cup (G - Q^{cl})$, Qx^{-1} is closed in $Qx^{-1} \cup (G - Q^{cl}x^{-1})$, whence $\omega \cap Qx^{-1}$ is compact $\Rightarrow \pi(\omega \cap Qx^{-1})$ is compact $\Rightarrow \pi(\omega) \cap \pi(H_2) = \pi(\omega \cap Qx^{-1})$ is compact, as we wished to show.

(4) Since ϕ is a homeomorphism of $H_2/(xH_1x^{-1}) \cap H_2$ onto its range $\pi(H_2)$ and since $\pi(\omega) \cap \pi(H_2)$ is compact, it follows that $\phi^{-1}(\pi(\omega) \cap \pi(H_2))$ is compact in $H_2/(xH_1x^{-1}) \cap H_2$. Therefore, letting μ denote the natural map $H_2 \rightarrow H_2/(xH_1x^{-1}) \cap H_2$, we see that there exists a compact subset $\tilde{\omega}$ of H_2 such that

$$\phi^{-1}(\pi(\omega) \cap \pi(H_2)) \subset \mu(\tilde{\omega}).$$

If $y \in (\omega(xH_1x^{-1})) \cap H_2$, then

$$\begin{aligned}\phi(\mu(y)) = \pi(y) \in \pi(\omega) \cap \pi(H_2) &\Rightarrow \mu(y) \in \phi^{-1}(\pi(\omega) \cap \pi(H_2)) \subset \mu(\tilde{\omega})\\ &\Rightarrow y \in \tilde{\omega}((xH_1x^{-1}) \cap H_2).\end{aligned}$$

Hence the lemma. □

Corollary 5.3.2.8 Let ω_1, ω_2 be compact subsets of G such that

$$\omega_2^{-1}\omega_1 \cap (Q^{cl} - Q)x^{-1} = \varnothing.$$

Then there exists a compact subset ω of G such that

$$\omega_1(xH_1x^{-1}) \cap \omega_2 H_2 \subset \omega((xH_1x^{-1}) \cap H_2).$$

Proof Taking $\omega_2^{-1}\omega_1$ to be the ω of the preceding lemma, choose $\tilde{\omega}$ accordingly. Let $y \in \omega_1(xH_1x^{-1}) \cap \omega_2 H_2$ – then we can write

$$y = z_1\xi_1 = z_2\xi_2 \qquad (z_1 \in \omega_1, \xi_1 \in xH_1x^{-1}, z_2 \in \omega_2, \xi_2 \in H_2).$$

Plainly

$$z_2^{-1}y \in (\omega_2^{-1}\omega_1)(xH_1x^{-1}) \cap H_2 \subset \tilde{\omega}((xH_1x^{-1}) \cap H_2).$$

The corollary follows upon setting $\omega = \omega_2\tilde{\omega}$. □

Notation Let H be a closed subgroup of G, $d\mu_H(\dot{x})$ a quasi-invariant measure on G/H – then we shall often write more briefly $d\mu_H(x)$ (similarly in other cases too).

As usual, let $H_x = H_1 \cap (x^{-1}H_2x)$; let σ_x be a rho-function on G for the subgroup xH_xx^{-1}, ν_x the corresponding quasi-invariant measure on G/xH_xx^{-1}.

Proposition 5.3.2.9 Suppose there exists a positive H_2-invariant measure ν_2 on H_2/xH_xx^{-1}. Let

$$f_1 \in C_c^\infty(G; E^1), \quad f_2 \in C_c^\infty(G; E^2)$$

and assume that $\operatorname{spt}(f_2)^{-1}\operatorname{spt}(f_1) \cap (Q^{cl} - Q) = \varnothing$; define a map $f: G \to E^1 \hat{\otimes} E^2$ by the rule $f(z) = \int_G f_1(yz) \otimes f_2(y)\, d_G(y)$ $(z \in G)$ – then $f \in C_c^\infty(\mathcal{O}; E^1 \hat{\otimes} E^2)$. Moreover, if $\mathbf{z}, S, T$, are entities arising per Proposition 5.3.2.5 and ensuing discussion, then, for a suitable choice of σ_x and ν_x, we have

$$\langle f, T\rangle = \int_{G/xH_xx^{-1}} \mathbf{z}(\pi^1(f_1)(yx), \pi^2(f_2)(y))\sigma_x(y)^{-1}\, d\nu_x(y).$$

[The integrand appearing here is well-defined, continuous, and has compact support in G/xH_xx^{-1} (cf. infra).]

Proof Obviously $f \in C_c^\infty(\mathcal{O}; E^1 \hat{\otimes} E^2)$. Consider now the integrand above; an elementary computation shows that it is well-defined and it is certainly continuous. Let us verify that it has compact support in G/xH_xx^{-1}. Thus $\pi^1(f_1)$ and $\pi^2(f_2)$ vanish outside of $\operatorname{spt}(f_1)H_1$ and $\operatorname{spt}(f_2)H_2$, respectively, and so the integrand vanishes unless

$$\begin{aligned} y &\in \operatorname{spt}(f_1)H_1x^{-1} \cap \operatorname{spt}(f_2)H_2 \\ &= \operatorname{spt}(f_1)x^{-1}(xH_1x^{-1}) \cap \operatorname{spt}(f_2)H_2. \end{aligned}$$

By hypothesis $\operatorname{spt}(f_2)^{-1}\operatorname{spt}(f_1) \cap (Q^{cl} - Q) = \varnothing$ – hence

$$\operatorname{spt}(f_2)^{-1}\operatorname{spt}(f_1)x^{-1} \cap (Q^{cl} - Q)x^{-1} = \varnothing.$$

Owing to Corollary 5.3.2.8, then, there is a compact subset $\omega \subset G$ having the property that

$$\operatorname{spt}(f_1)x^{-1}(xH_1x^{-1}) \cap \operatorname{spt}(f_2)H_2 \subset \omega(xH_xx^{-1}).$$

From this we deduce that the integrand in question vanishes unless $y \in \omega(xH_xx^{-1})$, which shows that it has compact support in G/xH_xx^{-1}.

(1) The hypothesis of an H_2-invariant measure ν_2 on H_2/xH_xx^{-1} permits one to choose a rho-function ρ on $H_2 \times H_1$ relative to $(H_2 \times H_1)_x$ (the stability subgroup of $H_2 \times H_1$ for x) in a special way, namely $\rho(\xi_2, \xi_1) = \delta_1(\xi_1^{-1})$ (cf. the discussion revolving around Lemma A 1.4 (Appendix 1)). If μ denotes the corresponding quasi-invariant measure on $H_2 \times H_1/(H_2 \times H_1)_x \sim Q$, then Theorem 5.2.1.4 tells us that

$$\langle \phi, S\rangle = \int_Q \delta_1(\xi_1)\mathbf{z}(A^{-1}((\xi_2, \xi_1)^{-1}, \xi_2 x\xi_1^{-1})\phi(\xi_2 x\xi_1^{-1}))\, d\mu(\xi_2 x\xi_1^{-1})$$
$$(\phi \in C_c^\infty(Q; E^1 \hat{\otimes} E^2)).$$

An application of Lemma A 1.4 to the integral at hand leads to

$$\langle \phi, S\rangle = \int_{H_2/xH_xx^{-1}} d\nu_2(\xi_2) \int_{H_1} (\delta_G(\xi_1\xi_2))^{1/2}(\delta_1(\xi_1)\delta_2(\xi_2))^{-1/2}$$
$$\times\, \mathbf{z}(L^1(\xi_1) \hat{\otimes} L^2(\xi_2^{-1})\phi(\xi_2 x\xi_1))\, d_1(\xi_1)$$
$$(\phi \in C_c^\infty(Q; E^1 \hat{\otimes} E^2)).$$

Of course the same formula is also valid for T (ϕ then being an element of $C_c^\infty(\mathcal{O}; E^1 \hat{\otimes} E^2)$) – therefore, in particular,

$$\langle f, T\rangle = \int_G d_G(y) \int_{H_2/xH_xx^{-1}} d\nu_2(\xi_2) \int_{H_1} (\delta_G(\xi_1\xi_2))^{1/2}(\delta_1(\xi_1)\delta_2(\xi_2))^{-1/2}$$
$$\times\, \mathbf{z}(L^1(\xi_1)f_1(y\xi_2 x\xi_1) \otimes L^2(\xi_2^{-1})f_2(y))\, d_1(\xi_1)$$
$$= \int_G d_G(y) \int_{H_2/xH_xx^{-1}} d\nu_2(\xi_2)\left(\frac{\delta_G(\xi_2)}{\delta_2(\xi_2)}\right)^{1/2}$$
$$\times\, \mathbf{z}(\pi^1(f_1)(y\xi_2 x) \otimes L^2(\xi_2^{-1})f_2(y)).$$

(2) Let ρ_2 be a rho-function on G with respect to H_2, μ_2 the corresponding quasi-invariant measure on G/H_2 – then, for all $f \in C_c(G)$,

$$\int_G f(y)\, d_G(y) = \int_{G/H_2} d\mu_2(y) \int_{H_2} \frac{f(y\eta_2)}{\rho_2(y\eta_2)}\, d_2(\eta_2)$$
$$= \int_{G/H_2} d\mu_2(y)\rho_2(y)^{-1} \int_{H_2} \delta_G(\eta_2)\delta_2(\eta_2^{-1})f(y\eta_2)\, d_2(\eta_2).$$

Hence

$$\langle f, T\rangle = \int_{G/H_2} d\mu_2(y)\rho_2(y)^{-1} \int_{H_2} \delta_G(\eta_2)\delta_2(\eta_2^{-1})\, d_2(\eta_2)X,$$

where

$$X = \int_{H_2/xH_xx^{-1}} d\nu_2(\xi_2)\left(\frac{\delta_G(\xi_2)}{\delta_2(\xi_2)}\right)^{1/2} \mathbf{z}(\pi^1(f_1)(y\eta_2\xi_2 x) \otimes L^2(\xi_2^{-1})f_2(y\eta_2)).$$

Making the substitution $\xi_2 \mapsto \eta_2^{-1}\xi_2$ and bearing in mind that ν_2 is H_2-invariant, we find that

$$\langle f, T\rangle = \int_{G/H_2} d\mu_2(y)\rho_2(y)^{-1} \int_{H_2} \left(\frac{\delta_G(\eta_2)}{\delta_2(\eta_2)}\right)^{1/2} d_2(\eta_2)X,$$

where

$$X = \int_{H_2/xH_xx^{-1}} d\nu_2(\xi_2)\left(\frac{\delta_G(\xi_2)}{\delta_2(\xi_2)}\right)^{1/2} \mathbf{z}(\pi^1(f_1)(y\xi_2 x) \otimes L^2(\xi_2^{-1}\eta_2)f_2(y\eta_2)).$$

Our next objective is to interchange the order of the integrations over H_2 and H_2/xH_xx^{-1}. In order to justify such a maneuver, it will be sufficient to verify that the (continuous) integrand in question has compact support in $H_2 \times (H_2/xH_xx^{-1})$ (cf. (3) below).

(3) Fix $y \in G$ – then our integrand is certainly continuous on $H_2 \times (H_2/xH_xx^{-1})$. To verify that it has compact support in $H_2 \times (H_2/xH_xx^{-1})$, first observe that it vanishes unless $y\eta_2 \in \operatorname{spt}(f_2)$ and $y\xi_2 x \in \operatorname{spt}(f_1)H_1$. The condition $y\eta_2 \in \operatorname{spt}(f_2)$ restricts η_2 to the compact set $y^{-1}\operatorname{spt}(f_2)$; this condition also tells us that $y^{-1} \in H_2 \operatorname{spt}(f_2)^{-1}$ so there is $\zeta_2 \in H_2$ such that $y^{-1} \in \zeta_2 \operatorname{spt}(f_2)^{-1}$. The condition $y\xi_2 x \in \operatorname{spt}(f_1)H_1$, in conjunction with the relation $y^{-1} \in \zeta_2 \operatorname{spt}(f_2)^{-1}$, then implies that

$$\begin{aligned}\xi_2 &\in y^{-1}\operatorname{spt}(f_1)x^{-1}(xH_1x^{-1}) \cap H_2 \\ &\subset \zeta_2 \operatorname{spt}(f_2)^{-1}\operatorname{spt}(f_1)x^{-1}(xH_1x^{-1}) \cap H_2.\end{aligned}$$

By hypothesis $\operatorname{spt}(f_2)^{-1}\operatorname{spt}(f_1) \cap (Q^{cl} - Q) = \varnothing$; since $Q^{cl} - Q$ is stable under left multiplication by H_2, we deduce that

$$\zeta_2 \operatorname{spt}(f_2)^{-1}\operatorname{spt}(f_1) \cap (Q^{cl} - Q) = \varnothing,$$

whence

$$\zeta_2 \operatorname{spt}(f_2)^{-1}\operatorname{spt}(f_1)x^{-1} \cap (Q^{cl} - Q)x^{-1} = \varnothing.$$

Lemma 5.3.2.7 now gives us a compact subset $\omega \subset G$ with the property that

$$(\zeta_2 \operatorname{spt}(f_2)^{-1}\operatorname{spt}(f_1)x^{-1}(xH_1x^{-1})) \cap H_2 \subset \omega(xH_xx^{-1}).$$

Consequently our integrand vanishes unless $\eta_2 \in y^{-1}\operatorname{spt}(f_2)$ and $\xi_2 \in \omega(xH_xx^{-1})$, hence has compact support in $H_2 \times (H_2/xH_xx^{-1})$ as desired.

(4) The discussion in (3), together with Fubini's Theorem, allows us to switch the integrations over H_2 and H_2/xH_xx^{-1} – thus we get

$$\begin{aligned}\langle f, T\rangle = \int_{G/H_2} d\mu_2(y)\rho_2(y)^{-1} \int_{H_2/xH_xx^{-1}} d\nu_2(\xi_2)\left(\frac{\delta_G(\xi_2)}{\delta_2(\xi_2)}\right)^{1/2} \\ \times \mathbf{z}(\pi^1(f_1)(y\xi_2 x) \otimes L^2(\xi_2^{-1})\pi^2(f_2)(y)).\end{aligned}$$

Since ν_2 is H_2-invariant, the modular functions of H_2 and xH_xx^{-1} coincide on xH_xx^{-1} – hence ρ_2 is also a rho-function on G for the subgroup xH_xx^{-1}, call it σ_x. Let ν_x denote the corresponding quasi-invariant measure on G/xH_xx^{-1}. Keeping in mind that

$$\rho_2(y\xi_2) = \rho_2(y)\left(\frac{\delta_2(\xi_2)}{\delta_G(\xi_2)}\right), \pi^2(f_2)(y\xi_2) = \rho_2(\xi_2)^{1/2}L^2(\xi_2^{-1})\,\pi^2(f_2)(y)$$
$$(\xi_2 \in H_2, y \in G),$$

an application of Lemma A 1.2 (Appendix 1) to the function

$$(\sigma_x(\,.\,))^{-1}\mathbf{z}(\pi^1(f_1)(\,.\,x) \otimes \pi^2(f_2)(\,.\,))$$

(which, as has been noted above, is an element of $C_c(G/xH_xx^{-1})$) now leads immediately to the sought for expression for $\langle f, T\rangle$.

The proof of Proposition 5.3.2.9 is complete. □

The following result is nothing more than an explication of the preceding proposition in a particular case.

Proposition 5.3.2.10 Let us suppose that the H_2, H_1 double coset Q containing x is closed and that, moreover, the homogeneous space H_2/xH_xx^{-1} carries a positive H_2-invariant measure ν_2. Let **B** be a U^{L^1}, U^{L^2} intertwining form, whose G-trace T has its support contained in Q, with transversal order zero there (i.e, T has transversal order zero relative to Q) – then there exists an element $\mathbf{z} \in (E^1 \hat{\otimes} E^2)^*$, which intertwines the representations

$$\zeta \mapsto L^1(\zeta) \hat{\otimes} L^2(x\zeta x^{-1}), \quad \zeta \mapsto \left(\frac{\delta_2(x\zeta x^{-1})}{\delta_1(\zeta)}\right)^{1/2} \qquad (\zeta \in H_x)$$

of H_x, such that

$$\mathbf{B}(f_1, f_2) = \int_{G/xH_xx^{-1}} \mathbf{z}(f_1(yx) \otimes f_2(y))\sigma_x(y)^{-1}\, d\nu_x(y)$$

for all $f_1 \in {}^{L^1}C^\infty(G; E^1)$, $f_2 \in {}^{L^2}C^\infty(G; E^2)$.

[It should, perhaps, be observed that $\delta_{xH_xx^{-1}} = \delta_2$ on xH_xx^{-1} (since H_2/xH_xx^{-1} admits an H_2-invariant measure).]

Here is another consequence of Proposition 5.3.2.9.

Proposition 5.3.2.11 Let us suppose that the H_2, H_1 double coset Q containing x is closed and that, moreover, the homogeneous space H_2/xH_xx^{-1} carries a positive H_2-invariant measure ν_2. Let L^i be a continuous representation of H_i on a Fréchet space E^i, L^i_∞ the associated differentiable representation of H_i on E^i_∞ ($i = 1, 2$). Let **z** be an element of $(E^1 \hat{\otimes} E^2)^*$ which intertwines the representations

$$\zeta \mapsto L^1(\zeta) \hat{\otimes} L^2(x\zeta x^{-1}), \quad \zeta \mapsto \left(\frac{\delta_2(x\zeta x^{-1})}{\delta_1(\zeta)}\right)^{1/2} \qquad (\zeta \in H_x)$$

of H_x; let S be the quasi-invariant $(E^1_\infty \hat{\otimes} E^2_\infty)$-distribution on Q with multiplier A corresponding, via Proposition 5.3.2.5, with the restriction of **z** to $E^1_\infty \hat{\otimes} E^2_\infty$, T the extension of S to an $E^1_\infty \hat{\otimes} E^2_\infty$-distribution on G (recall that T is also quasi-invariant with multiplier A) – then there is a unique intertwining form **B** for the continuously induced representations U^{L^1} and U^{L^2} of G with the property that its restriction to ${}^{L^1_\infty}C^\infty(G; E^1_\infty) \times {}^{L^2_\infty}C^\infty(G; E^2_\infty)$ has G-trace equal to T. In fact **B** is given by

$$\mathbf{B}(f_1, f_2) = \int_{G/xH_xx^{-1}} \mathbf{z}(f_1(yx) \otimes f_2(y))\sigma_x(y)^{-1}\, d\nu_x(y)$$

for all $f_1 \in {}^{L^1}C(G; E^1)$, $f_2 \in {}^{L^2}C(G; E^2)$.

Proof The same argument as was given at the beginning of the proof of Proposition 5.3.2.9 shows that the integrand appearing on the right above vanishes outside of $\omega(xH_x x^{-1})$, ω a certain compact subset of G. Since E^1 and E^2 are both Fréchet spaces, $\mathbf{z}$ is continuous on the product $E^1 \times E^2$ – therefore $\mathbf{B}$, as defined by the integral supra, is bilinear and separately continuous on ${}^{L^1}C(G; E^1) \times {}^{L^2}C(G; E^2)$. In addition $\mathbf{B}$ intertwines U^{L^1} and U^{L^2}; thus, if $f_1 \in {}^{L^1}C(G; E^1)$, $f_2 \in {}^{L^2}C(G; E^2)$, $z \in G$, then

$$\begin{aligned}\mathbf{B}(U^{L^1}(z)f_1, U^{L^2}(z)f_2) &= \int_{G/xH_x x^{-1}} \mathbf{z}(f_1(z^{-1}yx) \otimes f_2(z^{-1}y))\sigma_x(y)^{-1}\, d\nu_x(y)\\ &= \int_{G/xH_x x^{-1}} \mathbf{z}(f_1(yx) \otimes f_2(y))\sigma_x(zy)^{-1}\, d\nu_x(zy)\\ &= \mathbf{B}(f_1, f_2),\end{aligned}$$

the last equality following from the fact that $d\nu_x(zy) = (\sigma_x(zy)/\sigma_x(y))d\nu_x(y)$. Of course this shows that the restriction of $\mathbf{B}$ to ${}^{L^1\infty}C^\infty(G; E^1_\infty) \times {}^{L^2\infty}C^\infty(G; E^2_\infty)$ is $U^{L^1\infty}$, $U^{L^2\infty}$ intertwining; Proposition 5.3.2.9 then implies that T is its G-trace. □

Corollary 5.3.2.12 Under the assumptions of Proposition 5.3.2.11,

$$i(U^{L^1}, U^{L^2}) \geqslant \sum_Q i(L^1, L^2; Q, 0),$$

the sum $\sum_Q$ running over all closed H_2, H_1 double cosets Q such that $H_2/xH_x x^{-1}$ admits an H_2-invariant measure for some $x \in Q$ (and so for all $x \in Q$).

Before giving the proof of the promised Proposition 5.3.2.4 we shall derive one more technical fact.

Lemma 5.3.2.13 Let E^i be a Fréchet space ($i = 1, 2$), ω a closed subset of G; let $\mathscr{S}$ be the linear span of the set of all functions $f: G \to E^1 \hat{\otimes} E^2$, $f(z) = \int_G f_1(yz) \otimes f_2(y) d_G(y)$ ($z \in G$), where $f_1 \in C_c^\infty(G; E^1)$, $f_2 \in C_c^\infty(G; E^2)$ and $\operatorname{spt}(f_2)^{-1} \operatorname{spt}(f_1) \cap \omega = \varnothing$ – then $\mathscr{S}$ is a dense subset of $C_c^\infty(G - \omega; E^1 \hat{\otimes} E^2)$.

Proof Obviously any such f is a C^∞ function on G with support contained in $\operatorname{spt}(f_2)^{-1} \operatorname{spt}(f_1) \subset G - \omega$, whence $\mathscr{S} \subset C_c^\infty(G - \omega; E^1 \hat{\otimes} E^2)$. As for the density of $\mathscr{S}$ in $C_c^\infty(G - \omega; E^1 \hat{\otimes} E^2)$, we begin by observing that, in view of the existence of C^∞ partitions of unity, it will be enough to show that each $z_0 \in G - \omega$ admits a compact neighborhood $\mathcal{O} = \mathcal{O}(z_0)$ having the property that the closure of $\mathscr{S}$ in $C_c^\infty(G - \omega; E^1 \hat{\otimes} E^2)$ contains $C_{\mathcal{O}}^\infty(G; E^1 \hat{\otimes} E^2)$. So let $\mathcal{O}_1$ and $\mathcal{O}_2$ be compact neighborhoods of z_0 and 1, respectively, such that $\mathcal{O}_2^{-1}\mathcal{O}_1 \cap \omega = \varnothing$; let ϕ denote the manifold automorphism of $G \times G$ given by $(x_1, x_2) \mapsto (x_2^{-1}x_1, x_2)$ – then $\phi(\mathcal{O}_1 \times \mathcal{O}_2)$ is a compact neighborhood of $\phi(z_0, 1) = (z_0, 1)$. Choose compact neighborhoods $\mathscr{P}_1$ and $\mathscr{P}_2$ of z_0 and 1, respectively, such that $\mathscr{P}_1 \times \mathscr{P}_2 \subset \phi(\mathcal{O}_1 \times \mathcal{O}_2)$ – we shall now prove that the closure of $\mathscr{S}$ in $C_c^\infty(G - \omega; E^1 \hat{\otimes} E^2)$ contains $C_{\mathscr{P}_1}^\infty(G; E^1 \hat{\otimes} E^2)$. To this end, let

$$I: C_c^\infty(G \times G; E^1 \hat{\otimes} E^2) \to C_c^\infty(G; E^1 \hat{\otimes} E^2)$$

be the (continuous) map defined by

$$I_h(z) = \int_G h(z, y)\, d_G(y) \qquad (h \in C_c^\infty(G \times G; E^1 \hat{\otimes} E^2)).$$

Plainly

$$C_{\mathscr{O}_1}^\infty(G; E^1 \hat{\otimes} E^2) \subset I(C_{\phi(\mathcal{O}_1 \times \mathcal{O}_2)}^\infty(G \times G; E^1 \hat{\otimes} E^2)) \subset C_{\mathcal{O}_2^{-1}\mathcal{O}_1}^\infty(G; E^1 \hat{\otimes} E^2).$$

Because $C_{\mathcal{O}_1}^\infty(G; E^1) \otimes_C C_{\mathcal{O}_2}^\infty(G; E^2)$ is dense in $C_{\mathcal{O}_1 \times \mathcal{O}_2}^\infty(G \times G; E^1 \hat{\otimes} E^2)$,

$$\phi(C_{\mathcal{O}_1}^\infty(G; E^1) \otimes_C C_{\mathcal{O}_2}^\infty(G; E^2))$$

is dense in $C_{\phi(\mathcal{O}_1 \times \mathcal{O}_2)}^\infty(G \times G; E^1 \hat{\otimes} E^2)$. Furthermore, if $f_i \in C_{\mathcal{O}_i}^\infty(G; E^i)$ $(i = 1, 2)$, then

$$(I \circ \phi)(f_1 \otimes f_2)(z) = \int_G f_1(yz) \otimes f_2(y)\, d_G(y).$$

Hence

$$I(\phi[C_{\mathcal{O}_1}^\infty(G; E^1) \otimes_C C_{\mathcal{O}_2}^\infty(G; E^2)]) \subset \mathscr{S} \cap C_{\mathcal{O}_2^{-1}\mathcal{O}_1}^\infty(G; E^1 \hat{\otimes} E^2).$$

The continuity of I now leads at once to the contention stated above. □

Proof of Proposition 5.3.2.4 Fix $x \in Q \in \Phi$; our hypotheses entail that $\delta_x (= \delta_{H_x}) = \delta_1$ on $H_x = H_1 \cap (x^{-1}H_2x)$ and that $\delta_{xH_xx^{-1}} = \delta_2$ on $xH_xx^{-1} = (xH_1x^{-1}) \cap H_2$, whence the representation A_0 of H_x is, in the case at hand, the trivial one dimensional representation of H_x (for the definition of A_0, see the discussion preceding Theorem 5.3.2.3). This means that we shall want to consider the space Z_Q comprised of those $\mathbf{z}$ in the dual of $E^1 \hat{\otimes} E^2$ which verify the intertwining condition

$$^t(L^1(\zeta) \hat{\otimes} L^2(x\zeta x^{-1}))\mathbf{z} = \mathbf{z} \qquad (\zeta \in H_x).$$

As will be seen below (cf. (3)), to each such $\mathbf{z}$ one may assign, in a natural fashion, a U^{L^1}, U^{L^2} intertwining form $\mathbf{B}_{\mathbf{z}}$; with the restriction of $\mathbf{B}_{\mathbf{z}}$ to ${}^{L^1\infty}C^\infty(G; E_\infty^1) \times {}^{L^2\infty}C^\infty(G; E_\infty^2)$ there is associated its G-trace $T_{\mathbf{z}}$. A key step in the argument is to show that the (linear) map $\mathbf{z} \mapsto T_{\mathbf{z}}$ is one-to-one on Z_Q (see (4)). It being evident that the correspondence $\mathbf{B}_{\mathbf{z}} \mapsto T_{\mathbf{z}}$ is one-to-one, the proof of the proposition is then completed by verifying that the various linear spaces $\Sigma_Q = \{T_{\mathbf{z}} : \mathbf{z} \in Z_Q\}$ $(Q \in \Phi)$ are themselves linearly independent (cf.(5)).

(1) Let ρ_1 be a rho-function on G with respect to H_1, μ_1 the corresponding quasi-invariant measure on G/H_1; since δ_1 and δ_{H_x} coincide on H_x, ρ_1 is also a rho-function on G for the subgroup H_x – call it ρ_x. Let μ_x denote the corresponding quasi-invariant measure on G/H_x. On the other hand, let ρ_2 be a rho-function on G with respect to H_2, μ_2 the corresponding quasi-invariant measure on G/H_2; since δ_2 and $\delta_{xH_xx^{-1}}$ coincide on xH_xx^{-1}, ρ_2 is also a rho-function on G for the subgroup xH_xx^{-1} – call it σ_x. Let ν_x denote the corresponding quasi-invariant measure on G/xH_xx^{-1} (cf. part (4) of the proof of Proposition 5.3.2.9). The inner automorphism $y \mapsto x^{-1}yx$ carries G/xH_xx^{-1} onto G/H_x; the image of ν_x under this map is a quasi-invariant measure on G/H_x, namely $(\rho_2(x.x^{-1})/\rho_1(.))d\mu_x(.)$.

(2) Let $\mathbf{z} \in Z_Q, f_i \in {}^{L^i\infty}C^\infty(G; E^i_\infty)$ $(i = 1, 2)$ – our immediate objective is to investigate the summability of $\mathbf{z}(f_1(.x) \otimes f_2(.))\sigma_x(.)^{-1}$ with respect to ν_x. Let k be a positive constant having the property that $|\mathbf{z}(a_1, a_2)| \leqslant k\|a_1\|\|a_2\|$ $(a_i \in E^i$ $(i = 1, 2)$, $\|.\|$ the norm in the Hilbert space under consideration). Owing to the Schwarz inequality, then, the integral

$$\int_{G/xH_zx^{-1}} |\mathbf{z}(f_1(yx) \otimes f_2(y))|\sigma_x(y)^{-1}\, d\nu_x(y)$$

is majorized by

$$k\left[\int_{G/xH_zx^{-1}} \|f_1(yx)\|^2\sigma_x(y)^{-1}\, d\nu_x(y)\right]^{1/2} \times \left[\int_{G/xH_zx^{-1}} \|f_2(y)\|^2\sigma_x(y)^{-1}\, d\nu_x(y)\right]^{1/2}$$

The second integral appearing here can be transformed, via Lemma A 1.2 (Appendix 1), to read

$$\begin{aligned}
&\int_{G/xH_zx^{-1}} \|f_2(y)\|^2\sigma_x(y)^{-1}\, d\nu_x(y)\\
&= \int_{G/H_2} d\mu_2(y) \int_{H_2/xH_zx^{-1}} \|f_2(y\xi_2)\|^2\rho_2(y\xi_2)^{-1}\, d\nu_2(\xi_2)\\
&= \nu_2(H_2/xH_xx^{-1}) \int_{G/H_2} \|f_2(y)\|^2\rho_2(y)^{-1}\, d\mu_2(y)\\
&= \nu_2(H_2/xH_xx^{-1})\|f_2\|^2,
\end{aligned}$$

the last equality following from the definition of the norm in the representation space E^{L^2} for U^{L^2}. Bearing in mind the facts mentioned in (1) and using Lemma A 1.2 once again, the first integral can be written

$$\begin{aligned}
&\int_{G/xH_zx^{-1}} \|f_1(yx)\|^2\sigma_x(y)^{-1}\, d\nu_x(y)\\
&= \int_{G/H_z} \|f_1(xy)\|^2\rho_x(y)^{-1}\, d\mu_x(y)\\
&= \int_{G/H_1} d\mu_1(y) \int_{H_1/H_z} \|f_1(xy\xi_1)\|^2\rho_1(y\xi_1)^{-1}\, d\nu_1(\xi_1)\\
&= \nu_1(H_1/H_x)\|f_1\|^2.
\end{aligned}$$

Putting these remarks together, we finally obtain the estimate

$$\begin{aligned}
&\int_{G/xH_zx^{-1}} |\mathbf{z}(f_1(yx) \otimes f_2(y))|\sigma_x(y)^{-1}\, d\nu_x(y)\\
&\leqslant k\nu_1(H_1/H_x)^{1/2}\nu_2(H_2/xH_xx^{-1})^{1/2}\|f_1\|\|f_2\| < \infty.
\end{aligned}$$

[Let us bear in mind that ν_1 and ν_2 have finite total mass.]

(3) According to the calculations in (2), the bilinear form $\mathbf{B}_z$ on ${}^{L^1\infty}C^\infty(G; E^1_\infty) \times {}^{L^2\infty}C^\infty(G; E^2_\infty)$ defined by

$$\mathbf{B}_z(f_1, f_2) = \int_{G/xH_zx^{-1}} \mathbf{z}(f_1(yx) \otimes f_2(y))\sigma_x(y)^{-1}\, d\nu_x(y)$$

extends uniquely to a continuous bilinear form (which will still be called $\mathbf{B}_\mathbf{z}$) on $E^{L^1} \times E^{L^2}$. An argument given during the proof of Proposition 5.3.2.11 shows that $\mathbf{B}_\mathbf{z}$ is U^{L^1}, U^{L^2} intertwining. Thus, to each $\mathbf{z} \in Z_Q$, one may assign, in a natural way, a U^{L^1}, U^{L^2} intertwining form $\mathbf{B}_\mathbf{z}$. We shall now investigate the G-trace $T_\mathbf{z}$ corresponding to the restriction of $\mathbf{B}_\mathbf{z}$ to ${}^{L^1\infty}C^\infty(G; E^1_\infty) \times {}^{L^2\infty}C^\infty(G; E^2_\infty)$.

(4) Let S be the A-quasi-invariant $E^1_\infty \hat{\otimes} E^2_\infty$-distribution on Q associated with $\mathbf{z}$ per Proposition 5.3.2.5, T the extension of S to the (open) set $\mathcal{O} = Q \cup (G - Q^{cl})$ – then $T = T_\mathbf{z}$ on $\mathcal{O}$. To see this, it is sufficient, thanks to the preceding lemma, to verify that T and $T_\mathbf{z}$ coincide on functions $f: G \to E^1_\infty \hat{\otimes} E^2_\infty$ of the form

$$f(z) = \int_G f_1(yz) \otimes f_2(y)\, d_G(y) \qquad (z \in G)$$

where $f_i \in C_c^\infty(G; E^i_\infty)$ $(i = 1, 2)$ and $\operatorname{spt}(f_2)^{-1} \operatorname{spt}(f_1) \cap (Q^{cl} - Q) = \varnothing$. But, by definition of the G-trace, $\mathbf{B}_\mathbf{z}(\pi^1(f_1), \pi^2(f_2)) = \langle f, T_\mathbf{z}\rangle$; on the other hand, Proposition 5.3.2.9 tells us that

$$\begin{aligned}\langle f, T\rangle &= \int_{G/xH_\mathbf{z}x^{-1}} \mathbf{z}(\pi^1(f_1)(yx) \otimes \pi^2(f_2)(y))\sigma_x(y)^{-1}\, d\nu_x(y)\\ &= \mathbf{B}_\mathbf{z}(\pi^1(f_1), \pi^2(f_2)),\end{aligned}$$

and so T and $T_\mathbf{z}$ do indeed coincide on $\mathcal{O}$. Consequently, the map $\mathbf{z} \mapsto T_\mathbf{z}$ is one-to-one on Z_Q.

(5) (Fell) To complete the proof of our proposition, it will be enough to verify, in view of the last sentence in (4) above, that the various linear spaces $\Sigma_Q = \{T_\mathbf{z} : \mathbf{z} \in Z_Q\}$ $(Q \in \Phi)$ are linearly independent. This being so, let us now agree to use the notations centering around Lemma 5.2.4.4. [The roles of G and M there being played, in the present context, by $H_2 \times H_1$ and G; the space $\mathbf{T}$ by the $\mathbf{T}$ of Theorem 5.3.2.1.] The discussion in (4), combined with the fact that the support of T is contained in Q, implies that the support of $T_\mathbf{z}$ is contained in Q^{cl}; therefore if the H_2, H_1 double coset Q is associated with the ordinal α, then $T_\mathbf{z}$ has its support contained in S_α and so $T_\mathbf{z} \in \mathbf{T}_\alpha$. In addition, (4) shows that if Q' is another H_2, H_1 double coset, distinct from Q but associated with the same ordinal α, then $T_\mathbf{z}$ restricted to $\Omega_{Q'} = Q' \cup (G - S_\alpha)$ vanishes identically. Now suppose that $Q_1, \ldots, Q_r$ are pairwise distinct elements of Φ and that $\sum_1^r T_{\mathbf{z}_i} = 0$, the $\mathbf{z}_i$ being *non-zero* elements of Z_{Q_i}. Let Q_i be associated with the ordinal α_i where $\alpha_1 \leqslant \alpha_2 \leqslant \cdots \leqslant \alpha_r, \alpha_1 = \alpha_2 = \cdots = \alpha_s$, $\alpha_s < \alpha_{s+1}$(say) – then $T_{\mathbf{z}_i} \in \mathbf{T}_{\alpha_i}$ for all i. Let $f = f_{\alpha_1}$ be the map defined in part (1) of the proof of Lemma 5.2.4.4. Since $S_{\alpha_i} \cap \Omega_{Q_j} = \varnothing$ for $i > s$, $j \leqslant s$, $f(T_{\mathbf{z}_i}) = 0$ $(i > s)$; hence $\sum_1^r T_{\mathbf{z}_i} = 0 \Rightarrow \sum_1^s f(T_{\mathbf{z}_j}) = 0$. If Q' is any H_2, H_1 double coset associated with α_1 and if $j = 1, \ldots, s$, then, as has been observed above, $f(T_{\mathbf{z}_j})_{Q'} = 0$ unless $Q_j = Q'$, while $f(T_{\mathbf{z}_j})_{Q_j} = T_j$ on Ω_{Q_j} (T_j the extension of S_j to $\mathcal{O}_j$, S_j the $E^1_\infty \hat{\otimes} E^2_\infty$-distribution connected with $\mathbf{z}_j$ via Proposition 5.3.2.5). We see, therefore, that the relation $\sum_1^s f(T_{\mathbf{z}_j}) = 0$ necessarily entails $T_j = 0$ $(j = 1, \ldots, s)$;

this, in turn leads to the conclusion that $\mathbf{z}_j = 0$ $(j = 1, \ldots, s)$, which contradicts our original supposition on the $\mathbf{z}_j$.

This completes the proof of Proposition 5.3.2.4. ☐

Theorem 5.3.2.14 (Bruhat) Let H be a closed subgroup of G having the property that there exist in G at most countably many H, H double cosets; let L^i be a continuous representation of H on a Fréchet space E^i $(i = 1, 2)$. If for every H, H double coset Q distinct from H and for every non-negative integer n,

$$i(L_\infty^1, L_\infty^2; Q, n) = 0,$$

then

$$i(U^{L^1}, U^{L^2}) = i(L^1, L^2).$$

[Here U^{L^i} is the representation of G continuously induced by L^i $(i = 1, 2)$.]

Proof On the basis of the hypotheses of our theorem, the inequality $i(L^1, L^2) \leqslant i(U^{L^1}, U^{L^2})$ follows from Corollary 5.3.2.12 (take $Q = H$, $x = 1$) – hence our main task is to establish the reverse inequality $i(U^{L^1}, U^{L^2}) \leqslant i(L^1, L^2)$.

(1) Let $\mathbf{B}$ be a U^{L^1}, U^{L^2} intertwining form, T the G-trace of its restriction to ${}^{L^1\infty}C^\infty(G; E_\infty^1) \times {}^{L^2\infty}C^\infty(G; E_\infty^2)$ – then the support of T is contained in H. [Introduce the space $\mathbf{T}$ per Theorem 5.3.2.1; with each H, H double coset Q one may associate, via Lemma 5.2.4.4, a space $\mathbf{T}[Q]$; but here, thanks to Theorem 5.3.2.3 and supporting discussion, $\mathbf{T}[Q] = \{0\}$ for all H, H double cosets $Q \neq H$. Thus, if $Q \neq H$, then T vanishes on some open subset Ω_Q of G containing Q, and so T vanishes on $\bigcup_{Q \neq H} \Omega_Q \supset G - H$, which shows that the support of T is contained in H.] For $f_i \in C_c^\infty(G; E_\infty^i)$ $(i = 1, 2)$, we have, of course, the relation

$$(*) \quad \mathbf{B}(\pi^1(f_1), \pi^2(f_2)) = \int_G d_G(x_2) \int_G f_1(x_2 x_1) \otimes f_2(x_2)\, dT(x_1).$$

On the other hand, as we know, the map π^i may actually be defined on $C_c(G; E^i)$ $(i = 1, 2)$; since $\pi^i : C_c(G; E^i) \to {}^{L^i}C(G; E^i)$ $(i = 1, 2)$ is continuous and since $\mathbf{B}$ is separately continuous on ${}^{L^1}C(G; E^1) \times {}^{L^2}C(G; E^2)$, we see that the left hand side of $(*)$ above extends to a separately continuous bilinear form on $C_c(G; E^1) \times C_c(G; E^2)$. Lemma 2 in the Appendix to this number then tells us that T has transversal order zero relative to H (and so is the extension to G of an $E^1 \hat{\otimes} E^2$-distribution on H). Consequently (cf. Proposition 5.3.2.10) there exists an L_∞^1, L_∞^2 intertwining form $\mathbf{z}$ such that

$$\mathbf{B}(\pi^1(f_1), \pi^2(f_2)) = \int_{G/H} \mathbf{z}(\pi^1(f_1)(y), \pi^2(f_2)(y)) \rho_H(y)^{-1}\, d\mu_H(y)$$

for all $f_i \in C_c^\infty(G; E_\infty^i)$ $(i = 1, 2)$. Now the form $\mathbf{z}$ is, a priori, only defined on $E_\infty^1 \times E_\infty^2$; however, it will be seen in (2) below that $\mathbf{z}$ may be extended to a separately continuous bilinear form (call it $\mathbf{z}$ still) on $E^1 \times E^2$ and that, moreover, this extension intertwines L^1 and L^2. Because the map $\mathbf{z} \mapsto \mathbf{B}$ which assigns to each such extension $\mathbf{z}$ the corresponding U^{L^1}, U^{L^2} intertwining form $\mathbf{B}$ is well-defined, linear, and onto the space of all such $\mathbf{B}$, it follows that $i(U^{L^1}, U^{L^2}) \leqslant i(L^1, L^2)$, which serves to complete the proof of the present theorem.

(2) To verify that the $\mathbf{z}$ in (1) is necessarily the restriction to $E^1_\infty \times E^2_\infty$ of an L^1, L^2 intertwining form on $E^1 \times E^2$, one may proceed in the following way. Let $\dot{y} \mapsto s(\dot{y})$ be a C^∞ section of G fibered by H in a small neighborhood $\mathscr{P}$ of $\dot{1}$ in G/H. Fix a real valued element $\phi \in C_c^\infty(\mathscr{P})$; let $a_i \in E^i_\infty$ $(i = 1, 2)$; put

$$f_{a_i}(s(\dot{y})\xi) = \rho_H(\xi)^{1/2}\phi(\dot{y})(L^i(\xi^{-1})a_i) \qquad (\dot{y} \in \mathscr{P}, \xi \in H)$$

and take f_{a_i} to be zero outside of $s(\mathscr{P})H$ $(i = 1, 2)$ – then clearly

$$f_{a_i} \in {}^{L^i\infty}C^\infty(G; E^i_\infty) \qquad (i = 1, 2).$$

The formula for $\mathbf{B}$ obtained in (1) tells us that

$$\begin{aligned}\mathbf{B}(f_{a_1}, f_{a_2}) &= \int_{G/H} \mathbf{z}(f_{a_1}(y), f_{a_2}(y))\rho_H(y)^{-1}\, d\mu_H(y)\\ &= \mathbf{z}(a_1, a_2)\int_{\mathscr{P}} \phi(\dot{y})^2\rho_H(s(\dot{y}))^{-1}\, d\mu_H(\dot{y}).\end{aligned}$$

Let us suppose that ϕ has been chosen in such a way as to guarantee that

$$\int_{\mathscr{P}} \phi(\dot{y})^2\rho_H(s(\dot{y}))^{-1}\, d\mu_H(\dot{y}) \neq 0.$$

Because the map $a_i \mapsto f_{a_i}$ of E^i_∞ into ${}^{L^i\infty}C^\infty(G; E^i_\infty)$ extends to a continuous map of E^i into ${}^{L^i}C(G; E^i)$ $(i = 1, 2)$, due to the separate continuity of $\mathbf{B}$ on ${}^{L^1}C(G; E^1) \times {}^{L^2}C(G; E^2)$, it is now clear that $\mathbf{z}$ extends to a separately continuous bilinear form on $E^1 \times E^2$ which is surely L^1, L^2 intertwining.

Hence the theorem. □

Here is the main result of the present number.

Theorem 5.3.2.15 (Bruhat) Let H be a closed subgroup of G having the property that there exist in G at most countably many H, H double cosets; let L be an irreducible unitary representation of H on a Hilbert space E. If for every H, H double coset Q distinct from H and for every non-negative integer n,

$$i(L_\infty, \bar{L}_\infty; Q, n) = 0,$$

then the unitarily induced representation U^L of G is irreducible.

Proof We must prove that $i(U^L, \overline{U^L}) = i(U^L, U^{\bar{L}}) = 1$; but this is evident since $1 \leqslant i(U^L, \overline{U^L}) = i(U^L, U^{\bar{L}}) \leqslant i(L, \bar{L})$ (Theorem 5.3.2.14) $=$ 1 (by assumption). □

The Remark given after the statement of Theorem 5.3.2.3 shows that one may replace the condition 'for every integer $n \geqslant 0$' in Theorem 5.3.2.15 by 'for every integer n, $0 \leqslant n \leqslant 2 \dim (G) + 1$'.

Remarks The preceding theorem may be viewed as the analogue, for infinite groups, of the sufficiency assertion in the Mackey-Shoda criterion (Introduction, Corollary 1) for the irreducibility of unitarily induced representations in the case of finite groups (provided, of course, that G contains at most countably many H, H double cosets). If we give up the assumption that there exist in G at most countably many H, H double cosets (but impose instead alternative requirements on H), then it is still possible to obtain results along this line – cf. infra.

Let H be a closed subgroup of G.

(1) Let L be an irreducible unitary representation of H, U^L the corresponding unitarily induced representation of G; assume that there are in G but a finite number of left H cosets – then U^L is irreducible iff for every $x \in G - H$ the unitary representations $\zeta \mapsto L(\zeta)$, $\zeta \mapsto L(x\zeta x^{-1})$ of the subgroup $H_x = H \cap (x^{-1}Hx)$ are disjoint (cf. Mackey [9, p. 643]). [Notice that the assertion in question makes sense no matter how many left H cosets there are; unfortunately it is not true in general.]

(2) Let L be an irreducible unitary representation of H, U^L the corresponding unitarily induced representation of G; assume that H is normal in G and is a type I group – then U^L is irreducible iff for every $x \in G - H$, the unitary representations $\zeta \mapsto L(\zeta)$, $\zeta \mapsto L(x\zeta x^{-1})$ of H are disjoint (cf. Mackey, 'Induced representations and normal subgroups', Proceedings Symposium on Linear Spaces, Jerusalem, 1961). [This result should be compared with Corollary 2 in the Introduction.]

(3) Let L be a finite dimensional irreducible unitary representation of H, U^L the corresponding unitarily induced representation of G; assume that H is normal in G – then U^L is irreducible iff for every $x \in G - H$, the unitary representations $\zeta \mapsto L(\zeta)$, $\zeta \mapsto L(x\zeta x^{-1})$ of H are disjoint (cf. Bruhat [1, p. 166]). [The point here is, of course, that H need not be a type I group.] Furthermore it can be shown that two such unitary representations U^{L^1} and U^{L^2} are unitarily equivalent iff there exists an x in G such that the unitary representations $\zeta \mapsto L^1(\zeta)$, $\zeta \mapsto L^2(x\zeta x^{-1})$ of H are not disjoint (cf. with Corollary 4 in the Introduction).

(4) Let us suppose that G is a connected, simply connected nilpotent Lie group, H a closed connected subgroup of G; let L be a unitary character of H – then the unitarily induced representation U^L of G is irreducible iff the representations $\zeta \mapsto L(\zeta)$, $\zeta \mapsto L(x\zeta x^{-1})$ of $H \cap (x^{-1}Hx)$ are distinct for every $x \in G - H$ (cf. Pukanszky [2, p. 157]).

Let H be a closed subgroup of G with the property that there exists in G at most countably many H, H double cosets; let L be an irreducible unitary representation of H on a Hilbert space E – then Theorem 5.3.2.15 provides us with a condition sufficient to ensure the irreducibility of the unitarily induced representation U^L of G. In turn, it is clear, with a view toward the applications, that it would be desirable to have at hand a criterion which would guarantee us that $i(L_\infty, \bar{L}_\infty; Q, n) = 0$ in any particular case – along these lines, it is possible to offer the following observations.

Let us agree to return to the notations revolving around Theorem 5.3.2.3; thus we shall now be dealing with a differentiable representation L^i of the closed subgroup H_i on a Fréchet space E^i $(i = 1, 2)$. Given

$x \in G$, define the representations $\zeta \mapsto A_n(\zeta)$ of $H_x = H_1 \cap (x^{-1}H_2x)$ as there ($n \in \mathbf{N}$).

Lemma 5.3.2.16 Let $x \in G$, $Q = H_2xH_1$; let n be a non-negative integer. If the representation A_n has no irreducible component which is equivalent to a subquotient of the representation $\zeta \mapsto L^1(\zeta) \hat{\otimes} L^2(x\zeta x^{-1})$ of H_x on $E^1 \hat{\otimes} E^2$, then $i(L^1, L^2; Q, n) = 0$.

Proof Let $\mathbf{z}$ be a non-zero element in $(E^1 \hat{\otimes} E^2 \hat{\otimes} \mathbf{C}^s)^*$ verifying the intertwining condition

$$^t(L^1(\zeta) \hat{\otimes} L^2(x\zeta x^{-1}) \hat{\otimes} A_n^\vee(\zeta))\mathbf{z} = \mathbf{z} \qquad (\zeta \in H_x)$$

(cf. the discussion preceding Theorem 5.2.3.1). Alternatively, such a $\mathbf{z}$ may be viewed as a continuous linear map Z of $E^1 \hat{\otimes} E^2$ into $\mathbf{C}^s$ verifying for every $\zeta \in H_x : Z\circ(L^1(\zeta) \hat{\otimes} L^2(x\zeta x^{-1})) = A_n(\zeta)\circ Z$. Let I be the smallest A_n-stable subspace of $\mathbf{C}^s$ containing $Z(E^1 \hat{\otimes} E^2)$ (i.e. let $I = Z(E^1 \hat{\otimes} E^2)$), J a maximal A_n-stable subspace of I. The corresponding representation, say $\tilde{A}_n$, of H_x on I/J is an irreducible component of A_n. Let $\tilde{Z}$ denote the composition of Z with the canonical projection of I onto I/J – then $\tilde{Z}$ is non-zero and

$$\tilde{Z}\circ(L^1(\zeta) \hat{\otimes} L^2(x\zeta x^{-1})) = \tilde{A}_n(\zeta)\circ\tilde{Z} \quad (\zeta \in H_x).$$

Because $\tilde{A}_n$ is irreducible, $\tilde{Z}$ is surjective. The assertion of the lemma now follows at once. □

Corollary 5.3.2.17 Let $x \in G$, $Q = H_2xH_1$; let n be a non-negative integer. If the representations $\zeta \mapsto A_n(\zeta)$, $\zeta \mapsto L^1(\zeta) \hat{\otimes} L^2(x\zeta x^{-1})$ of H_x have no common irreducible components, then $i(L^1, L^2; Q, n) = 0$.

Corollary 5.3.2.18 Let $x \in G$, $Q = H_2xH_1$; let n be a non-negative integer. If L^i $(i = 1, 2)$ is one dimensional and if the representation $\zeta \mapsto L^1(\zeta)L^2(x\zeta x^{-1})$ of H_x is not contained in A_n, then $i(L^1, L^2; Q, n) = 0$.

Let us suppose that H_1 is open in G – then $H_x = H_1 \cap (x^{-1}H_2x)$ is open in $x^{-1}H_2x$. This means that $\delta_1 = \delta_G$ on H_1, $\delta_{H_x} = \delta_{x^{-1}H_2x}$ on H_x. If our representations L^1 and L^2 are, in addition, finite dimensional and unitary, then so is the representation $\zeta \mapsto L^1(\zeta) \hat{\otimes} L^2(x\zeta x^{-1})$ of H_x. Of course the irreducible components of the representation $\zeta \mapsto L^1(\zeta) \hat{\otimes} L^2(x\zeta x^{-1})$ are, in this case, also unitary. On the other hand, in view of what was said above about the modular functions,

$$A_0(\zeta) = \left(\frac{\delta_G(\zeta)}{\delta_2(x\zeta x^{-1})}\right)^{1/2} \qquad (\zeta \in H_x);$$

furthermore, A_0 cannot be equivalent to a unitary representation unless

$$\delta_G(\zeta) = \delta_2(x\zeta x^{-1}) \qquad (\zeta \in H_x).$$

Hence $i(L^1, L^2; Q, 0) = 0$ unless $\delta_G = \delta_2$ on $xH_xx^{-1} = (xH_1x^{-1}) \cap H_2$.

The preceding discussion shows that Theorem 5.3.2.15 may be restated as follows.

Theorem 5.3.2.19 Let H be a closed subgroup of G having the property that there exist in G at most countably many H, H double cosets; let L be an irreducible unitary representation of H on a Hilbert space E. If, for every $x \in G$ not in H and for every non-negative integer n, the representations

$$\zeta \mapsto A_n(\zeta), \zeta \mapsto L_\infty(\zeta) \hat{\otimes} \bar{L}_\infty(x\zeta x^{-1})$$

of $H_x = H \cap (x^{-1}Hx)$ have no common irreducible components, then the unitarily induced representation U^L is irreducible.

On the basis of Proposition 5.3.2.4, it is clear that Theorem 5.3.2.15 admits a partial converse, namely:

Proposition 5.3.2.20 Let H be as in Theorem 5.3.2.15; let L be an irreducible unitary representation of H on a Hilbert space E. If there exists an $x \in G - H$ having the property that $H/H \cap (x^{-1}Hx)$, $H/H \cap (xHx^{-1})$ both carry finite H-invariant measures and such that the representations $\zeta \mapsto L(\zeta)$, $\zeta \mapsto L(x\zeta x^{-1})$ of $H_x = H \cap (x^{-1}Hx)$ are not disjoint, then the unitarily induced representation U^L of G is not irreducible.

[Examples show that the mere non-disjointness of the representations $\zeta \mapsto L(\zeta)$, $\zeta \mapsto L(x\zeta x^{-1})$ is not sufficient to ensure the validity of Proposition 5.3.2.20.]

In passing, let us record the following simple result.

Proposition 5.3.2.21 Let H_1 and H_2 be closed subgroups of G per Theorem 5.3.2.3; let L^i be a unitary representation of H_i on a Hilbert space E^i ($i = 1, 2$). If for every double coset Q and every integer $n \geqslant 0$, $i(L^1_\infty, \overline{L^2_\infty}; Q, n) = 0$, then the unitarily induced representations U^{L^1} and U^{L^2} of G are disjoint.

Proof We must prove that $i(U^{L^1}, \overline{U^{L^2}}) = 0$; but $i(U^{L^1}, U^{\overline{L^2}}) \leqslant i(U^{L^1_\infty}, U^{\overline{L^2_\infty}}) \leqslant \sum_Q \sum_n i(L^1_\infty, \overline{L^2_\infty}; Q, n)$ (Theorem 5.3.2.3) $= 0$ (by hypothesis). □

Remark Suppose, in the preceding proposition, that we drop the assumption that there exists in G at most countably many H_2, H_1 double cosets; let us assume instead that H_1 and H_2 are compact. In this case Blattner [1, Th. 4] has obtained the following disjointness criterion. Let L^i be a finite dimensional unitary representation of H_i on a Hilbert space E^i ($i = 1, 2$); if, for almost all x in G, the unitary representations $\zeta \mapsto L^1(\zeta)$, $\zeta \mapsto L^2(x\zeta x^{-1})$ of $H_1 \cap (x^{-1}H_2x)$ are disjoint, then the unitarily induced representations U^{L^1} and U^{L^2} of G are disjoint (cf. Corollary 3 in the Introduction).

Appendix Our objective is to prove Lemma 2 below; we begin with a technical observation.

Lemma 1 Let r be a positive integer, $\mathcal{O}$ a non-empty open subset of $\mathbf{R}^r$; let $t_1, \ldots, t_r$ be the coordinates in $\mathbf{R}^r$,

$$D_i = \frac{\partial}{\partial t_i}, \quad D^I = D_1^{i_1} \cdots D_r^{i_r} \qquad (|I| = i_1 + \cdots + i_r);$$

let

$$D = \sum_{|I|<n} a_I D^I \qquad (a_I \in C^\infty(\mathcal{O})).$$

If, for every continuous complex valued function f on $\mathcal{O}$, Df (taken in the sense of distributions) is a measure, then D is of order zero (i.e. $a_I = 0$ for $|I| > 0$).

Proof (Fell) Proceeding by contradiction, assume that $a_I \neq 0$ for some $|I| > 0$; without loss of generality, it may be supposed that $i_1 > 0$ for some I such that $a_I \neq 0$. Let ν be the largest value of i_1 among those I possessing the property that $a_I \neq 0$ (hence $\nu > 0$); for definiteness, assume that $a_J \neq 0$, $j_1 = \nu$. Shrinking $\mathcal{O}$ if necessary, we can suppose that a_J never vanishes on $\mathcal{O}$. Now choose a function g of a real variable t such that $g, dg/dt, \ldots, d^{\nu-1}g/dt^{\nu-1}$ are all continuous, but such that $d^{\nu-1}g/dt^{\nu-1}$ is not of bounded variation in any real interval, so that $d^\nu g/dt^\nu$ (in the sense of distributions) is not a complex measure in any interval. Fix a point $t^0 = (t_1^0, \ldots, t_r^0)$ in $\mathcal{O}$; let h be a C^∞ function of $r-1$ real variables $t_2, \ldots, t_r$ having the property that for those $I \neq J$ with $a_I \neq 0$, $i_1 = \nu$

$$(D_2^{i_2} \cdots D_r^{i_r} h)(t_2^0, \ldots, t_r^0) = 0,$$

but such that

$$(D_2^{j_2} \cdots D_r^{j_r} h)(t_2^0, \ldots, t_r^0) = 1.$$

Put

$$f(t_1, \ldots, t_r) = g(t_1)h(t_2, \ldots, t_r).$$

Evidently $\sum_{|I|, i_1<\nu} a_I D^I f$ is a continuous function (hence a measure); by hypothesis $\sum_{|I|<n} a_I D^I f$ is a measure – therefore $\phi = \sum_{|I|, i_1=\nu} a_I D^I f$ is a measure. We can write

$$\begin{aligned}\phi &= \sum_{|I|, i_1=\nu} a_I (D_1^\nu g \otimes D_2^{i_2} \cdots D_r^{i_r} h) \\ &= \Phi(D_1^\nu g \otimes 1),\end{aligned}$$

Φ being the C^∞ function

$$\sum_{|I|, i_1=\nu} a_I (1 \otimes D_2^{i_2} \cdots D_r^{i_r} h).$$

The very definition of h ensures us that $\Phi(t_2^0, \ldots, t_r^0) \neq 0$; shrinking $\mathcal{O}$ further if need be, we can assume that Φ never vanishes on $\mathcal{O}$. Now ϕ is a measure on $\mathcal{O}$; hence so is $\psi = D_1^\nu g \otimes 1$. An elementary argument then tells us that $D_1^\nu g$ is a measure, in contradiction with our supposition on g. $\square$

Let H be a closed subgroup of the Lie group G; let E^i be a Fréchet space ($i = 1, 2$); let T be an $E^1 \hat{\otimes} E^2$-distribution on G, $\mathbf{B}^t$ the separately continuous bilinear form on $C_c^\infty(G; E^1) \times C_c^\infty(G; E^2)$ defined by

$$\mathbf{B}^t(f_1, f_2) = \int_G d_G(x_2) \int_G f_1(x_2 x_1) \otimes f_2(x_2)\, dT(x_1)$$
$$(f_i \in C_c^\infty(G; E^i), i = 1, 2).$$

Lemma 2. If the support of T is contained in H and if $\mathbf{B}^t$ extends to a separately continuous bilinear form on $C_c(G; E^1) \times C_c(G; E^2)$, then T has transversal order zero relative to H (and so is the extension to G of an $E^1 \hat{\otimes} E^2$-distribution on H).

Proof (1) First of all, if T satisfies the hypotheses of our lemma, then so does the right translate of T by any element of H – hence it will be sufficient to show that T has transversal order zero near 1. In all that follows, then, we shall work in a small, usually unspecified, neighborhood of 1 in G. Let $X_1, \ldots, X_r$ be a basis for the Lie algebra of G modulo the Lie algebra of H. Because the exponential map exp is a local diffeomorphism, the map

$$(t, \xi) \mapsto \exp\left(\sum_1^r t_i X_i\right)\xi \qquad (t \in \mathbf{R}^r, \xi \in H)$$

is a diffeomorphism from some neighborhood of $(0, 1)$ in $\mathbf{R}^r \times H$ onto a neighborhood of 1 in G. The differential operators $D_i = \partial/\partial t_i$ may now be defined (on such a neighborhood of 1 in G) by the usual prescription

$$(D_i f)\left(\exp\left(\sum_1^r t_j X_j\right)\xi\right) = \frac{d}{ds_i} f\left(\exp\left(\sum_1^r (t_j + \delta_{ij} s_i) X_j\right)\xi\right)\Bigg|_{s_i = 0}.$$

(2) Fix a vector $a_2 \in E^2$; let S be the E^1-distribution on G given by

$$\langle f_1, S\rangle = \langle f_1 \otimes a_2, T\rangle \qquad (f_1 \in C_c^\infty(G; E^1)).$$

Then S has its support contained in H. We shall prove below that S has transversal order zero relative to H. In view of the arbitrariness of a_2 and Proposition A 2.1.2, it then follows without difficulty that T too has transversal order zero relative to H.

(3) To establish that S has transversal order zero relative to H, we may, due to the observations in (1), confine our attention to a small neighborhood of 1 in G. This being so, write $S = \sum_{|I| \leq n} (-1)^{|I|} D^I S_I$, the S_I being distributions on a neighborhood of 1 in G which are extensions to G of E^1-distributions on a neighborhood of 1 in H, n the transversal order of S there. Assuming that $n > 0$ (so that the S_I, for $|I| = n$, do not all vanish on any neighborhood of 1), we shall proceed to derive a contradiction, thus proving the lemma.

(4) For any r-tuple $I = (i_1, \ldots, i_r)$ of non-negative integers, let $D^I \delta_x$ denote the distribution $f \mapsto (D^I f)(x)$ $(x \in G)$; we shall write $D^I \delta$ for $D^I \delta_1$. The group G operates on itself by left translation; hence, if $|I| = n$ and x is near 1, then we have $(D^I \delta)^x \equiv \sum_{|J| = n} \lambda_{JI}(x) D^J \delta_x$, the symbol $\equiv$ indicating that the relation in question is true modulo terms with derivatives of order $< |I|$ in $t_1, \ldots, t_r$. Here the functions $\lambda_{JI}(x)$ are defined and differentiable on a G-neighborhood of 1.

(5) We can write, under the obvious assumptions,

$$\langle f_1, S\rangle = \sum_{|I| \leq n} \int d(D^I \delta)(x_1) \int f_1(x_1 \xi)\, dS_I(\xi).$$

It will be convenient to consider functions f_1 of a special form, namely $f_1 = \phi_1 \psi_1$ where ϕ_1 has the property that $\phi_1(\exp(\sum t_i X_i)\xi)$ is independent of ξ and ψ_1 is such that $\psi_1(\exp(\sum t_i X_i)\xi)$ depends only on ξ. If x_2 is near 1, then

$$\begin{aligned}\langle \phi_1 \psi_1, S^{x_2}\rangle &= \int \phi_1(x_2 x_1) \psi_1(x_2 x_1)\, dS(x_1) \\ &= \sum_{|I| \leq n} \int d(D^I \delta)(x_1) \phi_1(x_2 x_1) \int \psi_1(x_2 x_1 \xi)\, dS_I(\xi).\end{aligned}$$

In view of (4), we have

$$\int \phi_1(x_2x_1)\, d(D^I\delta)(x_1) \equiv \sum_{|J|=n} \lambda_{JI}(x_2) D^J\phi_1(x_2)$$

and so we can transform $\langle \phi_1\psi_1, S^{x_2}\rangle$ to read

$$\langle \phi_1\psi_1, S^{x_2}\rangle \equiv \sum_{|I|=n} \sum_{|J|=n} \lambda_{JI}(x_2) D^J\phi_1(x_2) \int \psi_1(x_2\xi)\, dS_I(\xi).$$

Let g_2 be a C^∞ function on G with support concentrated near 1; let $f_2(.) = g_2(.) \otimes a_2$ – then, with these choices of f_i $(i = 1, 2)$,

$$\begin{aligned}\mathbf{B}^t(f_1, f_2) &= \int_G d_G(x_2) \int_G f_1(x_2x_1) \otimes f_2(x_2)\, dT(x_1)\\ &= \int_G d_G(x_2) \int_G f_1(x_2x_1) g_2(x_2)\, dS(x_1)\\ &\equiv \sum_{|I|=n} \sum_{|J|=n} \int_G d_G(x_2) \lambda_{JI}(x_2) D^J\phi_1(x_2) g_2(x_2) \int \psi_1(x_2\xi)\, dS_I(\xi).\end{aligned}$$

For each J such that $|J| = n$, put $\alpha_J(x_2) = \sum_{|I|=n} \lambda_{JI}(x_2) \int \psi_1(x_2\xi) dS_I(\xi)$; let $\Gamma(x_2) = \Gamma(t, \xi)$ $(x_2 = \exp(\sum t_iX_i)\xi)$ denote the density of the left Haar measure $d_G(x_2)$ with respect to $dt_1 \cdots dt_r d_H(\xi)$ on our small neighborhood of 1 in G ($d_H(\xi)$ being left Haar measure on H) – then

$$\begin{aligned}\mathbf{B}^t(f_1, f_2) &\equiv \sum_{|J|=n} \int_H \int_{\mathbf{R}^r} D^J\phi_1(x_2) g_2(x_2) \alpha_J(x_2) \Gamma(x_2)\, dt_1 \cdots dt_r d_H(\xi)\\ &\equiv \sum_{|J|=n} \int_{\mathbf{R}^r} D^J\phi_1(t) \beta_J(t)\, dt \qquad (dt = dt_1 \cdots dt_r),\end{aligned}$$

where we have set

$$\beta_J(t) = \int_H g_2(\exp(\textstyle\sum t_iX_i)\xi) \alpha_J(\exp(\textstyle\sum t_iX_i)\xi) \Gamma(t, \xi)\, d_H(\xi).$$

We shall, for the moment, regard ψ_1, g_2 and a_2 as being fixed but will allow ϕ_1 to vary over $C_c^\infty(\mathcal{O})$, $\mathcal{O}$ a small neighborhood of zero in $\mathbf{R}^r$. According to the hypotheses of our lemma, the map $\phi_1 \mapsto \mathbf{B}^t(f_1, f_2)$ is continuous in the topology of $C_c(\mathcal{O})$ relativised to $C_c^\infty(\mathcal{O})$. Consequently

$$\mathbf{B}^t(f_1, f_2) \equiv \sum_{|J|=n} \langle \beta_J, D^J\phi_1\rangle$$

for all $\phi_1 \in C_c(\mathcal{O})$ ($D^J\phi_1$ the derivative of ϕ_1 in the sense of distributions). [We are denoting by $\mathbf{B}^t$ the extension of $\mathbf{B}^t$ to $C_c(G; E^1) \times C_c(G; E^2)$.] Consider now those g_2 which can be written in the form $g_2(x_2) = \phi_2(t)\psi_2(\xi)$ $(x_2 = \exp(\sum t_iX_i)\xi)$, ϕ_2 and ψ_2 being C^∞ with compact support; fix ψ_2 – then by our extendability hypothesis on $\mathbf{B}^t$, it follows as above that for any $\phi_1 \in C_c(\mathcal{O})$, the map $\phi_2 \mapsto \mathbf{B}^t(f_1, f_2)$ is continuous in the topology of $C_c(\mathcal{O})$. In this case, $\beta_J(t) = \phi_2(t)\gamma_J(t)$ where

$$\gamma_J(t) = \int_H \psi_2(\xi) \alpha_J(\exp(\textstyle\sum t_iX_i)\xi) \Gamma(t, \xi)\, d_H(\xi)$$

and so, γ_J being C^∞, we deduce that $\mathbf{B}^t(f_1, f_2) \equiv \sum_{|J|=n} \langle \phi_2\gamma_J, D^J\phi_1\rangle$. This last expression for $\mathbf{B}^t$ tells us that

$$\mathbf{B}^t(f_1, f_2) = \langle \phi_2, D\phi_1\rangle,$$

D a certain differential operator on $\mathcal{O}$ of order $\leqslant n$. Because the map $\phi_2 \mapsto$

$\mathbf{B}'(f_1, f_2)$ is continuous in the topology of $C_c(\mathcal{O})$, it actually defines a measure on $\mathcal{O}$ – thus, for every $\phi_1 \in C_c(\mathcal{O})$, $D\phi_1$ is a measure and so, thanks to Lemma 1, D must be of order zero. [If $D\phi_1$ is a measure for every $\phi_1 \in C_c(\mathcal{O})$, then certainly $D\phi_1$ is a measure for every $\phi_1 \in C(\mathcal{O})$.] But we are assuming that n is greater than zero – hence $\gamma_J = 0$ (all J with $|J| = n$). The arbitrariness of ψ_2 then implies that $\alpha_J\Gamma = 0$ (all J with $|J| = n$); since the density function Γ is strictly positive, we infer from this that $\alpha_J = 0$; the matrix determined by the λ_{IJ} is non-singular near 1 – therefore the vanishing of α_J leads us to the conclusion that $\int \psi_1(x_2\xi)dS_I(\xi) = 0$ (all I with $|I| = n$, x_2 near 1 in G). From the arbitrariness of ψ_1, it follows that $S_I = 0$ near 1 (all I with $|I| = n$) – this, however, contradicts our assumption about n.

In so doing, it brings the proof of Lemma 2 to a close. □

5.3.3 Reciprocity Theorems

The prototype of the results we wish to establish in this number is: Let G be a compact topological group, H a closed subgroup. Let M be an irreducible unitary representation of G, L an irreducible unitary representation of H – then the number of times that the unitarily induced representation U^L contains M as a discrete direct summand is equal to the number of times that M restricted to H contains L as a discrete direct summand.

This theorem is Weil's [1, p. 83] formulation for compact groups of the classical Frobenius Reciprocity Theorem for induced representations of finite groups – it admits an evident interpretation in terms of intertwining numbers.

A partial generalization of the Frobenius Reciprocity Theorem may be obtained as follows. In Theorem 5.3.2.3, suppose that we take $H_1 = H$(say), $H_2 = G$, $x = 1$ – then there is only one H_2, H_1 double coset, namely the open set G itself. Theorem 5.3.2.1 then leads at once to:

Theorem 5.3.3.1 Let G be a Lie group, H a closed subgroup of G; let L and M be differentiable representations of H and G, respectively, on Fréchet spaces E and F – then the intertwining number $i(U^L, M)$ (U^L the representation of G differentiably induced by L) is equal to the intertwining number of the representations $\xi \mapsto L(\xi) \hat{\otimes} M(\xi)$, $\xi \mapsto (\delta_H(\xi)/\delta_G(\xi))^{1/2}$ of H on $E \hat{\otimes} F$ and $\mathbf{C}$, respectively, i.e., to the intertwining number of the representations $\xi \mapsto (\delta_H(\xi)/\delta_G(\xi))^{1/2}L(\xi)$, $\xi \mapsto M(\xi)$ of H on E and F, respectively.

Corollary 5.3.3.2 Retain the notations of Theorem 5.3.3.1 but assume now that L and M are unitary representations of H and G, respectively – then $i(U^L, M)$ (U^L the unitary representation of G unitarily induced by L) is less than or equal to $i(U^{L_\infty}, M_\infty)$, $i(U^{L_\infty}, M_\infty)$ being equal to the intertwining number of the representations $\xi \mapsto (\delta_H(\xi)/\delta_G(\xi))^{1/2}L_\infty(\xi)$, $\xi \mapsto M_\infty(\xi)$ of H.

If, in addition, L and M are finite dimensional unitary representations of H and G, respectively, then, of course $L = L_\infty$, $M = M_\infty$; the discussion following Corollary 5.3.2.18 shows that the intertwining number of the representations $\xi \mapsto L(\xi) \hat{\otimes} M(\xi)$, $\xi \mapsto (\delta_H(\xi)/\delta_G(\xi))^{1/2}$ of H cannot be non-zero unless $\delta_H = \delta_G$ on H – therefore, in view of Corollary 5.3.3.2, we have:

Corollary 5.3.3.3 If L is a finite dimensional unitary representation of H and if δ_H and δ_G do not coincide on H, then the unitarily induced representation U^L of G does not contain any (non-zero) finite dimensional subrepresentation.

In the special case when G/H carries a finite G-invariant measure, Corollary 5.3.3.2 can be improved substantially – thus:

Proposition 5.3.3.4 Let H be a closed subgroup of G; let L and M be unitary representations of H and G, respectively. If G/H possesses a finite G-invariant measure, then

$$i(L, M \mid H) \leqslant i(U^L, M) \leqslant i(L_\infty, M_\infty \mid H).$$

Proof The first inequality follows upon applying Proposition 5.3.2.4 to the case $H_1 = H$, $H_2 = G$; the second inequality is contained in Corollary 5.3.3.2 (δ_H being equal to δ_G on H here). □

Using this proposition, it is a simple matter to derive the following result due to Mackey.

Theorem 5.3.3.5 (Mackey) Let H be a closed subgroup of G. Let L and M be finite dimensional irreducible unitary representations of H and G, respectively. If G/H carries a finite G-invariant measure, then the unitarily induced representation U^L contains M as a discrete direct summand exactly as many times as M restricted to H contains L as a discrete direct summand. On the other hand, if G/H does not admit a G-invariant measure (finite or infinite), then, for any finite dimensional unitary representation L of H, U^L contains no (non-zero) finite dimensional subrepresentation.

Proof The second assertion of our theorem is just Corollary 5.3.3.3 above. As for the first, bear in mind that the multiplicity of M in U^L is given by the number $i(U^{\bar{L}}, M)$; since L and M are finite dimensional, this number is equal to, in view of Proposition 5.3.3.4, $i(\bar{L}, M \mid H) = i(L, \bar{M} \mid H) =$ multiplicity of L in $M \mid H$. □

Remark Actually it is possible to sharpen the second statement in Theorem 5.3.3.5 – thus one may show (without essential difficulty) that even if there exists a G-invariant measure of infinite total mass on G/H, then it is still true that U^L contains no (non-zero) finite dimensional subrepresentation (cf. Bruhat [1, p. 176]).

Corollary 5.3.3.6 (Weil) In the notations of Theorem 5.3.3.5, if G is compact, then U^L contains M exactly as many times as M restricted to H contains L.

Corollary 5.3.3.7 If G is not compact, then the left regular representation of G on $L^2(G)$ contains no (non-zero) finite dimensional subrepresentation.

[Here we have to keep in mind the Remark supra.]

5.3.4 Decomposition Theorems

In this number the Hilbert spaces under consideration are assumed to be separable; we shall then agree to realize all unitarily induced representations as 'multiplier' representations (5.1.1).

Let H_1 and H_2 be closed subgroups of the Lie group G; we shall say that H_1 and H_2 are *discretely related* if there exists a subset of G whose complement has Haar measure zero and which is itself the union of countably many H_1, H_2 double cosets.

The following result is known in the literature as Mackey's Subgroup Theorem.

Theorem 5.3.4.1 (Mackey) Let $\tilde{H}$ and H be (closed) discretely related subgroups of G; let L be a unitary representation of H on a Hilbert space E, U^L the corresponding unitarily induced representation of G. For each x in G, let $\tilde{V}_x$ denote the representation of $\tilde{H}$ unitarily induced by the unitary representation $\zeta \mapsto L(x^{-1}\zeta x)$ of the subgroup $\tilde{H} \cap (xHx^{-1})$ – then $\tilde{V}_x$ is determined to within unitary equivalence by the double coset $D(x) = \tilde{H}xH$ to which x belongs and we may write $\tilde{V}_D$ for $\tilde{V}_x$ where $D = D(x)$. Finally U^L restricted to $\tilde{H}$ is the (Hilbert space) direct sum of the $\tilde{V}_D$, the D ranging over those $\tilde{H}$, H double cosets which are not of Haar measure zero.

Proof By hypothesis, G is, up to a set of measure zero, the countable union of non-null $\tilde{H}$, H double cosets $D(x_i) = \tilde{H}x_iH$ (say). Such a decomposition of G leads immediately to a decomposition of E^L (the representation space for U^L) into mutually orthogonal subspaces $\mathscr{S}_i$, each of which is invariant under the restriction of U^L to $\tilde{H}$ (take for $\mathscr{S}_i$ the subspace of E^L comprised of those functions which vanish outside of $\tilde{H}x_iH$). To study the representation $(U^L|\tilde{H})|\mathscr{S}_i$, note that $\mathscr{S}_i$ may be regarded as the Hilbert space consisting of those E-valued Borel functions f_i on the double coset $\tilde{H}x_iH$ which verify the relation

$$f_i(x\xi) = L(\xi^{-1})f_i(x) \qquad (\xi \in H, x \in \tilde{H}x_iH)$$

and have the property that $\int_{Q_i} (f_i(x), f_i(x))d\mu_i(\dot{x}) < \infty$ (Q_i the image of $\tilde{H}x_iH$ in G/H under the canonical map π). [Here μ_i is the restriction to Q_i of the quasi-invariant measure μ_H on G/H; let us keep in mind that a subset of G/H is μ_H-null iff its π-counterimage in G is of Haar measure zero.] This being so, we shall now define a map from $\mathscr{S}_\iota$ to the representation space for $\tilde{V}_{x_\iota}$ which will serve to implement the required unitary equivalence between $(U^L|\tilde{H})|\mathscr{S}_\iota$ and $\tilde{V}_{x_\iota}$.

(1) Given an E-valued Borel function f_ι on $\tilde{H}x_\iota H$ which satisfies the relation $f_\iota(x\xi) = L(\xi^{-1})f_\iota(x)$ $(\xi \in H, x \in \tilde{H}x_\iota H)$, define a function $\tilde{f}_\iota$ on $\tilde{H}$ by the prescription $\tilde{f}_\iota(\tilde{\xi}) = f_\iota(\tilde{\xi}x_\iota)$ $(\tilde{\xi} \in \tilde{H})$; of course $\tilde{f}_\iota$ is a Borel function on $\tilde{H}$. Moreover, if $\zeta \in \tilde{H} \cap (x_\iota Hx_\iota^{-1})$ and $\zeta = x_\iota\xi x_\iota^{-1}$ $(\xi \in H)$, then

$$\begin{aligned}\tilde{f}_\iota(\tilde{\xi}\zeta) &= f_\iota(\tilde{\xi}x_\iota\xi x_\iota^{-1}x_\iota) = f_\iota(\tilde{\xi}x_\iota\xi) = L(\xi^{-1})f_\iota(\tilde{\xi}x_\iota)\\ &= L(x_\iota^{-1}\zeta^{-1}x_\iota)\tilde{f}_\iota(\tilde{\xi})\ (\tilde{\xi} \in \tilde{H}).\end{aligned}$$

Conversely, let $\tilde{f}_\iota$ be an E-valued Borel function on $\tilde{H}$ with the property that

$$\tilde{f}_\iota(\tilde{\xi}\zeta) = L(x_\iota^{-1}\zeta^{-1}x_\iota)\tilde{f}_\iota(\tilde{\xi}) \qquad (\zeta \in \tilde{H} \cap (x_\iota Hx_\iota^{-1}), \tilde{\xi} \in \tilde{H});$$

put

$$f_\iota(\tilde{\xi}x_\iota\xi) = L(\xi^{-1})\tilde{f}_\iota(\tilde{\xi}) \qquad (\tilde{\xi} \in \tilde{H}, \xi \in H).$$

Then f_ι is a well-defined, E-valued Borel function on $\tilde{H}x_\iota H$ such that

$$f_\iota(x\xi) = L(\xi^{-1})f_\iota(x) \qquad (\xi \in H, x \in \tilde{H}x_\iota H).$$

[If

$$\tilde{\xi}_1x_\iota\xi_1 = \tilde{\xi}_2x_\iota\xi_2 \qquad (\tilde{\xi}_1, \tilde{\xi}_2 \in \tilde{H}; \xi_1, \xi_2 \in H),$$

then $x_\iota^{-1}\tilde{\xi}_2^{-1}\tilde{\xi}_1x_\iota = \xi_2\xi_1^{-1}$; hence

$$\begin{aligned}\tilde{f}_\iota(\tilde{\xi}\tilde{\xi}_2^{-1}\tilde{\xi}_1) &= L(x_\iota^{-1}\tilde{\xi}_2^{-1}\tilde{\xi}_1x_\iota)^{-1}\tilde{f}_\iota(\tilde{\xi}) = L(\xi_1\xi_2^{-1})\tilde{f}_\iota(\tilde{\xi})\\ &\Rightarrow L(\xi_1^{-1})\tilde{f}_\iota(\tilde{\xi}\tilde{\xi}_2^{-1}\tilde{\xi}_1) = L(\xi_2^{-1})\tilde{f}_\iota(\tilde{\xi})\\ &\Rightarrow L(\xi_1^{-1})\tilde{f}_\iota(\tilde{\xi}_1) = L(\xi_2^{-1})\tilde{f}_\iota(\tilde{\xi}_2)\end{aligned}$$

and so f_ι is well-defined. We shall leave to the reader the mildly tedious task of verifying that f_ι is a Borel function having the stated property.] Thus: The map $f_\iota \mapsto \tilde{f}_\iota$ is a one-to-one map of E-valued Borel functions f_ι on $\tilde{H}x_\iota H$ which satisfy the condition

$$f_\iota(x\xi) = L(\xi^{-1})f(x) \qquad (\xi \in H, x \in \tilde{H}x_\iota H)$$

onto E-valued Borel functions $\tilde{f}_\iota$ on $\tilde{H}$ such that

$$\tilde{f}_\iota(\tilde{\xi}\zeta) = L(x_\iota^{-1}\zeta^{-1}x_\iota)\tilde{f}_\iota(\tilde{\xi}) \qquad (\zeta \in \tilde{H} \cap (x_\iota Hx_\iota^{-1}), \tilde{\xi} \in \tilde{H}).$$

(2) The obvious canonical mapping sets up a one-to-one Borel set preserving correspondence between the points of Q_ι and those of the coset space $\tilde{H}/\tilde{H} \cap (x_\iota Hx_\iota^{-1})$; we can use this map to transfer the measure μ_ι on Q_ι to a quasi-invariant measure $\tilde{\nu}_\iota$ on $\tilde{H}/\tilde{H} \cap (x_\iota Hx_\iota^{-1})$ – it is then clear that the map $f_\iota \mapsto \tilde{f}_\iota$ establishes the desired unitary equivalence between $(U^L|\tilde{H})|\mathscr{S}_\iota$ and $\tilde{V}_{x_\iota}$.

All the assertions of the theorem are now obvious. □

Corollary 5.3.4.2 Let $\tilde{H}$ and H be closed subgroups of G with the property that for some x in G the double coset $\tilde{H}xH$ is an open set with null complement. If L is a unitary representation of H, U^L the corresponding unitarily induced representation of G, then U^L restricted to $\tilde{H}$ is unitarily equivalent to $\tilde{V}_D$, $\tilde{V}_D$ the representation of $\tilde{H}$ unitarily induced by the unitary representation $\zeta \mapsto L(x^{-1}\zeta x)$ of the subgroup $\tilde{H} \cap (xHx^{-1})$ $(D = D(x) = \tilde{H}xH)$.

Example We shall leave it to the reader to explicate Corollary 5.3.4.2 in the following situation. Let G be a connected semi-simple Lie group with finite center, P a minimal parabolic subgroup of G; let L be a finite dimensional irreducible unitary representation of P (so that U^L is a representation in the principal P-series for G.– cf. 5.5.2 infra); finally, keeping in mind Proposition 1.2.3.4, take $H = \tilde{H} = P \ldots$.

It is an easy step from Theorem 5.3.4.1 to the following important result on the decomposition of Hilbert products of unitarily induced representations.

Theorem 5.3.4.3 (Mackey) Let H_i be a closed subgroup of G, L^i a unitary representation of H_i on a Hilbert space E^i, U^{L^i} the corresponding unitarily induced representation of G $(i = 1, 2)$; assume that H_1 and H_2 are discretely related. For each pair $(x_1, x_2) \in G \times G$, let M_{x_1,x_2} denote the Hilbert product of the unitary representations $\zeta \mapsto L^1(x_1^{-1}\zeta x_1)$, $\zeta \mapsto L^2(x_2^{-1}\zeta x_2)$ of the subgroup $(x_1H_1x_1^{-1}) \cap (x_2H_2x_2^{-1})$, V_{x_1,x_2} the unitary representation of G unitarily induced by M_{x_1,x_2} – then V_{x_1,x_2} is determined to within unitary equivalence by the double coset $D = D(x_1, x_2) = H_1x_1^{-1}x_2H_2$ to which $x_1^{-1}x_2$ belongs and we may write V_D for V_{x_1,x_2}. In addition the Hilbert product of U^{L^1} with U^{L^2} is unitarily equivalent to the (Hilbert space) direct sum of the V_D, D ranging over those double cosets which are not of Haar measure zero.

Proof The Hilbert product of U^{L^1} with U^{L^2} is the unitary representation of G obtained from the unitary representation $U^{L^1} \times U^{L^2}$ of $G \times G$ by restriction to the isomorphic replica $\tilde{G}$ of G consisting of those pairs (x_1, x_2) in $G \times G$ such that $x_1 = x_2$. On the other hand, in view of Proposition 5.1.3.3, $U^{L^1} \times U^{L^2}$ is unitarily equivalent to $U^{L^1 \times L^2}$; as $L^1 \times L^2$ is a unitary representation of $H_1 \times H_2$, we see that the setting for the present theorem is that of investigating what happens when a certain unitarily induced representation (viz. $U^{L^1 \times L^2}$) is restricted to a certain subgroup (viz. $\tilde{G}$) – thus we may try to apply Theorem 5.3.4.1.

(1) It will be necessary to verify that the subgroups $\tilde{G}$ and $H_1 \times H_2$ of $G \times G$ are discretely related. To this end, observe that the map $(x_1, x_2) \mapsto x_1^{-1}x_2$ of $G \times G$ onto G serves to establish a one-to-one correspondence between the $\tilde{G}$, $H_1 \times H_2$ double cosets of $G \times G$ and the H_1, H_2 double cosets of G (with $\tilde{G}(x_1, x_2)(H_1 \times H_2)$ corresponding to $H_1x_1^{-1}x_2H_2$). Moreover, under this mapping, double cosets of measure zero correspond

to double cosets of measure zero. [Note first that two points of $G \times G$ go into the same point of G iff they belong to the same right $\tilde{G}$ coset; this leads, then, in the obvious way, to a one-to-one map of G onto the right coset space $\tilde{G}\backslash G \times G$. As we know, a double coset in $G \times G$ is of measure zero iff its image in $\tilde{G}\backslash G \times G$ is of measure zero with respect to some (and hence all) quasi-invariant measures there; but the Haar measure on G, when transfered to $\tilde{G}\backslash G \times G$, is quasi-invariant . . . .] Therefore $\tilde{G}$ and $H_1 \times H_2$ are indeed discretely related.

(2) Because of (1), the hypotheses of Theorem 5.3.4.1 are met here. This theorem tells us that when $U^{L^1 \times L^2}$ is restricted to $\tilde{G}$ it breaks up into a direct sum of unitary representations over the double cosets $\tilde{G}(x_1, x_2)(H_1 \times H_2)$ which are not of measure zero, the summand associated with the double coset containing (x_1, x_2) being the representation of $\tilde{G}$ unitarily induced by the unitary representation

$$(\zeta, \zeta) \mapsto L^1 \times L^2((x_1, x_2)^{-1}(\zeta, \zeta)(x_1, x_2))$$

of the subgroup $\tilde{G} \cap ((x_1, x_2)(H_1 \times H_2)(x_1, x_2)^{-1})$. The natural isomorphism of $\tilde{G}$ onto G carries the subgroup

$$\tilde{G} \cap ((x_1, x_2)(H_1 \times H_2)(x_1, x_2)^{-1})$$

onto the subgroup $x_1 H_1 x_1^{-1} \cap x_2 H_2 x_2^{-1}$; the unitary representation

$$(\zeta, \zeta) \mapsto L^1 \times L^2((x_1, x_2)^{-1}(\zeta, \zeta)(x_1, x_2))$$

of $\tilde{G} \cap ((x_1, x_2)(H_1 \times H_2)(x_1, x_2)^{-1})$ then becomes the unitary representation $\zeta \mapsto L^1(x_1^{-1}\zeta x_1) \times L^2(x_2^{-1}\zeta x_2)$ of $x_1 H_1 x_1^{-1} \cap x_2 H_2 x_2^{-1}$.

The contentions of the theorem are now evident. □

Needless to say, Theorem 5.3.4.3 becomes particularly interesting when U^{L^1} and U^{L^2} are both irreducible.

It is only natural to wonder whether Theorems 5.3.4.1 and 5.3.4.3 are valid when the subgroups involved are no longer discretely related – the answer here is 'yes' provided the double cosets in question fill out G in a 'smooth' manner.

Let H_1 and H_2 be closed subgroups of G; let $\mathfrak{D}$ denote the set of H_1, H_2 double cosets in G; let π denote the mapping which assigns to each $x \in G$, the H_1, H_2 double coset to which x belongs. A subset $\mathfrak{B}$ of $\mathfrak{D}$ is said to be a Borel set provided $\pi^{-1}(\mathfrak{B})$ is a Borel set in G. If the points in $\mathfrak{D}$ are separated by countably many Borel sets, then we shall say that H_1 and H_2 are *regularly related;* the regular relatedness of H_1 and H_2 guarantees the existence of a unique measure class in $\mathfrak{D}$, the null sets of which being the sets $\mathfrak{B}$ such that $\pi^{-1}(\mathfrak{B})$ is a null set in G – this measure class will be referred to as the *canonical measure class* in $\mathfrak{D}$.

Theorem 5.3.4.4 (Mackey) Let $\tilde{H}$ and H be closed regularly related subgroups of G; let L be a unitary representation of H on a Hilbert space E, U^L the corresponding unitarily induced representation of G. For each x in G, let $\tilde{V}_x$ denote the representation of $\tilde{H}$ unitarily induced by the unitary representation $\zeta \mapsto L(x^{-1}\zeta x)$ of the subgroup $\tilde{H} \cap (xHx^{-1})$ – then $\tilde{V}_x$ is determined to within unitary equivalence by the double coset $D(x) = \tilde{H}xH$ to which x belongs and we may write $\tilde{V}_D$ for $\tilde{V}_x$ where $D = D(x)$. Finally U^L restricted to $\tilde{H}$ is a direct integral over $\mathfrak{D}$ (the set of $\tilde{H}$, H double cosets), with respect to any measure in the canonical measure class, of the representations $\tilde{V}_D$ ($D \in \mathfrak{D}$).

[The proof of this 'continuous' version of the Subgroup Theorem may be found in Mackey [3, p. 127].]

The following result is an easy consequence of Theorem 5.3.4.4.

Theorem 5.3.4.5 (Mackey) Let H_i be a closed subgroup of G, L^i a unitary representation of H_i on a Hilbert space E^i, U^{L^i} the corresponding unitarily induced representation of G ($i = 1, 2$); assume that H_1 and H_2 are regularly related. For each pair $(x_1, x_2) \in G \times G$, let M_{x_1, x_2} denote the Hilbert product of the unitary representations

$$\zeta \mapsto L^1(x_1^{-1}\zeta x_1), \quad \zeta \mapsto L^2(x_2^{-1}\zeta x_2)$$

of the subgroup

$$(x_1 H_1 x_1^{-1}) \cap (x_2 H_2 x_2^{-1}),$$

V_{x_1, x_2} the unitary representation of G unitarily induced by M_{x_1, x_2} – then V_{x_1, x_2} is determined to within unitary equivalence by the double coset $D = D(x_1, x_2) = H_1 x_1^{-1} x_2 H_2$ to which $x_1^{-1}x_2$ belongs and we may write V_D for V_{x_1, x_2}. Furthermore the Hilbert product of U^{L^1} with U^{L^2} is unitarily equivalent to a direct integral of the $V_D(D \in \mathfrak{D})$ with respect to any measure in the canonical measure class (in $\mathfrak{D}$).

Remark Needless to say, the preceding results are valid for locally compact groups which satisfy the second axiom of countability; on the other hand, as L. Robertson has shown by example, Mackey's Subgroup Theorem is false 'in general'.

Example (Robertson) Let G be the direct product $G = H \times \mathbf{R}$ of the discrete reals H and the ordinary reals $\mathbf{R}$; let $\tilde{H}$ be the diagonal subgroup $\tilde{H} = \{(t, t) : t \in \mathbf{R}\}$ – then $G = H\tilde{H}$. Let L be the trivial one dimensional representation of H – then, in particular, the restriction to $\tilde{H}$ of U^L operates on a separable Hilbert space. On the other hand, the representation of $\tilde{H}$ which is unitarily induced from the trivial one dimensional representation of $\{1\}$ operates on a non-separable Hilbert space . . . .

5.4 Systems of Imprimitivity

5.4.1 Mackey's Orbital Analysis

In the theory of unitary representations of finite groups, the notion of a system of imprimitivity plays an important role (cf. the Introduction). The first objective of the present number is to formulate an appropriate generalization of this concept for infinite groups; once this is done, we shall then indicate how it is possible to analyze, using (unitarily) induced representations, the unitary duals of certain important classes of groups.

As usual, G will denote a Lie group countable at infinity.

Definition (Mackey) Let M be a locally compact Hausdorff space, satisfying the second axiom of countability, on which G operates to the left via homeomorphisms; let U be a unitary representation of G on a Hilbert space E – then, by a *system of imprimitivity* P for U with *base* M, we shall understand a regular E-projection-valued measure P on the Borel subsets of M such that $U(x)\mathsf{P}(\mathfrak{B})U(x^{-1}) = \mathsf{P}(x\cdot\mathfrak{B})$ for all x in G and every Borel subset $\mathfrak{B}$ of M. If G operates transitively on M, then the system of imprimitivity P is said to be *transitive*.

Let U^i be a unitary representation of G on a Hilbert space E^i; let P_i be a system of imprimitivity for U^i based on the G-space M $(i = 1, 2)$ – then P_1 and P_2 are said to be *unitarily equivalent* provided there exists an isometry Q of E^1 onto E^2 such that

$$QU^1(x) = U^2(x)Q, \quad Q\mathsf{P}_1(\mathfrak{B}) = \mathsf{P}_2(\mathfrak{B})Q,$$

for all x in G and every Borel subset $\mathfrak{B}$ of M.

Example Every unitarily induced representation of G gives rise, in a canonical way, to a transitive system of imprimitivity. Thus let H be a closed subgroup of G, L a unitary representation of H on a Hilbert space E, U^L the corresponding unitarily induced representation of G; let M be the left coset space G/H considered as a G-transformation space. For each Borel subset $\mathfrak{B}$ of M, let $\mathsf{P}^L(\mathfrak{B})$ be the projection on E^L defined by

$$(\mathsf{P}^L(\mathfrak{B})f)(x) = \begin{cases} f(x) & \text{if } xH \in \mathfrak{B} \\ 0 & \text{if } xH \notin \mathfrak{B}, \end{cases} \qquad (x \in G, f \in E^L).$$

It is then clear that P^L is a regular E^L-projection-valued measure on the Borel subsets $\mathfrak{B}$ of M, verifying the condition

$$U^L(x)\mathsf{P}^L(\mathfrak{B})U^L(x^{-1}) = \mathsf{P}^L(x\cdot\mathfrak{B}) \qquad (x \in G)$$

– otherwise said, P^L is a system of imprimitivity for U^L with base $M = G/H$, the *canonical system of imprimitivity* for U^L. One can prove (and this is not obvious) that L is irreducible iff P^L is irreducible (i.e. iff E^L admits no non-trivial closed subspaces which are invariant simultaneously under all the $U^L(x)$ and all the $\mathsf{P}^L(\mathfrak{B})$ ($x \in G$, $\mathfrak{B}$ a Borel subset of G/H)).

Here is Mackey's celebrated Imprimitivity Theorem.

Theorem 5.4.1.1 (Mackey) Let H be a closed subgroup of G, U a unitary representation of G; if there exists a (transitive) system of imprimitivity P for U based on G/H, then there is a unitary representation L of H, unique to within unitary equivalence, such that P is unitarily equivalent to the canonical system of imprimitivity for the unitarily induced representation U^L.

[As far as transitive systems of imprimitivity are concerned, there is no loss of generality in assuming that they are based on a coset space G/H.]

Thus the Imprimitivity Theorem asserts that there are no systems of imprimitivity based on G/H other than the canonical ones.

One can use the Imprimitivity Theorem to analyze $\hat{G}$, at least if G contains normal subgroups of a suitable type. Thus let H be a closed normal abelian subgroup of G, $\hat{H}$ its dual group – then G operates to the left on $\hat{H}$ via the prescription

$$\hat{\xi} \mapsto x \cdot \hat{\xi}, \langle \xi, x \cdot \hat{\xi} \rangle = \langle x^{-1} \xi x, \hat{\xi} \rangle \qquad (\xi \in H, \hat{\xi} \in \hat{H}; x \in G).$$

[In passing observe that H, being a Lie group, is, according to our general agreements, an 'elementary group' of the form $\mathbf{R}^n \times \mathbf{T}^p \times \mathbf{Z}^q \times F$, F a finite group.] Given a character $\hat{\xi} \in \hat{H}$, let $H_{\hat{\xi}}$ denote its stabilizer in G; of course $H_{\hat{\xi}}$ is a closed subgroup of G containing H. Consider the following conditions on G and H.

(O_1) There exists a countable family $\mathfrak{B} = \{\mathfrak{B}_i\}$ of Borel subsets of $\hat{H}$ such that:

(i) Each set $\mathfrak{B}_i$ in $\mathfrak{B}$ is a union of G-orbits;

(ii) If Q_1 and Q_2 are any two distinct G-orbits in $\hat{H}$, then there is a $\mathfrak{B}_i$ in $\mathfrak{B}$ having the property that $Q_1 \subset \mathfrak{B}_i$, $Q_2 \cap \mathfrak{B}_i = \varnothing$.

(O_2) For every $\hat{\xi} \in \hat{H}$, the canonical continuous one-to-one map $xH_{\hat{\xi}} \mapsto x \cdot \hat{\xi}$ of $G/H_{\hat{\xi}}$ onto the orbit Q to which $\hat{\xi}$ belongs is a homeomorphism (Q being considered as equipped with the topology induced by $\hat{H}$).

[Condition (O_1) is usually expressed by saying that the orbit space of $\hat{H}$ is *countably separated;* it guarantees that the various orbits are Borel subsets of $\hat{H}$ (since any orbit can be written as the (necessary countable) intersection of the $\mathfrak{B}_i$ which contain it).]

We shall assume, in what follows, that conditions (O_1) and (O_2) hold for G and its closed normal abelian subgroup H.

Let U be an irreducible unitary representation of G on a Hilbert space E; consider the restriction of U to H – then, in view of well-known generalities, there exists a unique regular E-projection-valued measure P on the Borel subsets of $\hat{H}$ such that

$$U(\xi) = \int_{\hat{H}} \langle \xi, \hat{\xi} \rangle \, d\mathsf{P}(\hat{\xi}) \qquad (\xi \in H).$$

The measure P is usually referred to as the *spectral measure* of $U | H$. There is no difficulty in verifying that P is a system of imprimitivity for U with base $\hat{H}$. Moreover, using condition (O_1) supra, one can show that there is a unique orbit $Q \subset \hat{H}$ having the property that $\mathsf{P}(\hat{H} - Q) = 0$ (i.e. $\mathsf{P}(Q) = I$) – we shall say that U is *associated with* Q. Fix an element $\hat{\xi} \in Q$ – then, as is easily seen, the identification i, $xH_{\hat{\xi}} \mapsto x \cdot \hat{\xi}$, of $G/H_{\hat{\xi}}$ with Q carries Borel sets into Borel sets and so does its inverse. Therefore, using i, we can pull back $\mathsf{P} | Q$ to $G/H_{\hat{\xi}}$ and thus obtain a regular E-projection-valued measure P_Q (say) on the Borel subsets of $G/H_{\hat{\xi}}$ (the regularity of P_Q being a consequence of the fact that i^{-1} carries compact sets into compact sets) – it is then immediate that P_Q is a transitive system of imprimitivity for U with base $G/H_{\hat{\xi}}$. Owing to the Imprimitivity Theorem, there is an irreducible unitary representation L of $H_{\hat{\xi}}$ such that P_Q is unitarily equivalent to the canonical system of imprimitivity for the unitarily induced representation U^L – in particular U is unitarily equivalent to U^L. Furthermore, it can be shown that $L|H$ is a direct sum of copies of $\hat{\xi}$.

The preceding discussion isolates the main steps which are involved in the 'difficult half' of the proof of the following theorem.

Theorem 5.4.1.2 (Mackey) Let H be a closed normal abelian subgroup of G for which conditions (O_1) and (O_2) hold. Let Q be a G-orbit in $\hat{H}$, $\hat{\xi}$ an element of Q, $H_{\hat{\xi}}$ the stability subgroup of $\hat{\xi}$; let $\mathscr{L}(\hat{\xi})$ be the family of those elements L in $\hat{H}_{\hat{\xi}}$ such that $L(\xi) = \langle \xi, \hat{\xi} \rangle I$ $(\xi \in H)$ – then the map

$$L \mapsto U^L \qquad (L \in \mathscr{L}(\hat{\xi}))$$

is a one-to-one map of $\mathscr{L}(\hat{\xi})$ onto the family $\hat{G}(Q)$ of those elements in $\hat{G}$ which are associated with Q.

Note If Q is the one element orbit $\{1\}$, then the elements of $\hat{G}$ associated with Q are just the representations of the form $x \mapsto U(xH)$ where $U \in (G/H)\hat{}$ (i.e. those 'lifted' from irreducible unitary representations of G/H).

In general the classification of $\mathscr{L}(\hat{\xi})$ is a difficult matter – this problem can be reduced, however, to a simpler one whenever the orbit Q is 'accommodating'.

Definition Let Q be a G-orbit in $\hat{H}$ – then Q is said to be *accommodating* provided some $\hat{\xi}$ in Q can be extended to a unitary character of the stability subgroup $H_{\hat{\xi}}$ of $\hat{\xi}$.

[Of course if one element in Q can be extended to a unitary character of its stability subgroup, then the same is true of any other element in Q.]

Lemma 5.4.1.3 Let us suppose that the orbit Q is accommodating; let $\hat{\mathfrak{x}}$ be a unitary character of $H_{\hat{\xi}}$ which extends $\hat{\xi}$ – then there is a one-to-one correspondence $U \leftrightarrow L$ between $(H_{\hat{\xi}}/H)\hat{}$ and $\mathscr{L}(\hat{\xi})$, given (in one direction) by $U(\xi) = \langle \xi, \hat{\mathfrak{x}} \rangle L(\pi(\xi))$ $(\xi \in H_{\hat{\xi}})$ (π the natural homomorphism of $H_{\hat{\xi}}$ onto $H_{\hat{\xi}}/H$).

There is one important situation in which every orbit is accommodating, namely when G can be written as the semi-direct product of the closed normal abelian subgroup H and a closed subgroup K (say). [Thus let Q be any G-orbit in $\hat{H}$ – then one can extend $\hat{\xi}(\hat{\xi} \in Q)$ to $H_{\hat{\xi}}$ via the prescription $\langle k\xi, \hat{\mathfrak{x}} \rangle = \langle \xi, \hat{\xi} \rangle$ $(k \in H_{\hat{\xi}} \cap K,\ \xi \in H)$.]

Theorem 5.4.1.4 (Mackey) Let H be a closed normal abelian subgroup of G for which conditions (O_1) and (O_2) hold; assume in addition that G is the semi-direct product of H with a closed subgroup K. Let Q be a G-orbit in $\hat{H}$. For each L in $(H_{\hat{\xi}} \cap K)^\wedge$, let $\hat{\xi} \times L$ denote the element of $\hat{H}_{\hat{\xi}}$ given by $(\hat{\xi} \times L)(k\xi) = \langle \xi, \hat{\xi} \rangle L(k)$ $(k \in H_{\hat{\xi}} \cap K, \xi \in H)$ – then the map

$$L \mapsto U^{\hat{\xi} \times L}$$

is a one-to-one map of $(H_{\hat{\xi}} \cap K)^\wedge$ onto the family $\hat{G}(Q)$ of those elements in $\hat{G}$ which are associated with $Q(\hat{\xi} \in Q)$.

Thus, if G is the semi-direct product of a closed normal abelian subgroup H and a closed subgroup K, then there is a systematic procedure for obtaining the elements of $\hat{G}$ in terms of the characters of H and the irreducible unitary representations of certain subgroups of K: Every element U of $\hat{G}$ is associated with one and only one G-orbit in $\hat{H}$; if Q is a G-orbit in $\hat{H}$, $\hat{\xi}$ an element of Q, and $H_{\hat{\xi}}$ the stability subgroup of $\hat{\xi}$, then the elements of $\hat{G}$ associated with Q are precisely the $U^{\hat{\xi} \times L}$, where L runs over the elements of $(H_{\hat{\xi}} \cap K)^\wedge$.

Remark Although we have only formulated the preceding theory for Lie groups G, it is actually valid, with no essential change for general, locally compact G. A word or two about the conditions (O_1) and (O_2) above is now appropriate. Interestingly enough, if G is second countable (in particular, according to our conventions, if G is a Lie group), then conditions (O_1) and (O_2) are equivalent (cf. Glimm [2]); on the other hand, one can give examples where both (O_1) and (O_2) fail. This being so, consider now the conjunction (O_1') of the following two conditions: (i) Each G-orbit in $\hat{H}$ is a Borel subset of $\hat{H}$; (ii) If E is a Hilbert space and P is any regular E-projection-valued measure on the Borel subsets of $\hat{H}$ which is ergodic (i.e. $\mathsf{P}(\mathfrak{B}) = 0$ or I for every Borel subset $\mathfrak{B}$ of $\hat{H}$ such that $\mathsf{P}(x \cdot \mathfrak{B}) = \mathsf{P}(\mathfrak{B})$ (all $x \in G$)), then P is carried by a single orbit Q (that is $\mathsf{P}(\hat{H} - Q) = 0$). It is not difficult to prove that $(O_1) \Rightarrow (O_1')$; moreover the proof of Theorem 5.4.1.2 requires only (O_1') and (O_2). There is an important special case in which (O_1') and (O_2) hold automatically, namely when our closed normal abelian subgroup H has the property that G/H is compact. [Each G-orbit is then compact, and so (O_2) and (i) in (O_1') are valid; to get at (ii), observe that the relation of G-equivalence in $\hat{H}$ is closed, hence the space of G-orbits (in the quotient topology) is a locally compact *Hausdorff* space . . . .]

5.4.2 Examples

In this number we shall present some examples which should serve to illustrate the considerations outlined in 5.4.1.

Example 1 *The '$\alpha t + \beta$' Group*. Let G be the affine group of the real line – thus $G = \{(\alpha, \beta) : \alpha > 0, \beta \in \mathbf{R}\}$, the pairs multiplying according to the rule $(\alpha_1, \beta_1)(\alpha_2, \beta_2) = (\alpha_1\alpha_2, \alpha_1\beta_2 + \beta_1)$ (of course G may also be represented as the group of real matrices of the form $\left(\begin{smallmatrix} \alpha & \beta \\ 0 & 1 \end{smallmatrix}\right)$ $(\alpha > 0)$ under ordinary matrix multiplication). The group G has a Lie algebra $\mathfrak{g}$ with basis $\{X, Y\}$ and relation $[X, Y] = Y$; evidently $\exp(tX) = (e^t, 0)$, $\exp(tY) = (1, t)$ $(t \in \mathbf{R})$. Let H be the one parameter group determined by Y – then H is a closed normal abelian subgroup of G. If K is the one parameter group determined by X, then clearly G is the semi-direct product of H and K. The dual group $\hat{H}$ consists of the unitary characters $\hat{\xi}_r$, $(1, \beta) \mapsto e^{\sqrt{-1}\beta r}$ $(r \in \mathbf{R})$; the action of G on $\hat{H}$ is easily seen to be given by $(\alpha, \beta) \cdot \hat{\xi}_r = \hat{\xi}_{r/\alpha}$ $((\alpha, \beta) \in G, r \in \mathbf{R})$ – therefore $\hat{H}$ admits three G-orbits, namely $Q^0 = \{\hat{\xi}_0\}$, $Q^+ = \{\hat{\xi}_r : r > 0\}$, and $Q^- = \{\hat{\xi}_r : r < 0\}$. Thanks to Lemma 5.2.4.1 (iii) and the discussion in the Remark following Theorem 5.4.1.4, the theory developed in 5.4.1 is applicable here – in particular, Theorem 5.4.1.4 is available. The elements of $\hat{G}$ associated with the trivial orbit Q^0 are just the (unitary) characters U^r,

$$(\alpha, \beta) \mapsto \exp(\sqrt{-1}r \log \alpha) \qquad (r \in \mathbf{R});$$

the typical element $\hat{\xi}_1$ of the orbit Q^+ has for its stability subgroup H itself and so there is only one element of $\hat{G}$ associated with Q^+, namely $U^+ = U^{\xi_1}$; similarly there is just one element of G associated with Q^-, namely $U^- = U^{\xi_{-1}}$ (U^- is the contragredient to U^+). We conclude, therefore, that every irreducible unitary representation of G is unitarily equivalent to either U^+, U^- or to a unitary character U^r, a classical fact due to Gelfand and Naimark. The unitarily induced representations $U^\pm$ can be realized explicitly on $L^2(\mathbf{R})$; thus the mapping π, $(\alpha, \beta) \mapsto \log \alpha$, is a homomorphism of G onto the additive reals and so it is clear that

$$U^\pm(\alpha, \beta)f(t) = \exp(\pm\sqrt{-1}e^t\beta)f(t + \log \alpha) \qquad (f \in L^2(\mathbf{R}), t \in \mathbf{R})$$

(see the Remark immediately preceding the discussion on fiber bundles in 5.1.1 – in particular formula (†)).

Remarks (1) The reader will easily convince himself that $U^+(U^-)$ can be realized on $L^2_+ = L^2([0, +\infty))(L^2_- = L^2((-\infty, 0]))$, the explicit formula being

$$U^\pm(\alpha, \beta)f(t) = e^{\sqrt{-1}\beta t}\sqrt{\alpha}f(\alpha t) \qquad (f \in L^2_\pm, t \geqslant, \leqslant 0).$$

(2) The '$\alpha t + \beta$' group is type I (hence postliminaire) but not liminaire – see, for instance, Nelson and Stinespring [1, p. 556].

Example 2 Consider the three dimensional Lie algebra $\mathfrak{g}$ which has a basis $\{X, Y, Z\}$ and relations $[X, Z] = \sigma Y + Z$, $[X, Y] = Y - \sigma Z$, $[Y, Z] = 0$, σ a non-zero real number. The connected, simply connected solvable Lie group G corresponding to $\mathfrak{g}$ may be represented by the pairs $\{(\alpha, \beta) : \alpha \in \mathbf{R}, \beta \in \mathbf{C}\}$, multiplication being given by the rule

$$(\alpha_1, \beta_1)(\alpha_2, \beta_2) = (\alpha_1 + \alpha_2, e^{(1-\sqrt{-1}\sigma)\alpha_1}\beta_2 + \beta_1).$$

Our group G is the semi-direct product of the closed normal abelian subgroup $H = \{(0, \beta)\}$ with the one parameter group K determined by X. The dual group $\hat{H}$ consists of the unitary characters $\hat{\xi}_z$, $(0, \beta) \mapsto \exp[\sqrt{-1}\mathscr{R}(\beta z)]$ $(z \in \mathbf{C})$; the action of G on $\hat{H}$ is easily seen to be given by $(\alpha, \beta) \cdot \hat{\xi}_z = \hat{\xi}_{\tilde{z}}$, where $\tilde{z} = ze^{-(1-\sqrt{-1}\sigma)\alpha}$ – thus, for $z \neq 0$, the G-orbits in $\hat{H}$ are spirals in the z-plane and so it is clear that the theory outlined in 5.4.1 is applicable here. The stability subgroup for $\hat{\xi}_z$ $(z \neq 0)$ is just H itself while, of course, if $z = 0$, then it is all of G. A convenient cross section for the non-trivial orbits is the set of $\hat{\xi}_s$, s a real number lying between 1 and $e^{2\pi/\sigma}$; denoting the corresponding unitarily induced representation by U^s, we may now conclude that every irreducible unitary representation of G is unitarily equivalent to a U^s, with s in the half open interval whose end points are 1 and $e^{2\pi/\sigma}$, or is a one dimensional representation U^r, $(\alpha, \beta) \mapsto e^{\sqrt{-1}\alpha r}$ (r real). The unitarily induced representations U^s can be realized on $L^2(\mathbf{R})$ (cf. Example 1); thus the mapping π, $(\alpha, \beta) \mapsto \alpha$, is a homomorphism of G onto the additive reals and so it is clear that

$$U^s(\alpha, \beta)f(t) = \exp[\sqrt{-1}s\mathscr{R}(e^{(1-\sqrt{-1}a)t}\beta)]f(t + \alpha) \qquad (f \in L^2(\mathbf{R}), t \in \mathbf{R}).$$

Example 3 *The Heisenberg Group.* Let G be the three dimensional nilpotent Lie group consisting of all matrices of the form

$$(x, y, z) = \begin{pmatrix} 1 & x & z \\ 0 & 1 & y \\ 0 & 0 & 1 \end{pmatrix} \qquad (x, y, z \in \mathbf{R}) - G \text{ is called the } \textit{Heisenberg group.}$$

Evidently

$$(x_1, y_1, z_1)(x_2, y_2, z_2) = (x_1 + x_2, y_1 + y_2, z_1 + z_2 + x_1 y_2);$$

hence the center Z of G is the set $\{(0, 0, z) : z \in \mathbf{R}\}$. The group G is the semi-direct product of the closed normal abelian subgroup H with the closed subgroup K, where

$$H = \{(0, y, z) : y, z \in \mathbf{R}\}, K = \{(x, 0, 0) : x \in \mathbf{R}\}.$$

The dual group $\hat{H}$ consists of the unitary characters

$$\hat{\xi}_{r,s}, (0, y, z) \mapsto e^{\sqrt{-1}(ry+sz)} \qquad (r, s \in \mathbf{R});$$

the action of G on $\hat{H}$ is given by $(x, y, z)\cdot\hat{\xi}_{r,s} = \hat{\xi}_{r-xs,s}$ – thus the G-orbits in $\hat{H}$ fall into two classes, namely the one element orbits $\{\hat{\xi}_{r,0}\}$ $(r \in \mathbf{R})$ and the orbits Q_s $(s \neq 0)$, $Q_s = \{\hat{\xi}_{r,s} : r \in \mathbf{R}\}$. It is then clear that the orbit space of $\hat{H}$ is countably separated, hence Theorem 5.4.1.4 is now at our disposal. This theorem tells us that the elements of $\hat{G}$ associated with an orbit $\{\hat{\xi}_{r,0}\}$ are the unitary characters

$$U^{\rho,r}, (x, y, z) \mapsto e^{\sqrt{-1}(\rho x+ry)} \qquad (r, \rho \in \mathbf{R});$$

the stability subgroup for the typical element $\hat{\xi}_{0,s}$ $(s \neq 0)$ of Q_s is just H itself and so there is only one element of $\hat{G}$ associated with Q_s, namely the unitarily induced representation $U^s = U^{\xi_{0,s}}$. Hence, writing r for s, we now see that every irreducible unitary representation of the Heisenberg group G is unitarily equivalent to either a unitarily induced representation U^r $(r \neq 0)$ or to a unitary character $U^{r,\rho}$ (these representations are, evidently, pairwise inequivalent). The unitarily induced representations U^r can be realized on $L^2(\mathbf{R})$ (cf. Example 1); thus the mapping π, $(x, y, z) \mapsto x$, is a homomorphism of G onto the additive reals and so it is clear that

$$U^r((x, y, z))f(t) = \exp(\sqrt{-1}r(z + ty))f(t + x) \qquad (f \in L^2(\mathbf{R}), t \in \mathbf{R}).$$

Example 4 *The Euclidean Groups.* Let G be the n dimensional (real) Euclidean group $\mathbf{E}(n)$ – thus G consists of all pairs (k, a) with $k \in \mathbf{O}(n)$, $a \in \mathbf{R}^n$, the pairs multiplying according to the rule $(k_1, a_1)(k_2, a_2) = (k_1k_2, a_1 + k_1a_2)$. It is clear that G is the semi-direct product of the closed normal abelian subgroup H with the closed subgroup K, where

$$H = \{(1, a) : a \in \mathbf{R}^n\}, K = \{(k, 0) : k \in \mathbf{O}(n)\}.$$

The dual group $\hat{H}$ is comprised of the unitary characters $\hat{\xi}_a$, $b \mapsto e^{\sqrt{-1}(a,b)}$ $(a, b \in \mathbf{R}^n)$, $((\,.\,,\,.\,)$ the inner product in $\mathbf{R}^n)$; the G-orbit of a given character $\hat{\xi}_a$ consists of all $\hat{\xi}_b$ $(b \in \mathbf{R}^n)$ where $\|a\| = \|b\|$ – therefore the G-orbits in $\hat{H}$ are spheres Q_r, $Q_r = \{\hat{\xi}_a : \|a\| = r\}$ $(r \geqslant 0)$. Because G/H is compact, the theory outlined in 5.4.1 is automatically applicable in the present situation (cf. the Remark following Theorem 5.4.1.4). The elements of $\hat{G}$ associated with the trivial orbit Q_0 are those irreducible unitary representations of G which are 'lifted' from elements of $(G/H)^\wedge = (\mathbf{O}(n))^\wedge$; the stability subgroup $H_{\hat{\xi}_r}$ of the typical element $\hat{\xi}_r$ of Q_r ($r \sim (0, 0, \ldots, 0, r)$, $r > 0$) is, evidently, the set $\{(k, a) : k \in \mathbf{O}(n-1), a \in \mathbf{R}^n\}$, $\mathbf{O}(n-1)$ being identified with the subgroup of $\mathbf{O}(n)$ which leaves fixed the point $(0, 0, \ldots, 0, 1)$ – therefore the infinite dimensional elements of $\hat{G}$ are the unitarily induced representations $U^{\hat{\xi}_r \times L}$ ($r > 0, L \in (\mathbf{O}(n-1))^\wedge$), $\hat{\xi}_r \times L$ the element of $\hat{H}_{\hat{\xi}_r}$ given by

$$(\hat{\xi}_r \times L)(k\xi) = \langle \xi, \hat{\xi}_r \rangle L(k) \qquad (k \in H_{\hat{\xi}_r} \cap K = \mathbf{O}(n-1), \xi \in H = \mathbf{R}^n).$$

Example 5 *The Inhomogeneous Lorentz Group.* Let $\mathbf{L}(4)$ be the multiplicative group of 4 by 4 real matrices which, considered as operators on $\mathbf{R}^4$, leave invariant the form $a_1^2 + a_2^2 + a_3^2 - a_4^2$; equipped with the usual topology, $\mathbf{L}(4)$ is customarily referred to as the *homogeneous Lorentz group*. Let G be the group (with the direct product topology) comprised of all pairs (k, a) ($k \in \mathbf{L}(4)$, $a \in \mathbf{R}^4$), multiplication being given by $(k_1, a_1)(k_2, a_2) = (k_1k_2, a_1 + k_1a_2)$ – G is called the *inhomogeneous Lorentz group.* Evidently G is the semidirect product of the closed normal abelian subgroup H and the closed subgroup K, where

$$H = \{(1, a) : a \in \mathbf{R}^4\}, K = \{(k, 0) : k \in \mathbf{L}(4)\}.$$

The dual group $\hat{H}$ consists of the unitary characters $\hat{\xi}_a, b \mapsto e^{\sqrt{-1}(a,b)}$ ($a, b \in \mathbf{R}^4$), $((.,.)$ the Lorentz inner product in $\mathbf{R}^4$, i.e. $(a, b) = a_1b_1 + a_2b_2 + a_3b_3 - a_4b_4)$; the action of G on $\hat{H}$ is easily seen to be given by $(k, a)\cdot\hat{\xi}_b = \hat{\xi}_{k\cdot b}$ (therefore the orbit that contains $\hat{\xi}_b$ is itself contained in the set $\{\hat{\xi}_a : (a, a) = (b, b)\}$) – in fact it is clear that the G-orbits in $\hat{H}$ fall into three classes, namely the trivial orbit $\{\hat{\xi}_0\}$, the orbit $Q_0 = \{\hat{\xi}_a : a \neq 0\ (a \in \mathbf{R}^4), (a, a) = 0\}$, and the Q_r ($r \neq 0$), $Q_r = \{\hat{\xi}_a : a \in \mathbf{R}^4, (a, a) = r\}$. The theory developed in 5.4.1 is clearly applicable here; this being so, we now ask: What elements of $\hat{G}$ are associated with these orbits?

(i) The irreducible unitary representations of G determined by the trivial orbit $\{\hat{\xi}_0\}$ are, of course, those of the form $(k, a) \mapsto U(k)$, $U \in (\mathbf{L}(4))^\wedge$.

(ii) Let $r < 0$. A typical element in Q_r is $\hat{\xi}_r$ ($r \sim (0, 0, 0, \sqrt{|r|})$); because the stability subgroup $H_{\hat{\xi}_r}$ of $\hat{\xi}_r$ consists of those (k, a) in G such that k leaves fixed the point $(0, 0, 0, 1)$ (viewed as a column vector), it is clear that $H_{\hat{\xi}_r} \cap K = \mathbf{O}(3)$ – therefore the elements of $\hat{G}$ associated with Q_r are the unitarily induced representations $U^{\hat{\xi}_r \times L}$ ($L \in (\mathbf{O}(3))^\wedge$).

(iii) Let $r > 0$. A typical element in Q_r is $\hat{\xi}_r$ ($r \sim (\sqrt{r}, 0, 0, 0)$); its stability subgroup $H_{\hat{\xi}_r}$ is composed of the pairs (k, a) in G with the property that k leaves fixed the point $(1, 0, 0, 0)$ (viewed as a column vector); thus $H_{\hat{\xi}_r} \cap K$ may be regarded as consisting of all $(k, 0)$, with k belonging to the group $\mathbf{L}(3)$ of 3 by 3 (real) matrices which leave invariant the form $a_2^2 + a_3^2 - a_4^2$ (we are, of course, making the obvious identifications) – therefore the elements of $\hat{G}$ associated with Q_r are the unitarily induced representations $U^{\hat{\xi}_r \times L}$ ($L \in (\mathbf{L}(3))^\wedge$).

(iv) Let $r = 0$. A typical element in the orbit Q_0 is the character $\hat{\xi}_{(0,0,1,1)}$; one may then check that the intersection of the stability subgroup for $\hat{\xi}_{(0,0,1,1)}$ with K is isomorphic to $\mathbf{E}(2)$, the Euclidean group of the plane – we leave the (straightforward) details to the reader. Writing $\hat{\xi}_0$ for $\hat{\xi}_{(0,0,1,1)}$, we now conclude that the elements of $\hat{G}$ associated with Q_0 are the unitarily induced representations $U^{\hat{\xi}_0 \times L}$ ($L \in (\mathbf{E}(2))^\wedge$).

The investigation of the irreducible unitary representations of G has thus been reduced to studying the irreducible unitary representations of four smaller groups, viz. **L**(4), **L**(3), **O**(3), and **E**(2); the unitary duals of **O**(3) and **E**(2) are, of course, known; on the other hand, the representation theory for **L**(4) and **L**(3) is closely tied up with that of **SL**(2, **C**) and **SL**(2, **R**), respectively (groups to which Mackey's method does not apply).

Remark The inhomogeneous Lorentz group is not liminaire (cf. Nelson and Stinespring [1, p. 556]).

Example 6 *É. Cartan's Motion Group.* Let $\mathfrak{g}$ be a semi-simple Lie algebra over **R**, $\mathfrak{g} = \mathfrak{k} + \mathfrak{p}$ a Cartan decomposition for $\mathfrak{g}$; let K be the analytic subgroup of Int ($\mathfrak{g}$) corresponding to $\mathfrak{k}$ – then K operates on $\mathfrak{p}$ in the obvious way via orthogonal transformations (it being understood, of course, that $\mathfrak{p}$ is equipped with the Euclidean structure derived from the restriction of the Killing form B of $\mathfrak{g}$ to $\mathfrak{p}$). Let G be the group (with the direct product topology) comprised of all pairs (k, a) $(k \in K, a \in \mathfrak{p})$, multiplication being given by $(k_1, a_1)(k_2, a_2) = (k_1k_2, a_1 + k_1a_2)$. G will be referred to as *É. Cartan's motion group.* It is clear that G is the semi-direct product of the closed normal abelian subgroup H and the closed subgroup K, where

$$H = \{(1, a) : a \in \mathfrak{p}\}, K = \{(k, 0) : k \in K\}.$$

The dual group $\hat{H}$ consists of the unitary characters $\hat{\xi}_a$, $b \mapsto e^{\sqrt{-1}(a, b)}$ $(a, b \in \mathfrak{p})$, ((. , .) the Killing inner product on $\mathfrak{p}$). In order to describe the G-orbit structure of $\hat{H}$ in a convenient way, we proceed as follows. Fix a maximal abelian subspace $\mathfrak{a}_\mathfrak{p}$ of $\mathfrak{p}$; let $p_1, \ldots, p_l$ be a set of algebraically independent homogeneous generators for the set $I(\mathfrak{p}_c)$ of K-invariants in the symmetric algebra $S(\mathfrak{p}_c)$ $(l = \dim(\mathfrak{a}_\mathfrak{p}))$; we may assume that the p_i have real coefficients and, if desired, that $p_1 = B(. , .)$. By the *complex distance* of a point $a \in \mathfrak{p}$ from the origin, we shall understand the l-tuple $(p_1(a), \ldots, p_l(a))$. Two points in $\mathfrak{p}$ are K-conjugate (i.e. lie on the same K-orbit) iff they have the same complex distance from the origin. [Since the p_i are K-invariant, K-conjugate points have the same complex distance from the origin; conversely suppose that we have points a and b in $\mathfrak{p}$ such that $p_i(a) = p_i(b)$ $(i = 1, \ldots, l)$ – then we must prove that a and b are K-conjugate. Now $\mathfrak{p} = \bigcup_k k(\mathfrak{a}_\mathfrak{p})$ and so we may assume that a and b both lie in $\mathfrak{a}_\mathfrak{p}$; but the restrictions to $\mathfrak{a}_\mathfrak{p}$ of the p_i generate the set of W-invariants in $S(\mathfrak{a}_{\mathfrak{p}_c})$ (Chevalley's Restriction Theorem) (W denoting, as usual the Weyl group of the pair $(\mathfrak{g}, \mathfrak{a}_\mathfrak{p})$); therefore, on the basis of standard arguments, we see that a and b are W-conjugate, which serves to establish our contention.] Consequently the G-orbits in $\hat{H}$ are 'complex spheres'; this being so, fix a Weyl chamber $\mathscr{C}$ in $\mathfrak{a}_\mathfrak{p}$; the closure of $\mathscr{C}$ in $\mathfrak{a}_\mathfrak{p}$ is a fundamental domain (in $\mathfrak{a}_\mathfrak{p}$) for the action of W (see Theorem 1.1.2.7) – therefore a convenient cross section in $\hat{H}$ for the set of G-orbits is $\mathscr{C}^{cl}$. With this preparation, the reader will have no difficulty in analyzing the unitary dual $\hat{G}$ of G. [Notice, in particular, that the intersection with K of the stability subgroups corresponding to Σ-regular elements in $\mathscr{C}^{cl}$ (i.e. to elements of $\mathscr{C}$) is M, M the centralizer of $\mathfrak{a}_\mathfrak{p}$ in K.]

5.5 Applications to Semi-Simple Lie Groups

5.5.1 The Subquotient Theorem

Let G be a connected semi-simple Lie group with Lie algebra $\mathfrak{g}$ – assume that G has finite center. Let $\mathfrak{g} = \mathfrak{k} + \mathfrak{a}_\mathfrak{p} + \mathfrak{n}$ be an Iwasawa decomposition for $\mathfrak{g}$, $G = KA_\mathfrak{p}N^+$ the corresponding Iwasawa decomposition for G and $P = MA_{\mathfrak{p}}N^+$ the associated minimal parabolic subgroup (M the centralizer of $A_{\mathfrak{p}}$ in K). Following Harish-Chandra [5, p. 239], we shall first construct a family $\{U^\mu\}$ of (continuous) representations of G parameterized by complex valued linear functions μ on $\mathfrak{a}_{\mathfrak{p}}$ and realized on $L^2(K)$. Although these representations (which may be viewed as being induced from characters of $A_{\mathfrak{p}}N^+$) are, in general, non-unitary, it turns out that they are intimately related with the representations unitarily induced by the finite dimensional irreducible unitary representations of P; furthermore, an important theorem due to Harish-Chandra says that *every* TCI Banach representation of G is Naimark equivalent to a subquotient of some U^μ.

Note Everything that we shall have to say in this number applies, with no essential change, to the pairs (G, K) of number 1.1.5.

Given $x \in G$, $k \in K$, write $xk = \kappa(xk)\exp(H(x, k))n(x, k)$ with $\kappa(xk) \in K$, $H(x, k) \in \mathfrak{a}_{\mathfrak{p}}$, $n(x, k) \in N^+$; for fixed x in G, define maps $\kappa_x : K \to K$, $H_x : K \to \mathfrak{a}_{\mathfrak{p}}$ by the rules

$$\kappa_x(k) = \kappa(xk), \quad H_x(k) = H(x, k) \qquad (k \in K).$$

It is easy to check that:

$$\text{(a)} \quad \kappa_{xy} = \kappa_x \circ \kappa_y; \qquad \text{(b)} \quad H_{xy} = H_x \circ \kappa_y + H_y \qquad (x, y \in G).$$

In addition, a standard argument shows that

$$\int_K f(k)\, dk = \int_K f\circ\kappa_x(k)\exp(-2\rho\circ H_x(k))\, dk \; (\rho = 2^{-1}\sum_{\lambda>0} m(\lambda)\lambda)$$

where dk is normalized Haar measure on K, $f \in L^1(K)$.

Let μ be a (complex valued) linear function on $\mathfrak{a}_{\mathfrak{p}}$; given $x \in G$, define $U^\mu(x)$ by the prescription

$$U^\mu(x)f = \exp(-(\mu + 2\rho)\circ H_{x^{-1}})(f\circ\kappa_{x^{-1}}) \qquad (f \in L^2(K)).$$

Claim: U^μ is a continuous representation of G on $L^2(K)$. For this, let's first check that $U^\mu(x)f$ really is in $L^2(K)$. Thus, noting that $H_{x^{-1}} = -H_x\circ\kappa_{x^{-1}}$ $(x \in G)$, we have

$$\begin{aligned}\int_K |U^\mu(x)f|^2\, dk &= \int_K |\exp((\mu + 2\rho)\circ H_x\circ\kappa_{x^{-1}})(f\circ\kappa_{x^{-1}})|^2\, dk \\ &= \int_K |\exp((\mu + 2\rho)\circ H_x)(f)|^2 \exp(-2\rho\circ H_x)\, dk \\ &= \int_K |\exp((\mu + \rho)\circ H_x)(f)|^2\, dk \leqslant M_x^2 \|f\|^2 < \infty,\end{aligned}$$

where $M_x = \sup_K |\exp((\mu + \rho)\circ H_x)|$ – hence $U^\mu(x)f \in L^2(K)$ and $\|U^\mu(x)\| \leqslant M_x$ $(x \in G)$. It is evident that $U^\mu(xy) = U^\mu(x)U^\mu(y)$ (all $x, y \in G$). As for the continuity of U^μ, note first that for each compact neighborhood ω of 1 in G there exists a number $M_\omega > 0$ such that

$\| U^\mu(x) \| \leqslant M_\omega$ (all $x \in \omega$) – therefore the continuity of U^μ will follow when it is shown that $\lim_{x\to 1} \| U^\mu(x)f - f \| = 0$ (all $f \in L^2(K)$). To this end, fix $f \in L^2(K)$; given $\epsilon > 0$, choose a continuous function g on K such that $\| f - g \| \leqslant \epsilon$ and then fix a compact neighborhood ω of 1 in G such that $\| U^\mu(x)g - g \|_\infty \leqslant \epsilon$ (all $x \in \omega$) – we thus have:

$$\begin{aligned} \| U^\mu(x)f - f \| &= \| U^\mu(x)(f - g) - (f - g) + (U^\mu(x)g - g) \| \\ &\leqslant \| U^\mu(x) \| \epsilon + \epsilon + \epsilon \\ &\leqslant (M_\omega + 2)\epsilon \qquad \text{(all } x \in \omega). \ldots \end{aligned}$$

Conclusion: U^μ is a continuous representation of G on $L^2(K)$ (which is even *unitary* if $\mu + \rho$ assumes only pure imaginary values).

Note Fix $x \in G$ – then, as is easily seen, the Hilbert space adjoint of $U^\mu(x)$ is $U^{-\bar{\mu}-2\rho}(x^{-1})$.

We shall now show that the representations U^μ are induced from certain one dimensional representations (i.e. characters) of $A_\mathfrak{p}N^+$.

Lemma 5.5.1.1 The modular function δ_P is given by

$$\delta_P(mhn) = h^{-2\rho} \qquad (m \in \mathsf{M}, h \in A_\mathfrak{p}, n \in N^+).$$

Proof Since M is compact, $\delta_P | \mathsf{M}$ is one; since N^+ is the derived group of $A_\mathfrak{p}N^+$, $\delta_P | N^+$ is one. It remains to determine $\delta_P | A_\mathfrak{p}$ – thus, for all h in $A_\mathfrak{p}$, we have

$$\begin{aligned} \delta_P(\exp(\log h)) &= \det(\mathrm{Ad}_P(\exp(-\log h))) \\ &= e^{-\mathrm{tr}(\mathrm{ad}_\mathfrak{p}(\log h))} = h^{-2\rho}. \quad \square \end{aligned}$$

Corollary 5.5.1.2 The modular function $\delta_{A_\mathfrak{p}N^+}$ is given by

$$\delta_{A_\mathfrak{p}N^+}(hn) = h^{-2\rho} \qquad (h \in A_\mathfrak{p}, n \in N^+).$$

Let $\mu : \mathfrak{a}_\mathfrak{p} \to \mathbf{C}$ be a linear function on $\mathfrak{a}_\mathfrak{p}$, L^μ the one dimensional representation of $A_\mathfrak{p}N^+$ defined by the prescription $L^\mu(hn) = h^{\mu+\rho}$ ($h \in A_\mathfrak{p}, n \in N^+$). Consider the set ${}^{L^\mu}C(G)$ of those continuous functions f on G such that $f(xhn) = h^{-\rho}L^\mu((hn)^{-1})f(x)$ ($x \in G, hn \in A_\mathfrak{p}N^+$; recall that G is unimodular); letting G operate by left translation on ${}^{L^\mu}C(G)$ leads, in the natural way, to a continuous representation of G on ${}^{L^\mu}C(G)$, namely the representation U^{L^μ} continuously induced by L^μ. Let i denote the map which assigns to each $f \in {}^{L^\mu}C(G)$ its restriction to K – then $i : {}^{L^\mu}C(G) \to C(K)$ is a homeomorphism. Plainly the image of ${}^{L^\mu}C^\infty(G)$ under i in $L^2(K)$ is dense. Finally we claim that $U^\mu(x)\circ i = i \circ U^{L^\mu}(x)$ ($x \in G$). Thus, if $f \in {}^{L^\mu}C(G)$, then, for all $k \in K$, we have

$$U^\mu(x)\circ i(f)(k) = \exp(-(\mu + 2\rho)\circ H_{x^{-1}}(k))f(\kappa(x^{-1}k))$$

and

$$i \circ U^{L^\mu}(x)(f)(k) = f(x^{-1}k) \qquad (x \in G).$$

Write $x^{-1}k = \kappa(x^{-1}k)\exp(H(x^{-1}, k))n(x^{-1}, k)$ $(x \in G, k \in K)$ – then

$$f(x^{-1}k) = \exp(-(\mu + 2\rho)\circ H_{x^{-1}}(k))f(\kappa(x^{-1}k))$$

since $f \in {}^{L^\mu}C(G)$. We see, therefore, that U^μ is induced by L^μ in the sense of the definition given in 5.3.1.

Remark For the applications (cf. infra) it is sometimes convenient to realize the U^μ by 'right translation'; thus, inverting our definition leads to the consideration of the set of those continuous functions f on G such that $f(nhx) = h^\rho L^\mu(nh)f(x)$ $(nh \in N^+A_{\mathfrak{p}}, x \in G)$ . . . .

Let L denote the left regular representation of K on $L^2(K)$; clearly $L = U^\mu | K$. For any $\delta \in \hat{K}$, let $L^2(K; \delta)$ be the subspace of $L^2(K)$ consisting of those functions which transform under L according to δ — then $\dim(L^2(K; \delta)) = d(\delta)^2$. Consequently every vector in $\sum_{\delta \in \hat{K}} L^2(K; \delta)$ is analytic for U^μ (cf. Corollary 4.4.5.17).

Let R denote the right regular representation of K on $L^2(K)$ – then $R(m)U^\mu(x) = U^\mu(x)R(m)$ $(x \in G, m \in M)$. [This follows upon observing that

$$\begin{aligned} xkm &= (\kappa(xk)\exp(H(x, k))n(x, k))m \\ &= \kappa(xk)m\exp(H(x, k))n_m(x, k) \qquad (k \in K) \end{aligned}$$

where $n_m(x, k) = m^{-1}n(x, k)m \in N^+$.] Let $\hat{M}$ denote the set $\{\sigma\}$ of all equivalence classes of finite dimensional irreducible representations of M, ξ_σ the character of a class $\sigma \in \hat{M}$, $\chi_\sigma = d(\sigma)\xi_\sigma(d(\sigma)$ the degree of σ). Let μ_σ, $m \mapsto (a_{ij}(m))$ $(1 \leqslant i, j \leqslant d(\sigma))$, be an irreducible unitary matrix representation of M in the class σ. Put

$$Q_R(\sigma) = \int_M \chi_\sigma(m^{-1})R(m)\,dm, \quad Q_R(\sigma_{ij}) = d(\sigma)\int_M a_{ij}(m^{-1})R(m)\,dm,$$

dm being normalized Haar measure on the compact group M. The operators $Q_R(\sigma)$, $Q_R(\sigma_{ij})$ all commute with $U^\mu(\,.\,)$. Furthermore $Q_R(\sigma)$ and $Q_R(\sigma_{ii})$ $(1 \leqslant i \leqslant d(\sigma))$ are projections in $L^2(K)$; of course $Q_R(\sigma)L^2(K)$ is the isotypic M-submodule in $L^2(K)$ of type σ. Note too that $Q_R(\sigma_{ij})$ is a partial isometry mapping $Q_R(\sigma_{ii})L^2(K)$ onto $Q_R(\sigma_{jj})L^2(K)$ $(1 \leqslant i, j \leqslant d(\sigma))$. We have

$$L^2(K) = \sum_{\sigma \in \hat{M}} Q_R(\sigma)L^2(K), \quad Q_R(\sigma)L^2(K) = \sum_{i=1}^{d(\sigma)} Q_R(\sigma_{ii})L^2(K),$$

the (orthogonal) sums being taken in the sense of Hilbert space theory.

The representations U^μ are therefore highly reducible – in fact the closed subspaces $Q_R(\sigma_{ii})L^2(K)$ are U^μ-stable and the $d(\sigma)$ representations of G so obtained are all unitarily equivalent $(\sigma \in \hat{M})$. It will be convenient to refer to the U^μ as the *standard* representations of G; any of the various $U^\mu | Q_R(\check{\sigma}_{ii})L^2(K)$ ($\check{\sigma}$ the class contragredient to a given $\sigma \in \hat{M}$) is said to be an *elementary* representation of G (and will be denoted, indifferently, by

$U^{\sigma,\mu}$). [In passing let us record the following fact. Fix a class $\delta \in \hat{K}$ – then, on the basis of the Frobenius Reciprocity Theorem, $U^{\sigma,\mu} | K$ contains δ exactly $[\delta : \sigma]$ times ($[\delta : \sigma]$ the number of times that σ occurs in $\delta | \mathsf{M}$).]

Now suppose that $\mu + \rho$ assumes only *pure imaginary values* on $\mathfrak{a}_\mathfrak{p}$; as observed above U^μ is then a unitary representation of G. The representation $L^{\sigma,\mu}$ of P given by the prescription $mhn \mapsto h^{\mu+\rho}\mu_\sigma(m)$ ($m \in \mathsf{M}$, $h \in A_\mathfrak{p}$, $n \in N^+$) is a finite dimensional irreducible unitary representation of P and these are the only ones in view of the following lemma.

Lemma 5.5.1.3 Let L be a continuous finite dimensional irreducible representation (not necessarily unitary) of P on E – then $L(mhn) = \xi(h)L(m)$ ($m \in \mathsf{M}, h \in A_\mathfrak{p}, n \in N^+$), where $h \mapsto \xi(h)$ is a continuous homomorphism of $A_\mathfrak{p}$ into $\mathbf{C}^*$ and $m \mapsto L(m)$ is a continuous irreducible representation of M on E.

Proof The group $A_\mathfrak{p}N^+$ is connected and solvable – hence, by Lie's Theorem, there exists a non-zero vector $a \in E$ such that $L(hn)a = \xi(hn)a$ ($h \in A_\mathfrak{p}, n \in N^+$), ξ a one dimensional continuous representation of $A_\mathfrak{p}N^+$. Since N^+ is the derived group of $A_\mathfrak{p}N^+$, $\xi(n) = 1$ for all $n \in N^+$. On the other hand, $L(hn)L(m)a = \xi(h)L(m)a$ and so the span of the $L(m)a$ is L-stable; by the irreducibility of L, this span is E – thus $L(hn)$ reduces to the scalar $\xi(h)$. □

We propose now to show that $U^\mu | \mathsf{Q}_R(\check{\sigma}_{11})L^2(K)$ is unitarily equivalent to the unitarily induced representation $U^{L^{\sigma,\mu}}$ (which, for simplicity, will be denoted by $U^{\sigma,\mu}\ldots$).

It will be helpful to keep in mind the following well-known measure theoretic lemma (cf. Bourbaki [3, p. 66]).

Lemma 5.5.1.4 Let G be a locally compact group countable at infinity; let H_1 and H_2 be closed subgroups of G with the following properties: H_1H_2 is open in G, the complement of H_1H_2 in G has Haar measure zero, and $H_1 \cap H_2$ is compact. If d_{H_1} and d_{H_2} are left Haar measures on H_1 and H_2, respectively, then the left Haar measure d_G on G may be normalized in such a way that for any integrable or non-negative Borel function f on G

$$\int_G f(x)\, d_G(x) = \int_{H_1 \times H_2} f(\xi_1\xi_2)\frac{\delta_G(\xi_2)}{\delta_{H_2}(\xi_2)}\, d_{H_1}(\xi_1)\, d_{H_2}(\xi_2)$$

where δ_{H_2}, δ_G are the modular functions on H_2 and G, respectively. In addition, if $H_1 \cap H_2 = \{1\}$ and H_1H_2 is given the relative topology, then the map $(\xi_1, \xi_2) \mapsto \xi_1\xi_2$ is a homeomorphism of $H_1 \times H_2$ onto H_1H_2.

Let μ_σ, $m \mapsto (a_{ij}(m))$ ($1 \leqslant i, j \leqslant d(\sigma)$), be an irreducible unitary matrix representation in the class σ($m \in \mathsf{M}$) – then $\mu_\sigma(m)$ ($m \in \mathsf{M}$) operates on $\mathbf{C}^{d(\sigma)}$ by the rule

$$\begin{pmatrix} z_1 \\ \cdot \\ \cdot \\ \cdot \\ z_{d(\sigma)} \end{pmatrix} \mapsto \begin{pmatrix} a_{11}(m) & \cdots & a_{1,d(\sigma)}(m) \\ \cdot & & \cdot \\ \cdot & & \cdot \\ \cdot & & \cdot \\ a_{d(\sigma),1}(m) & \cdots & a_{d(\sigma),d(\sigma)}(m) \end{pmatrix} \begin{pmatrix} z_1 \\ \cdot \\ \cdot \\ \cdot \\ z_{d(\sigma)} \end{pmatrix} \qquad (z_i \in \mathbf{C}).$$

For $x \in G$, write $x = khn$ $(k \in K, h \in A_{\mathfrak{p}}, n \in N^+)$; put $\rho_P(x) = h^{-2\rho}$ – then ρ_P is a C^∞ rho-function on G relative to left invariant Haar measures on G and P. The representation space $E^{\sigma,\mu}$ for $U^{\sigma,\mu}$ can be regarded as consisting of those Borel functions $f: G \to \mathbf{C}^{d(\sigma)}$ such that

(a) $f(xmhn) = h^{-(\mu+2\rho)}\mu_\sigma(m^{-1})f(x)$ $(x \in G; mhn \in P = MA_{\mathfrak{p}}N^+)$;

(b) $\int_{G/P} \rho_P(x)^{-1} \| f(x) \|^2 \, d\mu_P(\dot{x}) < \infty$.

Note that the functions f in ${}^{\sigma,\mu}C(G; \mathbf{C}^{d(\sigma)})$ may be identified (by restriction) with the set $\mathscr{S}^{\sigma,\mu}$ of continuous functions $s: K \to \mathbf{C}^{d(\sigma)}$ such that $s(km) = \mu_\sigma(m^{-1})s(k)$ $(k \in K, m \in M)$. Now, since $G = KP$ and $M = K \cap P$ is compact, the left Haar measures d_G and d_P on G and P, respectively, can be normalized so that Lemma 5.5.1.4 is applicable. Fix $f \in {}^{\sigma,\mu}C(G; \mathbf{C}^{d(\sigma)})$; choose $\phi \in C_c^+(G)$ such that $\int_P \phi(xp)d_P(p) = 1$ for all $x \in G$ (this is possible since $G = KP$ and K is compact) – then

$$\begin{aligned} \| f \|^2 &= \int_{G/P} \rho_P(x)^{-1} \| f(x) \|^2 \, d\mu_P(\dot{x}) = \int_G \| f(x) \|^2 \phi(x) \, d_G(x) \\ &= \int_{K \times P} \| f(kp) \|^2 \phi(kp) \delta_P(p^{-1}) \, dk d_P(p) \\ &= \int_K \| f(k) \|^2 \, dk = \| s \|^2 \qquad (s = f | K). \end{aligned}$$

Finally we observe that the transport of $U^{\sigma,\mu}$ to $\mathscr{S}^{\sigma,\mu}$ is given by

$$U^{\sigma,\mu}(x)s = \exp(-(\mu + 2\rho)\circ H_{x^{-1}})(s \circ \kappa_{x^{-1}}) \qquad (s \in \mathscr{S}^{\sigma,\mu}).$$

Let $g \in \mathbf{Q}_R(\check{\sigma}_{11})C(K)$ – then g determines an element s_g of $\mathscr{S}^{\sigma,\mu}$ in the following way. Put

$$g_i(k) = d(\sigma) \int_M \overline{a_{1i}(m^{-1})} g(km) \, dm \qquad (i = 1, \ldots, d(\sigma))$$

and let s_g be defined by the rule

$$k \mapsto \frac{1}{\sqrt{d(\sigma)}} \begin{pmatrix} g_1(k) \\ \cdot \\ \cdot \\ \cdot \\ g_{d(\sigma)}(k) \end{pmatrix} \qquad (k \in K).$$

In order to check that $s_g \in \mathscr{S}^{\sigma,\mu}$ we must verify that $s_g(km) = \mu_\sigma(m^{-1})s_g(k)$ $(k \in K, m \in M)$, i.e. that $\sum_{j=1}^{d(\sigma)} a_{ij}(m^{-1})g_j(k) = g_i(km)$ $(i = 1, \ldots, d(\sigma))$ – but

$$\begin{aligned} \sum_j a_{ij}(m^{-1})g_j(k) &= d(\sigma) \int_M \Big(\sum_j a_{ij}(m^{-1})a_{j1}(\tilde{m}) \Big) g(k\tilde{m}) \, d\tilde{m} \\ &= d(\sigma) \int_M a_{i1}(m^{-1}\tilde{m}) g(k\tilde{m}) \, d\tilde{m} = g_i(km), \end{aligned}$$

as desired. Bearing in mind that $\mathsf{Q}_R(\check{\sigma}_{1i})$ maps $\mathsf{Q}_R(\check{\sigma}_{11})L^2(K)$ isometrically onto $\mathsf{Q}_R(\check{\sigma}_{ii})L^2(K)$, we find next that

$$\|s_g\|^2 = d(\sigma)^{-1}\sum_i \int_K g_i(k)\overline{g_i(k)}\,dk$$

$$= d(\sigma)^{-1}\sum_i \int_K \mathsf{Q}_R(\check{\sigma}_{1i})g(k)\overline{\mathsf{Q}_R(\check{\sigma}_{1i})g(k)}\,dk = \|g\|^2.$$

The map $g \mapsto s_g$ of $\mathsf{Q}_R(\check{\sigma}_{11})C(K)$ into $\mathscr{S}^{\sigma,\mu}$ is clearly injective. To see that it is bijective, take any

$$s = \begin{pmatrix} s_1 \\ \cdot \\ \cdot \\ \cdot \\ s_{d(\sigma)} \end{pmatrix} \in \mathscr{S}^{\sigma,\mu} \text{ -- then } s_1 \in \mathsf{Q}_R(\check{\sigma}_{11})C(K).$$

In fact, for $k \in K$, we have

$$\mathsf{Q}_R(\check{\sigma}_{11})s_1(k) = d(\sigma)\int_{\mathsf{M}} \overline{a_{11}(m^{-1})}s_1(km)\,dm$$

$$= d(\sigma)\sum_i \left(\int_{\mathsf{M}} a_{11}(m)a_{1i}(m^{-1})\,dm\right)s_i(k) = s_1(k).$$

Finally we claim that $s_i(k) = d(\sigma)\int_{\mathsf{M}} \overline{a_{1i}(m^{-1})}s_1(km)dm$ $(k \in K)$ – thus

$$d(\sigma)\int_{\mathsf{M}} \overline{a_{1i}(m^{-1})}s_1(km)\,dm = d(\sigma)\sum_j \left(\int_{\mathsf{M}} a_{i1}(m)\overline{a_{j1}(m)}\,dm\right)s_j(k)$$

$$= s_i(k) \qquad (k \in K).$$

The map $Q, g \mapsto s_g$, of $\mathsf{Q}_R(\check{\sigma}_{11})C(K)$ onto $\mathscr{S}^{\sigma,\mu}$ is, therefore, a linear isometry; but Q clearly commutes with the action of G and so, in view of the remarks in the previous paragraph, we conclude that $U^{\mu} | \mathsf{Q}_R(\check{\sigma}_{11})L^2(K)$ and $U^{\sigma,\mu}$ are indeed unitarily equivalent.

Remark Let us explicitly record the fact that, for an appropriate normalization of Haar measures, one has

$$\|f\|^2 = \int_K \|f(k)\|^2\,dk \qquad (f \in E^{\sigma,\mu}).$$

The question of the irreducibility of the $U^{\sigma,\mu}$ will be taken up in 5.5.2; at the moment, however, let us ask: How many times does the restriction of $U^{\sigma,\mu}$ to K contain a given class $\delta \in \hat{K}$? The answer to this is immediate provided we observe that the restriction $U^{\sigma,\mu} | K$ is the unitary representation of K unitarily induced by the irreducible unitary representation $m \mapsto \mu_\sigma(m)$ of M, so that, in view of the Frobenius Reciprocity Theorem, the number of times that $U^{\sigma,\mu} | K$ contains a given $\delta \in \hat{K}$ is precisely $[\delta : \sigma]$, $[\delta : \sigma]$ the number of times that σ occurs in $\delta | \mathsf{M}$. [Consequently the continuous 'parameter' μ has no influence on the restriction of $U^{\sigma,\mu}$ to K.]

Let N^- denote the analytic subgroup of G corresponding to $\mathfrak{n}^-$ – then the $U^{\sigma,\mu}$ can also be realized on $L^2(N^-; \mathbf{C}^{d(\sigma)})$. To see this, first recall that N^-P is a dense open submanifold of G whose complement has zero Haar measure (cf. Proposition 1.2.3.5) – moreover $N^- \cap P = \{1\}$. Therefore, if the Haar measures on G and P are as in the Remark supra, then, according to Lemma 5.5.1.4, we have

$$\int_G f(x)\, d_G(x) = \int_{N^- \times P} f(np)\delta_P(p^{-1})\, d_{N^-}(n)\, d_P(p)$$

for any integrable f provided the Haar measure on N^- is normalized in a suitable way. This being so, select now a $\phi \in C_c^+(G)$ such that $\int_P \phi(xp) d_P(p) = 1$ $(x \in G)$ – then, for any $f \in E^{\sigma,\mu}$, one has

$$\begin{aligned} \|f\|^2 &= \int_G \|f(x)\|^2\phi(x)\, d_G(x) \\ &= \int_{N^- \times P} \|f(np)\|^2\phi(np)\delta_P(p^{-1})\, d_{N^-}(n)\, d_P(p) \\ &= \int_{N^-} \|f(n)\|^2\, d_{N^-}(n) \Rightarrow f \mid N^- \in L^2(N^-; \mathbf{C}^{d(\sigma)}). \end{aligned}$$

On the other hand, given $g \in L^2(N^-; \mathbf{C}^{d(\sigma)})$ extend it to a function f on N^-P by writing

$$f(np) = \exp\left(-(\mu + 2\rho)(H_p(1))\right)\mu_\sigma(m^{-1})g(n)$$

and set $f(x) = 0$ for any x in G which is not in N^-P. Plainly $f|N^- = g$ and, of course, $f \in E^{\sigma,\mu}$. Thus restricting elements in $E^{\sigma,\mu}$ to N^- sets up an isometry between $E^{\sigma,\mu}$ and $L^2(N^-; \mathbf{C}^{d(\sigma)})$ by means of which $U^{\sigma,\mu}$ may be transported to a unitary representation of G on $L^2(N^-; \mathbf{C}^{d(\sigma)})$. It is not difficult to determine the form of the transport of $U^{\sigma,\mu}$ to $L^2(N^-;$ $\mathbf{C}^{d(\sigma)})$. Thus let π_{N^-} and π_P denote the coordinate projections of N^-P onto N^- and P, respectively. For each $x \in G$, the set $N_x^- = \{n \in N^- : x^{-1}n \in N^-P\}$ is open in N^- and its complement has Haar measure zero. The transport of $U^{\sigma,\mu}(x)$ $(x \in G)$ to $L^2(N^-; \mathbf{C}^{d(\sigma)})$ is then seen to be given by

$$U^{\sigma,\mu}(x)f(n) = \delta_P^{1/2}(\pi_P(x^{-1}n))L^{\sigma,\mu}(\pi_P(x^{-1}n)^{-1})f(\pi_{N^-}(x^{-1}n))$$
$$(N^- \text{ almost everywhere}).$$

Example Let us explicate the preceding considerations in the special case when $G = \mathbf{SL}(2, \mathbf{R})$. We may take for M the set $\{\left(\begin{smallmatrix} 1 & 0 \\ 0 & 1 \end{smallmatrix}\right), \left(\begin{smallmatrix} -1 & 0 \\ 0 & -1 \end{smallmatrix}\right)\}$, for $A_\mathfrak{p}$ the set $\{\left(\begin{smallmatrix} a & 0 \\ 0 & a^{-1} \end{smallmatrix}\right) : a \in \mathbf{R}^+\}$, for P the set $\{\left(\begin{smallmatrix} a & b \\ 0 & a^{-1} \end{smallmatrix}\right) : a, b \in \mathbf{R}; a \neq 0\}$ and for N^- the set $\{\left(\begin{smallmatrix} 1 & 0 \\ c & 1 \end{smallmatrix}\right) : c \in \mathbf{R}\}$. Suppose that $x = \left(\begin{smallmatrix} a & b \\ c & d \end{smallmatrix}\right)$, $n = \left(\begin{smallmatrix} 1 & 0 \\ t & 1 \end{smallmatrix}\right)$ – then, by a matrix computation, we find that

$$\pi_{N^-}(x^{-1}n) = \begin{pmatrix} 1 & 0 \\ \dfrac{at - c}{-bt + d} & 1 \end{pmatrix}, \quad \pi_P(x^{-1}n) = \begin{pmatrix} (-bt + d) & -b \\ 0 & (-bt + d)^{-1} \end{pmatrix}$$
$$(t \neq d/b).$$

The two element group M has two finite dimensional irreducible unitary representations say '+' and '−'; here '+' denotes the trivial $(m \mapsto 1)$ one dimensional representation while '$-$' denotes the identity ($m \mapsto m$) representation. Consequently the finite dimensional irreducible unitary representations of P fall into two classes, namely

$$\begin{pmatrix} a & b \\ 0 & a^{-1} \end{pmatrix} \mapsto \begin{cases} |a|^{\sqrt{-1}r} \\ \operatorname{sign}(a)|a|^{\sqrt{-1}r} \end{cases} \qquad (r \in \mathbf{R}).$$

We may identify $L^2(N^-)$ with $L^2(\mathbf{R})$ – thus, for $f \in L^2(\mathbf{R})$, the resulting unitary representations of G are

$$U^{+,r}(x)f(t) = |-bt + d|^{-1-\sqrt{-1}r} f\left(\frac{at - c}{-bt + d}\right)$$

and

$$U^{-,r}(x)f(t) = \operatorname{sign}(-bt + d)|-bt + d|^{-1-\sqrt{-1}r} f\left(\frac{at - c}{-bt + d}\right)$$

(**R** almost everywhere).

It will be seen below (cf. too 5.5.2) that the representations in the first class are irreducible for every r; those in the second class are irreducible if $r \neq 0$ (if $r = 0$, then $U^{-,0}$ is a direct sum of two irreducibles). Moreover U^{+,r_1} is unitarily equivalent with U^{+,r_2} iff $r_1 = r_2$ or $r_1 = -r_2$; similarly for $U^{-,r}$. [If we use right translation (and hence right Haar measure) instead of left translation in the definition of the $U^{\pm,r}$, then our formulae assume a somewhat more symmetrical form: A direct computation shows that for $f \in L^2(\mathbf{R})$, one has

$$U^{+,r}(x)f(t) = |bt + d|^{-1-\sqrt{-1}r} f\left(\frac{at + c}{bt + d}\right)$$

and

$$U^{-,r}(x)f(t) = \operatorname{sign}(bt + d)|bt + d|^{-1-\sqrt{-1}r} f\left(\frac{at + c}{bt + d}\right)$$

(**R** almost everywhere).]

We shall leave it to the reader to work out for himself the explicit form the formulae take when $G = \mathbf{SL}(2, \mathbf{C})$ or when G is the universal covering group of the DeSitter group.

We shall now drop the assumption that $\mu + \rho$ assumes pure imaginary values on $\mathfrak{a}_\mathfrak{p}$ and return to the U^μ in general. Our objective will be to show that every TCI Banach representation of G is Naimark equivalent to an irreducible component of some elementary representation of G. This important result (which may be viewed as a first step in the classification of $\hat{G}$) shows, among other things, that K actually is large in G, thus serving to complete the discussion in number 4.5.2 (and justifying, incidentally, our use of the symbol $\hat{G}$).

Theorem 5.5.1.5 (Harish-Chandra) Let G be a connected semi-simple Lie group with finite center; let U be a TCI Banach representation of G on E – then U is Naimark equivalent to an irreducible component of some elementary representation of G.

Corollary 5.5.1.6 The standard representations $\{U^\mu\}$ form a complete system of representations of G (i.e. of $C_c(G)$).

[Since Naimark equivalent TCI Banach representations of G have the same kernel (in $C_c(G)$), our contention is an immediate consequence of Theorem 5.5.1.5 (modulo Gelfand-Raikov . . .).]

In view of the Frobenius Reciprocity Theorem, the preceding theorem will lead without difficulty to:

Theorem 5.5.1.7 (Harish-Chandra) Let G be a connected semi-simple Lie group with finite center; let U be a TCI Banach representation of G on E – then there exists a class $\sigma \in \hat{M}$ such that the multiplicity of any given $\delta \in \hat{K}$ which occurs in $U|K$ does not exceed $[\delta : \sigma]$, $[\delta : \sigma]$ the number of times that σ occurs in $\delta | M$.

Corollary 5.5.1.8 (Harish-Chandra) Let G be a connected semi-simple Lie group with finite center – then the maximal compact subgroup K is uniformly large in G. In fact a given $\delta \in \hat{K}$ occurs no more than $d(\delta)$ times in the restriction to K of any TCI Banach representation of G.

[On the basis of Theorem 5.5.1.7, this is obvious.]

Remark Let U be a TCI Banach representation of G on E; Theorem 5.5.1.7 gives an upper bound on the multiplicity with which a given $\delta \in \hat{K}$ occurs in $U|K$. Examples show that these multiplicities may very well be *strictly less* than the upper bound in question (in this connection, recall the situation as concerns the unitarily induced representations $U^{\sigma,\mu}$ supra).

Historically Harish-Chandra [6, Theorem 4] first gave a proof of Theorem 5.5.1.5 in the case when G admits a faithful finite dimensional representation; later, using the theory of spherical functions, he was able to prove the theorem in complete generality. [Accordingly, since we shall have to use a couple of facts which will not be established until later, the reader may wish to put off the detailed study of the proof for the time being.] Recently, a different proof was given by Rader [1].

Let $\mathfrak{X}$ denote the centralizer of $\mathfrak{k}_c$ in $\mathfrak{G}$ ($\mathfrak{G}$ the universal enveloping algebra of $\mathfrak{g}_c$).

The present investigation rests on the following lemma.

Lemma 5.5.1.9 Let U be a TCI Banach representation of G on a Banach space E; fix a class $\delta_0 \in \hat{K}$ which occurs in $U|K$ – then there exists a linear function $\mu : \mathfrak{a}_\mathfrak{p} \rightarrow \mathbf{C}$ with the property that $\omega(D : \mu + \rho) \in \operatorname{Ker}(\mu_{\delta_0}) \Rightarrow D \in \operatorname{Ker}(U_{\delta_0})$ $(D \in \mathfrak{X})$.

[The reader will find the proof of this lemma (together with an explanation of the notation) in Vol. II, number 9.1.3.]

So fix a TCI Banach representation U of G on a space E, say; fix a class $\delta_0 \in \hat{K}$ which occurs in $U|K$ and let μ be the linear function on $\mathfrak{a}_\mathfrak{p}$ given to us by the preceding lemma. Let $f_0 = \bar{\chi}_{\delta_0}$ – then it is clear that $f_0 \in L^2(K; \delta_0)$ (the space of square summable functions on K which transform under the left regular representation according to δ_0) and, in fact, $U_K^\mu(\mathfrak{K})f_0 = L^2(K; \delta_0)$ ($\mathfrak{K}$ the universal enveloping algebra of $\mathfrak{k}_c$). Now

$$U^\mu({}^{k_0}(nhk))f_0(k_0) = d(\delta_0)\,\mathrm{tr}\,(\mu_{\delta_0}(k_0^{-1}k))h^{(\mu+2\rho)}$$
$$(k, k_0 \in K; h \in A_{\mathfrak{p}}; n \in N^+)$$

and so we see that

$$U_K^\mu(D)f_0(k) = U_K^\mu(kD)f_0(k) = d(\delta_0)\,\mathrm{tr}\,(\mu_{\delta_0}(k^{-1})\mu_{\delta_0}(\omega(D:\mu+\rho)))$$
$$(k \in K)$$

for all $D \in \mathfrak{X}$. Suppose that D is an element of $\mathfrak{X}$ with the property that $U_K^\mu(D)f_0 = 0$ – then it must necessarily be the case that

$$\mathrm{tr}\,(\mu_{\delta_0}(k)\mu_{\delta_0}(\omega(D:\mu+\rho))) = 0 \qquad (\text{all } k \in K),$$

whence $\mu_{\delta_0}(\omega(D:\mu+\rho)) = 0$ (μ_{δ_0} being irreducible). Taking into account Lemma 5.5.1.9 then leads us to the conclusion that $U_K^\mu(D)f_0 = 0 \Rightarrow U_{\delta_0}(D)$ $(= \mathrm{P}(\delta_0)U_K(D)\mathrm{P}(\delta_0)\ldots) = 0$ $(D \in \mathfrak{X})$. In order to exploit this observation, we shall utilize the following simple result.

Lemma 5.5.1.10 Let π^0 and π_0 be finite dimensional representations of a subalgebra $\mathfrak{B}$ of $\mathfrak{G}$ which contains $\mathfrak{K}\mathfrak{X}$; suppose that π_0 is semi-simple and that $D \in \mathrm{Ker}\,(\pi^0) \Rightarrow D \in \mathrm{Ker}\,(\pi_0)$ $(D \in \mathfrak{X})$ – then $B \in \mathrm{Ker}\,(\pi^0) \Rightarrow B \in \mathrm{Ker}\,(\pi_0)$ $(B \in \mathfrak{B})$.

Proof We may represent $\mathfrak{k}_c$ (and hence $\mathfrak{K}$) on $\mathfrak{B}$ via the prescription $d(X)B = [X, B]$ $(X \in \mathfrak{k}_c, B \in \mathfrak{B})$; this being so, fix C in $\mathfrak{B}$ – then the degree of any element in $d(\mathfrak{K})C$ cannot exceed $\deg(C)$, whence $\dim(d(\mathfrak{K})C) < \infty$. It therefore follows from Lemma 2.1.2.1 that C can be written in the form

$$C = D + \sum_{i=1}^{r} [X_i, C_i]$$

where $D \in \mathfrak{X} \cap (d(\mathfrak{K})C)$, $X_i \in \mathfrak{k}_c$ and $C_i \in d(\mathfrak{K})C$ $(1 \leqslant i \leqslant r)$. Now fix an element B in $\mathfrak{B}$ such that $\pi^0(B) = 0$; we are to show that $\pi_0(B) = 0$. For this purpose let A be an arbitrary element of $\mathfrak{B}$ – then the preceding remarks are applicable to $C = AB$. Let $\mathfrak{B}^0$ be the kernel of π^0 in $\mathfrak{B}$ – then $C \in \mathfrak{B}^0$ and since $\mathfrak{B}^0$ is an ideal in $\mathfrak{B}$, it follows that $d(\mathfrak{K})C \subset \mathfrak{B}^0$. This implies that $\pi^0(D) = 0$, hence $\pi_0(D) = 0$ (from our hypotheses); on the other hand it is obvious that $\mathrm{tr}\,(\pi_0([X_i, C_i])) = 0$ and so $\mathrm{tr}\,(\pi_0(C)) = \mathrm{tr}\,(\pi_0(D)) = 0$. Consequently $\mathrm{tr}\,(\pi_0(AB)) = 0$ for every $A \in \mathfrak{B}$; but π_0 is a semi-simple representation – therefore $\pi_0(B) = 0$, as desired. □

Returning to our TCI Banach representation U, let us agree to denote by $\mathfrak{B}$ the subalgebra of $\mathfrak{G}$ comprised of those elements B in $\mathfrak{G}$ such that 1) $U_K^\mu(B)(L^2(K; \delta_0)) \subset L^2(K; \delta_0)$ and 2) $U_K(B)(E(\delta_0)) \subset E(\delta_0)$ – then $\mathfrak{B} \supset \mathfrak{K}\mathfrak{X}$. Let π^0 and π_0 be the representations of $\mathfrak{B}$ defined on $L^2(K; \delta_0)$ and $E(\delta_0)$ under U_K^μ and U_K, respectively. Since $E(\delta_0)$ is irreducible under

$U_K(\mathfrak{K}\mathfrak{X})$, π_0 is irreducible. [To verify that $E(\delta_0)$ is irreducible under $U_K(\mathfrak{K}\mathfrak{X})$, it would clearly be enough to establish the following statement: If E_1 and E_2 are any two subspaces of $E(\delta_0)$ which are stable and simple under $U_K(\mathfrak{K})$, then $E_2 \subset U_K(\mathfrak{K}\mathfrak{X})(E_1)$. Take E_1 and E_2 per our assertion – then E_1 and E_2 are both simple $\mathfrak{K}$-modules in the class δ_0 and so there exists a $\mathfrak{K}$-module isomorphism Q which maps E_1 onto E_2. Because the representation U_K of $\mathfrak{G}$ on $E_K = \sum_{\delta \in \hat{K}} E(\delta)$ is algebraically completely irreducible (cf. Theorem 4.5.5.4),

$$\mathrm{Hom}_{\mathbf{C}}(E(\delta_0), E(\delta_0)) = \{\mathsf{P}(\delta_0)U_K(D)\mathsf{P}(\delta_0) : D \in \mathfrak{G}\}.$$

Therefore there exists a $T \in \mathfrak{G}$ such that $Q = \mathsf{P}(\delta_0)U_K(T)\mathsf{P}(\delta_0) \mid E_1$ and we claim that actually $Q = \mathsf{P}(\delta_0)U_K(T_K)\mathsf{P}(\delta_0) \mid E_1$ (T_K the image of T in $\mathfrak{X}$ under the 'canonical projection' of $\mathfrak{G}$ onto $\mathfrak{X}$ – cf. Lemma 2.1.2.1). In fact, for all $D \in \mathfrak{K}$, we have $U_K(D)Q = QU_K(D)$, whence, for all $D \in \mathfrak{K}$, $\mathsf{P}(\delta_0)U_K([D, T])\mathsf{P}(\delta_0) = 0$ on E_1; owing to the Extended Burnside Theorem, we can write $T = T_K + \sum_i T_i$ ($T_i \in \mathfrak{G}$) where each T_i has the form

$$T_i = \sum_j c_j^i [D_{n_j}^{i,j}, [D_{n_j-1}^{i,j}, \ldots [D_1^{i,j}, T] \ldots] \qquad (c_j^i \in \mathbf{C}),$$

the $D^{i,j}$ lying in $\mathfrak{K}$ – but then it is clear that Q and $\mathsf{P}(\delta_0)U_K(T_K)\mathsf{P}(\delta_0)$ agree on E_1, as claimed. Consequently $E_2 = Q(E_1) = \mathsf{P}(\delta_0)U_K(T_K)(E_1) = U_K(T_K)(E_1) \subset U_K(\mathfrak{K}\mathfrak{X})(E_1) \ldots$.] Furthermore, as has been noted above, if $D \in \mathrm{Ker}(\pi^0)$ for some $D \in \mathfrak{X}$, then, of necessity, $D \in \mathrm{Ker}(\pi_0)$ – therefore, in view of Lemma 5.5.1.10, we conclude that $B \in \mathrm{Ker}(\pi^0) \Rightarrow B \in \mathrm{Ker}(\pi_0)$ ($B \in \mathfrak{B}$).

The elements in $L^2(K; \delta_0)$ transform to the left, of course, according to δ_0 – therefore they transform to the right according to $\check{\delta}_0$. Let $\check{\sigma}_1, \ldots, \check{\sigma}_r$ (say) be all the distinct classes in $\hat{M}$ which occur in $\check{\delta}_0 \mid M$ – then

$$L^2(K; \delta_0) = \sum_{i=1}^{r} \mathsf{Q}_R(\check{\sigma}_i)L^2(K; \delta_0)$$

and each of the spaces $\mathsf{Q}_R(\check{\sigma}_i)L^2(K; \delta_0)$ is π^0-stable. Let π_i^0 be the representation of $\mathfrak{B}$ on $\mathsf{Q}_R(\check{\sigma}_i)L^2(K; \delta_0)$ obtained by restricting π^0 to $\mathsf{Q}_R(\check{\sigma}_i)L^2(K; \delta_0)$ – then, in view of the generality which appears in the Appendix to the present number, we conclude that there exists an index i (say $i = 1$) such that $B \in \mathrm{Ker}(\pi_1^0) \Rightarrow B \in \mathrm{Ker}(\pi_0)$. We shall now drop the index 1 and write $\check{\sigma}$, $\mathsf{Q}_R(\check{\sigma})$, and π^0 instead of $\check{\sigma}_1$, $\mathsf{Q}_R(\check{\sigma}_1)$, and π_1^0, respectively. The space $\mathsf{Q}_R(\check{\sigma})L^2(K; \delta_0)$ breaks up according to the rule

$$\mathsf{Q}_R(\check{\sigma})L^2(K; \delta_0) = \sum_{i=1}^{d(\sigma)} \mathsf{Q}_R(\check{\sigma}_{ii})L^2(K; \delta_0)$$

and each of the $\mathsf{Q}_R(\check{\sigma}_{ii})L^2(K; \delta_0)$ is π^0-stable. We then conclude once again that there exists an index i (say $i = 1$) such that $B \in \mathrm{Ker}(\pi_1^0) \Rightarrow B \in \mathrm{Ker}(\pi_0)$ ($\pi_1^0 = \pi^0 \mid \mathsf{Q}_R(\check{\sigma}_{11})L^2(K; \delta_0)$). Let us agree to write π^0 for π_1^0. Choose now π^0-stable subspaces $\mathscr{S}^0$ and $\mathscr{S}_0$ of $\mathsf{Q}_R(\check{\sigma}_{11})L^2(K; \delta_0)$ with the following properties: (1) $\mathscr{S}^0 \supset \mathscr{S}_0$; (2) The representation of $\mathfrak{B}$ defined on the factor space $\mathscr{S}^0/\mathscr{S}_0$ under π^0 is equivalent to the irreducible representation π_0 of $\mathfrak{B}$ on $E(\delta_0)$. Fix an element $f^0 \in \mathscr{S}^0$ which is not in $\mathscr{S}_0$; let $\mathfrak{B}_0$ be the set of all $B \in \mathfrak{B}$ such that $\pi^0(B)f^0 \in \mathscr{S}_0$ – then $\mathfrak{B}_0$ is a maximal left ideal in $\mathfrak{B}$ and, moreover, $\mathfrak{G}\mathfrak{B}_0 \neq \mathfrak{G}$. [For otherwise we could choose $D_i \in \mathfrak{G}$ and $B_i \in \mathfrak{B}_0$ $(1 \leqslant i \leqslant r)$ such that $f^0 = \sum_{i=1}^r U_K^\mu(D_i B_i)f^0$. Let $\delta_1, \ldots, \delta_s$ be a finite number of distinct classes in $\hat{K}$, none of which is equal to δ_0; utilizing the Extended Burnside Theorem, select an element $D_0 \in \mathfrak{K}$ such that $U_K^\mu(D_0)\mathsf{P}^\mu(\delta_0) = \mathsf{P}^\mu(\delta_0)$, $U_K^\mu(D_0)\mathsf{P}^\mu(\delta_j) = 0$ $(1 \leqslant j \leqslant s)$; it is then clear that we also have

$$U_K(D_0)\mathsf{P}(\delta_0) = \mathsf{P}(\delta_0), \quad U_K(D_0)\mathsf{P}(\delta_j) = 0 \qquad (1 \leqslant j \leqslant s).$$

It can be assumed that the set $\delta_1, \ldots, \delta_s$ has been chosen so large that

$$U_K^\mu(D_i)L^2(K; \delta_0) \subset (\mathsf{P}^\mu(\delta_0) + \sum_j \mathsf{P}^\mu(\delta_j))L^2(K)$$

and

$$U_K(D_i)E(\delta_0) \subset (\mathsf{P}(\delta_0) + \sum_j \mathsf{P}(\delta_j))E \qquad (1 \leqslant i \leqslant r).$$

Then we have $U_K^\mu(D_0 D_i)L^2(K; \delta_0) \subset L^2(K; \delta_0)$ and $U_K(D_0 D_i)E(\delta_0) \subset E(\delta_0)$ which proves that $D_0 D_i \in \mathfrak{B}$ and therefore $D_0 D_i B_i \in \mathfrak{B}_0$ $(1 \leqslant i \leqslant r)$. On the other hand $f^0 \in L^2(K; \delta_0)$ and so

$$f^0 = U_K^\mu(D_0)f^0 = \sum_{i=1}^r U_K^\mu(D_0 D_i B_i)f^0 \in U_K^\mu(\mathfrak{B}_0)f^0 \subset \mathscr{S}_0,$$

a contradiction . . . .]

Invoking Zorn's Lemma, choose a maximal left ideal $\mathfrak{G}_0$ in $\mathfrak{G}$ containing $\mathfrak{G}\mathfrak{B}_0$ – then $\mathfrak{G}_0 \cap \mathfrak{B} \supset \mathfrak{B}_0$ and $1 \notin \mathfrak{G}_0 \cap \mathfrak{B}$, whence $\mathfrak{G}_0 \cap \mathfrak{B} = \mathfrak{B}_0$ ($\mathfrak{B}_0$ being a maximal left ideal in $\mathfrak{B}$). Let E^0 and E_0 be the closures in $L^2(K)$ of the spaces $U_K^\mu(\mathfrak{G})f^0$ and $U_K^\mu(\mathfrak{G}_0)f^0$, respectively – then, since f^0 is U^μ-analytic, it follows from Corollary 4.4.5.5 that both E^0 and E_0 are U^μ-stable. We have, moreover, $E_0 \subset E^0 \subset \mathsf{Q}_R(\check{\sigma}_{11})L^2(K)$. The restriction of U^μ to E^0 passes to the quotient to define a representation of G on E^0/E_0; since $\mathfrak{G}_0$ is a maximal left ideal in $\mathfrak{G}$ containing $\mathfrak{B}_0$, this representation of G is TCI – therefore, to finish the proof of Theorem 5.5.1.5, we need only show that it is Naimark equivalent to U. On the basis of Proposition 6.1.2.2 infra this is, however, clear (evidently $\mathscr{S}^0 = E^0 \cap L^2(K; \delta_0)$, $\mathscr{S}_0 = E_0 \cap L^2(K; \delta_0) \ldots$).

Let us pass now to Theorem 5.5.1.7. Owing to the Frobenius Reciprocity Theorem, the number of times that δ appears in the restriction of U^μ to $\mathsf{Q}_R(\check{\sigma}_{11})L^2(K)$ is precisely $[\delta : \sigma]$; a fortiori, then, $\dim(E(\delta)) \leqslant d(\delta)[\delta : \sigma] \leqslant d(\delta)^2$ (all $\delta \in \hat{K}$). . . .

The proofs of Theorems 5.5.1.5 and 5.5.1.7 are therefore complete.

In order to obtain a complete classification of $\bar{\bar{G}}$ it is necessary to 'decompose' each of the U^μ (or even the $U^{\sigma,\mu}$), that is, it is necessary to find a Jordan-Hölder series for the various standard (or elementary) representations of G; this is by no means an easy problem and in fact has only been seriously investigated in the case when G is in addition complex (cf. the papers of Želobenko). The discussion in the following Example gives a complete answer to this question when $G = \mathbf{SL}(2, \mathbf{R})$.

Example Let us take $G = \mathbf{SL}(2, \mathbf{R})$; in $\mathfrak{g} = \mathfrak{sl}(2, \mathbf{R})$, put $H = \left(\begin{smallmatrix} 1 & 0 \\ 0 & -1 \end{smallmatrix}\right)$, $X = \left(\begin{smallmatrix} 0 & 1 \\ 0 & 0 \end{smallmatrix}\right)$, $Y = \left(\begin{smallmatrix} 0 & 0 \\ 1 & 0 \end{smallmatrix}\right)$ – then the triple $\{H, X, Y\}$ forms a basis for $\mathfrak{sl}(2, \mathbf{R})$ and we have $[X, Y] = H, [H, X] = 2X, [H, Y] = -2Y$. We shall then agree to write $G = N^+A_{\mathfrak{p}}K$ where

$$K = \{\exp(\theta(X - Y)) : 0 \leqslant \theta < 2\pi\},$$
$$A_{\mathfrak{p}} = \{\exp(tH) : -\infty < t < \infty\},$$

and

$$N^+ = \{\exp(tX) : -\infty < t < \infty\}.$$

In what follows it will be convenient to realize the standard representations of G on $L^2(K)$ via *right* translation (cf. supra); furthermore we shall write U^μ for the standard representation that would ordinarily be denoted by $U^{\mu-2\rho}$ (μ a given linear form on $\mathfrak{a}_{\mathfrak{p}}$). [In passing observe that here $\rho(H) = 1$; therefore, in the new notations, U^μ is unitary whenever the number

$$\mu(H) \in \{1 + \sqrt{-1}y : y \in \mathbf{R}\}.]$$

Fix μ– then, for any C^∞ function f on K, we have the relations

$$(U_\infty^\mu(X - Y)f)(\theta) = \frac{\partial f(\theta)}{\partial \theta},$$
$$(U_\infty^\mu(H \pm \sqrt{-1}(X + Y))f)(\theta) = e^{\pm 2\sqrt{-1}\theta}\left(\mu(H) \mp \sqrt{-1}\frac{\partial}{\partial \theta}\right)f(\theta)$$
$$(0 \leqslant \theta < 2\pi).$$

[The first relation is obvious; as for the second, let F denote the unique extension of f to G which is determined by the requirement $F(nhk) = e^{\mu(\log h)}F(k)$ $(nh \in N^+A_{\mathfrak{p}}, k \in K)$ – then we have

$$(U_\infty^\mu(H \pm \sqrt{-1}(X + Y))f)(\theta)$$
$$= e^{\pm 2\sqrt{-1}\theta}(-H \mp \sqrt{-1}(X + Y))F(\exp(\theta(X - Y)))$$
$$= e^{\pm 2\sqrt{-1}\theta}\left(\mu(H) \mp \sqrt{-1}\frac{\partial}{\partial \theta}\right)f(\theta) \qquad (0 \leqslant \theta < 2\pi)$$

where, in going from the second equality to the last, we observed that $X + Y = 2X - (X - Y)$ and then utilized the fact that X annihilates F.] The set $\hat{K}$ may be identified with the set of integers; the mth isotypic component of $L^2(K)$ $(m \in \mathbf{Z})$ is one dimensional; the function $\theta \mapsto e^{\sqrt{-1}m\theta}$ determines a basis for $L^2(K; m)$ $(m \in \mathbf{Z})$. The two element group M (the centralizer of $A_{\mathfrak{p}}$ in K) has two finite dimensional irreducible unitary representations, say '+' and '−'; here '+' denotes the trivial $(m \mapsto 1)$ one dimensional representation while '−' denotes the identity $(m \mapsto m)$ representation. Let $L_+^2(K)$ denote the Hilbert space sum of the $\theta \mapsto e^{\sqrt{-1}m\theta}$ (m an even integer); let $L_-^2(K)$ denote the Hilbert space sum of the $\theta \mapsto e^{\sqrt{-1}m\theta}$ (m an odd integer) – then $L^2(K) = L_+^2(K) + L_-^2(K)$ (orthogonal direct sum) and $L_+^2(K)$ (respectively $L_-^2(K)$) is U^μ-stable (in fact is the representation space for the elementary representation $U^{+,\mu}$ (respectively $U^{-,\mu}$)). The following statement gives a complete analysis of the $U^{\pm,\mu}$ (and hence serves to classify $\bar{G}$).

(1) If $\mu(H)$ is not an integer, then both $U^{+,\mu}$ and $U^{-,\mu}$ are topologically irreducible.

(2) For M a positive integer, let $\mathscr{F}_M$ denote the subspace of $L^2(K)$ spanned by the $\theta \mapsto e^{\sqrt{-1}m\theta}$ ($m = -M, -M+2, \ldots, M-2, M$) – then, if $\mu(H) = -M$, $\mathscr{F}_M$ is U^μ-stable. For M an arbitrary integer, let $\mathscr{S}_M$ (respectively $\mathscr{T}_M$) denote the (closed) subspace of $L^2(K)$ spanned by the $\theta \mapsto e^{\sqrt{-1}m\theta}$ ($m = M + 2i$; $i \geqslant 0$) (respectively by the $\theta \mapsto e^{\sqrt{-1}m\theta}$ ($m = -M - 2i$; $i \geqslant 0$)) – then, if $\mu(H) = M$, both $\mathscr{S}_M$ and $\mathscr{T}_M$ are U^μ-stable.

(a) Suppose that $\mu(H) = M$ is an *odd* integer – then the elementary representation $U^{+,\mu}$ is topologically irreducible.

(α) If $\mu(H) = M$ (M odd and positive), then $\mathscr{S}_M \cap \mathscr{T}_M = \{0\}$ and $L^2_-(K)$ contains three $U^{-,\mu}$-stable subspaces, namely $\mathscr{S}_M$, $\mathscr{T}_M$, and $\mathscr{R}_M = \mathscr{S}_M + \mathscr{T}_M$ (whence the restriction of $U^{-,\mu}$ to $\mathscr{S}_M$ (or $\mathscr{T}_M$) is topologically irreducible as is the quotient representation on $L^2_-(K)/\mathscr{R}_M$).

(β) If $\mu(H) = -M$ (M odd and positive), then $\mathscr{S}_{-M} \cap \mathscr{T}_{-M} = \mathscr{F}_M$ and $L^2_-(K)$ contains three $U^{-,\mu}$-stable subspaces, namely $\mathscr{S}_{-M}$, $\mathscr{T}_{-M}$, and $\mathscr{F}_M$.

(b) Suppose that $\mu(H) = M$ is an *even* integer – then the elementary representation $U^{-,\mu}$ is topologically irreducible.

(α) If $\mu(H) = M$ (M even and positive), then $\mathscr{S}_M \cap \mathscr{T}_M = \{0\}$ and $L^2_+(K)$ contains three $U^{+,\mu}$-stable subspaces, namely $\mathscr{S}_M$, $\mathscr{T}_M$, and $\mathscr{R}_M = \mathscr{S}_M + \mathscr{T}_M$ (whence the restriction of $U^{+,\mu}$ to $\mathscr{S}_M$ (or $\mathscr{T}_M$) is topologically irreducible as is the quotient representation on $L^2_+(K)/\mathscr{R}_M$).

(β) If $\mu(H) = -M$ (M even and non-negative), then $\mathscr{S}_{-M} \cap \mathscr{T}_{-M} = \mathscr{F}_M$ and $L^2_+(K)$ contains three $U^{+,\mu}$-stable subspaces, namely $\mathscr{S}_{-M}$, $\mathscr{T}_{-M}$, and $\mathscr{F}_M$. [Since the triple $\{H, X - Y, X + Y\}$ forms a basis for $\mathfrak{g}$, all contentions concerning invariance and irreducibility follow from the formula

$$U^\mu_\infty(H \pm \sqrt{-1}(X+Y))^{n+1} \cdot e^{\sqrt{-1}m\theta} = (\mu(H) \pm m)(\mu(H) \pm m + 2) \cdots (\mu(H) \pm m + 2n) e^{\sqrt{-1}(m \pm 2n \pm 2)\theta}$$

valid for an arbitrary integer m ($n \in \mathbf{N}$).] Several remarks are in order here. First of all it is clear that all finite dimensional irreducible representations of G are picked up via the above procedure in the obvious way. Secondly our considerations show that the 'Theorem of Complete Irreducibility' is false for infinite dimensional $\mathfrak{G}$-modules (cf. supra . . .). Finally, if $\mu(H) = 1 + \sqrt{-1}y$ ($y \in \mathbf{R}$), then the $U^{\pm,\mu}$ are the unitary representations of the principal P-series for G (cf. 5.5.2); if $y \neq 0$, then a theorem due to Bruhat (cf. 5.5.2) automatically guarantees us that the $U^{\pm,\mu}$ are topologically irreducible; however if $y = 0$, his theorem is no longer applicable – but in this case it follows from what has been said above that $U^{+,\mu}$ is still topologically irreducible while, mysteriously enough, $U^{-,\mu}$ splits into a direct sum of *two* distinct irreducibles. [There are, of course, a number of equivalences among the various representations listed above, the explicit description of which may be left to the reader.]

Appendix We intend to establish here the following result.

Lemma Let A be an associative algebra over $\mathbf{C}$; let U and $V_1, \ldots, V_r$ be a finite set of finite dimensional representations of A; let V be the direct sum of the V_i ($1 \leqslant i \leqslant r$). Suppose that U is irreducible and $x \in \operatorname{Ker}(V) \Rightarrow x \in \operatorname{Ker}(U)$ ($x \in A$) – then there exists an index i such that $x \in \operatorname{Ker}(V_i) \Rightarrow x \in \operatorname{Ker}(U)$ ($x \in A$).

Proof Because $V(A)$ is a finite dimensional associative algebra, we can write $V(A) = \mathscr{S}_0 + \mathscr{N}_0$ where $\mathscr{N}_0$ is the radical and $\mathscr{S}_0$ is a semi-simple subalgebra of $V(A)$; let $\mathscr{S}$ and $\mathscr{N}$ be the complete inverse images of $\mathscr{S}_0$ and $\mathscr{N}_0$, respectively, under V. Since the assignment $V(x) \mapsto U(x)$ ($x \in A$) is an irreducible representation of $V(A)$, it is clear that $U(\mathcal{N}) = \{0\}$. Consider now the representation π of $V(\mathcal{S})$ which is defined by the rule $V(s) \mapsto U(s)$ $(s \in \mathcal{S})$; because $V(\mathcal{S})$ is semi-simple and U is irreducible, π must be equivalent to some irreducible constituent of the identity representation $V(s) \mapsto V(s)$ $(s \in \mathcal{S})$, that is π must be equivalent to some irreducible constituent of one of the representations $V(s) \mapsto V_i(s)$ $(s \in \mathcal{S})$ – hence there exists an index $i(1 \leqslant i \leqslant r)$ such that $s \in \operatorname{Ker}(V_i) \Rightarrow s \in \operatorname{Ker}(U)$ $(s \in \mathcal{S})$. This being so, let us suppose that $x \in \operatorname{Ker}(V_i)$; write $x = s + n$ $(s \in \mathcal{S}, n \in \mathcal{N})$ – then $V_i(s) = -V_i(n) \in V_i(\mathcal{N})$. However $V_i(\mathcal{S})$ is a semi-simple algebra and $V_i(\mathcal{N})$ is a nilpotent ideal in $V_i(A)$ – therefore $V_i(\mathcal{S}) \cap V_i(\mathcal{N}) = \{0\}$. Thus $V_i(s) = 0$ and so $U(x) = U(s + n) = U(s) = 0$, whence $x \in \operatorname{Ker}(V_i) \Rightarrow x \in \operatorname{Ker}(U)$ $(x \in A)$, as desired. □

5.5.2 Irreducibility of the Principal P-Series – P Minimal

Let G be a connected semi-simple Lie group with finite center, $G = KA_{\mathfrak{p}}N^+$ an Iwasawa decomposition for G, $P = MA_{\mathfrak{p}}N^+$ the corresponding minimal parabolic subgroup of G. We are now going to consider, in more detail, the unitary representations U^L of G unitarily induced by finite dimensional irreducible unitary representations L of P. As has been noted already (cf. Lemma 5.5.1.3), such an L is of the form $mhn \mapsto \xi(h)L(m)$ $(mhn \in P = MA_{\mathfrak{p}}N^+)$ with ξ a unitary character of $A_{\mathfrak{p}}$ and $L|M$ an irreducible unitary representation of M.

Definition The unitary representations U^L of G unitarily induced by finite dimensional irreducible unitary representations L of P will be referred to as the *principal P-series* of unitary representations for G.

Let W be the Weyl group of the pair $(\mathfrak{g}, \mathfrak{a}_{\mathfrak{p}})$ – thus $W = M^*/M$, M^* the normalizer of $A_{\mathfrak{p}}$ in K. The group W operates, in the obvious manner, on the set of equivalence classes of (continuous) finite dimensional irreducible representations of P. [Let $w \in W$; if L is a continuous finite dimensional irreducible representation of $MA_{\mathfrak{p}}$, then we shall let wL denote the representation of $MA_{\mathfrak{p}}$ given by

$$mh \mapsto \xi(m_w^{-1}hm_w)L(m_w^{-1}mm_w) \qquad (m \in M, h \in A_{\mathfrak{p}}),$$

m_w any element in M^* associated with w.]

The main question to be answered is: When are the U^L of the principal P-series irreducible? Here is a useful sufficient condition.

Theorem 5.5.2.1 (Bruhat) Let L be a finite dimensional irreducible unitary representation of P. If, for every element $w \neq 1$ of W, the representations L and wL of $MA_{\mathfrak{p}}$ are not equivalent, then the unitarily induced representation U^L of G is irreducible.

The proof of Theorem 5.5.2.1 (which depends on the results obtained in 5.3.2) will emerge during the course of the following discussion.

It will be necessary to consider the representations

$$\zeta \mapsto A_n(\zeta) = (\delta_P(\zeta)\delta_P(x\zeta x^{-1}))^{1/2}\delta_{P_x}(\zeta^{-1})\Lambda_n((\zeta,\zeta),\dot{1}) \quad (\zeta \in P_x = P \cap (x^{-1}Px))$$

with x running through a system of representatives for the double cosets in G mod P. The Bruhat Lemma (cf. Theorem 1.2.3.1) tells us that W parameterizes the P, P double cosets in G; this being so, fix a $w \in W$ and let m_w be any element in M^* associated with w – then, employing an evident change in notation, we have that $P_w = P \cap m_w P m_w^{-1} = \mathsf{M} A_{\mathfrak{p}} N_w^+$ ($N_w^+ = \exp(\mathfrak{n}_w^+)$, $\mathfrak{n}_w^+ = \mathfrak{n}^+ \cap \mathrm{Ad}(m_w)\mathfrak{n}^+$; cf. Proposition 1.1.4.5).

We shall start off by verifying that $A_0(\zeta) = 1$ ($\zeta \in P_x$; all $x \in G$). Thus let Σ^+ (Σ^-) denote the positive (negative) roots of the pair $(\mathfrak{g}, \mathfrak{a}_{\mathfrak{p}})$; given a $w \in W$, put

$$\Sigma^+(w) = \{\lambda \in \Sigma^+ : w^{-1}(\lambda) > 0\}, \quad \Sigma^-(w) = \{\lambda \in \Sigma^+ : w^{-1}(\lambda) < 0\}.$$

Now fix a $w \in W$ – then Lemma 5.5.1.1 tells us that

$$\delta_P(mhn) = \prod_{\lambda \in \Sigma^+} h^{-m(\lambda)\lambda} \qquad (mhn \in P)$$

whereas the reasoning employed during the *proof* of Lemma 5.5.1.1 implies that

$$\delta_{P_w}(mhn) = \prod_{\lambda \in \Sigma^+(w)} h^{-m(\lambda)\lambda} \qquad (mhn \in P_w).$$

On the other hand, for $mhn \in P_w$, we have

$$\delta_P(m_w^{-1} mhn m_w) = \delta_P(m_w^{-1} m m_w m_w^{-1} h m_w m_w^{-1} n m_w) = \prod_{\lambda \in w\Sigma^+} h^{-m(\lambda)\lambda}$$

and so, in view of the relation $w\Sigma^+ = \Sigma^+(w) \cup (-\Sigma^-(w))$, it follows that

$$\delta_P(mhn)\delta_P(m_w^{-1} mhn m_w) = \prod_{\lambda \in \Sigma^+(w)} h^{-2m(\lambda)\lambda} \qquad (mhn \in P_w),$$

whence

$$\frac{(\delta_P(mhn)\delta_P(m_w^{-1} mhn m_w))^{1/2}}{\delta_{P_w}(mhn)} = 1 \qquad (mhn \in P_w).$$

Because w is an arbitrary element of W, there can now be no question that $A_0(\zeta) = 1$ ($\zeta \in P_x$; all $x \in G$), as contended.

Our next task will be to determine the irreducible components of the representation $\zeta \mapsto A_n(\zeta)$ ($n \geqslant 1$) restricted to $A_{\mathfrak{p}}$ ($\zeta \in P_w$; $w \in W$). Let us begin with the case $n = 1$. Owing to the discussion following Theorem 5.3.2.3 and the fact that $A_0 = 1$, it is clear that the assignment $\zeta \mapsto A_1(\zeta) = \Lambda_1((\zeta, \zeta), \dot{1})$ ($\zeta \in P_w$) is the representation of P_w on $\mathfrak{g}_c/(\mathfrak{p}_c + \mathrm{Ad}(m_w)\mathfrak{p}_c)$ obtained from the adjoint representation of P_w on $\mathfrak{g}_c$ by passage to the quotient ($\mathfrak{p}_c$ the complexification of the Lie algebra of $P \ldots$); a complement to $\mathfrak{p}_c + \mathrm{Ad}(m_w)\mathfrak{p}_c$ in $\mathfrak{g}_c$ is spanned by the $X_{-\alpha}$ where α is an element of P_+ whose restriction to $\mathfrak{a}_{\mathfrak{p}}$ is in $\Sigma^+(w)$; plainly, then, the irreducible components of A_1 restricted to $A_{\mathfrak{p}}$ are the representations $h \mapsto h^{-\alpha}$ ($h \in A_{\mathfrak{p}}$) where $\alpha \in P_+(w)$ $(= \{\alpha \in P_+ : \tilde{\alpha} = \alpha | \mathfrak{a}_{\mathfrak{p}} \in \Sigma^+(w)\})$. As for the general case, recall that A_n is a symmetric power of A_1; consequently the irreducible components of A_n restricted to $A_{\mathfrak{p}}$ are the representations

$$h \mapsto \prod_{\alpha \in P_+(w)} h^{-n_\alpha \alpha} \qquad (h \in A_{\mathfrak{p}})$$

where the n_α are non-negative integers and $\Sigma\, n_\alpha = n$.

Let L^i, $mhn \mapsto \xi^i(h)L^i(m)$ $(mhn \in P)$, be a finite dimensional irreducible unitary representation of $P(i = 1, 2)$; let U^{L^i} be the unitary representation of G unitarily induced by $L^i(i = 1, 2)$. In order to estimate $i(U^{L^1}, U^{\bar{L}^2})$ it is necessary to consider the numbers $i(L^1, \bar{L}^2; Q, n)$, $Q = PxP$ an orbit in G for the action of $P \times P$ on $G(x \in G)$. Since the orbits Q are parameterized by W, we shall write $i(L^1, \bar{L}^2; w, n)$ for $i(L^1, \bar{L}^2; Q, n)$ $(Q = Pm_w^{-1}P, m_w$ an element in M^* associated with $w \in W)$. [Here the change in notation is in conformity with the one agreed to above. . . .] Suppose that $n > 0$ – then, in view of what was said in the preceding paragraph, the irreducible components of A_n (and hence of $A_n^{\cdot}$) restricted to $A_\mathfrak{p}$ are representations $\neq 1$ of $A_\mathfrak{p}$ on the multiplicative group $\mathbf{R}^+$ of positive real numbers. On the other hand, the restriction to $A_\mathfrak{p}$ of the representation $\zeta \mapsto L^1(\zeta) \hat{\otimes} \bar{L}^2(m_w^{-1}\zeta m_w)$ $(\zeta \in P_w)$ is unitary and therefore certainly has no irreducible components $\neq 1$ which are representations of $A_\mathfrak{p}$ on $\mathbf{R}^+$ Thus $i(L^1, \bar{L}^2; w, n) = 0$ for $n > 0$ (all $w \in W$) and so

$$i(U^{L^1}, U^{\bar{L}^2}) \leqslant \sum_{w \in W} i(L^1, \bar{L}^2; w, 0).$$

If $n = 0$, then $A_0 = 1$ – hence $i(L^1, \bar{L}^2; w, 0)$ is the intertwining number of the representations $\zeta \mapsto L^1(\zeta)$, $\zeta \mapsto \bar{L}^2(m_w^{-1}\zeta m_w)$ of P_w. Now $\mathsf{M}A_\mathfrak{p} \subset P_w$ and L^1, wL^2 are irreducible representations of $\mathsf{M}A_\mathfrak{p}$; since $L^1 | N_w^+ = wL^2 | N_w^+ = 1$, Schur's Lemma tells us that $i(L^1, \bar{L}^2; w, 0) = 1$ if the representations L^1 and wL^2 of $\mathsf{M}A_\mathfrak{p}$ are equivalent while $i(L^1, \bar{L}^2; w, 0) = 0$ if they are not. Thus: The intertwining number $i(U^{L^1}, U^{\bar{L}^2})$ is less than or equal to the number of elements $w \in W$ such that the representations L^1 and wL^2 are equivalent.

The preceding remarks serve to prove Theorem 5.5.2.1 and also give:

Corollary 5.5.2.2 Let L be a finite dimensional irreducible unitary representation of P, U^L the corresponding unitarily induced representation of G; suppose that U^L is not irreducible – then U^L is discretely decomposable into a direct sum of irreducible unitary representations of G, the number of which cannot exceed the order of W.

In connection with this result, let us make the following observation. Suppose that our principal P-series representation U^L is not irreducible; write $U^L = \sum_i U^i$ with the U^i irreducible – then no U^i is unitarily equivalent to any member of the *irreducible* principal P-series for G. [In fact suppose to the contrary that $U^i \sim U^{\tilde{L}}$ for some i and some $\tilde{L}$ with $U^{\tilde{L}}$ irreducible. If there exists a $w \in W$ such that $wL \sim \tilde{L}$, then $U^L \sim U^{w\tilde{L}}$ (cf. infra), whence U^L is irreducible, a contradiction; on the other hand, if no such w exists, then U^L and $U^{\tilde{L}}$ are evidently disjoint, a contradiction to the occurrence of $U^{\tilde{L}}$ as a direct summand in U^L . . . .] This being so, one may then ask: Are the U^i necessarily pairwise inequivalent? At the time of this writing, the question had not been resolved.

Keeping to the above notations, fix an irreducible unitary representation L of $\mathsf{M}A_\mathfrak{p}$; let W_L denote the subgroup of W consisting of those w

which have the property that wL is equivalent to L – then W_L is called the *ramification group* of L and its order $[W_L]$ is called the *ramification index* of L. We shall say that L is *unramified* if its ramification group is trivial. In this terminology, then, the preceding theorem asserts that the unitarily induced representation U^L is irreducible whenever L is unramified; on the other hand, if L is ramified, then it may or may not be the case that U^L is irreducible (cf. the situation for $G = \mathbf{SL}(2, \mathbf{R})$ supra . . .).

Note In 5.5.3 we shall establish the following fact: Let L^1 and L^2 be two finite dimensional irreducible unitary representations of P – then the unitarily induced representations U^{L^1} and U^{L^2} are unitarily equivalent iff there exists a $w \in W$ such that the representations L^1 and wL^2 of $\mathsf{M}A_{\mathfrak{p}}$ are equivalent.

By the *class one principal P-series* for G (P minimal), we shall understand the collection comprised of those principal P-series representations $U^{\sigma,\mu}$ where σ is the trivial one dimensional representation of M.

Theorem 5.5.2.3 (Kostant) Let G be a connected semi-simple Lie group with finite center, P a minimal parabolic subgroup of G – then every member of the class one principal P-series for G is irreducible.

[The reader will find a discussion of this result in Kostant [5]; the complete proof, which at the time of this writing had not been published, is very long and difficult.]

Using a simple but ingenious argument, Wallach [3] employed Kostant's Theorem to deduce:

Theorem 5.5.2.4 (Wallach) Let G be a complex connected semi-simple Lie group, P a minimal parabolic subgroup of G – then every member of the principal P-series for G is irreducible.

Examples (1) (Knapp and Stein) Let G be a connected semi-simple Lie group with finite center, $G = KA_{\mathfrak{p}}N^+$ an Iwasawa decomposition for G; assume that the split-rank of G is 1, i.e. that $\dim(A_{\mathfrak{p}}) = 1$. Write $P = \mathsf{M}A_{\mathfrak{p}}N^+$ and, in the notations of 5.5.1, let $L = L^{\sigma,\mu}$ be an irreducible unitary representation of P – then, since $W(= \mathsf{M}^*/\mathsf{M})$ has but two elements in the case at hand, Theorem 5.5.2.1 tells us that the unitarily induced representation $U^{\sigma,\mu}$ of G is irreducible unless $\mu = -\rho$ and $w\sigma = \sigma$ (w the unique non-trivial element in W). Let us suppose, therefore, that $\mu = -\rho$ and $w\sigma = \sigma$ – then it is possible to define $\mu_\sigma(w)$ uniquely except for a minus sign in such a way that μ_σ extends to a representation of M^*. [We recall that μ_σ is a fixed irreducible unitary representation of M in the class σ.] Let π_P and π_{N^-} denote the coordinate projections of PN^- onto P and N^-, respectively – then Knapp and Stein [1] have established the following result: Suppose that $w\sigma = \sigma$ – then the unitarily induced representation $U^{\sigma,-\rho}$ is reducible iff

$$\int_{\Omega_{r,1}} \operatorname{tr}(\mu_\sigma(w)\mu_\sigma^{-1}(\pi_P(nw)))\, d_{N^-}(n) = 0$$

for all r, s such that $0 < r < s$ where

$$\Omega_{r,s} = \{n \in N^- : r \leqslant \delta_P^{-1/2}(\pi_P(nw)) \leqslant s\}.$$

[Here we have tacitly extended μ_σ to all of P by rendering it trivial on $A_{\mathfrak{p}}N^+$.] As an example, one may explicate this criterion in the case when G is the twofold covering group of $\mathbf{SO}_e(n, 1)$ $(n \geqslant 3)$. Thus, relative to the 'obvious' Iwasawa decomposition for G (cf. Vol. II, number 6.2.2), $\mathsf{M} = \mathbf{Spin}\,(n-1)$ is a double covering of $\mathbf{SO}(n-1)$ while M^* is a double covering of $\mathbf{O}(n-1)$. Suppose now that $w\sigma = \sigma$ – then our criterion becomes

$$\operatorname{tr}\left(\mu_\sigma\left(\begin{matrix} -1 & & & & \\ & 1 & & & \\ & & \cdot & & \\ & & & \cdot & \\ & & & & \cdot \\ & & & & & 1\end{matrix}\right)\right) = 0$$

where the notation refers to either preimage in M^* of the diagonal orthogonal matrix. When n is even, this condition can be shown to mean that reducibility occurs iff μ_σ is non-trivial on the two-element center of $\mathbf{Spin}\,(n-1)$ while if n is odd, the condition means that *every* member of the principal P-series for G is irreducible. [On the other hand, every representation in the principal P-series for $\mathbf{SO}_e(n, 1)$ itself (all $n \geqslant 3$) is irreducible; cf. Hirai [2].]

(2) Let us take $G = \mathbf{SL}(n, \mathbf{R})$, $K = \mathbf{SO}(n)$; taking for $A_{\mathfrak{p}}$ the group of all diagonal elements of G with positive entries, M becomes the group of all diagonal elements in K with entries ± 1. Given $m = D(m_1, \ldots, m_n) \in \mathsf{M}$, write $\xi_i(m) = m_i$ $(i = 1, \ldots, n-1)$ – then every non-trivial unitary character of M is of the form $\xi_{i_1} \ldots \xi_{i_r}$ $(1 \leqslant i_1 < \cdots < i_r \leqslant n-1)$. Set $\xi_n = \xi_1 \ldots \xi_{n-1}$ – then W acts by permuting $\xi_1, \ldots, \xi_n$. This being so, suppose first that n is odd – then Wallach [3] has shown that every member of the principal P-series for $\mathbf{SL}(n, \mathbf{R})$ is irreducible. However, if n is even, then certain members of the principal P-series for $\mathbf{SL}(n, \mathbf{R})$ are always reducible (cf. Knapp and Stein [3]). [Suppose that n is even; assuming that $\sigma \sim \xi_1\xi_3\xi_5 \ldots \xi_{n-1}$ and that $w \sim (1, 2)(3, 4) \ldots (n-1, n)$, we have $w\sigma = \sigma$ and it can be shown that $U^{\sigma,\mu}$ is reducible whenever $w(\mu + \rho) = \mu + \rho$.]

Remark As in Example 1 supra, take for G the twofold covering group of $\mathbf{SO}_e(n, 1)$ $(n \geqslant 3)$ with n odd (or $\mathbf{SO}_e(n, 1)$ itself, n odd) – then we have seen that every member of the principal P-series for G is irreducible. Since such a G has but *one* conjugacy class of Cartan subgroups, in view of Theorem 5.5.2.4 it is easy to conjecture: Let G be a connected semi-simple Lie group with finite center, P a minimal parabolic subgroup of G; suppose that G has one conjugacy class of Cartan subgroups – then every member of the principal P-series for G is irreducible. [Wallach has shown that this conjecture is a theorem.]

5.5.3 The Characters of the Principal *P*-Series – *P* Minimal

Let G be an acceptable (cf. Vol. II, number 8.1.1) connected semi-simple Lie group with finite center, $G = KA_{\mathfrak{p}}N^+$ an Iwasawa decomposition for G; extend $\mathfrak{a}_{\mathfrak{p}}$ to a Cartan subalgebra $\mathfrak{a}$ of $\mathfrak{g}$ and let A be the Cartan subgroup of G associated with $\mathfrak{a}$; let $\mathsf{J}^1, \ldots, \mathsf{J}^r = \mathsf{A}$ be a maximal collection of mutually non-conjugate θ-stable Cartan subgroups of G. Let M be the centralizer of $A_{\mathfrak{p}}$ in K and put $P = \mathsf{M}A_{\mathfrak{p}}N^+$ – then P is a minimal parabolic subgroup of G. The objective of the present number is to calculate the characters $T^{\sigma,\mu}$ of the unitary representations $U^{\sigma,\mu}$ which comprise the principal P-series for G.

The main result can be formulated as follows.

Theorem 5.5.3.1 (Harish-Chandra) Let $U^{\sigma,\mu}$ be a member of the principal P-series for G – then the character $T^{\sigma,\mu}$ of $U^{\sigma,\mu}$ agrees on G, in the sense of distribution theory, with a locally summable function $F^{\sigma,\mu}$ which is even analytic on G'. The G-central function $F^{\sigma,\mu}$ vanishes on $G(\mathsf{J}^i)$ $(i = 1, \ldots, r-1)$; on the other hand, for a suitable normalization of the relevant Haar measures, $F^{\sigma,\mu}$ is determined on $G(\mathsf{A})$ via the prescription

$$F^{\sigma,\mu}(h) = ([\mathsf{A} : \mathsf{A}_0][W(M, A_K)]\,|\,\Delta_{R,C}(h)\,|)^{-1} \sum_{w \in W(G, \mathsf{A}_0)} \xi_{w\sigma}(h_K) h_{\mathfrak{p}}^{w(\mu+\rho)}$$

$$(h = h_K h_{\mathfrak{p}} \in \mathsf{A}).$$

[The meaning of the various symbols will be explained during the course of the proof.]

Note In reality our theorem gives the formula for the character of *any* elementary representation of G.

Owing to Theorem 4.5.8.7 we know a priori that the character $T^{\sigma,\mu}$ of $U^{\sigma,\mu}$ is a function $F^{\sigma,\mu}$ (say) on G; accordingly our main problem is to 'identify' $F^{\sigma,\mu}$, i.e. to establish the aforestated explicit formula for $F^{\sigma,\mu}$.

The ensuing discussion relies rather heavily on certain integral formulae which are to be found in Vol. II, Chapter 8; therefore, the reader may wish to defer the detailed study of the proof of Theorem 5.5.3.1 for the time being.

Conventions The Haar measure on G is fixed per the agreements in Vol. II, number 8.1.2; the Haar measure on $A_{\mathfrak{p}}$ (respectively N^+) is the transport via the exponential map of the 'privileged' Euclidean measure on $\mathfrak{a}_{\mathfrak{p}}$ (respectively $\mathfrak{n}^+$) ('privileged' relative to the Euclidean structure $(\,.\,,\,.\,)_\theta$ – cf. Vol. II, numbers 8.1.2 and 8.1.3) – then, for a certain positive explicitly computable constant $\mathbf{c}_G$, we have

$$\int_G f(x)\, d_G(x) = \mathbf{c}_G \iiint_{K \times A_{\mathfrak{p}} \times N^+} f(kh_{\mathfrak{p}}n) h_{\mathfrak{p}}^{2\rho}\, dk\, d_{A_{\mathfrak{p}}}(h_{\mathfrak{p}})\, d_{N^+}(n)$$

$$(f \in C_c(G)).$$

Let A_0 be the center of A; since $\mathsf{A}_0 = (\mathsf{A}_0 \cap K)A_{\mathfrak{p}}$, the Haar measure on A_0 is determined when we require that $\mathsf{A}_0 \cap K$ be assigned mass 1; the Haar measure on A is then fixed upon stipulating that $d_{\mathsf{A}} = d_{\mathsf{A}_0}$ on A_0 (this will account for the appearance of the factor $[\mathsf{A} : \mathsf{A}_0]$ in the calculations infra). Because the normalization of the G-invariant measure on G/A_0 can be forced through the relation $\int_G = \int_{G/\mathsf{A}_0} \int_{\mathsf{A}_0}$, we have (cf. Vol. II, number 8.1.3)

$$\int_{G(A)} f(x)\,d_G(x) = [W(G, A_0)]^{-1} \int_A |\Delta(h)|^2\,d_A(h) \int_{G/A_0} f({}^{\dot{x}}h)\,d_{G/A_0}(\dot{x})$$
$$(f \in C_c(G(A))).$$

Note In what follows we shall frequently use the well-known rule which asserts that for any $f \in C_c(N^+)$

$$\int_{N^+} f(n)\,d_{N^+}(n) = h_{\mathfrak{p}}^{2\rho} \int_{N^+} f(h_{\mathfrak{p}} n h_{\mathfrak{p}}^{-1})\,d_{N^+}(n) \qquad (\text{all } h_{\mathfrak{p}} \in A_{\mathfrak{p}})$$

where $\rho = 2^{-1} \sum_{\lambda>0} m(\lambda)\lambda$.

Fix a linear function $\mu : \mathfrak{a}_{\mathfrak{p}} \to \mathbf{C}$ and let U^μ be the corresponding standard representation of G on $L^2(K)$; we recall that U^μ operates on elements ϕ in $L^2(K)$ by means of the rule

$$U^\mu(x)\phi = \exp(-(\mu + 2\rho)\circ H_{x^{-1}})(\phi\circ\kappa_{x^{-1}}) \qquad (x \in G).$$

This being so, fix an $f \in C_c^\infty(G)$ – then the operator

$$U^\mu(f) = \int_G f(x) U^\mu(x)\,d_G(x)$$

is of the trace class. [Since a given δ in $\hat{K}$ occurs exactly $d(\delta)$ times in $U^\mu | K$, our assertion is seen to be a consequence to Theorem 4.5.8.5.] Let R denote the right regular representation of K on $L^2(K)$; fix an element $g \in C(M)$ and set $R(g) = \int_M g(m)R(m)dm$ (dm being normalized Haar measure on M) – then, since $R(g)$ is a bounded linear operator on $L^2(K)$, it must be the case that $U^\mu(f)R(g)$ is also of the trace class and, as we shall now show,

$$\operatorname{tr}(U^\mu(f)R(g))$$
$$= \mathbf{c}_G \iiint_{M\times A_{\mathfrak{p}}\times N^+} f_K(mh_{\mathfrak{p}}n)\left[\int_M g(\tilde{m}m\tilde{m}^{-1})\,d\tilde{m}\right] h_{\mathfrak{p}}^{\mu+2\rho}\,dm\,d_{A_{\mathfrak{p}}}(h_{\mathfrak{p}})\,d_{N^+}(n).$$

Thus fix a $\phi \in L^2(K)$ – then, for all $k \in K$ and $m \in M$, we have

$$(U^\mu(f)R(m)\phi)(k) = \int_G f(kx^{-1})(U^\mu(kx^{-1})R(m)\phi)(k)\,d_G(x)$$
$$= \mathbf{c}_G \iiint_{K\times A_{\mathfrak{p}}\times N^+} f(k(h_{\mathfrak{p}}n)^{-1}\tilde{k}^{-1})h_{\mathfrak{p}}^{-\mu}\phi(\tilde{k}m)d\tilde{k}\,d_{A_{\mathfrak{p}}}(h_{\mathfrak{p}})\,d_{N^+}(n)$$
$$= \mathbf{c}_G \iiint_{K\times A_{\mathfrak{p}}\times N^+} f(kh_{\mathfrak{p}}n\tilde{k}^{-1})h_{\mathfrak{p}}^{\mu+2\rho}\phi(\tilde{k}m)d\tilde{k}\,d_{A_{\mathfrak{p}}}(h_{\mathfrak{p}})\,d_{N^+}(n),$$

whence, for all $k \in K$,

$$(U^\mu(f)R(g)\phi)(k) = \int_M g(m)(U^\mu(f)R(m)\phi)(k)\,dm$$
$$= \int_K \left[\int_M K_f(k, \tilde{k}m^{-1})g(m)dm\right]\phi(\tilde{k})\,d\tilde{k}$$

where

$$K_f(k, \tilde{k}) = \mathbf{c}_G \iint_{A_\mathfrak{p}\times N^+} f(kh_\mathfrak{p}n\tilde{k}^{-1})h_\mathfrak{p}^{\mu+2\rho} d_{A_\mathfrak{p}}(h_\mathfrak{p})\, d_{N^+}(n) \qquad (k, \tilde{k} \in K).$$

This shows that $U^\mu(f)R(g)$ is an integral operator on $L^2(K)$ with kernel

$$K_{f,g}(k, \tilde{k}) = \int_{\mathsf{M}} K_f(k, \tilde{k}m^{-1})g(m)\, dm \qquad (k, \tilde{k} \in K).$$

Owing to a standard generality (see the Appendix to the present number), it therefore follows that

$$\begin{aligned}
&\operatorname{tr}(U^\mu(f)R(g)) \\
&= \int_K K_{f,g}(k, k)\, dk = \int_K \left[\int_{\mathsf{M}} K_f(k, km^{-1})g(m)\, dm\right] dk \\
&= \mathbf{c}_G \iiint_{\mathsf{M}\times A_\mathfrak{p}\times N^+} f_K(h_\mathfrak{p}nm)g(m)h_\mathfrak{p}^{\mu+2\rho} dm\, d_{A_\mathfrak{p}}(h_\mathfrak{p})\, d_{N^+}(n) \\
&= \mathbf{c}_G \iiint_{\mathsf{M}\times A_\mathfrak{p}\times N^+} f_K(mh_\mathfrak{p}n)g(m)h_\mathfrak{p}^{\mu+2\rho} dm\, d_{A_\mathfrak{p}}(h_\mathfrak{p})\, d_{N^+}(n) \\
&= \mathbf{c}_G \iiint_{\mathsf{M}\times A_\mathfrak{p}\times N^+} f_K(mh_\mathfrak{p}n)\left[\int_{\mathsf{M}} g(\tilde{m}m\tilde{m}^{-1})d\tilde{m}\right] h_\mathfrak{p}^{\mu+2\rho} dm\, d_{A_\mathfrak{p}}(h_\mathfrak{p})\, d_{N^+}(n),
\end{aligned}$$

the Haar measure on N^+ being invariant under the transformations $n \mapsto mnm^{-1}$ $(m \in \mathsf{M})$.

Consider now a representation $U^{\sigma,\mu}$ in the principal P-series for G (thus, in particular, $\mu + \rho$ assumes pure imaginary values on $\mathfrak{a}_\mathfrak{p} \ldots$); adopting the notations of 5.5.1, we clearly have

$$T^{\sigma,\mu}(f) = \operatorname{tr}(U^{\sigma,\mu}(f)) = \operatorname{tr}(U^\mu(f)\mathsf{Q}_R(\check{\sigma}_{11})) \qquad (f \in C_c^\infty(G)).$$

By definition $\mathsf{Q}_R(\check{\sigma}_{11}) = d(\sigma)\int_{\mathsf{M}} \overline{a_{11}(m^{-1})}R(m)dm$; so, in view of the elementary relation

$$\xi_\sigma(m) = d(\sigma)\int_{\mathsf{M}} a_{11}(\tilde{m}m\tilde{m}^{-1})\, d\tilde{m} \qquad (m \in \mathsf{M}),$$

it follows from what has been said above that

$$T^{\sigma,\mu}(f) = \mathbf{c}_G \iiint_{\mathsf{M}\times A_\mathfrak{p}\times N^+} f_K(mh_\mathfrak{p}n)\xi_\sigma(m)h_\mathfrak{p}^{\mu+2\rho} dm\, d_{A_\mathfrak{p}}(h_\mathfrak{p})\, d_{N^+}(n) \qquad (f \in C_c^\infty(G)).$$

From here on we shall feel free to use the notations and conventions of Vol. II, Chapter 8 in so far as they are applicable to the Cartan subgroup A $(= \mathsf{A}_K A_\mathfrak{p})$; in particular, then, the symbols Δ, Δ_I, and $\Delta_{R,C}$ are to be taken relative to A. [Let us recall that $\Delta = \Delta_I \cdot \Delta_{R,C}$.]

The identity component A_K of A_K is a maximal torus in M (the identity component of M); agreeing to assign to both A_K and M total mass 1, the well-known integral formula of H. Weyl tells us that

$$\int_M f(m)\,dm = [W(M, A_K)]^{-1} \int_{A_K} |\Delta_I(h_K)|^2 f(h_K)\,dh_K$$

for any continuous class function f on M. [Here, of course, $W(M, A_K)$ denotes the Weyl group of the pair (M, A_K), i.e. $W(M, A_K)$ is the quotient of the normalizer of A_K in M by A_K itself.] Using Lemma 1.1.3.8, select elements $a_1 = 1, a_2, \ldots, a_r \in \mathsf{Z}(A_{\mathfrak{p}})$ $(r = [\mathsf{M} : M])$ such that $\mathsf{M} = \bigcup_i a_i M$ (disjoint union) – then it is easy to see that $\mathsf{A}_K = \bigcup_i a_i A_K$ (disjoint union). The elements of $\mathsf{Z}(A_{\mathfrak{p}})$ *commute* with the elements of M; so, if the total volume of A_K is 1, then, for any continuous class function f on M, we have

$$\begin{aligned}
&\int_{\mathsf{M}} f(m)\,dm \\
&= r^{-1} \sum_{i=1}^{r} \int_M f(a_i m)\,dm \\
&= r^{-1}[W(M, A_K)]^{-1} \sum_{i=1}^{r} \int_{A_K} |\Delta_I(h_K)|^2 \left[\int_M f(a_i m h_K m^{-1})\,dm\right] dh_K \\
&= r^{-1}[W(M, A_K)]^{-1} \sum_{i=1}^{r} \int_{A_K} |\Delta_I(h_K)|^2 f(a_i h_K)\,dh_K \\
&= [W(M, A_K)]^{-1} \int_{\mathsf{A}_K} |\Delta_I(h_K)|^2 f(h_K)\,dh_K,
\end{aligned}$$

it being clear that $|\Delta_I(a_i h_K)| = |\Delta_I(h_K)|$ $(h_K \in A_K)$ (since $\xi_\alpha(a_i) = 1$ for every imaginary root α of the pair $(\mathfrak{g}_c, \mathfrak{a}_c)\ldots$).

Returning to our principal P-series representation $U^{\sigma,\mu}$, we thus see that its character $T^{\sigma,\mu}$ admits the representation

$$T^{\sigma,\mu}(f) = \mathbf{c}_G[W(M, A_K)]^{-1} \iiint_{\mathsf{A}_K \times A_{\mathfrak{p}} \times N^+} f_K(h_K h_{\mathfrak{p}} n)\xi_\sigma(h_K)|\Delta_I(h_K)|^2 \times h_{\mathfrak{p}}^{\mu+2\rho} dh_K d_{A_{\mathfrak{p}}}(h_{\mathfrak{p}})\, d_{N^+}(n),$$

valid for all $f \in C_c^\infty(G)$. Let A_0 be the center of A – then, according to Proposition 8.1.3.5, for any $f \in C_c^\infty(G)$ we have

$$\epsilon_R(h)\Delta(h) \int_{G/\mathsf{A}_0} f(\dot{x}h)\, d_{G/\mathsf{A}_0}(\dot{x}) = \mathbf{c}_G \Delta_I(h) h_{\mathfrak{p}}^{\rho} \int_{N^+} f_K(hn)\, d_{N^+}(n) \qquad (h \in \mathsf{A}'),$$

the normalization of the G-invariant measure on G/A_0 being that agreed to in Vol. II, number 8.1.3 (cf. supra). Hence, upon introducing the invariant integral calculated relative to A (cf. Vol. II, section 8.5), the integral representation for the character $T^{\sigma,\mu}$ takes the form

$$T^{\sigma,\mu}(f) = [W(M, A_K)]^{-1} \iint_{\mathsf{A}_K \times A_{\mathfrak{p}}} \overline{\Delta_I(h_K)}\xi_\sigma(h_K) h_{\mathfrak{p}}^{\mu+\rho} \Phi_f(h_K h_{\mathfrak{p}}) dh_K d_{A_{\mathfrak{p}}}(h_{\mathfrak{p}}) \qquad (f \in C_c^\infty(G)).$$

The following points are clear: (1) The group $W(G, \mathsf{A}_0)$ operates on $\hat{\mathsf{M}}$, A, A_K, $A_{\mathfrak{p}}$ (and hence $\mathfrak{a}_{\mathfrak{p}}$); (2) $|w\cdot\Delta_{R,C}| = |\Delta_{R,C}|$ (all $w \in W(G, \mathsf{A}_0)$);

(3) $|\Delta_{R,C}(h)| = \epsilon_R(h)\Delta_{R,C}(h)$ (all $h \in \mathsf{A}'$) (cf. Lemma 8.1.3.8); (4) Fix $f \in C_c^\infty(G)$ – then, for any $w \in W(G, \mathsf{A}_0)$,

$$\epsilon_R(w\cdot h)\Phi_f(w\cdot h) = \det(w)\epsilon_R(h)\Phi_f(h) \qquad (\text{all } h \in \mathsf{A}')$$

(cf. Remark 4 in Vol. II, number 8.5.1). This being so, consider now the G-central function $F^{\sigma,\mu}$ on G defined by the conditions:

(i) $F^{\sigma,\mu}(x) = 0 \quad (x \in G - G(\mathsf{A}))$;

(ii) $F^{\sigma,\mu}(xyx^{-1}) = F^{\sigma,\mu}(y) \quad (x \in G, y \in G(\mathsf{A}))$;

(iii) $F^{\sigma,\mu}(h) = ([\mathsf{A}:\mathsf{A}_0][W(M, A_K)]|\Delta_{R,C}(h)|)^{-1} \sum_{w\in W(G,\mathsf{A}_0)} \xi_{w\sigma}(h_K)h_{\mathfrak{p}}^{w(\mu+\rho)}$ $(h = h_K h_{\mathfrak{p}} \in \mathsf{A}')$.

Then we claim that, in the sense of distribution theory, $T^{\sigma,\mu} = F^{\sigma,\mu}$ on G. Thus fix an f in $C_c^\infty(G)$ – then, bearing in mind remarks (1)–(4) supra, we find that

$$\int_G f(x)F^{\sigma,\mu}(x)\,d_G(x)$$

$$= [W(G, \mathsf{A}_0)]^{-1}\int_{\mathsf{A}} |\Delta(h)|^2 F^{\sigma,\mu}(h)\,d_{\mathsf{A}}(h)\int_{G/\mathsf{A}_0} f({}^{\dot{x}}h)\,d_{G/\mathsf{A}_0}(\dot{x})$$

$$= [W(G, \mathsf{A}_0)]^{-1}\int_{\mathsf{A}} \overline{\Delta(h)}F^{\sigma,\mu}(h)\epsilon_R(h)\Phi_f(h)\,d_{\mathsf{A}}(h)$$

$$= [W(M, A_K)]^{-1}\iint_{\mathsf{A}_K\times A_{\mathfrak{p}}} \overline{\Delta_I(h_K)}\xi_\sigma(h_K)h_{\mathfrak{p}}^{\mu+\rho}\Phi_f(h_K h_{\mathfrak{p}})dh_K d_{A_{\mathfrak{p}}}(h_{\mathfrak{p}})$$

$$= T^{\sigma,\mu}(f),$$

thereby establishing Theorem 5.5.3.1.

Remark We remind the reader that the Haar measure on A has been normalized in such a way as to agree, upon restriction, with the Haar measure on A_0; since the mass of $\mathsf{A}_0 \cap K$ is 1 and since we want the mass of A_K to also be 1, it is necessary to insert the factor $[\mathsf{A}:\mathsf{A}_0]^{-1}$ into the definition of $F^{\sigma,\mu}$ . . . .

Keeping to the above notations, write $\mu + \rho = \sqrt{-1}\nu$ and put $T^{\sigma,\nu} = T^{\sigma,\mu}$ – then, on the basis of Theorem 5.5.3.1, it is clear that $T^{w\sigma,w\nu} = T^{\sigma,\nu}$ for all $w \in W(G, \mathsf{A}_0)$. On the other hand suppose that $T^{\sigma_1,\mu_1} = T^{\sigma_2,\mu_2}$ – then we claim that there exists a $w \in W(G, \mathsf{A}_0)$ such that $w\sigma_1 = \sigma_2$, $w(\mu_1 + \rho) = \mu_2 + \rho$. In fact, thanks to our hypotheses and Theorem 5.5.3.1, it must be the case that

$$\sum_{w\in W(G,\mathsf{A}_0)} \xi_{w\sigma_1}(h_K)h_{\mathfrak{p}}^{w(\mu_1+\rho)} = \sum_{w\in W(G,\mathsf{A}_0)} \xi_{w\sigma_2}(h_K)h_{\mathfrak{p}}^{w(\mu_2+\rho)} \qquad (h = h_K h_{\mathfrak{p}} \in \mathsf{A}').$$

Because both sides of this relation are continuous functions while A' is dense in A, the equation in question is actually valid on all of A; since the exponentials of distinct linear functions are linearly independent, it follows on putting $h_K = 1$ that $\mu_2 = w_0(\mu_1 + \rho) - \rho$ for some $w_0 \in W(G, \mathsf{A}_0)$. Let $\tilde{W}$ be the subgroup of $W(G, \mathsf{A}_0)$ consisting of those $\tilde{w}$ such that $\mu_2 = \tilde{w}(\mu_2 + \rho) - \rho$ – then it is clear that

$$\sum_{\tilde{w}\in\tilde{W}} \xi_{\tilde{w}w_0\sigma_1}(h_K) = \sum_{\tilde{w}\in\tilde{W}} \xi_{\tilde{w}\sigma_2}(h_K) \qquad (\text{all } h_K \in \mathsf{A}_K),$$

or, since every element of M is conjugate to some element of A_K (cf. supra),

$$\sum_{\tilde{w} \in \tilde{W}} \xi_{\tilde{w}w_0\sigma_1}(m) = \sum_{\tilde{w} \in \tilde{W}} \xi_{\tilde{w}\sigma_2}(m) \qquad (\text{all } m \in \mathsf{M}).$$

However the characters corresponding to distinct irreducible classes are linearly independent, whence $\sigma_2 = \tilde{w}w_0\sigma_1$ for some $\tilde{w} \in \tilde{W}$; now put $w = \tilde{w}w_0$ – then $w\sigma_1 = \sigma_2$, $w(\mu_1 + \rho) = \mu_2 + \rho$, as claimed. Thus, taking into account Theorem 4.5.8.1, we have proved the following lemma.

Lemma 5.5.3.2 Let G be an acceptable connected semi-simple Lie group with finite center; let $G = KA_{\mathfrak{p}}N^+$ be an Iwasawa decomposition for G, $P = \mathsf{M}A_{\mathfrak{p}}N^+$ the associated minimal parabolic subgroup – then two members U^{σ_1,μ_1}, U^{σ_2,μ_2} of the principal P-series for G are unitarily equivalent iff there exists a $w \in W(G, \mathsf{A}_0)$ such that $w\sigma_1 = \sigma_2$, $w(\mu_1 + \rho) = \mu_2 + \rho$.

Let W denote the Weyl group of the pair $(\mathfrak{g}, \mathfrak{a}_{\mathfrak{p}})$ – then we claim that $F^{\sigma,\mu}$ can actually be written in terms of W (rather than $W(G, \mathsf{A}_0)$). To see this, note first that $W(G, \mathsf{A}) = \mathsf{A}^*/\mathsf{A} \sim W(G, \mathsf{A}_0)/(\mathsf{A}/\mathsf{A}_0)$; therefore, since A/A_0 operates trivially on both $\hat{\mathsf{M}}$ and $A_{\mathfrak{p}}$, it follows that

$$F^{\sigma,\mu}(h) = ([W(M, A_K)]\,|\Delta_{R,C}(h)|)^{-1} \sum_{w \in W(G,\mathsf{A})} \xi_{w\sigma}(h_K)h_{\mathfrak{p}}^{w(\mu+\rho)} \qquad (h = h_K h_{\mathfrak{p}} \in \mathsf{A}').$$

Let M^* denote the normalizer of $A_{\mathfrak{p}}$ in K – then $\mathsf{M}^*A_{\mathfrak{p}}$ is the normalizer of $A_{\mathfrak{p}}$ in G and it is easy to see that $\mathsf{M}^*A_{\mathfrak{p}} = (\mathsf{M}A_{\mathfrak{p}})(\mathsf{A}_K^*\mathsf{A}_{\mathfrak{p}})$ (cf. the Remark following the proof of Proposition 1.4.2.4). On the other hand, suppose that we let $W(\mathsf{M}, \mathsf{A}_K)$ denote the quotient of the normalizer of A_K in M by A_K itself – then, fortuitously enough, $W(\mathsf{M}, \mathsf{A}_K) \sim W(M, A_K)$. [Let $N(\mathsf{A}_K)$ (respectively $N(A_K)$) denote the normalizer of A_K (respectively A_K) in M (respectively M) – then $N(\mathsf{A}_K) = \mathsf{Z}(A_{\mathfrak{p}})N(A_K)$, whence

$$\begin{aligned} &N(\mathsf{A}_K)/\mathsf{A}_K \\ &\quad = \mathsf{Z}(A_{\mathfrak{p}})N(A_K)/\mathsf{Z}(A_{\mathfrak{p}})A_K \sim N(A_K)/\mathsf{Z}(A_{\mathfrak{p}}) \cap N(A_K)\big/A_K/\mathsf{Z}(A_{\mathfrak{p}}) \cap A_K \\ &\qquad \sim N(A_K)/A_K, \end{aligned}$$

it being clear that $\mathsf{Z}(A_{\mathfrak{p}}) \cap N(A_K) = \mathsf{Z}(A_{\mathfrak{p}}) \cap A_K$ (since A_K is a maximal torus in M . . .).] It therefore follows that

$$\begin{aligned} W = \mathsf{M}^*A_{\mathfrak{p}}/\mathsf{M}A_{\mathfrak{p}} &\sim \mathsf{A}^*/\mathsf{A}\big/(\mathsf{M} \cap \mathsf{A}_K^*)A_{\mathfrak{p}}/\mathsf{A}_K A_{\mathfrak{p}} \\ &\sim W(G, \mathsf{A})/W(M, A_K). \end{aligned}$$

Since the group $W(M, A_K)$ operates trivially on both $\hat{\mathsf{M}}$ and $A_{\mathfrak{p}}$, we can now write

$$F^{\sigma,\mu}(h) = |\Delta_{R,C}(h)|^{-1} \sum_{w \in W} \xi_{w\sigma}(h_K)h_{\mathfrak{p}}^{w(\mu+\rho)} \qquad (h = h_K h_{\mathfrak{p}} \in \mathsf{A}').$$

Theorem 5.5.3.3 (Bruhat) Let G be an acceptable connected semi-simple Lie group with finite center, $G = KA_{\mathfrak{p}}N^+$ an Iwasawa decomposition for G, $P = MA_\mathfrak{p}N^+$ the associated minimal parabolic subgroup – then two members U^{σ_1,μ_1}, U^{σ_2,μ_2} of the principal P-series for G are unitarily equivalent iff there exists a $w \in W$ such that $w\sigma_1 = \sigma_2$, $w(\mu_1 + \rho) = \mu_2 + \rho$.

[In view of the discussion leading up to Lemma 5.5.3.2 and the immediately preceding observations, our contention is clear (modulo Theorem 4.5.8.1, of course).]

Remark Let us drop the assumption that G is acceptable; write $G = KA_\mathfrak{p}N^+$ and put $P = MA_\mathfrak{p}N^+$; let L^i be a finite dimensional irreducible unitary representation of P – then it is still the case that $U^{L^1} \sim U^{L^2}$ iff there exists a $w \in W$ such that $wL^1 \sim L^2$. [This is a ready consequence of the formula for the spherical trace functions associated with the elements of the principal P-series for G (cf. Vol. II, number 6.2.2); it is necessary to bear in mind that the function $h_\mathfrak{p} \mapsto h_\mathfrak{p}^\rho \int_{N^+} f_K(h_\mathfrak{p} n) d_{N^+}(n)$ $(h_\mathfrak{p} \in A_\mathfrak{p})$ is invariant under the action of W for any $f \in C_c(G)$.]

Appendix Here we intend to establish the following result.

Lemma Let K be a compact Lie group; fix a $\lambda \in C^\infty(K \times K)$ and define a bounded linear operator Λ on $L^2(K)$ by the rule

$$\Lambda f(k) = \int_K \lambda(k, \bar{k}) f(\bar{k})\, d\bar{k} \qquad (f \in L^2(K)).$$

Then Λ is of the trace class with

$$\operatorname{tr}(\Lambda) = \int_K \lambda(k, k)\, dk.$$

Proof Since λ is a C^∞ function, an argument entirely analogous to that of Theorem 4.5.8.5 shows that Λ is of the trace class .... To compute $\operatorname{tr}(\Lambda)$, we proceed as follows. Given a class $\delta \in \hat{K}$, fix an irreducible unitary matrix representation μ_δ (say) in δ and let ϕ_{ij}^δ $(1 \leqslant i, j \leqslant d(\delta))$ be the corresponding coefficients for μ_δ relative to an orthonormal basis in the representation space of μ_δ – then, as is well-known, the functions $d(\delta)^{1/2}\phi_{ij}^\delta$ $(\delta \in \hat{K})$ form an orthonormal basis for $L^2(K)$, whence

$$\operatorname{tr}(\Lambda) = \sum_\delta d(\delta) \sum_{i,j} (\Lambda\phi_{ij}^\delta, \phi_{ij}^\delta).$$

Let ξ_δ be the character of δ $(\delta \in \hat{K})$ – then it is clear that

$$\xi_\delta(\bar{k}k^{-1}) = \sum_i \phi_{ii}^\delta(\bar{k}k^{-1}) = \sum_{i,j} \phi_{ij}^\delta(\bar{k})\phi_{ji}^\delta(k^{-1}) \qquad (k, \bar{k} \in K)$$

and so

$$\begin{aligned}\sum_{i,j} (\Lambda\phi_{ij}^\delta, \phi_{ij}^\delta) &= \sum_{i,j} \int_K \int_K \lambda(k, \bar{k})\phi_{ij}^\delta(\bar{k})\phi_{ji}^\delta(k^{-1})\, dk\, d\bar{k} \\ &= \int_K \int_K \lambda(k, \bar{k})\xi_\delta(k^{-1}\bar{k})\, dk\, d\bar{k}.\end{aligned}$$

Consequently

$$\operatorname{tr}(\Lambda) = \sum_\delta d(\delta) \int_K \xi_\delta(k)\lambda_0(k)\, dk$$

where

$$\lambda_0(k) = \int_K \lambda(\tilde{k}, \tilde{k}k)\, d\tilde{k} \qquad (k \in K).$$

Because λ_0 is a C^∞ function on K, its Fourier series formed relative to the left regular representation of K on $C^\infty(K)$ converges to λ_0 absolutely and uniformly (cf. Proposition 4.4.3.2); therefore, in particular,

$$\sum_\delta d(\delta) \int_K \xi_\delta(k)\lambda_0(k)\, dk (= \mathrm{tr}\,(\Lambda)) = \lambda_0(1)\Big(= \int_K \lambda(k, k)\, dk\Big),$$

as we wished to prove. □

5.5.4 The Riemann-Lebesgue Lemma for the Principal *P*-Series – *P* Minimal

Let G be an acceptable (cf. Vol. II, number 8.1.1) connected semi-simple Lie group with finite center, $G = KA_\mathfrak{p}N^+$ an Iwasawa decomposition for G; let M be the centralizer of $A_\mathfrak{p}$ in K and put $P = \mathsf{M}A_\mathfrak{p}N^+$ – then P is a minimal parabolic subgroup of G. Let L be the centralizer of $A_\mathfrak{p}$ in G – then $\mathsf{L} = \mathsf{M}A_\mathfrak{p}$. For every f in $C_c^\infty(G)$, set

$$f_{K,P}(l) = \mathbf{c}_G \mathrm{d}_P(l) \int_{N^+} f_K(ln)\, d_{N^+}(n) \qquad (l \in \mathsf{L})$$

where $\mathrm{d}_P(p) = \det\,(\mathrm{Ad}\,(p)\,|\,\mathfrak{n}^+)^{1/2}$ $(p \in P)$ – then $f_{K,P} \in C_c^\infty(\mathsf{L})$ (cf. Vol. II, number 8.5.3), and, in fact, the assignment $f \mapsto f_{K,P}$ defines a continuous mapping of $C_c^\infty(G)$ into $C_c^\infty(\mathsf{L})$. Suppose now that $U^{\sigma,\mu}$ is a member of the principal P-series for G with character $T^{\sigma,\mu}$; agreeing to write $\mu + \rho = \sqrt{-1}\nu$ (and hence $U^{\sigma,\nu}$ for $U^{\sigma,\mu}$; $T^{\sigma,\nu}$ for $T^{\sigma,\mu}$ etc.), in view of what has been said in 5.5.3, we have

$$T^{\sigma,\nu}(f) = \iint\limits_{\mathsf{M}\times A_\mathfrak{p}} f_{K,P}(mh_\mathfrak{p})\xi_\sigma(m)h_\mathfrak{p}^{\sqrt{-1}\nu}\, dm\, d_{A_\mathfrak{p}}(h_\mathfrak{p}) \qquad (\text{all } f \in C_c^\infty(G)).$$

Definition Fix an f in $C_c^\infty(G)$ – then the *Fourier transform* of f with respect to P is the function $\hat{f}_P$ on $\hat{\mathsf{M}} \times \hat{A}_\mathfrak{p}$ defined by the rule

$$\hat{f}_P(\sigma, \nu) = T^{\sigma,\nu}(f) \qquad ((\sigma, \nu) \in \hat{\mathsf{M}} \times \hat{A}_\mathfrak{p}).$$

Because M is compact, the hull-kernel topology on $\hat{\mathsf{M}}$ is the same as the discrete topology; on the other hand, the hull-kernel topology on $\hat{A}_\mathfrak{p}$ is the same as the usual topology which it inherits as the character group of a locally compact abelian group – we shall therefore agree to endow the space $\hat{\mathsf{M}} \times \hat{A}_\mathfrak{p}$ with the corresponding product topology. [The reader will find a résumé of these notions in Vol. II, Chapter 7.]

The main result of the present number can now be stated as follows.

Theorem 5.5.4.1 (Riemann-Lebesgue) Keeping to the above notations, suppose that $\{(\sigma_n, \nu_n)\}$ is a sequence which converges to infinity in $\hat{\mathsf{M}} \times \hat{A}_\mathfrak{p}$ – then

$$\hat{f}_P(\sigma_n, \nu_n) \to 0$$

for every f in $C_c^\infty(G)$.

Proof Fix an f in $C_c^\infty(G)$; let Φ_f denote the invariant integral of f calculated relative to the Cartan subgroup A – then, since $\mathfrak{a}_\mathfrak{p}$ is maximal abelian in $\mathfrak{p}$, $\mathsf{A}'(SI) = \mathsf{A}$, whence $\Phi_f \in \mathscr{C}(\mathsf{A})$ (cf. Corollary 8.5.1.2). We shall now divide the proof into two cases depending on whether M is connected or not.

(1) Suppose first that M is connected – then, in view of what has been said in 5.5.3, A_K is also connected. Fix $\sigma \in \hat{M}$, $\nu \in \hat{A}_\mathfrak{p}$; let Λ_σ be the highest weight of σ (or rather μ_σ); put $\rho_I = 2^{-1}\sum_{\alpha\in\Phi_I^+}\alpha$ (of course $\Phi_I^+ = P_- \ldots$) – then, taking into account the results obtained in 5.5.3, along with Weyl's character formula, for any $f \in C_c^\infty(G)$, we have

$$
\begin{aligned}
&[W(M, A_K)]\hat{f}_P(\sigma, \nu)\\
&= (-1)^{r_I} \iint\limits_{A_K\times A_\mathfrak{p}} \Delta_I(h_K)\xi_\sigma(h_K)h_\mathfrak{p}^{\sqrt{-1}\nu}\Phi_f(h_Kh_\mathfrak{p})\,dh_K\,d_{A_\mathfrak{p}}(h_\mathfrak{p})\\
&= (-1)^{r_I} \sum_{w\in W(M,A_K)} \det(w) \iint\limits_{A_K\times A_\mathfrak{p}} \xi_{w(\Lambda_\sigma+\rho_I)}(h_K)h_\mathfrak{p}^{\sqrt{-1}\nu}\Phi_f(h_Kh_\mathfrak{p})\,dh_K\,d_{A_\mathfrak{p}}(h_\mathfrak{p})\\
&= [W(M, A_K)](-1)^{r_I} \iint\limits_{A_K\times A_\mathfrak{p}} \xi_{\Lambda_\sigma+\rho_I}(h_K)h_\mathfrak{p}^{\sqrt{-1}\nu}\Phi_f(h_Kh_\mathfrak{p})\,dh_K\,d_{A_\mathfrak{p}}(h_\mathfrak{p})
\end{aligned}
$$

where, in passing from the second equality to the third, we used the fact that $\Phi_f(w\cdot h) = \det(w)\Phi_f(h)$ (all $w \in W(M, A_K)$). [Here we have to bear in mind Remark 4 in Vol. II, number 8.5.1; it is clear that $W(M, A_K) \subset W(G, A)$ and, moreover, in the present case $\epsilon_R(w) = \epsilon_w = 1$ (all $w \in W(M, A_K)$).] It therefore follows that

$$|\hat{f}_P(\sigma, \nu)| \leq |\hat{\Phi}_f(\Lambda_\sigma + \rho_I, \nu)| \qquad (\text{all } f \in C_c^\infty(G)).$$

Because $\Phi_f \in \mathscr{C}(A)$ (any f in $C_c^\infty(G)$), classical Fourier analysis tells us that its Fourier transform $\hat{\Phi}_f$ vanishes at infinity – hence

$$\hat{f}_P(\sigma_n, \nu_n) \to 0 \qquad (f \in C_c^\infty(G))$$

when $(\sigma_n, \nu_n) \to \infty$ in $\hat{M} \times \hat{A}_\mathfrak{p}$, as we wished to prove.

(2) To reduce the general case to the special one treated above, we shall employ the following artifice. Because $\mathsf{M} = \mathsf{Z}(A_\mathfrak{p})M$ with M and $\mathsf{Z}(A_\mathfrak{p})$ commuting, M is isomorphic to $(\mathsf{Z}(A_\mathfrak{p}) \times M)/\tilde{\mathsf{Z}}(A_\mathfrak{p})$ where

$$\tilde{\mathsf{Z}}(A_\mathfrak{p}) = \{(a, a^{-1}) : a \in M \cap \mathsf{Z}(A_\mathfrak{p})\};$$

thus

$$\hat{\mathsf{M}} = [(\mathsf{Z}(A_\mathfrak{p}) \times M)/\tilde{\mathsf{Z}}(A_\mathfrak{p})]\hat{} = \{\sigma \in (\mathsf{Z}(A_\mathfrak{p}) \times M)\hat{} : \mu_\sigma|\tilde{\mathsf{Z}}(A_\mathfrak{p}) = 1\}$$

and so, for a given σ in $\hat{\mathsf{M}}$, there exist uniquely determined elements $\tau \in \hat{M}$, $\zeta \in \mathsf{Z}(A_\mathfrak{p})\hat{}$ such that $\mu_\tau|M \cap \mathsf{Z}(A_\mathfrak{p})$, $\mu_\zeta|M \cap \mathsf{Z}(A_\mathfrak{p})$ act by the same scalar operator with $\mu_\sigma(am) = \mu_\zeta(a) \otimes \mu_\tau(m)$ (all $a \in \mathsf{Z}(A_\mathfrak{p})$, $m \in M$). Therefore

$$\xi_\sigma(am) = \xi_\zeta(a)\xi_\tau(m) \qquad (\text{all } a \in \mathsf{Z}(A_\mathfrak{p}), m \in M).$$

This being the case, select elements

$$a_1 = 1, \quad a_2, \ldots, a_r \in \mathsf{Z}(A_\mathfrak{p}) \qquad (r = [\mathsf{A}_K : A_K])$$

such that $A_K = \bigcup_i a_i A_K$ (disjoint union); fix elements $\sigma \in \hat{M}$, $\nu \in \hat{A}_{\mathfrak{p}}$ – then, for any $f \in C_c^\infty(G)$, we have

$$
\begin{aligned}
&[W(M, A_K)]\hat{f}_P(\sigma, \nu) \\
&= \iint_{A_K \times A_{\mathfrak{p}}} \overline{\Delta_I(h_K)}\xi_\sigma(h_K)h_{\mathfrak{p}}^{\sqrt{-1}\nu}\Phi_f(h_K h_{\mathfrak{p}})\, dh_K\, d_{A_{\mathfrak{p}}}(h_{\mathfrak{p}}) \\
&= r^{-1}\sum_{i=1}^{r} \iint_{A_K \times A_{\mathfrak{p}}} \overline{\Delta_I(a_i h_K)}\xi_\sigma(a_i h_K)h_{\mathfrak{p}}^{\sqrt{-1}\nu}\Phi_f(a_i h_K h_{\mathfrak{p}})\, dh_K\, d_{A_{\mathfrak{p}}}(h_{\mathfrak{p}}) \\
&= r^{-1}(-1)^{r_I}\sum_{i=1}^{r} \overline{\xi_\rho(a_i)}\xi_\tau(a_i) \iint_{A_K \times A_{\mathfrak{p}}} \Delta_I(h_K)\xi_\tau(h_K)h_{\mathfrak{p}}^{\sqrt{-1}\nu}\Phi_f^i(h_K h_{\mathfrak{p}})\, dh_K\, d_{A_{\mathfrak{p}}}(h_{\mathfrak{p}})
\end{aligned}
$$

where $\Phi_f^i(h) = \Phi_f(a_i h)$ $(h \in A = A_K A_{\mathfrak{p}};\ i = 1, \ldots, r)$. [Here ξ_ρ is to be taken per Vol. II, number 8.1.1.] It therefore follows that

$$
|\hat{f}_P(\sigma, \nu)| \leq \sum_{i=1}^{r} |\hat{\Phi}_f^i(\Lambda_\tau + \rho_I, \nu)| \qquad (\text{all } f \in C_c^\infty(G))
$$

from which we can read off once again the desired assertion.

This completes the proof of the theorem. □

Remark Because the characters of the principal P-series (P minimal) are tempered distributions on G (cf. Vol. II, number 8.3.8), the preceding theorem is actually valid for any f in $\mathscr{C}(G)$.

Appendices

1 Quasi-Invariant Measures

Let G be a locally compact group subject to our general conventions (countable at infinity). Let d_G denote a left Haar measure on G. Let δ_G denote the *right modular function* for G (or more briefly the modular function for G). Thus one has the relations

$$\int_G f(xy^{-1})\, d_G(x) = \delta_G(y) \int_G f(x)\, d_G(x),$$

$$\int_G f(x^{-1})\, d_G(x) = \int_G \frac{f(x)}{\delta_G(x)}\, d_G(x)$$

for all $f \in C_c(G)$. We remind the reader that δ_G is a continuous homomorphism of G into the multiplicative group $\mathbf{R}^+$ of strictly positive real numbers; if G is a Lie group, then

$$\delta_G(x) = |\det(\mathrm{Ad}_G(x^{-1}))| \qquad (x \in G).$$

Let H be a closed subgroup of G, the elements of which will be denoted by $\xi, \eta, \ldots$. The canonical projection of G onto the space G/H of left cosets mod H will be denoted by $x \mapsto \pi(x)$ or $x \mapsto \dot{x}$; G/H is paracompact. Let μ be a Radon measure on G/H. The transform μ^x of μ by an element $x \in G$ is the measure

$$f \mapsto \mu(f^{x^{-1}})$$

where $f \in \dot{C}_c(G/H)$ and

$$f^x(\dot{y}) = f(x^{-1} \cdot \dot{y}) \qquad (x, y \in G; x \cdot \dot{y} = xyH).$$

Definition A positive Radon measure μ on G/H is said to be *quasi-invariant* if μ and μ^x are equivalent for all $x \in G$ (i.e. μ and μ^x have the same null sets, i.e. same sets of measure zero).

Although in general positive G-invariant Radon measures do not exist on G/H, nevertheless quasi-invariant measures are always present (Dieudonné [1]). In fact if ν is a non-trivial finite positive Radon measure on G whose null sets are those of Haar measure zero, then, putting

$$\mu(f) = \nu(f \circ \pi) \qquad (f \in C_c(G/H)),$$

one easily checks that μ is quasi-invariant. In turn, the existence of ν follows from the fact that G is countable at infinity.

We shall now give a brief description of the main results on quasi-invariant measures (for more details consult Bourbaki [3, Chapter VII, §2, n°5]). Let δ_H denote the modular function for H.

In the first place quasi-invariant measures on G/H always exist and any two quasi-invariant measures are equivalent. A way to manufacture quasi-invariant measures is this. Let ρ be a strictly positive Borel function on G bounded above and below on compact subsets and verifying for every $\xi \in H$

$$\rho(x\xi) = \frac{\delta_H(\xi)}{\delta_G(\xi)}\rho(x) \quad (x \in G).$$

A function with these properties is called a *rho-function.* Fix now a rho-function ρ; associated with this ρ is a quasi-invariant measure μ_ρ defined by

$$\int_G f(x)\rho(x)\,d_G(x) = \int_{G/H} d\mu_\rho(\dot{x})\int_H f(x\xi)\,d_H(\xi), \quad f \in C_c(G).$$

[We recall that the map

$$f \mapsto \dot{f}, \quad \dot{f}(\dot{x}) = \int_H f(x\xi)\,d_H(\xi),$$

is a continuous surjection of $C_c(G)$ onto $C_c(G/H)$. Hence the main thing to prove is that

$$\int_H f(x\xi)\,d_H(\xi) = 0$$

for all $x \in G$ implies

$$\int_G f(x)\rho(x)\,d_G(x) = 0.$$

This is done by using the following classical argument due to Weil. Fix an arbitrary $\phi \in C_c(G)$; noting that $\dot{f} = 0 \Rightarrow$

$$\int_H f(x\xi)\,d_H(\xi) = \int_H f(x\xi^{-1})\delta_H(\xi^{-1})\,d_H(\xi) = 0,$$

we have

$$\begin{aligned} 0 &= \int_G\int_H \rho(x)\phi(x)f(x\xi^{-1})\delta_H(\xi^{-1})\,d_H(\xi)\,d_G(x) \\ &= \int_H\int_G \rho(x)\phi(x)f(x\xi^{-1})\delta_H(\xi^{-1})\,d_G(x)\,d_H(\xi) \\ &= \int_H\int_G \rho(x\xi)\phi(x\xi)f(x)\delta_G(\xi)\delta_H(\xi^{-1})\,d_G(x)\,d_H(\xi) \\ &= \int_G f(x)\rho(x)\left[\int_H \phi(x\xi)\,d_H(\xi)\right]d_G(x). \end{aligned}$$

All that remains to do is choose ϕ so that the function

$$x \mapsto \int_H \phi(x\xi)\,d_H(\xi)$$

is 1 on the compact support of f.]

For each $f \in C_c(G/H)$ one has the relation

$$\int_{G/H} f(x\cdot\dot{y})\,d\mu_\rho(\dot{y}) = \int_{G/H} s_\rho(x^{-1}, \dot{y})f(\dot{y})\,d\mu_\rho(\dot{y}).$$

The function s_ρ is obtained in the following way. Let

$$s_\rho(x, y) = \frac{\rho(xy)}{\rho(y)} \qquad (x, y \in G).$$

Then s_ρ passes to a function on $G \times G/H$, which again will be called s_ρ, such that:

(i) $s_\rho(xy, \dot{z}) = s_\rho(y, \dot{z}) s_\rho(x, y \cdot \dot{z})$ $(x, y, z \in G)$;

(ii) $s_\rho(\xi, \dot{1}) = \delta_H(\xi)/\delta_G(\xi)$ $(\xi \in H)$;

(iii) $s_\rho(x, \dot{1})$ is bounded on compact sets as a function of x.

The function s_ρ will be called a *multiplier* (cf. number 5.2.1).

On the other hand every quasi-invariant measure μ gives rise in a canonical manner to a rho-function so that all quasi-invariant measures are obtained in this way.

[In passing, note that if G is a Lie group, then every positive Radon measure on G/H which, on every local chart of G/H, is equivalent to Lebesgue measure is, in fact, quasi-invariant.]

Lemma A 1.1 (Bruhat) If G is a Lie group, then there exist C^∞ rho-functions.

Proof As mentioned above G/H is paracompact. So let $\{\mathcal{O}_i\}$ be a locally finite covering of G/H by relatively compact open sets $\mathcal{O}_i$ and let $\{\phi_i\}$ be a C^∞ partition of unity subordinate to the covering $\{\mathcal{O}_i\}$. For each i there exists a positive C^∞ function ψ_i on G such that the compact support of ψ_i is contained in $\pi^{-1}(\mathcal{O}_i)$ and

$$\phi_i(\dot{x}) = \int_H \psi_i(x\xi)\, d_H(\xi) \qquad (x \in G).$$

Consider the sum $\psi = \sum \psi_i$. Since each $x \in G$ belongs to only a finite number of the $\pi^{-1}(\mathcal{O}_i)$, ψ exists. In fact for each $x \in G$ there exists a neighborhood in G of the coset xH such that only finitely many terms in the expression for ψ are non-zero in this neighborhood. Hence ψ is a positive C^∞ function on G. In addition, for each $x \in G$, the function $\xi \mapsto \psi(x\xi)$ is compactly supported in H and is non-zero since

$$\int_H \psi(x\xi)\, d_H(\xi) = 1.$$

Put

$$\rho(x) = \int_H \delta_H(\xi)^{-1}\delta_G(\xi)\psi(x\xi)\, d_H(\xi) \qquad (x \in G).$$

Then ρ is a C^∞ rho-function on G (cf. Proposition A2.4.1). □

Convention Suppose that G is a Lie group – then, unless specified to the contrary, we shall assume $\rho = \rho_H$ chosen so that $\rho_H \in C^\infty(G)$ and normalized so that $\rho_H(1) = 1$ (hence

$$\rho_H(\xi) = \frac{\delta_H(\xi)}{\delta_G(\xi)}$$

for all $\xi \in H$). The quasi-invariant measure associated with ρ_H will be denoted by μ_H. Observe that now the multiplier $s_{\rho_H} = s_H$ is in $C^\infty(G \times G/H)$.

Notation Let H be a closed subgroup of G, $d\mu_H(\dot{x})$ a quasi-invariant measure on G/H – then we shall often write more briefly $d\mu_H(x)$ (similarly in other cases too).

For reference, we shall list here a number of technical lemmas concerning rho-functions and the like.

Let H_1, H_2 be closed subgroups of our locally compact group G – assume that $H_1 \subset H_2$. Let d_i $(= d_{H_i})$ denote left Haar measure on H_i, δ_i the corresponding modular function $(i = 1, 2)$.

Lemma A 1.2 Let us suppose that the homogeneous space H_2/H_1 admits a positive H_2-invariant measure ν_2; let ρ_2 be a rho-function on G for the subgroup H_2 (and hence also for the subgroup H_1, δ_1 being equal to δ_2 on H_1); let μ_1 and μ_2 be the quasi-invariant measures on G/H_1 and G/H_2 corresponding to ρ_2, respectively (considered relative to the subgroups H_1 and H_2) – then, for a suitable normalization of the Haar measures, we have

$$\int_{G/H_1} \phi(x)\, d\mu_1(x) = \int_{G/H_2} d\mu_2(x) \int_{H_2/H_1} \phi(x\xi_2)\, d\nu_2(\xi_2) \qquad (\phi \in C_c(G/H_1)).$$

[The notation is, of course, being abused, in the obvious way.]

Proof In view of the hypotheses, we have

$$\int_{H_2} f_2(\xi_2)\, d_2(\xi_2) = \int_{H_2/H_1} d\nu_2(\xi_2) \int_{H_1} f_2(\xi_2\xi_1)\, d_1(\xi_1) \qquad (f_2 \in C_c(H_2)).$$

If $f \in C_c(G)$, then

$$\begin{aligned}\int_G f(x)\rho_2(x)\, d_G(x) &= \int_{G/H_1} d\mu_1(x) \int_{H_1} f(x\xi_1)\, d_1(\xi_1)\\ &= \int_{G/H_2} d\mu_2(x) \int_{H_2} f(x\xi_2)\, d_2(\xi_2)\\ &= \int_{G/H_2} d\mu_2(x) \int_{H_2/H_1} d\nu_2(\xi_2) \int_{H_1} f(x\xi_2\xi_1)\, d_1(\xi_1).\end{aligned}$$

Our lemma now follows from the fact that all functions in $C_c(G/H_1)$ are of the form

$$\int_{H_1} f(.\xi_1)\, d_1(\xi_1) \qquad (f \in C_c(G)). \quad \square$$

Lemma A 1.3 Let us suppose that the homogeneous space G/H_2 admits a positive G-invariant measure ν_2; let ρ_1 be a rho-function on G for the subgroup H_1 – then $\rho_1 | H_2$ is a rho-function on H_2 relative to H_1 (since $\delta_2 = \delta_G$ on H_2). Let μ_1 and μ_2 be the quasi-invariant measures on G/H_1 and H_2/H_1 corresponding, respectively, to ρ_1 and $\rho_1 | H_2$ – then, for a suitable normalization of the Haar measures, we have

$$\int_{G/H_1} \phi(x)\, d\mu_1(x) = \int_{G/H_2} d\nu_2(x) \int_{H_2/H_1} (\rho_1(x\xi_2)/\rho_1(\xi_2))\phi(x\xi_2)\, d\mu_2(\xi_2)$$
$$(\phi \in C_c(G/H_1)).$$

Proof Let $f_2 \in C_c(H_2)$ – then

$$\int_{H_2} f_2(\xi_2)\, d_2(\xi_2) = \int_{H_2/H_1} d\mu_2(\xi_2) \int_{H_1} f_2(\xi_2\xi_1)/\rho_1(\xi_2\xi_1)\, d_1(\xi_1).$$

Let $f \in C_c(G)$ – then

$$\begin{aligned}\int_G f(x)\, d_G(x) &= \int_{G/H_2} d\nu_2(x) \int_{H_2} f(x\xi_2)\, d_2(\xi_2)\\ &= \int_{G/H_2} d\nu_2(x) \int_{H_2/H_1} d\mu_2(\xi_2) \int_{H_1} f(x\xi_2\xi_1)/\rho_1(\xi_2\xi_1)\, d_1(\xi_1).\end{aligned}$$

But we also have

$$\int_G f(x)\,d_G(x) = \int_{G/H_1} d\mu_1(x) \int_{H_1} f(x\xi_1)/\rho_1(x\xi_1)\,d_1(\xi_1).$$

Hence our lemma follows upon remarking that every function in $C_c(G/H_1)$ is of the form

$$\int_{H_1} f(.\xi_1)/\rho_1(.\xi_1)\,d_1(\xi_1) \qquad (f \in C_c(G)). \quad \square$$

Let us drop the assumption that H_1 is contained in H_2 – we shall, however, agree to retain the other conventions which were laid down above. The group $H_2 \times H_1$ operates to the left on G via the prescription

$$x \mapsto (\xi_2, \xi_1)\cdot x = \xi_2 x \xi_1^{-1} \qquad (x \in G; \xi_i \in H_i, i = 1, 2)$$

(cf. A2.4). Fix an element x in G and let Q be the $H_2 \times H_1$-orbit of x, that is, the H_2, H_1 double coset $H_2 x H_1$. The stability subgroup $(H_2 \times H_1)_x$ of $H_2 \times H_1$ for the element $x \in Q$ is

$$(H_2 \times H_1)_x = \{(x\zeta x^{-1}, \zeta) : \zeta \in H_x\}$$

where $H_x = H_1 \cap (x^{-1}H_2 x)$. Thus there is a natural identification

$$(\xi_2, \xi_1)(H_2 \times H_1)_x \leftrightarrow \xi_2 x \xi_1^{-1}$$

between $H_2 \times H_1/(H_2 \times H_1)_x$ and Q; as is known (A 2.4), Q may be equipped with a topology which renders this identification a homeomorphism.

Let us now suppose that the homogeneous space $H_2/xH_x x^{-1}$ carries a positive H_2-invariant measure ν_2. This assumption implies that $\delta_2 = \delta_{xH_x x^{-1}}$ on $xH_x x^{-1}$; consequently, if we put

$$\rho(\xi_2, \xi_1) = \delta_1(\xi_1^{-1}) \qquad (\xi_i \in H_i, i = 1, 2),$$

then ρ is a rho-function on $H_2 \times H_1$ relative to $(H_2 \times H_1)_x$. Consider the group $xH_x x^{-1} \times H_1$ – then

$$H_2/xH_x x^{-1} \sim H_2 \times H_1/(xH_x x^{-1}) \times H_1$$

and the image of ν_2 under this identification is an $H_2 \times H_1$-invariant measure on $H_2 \times H_1/(xH_x x^{-1}) \times H_1$ (in particular $\rho\,|(xH_x x^{-1}) \times H_1$ is a rho-function for $(H_2 \times H_1)_x$). We may identify $(xH_x x^{-1}) \times H_1/(H_2 \times H_1)_x$ with H_1:

$$(1, \xi_1)(H_2 \times H_1)_x \leftrightarrow \xi_1, \xi_1 \in H_1.$$

The image, under this identification, of the quasi-invariant measure on

$$(xH_x x^{-1}) \times H_1/(H_2 \times H_1)_x$$

corresponding to $\rho\,|(xH_x x^{-1}) \times H_1$ is, evidently, *right* Haar measure on H_1. Let μ denote the quasi-invariant measure on $H_2 \times H_1/(H_2 \times H_1)_x \sim Q$ corresponding to ρ; substituting $H_2 \times H_1$, $xH_x x^{-1} \times H_1$, $(H_2 \times H_1)_x$ for G, H_2, and H_1 in Lemma A 1.3 supra then leads to the following result.

Lemma A 1.4 Under the above hypotheses, we have, for a suitable normalization of the Haar measures,

$$\int_Q \phi(q)\,d\mu(q) = \int_{H_2/xH_x x^{-1}} d\nu_2(\xi_2) \int_{H_1} \phi(\xi_2 x \xi_1)\,d_1(\xi_1) \qquad (\phi \in C_c(Q)).$$

2 Distributions on a Manifold

2.1 Differential Operators and Function Spaces

Let M be a C^∞ manifold countable at infinity (i.e. M may be written as a countable union of compact subsets).

Notations $C^\infty(M)$ will denote the set of C^∞ complex valued functions on M, $C_c^\infty(M)$ the set of C^∞ functions with compact support, $C_\omega^\infty(M)$ the subspace of $C_c^\infty(M)$ consisting of those functions with support contained in the compact set ω, $C^\infty(p)$ $(p \in M)$ the functions on M which are differentiable at p.

With each C^∞ vector field X on M one may associate an endomorphism D_X of $C^\infty(M)$, viz. the map $f \mapsto Xf \, (f \in C^\infty(M))$. Likewise, with each $f \in C^\infty(M)$, there is associated an endomorphism D_f of $C^\infty(M)$, viz. the map $g \mapsto fg \, (g \in C^\infty(M))$. The algebra $\mathfrak{D}_F = \mathfrak{D}_F(M)$ generated by all the D_X and D_f is called the *algebra of differential operators of finite order* on M. The composition of two elements $D_1, D_2 \in \mathfrak{D}_F$ will be written $D_1 \circ D_2$; put

$$\{D_1, D_2\} = D_1 \circ D_2 - D_2 \circ D_1.$$

If $f \in C^\infty(M)$, $D \in \mathfrak{D}_F$, $p \in M$, we shall often write $f(p; D)$ instead of $(Df)(p)$ to denote the value of Df at p.

We shall now describe the topologies with which $C^\infty(M)$ and $C_c^\infty(M)$ will be equipped.

Definition Let $D \in \mathfrak{D}_F$ – then the *support* of D (written spt (D)) is the complement of the set of all $p \in M$ such that $p \notin$ spt (Df) for every $f \in C^\infty(M)$.
Evidently spt (D) is a closed subset of M.

Definition An indexed family $\{S_\alpha : \alpha \in \mathfrak{A}\}$ of subsets of M is said to be *scattered* if for every compact subset ω of M, $\omega \cap S_\alpha$ is empty for all but a finite number of α.

A subset Φ of $\mathfrak{D}_F$ is said to be scattered if the family $\{\text{spt}\,(D) : D \in \Phi\}$ is scattered.

For each $D \in \mathfrak{D}_F$ put $|f|_D = \sup_M |Df|$ and for any scattered subset Φ of $\mathfrak{D}_F$ put $|f|_\Phi = \sum_{D \in \Phi} |f|_D \, (f \in C_c^\infty(M))$ – then each $|\,.\,|_\Phi$ is a semi-norm on $C_c^\infty(M)$. Letting Φ range through the set $\mathscr{S}$ of all scattered subsets of $\mathfrak{D}_F$, the corresponding collection of semi-norms $\{|\,.\,|_\Phi : \Phi \in \mathscr{S}\}$ equips $C_c^\infty(M)$ with the structure of a locally convex, complete, Hausdorff, topological vector space. This structure (topology) on $C_c^\infty(M)$ is called the *Schwartz structure* (*topology*) for $C_c^\infty(M)$.

Let ω be a compact subset of M. The collections of semi-norms

$$\{|\,.\,|_D : D \in \mathfrak{D}_F\} \quad \text{and} \quad \{|\,.\,|_\Phi : \Phi \in \mathscr{S}\}$$

define the same locally convex structure on $C_\omega^\infty(M)$ – in fact it is easy to see that $C_\omega^\infty(M)$ then becomes a Fréchet space. On the other hand, owing to our assumption on M, we may find a sequence $\{\omega_j\}$ of compact subsets of M such that $\omega_1 \subset \omega_2 \subset \ldots, \bigcup_j \omega_j = M$, and every compact subset ω of M is contained in some ω_j.

One may then describe the Schwartz topology on $C_c^\infty(M)$ in an alternative way. Thus $C_c^\infty(M)$ is the inductive limit of the spaces $C_\omega^\infty(M)$ (ω a compact subset of M). Moreover, $C_c^\infty(M)$ is the inductive limit of the sequence

$$\{C_{\omega_j}^\infty(M) : j = 1, 2, \ldots\}$$

and since $\{C^\infty_{\omega_j}(M) : j = 1, 2, \ldots\}$ is a sequence of Fréchet spaces, $C_c^\infty(M)$ is an LF-space (see A 2.2). The consequences of the fact that $C_c^\infty(M)$ is an LF-space are frequently useful and will be cited as needed.

By a generalized semi-norm on $C_c^\infty(M)$ we shall understand a semi-norm μ on $C_c^\infty(M)$ in the usual sense (although not necessarily continuous) assuming, possibly, the value $+\infty$. Let us call such a semi-norm *locally continuous* if every point $p \in M$ admits an open neighborhood $\mathcal{O}_p$ (in M) on which μ is continuous (i.e. assumes only finite values there, and, regarded as a mapping of $C_c^\infty(\mathcal{O}_p)$ into $\mathbf{R}$, is continuous) – then one has the following simple result.

Lemma A 2.1.1 A generalized semi-norm μ is continuous iff it is locally continuous.

Definition A *distribution* T on M is a continuous linear mapping of $C_c^\infty(M)$ into $\mathbf{C}$.

The value of a distribution T at an element $f \in C_c^\infty(M)$ will be written $T(f)$, $\langle f, T\rangle$, or $\int_M f(p)dT(p)$.

Criterion (Schwartz [1, p. 85]) Let $T: C_c^\infty(M) \to \mathbf{C}$ be linear – then T is a distribution iff given any open relatively compact subset $\mathcal{O}$ of M there exists a finite number of differential operators $D_i \in \mathfrak{D}_F$ $(i = 1, \ldots, n)$ such that

$$|\langle f, T\rangle| = |T(f)| \leqslant \sum_{i=1}^{n} \sup |D_i f| \qquad (f \in C_c^\infty(\mathcal{O})).$$

The space $C^\infty(M)$ will be equipped with the topology of uniform convergence on compact subsets of a function and each of its derivatives. More precisely for each $D \in \mathfrak{D}_F$ and each compact subset $\omega \subset M$, set

$$|f|_{D,\omega} = \sup_{\omega} |Df| \qquad (f \in C^\infty(M)).$$

The collection of semi-norms $\{|\cdot|_{D,\omega}\}$ equips $C^\infty(M)$ with the structure of a locally convex, complete, Hausdorff, topological vector space. Since M is countable at infinity, $C^\infty(M)$ is a Fréchet space. Recall that the distributions with compact support (the support of a distribution being defined in the usual way) constitute the space of continuous linear forms on $C^\infty(M)$ (Schwartz [1, p. 89]).

Let $C_c^p(M)$ $(0 \leqslant p \leqslant \infty)$ denote the space of p-times continuously differentiable complex valued functions on M with compact support equipped with the usual (i.e. Schwartz-like) topology. A distribution T on M is said to be of *finite order* $\leqslant p$ if T is continuous when $C_c^\infty(M)$ is equipped with the topology induced by $C_c^p(M)$. A distribution of order zero is a measure.

In general, by a *differential operator* on M (not necessarily of finite order) we mean a continuous endomorphism of $C^\infty(M)$ with the property that on every local chart $(\mathcal{O}, \phi)$ of M the mapping $f \mapsto Df$ of $C_c^\infty(\mathcal{O})$ into $C^\infty(M)$ maps $C_c^\infty(\mathcal{O})$ into itself and defines a differential operator on $\mathcal{O}$ (i.e. if $\{x_1, \ldots, x_m\}$ are coordinates on $\mathcal{O}$, then

$$D = \sum_I a_I(x_1, \ldots, x_m)\partial^{|I|}/\partial x_1^{i_1} \cdots \partial x_m^{i_m}$$

where the $a_I \in C^\infty(\mathcal{O})$ and the sets $\{\mathrm{spt}(a_I)\}$ are scattered in $\mathcal{O}$). It is an immediate consequence of this definition that D decreases supports. Conversely, it can be shown that every endomorphism of $C^\infty(M)$ which decreases supports is necessarily a differential operator on M (see Peetre [1]). The set of all differential operators on M, with the evident operations, constitutes an algebra $\mathfrak{D}(M)$, the *algebra of differential operators* on M. The composition of two elements $D_1, D_2 \in \mathfrak{D}(M)$ will be written $D_1 \circ D_2$; put

$$\{D_1, D_2\} = D_1 \circ D_2 - D_2 \circ D_1.$$

The space of distributions on M is easily turned into a module over $C^\infty(M)$. Thus let T be a distribution, $f \in C^\infty(M)$ and define $f \cdot T$ by

$$\langle g, f \cdot T\rangle = \langle fg, T\rangle \qquad (g \in C_c^\infty(M)).$$

Let us recall a useful result concerning the structure of distributions, the support of which is contained in a closed submanifold Q of M.

Let $i : Q \to M$ denote the (regular) injection of Q into M. If S is a distribution on Q, then S extends to a distribution (the *extension* of S) on M (which will still be denoted by S) by the rule

$$\langle f, S\rangle = \langle f \circ i, S\rangle, \quad f \in C_c^\infty(M).$$

Let $p_0 \in Q$; on a sufficiently small open relatively compact neighborhood $\mathcal{O}_M(p_0)$ of p_0 in M there exists a local coordinate system $\{u_1, \ldots, u_q, v_1, \ldots, v_r\}$ such that $\mathscr{P} = Q \cap \mathcal{O}_M(p_0)$ is defined by the equations $v_i = 0$. For every system $I = (i_1, \ldots, i_r)$ of r non-negative integers put

$$D^I = D_1^{i_1} \cdots D_r^{i_r} = \partial^{|I|}/\partial v_1^{i_1} \cdots \partial v_r^{i_r}$$

where $|I| = i_1 + \cdots + i_r$, $D_i = \partial/\partial v_i$. If S is a distribution on $\mathscr{P}$ the symbol $D^I S$ will stand for the application of the differential operator D^I to the extension of S to $\mathcal{O}_M(p_0)$ – thus, by definition, one has

$$\langle f, D^I S\rangle = (-1)^{|I|}\langle D^I(f \circ i), S\rangle \qquad (f \in C_c^\infty(\mathcal{O}_M(p_0))).$$

The distribution $D^I S$ is called a *transversal derivative* of S.

Proposition A 2.1.2 (Schwartz [1, p. 102]) Every distribution T on M with support contained in Q, when viewed as a distribution on $\mathcal{O}_M(p_0)$, admits a *unique* decomposition into a finite linear combination of transversal derivatives of extensions to $\mathcal{O}_M(p_0)$ of distributions on $\mathscr{P}$,

$$\text{i.e.} \quad T = \sum (-1)^{|I|} D^I T_I,$$

T_I a distribution on $\mathscr{P}$.

The greatest of the orders $|I| = i_1 + \cdots + i_r$ such that $T_I \neq 0$ is independent of the choice of coordinates and is called the *transversal order* of T on $\mathcal{O}_M(p_0)$ (Schwartz [1, p. 103]).

Suppose in addition that M is orientable, the orientation being determined by a positive differential form $\omega > 0$. Let μ be the Radon measure on M canonically associated with ω (cf. Helgason [2, Chapter 10, § 1]) – then for every $D \in \mathfrak{D}(M)$, there exists a unique differential operator D^t, called the *adjoint* of D (with respect to ω), such that

$$\int_M (Df)(g)\, d\mu = \int_M f(D^t g)\, d\mu$$

whenever f and g are two C^∞ functions on M, at least one of which has compact support. The map $D \mapsto D^t$ is an anti-automorphism of $\mathfrak{D}(M)$ of order 2.

Consider the space $\mathfrak{L}(C_c^\infty(M); \mathbf{C})$ of scalar distributions on M (similar remarks will apply to vector distributions – see A2.3). One may equip $\mathfrak{L}(C_c^\infty(M); \mathbf{C})$ with the structure of a $\mathfrak{D}(M)$-module as follows: If $T \in \mathfrak{L}(C_c^\infty(M); \mathbf{C})$, $D \in \mathfrak{D}(M)$, then define the distribution $D \cdot T$ by the rule $f \mapsto T(D^t f)$ $(f \in C_c^\infty(M))$.

Take $M = E$, E a finite dimensional vector space over $\mathbf{R}$.

Lemma A 2.1.3 Let T be a distribution on an open set $\mathcal{O}$ in E and let X_0 be a point in $\mathcal{O}$. Let f be a function in $C^\infty(\mathcal{O})$ such that $f(X) = 0$ for every $X \in$ spt (T) – then there exists an integer $m \geqslant 0$ such that $X_0 \notin \operatorname{spt}(f^m T)$ (m depending on X_0).

[A proof of this elementary fact may be given, for instance, on the basis of the results in Schwartz [1, pp. 99–100].]

2.2 Tensor Products of Topological Vector Spaces

Let E and F be two locally convex, Hausdorff, topological vector spaces over $\mathbf{C}$ – this assumption will not be repeated. We shall recall without proof a few fundamental facts concerning the locally convex structures which one may place on the algebraic tensor product $E \otimes_{\mathbf{C}} F$. [For details see the fundamental work of Grothendieck [1]; the exposition of Treves [2] is also a convenient reference.]

Let λ be the canonical bilinear map from $E \times F$ (the Cartesian product) into $E \otimes_{\mathbf{C}} F$ (hence $\lambda(a, b) = a \otimes b$, for all $a \in E$, $b \in F$). If G is any other linear space over $\mathbf{C}$ and if ϕ is a bilinear map from $E \times F$ into G, then there exists a unique linear map ψ from $E \otimes_{\mathbf{C}} F$ into G such that $\psi \circ \lambda = \phi$. That is we have the commutative diagram

$$\begin{array}{ccc} E \times F & \xrightarrow{\ \lambda\ } & E \otimes_{\mathbf{C}} F \\ & \underset{\phi}{\searrow} & \downarrow \psi \\ & & G \end{array}$$

Topological tensor products are formed by placing topological requirements on the above diagram.

Projective Tensor Product Suppose the topologies of E and F are described by families $\{|\cdot|_\alpha : \alpha \in \mathfrak{A}\}$, $\{|\cdot|_\beta : \beta \in \mathfrak{B}\}$ of continuous semi-norms, respectively – then the algebraic tensor product $E \otimes_{\mathbf{C}} F$ can be given a locally convex structure as follows. For each $u \in E \otimes_{\mathbf{C}} F$ let

$$|u|^{\wedge}_{\alpha,\beta} = \inf \sum_{i=1}^{n} |a_i|_\alpha |b_i|_\beta$$

where the inf is taken over all finite sequences

$$\{(a_i, b_i) : a_i \in E, b_i \in F, 1 \leqslant i \leqslant n\}$$

such that

$$u = \sum_{i=1}^{n} a_i \otimes b_i.$$

We shall write

$$\left| \sum_{i=1}^{n} a_i \otimes b_i \right|^{\wedge}_{\alpha,\beta} = |u|^{\wedge}_{\alpha,\beta}$$

for each sum such that

$$\sum_{i=1}^{n} a_i \otimes b_i = u.$$

It is easy to prove:

Proposition A 2.2.1 The semi-norms $|\,.\,|^{\wedge}_{\alpha,\beta}$ have the following properties.

(i) For each $u \neq 0$ in $E \otimes {}_C F$ there are $\alpha \in \mathfrak{A}$, $\beta \in \mathfrak{B}$ such that $|u|^{\wedge}_{\alpha,\beta} \neq 0$.

(ii) For any two pairs (α', β'), (α'', β'') there exists a pair (α, β) such that for all $u \in E \otimes {}_C F$

$$|u|^{\wedge}_{\alpha,\beta} \geqslant |u|^{\wedge}_{\alpha',\beta'}; \quad |u|^{\wedge}_{\alpha,\beta} \geqslant |u|^{\wedge}_{\alpha'',\beta''}.$$

(iii) For all $a \in E$, $b \in F$

$$|a \otimes b|^{\wedge}_{\alpha,\beta} = |a|_{\alpha} |b|_{\beta}$$

for each pair (α, β).

Hence the semi-norms $|\,.\,|^{\wedge}_{\alpha,\beta}$ are a filtrant set and thereby define a locally convex structure on $E \otimes {}_C F$. The Hausdorff topology so obtained is called the *projective topology* for $E \otimes {}_C F$.

Definition The completion of the locally convex space $E \otimes {}_C F$ equipped with the projective topology is called the *projective tensor product* of the locally convex spaces E and F. It is denoted by $E \hat{\otimes} F$.

Proposition A 2.2.2 If E and F are metrizable, then $E \hat{\otimes} F$ is a Fréchet space.

Equip the Cartesian product $E \times F$ with the usual topology.

Theorem A 2.2.3 Let ϕ be a bilinear map from $E \times F$ into a locally convex space G, ψ the unique linear extension of ϕ to $E \otimes {}_C F$ (i.e. $\psi \circ \lambda = \phi$). In order that ψ be continuous in the projective topology, it is necessary and sufficient that ϕ be continuous. The projective topology is the only topology on $E \otimes {}_C F$ with this property.

Thus this theorem identifies the space of continuous bilinear maps of $E \times F$ into G with the set $\mathfrak{L}(E \otimes {}_C F; G)$ of continuous linear maps of $E \otimes {}_C F$ (equipped with the projective topology) into G. In addition it may be shown that the equicontinuous subsets of these spaces correspond under this identification.

Inductive Tensor Product Let G be a locally convex space, $\phi : E \times F \to G$ a bilinear function: ϕ is said to be *separately continuous* iff all the linear maps $a \mapsto \phi(a, b)$, $b \mapsto \phi(a, b)$ are continuous ($a \in E$, $b \in F$).

Theorem A 2.2.4 There exists one and only one Hausdorff locally convex topology $\mathfrak{T}$ on the algebraic tensor product $E \otimes {}_C F$ with the following property. Let ϕ be a bilinear map from $E \times F$ into a locally convex space G, ψ the unique linear extension of ϕ to $E \otimes {}_C F$ (i.e. $\psi \circ \lambda = \phi$). In order that ψ be continuous in the topology $\mathfrak{T}$ it is necessary and sufficient that ϕ be separately continuous.

The topology $\mathfrak{T}$ is called the *inductive topology* for $E \otimes {}_C F$.

Definition The completion of the locally convex space $E \otimes {}_C F$ equipped with the inductive topology is called the *inductive tensor product* of the locally convex spaces E and F. It is denoted by $E \bar{\otimes} F$.

Theorem A 2.2.4 identifies the space of separately continuous bilinear maps of $E \times F$ into G with the set $\mathfrak{L}(E \otimes {}_C F; G)$ of continuous linear maps of $E \otimes {}_C F$ (equipped with the inductive topology) into G.

The inductive topology on $E \otimes {}_C F$ is finer than the projective topology. On the other hand if E and F are both Fréchet spaces, then every separately continuous bilinear map is continuous (Kelley and Namioka [1, p. 107]) and so in this case the inductive and projective topologies are the same.

Theorem A 2.2.5 Suppose that E is the inductive limit of locally convex spaces $\{E_i : i \in I\}$, F is the inductive limit of locally convex spaces $\{F_j : j \in J\}$ – then $E \bar{\otimes} F$ is the inductive limit of the locally convex spaces

$$\{E_i \bar{\otimes} F_j : i \in I, j \in J\}.$$

Let (E_1, E_2), (F_1, F_2) be two pairs of locally convex, Hausdorff, topological vector spaces, A_i a continuous map of E_i into F_i $(i = 1, 2)$. Equip $E_1 \otimes_{\mathbf{C}} E_2$, $F_1 \otimes_{\mathbf{C}} F_2$ with the projective topology. The map $(a_1, a_2) \mapsto A_1 a_1 \otimes A_2 a_2$ of $E_1 \times E_2$ into $F_1 \otimes_{\mathbf{C}} F_2$ is plainly continuous, therefore the corresponding map of $E_1 \otimes_{\mathbf{C}} E_2$ into $F_1 \otimes_{\mathbf{C}} F_2$ is continuous and extends by continuity to a continuous linear map of $E_1 \hat{\otimes} E_2$ into $F_1 \hat{\otimes} F_2$. This latter map will be denoted by $A_1 \hat{\otimes} A_2$ and referred to as the (projective) *tensor product* of the continuous linear maps A_1 and A_2. Elementary properties are:

(i) If the spaces under consideration are normed, then

$$\| A_1 \hat{\otimes} A_2 \| \leqslant \| A_1 \| \cdot \| A_2 \|.$$

(ii) Under the obvious assumptions,

$$(A_1 \hat{\otimes} A_2) \circ (B_1 \hat{\otimes} B_2) = (A_1 \circ B_1) \hat{\otimes} (A_2 \circ B_2).$$

Examples Let us try to clarify the preceding considerations. First we need some notations and a definition. It is presupposed that the reader is familiar with inductive limits and strict inductive limits. A space $(E, \mathfrak{T})$ is called a *strict LF-space* if $(E, \mathfrak{T})$ is the strict inductive limit of Fréchet spaces ($\mathfrak{T}$ being the topology on E).

Let M be a non-compact C^∞ manifold countable at infinity, ω a compact subset of M.

Example 1 The space $C^\infty_\omega(M)$ is a Fréchet space, while $C^\infty_c(M)$ is a strict LF-space.

Assume in what follows that E and F are also complete.

Consider the set $C^\infty(M; E)$ of E-valued C^∞ functions on M. We shall equip $C^\infty(M; E)$ with the topology of uniform convergence on compact subsets of a function and each of its derivatives. More precisely if $\{| . |_\alpha : \alpha \in \mathfrak{A}\}$ is a family of continuous semi-norms describing the topology of E, if $D \in \mathfrak{D}_F$, and if ω is a compact subset of M, put

$$| f |^\alpha_{D,\omega} = \sup_\omega | Df |_\alpha \qquad (f \in C^\infty(M; E)).$$

Then the collection $\{| . |^\alpha_{D,\omega}\}$ of semi-norms equips $C^\infty(M; E)$ with the structure of a locally convex, complete, Hausdorff, topological vector space.

Remark It is due to Grothendieck that a function f on M with values in E is C^∞ iff it is weakly C^∞ (i.e. for all $\mathbf{z} \in E^*$, $p \mapsto \langle f(p), \mathbf{z} \rangle$ is a complex valued C^∞ function on M). For a discussion and references see Neto [1].

Similar considerations apply to the set $C^\infty_c(M; E)$ of E-valued C^∞ functions with compact support on M. In the notations of A2.1, the semi-norms

$$|f|_{\Phi}^{\alpha} = \sum_{D \in \Phi} |f|_D^{\alpha}$$

where $|f|_D^{\alpha} = \sup_M |Df|_{\alpha}$ $(f \in C_c^{\infty}(M; E))$ describe the locally convex, complete, Hausdorff topology of $C_c^{\infty}(M; E)$.

If $C_{\omega}^{\infty}(M; E)$ denotes the set of E-valued C^{∞} functions on M with support contained in the compact set ω, and if $C_{\omega}^{\infty}(M; E)$ is equipped with the topology inherited from $C^{\infty}(M; E)$, then $C_c^{\infty}(M; E)$ is the inductive limit of the $C_{\omega}^{\infty}(M; E)$.

Example 2 Evidently it is true that $C^{\infty}(M) \otimes_{\mathbf{C}} E \subset C^{\infty}(M; E)$. [Let $f \in C^{\infty}(M)$, $a \in E$ – then the function $f \otimes a$, $p \mapsto f(p)a$ $(p \in M)$, lies in $C^{\infty}(M; E)$.] In addition (Grothendieck [1, Chapter 2, p. 81])

$$C^{\infty}(M) \hat{\otimes} E = C^{\infty}(M; E).$$

Example 3 Let ω be a compact subset of M – then (Grothendieck [1, Chapter 2, p. 81])

$$C_{\omega}^{\infty}(M) \hat{\otimes} E = C_{\omega}^{\infty}(M; E).$$

Example 4 Suppose that E is a Banach space, μ a positive Radon measure on M. Let $L_{\mu}^1(M; E)$ denote the set of μ-summable functions on M with values in E – then (Grothendieck [1, Chapter 1, p. 59])

$$L_{\mu}^1(M; E) = L_{\mu}^1(M) \hat{\otimes} E.$$

Example 5 If E is a Fréchet space, then (Grothendieck [1, Chapter 2, p. 84])

$$C_c^{\infty}(M; E) = C_c^{\infty}(M) \bar{\otimes} E.$$

Example 6 If E is a Fréchet space, then

$$C_{\omega}^{\infty}(M) \bar{\otimes} E = C_{\omega}^{\infty}(M) \hat{\otimes} E = C_{\omega}^{\infty}(M; E)$$

(recall that $C_{\omega}^{\infty}(M)$ is a Fréchet space). Hence in this case $C_{\omega}^{\infty}(M; E)$ is also a Fréchet space. Moreover here $C_c^{\infty}(M; E)$ is a strict LF-space (M is countable at infinity).

Example 7 Let M_1, M_2 be two C^{∞} manifolds both countable at infinity. If $f_i \in C_c^{\infty}(M_i)$ $(i = 1, 2)$, then $f_1 \otimes f_2$ is defined by $(p_1, p_2) \mapsto f_1(p_1)f_2(p_2)$. It is well-known (Schwartz [1, p. 108]) that $C_c^{\infty}(M_1) \otimes_{\mathbf{C}} C_c^{\infty}(M_2)$ is dense in $C_c^{\infty}(M_1 \times M_2)$; in fact (Grothendieck [1, Chapter 2, p. 84])

$$C_c^{\infty}(M_1) \bar{\otimes} C_c^{\infty}(M_2) = C_c^{\infty}(M_1 \times M_2).$$

Example 8 (Schwartz's Kernel Theorem) Let E and F be Fréchet spaces – then

$$C_c^{\infty}(M; E) \bar{\otimes} C_c^{\infty}(M; F) = C_c^{\infty}(M \times M; E \hat{\otimes} F).$$

Let us prove this result. In the first place $C_c^{\infty}(M; E)$ (resp. $C_c^{\infty}(M; F)$) is the strict inductive limit of a sequence $C_{\omega_i}^{\infty}(M; E)$ (resp. $C_{\omega_j}^{\infty}(M; F)$) of subspaces, ω_i (resp. ω_j) a compact subset of M (Example 6). Since E (resp. F) is a Fréchet space, $C_{\omega_i}^{\infty}(M; E)$ is a Fréchet space (resp. $C_{\omega_j}^{\infty}(M; F)$ is a Fréchet space). Theorem A 2.2.5 implies that $C_c^{\infty}(M; E) \bar{\otimes} C_c^{\infty}(M; F)$ is the strict inductive limit of the sequence

$$C^{\infty}_{\omega_i}(M; E) \mathbin{\bar{\otimes}} C^{\infty}_{\omega_j}(M; F) = (C^{\infty}_{\omega_i}(M) \mathbin{\hat{\otimes}} E) \mathbin{\hat{\otimes}} (C^{\infty}_{\omega_j}(M) \mathbin{\hat{\otimes}} F)$$
$$= (C^{\infty}_{\omega_i}(M) \mathbin{\hat{\otimes}} C^{\infty}_{\omega_j}(M)) \mathbin{\hat{\otimes}} (E \mathbin{\hat{\otimes}} F).$$

Now

$$C^{\infty}_{\omega_i}(M) \mathbin{\hat{\otimes}} C^{\infty}_{\omega_j}(M) = C^{\infty}_{\omega_i}(M) \mathbin{\bar{\otimes}} C^{\infty}_{\omega_j}(M)$$

(both are Fréchet spaces) $= C^{\infty}_{\omega_i \times \omega_j}(M \times M)$ (Example 7), whence

$$C^{\infty}_{c}(M; E) \mathbin{\bar{\otimes}} C^{\infty}_{c}(M; F)$$

is the strict inductive limit of the sequence

$$C^{\infty}_{\omega_i \times \omega_j}(M \times M) \mathbin{\hat{\otimes}} (E \mathbin{\hat{\otimes}} F)$$

and by definition the latter inductive limit is precisely

$$C^{\infty}_{c}(M \times M; E \mathbin{\hat{\otimes}} F).$$

Example 9 As pointed out in Example 5, if E is a Fréchet space, then $C^{\infty}_{c}(M; E) = C^{\infty}_{c}(M) \mathbin{\bar{\otimes}} E$. In general all one can assert is (Grothendieck [1, Chapter 2, p. 83]): The topology on $C^{\infty}_{c}(M) \otimes_{C} E$ induced by $C^{\infty}_{c}(M; E)$ is finer than that induced by $C^{\infty}_{c}(M) \mathbin{\hat{\otimes}} E$ and less fine than that induced by $C^{\infty}_{c}(M) \mathbin{\bar{\otimes}} E$.

2.3 Vector Distributions

The theory of vector distributions has had a rather extensive development (cf. Schwartz [3], [4]). Nevertheless only the most elementary definitions and results will be needed in this book.

Throughout E will denote a locally convex, complete, Hausdorff, topological vector space, M a C^{∞} manifold countable at infinity.

Definition A continuous linear map T of $C^{\infty}_{c}(M)$ into E will be called a *distribution on M with values in E*.

Hence a distribution T on M with values in E is simply an element of $\mathfrak{L}(C^{\infty}_{c}(M); E)$. The space $\mathfrak{L}(C^{\infty}_{c}(M); E)$ will be equipped with the topology of uniform convergence on bounded subsets of $C^{\infty}_{c}(M)$.

Definition A continuous linear map T of $C^{\infty}_{c}(M; E)$ into **C** will be called an *E-distribution on M*.

Hence an E-distribution T on M is simply an element of $\mathfrak{L}(C^{\infty}_{c}(M; E); \mathbf{C})$. The space $\mathfrak{L}(C^{\infty}_{c}(M; E); \mathbf{C})$ will be equipped with the topology of uniform convergence on bounded subsets of $C^{\infty}_{c}(M; E)$.

Let F be a locally convex, complete, Hausdorff, topological vector space. We shall then agree to call the elements of $\mathfrak{L}(C^{\infty}_{c}(M; E); F)$ *E-distributions on M with values in F* – of course this definition encompasses the two preceding ones. [Let us note, in passing, that the completeness of F is not a crucial requirement for many considerations.] The space $\mathfrak{L}(C^{\infty}_{c}(M; E); F)$ will be equipped with the topology of uniform convergence on bounded subsets of $C^{\infty}_{c}(M; E)$

The distributions T appearing in these definitions will be called *vector distributions*. The value of T at an element f will be written $T(f)$, $\langle f, T \rangle$, or $\int_M f(p) dT(p)$.

Examples (1) Assume that E is a Fréchet space. Let T be an E-distribution on M – then T determines canonically a distribution S on M with values in

E_b^* (the strong dual of E). [If $f \in C_c^\infty(M)$, then $S(f)$ is that element of E^* which maps a to $\langle f \otimes a, T\rangle$ $(a \in E)$; it is easy to see that the map $S : C_c^\infty(M) \to E_b^*$ is continuous.] On the other hand if S is a distribution on M with values in E_b^*, then S determines a separately continuous bilinear map T of $C_c^\infty(M) \times E$ into $\mathbf{C}$, viz. $(f, a) \mapsto \langle a, S(f)\rangle$. That is to say, S determines a continuous linear map of $C_c^\infty(M) \bar{\otimes} E$ into $\mathbf{C}$. Since E is a Fréchet space,

$$C_c^\infty(M) \bar{\bar{\otimes}} E = C_c^\infty(M; E)$$

(A2.2, Example 5). Hence S does indeed determine an E-distribution T. Evidently the correspondence $T \leftrightarrow S$ is bijective.

(2) Let T be an E-distribution on M with values in F; let X be another locally convex, complete, Hausdorff, topological vector space – then T gives rise, in a natural way, to an $E \hat{\otimes} X$-distribution S on M with values in $F \hat{\otimes} X$. [Thus, for each compact set $\omega \subset M$, the restriction of T to $C_\omega^\infty(M; E)$, when tensored with the identity map $X \to X$, leads to a continuous map

$$S_\omega : C_\omega^\infty(M; E) \hat{\otimes} X \to F \hat{\otimes} X.$$

But, in view of Example 3 in 2.2,

$$\begin{aligned} C_\omega^\infty(M; E) \hat{\otimes} X &= (C_\omega^\infty(M) \hat{\otimes} E) \hat{\otimes} X = C_\omega^\infty(M) \hat{\otimes} (E \hat{\otimes} X) \\ &= C_\omega^\infty(M; E \hat{\otimes} X). \end{aligned}$$

Therefore we can regard S_ω as a continuous linear map on $C_\omega^\infty(M; E \hat{\otimes} X)$ with values in $F \hat{\otimes} X$. It is now clear how to define the vector distribution S.]

Most of the basic definitions and theorems of the scalar case extend with little or no change to the vector distributions. Thus let T be a vector distribution. It makes sense to speak of the support of T, to speak of the restriction of T to an open subset of M, to multiply T by an element of $C^\infty(M)$, and to differentiate T by differential operators $D \in \mathfrak{D}(M)$ (if M is orientable). If T is an E-distribution on M with values in F and with compact support, then T extends uniquely to an element of $\mathfrak{L}(C^\infty(M; E); F)$. More generally, if T does not have compact support, then it is at least true that T extends to the set $\mathscr{S}_T$ consisting of those f in $C^\infty(M; E)$ such that spt $(T) \cap$ spt (f) is compact – the extension of T to $\mathscr{S}_T$ will be called the *natural extension* of T. [Thus suppose that $f \in \mathscr{S}_T$; choose $\phi \in C_c^\infty(M)$ such that $\phi = 1$ on a neighborhood of spt $(T) \cap$ spt (f) – then $\phi f \in C_c^\infty(M; E)$ and, as is easily checked, the value of $\langle \phi f, T\rangle$ is independent of the choice of ϕ. We may therefore legitimately write $\langle f, T\rangle$ for $\langle \phi f, T\rangle$ etc.]

The order of a vector distribution T is defined as usual. One proves just as in Schwartz [1, p. 85] that every E-distribution T on M is of finite order on every open relatively compact subset of M. [It must be observed that this result is not true for distributions on M with values in E: Take the identity map of $C_c^\infty(M)$ into itself.]

Let M_1, M_2 be two C^∞ manifolds both countable at infinity; put $M = M_1 \times M_2$; let T_i be an E_i-distribution on M_i with values in F_i $(i = 1, 2)$ – then the *Cartesian product* T of T_1 and T_2, denoted by $T_1 \times T_2$, is the $E_1 \hat{\otimes} E_2$-distribution on M with values in $F_1 \hat{\otimes} F_2$ which is constructed in the following way. Observing that, for compact $\omega_i \subset M_i$ $(i = 1, 2)$,

$$C_{\omega_1}^\infty(M_1; E_1) \hat{\otimes} C_{\omega_2}^\infty(M_2; E_2) = C_{\omega_1 \times \omega_2}^\infty(M; E_1 \hat{\otimes} E_2)$$

(cf. A2.2, Example 7), we see that the tensor product map formed from the restriction to $C_{\omega_i}^\infty(M_i; E_i)$ of T_i $(i = 1, 2)$ may be identified with a mapping of $C_{\omega_1 \times \omega_2}^\infty(M; E_1 \hat{\otimes} E_2)$ into $F_1 \hat{\otimes} F_2$; taking the union of these maps over the $\omega_1 \times \omega_2$ $(\omega_i \subset M_i$, ω_i compact) leads to a continuous linear map on

$$C_c^\infty(M; E_1 \hat{\otimes} E_2)$$

with values in $F_1 \hat{\otimes} F_2$ – this, by definition, is the distribution T referred to above. Evidently the support of $T_1 \times T_2$ is equal to spt $(T_1) \times$ spt (T_2). The formation of the Cartesian product of distributions is associative (in the obvious sense). Moreover, just as in the scalar case, one may evaluate the Cartesian product $T = T_1 \times T_2$ of distributions T_1 and T_2 on functions $f \in C_c^\infty(M_1 \times M_2; E_1 \hat{\otimes} E_2)$ in an iterative way. Thus, as in Example 2 supra, let S_2 be the $E_1 \hat{\otimes} E_2$-distribution on M_2 with values in $E_1 \hat{\otimes} F_2$ obtained from T_2 by tensoring with E_1; similarly let S_1 be the $E_1 \hat{\otimes} F_2$-distribution on M_1 with values in $F_1 \hat{\otimes} F_2$ obtained from T_1 by tensoring with F_2 – then, for each $f \in C_c^\infty(M_1 \times M_2; E_1 \hat{\otimes} E_2)$, the expression $(S_2)_{p_2} f(p_1, p_2)$ (as a function of $p_1 \in M_1$) belongs to $C_c^\infty(M; E_1 \hat{\otimes} F_2)$ and

$$(T_1 \times T_2)(f) = (S_1)_{p_1}(S_2)_{p_2} f(p_1, p_2)$$

($(S_2)_{p_2}$ means S_2 acting on the expression which follows considered as a function of $p_2 \in M_2$; similarly for $(S_1)_{p_1}$). [Let ω_i be a compact subset of M_i $(i = 1, 2)$. The distribution S_2 gives rise to a continuous linear map R_2 of $C_{\omega_1 \times \omega_2}^\infty(M_1 \times M_2; E_1 \hat{\otimes} E_2)$ into $C_{\omega_1}^\infty(M_1; E_1 \hat{\otimes} F_2)$ (assign to

$$f \in C_{\omega_1 \times \omega_2}^\infty(M_1 \times M_2; E_1 \hat{\otimes} E_2)$$

the map $p_1 \mapsto (S_2)_{p_2} f(p_1, p_2)$); the composition $S_1 \circ R_2$ is a continuous map of $C_{\omega_1 \times \omega_2}^\infty(M_1 \times M_2; E_1 \hat{\otimes} E_2)$ into $F_1 \hat{\otimes} F_2$ – to conclude that $T_1 \times T_2 = S_1 \circ R_2$, one need only establish equality on

$$C_{\omega_1}^\infty(M_1; E_1) \otimes_C C_{\omega_2}^\infty(M_2; E_2)$$

and this is evident.] Making the obvious identifications, it is clear that the iteration may be reversed (that is can be performed in either order).

Let M, N be two C^∞ manifolds and $\pi : M \to N$ a C^∞ map. Let T be an E-distribution on M with values in F – assume: For every compact subset $\omega \subset N$, $\pi^{-1}(\omega) \cap$ spt (T) is compact. [Note, in particular, that this assumption is met whenever T itself has compact support.] Let $f \in C_c^\infty(N; E)$ – then our assumption has the consequence that the element $\pi^* f = f \circ \pi$ of $C^\infty(M; E)$ lies in the domain $\mathscr{S}_T$ of the natural extension of T. This means that we can define a linear map $\pi_* T : C_c^\infty(N; E) \to F$ by writing

$$\pi_* T(f) = T(\pi^* f) \qquad (f \in C_c^\infty(N; E)).$$

An elementary argument now shows that $\pi_* T$ is an E-distribution on N with values in F. We shall call $\pi_* T$ the *image of* T under π. If the E-distribution T has compact support, then so does its image $\pi_* T$ – this follows from the observation that spt $(\pi_* T) \subset \pi(\text{spt}\,(T))^{cl}$ (which is true regardless of whether T has compact support or not).

If $\pi_i : M_i \to N_i$ is a C^∞ map of manifolds, and if T_i is an E_i-distribution on M_i with values in F_i verifying the assumption of the previous paragraph relative to π_i $(i = 1, 2)$, then the Cartesian product $T = T_1 \times T_2$ also meets this requirement (relative to $\pi = \pi_1 \times \pi_2 : M_1 \times M_2 \to N_1 \times N_2$) and we have $\pi_* T = (\pi_1)_* T_1 \times (\pi_2)_* T_2$.

2.4 Distributions on a Lie Group

Let G be a Lie group countable at infinity. Let E be a locally convex, complete, Hausdorff, topological vector space. Let d_G denote left Haar measure on G. The existence on G of this canonical element of volume permits one to identify, whenever it is convenient, a function f, locally summable with respect to d_G, with the distribution $g \mapsto \int_G g(x)f(x)d_G(x)$ $(g \in C_c^\infty(G))$. In such situations we shall refer to T as a locally summable function, or a C^∞ function, or a real analytic function etc. [Note too that every $f \in C(G; E)$ gives rise in a natural manner to a distribution on G with values in E via the prescription

$$g \mapsto \int_G g(x)f(x)\, d_G(x) \qquad (g \in C_c^\infty(G))$$

(the integral being taken in the Bochner sense) – again we shall frequently permit ourselves to say that this distribution 'is' a continuous function on G etc.]

For $i = 1, 2$, let T_i be an E_i-distribution on G with values in F_i (E_i, F_i as always). We shall suppose that at least one of the T_i has compact support. Let π: $G \times G \to G$ denote the mapping $(x, y) \mapsto xy$ $(x, y \in G)$. Our assumption then implies that it makes sense to consider the image $\pi_*(T_1 \times T_2)$ of $T_1 \times T_2$ under π (see A 2.3) – this image is called the *convolution* of T_1 and T_2 and is denoted by $T_1 * T_2$. Thus the convolution $T_1 * T_2$ is an $E_1 \hat{\otimes} E_2$-distribution on G with values in $F_1 \hat{\otimes} F_2$. This definition of convolution is the same as that given in Schwartz [2, Chapter 6] for the scalar case and possesses most of the expected formal properties (with the exception, perhaps, of commutativity). In particular the convolution product $T_1 * \cdots * T_n$ can be defined when all but possibly one of the factors have compact supports and the associative rule holds in this case.

Proposition A 2.4.1 Let T be a distribution on G belonging to $\mathfrak{L}(C_c^\infty(G); E)$ (resp. $\mathfrak{L}(C^\infty(G); E)$, $\mathfrak{L}(C_c^\infty(G); \mathbf{C})$, $\mathfrak{L}(C^\infty(G); \mathbf{C})$) and f a function belonging to $C_c^\infty(G)$ (resp. $C^\infty(G)$, $C_c^\infty(G; E)$, $C^\infty(G; E)$) – then the convolutions $f * T$ and $T * f$ are C^∞ functions on G with values in E given by

$$f*T(x) = \int_G f(xy^{-1})\delta_G(y^{-1})\, dT(y), \quad T*f(x) = \int_G f(y^{-1}x)\, dT(y),$$

δ_G the modular function on G. Moreover the bilinear maps $(f, T) \mapsto f*T$, $(f, T) \mapsto T*f$ are separately continuous from $C_c^\infty(G) \times \mathfrak{L}(C_c^\infty(G); E)$ (resp. $C^\infty(G) \times \mathfrak{L}(C^\infty(G); E)$, $C_c^\infty(G; E) \times \mathfrak{L}(C_c^\infty(G); \mathbf{C})$, $C^\infty(G; E) \times \mathfrak{L}(C^\infty(G); \mathbf{C})$) into $C^\infty(G; E)$.

[The proofs of these facts do not differ in any essential way from the ones in the scalar case.]

Let G be a Lie group, M a C^∞ manifold.

Definition G is said to be a *differentiable Lie transformation group* of M if with each $x \in G$ there is associated a diffeomorphism $p \mapsto x \cdot p$ of M onto itself such that:

(i) $(xy) \cdot p = x \cdot (y \cdot p)$ $(x, y \in G)$;

(ii) The map $(x, p) \mapsto x \cdot p$ is a differentiable map of the product manifold $G \times M$ onto M.

[Under these circumstances, one frequently says that G operates to the *left* on M; to get a differentiable *right* action, one demands that there be associated with each $x \in G$ a diffeomorphism $p \mapsto p \cdot x$ of M onto itself such that:

(i) $p \cdot (xy) = (p \cdot x) \cdot y$ $(x, y \in G)$;

(ii) The map $(p, x) \mapsto p \cdot x$ is a differentiable map of the product manifold $M \times G$ onto M.]

Suppose then that G is a differentiable Lie transformation group of M – briefly, suppose that G operates on M. The *stabilizer* of a point $p \in M$, denoted by G_p, is the subgroup of G consisting of those $x \in G$ such that $x \cdot p = p$; the *orbit* of a point $p \in M$ is the set $\{x \cdot p : x \in G\}$. [Let us recall that the orbit of a point $p \in M$ under G can always be given the structure of a C^∞ submanifold of M. In fact the points on the orbit are in a one-to-one correspondence with the points in G/G_p; but G/G_p is a manifold (G_p is closed) and the differential of the map $xG_p \mapsto x \cdot p$ at any point of G/G_p is non-singular – therefore the manifold structure on G/G_p may be transported to the orbit so as to convert the orbit into a submanifold of M (the submanifold topology being, in general, finer than the induced topology).]

The action of G on M lifts to a linear action of G on the various linear spaces canonically associated with M (that is to say, we may represent G on these linear spaces). Thus G operates linearly on $C^\infty(M; E)$ by associating with each $f \in C^\infty(M; E)$ the function f^x $(= xf)$ defined by $p \mapsto f(x^{-1} \cdot p)$ $(x \in G, p \in M)$; G operates linearly on $\mathfrak{D}(M)$ by associating with each $D \in \mathfrak{D}(M)$ the differential operator xD defined by $f \mapsto (Df^{x^{-1}})^x$ $(x \in G, f \in C^\infty(M))$; G operates linearly on the space of distributions on M by associating with each such T the distribution T^x $(= xT)$ defined by $f \mapsto T(f^{x^{-1}})$ (evidently the map $f \mapsto f^x$ is a continuous endomorphism of $C_c^\infty(M)$). [Of course G also operates on the space of E-distributions on M with values in F.]

Notation Let T be an E-distribution on M with values in F; let $x \in G$ – then we shall frequently write

$$\langle f, T^x \rangle = \int_M f(p)\, dT^x(p) = \int_M f(x \cdot p)\, dT(p) = \int_M f(p)\, dT(x^{-1} \cdot p)$$
$$(f \in C_c^\infty(M; E)),$$

that is $dT^x(p) = dT(x^{-1} \cdot p)$.

A distribution T (or a vector distribution) on M is said to be *invariant* (relative to the given action of G on M) if $T^x = T$ for all $x \in G$. A differential operator D on M is said to be *invariant* if $xD = D$ for all $x \in G$.

Let $\mathfrak{g}$ denote the Lie algebra of G – then we may represent $\mathfrak{g}$ (and hence $\mathfrak{G}$, $\mathfrak{G}$ the universal enveloping algebra of $\mathfrak{g}_c$) on the algebra of differential operators on M as follows. Let $X \in \mathfrak{g}$; define the vector field $\bar{X}$ on M by the prescription

$$f(p; \bar{X}) = df^{\exp(tX)}(p)/dt\,|_{t=0} \qquad (p \in M, f \in C^\infty(M)).$$

It is well-known that $[\bar{X}, \bar{Y}] = \overline{[X, Y]}$ $(X, Y \in \mathfrak{g})$ (cf. Helgason [2, p. 112]) and so the map $X \mapsto \bar{X}$ does indeed give rise to a representation of $\mathfrak{G}$ on $\mathfrak{D}(M)$. [In particular suppose that $M = G$; letting G operate on itself by left translation $(y \mapsto xy)$ leads to a representation of $\mathfrak{G}$ on the algebra of right invariant differential operators on G, while letting G operate on itself by right translation $(y \mapsto yx^{-1})$ leads to a representation of $\mathfrak{G}$ on the algebra of left invariant differential operators on G.]

Suppose, in addition, that M is orientable, the orientation being determined by a positive differential form $\omega > 0$ (cf. A2.1). A distribution T on an open subset $\mathcal{O}$ of M is called *locally invariant* if $\bar{X} \cdot T$ vanishes on $\mathcal{O}$ for all $X \in \mathfrak{g}$. A differential operator D on $\mathcal{O}$ is said to be *locally invariant* if $\bar{X} \circ D = D \circ \bar{X}$ for all $X \in \mathfrak{g}$. Since

$$t^{-1}(f^{\exp(tX)} - f) \to \bar{X}f$$

in $C_c^\infty(M)$ as $t \to 0$, every invariant distribution is locally invariant. Likewise an invariant differential operator is locally invariant.

Let $\mathfrak{g}$ be the Lie algebra of G, $\mathfrak{G}$ the universal enveloping algebra of $\mathfrak{g}_c$. In the well-known way, we may regard $\mathfrak{G}$ as the algebra of right invariant differential operators on G; on the other hand there is also a natural isomorphism between this realization of $\mathfrak{G}$ and the algebra of scalar distributions on G with support the identity $\{1\}$ – thus assign to each $D \in \mathfrak{G}$ (viewed as a right invariant differential operator on G) the distribution T_D defined by

$$\langle f, T_D \rangle = Df'(1) \qquad (f \in C_c^\infty(G)),$$

f' denoting the map $x \mapsto f(x^{-1})$ $(x \in G)$. [If we were regarding the elements of $\mathfrak{G}$ as left invariant differential operators on G, then we would assign to each $D \in \mathfrak{G}$ the distribution T_D defined by

$$\langle f, T_D \rangle = Df(1) \qquad (f \in C_c^\infty(G)).]$$

It is straightforward to verify that the map $D \mapsto T_D$ is a homomorphism (the algebra product (in $\mathfrak{G}$) corresponding to the convolution product); the fact that our mapping is bijective follows quickly from Helgason [2, p. 98] and the well-known result of Schwartz [1, p. 100] which asserts that, relative to a local coordinate system at 1, every distribution on G whose support is $\{1\}$ must actually be a finite linear combination of derivatives of the Dirac measure δ_1. Notice too that we now have the convenient formula

$$Df = T_D * f \qquad (D \in \mathfrak{G}, f \in C^\infty(G)).$$

[Thus let $X \in \mathfrak{g}$; then, on the one hand,

$$Xf(x) = \frac{d}{dt} f(\exp(-tX)x)\bigg|_{t=0},$$

while, on the other,

$$T_X * f(x) = \int_G f(y^{-1}x)\, dT_X(y) = \langle (f \circ R_x)', T_X \rangle$$
$$= X(f \circ R_x)(1) = \frac{d}{dt} f(\exp(-tX)x)\bigg|_{t=0} \quad \text{etc.}]$$

It is equally clear that $T_{\mathrm{Ad}(x)D} = \delta_x * T_D * \delta_{x^{-1}}$ $(x \in G, D \in \mathfrak{G})$.

Let $f \in C_c^\infty(G)$; put $f'(x) = f(x^{-1})$, $f''(x) = \delta_G(x^{-1}) f(x^{-1})$ $(x \in G)$. If $T \in \mathfrak{L}(C_c^\infty(G); E)$, then

$$\int_G f(x)\, dT(x) = T * f'(1) = f'' * T(1).$$

Moreover it is clear that $f \mapsto f'$, $f \mapsto f''$ are involutive automorphisms of $C_c^\infty(G)$ and

$$\int_G f_1'(x) f_2''(x)\, d_G(x) = \int_G f_1''(x) f_2'(x)\, d_G(x).$$

If $T \in \mathfrak{L}(C_c^\infty(G); E)$, define T' and T'' by $T'(f) = T(f'')$, $T''(f) = T(f')$; plainly $(T_1 * T_2)' = T_2' * T_1'$, $(T_1 * T_2)'' = T_2'' * T_1'' \ldots$.

Let $X \in \mathfrak{g}$, T_X the corresponding distribution on G with support $\{1\}$. If $f \in C_c^\infty(G)$, $T \in \mathfrak{L}(C_c^\infty(G); E)$, then

$$(X \cdot T)(f) = T(X^{\mathfrak{t}}(f)) = -T(X(f)) = -T(T_X * f) = -(T_X * f)'' * T(1)$$
$$= -f'' * T_X'' * T(1) = (T_X * T)(f).$$

Thus $X \cdot T = T_X * T$ and so we see that the action of elements in $\mathfrak{G}$, regarded as right invariant differential operators, lifts in the expected manner to elements of $\mathfrak{L}(C_c^\infty(G); E)$. [Needless to say, the adjoint $D^{\mathfrak{t}}$ of $D \in \mathfrak{G}$ is calculated relative to the left invariant Haar measure on G – it is clear that $D^{\mathfrak{t}} = D^+$, $D \mapsto D^+$ the principal anti-automorphism of $\mathfrak{G}$ over $\mathbf{C}$.]

General Notational Conventions

(1) By $\mathbf{Z}$, $\mathbf{N}$, $\mathbf{N}^+$, $\mathbf{Q}$, $\mathbf{R}$, $\mathbf{R}^+$, $\mathbf{R}'$, $\mathbf{C}$, $\mathbf{C}^*$, $\mathbf{T}$, $\mathbf{H}$, $\mathbf{Z}_n$ $(n \geqslant 1)$ we shall understand, respectively, the integers, the non-negative integers, the strictly positive integers, the rational numbers, the real numbers, the strictly positive real numbers, the non-zero real numbers, the complex numbers, the non-zero complex numbers, the complex numbers of absolute value 1, Hamilton's quaternions, the cyclic group of order n.

(2) If z is a complex number, then $\mathscr{R}(z)$ (respectively $\mathscr{I}(z)$) denotes the real (respectively imaginary) part of z.

(3) By $\varnothing$ we shall understand the empty set. If S and T are sets, then we shall sometimes write $S \subset T$ when S is a proper subset of T. If F is a finite set, then $[F]$ denotes the cardinality of F. Set theoretic differences are indicated by a minus sign, set theoretic complements by $\complement$.

(4) If E is a locally convex topological vector space, then, unless stated to the contrary, the base field is $\mathbf{C}$; this convention is applicable, in particular, to all Banach spaces and Hilbert spaces. The symbol $\mathfrak{L}(E; E)$ stands for the space of continuous linear operators on E; we write $\mathfrak{L}_s(E; E)$ (respectively $\mathfrak{L}_b(E; E)$) when we wish to emphasize that $\mathfrak{L}(E; E)$ is carrying the topology of simple convergence (respectively the topology of bounded convergence). [If F is another space satisfying the same general conditions as E, then the symbol $\mathfrak{L}(E; F)$ is to be assigned the obvious connotation; similarly in other cases.] By E^* we shall understand the topological dual of E (or simply dual); the symbols E_s^*, E_c^*, or E_b^* stand, respectively, for E^* equipped with the topology of simple, compact, or bounded convergence.

(5) If $\{E_i : i \in I\}$ is a collection of Hilbert spaces, then the orthogonal direct sum of the E_i is denoted by $\sum_{i \in I} \oplus E_i$.

(6) Let X be a topological space – then X is said to be *locally compact* if every point in X has a neighborhood basis comprised of compact subspaces. [Note: It is not assumed that the topology on X is Hausdorff.] According to this convention, a compact space X may not be locally compact (since X need not be Hausdorff). A locally compact Hausdorff space X is said to be *countable at infinity* if X can be written as the countable union of compact subspaces.

(7) Let G be a group. If $x \in G$, then L_x (respectively R_x) denotes the left (respectively right) translation by x; the effect of applying the inner automorphism of G per x to the element y $(y \in G)$ is denoted either by ${}^x y$ or xyx^{-1}.

(8) For us the term 'locally compact topological group G' means: G is a topological group whose underlying topology is Hausdorff and which, as a locally compact topology, is countable at infinity. [In making this latter assumption, we thus avoid certain measure theoretic pathologies; in certain instances, it will be dropped.] Let, then, G be a locally compact topological group. By $C(G)$, $C_c(G)$, $C^+(G)$, and $C_c^+(G)$ we shall understand, respectively, the space of continuous complex valued functions on G, the space of continuous compactly supported complex valued functions on G, the space of continuous functions on G whose range lies in the set of non-negative real numbers, and the space of continuous compactly supported functions on G whose range lies in the set of non-negative real numbers. More generally, if E is a space per (4) supra, then $C(G; E)$ (respectively $C_c(G; E)$) denotes the space of continuous E-valued func-

tions on G (respectively the space of continuous compactly supported E-valued functions on G); if $f \in C_c(G; E)$, then spt(f) is the support of f. The space of complex Radon measures on G is denoted by $M(G)$ while the space of compactly supported complex Radon measures on G is denoted by $M_c(G)$; if $x \in G$, then δ_x is the Dirac measure concentrated at x.

(9) Let G be a Lie group – then, according to the convention laid down in (8) above, G is countable at infinity. Typically Lie groups are denoted by upper case Roman letters $G, H, \ldots$, their (real) Lie algebras by lower case German letters $\mathfrak{g}, \mathfrak{h}, \ldots$, the complexifications of the latter by $\mathfrak{g}_c, \mathfrak{h}_c, \ldots$, and their universal enveloping algebras by $\mathfrak{G}, \mathfrak{H}, \ldots$ (so we drop the subscript 'c'). The adjoint group of a Lie group G is denoted by Ad(G) or Int($\mathfrak{g}$) (if G is connected); G operates on its Lie algebra $\mathfrak{g}$ via the adjoint action which we shall agree to write as: $\mathrm{Ad}(x)X = xX$ ($x \in G$, $X \in \mathfrak{g}$). The word 'analytic', when used without qualification, means 'real analytic'.

(10) If G is a connected semi-simple Lie group with finite center, $G = KA_{\mathfrak{p}}N^+$ an Iwasawa decomposition for G, then, per a given x in G, we shall denote by $\kappa(x)$ and $H(x)$ the unique elements in K and $\mathfrak{a}_{\mathfrak{p}}$ (the Lie algebra of $A_{\mathfrak{p}}$), respectively, such that $x \in \kappa(x) \exp(H(x))N^+$. If λ is a (complex valued) linear function on $\mathfrak{a}_{\mathfrak{p}}$, then we shall often write h^λ for $\exp(\lambda(\log h))$ ($h \in A_{\mathfrak{p}}$, $\log h$ the unique element in $\mathfrak{a}_{\mathfrak{p}}$ such that $h = \exp(\log h)$). [A similar mode of writing will be used in other cases as well.]

(11) Unless specifically stated to the contrary, it is understood that all Lie algebras are finite dimensional over the base field; thus, whenever it is a question of a semi-simple Lie algebra, or a solvable Lie algebra, or a nilpotent Lie albegra, etc., then the dimension of this Lie algebra is tacitly finite.

(12) A *character* of a topological group G is any continuous homomorphism of G into $\mathbf{C}^*$; if the range of this character is a subset of $\mathbf{T}$ (the complex numbers of absolute value 1), then the character is said to be *unitary*. Unless stated to the contrary, all representations of G are on complex vector spaces; all finite dimensional representations of G are understood to be continuous (and hence analytic if G is a Lie group). If G is a locally compact abelian group, then $\hat{G}$ is its dual group (in the sense of Pontryagin).

(13) Let f, g be two functions defined on the topological space X with values in the set of non-negative real numbers.

(i) We write $f \asymp g$ if there exist $r_1, r_2 > 0$ such that $r_1 f(x) \leqslant g(x) \leqslant r_2 f(x)$ for all $x \in X$.

(ii) We write $f \preceq g$ (or $g \succeq f$) if there exists $r > 0$ such that $f(x) \leqslant rg(x)$ for all $x \in X$.

(14) In so far as the classical groups are concerned, we shall agree to use the customary notations as set forth, e.g. in Helgason [2]. The following notations will be employed in particular.

I_n: Identity matrix of degree n.
$D(a_1, \ldots, a_n)$: Diagonal matrix with diagonal elements $a_1, \ldots, a_n$.
$\mathfrak{gl}(n,\mathbf{C})(\mathfrak{gl}(n,\mathbf{R}))$: The set of all complex (real) n by n matrices.
$\mathfrak{sl}(n,\mathbf{C})(\mathfrak{sl}(n,\mathbf{R}))$: The set of all complex (real)n by n matrices of zero trace.
$\mathbf{S}(n,\mathbf{C})(\mathbf{S}(n,\mathbf{R}))$: The set of all complex (real) n by n symmetric matrices.
$\mathfrak{o}(n,\mathbf{C})(\mathfrak{o}(n,\mathbf{R}))$: The set of all complex (real) n by n skew symmetric matrices.
$\mathbf{H}(n)$: The set of all n by n Hermitian matrices.
$\mathfrak{u}(n)$: The set of all n by n skew Hermitian matrices.
$\mathfrak{su}(n)$: The set of all n by n skew Hermitian matrices of zero trace.

Note Other commonly used notations are introduced in Appendices 1 and 2.

List of Notations

Chapter 1

1.1.1 $\mathfrak{g}, \mathfrak{k}, \mathfrak{p}, \mathfrak{a}_{\mathfrak{k}}, \mathfrak{a}_{\mathfrak{p}}, \mathfrak{a}, \mathfrak{m}, \mathfrak{n}^+; \mathfrak{g}_c, \mathfrak{g}_c^{\alpha}, \Phi, B, \mathfrak{j}_c, \mathfrak{j}_c^{\vee}, \mathfrak{j}^*; H_{\alpha}, X_{\alpha}, N_{\alpha,\beta}, \{\alpha_1, \ldots \alpha_l\}, X_i, Y_i, H_i, A_{ij}(1 \leqslant i, j \leqslant l); \mathrm{Int}(\mathfrak{g}); \sigma, \tau, \theta, P_+, P_-$

1.1.2 $\Phi, W(=W_{\Phi}), l(=l_{\Phi}), \mathfrak{W}(=\mathfrak{W}_{\Phi}), \mathfrak{w}(=\mathfrak{w}_{\Phi}); \alpha, \beta, \alpha^{\vee}, \beta^{\vee}, A_{\alpha,\beta}; \Phi^+, \Psi, \Theta, \langle\Theta\rangle, \Xi, \Xi_s, \Xi_u, W_s, W_u, \Phi_w^+(w \in W); \mathscr{C}, w^{\dagger}, w_{\alpha}, \Delta(w), |\alpha|$

1.1.3 $(\Phi, \sigma), \Phi_+, \Phi_-, E_+, E_-, l_+, l_-; \Sigma, \Sigma^+, \tilde{\alpha} = \lambda, \Sigma(\lambda), m(\lambda), Q_{\lambda}, W_{\Sigma} (= W_{\sigma}/W_-), l_{\Sigma}, \Upsilon = \{\lambda_1, \ldots, \lambda_l\}(l = l_{\Sigma}); \mathfrak{g}^{\lambda}, G, K, A_{\mathfrak{p}}, N^+, N^-, G = KA_{\mathfrak{p}}N^+, M, \mathsf{M}, \mathsf{M}^*, \mathsf{Z}(A_{\mathfrak{p}}), W = \mathsf{M}^*/\mathsf{M}$

1.1.4 $\mathfrak{a}_{\mathfrak{p}}', m_w, \mathfrak{n}_w^+, \mathfrak{u}_w^+, \mathfrak{u}_w^-, N_w^+, U_w^+, U_w^-; \Sigma_0, \Delta_0(w)$

1.1.5 $(\mathfrak{g}, \mathfrak{k}), \bar{\mathfrak{g}}, \hat{\mathfrak{g}}, \mathfrak{c}, \mathfrak{c}_{\mathfrak{k}}, \mathfrak{c}_{\mathfrak{p}}; (G, K), \bar{G}, \widehat{G}, C, C_{\mathfrak{k}}, C_{\mathfrak{p}}, \bar{Z}, \widehat{Z}, Z = \widehat{Z} \times C_{\mathfrak{p}}; i_c, G_c$

1.2.1 $(G, P, M), P_{\Theta}, W, W_{\Theta}, L(w)$

1.2.2 $\mathfrak{g}_c, \mathfrak{j}_c, \mathfrak{b}_c, \mathfrak{p}_c, \mathfrak{l}_c, \mathfrak{n}_c^{\pm}; G_c, J_c, B_c, P_c, L_c, N_c^{\pm}; \mathfrak{w}, \mathfrak{W}, \hat{\mathfrak{w}}, \hat{\mathfrak{W}}, \mathfrak{L}(G_c)$

1.2.3 $P = \mathsf{M}A_{\mathfrak{p}}N^+, \quad \mathfrak{p} = \mathfrak{m} + \mathfrak{a}_{\mathfrak{p}} + \mathfrak{n}^+, \quad G = \bigcup_{w \in W} PwP = \bigcup_{w \in W} U_w^+ wP, N^-P, w^{\dagger}, (G, P, \mathsf{M}^*), P_{\Theta}; \mathfrak{g}(\Theta), \mathfrak{k}(\Theta), \mathfrak{p}(\Theta), \mathfrak{a}_{\mathfrak{p}}(\Theta), \mathfrak{n}^+(\Theta), \mathfrak{n}^-(\Theta), \mathfrak{a}_{\Theta}$

1.2.4 $\mathfrak{p}_{\Theta}, \mathfrak{l}_{\Theta}, \mathfrak{m}_{\Theta}, \mathfrak{n}_{\Theta}^+, \mathfrak{n}_{\Theta}^-, \Sigma_{\Theta}; P_{\Theta}, \mathsf{L}_{\Theta}, L_{\Theta}, \mathsf{M}_{\Theta}, M_{\Theta}, \mathsf{M}_{\Theta}(K), M_{\Theta}(K), A_{\mathfrak{p}}(\Theta), A_{\Theta}, N_{\Theta}^+, N_{\Theta}^-; P_{\Theta} = \mathsf{M}_{\Theta}(K)A_{\mathfrak{p}}N^+ = \mathsf{M}_{\Theta}A_{\Theta}N_{\Theta}^+; P, \mathrm{prk}\,(P), R_u(P), A, \mathsf{L}, \mathsf{M}, P = \mathsf{M}AR_u(P), W(A), \mathfrak{a}, \Sigma^+(\mathfrak{g}, \mathfrak{a}); (P_1, A_1) \succ (P_2, A_2), W(A_1, A_2); \mathfrak{P}, \mathfrak{C}_{\mathfrak{P}}, \mathfrak{P}_A$

1.3.1 $\mathfrak{j}, \mathfrak{j}_I, \mathfrak{j}_R, \mathfrak{j}_{\mathfrak{k}}, \mathfrak{j}_{\mathfrak{p}}; \mathfrak{j}^1, \ldots, \mathfrak{j}^r$

1.3.2 $\Phi, \Phi^+, \Phi_R^+, \Phi_C^+, \Phi_{CP}^+, \Phi_{SI}^+, \Phi_I^+; E_{\alpha}, E_{-\alpha}(\alpha \in \Phi^+); \mathfrak{l}_c^{\alpha}, \mathfrak{l}^{\alpha}(\alpha \in \Phi^+)$

1.3.3 $l_-(\mathfrak{j}), l_-; \omega, X \mapsto \mathrm{tr}(\mathrm{ad}(X))^2 (X \in \mathfrak{g})$

1.3.4 $l, d_l, \mathfrak{g}', \mathfrak{j}' = \mathfrak{j} \cap \mathfrak{g}', \mathfrak{g}(\mathfrak{j}) = \bigcup_{x \in G} x(\mathfrak{j}')(G = \mathrm{Int}(\mathfrak{g})), \pi = \prod_{\alpha > 0} \alpha; \Gamma, \mathfrak{g}_{\Gamma} = \mathfrak{l}_{\Gamma} + \mathfrak{c}_{\Gamma}, \mathfrak{h}_{\Gamma}^+, \mathfrak{h}_{\Gamma}^-, \{H^*, X^*, Y^*\}, \nu = \exp\{-\sqrt{-1}\,(\pi/4)\,\mathrm{ad}(X^* + Y^*)\}, \nu(\mathfrak{h}_{\Gamma}^+) = \mathfrak{h}_{\Gamma}^-, \alpha_+, \alpha_- = \alpha_+^{\nu}, \mathfrak{g}^s$

1.3.5 $\mathscr{S}, \mathscr{N}, X = X_s + X_n(X_s \in \mathscr{S}, X_n \in \mathscr{N}; [X_s, X_n] = 0), \Gamma \in \mathscr{S}, \mathfrak{g}_{\Gamma}$

1.4.1 $\mathsf{J}, \mathsf{J}_0, J, \mathsf{J}_K, J_K, J_{\mathfrak{p}}, \mathsf{Z}(J_{\mathfrak{p}}); D_l, G', \mathsf{J}' = \mathsf{J} \cap G', G(\mathsf{J}) = \bigcup_{x \in G} x\mathsf{J}'x^{-1}; \mathsf{J}^1, \ldots, \mathsf{J}^r$

1.4.2 $\mathsf{J}^*, \mathsf{J}_K^*, W(G, \mathsf{J}), W(G, \mathsf{J}_0)$

1.4.3 $\mathscr{N}_G, x = x_s x_u, \gamma, \mathfrak{g}_{\gamma}, G_{\gamma}, \mathsf{G}_{\gamma}$

Chapter 2

2.1.1 $E, X_E(X \in \mathfrak{g}), U, \delta, E(\delta), \hat{\mathfrak{g}}, S = S(E), T = T(E), \Lambda = \Lambda(E); F(\mathfrak{g}), R(\mathfrak{g}), K(\mathfrak{g}) = F(\mathfrak{g})/R(\mathfrak{g}), \mathfrak{R}(\mathfrak{g}), \mathrm{ch}_{\mathfrak{g}}(E), \lambda_t, \mathrm{Aug}$

2.1.2 $S(E), P(E), E_I, S, S_I, S^+, S_I^+$

2.1.3 $P, Q(P), P_I, P_I^+, Q(P_I), [\mathbf{k}_1 : \mathbf{k}_2]$

2.1.4 $\mathfrak{T}(E), \mathfrak{P}(E), \mathbf{e}^b, S^*(F)$

2.1.5 $I(\mathfrak{g}_c), I(\mathfrak{p}_c), I(\mathfrak{a}_c), I(\mathfrak{a}_{\mathfrak{p}_c}), I(\mathfrak{g}_c; \mathfrak{p}_c), I(\mathfrak{g}_c; \mathfrak{p}_c; \mathfrak{a}_{\mathfrak{p}_c}), I(\mathfrak{g}_c; \mathfrak{a}_c), I(\mathfrak{g}_c; \mathfrak{a}_c; \mathfrak{a}_{\mathfrak{p}_c}) (= I(\mathfrak{a}_c; \mathfrak{a}_{\mathfrak{p}_c})); W(\mathfrak{g}_c, \mathfrak{j}_c), W(\mathfrak{h}_c, \mathfrak{j}_c), I(\mathfrak{g}_c; \mathfrak{h}_c); H(\mathfrak{g}_c); \pi_{\Phi}, \pi_{\Sigma}$

2.2.1 $\mathfrak{k}_c, \mathfrak{G}, \mathfrak{K}, \mathfrak{Z}$

2.2.2 $m_U, \mathfrak{P} = \lambda(S(\mathfrak{p}_c)), \mathfrak{P}^m = \lambda(S^m(\mathfrak{p}_c)), H(\mathfrak{p}_c)$

2.3.1 $\mathfrak{g}, \mathfrak{G}, \mathfrak{G}^+, \nabla, \lambda : S = S(\mathfrak{g}) \to \mathfrak{G}, \mathfrak{G}^m, \mathfrak{G}_m, \mathfrak{X}$

2.3.2 $\mathfrak{sl}(E)$

2.3.3 $\epsilon^-, \epsilon^+, \epsilon = \epsilon^- \otimes 1 \otimes \epsilon^+, \gamma^*, \gamma_1, \gamma_2, \gamma_3, \gamma = \gamma_1^{-1} \circ \gamma^*; \mathfrak{n}^-, \mathfrak{n}^+, \mathfrak{N}^-, \mathfrak{N}^+, \mathfrak{Q}^+ = \mathfrak{G}\mathfrak{n}^+; \omega = \sum_{i,j} g^{ij} X_i X_j = -\omega_- + \omega_+; \mu_{\mathfrak{g}/\mathfrak{h}} : \mathfrak{Z} \to \mathfrak{Z}_{\mathfrak{h}}$

2.4.1 $E^{\lambda}, a_{\Lambda}, E_{\Lambda}, \{\Lambda_1, \ldots, \Lambda_l\}, \{E_1, \ldots, E_l\}, \rho = \frac{1}{2} \sum_{\alpha > 0} \alpha$

2.4.2 $\mathbf{Z}[\mathfrak{W}], \mathbf{Z}[\mathfrak{j}^{\vee}] \sim \mathfrak{R}(\mathfrak{b}), \mathfrak{W}^+(\rho), \mathfrak{W}_s^+(\rho), \rho_s, \rho_u, \rho = \rho_s + \rho_u, \Delta = \prod_{\alpha > 0} (e^{\alpha/2} - e^{-\alpha/2})$

2.4.3 κ, κ_Λ, $m(\lambda)$, $\mathbf{P}$, $\mathbf{e}^\lambda$, f_λ

2.5.1 $C^q(\mathfrak{g}, E)$, $C(\mathfrak{g}, E)$, $H^q(\mathfrak{g}, E)$, $H(\mathfrak{g}, E)$, $\theta = \theta_\mathfrak{g}$, $i = i_E \sim i_\mathbf{k} \otimes 1$, $d = d_\mathfrak{g}$, $\mathbf{\Omega}_E$, $\mathbf{X}_\mathfrak{g}(\mathrm{t}; E)$, $\chi_\mathfrak{g}(\mathrm{t}; \mathrm{ch}_\mathfrak{g}(E))$

2.5.2 $W(q)$, $W_u(q)$, $\square$, $>F<$

Chapter 3

3.1.1 $\mathfrak{g}_c^+$, $\mathfrak{g}_c^-$, $\mathfrak{u}_c^+$, $\mathfrak{u}_c^-$, $N(G)$, $\mathsf{L}(G_c)$, $\hat{J}_c$

3.1.2 $\mathsf{W}_s^+ = \mathfrak{W}/W_s$

3.2.1 ϕ_f, $\langle\pi, \pi\rangle = [W] \prod_{\alpha>0} (\rho, \alpha)$

3.2.2 χ_Λ, d_Λ, $d\mu(\Lambda)$, Δ_T, Φ_f, $\check{f}$, Θ_τ, ξ_τ

3.3.1 L_0, $\mathfrak{L}$

3.3.2 $\mathfrak{a}_\mathfrak{p}^+$, $\rho = 2^{-1} \sum_{\lambda>0} m(\lambda)\lambda$

Chapter 4

4.1.1 E, $\{|\cdot|_\alpha : \alpha \in \mathfrak{A}\}$, $M_c(G)$, $|\mu|$, $\|\mu\|$, $\{g_\mathcal{O}\}$, d_G, U, $U(x)$, $U(f)$, $U(\mu)$

4.1.2 E^*, E_s^*, E_c^*, E_b^*, $E^\vee$, $\bar{E}$, $E_1 \hat{\otimes} E_2$, $E_1 \times E_2$; U^*, $U^\vee$, $\bar{U}$, $U_1 \hat{\otimes} U_2$, $U^1 \times U^2$

4.1.3 $\mathbf{B}$, $I(U, V)$, $\mathfrak{J}(U, V)$, $i(U, V)$, $\mathrm{i}(U, V)$

4.2.1 TCI, FDS; A, $\tilde{A}$, $I(U)$, $\mathrm{Ker}(U)$, E_0, E_0^*, U_0, U_0^*, $\mathscr{E}$

4.2.2 ζ_U, ρ, $C_\rho(G)$, $\|\mu\|_\rho$, $\|f\|_\rho$

4.3.1 $\check{f} = \delta_G^{-1/2} f'$, $f^* = \delta_G^{-1} \bar{f}'$; $\phi_{a,b}$, μ^*

4.3.2 $\mu_\delta(\delta \in \hat{K})$, sp

4.4.1 E_∞, E_α, U_∞, $a \mapsto \tilde{a}\ (a \in E_\infty)$, $D_c(G)$, κ_U

4.4.2 $\delta \in \hat{K}$, $d(\delta)$, ξ_δ, χ_δ, $\mathsf{P}(\delta)$, $[\delta : \delta_0]$, $D_\infty \in \mathfrak{K}$

4.4.3 $\mathsf{P}_L(\delta)f = \bar{\chi}_\delta * f$, $\mathsf{P}_R(\delta)f = f * \chi_\delta$, $T_\delta = T * \chi_\delta$

4.4.4 $D^*(D \in \mathfrak{G})$, $\Delta = X_1^2 + \ldots + X_n^2$, $dU(X)$, $\sum_{(m)}$, $\mathrm{Dom}(T)$

4.4.5 $C^\omega(M; E)$, E_ω, U_ω, $\mathfrak{G}a$, $p_t = e^{t\Delta} f$, $d(x, y)$, D^R, D^L, ∇f, $E_\omega(\delta)(\delta \in \hat{K})$, E_K, T_K, D_K

4.4.6 $A = \overline{U_\infty(1 - \Delta)}$, $B = A^{1/2}$, $|a|_m$

4.4.7 $E_{PW}(T)$, $E_n(T) = \mathsf{P}([-n, n])(E)\ (n \in \mathbf{N})$

4.5.1 ${}^\delta C_c(G)$, $C_c^\delta(G)$, $C_{c,\delta}(G)$, $I_{c,\delta}(G)$, $I(\delta)$, ${}_\delta f$, f_δ, f_K, U_δ, $m(\delta)$; $\mathfrak{J}_c(G)$, $I_c(G)$; $\bar{G}$, $\bar{G}(\delta)$, $\bar{C}_{c,\delta}(G)$, $\bar{I}_{c,\delta}(G)$; $C_{\rho,\delta}(G)$, $I_{\rho,\delta}(G)$; $\mu_K(f) = \mu(f_K)$, $\phi(\mu) = \mu(\phi)$, $\phi(f) = f(\phi)$

4.5.2 $\hat{\xi}$, ${}^\xi C(G)$, U^ξ

4.5.3 m_U, n_U

4.5.4 $C_{c,F}(G)\ (F \subset \hat{K},\ F \text{ finite})$, $C_*(G)$, U_F, E_F, $\bar{\chi}_F$, U_K, $E_K = \sum_{\delta \in \mathfrak{K}} E(\delta)$

4.5.5 $U_K = U_\omega | E_K$, $I_\infty = \mathrm{Ker}(U_\infty)$, $\mathrm{Prim}\ (\mathfrak{G})$

4.5.6 $m(\mathscr{E}; \bar{V})$, $m(U; \bar{V})$, $m(U; V)$

4.5.7 $C^*(G)$, $\|\ \|_{HS}$

4.5.8 $C_*^\infty(G)$, T_U, $T_{\bar{U}}$, $T_{\bar{U}}$, F_U, $\mathfrak{S}(\kappa, \zeta)$

4.5.9 d_U, $\mathscr{E}_U$, $d_{\bar{U}}$, $\mathscr{E}_{\bar{U}}$, $L^2(G)_d$

Chapter 5

5.1.1 $\mathscr{E}^L$, E^L, ${}_G E^L$, U^L, ${}_G U^L$, $\mu_{f,g}$; ${}^L C^\infty(G; E)$, ${}^{L\infty} C^\infty(G; E_\infty)$, ${}^L C(G; E)$, π^L, f^L, ${}_L h$; $\mathfrak{B} = (B, \pi)$, $\mathfrak{B}_x(x \in X)$, $L^2(\mathfrak{B}; \mu)$; $\mathbf{E} = \mathbf{E}(G, H, L)$, $\mathfrak{E}(\rho_H) = (\mathbf{E}, \pi_\mathbf{E})$, ρ_H, μ_H, s_H, $\dot{\phi}$, r_ω

5.1.2 α, β, I_α, I_β, E_α, E_β, U^A, B, μ_ρ, ϕ_x^f

5.1.3 None

5.2.1 $A : G \times M \to \mathfrak{L}(E; E)$, $T^x = {}^t A(x^{-1}, .)T$

5.2.2 $A(x, \dot{y})$, $\mathscr{P}$

5.2.3 Q, $\mathbf{T}(Q)$, $\mathbf{T}^n(Q)$, $D^I \delta_p(p \in M)$, $\Lambda_n(x, p)$, $i(E; Q, n)$

5.2.4 $\{S_\alpha\}$, Q, Ω_Q; $\mathbf{T}$, $\mathbf{T}_\alpha$, $\mathbf{T}[Q]$, T_Q, $\mathbf{Q}_\alpha$

5.3.1 U^{L_∞}, U^L

5.3.2 $Q = H_2 x H_1(x \in G)$, $\mathbf{B}$, $\mathbf{B}^t$, T^t, T; $\mathbf{T}$, $\mathbf{T}(Q)$, $\mathbf{T}(Q_x)$; H_1, H_2, δ_1, δ_2, d_1, d_2, ρ_1, ρ_2, $H_x = H_1 \cap (x^{-1} H_2 x)(x \in G)$, σ_x, ν_x, δ_x; A_n, $i(L^1, L^2; Q, n)$, Φ, S, $\mathcal{O} = Q \cup (G - Q^{cl})$, $\mathbf{B}_z$, T_z; $i(L_\infty, \bar{L}_\infty; Q, n)$

5.3.3 $i(U^L, M)$, $i(U^{L_\infty}, M_\infty)$

5.3.4 $\tilde{V}_x$, $\tilde{V}_D$, V_{x_1,x_2}, V_D, M_{x_1,x_2}, $\mathfrak{D}$
5.4.1 P, P^L, $\mathfrak{B}$, $\hat{\xi} \in \hat{H}$, H_ξ, $\mathscr{L}(\hat{\xi})$, P_Q, $\hat{G}(Q)$, $\hat{\xi} \times L$
5.4.2 '$\alpha t + \beta$', U^+, U^-; U^r; $\mathbf{E}(n)$, $\mathbf{L}(4)$
5.5.1 κ_x, H_x, $H(x, k)$, $\rho = 2^{-1} \sum_{\lambda>0} m(\lambda)\lambda$, δ_P, ρ_P, $L^\mu(\mu: \mathfrak{a}_\mathfrak{p} \to \mathbf{C})$, ${}^{L^\mu}C(G)$, U^μ, $L^2(K; \delta)(\delta \in \hat{K})$, $\hat{\mathsf{M}} = \{\sigma\}$, μ_σ, χ_σ, ξ_σ, $d(\sigma)$, $\mathbf{Q}_R(\sigma)$, $\mathbf{Q}_R(\sigma_{ij})$, $L^{\sigma,\mu}$, $U^{\sigma,\mu}$, $E^{\sigma,\mu}$, ${}^{\sigma,\mu}C(G; \mathbf{C}^{d(\sigma)})$, $\mathscr{S}^{\sigma,\mu}$, $[\delta : \sigma]$, π_P, π_{N^-}, N_x^-, $\mathfrak{X}$
5.5.2 wL, $P_w = P \cap m_w P m_w^{-1}$, $i(L^1, \bar{L}^2; w, n)$, W_L
5.5.3 $U^{\sigma,\mu}$, $T^{\sigma,\mu}$, $F^{\sigma,\mu}$, $W(M, A_K)$, Φ_f
5.5.4 d_P, $f_{K,P}$, $U^{\sigma,\nu}$, $T^{\sigma,\nu}(\mu + \rho = \sqrt{-1}\nu)$, $\hat{f}_P$, Λ_σ, ρ_I, $\tilde{\mathbf{Z}}(A_\mathfrak{p})$

Appendices

A1 δ_G, δ_H, ρ, μ_ρ, s_ρ, ρ_H, μ_H, s_H, $\mu^x(x \in G)$
A2.1 $C^\infty(M)$, $C^\infty(p)(p \in M)$, $C_c^\infty(M)$, $C_\omega^\infty(M)$, $C_c^p(M)(0 \leqslant p \leqslant \infty)$, $\mathfrak{D}_F = \mathfrak{D}_F(M)$, $\mathfrak{D}(M)$, $D_1 \circ D_2$, $\{D_1, D_2\}$, $T(f)$, $\langle f, T\rangle$, $\int_M f(p)\, dT(p)$, $|\cdot|_D$, $|\cdot|_{D\,\omega}$, $f \cdot T$, D^t, $D \cdot T$
A2.2 $E \hat{\otimes} F$, $E \bar{\otimes} F$, $C^\infty(M; E)$, $|\cdot|_{D,\omega}^\alpha$
A2.3 $\mathscr{S}_T$, $\pi^* f$, $\pi_* T$
A2.4 $f * T$, $T * f$, $T_1 * T_2$, $f^x = xf$, $T^x = xT$, $D^x = xD$, $\bar{X}$, T_D, f', f'', T', T'', $G_p(p \in M)$

Guide to the Literature

Chapter 1

1.1.1 Adams [1], Helgason [2], Hochschild [1], Jacobson [3], Loos [1], Serre [3], [4]

1.1.2 Aribaud [1], Bourbaki [5], Serre [4], Steinberg [5]

1.1.3 Araki [1], Goto and Kobayashi [1], Harish-Chandra [15], [27], Helgason [2], Kunze and Stein [3], Murakami [1], Satake [1], Schattschneider [1]

1.1.4 Borel and Harish-Chandra [1], Harish-Chandra [8], [17], [29], Kunze and Stein [3], Langlands [5], Schiffmann [2]

1.1.5 Bourbaki [4], Helgason [2], Jacobson [3]

1.2.1 Bourbaki [5], Carter [1], Coxeter and Moser [1], Dornhoff [1], Tits [1], [2]

1.2.2 Aribaud [1], Borel [4], Griffiths and Schmid [1], Helgason [2], Serre [4], Wang [1]

1.2.3 Aomoto [1], Borel and Tits [1], Bruhat [1], Chevalley [3], Furstenberg [1], Harish-Chandra [12], Helgason [12], Iwahori and Matsumoto [1], Kunze and Stein [3], C. C. Moore [2], Satake [1]

1.2.4 Borel and Tits [1], Bruhat [1], Harish-Chandra [15], [31], Langlands [2], C. C. Moore [2], Oniščik [1], [2], Satake [1], Takeuchi [1]

1.3.1 Harish-Chandra [13], Kostant [1], Sugiura [1], Rothschild [1]

1.3.2 Harish-Chandra [15], [25]

1.3.3 Harish-Chandra [15]

1.3.4 Harish-Chandra [15], [16], [25], Serre [4]

1.3.5 Borel and Harish-Chandra [1], Bourbaki [4], Harish-Chandra [26], Jacobson [3], Kostant [2], Rallis [1], Richardson [1], Vinberg [1]

1.4.1 Harish-Chandra [13], [25], Steinberg [4]

1.4.2 Harish-Chandra [15], [16], [26], [27], Hirai [7], [8]

1.4.3 Borel [4], Borel and Harish-Chandra [1], Borel and Mostow [1], Harish-Chandra [27], [29], Steinberg [4]

Chapter 2

2.1.1 Aribaud [1], Bourbaki [4], Harish-Chandra [6], Jacobson [3]

2.1.2 Bourbaki [4], Dixmier [1], Hadziev [1], Harish-Chandra [6], Nagata [1]

2.1.3 Bourbaki [5], Chevalley [5], Dress [1], Flatto [1], [2], Flatto and Wiener [1], Harish-Chandra [18], Helgason [2], Séminaire Sophus Lie [1], Shephard and Todd [1], Solomon [1], Steinberg [3], [5]

2.1.4 Harish-Chandra [24], Séminaire Sophus Lie [1]

2.1.5 Borel and Chevalley [1], Bourbaki [5], Chevalley [2], Coleman [1], Coxeter [1], Harish-Chandra [5], [15], [18], [24], Helgason [2], [3], [5], Kostant [2], [4], Rallis [1], Steinberg [5], Varadarajan [1]

2.2.1 Harish-Chandra [3], [5]

2.2.2 Dixmier [1], Harish-Chandra [3], [5], [6], Rallis [1]

2.3.1 Bourbaki [4], Harish-Chandra [1], [2], [5], Jacobson [3], Séminaire Sophus Lie [1]

2.3.2 Harish-Chandra [1]

2.3.3 Gelfand [3], Harish-Chandra [3], [13], [27], Kostant [4], Séminaire Sophus Lie [1]
2.4.1 Bouwer [1], Harish-Chandra [3], Jacobson [3], Kimura [1], Lemire [1], Séminaire Sophus Lie [1]
2.4.2 Aribaud [1], Jacobson [3], Kostant [3], Séminaire Sophus Lie [1], Serre [4]
2.4.3 Harish-Chandra [3], [6], Séminaire Sophus Lie [1]
2.5.1 Aribaud [1], Bourbaki [4], Hattori [1], Hochschild and Serre [1], Koszul [1]
2.5.2 Aribaud [1], Bott [1], Cartier [3], Kostant [3]

Chapter 3

3.1.1 Harish-Chandra [2], Hochschild [1], Séminaire Sophus Lie [1], Serre [4]
3.1.2 Aribaud [1], Bott [1], Griffiths and Schmid [1], Hirzebruch [3], [4], Kostant [3], G. Segal [1], Wallach [1], [2]
3.2.1 Harish-Chandra [14], [27]
3.2.2 Harish-Chandra [30], Séminaire Sophus Lie [1]
3.3.1 Harish-Chandra [18], Helgason [14], Sugiura [2], [3]
3.3.2 Harish-Chandra [18], [30]

Chapter 4

4.1.1 Bourbaki [3], Bruhat [1], R. T. Moore [2]
4.1.2 Bourbaki [3], Bruhat [1], R. T. Moore [2]
4.1.3 Bruhat [1]
4.2.1 Dixmier [10], Fell [3], [4], [8], Godement [7], Jacobson [2], Naimark [2]
4.2.2 Berezin [1], Fell [4], Godement [7], de Leeuw and Glicksberg [1], Shiga [1]
4.3.1 Dixmier [15], Hewitt and Ross [1], Kaplansky et al. [1], Langlands [2], Loebel [1], Loomis [1], Mackey [1], [5], [9], [11], Naimark [1], [2]
4.3.2 Bichteler [1], Dixmier [15], Segal and von Neumann [1], van der Waerden [1]
4.4.1 Blattner [1], Bruhat [1], Fell [3], [4], Gårding [1], Harish-Chandra [30], Segal [6], Stein [1]
4.4.2 Harish-Chandra [30], Taylor [1]
4.4.3 Harish-Chandra [29], [30]
4.4.4 Bers, John, and Schechter [1], Goodman [2], [4], [7], John [1], Nelson [1], Nelson and Stinespring [1], Segal [5], [11], Sternheimer [1]
4.4.5 Browder [1], Cartier and Dixmier [1], Gaffney [1], Gårding [2], Godement [7], Harish-Chandra [5], Nelson [1]
4.4.6 Goodman [1], [2], Kotake and Narasimhan [1], R.T. Moore [1], Nelson [1], Tits and Waelbroeck [1]
4.4.7 Goodman (unpublished; cf. Zeitlin [1]), Nelson and Stinespring [1]
4.5.1 Dieudonné [3], Dixmier [15], Fell [4], [8], Godement [7], Harish-Chandra [7], Kaplansky [2], Mackey [9]
4.5.2 Fell [8], Godement [7], Harish-Chandra [5], [6], [7], [21], Rader [1]
4.5.3 Fell [2], [4], [6], [8]
4.5.4 Fell [8], Harish-Chandra [11]
4.5.5 Dixmier [11], [17], Fell [4], Godement [7], Harish-Chandra [5]
4.5.6 Fell [8]
4.5.7 Dixmier [15], Godement [7], Harish-Chandra [5], [7], Nelson and Stinespring [1], Stinespring [1]
4.5.8 Fell [8], Godement [6], [13], [14], Harish-Chandra [6], [7], [8], [13], [27]
4.5.9 Bargmann [1], Dixmier [15], Godement [1], [2], Harish-Chandra [10], Kunze [3], Kunze and Stein [1], Mackey [9], Rieffel [2]

Chapter 5

5.1.1 Auslander, Green, and Hahn [1], Blattner [1], [2], Bruhat [1], Dauns and Hoffmann [1], Fell [7], [8], Kleppner [1], Mackey [2], [3], [4], [5], [9], Mautner [1], [2], [3], [4], Poulsen [1], [2], Rieffel [1], Rigelhof [1]

5.1.2 Blattner [3], Fell [8]

5.1.3 Mackey [3], Fell [8]

5.2.1 Bruhat [1]

5.2.2 Bruhat [1]

5.2.3 Bruhat [1]

5.2.4 Bruhat [1], Fell [8]

5.3.1 Bruhat [1]

5.3.2 Bruhat [1], Fell [8]

5.3.3 Bruhat [1], Mackey [3], [4], Mautner [4], C.C. Moore [1], Weil [1]

5.3.4 Mackey [3]

5.4.1 Fell [8], Mackey [9], [11]

5.4.2 Auslander, Green, and Hahn [1], Fell [8], Gindikin [2], Kohari [1], Mackey [9], [11], Naimark [1], Nelson and Stinespring [1], Pukanszky [2]

5.5.1 Gross [1], Harish-Chandra [5], [6], [21], Kunze and Stein [3], Parthasarathy, K.R., Ranga Rao, R. and Varadarajan, V.S. [1], Rader [1], Sakai [1], [2], Takahashi [1]

5.5.2 Arthur [1], Bruhat [1], Harish-Chandra [8], [32], [36], Knapp and Stein [1], [4], Kostant [5], Wallach [3]

5.5.3 Harish-Chandra [8], Hirai [6], Lipsman [8]

5.5.4 Harish-Chandra [34], Lipsman [6]

Appendices

A1 Bourbaki [3], Bruhat [1], Helgason [2], Mackey [3]

A2.1 Harish-Chandra [13], [23], Helgason [2], Peetre [1], Schwartz [1], Treves [1], [2]

A2.2 Grothendieck [1], Treves [2]

A2.3 Bruhat [1], Schwartz [3], [4]

A2.4 Bruhat [1], Godement [7], Harish-Chandra [13], [23], [24], Helgason [2]

Bibliography

Adams, J. F.: [1] Lectures on Lie groups, Benjamin Inc., New York, 1969.

Aomoto, K.: [1] On some double coset decompositions of complex semi-simple Lie groups, J. Math. Soc. Japan, vol. 18 (1966), pp. 1–44.

— [2] L'analyse harmonique sur les espaces riemanniens à courbure riemannienne negative I, J. Fac. Sci. Univ. of Tokyo, vol. 13 (1966), pp. 85–107.

— [3] Sur les transformations d'horisphere et les équations integrales qui s'y rattachent, J. Fac. Sci. Univ. of Tokyo, vol. 14 (1967), pp. 1–23.

Araki, S.: [1] On root systems and an infinitesimal classification of irreducible symmetric spaces, J. of Math., Osaka City Univ., vol. 13 (1962), pp. 1–34.

Aribaud, F.: [1] Une nouvelle démonstration d'un théorème de R. Bott et B. Kostant, Bull. Math. Soc. France, vol. 95 (1967), pp. 205–242.

Arthur, J. G.: [1] Harmonic analysis of tempered distributions on semi-simple Lie groups of real rank one, Ph.D. Thesis, Yale University, 1970.

Atiyah, M. and Bott, R.: [1] A Lefschetz fixed point formula for elliptic complexes I, Ann. of Math., vol. 86 (1967), pp. 374–407.

— [2] A Lefschetz fixed point formula for elliptic complexes II, Ann. of Math., vol. 88 (1968), pp. 451–491.

Atiyah, M. and Segal, G.: [1] The index of elliptic operators II, Ann. of Math., vol. 87 (1968), pp. 531–545.

Atiyah, M. and Singer, I.: [1] The index of elliptic operators I, Ann. of Math., vol. 87 (1968), pp. 484–530.

— [2] The index of elliptic operators III, Ann. of Math., vol. 87 (1968), pp. 546–604.

Auslander, L., Green, L. and Hahn, F.: [1] Flows on homogeneous spaces, Princeton University Press, Princeton, New Jersey, 1963.

Auslander, L. and Kostant, B.: [1] Quantization and representations of solvable Lie groups, Bull. Amer. Math. Soc., vol. 73 (1967), pp. 692–695.

Auslander, L. and Moore, C. C.: [1] Unitary representations of solvable Lie groups, Memoirs Amer. Math. Soc., n°62, Providence, Rhode Island, 1966.

Baggett, L.: [1] Hilbert-Schmidt representations of groups, Proc. Amer. Math. Soc., vol. 21 (1969), pp. 502–506.

Bargmann, V.: [1] Irreducible unitary representations of the Lorentz group, Ann. of Math., vol. 48 (1947), pp. 568–640.

— [2] On unitary ray representations of continuous groups, Ann. of Math., vol. 59 (1954), pp. 1–46.

Berezin, F. A.: [1] Laplace operators on semi-simple Lie groups, Amer. Math. Soc. Translations (2), vol. 21 (1962), pp. 239–339.

Bernat, P.: [1] Sur les représentations unitaires des groupes de Lie résolubles, Ann. Sci. École. Norm. Sup., vol. 82 (1965), pp. 37–99.

Bernat, P. and Dixmier, J.: [1] Sur le dual d'un groupe de Lie, C. R. Acad. Sc. Paris, vol. 250 (1960), pp. 1778–1779.

Bers, L., John, F. and Schechter, M.: [1] Partial differential equations, Lectures in Applied Mathematics, vol. III, Interscience, New York, 1964.

Bhanu, Murthy, T. S.: [1] Plancherel's measure for the factor space **SL**(n, **R**)/**SO**(n, **R**), Doklady Akad. Nauk. SSSR, vol. 133 (1960), pp. 503–506.
— [2] The asymptotic behavior of zonal spherical functions on the Siegel upper half-plane, Doklady Akad. Nauk. SSSR, vol. 135 (1960), pp. 1027–1029.
Bichteler, K.: [1] On the existence of noncontinuous irreducible representations of a locally compact group, Inventiones Math., vol. 6 (1968), pp. 159–162.
Blattner, R. J.: [1] On induced representations, Amer. J. Math., vol. 83 (1961), pp. 79–98.
— [2] On induced representations II: Infinitesimal induction, Amer. J. Math., vol. 83 (1961), pp. 499–512.
— [3] Positive definite measures, Proc. Amer. Math. Soc., vol. 14 (1963), pp. 423–428.
— [4] Group extension representations and the structure space, Pacific J. Math., vol. 15 (1965), pp. 1101–1113.
Borel, A.: [1] Les fonctions automorphes de plusieurs variables complexes, Bull. Soc. Math. France, vol. 80 (1952), pp. 167–182.
— [2] Kählerian coset spaces of semi-simple Lie groups, Proc. Nat. Acad. Sci. U.S.A., vol. 40 (1954), pp. 1147–1151.
— [3] Topology of Lie groups and characteristic classes, Bull. Amer. Math. Soc., vol. 61 (1955), pp. 397–432.
— [4] Groupes linéaires algébriques, Ann. of Math., vol. 64 (1956), pp. 20–82.
— [5] On the curvature tensor of the Hermitian symmetric manifolds, Ann. of Math., vol. 71 (1960), pp. 508–521.
— [6] Density properties for certain subgroups of semi-simple groups without compact components, Ann. of Math., vol. 72 (1960), pp. 179–188.
— [7] Ensembles fondamentaux pour les groupes arithmétiques, Colloq. des Groupes Algébriques, Bruxelles, 1962, pp. 23–40.
— [8] Arithmetic properties of linear algebraic groups, Proc. Int. Congr. of Math., pp. 10–22, Stockholm, 1962.
— [9] Compact Clifford-Klein forms of symmetric spaces, Topology, vol. 2 (1963), pp. 111–122.
— [10] Introduction aux groupes arithmétiques, Publications de l'Institut Mathématique de l'Université de Strasbourg XV, Hermann, Paris, 1969.
Borel, A. and Chevalley, C.: [1] The Betti numbers of the exceptional groups, Memoirs Amer. Math. Soc., n°14, pp. 1–9, Providence, Rhode Island, 1955.
Borel, A. and Harish-Chandra: [1] Arithmetic subgroups of algebraic groups, Ann. of Math., vol. 75 (1962), pp. 485–535.
Borel, A. and Mostow, G. D.: [1] On semi-simple automorphisms of Lie algebras, Ann. of Math., vol. 61 (1955), pp. 389–405.
Borel, A. and de Siebenthal, J.: [1] Les sous-groupes fermés de rang maximum des groupes de Lie clos, Comment. Math. Helv., vol. 23 (1949), pp. 200–221.
Borel, A. and Tits, J.: [1] Groupes réductifs, Inst. Hautes Études Sci. Publ. Math., vol. 27 (1965), pp. 55–152.
Bott, R.: [1] Homogeneous vector bundles, Ann. of Math., vol. 66 (1957), pp. 203–248.
Bourbaki, N.: [1] Éléments de mathématique, vol. VI, Intégration, Chapters I–IV, Hermann, Paris, 1952.
— [2] Éléments de mathématique, Vol. VI, Intégration, Chapter VI, Hermann, Paris, 1959.
— [3] Éléments de mathématique, Vol. VI, Intégration, Chapters VII–VIII, Hermann, Paris, 1963.
— [4] Éléments de mathématique, Groupes et algèbres de Lie, Chapter I, Hermann, Paris, 1960.

— [5] Éléments de mathématique, Groupes et algèbres de Lie, Chapters IV–VI, Hermann, Paris, 1968.

Bouwer, I. Z.: [1] Standard representations of simple Lie algebras, Canad. J. Math., vol. 20 (1968), pp. 344–361.

Brezin, J.: [1] Unitary representation theory for solvable Lie groups, Memoirs Amer. Math. Soc., n°79, Providence, Rhode Island, 1968.

— [2] Harmonic analysis on nilmanifolds, Trans. Amer. Math. Soc., vol. 150 (1970), pp. 611–618.

Browder, F. I.: [1] Analyticity and partial differential equations I, Amer. J. Math., vol. 84 (1962), pp. 666–710.

Bruhat, F.: [1] Sur les représentations induites des groupes de Lie, Bull. Soc. Math. France, vol. 84 (1956), pp. 97–205.

Butzer, P. L. and Berens, H.: [1] Semi-groups of operators and approximation, Springer-Verlag, New York, 1967.

Carmona, J.: [1] Représentations unitaires induites holomorphes des groupes de Lie; application aux groupes de Lie semi-simples connexes reels, Ph.D. Thesis, Marseille, 1968.

Cartan, H.: [1] Fonctions automorphes, Séminaire, 1957–1958, Paris.

Carter, R.: [1] Simple groups and simple Lie algebras, J. London Math. Soc., vol. 40 (1965), pp. 193–240.

Cartier, P.: [1] Développements de fonctions arbitraires suivant les fonctions propres d'un opérateur différential, Séminaire Bourbaki, n° 102 (1954), Paris.

— [2] On H. Weyl's character formula, Bull. Amer. Math. Soc., vol. 67 (1961), pp. 228–230.

— [3] Remarks on "Lie algebra cohomology and the generalized Borel-Weil theorem," by B. Kostant, Ann. of Math., vol. 74 (1961), pp. 388–390.

Cartier P. and Dixmier, J.: [1] Vecteurs analytiques dans les représentations des groupes de Lie, Amer. J. Math., vol. 80 (1958), pp. 131–145.

Chevalley, C.: [1] Theory of Lie groups, vol. 1, Princeton University Press, Princeton, New Jersey, 1946.

— [2] The Betti numbers of the exceptional simple Lie groups, Proc. Int. Congr. of Math., vol. 2, pp. 21–24, Harvard, 1950.

— [3] Sur certains groupes simples, Tôhoku Math. J., vol. 7 (1955), pp. 14–66.

— [4] Theorie des groupes de Lie, vol. 3, Hermann, Paris, 1955.

— [5] Invariants of finite groups generated by reflections, Amer. J. Math., vol. 77 (1955), pp. 778–782.

— [6] Fundamental concepts of algebra, Academic Press, New York, 1956.

Cohn, L.: [1] The dimension of spaces of automorphic forms on a certain two dimensional complex domain, Ph.D. Thesis, University of Chicago, 1969.

Coifman, R. R. and Weiss, G.: [1] Representations of compact groups and spherical harmonics, L'Enseignement Mathématique, vol. 14 (1969), pp. 121–173.

Coleman, A. J.: [1] The Betti numbers of the simple Lie groups, Canad. J. Math., vol. 10 (1958), pp. 349–356.

Coxeter, H. S. M.: [1] The product of the generators of a finite group generated by reflections, Duke Math. J., vol. 18 (1951), pp. 765–782.

Coxeter, H. S. M. and Moser, W.: [1] Generators and relations for discrete groups, Berlin, 1965.

Curtis, C. and Reiner, I.: [1] Representation theory of finite groups and associative algebras, Interscience, New York, 1962.

Dauns, J. and Hofmann, K.: [1] Representations of rings by sections, Memoirs Amer. Math. Soc., n°83, Providence, Rhode Island, 1968.

Dieudonné, J.: [1] Sur le théorème de Lebesgue-Nikodym (III), Ann. Univ. Grenoble (NS), vol. 23 (1948), pp. 25–53.
— [2] Foundations of modern analysis, Academic Press, New York, 1960.
— [3] Note sur les fonctions sphériques, J. Math. Pures Appl., vol. 41 (1962), pp. 233–240.
— [4] Éléments d'analyse, Tome II, Gauthier-Villars, Paris, 1968.

Dixmier, J.: [1] Sur un théorème d'Harish-Chandra, Acta Sci. Math. Szeged., vol. 14 (1952), pp. 145–156.
— [2] Les algèbres d'operateurs dans l'espace Hilbertien (algèbres de von Neumann), Gauthier-Villars, Paris, 1957.
— [3] Sur les représentations unitaires des groupes de Lie nilpotents II, Bull. Soc. Math. France, vol. 85 (1957), pp. 325–388.
— [4] Sur les représentations unitaires des groupes de Lie algébriques, Ann. Inst. Fourier (Grenoble), vol. 7 (1957), pp. 315–328.
— [5] Sur les représentations unitaires des groupes de Lie nilpotents III, Canad. J. Math., vol. 10 (1958), pp. 321–348.
— [6] Sur les représentations unitaires des groupes de Lie nilpotents I, Amer. J. Math., vol. 81 (1959), pp. 160–170.
— [7] Sur les représentations unitaires des groupes de Lie nilpotents IV, Canad. J. Math., vol. 11 (1959), pp. 321–344.
— [8] Sur les représentations unitaires des groupes de Lie nilpotents V, Bull. Soc. Math. France, vol. 87 (1959), pp. 65–79.
— [9] Sur les représentations unitaires des groupes de Lie nilpotents VI, Canad. J. Math., vol. 12 (1960), pp. 324–352.
— [10] Opérateurs de rang fini dans les représentations unitaires, Inst. Hautes Études Sci. Publ. Math., vol. 6 (1960), pp. 13–25.
— [11] Représentations intégrables du groupe de De Sitter, Bull. Soc. Math. France, vol. 89 (1961), pp. 9–41.
— [12] Points isolés dans le dual d'un groupe localement compact, Bull. Sc. Math., vol. 85 (1961), pp. 91–96.
— [13] Sur les représentations unitaires des groupes de Lie résolubles, Math. J. Okayama Univ., vol. 11 (1962), pp. 1–18.
— [14] Représentations irréductibles des algèbres de Lie nilpotentes, Anais. Acad. Brasil. Ciencias, vol. 35 (1963), pp. 491–519.
— [15] Les C^*-algèbres et leurs représentations, Gauthier-Villars, Paris, 1964.
— [16] Représentations irréductibles des algèbres de Lie résolubles, J. Math. Pures et Appl., vol. 45 (1966), pp. 1–66.
— [17] Sur le noyau infinitesimal d'une représentation unitaire d'un groupe résoluble, C. R. Acad. Sc. Paris, vol. 262 (1966), Serie A, pp. 483–486.
— [18] Représentations induites holomorphes des groupes résolubles algébriques, Bull. Soc. Math. France, vol. 94 (1966), pp. 181–206.
— [19] Espace dual d'une algèbre ou d'un groupe localement compact, Proc. Int. Congr. of Math., pp. 357–366, Moscow, 1966.
— [20] Sur le dual d'un groupe de Lie nilpotent, Bull. Sc. Math., vol. 90 (1966), pp. 113–118.
— [21] Sur le centre de l'algèbre enveloppante d'une algèbre de Lie, C. R. Acad. Sc. Paris, vol. 265 (1967), Serie A, pp. 408–410.
— [22] Représentations induites des algèbres de Lie, L'Enseignement Mathématique, vol. 16 (1970), pp. 169–175.

Dopplicher, S., Kastler, D. and Robinson, D.: [1] Covariance algebras in field theory and statistical mechanics, Comm. Math. Phys., vol. 3 (1966), pp. 1–28.

Dornhoff, L.: [1] Linear and reflection groups, Lecture Notes, Yale University, 1967.

Dress, A.: [1] On finite groups generated by pseudoreflections, J. of Algebra, vol. 11 (1969), pp. 1–5.

Duflo, M.: [1] Caractères des groupes et des algèbres de Lie résolubles, Ann. Scientifiques L'École Normale Supérieure, vol. 3 (1970), pp. 23–74.

— [2] Fundamental series representations of a semi-simple Lie group, Func. Analysis and Applic., vol. 4 (1970), pp. 122–126.

Dunford, N. and Schwartz, J. T.: [1] Linear operators, Part I, Interscience, New York, 1958.

— [2] Linear operators, Part II, Interscience, New York, 1963.

Dunkl, C. F.: [1] Functions that operate in the Fourier algebra of a compact group, Proc. Amer. Math. Soc., vol. 21 (1969), pp. 540–544.

Effros, E.: [1] Transformation groups and C^*-algebras, Ann. of Math., vol. 81 (1965), pp. 38–55.

Effros, E. and Hahn, F.: [1] Locally compact transformation groups and C^*-algebras, Memoirs Amer. Math. Soc., n°75, Providence, Rhode Island, 1967.

Ehrenpreis, L. and Mautner, F. I.: [1] Some properties of the Fourier transform on semi-simple Lie groups I, Ann. of Math., vol. 61 (1955), pp. 406–439.

— [2] Some properties of the Fourier transform on semi-simple Lie groups II, Trans. Amer. Math. Soc., vol. 84 (1957), pp. 1–55.

— [3] Some properties of the Fourier transform on semi-simple Lie groups III, Trans. Amer. Math. Soc., vol. 90 (1959), pp. 431–484.

Eymard, P.: [1] L'algèbre de Fourier d'un groupe localement compact, Bull. Soc. Math. France, vol. 92 (1964), pp. 181–236.

Fell, J. M. G.: [1] The dual spaces of C^*-algebras, Trans. Amer. Math. Soc., vol. 94 (1960), pp. 365–403.

— [2] Weak containment and induced representations of groups, Canad. J. Math., vol. 14 (1962), pp. 237–268.

— [3] The dual spaces of Banach algebras, Trans. Amer. Math. Soc., vol. 114 (1965), pp. 227–250.

— [4] Non-unitary dual spaces of groups, Acta Math., vol. 114 (1965), pp. 267–310.

— [5] Conjugating representations and related results on semi-simple Lie groups, Trans. Amer. Math. Soc., vol. 127 (1967), pp. 405–426.

— [6] An extension of Mackey's method to algebraic bundles over finite groups, Amer. J. Math., vol. 91 (1969), pp. 203–238.

— [7] An extension of Mackey's method to Banach *-algebraic bundles, Memoirs Amer. Math. Soc., n°90, Providence, Rhode Island, 1969.

— [8] Group representation theory, To Appear.

Flatto, L.: [1] Basic sets of invariants for finite reflection groups, Bull. Amer. Math. Soc., vol. 74 (1968), pp. 730–734.

— [2] Invariants of finite reflection groups and mean value problems II, Amer. J. Math., vol. 92 (1970), pp. 552–561.

Flatto, L. and Wiener, M. M.: [1] Invariants of finite reflection groups and mean value problems, Amer. J. Math., vol. 91 (1969), pp. 591–598.

Flensted-Jensen, M.: [1] On the Fourier transform on a symmetric space of rank one, Institut Mittag-Leffler, 1970.

Freudenthal, H.: [1] Zur Berechnung der Charaktere der halbeinfachen Lieschen Gruppen I, II and III. I: Indag. Math., vol. 16 (1954), pp. 369–376. II: ibid., pp. 487–491. III: ibid., vol. 18 (1956), pp. 511–514.

Frobenius, G.: [1] Über Relationen zwischen den Charakteren einer Gruppe und denen ihrer Untergruppen, Sitz. Preus. Akad. Wiss., 1898, pp. 501–515.

Furstenberg, H.: [1] A Poisson formula for semi-simple Lie groups, Ann. of Math., vol. 77 (1963), pp. 335–386.

— [2] Translation-invariant cones of functions on semi-simple Lie groups, Bull. Amer. Math. Soc., vol. 71 (1965), pp. 271–326.
— [3] Poisson boundaries and envelopes of discrete groups, Bull. Amer. Math. Soc., vol. 73 (1967), pp. 350–356.

Gaffney, M.: [1] The conservation property of the heat equation on Riemannian manifolds, Comm. Pure and Applied Math., vol. 12 (1959), pp. 1–11.

Gangolli, R.: [1] Isotropic infinitely divisible measures on symmetric spaces, Acta Math., vol. 111 (1964), pp. 213–246.
— [2] Asymptotic behaviour of spectra of compact quotients of certain symmetric spaces, Acta Math., vol. 121 (1968), pp. 151–192.
— [3] On the Plancherel formula and the Paley-Wiener theorem for spherical functions on semi-simple Lie groups, Ann. of Math., vol. 93 (1971), pp. 150–165.

Gårding, L.: [1] Note on continuous representations of Lie groups, Proc. Nat. Acad. Sci. U.S.A., vol. 33 (1947), pp. 331–332.
— [2] Vecteurs analytiques dans les représentations des groupes de Lie, Bull. Soc. Math. France, vol. 88 (1960), pp. 73–93.
— [3] Applications of the theory of direct integrals of Hilbert spaces to some integral and differential operators, Institute for Fluid Dynamics and Applied Mathematics Lecture Series, n° 11, University of Maryland.
— [4] Transformation de Fourier des distributions homogènes, Bull. Soc. Math. France, vol. 89 (1961), pp. 381–428.

Garland, H.: [1] The spectrum of non-compact G/Γ and the cohomology of arithmetic groups, Bull. Amer. Math. Soc., vol. 75 (1969), pp. 807–811.

Garland, H. and Goto, M.: [1] Lattices and the adjoint group of a Lie group, Trans. Amer. Math. Soc., vol. 124 (1966), pp. 450–460.

Garland, H. and Raghunathan, M. S.: [1] Fundamental domains for lattices in (**R**)-rank 1 semi-simple Lie groups, Ann. of Math., vol. 92 (1970), pp. 279–326.

Gelbart, S.: [1] Fourier analysis on matrix space, Memoirs Amer. Math. Soc., n° 108, Providence, Rhode Island, 1971.

Gelfand, I.: [1] On one-parameter groups of operators in a normed space, Doklady Akad. Nauk. SSSR (N. S.), vol. 25 (1939), pp. 713–718.
— [2] Spherical functions on symmetric spaces, Doklady Akad. Nauk. SSSR, vol. 70 (1950), pp. 5–8.
— [3] The center of an infinitesimal group algebra, Mat. Sb., vol. 26 (1950), pp. 103–112.
— [4] Integral geometry and its relation to group representations, Russ. Math. Surveys, vol. 15 (1960), pp. 143–151.
— [5] Automorphic functions and the theory of representations, Proc. Int. Congr. of Math., pp. 74–85, Stockholm, 1962.

Gelfand, I. M. and Graev, M. I.: [1] The geometry of homogeneous spaces, group representations in homogeneous spaces and questions in integral geometry related to them, Amer. Math. Soc. Translations (2), vol. 37 (1964), pp. 351–429.

Gelfand, I. M., Graev, M. I. and Pyateckii-Shapiro, I. I.: [1] Representation theory and automorphic functions, W. B. Saunders Co., Philadelphia, 1969.

Gelfand, I. M., Graev, M. I. and Shapiro, S. J.: [1] Differential forms and integral geometry, Func. Analysis and Applic., vol. 3 (1969), pp. 24–40.

Gelfand I. M. and Kirillov A. A.: [1] Sur les corps Liés aux algèbres enveloppantes des algèbres de Lie, Inst. Hautes Études Sci. Publ. Math., vol. 31 (1966), pp. 5–20.

Gelfand, I. M. and Naimark, M. A.: [1] Unitäre Darstellungen der klassischen Gruppen, Akademie Verlag, Berlin, 1957.

Gelfand, I. M. and Pyateckii-Shapiro, I. I.: [1] Theory of representations and theory of automorphic functions, Amer. Math. Soc. Translations (2), vol. .26 (1963), pp. 173–200.

Gindikin, S.: [1] Analysis in homogeneous domains, Russ. Math. Surveys, vol. 19 (1964), pp. 1–89.

— [2] Unitary representations of groups of automorphisms of Riemannian symmetric spaces, Func. Analysis and Applic., vol. 1 (1967), pp. 28–32.

Gindikin, S. G. and Karpelevič, F. I.: [1] Plancherel measure of Riemannian symmetric spaces of non-positive curvature, Doklady Akad. Nauk. SSSR, vol. 145 (1962), pp. 252–255.

— [2] On a problem in integral geometry, Chebotarev Memorial Volume, Kazan University, 1964.

— [3] On an integral connected with symmetric Riemannian spaces of negative curvature, Amer. Math. Soc. Translations (2), vol. 85 (1969), pp. 249–258.

Glimm, J.: [1] Type I·C^*-algebras, Ann. of Math., vol. 73 (1961), pp. 572–612.

— [2] Locally compact transformation groups, Trans. Amer. Math. Soc., vol. 101 (1961), pp. 124–138.

— [3] Families of induced representations, Pacific J. Math., vol. 12 (1962), pp. 885–911.

Godement, R.: [1] Sur les relations d'orthogonalité de V. Bargmann I, Resultats préliminaires, C. R. Acad. Sc. Paris, vol. 225 (1947), pp. 521–523.

— [2] Sur les relations d'orthogonalité de V. Bargmann II, Demonstration générale, C. R. Acad. Sc. Paris, vol. 225 (1947), pp. 657–659.

— [3] Les fonctions de type positif et la théorie des groupes, Trans. Amer. Math. Soc., vol. 63 (1948), pp. 1–84.

— [4] Sur la transformation de Fourier dans les groupes discrets, C. R. Acad. Sc. Paris, vol. 228 (1949), pp. 627–628.

— [5] Sur la théorie des représentations unitaires, Ann. of Math., vol. 53 (1951), pp. 68–124.

— [6] Mémoire sur la théorie des caractères dans les groupes localement compacts unimodulaires, J. Math. Pures Appl., vol. 30 (1951), pp. 1–110.

— [7] A theory of spherical functions I, Trans. Amer. Math. Soc., vol. 73 (1952), pp. 496–556.

— [8] Les travaux de E. Hecke I, Séminaire Bourbaki, n° 51 (1951), Paris.

— [9] Les travaux de E. Hecke II, Séminaire Bourbaki, n° 59 (1952), Paris.

— [10] Les travaux de E. Hecke III, Séminaire Bourbaki, n° 74 (1953), Paris.

— [11] Les travaux de E. Hecke IV, Séminaire Bourbaki, n° 80 (1953), Paris.

— [12] Cohomologie des groupes discontinus, Séminaire Bourbaki, n° 90 (1954), Paris.

— [13] Théorie des caractères I: Algèbres unitaires, Ann. of Math., vol. 59 (1954), pp. 47–62.

— [14] Théorie des caractères II: Définition et propriétés générales des caractères, Ann. of Math., vol. 59 (1954), pp. 63–85.

— [15] Introduction aux travaux de A. Selberg, Séminaire Bourbaki, n° 144 (1957), Paris.

— [16] Les fonctions ζ des algèbres simples I, Séminaire Bourbaki, n° 171 (1958), Paris.

— [17] Les fonctions ζ des algèbres simples II, Séminaire Bourbaki, n° 176 (1959), Paris.

— [18] Articles in H. Cartan [1].

— [19] Groupes linéaires algébriques sur un corps parfait, Séminaire Bourbaki, n° 206 (1960), Paris.

— [20] La formule des traces de Selberg, Séminaire Bourbaki, n° 244 (1962), Paris.

— [21] Domaines fondamentaux des groupes arithmétiques, Séminaire Bourbaki, n° 257 (1963), Paris.

— [22] Analyse spectrale des fonctions modulaires, Séminaire Bourbaki, n° 278 (1964), Paris.
— [23] The decomposition of $L^2(G/\Gamma)$ for $\Gamma = SL(2, Z)$, Proc. Sympos. Pure Math., vol. 9, Amer. Math. Soc., Providence, Rhode Island, 1966, pp. 211–224.
— [24] The spectral decomposition of cusp forms, Proc. Sympos. Pure Math., vol. 9, Amer. Math. Soc., Providence, Rhode Island, 1966, pp. 225–234.
— [25] Introduction a la théorie de Langlands, Séminaire Bourbaki, n° 321 (1967), Paris.
— [26] Formes automorphes et produits Euleriens, Séminaire Bourbaki, n° 349 (1968), Paris.

Goodman, R.: [1] Analytic domination by fractional powers of a positive operator, J. of Functional Analysis, vol. 3 (1969), pp. 246–264.
— [2] Analytic and entire vectors for representations of Lie groups, Trans. Amer. Math. Soc., vol. 143 (1969), pp. 55–76.
— [3] Differential operators of infinite order on a Lie group I, J. Math. and Mechanics, vol. 19 (1970), pp. 879–894.
— [4] Some regularity theorems for operators in an enveloping algebra, To Appear.
— [5] One parameter groups generated by operators in an enveloping algebra, J. of Functional Analysis, vol. 6 (1970), pp. 218–236.
— [6] Complex Fourier analysis on nilpotent Lie groups, To Appear.
— [7] Differential operators of infinite order on a Lie group II, To Appear.

Goto, M. and Kobayashi, E.: [1] On the subgroups of the centers of simply connected simple Lie groups—classification of simple Lie groups in the large, Osaka J. Math., vol. 6 (1969), pp. 251–281.

Griffiths, P. and Schmid, W.: [1] Locally homogeneous complex manifolds, Acta Math., vol. 123 (1969), pp. 253–302.

Gross, K.: [1] The Plancherel transform on the nilpotent part of $\mathbf{G}_2$ and some applications to the representation theory of $\mathbf{G}_2$, Tulane University, 1968.

Gross, K. and Kunze, R. A.: [1] Fourier decompositions of certain representations, To Appear.

Grothendieck, A.: [1] Produits tensoriels topologiques et espaces nucléaires, Memoirs Amer. Math. Soc., n° 16, Providence, Rhode Island, 1955.

Gutkin, E. A.: [1] Representations of the principal series of a complex semi-simple Lie group, Func. Analysis and Applic., vol. 4 (1970), pp. 117–121.

Hadziev, Dž.: [1] Some questions in the theory of vector invariants, Math. of the U.S.S.R.—Sbornik, Amer. Math. Soc. Translation, vol. 1 (1967), pp. 383–396.

Harish-Chandra: [1] On representations of Lie algebras, Ann. of Math., vol. 50 (1949), pp. 900–915.
— [2] Lie algebras and the Tannaka duality theorem, Ann. of Math., vol. 51 (1950), pp. 299–330.
— [3] On some applications of the universal enveloping algebra of a semi-simple Lie algebra, Trans. Amer. Math. Soc., vol. 70 (1951), pp. 28–96.
— [4] Plancherel formula for the 2×2 real unimodular group, Proc. Nat. Acad. Sci. U.S.A., vol. 38 (1952), pp. 337–342.
— [5] Representations of semi-simple Lie groups I, Trans. Amer. Math. Soc., vol. 75 (1953), pp. 185–243.
— [6] Representations of semi-simple Lie groups II, Trans. Amer. Math. Soc., vol. 76 (1954), pp. 26–65.
— [7] Representations of semi-simple Lie groups III, Trans. Amer. Math. Soc., vol. 76 (1954), pp. 234–253.
— [8] The Plancherel formula for complex semi-simple Lie groups, Trans. Amer. Math. Soc., vol. 76 (1954), pp. 485–528.

— [9] Representations of semi-simple Lie groups IV, Amer. J. Math., vol. 77 (1955), pp. 743–777.
— [10] Representations of semi-simple Lie groups V, Amer. J. Math., vol. 78 (1956), pp. 1–41.
— [11] Representations of semi-simple Lie groups VI, Amer. J. Math., vol. 78 (1956), pp. 564–628.
— [12] On a lemma of F. Bruhat, J. Math. Pures Appl., vol. 35 (1956), pp. 203–210.
— [13] The characters of semi-simple Lie groups, Trans. Amer. Math. Soc., vol. 83 (1956), pp. 98–163.
— [14] Differential operators on a semi-simple Lie algebra, Amer. J. Math., vol. 79 (1957), pp. 87–120.
— [15] Fourier transforms on a semi-simple Lie algebra I, Amer. J. Math., vol. 79 (1957), pp. 193–257.
— [16] Fourier transforms on a semi-simple Lie algebra II, Amer. J. Math., vol. 79 (1957), pp. 653–686.
— [17] A formula for semi-simple Lie groups, Amer. J. Math., vol. 79 (1957), pp. 733–760.
— [18] Spherical functions on a semi-simple Lie group I, Amer. J. Math., vol. 80 (1958), pp. 241–310.
— [19] Spherical functions on a semi-simple Lie group II, Amer. J. Math., vol. 80 (1958), pp. 553–613.
— [20] Automorphic forms on a semi-simple Lie group, Proc. Nat. Acad. Sci. U.S.A., vol. 45 (1959), pp. 570–573.
— [21] Some results on differential equations and their applications, Proc. Nat. Acad. Sci. U.S.A., vol. 45 (1959), pp. 1763–1764.
— [22] Invariant eigendistributions on semi-simple Lie groups, Bull. Amer. Math. Soc., vol. 69 (1963), pp. 117–123.
— [23] Invariant distributions on Lie algebras, Amer. J. Math., vol. 86 (1964), pp. 271–309.
— [24] Invariant differential operators and distributions on a semi-simple Lie algebra, Amer. J. Math., vol. 86 (1964), pp. 534–564.
— [25] Some results on an invariant integral on a semi-simple Lie algebra, Ann. of Math., vol. 80 (1964), pp. 551–593.
— [26] Invariant eigendistributions on a semi-simple Lie algebra, Inst. Hautes Études Sci. Publ. Math., vol. 27 (1965), pp. 5–54.
— [27] Invariant eigendistributions on a semi-simple Lie group, Trans. Amer. Math. Soc., vol. 119 (1965), pp. 457–508.
— [28] Discrete series for semi-simple Lie groups I, Acta Math., vol. 113 (1965), pp. 241–318.
— [29] Two theorems on semi-simple Lie groups, Ann. of Math., vol. 83 (1966), pp. 74–128.
— [30] Discrete series for semi-simple Lie groups II, Acta Math., vol. 116 (1966), pp. 1–111.
— [31] Automorphic forms on semi-simple Lie groups, Springer-Verlag, Lecture Notes in Mathematics, vol. 62, 1968.
— [32] Eisenstein series over finite fields, Functional Analysis and Related Fields, pp. 76–88, Springer-Verlag, Berlin, 1970.
— [33] Some applications of the Schwartz space of a semi-simple Lie group, Lectures in Modern Analysis and Applications, Georgetown, 1969. [Lecture Notes in Mathematics, Springer-Verlag, vol. 140, 1970.]
— [34] Harmonic analysis on semi-simple Lie groups, Bull. Amer. Math. Soc., vol. 76 (1970), pp. 529–551.
— [35] Harmonic analysis on reductive $\mathfrak{p}$-adic groups, Springer-Verlag, Lecture Notes in Mathematics, vol. 162, 1970.
— [36] On the theory of the Eisenstein integral, To Appear.

Hattori, A.: [1] On 1-cohomology groups of infinite dimensional representations of semi-simple Lie algebras, J. Math. Soc. Japan, vol. 16 (1964), pp. 226–229.

Helgason, S.: [1] Differential operators on homogeneous spaces, Acta Math., vol. 102 (1959), pp. 239–299.

— [2] Differential geometry and symmetric spaces, Academic Press, New York, 1962.

— [3] Invariants and fundamental functions, Acta Math., vol. 109 (1963), pp. 241–258.

— [4] Duality and Radon transform for symmetric spaces, Amer. J. Math., vol. 85 (1963), pp. 667–692.

— [5] Fundamental solutions of invariant differential operators on symmetric spaces, Amer. J. Math., vol. 86 (1964), pp. 565–601.

— [6] A duality in integral geometry; some generalizations of the Radon transform, Bull. Amer. Math. Soc., vol. 70 (1964), pp. 435–446.

— [7] The Radon transform on Euclidean spaces, compact two-point homogeneous spaces and Grassmann manifolds, Acta Math., vol. 113 (1965), pp. 153–180.

— [8] Radon-Fourier transforms on symmetric spaces and related group representations, Bull. Amer. Math. Soc., vol. 71 (1965), pp. 757–763.

— [9] A duality in integral geometry on symmetric spaces, Proc. U.S.-Japan Seminar in Differential Geometry, Kyoto, 1965.

— [10] An analogue of the Paley-Wiener theorem for the Fourier transform on certain symmetric spaces, Math. Ann., vol. 165 (1966), pp. 297–308.

— [11] Totally geodesic spheres in compact symmetric spaces, Math. Ann., vol. 165 (1966), pp. 309–317.

— [12] Lie groups and symmetric spaces, Battelle Rencontres, pp. 1–71, Benjamin, New York, 1968.

— [13] Applications of the Radon transform to representations of semi-simple Lie groups, Proc. Nat. Acad. Sci. U.S.A., vol. 63 (1969), pp. 643–647.

— [14] A duality for symmetric spaces with applications to group representations, Advances in Math., vol. 5 (1970), pp. 1–154.

Helgason, S. and Johnson, K.: [1] The bounded spherical functions on symmetric spaces, Advances in Math., vol. 3 (1969), pp. 586–593.

Helgason, S. and Korányi, A.: [1] A Fatou-type theorem for harmonic functions on symmetric spaces, Bull. Amer. Math. Soc., vol. 74 (1968), pp. 258–263.

Herz, C. S.: [1] Bessel functions of matrix argument, Ann. of Math., vol. 61 (1955), pp. 474–523.

— [2] Sur le phénomène de Kunze-Stein, C. R. Acad. Sc. Paris, vol. 271 (1970), Serie A, pp. 491–493.

Hewitt, E.: [1] A certain algebra of functions on a compact group, Queen's Papers on Pure and Applied Math., Queen's University, Kingston, Ontario, 1967.

Hewitt, E. and Ross, K. A.: [1] Abstract harmonic analysis, vol. I, Springer-Verlag, Berlin, 1963.

— [2] Abstract harmonic analysis, vol. II, Springer-Verlag, Berlin, 1970.

Heyer, H.: [1] Dualität lokalkompakter Gruppen, Springer-Verlag, Lecture Notes in Mathematics, vol. 150, 1970.

Hirai, T.: [1] On infinitesimal operators of irreducible representations of the Lorentz group of n-th order, Proc. Japan Acad., vol. 38 (1962), pp. 83–87.

— [2] On irreducible representations of the Lorentz group of n-th order, Proc. Japan Acad., vol. 38 (1962), pp. 258–262.

— [3] The characters of irreducible representations of the Lorentz group of n-th order, Proc. Japan Acad., vol. 41 (1965), pp. 526–531.

— [4] The Plancherel formula for the Lorentz group of n-th order, Proc. Japan Acad., vol. 42 (1965), pp. 323–326.

— [5] Classification and the characters of irreducible representations of SU(p, 1), Proc. Japan Acad., vol. 42 (1966), pp. 907–912.
— [6] The characters of some induced representations of semi-simple Lie groups, J. Math. Kyoto Univ., vol. 8 (1968), pp. 313–363.
— [7] Invariant eigendistributions of Laplace operators on real simple Lie groups I, To Appear.
— [8] The Plancherel formula for SU(p, q), J. Math. Soc. Japan, vol. 22 (1970), pp. 134–179.

Hirzebruch, F.: [1] Characteristic numbers of homogeneous domains, Seminars on Analytic Functions, vol. 2, pp. 92–104, Institute for Advanced Study, 1957.
— [2] Automorphe Formen und der Satz von Riemann-Roch, Symp. Intern. Top. Alg., 1956, pp. 129–144, Universidad de Mexico, 1958.
— [3] Topological methods in algebraic geometry, Third Edition, Springer-Verlag, New York, 1966.
— [4] Elliptische Differentialoperatoren auf Mannigfaltigkeiten, Weierstrass Festband, Westdeutscher Verlag, Opladen, 1966.

Hochschild, G.: [1] The structure of Lie groups, Holden-Day Inc., San Francisco, 1965.

Hochschild, G. and Serre, J-P.: [1] Cohomology of Lie algebras, Ann. of Math., vol. 57 (1953), pp. 591–603.

Hörmander, L.: [1] Linear partial differential operators, Springer-Verlag, Berlin, 1963.
— [2] Complex analysis in several variables, Van Nostrand, New York, 1966.

Hotta, R.: [1] Elliptic complexes on some homogeneous spaces, Osaka J. Math., vol. 7 (1970), pp. 117–160.
— [2] On realization of the discrete series for semi-simple Lie groups, Proc. Japan Acad., vol. 46 (1970), pp. 993–996.

Howe, R.: [1] On Frobenius reciprocity for unipotent algebraic groups over **Q**, To Appear.
— [2] On representations of discrete, finitely generated, torsion-free, nilpotent groups, To Appear.

Ise, M.: [1] Generalized automorphic forms and certain holomorphic vector bundles, Amer. J. Math., vol. 86 (1964), pp. 70–108.

Iwahori, N.: [1] On real irreducible representations of Lie algebras, Nagoya Math. J., vol. 14 (1959), pp. 59–83.
— [2] On the structure of a Hecke ring of a Chevalley group over a finite field, J. Fac. Sci. Univ. Tokyo, vol. 10 (1964), pp. 215–236.

Iwahori, N. and Matsumoto, H.: [1] On some Bruhat decompositions and the structure of the Hecke rings of $\mathfrak{p}$-adic Chevalley groups, Inst. Hautes Études Sci. Publ. Math., vol. 25 (1965), pp. 5–48.

Jacobson, N.: [1] Lectures in abstract algebra, vol. I, Van Nostrand, New York, 1951.
— [2] Lectures in abstract algebra, vol. II, Van Nostrand, New York, 1953.
— [3] Lie algebras, Interscience, New York, 1962.
— [4] Structure of rings, Amer. Math. Soc. Colloquium Publication, vol. 37, Providence, Rhode Island, 1964.

Jacquet, H.: [1] Fonctions de Whittaker associées aux groupes de Chevalley, Bull. Soc. Math. France, vol. 95 (1967), pp. 243–309.

Jeanquartier, P.: [1] Distributions et opérateurs différentiels homogènes et invariants, Comment. Math. Helv., vol. 39 (1964), pp. 205–252.

John, F.: [1] Plane waves and spherical means, Interscience, New York, 1955.

Kaneyuki, S. and Nagano, T.: [1] On the first Betti numbers of compact quotient spaces of complex semi-simple Lie groups by discrete groups, Sci. Papers Coll. Gen. Ed. Univ. Tokyo, vol. 12 (1962), pp. 1–11.
— [2] On certain quadratic forms related to symmetric Riemannian spaces, Osaka Math. J., vol. 14 (1962), pp. 1–20.

Kaplansky, I.: [1] Primary ideals in group algebras, Proc. Nat. Acad. Sci. U.S.A., vol. 35 (1949), pp. 133–136.
— [2] Groups with representations of bounded degree, Canad. J. Math., vol. 1 (1949), pp. 105–112.

Kaplansky, I. et al.: [1] Operator theory and group representations, National Academy of Sciences – National Research Council, Publication 387, Washington, D. C., 1955.

Karpelevič, F. I.: [1] Orispherical radial parts of Laplace operators on symmetric spaces, Soviet Mathematics, vol. 3 (1962), pp. 528–531.
— [2] The geometry of geodesics and the eigenfunctions of the Laplace-Beltrami operator on symmetric spaces, Trans. Moscow Math. Soc., vol. 14 (1965), pp. 48–185.

Kazdan, D.: [1] Connection of the dual space of a group with the structure of its closed subgroups, Func. Analysis and Applic., vol. 1 (1967), pp. 63–65.

Kazdan, D. A. and Margolies, G. A.: [1] A proof of Selberg's hypothesis, Mat. Sb. (N. S.), vol. 75 (1968), pp. 163–168.

Kelley, J. L. and Namioka, I.: [1] Linear topological spaces, Van Nostrand, Princeton, New Jersey, 1963.

Kimura, H.: [1] On some infinite dimensional representations of semi-simple Lie algebras, Nagoya Math. J., vol. 25 (1965), pp. 211–220.

Kirillov, A. A.: [1] Unitary representations of nilpotent Lie groups, Russ. Math. Surveys, vol. 17 (1962), pp. 53–104.
— [2] Plancherel measure for nilpotent Lie groups, Func. Analysis and Applic., vol. 1 (1967), pp. 330–331.
— [3] Method of orbits in the theory of unitary representations of Lie groups, Func. Analysis and Applic., vol. 2 (1968), pp. 90–93.
— [4] The characters of unitary representations of Lie groups, Func. Analysis and Applic., vol. 2 (1968), pp. 133–146.

Kleppner, A.: [1] Intertwining forms for summable induced representations, Trans. Amer. Math. Soc., vol. 112 (1964), pp. 164–183.

Klimyk, A.: [1] Decomposition of a direct product of irreducible representations of a semi-simple Lie algebra into a direct sum of irreducible representations, Amer. Math. Soc. Translations (2), vol. 76 (1968), pp. 63–73.

Knapp, A. W. and Okamoto, K.: [1] Limits of holomorphic discrete series, To Appear.

Knapp, A. W. and Stein, E. M.: [1] Singular integrals and the principal series, Proc. Nat. Acad. Sci. U.S.A., vol. 63 (1969), pp. 281–284.
— [2] Existence of complementary series, Problems in Analysis: Symposium in Honor of Salomon Bochner, Princeton University Press, Princeton, New Jersey, 1970.
— [3] Singular integrals and the principal series II, Proc. Nat. Acad. Sci. U.S.A., vol. 66 (1970), pp. 13–17.
— [4] Intertwining operators for semi-simple groups, Ann. of Math., vol. 93 (1971), pp. 489–578.

Kohari, A.: [1] Harmonic analysis on the group of linear transformations of the straight line, Proc. Jap. Acad., vol. 37 (1962), pp. 250–254.

Korányi, A. and Wolf, J. A.: [1] Realization of Hermitian symmetric spaces as generalized half-planes, Ann. of Math., vol. 81 (1965), pp. 265–288.

Kostant, B.: [1] On the conjugacy of real Cartan subalgebras I, Proc. Nat. Acad. Sci. U.S.A., vol. 41 (1955), pp. 967–970.
— [2] The principal three-dimensional subgroup and the Betti numbers of a complex simple Lie group, Amer. J. Math., vol. 81 (1959), pp. 973–1032.
— [3] Lie algebra cohomology and the generalized Borel-Weil theorem, Ann. of Math., vol. 74 (1961), pp. 329–387.
— [4] Lie group representations on polynomial rings, Amer. J. Math., vol. 85 (1963), pp. 327–404.
— [5] On the existence and irreducibility of certain series of representations, Bull. Amer. Math. Soc., vol. 75 (1969), pp. 627–642.
— [6] Quantization and unitary representations, Lectures in Modern Analysis and Applications, Georgetown, 1969. [Lecture Notes in Mathematics, Springer-Verlag, vol. 170, 1970.]

Kostant, B. and Rallis, S.: [1] On orbits associated with symmetric spaces, Bull. Amer. Math. Soc., vol. 75 (1969), pp. 879–883.
— [2] On representations associated with symmetric spaces, Bull. Amer. Math. Soc., vol. 75 (1969), pp. 884–888.

Koszul, J-L.: [1] Homologie et cohomologie des algèbres de Lie, Bull. Soc. Math. France, vol. 78 (1950), pp. 65–127.

Kotake, T. and Narasimhan, M. S.: [1] Regularity theorems for fractional powers of a linear operator, Bull. Soc. Math. France, vol. 90 (1962), pp. 449–471.

Kubota, T.: [1] Introduction to Selberg's theory, Lecture Notes, Nagoya Univ., 1966.
— [2] An elementary theory of Eisenstein series, Tokyo Univ. Seminar Lecture Notes, vol. 18, 1968.
— [3] On automorphic functions and the reciprocity law in a number field, Kinokuniya Book Store Co., Tokyo, Japan, 1969.

Kuga, M.: [1] Topological analysis and its applications in weakly symmetric Riemannian spaces, Sûgaku, vol. 9 (1957/58), pp. 166–185.
— [2] On a uniformity of distribution of O-cycles and the eigenvalues of Hecke's operators I, Sci. Papers Coll. Gen. Ed. Univ. Tokyo, vol. 10 (1960), pp.1–16.
— [3] On a uniformity of distribution of O-cycles and the eigenvalues of Hecke's operators II, Sci. Papers Coll. Gen. Ed. Univ. Tokyo, vol. 10 (1960), pp. 171–186.
— [4] Fiber varieties over a symmetric space whose fibers are abelian varieties, University of Chicago Lecture Notes, 1963-1964.

Kuga, M. and Shimura, G.: [1] On vector differential forms attached to automorphic forms, J. Math. Soc. Japan, vol. 12 (1960), pp. 258–270.

Kunze, R. A.: [1] Positive definite operator-valued kernels and unitary representations, Functional Analysis (Proc. Conf., Irvine, Calif., 1966), pp. 235–247, Thompson Book Co., Washington, D. C., 1967.
— [2] Analytic continuation of intertwining operators, To Appear.
— [3] A note on square integrable representations, J. Functional Analysis, vol. 6 (1970), pp. 454–459.

Kunze, R. A. and Stein, E. M.: [1] Uniformly bounded representations and harmonic analysis on the 2×2 real unimodular group, Amer. J. Math., vol. 82 (1960), pp. 1–62.
— [2] Uniformly bounded representations II: Analytic continuation of the principal series of representations of the $n \times n$ complex unimodular group, Amer. J. Math., vol. 83 (1961), pp. 723–786.
— [3] Uniformly bounded representations III: Intertwining operators for the principal series on semi-simple groups, Amer. J. Math., vol. 89 (1967), pp. 385–442.

— [4] Uniformly bounded representations IV: Analytic continuation of the principal series for complex classical groups of types $\mathbf{B}_n$, $\mathbf{C}_n$, $\mathbf{D}_n$, To Appear.
Langlands, R. P.: [1] The dimension of spaces of automorphic forms, Amer. J. Math., vol. 85 (1963), pp. 99–125.
— [2] On the functional equations satisfied by Eisenstein series, unpublished.
— [3] Eisenstein series, Proc. Sympos. Pure Math., vol. 9, Amer. Math. Soc., Providence, Rhode Island, 1966, pp. 235–252.
— [4] Dimension of spaces of automorphic forms, Proc. Sympos. Pure Math., vol. 9, Amer. Math. Soc., Providence, Rhode Island, 1966, pp. 253–257.
— [5] Euler products, Lecture Notes, Yale University, 1967.
— [6] Representations of abelian algebraic groups, Lecture Notes, Yale University, 1968.
— [7] Problems in the theory of automorphic forms, Lectures in Modern Analysis and Applications, Georgetown, 1969. [Lecture Notes in Mathematics, Springer-Verlag, vol. 170, 1970.]
— [8] On the functional equation of the Artin L-functions, To Appear.
de Leeuw, K. and Glicksberg, I.: [1] The decomposition of certain group representations, J. d'Analyse Math., vol. 15 (1965), pp. 135–192.
Lemire, F. W.: [1] Irreducible representations of a simple Lie algebra admitting a one-dimensional weight space, Proc. Amer. Math. Soc., vol. 19 (1968), pp. 1161–1164.
Lipsman, R.: [1] Uniformly bounded representations of SL(2, C), Amer. J. Math., vol. 91 (1969), pp. 47–66.
— [2] An indicator diagram for locally compact unimodular groups, Duke Math. J., vol. 36 (1969), pp. 765–780.
— [3] Harmonic analysis on SL(n,C), J. of Functional Analysis, vol. 3 (1969), pp. 126–155.
— [4] Uniformly bounded representations of the Lorentz group, Amer. J. Math., vol. 91 (1969), pp. 938–962.
— [5] An indicator diagram for locally compact unimodular groups II, Duke Math. J., vol. 37 (1970), pp. 459–466.
— [6] The dual topology for the principal and discrete series on semi-simple groups, Trans. Amer. Math. Soc., vol. 152 (1970), pp. 399–417.
— [7] An explicit realization of Kostant's complementary series with applications to uniformly bounded representations, To Appear.
— [8] On the characters and equivalence of continuous series representations, To Appear.
Loebel, E. et. al.: [1] Group theory and its applications, Academic Press, New York, 1968.
Loomis, L. H.: [1] Abstract harmonic analysis, Van Nostrand, New York, 1953.
— [2] Positive definite functions and induced representations of groups, Duke Math. J., vol. 27 (1960), pp. 569–580.
Loos, O.: [1] Symmetric spaces, vols. I–II, Benjamin, New York, 1969.
Maas, H.: [1] Über eine neue Art von nichtanalytischen automorphen Funktionen, Math. Ann., vol. 121 (1949), pp. 141–183.
— [2] Spherical functions and quadratic forms, J. Indian Math. Soc., vol. 20 (1956), pp. 117–162.
— [3] Zur Theorie der Kugelfunktionen einer Matrixvariablen, Math. Ann., vol. 135 (1958), pp. 391–416.
MacDonald, I.: [1] Spherical functions on a $\mathfrak{p}$-adic Chevalley group, Bull. Amer. Math. Soc., vol. 74 (1968), pp. 520–525.
Mackey, G. W.: [1] Functions on locally compact groups, Bull. Amer. Math. Soc., vol. 56 (1950), pp. 385–412.
— [2] On induced representations of groups, Amer. J. Math., vol. 73 (1951), pp. 576–592.

— [3] Induced representations of locally compact groups I, Ann. of Math., vol. 55 (1952), pp. 101–139.
— [4] Induced representations of locally compact groups II, Ann. of Math., vol. 58 (1953), pp. 193–221.
— [5] Mimeographed notes on group representations, Univ. of Chicago, 1956.
— [6] Borel structure in groups and their duals, Trans. Amer. Math. Soc., vol. 85 (1957), pp. 134–165.
— [7] Unitary representations of group extensions I, Acta Math., vol. 99 (1958), pp. 265–311.
— [8] Point realizations of transformation groups, Illinois J. Math., vol. 6 (1962), pp. 327–335.
— [9] Infinite dimensional group representations, Bull. Amer. Math. Soc., vol. 69 (1963), pp. 628–686.
— [10] Ergodic transformation groups with a pure point spectrum, Illinois J. Math., vol. 8 (1964), pp. 593–600.
— [11] Mimeographed notes on group representations and non-commutative harmonic analysis, University of California (Berkeley), 1965.
— [12] Ergodic theory and virtual groups, Math. Ann., vol. 166 (1966), pp. 187–207.

Magnus, W. and Oberhettinger, F.: [1] Formeln und Sätze für die Speziellen Funktionen der Mathematischen Physik, Springer-Verlag, Berlin, 1948.

Matsumoto, H.: [1] Fonctions sphériques sur un groupe semi-simple $\mathfrak{p}$-adic, C. R. Acad. Sc. Paris, vol. 269 (1969), Serie A, pp. 829–832.

Matsushima, Y.: [1] On the first Betti number of compact quotient spaces of higher dimensional symmetric spaces, Ann. of Math., vol. 75 (1962), pp. 312–330.
— [2] On Betti numbers of compact, locally symmetric Riemannian manifolds, Osaka Math. J., vol. 14 (1962), pp. 1–20.
— [3] A formula for the Betti numbers of compact locally symmetric Riemannian manifolds, J. Diff. Geom., vol. 1 (1967), pp. 99–109.

Matsushima, Y. and Murakami, S.: [1] On vector bundle valued harmonic forms and automorphic forms on symmetric Riemannian manifolds, Ann. of Math., vol. 78 (1963), pp. 365–416.
— [2] On certain cohomology groups attached to Hermitian symmetric spaces, Osaka J. Math., vol. 2 (1965), pp. 1–35.
— [3] On certain cohomology groups attached to Hermitian symmetric spaces II, Osaka J. Math., vol. 5 (1968), pp. 223–241.

Matsushima, Y. and Shimura, G.: [1] On the cohomology groups attached to certain vector-valued differential forms on the product of upper half-planes, Ann. of Math., vol. 78 (1963), pp. 417–449.

Mautner, F. I.: [1] Unitary representations of locally compact groups I, Ann. of Math., vol. 51 (1950), pp. 1–25.
— [2] Unitary representations of locally compact groups II, Ann. of Math., vol. 52 (1950), pp. 528–556.
— [3] On the decomposition of unitary representations of Lie groups, Proc. Amer. Math. Soc., vol. 2 (1951), pp. 490–496.
— [4] A generalization of the Frobenius reciprocity theorem, Proc. Nat. Acad. Sci. U.S.A., vol. 37 (1951), pp. 431–435.
— [5] Fourier analysis and symmetric spaces, Proc. Nat. Acad. Sci. U.S.A., vol. 37 (1951), pp. 529–533.
— [6] Induced representations, Amer. J. Math., vol. 74 (1952), pp. 737–758.
— [7] Geodesic flows and unitary representations, Proc. Nat. Acad. Sci. U.S.A., vol. 40 (1954), pp. 33–36.
— [8] Geodesic flows on symmetric Riemann spaces, Ann. of Math., vol. 65 (1957), pp. 416–431.

— [9] Spherical functions over $\mathfrak{p}$-adic fields I, Amer. J. Math., vol. 80 (1958), pp. 441–457.
— [10] Spherical functions over $\mathfrak{p}$-adic fields II, Amer. J. Math., vol. 86 (1964), pp. 171–200.
— [11] The trace of Hecke operators, Monat. für Math., vol. 72 (1968), pp. 137–143.
— [12] Fonctions propres des opérateurs de Hecke, C. R. Acad. Sc. Paris, vol. 269 (1969), Serie A, pp. 940–943.
— [13] Fonctions propres des opérateurs de Hecke, C. R. Acad. Sc. Paris, vol. 270 (1970), Serie A, pp. 89–92.

McKean, P. and Singer, I.: [1] Curvature and the eigenvalues of the Laplacian, J. Diff. Geom., vol. 1 (1967), pp. 43–70.

Methée, P-D.: [1] Sur les distributions invariantes dans le groupe des rotations de Lorentz, Comment. Math. Helv., vol. 28 (1954), pp. 225–269.

Miller, J. and Simms, D. J.: [1] Radial limits of the rational functions $'\Gamma_\mu$ of Harish-Chandra, Proc. Royal Irish Acad. (Sec. A), vol. 68 (1969), pp. 41–47.
— [2] A difference equation satisfied by the functions $'\Gamma_\mu$ of Harish-Chandra, Amer. J. Math., vol. 92 (1970), pp. 362–368.

Minakshisundaram S. and Pleijel, A.: [1] Eigenfunctions of the Laplace operator on Riemann manifolds, Canad. J. Math., vol. 1 (1949), pp. 242–256.

Moore, C. C.: [1] On the Frobenius reciprocity theorem for locally compact groups, Pacific J. Math., vol. 12 (1962), pp. 359–365.
— [2] Compactifications of symmetric spaces, Amer. J. Math., vol. 86 (1964), pp. 201–218.
— [3] Compactifications of symmetric spaces, Amer. J. Math., vol. 86 (1964), pp. 358–378.

Moore, R. T.: [1] Exponentiation of operator Lie algebras on Banach spaces, Bull. Amer. Math. Soc., vol. 71 (1965), pp. 903–908.
— [2] Measurable, continuous and smooth vectors for semigroups and group representations, Memoirs Amer. Math. Soc., n°78, Providence, Rhode Island, 1968.

Morimoto, M: [1] Sur les transformations horosphériques généralisées dans les espaces homogènes, J. Fac. Sci. Univ. of Tokyo, vol. 13 (1966), pp. 65–83.

Mostow, G. D.: [1] Some new decomposition theorems for semi-simple groups, Memoirs Amer. Math. Soc., n° 14, pp. 31-54, Providence, Rhode Island, 1955.
— [2] Self-adjoint group, Ann. of Math., vol. 62 (1955), pp. 44–55.
— [3] Fully reducible subgroups of algebraic groups, Amer. J. Math., vol. 78 (1956), pp. 200–221.
— [4] Equivariant embeddings in Euclidean space, Ann. of Math., vol. 65 (1957), pp. 432–446.
— [5] On maximal subgroups of real Lie groups, Ann. of Math., vol. 74 (1961), pp. 503–517.
— [6] Covariant fiberings of Klein spaces II, Amer. J. Math., vol. 84 (1962), pp. 466–474.
— [7] Homogeneous spaces with finite invariant measure, Ann. of Math., vol. 75 (1962), pp. 17–37.
— [8] Representative functions on groups, Advanced Science Seminar, pp. 1–37, Bowdoin College, 1968.
— [9] Quasi-conformal mappings in n-space and the rigidity of hyperbolic space forms, Inst. Hautes Études Sci. Publ. Math., vol. 34 (1968), pp. 53–104.

Mostow, G. D. and Tamagawa, T.: [1] On the compactness of arithmetically defined homogeneous spaces, Ann. of Math., vol. 76 (1962), pp. 446-463.

Murakami, S.: [1] Sur la classification des algèbres de Lie réelles et simples, Osaka J. Math., vol. 2 (1965), pp. 291–307.

Nachbin, L.: [1] The Haar integral, Van Nostrand, New York, 1965.

Nagata, M.: [1] On the 14th problem of Hilbert, Amer. J. Math., vol. 81 (1959), pp. 766–772.

Naimark, M. A.: [1] Normed rings, Noordhoff, Groningen—The Netherlands, 1964.

— [2] Linear representations of the Lorentz group, Macmillan, New York, 1964.

Narasimhan, M. S. and Okamoto, K.: [1] An analogue of the Borel-Weil-Bott theorem for Hermitian symmetric pairs of non-compact type, Ann. of Math., vol. 91 (1970), pp. 486–511.

Nelson, E.: [1] Analytic vectors, Ann. of Math., vol. 70 (1959), pp. 572–615.

Nelson, E. and Stinespring, W. F.: [1] Representation of elliptic operators in an enveloping algebra, Amer. J. Math., vol. 81 (1959), pp. 547–560.

Neto, J. B.: [1] Spaces of vector valued real analytic functions, Trans. Amer. Math. Soc., vol. 112 (1964), pp. 381–391.

Okamoto, K.: [1] On the Plancherel formulas for some types of simple Lie groups, Osaka J. Math., vol. 2 (1965), pp. 247–282.

— [2] On induced representations, Osaka J. Math., vol. 4 (1967), pp. 85–94.

— [3] On square integrable $\bar{\partial}$-cohomology spaces attached to homogeneous symplectic manifolds, To Appear.

Okamoto, K. and Ozeki, H.: [1] On square integrable $\bar{\partial}$-cohomology spaces attached to Hermitian symmetric spaces, Osaka J. Math., vol. 4 (1967), pp. 95–110.

Oniščik, A. L.: [1] Lie groups transitive on compact manifolds, Amer. Math. Soc. Translations (2), vol. 73 (1968), pp. 59–72.

— [2] Lie groups transitive on compact manifolds II, Math. U.S.S.R.-Sbornik, vol. 3 (1967), pp. 373–388.

Osima, M.: [1] Note on the Kronecker product of representations of a group, Proc. Imperial Acad. Tokyo, vol. 17 (1941), pp. 411–413.

Palais, R.: [1] Imbedding of compact, differentiable transformation groups in orthogonal representations, J. Math. Mech., vol. 6 (1957), pp. 673–678.

— [2] Seminar on the Atiyah-Singer Index Theorem, Princeton University Press, Princeton, New Jersey, 1965.

Paley, R. and Wiener, N : [1] Fourier transforms in the complex domain, Amer. Math. Soc., New York, 1934.

Parthasarathy, K. R., Ranga Rao, R. and Varadarajan, V. S.: [1] Representations of complex semi-simple Lie groups and Lie algebras, Ann. of Math., vol. 85 (1967), pp. 383–429.

Peetre, J : [1] Rectification à l'article "Une charactérisation abstraite des opérateurs differentials," Math. Scand., vol. 8 (1960), pp. 116–120.

Poulsen, N.: [1] Regularity aspects of the theory of infinite dimensional representations of Lie groups, Ph.D. Thesis, Massachusetts Institute of Technology, 1970.

— [2] On C^∞-vectors and intertwining bilinear forms for representations of Lie groups, To Appear.

Pukanszky, L.: [1] The Plancherel formula for the universal covering group of **SL**(2, **R**), Math. Ann., vol. 156 (1964), pp. 96–143.

— [2] Leçons sur les représentations des groupes, Dunod, Paris, 1967.

— [3] On the theory of exponential groups, Trans. Amer. Math. Soc., vol. 126 (1967), pp. 487–507.

— [4] On the characters and the Plancherel formula of nilpotent groups, J. of Functional Analysis, vol. 1 (1967), pp. 255–280.

— [5] On the unitary representations of exponential groups, J. of Functional Analysis, vol. 2 (1968), pp. 73–113.
— [6] Characters of algebraic solvable groups, J. of Functional Analysis, vol. 3 (1969), pp. 435–494.

Pyateckii-Shapiro, I. I.: [1] The geometry of classical domains and the theory of automorphic functions, Moscow, 1961.

Rader, C.: [1] Spherical functions on a semi-simple Lie group, Ph.D. Thesis, University of Washington, 1971.

Rallis, S.: [1] Lie group representations associated to symmetric spaces, Ph.D. Thesis, Massachusetts Institute of Technology, 1968.

de Rham, G.: [1] Solution élémentaire d'opérateurs différentiels du second ordre, Ann. de l'Institut Fourier, vol. 8 (1958), pp. 337–366.

Rhie, Y : [1] A density theorem on spectra of discrete subgroups of semi-simple Lie groups, Illinois J. Math., vol. 10 (1966), pp. 147–158.

Richardson Jr., R. W.: [1] Conjugacy classes in Lie algebras and algebraic groups, Ann. of Math., vol. 86 (1967), pp. 1–15.

Rickart, C. E.: [1] General theory of Banach algebras, Van Nostrand, New York, 1960.

Rieffel, M. A.: [1] Induced Banach representations of Banach algebras and locally compact groups, J. of Functional Analysis, vol. 1 (1967), pp. 443–491.
— [2] Square integrable representations of Hilbert algebras, J. of Functional Analysis, vol. 3 (1969), pp. 265–300.

Riesz, F. and Sz-Nagy, B.: [1] Functional analysis, Ungar, New York, 1955.

Rigelhof, R.: [1] Induced representations of locally compact groups, Acta Math., vol. 125 (1970), pp. 155–187.

Roelcke, W.: [1] Über die Wellengleichung bei Grenzkreisgruppen erster Art, Sitz. Ber. Heidelberger Ak. d. Wiss., Math.-nat. Kl. 1955, 4. Abh.
— [2] Das Eigenwertproblem der automorphen Formen in der hyperbolischen Ebene I, Math. Ann., vol. 167 (1966), pp. 292–337.
— [3] Das Eigenwertproblem der automorphen Formen in der hyperbolischen Ebene II, Math. Ann., vol. 168 (1967), pp. 261–324.

Romm, B. D.: [1] Analogue to the Plancherel formula for the real unimodular group of the *n*th order, Amer. Math. Soc. Translations (2), vol. 58 (1966), pp. 155–215.
— [2] Fully reducible representations of a semi-simple Lie algebra, Soviet Math., vol. 175 (1967), pp. 861–864.

Rothschild, L.: [1] Conjugacy classes of real Cartan subalgebras revisited, A.M.S. Notices, vol. 18 (1971), Abstract 682-22-9.

Sakai, S.: [1] On the representations of semi-simple Lie groups, Proc. Japan Acad., vol. 30 (1954), pp. 14–18.
— [2] On infinite-dimensional representations of semi-simple Lie algebras and some functionals on the universal enveloping algebras I, Proc. Japan Acad., vol. 30 (1954), pp. 305–312.

Sally, P. and Shalika, J. [1] Characters of the discrete series of representations of **SL**(2) over a local field, Proc. Nat. Acad. Sci. U.S.A., vol. 61 (1968), pp. 1231–1237.
— [2] The Plancherel formula for **SL**(2) over a local field, Proc. Nat. Acad. Sci. U.S.A., vol. 63 (1969), pp. 661–667.

Satake, I.: [1] On representations and compactifications of symmetric Riemannian spaces, Ann. of Math., vol. 71 (1960), pp. 77–110.
— [2] On compactifications of the quotient space for arithmetically defined discontinuous groups, Ann. of Math., vol. 72 (1960), pp. 555–580.

— [3] Theory of spherical functions on reductive algebraic groups over p-adic fields, Inst. Hautes Études Sci. Publ. Math., vol. 18 (1963), pp. 5–69.

Schattschneider, D.: [1] On restricted roots of semi-simple algebraic groups, J. Math. Soc. Japan, vol. 21 (1969), pp. 94–115.

Schiffmann, G.: [1] Distributions centrales de type positif sur un groupe de Lie nilpotent, Bull. Soc. Math. France, vol. 96 (1968), pp. 347–355.

— [2] Intégrales d'entrelacement et fonctions de Whittaker, Bull. Soc. Math. France, vol. 99 (1971), pp. 3–72.

Schmid, W.: [1] Homogeneous complex manifolds and representations of semi-simple Lie groups, Ph.D. Thesis, University of California—Berkeley, 1967.

— [2] Die Randwerte holomorpher Funktionen auf hermitesch symmetrischen Räumen, Inventiones Math., vol. 9 (1969), pp. 61–80.

— [3] On a conjecture of Langlands, Ann. of Math., vol. 93 (1971), pp. 1–42.

— [4] Some properties of square integrable representations of semi-simple Lie groups, To Appear.

Schochetman, I.: [1] Topology and the duals of certain locally compact groups, Trans. Amer. Math. Soc., vol. 150 (1970), pp. 477–489.

Schwartz, L : [1] Théorie des distributions, vol. I, Hermann, Paris, 1957.

— [2] Théorie des distributions, vol. II, Hermann, Paris, 1957.

— [3] Distributions à valeurs vectorielles I, Ann. Inst. Fourier Grenoble, vol. 7 (1957), pp. 1–141.

— [4] Distributions à valeurs vectorielles II, Ann. Inst. Fourier Grenoble, vol. 8 (1959), pp. 1–207.

Segal, G.: [1] The representation ring of a compact Lie group, Inst. Hautes Études Sci. Publ. Math., vol. 34 (1968), pp. 113–128.

Segal, I. E.: [1] The group algebra of a locally compact group, Trans. Amer. Math. Soc., vol. 61 (1947), pp. 69–105.

— [2] Irreducible representations of operator algebras, Bull. Amer. Math. Soc., vol. 53 (1947), pp. 73–88.

— [3] The two-sided regular representation of a unimodular locally compact group, Ann. of Math., vol. 51 (1950), pp. 293–298.

— [4] An extension of Plancherel's formula to separable unimodular groups, Ann. of Math., vol. 52 (1950), pp. 272–292.

— [5] A class of operator algebras which are determined by groups, Duke Math. J., vol. 18 (1951), pp. 221–265.

— [6] Hypermaximality of certain operators on Lie groups, Proc. Amer. Math. Soc., vol. 3 (1952), pp. 13–15.

— [7] A non-commutative extension of abstract integration, Ann. of Math., vol. 57 (1953), pp. 401–457.

— [8] A theorem on the measurability of group-invariant operators, Duke Math. J., vol. 26 (1959), pp. 549–552.

— [9] Infinite dimensional irreducible representations of compact semi-simple groups, Bull. Amer. Math. Soc., vol. 70 (1964), pp. 155–160.

— [10] Algebraic integration theory, Bull. Amer. Math. Soc., vol. 71 (1965), pp. 419–489.

— [11] An extension of a theorem of L. O'Raifeartaigh, J. of Functional Analysis, vol. 1 (1967), pp. 1–21.

Segal, I. E. and von Neumann, J.: [1] A theorem on unitary representations of semi-simple Lie groups, Ann. of Math., vol. 52 (1950), pp. 509–517.

Selberg, A.: [1] Harmonic analysis and discontinuous groups in weakly symmetric Riemannian spaces with applications to Dirichlet series, J. Indian Math. Soc., vol. 20 (1956), pp. 47–87.

— [2] Automorphic functions and integral operators, Seminars on Analytic Functions, Institute for Advanced Study, Princeton, New Jersey, vol. 2 (1957), pp. 152–161.

— [3] A new type of zeta function connected with quadratic forms, Report of the Institute in the Theory of Numbers, pp. 207–210, Boulder, Colorado, 1959.
— [4] On discontinuous groups in higher dimensional symmetric spaces, Contributions to Function Theory, pp. 147–164, Bombay, 1960.
— [5] Discontinuous groups and harmonic analysis, Proc. Int. Congr. of Math., pp. 177–189, Stockholm, 1962.
— [6] On the estimation of Fourier coefficients of modular forms, Proc. Sympos. Pure Math., vol. 8, Amer. Math. Soc., Providence, Rhode Island, 1965, pp. 1–15.
— [7] Recent developments in the theory of discontinuous groups of motions of symmetric spaces, Proceedings of the 15th Scandinavian Congress, Oslo, 1968. [Lecture Notes in Mathematics, vol. 118, 1970, Springer-Verlag.]

Séminaire Henri Cartan: [1] Fonctions automorphes, Paris, 1957–58.

Séminaire Sophus Lie: [1] Théorie des algèbres de Lie, Topologie des groupes de Lie, Paris, 1955.

Semyanistyi, V. I.: [1] Homogeneous functions and some problems in integral geometry in spaces of constant curvature, Soviet Math., vol. 2 (1961), pp. 59–62.

Serre, J-P.: [1] Faisceaux algébriques cohérents, Ann. of Math., vol. 61 (1955), pp. 197–278.
— [2] Une théorème de dualité, Comment. Math. Helv., vol. 29 (1955), pp. 9–26.
— [3] Lie algebras and Lie groups, Benjamin, New York, 1965.
— [4] Algèbres de Lie semi-simples complexes, Benjamin, New York, 1966.

Shephard, G. and Todd, J.: [1] Finite unitary reflection groups, Canad. J. Math., vol. 6 (1954), pp. 274–304.

Sherman, T.: [1] A weight theory for unitary representations, Canad. J. Math., vol. 18 (1966), pp. 159–168.

Shiga, K.: [1] Representations of a compact group on a Banach space, J. Math. Soc. Japan, vol. 7 (1955), pp. 224–248.

Shimizu, H.: [1] On discontinuous groups operating on the product of upper half planes, Ann. of Math., vol. 77 (1963), pp. 33–71.
— [2] On traces of Hecke operators, J. Fac. of Sci. Univ. of Tokyo, Sec. 1, vol. 10 (1963), pp. 1–19.
— [3] On zeta functions of quaternion algebras, Ann. of Math., vol. 81 (1965), pp. 166–193.

Shimizu, Y.: [1] An analogue of the Paley-Wiener theorem for certain function spaces on the generalized Lorentz group, J. Fac. of Sci. Univ. of Tokyo, vol. 16 (1969), pp. 13–51.

Shimura, G.: [1] Sur les intégrales attachées aux formes automorphes, J. Math. Soc. Japan, vol. 11 (1959), pp. 291–311.

Shoda, K.: [1] Über die monomialen Darstellungen einer endlichen Gruppe, Proc. Physico-Mathematical Soc. Japan, vol. 15 (1933), pp. 249–257.

de Siebenthal, J.: [1] Sur certains modules dans une algèbre de Lie semi-simple, Comment. Math. Helv., vol. 44 (1969), pp. 1–44.

Sobolev, S.: [1] Applications of functional analysis in mathematical physics, Translations of Math. Mono., vol. 7, Amer. Math. Soc., Providence, Rhode Island, 1963.

Solomon, L.: [1] Invariants of finite reflection groups, Nagoya Math. J., vol. 22 (1963), pp. 57–64.

Solomon, L. and Verma, D. N.: [1] Sur le corps des quotients de l'algèbre enveloppante d'une algèbre de Lie, C. R. Acad. Sc. Paris, vol. 264 (1967), Serie A, pp. 985–986.

Springer, T. A.: [1] Weyl's character formula for algebraic groups, Inventiones Math., vol. 5 (1968), pp. 85–105.

Stein, E. M.: [1] A survey of representations of non-compact groups, Lect. Sem. on High Energy Physics and Elementary Particles (Trieste, 1965), International Atomic Energy Agency, Vienna, 1965, pp. 563–584.

— [2] Analysis in matrix spaces and some new representations of **SL**(n, **C**), Ann. of Math., vol. 86 (1967), pp. 461–490.

— [3] Analytic continuation of group representations, Advances in Math., vol. 4 (1970), pp. 172–207.

Stein, E. M. and Weiss, G.: [1] Generalization of the Cauchy-Riemann equations and representations of the rotation group, Amer. J. Math., vol. 90 (1968), pp. 163–196.

Steinberg, R.: [1] A general Clebsch-Gordan theorem, Bull. Amer. Math. Soc., vol. 67 (1961), pp. 406–407.

— [2] Representations of algebraic groups, Nagoya Math. J., vol. 22 (1963), pp. 33–56.

— [3] Differential equations invariant under finite reflection groups, Trans. Amer. Math. Soc., vol. 112 (1964), pp. 392–400.

— [4] Regular elements of semi-simple algebraic groups, Inst. Hautes Études Sci. Publ. Math., vol. 25(1965), pp. 49–80.

— [5] Lectures on Chevalley groups, Lecture Notes, Yale University, 1967.

Stern, A. I.: [1] Completely irreducible class I representations of real semi-simple Lie groups, Soviet Math., vol. 10 (1969), pp. 1254–1257.

Sternheimer, D.: [1] Propriétés spectrales dans les représentations de groupes de Lie, J. Math. Pures et Appl., vol. 47 (1968), pp. 289–319.

Stinespring, W. F.: [1] A semi-simple matrix group is of type I, Proc. Amer. Math. Soc., vol. 9 (1958), pp. 965–967.

— [2] Integrability of Fourier transforms for unimodular Lie groups, Duke Math. J., vol. 26 (1959), pp. 123–131.

Stone, M.: [1] Linear transformations in Hilbert space, Amer. Math. Soc. Colloquium Publication, vol. 15, New York, 1932.

Sugiura, M.: [1] Conjugate classes of Cartan subalgebras in real semi-simple Lie algebras, J. Math. Soc. Japan, vol. 11 (1959), pp. 374–434.

— [2] Spherical functions and representation theory of compact Lie groups, Sci. Papers Coll. Gen. Ed. Univ. Tokyo, vol. 10 (1960), pp. 187–193.

— [3] Representations of compact groups realized by spherical functions on symmetric spaces, Proc. Japan Acad., vol. 38 (1962), pp. 111–113.

Takahashi, R.: [1] Sur les fonctions sphériques et la formule de Plancherel dans le groupe hyperbolique, Jap. J. Math., vol. 31 (1961), pp. 55–90.

— [2] Sur les représentations unitaires des groupes de Lorentz généralisés, Bull. Soc. Math. France, vol. 91 (1963), pp. 289–433.

Takenouchi, O.: [1] Sur une classe de fonctions continues de type positif sur un groupe localement compact, Math. J. Okayama U., vol. 4 (1955), pp. 143–173.

— [2] Sur les facteurs représentations d'un groupe de Lie résoluble de type E, Math. J. Okayama U., vol. 7 (1957), pp. 151–161.

Takesaki, M.: [1] Covariant representations of C^*-algebras and their locally compact automorphism groups, Acta Math., vol. 119 (1967), pp. 273–303.

Takeuchi, M.: [1] Cell decompositions and Morse equalities on certain symmetric spaces, J. Fac. of Sci. Univ. of Tokyo, vol. 12 (1965), pp. 81–191.

— [2] On orbits in a compact Hermitian symmetric space, Amer. J. Math., vol. 90 (1968), pp. 657–680.

Tamagawa, T.: [1] On Selberg's trace formula, J. Fac. of Sci. Univ. of Tokyo, vol. 8 (1960), pp. 363–386.

— [2] On the ζ-functions of a division algebra, Ann. of Math., vol. 77 (1963), pp. 387–405.
— [3] Harmonic analysis on adele groups, Advanced Science Seminar, pp. 1–39, Bowdoin College, 1968.

Tanaka, S.: [1] Selberg's trace formula and spectrum, Osaka J. Math., vol. 3 (1966), pp. 205–216.

Tatsuuma, N.: [1] Regularity of orbit spaces on semi-simple Lie groups, Proc. Japan Acad., vol. 42 (1966), pp. 84–86.

Taylor, M.: [1] Fourier series on compact Lie groups, Proc. Amer. Math. Soc., vol. 19 (1968), pp. 1103–1105.

Tengstrand, A.: [1] Distributions invariant under an orthogonal group of arbitrary signature, Math. Scand., vol. 8 (1960), pp. 201–218.

Tirao, J. and Wolf, J.: [1] Homogeneous holomorphic vector bundles, J. Math. and Mechanics, vol. 20 (1970), pp. 15–31.

Titchmarsh, E. C.: [1] Eigenfunction expansions, Oxford University Press, London, 1946.

Tits, J.: [1] Théorème de Bruhat et sous-groupes paraboliques, C. R. Acad. Sci. Paris, vol. 254 (1962), pp. 2910-2912.
— [2] Algebraic and abstract simple groups, Ann. of Math., vol. 80 (1964), pp. 313–329.
— [3] Tabellen zu den einfachen Lie Gruppen und ihren Darstellungen, Springer-Verlag, Lecture Notes in Mathematics, vol. 40, 1967.

Tits, J. and Waelbroeck, L.: [1] The integration of a Lie algebra representation, Pacific J. of Math., vol. 26 (1968), pp. 595–600.

Treves, F.: [1] Linear partial differential equations with constant coefficients, Gordon and Breach, New York, 1966.
— [2] Topological vector spaces, distributions and kernels, Academic Press, New York, 1967.

Trombi, P. C. and Varadarajan, V. S.: [1] Spherical transforms on semi-simple Lie groups, Ann. of Math., vol. 94 (1971), pp. 246–303.
— [2] Asymptotic behaviour of eigenfunctions on a semi-simple Lie group: The discrete spectrum, To Appear.

Tsuchikawa, M.: [1] On the representations of SL(3, C) I, Proc. Japan Acad., vol. 43 (1967), pp. 852–855.
— [2] On the representations of SL(3, C) II, Proc. Japan Acad., vol. 44 (1968), pp. 127–129.
— [3] On the representations of SL(3, C) III, Proc. Japan Acad., vol. 44 (1968), pp. 130–132.

van Dijk, G.: [1] On symmetry of group algebras of motion groups, Math. Ann., vol. 179 (1969), pp. 219–226.

van der Waerden, B. L.: [1] Stetigketissätze der halbeinfachen Lieschen Gruppen, Math. Zeit., vol. 36 (1933), pp. 780–786.

Varadarajan, V. S.: [1] On the ring of invariant polynomials on a semi-simple Lie algebra, Amer. J. Math., vol. 90 (1968), pp. 308–317.

Vilenkin, N.: [1] On the theory of associated spherical functions on Lie groups, Mat. Sb., vol. 42 (1957), pp. 485–496.
— [2] Special functions and the theory of group representations, Translations of Math. Mono., vol. 22, Amer. Math. Soc., Providence, Rhode Island, 1968.

Vinberg, È. B.: [1] The Borel-Morosow theorem for real Lie groups, Soviet Math., vol. 2 (1961), pp. 1416–1419.

Wallach, N.: [1] Induced representations of Lie algebras and a theorem of Borel-Weil, Trans. Amer. Math. Soc., vol. 136 (1969), pp. 181–187.

— [2] Induced representations of Lie algebras II, Proc. Amer. Math. Soc., vol. 21 (1969), pp. 161–166.

— [3] Cyclic vectors and irreducibility for principal series representations, To Appear.

Wang, H. C.: [1] Closed manifolds with homogeneous complex structure, Amer. J. Math., vol. 76 (1954), pp. 1–32.

— [2] On a maximality property of discrete subgroups with fundamental domain of finite measure, Amer. J. Math., vol. 89 (1967), pp. 124–132.

Wang, S. P.: [1] The dual space of semi-simple Lie groups, Amer. J. Math., vol. 91 (1969), pp. 921–937.

Weil, A.: [1] L'intégration dans les groupes topologiques et ses applications, Hermann, Paris, 1940.

— [2] On discrete subgroups of Lie groups I, Ann. of Math., vol. 72 (1960), pp. 369–384.

— [3] On discrete subgroups of Lie groups II, Ann. of Math., vol. 75 (1962), pp. 578–602.

— [4] Remarks on the cohomology of groups, Ann. of Math., vol. 80 (1964), pp. 149–157.

Weyl, H.: [1] The classical groups, Princeton University Press, Princeton, New Jersey, 1939.

Whitney, H.: [1] Elementary structure of real algebraic varieties, Ann. of Math., vol. 66 (1957), pp. 545–556.

Whittaker, E. T. and Watson, G. N.: [1] Modern analysis, Cambridge Press, London, 1927.

Wolf, J. A.: [1] The geometry and structure of isotropy irreducible homogeneous spaces, Acta Math., vol. 120 (1968), pp. 59–148.

— [2] The action of a real semi-simple group on a complex flag manifold I: Orbit structure and holomorphic arc components, Bull. Amer. Math. Soc., vol. 75 (1969), pp. 1121–1237.

— [3] The action of a real semi-simple group on a complex flag manifold II: Unitary representations on partially holomorphic cohomology spaces, To Appear.

Wolf, J. A. and Gray, A.: [1] Homogeneous spaces defined by Lie group automorphisms, J. Diff. Geom., vol. 2 (1968), pp. 77–159.

Wolf, J. A. and Korányi, A.: [1] Generalized Cayley transformations of bounded symmetric domains, Amer. J. Math., vol. 87 (1965), pp. 899–939.

Zeitlin, J.: [1] Correspondence between Lie algebra invariant subspaces and Lie group invariant subspaces of representations of Lie groups, Ph. D. Thesis, University of California-Los Angeles, 1969.

Želobenko, D. P.: [1] Infinitely differentiable vectors in the theory of representations, Vestnik. Moskov. Univ. Ser I. Math. Meh. 1965, pp. 3–10.

— [2] Harmonic analysis of functions on semi-simple Lie groups I, Amer. Math. Soc. Translations (2), vol. 54 (1966), pp. 177–230.

— [3] Analogue of the Cartan-Weyl theory for infinite dimensional representations of a complex semi-simple Lie group, Soviet Math., vol. 8 (1967), pp. 798–802.

— [4] Symmetry in a class of elementary representations of a semi-simple complex Lie group, Func. Analysis and Applic., vol. 1 (1967), pp. 15–38.

— [5] The analysis of irreducibility in the class of elementary representations of a complex semi-simple Lie group, Math. U.S.S.R.-Izvestija, vol. 2 (1968), pp. 105–128.

— [6] Operational calculus on a complex semi-simple Lie group, Math. U.S.S.R.-Izvestija, vol. 3 (1969), pp. 881–915.

— [7] On the harmonic analysis of functions on semi-simple Lie groups II, Math. U.S.S.R.-Izvestija, vol. 3 (1969), pp. 1183–1217.

— [8] On the irreducible representations of a complex semi-simple Lie group, Func. Analysis and Applic., vol. 4 (1970), pp. 163–165.

Želobenko, D. P. and Naimark, M. A.: [1] Description of the completely irreducible representations of a complex semi-simple Lie group, Math. U.S.S.R.-Izvestija, vol. 4 (1970), pp. 59–83.

Subject Index

Die Grundlehren der mathematischen Wissenschaften in Einzeldarstellungen mit besonderer Berücksichtigung der Anwendungsgebiete

Eine Auswahl

146. Treves: Locally Convex Spaces and Linear Partial Differential Equations
147. Lamotke: Semisimpliziale algebraische Topologie.
148. Chandrasekharan: Introduction to Analytic Number Theory
149. Sario/Oikawa: Capacity Functions
150. Iosifescu/Theodorescu: Random Processes and Learning
151. Mandl: Analytical Treatment of One-dimensional Markov Processes
152. Hewitt/Ross: Abstract Harmonic Analysis. Vol. 2: Structure and Analysis for Compact Groups. Analysis on Locally Compact Abelian Groups
153. Federer: Geometric Measure Theory
154. Singer: Bases in Banach Spaces I
155. Müller: Foundations of the Mathematical Theory of Electromagnetic Waves
156. van der Waerden: Mathematical Statistics
157. Prohorov/Rozanov: Probability Theory
159. Köthe: Topological Vector Spaces I
160. Agrest/Maksimov: Theory of Incomplete Cylindrical Functions and Their Applications
161. Bhatia/Szegö: Stability Theory of Dynamical Systems
162. Nevanlinna: Analytic Functions
163. Stoer/Witzgall: Convexity and Optimization in Finite Dimensions I
164. Sario/Nakai: Classification Theory of Riemann Surfaces
165. Mitrinović/Vasić: Analytic Inequalities
166. Grothendieck/Dieudonné: Eléments de Géométrie Algébrique I
167. Chandrasekharan: Arithmetical Functions
168. Palamodov: Linear Differential Operators with Constant Coefficients
170. Lions: Optimal Control Systems Governed by Partial Differential Equations
171. Singer: Best Approximation in Normed Linear Spaces by Elements of Linear Subspaces
172. Bühlmann: Mathematical Methods in Risk Theory
173. F. Maeda/S. Maeda: Theory of Symmetric Lattices
174. Stiefel/Scheifele: Linear and Regular Celestial Mechanics. Perturbed Two-body Motion—Numerical Methods—Canonical Theory
175. Larsen: An Introduction of the Theory of Multipliers
176. Grauert/Remmert: Analytische Stellenalgebren
177. Flügge: Practical Quantum Mechanics I
178. Flügge: Practical Quantum Mechanics II
179. Giraud: Cohomologie non abélienne
180. Landkoff: Foundations of Modern Potential Theory
181. Lions/Magenes: Non-Homogeneous Boundary Value Problems and Applications I
182. Lions/Magenes: Non-Homogeneous Boundary Value Problems and Applications II
183. Lions/Magenes: Non-Homogeneous Boundary Value Problems and Applications III. In preparation
184. Rosenblatt: Markov Processes. Structure and Asymptotic Behavior
185. Rubinowicz: Sommerfeldsche Polynommethode
186. Wilkinson/Reinsch: Handbook for Automatic Computation II, Linear Algebra
187. Siegel/Moser: Lectures on Celestial Mechanics
188. Warner: Harmonic Analysis on Semi-Simple Lie Groups I
189. Warner: Harmonic Analysis on Semi-Simple Lie Groups II
190. Faith: Algebra: Rings, Modules, and Categories I. In preparation
191. Faith: Algebra: Rings, Modules, and Categories II. In preparation
192. Maltsev: Algebraic Systems. In preparation
193. Pólya/Szegö: Problems and Theorems in Analysis. Vol. 1. In preparation
194. Igusa: Theta Functions

MIX
Papier aus verantwortungsvollen Quellen
Paper from responsible sources
FSC® C105338

FSC
www.fsc.org

If you have any concerns about our products,
you can contact us on
ProductSafety@springernature.com

In case Publisher is established outside the EU,
the EU authorized representative is:
Springer Nature Customer Service Center GmbH
Europaplatz 3, 69115 Heidelberg, Germany

Printed by Libri Plureos GmbH
in Hamburg, Germany